Invertebrate Zoology

Invertebrate

Zoology

THIRD EDITION

Paul A. Meglitsch

Frederick R. Schram

NEW YORK OXFORD **OXFORD UNIVERSITY PRESS** 1991

Oxford University Press

Oxford New York Toronto
Delhi Bombay Calcutta Madras Karachi
Petaling Jaya Singapore Hong Kong Tokyo
Nairobi Dar es Salaam Cape Town
Melbourne Auckland

and associated companies in
Berlin Ibadan

Published by Oxford University Press, Inc.,
200 Madison Avenue, New York, New York 10016

Library of Congress Cataloging-in-Publication Data
Meglitsch, Paul A. (Paul Allen), 1914–
Invertebrate zoology / Paul A. Meglitsch, Frederick R. Schram.—3rd ed.
p. cm. Includes bibliographical references.
ISBN 0-19-504900-4
1. Invertebrates. I. Schram, Frederick R., 1943– II. Title.
QL362.M4 1991 592—dc20 89-26592

9 8 7 6 5 4 3 2 1

Printed in the United States of America
on acid-free paper

Preface to the Third Edition

The late Paul Meglitsch was in the process of planning a third edition of this book when he died some 14 years ago. At that time this volume was one of the major textbooks in invertebrate zoology, used in many countries around the world. Since then, with time and advances in the field, the text eventually went out of date. Yet *Invertebrate Zoology* has enjoyed steady use through those years, and the publisher received a stream of inquiries as to when a new edition might be forthcoming. As stocks were depleted, effort was made to find another author to carry on for Professor Meglitsch.

When Oxford University Press approached me, they expressed their desire for a major revision. They wanted the text made more readable and the total volume reduced by about 25 percent to maintain its affordability as a text for students. Several advances in the field since 1971 had to be included in the new edition. These advances involved the discovery of two new phyla and many new classes of invertebrates. In addition, there is increased information about all groups, derived from physiological and field studies, and new and improved techniques such as computer-based cladistic analysis and electron microscopy.

The needed reductions were achieved in several ways. The old photographic figures were eliminated. This text is designed for use with another Oxford University Press book, *Illustrated Invertebrate Anatomy* by J. K. Pierce and T. K. Mangel, an amply and beautifully illustrated laboratory manual. I felt that given the method of photo reproduction that has to be employed in a textbook of this kind to maintain price, and the low amount of information conveyed in such figures, keeping them could not be justified. Then, a close look at the line drawings of the Second Edition revealed that some of them conveyed little meaningful information or were actually repetitious of what was in other figures. A figure is supposed to convey information to supplement the text, and if one figure does that effectively, there is little need to include one or two more figures that repeat the process. The text itself was literally pulled apart line by line in a serious effort to remove repetitious material, as well as to make what remained terse and with increased clarity. In this process an attack was made on the old complaint of students about courses in invertebrate zoology—"too many terms." Although it may not be evident to students, a reduction in the number and a simplification in the use of anatomical terms has been made; and this is a process that will be extended in future editions. Furthermore, terms are boldfaced where their first use and/or definition occurs. The result, I feel, is a "leaner and meaner" textbook—a volume that is concise and more interesting than earlier editions and, as a result, more usable for teachers and students alike.

It is impossible to adequately express thanks to everyone who should be acknowledged in the preparation of a book like this. However, several need special recognition. My editor at Oxford University Press, William F. Curtis, has been a steady source of support with his continuous interest in my progress and his constant encouragement to proceed. Numerous anonymous reviewers helped in finding errors in the text and figures. Michael J. Emerson prepared the new figures, and John M. Simpson and Brian Kirkpatrick helped with recasting many of the old figures. Marjorie Rea prepared readable drafts from my sometimes almost incomprehensible editings and re-editings, and organized the indexing. A host of anonymous reviewers have helped to point the way when we have strayed from the paths of accuracy and clarity. And finally, the facilities of the Friday Harbor Laboratories and the biological libraries of the University of Washington were invaluable to me during two research leaves from my home institution, and the library and staff of the San Diego Natural History Museum were most effective in their service as well.

San Diego
January 1991

F.R.S.

Contents

Say not that this is so,
but that this is how it seems to me to be
as I now see the things I think I see.

J. Brooks Knight

Invertebrate Zoology

1

Introduction

Professor Meglitsch termed courses in invertebrate zoology an exploratory trip, and his text as a guidebook to an adventure. Indeed, the fascinating variety seen among invertebrate animals makes for a compelling area of study, both for the professional as well as the student. However, rather than enjoy the animals, students all too often get caught up in trying to memorize terms. Students forget that studying the terminology is like learning a foreign language. Once mastered, even in its rudiments, it becomes a gateway to a wonderful new world of ideas and concepts closed off from study without it. Professor Meglitsch carried his guidebook analogy along these lines when he pointed out that using a textbook like this is like using a guidebook to plan a trip; that is, the wise traveler plans an excursion that interests him or her and that, without the book, would result in a trip devoid of the interesting and fascinating insights that would have gone unseen. The terminology gets you going. Use what you want, retain what you will, and when you get ready to investigate a subject in greater detail, you will know how to begin enlarging on the vocabulary to become a specialist.

Courses in invertebrate zoology have been disappearing from curricula at many universities. Mastering even a small part of the total available information does require some effort, and as a result of the "relevancy reforms" in education in the 1970s, invertebrate zoology as a subject passed out of fashion. In this, it is no different than foreign language and English literature courses. And just as in the humanities, we have found that as courses such as invertebrate zoology decreased in enrollment, our biology graduates began to lose the vocabulary associated with being an educated biologist.

This loss was due to an assumption that faunistics courses like invertebrate zoology were not relevant. This, we now realize, is a naive and ill-founded notion. Invertebrate animals currently constitute about 98 percent of the known species of animals (May, 1988), and invertebrates probably constitute closer to 99.99 percent of the total number of animal species in the world. Indeed, as the information in Table 1.1 shows, there are a great many more invertebrates yet to be discovered and described.

Furthermore, most of the basic discoveries about animal biology were based on work with invertebrates. One need only try and conceive of genetics without the work on *Drosophila*, nerve physiology without recourse to the information derived from giant axons of squids, muscle–nerve interactions without input from barnacles, or understanding early phases of embryonic differentiation without the models provided by sea urchins and starfish. These are but a few examples. Nor is this importance about to end. With the increasing stridency of animal rights activists and antivivisectionists, more and more basic work in biological research will be shifted to the use of invertebrates, as academic administrators and granting agencies may come to view experimental use of vertebrates as "too politically hot." Yet even as this renaissance is blooming, we are faced with the fact that there appears to be a basic illiteracy about the invertebrates among the majority of the current generation of biology graduates.

In the face of this pessimism, there is evidence that the pendulum is swinging back. Many new textbooks in invertebrate studies have been published in the past few years. Indeed, many colleagues expressed pleasure that a new edition of Professor Meglitsch's book would be available. So there are teachers still ready and able to try and stimulate students.

Despite the voluminous information content of all these books, so much remains to be discovered. Invertebrate zoology is not a dead field; major questions still must be answered. Textbook treatments of invertebrates often distinguish between "major" and "minor" phyla. However, in a very real sense there are no minor phyla, just phyla for which little is known. In addition, much of the information that we do have about invertebrates needs to be recast. Ad hoc explanations are no longer acceptable. We need to develop testable hypotheses about invertebrate function and relationships, and then turn to

Table 1.1. Estimates for some phyla of how many species are known versus how many there may actually be

Phylum	Estimated species known	Estimated species living
Mesozoa	85	500+
Ctenophora	80	130–500
Gnathostomulida	80	1000
Gastrotricha	450	1000+
Rotifera	1800	2500–3000
Kinorhyncha	125	500
Priapulida	10	15–25
Nemotoda	1,000,000	?
Chaetognatha	70	100
Nemertinea	800	3000+
Pogonophora	120	500
Sipuncula	320	330
Echiura	130	140
Tardigrada	550	1000+
Insecta	750,000	1,000,000–10,000,000
Crustacea	75,000	90,000
Phoronida	13	20
Brachiopoda	330	400–500
Hemichordata	100	150+

Derived and expanded from Strathman, 1983, *Paleobiology* 9:99.

the animals to check our assumptions. We are scientists, and as such we must obey certain rules implicit in the scientific method. Chief of these is that we must try to falsify our hypotheses about animals. We should not seek to gather mountains of information to prove our positions, because that is an impossible task. Proof is relative, confirmed only by changeable levels of probability. Disproof, when it occurs, is absolute. Animals can provide us with definitive answers to our questions, if we know how to ask them. The admonitions of Professor Meglitsch to use a textbook merely as a guide to the actual animals, not as a replacement for them, seem as valid today as when they were first written two decades ago. Textbooks like this are temporary, constantly changing tools to encourage study of animals, not a shrine to some misplaced faith in eternal verities.

To facilitate this, in this edition each phylum has been given its own chapter. And each chapter is organized along similar lines. There is a definition of the group (focusing on unique advanced or derived features) followed by an outline of the basic anatomical plan for that group. Important terms are in boldface in or near the text that defines their meaning. Then, interesting aspects of the basic biology of the phylum are reviewed, with emphasis placed on aspects of ecology, behavior, physiology, and development that relate to problems of survival in nature. In this regard, it has unfortunately been necessary to leave out a great deal of the total available information because of the constraints of space. In the larger phyla, a review of the individual constituent groups is presented, again with emphasis on basic body plan and biology of adaptation. Then, some comments are given about fossil history, and current consensus is advanced about relationships in the phylum, using cladistic analysis when available (see Chapter 2). Finally, a current taxonomy is presented. Because this book is only a guide and introduction, interested students should seek to amplify the information given here. To this end, selected references are supplied after every chapter, but these references are not meant to be exhaustive. The sources included will help students begin a more thorough pursuit.

Many new terms will be encountered in every chapter; however, every specialist requires a far more extensive vocabulary than is used here. Those who are studying invertebrates as a basis for future specialization will, of course, want to learn as much of the terminology as possible, with the idea of adding new terms later. Those who are studying invertebrates for general background or to establish a basis for specialty in another field should work with the vocabulary only long enough to understand the main ideas and adaptational trends.

The Evolutionary Viewpoint

Every scientist approaches his or her subject in the light of some basic paradigm (Kuhn, 1970). A paradigm is a theoretical superstructure that serves to hold a science together, to relate all the various subdisciplines, to explain past findings, and to point the way along which the most profitable lines of future work can be pursued. For biologists, that paradigm is Evolutionary Theory. Note, we did not use the term "Theory of Evolution." Evolution itself is *not* a theory; it is a phenomenon of nature! It can be simply defined as "descent with change." Furthermore, it is a phenomenon that can be verified easily in any one of a number of ways. We can look at our children, our parents, and grandparents and see for ourselves what descent with change has wrought through several generations. Or we can reflect on what changes have occurred in many lines of microbes in the face of continued attack by antibiotics, or insect pest lineages after decades of pesticide use. The immunities and resistances we see in these creatures are the result of descent with change. Or we can examine the fossil record and see actual changes that have occurred through long periods of time.

The zoologist is concerned in part with animal form, the characteristic structural plan through which an animal type can be recognized. However, form is not a study of structure alone. Structure is the result of a complex series of interactions that involve development and evolution. Development is determined both by topographic as well as genetic factors. One example of how these factors interact is given by the nature of gastrulation and mesoderm formation in echinoderms. The mesoderm is a cell layer formed by a topographic folding process identical to the folding that produced the gastrula. A two-dimensional surface is extended into three-dimensional space when folds and pockets define subsurfaces of the whole. This distinctive process is in large part constrained by the sticky hyaline layer that holds the cells in a sheet that can only be folded to achieve differentiation. Although genes control the presence of a mesoderm, the topologic relationships of the cells in that membrane govern just how that mesoderm comes to be expressed.

So form and, by extension, function are not merely just phenotypic manifestations of genes. Genes control some aspects of the structure and biochemical functions of organisms, but not all of these are necessarily the result of natural selection. Charles Darwin and Alfred Russel Wallace founded Evolutionary Theory based on the concept of change achieved through the agency of natural selection. They realized that organisms changed through time; the fossil record indicated that. Why did they change? Resources are limited in nature. Organisms vary. The variations most effective in garnering resources are those that survive. It was an a posteriori explanation of what happened in nature, formulated in a particular time, place, and sociocultural context. The outlook of their time predisposed Darwin and Wallace to view the changes as progressive and, therefore, as part of a continuum of unfolding improvement. To achieve that improvement, that increasing level of fitness, the agency of natural selection was conceived. Darwin drew parallels to the a priori approaches of artificial selection in barnyard and greenhouse hybridization.

Abundant examples in nature and the laboratory were developed in the succeeding century to prove the idea of evolution by natural selection, and much insight was gained in this process. Natural selection stimulated the foundation and growth of the science of genetics, it fostered the modernization of ecology, and it encouraged study of the fossil record. Its effectiveness as a paradigm cannot be disputed. All this proof, however, only set a level of probability for truth—but it did not really *prove* anything.

Recent advances in our understanding of Evolutionary Theory have allowed us to see that there is much in evolution that is the result of pure chance rather than just natural selection. Of the 31 phyla recognized in this book, some are larger than others and some are more diverse in their variant forms than others. Traditionally, these differences would have been explained on the basis of the larger and more diverse phyla being somehow more fit than the small, structurally confined phyla.

To be sure, selection can and has taken place. An element of fitness can be interjected into the process of trying to understand form and function. It is a mistake, however, to assume that everything about animals can or should be explained in this way. There may be an element of adaptational success that helps determine phylum size and diversity. Nevertheless, modern Evolutionary Theory tells us that we are not entitled to look for such a causal explanation of something in nature until we have eliminated the possibility that what we see cannot be explained more easily by chance alone. Is the crab body form, unique as it is, so abundant in nature because being a crab is somehow more fit than being a noncrab? Or is it that we have so many crabs in the sea because crabs happen to be in a lineage that actively generates new species? Success cannot be measured by complexity alone; part of what we commonly recognize as success may just be luck. That is, animals are the way they are; they are not necessarily more fit than some other possibility.

This is a difficult outlook to accept. Our training and culture condition us to think of struggle, success, progress, and survivorship in terms of a history of "nature red in tooth and claw." Indeed, Spencer's classic characterization of Darwin's and Wallace's concept of evolution by natural selection as "survival of the fittest" has formed a powerful constraint on the way we have come to think of descent with change. However, there is another way of looking at the variations we note in the form and relative numbers of animal phyla—one form is not necessarily better than another, merely different. In this context, the variety we see in the invertebrates is not due to a mechanism that always rewarded a winner. Rather, nature has a more poetic beauty. Things survived because they were adequate to the tasks at hand. Thus, we are free to admire how the "little" phyla, like the nemertines, echiurans, and sipunculans, handle the problems of living, and not be forced to rationalize why their "big sister," the molluscs, have hogged the limelight. There are no phyla out in nature, just a lot of individuals trying to garner enough to eat and stay alive until they can reproduce in some way.

To survive, certain things have to be done. Food has to be gathered and broken down to its constituent parts and then reassembled into usable forms. Oxygen is needed to derive energy from that food, and if oxygen is not available, then energy has to be derived in other ways. Waste products are generated in this process, and this waste has to be disposed of in some way. And in the end, when it comes time to reproduce, there are certain ways this has to be done, given how life on this planet is organized. In all of this, it may just be sufficient for animals to muddle through; they do not necessarily have to beat nature or some competitor into submission to leave progeny in the next generation.

Does all this make evolution a passive phenomenon? No, just an inevitable one. Why should we not think about evolution more aggressively? Because what is needed is some effort to free ourselves of our Victorian, mid-nineteenth-century way of looking at life, and inculcate in ourselves a kind of Zen. Nature is the way it is—so let us seek to understand what is there and how it works. We do not need to understand why so much as we need to affirm that it is, and in so doing, we affirm our own role in the scheme of things. Ours is not to dominate, because we are better or more advanced; ours is to recognize what is, and to affirm that it is good. "Beauty is truth, truth beauty,—that is all ye know on earth, and all ye need to know."

Use of the Library

The literature on invertebrates is far too extensive to permit the inclusion of an exhaustive bibliography in any one book. The references included with each chapter are merely meant to serve to introduce students to the available literature. Listings of review papers or monographs and recent important publications are provided toward this end.

Invertebrate zoology as a discipline, however, has benefited from monographic, often multivolume compilations that have served to summarize the intimidating mountain of available information. A good and stimulating course in invertebrate zoology needs two things: (1) access to abundant and diverse materials in the laboratory, the animals themselves; and (2) a basic library. A major part of the pleasure of learning about invertebrates is to be able to deal with the rich literature available.

The annotated listing provided below includes some of the important works with which every student in an invertebrate zoology course should be familiar. One thing you will note is that some of them are in foreign languages. For those who are able to read some of these languages, the available data base will be increased. For those not yet able to handle German or French, fear not. At the very least, the classic illustrations and reference lists of these treatises can help provide a modicum of information to augment the works in English.

Selected Readings

Beklemishev, W. N. 1969. *Principles of Comparative Anatomy of the Invertebrates.* Univ. Chicago Press, Chicago. [A two-volume translation from the Russian that presents a rather different way of looking at invertebrates not encountered in Western-based literature.]

Bronn, H. 1873—. *Klassen und Ordnungen des Thierreiches.* Friedlander & Sohn, Leipzig. [In German, but an absolutely invaluable source book. Even the older editions on various groups are classic reference works.]

Bullock, T. H., and G. A. Horridge. 1965. *Structure and Function of the Nervous System of Invertebrates.* Freeman, San Francisco. [Two volumes, with the last word on neuroanatomy and biology.]

Giese, A. C., and J. S. Pearse. 1974—. *Reproduction of Marine Invertebrates.* Academic Press, New York. [An edited ongoing multivolume series dealing with aspects of reproductive anatomy and biology, as well as development of various major groups and minor phyla.]

Grassé, P.-P. 1948—. *Traité de Zoologie.* Masson, Paris. [Many classic illustrations, although the text is not as comprehensive as other available works. Updated volumes are now beginning to appear.]

Hyman, L. H. 1940–1967. *The Invertebrates.* McGraw-Hill, New York. [An incomplete treatise, yet the six volumes published remain examples of the finest single-author survey of invertebrates of all time.]

Kaestner, A. 1967–1969. *Invertebrate Zoology.* Wiley-Interscience, New York. [A fine three-volume partial English translation of a more extensive German work. Deals with the lower phyla and the noninsect arthropods.]

Kuhn, T. S. 1970. *The Structure of Scientific Revolutions,* 2nd edit. Univ. Chicago Press, Chicago.

Kükenthal, W., and T. Krumbach. 1923–1928. *Handbuch der Zoologie.* de Gruyter, Berlin. [An excellent treatise, on a par with Hyman, and written in an easily comprehensible German.]

Lankester, E. R. 1900–1909. *A Treatise on Zoology.* Adam and Charles Black, London. [An old series, but contains individually authored volumes by experts of the time that became benchmark works in the field.]

May, R. M. 1988. How many species are there on earth? *Science* 241:1441–1450.

Mayr, E. 1963. *Animal Species and Evolution.* Belknap Press, Cambridge. [A classic overview of traditional evolutionary theory, the Synthetic Theory of Evolution.]

Milkman, R. 1982. *Perspective on Evolution.* Sinauer Assc., Sunderland, Mass. [A concise overview of current thinking on evolutionary theory, the Stochastic Theory of Evolution.]

Moore, R. C., et al. 1957—. *Treatise on Invertebrate Paleontology.* Geol. Soc. Amer. and Univ. Kansas Press, Lawrence. [An ongoing series, with addendum and revised volumes now appearing. Each volume contains one or more introductory chapters on basic anatomy, biology, and paleontology of that group. Invaluable for biologists as well as paleontologists.]

Stanley, S. M. 1979. *Macroevolution. Pattern and Process.* Freeman, San Francisco. [A paleontological perspective on current evolutionary theory.]

2

Classification

All of us are in the habit of referring to the animals and plants that we see about us with common names. We refer to creatures with names like "oysters," "garden snails," "American lobsters," "crayfish," and "jellyfish," and these names try to convey to a listener some idea of what is being talked about. Some names, like "American lobster," are relatively informative and make reference to a specific kind of lobster found in a particular place in the Atlantic Ocean. Other names, like "jellyfish," are quite uninformative, as that common name could be used to refer to a myriad of creatures that belong to two different phyla. Common names can be used in everyday conversations, but are nearly useless as scientific tools. A sophisticated physiological study, accurate as to technique and materials, would completely fail if, after specifying the exact chemicals and instruments used, it went on to say "the work was done on a snail." No one could ever confirm the study. Was the snail freshwater, marine, terrestrial? Was it herbivorous or carnivorous? Was it laboratory-reared or collected in nature? Where was it found? A physiologist must either provide an extremely detailed and accurate description of the material used, or use the scientific name. For example, if the physiologist says *Thais emarginata* was used, that name would immediately tell the informed reader that a specific kind of intertidal snail from the western coast of California was studied. Accurate scientific names are a crucial part of effective biological communication. Without them, the biologist is illiterate.

Nomenclature

Nomenclature is so important that the creation and use of scientific names is heavily regulated. Scientific names must mean something. Taxonomy, or Systematics, is the science of naming organisms, or classification. Scientists who are not systematists often belittle the work of taxonomists. Yet, if a particular scientific name meant nothing more than a common name like "June bug," a term applied to hundreds of beetles in North America, there would be no advantage to creating and using scientific names. Every name must stand for one kind of organism, must apply to that creature everywhere it is found, and must be described in a way that anyone can make reference to it with the certainty that a scientific standard has been applied in its creation.

When animals and plants were first named, the idea of evolution was neither clearly described nor universally held. Species were thought for the most part as being changeless, or fixed, their form having been determined at the time of their creation. The modern viewpoint is broader. We understand that there is considerable diversity of form within a species; and we know that species are not immutable, that they change through time.

Modern systematists describe species as populations that are characterized by a distinctive form and behavior, live within a fairly definite geographic area, and are reproductively isolated from related species. The species then is not an arbitrary construct; it exists as a natural entity. Recognizing such populations is not always easy. Observations in the field must be combined with morphologic studies in the museum or laboratory. Attention to morphology alone has sometimes led to incorrect definitions of species. Sometimes reproductively isolated species were placed within one taxon, or a single interbreeding group was split into two or more taxa. So even with the most careful taxonomic work, improving techniques of analysis, more accurate perception derived from field work, or the development of new methods of study sometimes results in changes in names. This does not mean that the old work was invalid or that systematics is not a science. Rather, such changes merely reflect the operation of the scientific method. Species names and descriptions are testable; that is, they are valid only until they are found to be wrong in some way. When that occurs, as scientists we must be free to amend, correct, or reject a particular taxonomic work.

NAMES

The basis of scientific classification is the **binomen**. The system began in 1758 when Karl Linné, a.k.a. Linneus, in his *Systema Natura*, 10th ed., designated animals with two names, a genus and a species. For example, *Homarus americanus* is the binomen of the American lobster. The first name, *Homarus*, is the genus. It is always capitalized and underlined or italicized. It is equivalent to a family or surname. The second name, *americanus*, is the species. It is lowercase and also underlined or italicized. It is equivalent to a Christian or given name. Sometimes one will see a third name after these two; an example might be *H. americanus americanus*. This is a subspecies or varietal name, and not all species have them.

You will also note that in the second use of the genus previously mentioned, *H.* was substituted for *Homarus*. This is an accepted and common way of abbreviating a genus after its first use in a scientific text. However, care must be taken in a scientific paper not to abbreviate in this way if another genus is used that results in the same abbreviation. For example, if a paper on *Homarus americanus* also included references to the hydroid, *Hydra oligactis*, use of the *H.* abbreviation for both genera would have to be used carefully or avoided altogether. Sometimes a worker, usually not an expert in a group, will not be able to specify the species, in which case the form *Homarus* sp. will be used, or, if several species are possible, *Homarus* spp.

Where such information is available, the first reference to a species should include the name of the person who first described it. *Plasmodium ovale* Stevens is the correct form of such a reference. It is almost impossible to be sure that Stevens is the only person who has given the name *ovale* to a species of *Plasmodium*. A later reader would not then be able to discern which species might be meant. Furthermore, the name of the author helps to locate the original description. Sometimes you will see the authors' names in parentheses. *Plasmodium vivax* (Grassi and Faletti) means that the species *vivax* was originally assigned by Grassi and Faletti to some genus other than *Plasmodium*, and was later reassigned to *Plasmodium*.

THE CODE

Species have been described in many languages and in journals all over the world. Keeping up with the names is an impossible task for one person. Taxonomists who work with one particular group of organisms are specialists in that field, and often these people are the ultimate arbiters of taxonomic disputes. Even so, mistakes can still be made. Several workers in different places can describe the same species under different names. Sometimes a name is unwittingly used to identify an entirely different species. A worker might use one name to describe what is actually more than one species, or variable species might be given more than one name. To assist in sorting out these problems certain practices and rules are used.

To govern the creation and use of scientific names, the Fifth International Zoological Congress, in 1901, adopted the *International Code for Zoological Nomenclature* (1985). The code establishes rules and policies that stabilized nomenclature. Responsibility for the code passed to the International Congress of Systematic and Evolutionary Biology in 1973, but is supervised on a continuing basis by the International Commission on Zoological Nomenclature, currently headquartered in the British Museum (Natural History). The objectives of the code are to ensure that no two species share the same name, no one species has more than one name, and only valid names are permanently applied to a species.

If two species in the same genus come to have the same name accidently, those names are called **homonyms**. In that case, the first species given the name has priority and retains it. The second species must be given a new name. Not all names arise from inadequate literature searches. Sometimes two genera thought to be distinct prove to be the same. Thus, when all or parts of genera are combined, homonyms may be created.

Sometimes a species inadvertently gets described more than once, with the result that the species comes to have more than one name. In this case, the rule of priority applies. The first adequate description takes priority, and the name used in that description is the **valid name**. All other names are termed **synonyms**. The blood flukes were known for a long time by the generic name *Schistosoma*, but a name change due to priority has resulted in their now being known as *Bilharzia*. Sometimes very common animals used in dozens or hundreds of studies turn out to have been described long ago by some obscure worker in an obscure journal. The valid name, unknown to most zoologists, would cause difficult adjustments if forced to be used. Confusion would be increased instead of reduced. In such a case, the International Commission can exercise plenary powers to suspend the code and preserve the more widely familiar name.

It is not always easy to be sure that the population of animals will always be associated with the same name. Abundant, frequently encountered species have little problem in this regard. Rare species, however, especially in little-studied groups, may not be seen for decades after their description. Many species have never been collected after they were first described, or at least never mentioned in a second publication. If one collects something resembling a rare species, how can one be sure that the material in question is in fact that species? Word descriptions and illustrations of specimens are subject to different interpretations. This problem is solved in the code by insisting on the designation of a **holotype**.

When a species is first described, one specimen must be deposited in a museum repository so that it will be available for later comparison. The holotype specimen thus becomes the bearer of the name. The type is the international standard for what that species is. Several categories of types are recognized, but what these are is not important here. Similarly, the locality from which the species is originally collected is the type locality. The first species to be assigned to a genus is the type species. If a genus or species is later to be divided into parts or merged with another preexisting genus or species, the type specimen or type species determines which parts shall retain the original name.

There are standards that mark valid descriptions. Names used in a letter, newsletter, or newspaper article are not considered valid. Adequate publication requires that the description be in any of several languages, use reasonably permanent paper, and be offered for general sale. The article must clearly designate a holotype specimen and its repository, and contain a reasonable illustration(s) of that material. Not all of these rules have been in effect since the time of Linneus, and thus there is some confusion possible when dealing with the older literature. Rules, however, exist for handling these problems when they arise, and if the code fails for some reason to solve a problem, the International Commission is the ultimate judicial authority in these matters.

HIGHER TAXA

The code governs the creation of genus and species names. However, other taxonomic categories exist and indeed are the ones that will be most often encountered in this book: kingdom, phylum, class, order, and family. Many groups, especially those that are very diverse, for example, crustaceans, have supercategories and subcategories of this hierarchy. Except for the species, all higher categories are quite artificial. They are convenient devices to help sort species, and these groups are commonly used in discussions of animal relationships. These categories help taxonomists to catalogue, which is the first step in systematizing.

Even the genus is subject to this uncertainty. Because the exact course of evolution is seldom known in detail, the genus may or may not include species that are in fact closely related. The best that can be said is that species placed in the same genus appear to be more closely related to each other than they are to anything else.

Families include similar genera. Family names are governed by the code only insofar as the code regulates generic names, as a family name is usually based on a type genus. Family names are easily recognized, as in animal nomenclature they must end in *idae*. Many families are easily recognized and indeed are common vernacular concepts, for example, the dog and cat families, Canidae and Felidae.

Orders include similar families. They are typically based on stable, very general characters. Many orders correspond to common names; for example, beetles are an order of insects, the Coleoptera, and termites are members of the order Isoptera. Differences between orders are usually so striking, one hardly needs to be a zoologist to recognize them.

Classes are still higher groups. The traits that unite orders in a class are typically few in number and quite generalized. All crabs, lobsters, and shrimps belong to the crustacean class Malacostraca; that is, they have a thorax of eight segments, an abdomen of six, with the last pair of limbs combining with the terminal telson to form a tailfan, and with the gonopores on specific segments in the thorax. In many cases, an untrained observer might have difficulty in recognizing all members of a class. For example, one might be surprised to learn that lobsters are Malacostraca as are garden pill bugs. Another example, class Gastropoda includes the snails, and everyone knows what a snail is. Gastropoda, however, also includes the not very snail-like slugs and nudibranchs. Or the similarities that place corals and sea anemones in the class Anthozoa are not obvious unless one is familiar with the anatomical details.

Phyla are the next higher of the major subdivisions. Each phylum represents a distinct line of animal evolution distinguished by fundamental, but sometimes very subtle, features. Each phylum has a basic body plan or structural type. Sometimes phyla can be clustered into groups that are called branches, or grades, or divisions, or whatever. These clusterings may or may not have any phylogenetic validity, but may be treated simply as convenient handles. Indeed, some of these categories get vernacular names and are frequently used in general discussions, for example, terms like coelomates, protostomes, and lophophorates. Some of these categories overlap, like coelomate, gastroneuralian, and protostome.

This system of categorizing by hierarchy is remarkably efficient. If there were just ten phyla, and each had ten classes, and each of these had ten orders, and so on, we could categorize one million species and trace their relationships with just six names for each. Actually, some phyla are small, others are huge, but the fact remains that the system of classification based on binomial nomenclature is a very effective tool that enables biologists to collate information easily and efficiently.

Cladistics

Nomenclature is only one aspect of classification, and some have said it is more art than science. Indeed, there is a considerable element that is subjective and intuitive in the process of classifying. There is, however, another aspect to taxonomy that attempts to be more objective and scientific. This part of systematics frames taxonomic statements in terms of testable hypotheses that we attempt to falsify, and that we are free to accept, reject, or modify in an unceasing search to arrive at a more accurate description of nature. This approach has various names such as **phylogenetic systematics**, **Hennigian systematics** (after its first formulator, Willi Hennig), **cladism**, or **cladistics** (Hennig, 1966; Wiley, 1981). This last term will be used here.

CHARACTERS

Cladistics is a method to analyze characters objectively. In this process we seek to distinguish between two types of characters: those that are **apomorphic**, that is to say, **advanced** or **derived**; and those that are **plesiomorphic**, those that are **primitive**. (The student can keep these Greek-derived terms straight by remembering that the first letters of the Anglo-Saxon equivalents are the same—*a*pomorphies are *a*dvanced and *p*lesiomorphies are *p*rimitive.)

These characters have uses in determining relationships. A unique advanced character is an **autapomorphy** and is central to defining a group as distinct. For example, although the radula is an autapomorphy of the Mollusca, it does not serve to define the relationship of Mollusca to any other phylum. A shared advanced character is a **synapomorphy** and is crucial in linking groups. For example, the trochophore larva of Mollusca is a synapomorphy that serves to link Mollusca with other phyla that have such a larva, such as the Annelida. A shared primitive character is a **symplesiomorphy**. It is not useful in unambiguously defining groups or determining relationships. However, what is plesiomorphic at one level must be apomorphic at another level. For example, coelomates have a coelom, a lined cavity in mesoderm tissues. This is a primitive feature and one not particularly characteristic of annelids and molluscs, or effective in defining their position to the other coelomate phyla that also share this feature, such as arthropods and chordates. However, at the kingdom or superphylum level, a coelom is an advanced character that serves to separate those phyla with such a cavity from those phyla that lack such a cavity, such as the pseudocoelomate or acoelomate phyla.

Groups that are united by their sharing one or more advanced characters are said to be **sister groups**. (There is no particular significance to the gender of the noun; they might just as well have been called brother groups.) All classifications should be based on unfolding dichotomies of sister groups, each group sharing one or more advanced features. The more such **congruent features** are shared, the more probable the link is viewed to be as something actually reflecting events in the history of that group. Sister groups can be recognized at any level: species, order, phylum, whatever.

The concept of synapomorphies is closely linked with another concept, that of **homology**. Shared advanced features are homologues, that is, they share identical evolutionary origins and developmental sequences and are located in similar positions of the body. Homologous characters must be carefully distinguished from features that are **homoplasies**, characters that are similar to features found in nonsister groups but that are really convergences or parallelisms. **Convergences** are similarities that develop in lines that lack any close common ancestry. The development of a calcareous shell in Mollusca and Brachiopoda is said to be convergent, as these two phyla share only the most distant of ancestries. **Parallelisms** are similarities that develop in lineages that do share some common ancestry and thus have a common genetic or developmental potential for expressing similar features. The development of Malpighian tubules in insects and some arachnids is a parallelism because both the Uniramia and the Cheliceriformes apparently share some common ancestor. Sometimes it is difficult to decide whether characters are homologous or homoplastic.

GOAL OF CLADISTICS

The purpose of cladistic analyses and of the classifications that result from such studies is to identify groups that are **monophyletic** (Fig. 2.1A), that is, groups that include all descendants of a common ancestor, a form that possessed the derived feature(s) by itself. Thus, Crustacea is a monophyletic group,

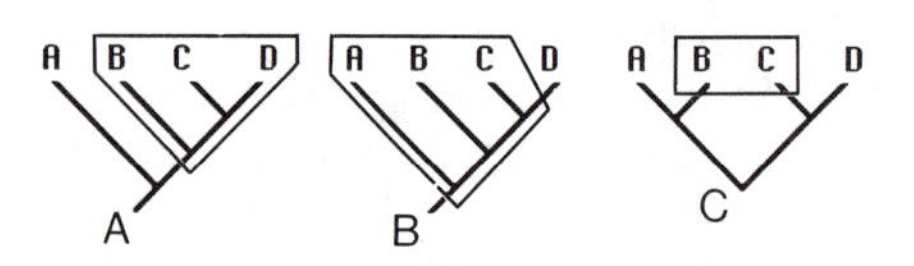

FIGURE 2.1. Types of groups encountered in phylogenetic studies. **A.** Monophyletic; taxa B, C, and D descended from one ancestor, all descendents of that ancestor within taxon BCD. **B.** Paraphyletic; taxa A, B, and C descended from one ancestor, but all descendents of that ancestor are included in two groups, taxon ABC and taxon D. **C.** Polyphyletic; taxon BC contains members descended from two different ancestors.

as all crustaceans are hypothesized to be derived from a common ancestor that had a series of unique features including two sets of antennae, gnathobasic jaws, and nauplius larvae. Monophyletic groups are to be distinguished from two other types of groups. A **paraphyletic** group (Fig. 2.1B) is one that contains only some descendents of a common ancestor. In the Platyhelminthes (see Fig. 10.28), the old taxonomic names Turbellaria and Trematoda designate groups that are paraphyletic, as not all the descendents of the ancestors are included in those taxa; that is to say, the turbellarian ancestor also gave rise to trematodes and cestodes, and the trematode ancestor also gave rise to the Cestoda. A **polyphyletic** group (Fig. 2.1C) is one that is descended from two or more ancestors. Early in the last century, the Mollusca also contained brachiopods, and together these were sometimes referred to as the Molluscoidea. That arrangement is now viewed as polyphyletic.

CHARACTER POLARITY

To arrive at a point where one has identified monophyletic groups defined by many synapomorphies with the fewest number of homoplasies held with other groups entails a detailed assessment of each character. To decide which expression of the character is primitive and which is advanced is to determine **character polarity**, that is, the direction of evolutionary change of that character. This is a very important yet often difficult step in cladistic analyses. Several approaches can be taken to polarize a character.

The most important and powerful is **outgroup analysis** (Watrous and Wheeler, 1981). Characters are advanced if they do not appear in a related group, and are primitive if they do. For example, if one is trying to determine if a segmented cuticle is advanced or primitive within the clade of annelids and arthropods, and you are comparing these two groups to the molluscs, one would conclude that a segmented cuticle is an advanced or synapomorphic feature in these two groups as molluscs do not share the character. On the other hand, if you are trying to decide whether a schizocoel in the annelid–arthropod clade is advanced or primitive as compared with molluscs, we would conclude that that character is primitive or symplesiomorphic as it is shared with molluscs. The use of multiple outgroups in this kind of analysis increases the chances of accuracy of the polarization (Maddison et al., 1984).

Outgroup analysis is not the only way to polarize characters. Other, less powerful sources of evidence can be used. **Ingroup analysis** looks at the distribution of a feature within a group. The idea here is that a rarely found character state is likely to be advanced, a common one primitive. However, this is not always so. For example, very few crustaceans actually have a free nauplius larva, yet outgroup analysis tells us that such a larva is an apomorphy of Crustacea. **Ontogenetic sequencing** is another source of evidence. Early appearance of a feature in an embryologic series often reflects an early appearance in the phylogeny of the group. Thus, the presence of an egg–nauplius stage in all crustaceans that lack a free larva helps to polarize the naupliar feature. Of course, paedomorphic processes (Gould, 1977), such as progenesis, where gonads develop in larva-like animals, or neoteny, where larval features are continued on into an adult stage, result in altered ontogenetic sequences. **Fossil history** is yet another source of information. The idea is that early appearance in a fossil sequence bespeaks a primitive feature, late appearance is advanced. However, the degree of completeness of the fossil record varies between groups, and one of the first lessons the junior author had in his first paleontology class was that "things are always older than you think they are."

To guard against the pitfalls inherent in the analysis, as many lines of evidence as possible should be used, and as many characters as possible should be used to sort the taxa in question. The latter is most important. If you are trying to sort out relationships between ten taxa and are only using seven characters, the appearance of convergences is unavoidable. To sort *n* taxa, one should have at least n + 1 characters. In this example, ten taxa would need at least 11 characters, and the more the better. The more characters that verify each sister group, termed **character congruence**, the more highly corroborated is the phylogenetic hypothesis.

CLADOGRAMS

A few characters for a handful of taxa can be easily evaluated with paper and pencil. However, large arrays of characters for many taxa are not likely to be analyzed accurately this manual way. Data processing must be used. The character states in each taxon are arranged tabularly in a matrix (e.g., Table 38.1) and fed into a computer for analysis. The goal of any such analysis is to produce a cladogram, that is, a tree of genealogical relationships. Ideally, this cladogram should exhibit **parsimony**; that is, no matter what the size of the matrix, the computer program produces trees with the fewest number of branchings, with the least amount of parallelism or convergence (homoplasy), and with the greatest degree of character congruence.

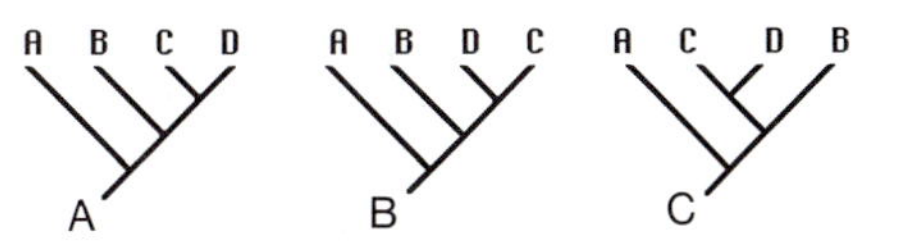

FIGURE 2.2. Three cladograms with identical information content. A taxon is not "better" or "more advanced" than its sister, only different.

One interesting feature of cladograms is that they can be rearranged in different ways and still portray the same information. In Figure 2.2 cladograms A, B, and C are identical, as sister groups can be rotated about nodes with no effect on information content. Any problems with interpreting these three cladograms resides in the observer. Cladogram A *does not* say that taxon D or taxon CD is the most highly evolved. To say that would be to succumb to the

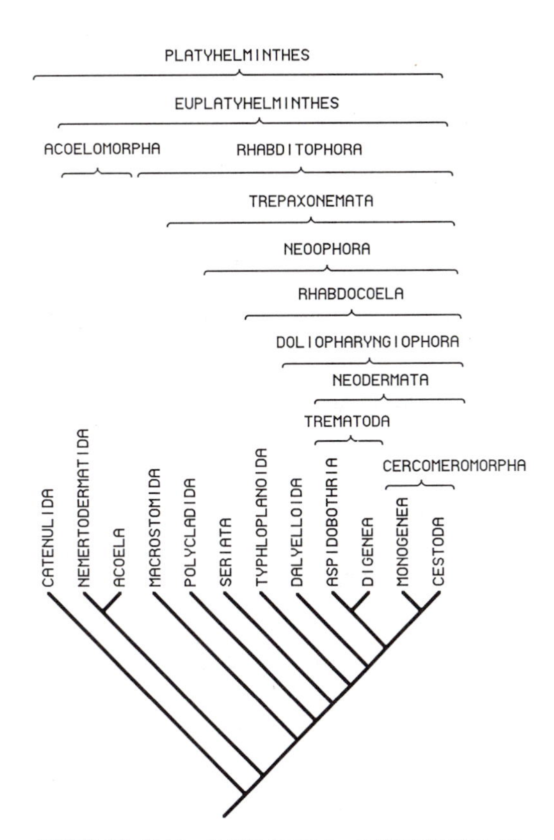

FIGURE 2.3. Strictly cladistic taxonomy of Platyhelminthes proposed by Ehlers. Consult Figure 10.28 for derived characters, and taxonomy at end of Chapter 10 for a more parsimonious approach to the taxonomy of this cladogram.

same fallacy that the Victorian naturalists did (see Chapter 1), namely, that evolution is progressive. In fact, all we can say about taxon CD is that it is the sister group of taxon B, and that these two groups are more closely related to each other than either is to taxon A.

Once a cladogram is obtained, a classification can be derived from it. Pure cladists would apply a taxonomic name to each and every branching in the cladogram (Fig. 2.3). This is not necessary, however, if one can be satisfied with identifying the basic structural plans of the group being analyzed and giving these higher taxonomic status (compare Fig. 2.3 with Fig. 10.28 and the taxonomy of flatworms at the end of Chapter 10).

Other Methods of Analysis

Although cladistics has developed into one of the more effective methods to analyze animal relationships, there are other approaches. **Phenetics**, or **numerical taxonomy**, seeks to measure as many characters as possible and then analyze them numerically to produce a phenogram or tree in which organisms are grouped by overall similarity. The underlying assumption of the phenetic school is that one can never be sure which of several phylogenies is correct; therefore, overall similarity is as good a criterion as any upon which to sort (Sneath and Sokal, 1973). Pure numerical taxonomists do not assess character polarity, or weigh characters for importance. They analyze their data by computer with programs that produce similarity/difference indices. The implicit hope in this approach is that a large enough sample of characters will begin to resemble the actual genomes of the creatures being analyzed.

One strength of the method is that it does present a way to handle strictly numerical data bases such as protein chemistry, cytochrome c differences, and RNA and DNA pairings. Unlike cladistics, however, it does not have a way to take into account convergent and parallel evolution. Proteins, cytochromes, and nucleotides, just like morphologic features, can be homoplastic.

Taxa, however, are grouped in a classification because they share a common ancestor, not because they look alike. This is the fundamental assumption of evolution. Two sisters, although they may not look alike, are more closely related than two cousins who may happen to look like identical twins. A sea cucumber, or holothurian, is more closely related to a sea lily, a crinoid, than it is to a peanut worm, a sipunculid. The holothurian and sipunculid may look alike—big fleshbags with retractable tentacles around the mouth. The holothurian, however, shares with the crinoid a common ancestry as members of the phylum Echinodermata.

Despite its deficiencies, phenetics has made some important contributions to systematics. It has stimulated a more quantitative approach to data collection. This can be good when it moves us away from purely subjective analyses of characters. It also has led the way in the use of computers in analyzing data, which has allowed larger data sets to be used than was heretofore possible.

The other approach to analyzing animal relationships is that of **evolutionary systematics** (see Ross, 1974). This is an unfortunate term in that it implies that other schools are not founded on evolutionary theory. Perhaps this should be referred to as **traditional taxonomy**. Evolutionary systematists point out that the character gaps between taxa vary, and that these differences are often linked to differences in niche occupation. Thus, this school attempts to qualify rates of evolution and will often engage in **character weighting**.

Traditional taxonomy considers all aspects of the biology of taxa when producing a classification, such as ecology, physiology, geographic distribution, in addition to structure. The result is that large character shifts or niche are reflected in classifications. For example, the traditional taxonomists in the eighteenth century were responsible for segregating amniote vertebrates into the reptiles, mammals, and birds. We now realize, however, that this scheme does not reflect the actual genealogies of the animals. Birds are now understood as effectively the last survivors of the dinosaur clade and, as such, are closely related to crocodiles and pterosaurs in a taxon Archosauria.

Cladists can point to specific characters in fossil and living forms as evidence to disprove the suggestion that Aves (birds) is a separate class. Traditional taxonomists cannot disprove the hypothesis that birds are more closely related to dinosaurs than to anything else, but only point to their intuition as justification for their position. As a result, pure traditional taxonomy is not particularly scientific. It does not set up hypotheses that can be disproved by anything other than the subjective opinion of another traditional taxonomist. Their idea of weighting characters cannot be defined by rules or procedures that would allow independent replication.

Traditional taxonomy has made valuable contributions toward understanding the history of life. The idea of character weighting can be used in converting a cladogram into a reasonable classification. Not every branch in a cladogram deserves a name, so the location of "important" or "major" advances on a tree can indicate where breaks between higher taxa should occur. In addition, the development of scenarios from the phylogenetic tree, can give cognizance to weighted features.

What has emerged from the debate between these three approaches to systematics has been a synthesis. The rigor and testability of cladistics has been tempered in the actual derivaton of a classification from a cladogram by allowing character weightng to be acknowledged. The use of computer techniques has allowed the use of larger data sets than would have been otherwise possible. The most productive approach to the problems of taxonomy would seem to be ones grounded in the judicious use of cladistic methods.

As will be discussed in Chapter 38, much of what has been expressed in the old literature about invertebrate evolution has lacked rigor and been quite subjective. Little attempt was made until recently to distinguish between primitive and advanced characters, to use as many characters as possible, or to produce the shortest phylogenetic trees possible with the most number of characters verifying each branch on the tree. The phylogenetic discussions in each chapter and at the end of this book have made attempts at doing that. However, discerning the history of invertebrates is still an important and interesting area of analysis. Much remains to be discovered about all the animal phyla, and as this information becomes available, it will become "grist for the mill" of continuing phylogenetic analyses.

Cladistics does not produce history writ in stone (Schram, 1983). It is merely one of many tools of analysis in the science of systematics.

Selected Readings

Gould, S. J. 1977. *Ontogeny and Phylogeny.* Harvard Univ. Press, Cambridge. [A scholarly but eminently readable survey of developmental factors in phylogenetic analysis, with important considerations of recapitulation, neoteny, and progenesis.]

Hennig, W. 1966. *Phylogenetic Systematics.* Univ. Illinois Press, Urbana. [A translation of original German text, it remains an effective introduction to cladistics even though the method has changed from Hennig's original concepts.]

International Code of Zoological Nomenclature, 3rd ed. 1985. International Trust for Zool. Nomenclature in assoc. with Br. Museum (Nat. Hist.) and Univ. California Press, Berkeley.

Maddison, W. P., M. J. Donoghue, and D. C. Maddison. 1984. Outgroup analysis and parsimony. *Syst. Zool.* 33:83–103.

Ross, M. H. 1974. *Biological Systematics.* Addison-Wesley Publ., Reading. [A fine practical introduction to evolutonary systematics, and more effective for the student than more famous but more ponderously theoretical volumes.]

Schram, F. R. 1983. Methods and madness in phylogeny. *Crust. Issues* 1:331–50.

Sneath, P. H. A., and R. R. Sokal. 1973. *Numerical Taxonomy. The Principles and Practice of Numerical Classification.* Freeman, San Francisco. [The latest compendium on the phenetic approach.]

Watrous, L. E., and Q. D. Wheeler. 1981. The outgroup comparison method of character analysis. *Syst. Zool.* 30:1–11.

Wiley, E. O. 1981. *Phylogenetics: The Theory and Practise of Phylogenetic Systematics.* John Wiley & Sons, New York. [An excellent beginning for the student, geared strongly toward practical application and consequently more interesting than the numerous more theoretical books available.]

3

Protozoa

Living things can be organized for conceptual purposes into five kingdoms (Whittaker, 1969). **Monera** are prokaryotic organisms, that is, have no organized nucleus and include bacteria, blue-green algae, and similar creatures. **Protista** are essentially unicellular eukaryotic organisms, that is, with an organized nucleus but not developed beyond a cellular level of organization and include protozoa, protophytans, and protofungi. **Plantae** are multicellular organisms that have cell walls of cellulose and possess photo-autotrophic modes of nutrition. **Fungi** are multicellular organisms that have cell walls of chitin and use osmotropic heterotrophy to feed, that is, they derive nourishment by means of chemical breakdown of materials. **Animalia** are multicellular organisms that do not have cell walls and generally feed by means of phagotropic or pinocytotic heterotrophy, that is, they generally consume other organisms or take in dissolved nutrients. The last three of these kingdoms are all eukaryotic, multicellular, and generally organized into tissues, that is, specialized associations of cells within a body to achieve some specific function. The species in the last of these kingdoms are, of course, the subject of this book.

Although the multicellular kingdoms are viewed as natural clades, that is, monophyletic lineages (see Chapter 2), for the most part the two "lower" kingdoms, the Monera and Protista, are better viewed as grades of evolution. Older textbooks separated a phylum Protozoa as an ancestral stock for higher animals. However, there has never been agreement as to what groups of unicellular organisms should be placed within such a phylum, nor which protozoan protists are the closest to the root stock of the metazoans (see Chapter 4). No unambiguous, unique, derived features define the Protista, that is, they are not defined as a whole by autapomorphic characters. If we cannot produce such a description at the kingdom level, there should be little wonder that there has not been agreement on the status of protozoan protists.

In this text, only a summary discussion of the animallike (that is, phagotropic) protist body plan will be presented along with an overview of protozoan classification. Invertebrate zoology is a discipline that deals with nonvertebrate multicellular animals; if further information on protists is desired, students should consult textbooks on protozoology (see Selected Readings list at the end of this chapter).

The concept of grades of evolution is an important one. Monera are specialized at the subcellular or biochemical level of organization. Higher organisms (plants, fungi, and animals) are specialized at the tissue level of organization. Between these two levels are the Protista who are specialized at the cellular level. It is not proper to label them "unicellular," "acellular," or "single-celled" forms. There are single-celled Protista, that is, creatures with one cell containing one nucleus. Protists are also encountered with many nuclei (**syncytia**) or with many cells (**colonies**). Ordinarily we think of a cell as an undivided protoplasmic mass containing a single nucleus and surrounded by a cell membrane. *Paramecium* is a single-celled ciliate, but it has a macronucleus and one or more micronuclei, whereas the amoeba *Arcella* contains two similar nuclei. At one stage, a malarial parasite has up to two dozen nuclei, and a large myxosporidian cyst may contain thousands of nuclei. Somewhere between the many uninucleate protists and the large multinucleate forms, the resemblance to a classic cell concept ends. It is important that we do not interpret "one-celled" naively, or overlook the importance of protists in the development of cell structure and function. Mitochondria, microsomes, Golgi material, endoplasmic reticulum, nuclei, and centrosomes are cellular components that first appeared in the course of evolution in organisms at the protistan grade of organization.

Are "protozoa" animals? Certainly there are protists that exhibit animallike modes of nutrition, that is, phagocytic heterotrophy. Their structural plan, however, is fundamentally different from that of "animals" (Metazoa) sensu stricto. In protozoa, cell organelles achieve specific ends, rather than specialized cells or groups of cells. A number of Protista form colonies consisting of a few to several thousand members. Differentiated vegetative and reproductive members are found in the more highly organized colonies. In such organisms, Protista approach the multicellular organization of Animalia and Plantae, and the large, highly integrated colonies look and behave very much as though they were multicellular organisms. When does an integrated group of individual protists emerge as a multicellular creature? A line must be drawn, and an arbitrary one has been selected. If more than one kind of nonreproducing cell type is found, the creature belongs to the metazoans; if not, it is a protist. Thus, a *Volvox* colony with several thousand members is protistan because it has only one kind of nonreproducing member, whereas a mesozoan with far fewer cells is an animal because there are several kinds of cells that do not enter into the reproductive process.

The protistan level of organization has proved extremely successful, adapting to permit invasion of many different habitats. Protista occur in soil, in all kinds of freshwater and marine habitats, and in all climatic regions, wherever temperatures rise above the freezing point. The modification of a successful organizational plan to produce organisms capable of occupying many kinds of habitats and assuming many life-styles is adaptive radiation, an evolutionary phenomenon seen in all groups of organisms. As an object lesson of just what adaptive radiation can achieve, we consider in this chapter the ways in which the protistan body plan is used by its "protozoan" members to achieve their ends.

Fundamentals of Form and Function

The best approach to the protozoa is to establish a basic understanding of how they are built and how they function, and to build enough of a vocabulary to understand the fundamentals of classification. After that, one can examine functional and morphological problems in greater detail and move to an understanding of their phylogenetic relationships. Even the simplest animals require division of labor among specialized parts that must interact to form a harmonious system serving all the necessary functions. The nature and function of one part, therefore, tends to influence adaptation of other parts, just as the interplay between form and function in the organism helps to determine its course of evolution as well as its ecological relationships. For example, food acquisition determines an organism's trophic niche and defines much of its relationship to its environment.

FEEDING AND DIGESTION

Protozoa acquire food in a variety of ways that lead to one of several distinctive adaptational lines. One can distinguish between two general kinds of feeding: hunting, which involves seeking and capturing prey; and trapping or suspension feeding, which is often used by sessile organisms. The two modes are not sharply separated, but at the extremes of both adaptations, marked differences are observed.

It is difficult to generalize about the protozoan hunter, for it may move by flagella, cilia, or pseudopodia, and may either have a definite mouth or engulf food at any point on the cell surface. An amoeba can capture the ciliate *Paramecium*, which moves much more rapidly than itself, not by pursuit but by protoplasmic engulfment of its unwary prey, such that a few hungry *Pelomyxa* in a culture that contains *Paramecium* are surprisingly successful at gathering meals. On the other hand, the ciliate *Didinium* actively pursues and captures *Paramecium*, and provides a spectacular display of hunting effectiveness. The best one can say is that the hunter is actively motile, moving about to sample its environment, and that it is either equipped with a large, often distensible **cytostome**, or cell mouth, or can form engulfing pseudopodia.

Suspension feeding and hunting become indistinguishable in motile ciliates that possess extensive buccal ciliatures that sweep particles into the mouth. Peritrichous ciliates use a buccal ciliature adapted to create effective currents that bring food into the buccal area. The sessile choanoflagellates also show an interesting adaptation to this type of feeding in that their flagella sweep food organisms against the collar that is then ingested by the cytoplasm at its base.

Some of the sessile, floating, and slow-moving protozoa have become adapted to trapping. The trapper increases its catch potential by sweeping an area greater than the effective cell surface. The tentacles of suctorians and radiating pseudopods (axopodia) of heliozoans and radiolarians, as well as the long, branching, and sometimes anastomosing pseudopods (filopodia and rhizopodia) of foraminiferans and some arcellinids, accomplish this purpose. Once contact with the prey is made, the trapper must subdue and ingest it. Toxic substances to stun or kill prey occur, for example, in the Suctoria, Heliozoa, and Radiolaria. The ephelote Suctoria have sharp, raptorial tentacles that can capture prey and separate suctorial tentacles for the actual ingestion. When protozoa capture large prey organisms, several tentacles or pseudopodia may cooperate in their ingestion, and digestion occurs in the endoplasm; but where prey organisms are small, digestion may happen within the pseudopodia. In some cases filtering adaptations occur in protozoa; for example, some Sarcodina use protoplasmic nets formed by anastomosing rhizopodia to serve as seines for trapping small organisms.

Once food is ingested, all holozoic protozoa use **food vacuoles**, the characteristic organelle for intracellular digestion. In many organisms, digestion is a highly organized process, involving the sequential use of enzymes that act best at different pH values. First, the vacuolar pH is usually neutral to slightly alkaline. Within the first ten minutes or so, the pH begins to fall. Maximal acidity reported by most investigators ranges between pH 3.0 and 4.5. From this point, pH rises to neutral or slightly alkaline and often falls again by the time of waste residue egestion. The range of foods that can be used is, of course, determined by the suite of enzymes produced. The ability to use protein is widespread, whereas the ability to use carbohydrates varies and is more limited.

Egestion is a simple matter in naked protozoa. Amoebae generally accumulate exhausted food vacuoles and eventually leave the contents behind as they move about. Where there is a definite cell membrane covering, some exact point for egestion is required, the cell anus or **cytoproct**. In a few ciliates, a definite canal leads to a permanent opening, the **cytopyge**.

RESPIRATION

The small size of protozoa makes respiration easy. Respiratory exchange occurs at the body surface by diffusion, no doubt facilitated by currents created by cilia and flagella.

EXCRETION AND OSMOREGULATION

Disposal of excretory wastes occurs at the body surface. On the basis of our present information, protozoa appear to be ammonotelic; that is, they excrete most of their nitrogen as ammonia, as do most aquatic invertebrates. *Paramecium aurelia* has been shown to liberate ammonia, presumably as a result of amino acid deamination, but not from purine metabolism. Purines are probably excreted as hypoxanthine, with some guanine and adenine. Early investigators often indicated some excretion of urea and uric acid; these results seem questionable in the light of recent work. Certainly, if they are formed, they are not a very important part of protozoan excretion. It should be noted that brownish inclusions in some radiolarians and foraminiferans, and various crystals seen in amoebae and other protozoa, may be storage excretion wastes, that is, the waste is converted to nontoxic substances and stored for a time. The stored wastes may be eventually expelled, either through the cell anus or by other methods used for voiding food vacuoles.

Soluble waste materials simply are released from the body surface without the intervention of special excretory organelles. Some soluble wastes may leave with the water expelled by the contractile vacuole system, but there is no evidence that the contractile vacuoles have any specific excretory function.

In many metazoan organisms, excretion and osmoregulation are closely related. However, the primary function of the contractile vacuole (Fig. 3.1) is to expel water taken in from a hypotonic environment. Contractile vacuoles are missing or poorly developed in protozoa living in isotonic marine or parasitic environments, but are well developed in protozoa from freshwater habitats. The contractile vacuole system may be simple or quite complex. In simple contractile vacuoles, two or more small accessory vacuoles unite or empty into a larger vacuole, which discharges after full dilation has occurred. In some ciliates, a system of canals discharges into the contractile vacuole (Fig. 3.1). In full dilation these canals are inconspicuous. As the main vacuole discharges, the outer regions of the canals, the ampullae, dilate with incoming fluid. The ampullar contents are injected into the forming contractile vacuole through injection canals causing the vacuole to dilate again.

INTERNAL TRANSPORT

Even the largest protozoa find internal transport no problem. Cytoplasmic streaming distributes materials throughout the continuous protoplasmic mass of the organism. This cyclosis follows no set pattern in most protozoa but in ciliates tends to become established as a definite cyclical movement of the endoplasm. The mechanics of cyclosis have yet to be satisfactorily explained. In ciliates it may be due to food vacuole movements along microtubules extending from the cytostome.

PROTECTIVE DEVICES

Life is hazardous, even for small creatures. Every organism faces so many threats that special protective devices have evolved. It is evident that much of protozoan behavior results in avoidance of toxic environmental conditions, and special protective structures are common. Most of these are useful in preventing mechanical injury or in providing protection against predators, drying, and excessive water intake.

Some protozoa, in particular ciliates, use active defense devices. It has been suggested that some **trichocysts** are organelles just beneath the pellicle that are discharged either to stun the prey or attacker or to ensnare and entangle it (see Fig. 3.15A,B).

Surface envelopes provide protection against mechanical injury and drying, as well as against some toxic chemicals. In a sense, the **plasmalemma** is the universal protective envelope, maintaining the integrity of the protoplasmic mass. A **pellicle** is an outside sheathing. Some Sarcodina, many flagellates, and all ciliates have a pellicle. Where pellicles are thin, considerable flexibility remains; where they become thick, the body form remains constant. A pellicle is usually pierced by openings for ingestion, egestion, and water expulsion.

Unfortunately, there has been no consistent terminology for coverings. Many sessile and some free-swimming protozoa inhabit a loose-fitting container called a **lorica**. Some protozoa, like *Difflugia*, build a **test** of sand grains or other foreign particles cemented together, and still others, like Foraminifera, secrete a heavy shell of silica or calcium carbonate. The terms lorica, shell, test, and case are often used synonymously.

Resistant coverings are sometimes formed for brief parts of a life cycle. This is especially true of parasites, which usually pass from one host to another as cysts or spores, covered by a resistant membrane that protects them while out of the host. Many free-living protozoa also encyst to survive unfavorable environmental conditions and to be dispersed by air or water currents in a protected stage.

MECHANISMS OF RESPONSE

Protozoa are sensitive to many kinds of stimuli including touch, temperature changes, light, and many chemicals (Table 3.1). How stimuli are received is not clear. Amoebae, with no visible organelles, are constantly changing and have no fixed ref-

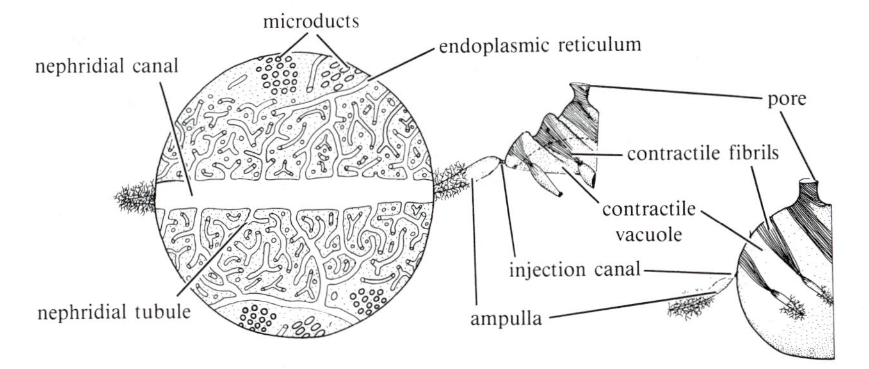

FIGURE 3.1. *Paramecium* contractile vacuole based on studies of ultrastructure in sections. Tiny canals connect endoplasmic reticulum to radial nephridial canals. Contraction of fibers in radial canal ampullae forces water through canal into main vacuole. Discharge of vacuole caused by contraction of fibers in its wall. (After Schneider.)

Table 3.1. Types of tactic behavior

Stimulus	Type of taxis
Temperature	Thermotaxis
Light	Phototaxis
Electric	Galvanotaxis
Pressure	Barotaxis
Gravity	Geotaxis
Contact	Thigmotaxis
Currents	Rheotaxis
Chemicals	Chemotaxis
Moisture	Hydrotaxis

erence points, yet their responses indicate that the general protoplasmic mass can receive stimuli as well as conduct them. Cilia and flagella appear to be highly sensitive, especially to touch. Some ciliates have motionless bristles that appear to act as receptors. For example, in *Euplotes*, dorsal bristles are surrounded by delicate fibrils that may be used in sensory conduction.

One special sensory organelle is the stigma or eyespot (Fig. 3.2). The stigma is a reddish body, commonly found in photosynthetic flagellates and usually located near the flagellar root system. The fine structure of the stigma resembles that of a chromoplast but with eyespot chambers adjacent to the flagellum. The stigma is often cupped, shading the light-sensitive material from one side, and, in a few dinoflagellates, has developed into an ocellus with a starchy lens.

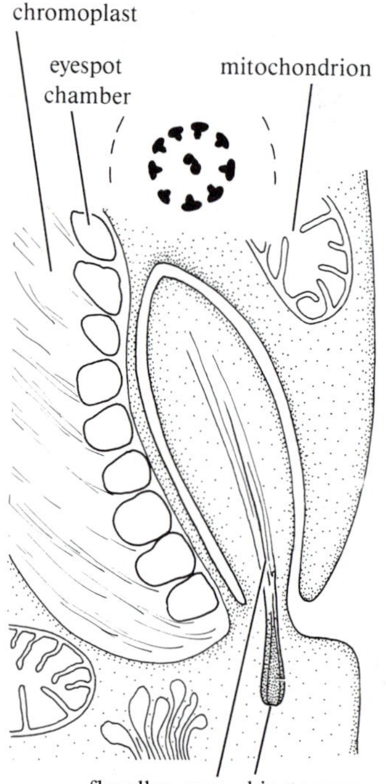

FIGURE 3.2. Diagram of optic stigma and flagellar base of *Chromulina*, a chrysomonad. Note 9 + 2 arrangement of fibrils in flagellum, as seen in cross section, at top. (After Fauré-Fremiet.)

It is obvious that stimuli are conducted from the point of reception to another point, where activity may change. How conduction occurs remains uncertain. In amoeboid organisms, generalized protoplasmic conduction is apparently used. In organisms with specialized, permanent locomotor organelles, it is not unreasonable to expect that some kind of permanent conduction organelles should evolve. However, what were once thought to be conducting fibrils in some ciliates, used to coordinate locomotory organelles, are now believed to be only supportive structures. So the issue of conduction in protozoa remains open.

LOCOMOTOR ORGANELLES AND COORDINATION

Three major types of locomotor organelles occur among protozoa. A considerable part of protozoan classification depends on the kinds and distribution of these locomotor parts. **Pseudopodia**, characteristic of Sarcodina, although they occur in flagellates and other groups as well, are semipermanent to transitory extensions of protoplasm of the body surface. **Flagella** are permanent or semipermanent locomotor organelles characteristic of the Mastigophora, although they occur on the swarmers or gametes of some Sarcodina and Sporozoa. Most flagellates bear flagella throughout life, but some may reabsorb their flagella and move about with amoeboid actions or become temporarily incapable of movement. Flagella are usually larger than cilia and occur singly or in small numbers. Cilia and flagella differ in the nature of their root systems and associated parts. **Cilia** are built like flagella but have a complex system of kinetosomes and fibers composing a characteristic infraciliature. Free cilia occur in rows on the body surface. Cilia are sometimes closely associated to produce compound ciliary organelles.

Pseudopodia Five kinds of pseudopodia can be distinguished among Sarcodina. In some cases the amoeba flows along without forming definite extensions of the body, and is said to be a limax form because of its sluglike shape (Fig. 3.3B). The broad, advancing margins at the tip of the body are protoplasmic waves. Most of the large amoebae and many smaller ones move by lobopodia, finger-shaped, round-tipped pseudopodia containing both ectoplasm and endoplasm (Fig. 3.3A). Filopodia are slender ectoplasmic pseudopodia, pointed at the tips, and sometimes branching. They never anastomose to form networks (Fig. 3.3C). Reticulopodia are endoplasmic, and branch extensively and fuse together to form meshworks that act as food traps. In some cases the reticulopodia of several individuals unite to form a multicellular reticulum (Fig. 3.3D). Axopodia are also slender and composed of granular ectoplasm. They contain a slender axial filament (Fig. 3.3E), composed of parallel microtubules rolled into a spiral when seen in the electron microscope. The fibrils are like flagellar fibrils and so cause axopodial movement.

A delicate membrane covers the surface of an amoeba. Beneath it is a clear zone free of granules, the ectoplasm, differentiated from the inner, more granular endoplasm. The ectoplasm and the outermost part of the endoplasm is in the gel state and makes up the **plasmagel**. The more fluid inner endoplasm is the **plasmasol**. The plasmagel exercises a constant, slight pressure on the plasmasol and can contract to increase this pressure. If for any reason the restraining influence of the plasmagel is locally insufficient, the plasmasol erupts, flowing into an extending pseudopodium (Fig. 3.3F,G). As the plasmasol reaches the advancing tip, it spreads fountain-like, turning outward and backward to be converted into plasmagel and thus extend the plasmagel tube of the pseudopod. If plasmagel forms too rapidly at the tip, a continuous retaining layer forms and stops further movement. If this does not happen and if new plasmasol is continuously formed at the opposite end of the organism, movement can continue indefinitely. The change of plasmasol to plasmagel involves a loss of volume; therefore, the protoplasm may stream forward to occupy the space made available in this way. To be effective, a pseudopod must keep contact with the substrate. A fringe of filaments

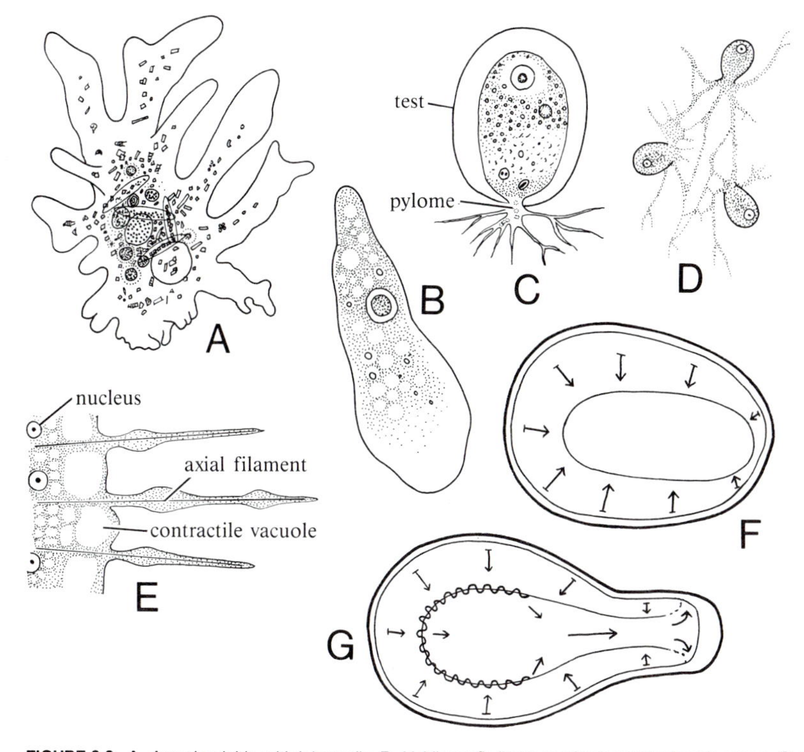

FIGURE 3.3. **A**. *Amoeba dubia,* with lobopodia. **B**. *Vahlkampfia limax,* moving by protoplasmic waves. **C**. *Chlamydophorus stercorea,* moving by filopodia. **D**. *Microgromia socialis,* with anastomosing reticulopodia. **E**. Axopodia in part of surface of *Actinosphaerium eichorni.* **F**,**G**. Two diagrams showing pseudopodial formation. *Arrows with square tails* show pressure of plasmagel on plasmasol; *straight arrows* show flow of protoplasm. *Wavy line* in G indicates site of conversion of plasmagel to plasmasol; *broken lines* at right shows region of reverse conversion where macromolecular folding reduces protoplasmic volume to facilitate its flow forward. (**A** after Schaffer; **B** after Kudo; **C** after Wenyon; **D** after Cash, from Kudo; **E**,**F**,**G** after Jahn.)

in the active region seen in electron micrographs may represent adhesive material used for this purpose.

In amoeboid movement, there can be no doubt that pseudopodia are produced in different ways in different organisms and that the mechanics of lobopodia, the best understood, differ from the mechanics of filopodia. Lobopodial movements are complex, involving several activities (Fig. 3.3F,G): (1) change of plasmagel into plasmasol at the "posterior" end, and of plasmasol to plasmagel at the "anterior" end, (2) contraction of the gelated protoplasm, and (3) streaming of the plasmasol.

Flagella Flagella are delicate, whiplike structures that beat to propel flagellates through the water. It is interesting to find that the detailed structure of flagella is approximately the same no matter what kind of cell or organism they are part of. The outer covering is continuous with the plasmalemma of the body surface, and contains a matrix in which a ring of nine doublet microtubules surrounds two central microtubules (Fig. 3.4A), an arrangement also seen in cilia. The 9 + 2 pattern is enclosed in an inner sheath. The two central fibers confer bilaterality, and the plane of flagellar beat may be associated with the orientation of the central filaments.

The flagellum is able to make a variety of movements, rapid or slow, and forward, backward, or lateral. One motion uses a spiral effective stroke to bring the cupped flagellum downward, forcing the animal forward and imparting a spiral spin. The recovery stroke consists of a wave passing along the flagellum to position it for a second effective stroke (Fig. 3.4B,C,D). In addition, undulating movements can be used to produce forward or backward movement, as well as movement to the side. It is not clear how controls for such varied movements are associated with the rather simple basal structure.

Flagellar movements cease if the flagellum is detached from the flagellar root system. Flagellar roots (Fig. 3.5) are extremely diverse. In the very simple flagellar substructure of trypanosomes, a small dark-staining body occurs at the base of the flagellum, and a larger DNA-containing body lies near it. The small body is called a **basal body** or **kinetosome**; the large body has been termed a **parabasal body** or **kinetoplast**.

Some of the most complex flagellar root systems are found in endozoic protozoa. In *Trichomonas* (Fig. 3.5), three flagella extend forward, and one trailing flagellum is attached to the body by a delicate undulating membrane. A prominent **costa** lies at the base of the undulating membrane and is also connected to the root system. The body is supported by an axostyle, often closely associated with the flagellar roots or nucleus.

Where flagella occur in rows or in other patterns, division of the body occurs between the rows rather than across them. It is thus a longitudinal division insofar as the flagellar substructure is concerned. Ciliates characteristically divide across their rows of cilia, providing an important criterion for distinguishing between flagellates with numerous flagella and ciliates.

Cilia Like flagella, cilia are composed of nine peripheral and two central microtubules enclosed in a sheath continuous with the plasmalemma. Each cilium terminates in a kinetosome (Fig. 3.6A,B,C). A fiber arises from the kinetosome. It joins other fibers from cilia of the same row. The compound fiber is the kinetodesmose. The row of cilia kinetosomes and their kinetodesmose is a **kinety** (Fig. 3.6D).

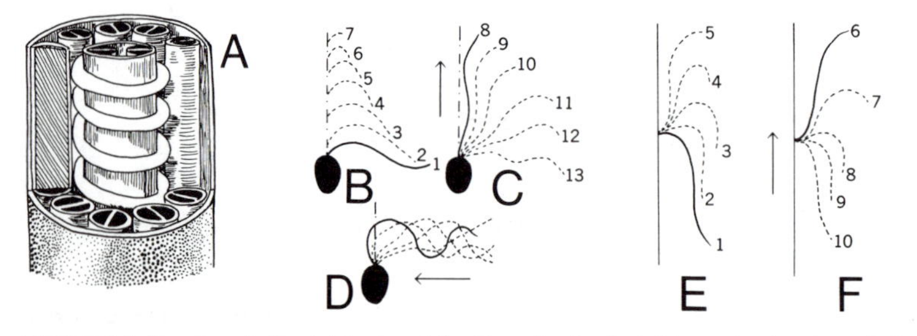

FIGURE 3.4. Flagella and cilia. **A.** Diagram of flagellum, showing 9 + 2 inner structure from electron micrographs, two central and nine peripheral fibers characteristic. **B,C,D.** Flagellar movements of *Monas,* a protomonad; **B.** preparatory stroke; **C.** effective stroke; **D.** lateral movement. **E,F.** Ciliary action; **E.** preparatory stroke; **F.** effective stroke. *Arrows* in **C**, **D**, and **F** show direction of movement, and numbers show successive positions of cilium or flagellum. (**A** after Manton and Clarke, from Berrill; **B,C,D** after Krijgsman; **E,F** after Verworn, from Kudo.)

These bodies and fibers make up the infraciliature, the details of which are extremely useful in determining ciliate relationships.

The infraciliature provides support (see Fig. 3.4E,F), as well as playes a decisive role during the differentiation of organisms in reproduction. A wave of effective beat passes along adjacent kineties in unison, causing in some a spiral movement of the organism as a whole. The details of coordination, however, remain uncertain in forms with uniform body ciliature.

Ciliary adaptation has followed several different lines. Somatic cilia occur at the body surface and have an infraciliature distinctively separated from buccal or oral cilia located in or near the oral region. In some ciliates, somatic cilia have been greatly reduced. The most important somatic organelles are cirri, tuftlike brushes of functionally fused cilia. Their kinetosomes form a basal fiber plate.

The oral ciliature is variously modified, depending on the nature of the oral apparatus. The most important compound ciliary organelles in the oral region are undulating membranes, formed of a row of cilia (see Fig. 3.16A), and membranelles. Each membranelle is a flat plate of cilia with a basal fiber

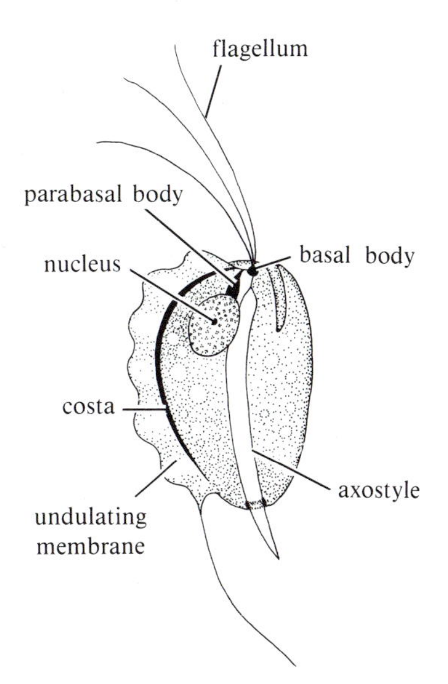

FIGURE 3.5. Diagram of flagellar root system of *Trichomonas*. (After Grell, 1973.)

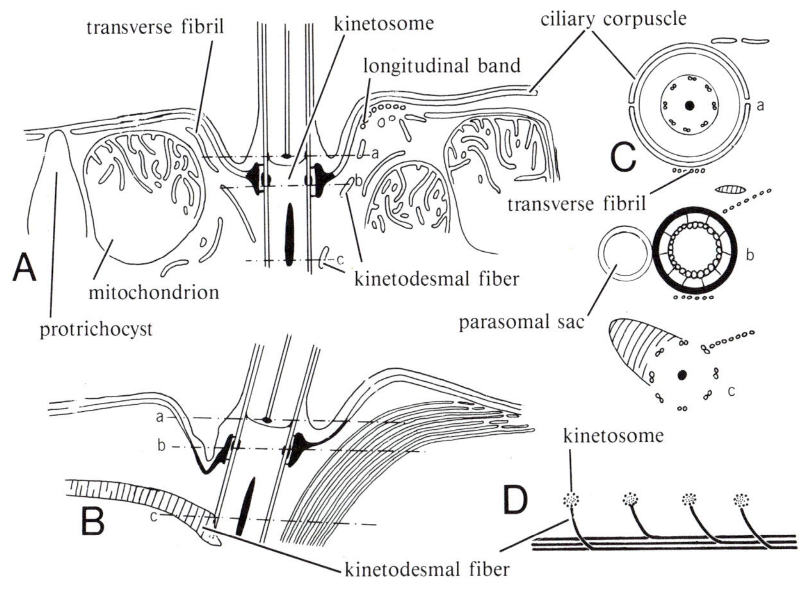

FIGURE 3.6. A,B,C. Diagram of cortex of *Colpoda campanula,* tetrahymenid ciliate, derived from electron micrographs; **A**. transverse section; **B**. longitudinal section; **C**. cross sections of cilium at levels marked (*a*), (*b*), and (*c*) in preceding figures. **D**. Diagram of kinetosome and kinetodesmose system. Only kinetosome and kinetodesmose system shown at **D**. Thread from each cilium passes to kinetodesmose. This basal apparatus constitutes kinety that extends entire length of ciliary row. **A**, **B**, and **C** show relationship of cilia and kinetosome to other cortical structures. Parasomal sac lies in front of each cilium. Flattened sacs (ciliary corpuscles) curve down from surface to connect with kinetosomes. Longitudinal and transverse fibers form network, distinct from kinetodesmose systems, associated with ciliary base. (**A,B,C** after Pitelka; **D** after Metz, from Grimstone.)

plate. They are arranged in a longitudinal series and beat in a coordinated fashion.

THE NUCLEUS

Protozoan nuclei are extremely interesting because of their diversity in structure and methods of division. Two basic types are recognized: vesicular, containing considerable nucleoplasm, and compact, containing little nucleoplasm. The distribution of the DNA material within the nucleus varies considerably.

When more than one nucleus is present, usually all are alike. Exceptions occur in ciliates, where a massive compact macronucleus and one or more small vesicular or compact micronuclei are found. In sporozoans, nuclei for vegetative and generative functions may occur.

When protozoa are cut into parts, pieces with an intact nuclear apparatus may recover, whereas those without nuclei soon deteriorate and die. The pieces that die show evidences of inability to produce shells and digestive enzymes.

METHODS OF ASEXUAL REPRODUCTION

Both sexual and asexual reproduction occur among protozoa; however, some reproduce asexually only. Asexual reproduction involves the division of the parent body, either equally or unequally, to produce one or more young individuals that develop into mature organisms. It always involves a single parent, and neither meiosis nor fertilization occurs.

Binary Fission Binary fission (Fig. 3.7A,B,C,D) is the most common form of asexual reproduction. The parent body divides equally to produce two daughter cells that regain adult size and form. Where there is a definite body axis, fission may be transverse, as in ciliates, or longitudinal, as in flagellates. Fission typically involves karyokinesis, or nuclear division, and cytokinesis, or cytoplasmic division. Preparation for division may be simple or involve complex reorganization. Extensive macronuclear reorganization accompanies fission of ciliates in many cases, and the infraciliature and compound ciliary organelles may also participate in predivision reorganization. Some protozoa undergo fission in the encysted state. In this case extensive dedifferentiation may precede fission. After or during the latter part of karyokinesis, cytoplasmic constriction begins. The two daughters then enter a period of growth and differentiation following cytokinesis.

Multiple Fission Multiple fission (Fig. 3.7E,F) is common in some groups of protozoa, especially the sporozoan parasites (**sporogony**), Foraminiferida, and radiolarians. Preparation for division involves repeated nuclear division to form a multinucleate

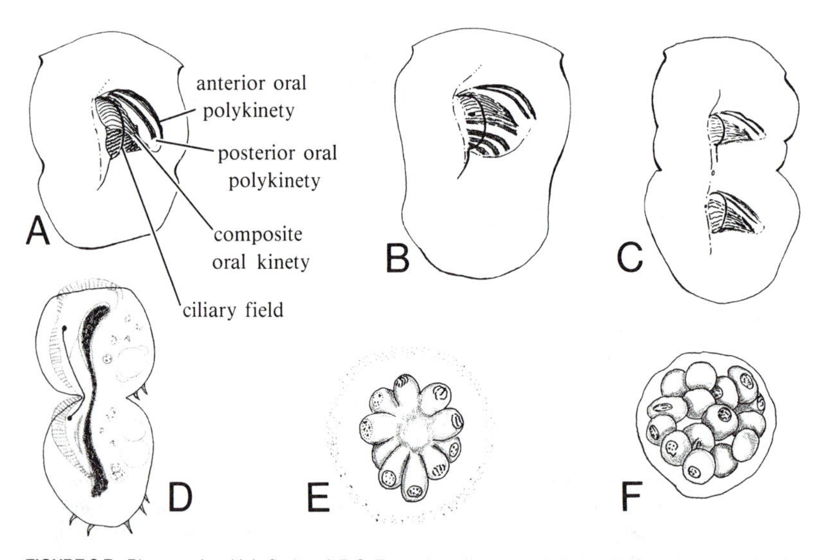

FIGURE 3.7. Binary and multiple fission. **A,B,C.** Formation of new mouth during division of *Urocentrum.* Note four special ciliary fields. Division of ciliary fields provides new ciliary field for anterior mouth and four ciliated fields for posterior mouth. **D.** Division of *Euplotes patella,* showing simultaneous division of micronucleus and macronucleus. **E,F.** Multiple fission during sporoblast formation in sporozoan, *Ovivora thalassemae.* Cytoplasm divides to form number of daughters, leaving small mass of residual protoplasm behind. (**A,B,C** after Fauré-Fremiet; **D** after Turner; **E,F** after MacKinnon and Ray, from Hall.)

organism with no cytokinesis. The cytoplasm then cleaves to form many uninucleated individuals. Although there are exceptions, the products of multiple fission often enter into a new phase of the life cycle of the organism, serving as swarmers or infective stages for new host cells. It is not uncommon for multiple fission to leave a mass of residual protoplasm that degenerates.

Budding Budding is a type of asexual reproduction in which one or more small daughters are produced by the parent. It may be considered a form of unequal fission and, like fission, involves preparatory phases before budding and a later period of growth and differentiation. Budding may occur at the body surface or in an internal cavity; the former is termed exogenous, the latter endogenous.

METHODS OF SEXUAL REPRODUCTION

Sexual reproduction involves meiotic nuclear divisions that result in a change from diploid to haploid, and the union of gametes to restore diploidy. It may be **amphimictic**, involving the union of gametes from different parents, or **automictic**, in which the gametes arise from the same parent. In either case, the uniting gametes may be whole organisms or nuclei only. Where whole organisms unite, the union is termed **syngamy**. Where only nuclei unite, the process is **conjugation**. Conjugation occurs only among ciliates, whereas syngamy occurs in all other groups where sexual reproduction occurs.

Meiosis Meiosis evidently occurs among sexually reproducing protozoa, but the details are not well known in most cases. Meiosis occurs in *Paramecium aurelia*, and chromosome counts showing haploid and diploid numbers have been reported from all groups of protozoa. Among ciliates, the second division of the micronucleus during conjugation is reductional, insofar as chromosome number is concerned, although the matter is somewhat complicated by the absence of a definite tetrad stage. The absence of distinctive tetrads suggests that the distinctive meiotic process seen in all of the metazoans may have been evolved independently. Nevertheless, there remains a strong tendency for protozoan meiosis to require two divisions, as in metazoans. Among some of the photosynthetic flagellates, meiosis immediately follows zygote formation, as often happens in algae.

Syngamy The gametes may be morphologically similar (**isogametes**) or dissimilar (**anisogametes**). In at least some isogamous species, the gametes are chemically differentiated. In this case a gamete belongs to a mating type and can unite only with a gamete from a different type, thus ensuring outbreeding. Gametes vary widely in form. They may be flagellate or amoeboid and in some cases are highly differentiated, especially in the case of motile microgametes. The zygote commonly enters into a quiescent phase. For example, in colonial flagellates the zygote secretes a heavy cyst membrane and completes its development after winter or when conditions have become more favorable and thus resembles algal zygospores. In some of the sporozoans, the zygote becomes an oöcyst, in which the spores, quiescent until they reach a new host, are formed.

Conjugation Conjugation is characteristic of ciliates. Its details vary with the species and with the number of macronuclei and micronuclei present. In all cases, however, the old macronucleus is replaced, a new genome with genes from both parents is established, and the reproductive capacity of the organism is increased. The general features of conjugation are best seen in a species with a single macronucleus and a single micronucleus. Two ciliates, ready for conjugation, partially unite, the pellicle and body surface undergoing extensive local changes during union. As conjugation continues, the macronucleus has, or will soon, disintegrate. The micronucleus divides, usually twice, to produce four haploid micronuclei (or pronuclei), all but one of which disintegrate. The persisting nucleus divides once again, to form a stationary pronucleus and a migratory pronucleus (Fig. 3.8), the former remaining within the original parent and the other passing to the other conjugant. The two organisms now separate, and the pronuclei unite to form a zygote nucleus that divides repeatedly, giving rise to micronuclei and to macronuclear primordia that grow into macronuclei. Several postconjugation fissions, differing somewhat with the species, restore the normal nuclear complement to each daughter organism. Among some ciliates, especially stalked or sessile forms, conjugation has evolved in the same direction as syngamy, with the appearance of macroconjugants and microconjugants that fuse to form a single zygote.

It has been shown in a number of cases that substances stimulating syngamy or conjugation are liberated by gametes or by organisms ready for conjugation. These substances may play a role in favoring the meeting of the gametes or conjugants, or in establishing mating types. The appearance of mating types is important, for it tends to prevent inbreeding. In some mating-type systems, a gamete or conjugant can unite with individuals belonging to any

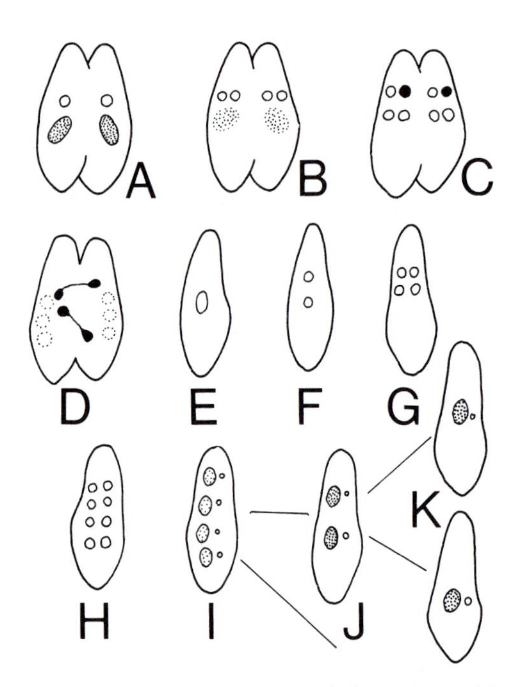

FIGURE 3.8. Diagram of conjugation, as in *Paramecium caudatum*. **A**. Joining of two animals. **B**. First meiotic division of micronucleus and degeneration of macronucleus. Macronuclear degeneration hastened in diagrams; in reality, macronuclear fragments still present when conjugants first separate. **C**. Second meiotic division of micronuclei. One haploid nucleus (pronucleus) remains viable in each organism; others degenerate. **D**. Degeneration of three nuclei and division of viable pronucleus into migratory and stationary pronuclei. **E**. Fusion of pronuclei in each conjugant, and separation of two animals with single, diploid, zygote nucleus formed. **F**. First division of zygote nucleus. **G**. Second nuclear division in postconjugant. **H**. Third nuclear division in postconjugant. **I**. Differentiation of micronuclear and macronuclear primordia. **J**. After first postconjugation fission. **K**. After second postconjugation fission, with normal nuclear complement restored. Eventual outcome is recombination of genes through meiosis and zygote formation, and replacement of old macronucleus with one derived from zygote nucleus.

mating type but its own. In others, especially well represented by *Paramecium aurelia*, the species is divided into a group of mating types, each consisting of two mating strains. Mating occurs only between the two strains of each type.

Autogamy Autogamy is a type of sexual reproduction involving a single individual. It occurs in several groups but is best known in ciliates. It is a modification of conjugation and results in some of the same benefits, the replacement of the old macronucleus, the restoration of vigor, and occasionally gene recombination. The macronucleus breaks down while the micronucleus undergoes meiosis. Two pronuclei from the same organism unite to form a zygote nucleus. In *Paramecium aurelia* the zygote nucleus is formed by the union of two nuclei derived from a single pronucleus, and the result is a homozygous organism.

METAGENETIC LIFE CYCLES

In a few groups of protozoa, an alternation of sexually reproducing and asexually reproducing generations occurs, each with a characteristic form. This is best seen in the Foraminiferida. The alternation of asexual and sexual reproduction has become formalized into a system in which the offspring of a sexually reproducing generation are obliged to reproduce asexually.

Among the Apicomplexa an alternation of sexual and asexual generations is almost universal. The general features of this kind of life cycle are not always apparent because of differences in the appearance of the various stages in different groups. **Schizogony** is a period of asexual reproduction, usually occurring by multiple fission, producing infective **merozoites**. These grow into **schizonts** that again undergo schizogony. The process continues until a **gamont** appears that develops into gametes, from which a zygote is produced. After many divisions, the zygote either directly, or indirectly by sporogony, produces **sporonts** that differentiate into spores. Each spore contains one or more **sporozoites** that reach a new host and here become schizonts.

Modifications of this type of cycle are numerous. In some cases two kinds of hosts are used, one for the schizogony part of the cycle and another for the sporogony part of the cycle, as in *Lankestrella* and *Plasmodium* (see Fig. 3.14). In other cases, as in *Plasmodium*, there may be two kinds of schizogony cycles in different host cells.

Representative Forms and Classification

Rather than reproduce here a group-by-group review of protozoan Protista, the student is advised to consult earlier editions of this book or any of several texts on protozoology. A summary taxonomy is provided here of only the protozoan protists. It is based on the widely accepted effort of Levine et al. (1980), and there is appended throughout this list some illustrations of representative types. Just which of these might in fact be ancestral to the Animalia sensu stricto is still a subject of great debate (e.g., see Chapter 4). A complete taxonomy of Protista would encompass some 45 phyla. If the student is interested in pursuing that classification further, Corliss (1984) can be consulted.

Phylum Sarcomastigophora. Single type of nucleus, except in Foraminiferida; sexuality, when present, essentially syngamy; flagella, pseudopodia, or both types of locomotor organelles.

Subphylum Mastigophora. One or more flagella typically present; asexual reproduction basically by binary fission; sexual reproduction known in some groups.

Class Phytomastigophorea. Typically with chloroplasts; mostly free-living.

Order Cryptomonadida. Two subequal flagella arising subapically in ventral groove; chloroplasts brown, red, olive-green, blue, or yellow; storage products starch and fat; cells flattened, naked, without wall or pellicle; sexual reproduction unknown.

Order Dinoflagellida. Two heterodynamic flagella, inserted apically or laterally, one ribbon-shaped with paraxial rod and single row of fine hairs, other smooth or with two rows of stiffer hairs; chloroplasts typically golden brown or green; storage products starch and fat; cells flattened or of complex symmetry with transverse and ventral grooves and often armor of cellulose plates; nucleus unique among eukaryotes in having chromosomes that consist primarily only of nonprotein-complexed DNA; mitosis intranuclear; sexual reproduction present.

Order Euglenida. Two (rarely more) flagella, one or both emerging from an anterior invagination of the cell; flagella with paraxial rods; chloroplasts grass-green, but absent in many genera; storage products paramylon, fat, and cyclic metaphosphates; cell with helical symmetry, naked but with complex pellicle of interlocking proteinaceous strips; nonspindle intranuclear mitosis; solitary or colonial.

Order Chrysomonadida. Two unequal flagella, one directed anteriorly, other trailing and smooth; chloroplasts golden brown or absent; storage products chrysolaminarin and fat; cells naked, with richly patterned silicified scales or with platelike lorica; sexual reproduction present.

Order Heterochlorida. Two unequal flagella; chloroplasts yellow-green; storage products oil and possibly chrysolaminarin; supposedly related to xanthophycean algae.

Order Chloromonadida. Two flagella; chloroplasts green; storage product oil; characteristic ring of Golgi bodies at anterior end; sexual reproduction by fusion of two individuals.

Order Prymnesiida. Two equal or subequal smooth flagella inserted laterally or anteriorly, with unique third appendage, the haptonema, between them; chloroplasts golden brown; storage products fat and possibly chrysolaminarin; cells covered with delicate organic scales of diagnostic pattern, scales sometimes calcified; sexual reproduction present.

Order Volvocida. Two or four equal, smooth, apical flagella; chloroplasts grass-green; storage products starch and fat; sexual reproduction present.

Order Prasinomonadida. One, two, four, or eight flagella, typically covered with rows of finely patterned scales; chloroplasts grass-green; storage product starch; cells typically covered with one or more layers of intricately patterned Golgi-derived scales; sexual reproduction present.

Order Silicoflagellida. One flagellum; chloroplasts golden brown or green-brown; storage product apparently chrysolaminarin; with star-shaped siliceous skeleton composed of tubular elements; sexual reproduction unknown.

Class Zoomastigophorea. Chloroplasts absent; one to many flagella; amoeboid forms, with or without flagella in some groups; sexual reproduction known in few groups; (probably a polyphyletic taxon).

Order Choanoflagellida. One flagellum, inserted apically, with proximal part surrounded by ring of microvilli (collar); with membranous sheath or basketlike siliceous lorica; stalked or free-swimming; free-living (Fig. 3.9C).

Order Kinetoplastida. One or two flagella arising from depression; single mitochondrion (nonfunctional in some forms) extending length of body as single tube, hoop, or network of branching tubes; parasitic (majority of known species) and free-living (Figs. 3.9B and 3.10).

Order Proteromonadida. One or two pairs of flagella without paraxial rods; single mitochondrion, curving around nucleus; cysts present; parasitic.

Order Retortamonadida. Two to four flagella, one turned posteriorly and associated with ventrally located cytostomal area bordered by fibril; mitochondria and Golgi apparatus absent; cysts present; parasitic.

Order Diplomonadida. One or two karyomastigonts; genera with two karyomastigonts with twofold rotational symmetry or, in one genus, primarily mirror symmetry; individuals with one to four flagella, typically one of them directed back over body and associated with cytostome or, in more advanced genera, with organelles forming cell axis; mitochondria and Golgi apparatus absent; cysts present; free-living or parasitic (Fig. 3.11B).

Order Oxymonadida. One or more karyomastigonts, each containing four flagella typically arranged in two pairs in motile stages; one or more flagella adhering to body surface for greater or lesser distance; mitochondria and Golgi apparatus absent; cysts in some species; sexuality in some species; parasitic (Fig. 3.11A).

Superorder Parabasalidea. Typically at least some kinetosomes, bearing flagella or barren, arranged in pattern; mitochondria absent; division spindle extranuclear.

Order Trichomonadida. Typically karyomastigonts with four to six flagella, but in one genus only one flagellum and no flagella in another; typically one flagellum recurrent, free, or with

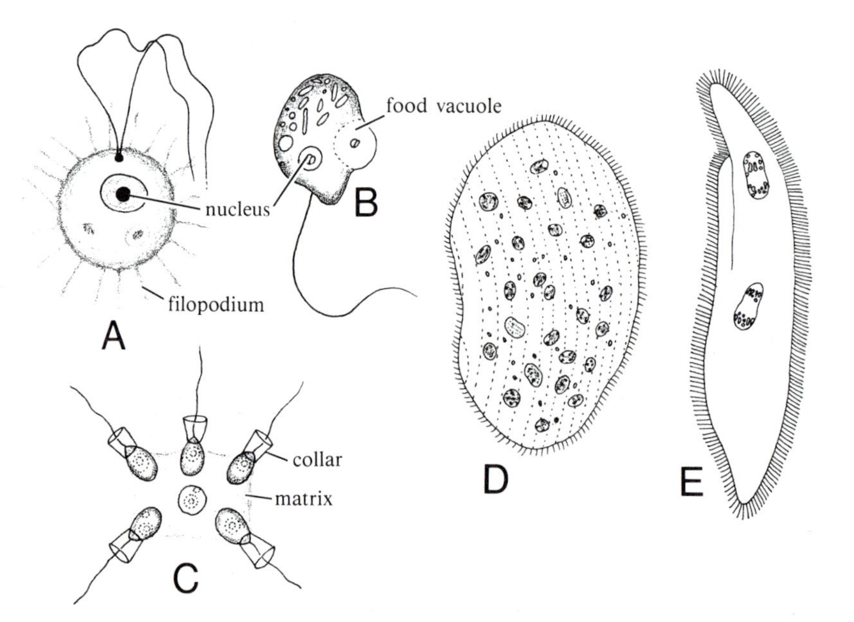

FIGURE 3.9. A,B,C. Some representative zooflagellates; **A.** *Heliobodo radians,* rhizomastigid with flagella and radiating filopodia; both flagella and some kind of pseudopodia occur in rhizomastigids; **B.** *Oikomonas termo,* minute kinetoplastid; **C.** *Protospongia haeckelii,* choanoflagellate, forming colonies with gelatinous matrix. **D,E.** Typical Opalinata; **D.** *Opalina ranarum,* flattened, multinucleated form common in frogs; **E.** *Protoopalina mitotica,* cylindrical, binucleate form. (**A** after Valkanov; **B,C** after Lemmermann; **D** after Ghatia and Gulati; **E** after Metcalf.)

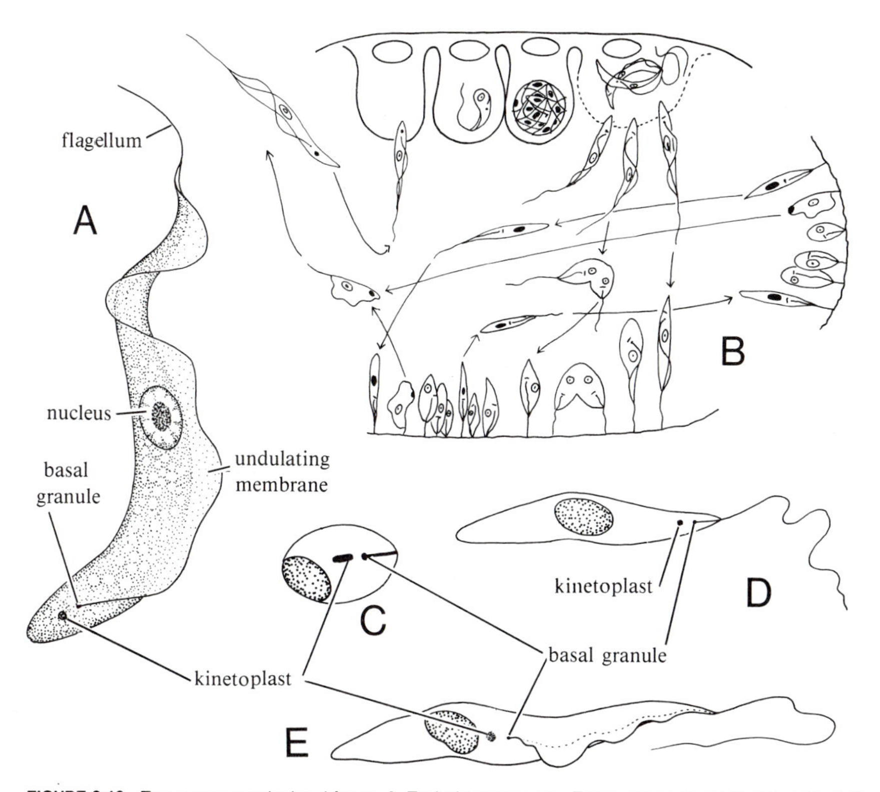

FIGURE 3.10. *Trypanosoma* and related forms. **A**. Typical trypanosome, *Trypanosoma brucei.* **B**. Life cycle of *Trypanosoma lewisi,* living in rats and fleas, showing transformation in form during life cycle. Trypanosomes, at top, enter flea stomach in rat blood; here they infect cells and reproduce. Daughters may continue fission cycle in stomach, or may enter rectum where they become leptomonads and approach becoming leishmania forms, at bottom. Leptomonads from rectum may enter pyloric part of hindgut, at right. Crithidia may appear in rectum or pylorus, and can infect rectal wall or pass out with feces and infect rats. **C,D,E**. Forms assumed by trypanosomes during their life cycles; **C**. leishmanis form; **D**. leptomonad form; **E**. crithidia form. (**A** after Grell; **B** based on Minchin and Thomson; **C,D,E** from Med. Protozoa and Helminth., U.S. Navy *Medical School Bull.*)

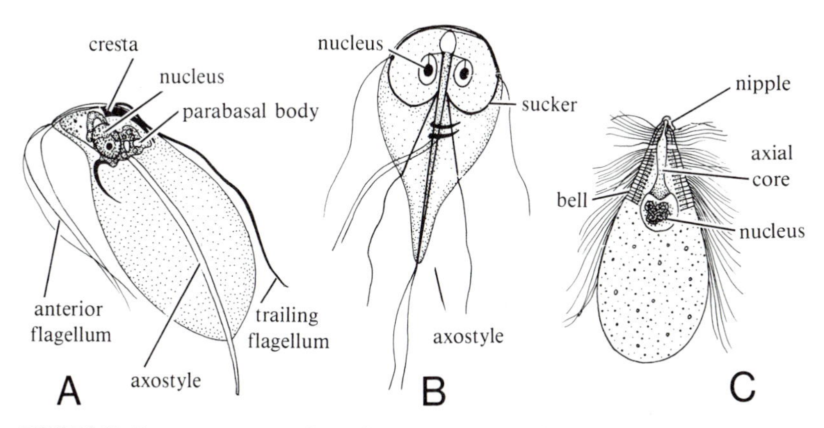

FIGURE 3.11. Some more representative zooflagellates. **A**. *Foaina inflata,* oxymonad from termite, typical monokaryomastigont, with single nucleus and flagellar root system. **B**. *Giardia intestinalis,* from human intestine, typical diplomonad, with dikaryomastigont structure with two nuclei and two flagellar root systems. **C**. *Trichonympha,* common hypermastigid symbiont in termites and wood roaches, polykaryomastigont with many nuclei and flagellar root systems. (**A** after Kirby; **B** after Kofoid and Swezy; **C** after Cleveland.)

proximal or entire length adherent to body surface as undulating membrane; true cysts infrequent, known in very few species; all or nearly all parasitic.

Order Hypermastigida. Numerous flagella; flagella-bearing kinetosomes distributed in complete or partial circle, in plate or plates, or in longitudinal or spiral rows meeting in a centralized structure; one nucleus per cell; cysts in some species; sexuality in some species; all parasitic (Fig. 3.11C).

Subphylum Opalinata. Numerous cilia in oblique rows over entire body surface; cytostome absent; known life cycles involve syngamy with anisogamous flagellated gametes; all parasitic.

Class Opalinatea. With characters of the subphylum.

Order Opalinida. With characters of the class (Fig. 3.9D,E).

Subphylum Sarcodina. Pseudopodia, or locomotive protoplasmic flow without discrete pseudopodia; flagella, when present, usually restricted to developmental or other temporary stages; body naked or with external or internal test or skeleton; asexual reproduction by fission; sexuality, if present, associated with flagellate or, more rarely, amoeboid gametes; most species free-living.

Superclass Rhizopoda. Locomotion by lobopodia, filopodia, or reticulopodia, or by protoplasmic flow without production of discrete pseudopodia.

Class Lobosea. Pseudopodia lobose or more or less filiform but produced from broader hyaline lobe; usually uninucleate; multinucleate forms not flattened or much-branched plasmodia; no sporangia or similar fruiting bodies.

Subclass Gymnamoebia. Without test.

Order Amoebida. Typically uninucleate; mitochondria typically present; no flagellate stage (Fig. 3.3A).

Order Schizopyrenida. Body with shape of cylinder, usually moving with more or less eruptive bulges; typically uninucleate; temporary flagellate stages in most species.

Order Pelobiontida. Body with shape of thick cylinder; monopodial, with true bidirectional fountain flow of cytoplasm common; typically multinucleate; lacking mitochondria but with symbiotic bacteria; no flagellate stage known.

Subclass Testacealobosia. Body enclosed by test, tectum, or other complex membrane external to plasma membrane.

Order Arcellinida. Test, tectum, or other external membrane with single aperture and composed of either organic or inorganic material or both (Fig. 3.12E,F).

Order Trichosida. Test composed of fibrous sheath and, in at least one extensive stage of life cycle, with calcareous spicules, and multiple apertures through which short, conical pseudopodia extend; locomotion by broad lobopodium; marine.

Class Acarpomyxea. Small plasmodia or much-expanded similar uninucleate forms, usually branching, sometimes forming reticulum of coarse branches; advancing tips lobose; no regular reversal of streaming; no test; no spores or fruiting bodies known.

Order Leptomyxida. Typically thin sheets; sometimes more cylindrical; cysts produced by soil and freshwater species.

Order Stereomyxida. Marine amoeboid organisms with more or less branched pseudopodia producing only very slow motion or serving as flotation organelles.

Class Acrasea. Uninucleate amoebae with eruptive, lobose pseudopodia; amoebae aggregating to form pseudoplasmodium that gives rise to fruiting bodies without stalk tube; flagellate cells known in only one species; sexuality unknown.

Order Acrasida. With characters of the class.

Class Eumycetozoea. Myxamoebae with filiform subpseudopodia; flagella sometimes present;

producing aerial fruiting bodies; stalk tube typically present in fruiting bodies of first two subclasses and in some members of third.

Subclass Protosteliia. Trophic stage varying from single amoebae to plasmodia that lack shuttle streaming; flagellate cells present or absent; fruiting bodies consisting of one to several spores on narrow, hollow stalk; sexuality known in one species.

Order Protosteliida. With characters of the subclass.

Subclass Dictyosteliia. Amoebae aggregate to form multicellular pseudoplasmodium that gives rise to multispored fruiting body; stalk tube present; no flagellate cells; sexuality indicated in some species.

Order Dictyosteliida. With characters of the subclass.

Subclass Myxogastria. Major trophic stage multinucleate plasmodium typically with shuttle streaming; fruiting bodies multispored; flagellate cells present; syngamy and meiosis in life cycle.

Order Echinosteliida. Sporangia stalked, minute; plasmodium small, amoebalike, nonreticulate.

Order Liceida. Spore mass usually light-colored.

Order Trichiida. Spore mass usually light-colored.

Order Stemonitida. Spore mass usually dark-colored.

Order Physarida. Spore mass usually dark-colored; calcareous.

Class Plasmodiophorea. Obligate intracellular parasites with minute plasmodia; zoospores produced in zoosporangia and bearing anterior pair of unequal flagella; resting spores formed in compact sori or loose clusters within host cells; sexuality reported in some species.

Order Plasmodiophorida. With characters of the class.

Class Filosea. Filiform pseudopodia, often branching, sometimes anastomosing; no spores or flagellate stages known.

Order Aconchulinida. Without external skeletal material; filopodia produced from main mass of cell, not from hyaline lobe.

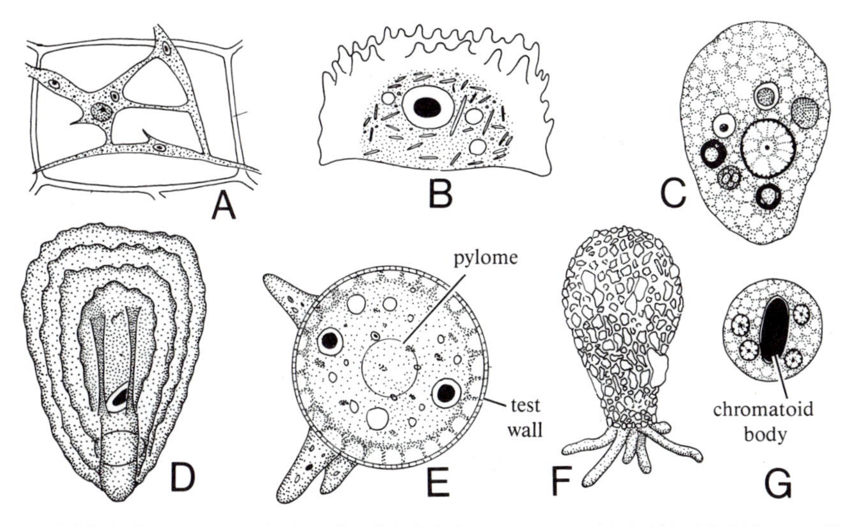

FIGURE 3.12. Some representative Sarcodina. **A.** *Labyrinthomyxa sauvageaui,* typical labyrinthulid that parasitizes algal cells. **B.** *Hartmanella klitzkei,* amoeba with conspicuous ectoplasmic zone. **C,G.** Trophic and encysted stages of *Entamoeba histolytica,* amoeba that parasitizes human intestine and causes amoebic dysentery. **D.** *Thecamoeba verrucosa,* amoeba with heavy pellicle, as it appears when moving slowly. **E.** *Arcella,* arcellinid with transparent, secreted test. **F.** *Difflugia pyriformis,* arcellinid with test formed of foreign particles. **A** after Duboscq; **B** after Arndt; **C,G** after Kudo; **D** after Jahn; **F** after Pénard, from Hall.)

Order Gromiida. Body enclosed by test or rigid external membrane with distinct aperture.

Class Granuloreticulosea. Delicate, finely granular reticulopodia, or, rarely, finely pointed, granular, but nonanastomosing pseudopodia.

Order Athalamida. Naked.

Order Monothalamida. With single-chambered organic or calcareous test, sometimes including foreign matter; no alternation of generations.

Order Foraminiferida. Test with one to many chambers; pseudopodia protruding from aperture, wall perforations, or both; reproduction with alternation of sexual and asexual generations, of which one may be secondarily repressed; gametes usually flagellate, rarely amoeboid; nuclear dimorphism in developmental stages of some species (Fig. 3.13D,E,F,G,H).

Class Xenophyophorea. Multinucleate plasmodium enclosed in branched-tube system composed of transparent organic substance; numerous barium sulfate crystals in cytoplasm; fecal pellets retained outside organic tube system as conspicuous dark masses; test of foreign matter surrounding tube system and fecal-pellet masses; marine.

Order Psamminida. Without threads forming part of test; body more or less rigid.

Order Stannomida. With threads forming part of test; body flexible.

Superclass Actinopoda. Often spherical, usually planktonic; axopodia; skeleton, when present, composed of organic matter and/or silica, or of strontium sulfate; reproduction asexual and/or sexual; trophic cells rarely flagellated; in many species small flagellated stages whose exact nature (gametes or spores) is still uncertain.

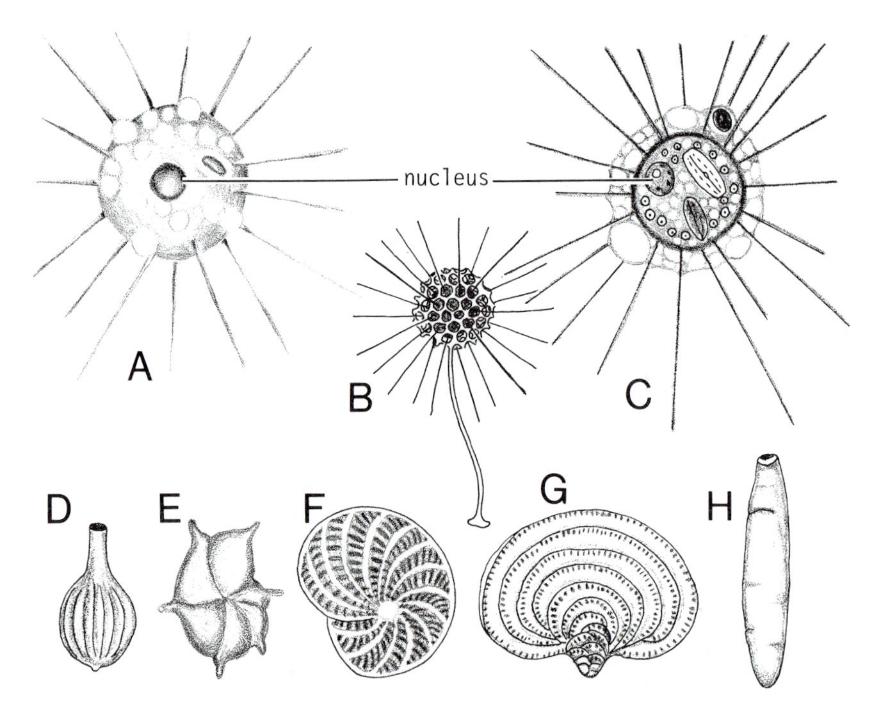

FIGURE 3.13. **A,B,C.** Representative Heliozoa; **A.** *Actinophrys sol,* common uninucleate heliozoan; **B.** *Clathrulina elegans,* stalked heliozoan with perforate test, through which axopodia protrude; **C.** *Actinosphaerium eichorni,* common multinucleate heliozoan. **D,E,F,G,H.** Foraminifera; some examples of tests to illustrate range of shapes; **D.** *Lagenastriata;* **E.** *Hantkenia alabamensis;* **F.** *Elphidium crispa;* **G.** *Pavonina flabelliformis;* **H.** *Bathysiphon humilis.* (**B** after Leidy; **C** after Kudo; **D** after Rhumbler; **E** after Cushman, from Kudo; **G** after Parr; **H** after Calvez, from Hall.)

Class Acantharea. Strontium sulfate skeleton, usually composed of 20 radial or 10 diametrical spines, rarely 16 diametrical or 32 radial spines; sometimes many more spines randomly oriented; spines more or less joined in cell center; extracellular outer (cortex) and inner envelopes usually present; inner envelope (called capsular membrane) often closely lining central cell mass; marine, usually planktonic.

Order Holacanthida. Usually 10, sometimes 16 diametrical spines, crossing in center; inner envelope far outside central cell mass, or absent; encystment before sporogenesis, at least in several species.

Order Symphyacanthida. Twenty radial spines totally fused in cell center or forming small sphere by apposition of their basal pyramids; inner envelope far outside central cell mass; encystment before sporogenesis, at least in some species.

Order Chaunacanthida. Twenty radial spines with bases more or less loosely articulated; inner envelope at some distance outside central cell mass, or absent; encystment before sporogenesis in most or perhaps all species.

Order Arthracanthida. Usually 20 radial spines joined at cell center by apposition of bases; inner envelope usually closely lining central cell mass; no cysts.

Order Actineliida. Variable number of radial spines; mostly planktonic, one benthic genus.

Class Polycystinea. Siliceous skeleton present in most species; made up usually of solid elements, consisting of one or more latticed shells with or without radial spines, or of one or more isolated spicules; capsular membrane composed usually of grossly polygonal plates and containing many more than three pores; marine, planktonic.

Order Spumellarida. Capsular membrane with uniformly distributed pores.

Order Nassellarida. Capsular membrane with pores gathered at a single pole; skeleton of one piece, often basket-shaped.

Class Phaeodarea. Skeleton (sometimes absent) of mixed silica and organic matter, consisting of usually hollow spines and/or shells; very thick capsular membrane; marine, planktonic.

Order Phaeocystida. Skeleton absent, or consisting of spicules either free or radiating from common junction point.

Order Phaeosphaerida. Skeleton consisting mainly of very large latticed shell with wide polygonal meshes.

Order Phaeocalpida. Skeleton consisting mainly of small shell, usually with numerous pores, often with one large opening; shell texture usually porcellanous; radial spines often present.

Order Phaeogromida. Skeleton consisting mainly of small diatomaceous or alveolar shell with one large opening; shell, sometimes greatly reduced, may bear spines.

Order Phaeoconchida. Skeleton consisting of two thick, usually hemispherical valves pressed against each other.

Order Phaeodendrida. Skeleton consisting of two noncontiguous valves, from which originate long, branching spines with ramifications that may produce enormous external latticed spongious shells.

Class Heliozoa. Without central capsule; skeletal structures, if present, siliceous or organic; axopodia radiating on all sides; most species freshwater, some marine (Fig. 3.13A,B,C).

Order Desmothoracida. Cell enclosed in usually spherical, latticed organic capsule stalked in most species; uniflagellate or diflagellate zoospores.

Order Actinophryida. No skeleton; some with flagella or flagellated stage; sexuality known in some genera.

Order Taxopodida. Bilaterally symmetrical, planktonic cells with siliceous spines; swimming by rowing action of axopodia arranged in parallel longitudinal rows; small biflagellated species; marine.

Order Centrohelida. Frequently with a skeleton of siliceous plates and/or spines or of organic spicules; some species with flagella or flagellated stages.

Phylum Labyrinthomorpha. Trophic stage, ectoplasmic network with spindle-shaped or spherical, nonamoeboid cells; in some genera amoeboid cells move within network by gliding; zoospores produced by most species; saprobic and parasitic on algae, mostly in marine and estuarine waters.

Class Labyrinthulea. With characters of the phylum.

Order Labyrinthulida. With characters of the class (Fig. 3.12A).

Phylum Apicomplexa. Apical complex (visible with electron microscope); cilia absent; sexuality by syngamy; all species parasitic.

Class Perkinsea. Conoid forming incomplete cone; "zoospores" (sporozoites?) flagellated, with anterior vacuole; no sexual reproduction.

Order Perkinsida. With characters of the class.

Class Sporozoa. Conoid, if present, forming complete cone; reproduction generally both sexual and asexual; oöcysts generally containing infective sporozoites that result from sporogony; locomotion of mature organisms by body flexion, gliding, or undulation of longitudinal ridges; flagella present only in microgametes of some groups; pseudopods ordinarily absent, but if present, used for feeding, not locomotion.

Subclass Gregarinia. Mature gamonts large, extracellular; gametes usually similar (isogamous) or nearly so, with similar numbers of male and female gametes produced by gamonts; zygotes forming oöcysts within gametocytes; life cycle characteristically consisting of gametogony and sporogony; in digestive tract or body cavity of invertebrates or lower chordates (Fig. 3.14).

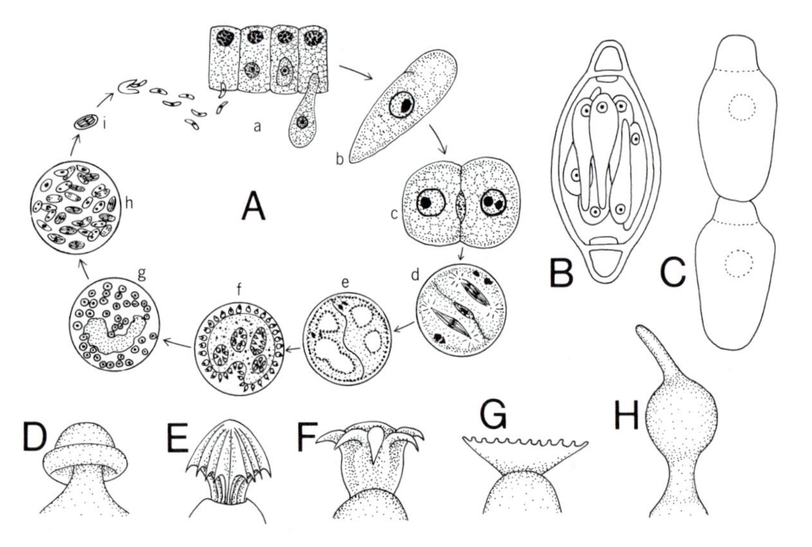

FIGURE 3.14. Gregarines. **A.** Life cycle of *Lankestrella culicis.* At (*a*) sporozoites from germinating spore enter epithelial cells for period of intracellular growth and development, and emerge to develop into mature trophic forms (sporadins) (*b*) in gut lumen. Two sporadins come together (*c*) and secrete common cyst membrane (*d*). Repeated nuclear divisions and multiple cleavage *(e,f)* produce gametes that undergo syngamy to form zygotes (*g*). Each zygote develops into spore (*h*) containing eight sporozoites. Sporozoites escape when appropriate host eats spore. **B.** Spore of *Rhynchocystis porrecta,* with sporozoites. **C.** Two sporadins of *Gregarina blattarum.* **D,E,F,G,H.** Epimerites of several cephaline gregarines, illustrating diversity of form and variety of attachment organs; **D.** *Discorhynchus truncatus;* **E.** *Sciadiophora phalangii;* **F.** *Corycella armata;* **G.** *Dactylophorus robusta;* **H.** *Bulbocephalus elongatus.* (**A** after Wenyon; **B** after Hesse; **C** after Kudo, from Kudo; **D,E,F,G** after Léger; **H** after Watson.)

Order Archigregarinida. Life cycle apparently primitive, characteristically with merogony, gametogony, sporogony, gamonts (trophozoites) aseptate; parasitic in annelids, sipunculids, hemichordates, or ascidians.

Order Eugregarinida. Merogony absent; gametogony and sporogony present; locomotion by gliding or undulation of longitudinal ridges; typically parasites of annelids and arthropods, but some species in other invertebrates.

Order Neogregarinida. Merogony, presumably acquired secondarily; in Malpighian tubules, intestine, hemocoel, or fat tissues of insects.

Subclass Coccidia. Gamonts ordinarily present; mature gamonts small, typically intracellular; syzygy generally absent, if present involves markedly anisogamous gametes; life cycle characteristically consisting of merogony, gametogony, and sporogony; most species in vertebrates.

Order Agamococcidiida. Merogony and gametogony absent.

Order Protococcidiida. Merogony absent; in invertebrates.

Order Eucoccidiida. Merogony present; in vertebrates and/or invertebrates.

Subclass Piroplasmia. Piriform, round, rod-shaped, or amoeboid; no oöcysts; spores and pseudocysts; flagella absent; locomotion by body flexion, gliding, or, in sexual stages (in Babesiidae and Theileriidae, at least), by large axopodiumlike rays; asexual and probably sexual reproduction; parasitic in red blood cells; with merogony in vertebrates and sporogony in invertebrates; sporozoites with single-membraned wall; so far as known, vectors are ticks.

Order Piroplasmida. With characters of the subclass.

Phylum Microspora. Unicellular spores, each with imperforate wall, containing one uninucleate or dinucleate sporoplasm and simple or complex extrusion apparatus; without mitochondria; often, if not usually, dimorphic in sporulation sequence; obligatory intracellular parasites in nearly all major animal groups.

Class Rudimicrosporea. Spore with simple extrusion apparatus; spore spherical or subspherical; sporulation sequence with dimorphism, occurring either in vacuole or in thick-walled cyst; hyperparasites of gregarines in annelids.

Order Metchnikovellida. With characters of the class.

Class Microsporea. Spore with complex extrusion apparatus of Golgi origin; spore shape various, depending largely on structure of extrusion apparatus; spore wall with three layers; proteinaceous exospore, chitinous endospore, and membranous inner layer; outer two layers varying considerably in structure; sporocyst present or absent; often dimorphic in sporulation sequence.

Order Minisporida. General tendency toward minimum development of components of extrusion apparatus and spore wall and accompanying tendency toward maximum development of sporocysts; shape spherical or slightly ovoid; merogony present or absent; sporulation stages usually separated from host cell cytoplasm by intracellular sporocyst.

Order Microsporida. General tendency toward maximum development and varied specialization of components of extrusion apparatus and spore wall with accompanying reduction of sporocysts; sporocysts inside host cell present or absent; cysts of other kinds sometimes formed from host cell membrane or other host material; merogony present; spore shape variable.

Phylum Ascetospora. Spore apparently multicellular (or unicellular?); all parasitic.

Class Stellatosporea. Spore with one or more sporoplasms.

Order Occlusosporida. Sporulation involving series of endogenous buddings that produce sporoplasm(s) within sporoplasm(s); spore wall entire.

Order Balanosporida. Spore wall interrupted anteriorly by orifice; orifice covered externally by operculum or internally by diaphragm.

Class Paramyxea. Spore bicellular, consisting of parietal cell and one sporoplasm; spore without orifice.

Order Paramyxida. With characters of the class.

Phylum Myxozoa. Spores of multicellular origin; with one, two, or three (rarely more) valves; all species parasitic.

Class Myxosporea. Spore membrane generally with two, occasionally up to six valves; trophozoite stage well developed, main site of proliferation; in cold-blooded vertebrates.

Order Bivalvulida. Spore wall with two valves.

Order Multivalvulida. Spore wall with three or more valves.

Class Actinosporea. Spore membrane with three valves; trophozoite stage reduced, proliferation mainly during sporogenesis; in invertebrates, especially annelids.

Subclass Actinomyxia. With characters of the class.

Order Actinomyxida. With characters of the class.

Phylum Ciliophora. Simple cilia or compound ciliary organelles typical in at least one stage of life cycle; with subpellicular cilia present even when surface cilia absent; two types of nuclei, with rare exception; binary fission transverse, but budding and multiple fission also occur; sexuality involving conjugation, autogamy, and cytogamy; nutrition heterotrophic; contractile vacuole typically present; most species free-living, but many commensal, some truly parasitic, and large number found as symphorionts on variety of hosts.

Class Kinetofragminophorea. Oral cilia only slightly distinct from somatic cilia; cytostome often apical (or subapical) or midventral, on surface of body or at bottom of atrium or vestibulum; cytopharyngeal apparatus commonly prominent; compound cilia typically absent.

Subclass Gymnostomatia. Cytostomal area superficial, apical, or subapical; circumoral cilia without kinetosomal differentiation other than closer packing of kinetosomes; toxicysts common; somatic ciliation usually uniform (Fig. 3.15C,D).

Order Prostomatida. Cytostome apical or subapical; circumoral cilia involving anterior parts of all somatic kineties; typical polyploid independent macronucleus; body often large; commonly carnivorous.

Order Pleurostomatida. Cytostome slitlike, lateral; circumoral cilia including anterior parts of only few somatic kineties and showing differentiation into left and right components; body often large, laterally compressed; macronucleus possibly of low ploidy number; voracious carnivores.

Order Primociliatida. Nuclei with prominent RNA-rich nucleolus; cytostome apical, slitlike; somatic cilia sparse, ventral; small marine benthic forms, thigmotactic, often algivorous.

Order Karyorelictida. Macronucleus usually diploid and nondividing; fragile, highly thigmotactic; oral area apical or ventral slit; contractile vacuoles absent; mainly interstitial sand-dwelling forms, often carnivorous.

Subclass Vestibuliferia. Usually apical or near-apical cytostome; mouth formation sometimes involving two anlagen; free-living or parasitic, especially in digestive tract of vertebrates and invertebrates.

Order Trichostomatida. Endocommensal in vertebrate hosts (Fig. 3.15E).

Order Entodiniomorphida. Somatic cilia in form of unique ciliary tufts or bands, otherwise body naked; oral area sometimes retractable; pellicle generally firm, sometimes drawn out into processes; skeletal plates in many species; commensal in mammalian herbivores, including anthropoid apes.

Order Colpodida. Mouth formation sometimes involving two anlagen; body often contorted, rendering morphogenetics of division complex; cysts common; mostly free-living, often in edaphic habitats.

Subclass Hypostomatia. Cytostome on ventral surface; body cylindric or flattened dorsoventrally, often with reduction of somatic cilia; oral area may be sunk into atrium, with atrial cilia present; some species with mouth; free-living, ectocommensal, or endocommensal, principally of invertebrates.

Superorder Nassulidea. Body often cylindrical, with complete somatic cilia; free-living, most often in freshwater habitats.

Order Synhymeniida. Body often cylindrical, with complete ciliation.

Order Nassulida. Distinct preoral suture.

Superorder Phyllopharyngidea. Circumoral cilia restricted to three short rows of kinetosomes near oral opening; somatic cilia only on ventral surface, in two dissymmetric fields; preoral suture skewed to left.

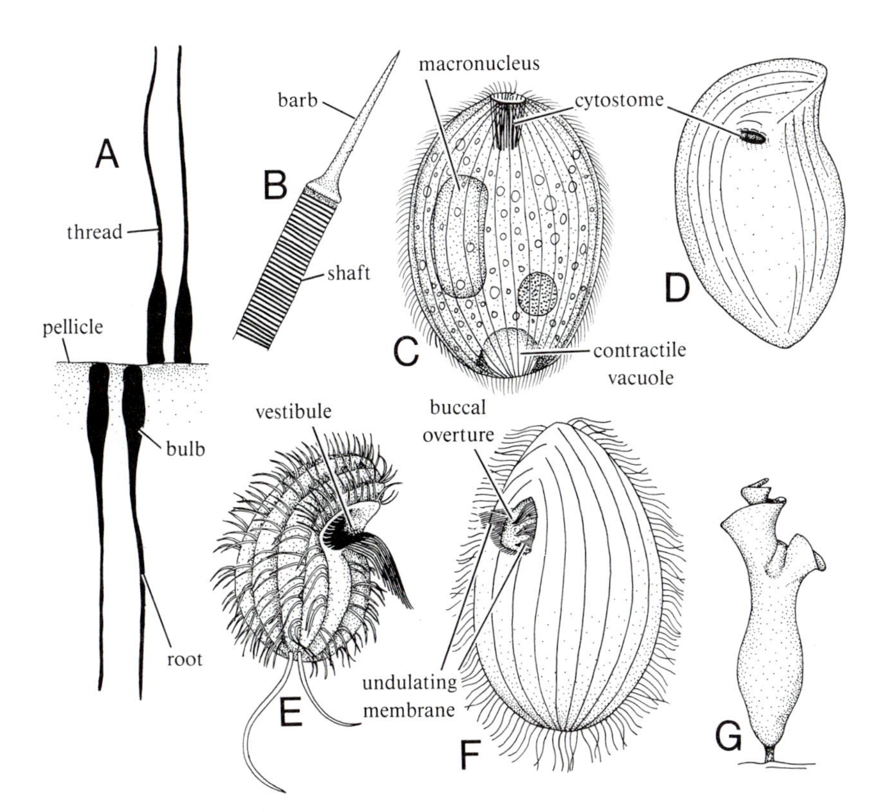

FIGURE 3.15. Trichocysts and some representative ciliates. **A**. Trichocysts of *Dileptus,* before and after discharge, suggesting they evert during discharge. **B**. Diagram of discharged trichocyst of *Paramecium,* from electron micrographs, slowing solid barb and striated shaft. **C**. *Prorodon discolor,* typical rhabdophorine gymnostome. **D**. *Chilodonella labiata,* typical cyrtophorine gymnostome, without specialized oral ciliature but with rather complex oral structure. **E**. *Colpoda steinii,* fairly complex trichostome with vestibule containing modified cilia, laterally placed. **F**. *Tetrahymena pyriformis,* hymenostome with buccal cavity containing specialized ciliature, including undulating membrane on left and three membranellelike membranes on right. **G**. *Spirochona gemmipara,* showing absence of somatic cilia and extension of vestibular region to form complex funnel, characteristic of group. (**A** after Hayes; **B** based on Jakus and Hall; **C**,**D**,**E**,**F**,**G** after Corliss.)

Order Cyrtophorida. Body dorsoventrally flattened or laterally compressed; ventral cilia often thigmotactic; many species with "glandular" adhesive organelle near posterior end.

Order Chonotrichida. Variously vase-shaped, sessile and sedentary forms; naked, except for cilia of ventral surface (displaced to apical end of body); adhesive organelle active in stalk production; reproduction by budding; marine and freshwater species, ectocommensal principally on crustaceans.

Superorder Rhynchodea. Aberrant forms, with sucking tube; body of mature stage often nearly naked or with somatic cilia limited to thigmotactic field; buds or "larvae" typically ciliated (in two fields); commensal or pathogenic, most commonly on gills of marine bivalves.

Order Rhynchodida. With characters of the superorder.

Superorder Apostomatidea. Cytostome inconspicuous or, in certain stages of polymorphic life cycle, absent; glandular complex typically near oral area; in mature forms somatic cilia spiraled, often widely spaced; commonly anterior thigmotactic ciliary field; life cycle complex, sometimes involving alternation of hosts (unique in phylum); most species associated with marine crustaceans.

Order Apostomatida. With characters of the superorder (Fig. 3.16F).

Subclass Suctoria. Suctorial tentacles, generally multiple; adult body sessile and sedentary, seldom with cilia; reproduction by budding stalk commonly present; conjugation often involving

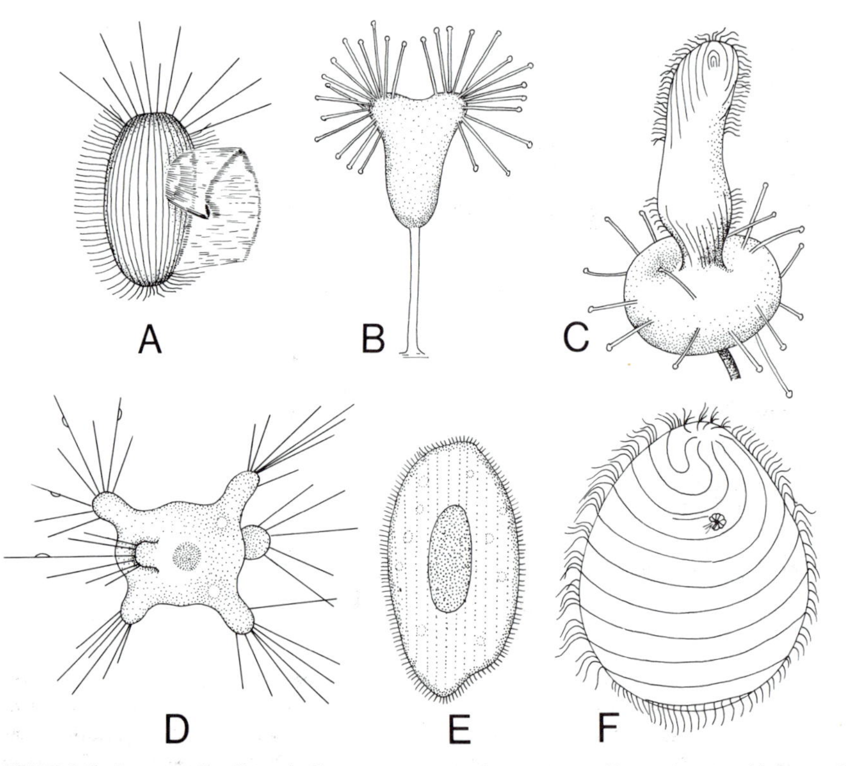

FIGURE 3.16. Representative ciliates. **A.** *Pleuronema coronatum,* hymenostome with conspicuous undulating membrane on right side of buccal overture. **B.** *Acineta tuberosa,* suctorian with two clusters of capitate tentacles. **C.** *Podophrya collina,* suctorian, reproduction by budding. Note holotrichlike ciliation of bud. **D.** *Staurophrya elegans,* asymmetrical suctorian with raptorial tentacles. **E.** *Anoplophrya orchestii,* astome ciliate. **F.** *Foettingeria actinarium,* apostome ciliate, showing rosette-shaped mouth. (**A,B,C,F** and Corliss; **D** after Zacharias, from Kudo; **E** after Summers and Kidder.)

microconjugants and macroconjugants; migratory larva ciliated (with right field and possibly vestigial left field), without tentacles or stalk; widespread on marine and freshwater organisms, occasionally endocommensal.

Order Suctorida. With characters of the subclass (Fig. 3.16B,C,D).

Class Oligohymenophorea. Oral apparatus generally well defined, although absent in one group; oral cilia, clearly distinct from somatic cilia; cytostome usually ventral and/or near anterior end, present at bottom of buccal or infundibular cavity; cysts not uncommon; various species loricate; colony formation common in some groups.

Subclass Hymenostomatia. Body ciliation often uniform and heavy; buccal cavity, when present, ventral; sessile forms and stalk, colony, and cyst formation relatively rare; freshwater forms predominant (Fig. 3.16A).

Order Hymenostomatida. Buccal cavity well defined; oral area on ventral surface, usually in anterior half of body (Fig. 3.15F).

Order Scuticociliatida. Body uniformly to sparsely ciliated; thigmotactic area common in many species; buccal cilia often dominated by tripartite (anterior, middle, and posterior segments); mitochondria long, interkinetal, sometimes fused to form gigantic "chondriome"; probably no trichocysts; cysts common (Fig. 3.15G).

Order Astomatida. Body usually large or long, uniformly ciliated; mouth absent; complex infraciliary endoskeleton and often elaborate holdfast organelles (hooks, spines, or sucker) may be present at anterior end; fission may be by budding, with chain formation; cytoproct absent; contractile vacuoles present; all endoparasitic, mostly in oligochaetes (soil, freshwater, marine); few species in other annelids, molluscs, and turbellarian flatworms; one major group in caudate amphibians (Fig. 3.16E).

Subclass Peritrichia. Oral ciliary field prominent; somatic cilia reduced to temporary posterior circlet of locomotor cilia; widely distributed species, many-stalked and sedentary, others mobile; dispersal by migratory telotroch (larval form); conjugation total, involving fusion of microconjugants and macroconjugants.

Order Peritrichida. With characters of the subclass.

Class Polymenophorea. Somatic cilia complete or reduced, or appearing as cirri; cytostome at bottom of buccal cavity or infundibulum; cytoproct often absent; cysts, and especially loricae, very common in some groups; often large and commonly free-living, free-swimming forms in great variety of habitats.

Subclass Spirotrichia. With characters of the class.

Order Heterotrichida. Generally large to very large forms, often highly contractile, sometimes pigmented; macronucleus oval or often beaded; parasitic and free-living species.

Order Odontostomatida. Laterally compressed, wedge-shaped, with armorlike plates and often posterior spines; somatic cilia reduced; cytoproct absent; several small species, chiefly in putrefying organic matter in freshwater habitats, few marine.

Order Oligotrichida. Body ovoid to elongate, sometimes with tail; pellicle thickened; somatic cilia reduced; macronuclear reorganization bands present; cytoproct absent; free-swimming, macrophagous, mainly pelagic.

Order Hypotrichida. Dorsoventrally flattened, highly mobile (yet often thigmotactic), with unique cursorial type of locomotion; body dominated by compound ciliary structures; rows of widely spaced "sensory-bristle" cilia common on dorsal surface; some species loricate, few colony-forming; macronuclear reorganization bands common; species numerous and very widespread.

Selected Readings

Anderson, O. R. 1988. *Comparative Protozoology*. Springer, New York.

Cavalier-Smith, T. 1987. Eukaryotes with no mitochondria. *Nature* 326:332–33.

Corliss, J. O. 1979. *The Ciliated Protozoa*, 2nd ed. Pergamon Press, New York.

Corliss, J. O. 1984. The Kingdom Protista and its 45 phyla. *BioSystems* 17:87–126.

Corliss, J. O. 1986. Advances in studies on phylogeny and evolution of protists. *Insect. Sci. Applic.* 7:305–312.

Fenchel, T. 1987. *Ecology of Protozoa*. Science Tech. Publ., Madison.

Grell, K. G. 1973. *Protozoology*. Springer, New York.

Kudo, R. R. 1966. *Protozoology*, 5th ed. Charles Thomas, Springfield.

Lee, J. J., S. H. Hutner, and E. C. Bovee. 1985. *An Illustrated Guide to the Protozoa*. Soc. Protozool., Lawrence.

Levine, N. D., et al. 1980. A newly revised classification of the Protozoa. *J. Protozool.* 27:37–58.

Sleigh, M. 1973. *The Biology of Protozoa*. Edward Arnold, London.

Whittaker, R. H. 1969. New concepts of kingdoms of organisms. *Science* 163:150–159.

4

Metazoans

The highly plastic and undeniably successful protozoans have one serious limitation—size. An undivided mass of protoplasm becomes physiologically and structurally ineffective if too large. It suffers from lack of mechanical strength if flattened or elongated, and from lack of surface area if it approaches a spherical shape. Physiological efficiency is limited by constraints imposed by subcellular structure. Animalia (metazoans) are of necessity multicellular.

Of course, some metazoans are smaller than the largest protozoans. Protozoans can form colonies, some of which have thousands of members, whereas some of the smallest metazoans consist of less than a hundred cells. Among those relatively few cells, however, several kinds of nonreproductive cell types are found. This feature is the main distinction between protozoans and metazoans, somatic cell differentiation; that is, cells in metazoans are specialized to achieve specific ends. This is something that can be done *only* at the subcellular level in protozoans.

Structural Plans

Animals are organized along specific anatomical and developmental lines that place constraints on how the individual functions. Several major types of metazoans can be easily outlined.

The simplest arrangement is found in Mesozoa and Placozoa. The outer covering of these small animals is formed of a single layer of cells, but no well-organized inner layer is present. One or more reproductive cells occur within the body; in this they could resemble some protozoan colonies. However, their organizational level is above that of protozoans in that there *is* differentiation of the somatic cells of the covering layer.

Porifera, sponges, exhibit a different plan. A single layer of cells forms the outer covering, but inside the animal there is great differentiation of cell types. There is an internal cavity, or system of cavities, lined by a layer of flagellated cells that resemble choanoflagellates (see Figs. 3.9C and 6.1B,E). They propel water through the cavity system and ingest food. Between the inner cells and the outer layer is a mesogloea that contains a number of different cell types as well as skeletal elements. Sponges do not have the organ systems that are found in other metazoans.

All of the animals above the poriferan level are referred to as eumetazoans. The body is covered by cells derived from the embryonic layer called ectoderm. Typically, eumetazoans have a mouth, a digestive cavity lined by cells derived from the embryonic endoderm, and a third region derived from the mesoderm, a layer that is organized between the inner and outer layers. The mesoderm is derived from either the outer or inner layer of cells during embryonic development, but establishes itself as an independent, third embryonic layer.

Mesozoa, Placozoa, and Porifera suffer under great structural constraints, although they exhibit individual developmental versatility. Eumetazoans exhibit infinite structural versatility as a group, but individually suffer considerable developmental constraints. The secret of metazoan evolutionary success is the division of labor among its cells. The lower metazoans never effectively exhibit this, as the function of individual cells are generally indeterminant; that is, cell function can vary with circumstances. The evolution of embryonic differentiation channels cells into specific functions. Although specific individual cells in a eumetazoan body are quite limited in what each type can do, the animal itself is potentially more versatile in what it can do by virtue of the wide variety of cell types that embryonic differentiation allows to be expressed. In a real sense, less is more.

Histogenesis, the embryonic differentiation of specialized cell and tissue types, is, therefore, characteristic of metazoans. Although a commonwealth of cells, each metazoan is able to function as an integrated organism, and the masses of cells are neither random nor chaotic in their spatial relationships or function. Embryogenesis, the specific pattern of differentiation, is important not only for imposing constraints on function, but it also allows us to consider historic patterns in the development of form and function and is, therefore, useful in helping us to determine phylogenetic relationship.

Metazoan embryology involves an initial period of cleavage, or cell divisions. Cleavage is then followed by cell movements. These movements produce changes in the form of the embryo and the relationships of masses and layers of cells as they are transformed into the various structures of the adult. The first movements result in the formation of distinctive embryonic cell layers, called germ layers. The layered origin of the adult may be almost completely hidden by profound changes in the germ layers during later development, or may be clearly visible in the adult, with the layers retaining something of their original embryonic relationships. Histogenesis may start during early ontogeny, but the more specialized changes usually occur during later organ development, after the organ rudiments have been shaped by the early cell and germ layer movements. These morphogenetic movements can produce certain body cavities during the course of development. These cavities, and their spatial relationship to the germ layers, are also important factors in recognizing phylogenetic relationships.

Cleavage

Cleavage is a period of development during which the zygote is divided into a number of smaller cells called blastomeres, or cleavage cells. Cleavage leads to the formation of a blastula. The form of the blastula is a direct outgrowth of the pattern of cell divisions occurring during cleavage.

Cleavage divisions follow a definite pattern, characteristic of the organism. They are associated with the form of the zygote and the distribution of yolk material in it. Three basic types of ova may be recognized: isolecithal, telolecithal, and centrolecithal (Fig. 4.1). **Isolecithal ova** have little yolk equally distributed throughout the cytoplasm. **Telolecithal ova** have yolk concentrated at the vegetal pole. Ova may be moderately or strongly telolecithal, depending on the quantity of yolk present. **Centrolecithal ova** have the yolk formed as a central mass.

Isolecithal ova usually cleave in a characteristic manner (Fig. 4.1A). The first cleavage follows the primary axis of the ovum, passing from animal to vegetal pole, producing a two-celled embryo. The whole ovum divides, so cleavage is said to be total, or **holoblastic**. The second division is at right angles to the first, and also passes from the animal to the vegetal poles, producing a four-celled embryo (Fig. 4.1A2). The third cleavage is at right angles to the first two, therefore cutting across the primary axis. This is a transverse cleavage, and produces an embryo consisting of a quartet of blastomeres in the animal half and a similar quartet in the vegetal half (Fig. 4.1A3). If the two quartets are approximately equal in size, cleavage is said to be equal; if they are markedly dissimilar in size, it is termed unequal. Moderately telolecithal ova usually develop like isolecithal ova, undergoing holoblastic cleavage (Fig. 4.1B). However, most isolecithal ova undergo equal cleavage, whereas telolecithal ova usually undergo unequal cleavage. The difference in size of the two quartets of cells increases with increasing yolk concentration at the vegetal pole.

Whether cleavage is equal or unequal, the third cleavage spindles may parallel the primary axis (Fig. 4.1B11) or be at an angle to it (Fig. 4.1B9). If the spindles parallel the main axis, the upper quartet of cells lies directly above the lower ones, and cleavage is **radial** (Fig. 4.1B12). If the spindles are canted (Fig. 4.2), the cells of the upper quartet lie over the furrows between the cells of the lower quartet, and cleavage is said to be **spiral** (Fig. 4.1B10).

The cytoplasm of strongly telolecithal ova is concentrated in a small area at the animal pole (Fig. 4.1C). Only the cytoplasm divides; the large mass of yolk remains undivided (Fig. 4.1C16,17). Such cleavage is termed partial or **meroblastic**. Centrolecithal ova have cytoplasm concentrated at the periphery of the ovum, and generally the only part of the ovum to divide is the peripheral layer (Fig. 4.1D). Cleavage of this type is **superficial**.

In echinoderms, chordates, and some other kinds of animals, cleavage is generally **indeterminate**; that is, the first blastomeres are equally potent in being able to develop a complete organism. If the first two or four blastomeres are separated, each continues its cleavage to form a diminutive blastula, which in some cases continues its development to form di-

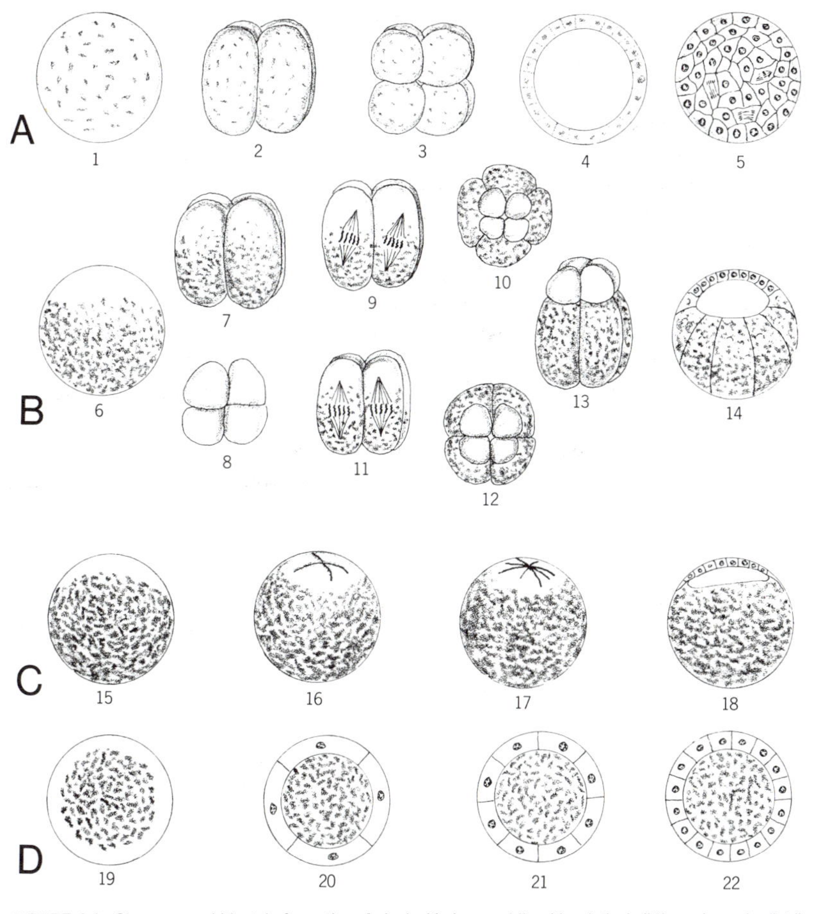

FIGURE 4.1. Cleavage and blastula formation. **A**. Isolecithal ovum (*1*), with relatively little and evenly distributed yolk, undergoes total (holoblastic) and equal cleavage. Four-celled embryo (*2*) divides to form equal quartets of cells (*3*). Further equal divisions produce coeloblastula with central blastocoel (*4*) or stereoblastula (*5*) without blastocoel. **B**. Moderately telolecithal ovum (*6*), yolk concentrated in vegetal hemisphere, undergoes total but unequal cleavage. Four-celled embryo (*7,8*) shown in side and polar views. Protostomes typically undergo spiral cleavage (*9*); canting spindles result in micromere quartets that lie over furrows between four macromeres (*10*). Deuterostomes typically undergo radial cleavage (*11*); as spindles parallel primary axis, micromere quartet lies immediately over macromere quartet (*12,13*). Either spiral or radial cleavage of this kind may lead to coeloblastula (*14*) with excentric blastocoel, or to stereoblastula with unequal blastomeres. **C**. Strongly telolecithal ovum (*15*), yolk heavily concentrated in vegetal hemisphere, undergoes partial (meroblastic) cleavage. Cleavage furrows do not wholly divide zygote (*16,17*), and blastula (*18*) consists of cells resting on undivided yolk mass with space between yolk and cells comparable with blastocoel. **D**. Centrolecithal ovum, yolk concentrated in central mass (*19*), undergoes superficial cleavage. Cleavage divides surface protoplasm but not yolk mass (*20,21*). Resulting blastula (*22*) solid, with undivided yolk mass filling blastocoel.

minutive larvae. Twins, triplets, or quadruplets can be obtained from a single zygote. In many phyla cleavage is **determinate**, and the exact destiny of the early blastomeres (Fig. 4.2) is determined during cleavage when they are formed. When a two-celled embryo of such is separated into isolated cells, each produces a partial larva.

Blastula

The blastula varies considerably in different kinds of animals. Many blastulae have a blastocoel, located centrally if the ovum was isolecithal (Fig. 4.1A4) or displaced toward the animal pole if the ovum was telolecithal (Fig. 4.1B14). Such hollow blastulae are called **coeloblastulae**. Others are solid, without a blastocoel, and are termed **stereoblastulae**. Stereoblastulae arise from both isolecithal and moderately telolecithal ova. Where cleavage is meroblastic, only the upper part of the zygote is divided, so that the blastula consists of a disc of cells lying on top of an undivided yolk mass (Fig. 4.1C18). The blastocoel is reduced to the space separating cells from the yolk mass. Centrolecithal ova, undergoing superficial cleavage, produce blastulae that consist of a continuous layer of cells surrounding a central, undivided yolk mass (Fig. 4.1D22).

Gastrulation: The Beginning of Germ Layer Formation

Until the blastula is fully formed, embryonic development is dominated by cleavage divisions. The blastula is transformed into a gastrula by cell movements known as **gastrulation**. Gastrulation is accomplished in a variety of ways, but in all cases it produces the germ layers. The most important methods are invagination, epiboly, involution, delamination, and ingression.

Gastrulation by invagination (Fig. 4.3A) occurs in many coeloblastulae and is the type most commonly emphasized in introductory texts. The vegetal hemisphere of the blastula grows inward to form a pouch-like space, the **gastrocoel**, also known as the **archenteron** or the **primitive gut**. The gastrocoel opens to the exterior through the **blastopore**. As invagination proceeds, the gastrocoel grows at the expense of the blastocoel. The blastocoel may be either completely obliterated or may persist as a fairly large cavity.

Gastrulation by **epiboly** and **involution** (Fig. 4.2B,C) occurs in animals with telolecithal ova, where the large amount of yolk in the vegetal hemisphere makes inward growth of the blastomeres difficult or impossible. Where unequal cleavage has produced large cells, or macromeres, at the vegetal pole and small cells, and micromeres, at the animal pole, then the small, more active cells divide and

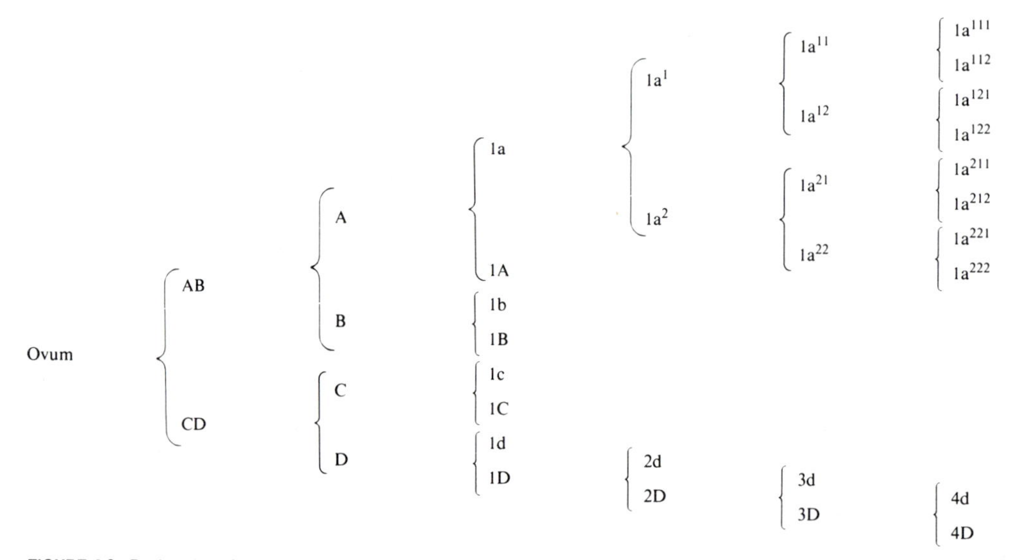

FIGURE 4.2. Designation of blastomeres in spiral cleavage. Scheme of divisions, with letter and number designations of blastomeres, shown up to first micromere quartet formation (1a, 1b, 1c, 1d). Beyond this point, only divisions of micromere 1a and macromere 1D shown. All other micromeres and macromeres divide in similar manner.

grow, and move down over the larger, more inert, yolk-filled cells. This movement is epiboly (Fig. 4.3C). It results in a gastrula with a blastopore at the vegetal pole and with at least a vestige of a gastrocoel. Gastrulation by involution (Fig. 4.3B) is most commonly seen in disclike blastulae produced by meroblastic cleavage. At some point, usually related to the future symmetry of the organism, cells turn inward and move back under the surface cells and over the yolk mass. This produces a doubled surface layer separated from the yolk by a small space, the rudiment of the gastrocoel. Evidently this type of gastrula is highly modified and lacks a typical blastopore or gastrocoel.

Gastrulation by invagination, epiboly, or involution produces a gastrula with a central cavity or some vestige of it, usually opening to the outside by a blastopore. This type of gastrula is called a **coelogastrula**.

When gastrulation occurs by delamination or ingression, however, there is nothing comparable to a blastopore and there may be no evidence of a gastrocoel. This type of gastrula is a **stereogastrula**. Gastrulation by **delamination** may occur in either of two ways (Fig. 4.3D,E). A coeloblastula may produce a two-layered gastrula by tangential cell divisions, cutting off the inner portions of the blastomeres as an inner mass of cells (Fig. 4.3D). This obscures the central cavity or produces a gastrula without an opening to the interior cavity. Stereoblastulae can gastrulate by delamination through the separation of the outer cells from an inner cell mass. This usually involves a sorting-out process, accompanied by flattening of the outer cells; it produces a solid gastrula without gastrocoel or blastopore. Gastrulation by **ingression** occurs when surface cells migrate to the interior. Ingression may be **unipolar**, occurring only at one pole, or **multipolar**, occurring at various points on the blastula surface. Ingression forms a gastrula without gastrocoel or blastopore (Fig. 4.3E). Unipolar ingression may be a modification of invagination.

Gastrulation produces a two-layered embryo; the outer layer is the ectoderm, and the inner layer the endoderm. If present, the gastrocoel is often the primordium of the digestive cavity, and the blastopore may be thought of as the primitive, embryonic mouth. For reasons that will become evident later,

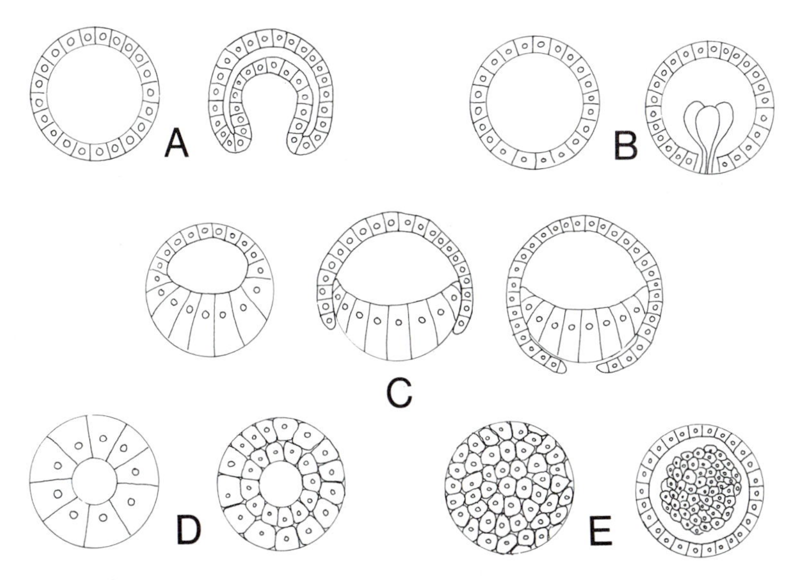

FIGURE 4.3. Blastulae and gastrulation. **A.** Coeloblastula with central blastocoel, as might result from total, equal cleavage of isolecithal ovum, undergoing invagination. Primitive gut (archenteron) opens to outside by way of blastopore. **B.** Coeloblastula undergoing gastrulation by polar ingression. No blastopore formed; endoderm cells may be solid mass or hollowed out to form primitive gut. **C.** Coeloblastula with blastocoel displaced toward animal pole, as commonly results from total, unequal cleavage of moderately telolecithal ova, undergoing gastrulation by epiboly, downward movement of micromeres over macromeres. Such gastrulae may lack primitive gut, although blastopore forms at vegetal pole. **D.** Blastula with large cells undergoing gastrulation by tangential cell division. No blastopore formed, and endoderm may be solid or hollowed out to form primitive gut. **E.** Stereoblastula gastrulating by delamination, resulting in no primitive gut or blastopore, although endoderm may later hollow out to form primitive gut.

the gastrula is of considerable interest in speculation about evolutionary relationships. The blastopore is sometimes retained as the adult mouth, sometimes it is closed off and a secondary mouth is formed some distance away; and some would divide the animal kingdom into two main branches, the Protostomia, which retain the blastopore as a mouth, and the Deuterostomia, which form a new mouth some distance away from the closed blastopore, which may persist as or be close to the anus.

Mesoderm: The Third Germ Layer

Despite variations in the form of gastrulae and in the methods of gastrulation, gastrulae have some remarkably constant features. That a two-layered embryo should have inner and outer layers is scarcely surprising. What is fascinating is that these layers should tend to form the same body parts, even in animals with dissimilar gastrulae that arise by diverse methods of gastrulation. The ectoderm of most animals gives rise to the outer, epithelial covering of the body and its derivatives, to the lining of the most anterior and posterior parts of the digestive tract, and to the nervous system and sense organs. The endoderm gives rise to the epithelial lining of the remainder of the digestive tract, and the glands and other derivatives arising from it. Because all the cells of a typical gastrula are a part of either the inner or outer layers, it is evident that all other tissues also must arise from them. This is brought about, however, by forming a third germ layer, the **mesoderm**. Mesoderm forms in a variety of ways and at different times during development. Ectoderm and endoderm are more or less homologous throughout the animal kingdom, except possibly in the sponges. Mesoderm, on the other hand, is variable in its origins.

Mesoderm or its equivalent arises either as a loose connective tissue or as an epithelial layer. The loose type of mesoderm is called **mesenchyme**. Mesoderm is, for example, mesenchymal in ctenophores, derived from the ectoderm, and so may be called **ectomesoderm**. In all advanced metazoans, the bulk of the mesoderm arises generally from the endoderm and is called **endomesoderm**. However, endomesoderm may be either epithelial or mesenchymal, and can arise at different times and in varied ways.

Ectomesoderm arises by simple ingression of cells from the ectoderm. The space between the ectoderm and endoderm is occupied by a jellylike matrix, the **mesoglea**, which comes to have a greater or lesser population of cells depending on the extent of cellular ingression.

Endomesoderm formation is too diverse to discuss in detail at this point. In some animals, all of the mesoderm arises from a special blastomere, formed during early cleavage. In others, it arises as a mesenchymal mass, and in still others it appears as an epithelial tissue stemming from the endoderm (Fig. 4.4).

Body Cavities

An important aspect of metazoan architecture is the relationship between important body cavities of the germ layers and their derivatives. The blastocoel is the first cavity to appear, and parts of it may persist in adults. The gastrocoel is generally permanent, forming all or part of the digestive cavity. Sponges are exceptional, in that the system of cavities is entirely unlike those found in other animals, in both function and derivation. In the most primitive eumetazoans, the gastrocoel is the only embryonic cavity, and the digestive cavity is the only body cavity of the adult. This is true of the Cnidaria and Ctenophora, and the acoelomate bilaterians.

In higher invertebrates, the mesoderm tissues do not completely fill the space between the digestive tube and body wall. A body cavity is present, separating the body wall from the digestive tube. There are two kinds of body cavities in this regard, the pseudocoel and the coelom. In animals with a **pseudocoel**, the mesoderm is mesenchymal and eventually disappears from near the gut as the tissue restricts its development to near the body wall. In these animals the body wall is lined with mesodermal tissue, but this is separated from the digestive tube by the pseudocoel (Fig. 4.4C). In animals with a **coelom**, the coelomic space appears within the mesoderm. As a result, the body wall is lined with mesoderm and the digestive tract is also surrounded by it (Fig. 4.4F).

Although usually thought of as steps in a transition series, lack of a cavity, pseudocoels, and coeloms are each unique methods of solving basic problems that animals encounter. The presence of a cavity profoundly alters body organization. Except at openings, the body wall and viscera are disconnected and can move or adapt independently. The perivisceral fluid in a cavity permits a more efficient distribution of materials in the body, and, in addition, the cavity and its contained fluid can have an extremely important function as a hydrostatic skeleton. Contractions of the body wall musculature must work in opposition to some rigid structural element if any movement is to be achieved. For example, the fluid pressure in the pseudocoel everts the corona of a rotifer. Because muscles cannot push but can only contract, a type of hydrostatic mechanism is needed whenever a part must be elevated, everted, or otherwise moved in the absence of a system of firmly anchored levers. The parenchyme of acoelomates also serves as a hydrostatic skeleton, but the free-moving fluids in the pseudocoel and coelom are considerably more effective in this regard.

The physically confining aspects of mesenchyme or parenchyme in the acoelomate condition limits the motions muscles can achieve, as well as the size of the body due to reliance on diffusion of nourishment into and waste products out of solid masses of tissues. Some acoelomates, like gnathostomulids, have secondarily developed cavities in parts of the body. The pseudocoelomate phyla have effective locomotory powers, but the restriction of muscles to the body wall imposes constraints on the structural plan. Gut musculature remains delicate, and peristaltic motions of the body wall are usually less important than the beating of cilia for food conduction. The digestive system of pseudocoelomates is a complete digestive system, but does not have the extensive diverticula for food distribution seen in acoelomates, indicating the effectiveness of the pseudocoel fluid itself in food transport. However, none of the pseudocoelomates has a circulatory system,

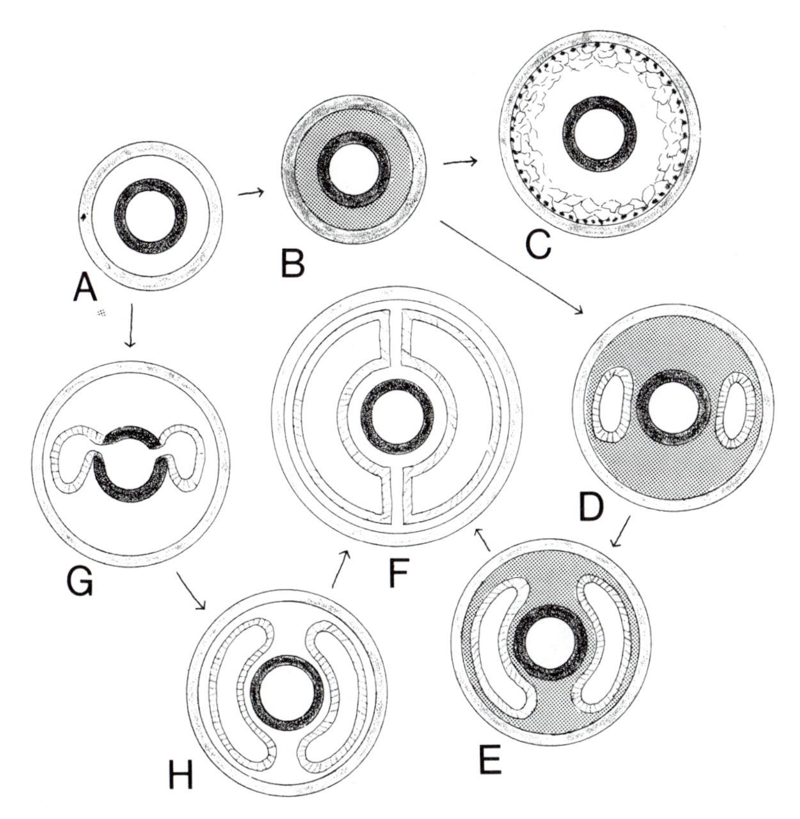

FIGURE 4.4. Germ layers and body cavities (in schematic transverse section). Endoderm as *dark layer,* ectoderm intermediate in *shade,* and mesoderm *cross hatched.* Cnidarian level of organization shown at **A**. Development of true mesoderm converts **A** into acoelomate (**B**). Pseudocoelomates separate mesoderm from endoderm, leaving pseudocoel between endoderm and body wall (**C**), with no true peritoneal linings. Typical coelomate construction shown at **F**, with only peritoneal mesoderm included with coelom, lined with peritoneum, between body wall and gut, divided into lateral compartments by dorsal and ventral, double-walled mesenteries. Most protostomes have schizocoels, in which spaces appear in mesoderm on each side of body (**D**), and grow (**E**), eventually achieving form shown at **F**. Mesoderm can arise as lateral outpocketings of primitive gut (**G**), which grow and detach (**H**), eventually meeting above and below gut to form dorsal and ventral mesenteries (**F**).

and they are almost all small. It is likely that size and the effectiveness of the pseudocoelomate plan have been the most significant factors in mitigating against a circulatory system. It is clear that the pseudocoel has imposed limits on the body plan.

In those animals that exhibit a high degree of organization of the pseudocoelomate body plan (viz., priapulids, kinorhynchs, and possibly chaetognaths), curious convergences to true coelomates appear. For example, the greater independence of the body wall and the presence of a cuticle in some of these forms have favored the development of pseudosegmentation. This is frequently accompanied by segmentation of the body wall musculature and a tendency for the nervous system to develop ganglia associated with the muscular segments. Although it is improbable that any modern pseudocoelomate groups are directly ancestral to metameric, coelomate animals, the manifestation of pseudosegmentation may reveal the kind of changes that produced truly segmented animals. Priapulida and chaetognaths also present curious attempts to line the pseudocoel akin to what is seen in a coelom.

A true coelom may be formed in either of two ways. When the mesoderm arises as mesenchyme, it is then at first a solid mass filling the space between ectoderm and endoderm. Delamination occurs to form a **schizocoel** that results in the separation of an outer somatic layer and an inner splanchnic layer (Fig. 4.4D,E,F) within the mesoderm. The somatic layer adheres to the body wall, and the splanchnic layer to the digestive tube. The coelomic space that is formed is thus completely surrounded by mesodermal tissue. Alternatively, where the mesoderm arises as a pouch or double epithelial layer from an outpocketing of the endoderm, an **enterocoel** is formed, and the coelom is a derivative of the embryonic digestive cavity (Fig. 4.4F,G,H).

The Origin of Metazoans

Zoologists disagree about the origin of metazoans. Inasmuch as fossil evidence cannot be obtained, our conclusions must be based on other kinds of information. Elaborate theories have been based on analyses of developmental patterns and the erection of hypothetical animals to explain animal history. Although many of these theories are interesting—some approach a type of biopoetry—very few are based on consideration of the meaning of form and function (see review in Chapter 38). It is far more effective for the student of invertebrate zoology to consider the animals themselves: which characters they share and which are unique, and what animals do and how they do it, to discern possible paths of animal history. One should not try to analyze animals within the confines of the many old theories of invertebrate evolution where fancy often had to make up for lack of facts.

Selected Readings

Barrington, E. J. W. 1979. *Invertebrate Structure and Function*, 2nd ed. John Wiley, New York.

Beklemishev, W. N. 1969. *Principles of Comparative Anatomy of Invertebrates*, vol. 1 and 2. Univ. Chicago Press, Chicago.

Clark, R. B. 1964. *Dynamics of Metazoan Evolution*. Clarendon Press, Oxford.

5

Mesozoa and Placozoa

The Mesozoa and Placozoa comprise a diverse array of microscopic animals that have intrigued invertebrate zoologists with their apparent simplicity of body plan. These forms have been placed at various times in different combinations in a number of different phyla, Rhombozoa (= Dicyemida), Orthonectida, and Placozoa, but are treated here together for convenience. These are animals that exhibit a simple multicellular body plan, with minimal cell and tissue specializations. These groups appear to be primitive (see Chapter 38) and are placed at the base of the animal family tree.

The Rhombozoa and Orthonectida are parasitic forms found in a number of invertebrate types. Because they are obligate parasites, it has been suggested by some that they cannot be considered primitive, but rather are more likely degenerate flatworms. Although this suggestion may have some merit, structurally these groups seem to share little in common with platyhelminths, and seem better retained as a separate group. The Placozoa are enigmatic, known from a single free-living species, *Trichoplax adhaerens.*

The rhombozoans and orthonectids are often treated as the Mesozoa, other times as two separate phyla; *Trichoplax* has typically been placed in a separate phylum. These groups, however, all share a general body plan, although we need to learn more about the life cycles of the parasitic forms and the overall biology and details of structure of all these animals.

A fourth form, *Salinella salina*, has been in taxonomic limbo for decades. A microscopic form, it reportedly was obtained in cultures made from saltbed material collected in Argentina. *Salinella* was supposed to have a single layer of cells, surrounding a central "digestive cavity" with a "mouth" and an "anus." The ventral body surface was ciliated, and bristles occurred on the dorsal surface around the mouth and the anus. The cell surface of the digestive cavity was ciliated. *Salinella* supposedly reproduced by transverse fission, as well as sexually when two organisms formed a common cyst. Blackwelder suggested a phylum Monoblastozoa for *Salinella.* Despite the best efforts by many people for decades to re-collect this species, these animals have never been found again. Grell actually expressed the opinions of many when he observed that the species had not been seen since its description and doubted it had even been seen then. In light of no confirmatory observations of *Salinella* in almost a century, it seems better to delete this "creature" from the invertebrate pantheon and concentrate our future attentions on observable beasts.

Definition

These groups are microscopic forms, with a simple, ciliated, epithelial cell layer surrounding an inner mass composed of either a single reproductive cell, or many gametes and sometimes other weakly dif-

*ferentiated fibrous cells. The mesozoans alternate between a dispersive/infective **infusoriform** stage and a reproductive **vermiform** stage.*

Rhombozoa

Rhombozoa are common parasites found in the nephridia of squids and octopi. Two orders are recognized: the Dicyemida and the Heterocyemida. The dicyemids are the better-understood group. The **vermiform** or **primary nematogen** (Fig. 5.1A) is a small, ciliated animal with a constant number of cells. The outer layer of cells, the **somatoderm**, surrounds an internal, elongated **axial cell**. The anterior end is composed of eight or nine polar cells forming a **polar cap**, and two parapolar cells immediately behind. The trunk somatoderm contains from 10 to 15 cells, depending on the species. Dicyemid vermiforms feed by taking in particulate and dissolved nutrients from the host urine using phagocytosis or pinocytosis by the somatoderm cells. As the primary nematogens mature, the internal axial cell divides to form axoblasts, or germinative cells. These undergo divisions within the parent to produce daughter **vermiforms**, which leave the parent to take up life on their own within the host nephridium.

When the population density of vermiforms reaches a critical level, they become sexually mature, and are sometimes referred to as **rhombogens**. This transformation may be linked to sexual maturation in the host. The vermiform axial cells again form an axoblast, and this divides to form a kind of hermaphroditic structure, the **infusorigen**, that produces both egg and sperm daughter cells. The gametes fuse to form a zygote, or cross-fertilization may occur from some adjacent individual. The zygote develops into a larva.

The resulting **infusoriform larva** (Fig. 5.1B) has two unciliated apical cells and several large, ciliated cells that cover the posterior surface of the short, oval larva. The interior of the larva is termed the **urn.** It is made up of four central cells, covered posteriorly and laterally by two large, flat capsule cells. Just anterior to the central cells is a small urn cavity, bounded anteriorly by two small, ciliated cells located just behind the apical cells, and a quartet of small cover cells. The infusoriform larvae escape from the cephalopod host, and it is suspected that these then are infective and are taken up by a new host. After possibly passing through the host's circulatory system, the parasite takes up residence in the host's nephridium.

There are about 65 species of dicyemids known at

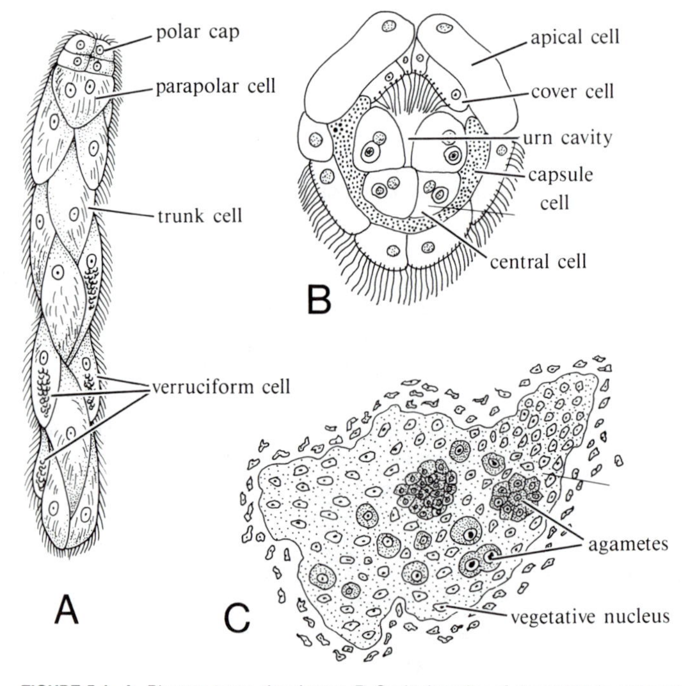

FIGURE 5.1. A. *Dicyema typus,* rhombogen. **B.** Sagittal section of dicyemid infusoriform larva. **C.** *Rhopalura ophiocomae,* male plasmodium with agametes. (**A** after Whitman; **B** after Neuvel; **C** after Caullery and Mesnil.)

present, and it is estimated that there may be as many as 500 more species to be described.

The heterocyemids consist of only two species in two separate genera. Their chief distinguishing characters from dicyemids are the lack of both surface cilia and polar caps in the vermiform/nematogen stage, although their infusoriform larvae are ciliated. *Conocyema* occurs in octopus nephridia, and little is known of its life cycle. *Microcyema* is found in cuttlefish, and has a syncytial outer somatic layer. Two alternative life cycles are known for *Microcyema*. In one, a larva develops not unlike the infusoriform larva of the dicyemids. In the other, a ciliated Wegener's larva is formed, a much reduced and densely ciliated variant of the infusoriform type.

Orthonectida

Orthonectida is an order of parasites found in a variety of invertebrate hosts such as platyhelminths, nemertines, brittle stars, pelecypods, and polychaetes. The parasitic asexual stages are completely unlike rhombozoans. They form a multinucleated, amoeboid plasmodium that grows and fragments into agametes (Fig. 5.1C). Infestations can become so bad that parasitic castration of the hosts can result.

These agametes ultimately give rise to males or females that resemble dicyemid vermiforms or nematogens (Fig. 5.2). The adult outer somatoderm, unlike that of dicyemids, is arranged into distinct rings of cells. Sexes are separate, and the number and size of rings in each sex and in different species vary. For example, in *Rhopalura ophiocomae* (Kozloff, 1969) the rings alternate between wide and narrow bands (Fig. 5.2). In *Rhopalura* the internal male reproductive cells are encased in a sheath of contractile cells, whereas in the females the contractile cells are dispersed. In the genus *Ciliocincta*, however, the contractile cells are restricted to around the genital pores (Kozloff, 1971). The possession of these contractile cells is unlike that of dicyemids. When sexually mature, the adult parasites leave their hosts and mate as free-swimming organisms. Zygotes develop within the female parent, and cili-

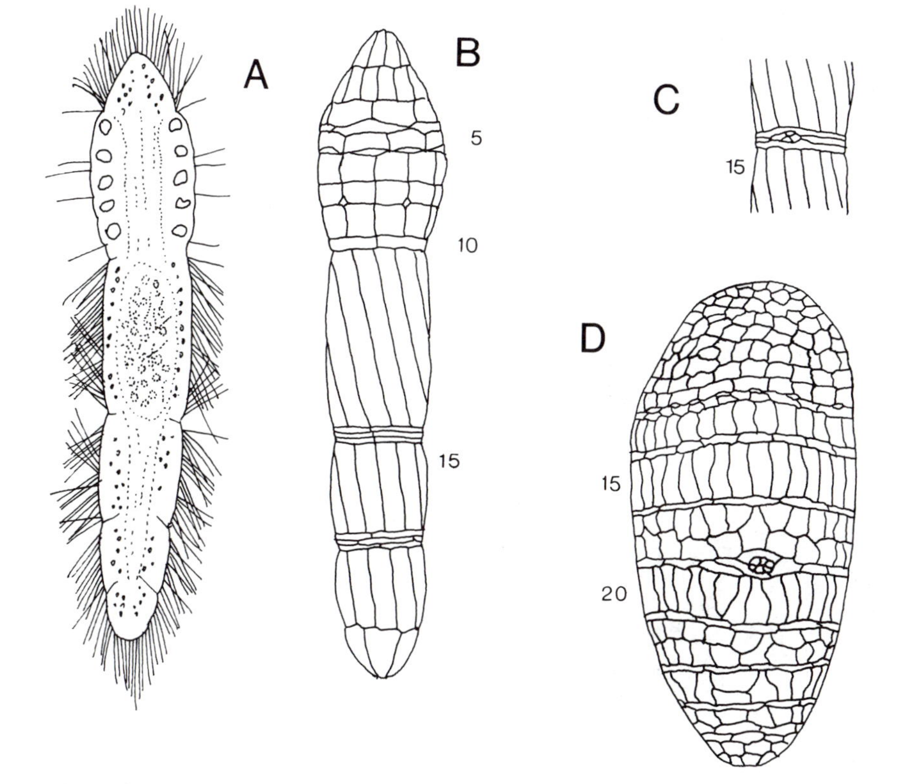

FIGURE 5.2. *Rhopalura ophiocomae.* **A**. Male, optical section showing cilia, lipid inclusions, and developing testis. **B,C,D**. Surface cell patterns with rows numbered; **B**. male; **C**. male gonopore; **D**. female, ovoid type. (After Kozloff.)

ated larvae result that resemble the infusoriform larvae of dicyemids in that the outer, ciliated layer surrounds internal germ cells. These larvae in turn escape from the female parent and, after a migratory period, become infective for new hosts. Once in the host, the outer cells are shed and the inner mass produces the plasmodial stage.

Currently only about 20 species of orthonectids are known, and because of their widespread and sporadic occurrence in all kinds of invertebrates, there is no way of estimating just how many species remain to be discovered.

Placozoa

The single representative of this group, *Trichoplax adhaerens* (Fig. 5.3A) has been studied extensively in laboratory aquarium culture (Rassat and Ruthmann, 1979). It occurs naturally in marine waters, but has been only rarely collected directly from nature. The animal is ciliated, essentially flattened, and somewhat discoidal, although the surface is irregularly folded. The body is composed of outer and inner layers (Fig. 5.3B). The dorsal epithelium is composed of thin, ciliated cells whose nuclei project inwardly, and these are interspersed with refractile bodies thought to be composed of lipid. The ventral epithelium is columnar, ciliated, and marked with many microvilli. Interspersed among the ciliated cells are gland cells and degenerating cells. The inner layer between the epithelia contains an intercellular space with contractile fiber cells whose nuclei are rather unusual in being tetraploid.

The gland cells in the ventral epithelium are assumed to produce secretions used in extracellular digestion, the food being taken up by pinocytosis into the ciliated cells. Symbiotic bacteria have been noted within the fiber cells, and it is conceivable that these may provide another source of nourishment. Changes in body shape are facilitated by contractions of the fiber cells, but actual locomotion is achieved by the ventral cilia. Asexual reproduction occurs by budding or binary fission. A possible sexual mode has been postulated (Grell, 1971); however, the details of fertilization and development are

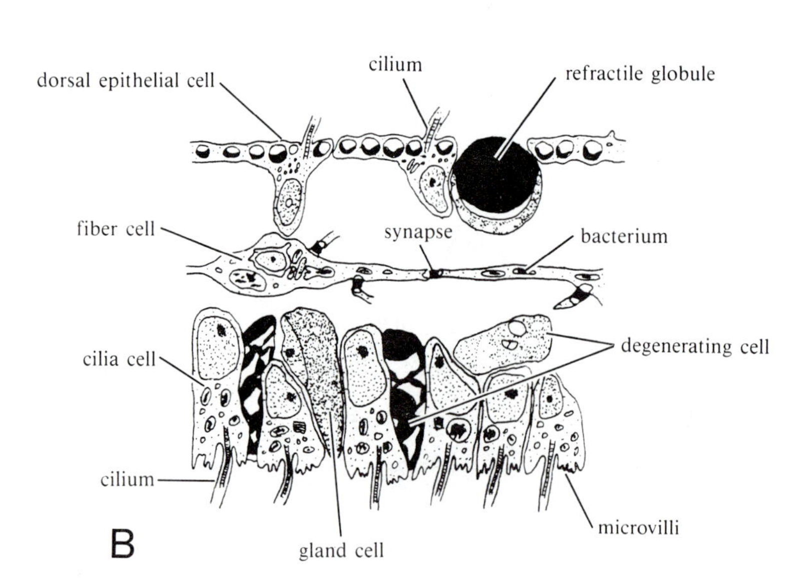

FIGURE 5.3. *Trichoplax adhaerens.* **A.** Cross section through margin. **B.** Schematic cross section of body layers. (**A** after Schulze; **B** after Rassat and Ruthmann.)

not clear at this time. The inner layer can develop large egglike cells that, as the parent degenerates, undergo early cleavage. In addition, nonflagellated cells develop in the inner layer, and these are presumed to be sperm.

Phyletic Position

The rhombozoan mesozoans were once thought to have possessed more complex life cycles than has proven to be the case, direct infection of hosts by infusorigens having been reported from two laboratories (Lapan and Morowitz, 1972). Orthonectids have long been known to complete their life cycle without an intermediate host. The postulated degeneration of mesozoans from flatworm stock is possibly overstated, and their slight resemblance to platyhelminth miracidian larvae is probably convergent. The simplicity of the mesozoan body plan is thus most likely primitive.

Of special note in this regard is the DNA content of these groups. In dicyemids, the guanine–cytosine content is only 23 percent of total DNA. This percentage is closer to that noted in ciliate and amoeboid protozoa (22 to 30 percent) than it is to that of flatworms (35 to 50 percent), whose percentage is more characteristic of higher metazoans. In this regard, the placozoan *Trichoplax* has been recorded as having the lowest total content of DNA of any metazoan, about two-thirds that seen in sponges and only about ten times that of bacteria. This DNA data in connection with the simple structural plan of the placozoans most likely indicates a truly primitive position for the group (Grell, 1981).

Selected Readings

Grell, K. G. 1971. Embryonalentwicklung bei *Trichoplax adhaerens*. *Naturwiss.* 58:571.

Grell, K. G. 1981. *Trichoplax adhaerens* and the origin of Metazoa. *Atti Acad. Naz. Lincei* 49:107–122.

Kozloff, E. N. 1969. Morphology of the orthonectid *Rhopalura ophiocomae*. *J. Parasit.* 55:171–195.

Kozloff, E. N. 1971. Morphology of the orthonectid *Ciliocincta sabellariae*. *J. Parasit.* 57:585–597.

Lapan, E. A., and H. Morowitz. 1972. The Mesozoa. *Sci. Am.* 227(6):94–101.

McConnaughey, B. H. 1963. The Mesozoa. In *The Lower Metazoa* (E. C. Dougherty, ed.), pp. 151–165. Univ. California Press, Berkeley.

Rassat, J., and A. Ruthmann. 1979. *Trichoplax adhaerens* (Placozoa) in the scanning electron microscope. *Zoomorph.* 93:59–72.

Stunkard, H. W. 1972. Clarification of the taxonomy in the Mesozoa. *Syst. Zool.* 21:210–214.

6

Porifera

Sponges have a way of life so unlike that of other metazoans that they were long believed to be plants and were only recognized as animals in 1825. Adult sponges are always attached and motionless. Even when touched they do not draw away, although some will either round up or contract the surface openings when manhandled. About 5000 poriferan species are currently recognized. They occur most abundantly in shallow, coastal waters, attached to the bottom or to submerged objects, but also are found in deep water, and some 200 species are adapted to fresh water.

Definition

Sponges are aquatic, sessile, and either multicellular or multinucleate-syncytial form. They do not have columnar tissues supported on a basal membrane, organ systems in the usual sense, or special sensory structures. The body is pierced by pores, with smaller and more numerous pores for incoming water, and larger and fewer pores for outgoing water. The pores are connected by means of a canal system, in part lined with a single layer of flagellated choanocytes or a choanosyncytium that generates and maintains the water currents. The internal skeleton is composed of an organic matrix of collagen, with or without compact fibers of spongin, and with or without siliceous or calcareous spicules.

Body Plan

Sponges have many unusual features, but the most obvious characteristic is the porous nature of the body, from which the name Porifera comes. Two kinds of pores are found (Fig. 6.1A). Particles suspended in the water near a living sponge enter the many small dermal or **incurrent pores**, or **ostia**, and emerge from the fewer large **excurrent pores**, or **oscula**. A system of passageways and chambers, differing in complexity with the species, connects the ostia and oscula. As water passes along these channels, suspended food particles are removed and ingested, and respiratory exchange occurs due to the flagellar action of the choanocytes or choanosyncytia. A unique skeleton of many needlelike **spicules**, sometimes in association with a meshwork of organic fibers composed of **spongin**, prevents the collapse of the body and the closure of its passageways and openings. Spongin can occur alone, and some poriferans have a matrix with massive deposits of calcium carbonate.

The sponge structural plan essentially is focused at a cellular level. Sponge biology must be understood in terms of individually specialized cells; each is an independent functional unit. Sponge cells do not differentiate into the many distinctive types generally found in higher Metazoa, and each cell is left with quite broad functional capacities. Sponges are best considered to lack well-organized tissues.

In cellularian sponges, highly flattened **pinaco-**

cytes cover the outer surface and line all internal canals or spaces that are not equipped with choanocytes (Fig. 6.1E). Pinacocytes sometimes form a syncytial epithelioid membrane, especially in hexactinellid sponges. Internal pinacocytes are called **endopinacocytes**. Some pinacocytes are contractile, although they contain no contractile fibers. They contract by rounding up and, thus, reduce the surface area they can cover. Since pinacocytes are attached tightly to one another, they squeeze the body and, if the skeleton is not too rigid, cause the body to round up slightly.

Asconoid sponges have **porocytes** scattered among the pinacocytes (Fig. 6.1E). These large, tubular cells contain the microscopic ostia. A cytoplasmic flange, the **prosopyle** (Fig. 6.1A,B,C), can close the ostium. Porocytes are able to ingest and digest food, and may sometimes metamorphose into amoebocytes. Hexactinellids and adult syconoid and leuconoid sponges have no porocytes; in these groups, the ostia are simple gaps in the pinacocyte layer.

Choanocytes are the most characteristic of sponge cells (Fig. 6.1B,E) and also the most important of the sponge cell types. Choanocyte flagella are responsible for producing water currents that bring in food and oxygen and expel wastes. The choanocyte **collars** are filtering mechanisms for trapping and engulfing food particles, and in many sponges the choanocytes are also responsible for digestion. The electron microscope reveals that the collar is not a solid structure, but is composed of many delicate, linear pseudopodialike **microvilli** (Fig. 6.2), strengthened by exceedingly fine **microfibrils** that pass from one microvillus to the next. The beating flagellum supposedly pulls water through the interstices of the collar, and allows the filtering out of particles larger than the opening. (Whether this is actually so or not is doubtful without detailed study of choanocyte function and the consideration of the physics of water flow at that scale, which would be a viscous medium with laminar flow with very small Reynolds numbers.) Food particles typically end up near the base of the collar to be ingested into food vacuoles. In calcareous sponges with large choanocytes, the food is digested in choanocytes; in other sponges the food particles are passed on to amoebocytes for digestion.

No mechanism for coordinating the beat of the flagella of the choanocytes has been found. Apparently each works independently, and efficiency in water flow depends on the contours of the chambers and canal system. This proves to be ample. A *Leuconia*, 10 cm high and 1 cm in diameter, has about

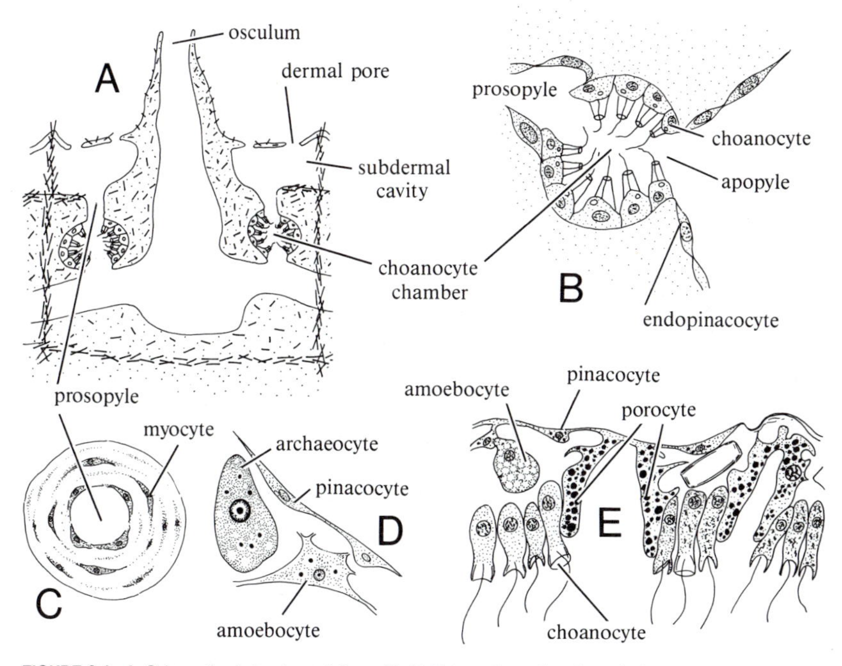

FIGURE 6.1. A. Schematized structure of *Spongilla*. **B.** Schematic section through choanocyte chamber. **C.** Prosopyle of calcareous sponge, showing concentrally arranged myocytes. **D.** Epidermis and amoebocytes of *Spongilla*. **E.** Cross section through wall of asconoid sponge. (**B** adapted from Brien; **C** after Dendy; **D** after Hyman; **E** after Prenant.)

2,250,000 choanocytes and pumps 22.5 liters of water through its body in a day. The position of choanocyte and beat of flagella help to direct the water toward excurrent apopyles, and pumping efficiency is increased by the larger size of apopyles over that of prosopyles (Fig. 6.1B). The large, open, internal cavity, or **spongocoel** (see Fig. 6.4C) of asconoid and syconoid sponges makes ejection of water through the osculum less efficient; in leuconoids the spongocoel is reduced or absent. As the excurrent canals join, the total cross-sectional area of the canal system decreases, and water flows at an ever-increasing pace as it approaches the osculum.

Other sponge cells are embedded in the mesogleal matrix. Nearly all are amoeboid. These **amoebocytes** carry out all of the processes in most sponges not accounted for by pinacocytes and choanocytes. This versatility demands a greater functional plasticity than is customary in metazoan cells. Sponges are difficult histological subjects because other cells may assume an amoeboid shape and wander off into the mesoglea. Porocytes and choanocytes are said to transform into amoebocytes in some sponges. Some amoebocytes develop slender, branching pseudopodia and often unite to form syncytial units; these are **collenocytes**. **Chromocytes** contain pigments, and **thesocytes** are filled with reserve food. **Archeocytes** (see Fig. 6.1D) are particularly important. They are undifferentiated, embryonic cells and usually form sperm or ova. An important function of amoebocytes is the secretion of spicules and spongin fibers of the skeleton. When an amoebocyte becomes active in skeleton formation it is called a **scleroblast**. Scleroblasts are named for the kind of skeleton they form: **calcoblasts** form calcareous spicules, **silicoblasts** form siliceous spicules, and **spongioblasts** form spongin fibers.

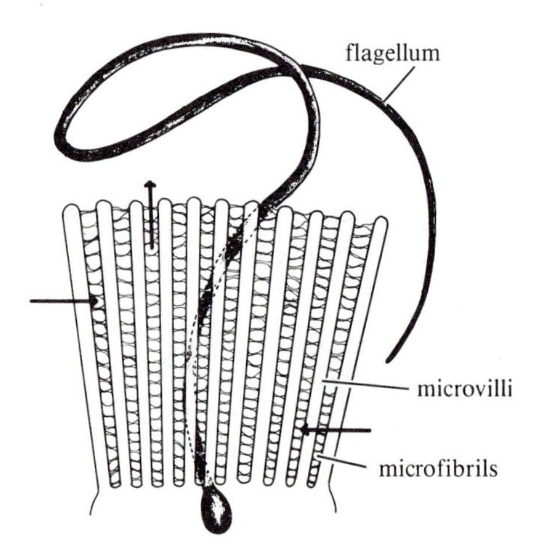

FIGURE 6.2. Details of choanocyte collar revealed by electron microscopy. Fine protoplasmic processes or microvilli connected by delicate microfibrils. *Arrows* indicate assumed flow of water. (Based on Fjerdingstad.)

A few other kinds of cells are found in some sponges. Slime-secreting gland cells occur in some. Two kinds of elongated cells also occur. **Desmacytes** are found in the outer cortex, providing fibrous strength as well as aiding in cortex contraction. **Myocytes** (Fig. 6.1C) are rather like smooth muscle cells and are arranged in sphincterlike bands around the pores of some sponges. They alter the diameter of the pores and regulate the amount of fluid flowing through the organism.

The above-mentioned cell types perform similar functions in all cellularian sponges, which are organized into one of essentially three major architectural body plans discussed below. However, analysis is beginning to indicate that these body plans are not useful in taxonomy. Consideration of these body types still retains instructional value, in that it outlines ways that sponges adapt to the problems of water movement.

The most simply organized sponges are known as **asconoids** because of their saclike form. The more complex sponges can be derived conceptually from the asconoid form, and the olynthus embryo is asconoid in character.

Leucoselenia is most commonly studied as an example of an asconoid sponge. It grows as groups of slender, tubular individuals attached by a common stolon to objects in shallow, wave-washed water. Water enters through microscopic incurrent openings and is expelled through a single large osculum (Fig. 6.3). Tubular buds without excurrent openings arise from the wall of the parent ascon.

The structure of the ascon wall is distinctive. An epidermis of close-fitting pinacocytes covers the surface. Scattered among the pinacocytes are tubular porocytes with the ostia. Collared choanocytes line the internal cavity or spongocoel. Their flagella force water toward the osculum and draw it into the spongocoel through the ostia. A jellylike matrix, the **mesoglea**, fills the space between the epidermis and the choanocytes. Wandering amoebocytes occur in the mesoglea. Calcoblasts secrete the spicules, which give skeletal support to the ascon tube. Most of the spicules are simple, elongated monaxons (see below) or three-rayed triaxons (see Fig. 6.6).

The next level of complexity in sponge structure is termed the **syconoid** type, for *Sycon*, a well-known example. Both asconoid and syconoid plans retain a tubular shape, and each individual ends in a single osculum. However, the syconoid body wall is thicker and more complex, and the spongocoel is lined with pinacocytes. The choanocytes have come to lie in a series of radially arranged choanocyte canals.

Scypha is usually studied as an example of syconoid structure. Each sponge is vase-shaped, with a terminal osculum protected by a fringe of monaxon spicules forming an oscular collar (Fig. 6.4A). Syconoids are never found in highly branched colonies, although one or more buds may be found on the side of the body.

Body organization is best seen in cross section. The spongocoel has evaginated to form the choanocyte-lined radial canals, or **choanocyte canals**. The spaces between them are filled with **incurrent canals**. Tiny openings, the **prosopyles**, connect the choanocyte and incurrent canals. Water flows through the external ostia, into the incurrent canals, through the prosopyles, and so into the choanocyte canals (Fig. 6.4B). Here food is ingested by the choanocytes. The choanocyte flagella force the water through internal ostia, which connect the choanocyte canals with the spongocoel. Water forced into the spongocoel emerges by way of the osculum.

Flattened pinacocytes cover the outer surface of the body and line the incurrent canals. The spongocoel is lined with similar endopinacocytes. Syconoid mesoglea resembles asconoid mesoglea, containing amoebocytes of various kinds as well as spicules. The spicules are arranged in a definite pattern, characteristic of the genus or species. Small monaxons project through the cortex. Triaxons lie along the walls of the canals, with the longest ramus directed toward the distal end. Small tetraxons or triaxons support the spongocoel wall.

Sycetta is a simpler syconoid with choanocyte canals only. Each choanocyte canal is a simple outpocketing of the spongocoel and is exposed directly to the surrounding water (Fig. 6.4C). However, the outer margins of the choanocyte canals of nearly all syconoids are attached, forming incurrent canals, as in *Scypha*. The most complicated syconoids, like *Grantia*, have a more highly developed cortex, pierced by external pores. Narrow, branching and anastomosing canals lead from the pores to the incurrent canals (Fig. 6.4D).

Leuconoids are the most complexly organized sponges, and most cellularian sponges are of this type. Most leuconoids form large masses, containing many oscula. Each osculum marks one member of the large colonial mass, but the limits of the members are so ill-defined that it is impossible to distinguish individuals clearly.

Leuconoid organization entails folding and outpocketing of the choanocyte canals to form clusters of round or oval **choanocyte chambers** (see Fig. 6.1A,B). The choanocyte chambers surround an excurrent canal into which they discharge water. Smaller excurrent canals merge to form larger ones; all eventually unite to form a major excurrent canal through which water reaches the osculum. The cortex contains a system of branching canals, delivering water to the choanocyte chambers by way of small pores, the prosopyles. Water leaves the choanocyte chambers through similar pores, the **apopyles**.

Spongilla, a freshwater sponge, may be used as an example of leuconoid structure. *Spongilla* forms

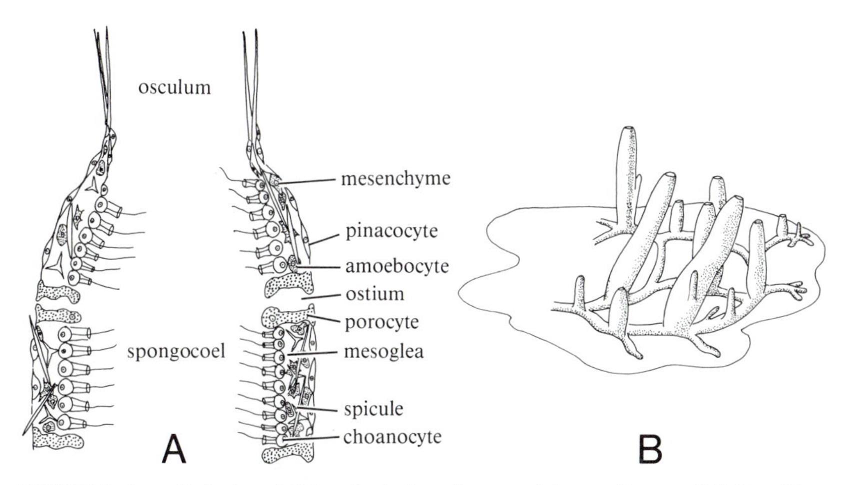

FIGURE 6.3. Asconoid structure. **A.** Schematic structure of upper part of asconoid sponge. **B.** Colony of *Leucosolenia*, typical ascon sponge. (**A** after Hyman.)

small masses attached to submerged objects, and is often greenish because of symbiotic flagellates, or **zoochlorellae**. A thin cortex forms the sponge surface (see Fig. 6.1A). It is pierced by many small dermal pores and propped up by columns of spicules extending from the lower mesenchymal mass, and covers the extensive subdermal cavity. Water flows from the subdermal cavity into short incurrent canals that open through wide prosopyles into the choanocyte chambers. Wide apopyles deliver the water into excurrent canals that eventually discharge into a small spongocoel terminating in an osculum. Unless very small, each *Spongilla* has many oscula.

Hexactinellida, the glass sponges, have the most beautiful of skeletons. Hexactinellids are well-formed, vase- to funnel-shaped individuals, attached to the substrate by a tuft of root spicules. They are typically deep-water forms with some 500 known species.

The skeletal framework is composed of six-rayed, or hexactine, spicules, from which the group gets its name. Small crossbars of silica may attach the large spicules together, forming a continuous network. Others have a solid skeleton, formed of spicules fused at their tips. The skeleton is pierced by many rather regularly arranged parietal spaces, and looks like a glass basket. In life, the solid framework of fused spicules is augmented by smaller spicules, generally missing in museum preparations.

Glass sponges have a unique structural plan (Mackie and Singla, 1983; Reiswig and Mackie, 1983). Pinacocytes are totally lacking. In their place is an all-pervading syncytium, the **trabecular net**, which literally drapes the entire network of spicules externally and internally. The hexactinellid body thus effectively is a single large, multinucleate cell (Fig. 6.5A). The trabecular net completely surrounds the choanoblasts, collar bodies, and collagenous mesolamella located in the **basal reticulum**; forms the **secondary reticulum** around the collars; and contributes to syncytia permeating the body wall (Fig. 6.5B).

The **choanosyncytium** is composed of choanoblasts that bud to form networks of **collar bodies** lining the flagellated chambers (Fig. 6.5C). This network does not contain nuclei.

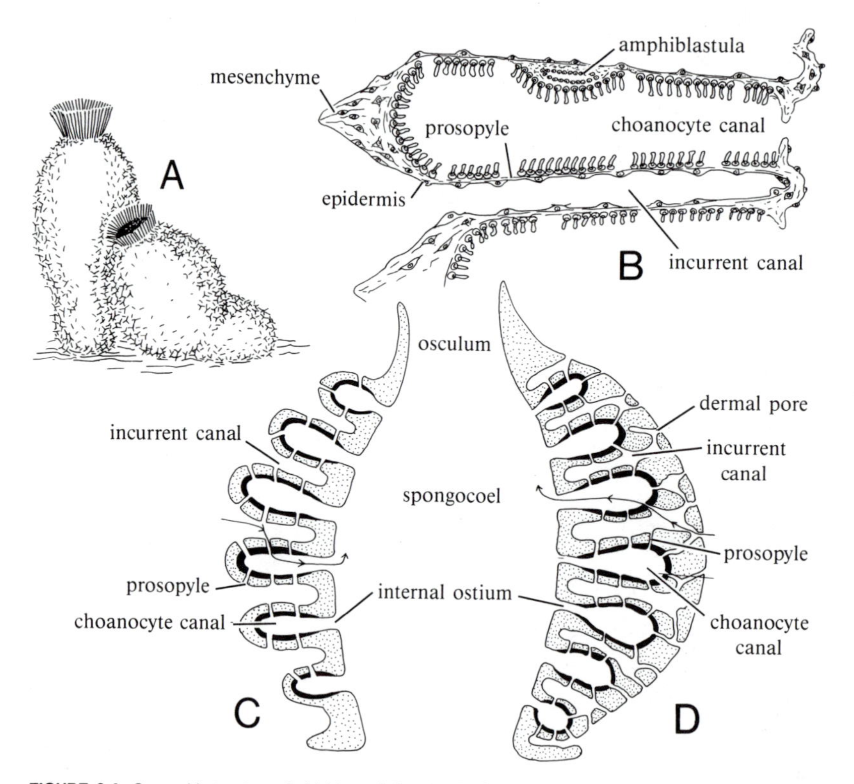

FIGURE 6.4. Syconoid structure. **A.** Habitus of *Scypha.* **B.** Schematic section of incurrent and choanocyte canal in syconoid sponge. **C.** Schematic section of wall of syconoid sponge without cortex, with choanocyte layer shown dark. **D.** Schematic section of wall of syconoid with cortex. (**A** after Brown; **B,C,D** after Hyman.)

The **mesolamella** is composed of sheets or strands, or both, of fibers resembling spongin A. These are enveloped by the trabecular net and serve to fortify the walls of the flagellated chambers as well as the trabecular net. In addition, a great deal of free mucopolysaccharide and glycoprotein is secreted to help hold and support the body framework.

Besides the above material, hexactinellids include many other cell types such as archaeocytes (which give rise to gametes, thesocytes and granulocytes for possible food storage), spherulous cells (with vacuoles and prominent Golgi bodies indicating a secretory role), and, of course, sclerocytes.

Thus, hexactinellidans are organized in a very different way than that seen in the cellularian sponges.

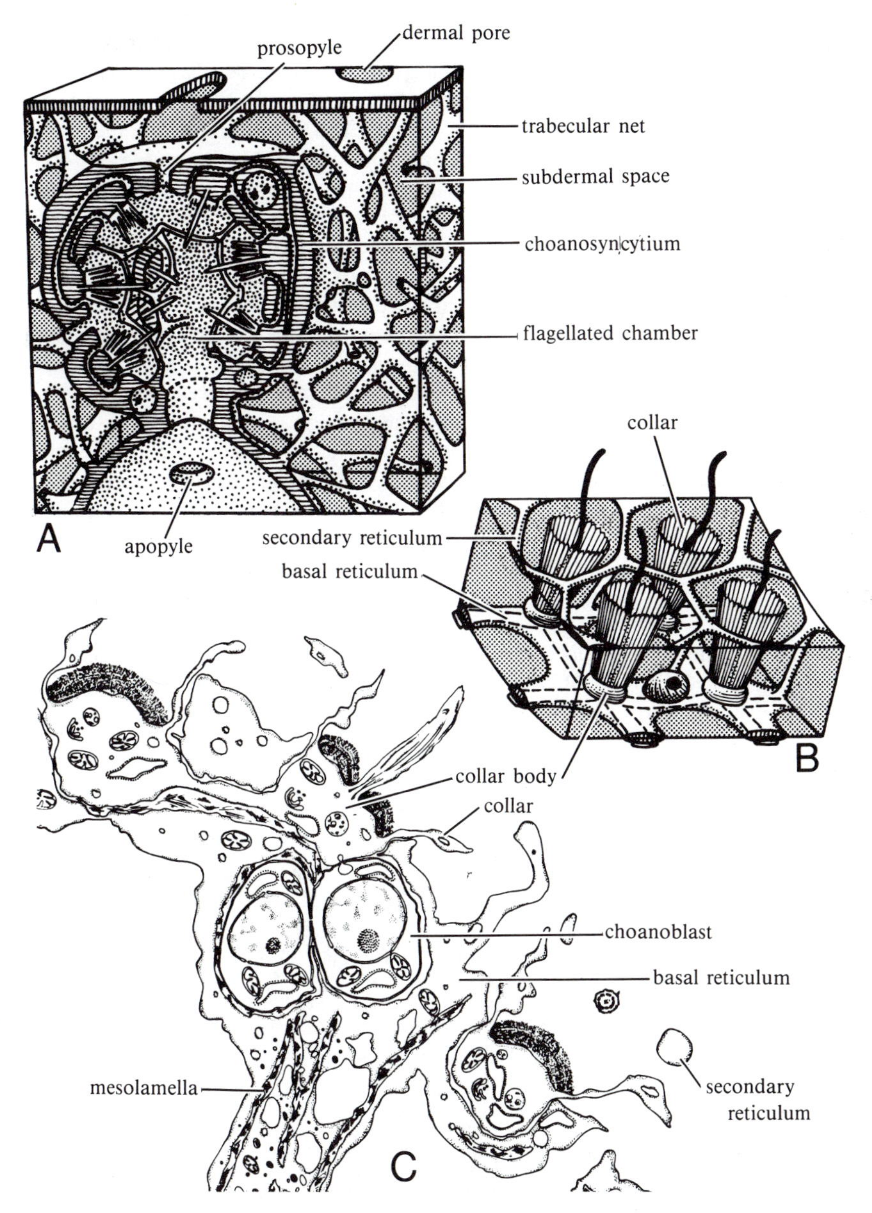

FIGURE 6.5. Hexactinellid structure, based on *Rhabdocalyptus dawsoni.* **A**. Body plan of idealized hexactinellid; water enters dermal pore, flows around trabecular net, goes through prosopyl into flagellated chamber, exits through apopyle into excurrent lacunae. **B**. Diagram of chamber wall with secondary reticulum around collars. **C**. Section through flagellated chamber; note role of basal reticulum in enclosing elements of trabecular mesolamella and choanosyncytium. (Adapted from Mackie and Singla, 1983.)

Biology

Sponges are sessile animals that remain attached to surfaces, although some very limited and slow shifting of position can occasionally be achieved at the holdfast.

Skeleton The skeletal system of sponges is formed from one or more distinctive elements: calcareous spicules composed of some form of calcium carbonate, either aragonite or magnesium calcite; siliceous spicules of silicon dioxide ("glass"); and/or organic fibers made of a proteinaceous substance called spongin. The spicules exhibit a wide variety of form (Fig. 6.6).

Calcareous spicule formation begins within the cytoplasm of a binucleate calcoblast. As it grows, the calcoblast divides into founder and thickener cells. Each axis of a spicule begins within a separate scleroblast; a triradiate spicule begins in three scleroblasts and develops eventually with three founders and three thickeners (Fig. 6.7A). The founder cell moves along the surface, apparently determining the length of the ray. It is followed by the thickener, which deposits additional material on the surface. When the spicule is fully formed, the founder and thickener move away and merge into the mesoglea.

The fully formed calcareous spicule is surrounded by a sheath, but apparently lacks any axial filament of an organic nature. Siliceous spicules, on the other hand, have an organic core in the form of an axial filament.

Siliceous spicules are formed within a single silicoblast unless they are very large, whereupon several cells may work together or form a syncytial mass within which the spicule is laid down.

Spongin fibers are formed in an entirely different manner. Each spongioblast develops a vacuole, within which spongin material collects. They line up so that the short sections are united into a single long fiber (Fig. 6.7B).

Sponge cells are so poorly intercoordinated in most matters that the relatively complex cellular cooperation observed in skeleton formation is amazing. Little is known of the factors that make it possible for a group of scleroblasts to lay down a spicule of complex form, with rays meeting in precisely determined angles. The factors resulting in the formation of different kinds of spicules in different body regions are also unknown. Intriguing questions are suggested, for example, by *Scypha* scleroblasts, which form giant monaxons in the oscular region and, in the walls of the choanocyte canals, lay down triradiate spicules with the long axis of each parallel to the canal. Evidently the scleroblasts receive some kind of information, genetic or otherwise, that results in spicule constancy in a given species.

Sponge classification and identification depend largely on skeletal details. Spicules are so varied that a large specialist vocabulary has been developed to permit their description. Only a few of the more basic terms are illustrated here (Fig. 6.6). The terms are introduced to reveal something of skeletal diversity and also to exemplify the kind of specialized vocabulary that develops in the course of detailed study of any animal group.

Some spicules make up the major skeletal ele-

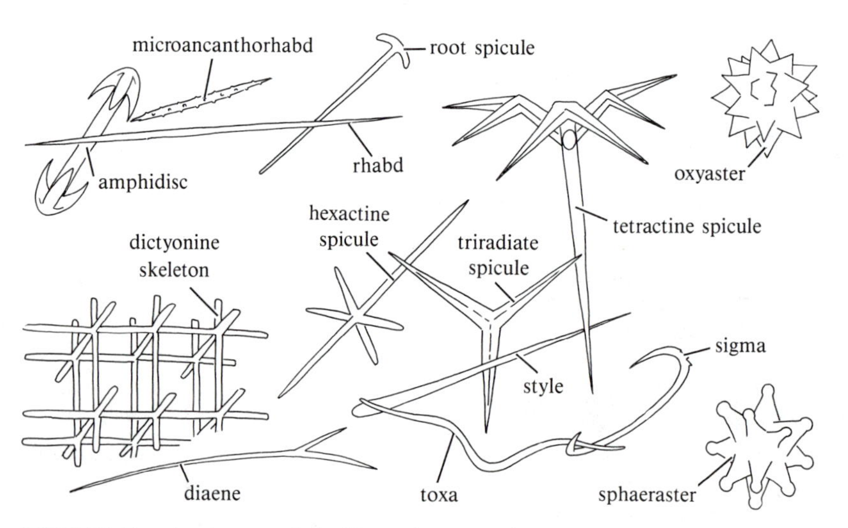

FIGURE 6.6. Examples of sponge spicules. (From various sources.)

ments, and others are more or less randomly scattered in the soft parts. Larger skeletal spicules are called **megascleres**, and smaller flesh spicules are **microscleres**. Some sponges lack flesh spicules, and in Calcarea there is no sharp distinction between megascleres and microscleres.

Spicules are named according to their form. A **monaxon** is a straight or curved spicule without branches, consisting of one or two rays. If growth has occurred in both directions and has similar ends, they are called **rhabds** (Fig. 6.6). **Tetraxons** are spicules with four rays, although some are simplified by the loss of one or more rays. Usually one arm is much longer than the others. The rays never meet at right angles, so tetraxon spicules are easily recognized, even when simplified by the loss of some rays. The most common spicule of calcareous sponges is the **triaene**, formed of three rays not meeting at right angles. It is thought to be a tetraxon with one ray missing. **Triaxon** spicules have three axes, meeting at right angles, and so have six rays. These hexactinal spicules are found only in hexactinellids and are diversely modified. Some rays branch, and sometimes some rays are reduced or lost. **Polyaxon** spicules, also called **asters**, have many rays meeting at a central point. Silica may be deposited irregularly on the spicules, which are often cemented together to form a solid meshwork, or **dictyonine skeleton**.

Something of the long evolutionary history of sponges is reflected in the great variety of spicule types. Evolutionary convergence and ecological constraint have often produced a similar body form in sponges that have quite different kinds of skeletons. The result is that the relationships of modern sponges are difficult to determine. Taxonomic work based on skeletal detail reflects diversity as well or better than anything that might be used, and is likely to reflect relationships.

Food Capture and Digestion It is only in obtaining food and other materials from the environment that sponges have capitalized on their multicellular organization. The activity of choanocytes in creating currents and ingesting food particles has been described; their effectiveness in this regard depends on the overall form of the organism and especially on the structure of the pore and canal system (Lagenbruch et al., 1985). Oxygen, silica, calcium salts, and other substances are absorbed as the water passes through the sponge canal system.

Food ingestion is entirely protozoan in character. Small particles are engulfed, food vacuoles are formed, and digestion is entirely intracellular. Choanocytes ingest by far the greater part of the food particles, but amoebocytes, porocytes, and even epidermal cells can ingest food. In addition, nutrition can be supplemented by both association of symbiotic bacteria and algae in the sponge and with uptake of dissolved organic matter from the water currents.

The amount of food ingested depends on the exposed surface and the volume of water filtered. Folding of the spongocoel to form choanocyte canals, and further folding to form choanocyte chambers, increases the available surface for food ingestion and also improves water circulation.

Digestion occurs within food vacuoles, and food is absorbed across the vacuolar membrane. However, in most sponges the cells that ingest food do not digest it. The small Demospongiae choanocytes take food in but pass it on immediately to amoebocytes for digestion. The ability to pass food from cell to cell is not unique to sponges but is unusually important in their physiology. Wandering amoebocytes transport food like postmen distributing mail. They may be transformed into storage cells or nurse cells (**trophocytes**) that build up food reserves for use by young oöcytes or by amoebocytes preparing to form gemmules. Embryos, for example amphiblastulae or gemmules, develop for a time within the parent body, lying adjacent to a trophic membrane

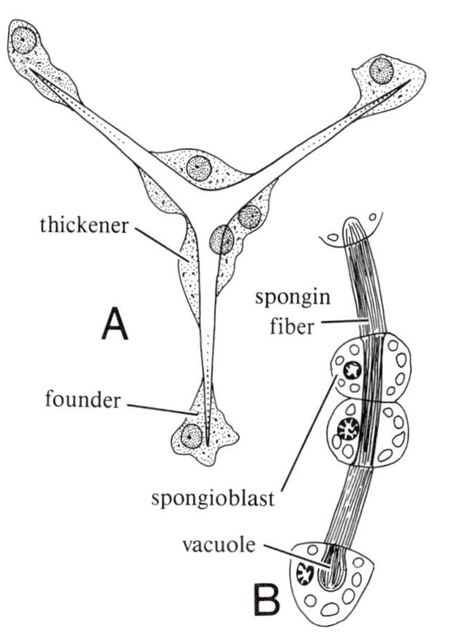

FIGURE 6.7. Skeleton formation. **A.** Triradiate spicule near completion of its development, showing founder cells and thickener cells. **B.** Spongioblasts assembled to form spongin fiber. (**A** after Woodland; **B** after Hyman.)

formed by the choanocytes, from which they directly receive food (Figs. 6.4B and 6.8B).

An important point must be made. Choanocyte-lined chambers are not comparable to the digestive cavities of other animals. In sponges, the ingestion, digestion, and absorption of food are far more diffuse cellular processes than exist in higher metazoans.

Excretion So little is known of the details of energy release and energy transfer in sponges that these processes cannot be discussed profitably. The metabolic wastes of sponges are also incompletely characterized. At least some sponges are ammonotelic, producing appreciable quantities of ammonia and little or no urea, uric acid, or other complex nitrogenous wastes. Crystals resembling wastes are said to accumulate in amoebocytes but have not been chemically identified. They are freed in the excurrent canals. Ammonotelic excretion is to be expected in sponges, as it is generally characteristic of aquatic animals.

Respiration The stream of water moving through the sponge body provides excellent conditions for respiratory exchange. A large surface is exposed to a constant supply of water. Oxygen withdrawal varies considerably with the species; *Aplysina aerophoba* absorbs 57 percent of the dissolved oxygen passing through its body, comparing favorably with many of the higher Metazoa, but some sponges absorb a very small percentage of available oxygen. Considering the quantity of water that sponges pump, this seems relatively unimportant. Where water flows rapidly over an extensive surface, low withdrawal rates can provide adequate oxygen.

Sponges are generally aerobic and sensitive to low oxygen availability. Closure of the osculum for a time results in an above-normal oxygen uptake during recovery. Evidently some kind of oxygen debt system operates in sponges. When metabolism is carried out under conditions of oxygen shortage, complex organic end products are formed. If these products accumulate in the body, they are oxidized when oxygen becomes available, and thus oxygen use rises.

Internal Transport It is evident that while the flow of water through the body is important in the transport of material, it is by no means comparable to the circulatory system of higher metazoans. Diffusion and active transport of materials across cell membranes and through syncytia account for most of the movement of materials in the body.

Cell-to-cell internal transport also cannot be ignored, however, for food is passed from choanocyte to amoebocyte, and motile cells play an important role in distribution of substances about the body. Food reserves in masses of amoebocytes are mobilized for use. Amoebocytes are the workhorses of sponge economy, functioning as instruments for internal transport and for a variety of other metabolic functions.

Protective Devices Sponges lack special sense organs and the ability to escape, and it may seem that they are helplessly exposed to damage. The facts are otherwise. Fishes tend to avoid sponges, and anyone who has handled them without protection understands why. The sharp spicules penetrate any soft tissue exposed to them. Some sponges apparently produce irritating substances.

Nevertheless, sponges do have enemies. Some fish, molluscs, and arthropods, such as sea spiders and insects, feed on them, and they are, of course, powerless to flee.

Many small animals apparently take up residence within sponges for protection. Some crabs decorate themselves with sponges, perhaps for camouflage or perhaps because so many predators avoid sponges. The crab *Dromia* holds a piece of live sponge over its back, and some hermit crabs occupy empty snail shells with attached sponges. The crab is apparently protected by the disgreeable nature of the sponge as it carries its partner about. This has become an intimate relationship, for some sponges are found almost exclusively in association with hermit crabs.

Mechanisms of Response Sensory stimuli transmitted by nerves evoke a response in almost all animals, but sponges are an exception. There is no nervous system. Any contractile parts of a sponge act independently in receiving stimuli and acting upon them.

Response to environmental stimuli is limited and centers about the direct response of individual cells. Myocytes, desmacytes, and pinacocytes can contract slowly, closing ostia and oscula and reducing body surface. A sharp stimulus given at a definite point slowly spreads and evokes responses in nearby cells. A needle prick near an osculum results in its closure, but a similar prick some distance away leaves the osculum virtually unaffected. Transmission of such a stimulus is slow, about 1 cm/min. This kind of conduction that requires no nervous system is termed **neuroid conduction**.

Growth and Reproduction Porifera possess a simple structural plan, with many cells retaining protozoalike characteristics. Their growth, reproduction, and morphogenesis pose tantalizing problems and contribute to our understanding of the nature of the earliest multicellular animal stocks.

Asexual reproduction occurs as the result of accidental fragmentation of the body, of exposure to unfavorable environmental conditions, or a part of the normal life cycle. Budding, essentially the differentiation of a small primordium on the parent body into a young individual, is the most common type of asexual reproduction. Small buds appear on the sides of the body and grow and differentiate into young sponges that detach to take up an independent life or remain as a new member of a colony. Freshwater and some marine sponges produce special kinds of buds, called **gemmules**. Gemmules of freshwater sponges have a complex, hard shell; they are produced most abundantly in the fall and germinate the next spring (Fig. 6.8A). Gemmules of marine Demospongiae are formed more or less continuously and undergo prompt development, giving rise to juveniles that pass through stages resembling the embryos produced during sexual reproduction.

Freshwater gemmules (Fig. 6.8A) begin as masses of amoebocytes, filled with food from special nurse cells. Other amoebocytes form a covering around the central mass and differentiate into a layer of columnar cells that secrete a hard inner membrane. Scleroblasts from the parent mesoglea form special gemmule spicules, depositing them into a cylindrical layer formed by the columnar cells. The gemmule wall is broken at the **micropyle**, a pore through which the cells emerge during early development.

Marine gemmules also begin as a mass of amoebocytes, surrounded by other cells. The outer layer of cells develops into a flagellated, surface epithelium. At maturity, gemmules have a flagellated and a nonflagellated pole. After a free-swimming period, they settle and attach at the nonflagellated end to develop into young sponges.

Sponges have a fabulous ability to regenerate. Tips of the branches of some sponges regularly break off and produce masses of cells that become young sponges. Many sponges form **reduction bodies** during unfavorable periods. The main mass of the body deteriorates, leaving a residual mass that develops into young sponges when favorable conditions return. Gemmules and reduction bodies are ideal stages for dissemination. Motile or long quiescent, they may be carried far away from the parent colony by water currents. The ability of the sexually or asexually reproduced young to reach areas remote from the parent is very important to sessile animals.

Sexual reproduction involves the production of gametes that generally arise from archaeocytes or choanocytes. Growth and food storage at the expense of nearby nurse cells transform an archeocyte into an ovum. Meiosis occurs at the end of the growth period.

Sperm development has been rarely seen, and there is uncertainty about the details. Archaeocytes and choanocytes have been observed in sperm formation. Spermatogonia form masses surrounded by layers of flattened cells, known as spermatocysts. The spermatocysts disappear as spermatids form. Mature sperm find their way to the excurrent canal system and are swept out with water currents to eventually reach the incurrent canal system of another sponge. Although most sponges are hermaphroditic, crossbreeding is ensured by different maturation times of sperm and ova.

Sperm are ingested by the choanocytes that in turn fuse with the ovum and set the engulfed sperm free. This unusual method of fertilizaton is probably common.

Development The embryonic development of sponges involves stages sometimes like and at other times quite unlike similar stages of other animals.

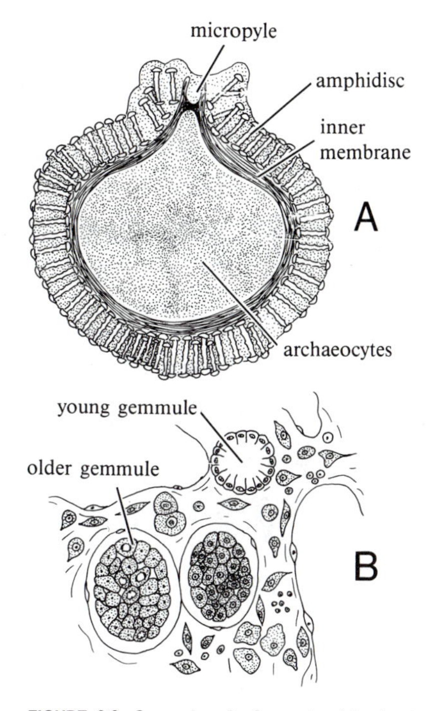

FIGURE 6.8. Gemmules. **A.** Gemmule of freshwater sponge. **B.** Early stages in formation of gemmules in marine sponge. (**A** after Evans; **B** after Wilson.)

The embryo remains within the mesenchyme of the parent during early stages, deriving nourishment from parent cells. It emerges as a solid **stereoblastula** or **stereogastrula**, and swims about for a time. Ultimately, it attaches and develops into an olynthus in the Calcarea, and a rhagon in the Demospongiae.

Calcarean zygotes undergo holoblastic cleavage. The young embryo lies embedded in the mesenchyme near the choanocytes. In the 16-cell stage, eight blastomeres are macromeres and eight are micromeres. The macromeres lie against the parent choanocytes. The micromeres appear to be in the embryo's animal hemisphere. The macromeres cease dividing, while the micromeres continue to divide and develop flagella that project into the blastocoel. Meanwhile, a functional mouth develops in the center of the macromeres, and ingests parent cells (Fig. 6.9B).

At this time inversion occurs, an unusual process during which the cells pass through the mouth to turn the embryo inside out and bring the flagella to the outer surface (Fig. 6.9A). After inversion, the parent choanocytes form a nutritive trophic membrane around the blastula. Both macromeres and micromeres now divide to form approximately equal masses when the **amphiblastula** escapes from the parent. The free-swimming amphiblastula swims with its flagellated micromeres forward (Fig. 6.9A,C).

The inversion of cells is not typical of most animals. The protist *Volvox* undergoes a similar inversion during development, and this has been used by some to support the idea that flagellated protozoan colonies were ancestral to sponges. The small cells at the animal pole are like the micromeres of other animals, and the large cells at the vegetal pole, with a history of being associated with the nourishment of the embryo, are not unlike the macromeres of most animals. However, the sponge embryo develops further along a totally unique path.

The amphiblastula usually attaches before gastrulation. Gastrulation may occur by invagination or by epiboly, but it is the micromeres that invaginate and the macromeres that grow over the micromeres in epiboly. What would have become the ectoderm of other metazoans is now inside, and what might have been the endoderm forms the outer covering. As development proceeds, the inner layer of cells produces choanocytes, archeocytes, and some amoebocytes, while the outer layer forms pinacocytes, porocytes, and scleroblasts.

After a period of swimming, the larva attaches at the blastopore and the central cavity forms the beginning of the spongocoel (Fig. 6.9B). As the oscu-

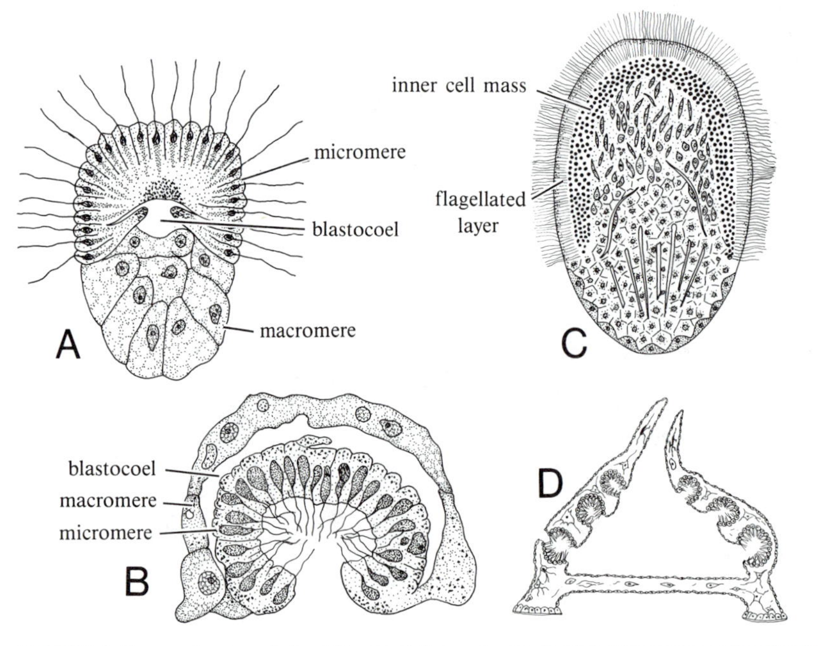

FIGURE 6.9. Spone embryology. **A.** Amphiblastula of *Sycon,* showing flagellated micromeres and nonflagellated macromeres. **B.** *Sycon* embryo after gastrulation and attachment with choanocytes developing from micromeres and outer wall arising from macromeres. **C.** Swimming demospongian larva as differentiating stereoblastula. **D.** Rhagon larva of tetractinellid sponge. (**A**,**B** after Hammer; **C** after Wilson; **D** after Sollas.)

lum breaks through at the free end, the embryo becomes a simple asconoid sponge, known as an **olynthus**. The spongocoel evaginates to form choanocyte canals in syconoids. *Leucoselenia* achieves the same result in a different manner. A solid gastrula is formed, with flagellated cells at the surface and nonflagellated cells within. When the larva attaches, the interior cells migrate to the surface to form the epidermis.

The embryo of primitive tetractinellid Demospongiae escapes from the parent as a coeloblastula, composed entirely of small flagellated cells, and so lacks macromeres. It attaches at its anterior end. Anterior cells invaginate and become the inner choanocytes, and posterior cells change into covering epithelium. The resulting **rhagon** stage (Fig. 6.9D) is somewhat conical. The rhagon choanocytes evaginate to lay down the primary choanocyte chambers and assume a syconoid form. Further outpocketings of the chambers later produce the typical choanocyte chambers of leuconoid sponges. An osculum eventually breaks through at the top.

The solid blastula of other Demospongiae gastrulates by delamination while still within the parent. The outer cells develop flagella and the inner cells become amoebocytes. A small area of nonflagellated cells occurs at the posterior pole of the larva. The stereogastrula (Fig. 6.9C) emerges from the parent and, after swimming about for a time, attaches at its anterior end. Inversion occurs, bringing the internal cells to the surface, where they form pinacocytes and mesenchyme, while surface cells lose flagella and end up inside to develop as choanocytes.

Regeneration Animals that reproduce asexually usually have high regenerative powers, and sponges are no exception. Small pieces of sponge detached from the parent organism can regenerate new sponges. Pieces grow slowly, often requiring months to reach adult size. Sponges also have the remarkable power of reconstitution. A sponge squeezed through silk bolting cloth is dissociated into isolated cells and small clusters of cells. The cells become amoeboid and proceed to reaggregate, at first into delicate meshworks and later into small spherules. They coalesce to form syncytial masses as they come together, with the more granular internal part of each cell retaining its individuality. The clear hyaloplasm of the whole mass thus becomes available to make large pseudopodia, hastening coalescence. Each spherule is composed of a covering layer of pinacocytes and a central mass of mesenchymal cells and choanocytes. The excurrent canal system is formed as internal cells die and are engulfed by amoebocytes that clear away lacunar spaces. The incurrent canal system is formed by collenocytes just under the covering layer. Choanocytes remain active and form choanocyte chambers.

Review of Groups

CLASS CALCAREA

Sponges of this group have skeletons of calcium carbonate, generally as calcite although sometimes as aragonite. (These are two distinct crystal forms of calcium carbonate.) All three forms of body organization (asconoid, syconoid, leuconoid) occur among calcareans.

Larvae can either be liberated to swim freely or be brooded in the parent tissues. Larval form can range from hollow coeloblastulae to amphiblastulae.

Calcarea include forms that are common in intertidal or shallow-water habitats. However, some are more typical of very protected conditions. The pharetronidans are often found in recessed niches or places like tunnels or caves. The sphinctozoans are common on reefs, but prefer the more protected sites such as crevices, overhangs, and tunnels.

CLASS DEMOSPONGIAE

The vast majority of living sponges, more than 95 percent, are members of this class. The skeletal composition varies, with various combinations of silica spicules, spongin, and/or collagenlike fibers. The spicules, when present, are divisible into the large megascleres or the smaller microscleres. The axonal variety of the spicules is also quite diverse (see Fig. 6.6).

The demosponges are often used for commercial purposes, and hold the greatest potential for further exploitation in these regards. The ceractinomorph order Dictyoceratida includes the bath sponges. These sponges produce skeletons of spongin that are characterized by a fine mesh of slender fibers with very few inclusions of foreign matter. In addition, several orders, again especially among ceractinomorphs, are noted to produce biochemically interesting compounds, such as cyclopropenes, sterols, and terpenes, that may prove useful in biomedical and other commercial uses.

Demosponges occur in marine habitats extending from intertidal to abyssal depths of 5000 m or more. One family, the myxillid ceractinomorphs, extends through this entire range. Most families, however, have a more restricted distribution. Demosponges, especially the spongillids, also occur in freshwater rivers, streams, and lakes.

CLASS SCLEROSPONGIAE

The coralline sponges build a skeleton that contains spicules of both silica and aragonitic calcite as well as spongin. This very ancient group, which includes extinct members of the order Stromatoporoida, bears so many resemblances to various demosponges that some authorities treat sclerosponges as merely a subclass of the Demospongiae.

The rarely encountered sclerosponges generally exhibit rather cryptic habits, and prefer caves, tunnels, crevices, and deep-water overhangs on coral reefs.

CLASS HEXACTINELLIDA

Until recently, these sponges were thought to be restricted to habitats that extended from moderately deep to abyssal depths. Because most were brought up from great depth by dredge collecting and, if they survived, were difficult or impossible to maintain in the laboratory, little had been known of their fine anatomy or biology. The identification of *Rhabdocalyptus dawsoni* (see Fig. 6.5) in shallower waters has already resulted in new insights concerning the glass sponges.

Evolution and Phylogenetic Relationships

Sponges have obviously evolved along a line different from other Metazoa. They may have arisen from similar, prostistan, choanoflagellate stock. However, if any more detailed statement is undertaken, difficulties arise.

Sessile life is associated with several important evolutionary trends, beautifully exhibited by sponges. Active animals benefit from streamlining and bilateral symmetry, whereas sessile animals tend to be radially symmetrical or asymmetrical. Selective pressures favoring improvements of muscular and nervous systems also benefit active animals. Since many sessile animals make few body movements, selective pressures favoring the development or maintenance of nervous, sensory, and locomotor parts may be lacking, so that these parts tend to degenerate in complex animals that assume a sessile habit. Sponges have apparently lived as sessile organisms from their earliest appearance, and have failed to acquire a nervous system or sense organs, and have very simple contractile elements.

Sessile creatures cannot pursue prey, and generally adaptations for trapping or filtering food from the water are associated with a sessile life. The pores and canals of sponges make an admirable filtering system, and major adaptations have centered around modifications in the complexity and arrangement of the canal system and of the chambers that contain the choanocytes. Their many unusual features emphasize sponge distinctiveness. That sponges have specialized along lines entirely unlike those followed by other animals does not mean that they are unsuccessful. In their own way they are exceedingly effective.

The sponge choanocytes resemble choanoflagellates. Indeed, a genus of choanoflagellates forming irregular masses of jelly containing scattered zooids was named *Protospongia*, as it seemed a possible design for a sponge ancestor. However, choanoflagellate colonies have no embryonic stages like sponges, except for the striking inversion of *Volvox*, so like the inversion of sponge embryos. Sponge ancestors might have been somewhat spherical, blastulalike colonial flagellates with embryonic inversion. Unfortunately, all of the choanoflagellates are photosynthetic. The unique use of macromeres and micromeres, which bears no similarity to the germ layers of other metazoan phyla, during sponge embryonic development emphasizes the phyletic isolation of Porifera. However, their basic biochemistry, microgamete ultrastructure, and isolation of reproductive cells places Porifera on the main animal stem (see Chapter 38).

The phylum is very old, with fossil representatives of all classes apparently known since the Cambrian. Relationships within the phylum are uncertain. Calcarea include the structurally simplest species, of an asconoid type. Some Calcarea have reached leuconoid organization but, in all probability, by several different avenues. Within the Demospongiae, available evidence points to homoscleromorphs as the most primitive, probably arising from ancestral stocks resembling the rhagon embryos, and to ceractinomorphs as the most advanced sponges. Hexactinellida are so unlike other sponges as to lack any suggestion of their possible relationship. Nevertheless, the recent recognition of the Sclerospongiae, with silica as well as calcareous spicules, has suggested to some that a possible link exists between hexactinellids and demosponges. However, the basically different cellular organization of glass sponges would seem to preclude this.

Classification of Sponges

Sponge classification is unusual in being of little value to anyone except a specialist (see Vacelet and Boury-Esnault, 1987). It is based almost entirely on microscopic skeletal structures, and while body organization is to some extent correlated with the skeletal parts, the body form of sponges classified in different orders is often quite similar. The taxonomy of the group is thus a pragmatic one and rather artificial, and probably does not reflect entirely naturally evolved groupings. Thus, for cellularians, the taxonomy given here is not carried below subclass level. The recognition of the distinct nature of the hexactinellids has delineated two clades within the Porifera, but some have gone so far as to suggest that these two groups should be elevated to separate phyla. Since both groups can be characterized with derived features in common (see "Definition" section), we retain a single phylum Porifera here.

Subphylum Cellularia. Cellular organization; epidermis with pinacocytes; mesoglea of collagenous matrix; amoeboid cells and skeletal elements various; choanocytes distinct and nucleated; myocytes present with limited contractility.

Class Calcarea. Calcareous sponges. Skeleton composed of discrete spicules of calcium carbonate, not differentiated into microscleres and megascleres.

Subclass Calcinia. Larva as flagellated, hollow ball.

Subclass Calcaronia. Larva with flagellated, anterior hemisphere and posterior, nonflagellated hemisphere. Examples: *Leucoselenia* (Fig. 6.3), *Scypha*, (Fig. 6.4).

Subclass Pharetronida. Skeleton with massive calcitic reinforcements.

Subclass Sphinctozoa. Solid skeleton of aragonite formed into discrete chambers.

Class Demospongiae. Skeleton composed of siliceous spicules, spongin fibers, or both; generally with small, microsclere, tissue spicules and large, megasclere, skeletal spicules; of leuconoid form.

Subclass Homoscleromorpha. Skeleton of equal-rayed tetraxons.

Subclass Tetractinomorpha. Skeleton with tetraxon and monaxon megascleres.

Subclass Ceractinomorpha. Skeleton of spicules and spongin, or spongin alone. Megascleres, when present, always monaxons. Example: *Spongilla* (Fig. 6.1).

Class Sclerospongiae. Coralline sponges. Skeleton composed of siliceous and aragonitic spicules, spongin fibers.

Subphylum Symplasma. Symplastic or syncytial organization; epidermis lacking pinacocytes; mesoglea as thin mesolamella; collar elements enucleate connected in reticular networks to nuclear choanoblasts, with secondary reticulum about choanoderm.

Class Hexactinellida. Glass sponges. Skeleton composed of six-rayed, siliceous spicules or triaxon spicules, often united to form networks; with syncytial trabecular net. Example: *Rhabdocalyptus* (Fig.6.5).

Order Hexasterophora. With hexasters, no amphidiscs.

Order Amphidiscophora. With "anchorlike" amphidiscs, no hexasters.

Selected Readings

Berquist, P. R. 1978. *Sponges*. Univ. California Press, Berkeley.

Bidder, G. P. 1923. The relation of the form of a sponge to its currents. *Q. J. Micro. Sci.* 67:293–323.

Fry, W. G. (ed.). 1970. *The Biology of the Porifera*. Academic Press, New York.

Lagenbruch, P.-E., T. L. Simpson, L. Scalera-Liaci. 1985. Body structure of marine sponges. III. The structure of the choanocyte chamber in *Petrosia filiformis*. *Zoomorph.* 105:383–387.

Laubenfels, M. W. de. 1936. Sponge fauna of the Dry Tortugas with material for a revision of families and orders. *Pap. Tortugas Lab.* 30:1–225.

Lévi, C. 1957. Ontogeny and systematics in sponges. *Syst. Zool.* 6:174–183.

Mackie, G. O., and C. L. Singla. 1983. Studies on hexactinellid sponges. I. Histology of *Rhabdocalyptus dawsoni*. *Phil. Trans. Roy. Soc. Lond.* (B)301:365–400.

Reiswig, H. M., and G. O. Mackie. 1983. Studies on hexactinellid sponges. III. The taxonomic status of Hexactinellida within the Porifera. *Phil. Trans. Roy. Soc. Lond.* (B)301:419–428.

Simpson, T. 1984. *The Cell Biology of Sponges*. Springer-Verlag, New York.

Vacelet, B., and N. Boury-Esnault. 1987. *Taxonomy of Porifera*. Springer-Verlag, New York.

7

Cnidaria

Hatschek, in 1888, was the first to distinguish sponges, cnidarians, and ctenophores, groups that had until then generally been placed together. The name Coelenterata was often used for the phylum.

Cnidaria includes jellyfish, sea anemones, corals, and various *Hydra*-like polyps. It is a very successful phylum, with about 10,000 living species and almost as many fossils. Cnidarians occur in oceans everywhere, but favor shallower, warmer waters. Only a few species have succeeded in fresh water. The largest jellyfish, *Cyanea arctica*, weighs nearly a ton and may possess tentacles up to a hundred feet long. Huge aggregations of coral polyps have formed reefs and atolls in the tropics, increasing the total surface available for life.

Definition

Cnidaria are radial, biradial, or radiobilateral in symmetry, and are structured about an oral–aboral axis derived from the primary gastrula axis. The body is composed of three layers: an outer epidermis derived from ectoderm, an inner gastrodermis derived from endoderm, and a cellular or noncellular mesoglea largely derived from ectoderm. There is but a single body cavity, or coelenteron, and a single mouth opening to the gut derived from the blastopore. Distinctive nematocysts are formed by interstitial cells. Typically, one or more whorls or fields of tentacles are arranged around the mouth. The neurosensory apparatus is composed of a nerve net of nonpolarized protoneurons, and various types of sensory cells sometimes assembled in special sense organs. There is a tendency toward polymorphism and an alternation of generations between sessile hydroid and swimming medusoid forms, and the body is often colonial with a variety of specialized individuals or zooids.

BODY PLAN

Two distinctive body styles are seen in Cnidaria: the medusoid or jellyfish form, for free-swimming life; and the polypoid or hydroid form, adapted to sessile life. The umbrella-shaped **medusa** is convex above and concave below, with the mouth centered in the concave surface. Tentacles lie around the margin of the body. The cylindrical **hydroid** is attached at its aboral end; the mouth is at the opposite end with a whorl of tentacles that typically occur around the margin at the oral end (Fig. 7.1A). The body is covered by ectodermal tissue, the **gastrovascular cavity**, or coelenteron, is lined with endodermal tissue, and mesoglea lies between. The mesoglea is thicker in medusae. The concave oral surface of a medusa (Fig. 7.1B) is comparable to the oral surface of a hydroid, extending from tentacles to mouth, and the convex upper surface is comparable to the aboral surface and the cylindrical body portion of a hydroid.

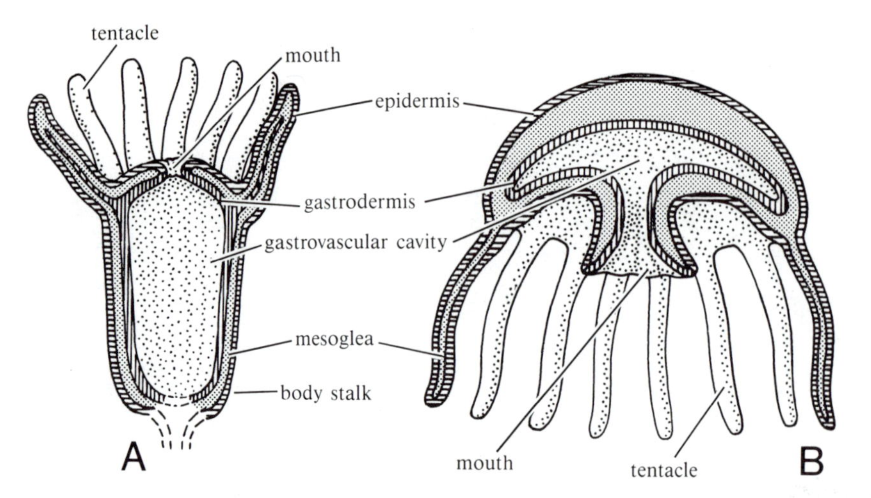

FIGURE 7.1. Comparison of polyp and medusa form; both composed of epidermal and gastrodermal layers separated by mesoglea, radially symmetrical with mouth centered in oral surface, with whorl of tentacles typically at outer margin of oral surface. Sessile polyp attached at aboral surface and normally oriented with mouth upwards; floating medusa lives with its mouth directed downward. Where both stages occur in life cycle, polyp reproduces asexually, whereas medusa reproduces sexually.

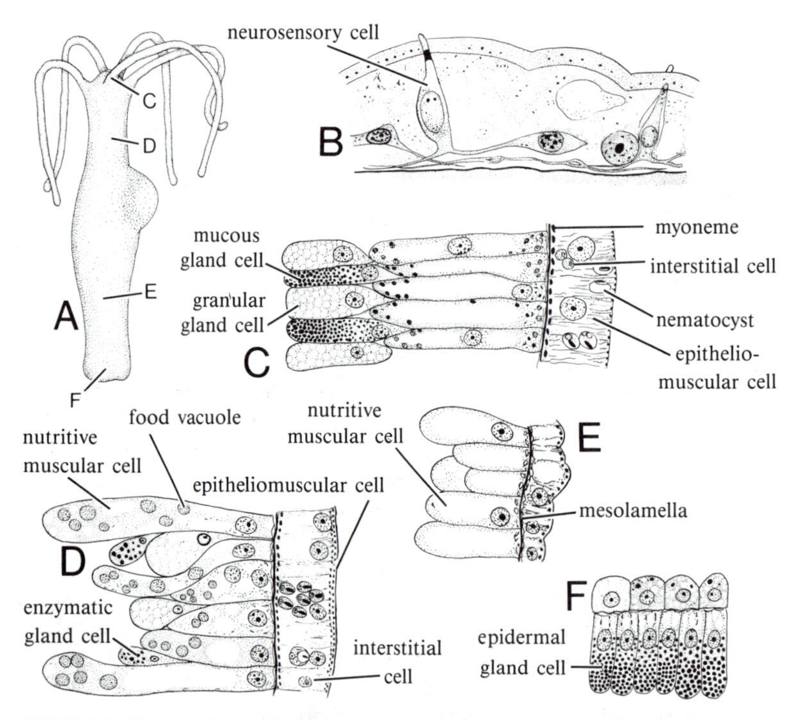

FIGURE 7.2. Histology of *Hydra*. **A**. Testis protruding on side of column. Lines *C* through *F* indicate levels at which corresponding sections have been cut. **B**. Epidermis of hypostome **C,D,E,F**. Sections through hypostome **C**, upper part of column in region of active digestion **D**, stalk region where digestion does not occur **E**, pedal disc **F**. Note highly developed glandular cells in gastrodermis of hypostome, used to lubricate mouth area. In gastric region many enzymatic gland cells occur and food vacuoles seen in nutritive–muscular cells. In stalk region enzymatic gland cells absent and gastrodermis vacuolated; epidermis composed of low, cuboidal cells. Gastrodermis of pedal disc inactive, and epidermis consists of glandular cells. Cellular specializations thus define oral, gastric, stalk, and basal regions in *Hydra,* although these may not be obvious externally. (**B** after Hadzi; **C,D,E** after Hyman.)

The familiar and readily available *Hydra* has become the classic type for the cnidarian structural plan (Burnett, 1973), and is used here to exemplify the simple level of tissue organization of the phylum (Tardent and Tardent, 1980).

Epidermis and **gastrodermis** are separated by a thin mesogleal plate, the **mesolamella**. Epidermis and gastrodermis are incipient tissues in *Hydra*, consisting predominantly of one cell type but containing a number of other cells. These cells have broad functions rather than the narrow uses typical of true tissues.

Histological structure of several regions of the body are shown in Figure 7.2. Large, vacuolated **epithelio-muscular cells** (Fig. 7.2D,E), with expanded bases containing two longitudinally oriented muscle threads, dominate the epidermis. The muscle threads or **myonemes** attach to the mesolamella and act to shorten the body when contracted. Narrow, elongated sensory cells (Fig. 7.2B), ending in fine processes or tiny bulbs, are scattered among the epithelio–muscular cells. Their basal fibers make contact with the nerve plexus. Multipolar neurons form a plexus in the gastrodermal as well as the epidermal layers. Their several neurites make connections with the neurites of adjacent protoneurons to form a nerve net. Adhesive secretions are formed by epidermal gland cells (Fig. 7.2F) with basal myonemes, situated on the pedal disc. Small interstitial cells (Fig. 7.2C) are packed in at the bases of the epithelio–muscular cells. They are unspecialized and can replace any of the other cell types. They are especially prominent in the growth zone at the upper end of the column and at points where buds or sex organs are developing. Many interstitial cells become cnidoblasts, forming nematocysts of various kinds.

Large **nutritive–muscular cells** (Fig. 7.2D) predominate in the gastrodermis. Their transverse myonemes constrict and lengthen the body when contracted, acting as antagonists to the epidermal myonemes. Mucous gland cells are abundant near the mouth, and enzymatic gland cells (Fig. 7.2C,D) secrete enzymes for the preliminary extracellular digestion of food. Nutritive–muscular cells have two flagella, and most gland cells have one. Their beating causes currents that aid in food distribution. Interstitial and sensory cells also occur in the gastrodermis but are less abundant than in the epidermis. The gastrodermis is regionally specialized, as shown by changes in nutritive–muscular cell form and by differing frequencies of gland cells (Fig. 7.2C,D,E,F). The oral region (Fig. 7.2C) secretes mucus to lubricate food to aid swallowing. The column (Fig. 7.2D) contains gastrodermis where most digestion and absorption occurs. The stem region (Fig. 7.2E) has a thin, inactive gastrodermis.

Cnidoblasts and Nematocysts When an interstitial cell begins to secrete a nematocyst, it is termed a **cnidoblast**. As the nematocyst matures, the cnidoblast attaches to the mesoglea and extends its distal tip to the epidermis surface. A bristlelike **cnidocil** develops that serves as a trigger when properly stimulated (Fig. 7.3A). Fibrils develop in the cytoplasm. Some are supportive, but others are thought to be contractile and aid in nematocyst discharge, at least in some Cnidaria. The nematocysts of *Physalia*, the Portuguese man-of-war, do not discharge when the animal is anesthetized, supporting the view that cnidoblast myonemes aid in discharge, and some fibers in *Hydra* cnidoblasts stain differentially like muscle fibers.

Each nematocyst consists of a capsule, its contents, a cap, and an inverted tube, sometimes expanded basally (Fig. 7.3E,F). The tube everts and the capsular contents are released when the nematocyst discharges. The tube and base are often covered with

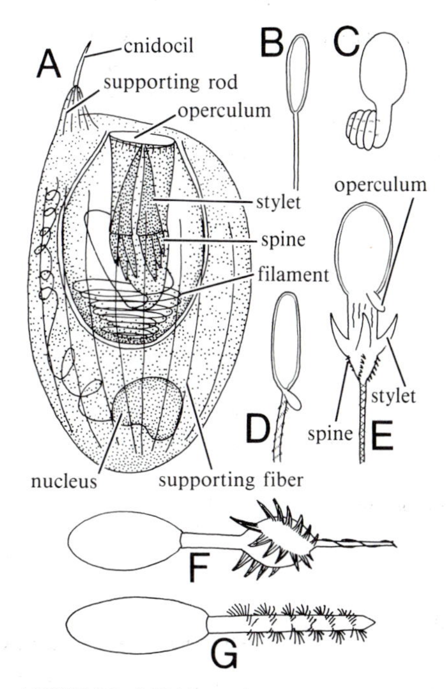

FIGURE 7.3. Cnidoblast and nematocyst types. **A.** Schematized cnidoblast containing undischarged nematocyst. **B,C,D,E,F,G.** Various discharged nematocysts; **B.** *Hydra* atrichous isorhiza (streptoline glutinant); **C.** *Hydra* desmoneme (volvent); **D.** *Hydra* holotrichous isorhiza (stereoline glutinant); **E.** *Hydra* stenotele (penetrant); **F.** eurytele; **G.** amastigophore. (**A** after Schulze; **B,C,D,E** after Hyman; **F,G** after Weill.)

bristles (Fig. 7.3F,G). At least 18 types of nematocysts have been described from Cnidaria, and differences of form occur in each type (Table 7.1; Fig. 7.3). Nematocysts stun, paralyze, or kill prey or enemies, entangle food and attach it to tentacles, and perform other, sometimes unknown, functions.

Nematocyst discharge is beginning to be understood (Holstein and Tardent, 1984). The discharge occurs in the space of 3 mμsec, that is, with a velocity of 2 m/sec or an acceleration of 40,000 *g*. With stimulation, there is an initial ten percent increase in the volume of the nematocyst, causing the operculum to open. Ultimately the capsule shrinks by about 75 percent of volume as its contents are discharged, apparently related to the release of mechanical energy stored in the fibrous structure of the wall. This mechanism appears to fit the evidence for nematocyst discharge in many cnidarians, but *Physalia* nematocysts appear to require muscle contraction.

Nematocysts do not typically require nerve stimuli for activation. There is much evidence for changing thresholds of stimulation. Few nematocysts discharge when they are touched with a glass rod, but more discharge if objects with different textures are used. Well-fed animals do not discharge nematocysts when they come in contact with food. Meat juices lower the threshold of activation, and glutathione is effective in lowering the threshold, such that a weaker mechanical stimulus evokes discharge.

Discharged nematocysts with open tubes expel the capsule contents. They generally release a toxic substance. Various toxins have been named and described by their effects, but the most toxic are those of the sea wasps (Cubomedusae) that form a monovalent cation channel in victim tissues that effectively depolarizes cardiac tissues (Skryock and Bianchi, 1983). Rapid anesthetization or death of the prey can follow a nematocyst attack. Some scyphomedusae (like sea wasps) and siphonophores can cause severe pain, serious illness, or even death in humans. Most cnidarians, however, cannot irritate the human skin.

Cnidarians lack the muscular and nervous equipment to grasp prey actively with their tentacles. Only by using an immobilizing agent and something to attach prey to the tentacles could the slow-moving cnidarians become successful carnivores. That this aptation is effective is shown by the fact that cnidarians are almost exclusively carnivorous.

Biology

Skeleton Cnidarians are soft-bodied creatures, but some groups do support themselves with skeletal material. Some hydrozoans use a periderm to afford some protection of the soft tissues below and offer support for the colony. However, the most noteworthy cnidarian skeletons are found among anthozoans that use both organic and calcareous materials, corals being the best examples of this (Wellington and Glynn, 1983).

Table 7.1. Types of nematocysts

Type	Characteristics	Function	Occurrence
Rhopaloneme	Tube, an elongated, closed sac	Entangling prey	Siphonophora
Desmoneme (or Volvent)	Tube, a closed thread, coiling on discharge	Engangling prey	Hydroida and Siphonophora
Isorhiza (or Glutinant)	Open tube of even diameter; without a butt	Uncertain	All classes
Anisorhiza	Open tube dilated at base; without a butt	Uncertain	Siphonophora
Mastigophore	Open tube with a cyclindrical butt; tube extends beyond the butt	Penetration, anchorage, toxic?	Anemones; Corals; Zoanthids
Amastigophore	Butt cylindrical; no tube beyond butt	Penetration, anchorage, toxic?	Anemones
Eurytele	Butt dilated at summit; tube long, open	Penetration, anchorage, toxic	Hydroida; Trachymedusae; Scyphozoa
Stenotele (or Penetrant)	Butt dilated at base, tube long, open	Penetration, anchorage, toxic	Hydrozoa

Locomotion Cnidarian polyps are mostly sessile. However, *Hydra* is known to move with somersault and/or inchwormlike movements that alternately use the oral tentacles and the aboral disc to adhere to the substrate and move the animal along. Many solitary anthozoan polyps are capable of ever so slowly gliding along surfaces on their pedal discs.

The siphonophores and the medusae display generally good, if rather gyratory, swimming abilities, achieved by rhythmic contractions of the muscles of the umbrellar margins (Gladfelter, 1973). The siphonophores often also use sails and floats to facilitate their movements. In medusae, a band of circular muscle lies between the epidermal layers of the velum and/or bell margin. Sudden contractions of the subumbrellar and velar muscles force a jet of water from the subumbrellar space, narrowed at its aperture by the velum, and this thrusts the animal forward. The muscles relax and the subumbrellar space fills again. Swimming movements are repeated rhythmically; the rate of pulsation changes with concentration of light, temperature, and the presence of food.

Food Capture and Digestion Most cnidarians are carnivores, the tentacles serving to capture food that comes in contact with the body surface. Tentacles are generally simple, although some Scyphozoa use elaborate oral arms instead. An extended tentacle with its nematocysts is a simple food-getting device, but it enables a small *Hydra* to capture a *Daphnia* as large as and a great deal faster than itself, and it enables a floating siphonophore to capture fish as powerful and as fast as mackerels. Food is either brought to the mouth by the tentacles, or the mouth is set on a movable extension of the body, the **manubrium**, and is extended partway toward the food. In addition, cilia or flagella on the tentacles may be used to direct food toward the mouth.

Flagellary and ciliary–mucus feeding occurs in some cnidarians. Food is caught in mucus, and food and mucus are then swept into the mouth. The feeding of *Aurelia* is an example. Mucus and food particles are conveyed to the mouth by flagellated grooves in the oral arms. Primitive anemones use ciliary currents in much the same manner, sweeping particles up along the side of the body, and onto the sticky, mucus-coated tentacles or oral disc, where more cilia propel food particles to the mouth.

Food enters at the mouth, and the mouth cavity typically is provided with lubricating slime glands; large organisms are ordinarily swallowed promptly. However, other methods may be used. The original mouth is lost in rhizostomes, and is replaced by many tiny "mouths" on the oral arms. Large fish may be caught and held against these suctorial mouths. Enzymes are secreted onto the prey; preliminary digestion takes place outside the body, and tiny, partially digested food particles are swept into the suctorial mouths by ciliary action.

Most Cnidaria lack a special swallowing region, but Anthozoa have a pharynx or stomodeum formed from in-turned epidermis at the mouth (see Figs. 7.14 and 7.18). It is equipped with many muscles and is important in swallowing.

The general features of digestion and absorption are the same in all forms. Food is rapidly attacked by extracellular enzymes to effect a preliminary digestion. The small particles that result are then ingested by gastrodermal cells, and final digestion and absorption occurs intracellularly. What is digested depends on the battery of enzymes available to hydrolyze foods. Cnidarians have little or no ability to digest carbohydrates and a very limited ability to digest fats. Proteases, usually most effective in neutral to alkaline conditions, are responsible for preliminary digestion.

Digestion tends to be a sequential process. The first phases of preliminary digestion produce intermediate products from proteins, and the pH of the coelenteron tends to shift toward the alkaline side. In the final, intracellular digestion, peptidases complete the process. As noted above, there is some tendency to differentiate the coelenteron into regions for preliminary digestion and for final, intracellular digestion. In colonial hydroids, preliminary digestion takes place in the hydranths, and much of final digestion occurs in the coenosarc or coenenchyme. In medusae, the system of radial and ring canals distributes the products of preliminary digestion to other parts of the body, where final digestion occurs. In many of the cnidarians, amoebocytes are also important in food transport.

The speed of digestion is limited by the available surface that bears enzymatic and nutritive cells. The consequences of surface-amplifying adaptations are seen in the larger cnidarians, for example, in the canal system of many medusae and the septa of Anthozoa. The unicellular glands of small polyps, like *Hydra*, tend to be replaced by patches of gland cells. In the gastric filaments of Scyphozoa and the septal filaments of Anthozoa, surface amplification is accompanied by grouping cells with a particular function.

Excretion As is generally true of aquatic invertebrates, Cnidaria appear to be ammonotelic, releasing most of their nitrogenous waste as ammonia.

Small quantities of uric acid, creatinine, xanthine, and guanidine have been recovered from cnidarian tissues. Inclusions that resemble complex nitrogen compounds have been described in cells located at various points, but no firm conclusions can be drawn. A dark region, sometimes called the "liver," is found below the float of some siphonophores. It contains guanine crystals, and may be an excretory structure.

Most Cnidaria have a very limited ability to control the water concentration of their tissues. This has been a factor in restricting most of them to marine habitats, where as osmoconformers they take up water or give off salts to match their environment. *Cyanea*, for example, reaches equilibrium with its environment within 36 hours, and corals respond similarly. Even in freshwater *Hydra*, exposure to distilled water soon leads to osmotic swelling and disintegration.

Most cnidarians that have been tested show a limited ability to control some specific ions. Medusae tend to have more potassium and less sulfate than the water in which they live, and the calcium content of anemones is below that of the surrounding medium.

If an animal cannot control its salt and water concentration, it can occur only in regions within the limits of its tolerance. The total salt concentration and the concentration of specific ions thus become specific, critical factors in affecting distribution. *Aurelia* survives well in brackish water because it has a high tolerance for low salt concentration, rather than because it has developed control mechanisms. The majority of the cnidarians have a narrow salt tolerance and are restricted to habitats that approach full ocean salinity. Thus, corals are killed by a few hours' to a few days' exposure to rain-diluted seawater.

Respiration Cnidarians have very low oxygen requirements; for example, red coral oxygen use is only 0.03 ml/g/hour, and *Aurelia* oxygen consumption is 0.02 ml/g/hour. Similar low figures are characteristic of other cnidarians. The large mass of mesoglea, containing few cells, and the passive sessile and floating habits are undoubtedly important in keeping oxygen requirements low. Activity changes the rate of oxygen consumption. Corals and anemones consume more oxygen when expanded, and *Aurelia* oxygen use varies with the oxygen content of the water.

The respiratory picture of many cnidarians is complicated by the presence of zooxanthellae or zoochlorellae. Whether they are necessary for their hosts or not, the algal cells liberate oxygen while carrying out photosynthesis and thus may elevate host activity in this way.

Internal Transport The gastrovascular cavity combines circulation with digestion. Circulatory problems are minimal in small animals like *Hydra*, since the surrounding water provides oxygen and a place to dump organic wastes, and no cell is far removed from the coelenteron or external medium. Circulation becomes important in large solitary individuals or in colonies with a large mesoglea. There is nothing in cnidarians comparable to blood. The gastrovascular fluid contains dissolved or particulate food, mucus, enzymes, and possibly other substances. In many instances the gastrovascular fluid resembles the sea outside the body, since the siphonoglyphs and septal flagella of Anthozoa (see Figs. 7.17 and 7.18) draw water into the cavity and discharge it, and the mouth is generally kept open in most ciliary feeders.

Body gyrations and the beating of gastrodermal flagella move the gastrovascular fluid through the coelenteron and its branches. Few muscular adaptations to move the fluid are found, but polyps of *Sertularia* may have a muscular gastric pouch that contracts and forces food into the coenosarc, and polyps of *Tubularia* use the basal part of the hydranth to pump food into the stem and roots.

Transport is a more severe problem in the more massive medusae. The challenge is met by canalization of the coelenteron (see Figs. 7.11B and 7.12) and a greater organization of flagellar propulsion of the coelenteric fluid. Canal systems tend to become truly circulatory in function, particularly in species with digestion restricted to a gastric cavity or gastric pouches with gastric filaments. In *Cyanea*, currents in the radial canals pass outward near the roof and inward along the floor. The flow is outward in the unbranched, and inward in the branched radial canals of *Aurelia* (see Fig. 7.12B).

The most effective circulatory devices are found in the Anthozoa, where siphonoglyph and septal flagella act to circulate the coelenteric fluid. That fluid circulation can be a critical problem is shown by the differentiation of some zooids as special circulatory individuals, siphonozooids that produce adequate currents to keep fluid moving in a complex system of tubes (see Fig. 7.16A).

Nervous System Cnidarians are animals without a well-organized central nervous system. There is never a structure comparable to a brain; the most coordinated forms have either a nerve ring that encircles the medusa bell, or a series of independent rhopalial ganglia. Radial symmetry appears to be correlated with poor nervous centralization (as will be noted in echinoderms).

The nervous system consists of multipolar or bipolar nerve cells that conduct stimuli in any direction. These protoneurons form a nerve net or plexus immediately below the epidermis that connects with neurosensory cells and muscle fibers as well as a similar, but more sparse, gastrodermal plexus. Fibers in the mesoglea connect the two plexi. The plexus is usually denser along the radial canals, forming radial nerves. Heavy concentrations of bipolar protoneurons form two nerve rings in hydrozoans, a larger upper ring just above the velar attachment, and a smaller ring just below the velar attachment. The two rings are separated by the thin mesogleal layer of the velum, the mesolamella, but are connected by fibers passing through it. The lower ring innervates the organs of balance and the velar and subumbrellar musculature. The upper ring innervates the tentacles, ocelli, tentacular bulbs, and tracts and patches of sensory epithelium at the bell margin.

Considerable difference of opinion exists about whether the neurites of the adjacent protoneurons are separated by synaptic spaces or are structurally united. It appears that both schools are correct and cnidarians have both kinds of nerve nets. In the chondrophore *Velella*, a nerve net of large syncytial fibers serves for rapid transmission of stimuli, and another, smaller, discontinuous net conducts slower, spreading stimuli. In netlike nerve plexi, stimuli spread in all directions, and the strength of the stimulus is correlated with the distance it travels. A stretched net conducts stimuli more rapidly, perhaps because it tends to modify the multipolar cells into bipolar cells. Generally, and perhaps for the same physiological reasons, impulses channeled along a constant pathway often follow aggregations of the nerve net cells composed of predominantly bipolar neurons.

Sense organs, nerves, and muscles are poorly developed in Cnidaria. Generalized neurosensory cells are found in the epidermis and, more sparsely, in the gastrodermis. Several morphologic types are recognized. Sensory cells are usually elongated (see Fig. 7.2B), with basal processes that make contact with nerve cells and with distal ends having bristles or bulbs. Cnidarians in which no other kinds of sensory receptors have been found are sensitive to touch, chemicals, and light.

Generalized or dermoptic light sensitivity is common. Coral polyps expand in diffuse, and contract in bright light. *Gonionemus* and some other medusae without eyespots change their swimming rates with changed light intensity. Hydras choose blue-green over red environments; some sea anemones are color-blind, whereas others respond differently to blue-green and red-yellow light. Dermoptic light sensitivity tends to disappear with the development of ocelli. *Aurelia* and *Sarsia* are photo-insensitive when their ocelli are removed.

Regions of high activity or parts more exposed to the environment tend to be better equipped with sensory cells. Concentration of sensory cells tends to occur near the mouth and on tentacles and pedal discs. Patches of sensory epithelia with long sensory hairs occur at the base and sides of scyphozoan rhopalia and in tracts associated with the nerve ring. A somewhat different sensory epithelium lines the sensory pits at the upper and lower surfaces of rhopalia in semaeostomes and rhizostomes.

Special light-sensitive organs occur in or near the tentacular bulbs of many hydromedusae and as a part of the rhopalia of some scyphomedusae. Scattered light-sensitive cells react to the intensity and quality of light, but are not well suited to respond to the direction from which it comes. The advantage of ocelli is that the combination of pigment cells and light-sensitive cells permits collecting of information necessary for a sense of direction. The simplest ocellus is a pigment spot, with pigment cells between the light-sensitive cells (Fig. 7.4A). Shading becomes more effective if the ocellus is invaginated as a cup (Fig. 7.4B), especially if a lens is added.

Ocelli may or may not be present in hydromedusae and scyphomedusae, and the complexity of ocelli varies independently of the level of body complexity. *Sarsia* ocelli are far more advanced than the other parts of its body, and otherwise closely related medusae may have simple ocelli or none at all.

Medusae usually orient with the oral surface down and the aboral surface up. This posture is favored by the weight distribution of the body, and many species tend to assume their normal position without balance organs. Many medusae, however, have organs of balance. They are characteristic of leptomedusae, but in most groups they are present in some and absent in others. Statocysts are variable in form. They contain heavy cells (lithocytes) weighted by inclusions (statoliths) contained in a cavity of the lithocyte. Leptomedusae lithocytes are specialized epidermal cells, lying in pockets of vesicles formed by the closure of pits (Fig. 7.4C). Generally, open statocysts have many lithocytes and closed statocysts have few. Narcomedusae have gastrodermal lithocytes, contained in club-shaped lithostyles surrounded by long sensory hairs. Trachymedusae have closed statocysts, each containing a lithostylelike club in a cavity lined by sensory epithelium. The rhopalium of *Aurelia* (see Fig. 7.12C) contains a cavity continuous with the ring canal, and ends in many lithocytes of gastrodermal origin, with sensory pits above and below. Balance organs tend to be held within rather narrow limits of design by the nature of the environmental factor they are intended to reflect. Gravity pulls a weighted body down. The weighted object may be a spheroid in a cavity or a weighted club. As vesicles or pits are overturned, the heavy object within stimulates different parts of the sensory epithelium and so provides information about the direction of gravitational pull.

Balance organs set up corrective "righting" reflexes that restore normal body orientation. If the bell of *Aurelia* is tilted, unequally strong swimming movements of the lower and raised side of the bell tend to return the animal to a normal position. If the rhopalia are removed, the ability to right is lost.

Reproductive System In the Hydrozoa, the polyp can reproduce asexually by budding off small, unciliated, planulalike frustules that creep about for a time, attach, and develop into new polyps. Eventually the mature polyps bud off medusae that detach and swim away. In the *Gonionemus* life cycle, it is clear that the polyp is little more than a larval form, although it can reproduce. The polyp is effectively absent in trachylines. Among hydroidans the polyp is far more important, and the medusa phase may be little more than a "gadget" for making eggs and sperm. The medusa stage is completely omitted in *Hydra*. Scyphozoan jellyfish have reduced polyps, and the sea anemones and corals have no medusa phase.

In most hydrozoans, the mature polyp or colony produces gonophores. The parent polyp may be a normal gastrozooid, as in most gymnoblasts, or may be a gonozooid, specialized for medusa formation. Gonophores are sometimes produced on stems or stolons. In any event, the young medusa bud is an evagination of the body wall, involving epidermis and gastrodermis (Fig. 7.5A). The tip epidermis develops into the primordium of the subumbrella. The gastrodermis below forms five growing tips: four marginal ones, which form the four radial canals, and a medial one, growing downward to form the manubrium. A single layer of gastroderm grows out from the radial canals to form the gastrodermal lamella. The ring canal is formed by lateral fusion of the radial canal tips. Meanwhile, a subumbrellar space has appeared, and a velar plate has formed at the apex of the bud. Rupture of the velar plate and growth of the tentacles and other marginal organs complete the medusa.

Hydrozoan gametes arise from interstitial cells or, more rarely, from normal epidermis or gastrodermis cells, aggregated to form a gonad. In anthozoans and scyphozoans, gametes form from gastrodermal interstitial cells. They gather as gonads, either on each side of the gastric septa or on the floor of the gastric pouches. Gametes are freed into the coelenteron and emerge through the mouth.

Cnidaria may be either monoecious or dioecious, and fertilization may be external or internal. Sper-

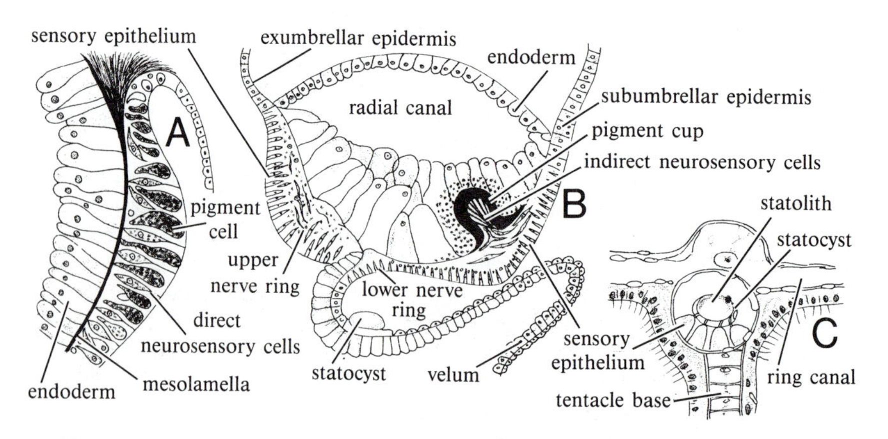

FIGURE 7.4. Sense organ cross sections. **A.** Pigment-spot ocellus from *Turris,* simplest type of ocellus with alternating pigment and neurosensory cells. **B.** Ocellus and adjacent region *Tiaropsis,* ocellus of advanced type with definite pigment cup to admit light only through cup opening, and with neurosensory cells inverted with sensory tips at bottom of cup. Note statocyst of open type at base of velum. **C.** Statocyst of closed type from *Obelia.* (**A,B** after Linko; **C** after the Hertwigs.)

matogenesis follows standard patterns. Many oöcytes form, but few mature, the others becoming nurse cells that fuse with the ovum. Ova are released or undergo early development in the medusa, sometimes in special brood chambers (Campbell, 1974).

Asexual reproduction is important in Hydrozoa and Anthozoa. Polyps are added to colonies as buds from older polyps. Some anemones reproduce asexually by longitudinal fission. *Metridium* reproduces asexually by fragmentation of the pedal disc to produce young with irregularities in septal and siphonoglyph arrangement. New mouths and whorls of tentacles appear in the coral coenenchyme. They develop septa continuous with the septa of adjacent polyps. Some corals form new mouths in the oral disc of a parent polyp. They may separate completely from the parent or remain attached to form a compound polyp of irregular form (Chia, in Mackie, 1976).

In Hydrozoa, especially in the siphonophores, asexual colonies lead to such specialization that individual polyps come to act as organs (discussed later). Indeed, cnidarians are an excellent example of a common phenomenon among invertebrates, the extensive use of asexual reproduction to form colonies (Fig. 7.5A). In the Cnidaria, asexual colony formation allows for the mimicking of organ systems, with individual polyps specializing for specific functions. In addition, asexual reproduction allows for exponential growth that allows quick utilization of ephemeral resources. The survival of the colony is size-independent; any number of individuals can ensure this, depending on the level of resources available. Asexual colony formation also reduces competition between individuals of a species by allowing them to share rather than fight over resources. Finally, asexual reproduction not only brings about rapid cloning of phenotypes proven effective by virtue of their survival, but it also imparts a high degree of morphologic flexibility and adaptability to allow the expression of seasonal and environmental polymorphism.

Development Patterns of cnidarian embryonic development vary in specifics from group to group (Campbell, in Muscatine and Lenhoff, 1974). Similarities do exist, however, to allow some generalizations to be made. Holoblastic cleavage usually produces a coeloblastula, which is sometimes freed as a ciliated larva in some hydrozoans, but alcyonarian anthozoans form a solid stereogastrula. Endoderm forms by multipolar ingression in most cnidarians (Fig. 7.6A), although this process is aborted in semaeostome scyphozoans in favor of invagination (Fig. 7.7A), and the alcyonarians differentiate layers by delamination. At this point in ontogeny, typically a gastrulated, mouthless, solid, ciliated larva is freed, the **planula** (Fig. 7.6B).

The fate of the planula larva is quite varied. It may attach and form a polyp (Fig. 7.7F,G,H,I), as in hydroidans and anthozoans. In the latter, a stomodeum is produced (Fig. 7.7E) and endodermal infoldings form the septa. In semaeostomes, the planula forms a polypoid larva, the **scyphistoma** (Fig. 7.7B), which later strobilates, that is, asexually divides horizontally (Fig. 7.7C) to form stacks of **ephyrae** (Fig. 7.7D) that become medusae. In most hydrozoans, the planula forms an **actinula** larva (Fig. 7.6C) that transforms into a medusa (Fig.

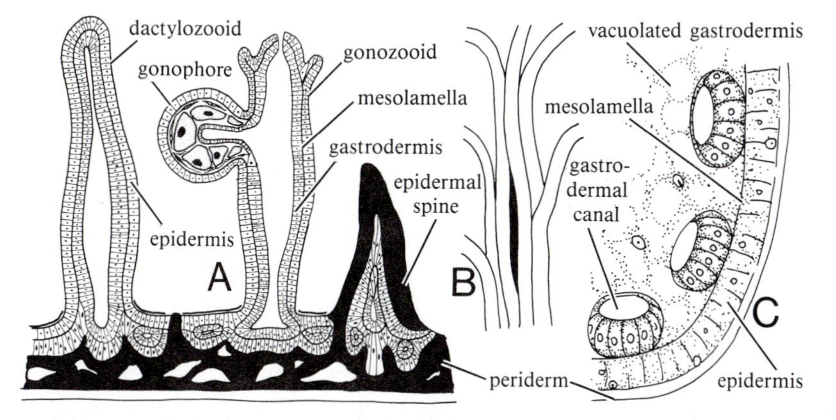

FIGURE 7.5. A. Schematic section through part of *Hydractinia* colony showing peridermal spine, dactylozoid, and gonozooid with sessile gonophore attached. Thin upper periderm interrupted, and gastrodermal canals pass through heavy basal periderm, forming matlike coenosarc. Colony results from hydrorhizal growth. **B**. Part of hydrocaulus of *Eudendrium*, consisting of parallel stolons. **C**. Cross section of *Tubularia* stem, showing vacuolated gastrodemis cells and gastrodermal canals. (**A** composite; **B** after Hyman; **C** after Warren.)

7.6E,F,G,H) or settles and creeps about before becoming a polyp.

Review of Groups

CLASS HYDROZOA

This traditional class contains a wide array of forms that appear to be only distantly related to each other (see section on classification and phylogeny section below). A few hydroids live in fresh water, notably *Hydra*, widely studied in introductory courses, and *Craspedacusta*, the freshwater jellyfish, but nearly all hydrozoans are marine. Some hydroids are solitary and others form colonies of thousands of zooids. A single polyp is usually small, but *Branchiocerianthus*, living nearly two miles below the surface, may approach 3 m in height. Hydrozoans generally prefer shallow water where their colonies contribute heavily to surface growth on submerged objects.

Actinulida These are minute forms resembling the actinula larva (Rees, 1966). Little is known concerning their biology. They are meiofaunal forms living in beach sands. Actinulids move by the beating of cilia rather than contractions of muscles.

Trachylina According to one view of cnidarian phylogeny, the trachyline medusae are the most primitive modern cnidarians. *Gonionemus* and *Craspedacusta* are often studied as representative trachylines.

Two suborders of trachylines are recognized, Trachymedusae and Narcomedusae. Trachymedusae are characterized by a smooth bell margin, gonads on the radial canal, and a manubrium (see Fig. 7.9A). Narcomedusae have a scalloped bell margin, gonads on the floor of the gastric cavity, and no manubrium (see Fig. 7.9B). The mouth opens directly into the gastric cavity. *Gonionemus* and *Craspedacusta* are good examples of Trachymedusae, showing a typical tetramerous radial symmetry in the four radial canals and quadrangular mouth. *Aglantha* (see Fig. 7.9A) possesses a very tall bell and a long **pseudomanubrium**, containing radial canals, that ends in a **true manubrium** without radial canals. As in some other Trachymedusae, there are more than four radial canals. *Cunina* (see Fig. 7.9B) is a typical narcomedusan that lacks the manubrium,

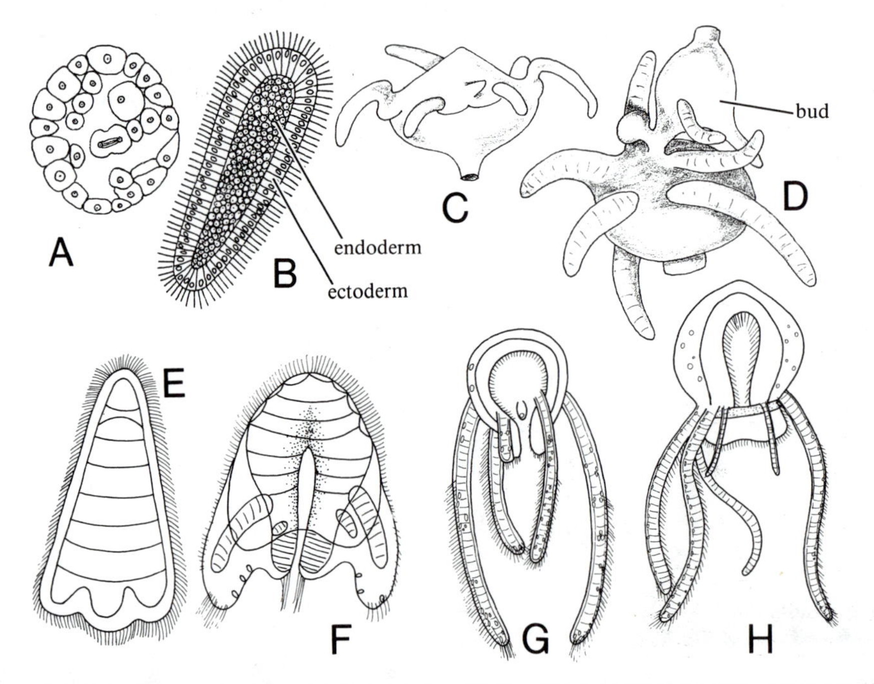

FIGURE 7.6. Hydrozoan development. **A.** Endoderm formation by multiple ingression in *Gonothyraea*. **B.** Planula larva. **C,D.** Stages in development of *Pegantha;* **C.** young actinula; **D.** older actinula budding off daughter actinulae (appearance of budding in developing actinulae larvae may have been first step in origin of colonial hydroid generation). **E,F,G,H.** Actinula developing into young medusa. (**A** after Wulfert; **C,D** after Bigelow; **E,F,G,H** after Metschnikoff.)

has a flattened form and scalloped bell margin, has scalloping of the gastric cavity, and possesses prominent sensory tracts. In general, Narcomedusae have lithostyles for balance organs, whereas Trachymedusae have statocysts.

Hydroida Hydroida have typically the best balance in their cycle between medusae and polyps, but the polyp is usually dominant. Polyps are sessile, whereas medusae float or swim feebly at the mercy of ocean currents and tides. The inactive polyp merely extends its tentacles to catch prey and retracts, often into a protective exoskeleton, to avoid enemies. Medusae orient themselves in daily cycles or with weather changes, alternately rising to the surface or swimming deep. Polyps typically reproduce asexually, and medusae sexually, although exceptions occur. Polyps often form large colonies, whereas medusae are typically solitary.

Typical hydroid polyps are composed of three

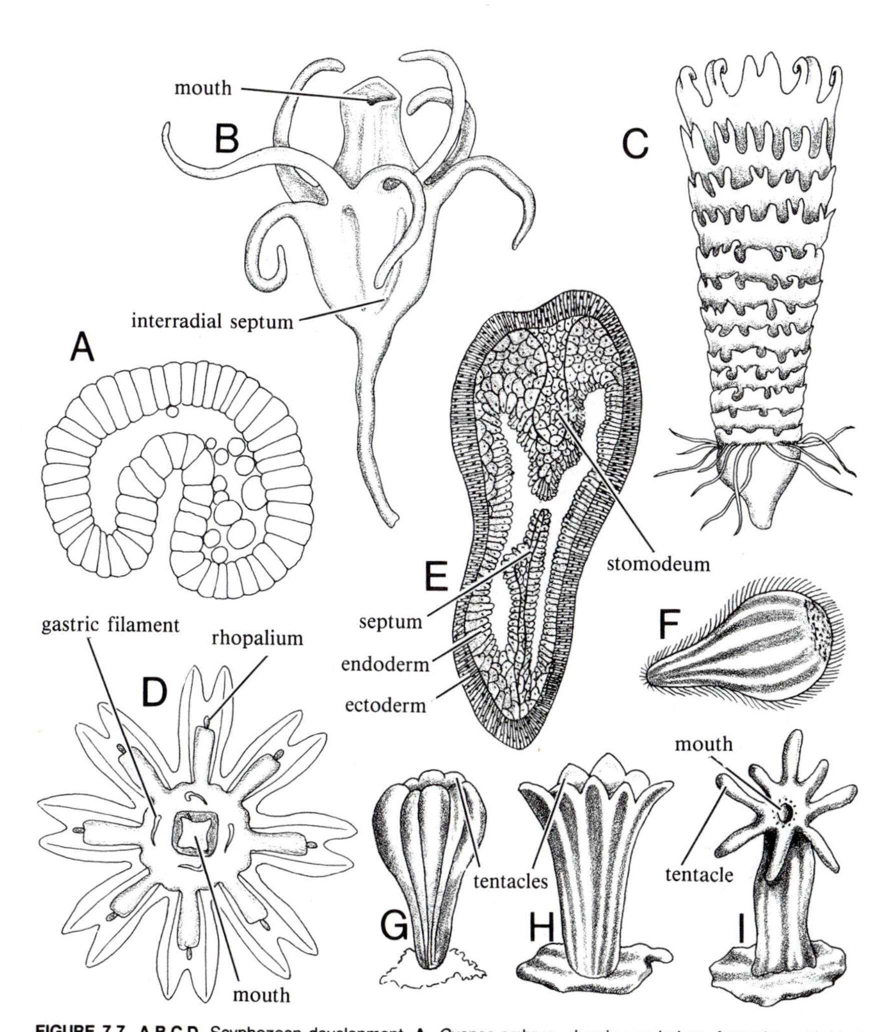

FIGURE 7.7. A,B,C,D. Scyphozoan development. **A.** *Cyanea* embryo, showing endoderm formation with blastocoel endoderm derived by ingression and second endoderm formed by invagination. **B,C,D.** *Aurelia* development, schematized. *Aurelia* zygote develops into planula that attaches and becomes scyphistoma **B**, with rudiments of four interradial septa characteristic of Scyphozoa. Scyphistomae live for several years, budding off new scyphistomae **C** during warm seasons and producing larval medusae known as ephyrae by transverse fission **D** during winter and spring. When released, ephyrae have eight rhopalia and rhopalial lappets, but undeveloped interradial areas. **E,F,G,H,I.** Anthozoan development. **E.** Larva of *Renilla,* in longitudinal section, showing stomodeum formation. At stage shown, stomodeum is solid; mouth opening formed by rupture of stomodeal region. **F,G,H,I.** Stages in development of sea anemone, *Lebrunia;* planula loses apical tuft when ready to settle **F**; after attachment, tentacles begin to form **G,H**, pedal disc develops **H**, and mouth breaks through **I**. (**A** after Okada; **B,C,D** after Schechter; **E** after Wilson; **F,G,H,I** after Duerden.)

parts: **base**, **stem**, and **hydranth**. Adaptations of the base center about attachment, with some provision for the protection of the living, basal tissues. Colonial forms usually arise from a **hydrorhiza**, a tangled mass of branching and anastomosing stolons (see Fig. 7.5B). Each **stolon** is essentially like the column, with a central coelenteron enclosed in a wall composed of epidermis, mesoglea, and gastrodermis, and covered by a protecting sheath of chitinous periderm secreted by epidermal cells. Solitary polyps usually attach by unbranched, solid holdfasts that die away to leave hollow peridermal tubes behind. *Hydractinia* commonly attaches to snail shells containing hermit crabs. Its stolons fuse into a network, from which continuous upper and lower sheets of periderm extend (see Fig. 7.5A). The lower epidermis secretes layer after layer of periderm that serves to attach the colony firmly. The upper periderm disappears, although scattered epidermal spines extend upward through the surface. *Hydra* is exceptional in showing no stolon branching and in having an adhesive pedal disc. Such species are somewhat motile, gliding on the pedal disc or creeping about with the aid of the tentacles.

The hollow stem and living basal parts make up the **coenosarc** of colonial hydroids. The coenosarc (Fig. 7.8C,D) houses a continuous coelenteron, distributing food throughout the colony. As a plant stem lifts leaves to the sun, the cnidarian stem lifts the hydranths above the bottom, increasing the number of animals that can occupy the available surface. The stem provides support as well as food

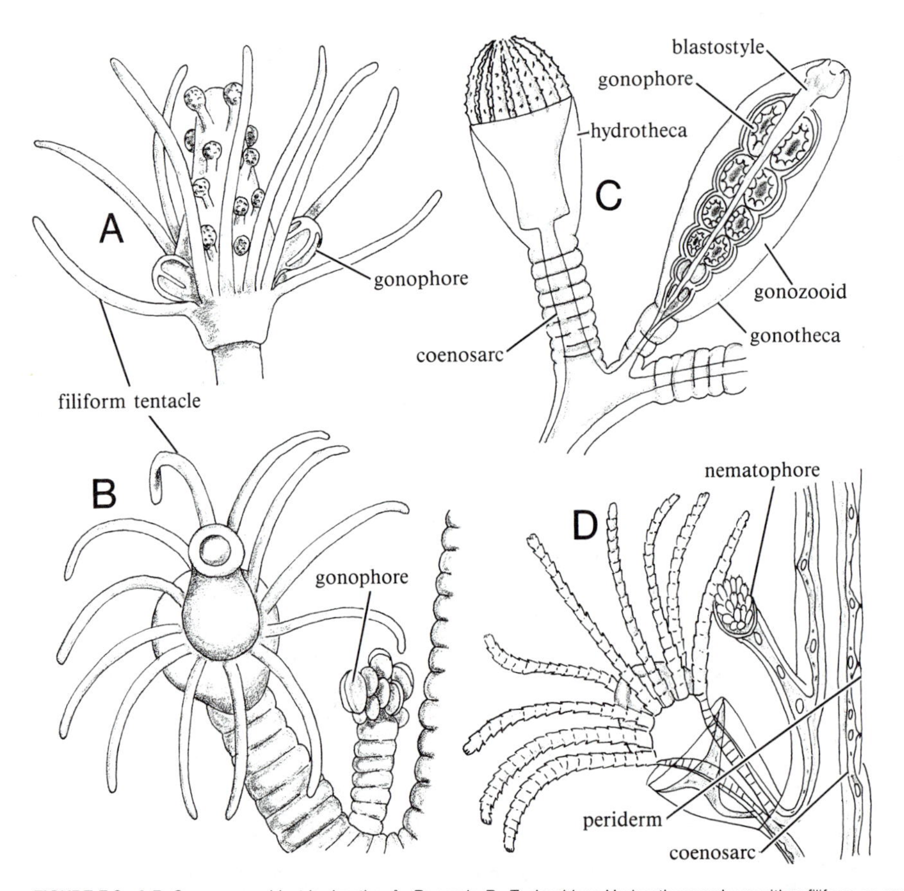

FIGURE 7.8. A,B. Some gymnoblast hydranths. **A.** *Pennaria.* **B.** *Eudendrium.* Hydranths may have either filiform or capitate tentacles. Latter have globular clusters of nematocysts at tip. Scattered tentacles considered primitive. In advanced forms, tentacles assume definite position, with formation of basal whorl, as in *Eudendrium.* Gonophores of most gymnoblasts arise from gastrozooids, often rising above basal whorl of tentacles. **C, D.** Some calyptoblast hydranths; **C.** *Obelia,* with retracted tentacles; **D.** *Plumularia.* Calyptoblast hydranths, characterized by their exoskeleton, have single whorl of filiform tentacles. Periderm continues beyond base of hydranth, forming protective cup. Gastrozoids sit in hydrotheca. Gonozooids, which consist of blastostyle surrounded by gonotheca, form many daughter gonophores. Term gonangium applied to gonozooid and its gonotheca. *Obelia* has simple sympodial growth pattern and relatively simple colony form. *Plumularia* colonies include specialized zooids, which contain many nematocysts termed nematophores. (**B** after Nutting; **D** after Hyman.)

transport and often buds off new hydranths or medusae. The upright main stem of colonial hydroids is the **hydrocaulus**. The stem and its periderm provide adequate support for most polyps, but if the hydranth is very large, as in *Tubularia* and *Corymorpha*, the stem is filled with vacuolated gastrodermal cells to provide support by turgor. In such stems, multiple gastrodermal canals extend to the base (see Fig. 7.5C).

Every hydroid colony has at least two kinds of zooids: **gastrozooids** and **gonophores**. Gastrozooids are ordinary feeding polyps, and gonophores are the buds from which medusae arise. In many hydroids, the medusae are reduced and permanently sessile; these gonophores are permanent zooids. Most calyptoblasts have reproductive polyps, gonozooids, that produce gonophores and have reduced tentacles, mouth, and coelenteron. Other kinds of specialized zooids with varied specialized functions occur among Hydrozoa and other Cnidaria (see Table 7.2).

The hydranths catch prey and carry out the preliminary digestion of food, in addition to other functions. The partly digested food is circulated through the colony to be digested in the gastrodermis of the coenosarc and hydranths. Hydranths may also bud off new hydranths or medusae (Fig. 7.8C). Adaptation of the hydranth has centered about its major activities. Tentacles vary in form and arrangement, and specializations for gonophore production are common.

Gymnoblastea hydranths are **athecate**; that is, the periderm ends at the base of the hydranth, leaving it naked. Several adaptational lines have appeared in gymnoblasts. *Syncoryne* retains the primitive characteristics of scattered capitate tentacles ending in nematocyst-filled knobs. *Pennaria* (Fig. 7.8A), with a basal whorl of fingerlike filiform and the primitive scattered capitate tentacles, has tended toward a definite pattern of tentacle placement. *Tubularia*, with distinct basal and oral whorls of filiform tentacles, would represent an end point of this adaptational line. A second line ends in forms like *Eudendrium*, with a single whorl of filiform tentacles (Fig. 7.8B). Some hydranths are highly specialized. *Hydractinia* colonies have spinelike dactylozooids, with the tentacles reduced or absent and interspersed among the gastrozooids and gonozooids (Fig. 7.5A).

Calyptoblastea hydranths are **thecate**, sitting in a **peridermal cup** into which they can withdraw. Calyptoblast hydranths usually have a single whorl of filiform tentacles. Adaptation has especially centered on skeletal modifications, colony growth, and specialization of gonozooids. Forms like *Obelia* are thought to be primitive because of the loose colony form and bell-shaped theca (Fig. 7.8C). In these

Table 7.2. Types of zooids

Type	Function	Characteristics	Occurrence
Gastrozooid (or Trophozooid)	Feeding polyps	Typical polyp form	Universal in hydroids
Gonophore	Immature or sessile medusae	Buds, often platelike or bell-shaped	Universal in hydroids
Gonozooid	Gonophore production	Usually a polyp with reduced tentacles, and often without mouth	Common
Dactylozooid	Protection, food capture	Modified polyp with reduced mouth, coelenteron and tentacles	Common in Hydrozoa
Tentaculozooid	Same as dactylozooid	Tentaclelike	Common in some hydrozoan groups
Nematophore (or Sarcostyle)	Food capture	Blunt ends, with nematocysts or adhesive cells	Plumulariidae
Pneumatophore	Float	Polyp, modified to form a gas-filled chamber	Siphonophora
Nectophore	Swimming	Modified medusoid, with a velum and four radial canals, but without a mouth	Siphonophora
Phyllozooid (or Bract)	Protective cover	Modified medusa, forming thick, leaflike, helmet-shaped or prismatic covers	Siphonophora

forms, gonozooids generally form in the axils of the branches and are surrounded by flask-shaped gonothecae. Sessile thecae and monopodial growth are characteristic of the more highly specialized calyptoblasts, as seen in *Plumularia* (Fig. 7.8D). Some calyptoblasts have a theca with a closing lid, the operculum, formed from one or several pieces of periderm.

The typical gonozooid takes the form of a blastostyle, a club-shaped zooid without mouth or tentacles. On its surface many medusa buds or gonophores develop (Fig. 7.8C). Plumularian colonies have an additional type of specialized polyp, the nematophore, whose club-shaped body is filled with adhesive glands and nematocysts.

Gymnoblasts produce **anthomedusae** (Fig. 7.9C,D), and calyptoblasts, **leptomedusae** (Fig. 7.9E,F,G). Anthomedusae are taller, retain the tetraradiate form in all cases, have gonads on the manubrium, and usually have ocelli. Leptomedusae are flatter, may have lost their tetraradiate appearance by branching of the radial canals, have gonads borne on these canals, and have statocysts on the bell margin. Only a few of the hydroid medusae are large. *Aequora* is an exception, and may reach several inches in diameter. Smaller medusae, like *Obelia* and *Pennaria*, show signs of reduction, but *Obelia* (Fig. 7.9G) retains the basic traits of its group and survives for some time as a free-swimming organism. *Pennaria* (Fig. 7.9D) is much further reduced. Its tentacles are mere vestiges, and it survives for only a brief time after its release. Many hydromedusae are reduced, remain permanently sessile on the parent hydranth, and function as the reproductive organs of their colonies.

Milleporina and Stylasterina These groups are often found among coral reefs in warm seas. Rather than a periderm, they form a heavy, calcareous exoskeleton, the **coenosteum**. The porous millepore coenosteum is permeated in its upper levels by living coenosarc tubes in deep canals, and has cups in

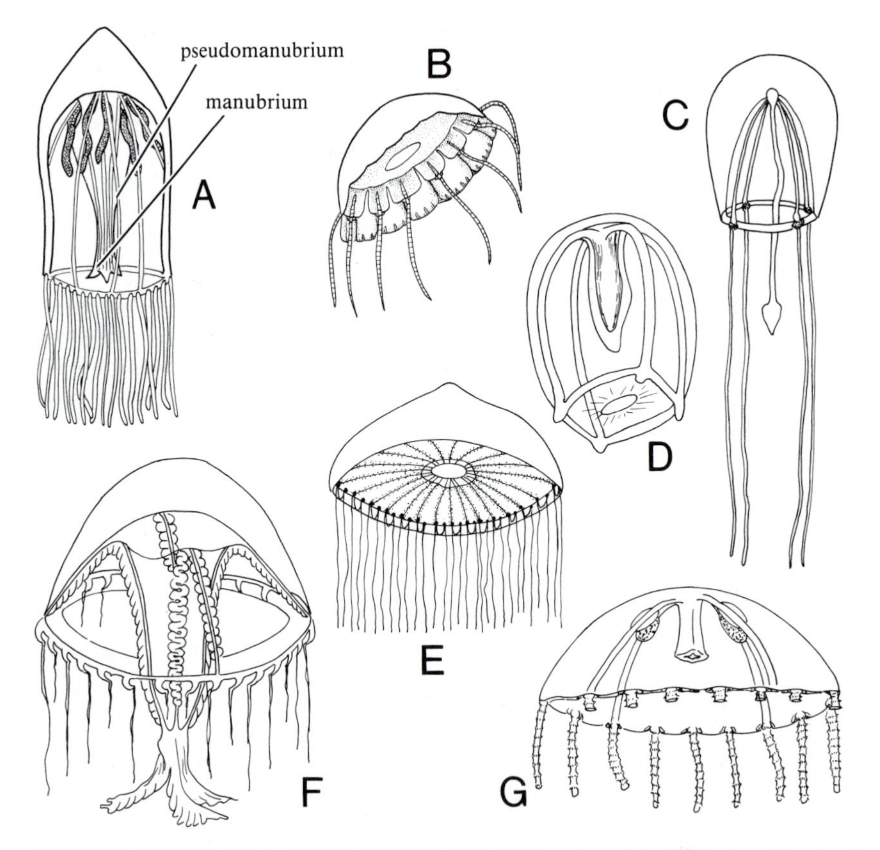

FIGURE 7.9. Medusae types. **A.** *Aglantha,* trachymedusa. **B.** *Cunina,* narcomedusa. **C.** *Sarsia,* well-developed anthomedusa. **D.** *Pennaria,* somewhat degenerate anthomedusa. **E.** *Aequora,* large leptomedusa. **F.** *Tima,* well-differentiated leptomedusa. **G.** *Obelia,* small but complete leptomedusa without velum. (**A** after Hargitt, from Pratt; **B** after Bigelow; **C,E** after Hyman; **D,F** after Mayer.)

which the polyps sit. As new coenosteum forms, the deeper coenosarc dies away, and cross partitions, or tabulae, form across the polyp cups. Millepores are usually massive and whitish or yellowish. Stylasters form more open, branched colonies and are often colorful, running to red and violet hues. Both have gastrozooids and dactylozooids. Millepore gastrozooids are short, with tentacles reduced to nematocyst-filled knobs. Stylaster gastrozooids are more elongate and have stumpy tentacles. Millepore dactylozooids are elongated and have scattered capitate tentacles. Stylaster dactylozooids have no tentacles and are slender. Both groups form much-reduced medusae within the coenosteum.

Siphonophora Siphonophores are striking creatures, of great interest because they are highly polymorphic colonies, pelagic rather than attached, and made up of many highly modified medusae and polyps. The colony is a true superorganism; the individual zooids are specialized to serve as organs of the superorganism. They are most common in warm seas, but ocean currents and storms sometimes bring large numbers of siphonophores into cooler waters. Many are attractively colored but are very dangerous because of their powerful nematocysts.

Various kinds of zooids bud from a common coenosarc. Three kinds of modified zooids occur: **gastrozooids**, **dactylozooids**, and **gonozooids**. Gastrozooids are feeding polyps, with a mouth and usually with a single, branching tentacle. Dactylozooids are long, tubular, tentaclelike zooids without a mouth, or shorter zooids with a long, unbranched tentacle. Gonozooids resemble gastrozooids but may not have a mouth. Their branches, the **gonodendra**, resemble calyptoblast blastostyles. Medusoid zooids are swimming bells, bracts, gonophores, and floats. Swimming bells (**nectophores**) lack mouth, tentacles, and sense organs, but are muscular and propel the colony. Bracts (**hydrophylla**) are protective, leaflike zooids, without resemblance to ordinary medusae. Gonophores vary from buds that live briefly after being freed to saclike, permanently sessile forms. Floats (**pneumatophores**), according to some recent studies, may be modified polyps.

Three suborders of Siphonophora are recognized: the Calycophorae, characterized by the absence of a float and the presence of nectophores; the Physonectae, characterized by small floats and a series of nectophores; and the Cystonectae, which have large floats but no nectophores.

Muggiaea is a typical calycophoran, with an apical swimming bell and a linear series of trailing zooids. The swimming bell is drawn out to form a protective sheath for the base of the coenosarc. The coenosarc trails as a long stem, to which a series of cormidia are attached. Each **cormidium** consists of a definite number and variety of individuals (Fig. 7.10A,B). New cormidia form at the coenosarc base, and increase in age and size distally. They eventually break off and take up a separate existence for a time.

The physonectan siphonophores develop a small float. With the appearance of a float, placed above the swimming bells, the linear stem becomes shorter, the members are more crowded, and cormidia lose their individuality. In forms like *Nectalia*, the float is surrounded by conspicuous nectophores. This adaptive line ends with all of the zooids crowded about a short, often massive coenosarc, and the complete absence of swimming bells, as in *Physalia*, the Portuguese man-of-war, a typical cystonectan form. In the absence of nectophores, these colonies are at the mercy of currents or winds carrying them into a lee shore. *Physalia* has a large, polypoid float with the aboral surface invaginated to form an inner air sac. Larger and smaller dactylozooids, gastrozooids, and gonozooids hang from the short coenosarc below the float. Food capture is the function of dactylozooids, armed with powerful nematocysts. They carry the prey upward and the gastrozooids fasten to it, accomplishing preliminary digestion. Partly digested food is distributed by way of the coenosarc.

Chondrophora Chondrophores differ from siphonophores in having the colonies formed entirely of polyps. *Velella* (Fig. 7.10C), the purple sailor, is a well-known example. The large pneumatophore is discoid and contains many chambers. A single large gastrozooid is surrounded by gonozooids and, at the margins of the disc, by dactylozooids. An upright sail extends from the top of the float.

CLASS SCYPHOZOA

The common jellyfish, so often seen in coastal waters and usually feared by local swimmers, are Scyphozoa. Although there are only about 200 species of Scyphozoa, they can become so abundant that they play an important ecologic role, and are found everywhere in oceans living at depths up to 3000 m or more.

Stauromedusae These are the most unusual scyphozoans. They are jellyfish, but attach to seaweeds and other objects by an aboral stalk. They are probably best understood as **paedomorphs**, that is, permanently juvenile animals that never quite complete medusa formation. In any event, they retain many primitive characteristics and may provide some idea of what scyphozoan stem animals were like when they diverged from the hydrozoan line.

Lucernaria and *Haliclystus* are the most commonly observed stauromedusans. They are more common in cool waters than in warm waters, and are from about 3 to 5 cm across. The aboral stalk expands sharply to form a sort of umbrella (Fig. 7.11). The mouth is centrally located, at the end of a short manubrium. It leads into a coelenteron divided by four large **septa** into four **gastric pockets**. The septa are not simple partitions, but are hollow and contain deep **subumbrellar funnels** that extend well into the stalk and open on the oral surface. The septa do not meet centrally; the coelenteron has a median open region. The oral surface is drawn out into lobes on which the tentacles are borne. Four gastric septa are characteristic of scyphozoans, although they are lost during embryonic development in some groups.

Semaeostomae *Aurelia* is a commonly studied example of a scyphozoan form (Fig. 7.12). The gastrovascular system is complex, with extensively branched radial canals and a coronal stomach composed of four gastric pouches. The symmetrical axes are easily determined. The margin of the bell is slightly lobated, with a complex sense organ, the **rhopalium**, at each indentation. The eight rhopalia

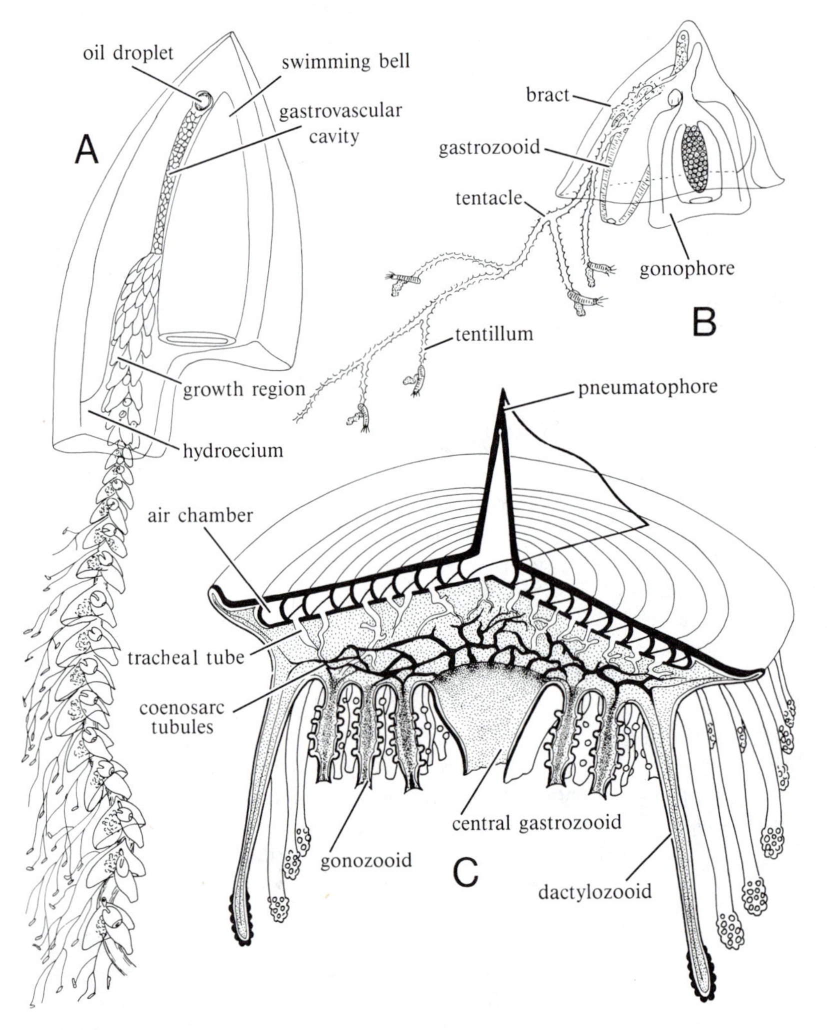

FIGURE 7.10. **A**. *Muggiaea,* calycophoran siphonophore, showing swimming bell and upper part of stem with long string of cormidia attached. **B**. *Muggiaea* cormidium. Gonozooid serves as swimming bell, with tentilla as branches of tentacle bearing nematocysts. **C**. *Velella,* Purple Sail, chondrophoran cut away to show inner structure. (**A,B** after Hyman; **C** composite.)

mark the four perradii and interradii. The four perradii are the major body axes; the four oral arms lie in the perradii, and the interradii are halfway between. Eight adradii lie between the perradii and interradii. In scyphomedusae, septa grow inward along the interradii to form four perradial gastric pouches, as in Stauromedusae. Gastric tentacles develop on the septa, and subumbrellar funnels, probably respiratory in function, push into the septa from the subumbrellar surface. In semaeostomes and rhizostomes, the septa and funnels disappear during development. The gastric cavity of *Aurelia* expands as the septa disappear, and the center of the adult gastric cavities lies where the gastric septa once were. Here, gastric tentacles cluster on the coelenteron floor. As the septa disappear, new depressions, the peristomial pits, appear at the site of the old subumbrellar funnels.

Scyphozoan nematocysts are relatively powerful, and many jellyfish feed on fish or other rather large prey. *Aurelia*, however, is a flagellary-mucus feeder. Small particles and plankton are caught in slime secreted on the exumbrella. These are licked off the lappets by the oral arms, and currents produced by flagella in grooves on the oral arms carry the particles into the stomach. Scyphozoa feeding on larger prey use the gastric tentacles to pull a victim into the gastric cavity and subdue it with nematocysts, if necessary. In addition, the filaments contain many gland cells that secrete enzymes for preliminary digestion.

Muscle cells lie below the epidermis. Longitudinal fibers are found in tentacles and manubrium, and form radial muscles on the subumbrella. Swimming movements are produced by a striated coronal muscle that encircles the subumbrella.

A subepidermal nerve plexus extends throughout the tentacles, the oral arms, and the manubrium. It

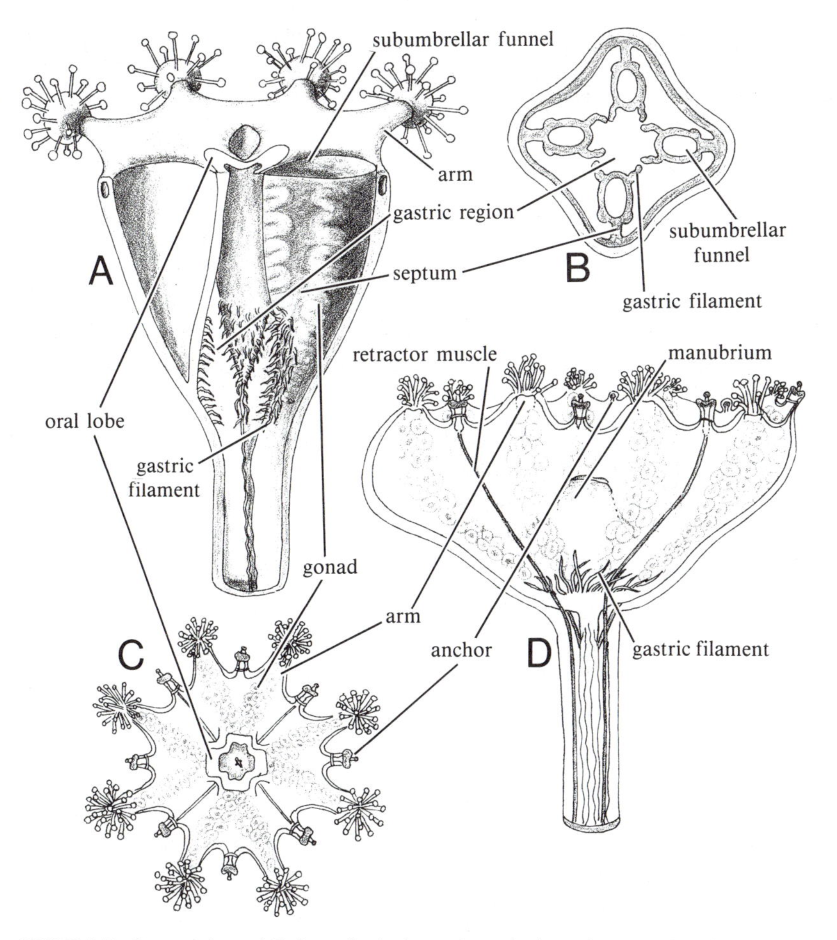

FIGURE 7.11. Stauromedusae. **A,B.** Generalized scheme of organization and cross section of stauromedusa. **C,D.** *Haliclystus salpinx* inside and oral view. **(C,D** after Berrill.)

is concentrated near the rhopalia to form the **rhopalial ganglia**. Protoneurons also aggregate to form radial strands along the main radii. Little is known of the **subgastrodermal plexus**, but it has been demonstrated in some scyphomedusae. Presumably it exercises control over the gastric tentacles.

Rhopalia are sense organs (Fig. 7.12C). In *Aurelia*, they occur in indentations between marginal lappets. Each rhopalium is a hollow club, growing at the base of the larval tentacles. One or more ocelli are usually present, as well as one or two sensory pits lined with sensory epithelium. Endodermal lithocytes, containing heavy, calcareous particles, weight the rhopalial tip. When the margin of the bell is lifted, the weighted rhopalium sinks, stimulating the sensory epithelium and initiating righting movements.

Rhizostomae These are modified semaeostomes, with the oral arms fused to close off the mouth opening, but leaving many tiny suctorial mouths as a result of the union. Enzymes are produced by the oral arms so that initial stages of digestion are technically outside the body. Partially digested material is then taken in with the help of cilia through the "mouths" for final processing inside the gut. Rhizostomes have no tentacles and often have a firm bell with dense mesoglea. The oral arms have nematocyst-filled appendages in place of tentacles. As in semaeostomes, the subumbrellar funnels are lost and replaced by subgenital pits. Rhizostomes are tropical forms, some becoming fairly large. Diameters of 35 cm or more are not uncommon.

Cubomedusae Cubomedusans are easily recognized by their cuboid shape and the four interradial tentacles or clusters of tentacles located at the corners. Each tentacle is flattened at its base into a bladelike pedalium. Septa divide the gastric cavity into perradial gastric pouches, each with a deep subumbrellar funnel (Fig. 7.13A). Most Cubomedusae are small, 4 to 5 cm across, and swim strongly. They are most common in littoral regions of warm seas. The affinities of Cubomedusae have been a subject of debate (Calder and Peters, 1975; Larson, in Mackie, 1976) and have recently been treated by some as a separate class, the Cubozoa, because of their supposedly intermediate position between Hydrozoa and Scyphozoa. However, the similarities with the latter are apomorphic, whereas those with

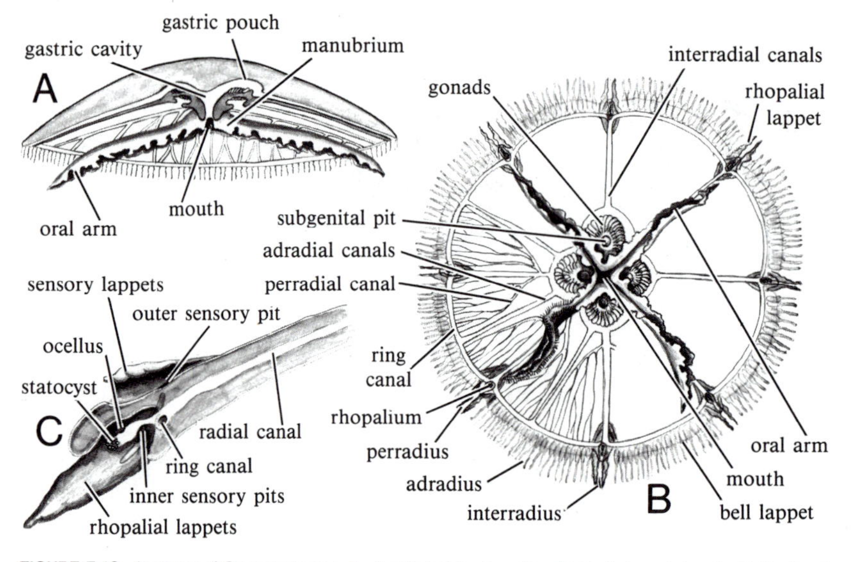

FIGURE 7.12. Anatomy of Semaeostomae. **A**. *Aurelia* in lateral section. **B**. Oral view of *Aurelia*. **C**. Section through rhopalium. Manubrium ends in quadrangular mouth, whose corners project as long, frilled oral arms. Complex sense organs, rhopalia, lie in shallow marginal indentations, flanked by conspicuous rhopalial lappets. Tentacles arise from bell margin divided into lappets by scalloping between rhopalia. Principal body axes: perradii, determined by oral arms and indentations between gastric pouches; interradii, midway between perradii; and adradii, end in middle of each marginal lappet. Gastric cavity indented at perradii to form gastric pouches; branched perradial canals arise between them, and branched interradial canals arise at centers. Eight unbranched adradial canals lie in adradii. All radial canals join ring canal near margin. Horseshoe-shaped gonads, with subgenital pits at center, empty into gastrovascular cavity. Each rhopalium composed of heavy statocyst at tip, which sags downward when bell margin tilts upward; ocellus; and inner and outer sensory pits. Extension of radial canal continues into base of rhopalium. Small sensory lappets lie on each side of rhopalial hood in *Aurelia*. (**A** modified from Schechter.)

the former are plesiomorphic. The polyps of the Scyphozoa and the Cubomedusae have muscles in the mesoglea and possess conspicuous nerve rings along the margins. The polyps of Cubomedusae and Hydrozoa share only the primitive features of pure radial symmetry (not tetraradial as in other scyphozoans), a lack of peristomial pits, and the ability to bud. The medusae of Scyphozoa and Cubomedusae share these advanced features: rhopalia, gastric filaments, peristomial pits, and statoliths of calcium sulfate (gypsum). The medusae of Hydrozoa and Cubomedusae share two problematic characters. Both have marginal nerves (Scyphozoa lack such altogether), although in Cubomedusae these do not act as pacemakers as they do in Hydrozoa. The Cubomedusae have a **velarium** around the edge of the subumbrella, similar to the Hydrozoan velum, but, unlike the hydrozoan structure, comes equipped with an extension of the gut. Cubomedusae, thus, would appear to be securely placed within the Scyphozoa.

Coronatae Coronatae are easily distinguished by the coronal groove, which divides the bell into an inflated upper part and a lower part with marginal lappets (Fig. 7.13B). They are more common in deep waters, range from about 5 to 15 cm across, and are often brightly colored. Their septa contain subumbrellar funnels.

CLASS ANTHOZOA

Anthozoa, literally the flower animals, include sea anemones and corals. In anthozoans medusae are never formed. The polyps have become large and complex. *Stoichactis*, from the Australian barrier reef, may reach 1 m in diameter. Septa partially divide the coelenteron, amplifying its surface. The epidermis invaginates at the oral end as a **stomodeum**, or **pharynx**, leading from the mouth to the coelenteron. One or two flagellated grooves, the **siphonoglyphs**, are typically located at the ends of the more or less slitlike stomodeum (Fig. 7.14). The lateral compression of mouth, stomodeum, and the siphonoglyphs convert the basic radial to a biradial symmetry. However, no anterior and posterior ends develop, and anthozoans live as radially symmetrical animals. The mesoglea is a richly cellular, dense mesenchyme or fibrous connective tissue.

Alcyonaria Alcyonarians make a good introduction to anthozoan form. All are colonial; the polyps project from stolons, fleshy mats, or vertical extensions of **coenenchyme**, the anthozoan equivalent of the coenosarc. The essentials of alcyonarian form are shown and described in Figure 7.14. The eight tentacles are branched and surround a mouth, with the body wall turned in as a pharynx. A flagellated groove, the siphonoglyph (Fig. 7.15C) pulls water into the coelenteron. Eight partitions, the septa, composed of gastrodermis and mesoglea, extend inward from the body wall, unite with the stomodeum, and end in free, thickened margins, the **septal filaments** (see Fig. 7.14). The two septa opposite the siphonglyph, the **asulcal septa**, differ from the others in that they have strong flagella that force water out of the pharynx (see Fig. 7.14). The polyp extends far below the colony surface, embedded in a common

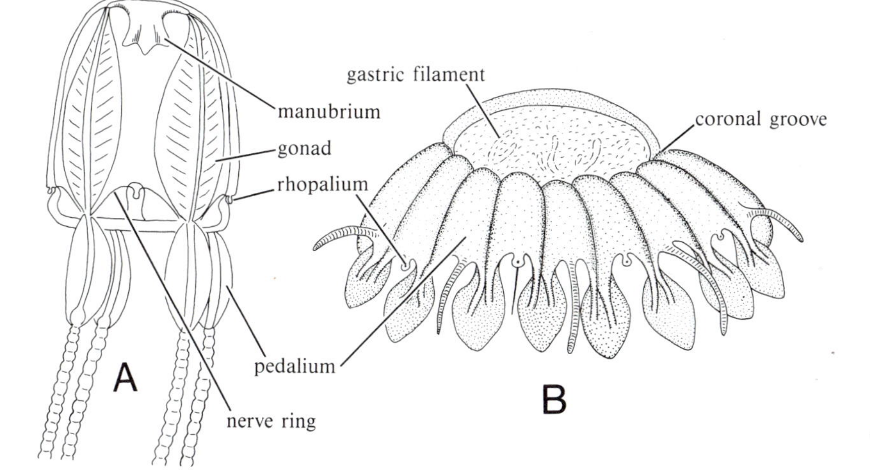

FIGURE 7.13. Some representative Scyphozoa. **A.** *Carybdea*, cubomedusa showing cubical shape and four tentacles with pedalia. **B.** *Nausithoë*, coronate showing characteristic coronal groove on exumbrella. (**A** after Hyman; **B** after Mayer.)

mesogleal mass in which the skeletal spicules develop, covered by a common sheet of epidermis. Endodermal canals, the **solenia**, branch throughout the living coenenchymal tissues.

The voracious polyps kill any food organisms of suitable size. Food is rapidly swallowed, lubricated by slime secreted in the pharynx. Septal filaments on all but the asulcal septa play the same role as scyphozoan gastric tentacles, grasping, immobilizing, and aiding in preliminary digestion of prey. When preliminary digestion is completed, food particles are engulfed and digested intracellularly by cells of the general gastrodermis as well as by septal filament cells. Egestion is by way of the mouth.

Water circulation in the coelenteron is maintained by the powerful flagella of the siphonoglyph and asulcal septa. Food and oxygen are thus distributed through the solenial tubes to the coenenchyme cells, and wastes are removed.

Muscles are poorly developed. The longitudinal epidermal muscles of the tentacles continue into the oral disc, and epidermal fibers of the disc sometimes follow the pharynx wall. A weak ring of circular muscle circumscribes the mouth. Gastrodermal fibers are antagonistic to epidermal fibers. They are weakly developed over most of the body and form transverse muscles on the asulcal face of each septum. More powerful longitudinal gastrodermal fibers on the sulcal face of the septa serve as retractors, drawing the polyps into their cup. The coenenchyme usually has no muscle fibers.

The nervous system is not elaborate. An epidermal plexus, concentrated near the mouth and tentacle bases, forms strands along the attachments of septa and oral disc. A gastrodermal nerve plexus lies in the septa. Stimulating one polyp causes contractions of others, indicating that a nerve plexus extends through the coenenchyme. Sensory cells are sparse; some species do not respond to touch. There are no special sense organs.

The elongated mouth and flattened pharynx with a single siphonoglyph confer a radiobilateral sym-

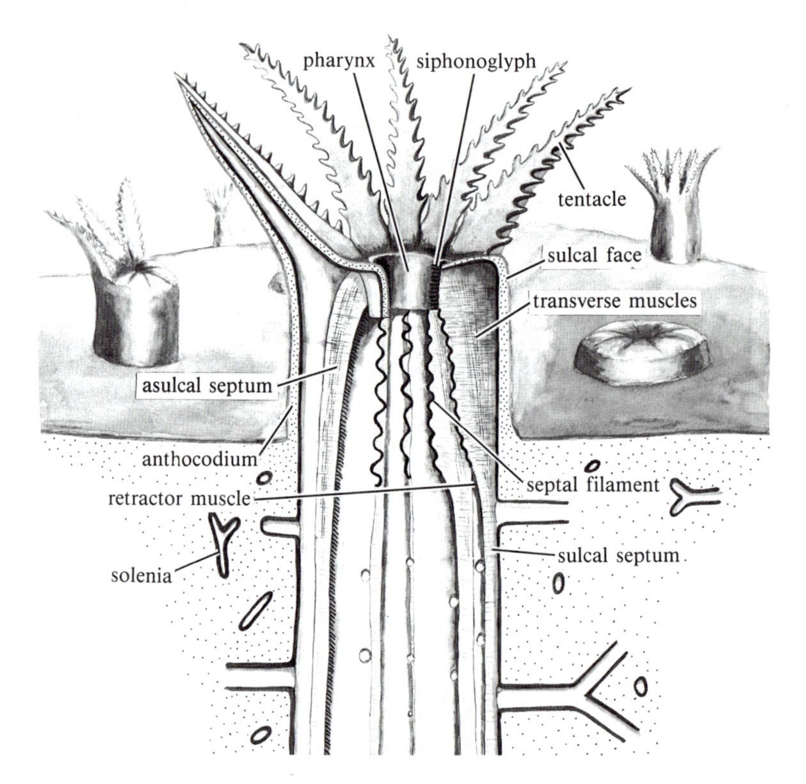

FIGURE 7.14. Alcyonarian anatomy. Schematic longitudinal section of alcyonarian polyp, passing through septum on right and between septa on left (based primarily on Hyman). Common basal coenenchyme contains mesoglea and branching gastrodermal tubes (solenia) continuous with coelenterons of polyps. Exposed distal part of polyp (anthocodium) bears eight pinnate tentacles. Mouth opens into pharynx, lined with ectoderm, with strongly flagellated furrow (siphonoglyph). Eight septa contain mesoglea covered with gastrodermis. At oral end, septa attach to pharynx, dividing coelenteron into compartments continuous with tentacle cavity. Septal margins thickened to form septal filaments below pharynx, where coelenteron incompletely divided by septa.

metry on the polyp and determine a sagittal plane, bisecting the siphonoglyph and passing between the two asulcal septa. This radiobilateral symmetry is superimposed on an octamerous radial symmetry. The siphonoglyph surface is sometimes called the "ventral surface." Inclined polyps are oriented with the siphonoglyph "below," taking water in "below" and discharging it "above," but it is preferable to refer to the sulcal and asulcal surfaces, as these terms reflect functional relationships.

Some polymorphism occurs in Alcyonaria. Some species have typical feeding **autozooids** and **siphonozooids**, specialized by powerful siphonoglyphs to circulate water through large, fleshy colonies. Some have incurrent and excurrent siphonozooids (Fig. 7.16A).

The basic form of the polyp is stable in Alcyonaria, but colony form differs among the orders. Stolonifera include the stolon corals and organ-pipe corals. They have no massive coenenchymal mass; each polyp rises from a stolon, as in *Clavularia* (see Fig. 7.15A), or a mat of tissue. Mesogleal spicules are usually present; they may be fused to form tubes, as in *Tubipora* (see Fig. 7.15B). The Telestacea are very similar but form erect, branching colonies. Alcyonacea includes the soft corals, with a massive coenenchymal mass containing spicules. A single genus, *Heliopora*, makes up the Coenothecalia, or blue corals. It, too, is characterized by a massive, calcareous skeleton that somewhat resembles the millepores. Gorgonacea are the sea fans, sea whips, and sea feathers. A proteinaceous, horny material, gorgonin, sometimes infiltrated with calcareous spicules, forms the axial core of the flexible skeleton. The coenenchyme spreads over the core, its spicules form the outer skeletal layer, and the individual polyps arise from it. Pennatulacea includes the sea pens, perhaps the most advanced of the Alcyonaria. *Renilla*, the sea pansy, is a well-known example (Fig. 7.16A,B). The colonies consist of several types of polyps. A very large axial polyp is usually divided into a peduncle, thrust into the soft sea bottom, and a rachis, from which the secondary polyps arise. Among the autozooids of the rachis are specialized incurrent and excurrent zooids that are responsible for maintaining water circulation in the colony. Solenia in the peduncle serve to connect an inferior and superior canal; water pressure inflates the colony and gives it hydraulic support.

Zoantharia Zoantharia contain over 1000 species of sea anemones and 3000 species of stony corals. Sea anemones are abundant along sea coasts, and corals, of course, form huge reefs in warmer waters. However, Zoantharia also occurs in deep waters, and some species are polar. It is a diverse group of animals, and because the method of asexual reproduction favors development of anomalies, members of the same species often differ in septal numbers and arrangement. Zoantheria is best characterized as anthozoans that never have eight single septa or eight branched tentacles like Alcyonaria. Most species are hexamerous, with pairs of septa occurring in multiples of six, but there are exceptions.

The order Actiniaria, sea anemones (Fig. 7.17), differ markedly from Alcyonaria, for their symmetry is usually hexamerous rather than octamerous, and biradial rather than radiobilateral, as most have two siphonoglyphs.

Sea anemones are quite diverse. The oral disc varies considerably in size and may bear one or more whorls of tentacles, or tentacles crowded over the

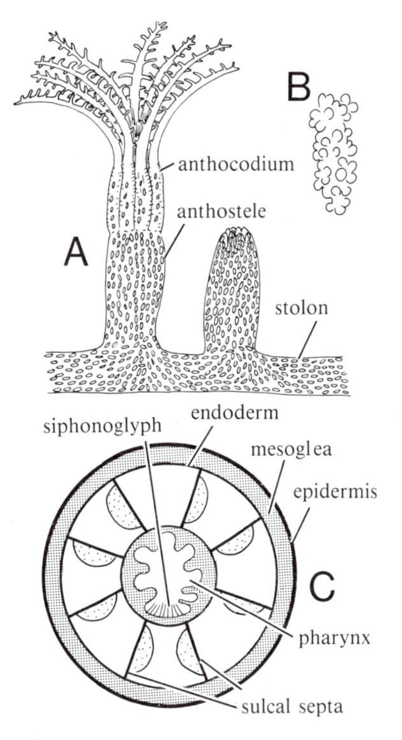

FIGURE 7.15. Alcyonaria. **A.** Part of stolon, expanded polyp, and contracted polyp of *Clavularia,* alcyonarian that forms simple colonies. **B.** Spicule from surface of *Clavularia* colony, enlarged. **C.** Scheme of alcyonarian polyp in cross section, through region of pharynx. Strong retractor muscle found on sulcal side of each septum. Sulcal septa, which lie on each side of siphonoglyph, have retractor muscles facing each other. Remaining, asulcal septa have strong flagella. Siphonoglyph flagella draw water into polyp and asulcal flagella drive water out, creating circulation of water through coelenteron and solenia. (**A**,**B** after Hyman.)

whole surface. Many sea anemones are particle feeders, with cilia on the body surface that cause upward currents to waft bits of food to the tentacles. The oral disc and tentacles are also ciliated. When a particle touches a tentacle, the tentacle bends and sweeps it toward the mouth. The tentacles have many nematocysts and adhesive glands, and when larger food organisms are eaten, food is pushed into the mouth.

Mouth and pharynx structures are much the same as in Alcyonaria, but most anemones have two siphonoglyphs; a few have one or none. Mucous glands in the pharynx aid in swallowing. Preliminary digestion is rapid and achieved by enzymes from gland cells in the septal filaments. The final, intracellular phase of digestion occurs in the general gastrodermis, greatly amplified by septa. Sea anemones have characteristic septal patterns. Primary septa extend from the body wall to the pharynx, as in the Alcyonaria. Most Zoantharia have additional, shorter, secondary septa between the primaries that reach only partway to the pharynx (Fig. 7.17A,B). Tertiary and quaternary cycles of septa, each decreasing in size, are often present. The septa occur in couples and pairs. A line passing through the sagittal plane and passing through the siphonoglyphs divides the organism into mirror-image halves; septa in corresponding positions on either side are couples. In most Zoantharia, the mesenteries are paired, that is, two lie together, side by side; in the two members of a pair, the retractor muscles face each other (Fig. 7.17B). The two septa that occur at each siphonoglyph are termed "directives." As they lie on opposite sides of the sagittal plane, they are actually couples rather than pairs, and differ from the pairs by having retractor muscles located back to back. Trilobed septal filaments (Fig. 7.17C) have tracts of nematocysts, gland cells, and flagella. In many anemones the filaments continue at the base as threadlike **acontia**, lacking the flagellated tracts found in septal filaments (Fig. 7.17D). Acontia are extruded from the mouth and from pores in the body wall, the **cinclides**, when anemones are disturbed, and these are probably protective as well as digestive.

The siphonoglyphs and flagellated tracts of the septal filaments cause a continual intake and circulation of water. The currents transport food particles and ventilate the internal surfaces, permitting more effective respiratory exchange. Food residues are eventually expelled from the mouth.

The anemone muscles are relatively complex. Epidermal muscles form longitudinal fibers in the tentacles and radial fibers in the oral disc. The latter fold in the disc upon contraction. The column and pharynx usually have no epidermal muscle. Circular muscle in all parts of the body is gastrodermal. Special circular muscle fibers form a sphincter beneath

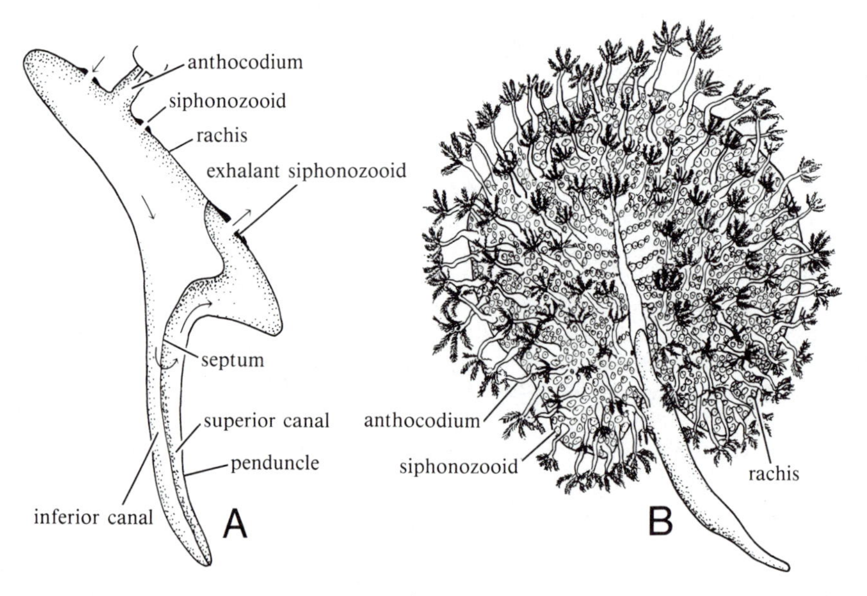

FIGURE 7.16. *Renilla,* pennatulacean sea pansy. **A**. Schematic section through colony, showing circulation of water in gastrovascular cavity. **B**. Colony of *Renilla,* showing broad rachis containing many polyps and siphonozooids and naked peduncle, thrust into soft bottom. (**A** after Parker, from Brown; **B** after Hyman.)

the oral disc, permitting it to be concealed when the animal contracts. The strong septal retractor muscles are gastrodermal. When the animal contracts, water escapes through the cinclides. Creeping movements of the pedal disc are largely made by the parietal and basilar muscles.

The body wall has an extensive epidermal nerve plexus, and the septa have a gastrodermal plexus. Fibers across the mesoglea connect the plexi. A double nerve net exists, a delicate one composed of small nerve cells, and a coarse one composed of larger nerve fibers, apparently most directly concerned with overall body contractions. The system is apparently a synaptic one, with transmission across synapses carried out without decrement and without fatigue. Stimuli spread from one tentacle to another and repeated stimuli fail to evoke responses. There are no special sense organs. Neurosensory cells are scattered throughout the epidermis but are concentrated in tentacles, oral disc, and pedal disc.

The order Madreporaria, or corals, are the most abundant Zoantharia and the most ecologically im-

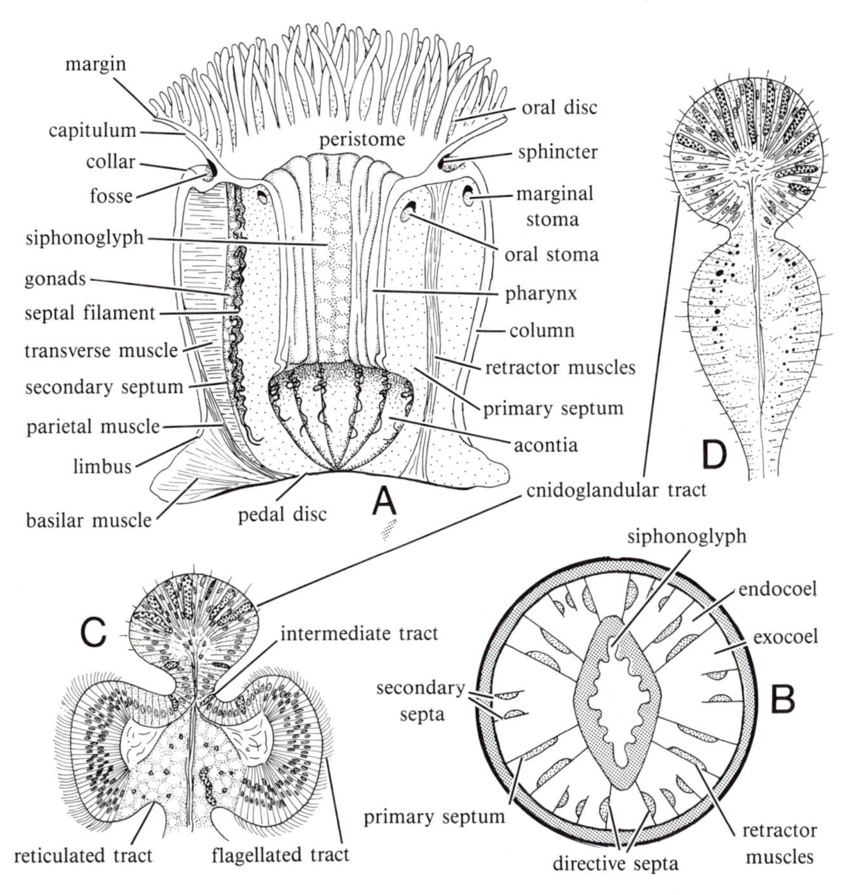

FIGURE 7.17. Sea anemone anatomy. **A**. Schematic hemisection showing endocoelic surface of primary septum and exocoelic surface of secondary septum. **B**. Schematic cross section through pharynx region. **C,D**. Cross sections through septal filament above and below pharynx. Body wall composed of outer epidermis, mesoglea, and inner gastrodermis. Pedal disc separated from column by groove, limbus. At upper end of column, collar bounds a groove, fosse, above which thin-walled capitulum is found. Sphincter contracts to protect withdrawn capitulum and oral disc at collar. Mouth separated from tentacles by peristome, and has opposing siphonoglyphs that extend to bottom of pharynx. Paired septa partially divide enteron into compartments. Primary septa reach from body wall to stomodeum. Directive septa are primaries at each side of siphonoglyphs. Shorter secondary septa do not reach pharynx. Still shorter tertiary septa found between secondaries. Space between two members of septal pair termed endocoel, space between pairs is exocoel. In pharynx region, marginal and oral stoma connect endocoels and exocoels. Septal filaments occur at free margin of septa. Beyond end of stomodeum, lateral lobes end and cnidoglandular tract continues. Beyond basal end of septum, septal filaments continue as acontia. Longitudinal muscles in septa form retractor muscles. Gastrodermal circular muscles extend into septa as transverse muscles. Parietal muscles extend from base of septa into pedal disc, where they continue as basilar muscles. Gonads located near base of septal filaments. (**A** composite; **C,D** after Hyman.)

portant. In shallow, warm waters they form extensive reefs. At night and on cloudy days, the polyps expand and feed on plankton.

Coral polyps are essentially colonial anemones that are connected by flat extensions of the body wall, the coenenchyme. A hard, calcareous exoskeleton is secreted by the lower surface of the coenenchyme. Each polyp sits in a depression, the **theca**. Ridges, the **sclerosepta**, project upward from the thecal base, and the body wall follows their contours, folding inward, usually between the two septa of a pair (Fig. 7.18). Septa, coelenteron, and pharynx are as in anemones, but the septal filaments lack flagellated tracts.

Corals do much to create a unique environment in shallow tropical seas. Many invertebrates and the bright reef fishes are found only in the special world created by corals. Reefs are limestone formations, formed of coral exoskeletons and heavily charged with organic material, especially in upper strata where growing corals occur. Calcareous algae, milleporines, and other calcareous Alcyonaria contribute to reefs.

Reef corals grow best at depths of 11 m or less and at 22°C or above, and contain **zooxanthellae** (actually dinoflagellates), as do many other cnidarians. Corals and zooxanthellae are in a mutualistic relationship; corals cannot live, or are handicapped, without them (Muscatine, in Muscatine and Lenhoff, 1974). This restricts coral distribution to areas where zooxanthellae can thrive. Corals die after 18 days in total darkness, but they are completely carnivorous and expel zooxanthellae when starved. Zooxanthellae probably seek corals to satisfy their own nitrate and phosphorous requirements, but although they produce oxygen, it is not clear how important this is to corals. Good evidence has been obtained that skeleton formation is facilitated by the presence of the zooxanthellae, as they help to dispose of H_2CO_3 released as a by-product of the formation of $CaCO_3$ (Rinkevich and Loya, 1984). Coral skeletons are often beautifully marked, with skeletal septa arranged in regular or irregular systems, depending on the growth patterns exhibited by the polyps. The many fossil corals can be recognized by these patterns and are of some importance in dating rock deposits.

The other orders of Zoantharia are distinguished by polyp form or details of colonial structure. Three of these orders are essentially like sea anemones.

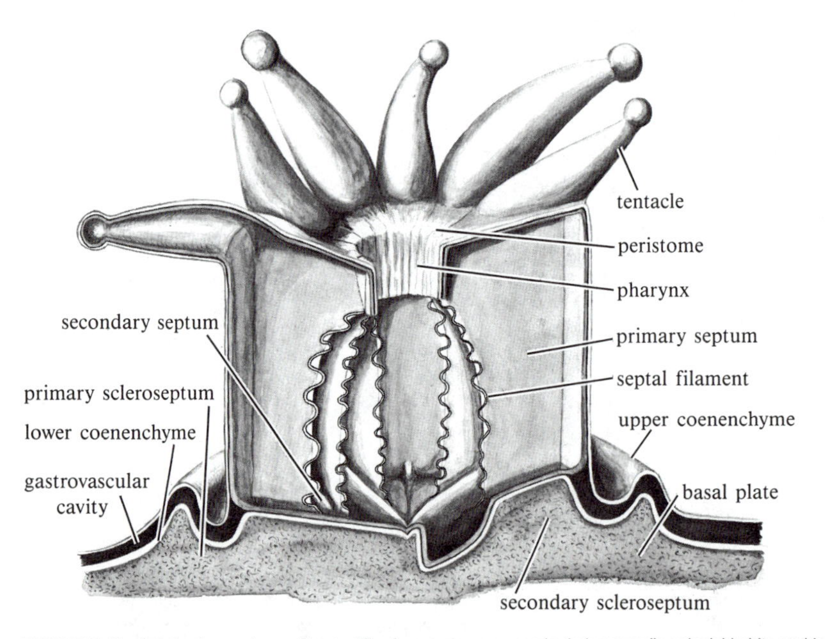

FIGURE 7.18. Coral polyp anatomy. Corals differ from sea anemones in their generally colonial habits and in having skeleton. Relationships of body wall, primary and secondary septa, pharynx, peristome, and tentacles essentially as in sea anemones, but septa interdigitate with sclerosepta of skeleton, pharynx has no siphonoglyph, and septal filaments lack flagellated tracts. Body sits in thecal cup, and coenencyme, which extends laterally as fold in body wall, extends to adjacent polyps. Lower layer of coenenchyme forms skeleton, therefore colony lies above skeleton.

The Corallimorpharia are corallike anemones, solitary or colonial, which have no skeleton but which tend to form buds, like some corals, on the oral disc, producing several mouths within a single whorl of tentacles. The Ptychodactiaria are another aberrant group of sea anemones without basal muscles and without flagellated tracts on the septal filaments. The Ceriantharia are the burrowing anemones (Fig. 7.19). The long body is thrust into the burrow, and the oral end, with a whorl of tentacles at the margin of the disc and one around the mouth, is extended. The Zoanthidea are the colonial anemones; polyps extend from basal stolons or mats as in the Alcyonaria, but they do not resemble the Alcyonaria or the Actiniaria in septal form and arrangements. Most of them live on the surface of other animals. The Antipatharia are the black corals; they have septa much like the cerianthids but have a skeleton with a horny axis, like the gorgonids, with the added feature of thorns projecting from it.

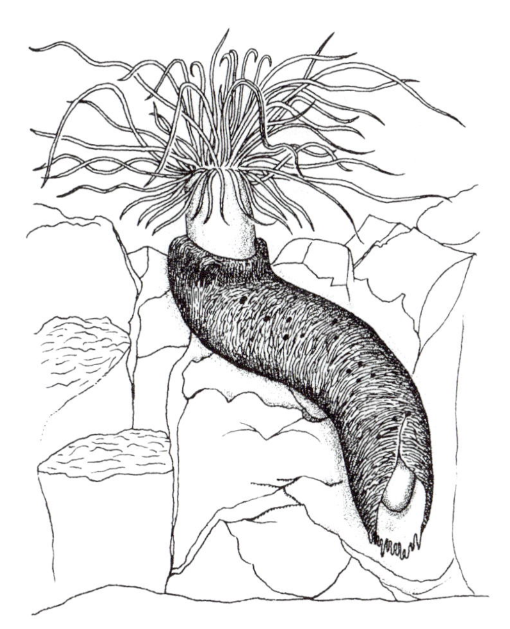

FIGURE 7.19. *Cerianthus,* member of order Ceriantharia. Note absence of pedal disc, and characteristic tube of this burrowing form. (After Andres, from Hickson.)

Fossil Record and Evolution

The Cnidaria have left an extensive fossil record, especially among the Anthozoa. There are close to 3000 living species of corals, and probably twice that number of fossil forms. The range of this phylum extends back to the Cambrian, possibly even into the Precambrian with the strange jellyfishlike fossils of the Ediacara fauna. Although most cnidarian fossils are corals, scyphozoan and hydrozoan forms are known. One particular problematic group that is usually assigned to the Scyphozoa is the Paleozoic conularids. These very strange animals, which resemble inverted pyramids, illustrate a common difficulty that occurs in trying to assign affinities for problematic groups. Typically, only the hard outer cortex is present on the fossils. Although such hard parts that are preserved are evocative of some living semaeostome scyphozoans, lack of preserved soft anatomy presents problems to the investigator in that it precludes definitive assignment.

Not so difficult is the placement of two other extinct groups within the Zoantharia: the Paleozoic Rugosa (or tetracorals) and the essentially Paleozoic (though possibly younger) Tabulata. These latter were once considered to be problematic, some authors suggesting that they even might be sponges rather than corals. However, the recent discovery of preserved soft parts clearly aligns tabulates within the zoantharian cnidarians (Cooper, 1985).

Though the fossil record of cnidarians is good, un-

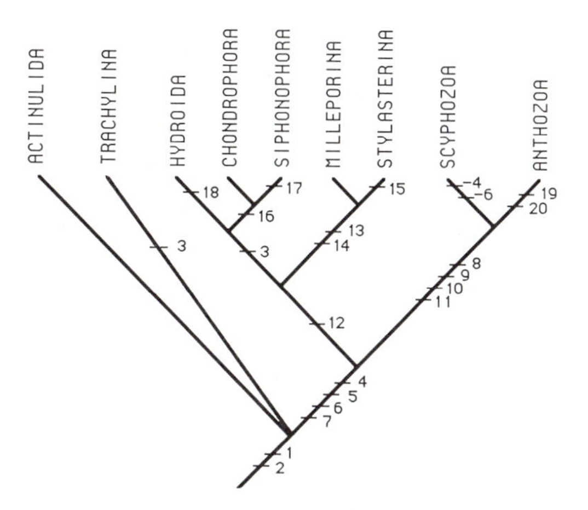

FIGURE 7.20. Cladogram depicting possible relationships among cnidarians. Derived characters are: (1) radial or biradial symmetry; (2) at least one type of nematocyst; (3) medusa with velum; (4) life cycle at least partially with polyps; (5) several kinds of nematocysts; (6) typically colonial; (7) alternation between sexual and asexual phases; (8) compartmentalized gastrovascular cavity; (9) gastric filaments; (10) septal gonads; (11) cells in mesoglea; (12) polyps polymorphic; (13) calcareous skeleton; (14) medusa stage degenerate; (15) medusa stage lost; (16) free-floating colony; (17) medusae polymorphic; (18) colony typically with organic perisarc; (19) exclusively polypoid; (20) epidermis forming stomodeum at mouth.

like that of some other phyla, it contributes at present little to our understanding of the evolution of Cnidaria. And little has been written on the evolution of living forms (Hand, 1959; Rees, 1966).

The issues of cnidarian phylogeny have been centered on consideration of living forms, and are vexing ones. Debates have occurred for decades as to whether the medusa or the polyp represents a more primitive condition, and whether cnidarians exhibit primary or secondary symmetry. These matters have bearing on the issue of which phylum is the closest relative of Cnidaria, as well as determining the relationships within Cnidaria. The former issue is taken up in the concluding chapter on phylogeny. An attempt at the latter issue is presented here (Fig. 7.20).

Cladistic analyses of all cnidarians have been rarely undertaken; therefore, the one presented here is a first approximation. Polarizing of characters was focused on outgroup analysis with Ctenophora and Porifera. On this basis, radial symmetry is viewed as derived, and the medusoid or free-swimming phase as more primitive. In addition, more complex life cycles and internal anatomies are evaluated as apomorphic. This analysis used the computer program PHYSYS, and produced no surprises (reproduced in Fig. 7.20 in a simplified form). It does suggest that the Hydrozoa as a class may be a paraphyletic group; that is, they do not appear to share a common ancestor unique to themselves and no other group. Scyphozoa and Anthozoa, on the other hand, seem to be good monophyletic taxa. However, as in all such analyses, different results can be obtained using a different data set *or* by analyzing the same set of characters with a different program (see Chapters 2 and 38). The important thing is that such analyses use as many characters as possible for all groups and do so in a way that ensures that others could confirm your results or try to challenge them.

Taxonomy

Three well-defined classes of the phylum Cnidaria traditionally are recognized, each containing several orders.

Class Hydrozoa. Solitary or colonial, with noncellular mesoglea, coelenteron without gastric tentacles or septa, without stomodeum. Medusae typically with velum (craspedote). Polypoid phases usually predominate in life cycle.

Order Actinulida. Body form resembles actinula larva.

Order Trachylina. Exclusively medusoid, or with rudimentary polypoid stages; with endodermal statoliths.

Suborder Trachymedusae. Bell margin not scalloped; with four, six, or eight radial canals, long manubrium or pseudomanubrium; gonads on radial canals. Examples: *Gonionemus*, *Aglantha* (Fig. 7.9A).

Suborder Narcomedusae. Bell margin scalloped; radial pouches and reduced radial canals; without manubrium; gonads on floor of gastric cavity. Example: *Cunina* (Fig. 7.9B).

Order Hydroida. Solitary or colonial polyps that bud off free medusae, produce sessile, degenerate medusae, or are exclusively hydroid.

Suborder Gymnoblastea. Gymnoblasts and anthomedusae. Stalk typically covered by perisarc; without thecal cups for zooids; medusae, if present, bell-shaped, with ocelli, and with gonads on manubrium. Examples: *Syncoryne*, *Pennaria*, *Tubularia*, *Eudendrium*, and *Sarsia* (Figs. 7.2A, 7.8A,B, 7.9C,D).

Suborder Calyptoblastea. Calyptoblasts and leptomedusae. Zooids contained in skeletal cups; medusae, if present, typically flattened, with statocysts, and with gonads on radial canals. Examles: *Campanularia*, *Obelia*, *Sertularia*, *Plumularia*, *Aequora*, and *Tima* (Figs. 7.8C,D, 7.9E,G).

Order Chondrophora. Free-floating polypoid colonies with high degree of polymorphism. Example: *Velella* (Fig. 7.10C).

Order Siphonophora. Free-swimming or floating colonies composed of both hydroid and medusoid members; highly polymorphic. Examples: *Physalia* and *Muggiaea* (Fig. 7.10A,B).

Order Milleporina. Stinging corals. Colonial hydroids forming massive, porous skeleton; with feeding gastrozooids and tactile dactylozooids; medusae degenerate, but free.

Order Stylasterina. Hydrocorals. Similar to millepores, but with smaller dactylozooids, arranged in systems, and with permanently sessile, degenerate medusae.

Class Scyphozoa. Medusa stage predominant, without velum (acraspedote); mesoglea cellular, coelenteron without stomodeum but with filamentous gastric tentacles, often divided into four compartments. Polypoid stage missing, or sometimes present, developing directly into medusae or giving rise to medusae by transverse fission.

Order Stauromedusae. Adults developing directly from scyphistoma, remaining permanently attached by aboral stalk; coelenteron divided by septa; without marginal sense organs other than modified tentacles. Example: *Haliclystus* (Fig. 7.11).

Order Cubomedusae. Cubical medusae with four tentacles or groups of tentacles on the perradii; four compound sense organs, rhopalia on interradii; margin of bell bent inward as velarium; coelenteron divided by septa. Example: *Carybdea* (Fig. 7.13A).

Order Coronatae. Medusae with scalloped margin, separated by furrow from bell margin; tentacles borne on flattened blades known as pedalia; coelenteron divided by septa. Example: *Nausithoë* (Fig. 7.13B).

Order Semaeostomae. Medusae with corners of mouth prolonged into frilly oral tentacles; without furrow on exumbrella or pedalia; eight to sixteen rhopalia and scalloped bell margin; without septa dividing gastric cavity. Example: *Aurelia* (Fig. 7.12).

Order Rhizostomae. Medusae with oral tentacles fused, containing many canals and small mouths; eight or more rhopalia and scalloped margin; without septa in coelenteron.

Class Anthozoa. Exclusively polypoid; coelenteron divided into compartments by radial septa with nematocysts along border; gametes develop from septal endoderm; body wall turns in at mouth to form pharynx or stomodeum lined with ectoderm; mesoglea richly cellular.

Subclass Alcyonaria (Octocorallia). Polyps with eight feathery tentacles and coelenteron divided by eight single, complete septa; single siphonoglyph; almost exclusively colonial.

Order Stolonifera. Simple polyps connected by basal stolons or mats; typically with skeleton of mesogleal spicules, sometimes fused to form tubes. Examples: *Clavularia* and *Tubipora* (Fig. 7.15).

Order Telestacea. Polyps connected by basal stolons, but producing lateral buds and erect colonies.

Order Alcyonacea. Soft corals. With lower end of polyps fused with fleshy coenenchyme and only oral ends protruding; skeleton of separate mesogleal spicules, often consolidated, without axial core (Fig.7.14).

Order Gorgonacea. Horny corals. Skeleton of mesogleal spicules and axial core of horny or calcareous material; polyps short. Examples: *Plexaurella* and *Gorgonia.*

Order Pennatulacea. Sea pens. With many small lateral polyps attached at sides of long, axial polyp; polyps always dimorphic; skeleton of separate calcareous spicules and axial rod. Example: *Renilla* (Fig. 7.16).

Subclass Zoantharia (= Hexacorallia). More than eight simple, usually unbranched tentacles; coelenteron divided by septa but not in alcyonarian pattern; solitary or colonial.

Order Actiniaria. Sea anemones. Solitary polyps, usually with pedal disc and one or more siphonoglyphs; septa paired, complete and incomplete, often in multiples of six (Fig. 7.17).

Order Madreporaria. Stony corals. Typically colonial polyps, lacking siphonoglyph and forming compact, calcareous skeleton. Examples: *Mancina* and *Astrangia* (Fig. 7.18).

Order Zoanthidea. Colonial anemones. Usually colonial, epizoic polyps, without skeleton and with one siphonoglyph; septal pairs usually consist of one complete and one incomplete septum.

Order Corallimorpharia. Coral-like anemones. Solitary or colonial polyps without skeleton, and without siphonoglyphs or basilar muscles.

Order Antipatharia. Thorny corals. Polyps with six, ten, or twelve single, complete septa, six simple or eight branched tentacles, and two siphonoglyphs; skeleton with thorny axial core of slender branching form.

Order Ceriantharia. Long, solitary polyps adapted for burrowing; oral and marginal whorls of tentacles; many single, complete septa and single siphonoglyph. Example: *Cerianthus* (Fig. 7.19).

Selected Readings

Burnett, A. L. (ed.). 1973. *Biology of Hydra.* Academic Press, New York.

Calder, D. R. and E. C. Peters. 1975. Nematocysts of *Chiropsalmus quadrummanus* with comments on the systematic status of Cubomedusae. *Helgoländer wiss. Meeresunters* 27:364–369.

Campbell, R. D. 1974. Cnidaria. In *Reproduction of Marine Invertebrates,* Vol. 1 (A. C. Giese and J. S. Pearse, eds.), pp. 133–200. Academic Press, New York.

Copper, P. 1985. Fossilized polyps in 430-myr-old *Favosites* corals. *Nature* 316:142–144.

Gladfelter, W. G. 1973. A comparative analysis of the locomotory system of medusoid Cnidaria. *Helgol. wiss. Meere.* 25:228–272.

Hand, C. 1959. On the origin and phylogeny of the coelenterates. *Syst. Zool.* 8:191–202.

Holstein, T., and P. Tardent. 1984. An ultra high-speed analysis of exocytosis: nematocyst discharge. *Science* 223:830–833.

Mackie, G. O. (ed.). 1976. *Coelenterate Ecology and Behavior.* Plenum Press, New York.

Muscatine, L., and H. M. Lenhoff (eds.). 1974. *Coelenterate Biology.* Academic Press, New York.

Rees, W. J. (ed.). 1966. *The Cnidaria and Their Evolution.* Academic Press, New York.

Rinkevich, B., and Y. Loya. 1984. Does light enhance calcification in hermatypic canals? *Mar. Biol.* 80:1–6.

Skryock, J. C., and C. P. Bianchi. 1983. Sea nettle nematocyst venom: mechanisms of action in muscle. *Toxicon* 21:81–95.

Tardent, P., and R. Tardent (eds.). 1980. *Developmental and Cellular Biology of Coelenterates.* Elsevier, Amsterdam.

Wellington, G. M., and P. W. Glynn. 1983. Environmental influences on skeletal bonding in eastern Pacific canals. *Coral Reefs* 1:215–222.

8

Ctenophora

Ctenophora are among the most striking of marine plankton and also can be found at great depths. A few are epibenthic in habits. Individual ctenophores are very delicate and are easily damaged by plankton nets. As a result, all too little is known about the biology and diversity of the not uncommon "comb jellies."

Definition

These marine forms have biradial symmetry, gastrovascular cavity, and gelatinous mesenchymal mesoglea with associated muscle cells. The principal distinctive characteristic is the radially arranged rows of cilia or ctenes. Tentacles are not arranged in whorls around the mouth. They have an aboral sense organ and special adhesive cells or colloblasts. Cleavage is determinate, and ontogeny culminates in a cydippid larva with virtually no planula.

Body Plan

There are seven currently recognized orders of ctenophores, each rather distinctive from the other. However, it is the cydippids that afford the best example of the basic ctenophore structural plan.

Cydippids, known as sea gooseberries or sea walnuts, are more or less oval creatures, whose biradial form is most evident in aboral view (Fig. 8.1A,D). Two long, trailing **tentacles** may be retracted into **tentacle sheaths**. The flattened pharynx is a stomodeum lined with epidermis. The stomach is flattened at right angles to the pharynx; it is lined with **gastrodermis**, as are the eight gastrovascular canals that radiate from it. The body is covered by an epidermis containing many gland cells, sensory cells, and pigment granules. Eight comb rows are composed of overlappng ciliated plates or **ctenes**. Each plate consists of hundreds of thousands of cilia that beat together as a unit (Fig. 8.1C). The ctenes beat strongly and propel the animal, sometimes very fast, mouth first, through the water. A gelatinous **mesoglea**, derived from the ectoderm, fills the space between the gastrodermis and the epidermis. It contains amoebocytes, connective tissue fibers, muscle fibers, and nerve fibers to the muscles (Siewing, 1977). Smooth muscles develop from the amoeboid mesogleal cells. Longitudinal, circular, and radial fibers encircle the mouth to form a sphincter. Longitudinal fibers in the tentacle core retract that structure into the sheath and manipulate it when it is covered with food. Striated muscles are present in the tentacles of the cydippid *Euplokamis*.

Biology

Food Capture and Digestion Prey includes copepods and phytoplankton. Food is usually caught by surface mucus and carried to the mouth by cilia, or ensnared by colloblasts on the tentacles (Reeve and Walter, 1978; Reeve et al., 1978). A **colloblast** is a highly specialized cell, lacking normal cell structures (Fig. 8.2A). Its terminus is covered by sticky globules that are tethered to a basement membrane by a straight filament. If the colloblast is pulled out of the surrounding tissue by a struggling prey, it is retracted back into position by action of a spiral contractile filament. The tentacles are effective food traps, and when they are contracted, food is wiped off onto the "lips" around the mouth. At least one ctenophore, *Haeckelia rubra*, was reported to have tentacular nematocysts rather than colloblasts, and nematocysts have been noted in the gastrodermis of other forms. However, these cells have now been shown to be secondarily acquired as kleptocnidae from prey cnidarians.

At high-prey densities, ctenophores can ingest up to ten times their body weight in a day. Food is initially digested rapidly in the flattened pharynx. The partly digested food enters the stomach and is circulated through the gastrovascular canals, where it is then digested intracellularly. Digestive efficiency exceeds 70 percent of total intake, but 80 percent of that is needed for basic energy requirements. Food residues are ejected back through the mouth or out of the small **anal pores** at the aboral end. To facilitate basic energy needs, branches of the gastrovascular system extend into the most important parts

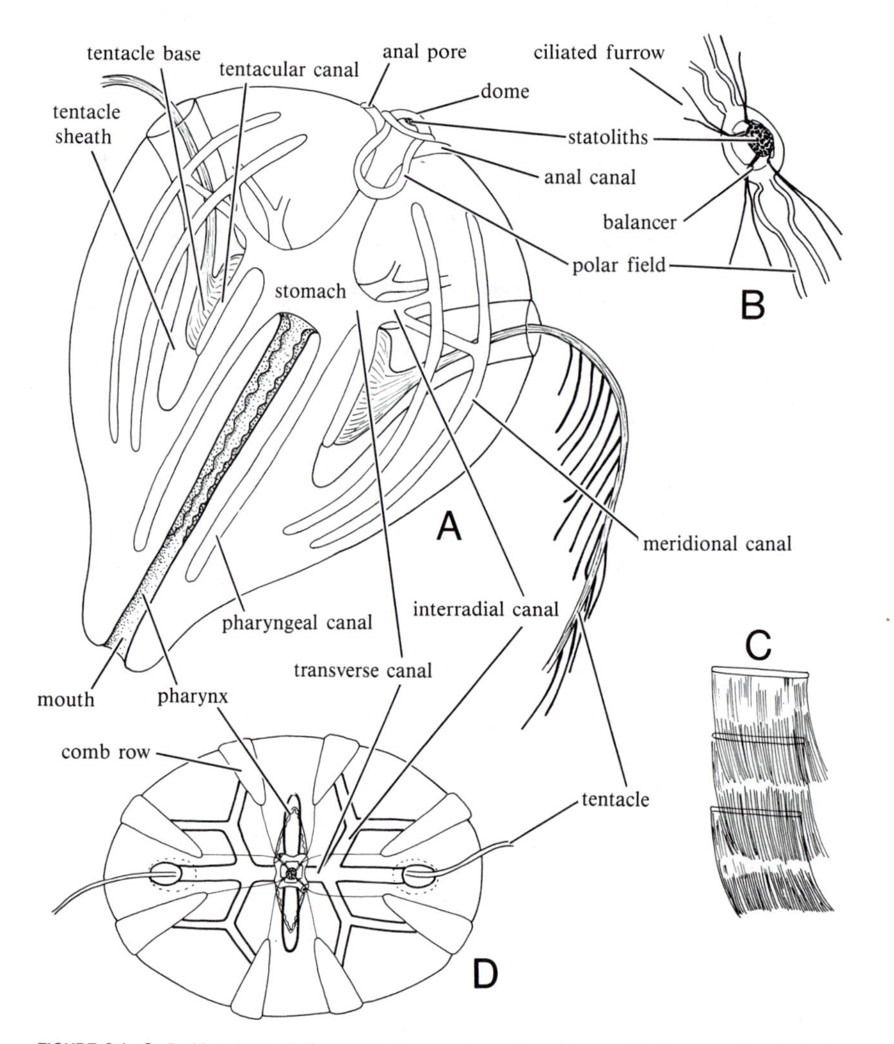

FIGURE 8.1. Cydippid anatomy. **A.** Schematic diagram of side view along sagittal plane. **B.** Aboral apical sense organ. **C.** Ctenes, highly magnified. **D.** Schematic diagram of cydippid from aboral pole. (**A** after Hyman.)

of the body; tentacular canals lie at the tentacle bases, and meridional canals lie beneath the eight comb rows.

The food-capture system described is modified in the lobates. In that order, mucus-covered lobes also serve to trap prey, while cilia sweep such tidbits toward the mouth. In the "cannibalistic" beroids, the mouth is appressed against a prey ctenophore. Then, with a single gulp, the victim is swallowed (Reeve et al., 1978).

Respiration and Excretion There are no special respiratory structures in ctenophores. **Cell rosettes** (Fig. 8.2B), consisting of a double circlet of ciliated gastrodermal cells, surround openings leading from the gastrovascular canals to the mesoglea. Their function is unknown. They may be excretory, may help regulate body fluids, or may help to distribute food to the mesoglea. In any case, most nitrogen is excreted as ammonia, presumably diffusing to the medium from body and gut surfaces.

Internal Transport Movement of materials is achieved by use of the radiating canals of the enterocoel. In at least some forms, for example, *Pleurobranchia*, gastrodermal cilia occur on the medial faces of these canals and facilitate the movement of materials in them. Given the effective bulk of the

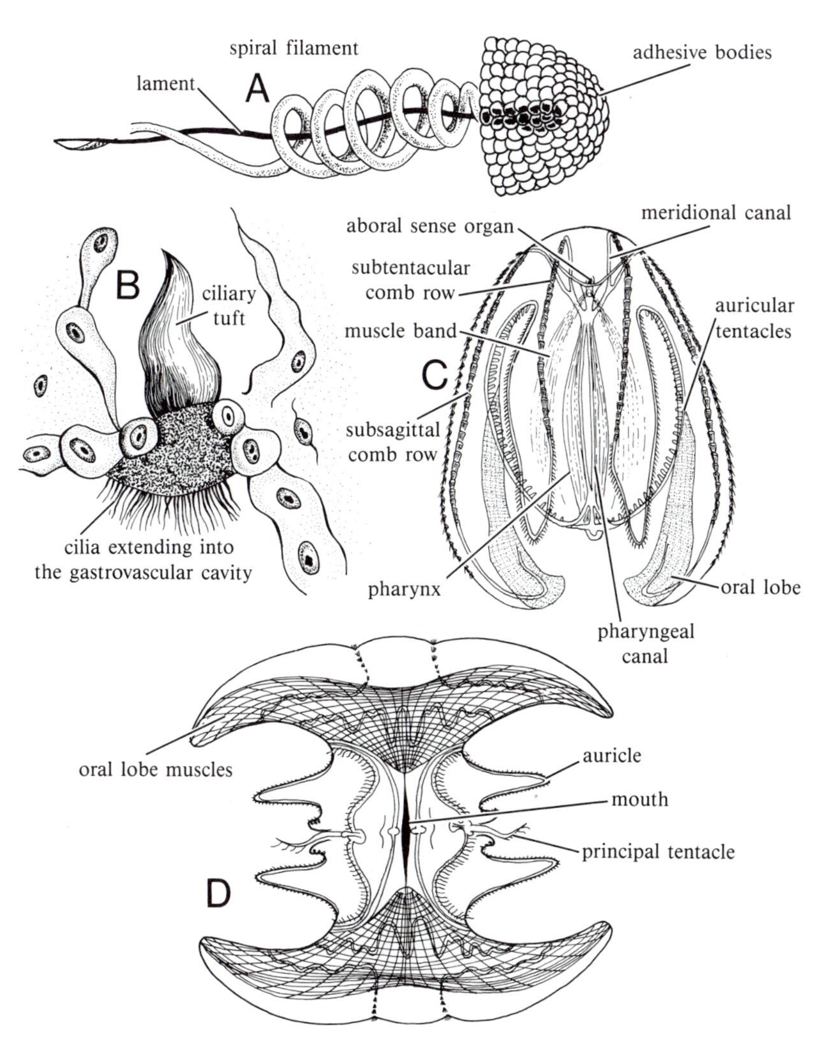

FIGURE 8.2. A. Colloblast. **B.** Rosette organ. **C,D.** Lobata, *Mnemiopsis* sp., sagittal and oral views. (**B** after Komai, from Andrew; **C** after Hyman; **D** after Mayer.)

mesoglea, these canals are important to the supply of tissues that are located at some distance from the central gut.

Nervous System The complex ctenophore **aboral sense organ** (see Fig. 8.1B) is a structure that comes close to being a true organ. It is basically a statocyst, but designed differently than the statocysts of cnidarians. The heavy body, also called a statolith, is formed of living cells containing large calcareous deposits. It is suspended on four sigmoid tufts of motile, mechanosensitive cilia (balancers) that are at the end of ciliated furrows leading to the comb rows. A dome of cilia covers the organ. Body orientation with respect to gravity determines which of the balancer tufts the statolith deforms, and this changes their beat frequency, resulting in excitation or inhibition of beating of the appropriate comb rows. The orientation of geotaxis, that is, up or down "mood," can reverse.

An epidermal nerve plexus is concentrated into a ring around the mouth and radial structures along the base of the comb rows. These neurons are not true nerves, but condensations of the nerve net. Neurosensory cells are scattered about the epidermis, with some concentration in the mouth region. The nervous system controls muscular movements and can reverse the ciliary beat or change the frequency (Tamm, 1982). If the statocyst or just the statolith cells are removed, the comb rows act independently and geotaxis disappears. Comb plates are coordinated mechanically, either by direct interaction (except in lobates) or by some action of the interplate ciliated groove (in lobates).

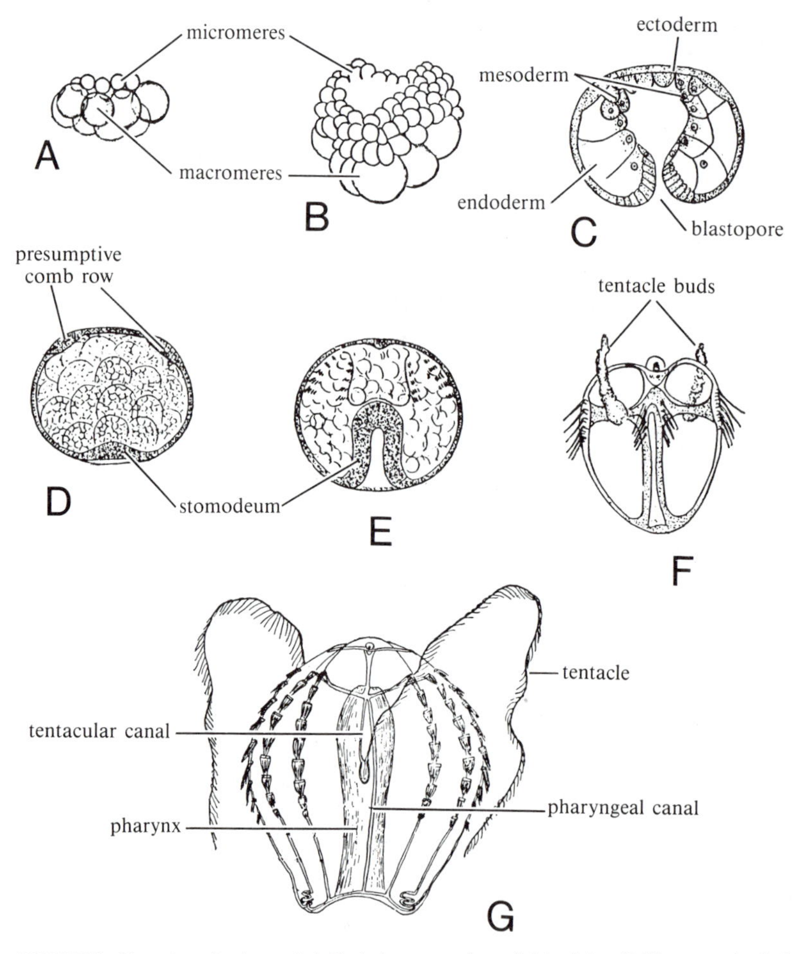

FIGURE 8.3. Ctenophore development. **A.** First micromeres given off, lateral view. **B.** Micromeres beginning to engulf macromeres. **C.** Invagination complete, with "mesoderm" on roof of archenteron. **D.** Later stage with presumptive comb rows now thickening. **E.** Stomodeal invagination and comb rows differentiating. **F.** Young cydippid larva. **G.** Late cydippid larva of the lobate *Mnemiopsis leidyi.* (Modified from Hyman.)

Reproduction Most ctenophores are simultaneous hermaphrodites. The gonads exist as testicular and ovarian bands located in each meridional canal. Gametes are typically voided through gonoducts. Fertilization is usually external. Some forms such as platyctenids brood their eggs. Platyctenids also exhibit protandry and in addition are capable of asexual reproduction by a type of budding.

Development Cleavage is total, and within a few divisions cells have differentiated in form and are determined in their fate. Cleavage produces micromeres, from which the ectoderm develops, and macromeres, which become the endoderm (Fig. 8.3A,B). After the micromeres cover the embryo to form the epidermis (Fig. 8.3C), four bands of rapidly dividing cells are formed (Fig. 8.3D). These bands differentiate into the comb rows, two from each band. The aboral ectoderm forms the apical sense organ, while the oral ectoderm invaginates as stomodeum (Fig. 8.3E). The growth of the stomodeum constricts the endoderm to cause it to form four pockets. The ectoderm then invaginates to form two tentacular sheaths, out of which the tentacles grow (Fig. 8.3F). At this stage, the embryo escapes as a **cydippid larva** (Fig. 8.3G), which is a characteristic larval type for all orders except the beroids.

Taxonomy and Review of Groups

Some texts divide ctenophores into two classes, the Tentaculata (six orders) and the Nuda (beroids only). However, since most of the groups are poorly known, any classification of the phylum is subject to revision.

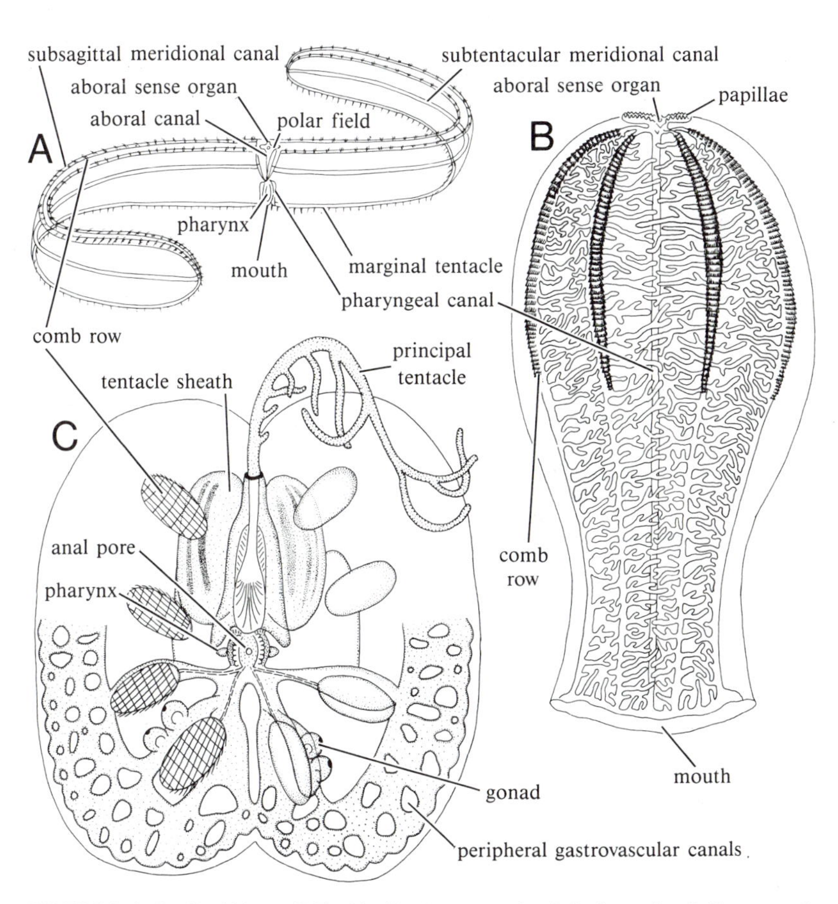

FIGURE 8.4. A. Cestida, *Velamen.* **B**. Beroida, *Beroë;* note complete lack of tentacles. **C**. Platyctena, *Ctenoplana.* (**A** after Mayer; **B** after Hyman; **C** after Komai.)

It seems best not to attempt any separation at the class level. The interplate ciliated groove would appear to distinguish the Lobata structurally and functionally from all other orders.

Order Cydippida. Simple spherical or ovoid forms with two retractile tentacles arising in sheaths; with gastrovascular branches ending blindly. The cydippids are generally regarded as primitive, since all orders (except the Beroida) have larval stages that resemble adult cydippids (Fig. 8.1A).

Order Platyctenea. Strongly flattened, with very short oral-aboral axis; creeping animals with two tentacles in sheaths; comb rows often missing in adult. Example: *Ctenoplana* (Fig. 8.4C).

Order Cestida. Body compressed and ribbonlike, with four rudimentary comb rows; with tentacular sheaths and reduced main tentacles and secondary tentacles along the oral margin of the body. Example: *Velamen* (Fig. 8.4A).

Order Thalassocalycida. Body compressed as in cydippids and formed as medusalike bell; tentacles lack sheaths; comb row short; blind meridional canals of gut long and rather complex. A single species, *Thalassocalyce inconstans*, composes this order. It is intermediate in form between the cydippids and the lobates.

Order Ganeshida. Tentacles complex with side branches; sheaths only at oral margins of comb rows; meridional and paragastric canals join at oral edge to form circumoral canal; mouth large and expanded in tentacular plane. The general body form and arrangement of gastric canals is similar to that of beroids, but development clearly involves a cydippid larva.

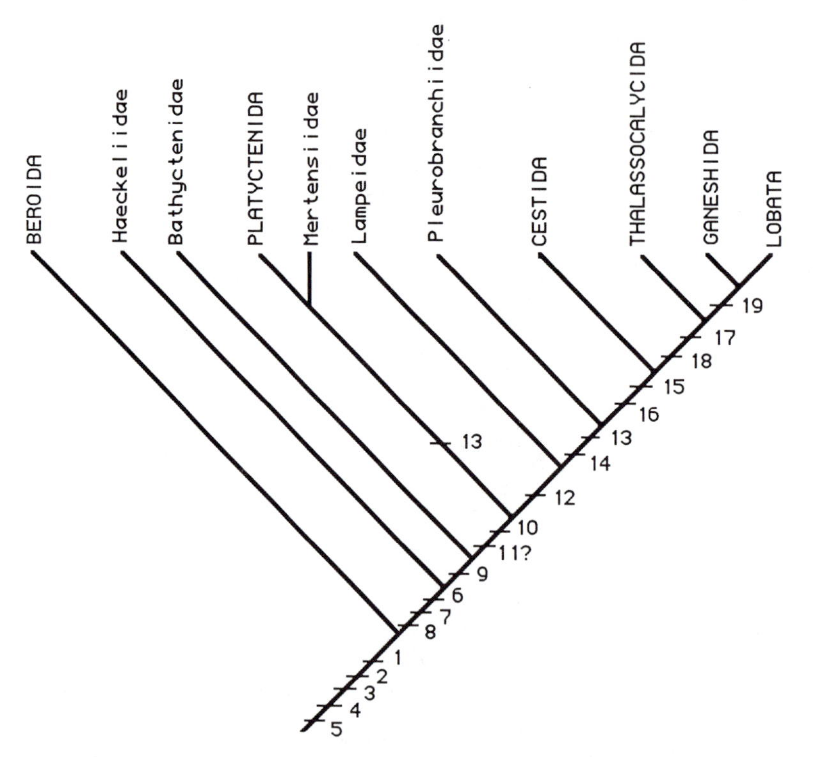

FIGURE 8.5. Possible interrelationships of ctenophore orders and families of Cydippida. Characters are as follows: (1) ctenes, (2) blind meridional canals. (3) pale plate with marginal papillae, (4) aradial canals arising from infundibulum, (5) statocyst, (6) simple tentacles in sheaths, (7) tentacular canals, (8) infundibular canal, (9) subtentacular aradial canals bifurcate, (10) tentacles branched (tentilla), (11) colloblasts?, (12) tentacular canal arising from infundibulum, (13) tentacles exit sheaths aborally, (14) interradial canals, (15) tentilla as fringe along oral margin, (16) interplate ciliated groove, (17) enlarged oral surface, (18) loss of tentacle sheaths, (19) oral surface divided into two lobes. (Modified from Harbison, 1985.)

Order Lobata. Body with two large oral lobes, and four folds, known as auricles, defining auricular grooves; interplate ciliated grooves; auricular grooves containing tentacles and used in feeding; tentacles variously arranged, without sheaths; with the oral ends of the gastrovascular canals jointed by a ring canal. Example: *Mnemiopsis* (Fig. 8.2C,D).

Order Beroida. Conical; with a wide mouth and voluminous stomodeum, with meridional, ramified, gastrovascular canals, without tentacles. These are predators on other ctenophores, often engulfing in one gulp animals as large as they are. Macrocilia on the inner lips can spread the lips over prey or act as teeth to bite chunks of "flesh" off prey. Example: *Beroë* (Fig. 8.4B).

Phylogeny

The Ctenophora have been subjected to some attempts at rigid character analysis (Harbison, 1985). As one might have suspected, the Cydippida, the group generally thought by most authorities to be primitive, is probably paraphyletic. The constituent families of that order can be sorted as monophyletic taxa, but not the order itself. Although we can clearly perceive that a revision of ctenophore higher taxonomy is inevitable, it seems best to avoid proposing drastic changes here. Most "orders" of the Ctenophora are too poorly known to provide an adequate data base for a meaningful revision.

In addition, the polarization of some features is in doubt, since there is no consensus as to the closest outgroup of Ctenophora. Characters that have been used to link Ctenophora and Cnidaria are: radial (biradial) symmetry; branching gastrovascular system; loosely organized nervous system; absence of a coelom, excretory system, and accessory genital structures, statocysts, and true organ systems. However, some of these features are primitive and thus of little use in indicating common ancestry. The platyctene ctenophores have been compared to polyclad platyhelminths because of their oval, flattened body form; presence of a pair of dorsal tentacles; central, ventral mouth; blind gut; radiating nerve net; epibenthic life-style; and often with swimming larvae that have eight ciliated lobes. While many of these features are derived, platyctenes seem to be rather specialized ctenophores, and thus these characters may be only examples of homoplasy rather than actual shared ancestry.

There is one fossil form (Stanley and Stürmer, 1983), known from the Devonian, and it leaves issues of polarity of characters within ctenophores unresolved. For example, it is still unclear whether the ctenophore ancestor had tentacles, whether those had side branches, and whether an infundibular canal was present. The cladogram in Figure 8.5 thus represents a useful, but only first, approximation of ctenophore interrelationships, which should be resolved as our knowledge of the Ctenophora improves.

Selected Readings

Harbison, G. R. 1985. On the classification and evolution of the Ctenophora. In *The Origins and Relationships of Lower Invertebrates* (S. Conway Morris, et al., eds.), pp. 78–100. Clarendon Press, Oxford.

Harbison, G. R., L. P. Modin, and N. R. Swanberg. 1978. On the natural history and distribution of oceanic ctenophores. *Deep-Sea Res.* 25:233–256.

Horridge, G. A. 1974. Recent studies on the Ctenophora. In *Coelenterate Biology* (L. Muscatine and H. M. Lenhoff, eds.), pp. 439–468. Academic Press, New York.

Reeve, M. R., and M. A. Walter. 1978. Nutritional ecology of Ctenophora—A review of recent research. *Adv. Mar. Biol.* 15:249–287.

Reeve, M. R., M. A. Walter, and T. Ikeda. 1978. Laboratory studies of ingestion and food utilization in lobate and tentaculate ctenophores. *Limnol. Oceanogr.* 23:740–751.

Siewing, R. 1977. Mesoderm in ctenophores. *Z. zool. Syst. Evolut.-forsch.* 15:1–8.

Stanley, G. D., and W. Stürmer. 1983. The first fossil ctenophore from the Lower Devonian of West Germany. *Nature* 303:518–520.

Tamm, S. L. 1982. Ctenophora. In *Electrical Conduction and Behavior in Simple Invertebrates* (G. A. B. Shelton, ed.), pp. 266–358. Clarendon Press, Oxford.

9

Gnathostomulida

First recognized in 1956, the Gnathostomulida is worldwide in distribution and contains about 80 described species within 20 genera. The position of the gnathostomulids in relation to other phyla has been a problem from the start; some ally them with pseudocoelomates, others with the acoelomates. Current consensus places them as a sister group to the platyhelminths.

Assessment of gnathostomulid anatomy is confused by their highly modified form related to their meiobenthic habits. They prefer to live in fine sand with a mean optimum grain size of 150 μm, but also occur in coarse sand provided it contains traces of organic material, as typical of sheltered bays and between coral reefs. The diversity and numbers of gnathostomulids increase as one proceeds deeper into their sediments and encounters increasingly anoxic and sulfurous conditions. They seem to represent a side branch of acoelomate evolution that has adapted to these extreme conditions that many other animals seem to avoid.

Definition

Gnathostomulids are bilateral acoelomates that are only slightly cephalized, with paired sense organs in some. They have monociliary epidermal cells. The parenchymal tissues are well developed in the rostral region, but weakly developed to absent in the trunk. The nervous system is epidermal and weakly developed. The gut has the mouth as the only opening, and possesses a well-developed pharynx that bears a complex jaw apparatus. Protonephridia have a flame cell bearing a single cilium surrounded by eight microvilli. Gnathostomulids are hermaphroditic, possibly protandric. Sperm is filiform in some, with internal fertilization. Cleavage is spiral, and development is direct without larval stages.

Body Plan

The body is essentially vermiform. Two orders are recognized, of which the more primitive filospermoids (Fig. 9.1A) have a very long and sinuous body and weakly delineated **rostrum**, whereas the bursovaginoids are shorter and wider, with a clearly delineated rostrum that bears sensory spines (Fig. 9.1C,E).

The epidermal cells bear a single cilium, a feature seen elsewhere only among some primitive gastrotrichs. The epidermis itself is composed of a single layer that rests on a very thin basal membrane. More advanced forms of gnathostomulids also develop **mucous gland** cells, and the most apomorphic forms, the conophoralians, have **rhabdite glands** as well.

The most obviously diagnostic feature of the phylum is the jawed pharynx. The apparatus consists of two sets of elements (Fig. 9.1B,D,F): an anterior basal plate and the jaws proper. These are cuticularized elements secreted by cells of the pharyngeal epidermis. The **basal plate** is a ribbed, occasionally toothlike, comb apparatus probably used for scraping. The **jaws** of filospermoids are simple compact structures, whereas in bursovaginoids they are complex, lamellar, toothed elements probably used for grabbing. One family, the Agnathiellidae, has lost the jaws altogether.

The **pharynx** is a muscular organ that in burso-

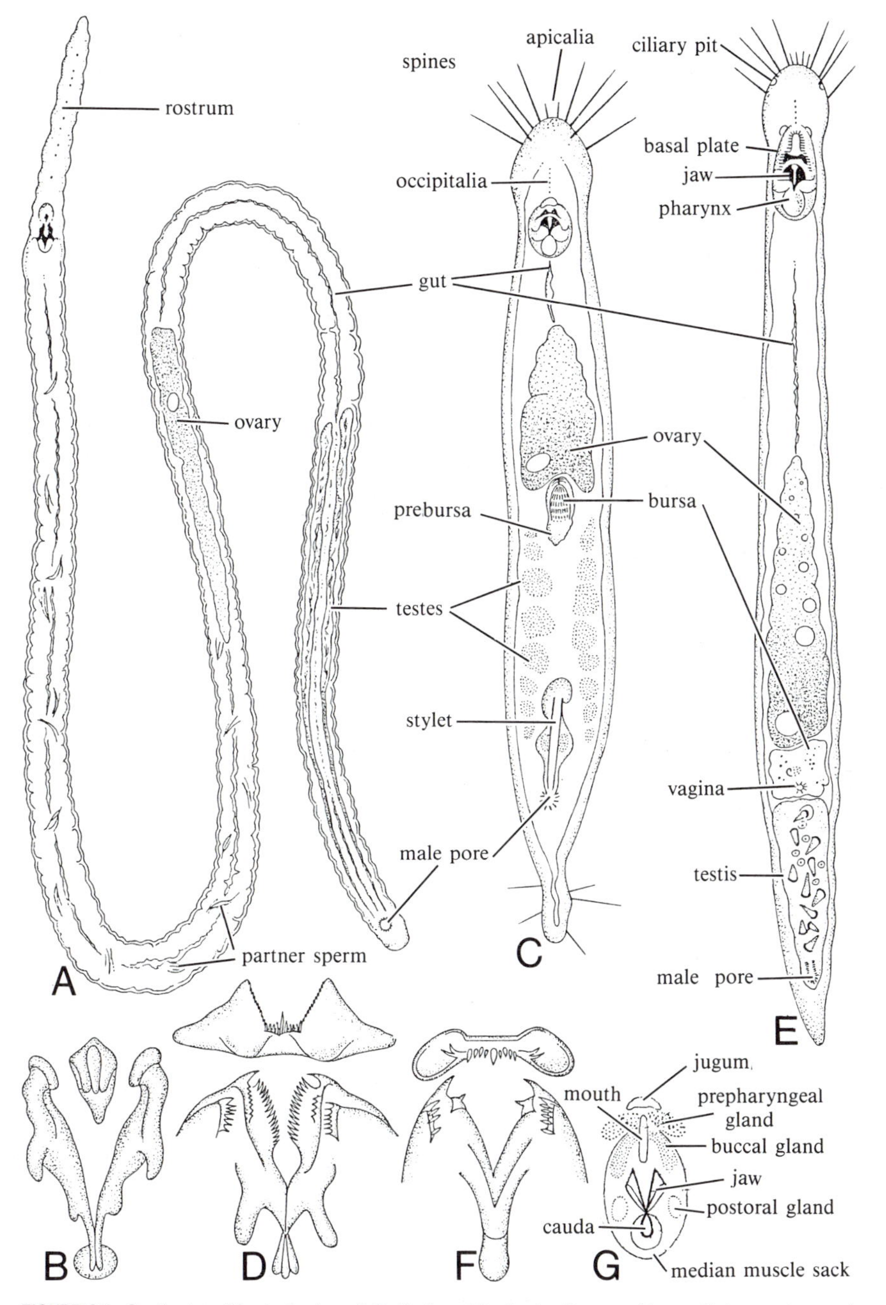

FIGURE 9.1. Gnathostomulidan body plans. A,B. *Haplognathia simplex,* filospermoidean, with long rostrum and trunk. C,D,E,F. Bursovaginoideans; C. *Gnathostomula jenneri* body; D. *G. mediterranea* jaw apparatus; E,F. *Austrognatharia kirsteueri.* G. Gnathostomulid mouth field and pharynx. (After Sterrer.)

vaginoids is tripartite, with median and lateral muscle compartments (Fig. 9.1G). Several glands are associated with the pharynx. The Gnathostomulidae have an additional "cartilaginous" crescentlike plate just in front of the mouth, the **jugum**. This apparently serves to help prevent the mouth opening from totally collapsing when the pharyngeal muscles contract. The gut is a tube formed by a simple layer of gastrodermis. The only major access to the gut is through the mouth. However, a strand of tissue has been identified that connects the posterior end of the gut with the dorsal epidermis in the elongate filospermoid *Haplognathia* and the shorter bursovaginoid *Gnathostomula.* This strand has been suggested as an **anal pore** and, if it proves to be consistently present in other genera, may be a diagnostic feature of the phylum.

The parenchyme is only well developed in the rostral area and around the reproductive organs. The rest of the body space between the epidermis and the gastrodermis is empty, but is not in any way a pseudocoel. The lack of trunk parenchyme is probably an adaptive response not only to the exigencies of tiny size and locomotory requirements in tightly packed spaces, but also to a need to supply some reservoir to store partner sperm after copulation.

The excretory system is protonephridial, with flame cells mostly located in the parenchyme around the gonads. The cell contains a single cilium surrounded by eight microvilli (Fig. 9.2A,B). This is a primitive morphological arrangement that is a precursor to the flame cell structures seen among flatworms (Fig. 9.2C,D,E,F). This flame cell monocilium, surrounded by eight microvilli, is homologous to the condition seen in the epidermal cells, that is, a discrete cilium situated in a pit surrounded by eight microvilli (Ax, 1985).

The gnathostomulid nervous system has been termed basiepithelial; that is, it is an integral part of the epidermis. However, in filospermoids, a subepidermal brain is noted. The **brain** is preoral, from which arises three pairs of **longitudinal nerves** that run along the base of the epidermis and that are closely associated with the basal membrane. In addition, a pair of **buccal nerves** connect the brain with a **buccal ganglion** near the posterior of the pharynx.

The male reproductive system is generally composed of paired **testes**, and in such forms that have a single dorsal testis, it is known to develop from paired embryonic anlagen that fused. The filospermoids have glandular tissues located around the ventral, male pore; bursovaginoids have a well-de-

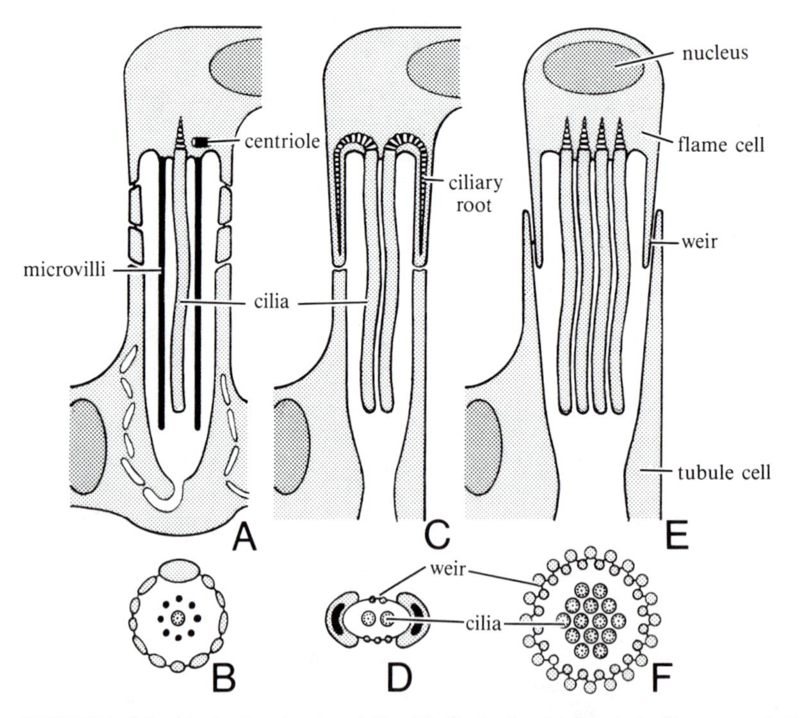

FIGURE 9.2. Schematic drawing of protonephridia. **A,B**. Gnathostomulid plan, single cilium surrounded by eight microvilli. **C,D**. Catenulidan platyhelminth, two cilia without microvilli. **E,F**. Rhabditiphoran platyhelminth, multiple cilia, two-cell weir. (After Ax.)

veloped, ventroposterior, muscular **penis** often armed with **stylets**. Sperm are of three types: **filiform sperm**, seen in filospermoids and similar to that seen in primitive flatworms, with a simple tail with a 9+2 microtubule pattern (Fig. 9.3A); **dwarf sperm**, which lack a tail but have nonciliary filaments (Fig. 9.3B); and a **conulus**, a huge structure that could be either a giant sperm or a spermatophore (Fig. 9.3C). Interestingly, the active filiform sperm are associated with a lack of a penis, whereas the less mobile dwarf sperm and conuli are found in forms with well-developed penes.

The female system centers on unpaired **ovaries**. Bursovaginoids have a vagina and a bursa. The **vagina** consists of a dorsal pore connected to a short canal that enters the bursa. The **bursa** is a large chamber behind the ovary that often has a posterior **prebursa**, an injected sperm packet from the last copulation. The ovary carries one mature egg at a time.

Biology

Locomotion One of the reasons gnathostomulids remained undiscovered for so long is their annoying propensity to stick to substrate particles. This is a reflection of their locomotory capacities. A matrix whose mean particle size is 150 μm has a minimum free pore space of about 900 μm^2. A typical gnathostomulid has a cross-sectional area of about 1900 μm^2. Thus, gnathostomulids must be quite deformable to virtually "flow" through the sediment pore spaces of their habitat (Sterrer, 1971). Actual movement is achieved by ciliary beating, and some genera have been reported capable of swimming, even backward by reversing ciliary beat.

Feeding Gnathostomulids feed on bacteria, fungi, and, on occasion, blue-green algae. All three have been found in their guts, although apparently the main food of preference is fungi. Food gathering is not well understood; it probably occurs as the basal plate scrapes material off sediment particles, allowing the jaws to grab and pull food items inward to be sucked into the pharynx. The jaws also may serve to prevent food from popping back out of the mouth when the pharyngeal muscles relax.

The amount of food in the gut is inversely proportional to the proximity of an individual to a reproductive peak. The animals apparently operate their life cycle on one or the other of two phases, nutritive or reproductive, but not both simultaneously.

Excretion The proximity of the protonephridia to the gonadal tissues apparently concentrates excretory activity near areas of highest metabolic activity.

Respiration The habitat in which gnathostomulids live is among the most anoxic known. Furthermore, gnathostomulids thrive as one proceeds deeper into the sediments as blacker and more sulfurous conditions prevail. The associated meiofauna of nematodes, turbellarians, and gastrotrichs under such conditions decrease in abundance and diversity as gnathostomulids increase. Their respiratory needs must, therefore, depend entirely on glycolysis. Oxygen consumption has been measured (Scheimer, 1973) at 282 and 413 mm^3 O_2/hour/g wet weight for *Gnathostomula* and *Haplognathia*, respectively. The former figure is the lowest ever recorded for oxygen consumption in meiobenthos, where requirements typically range from 370 to 3000 mm^3 O_2/hour/g wet weight.

Circulation The small size, extreme deformability, and lack of parenchyme undoubtedly make it easy for material to circulate and diffuse about the body.

Sense Organs Sense organs vary between groups, although all forms have a median dorsal row of cilia, termed the **occipitalia**. Filospermoids appear to have few sensory organs, although the large rostrum is undoubtedly tactile in nature. Bursovaginoids have a variety of sensory structures. A medial apical and paired lateral **ciliary pits** occur on the rostrum. **Apicalia**, sensory cilia, are found on the rostrum terminus. Pairs of sensory spines are characteristic of

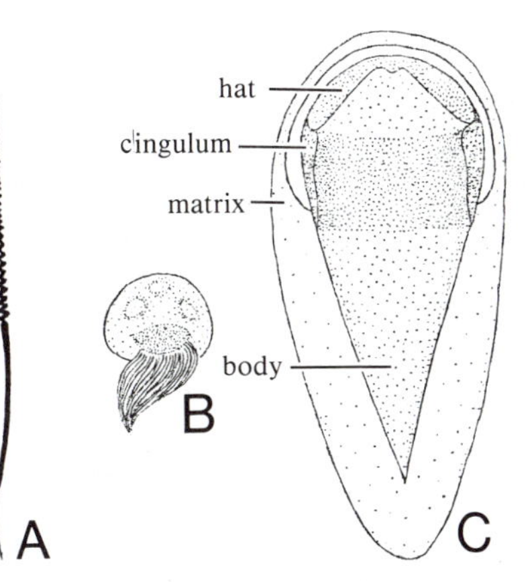

FIGURE 9.3. Gnathostomulid sperm types. **A.** Filiform type, *Haplognathia simplex*. **B.** Dwarf type, *Gnathostomaria* sp. **C.** Conulus of *Austrognathia riedli*. (After Ax.)

bursovaginoids, where up to five, but more typically four, pairs occur.

Reproduction Gnathostomulids are hermaphrodites, but all possible sexual combinations are known, from pure males, through male–female forms, to pure females, and even "senile" sexless forms (Sterrer, 1974). It has been suggested that at least some members of the group might be **protandric**, that is, sequential hermaphrodites with the male phase occurring first. Certainly such an option would make it possible to devote all available energy resources to the production of the most viable of sperm or ova.

Gnathostomulid mating has actually never been observed. Filospermoids swell the gland tissue around the male pore and then stick to a sex partner. Supposedly, the filiform sperm bore their way into the partner's body cavity. Bursovaginoids apparently deposit a sperm pocket in the bursa with their muscular penis. Venter-to-venter mutual insemination may be possible in filospermoids, but such is physically impossible in bursovaginoids in which male and female gonopores are located on opposite body surfaces. How long the sperm remain in the body before actual fertilization takes place is not now known. A single egg is laid. The body contracts in peristaltic waves, causing the egg to be forced posteriorly and lodged between the gut wall and bursa. If all goes well, the egg breaks through the epidermis and adheres to a sand grain. Occasionally, the egg breaks into the gut instead, whereupon it ruptures and is reabsorbed by the gastrodermis. In either case, the resultant wounds quickly heal.

Asexual reproduction may be possible. Some localities have produced anterior and posterior body halves. This might indicate the possibility of asexual reproduction by body fragmentation, but it has also been suggested that fragmentation may be a way to reduce food requirements during hibernation.

Development After being laid, the egg develops a rather hairy appearance (Fig. 9.4A,A'). Development conforms to a typical spiral pattern. Gastrulation occurs by combinations of invagination and epiboly. Early cleavages are rapid, but after gastrulation, development proceeds more slowly. Development is direct (Riedl, 1969; Sterrer, 1972). The hatchling (Fig. 9.4B,B') has a short tail, a pharynx, rudiments of a jaw apparatus, and a lumenless gut. As the juvenile grows, more organs appear and differentiate (Fig. 9.4C,D,E,F,G) until the adult form is finally achieved (Fig. 9.4H).

Fossil Record and Phylogeny

It has been suggested that the mysterious Paleozoic conodonts were extinct gnathostomulids. However, the discovery of fossilized conodont animals and the

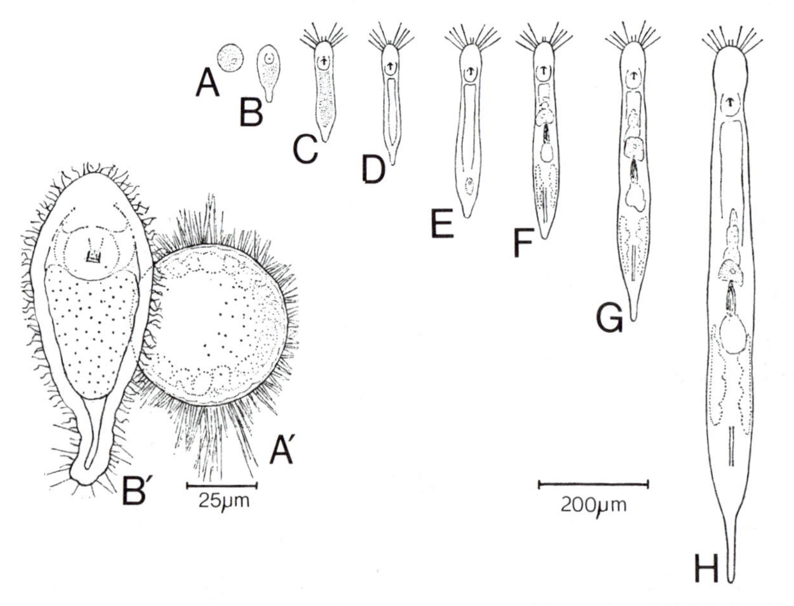

FIGURE 9.4. Developmental sequence of *Gnathostomula jenneri*. **A,A′**. Late embryo. **B,B′**. Tailed hatchling, with pharynx. **C,D,E,F**, Juveniles; **C**. with jaws; **D**. with gut lumen; **E**. with stylets; **F**. bursa and gonads. **G**. Small adult. **H**. Largest specimen. (After Riedl.)

recognition of their apparent affinities with the chordate lineage precludes any more serious consideration of this issue.

Relationships within gnathostomulids have been well documented. Those desiring more details should consult the interesting paper by Sterrer (1972). However, some serious problems remain since the Filospermoidea are defined largely by primitive features, and current understanding of the scleroperalian bursovaginoids reveals them to be paraphyletic. Study of new species and expanding use of scanning and transmission electron microscopy will undoubtedly help resolve these issues.

Current opinion places gnathostomulids in close relationship to platyhelminths (Sterrer et al., 1985). They are thus viewed as an offshoot of the acoelomate lineage that is specialized for interstitial or meiofaunal habitats. Although gnathostomulids have a few similarities to some pseudocoelomates, for example, the rotifer jaw apparatus and the gastrotrich monociliate epidermis, these similarities are now coming to be viewed either as convergent or as the retention of a primitive feature.

Taxonomy

In the most recent surveys of the group, ten families within two orders are recognized.

Order Filospermoidea. Filiform sperm; lacking vagina, bursa, injectory penis, and paired sense organs on rostrum; body and rostrum elongate; pharyngeal muscles loosely organized, jaws compact. Two families (Fig. 9.1A,B).

Order Bursovaginoidea. Sperm not filiform; with bursa, vagina, and injectory penis; paired sense organs on rostrum; body short, rostrum delineated from trunk; pharyngeal muscles well developed, jaws lamellar. Six families (Fig. 9.1C,D,E,F).

Suborder Scleroperalia. Dwarf sperm, testes paired, penis with stylet of cuticular rods; bursa well developed, sometimes a vagina; three to five pairs of sensory bristles and one pair of apicalia (Fig. 9.1C,D).

Suborder Conophoralia. Conical sperm (conulus), testis unpaired, muscular penis without stylets; soft bursa, vagina well developed; four pairs of sensory bristles, two pairs of apicalia (Fig. 9.1E,F).

Selected Readings

Ax, P. 1985. The position of Gnathostomulida and Platyhelminthes in the phylogenetic system of the Bilateria. In *The Origin and Relationshps of Lower Invertebrates* (S. Conway Morris et al., eds.), pp. 168–180. Clarendon Press, Oxford.

Riedl, R. J. 1969. Gnathostomulida from America. *Science* 163:445–452.

Schiemer, F. 1973. Respiration rates of two species of gnathostomulids. *Oecologia* 13:403–406.

Sterrer, W. 1971. On the biology of Gnathostomulida. *Vie Milieu Suppl.* 22:493–508.

Sterrer, W. 1972. Systematics and evolution within the Gnathostomulida. *Syst. Zool.* 21:151–173.

Sterrer, W. 1974. Gnathostomulida. In *Reproduction of Marine Invertebrates*, vol. 1 (A. C. Giese and J. S. Pearse, eds.), pp. 335–357. Academic Press, New York.

Sterrer, W. 1982. Gnathostomulida. In *Synopsis and Classification of Living Organisms*, vol. 1 (S. P. Parker, ed.), pp. 847–851. McGraw-Hill, New York.

Sterrer, W., M. Mainitz, R. M. Rieger. 1985. Gnathostomulida: enigmatic as ever. In *The Origins and Relationships of Lower Invertebrates* (S. Conway Morris, et al., eds.), pp. 181–199. Clarendon Press, Oxford.

10

Platyhelminthes

Platyhelminthes is a diverse group that remains the most representative example of a structural body plan that is based on bilateral symmetry and that lacks any body cavities outside the gut. Traditionally three groups are recognized, and these should be viewed as grades of evolution within the platyhelminths. The turbellarians comprise several groups of free-living flatworms. The trematodes are more advanced and are generally parasites. The cestodes are the most derived and comprise the parasitic tapeworms. Modern understanding of the interrelationships among the flatworms and their position with regard to other phyla is based on extensive electron microscopic studies of cell and tissue ultrastructure, a detailed presentation of which is not entirely appropriate for an introductory text. Students are referred to the Selected Readings to explore these matters more fully. It is traditionally felt that the most primitive Bilateria are acoelomate, that is, with the gut as the only body cavity. However, this may be an unwarranted assumption and alternative views are possible (e.g., see Chapter 38).

Eumetazoa that share a primary bilateral symmetry are sometimes grouped as the Bilateria. This symmetry can be lost in sessile or slow-moving adults but is almost always evident during early embryonic development. Bilateral symmetry is linked to a life-style of active movement. Locomotor organs are associated with the development of complex muscles, a tissue type almost wholly mesodermal in origin. As muscles develop, selection works on the sense organs and nervous system, which in turn further specialize by aggregating sense organs at the anterior end of a bilateral body with a nerve center or brain located nearby. Locomotor and sensory specializations polarize the animal body such that anterior/posterior, right/left, and dorsal/ventral orient the body to external stimuli.

As locomotor, sensory, and nervous systems evolve, additional metabolic requirements arise. Nutritive and excretory functions are amplified, and complex organ systems associated with serving these functions develop in more highly organized animals. Greater size and internal complexity require the development of mechanisms for internal transport. In short, bilaterality was a critical factor in determining the course of animal evolution.

Definition

Platyhelminthes are acoelomate. Although bilaterally symmetrical, they are cephalized only slightly. Typically they bear anterior eyespots, chemoreceptive organs, and a "brain." They possess multiciliary epidermal cells without accessory centrioles. There is a functional third germ layer, the mesoderm, that produces a mesenchymal parenchyme that fills the space between the epidermis and the gastrodermis. The digestive system has a single opening to a well-

developed gastrovascular cavity. A synaptic nerve net of unipolar, bipolar, and multipolar neurons tends to condense as longitudinal nerve trunks and a bilobed, anterior brain. There are no morphological distinctions between dendritic and axonic neuron endings, but neurons sometimes serve as definite sensory, associative, and motor elements. Protonephridial tubules with terminal flame cells are present. Reproductive systems are complex and typically hermaphroditic. Development occurs in free-living forms by means of determinate, spiral cleavage, with mesoderm and endoderm arising from a special mesendoblast. Sperm is filiform, threadlike in form, with an elongate head, cylindrical middle piece, and long uniciliary tail.

Body Plan

The basic flatworm body plan is best illustrated by the free-living forms, typically referred to as turbellarians (Fig. 10.1). In addition to providing a structural outline for study of the phylum, turbellarians will also be used here as a type for the presentation of flatworm biology. However, it will be necessary to note exceptions to the basic type that occur in the parasitic platyhelminths.

Small or primitive species have uniform ciliation with many cilia associated with each cell (Fig. 10.2D). Larger species, and especially terrestrial species, have strong ventral cilia and weak or no dorsal cilia. Large turbellarians have many **cyanophilous** or **slime glands** (Fig. 10.3B) on the ventral surface. Slime secretions lubricate the surface, increase water viscosity, and provide resistance against which the cilia can thrust. Many species cannot move without leaving a slime trail. The common planarian, *Dugesia*, uses its marginal adhesive zone to keep in contact with the substrate. Only when slime accumulates ventrally can it move effectively. Epidermal and subepidermal glands also form rhabdites (Figs. 10.2A,B and 10.4E). Rhabdites differ in form in different species, but appear in all cases to be secreted particles that dissolve rapidly when extruded. They probably do not contribute to movement, but serve to protect against irritating chemicals.

Some acoelomorphs have muscle fibers at the base of epidermal cells, as in cnidarians. Other than this, flatworm muscle arises from mesoderm and has a relatively simple structure in most flatworms. The elongated muscle cells have a persistent myoblast that clings to the surface and encloses the nucleus (Fig. 10.2C). Muscle fibers are usually unstriated but, when present, striae exhibit simple patterns.

Muscle fibers typically form layers below the epidermis and pass through the parenchyme. Subepidermal muscle usually consists of outer circular and inner longitudinal layers, separated by a few to many diagonal fibers (Fig. 10.2A,D). Land planarians have conspicuous bundles of longitudinal muscles. Pharyngeal muscles are modified subepidermal muscle (Fig. 10.3D). Parenchymal muscles are extremely variable, and large species may have interlacing patterns of transverse, longitudinal, dorsoventral, and diagonal fibers passing through the parenchyme. Primitive trematode parasites, monogeneans, have muscle layers like those of turbellarians, but other flukes have a diagonal muscle layer within the longitudinal layer (see Fig. 10.5).

Because external parasites are likely to be detached, attachment structures are among the most conspicuous and characteristic external features in parasitic flatworms. Monogeneans adapt to their precarious position by remarkable attachment organs. A **prohaptor**, usually consisting of a pair of **glandulomuscular organs**, is often present at the anterior end (see Fig. 10.25A). The posterior attachment organ or **opisthaptor** is larger and more conspicuous, consisting of a large disc with suckers, large hooks, and/or smaller hooklets (see Fig. 10.24A,C). The Aspidobothrea are no less remarkable, with the whole ventral surface being modified as a huge, compartmented sucker or a series of discrete suckers (see Fig. 10.24A). Digenean trematodes usually have a small oral sucker surrounding the mouth and a larger ventral sucker (or acetabulum) somewhere on the ventral midline (Fig. 10.17D). Cestodes use similar structures to attach to their hosts. The **scolex** or head is equipped with a variety of suckers and/or **bothridia**, suctorial slits. In addition, the anteriormost **rostellum** is armed with a battery of hooks.

Platyhelminths are characterized by a solid mass of mesenchymal cells, or parenchyme, that occupies the space between the epidermis and the gastrodermis. This mesenchyme occupies the same position as cnidarian mesoglea but is not the same. Mesogleal cells are scarce and wander into the matrix from the epidermis, and mesogleal formation is not a formal embryonic process. In flatworms and all higher phyla, the bulk of the mesoderm tissue arises from endoderm by discrete processes that are fixed as a definite part of embryonic development. The ectoderm and endoderm become more restricted in function as the mesoderm emerges as a functional germ layer. This **parenchyme** (Fig. 10.2A) is composed of large vacuolated cells, usually reported to be merged into a syncytium, but is in the form of discrete cells in some species. In any case, smaller discrete amoeboid cells are scattered among the parenchymal cells. These gather at a wound and are important in the formation of reproductive organs. Presumably these amoeboid cells are similar to the interstitial cells seen in radiates.

Digestive System Flatworms exhibit three types of pharynx construction: simple, bulbous, and plicate. A **simple pharynx** is merely a short length of invaginated epidermis, without specialized musculature (Fig. 10.3C). It is found in the more primitive turbellarians, such as acoelomorphs, and in some rhab-

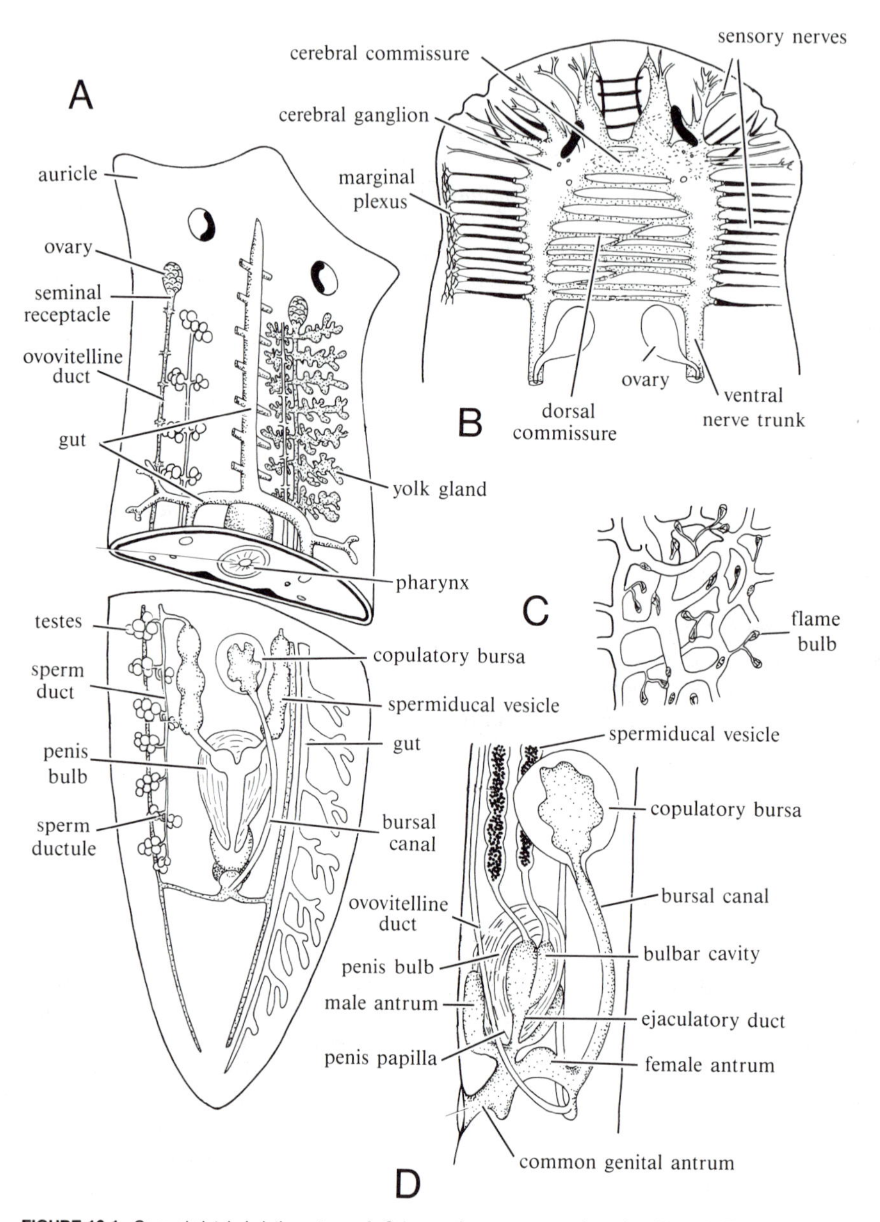

FIGURE 10.1. General platyhelminth anatomy. **A**. Scheme of organ systems, based on *Dugesia*. Digestive diverticula removed for clarity; male system featured on left side, female system on right. **B**. Brain and anterior part of nervous system of *Dugesia*. **C**. Part of protonephridial reticulum of branching and anastomosing primary and secondary longitudinal excretory canals, flame bulbs occurring as terminal twigs of reticulum. **D**. Copulatory apparatus of *Dugesia tigrina*. Common gonopore serves both male and female systems. (**B** after Micoletzky; **C** after Benham, from Hegner; **D** after Hyman, from Brown.)

ditophorans. A **bulbous pharynx** contains special muscles, and a thin-walled pharyngeal cavity lies between the mouth and the pharynx. A bulbous pharynx may be rosulate, doliiform, or variable. The spherical **rosulate pharynx** is set at right angles to the main body axis (Fig. 10.3D). The **doliiform pharynx** is cask-shaped and lies parallel to the body axis (see Fig. 10.23C). The circular and longitudinal muscles form a lattice that is often conspicuous. The **variable pharynx** is a modification of the doliiform, with modified muscle layers and a variable shape. A **plicate pharynx** is a fold of the body wall, projecting into the pharyngeal cavity. In its simplest form it is a tubular organ, as in *Dugesia* (see Fig. 10.1). The

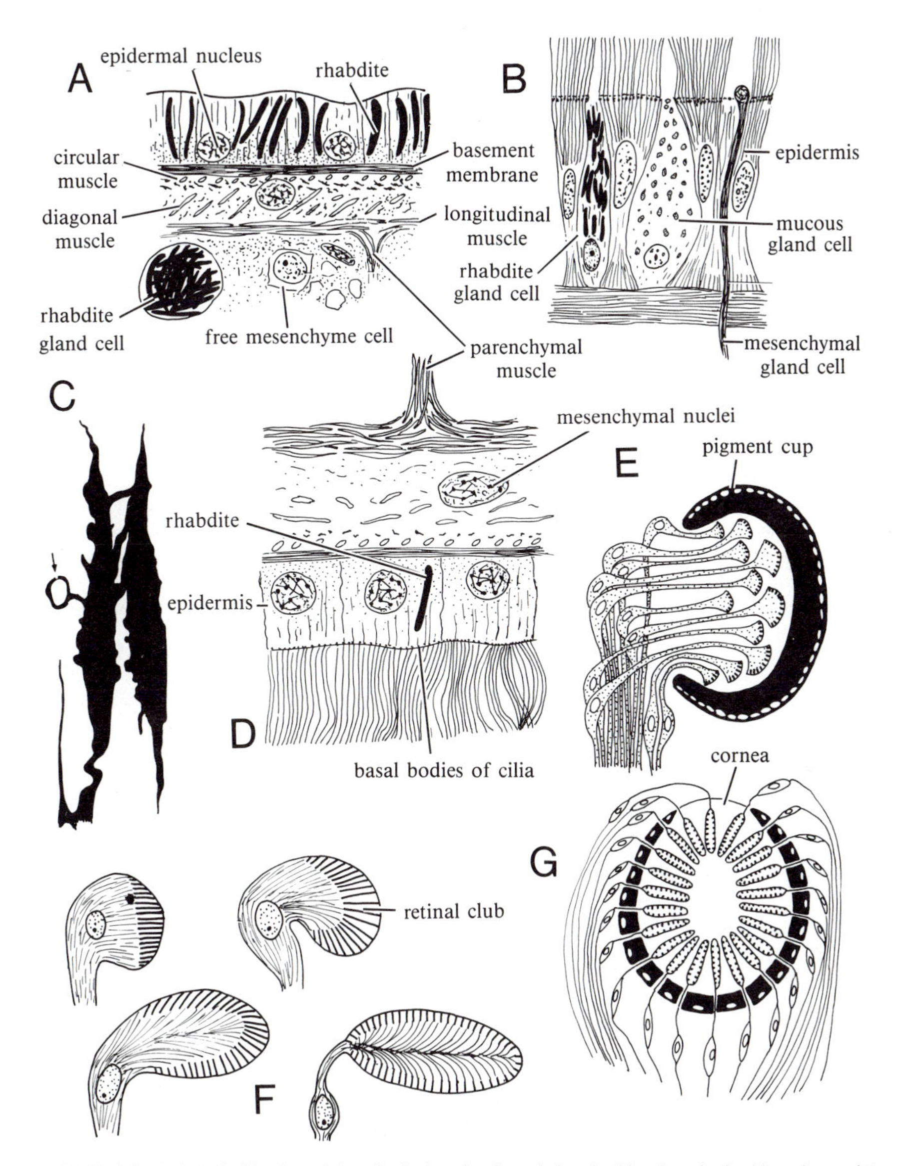

FIGURE 10.2. Turbellarian histology. **A.** Longitudinal section through dorsal epidermis and subepidermal musculature of freshwater planarian. **B.** Longitudinal section through epidermis of polyclad. **C.** Part of subepidermal longitudinal muscle fibers of planarian. Note bulge (*arrow*) containing nucleus on fiber on left. **D.** Longitudinal section through ventral surface of freshwater planarian. **E.** Scheme of inverse pigment-cup ocellus as seen in planarian; sensory cells enter through open side of cup, and sensory tips extend toward floor of cup. **F.** Four stages in the evolution of the retinal club, showing rod border spreading over tip of retinal cell and migration of nucleus away from sensory tip. **G.** Scheme of converse pigment-cup ocellus, as seen in land planarians; retinal cells pierce wall of pigment cup and extend toward opening. (**A,D** after Hyman; **B** after Pock; **C** after Gelei; **E** after Bresslau, in Kükenthal and Krumbach; **F** after Hofsten; **G** after Hesse.)

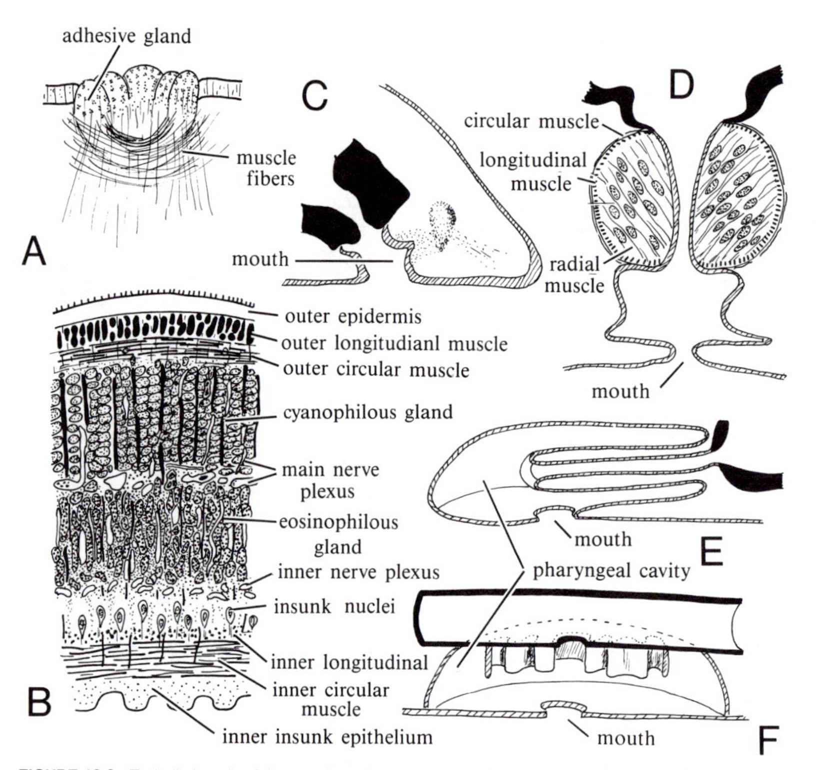

FIGURE 10.3. Turbellarian glandulomuscular adhesive organ and pharyngeal structure. **A**. Glandulomuscular organ of *Procotyla,* ventral view, with muscles not cut off from body by septum, as in true suckers. **B**. Segment of cross section through pharynx of *Dugesia,* showing component layers of tissue. **C**,**D**,**E**,**F**. Scheme of structure of various types of pharynx; **C**. simple, nonmuscular pharynx; **D**. rosulate pharynx, muscular and preceded by pharyngeal cavity that opens at mouth; **E**. cylindrical plicate pharynx; **F**. ruffled plicate pharynx. Plicate pharynges are extensions of body wall, containing muscle and other elements, and extend and thrust through mouth opening during feeding. Gastrodermis, area in black, does not enter into pharynx formation. (**A**,B,**C**,**D** after Hyman.)

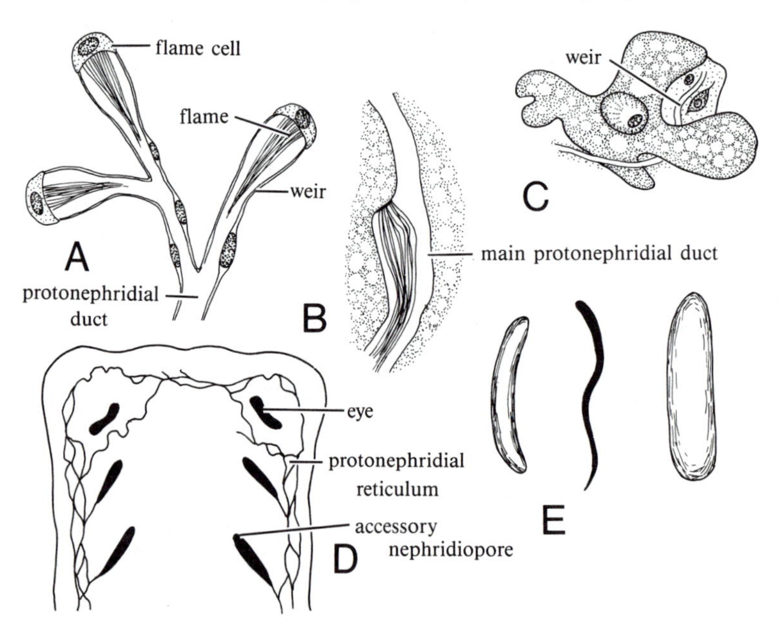

FIGURE 10.4. Protonephridial system and rhabdites. **A**. Three flame cells and weirs. Note upper end of each weir completely closed off by flame cells. **B**. Lateral flame of *Mesostoma* located in main protonephridial duct. Lateral flames agitate fluid in duct where relatively few flame cells occur. **C**. Athrocyte of *Rhynchomesostoma* around protonephridial weir probably picks up wastes and deposits them in tubule. **D**. Part of protonephridial reticulum of *Dendrocoelum,* somewhat magnified, showing part of protonephridial reticulum in head region and several lateral branches to accessory nephridiopores. **E**. Shapes of rhabdites. Ordinary rhabdite shown at *left,* rhammite in *center,* and chondrocyst on *right.* (**A** after Pennak; **B**,**C** after Luther; **D** after Wilhelmi; **E** after Hyman.)

plicate pharynx lies in a pharyngeal cavity lined with epidermis and is itself covered with epidermis (Fig. 10.3E,F). The plicate pharynx of many polyclads is a ruffled curtain that hangs from the roof of the pharyngeal cavity. The plicate pharynx, and sometimes the bulbous, is thrust out through the mouth during feeding, and food enters the gut through the pharyngeal opening. The histological construction of a plicate pharynx is complex (Fig. 10.3B). It is richly supplied with nerves, although its control is largely intrinsic as a detached pharynx can move about and will sometimes ingest food. A curious condition (polypharyngy) is seen in some planarians when two or more pharynges, sometimes associated with several mouths, are present.

The main body cavity is the gut itself. Turbellarians were once partly subdivided for purposes of taxonomy on the basis of gut shape. Acoels had no intestine. Rhabdocoels had a saccate or tubular gut without diverticula. Alloeocoels had lateral diverticula. Triclads had three main intestinal trunks, and polyclads had many. The main trunks of triclads and polyclads usually branch extensively. However, this is not a particularly effective system of classification as the form of the gut can be quite variable within groups.

Tall gastrodermal cells, ciliated in some macrostomid and polyclad turbellarians, line the gut. Preliminary digestion occurs in the lumen and precedes the final, intracellular phase of digestion. The relative importance of extracellular and intracellular phases differ; flatworms that have a ciliated gastrodermis and ingest prey organisms whole are more dependent on extracellular digestion. In some, no trace of intracellular digestion is seen.

In parasitic forms other than the gutless tapeworms, the stomodeal region consists of a mouth, a pharynx, and, usually, an oral sucker. The short buccal cavity is followed by a bulbous pharynx, rosulate in most trematodes but doliiform in some monogeneans. In Aspidobothrea and some Monogenea, the enteron is saccate, but most trematode forms have two intestinal caeca (**crura**), sometimes with **intestinal diverticula**.

Excretory System All but the primitive acoelomorph turbellarians have a **protonephridium** that consists of **protonephridial tubules** or **weirs** and terminal **solenocytes** or **flame cells**. The weir with terminal flame cells (Fig. 10.4A) is attached to the duct, so that the duct is closed off at the end. Each flame cell has at least two, in some cases several, flames or beating cilia. Systems with few flame cilia may have lateral cilia (Fig. 10.4B) on the walls of the protonephridia. The contents of the duct are discharged through a **nephridiopore**. The protonephridial system tends to vary in complexity among groups. Acoelomorphs lack protonephridia, but other turbellarians have one or two recurved protonephridia, and larger forms often have two or three protonephridia on each side with accessory nephridiopores. Triclad seriate turbellarians have the most complex system with as many as four protonephridia on each side that branch and anastomose to form a protonephridial reticulum with numerous nephridiopores (see Fig. 10.1C).

The protonephridial system is poorly developed in marine species and highly developed in freshwater forms. Accumulation of water in *Stenostomum* evokes formation of new flame cells; in blood flukes, flame cells that are inactive when the animal is in urine become active with dilution of that urine. Protonephridia evidently serve to discharge excess water. Acoelomorphs, which have no protonephridia, take up vital dyes into their mesenchymal cells and discharge these through the mouth. Some species have large cells, which take up vital dyes (Fig. 10.4C), and that are wrapped around the protonephridial tubules.

In parasitic trematodes, flame cells connect with paired protonephridial ducts. The system first appears in the miracidium larvae. Ducts and their branches have a constant arrangement, characteristic of various trematode groups. The number and arrangement of flame cells are also constant features, and can be of considerable value in taxonomy, helping to relate unknown larvae to adults. During the development of the larval cercaria stage, the two protonephridial ducts unite medially to form a bladder (Fig. 10.6). The tubules continue to the tail, where they open through separate nephridiopores. When the tail is shed, the posterior excretory pore, characteristic of Digenea trematodes, is formed. The

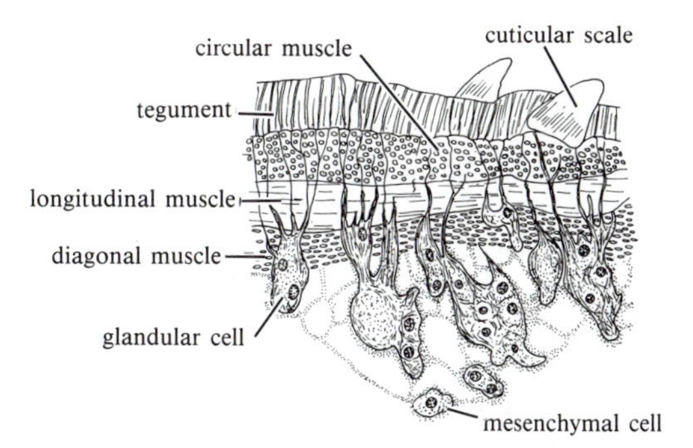

FIGURE 10.5. Longitudinal section through surface of *Fasciola hepatica*. (After Hein.)

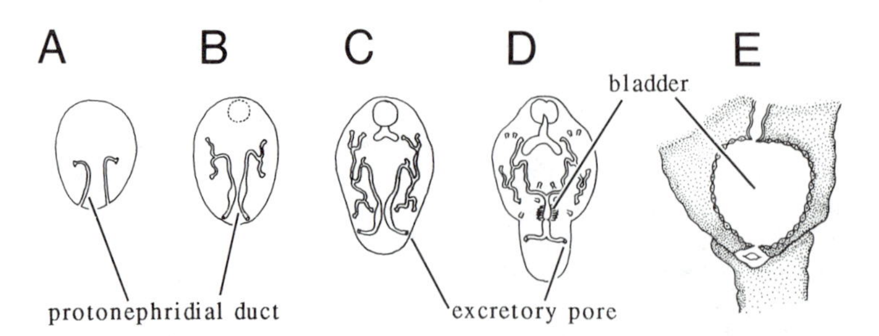

FIGURE 10.6. Development of protonephridial system in trematode cercaria, *Notocotylus urbanensis,* showing formation of bladder from union of two protonephridial ducts, early branching of ducts, and formation of excretory pores in cercarial tail. With loss of tail, original nephridiopores are lost, and opening from bladder serves to discharge urine. (After Kuntz.)

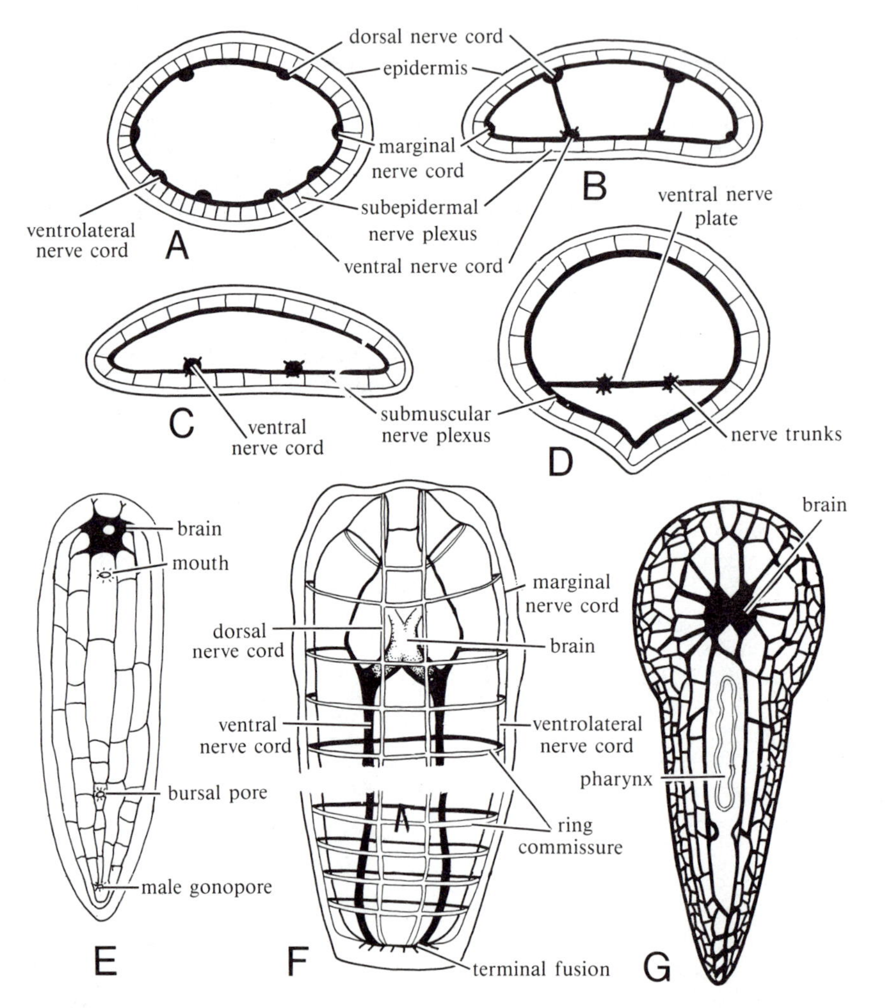

FIGURE 10.7. Turbellarian nervous systems. **A,B,C,D**. Scheme of nervous system of catenulids and triclads, showing relationship of nerve trunks and plexi; **A**. catenulid; **B**. marine triclad; **C**. freshwater triclad; **D**. land triclad. **E**. Nervous system of acoelomorph, *Convoluta,* showing anterior concentration of nerve plexus (brain) around statocyst, and tendency toward formation of longitudinal trunks. **F**. Anterior and posterior ends of nervous system of *Bothrioplana,* with ventral elements darkened and dorsal elements unshaded. **G**. Ventral, submuscular nerve plexus of polyclad, *Gnesioceros,* showing tendencies toward radial arrangement of trunks. (**A,B,C,D,G** after Hyman; **E** after Bresslau, in Kükenthal and Krumbach; **F** based on Reissinger.)

bladder develops with a thin wall at first, but may come to have a heavy epithelial lining. This ontogenetic trait is used to divide the Digenea into two groups.

The protonephridial system of tapeworms consists of a dorsal and a ventral excretory duct on each side of the body, from which secondary canals arise that end in flame cells. The ventral ducts are usually connected by a transverse excretory duct at the posterior end of each proglottid; dorsal ducts are rarely connected by a transverse duct. Other longitudinal canals may be present; some tapeworms, such as the Cyclophyllidea, have as many as twenty. The main longitudinal ducts continue into the anterior strobila, and are connected by a plexus of tiny tubules. Accumulation of vital dyes in granules in some cestodes suggests that storage excretion may also occur.

Circulatory System All platyhelminth parenchyme is essentially the same in lacking any special provisions for circulation, but some trematodes, like the Digenea, have a **lymphatic system**, the only special system for internal transport in the phylum! Canals, lined with flattened cells derived from the parenchyme, ramify extensively through the parenchyme and the muscular organs (see Fig. 10.9B). Free cells that resemble vertebrate hemocytoblasts float in the enclosed fluid and are formed in a center at the junction of the two intestinal crura.

Nervous System The most primitive flatworm nervous systems consist of a delicate nerve plexus between the epidermis and subepidermal muscle. The plexus is concentrated anteriorly near the statocyst and is thinner posteriorly, where vaguely defined longitudinal trunks can be seen. Except for the concentration at the statocyst and in the longitudinal trunks, it resembles the nervous system of cnidarians.

The main nervous layer in more complex flatworms is the submuscular plexus beneath the subepidermal muscle, although a delicate subepidermal plexus may also be present. The brain is a typically bilobed ganglionic mass, sometimes associated with a statocyst, but more often associated with a pair of ventral longitudinal nerve trunks. The plexus tends to be concentrated to form serially arranged ring commissures between the longitudinal trunks. More highly specialized flatworms tend to have a simplified system with fewer longitudinal trunks.

Most acoelomorph turbellarians have a submuscular nerve plexus, consisting of a fine and coarse net. Five pairs of longitudinal strands can be recognized, the dorsal, dorsolateral, marginal, ventrolateral, and ventral trunks (Fig. 10.7E). The plexus connects the trunks, and a brain surrounds the statocyst. The marginal trunk is sometimes missing, leaving a total of eight longitudinal trunks and creating a superficial resemblance to the eight strands associated with ctenophore comb rows.

The bilobed polyclad brain is enclosed in a capsule (Keenan et al., 1981), and bilateral but somewhat radially arranged nerve trunks arise from it (Figs. 10.7G and 10.8C). The two nerve trunks that parallel the pharynx may be homologous with the ventral nerve trunks of other turbellarians. All nerve trunks lie in the submuscular plexus. A delicate subepidermal plexus also is present.

In all other turbellarians, the nervous system is strongly bilateral and has little resemblance to nervous systems of radiate phyla. Primitive catenulids have four pairs of longitudinal nerve trunks (dorsal, marginal, ventrolateral, and ventral) connected by commissures tending to form rings around the body (Fig. 10.7A,F). Connections between the main tracts are located in the general submuscular plexus. Derived turbellarians, such as the rhabditophorans, show evidence of retrogressive changes. Some have the four pairs of longitudinal trunks found in primitive catenulids, but most have only two ventral nerve trunks (Fig. 10.7C,D), with few lateral branches or commissures.

The most complex brain is found in triclads. In *Dugesia*, the brain is formed by thickenings of the ventral trunks and a connecting commissure (see Fig. 10.1B). In some other triclads, the brain is composed of dorsal ganglia arising from the surface of the ventral trunks (Fig. 10.8A). Land planarians have developed along an entirely different line, perhaps in relation to the elongated body form. A nerve plate connects the two ventral margins of the submuscular plexus, and the longitudinal nerve trunks lie in the nerve plate (Fig. 10.7D). The nerve plate probably assumes the brain functions.

Unipolar, bipolar, and multiplar neurons occur in flatworms. The neuron arrangement is diffuse in primitive types but becomes organized in more complex forms. For example, the marginal nerve of the horseshoe crab leech, *Bdelloura*, consists both of a series of cells with tips ramifying into the epidermis and of fibers extending into the lateral nerves (Fig. 10.8B), evidently functioning as sensory neurons. Highly branched unipolar neurons in the nerve trunks innervate parenchymal muscles and serve as motor neurons. Other neurons lie entirely within the nerve trunks or commissures, and act as associational fibers. Although there is no evidence of the cytological differentiation of axons, functional differentiation of neurons into sensory, motor, and associational types is an important stage in nervous evolution.

The plan of the parasitic nervous system is similar to that of turbellarians. However, there is no subepidermal plexus, as the epidermis is missing. The surface layer is lost during development, and thereafter the protective tegument rests directly on the muscles and parenchyme (see Fig. 10.5). Unlike a cuticle, the tegument contains mitochondria. However, details of the formation of the tegument are not yet known. In-sunk gland cells and some subtegmental cells may be of epidermal origin. The submuscular plexus consists of a mesh in which three pairs of nerve trunks (dorsal, lateral, and ventral) are found. The trunks are connected by commissures that tend to form rings (Fig. 10.9A). The parasitic Monogenea lack dorsal trunks and may lack lateral trunks, thus resembling some of the simpler, less derived rhabditophoran turbellarians. Two dorsal cerebral ganglia lie between the oral sucker and the pharynx, and are connected by a broad, ganglionic commissure. Attachment organs, pharynx, and parts of the reproductive system are innervated by peripheral nerves.

Changes in the cestode musculature have apparently balanced the reduction of sense organs, for the nervous system is about as complex as in free-living flatworms. Two main longitudinal nerve trunks parallel the excretory ducts, and accessory longitudinal trunks are usually present (Fig. 10.10B). Each proglottid contains at least one ring commissure, whereas the brain lies between the two lateral trunks and is surrounded by a ring commissure, with large

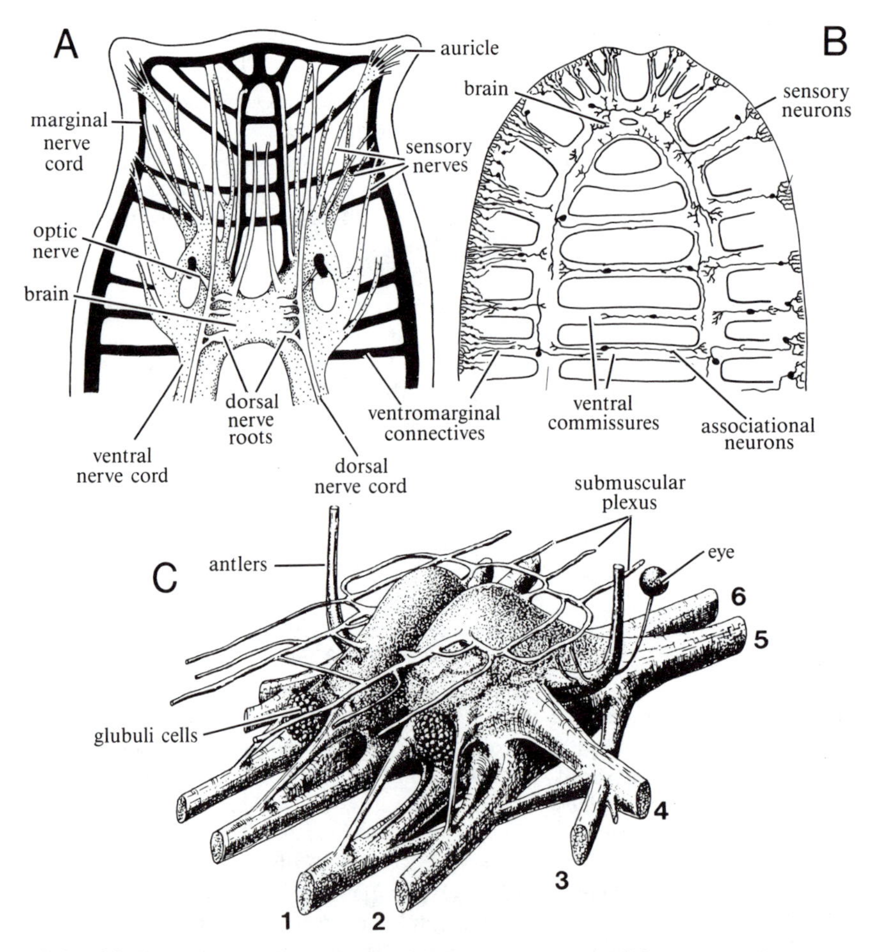

FIGURE 10.8. Turbellarian central nervous system. **A.** Brain and anterior nerves of *Crenobia alpina.* Dorsal nerves shown in *white,* ventral in *black.* Compare with *Dugesia* brain region (Fig. 10.1), *Crenobia,* unlike *Dugesia,* has full complement of longitudinal nerve trunks. **B.** Anterior end of *Bdelloura candida,* showing position of some associational and sensory neurons. **C.** Brain of polyclad, *Notoplana;* nerves 1–6 supply submuscular plexus. (**A** after Micoletzky; **B** after Hanström; **C** after Keenan et al.)

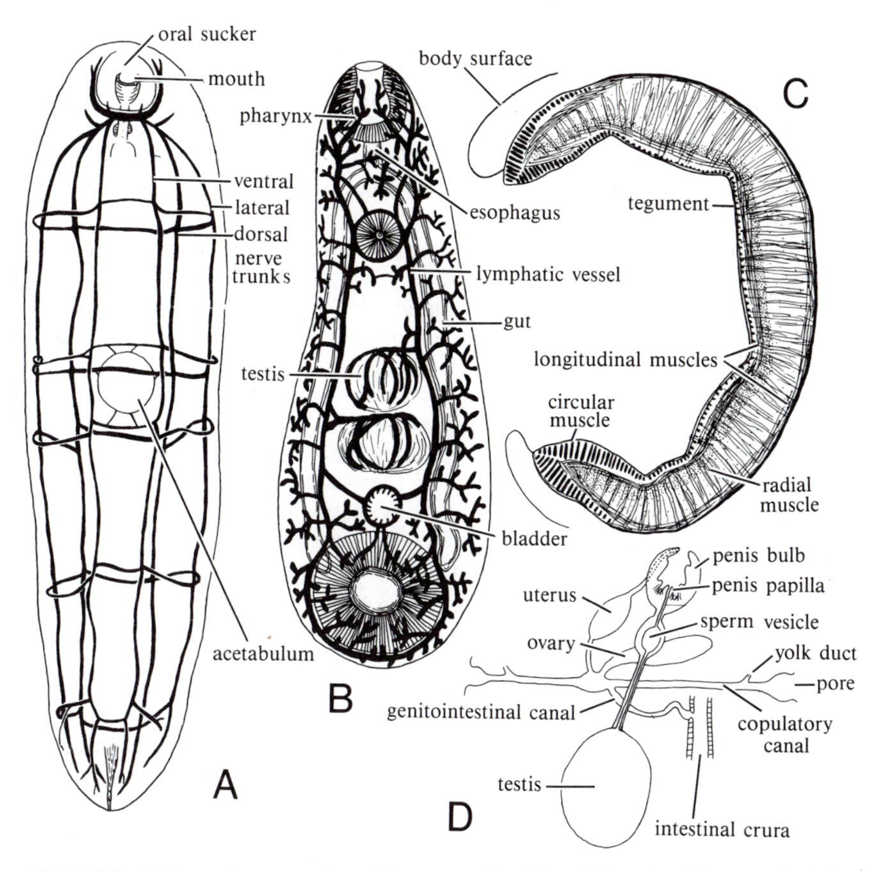

FIGURE 10.9. **A**. Scheme of nervous system of digenean parasites. **B**. Lymphatic system of digenean, *Cotylophoron cotylophorum*. **C**. Section through acetabulum of digenean. **D**. Scheme of reproductive system of monogenetic cercomeromorphan. (**A** after Looss; **B** after Wiley; **C** after Näsmark; **D** after Stunkard.)

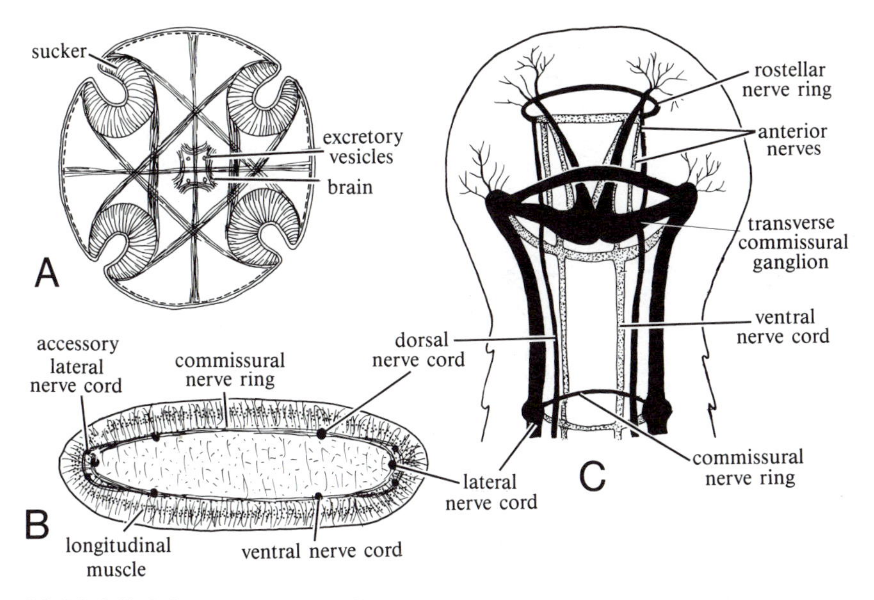

FIGURE 10.10. **A**. Section through scolex of taenioid tapeworm, showing arrangement of muscle fibers. **B**. Scheme of section through proglottid of taenioid tapeworm, showing distribution of nerve cords. **C**. Central nervous system of taenioid, *Moniezia*. (**A** after Fuhrmann; **B** after Becker; **C** after Tower.)

ganglia located at its junctions with the lateral trunks, and smaller ganglia at junctions with the other longitudinal trunks (Fig. 10.10C). A rostellar ring commissure in front of the brain connects the anterior projections of the longitudinal trunks.

Reproductive System Flatworms are almost exclusively hermaphrodites (Fig. 9D). The male and female systems are distinct, but may join terminally in a common chamber, the genital antrum, and may open through a common gonopore.

The acoelomorph **testis** is ill-defined. Free parenchymal cells aggregate in a region characteristic of the species and enter into spermatogenesis. In other turbellarians the testis is enclosed in a membrane and is sharply delimited from the parenchyme. Flatworms may have one, a pair, or many testes, depending on the species.

Sperm ducts are incomplete in some acoelomorphs, but are the most stable elements of the male system. Where many testes occur, each is served by a tiny **sperm ductule**; these unite to form a lateral sperm duct on each side (see Fig. 10.1A). Unless there is but one testis, the two lateral sperm ducts unite to form a common sperm duct, usually near the copulatory complex. The terminal part of the sperm duct is typically modified for sperm storage and passes through the copulatory complex as a muscular **ejaculatory duct** into which the prostatic apparatus opens, if one is present. Before mating takes place, sperm accumulate in paired or single muscular dilations of the sperm ducts or medial and unpaired structures of the ejaculatory duct (**seminal vesicles**).

The ejaculatory duct is the modified end of the common sperm duct, ending in a chamber, the **male antrum**, and serving as the center about which the copulatory complex is assembled. In addition to a seminal vesicle, the copulatory complex usually includes prostatic glands and an intromittent organ for mating. The most common intromittent organ is a **penis**, a muscular part protruded through the gonopore and inserted into the copulatory complex of the mate. Some have a **cirrus**, a muscular part of the tube that is everted during mating. Some turbellarians have no intromittent organ.

The prostatic apparatus consists of the **prostate glands** and whatever delivery system for the prostatic secretion the worm may have. In the simplest systems, the glands open directly into the ejaculatory duct or, more rarely, the sperm ducts. In most cases these glands empty into muscular **prostatic vesicles**, which accumulate the secretion and force it out when needed. The prostatic secretion contrib-

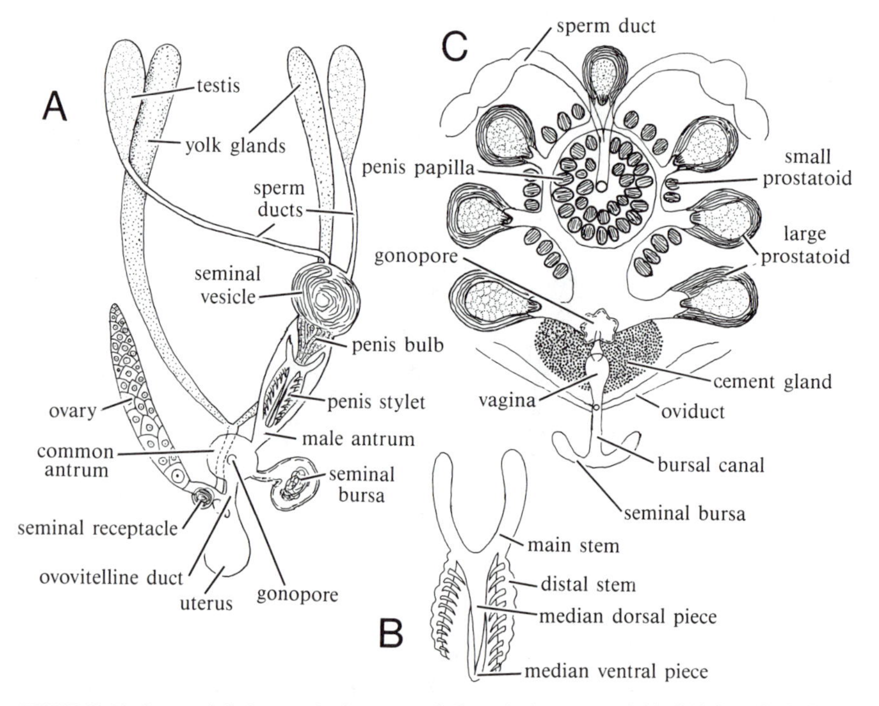

FIGURE 10.11. Some turbellarian reproductive organs. **A**. Reproductive system of dalyelloid, *Dalyellia*. **B**. Penis stylet of *Dalyellia rossi*. **C**. Copulatory complex of polyclad, *Coronadena*, with replicated male parts. (**A**,**B** after von Graff; **C** after Hyman.)

utes to the vehicle for sperm suspension and appears to be important for normal viability and activity of the sperm. Muscular contractions of the prostatic vesicle force secretions out during mating. Some turbellarians have **prostatoids**, pyriform, often cuticularized structures sometimes present in large numbers (Fig. 10.11C) whose exact function is unknown.

The penis lies beyond the prostatic apparatus. It is sometimes divided into a basal **penis bulb** and a **penis papilla** that projects into the male antrum (see Fig. 10.1D). The penis papilla may be muscular or often a specialized cuticular part consisting of hooks, bars, spines, or a cuticular tube known as the **penis stylet** (Fig. 10.11A,B). Sometimes the stylet completely replaces the muscular part of the penis papilla. Stylet form, exhibiting fantastic diversity, is important in identification of species.

Monogenean parasites have from one to many testes. Seminal vesicles sometimes occur near the copulatory complex. There is a muscular or armed penis. Aspidobothrea also have a turbellarian-like system, with one or two testes but with a cirrus as an intromittent organ. It should be noted again that parasitologists normally term both dilations of the sperm duct and the ejaculatory duct seminal vesicles. Their conventions are followed here.

Two compact, but sometimes greatly branched, testes are almost universal among digenean parasites. The two sperm ducts unite as a common sperm duct and enter the copulatory complex. The copulatory complex may include only an unspecialized ejaculatory duct, but in most Digenea it is contained in a large, muscular cirrus sac, and includes a seminal vesicle, prostatic glands, and sometimes a prostatic vesicle. The muscular cirrus is the terminal, often armed section of the ejaculatory duct that is everted during mating (see Fig. 10.12). A common gonopore is found. Blood flukes are exceptional in having separate sexes. The slender female is often found in a ventral groove on the body of the much larger male (see Fig. 10.24B).

Details of the tapeworm male system are essentially as in other flatworms. Detailed differences are useful in taxonomy, but contribute little to a further understanding of the reproductive process. Cestode testes are numerous and are strewn at random in the proglottid, or lie in lateral fields. Arrangement of the testes, and other reproductive organs, tends to be characteristic of cestode orders, and an examination of the details in representatives shown in Figures 10.26 and 10.27 will be helpful. Sperm ductules unite to form a sperm duct, usually much coiled, that enters the copulatory complex. Cestodes have an armed or unarmed cirrus, enclosed in a muscular cirrus sac. Sometimes a muscular dilation of the ejaculatory duct or sperm duct forms a propulsion vesicle. Prostatic glands usually open into the sperm duct or ejaculatory duct, but prostatic vesicles are rare.

In the female system, acoelomorph turbellarian **ovaries**, like the testes, are ill-defined, do not have a covering membrane, and consist of clusters of free parenchymal cells undergoing oögenesis. In all other turbellarians, the ovaries have definite boundary membranes. Two kinds of ova and two kinds of ovaries must be distinguished in turbellarians. In the more primitive forms, ovaries produce **endolecithal ova**, with yolk reserves stored in the egg cytoplasm. In more advanced forms, called neoophorans, **ectolecithal ova** are free of yolk but are encapsulated with special yolk cells containing the needed food reserves. Some of these have a **germovitellarium** that produces both ova and yolk cells, whereas others have a separate ovary and yolk gland (Fig. 10.13). The primitive arrangement seems to be an ovary with four arms that meet medially at the **female an-**

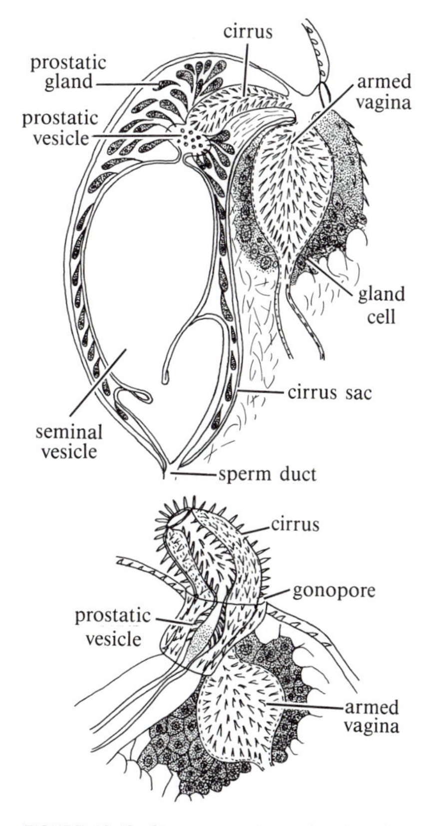

FIGURE 10.12. Cirrus sac and armed vagina of digenetic trematode, with cirrus withdrawn **A** and everted **B**. (After Looss.)

trum and **gonopore**. Each branch probably produced endolecithal ova (Fig. 10.13A). No living rhabditophorans have such an ovary, but *Gnosoñesima*, a primitive form, has a germovitellarium with four arms, each forming ova enclosed in a follicle of yolk cells. This may reflect the way the primitive ovary became a germovitellarium. If so, most turbellarians have secondarily acquired a different method of yolk cell formation. Part of the germovitellarium is converted into a **yolk gland**, and part into a region forming ectolecithal ova (Fig. 10.13B). Generally there are two yolk-producing arms and one ova-producing arm, with the remaining arm lost (Fig. 10.13C), although other patterns occur. Evolutionary adaptation has transformed a primitive ovary, producing endolecithal ova, through a germovitellarium, producing ectolecithal ova and yolk cells, to definitive secondary ovaries, producing ectolecithal ova, and separate yolk glands.

Except for acoelomorph turbellarians, which lack oviducts, delivery of ova and yolk cells is effected by a system of oviducts and yolk ducts, both derived from the oviduct system of the primitive ovary. Both ovaries and yolk glands are diverse in structure and position; either may occur as many discrete follicles, served by **yolk ductules** or **oviductules** that unite to form main yolk ducts or oviducts. Yolk ducts or yolk ductules eventually join the oviduct; beyond this point the oviduct should be termed an **ovovitelline duct** (see Fig. 10.1A). In more complex systems,

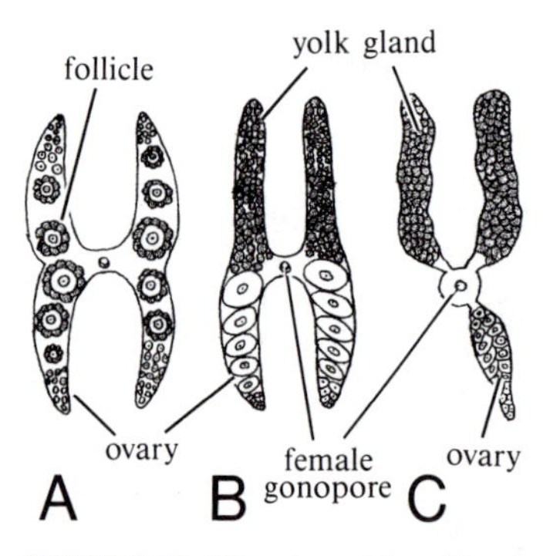

FIGURE 10.13. Stages in evolution of female system. Ovary develops from indistinct clusters of cells undergoig oogenesis found in most primitive turbellarians, originally tetrapartite, as in *Gnosonesima* **A**. Further development of germovitellarium has led to separation of function of yolk elaboration from gamete formation. In *Gnosonesima,* nurse cell follicles surround developing ova. More commonly, germovitellarium divides into two yolk-forming regions and two gamete-producing regions **B**. As regions become more discrete, often with disappearance of one ovum-producing arm, system with separate yolk glands and one or two ovaries develop **C**. (After Hyman.)

the ovovitelline duct system ends in a female antrum that may be preceded by a copulatory complex.

The trematode female system is more diverse than the male system, but is also turbellarian-like. As in turbellarians, detailed differences are important in taxonomy. Only essentials can be mentioned here; a text on parasitology should be consulted for details.

Monogeneans usually have one or two copulatory canals with independent copulatory pores (see Fig. 10.9D). The inner end of the copulatory canal opens into the oviduct or yolk duct system. There is no bursa, but a part of the copulatory canal may be modified as a seminal receptacle. Beyond the union of oviduct and yolk duct systems, the ovovitelline duct is slightly dilated as an **oötype**, surrounded by **Mehlis's gland (shell gland)**. Here, the ovum is encapsulated. The ovovitelline duct continues as a very short uterus; ova are released singly or a few at a time. An interesting feature is the **genitointestinal canal**, characteristic of one of the two monogenetic orders. When present, it connects the vagina and the right intestinal caecum, and is apparently identical with the genitointestinal canal of some turbellarians.

The basic pattern of the digenean female system is simple enough, but is subject to myriad modifications. The single ovary opens into an oviduct, down which ova pass. Scattered follicular yolk glands empty into yolk ductules, which unite to form the main yolk ducts on each side. The yolk ducts unite near the oviduct, and may expand as a yolk reservoir at the base of the common yolk duct. Ova are fertilized in the upper part of the ovovitelline duct, formed by union of common yolk duct and oviduct.

Laurer's duct (see Fig. 10.18A) connects the oviduct of some flukes with a dorsal pore; in others it ends blindly or is missing. It is apparently a remnant of a copulatory canal and is usually connected with the seminal receptacle, when present. However, the terminal end of the ovovitelline duct serves as the functional vagina and is often armed with hooks or has a special musculature (Fig. 10.12).

In cestodes, a single bilobed ovary, sometimes dorsoventrally forked to form four arms, lies in each proglottid. Yolk glands are almost universally follicular, but form a compact organ in *Taenia*. A yolk duct system empties into a single yolk duct and joins the oviduct shortly after its emergence from the ovary (see Fig. 10.26B). Except in one group, the cyclophyllideans, a muscular area pulls ova into the oviduct by strong peristaltic contractions, and nearby a **vagina** branches off from the oviduct. The vagina runs to the gonopores and often contains a dilated **seminal receptacle**, where sperm is stored after mating. Beyond the entrance of the vagina, the ovovitelline duct continues as a uterus, small in mature proglottids, but becoming coiled or branched in gravid proglottids filled with ova. Ova either are retained until the proglottid is shed, or exit by a uterine pore. A Mehlis's gland encloses the oviduct where the eggs are encapsulated, as in flukes.

The cestode and trematode female systems would correspond very well if the cestode vagina were the uterus and if the uterus were Laurer's canal. There must have been a shifting of external pores, or else the part known as Laurer's canal in trematodes must correspond to the cestode uterus. Both views have been held.

Biology

Locomotion Movement in turbellarians is largely achieved by the beating of cilia against a kind of carpet of mucus produced by the glands and rhabdites in the epidermis (Martin, 1978). Small turbellarians, half a millimeter or so long, flash across a microscope field in the laboratory like large ciliates, leaving an entirely different impression of locomotor capacities than large planarian worms like *Dugesia*. Muscles contribute little to the movement of small species. However, moving a large animal with cilia is rather like rowing an ocean liner. Wriggling movements become more important, and such large forms as the land planarians look superficially like earthworms as they move. Yet ciliary movement is never fully abandoned by turbellarians, and muscles are never fully exploited. The most effective muscular locomotor organ in the phylum is the cercarial tail that occurs in larval trematodes who are devoid of epidermis and external cilia (discussed later).

In the parasitic flatworms, with the development of a tegument and the loss of epidermis and external cilia, movement is achieved entirely by muscles. In fact, muscles reach their highest development in Platyhelminthes among the parasitic members. Parenchymal muscles are usually poorly developed, but are specialized for operating the hooks and anchors used for attachment. Suckers are formed of specialized subtegmental muscles, with modifications of the epidermal layers of the body wall (see Fig. 10.10A).

The tail of larval cercariae, the most effective muscular locomotor organ in the phylum, is a secondary adaptation for movement in a larval stage noted for its dispersal abilities. A layer of smooth muscle lies just beneath the cuticle, and beneath those lies a diagonal layer of striated muscle. A layer with dense concentrations of mitochondria surrounds the innermost parenchyme, presumably serving the high oxidative requirements of the tail muscles.

In cestodes, there are a thin layer of circular and one or two layers of longitudinal muscles present. The parenchymal muscles form one or two heavy layers of longitudinal fibers and dorsal or ventral sheets of transverse fibers. The transverse and longitudinal fibers may form a muscular septum enclosing the inner part of the parenchymal tissue. A few dorsoventral muscles are usually present. Parenchymal muscles are especially prominent in the scolex, where they are arranged to manipulate the scolex attachment organs (see Fig. 10.10A).

Food Capture and Digestion Many turbellarians contain zoochlorellae but, although some can subsist entirely by digesting their symbiotic partners if necessary, feeding is a major effort in most turbellarians. A few take in some plant material (Kozloff, 1972), but the majority are strict carnivores. Feeding habits have undoubtedly influenced sense organ development. Efficient chemoreceptors are sensitive to meat juices, and rheoreceptors enable some to detect nearby prey by water disturbances.

Food capture is not a simple matter. Ordinarily, special organs are useful to grasp the prey. Whorls of tentacles are suitable for sessile or floating animals, but not for actively moving forms. The flatworm capturing organs are built of gland cells, sometimes with the aid of muscle fibers. Two kinds of glands are available: **cyanophilous glands**, which secrete slime, and **eosinophilous glands**, which secrete adhesive substances. **Frontal glands**, formed of slime cells and sensory cells, are found in some turbellarians, such as acoelomorphs and some neoophorans (see Figs. 10.20A,B and 10.22C). **Glanduloepidermal adhesive organs** are formed only of adhesive glands and modified epidermis without muscles, the marginal adhesive zone of seriates being an organ of this kind. Caudal adhesive areas are often found that consist primarily of eosinophilous glands that project individually or form **adhesive papillae**. These parts, not used directly in food handling, enable the animal to cling to the substrate, often an important act during food capture, but more so in movement. Glandulomuscular adhesive organs include special musculature in addition to adhesive glands. The muscles are often arranged in a suckerlike pattern (see Fig. 10.3A), but true suckers, sharply set off from the parenchyme, are not common in turbellarians. In a few turbellarians, the glandulomuscular organ is developed into a grasping proboscis, as in *Kenkia* (see Fig. 10.22F).

Prey is grasped by the pharynx and adhesive organs. The worm clings to the substrate with whatever equipment it may have, and secretes mucus and an adhesive substance to entangle, sometimes paralyze, and immobilize the prey. The prey tries to pull away. Sometimes even the penis armature is used to injure the prey. The victim is usually swallowed whole, but species with a tubular pharynx sometimes thrust the pharynx into the tissue and secrete proteolytic enzymes, softening the tissue for ingestion; and some species with a bulbous pharynx also ingest small tissue fragments. *Stylochus* avoids the struggle by nipping off small bits of oyster tissue as it moves about within the shell.

In acoelomorph turbellarians without a pharynx, food goes directly to the solid mass of endodermal cells or syncytium where it is phagocytized. In *Convoluta*, the endodermal syncytium is thrust out through the mouth to permit phagocytosis. Generally, however, an embryonic stomodeum produces a pharynx, lined with cells derived from ectoderm, and a pharyngeal cavity between the mouth and the pharynx. The round or slit-shaped mouth appears at the site of the blastopore. It may occur anywhere on the midventral surface, but is usually located in the middle three-quarters of the body. Modified subepidermal muscles open and close the mouth.

Animals with a highly muscular, protrusible pharynx often use it to fragment the food mechanically or to suck in particles of soft tissues. *Dugesia* that are fed beef liver after a period of starvation ingest small bits of tissue rapidly. About a hour is needed to fill the intestinal diverticula. Food vacuole formation begins promptly, with little evidence of preliminary digestion (Jennings, in Riser and Morse, 1974). Gastrodermal cells imbibe water and swell to become syncytial. Food vacuoles form during the first eight hours and disappear slowly during the next five days, presumably as the result of proteolytic digestion. A different pattern is seen in *Leptoplana*, a polyclad turbellarian with a ruffled, plicate pharynx. The pharynx clings tightly to prey while secretions induce preliminary digestion outside the body. Partially digested tissues are sucked in and appear to be completely reduced in the intestine. Parenchymal muscles force the intestinal contents out to the tips of the diverticula and back to the main intestinal trunk repeatedly, presumably hastening food fragmentation and absorption by the gut epithelium.

Most noncestode parasites ingest detached body cells, blood cells, mucus, or tissue exudates. Host tissues are injured by the oral sucker and pharynx, sometimes with the aid of tegmental spines. Some species secrete enzymes that soften host tissues and prepare them for digestion. Food usually passes through an esophagus on its way to the main gut. Probably some food is absorbed through the body surface, but digestion is predominantly extracellular, and food is stored primarily as glycogen. Some fat is also stored.

All cestodes are internal parasites and absorb predigested food through the body surface. It is possible that where the head or scolex works itself deeply into the intestinal mucosa, it may absorb material not available to the segments. Cestodes resemble trematodes in that they are covered by a tegument. The tegument is covered by **microtrichs**, presumably important in absorption, and contains tiny vesicles, presumably pinocytic in character and thought to be important in absorption. Mitochondria also occur in the tegument.

Excretion It is reasonable to suppose that flatworms, like other aquatic invertebrates, are ammonotelic. Proteins can be stored as fats, indicating that deamination takes place, ammonia formation being a probable result.

The protonephridium has traditionally been thought of as a primitive precursor to more complex structures of excretion seen in higher invertebrates (Wilson and Webster, 1974). Recent work indicates, however, that rather than being primitive, protonephridia are a unique solution to the filtration of extracellular fluid in a body that has no coelom or blood-vascular system (Ruppert and Smith, 1985). In this case, the solenocytes must actively transport selective materials out of the body fluids into the protonephridial weir.

Respiration The mass of mesodermal tissue in which the organs of acoelomates are embedded is not an unmixed blessing, for it interferes with the internal transport of respiratory gases, food, and waste products. Spaces or body cavities that facilitate internal transport evolved in pseudocoelomate and coelomate animals, but the acoelomates are hampered in this regard, and because of this, flatworms suffer severe physiological handicaps. The mesodermal mesenchyme, or parenchyme, contains more cells than the mesoglea of lower phyla and has higher metabolic requirements. Planarians have a level of oxygen consumption about ten times that of the average cnidarian on a per-gram-per-hour basis, yet flatworms are scarcely better equipped than the radiate phyla to provide oxygen to tissues or to discharge carbon dioxide. Turbellarian energy release appears to be predominantly aerobic. It is largely cyanide-sensitive, which indicates that after dehydrogenation, hydrogen is transferred by a cytochrome–oxidase system to the final hydrogen acceptor, oxygen. A small cyanide-resistant fraction is probably maintained by iron-free oxidases.

No special organs for respiratory exchange or transport occur. The epidermal cilia ventilate the body surface and undoubtedly improve the rate of absorption. Hemoglobin has been reported from a few forms, but is not widely distributed. Available evidence, on the whole, supports the idea that for lack of a more effective system of respiratory exchange and transport, the respiratory requirements of internal cells are a critical limiting factor, favoring small size and a flattened shape.

Respiration in the parasitic platyhelminths is anaerobic, involving glycogen degradation and the formation of carbon dioxide and various organic acids. Different species release different combinations of acids, which indicate diversity among the array of respiratory enzymes. Available oxygen can be taken up and used. Cytochrome has been identified, and presumably a typical cytochrome–oxidase system exists. Under natural conditions, however, oxygen is rarely available in any appreciable amounts to the adults.

Circulation The mesodermal parenchyme around the internal organs is characteristic of all acoelomates. Parenchymal tissue fluid is typically circulated as muscular movements squeeze and distort the parenchyme. This aids in internal transport. Protoplasmic continuity of the parenchymal syncytium helps to distribute substances held in the cytoplasm. The systematic transport of materials, however, is generally poorly provided for, and the inefficient internal transport remains a major physiological weakness of the platyhelminth body plan.

Sense Organs Flatworms are by no means "brilliant," but experimental studies show that flatworm behavior can be modified on the basis of past experience. The learning process is laborious, requiring long conditioning, and retention is brief. *Dugesia* can be taught to make avoiding movements under conditions not ordinarily evoking them, but cannot be taught to turn toward conditions it normally avoids. Apparently flatworms have no ability, or a very limited one, to inhibit normal responses. A polyclad can be "taught" to remain more quiescent when exposed to light by touching its anterior margin whenever it begins to move, but even the most gifted worms never learn to remain wholly quiescent. Not too surprisingly, removal of the brain prevents the learning response, but does indicate its importance in associative learning.

Flatworm organization is not suitable for a dramatic development of sense organs, but active locomotion and an anteroposterior axis have favored some cephalization, and the widespread occurrence of light-sensitive organs indicates their great importance in flatworms. Receptors sensitive to chemicals, currents, and touch are somewhat more highly organized than in radiate phyla.

Photoreception. A simple pigment-spot ocellus is the primitive light-sensitive organ in flatworms, but occurs only in a few forms. Most flatworm eyes are pigment-cup ocelli. Animals may have inverse eyes (Fig. 10.2E), with the retina turned away from the light, or converse eyes (Fig. 10.2G), with the retina turned toward the light. The inverse pigment cup is most common, like the *Dugesia* eye, with retinal cells entering the open side of the pigment cup, and ending in a light-sensitive, striated border at the bottom of the cup. Land planarians have complex eyes of the converse type, with the retinal cells passing through the pigment cup wall. These eyes have an epidermal cornea and might be capable of forming a vague image.

Flatworm photoreceptors are neurosensory cells. They contain a light-sensitive striated border and a process that extends to the nervous system of a ganglion. A tendency to increase the area of the light-sensitive striated part of the cell and to separate it from the nucleus can be seen in flatworms (Fig. 10.2F). When the nucleus lies outside the light-sensitive region, the striated border tends to be differentiated as a retinal club, as in the eye of land planarians.

Many turbellarians react negatively to light. The pigment cup provides unilateral shading and so confers a direction sense. However, in *Dugesia* the retinal cells must be parallel to light rays to be stimulated, thus providing a mechanism for precise orientation to light. If all eyespots are removed, planarians select a dark background and eventually find the least highly illuminated region available. The choice of backgrounds depends on random movement rather than oriented response, but evidently a dermotropic light sense is also operating.

Chemoreception. Chemoreceptors are important in food recognition. Most chemoreceptors are ciliated pits or grooves containing glandular and chemoreceptive cells. Chemoreceptors of this type are common among flatworms; the ciliated auricular grooves at the base of planarian auricles and the frontal organs formed of slime cells and chemoreceptive cells are examples. Cilia circulate water over the chemoreceptive cells, presumably helping to give a directional sense. Each chemoreceptor has one or several bristles. Near the brain, chemoreceptors are elongated to form a sensory nerve (Fig. 10.14). Marginal chemoreceptors form a sensory epithelium with a ganglion at the base; the sensory nerve arises from the ganglion.

Planaria give highly positive, oriented responses to meat juices and snail blood, and can be made to follow blood in capillary tubes. The animal lifts its head and swings it to face directly toward the stimulus and moves directly toward it. Orientation is dependent on equal stimulation of the auricular sense organs, for if one auricle is removed, the animal turns continuously toward the intact auricle when in the presence of meat juices.

Touch Reception. Tactile cells, or tangoreceptors, are abundant throughout the body surface. They are somewhat concentrated anteriorly and sometimes cluster together to form definite tactile organs, as in the auricles of planarians. Touch receptors have one or more bristles protruding through or between epidermal cells, and processes extending to the nerve trunks or brain (Fig. 10.14).

Turbellarians usually respond negatively to strong and positively to weak mechanical stimuli. Planarians have a negatively tactile dorsal surface and a positively tactile ventral surface. When turned over, they right themselves by twisting the body. When the ventral tactile cells of the head touch the substrate, the worm moves off. This reaction is not dependent on gravity but on loss of contact at the ventral surface. It occurs, more slowly, in planarian pieces that might lack a brain. It is important, as contact with the substrate is necessary in locomotion.

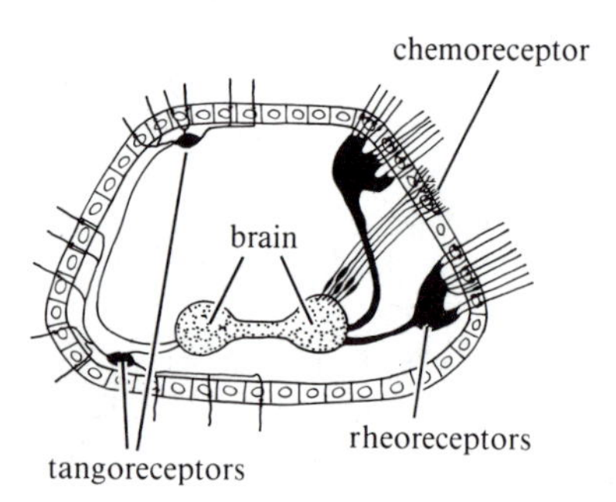

FIGURE 10.14. Schematic section through head of *Mesostoma*, showing relationship of neurosensory cells and brain. Tangoreceptors shown on *left* and rheoreceptors and chemoreceptors on *right*. (After Gelei.)

Rheoreception. Rheoreceptors are large cells, with sensory bristles projecting beyond the epidermis. In *Mesostoma* they occur along the sides of the body (Fig. 10.14).

Rheoreceptors are probably widespread in turbellarians and may aid in food capture. Oriented responses to water currents are common in animals from lotic (running water) environments but usually lacking in animals from still water. Lotic flatworms usually move upstream in weak currents and grasp the substrate in strong currents. Acclimatization in a current develops definite responses in some species that are normally indifferent to currents.

Reactions to Gravity. Many forms have statocysts. When present, a single median statocyst lies near the brain. It has been suggested that the accumulation of nerve tissue near the statocyst may have initiated brain formation in primitive flatworms. Flatworm statocysts resemble cnidarian statocysts. Each contains a large cavity, within which a lithocyte containing one or several statoliths is found.

Orientation to gravity is not important to many flatworms, for they right themselves by tactile responses. Flatworms with statocysts respond positively when oxygen is abundant, collecting at the bottom in a container. As oxygen levels fall, they react negatively, swimming to the surface. Apparently orientation to gravity is used to bring the animal to oxygen when oxygen supplies are limited.

Parasite Sense Organs. The sense organs of parasites are poorly developed. Some Monogenea have eyes, but otherwise eyes occur only in free-swimming larval stages of some Digenea. Eye structure is usually simple; sometimes only a single retinal cell occurs. Some trematode eyes, however, are quite well developed. Touch receptors are especially abundant in the attachment organs. Chemoreceptors are also present, aiding in orientation to hosts or host organs, as well as to other factors in free-living stages.

In the cestodes, many free nerve endings are probably sensory, but little is known of sensory reception in these forms. There are no special sense organs. Certain club-shaped neurosensory endings are known to be tactile.

Reproduction Reproductive habits in the platyhelminths are extremely diverse and fascinating, although certain broad themes can be discerned (Henley, 1974). With few exceptions, flatworms are hermaphrodites. Some turbellarians have no female copulatory organs, and the male stylet is used to inject sperm through the body surface into the parenchyme. Generally, however, it enters into the female copulatory duct. The sperm are of a threadlike, filiform shape, and may or may not possess a tail (aciliary condition).

A great range of problems arise in the female system, especially if the ovum is to be fertilized internally and any sort of protective cover is to be applied to the zygote. Protection of the developing zygote has many advantages; but the appearance of ectolecithal ova, dependent on adjacent yolk cells for food, makes encapsulation of ovum and yolk cells a requirement. Other problems arise unless the ovum is to be fertilized before it is surrounded by a protective layer. To a considerable extent, the developments that have occurred in the female reproductive system are to be understood as responses to the problems of (1) reception of sperm, (2) storage of sperm before fertilization (as the addition of a protective cover is most easily accomplished one at a time), (3) provisions for sperm and ova to meet just before encapsulation, (4) the mechanics of encapsulation of the zygote and the yolk cells, and (5) release of the product plus whatever maternal care may be useful.

A most interesting arrangement is seen among the acoelomorph turbellarians, which lack oviducts. Sperm are received anywhere, often by hypodermic impregnation, which, of course, imposes development of appropriate mechanisms in the male system. Sperm reach the ova without the advantage of formally prepared routes. Acoelomorphs have endolecithal ova; protective covers are not, therefore, a requirement, but if they are to be formed, it must be from epidermal gland cells not directly associated with the reproductive system.

If oviducts are present, the reception of sperm may occur by way of the terminal part of the oviduct, or by way of a separate copulatory or bursal pore (Fig. 10.23F) leading to an organ to retain the sperm, termed the copulatory bursa. Where discharge of zygotes and reception of sperm are carried out separately, three external reproductive openings may occur, a male pore, a female pore, and a copulatory pore, but as a rule, the male and female systems have tended to unite terminally in a common genital antrum, opening through a common gonopore. Most turbellarians use the female gonopore (or the common gonopore) as a copulatory pore, and the copulatory bursa is a diverticulum from the female antrum (Fig. 10.1D). Either one or two external reproductive openings are seen. It is not uncommon for a part of the female system to be adapted as a vagina (Fig. 10.11C), and sperm received in the bursa are sometimes more permanently stored in a dilated part of the oviduct system, termed the seminal receptacle. As newly released ova pass this point, they are fertilized.

In the turbellarians, fertilized ova are usually not retained for a long time; they are usually released one or a few at a time. In some species, a diverticulum from the female antrum, termed the uterus, is present (Fig. 10.11A), but zygotes remain here only temporarily while the capsule is applied. In polyclads, however, a true uterus for temporary storage of ova before encapsulation is found.

Glands associated with the terminal parts of the female system, sometimes with the cooperation of inclusions in the yolk cells, give rise to the capsule wall containing the zygote and the yolk cells. In some species, glands secrete substances used to cement the ova to submerged plants or other objects.

In trematodes, ova are fertilized soon after entering the ovovitelline duct, and pass into the oötype, a small chamber in which the capsule wall begins to form and that is surrounded by the loosely arranged cells of Mehlis's gland. Inclusions in the yolk cells that form the capsule wall are released immediately after the ova and the yolk cells have passed Mehlis's gland, and harden to form the capsule wall. The wall is thickened as additional material gathers on the inner surface. The hardening of the capsule wall is a quinone-tanning process, producing a scleroprotein. Mehlis's gland probably secretes a material that promotes the hardening of the wall and most likely stimulates the coalescence of the shell globules. The encapsulated ova move into a dilated part of the ovovitelline duct, the uterus. Here the capsule hardens and darkens, and ova usually undergo the first part of development.

The uterus is remarkably large in most trematodes, looping back and forth and sometimes making several circuits of the body. Thousands of ova may accumulate in such uteri. The uterus narrows sharply at the end and opens through the common gonopore.

Tapeworms leave no doubt that their major business is the formation of ova. The few unsegmented cestodes have a single hermaphroditic reproductive system, but other cestodes have at least one full set of reproductive organs in each proglottid—and little else.

The high regenerative capacities of freshwater planarians have conferred upon the turbellarians an undeserved reputation for great regenerative abilities and for common asexual reproduction. Some turbellarians can reproduce asexually (Moraczewski, 1977), but the ability to regenerate is relatively low in those groups that do not.

Land planarians reproduce asexually by fragmentation and regeneration. Asexual reproduction occurs in some forms by transverse fission. The site of the fission plane is predetermined; in planarians the regenerative pattern is modified at the site of the future head. Generally, fission precedes any visible differentiation of the daughter animal, but the eyes and pharynx of daughter *Dugesia paramensis* are visible before active fission occurs. This may be related to the fact that fission is controlled by an inhibitor that is photosensitive (Morita and Best, 1984). Macrostomid turbellarians are constricted at the site of future fissions, and considerable differentiation occurs before the daughters separate (Fig. 10.20G). Chains of zooids are sometimes formed, each in a different stage of differentiation.

The ability to regenerate in flatworms is limited, except in asexually reproducing forms where it is very high. The results of experiments repeatedly demonstrate the importance of the anteroposterior axis. Reasonably large pieces with anterior and posterior cut edges regenerate a head and tail but retain the polarity of the animal from which the piece was cut. Very short pieces produce two-headed or other abnormal forms, indicating insufficient polarity to evoke normal development. Anterior pieces regenerate more successfully than posterior pieces. The same principle applies in turbellarians with lesser regenerative capacities. The anterior pieces of a polyclad regenerate perfectly, but only parts anterior to the cerebral ganglia have this ability. Posterior pieces regenerate lost posterior parts, but not lost anterior parts. Factors responsible for axial organization are

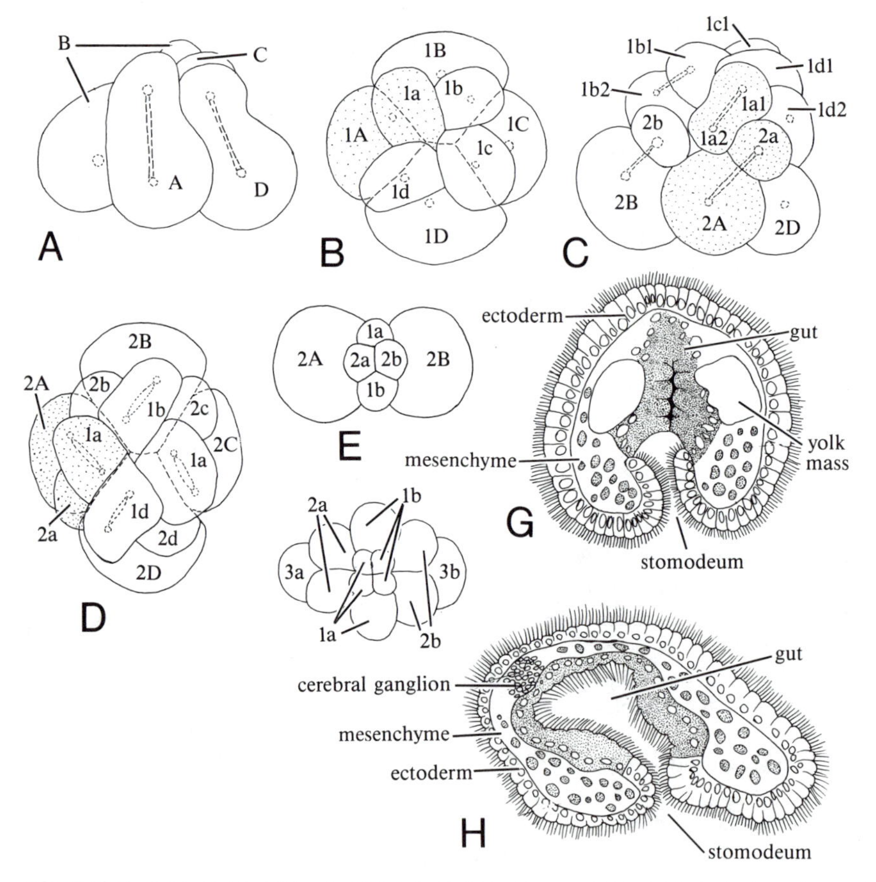

FIGURE 10.15. Endolecithal development. **A,B,C,D**. Spiral determinate cleavage in *Hoploplana,* with cell lineage traced by numbering (see Chapter 4). **E,F**. Spiral cleavage in acoelomorph, *Polychoerus,* resulting from transverse cleavages of two-celled rather than four-celled stages. **G,H**. Later stages in development of *Hoploplana,* showing shift in axis of embryo. (**A,B,C,D,G,H** after Surface; **E,F** after Gardiner.)

not yet clear, or how they influence free parenchymal cells, largely responsible for regeneration.

Development Endolecithal and ectolecithal ova develop in entirely different ways; the presence of external yolk cells leads to aberrant developmental patterns. The development of endolecithal ova is far more important for understanding relationships.

Endolecithal polyclad ova undergo holoblastic, unequal, spiral cleavage and produce an embryo with micromeres and macromeres. The spindles are set obliquely in dividing blastomeres, so the blastomeres lie over the furrows between blastomeres in the tier below (Fig. 10.15B). The fate of each cell is determined at the time of its formation, so cleavage is determinate. (For review of details, refer to Chapter 4.)

The first two divisions produce four equal blastomeres (Fig. 10.15A). The third cleavage division is unequal and oblique, producing a quartet of micromeres (1a, 1b, 1c, 1d) and a quartet of macromeres (1A, 1B, 1C, 1D) (Fig. 10.15B). The micromeres continue to divide, 1a dividing into cells $1a^1$ and $1a^2$, with $1a^1$ being the daughter closer to the animal pole. When $1a^1$ divides, it produces $1a^{11}$ and $1a^{12}$. This system of designation continues and can be expanded indefinitely.

The next division of the macromeres produces a second quartet of micromeres (2a, 2b, 2c, 2d) and a quartet of macromeres (2A, 2B, 2C, 2D) (Fig. 10.15C,D). The macromeres divide again to form micromeres 3a, 3b, 3c, 3d and macromeres 3A, 3B, 3C, 3D. The next division of macromeres forms the last quartet of micromeres, 4a, 4b, 4c, 4d, and macromeres 4A, 4B, 4C, 4D. After this division, macromeres 4A, 4B, 4C, 4D and micromeres 4a, 4b, 4c, which are filled with yolk material, move into the interior and are used as food supplies.

In polyclads, the first quartet of micromeres (1a, 1b, 1c, 1d) forms the anterior and dorsal ectoderm, the pigment cells of the eyes, the brain and nervous system, the frontal glands (when they are present), and in some cases an apical tuft of sensory hairs in the larva. This tuft corresponds to the apical sense organ of some planula larvae, and to a similar apical tuft in the trochophore larvae of annelids and molluscs. The second quartet of micromeres (2a, 2b, 2c, 2d) contributes to the surface ectoderm and forms most of the ectomesoderm of the pharyngeal musculature and associated parenchyme. The third quartet of micromeres (3a, 3b, 3c, 3d) contributes to the surface ectoderm and somewhat to the pharyngeal ectomesoderm. Cells 4a, 4b, 4c and 4A, 4B, 4C, 4D eventually deteriorate as their food content is exhausted. Cell 4d is the mesendoblast, giving rise to all of the endoderm and endomesoderm. It moves into the interior and at the first division forms cells $4d^1$, from which all endoderm is derived, and $4d^2$, which gives rise to the endomesoderm, eventually forming the bulk of the parenchyme, the parenchymal and subepidermal muscle, and the reproductive organs.

Gastrulation is by epiboly, and the blastopore soon closes. The ectoderm that will form the nervous system, the mesoderm from $4d^2$, and the endoderm from $4d^1$ are located to the interior. Yolk material coalesces to form oil droplets. The superficial ectoderm becomes ciliated and invaginates to form a stomodeum at the site of the blastopore, developing into a simple pharynx as it contacts the endodermal cells. Endodermal cells rearrange themselves to form a gut (Fig. 10.15G,H). The embryonic axis shifts, bringing the brain primordium forward and changing the gut position (Fig. 10.15H).

Usually the embryo flattens and develops into a tiny worm. Some pass through a free-swimming larval stage, known as **Müller's larva**, with eight ciliated ectodermal lobes. A few species produce **Gotte's larva**, with four ciliated lobes.

Early development of acoelomorphs is similar to the one described previously, but spiral cleavages begin at the two-cell stage, producing pairs instead of quartets of cells (Fig. 10.15E,F). The stereogastrula has no trace of a gut and develops without the intervention of larval stages.

Ectolecithal development is so aberrant that cleavage stages cannot be homologized with endolecithal cleavage, although traces of spiral cleavage have been reported. No definite germ layers can be distinguished, and organ development involves the differentiation of cell masses into larval or adult parts. Yolk cells do not contribute directly to embryo formation. They are surrounded during development, gradually disappearing as their contents are used.

Neoophoran turbellarians have the ovum divided into a mass of blastomeres that come to lie on the future ventral side of a syncytial mass of yolk cells (Benazzi and Gremigni, 1982). The blastomeres develop into an epithelium that gradually surrounds the yolk mass, and three inner cell masses (Fig. 10.16A). The anterior region becomes the brain, the nervous system, and the pigmented and sensory cells of the eyes. The middle region becomes the parenchyme and musculature of the pharynx. The posterior region forms the reproductive system and posterior end of the body. The blastomeres adjacent to the yolk digest it and eventually become the enteron, at first solid but later becoming hollow.

In seriates, several ova are often in the same capsule. Each divides to form a group of blastomeres. Some of these surround the fluid part of the yolk and the remaining blastomeres, and thus become the covering membrane of the embryonic sphere. Some internal blastomeres migrate to the surface and differentiate into a temporary embryonic pharynx and intestine (Fig. 10.16B). Yolk is drawn into the intestine through the embryonic pharynx, pushing the undifferentiated blastomeres against the surface membrane where they form the embryonic wall from which the body develops. Three masses (Fig. 10.16B), similar in position and destiny to those mentioned earlier, are formed, and the innermost blastomeres become the gastrodermis.

In general, eggs rich in yolk undergo modified development in all phyla. Yet beneath the aberrancy, patterns more or less comparable to the development of small, yolk-poor eggs can be seen. The three cell masses formed in ectolecithal ova correspond, to some extent, to the quartets of micromeres produced in endolecithal development. The anterior mass is somewhat like blastomeres 1a, 1b, 1c, 1d, and the middle mass like cells 2a, 2b, 2c, 2d.

The development of the ectolecithal ova of cestodes and trematodes is more highly modified and adds to the evidence of the profound influence of yolk on development (Bogitsh and Carter, 1982). For example, in Monogenea, a mass of blastomeres forms, and the outer cells become an epithelial membrane. Some internal cells deteriorate, forming an enteric cavity. This begins posteriorly, and for a time the enteron is open through an embryonic "anal pore" that pierces the epithelial covering (Fig. 10.16C). No explanation of the anal pore is available; it eventually closes. Two cell masses form; one develops into the pharynx, whereas the other, always in contact with the pharynx, becomes the brain.

Embryonic development provides evidence of turbellarian relationships. Vitelline cells of digenetic trematode embryos remain discrete and external to the embryo. The first division of the ovum produces one somatic and one propagative cell. Each subsequent division of the propagative cell produces a somatic and a propagative daughter (Fig. 10.17B); therefore, the embryo becomes a mass of somatic blastomeres and one propagative cell. An outer epithelial membrane eventually forms, enclosing the remaining blastomeres and the propagative cell. The propagative cell gives rise to the germ balls and propagative cells of the miracidium, and the remain-

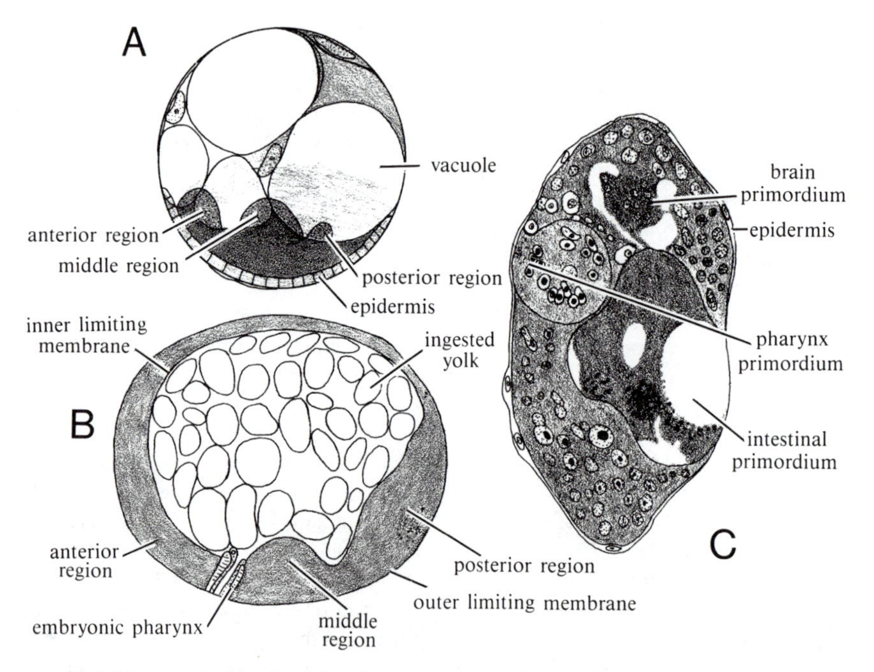

FIGURE 10.16. Ectolecithal development. **A.** Typical neoophoran with ectolecithal ovum, showing three embryonic regions. **B.** Seriate development, with yolk material already ingested through embryonic pharynx, and with three embryonic regions visible. **C.** Embryo of cestode, *Polystoma,* showing organs arising from cell masses reminiscent of those seen in ectolecithal turbellarians. (**A** Bresslau; **B** after Fulinski; **C** after Halkin.)

ing internal cells form masses that differentiate into the miracidial parts (Fig. 10.17A).

Digenetic trematodes always have a complex life cycle, involving several free-swimming and parasitic larval stages. The progression of larval stages is from **miracidium** to **sporocyst** to **redia** to **cercaria** (Fig. 10.18B,C,D,E,F,G). The cercaria becomes a metacercaria by encysting, which in turn develops into the adult in the final host. Life cycles may be lengthened by the addition of a generation of daughter sporocysts or rediae, or shortened by the omission of sporocysts or rediae.

Miracidium. The miracidium, although minute, is highly differentiated (Fig. 10.17A). Four or five tiers of ciliated epidermal cells cover the surface. Nerves that supply eyes, sensory papillae, and muscles arise from the brain. Each sinuous protonephridial tubule

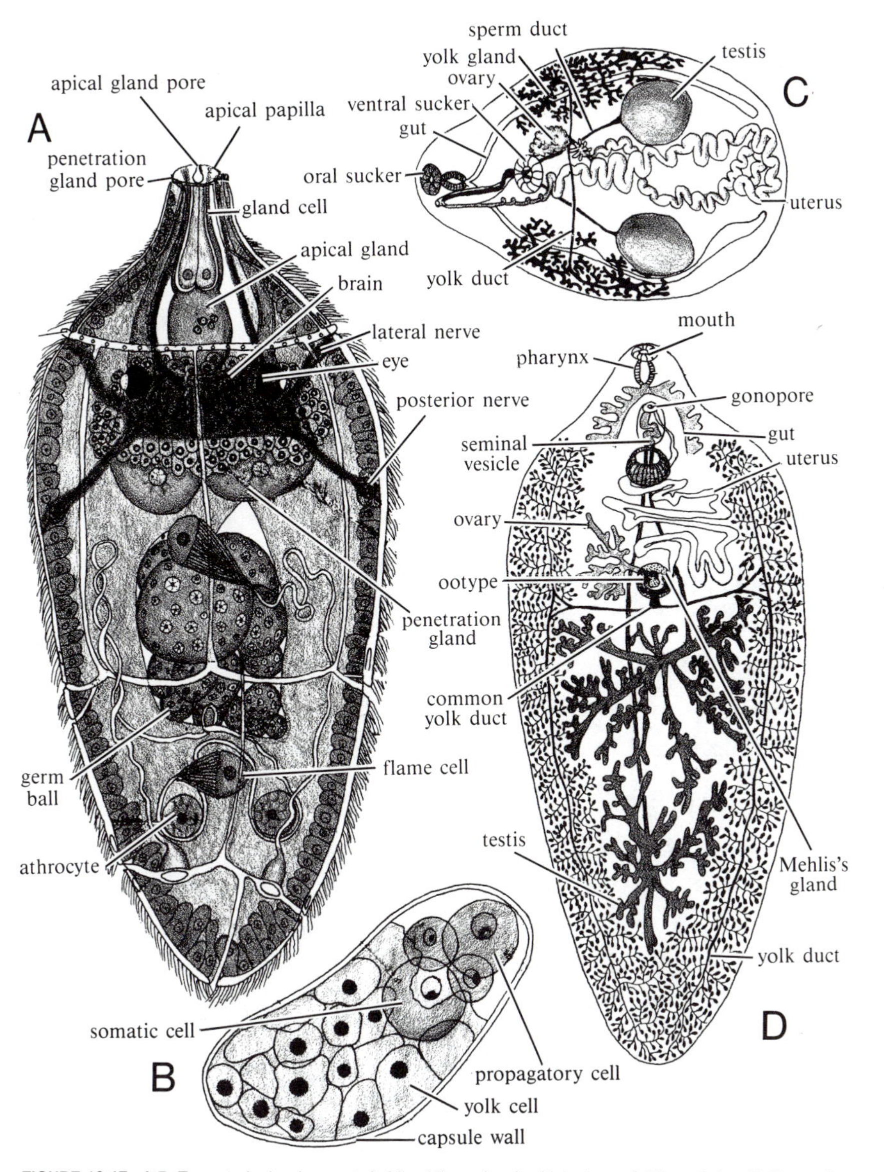

FIGURE 10.17. A,B. Trematode development. **A.** Miracidium, showing high stage of differentiation. **B.** Four-cell stage of *Parorchis.* Four blastomeres lie at upper end of capsule with mass of yolk cells below them. Propagatory cell distinguishable from other blastomeres. **C.** *Prosthogonimus macrorchis,* plagiorchid digenetic trematode. **D.** *Fasciola hepatica,* sheep liver fluke, digenetic trematode of echinostome type, somewhat schematized. Digestive tract shown only at anterior end. (**A** after Lynch; **B** after Rees; **C** after Macy; **D** after Chandler and Read.)

has one or two flame bulbs. An anterior apical papilla contains openings of the large central apical glands and somewhat more lateral penetration glands. Propagative cells and germ balls, derived from the embryonic propagative cell, fill the body space. Miracidia swim continuously and must find a host in a few hours. Some miracidia have a chemosensory mechanism to aid in host recognition, but others seem to find the host accidentally. The miracidium sheds its ciliated epidermal cells as it enters the host, with histolytic secretions of the penetration glands to aid entry. Most fluke species have miracidia that can successfully enter and develop in several different species of snails or clams.

Sporocyst. The miracidium metamorphoses into a sporocyst inside the host. Contact with the host induces metamorphosis, which continues even if the parasites are removed from the host after the initial contact and allowed to live in vitro. Metamorphosis involves extensive simplification, for the sporocyst is little more than a brood sac containing germ balls, propagative cells, and immature, developing larvae (Fig. 10.18D). Some trematode sporocysts give rise to a second generation of sporocysts that then produce cercariae. Some fluke sporocysts produce rediae. Thus, depending on the species, sporocysts may contain daughter sporocysts, rediae, or cercariae.

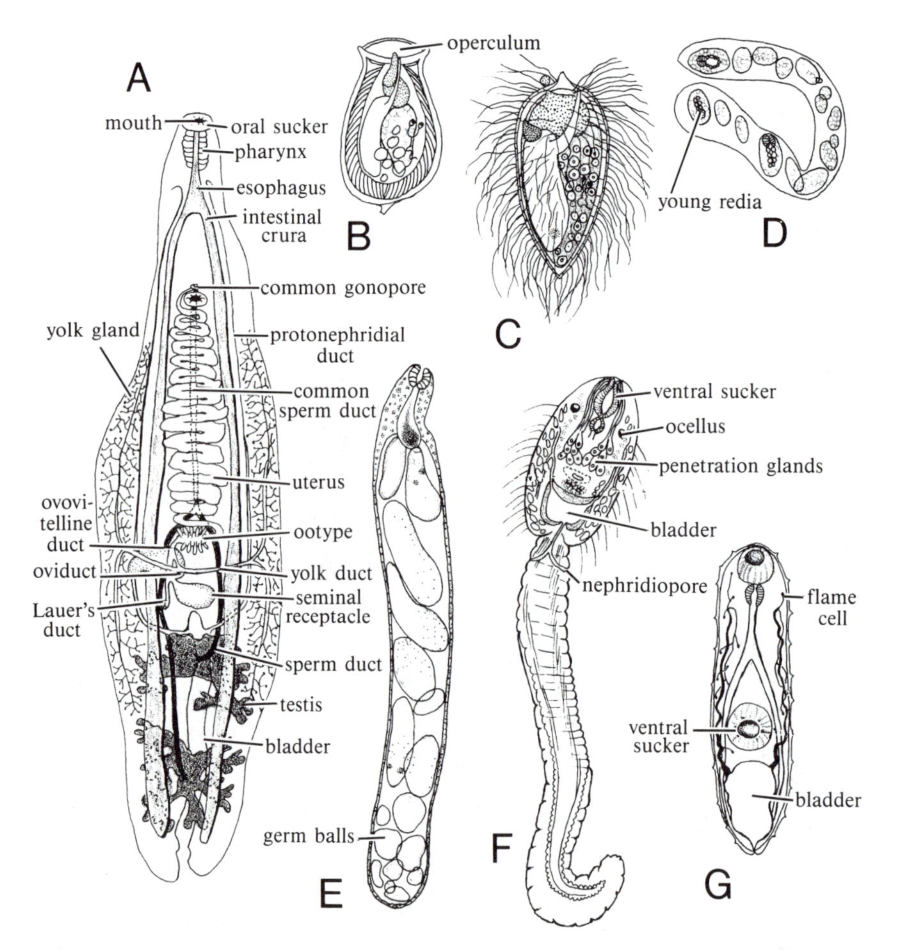

FIGURE 10.18. *Opisthorchis sinensis,* Chinese liver fluke, well-known digenean trematode that parasitizes humans. **A**. Scheme of adult. **B**. Capsule, containing miracidium. **C**. Miracidium, first larval stage, which emerges from capsule when eaten by snail to penetrate into hemocoelic lacunae where it develops into sporocyst. **D**. Mature sporocyst, containing daughter larvae known as rediae. **E**. Redia, containing developing germ balls. **F**. Cercaria that emerges from host snail and swims about for 1 to 2 days in search of acceptable fish to serve as second intermediate host. **G**. Metacercaria as it appears after encystment, which normally occurs when metacercaria is eaten by final host. (**A** after Belding; **C** after Hsu and Khaw, from Belding; **B,C,D,E** after Faust and Khaw; **F** after Komiya and Tajima, from Chandler and Read.)

Redia. The vermiform redia is considerably more differentiated than a sporocyst. The body wall is often pierced by a birth pore, and the body may have lateral or ventral appendages. Primordia of the digestive, nervous, and protonephridial systems are present (Fig. 10.18E). The central cavity is packed with propagative cells, germ balls, and immature larvae. Some rediae produce a second generation of rediae, but generally they produce cercariae.

Cercaria. The cercaria is a more advanced larva, with primordia of most of the adult organs, as well as larval organs that disappear during metamorphosis. The broad flukelike body ends in a muscular tail, variable in form and distinctive in the principal trematode groups. Digestive, nervous, and protonephridial systems are well developed (Fig. 10.18F). Propagative cells assemble to form the primordia of the reproductive organs. Many large, unicellular penetration glands are usually present, and some cercaria have an interior stylet that aids in host penetration. Many cercariae have eyespots. Cercariae emerge from the intermediate host and take up a short, free-swimming existence. Some encyst on vegetation likely to be eaten by herbivorous final hosts. Others penetrate a second intermediate host, encysting in its muscles or other organs. They develop into adults when the second intermediate host is eaten by the carnivorous final host. A few cercariae actively penetrate the skin of the final host, migrating to the appropriate host organ. Schistosomes (blood flukes) follow the last course, taking up final residence in mesenterial blood vessels. The encysted cercariae are known as metacercariae.

Metacercaria. Most flukes have tailless metacercariae, the tail being shed as cercariae enter the final host or as the cyst membrane is laid down. Metacercariae undergo a partial metamorphosis, the larval cercarial organs degenerating during and after encystment. The reproductive organs remain immature, and the final steps of metamorphosis occur in the final host.

Development in cestodes and monogeneans usually begins while the ova are in the uterus. Each ovum is encapsulated with one or several yolk cells. Two membranes form around the embryo during development of pseudophyllids and cyclophyllids. The outer membrane forms first; it is built from one or several blastomeres appearing during early cleavage.

The pseudophyllid outer membrane encloses the yolk, and the inner mass of cells produces a second, delicate inner membrane, apparently the embryonic ectoderm (Fig. 10.19A). The remaining cells develop into an **onchosphere**, or **hexacanth embryo**, an oval, hooked embryo (Fig. 10.19B). The capsules are liberated from the host, and eventually a ciliated larva known as a **coracidium** emerges (Fig. 10.19F) as the outer membrane is left behind and the inner ectodermal membrane bears cilia.

In the cyclophyllid *Taenia*, the ovum forms micromeres, mesomeres, and macromeres during cleavage. Outer vitelline cells form a syncytial nutritive membrane capsule around the embryo (Fig. 10.19C). Later, several mesomeres form an inner vitelline membrane around the remaining blastomeres. This inner membrane hardens into a heavy shell. The innermost cell mass forms the onchosphere. The capsule, and often the outer membrane, is lost (Fig. 10.19D), and the embryo is surrounded only by the shell when it leaves the host. Not all *Taenia* shed the capsule; a few retain it and have a thinner shell.

Pseudophyllid and cyclophyllid life cycles are best known. Pseudophyllids usually have two intermediate hosts, the first a crustacean, and the second a fish or some other vertebrate. Cyclophyllids have a single intermediate host, usually an arthropod, but sometimes a vertebrate or other kind of invertebrate. A few have no intermediate host, the final host being infected directly by the shelled larvae.

The larva emerging from the capsule is always hooked, but Cestodaria larvae (**decacanths**) have ten hooks, and Eucestoda larvae (**hexacanths**) have six hooks. The external ciliated layer of decacanths and some hexacanths is shed after the larva is eaten by a suitable host. The corresponding shell of cyclophyllid capsules is digested when the capsule is eaten.

Procercoid. In both Cestodaria and pseudophyllids, the embryo uses its hooks to work into the host tissues, and develops into a **procercoid**, an elongated larva covered by a thick cuticle and with a tail containing the hooks (Fig. 10.19E). Large frontal glands open at the anterior end. It remains quiescent until its host is eaten by a second intermediate host, usually a fish.

Plerocercoid. The procercoid penetrates the gut wall of the second intermediate host, apparently with the aid of frontal gland secretions. It becomes a **plerocercoid** (Fig. 10.19G) by elongating and forming the attachment organs of the scolex at the anterior end. Primordia of most of the organ systems are present except for the reproductive organs, which do not appear until proglottids are formed. Cestodaria, however, have only one intermediate host, and the procercoid develops into the adult. The adult does not strobilate and is plerocercoid-like. Cestodaria may be cestodes in which the larval plerocercoid has become sexually mature, instead of being primitive cestodes as was once believed.

When eaten by a suitable host, plerocercoids develop into adults. Little metamorphosis is required, for the scolex and major organs are already formed. Proglottids begin to develop and gradually mature.

Cysticercoid. The cyclophyllic onchosphere may develop into a **cysticercoid** larva in an invertebrate host (Fig. 10.19H). The middle of the body grows up around the anterior end to form a cavity containing the scolex, while the tail, containing the hooks, remains free. The scolex is not turned inside out. When the host containing the cysticercoid is eaten, the body is digested away from the scolex, which develops proglottids and becomes mature.

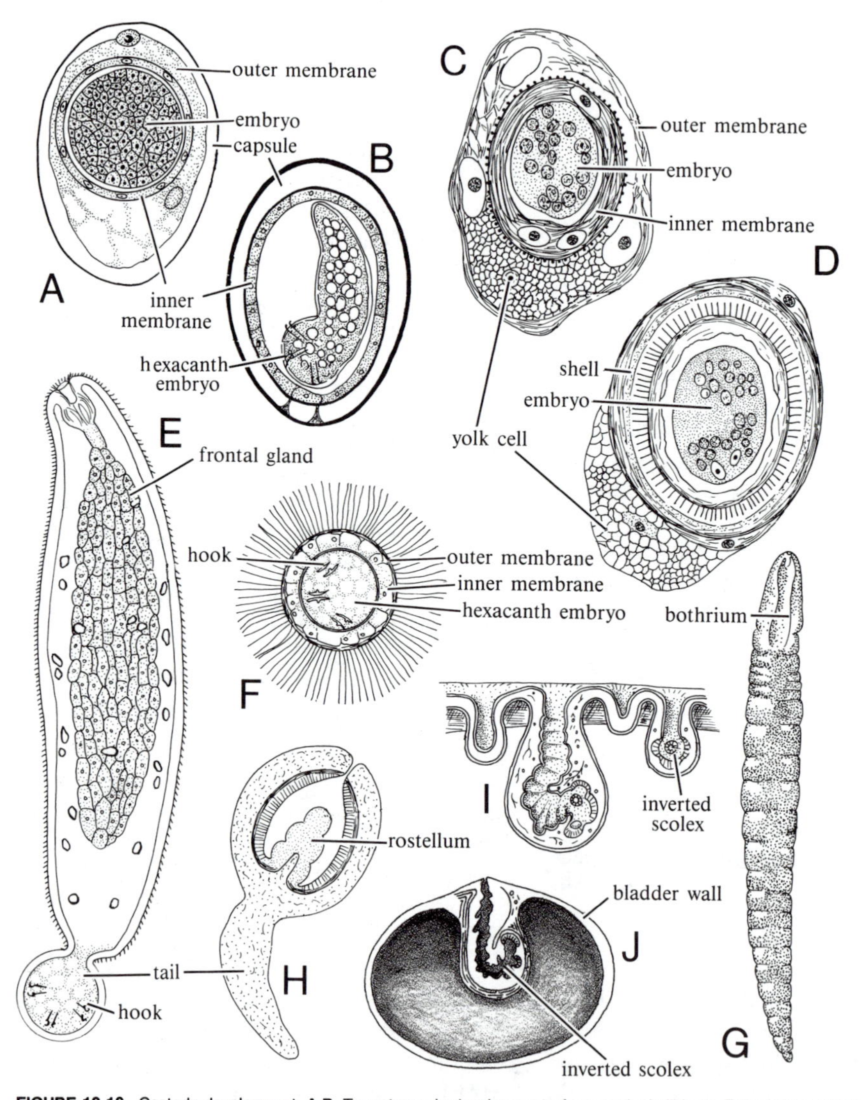

FIGURE 10.19. Cestode development. **A,B**. Two stages in development of a pseudophyllidean, *Eubothrium*, with inner and outer membrane laid down, and mature hexacanth. **C,D**. Two stages in development of *Taenia*, showing formation of inner membrane, and outer membrane disappearing while inner membrane forms striated shell. **E,F,G**. Larvae of Pseudophyllidea. **E**. Mature procercoid, showing hooks and large frontal gland. **F**. Ciliated coracidium, ready to be eaten by aquatic arthropod. **G**. Plerocercoid that develops when fish eats arthropod containing procercoid. **H,I,J**. Larvae of taenioid cestodes. **H**. Cysticercoid differing from cysticercus in its solid construction. **I**. Cysticercus with inverted scolex. **J**. Coenurus, cystlike larva containing a number of scolices. (**A,B** after Schauinsland; **C,D**. after Janicki; **E,F** after Rosen; **G** after Hamann; **H** after Belding; **I** after Leuckart and Belding; **J** after Fantham, Stephens, and Theobald, from Belding.)

Cysticercus. Some cyclophyllids use a vertebrate host and form a **cysticercus** instead of a cysticercoid. The scolex of a cysticercus is turned inside out (Fig. 10.19I) and hangs in a large cavity formed by the rest of the body. This larva, often known as a bladder worm, develops into an adult by eversion of the scolex and beginning of strobilization.

In some cases, cestode larvae reproduce by budding. The most noteworthy examples are the hydatid cysts of *Echinococcus granulosus*, a tiny tapeworm living in dogs. The eggs are eaten by herbivores, especially sheep, and the onchospheres are carried by portal blood to the liver, where hydatids are most commonly found. They grow into a cyst, with an inner, germinative layer that forms daughter cysts. Many inverted scolices form within the daughter cysts (Fig. 10.19J). The cysts may become very large and interfere with the normal functioning of infected organs, and multiple infections with hydatids prove fatal.

Review of Groups

All free-living flatworms traditionally were placed within a single class, Turbellaria. In turn, the parasitic groups were divided into two classes, Trematoda and Cestoda. The determination of the interrelationships of the various classic turbellarian orders was a vexing one, but recent analysis (e.g., see Ehlers, 1985) indicates that the turbellarians and trematodes are paraphyletic groups, with the monogenean trematodes more closely related to the cestodes than to the other trematodes. Although the concept of turbellarian and trematode is useful to conceptualize grades of evolution within the platyhelminths, and indeed have been employed in the preceding discussions of structure and biology, they probably should be abandoned as formal units of classification. A more up-to-date taxonomy is presented.

Turbellarians, the free-living flatworms, are predominantly marine, although there is a considerable freshwater fauna, and a few live in moist terrestrial habitats (Riser and Morse, 1974; Schockaert and Ball, 1980). They are about 0.5 mm to 60 cm long, with most less than 5 cm. Many are broad and leaf-shaped, and whereas some are nearly cylindrical, even these tend to be ventrally flattened. Freshwater species are white or uniformly dark-colored in most cases, but marine species have characteristic pigmentation, often bright in warm seas. *Dugesia tigrina* is the most common American species, and the larger *D. dorotocephala* is usually found in spring-fed waters. They are the most commonly studied species in student laboratories.

CLASS CATENULIDEA

These were formerly known as the catenulid rhabdocoels. Catenulids have a simple pharynx, a ciliated gastrodermis, a single protonephridium with a pair of cilia in the flame cell, four pairs of longitudinal nerve trunks, and a dorsal male pore and unarmed penis. They are often sexually immature and frequently reproduce asexually. Examples: *Stenostomum* (Fig. 10.20D) and *Catenula* (Fig. 10.20E).

CLASS ACOELOMORPHA

Small marine worms with a mouth but no gut, sometimes with a simple pharynx; without protonephridia; with a submuscular nerve plexus, concentrated anteriorly near a statocyst, and tending to form longitudinal trunks; usually without eyes; without definite ovaries or testes, gametes being produced in the parenchyme; usually free-swimming, occurring among algae and about rocks or in mud of the littoral zone; a few are endozoic in echinoderms. Examples: *Childia* (Fig. 10.20A), *Convoluta* (Fig. 10.20B), and *Polychoerus* (Fig. 10.20C).

CLASS RHABDITOPHORA

Flatworms that possess lamellated rhabdites, two types of adhesive glands, a double-cell wall to the protonephridial weir with multiciliary flame cells at the terminus. Three subclasses are recognized.

Subclass Macrostomida are marine and freshwater species, with a simple pharynx, ciliated gastrodermis, one pair of longitudinal nerve trunks, paired protonephridia, and complete male and female systems with separate gonopores and no yolk glands. Examples: *Macrostomum* (Fig. 10.20F) and *Microstomum* (Fig. 10.20G).

Subclass Polycladida are thin, often large, leaflike marine worms, with a many-branched enteron and plicate pharynx; many radiating nerve cords and usually with many eyes; many testes and ovaries and no yolk glands; usually dull but with some bright-colored species; some are epizoic or associated with hermit crabs or molluscs. There are two orders of polyclads. Acotylea have no sucker behind the female gonopore, and the eyes are not in paired clusters; tentacles, if present, are marginal rather than nuchal. Example: *Stylochus* (Fig. 10.21A). Cotylea have a posterior glandulomuscular disc, and tentacles are marginal when present; eyes are arrayed across the front margin, or in pairs of clusters in the tentacles or where the tentacles would occur; a pharynx ruffled or tubular; and no terminal bursa. They are characteristic of warmer waters. Examples: *Notoplana* (Fig. 10.21B) and *Euryleptus* (Fig. 10.21C).

Subclass Neoophora are flatworms that have ovaries that produce both ova and yolk cells that result in ectolecithal eggs. Several orders of neoophorans are recognized.

Seriata include the triclads and proseriate allocoels. These are moderate to large worms with a central to posterior mouth, and bulbous or plicate pharynx; enteron sometimes with lateral diverticula, and one anterior and two posterior trunks; reticulate protonephridia and numerous nephridiopores; one pair of ovaries and separate yolk glands; many testes or one pair; a penis papilla and a single gonopore; often with ciliated pits and grooves; and caudal adhesive organs. Examples: *Dugesia* (Fig. 10.1), *Monocelis* (Fig. 10.22A), *Otomesostoma* (Fig. 10.22B),

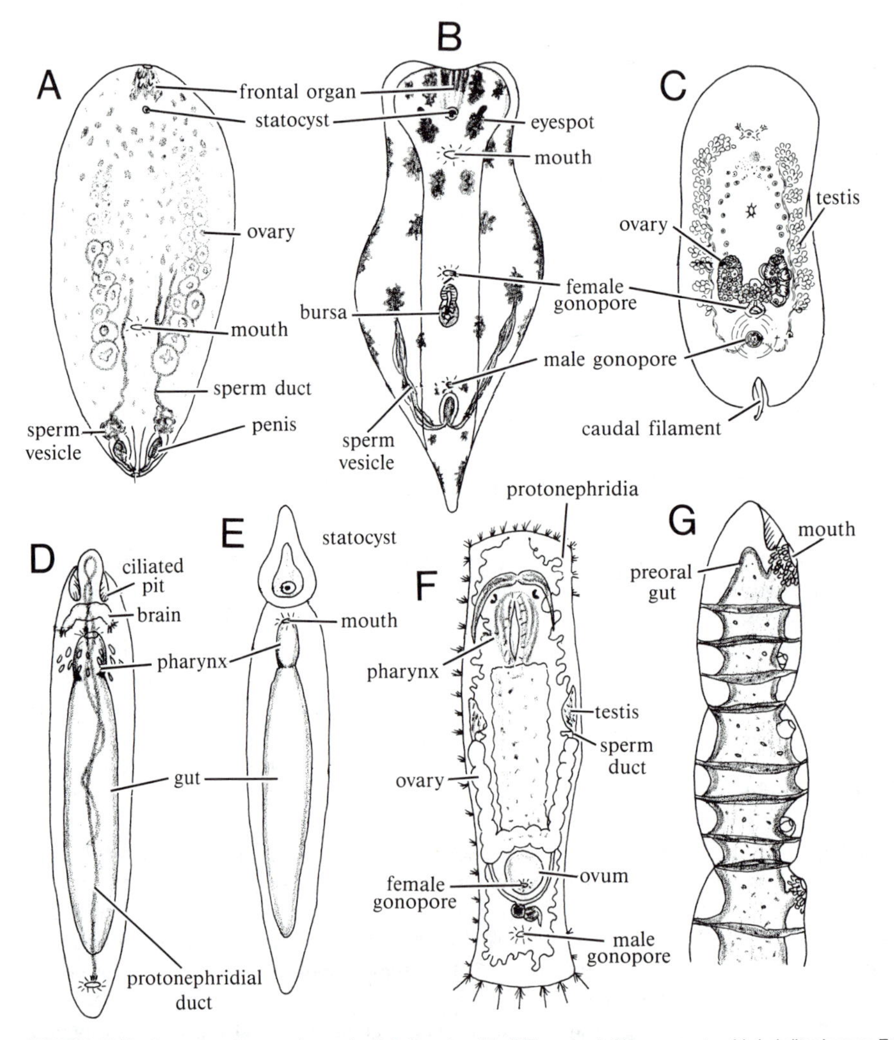

FIGURE 10.20. Representative acoelomorphs **A,B,C**, catenulids **D,E**, and primitive macrostomid rhabditophorans **F,G**. **A**. *Childia spinosa,* characterized by absence of female gonoducts. **B**. *Convoluta,* with at least one seminal bursa and gonopore. **C**. *Polychoerus,* with caudal filament. **D**. *Stenostomum,* characterized by simple pharynx, single medium protonephridium, and no female gonoducts. Stenostomidae have no circular groove in front of mouth. **E**. *Ctenula,* with circular preoral groove around body. **F**. *Macrostomum appendiculatum,* showing simple pharynx, complete male and female reproductive systems with separate gonopores, and no preoral extension of enteron. **G**. *Microstomum lineare,* with preoral extension of gut. Members of Microstomidae often reproduce asexually to form chains of zooids, as shown. Note differentiating mouths, indicating position of future zooids in chain. (**A,B,C,F** after Ferguson; **G** after von Graaf.)

Bothrioplana (Fig. 10.22C), *Bdelloura*, which lives on gills of horseshoe crabs (Fig. 10.22D), *Procotyla* (Fig. 10.22E), and *Kenkia* (Fig. 10.22F).

Typhloplanoida and Dalyelloida are flatworms not distinctly characterized but both having an unspecialized bulbous pharynx. Typhloplanoids tend to have a more posteriorly located mouth than the dalyelloids. Example: *Gyratrix* (Fig. 10.23A). Suborder Dalyelloida are marine, freshwater, or terrestrial, sometimes epizoic or endozoic rhabdocoels with a bulbous pharynx; two pairs of longitudinal nerve trunks; paired protonephridia; a germovitellarium that produces ova and yolk cells, or separate yolk glands. They do not reproduce asexually. Examples: *Paravortex* (Fig. 10.23B), *Dalyellia* (Fig. 10.23C), *Mesostoma* (Fig. 10.23D), and *Pseudostomum* (Fig. 10.2F). Suborder Temnocephalida are small, flattened, epizoic worms with a posterior adhesive disc and usually with five, six, or twelve tentacles; they are tropical or subtropical. Example: *Temnocephala* (Fig. 10.23E).

Parasitic Forms Important groups of flatworms are parasitic. Parasites live in environments as variable as free-living aquatic animals; each host species and host organ differs, and each presents peculiar adaptive problems. Many parasitic adaptations are invisible modifications of metabolic patterns, suitable for the oxygen levels, body defenses, and chemical peculiarities of the host organ and species. Others affect the visible form of the organ systems. Although the environments occupied by parasites are diverse, all must cope with some of the same problems, with a resulting tendency toward convergent changes in unrelated parasitic groups.

Some parasites live for several generations in the host body, but the host life span is limited and all parasites must often seek a new host. This discontinuity of the habitat a parasite calls "home" sparks several adaptive trends including the following, both of which are found in almost all parasites.

1. Parasites have a high reproductive potential. Sedentary animals often have migratory young or shelled eggs used as disseminules to spread the species. Parasites also depend primarily on young stages to locate new hosts. The chance that any one larva or egg will reach a suitable host may be very small. However, the improbable becomes probable if tried often enough! A parasite producing a million ova, each with one chance in ten thousand to be eaten by a suitable host, has a high probability to have some successful young. The two most common consequences are nicely exemplified by trematodes: reproductive organs tend to become large and complex, and larval stages often develop the ability to reproduce asexually.

2. Larval parasites tend to adapt, ecologically, to the habits of the host. An effective parasite evolves behavior patterns that conform to the habits of suitable host animals. This is usually achieved by larvae becoming part of the food chain of the final host. A larva that lives in a copepod, often eaten by fish, has an increased probability of eventually being consumed by a fish-eating bird or mammal. Or, larval

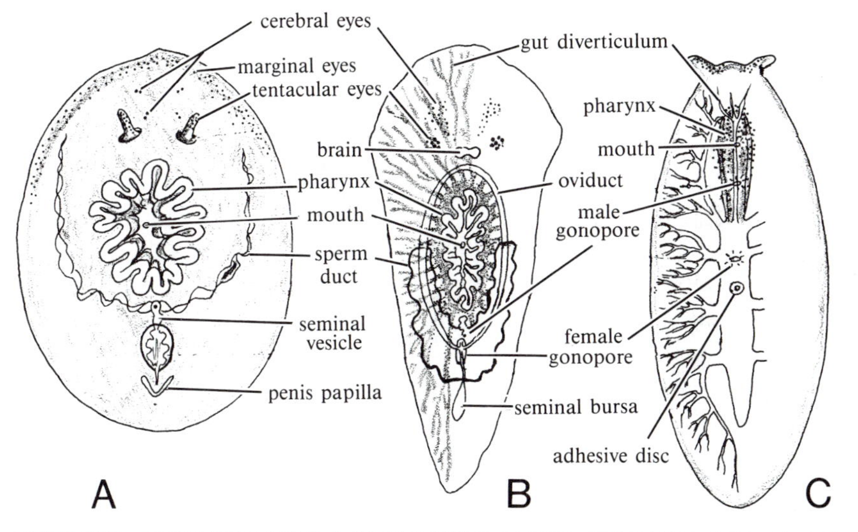

FIGURE 10.21. Representative polyclads. **A.** *Stylochus ellipticus*, acotylean, with band of marginal eyes in addition to cerebral and tentacular clusters of eyes. **B.** *Notoplana atomata*, acotylean, lacking marginal eyes and with four clusters of eyes near brain. **C.** Cotylean, with suckerlike glandulomuscular adhesive organ behind female gonopore and marginal tentacles with eyes. (After Hyman.)

parasites may live in blood-sucking animals that feed on the final host, and thus fit themselves into the food habits of the larval host. These adaptations are crucial for the maintenance of parasitic life cycles.

3. Parasites have little use for some anatomical parts that are important in free-living animals. Therefore, internal parasites rarely have eyes and usually have few sense organs of other kinds. Locomotor organs are poorly developed, and retention of the digestive tract may become useless if the parasite can absorb what is needed from the host through the body surface. Degenerative changes of this kind are almost universal consequences of a parasitic life. On

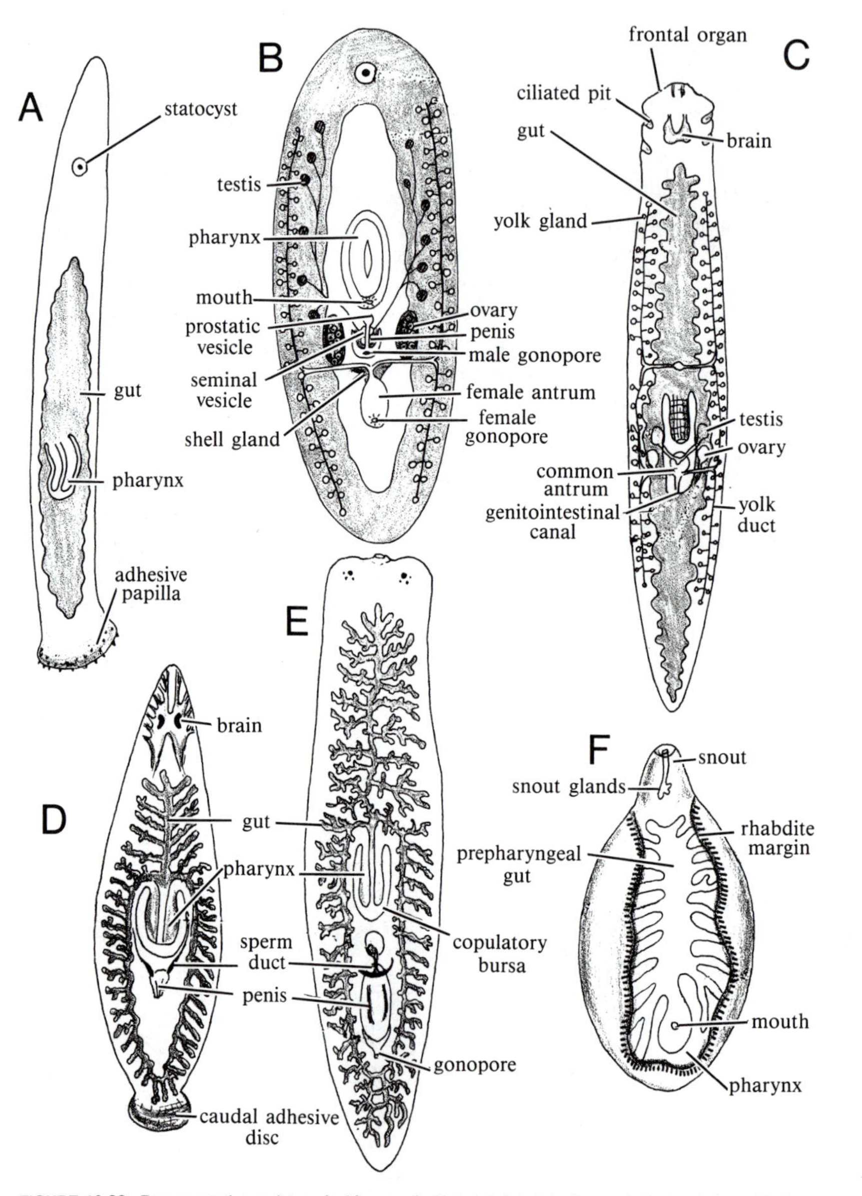

FIGURE 10.22. Representative seriates. **A.** *Monoscelis.* Note lobulate intestine and plicate pharynx. **B.** *Otomesostoma auditivum.* Note statocyst and absence of adhesive papillae. **C.** *Bothrioplana semperis,* with two ciliated pits but no statocyst. **D.** *Bdelloura candida,* which lives on horseshoe crabs, with caudal adhesive discs for host attachment. **E.** *Procotyla fluviatilis,* with anterior adhesive organ. **F.** *Kenkia rhynchida,* cave dweller, with snoutlike glanduloadhesive organ. (**A,C,D,E,F** after Hyman; **B** after Bresslau.)

the other hand, organs of attachment, which may not be important in free-living animals, are often very important in parasites.

The Order Trematoda includes the leaflike, unsegmented, parasitic flatworms with a gut, commonly known as flukes. Flukes are successful animals; nearly 6000 species have been described (Erasmus, 1972). Adults parasitize all vertebrate classes, with fishes the most common hosts. Trematodes typically have an indirect life cycle as they

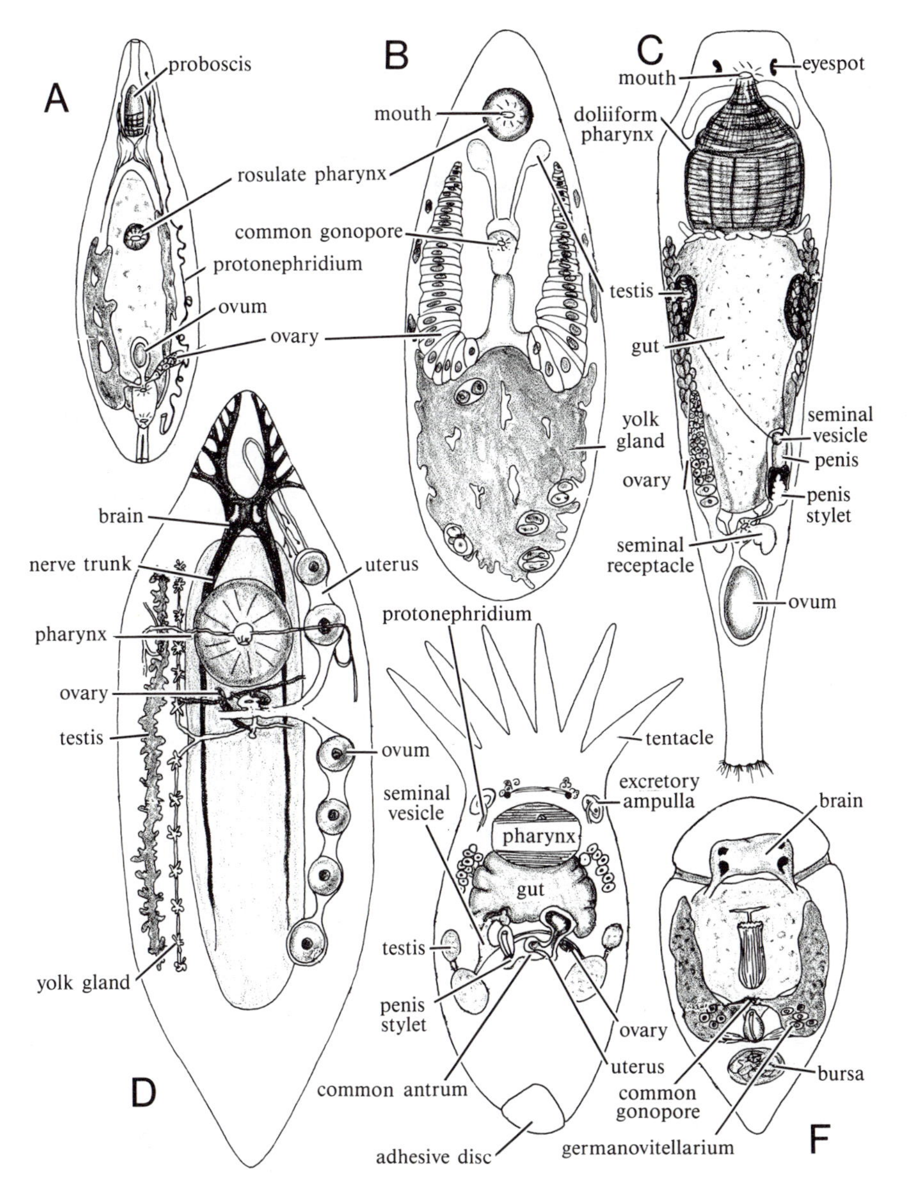

FIGURE 10.23. Representative typhloplanoids and dalyelloids. **A**. *Gyratrix hermaphroditicus,* typhloplanoid, with conspicuous glandulomuscular adhesive organ or proboscis. **B**. *Paravortex gemellipara,* dalyelloid, from mantle cavity and viscera of saltwater clam, *Modiolus.* Note gonopore set well forward. **C**. *Dalyellia,* one or two yolk glands that form yolk deposited in capsule rather than in ovum. Note doliiform pharynx. **D**. *Mesostoma,* freshwater dalyelloid, characterized by dorsal or lateral testes and protonephridia that open into pharyngeal cavity. **E**. *Temnocephala,* somewhat trematodelike dalyelloid, with posterior adhesive disc and characteristic tentacles. **F**. *Pseudostomum,* dalyelloid, with germovitellarium yolk glands partially distinct, and simple or slightly lobulated intestine. (**A**,**F** after von Graff; **B** after Ball; **C** after Ruebush and Hayes; **D** after Ruebush, from Pennak; **E** after Haswell.)

live as larvae in an intermediate host, and may live as larvae in a second intermediate host. Adults live in a final host. First, intermediate hosts are usually molluscs. Second, intermediate hosts, when present, are often fish or other cold-blooded vertebrates or, in some cases, invertebrates. The final hosts are vertebrates. *Opisthorchis sinensis*, the Chinese liver fluke (Fig. 10.18), makes an effective type for the group.

Knowledge of parasitic life cycles is of great importance, for relationships can often be determined more easily in immature than in mature forms, and the best methods of controlling parasites always depend on knowledge of the life cycle. The incidence of infection in a region can be cut down greatly by reducing the number of intermediate hosts or reducing their chances of being infected. Some wild or domestic animals serve as reservoir hosts, keeping an infection going even when sewage disposal or case treatment has reduced the likelihood of ova reaching intermediate hosts from human patients. Knowledge of the intermediate hosts and of the kinds of animals that feed upon them is often useful in locating the reservoir hosts in an area.

The ability of larval forms to multiply asexually adds greatly to the probability of successful completion of a life cycle. As many as 200,000 cercariae may be released by a *single* infected snail. Larval reproduction is not a true alternation of generations, however. It is more correct to think of it as a highly specialized form of polyembryony, fixed through embryonic development by the differentiation of propagative cells. Polyembryony, the formation of several embryos from a single ovum, occurs in several other animal phyla.

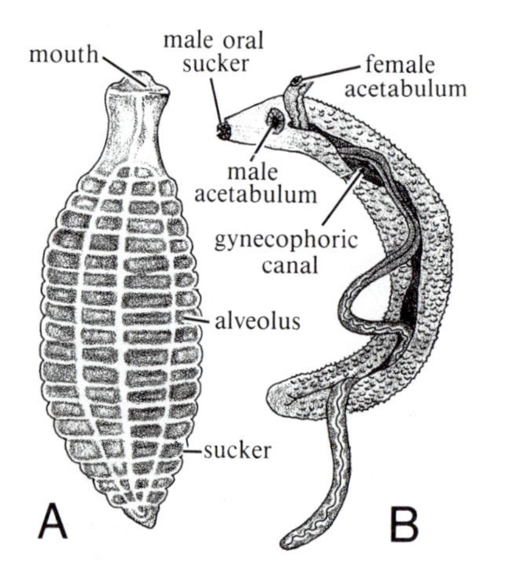

FIGURE 10.24. **A.** *Aspidogaster conchicola,* aspidobothrian trematode, characterized by huge ventral sucker. **B.** *Schistosoma mansoni,* strigeid blood fluke that parasitizes humans. Sexes separate, and smaller female carried in gynecophoric canal of male. (**A** after Monticelli; **B** after Looss, from Chandler and Reid.)

Trematode classification has been modified recently. The Monogenea, traditionally included, are now felt to be closer to cestodes. Only the suborders Digenea and Aspidobothrea are now considered to be Trematoda sensu stricto.

Aspidobothrea. Internal parasites of molluscs and cold-blooded vertebrates with a mouth lacking an oral sucker and, usually, a small pharynx and saccate enteron; a single posterior nephridiopore; oviducts divided by septa into chambers; a huge ventral sucker divided into compartments or a ventral row of suckers without hooks; life cycle direct, usually without an alternation of hosts. Example: *Aspidogaster* (Fig. 10.24A).

Digenea. Adults are internal parasites of vertebrates, with a mouth usually surrounded by an oral sucker, a bulbous pharynx, and a bifurcate gut; a single posterior excretory pore; a ventral sucker or acetabulum lacking hooks; a long uterus, usually filled with ova; a complex life cycle involving larval hosts, typically molluscs. Four superfamilies can be recognized. Strigeatoidea: With a fork-tailed cercaria, usually developing in sporocysts; miracidia primitively with two flame cells. Example: *Schistosoma* (Fig. 10.24B). Echinostomoidea: Cercaria with a strong, unforked tail; miracidia with one pair of flame cells. Example: *Fasciola* (Fig. 10.17D). Opisthorchoidea: Cercaria with excretory ducts in the tail and without a stylet. Example: *Opisthorchis* (Fig. 10.18). Plagiorchioidea: Cercaria without excretory ducts in the tail and usually with a stylet. Example: *Prosthogonimus* (Fig. 10.17C).

The order Cercomeromorpha are parasites in which the larvae and at least the young postlarval stages have up to 16 marginal hooks. There are two suborders: the Monogenea and the Cestoda.

Monogenea. External or semiexternal parasites of aquatic animals, especially of fishes and amphibia; with a mouth lacking an oral sucker or with a weak oral sucker, a pharynx and usually bifurcate gut; two anterior nephridiopores; a large posterior attachment organ, the opisthaptor, usually equipped with hooks and typically with two glandulomuscular attachment organs at the anterior end forming a prohaptor; rather short uterus containing few ova; direct life cycle, without alternation of hosts. There are two groups. Monopisthocotylea are Monogenea without a genitointestinal canal; opisthaptor as a single disc, usually with one to three pairs of large hooks and many marginal hooklets. Example: *Gyrodactylus* (Fig. 10.25A). Polyopisthocotylea are

Monogenea with a genitointestinal canal; opisthaptor with discrete suckers or clamps, on a single disc or free. Example: *Polystoma* (Fig. 10.25C).

Monogenea produce capsules in which development has already begun, and that hatch a few days to a few weeks after being freed. A free-swimming larva emerges (Fig. 10.25B), with eyes and usually with three or four transverse bands of cilia. The larval opisthaptor resembles the opisthaptor of *Gyrodactylus*. The larva dies in a day or two if it does not find a suitable host. It attaches to the host and metamorphoses into an adult. Metamorphosis often involves a profound change of the opisthaptor, accomplished either by shedding the larval attachment organ or by gradual metamorphosis. During metamorphosis, the ciliated epidermis is shed and the eyes are usually lost.

A curious feature of the *Polystoma* life cycle may hint at how a complex life cycle involving several host species might have developed. *Polystoma* lives in frog urinary bladders. Larvae attach to tadpoles and become sexually mature, but retain the larval form if the tadpoles are young. On older tadpoles they remain immature until the tadpole metamorphoses, whereupon they move into the urinary bladder, lose the larval traits, and become sexually mature.

Cestoda. Cestoda are internal parasites of vertebrates. The cestode body is composed of a head or scolex and a body usually divided into segments or

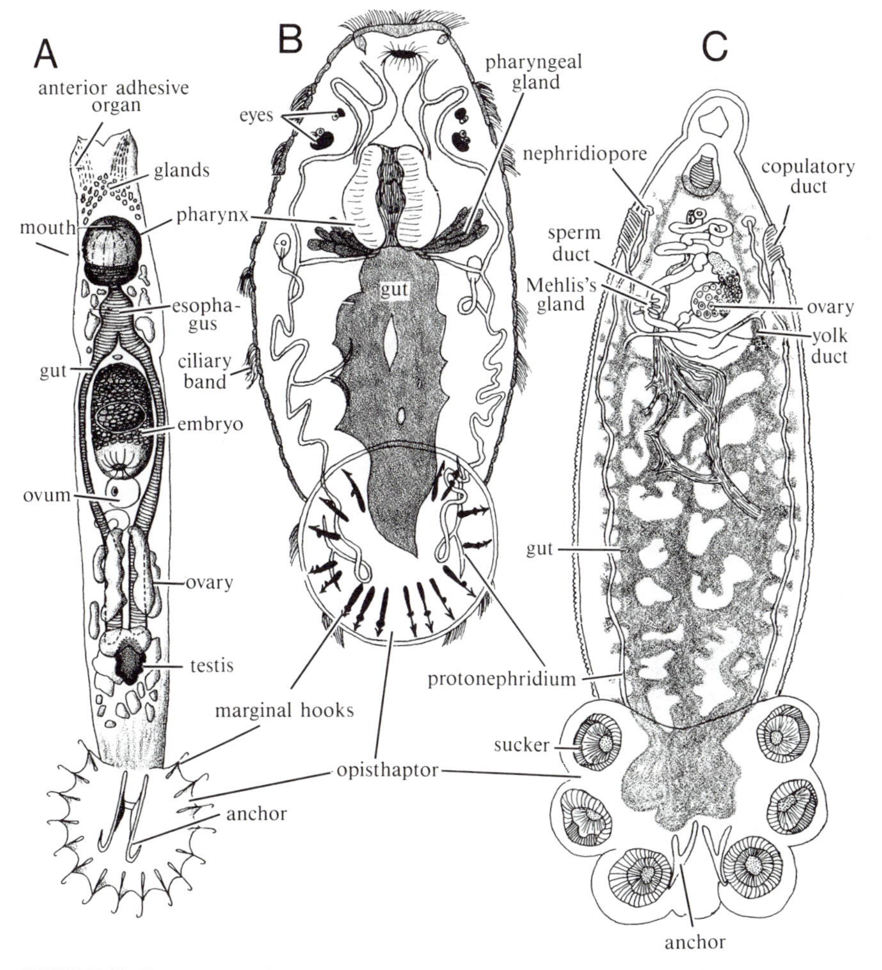

FIGURE 10.25. Monogenea. **A.** *Gyrodactylus,* with weak or no oral sucker, and simple posterior attachment organ or opisthaptor. No genitointestinal canal. **B.** *Polystoma* larva, with several bands of locomotor cilia. Note resemblance of opisthaptor to *Gyrodactylus,* although adult worm will have several suckers on opisthaptor. **C.** *Polystoma integerrinum,* with oral sucker, opisthaptor with several suckers and anchors, and genitointestinal canal. (**A** after Mueller and Van Cleve; **B** after Halkin; **C** after Paul.)

proglottids with one or two complete, hermaphroditic reproductive systems in each (Figs. 10.26 and 10.27). Cestoda completely lack a digestive system and have six to ten hooks in larval and postlarval stages. There are generally six groups recognized within the cestodes.

Cestodaria. Embryos with ten hooks; body unsegmented and without a scolex; a single reproductive system as in trematodes. Example: *Gyrocotyle* (Fig. 10.27E).

Tetraphyllidea. Embryos with six hooks; scolex with four lappetlike attachment organs known as bothridia and often with hooks; proglottids in various stages of development; yolk glands in lateral bands in the proglottids; genital pores lateral; live in elasmobranch fishes. Example: *Phyllobothrium* (Fig. 10.27C,D).

Trypanorhyncha. Embryos with six hooks; scolex with two or four bothridia and four protrusible, spiny proboscides in sheaths; yolk glands in cortical layer around the proglottid; genital pores lateral; live in elasmobranchs and some fishes. Example: *Tentacularia* (Fig. 10.27A,B).

Pseudophyllidea. Embryos with six hooks; usually segmented, with one terminal or two lateral groove-shaped attachment organs known as bothria; most proglottids in a strobila at the same stage of development; yolk glands in dorsal and ventral layers across proglottids; uterus opening at a uterine pore; genital pores usually midventral; live in bony fishes or terrestrial vertebrates. Example: *Diphyllobothrium* (Fig. 10.27F,G).

Proteocephaloidea. Embryos with six hooks; scolex with four cup-shaped suckers and with an apical sucker or glandular organ; yolk glands in lateral bands in each proglottid; uterus ruptures to form midventral uterine pores; genital pores lateral; live in cold-blooded vertebrates. Example: *Ophiotaenia* (Fig. 10.26D,E).

Cyclophyllidea. Embryos with six hooks; scolex

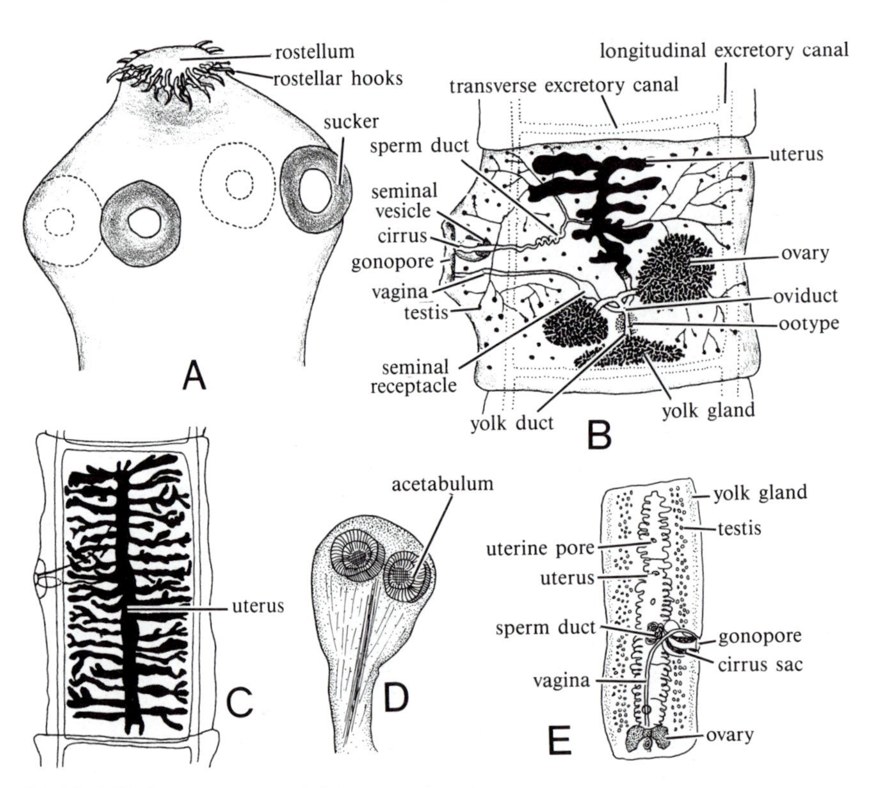

FIGURE 10.26. Cestode structure. **A,B,C.** *Taenia solium.* **A.** Scolex forms in larva that encysts in muscle tissue of intermediate host. Scolex attaches to host intestinal wall to bud off series of segments or proglottids. Youngest proglottids occur at budding zone, whereas older ones lie further down in chain. **B.** Mature proglottid, reproductive organs fully differentiated. Mature proglottids produce ova, encapsulated and stored in uterus, which branches repeatedly as ova accumulate. **C.** Gravid proglottid, with degenerated reproductive systems except for enlarged uterus filled with encapsulated ova, detach and emerge from final host with feces. Egg capsules liberated when proglottids break down. **D,E.** Scolex and proglottids of proteocephalid, *Ophiotaenia.* Note scolex with four acetabula, and proglottids with lateral bands of yolk glands and uterine pores formed by rupture of body wall. (**A** after Chandler and Reed; **B,C** after the Nobles; **D,E** after Van Cleave and Mueller.)

with four cup-shaped suckers and usually with an apical rostellum; proglottids in various stages of development; yolk gland compact and posterior; uterus without a uterine pore; genital pores usually lateral; typically live in warm-blooded vertebrates. Example: *Taenia* (Fig. 10.26A,B,C).

Phylogeny

The most recent phylogenetic analysis, that of Ehlers (Fig. 10.28), uses information from ultrastructural studies to augment some traditional features of gross anatomy. The scheme produced in Figure

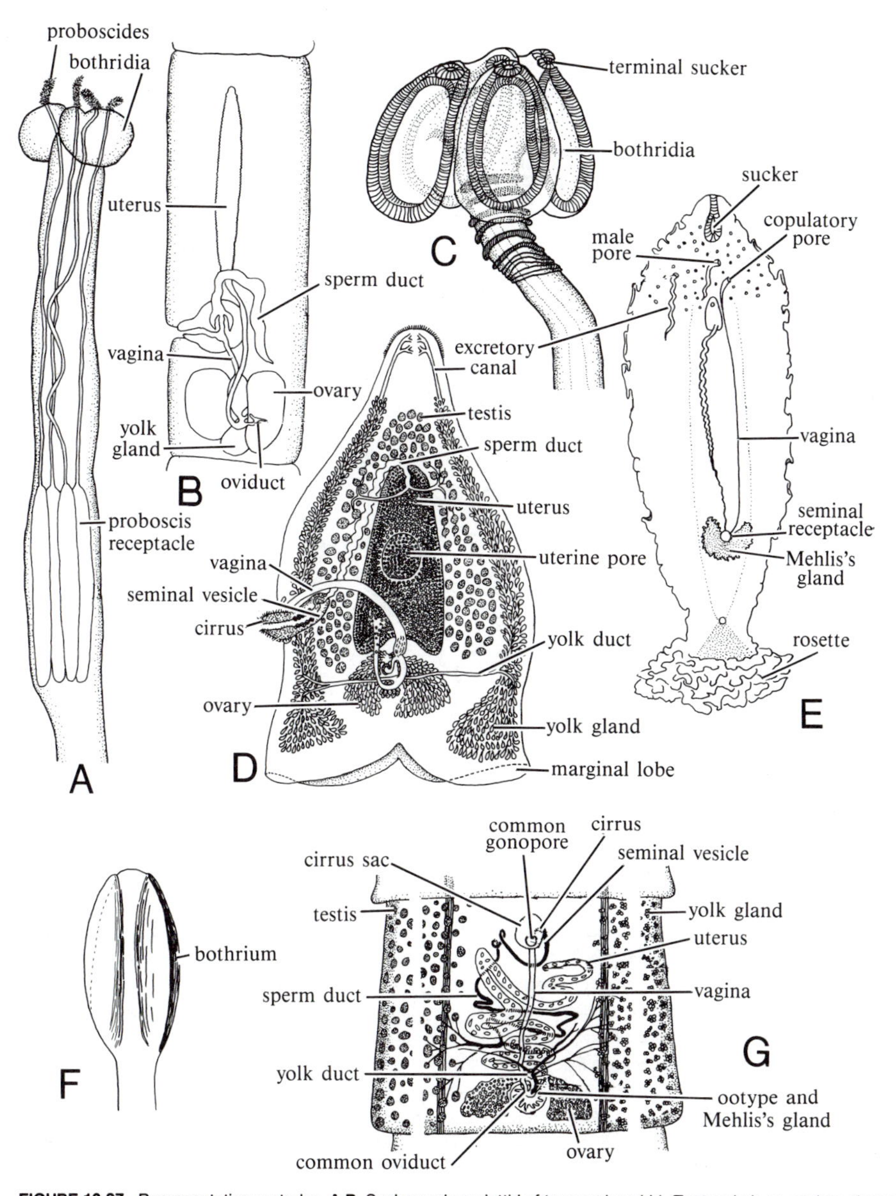

FIGURE 10.27. Representative cestodes. **A**,**B**. Scolex and proglottid of trypanorhynchid, *Tentacularia musculara,* characterized by four spiny, retractable proboscides on scolex. **C**,**D**. Scolex and proglottid of tetraphyllidean, characterized by four large bothridia on scolex and by proglottids with lateral bands of yolk glands and lateral gonopore. **E**. *Gyrocotyle urna,* cestodarian, with no scolex and body not segmented. **F**,**G**. Scolex and proglottid of pseudophyllidean, *Diphyllobothrium.* Note one or two sucking grooves or bothria on scolex and proglottids with discrete uterine pore. Testes shown only on left and yolk glands only right of proglottid. (**A**,**B** after Hart; **C**,**D** after Goodchild, from Brown; **E** after Lynch; **F** after Hyman; **G** after Goodchild, in Brown.)

10.28 is somewhat simplified, and also illustrates the grades of evolution that the traditional three-class system of platyhelminths represented. At the class level, the theme of evolution of this new classification centers on varying degrees of protonephridial development. Within the Rhabditophora, evolution is for the most part directed at increasing specializations in reproductive structures and developmental biology. However, it should be emphasized that the kind of analysis represented here is just beginning, and we can be sure that newer information will be forthcoming that will clarify relationships between and within these groups.

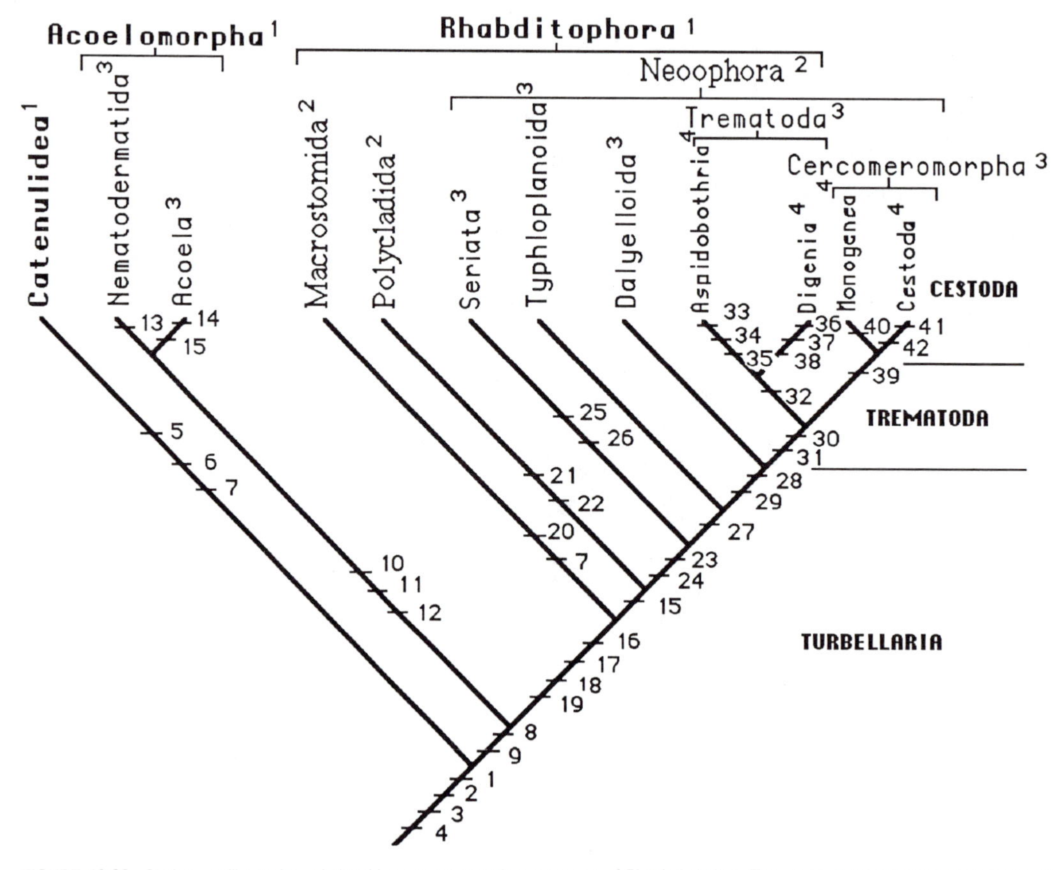

FIGURE 10.28. Cladogram illustrating relationships among constituent groups of Platyhelminthes. Taxonomic level indicated by superscript numbers as follows: Classes[1], Subclasses[2], Orders[3], Suborders[4]. Traditional classes that are now grades of evolution indicated on the right-hand side in bold, underlined script. Derived characters are: (1) multiciliary epidermis; (2) cilia with one basal body, without accessory centriole; (3) flame cells with at least two cilia; (4) flame cell without podlike microvilli; (5) protonephridium unpaired; (6) male gonopore dorsal; (7) aciliary sperm; (8) very dense cilia on epidermal cells; (9) frontal glands; (10) several ultrastructural modifications of epidermal cilia; (11) reduced protonephridia; (12) reduced or absent gut cavity; (13) statocyst with two statoliths; (14) statocyst with tubular body; (15) biciliary sperm; (16) lamellated rhabdites; (17) double-gland adhesive system; (18) double-cell weir in protonephridia; (19) multiciliary flame cells; (20) both adhesive glands emerge in one collar or microvilli; (21) intestine with branched diverticula; (22) reabsorption in development of 4A, 4B, 4C, 4D macromeres and 4a, 4b, 4c micromeres; (23) ova with both egg- and yolk-producing cells; (24) ectolecithal eggs; (25) tubiform pharynx; (26) serial arrangement of gonads; (27) bulbous pharynx; (28) doliiform pharynx; (29) reduction of double-gland adhesive system; (30) epidermis shed at end of larval stages; (31) syncytial neodermis; (32) neodermis separating epidermis from basement tissues; (33) catylocidium larva; (34) adult with huge ventral sucker; (35) oviducts divided into chambers; (36) early development stages (miracidium and sporocyst) without gut; (37) cercaria and adult with ventral sucker; (38) complex life cyles; (39) larvae and at least young postlarvae with up to 16 marginal hooks; (40) oncomiracidium larva epidermis with three zones of cilia; (41) at most ten marginal hooks in larval or postlarval phases; and (42) all stages of development lacking a gut. (Modified from Ehlers, 1985.)

Taxonomy

The classification used here is modified from that of Ehlers. It provides a framework to outline the relationships of structural types within the phylum.

Class Catenulidea. Unpaired protonephridia, biciliary flame cell, male genital pore dorsal, aciliary sperm. Order Catenulida (Fig. 10.20D,E).

Class Acoelomorpha. Loss of gut cavity with digestive parenchyme filling body, protonephridia reduced, ultrastructural specialization of epidermis. Orders: Acoela, Nematodermatida (Fig. 10.20A,B,C).

Class Rhabditophora. Rhabdites lamellated, two-gland adhesive system, two cells involved in forming walls of protonephridial ductules, flame cells with several cilia, protonephridia complex.

Subclass Macrostomida. All glands of the double-gland adhesive system emerging in one collar of microvilli, aciliary sperm. One order (Fig. 10.20F,G).

Subclass Polycladida. Gut highly branched with numerous diverticula. One order (Fig. 10.21).

Subclass Neoophora. Ovaries with yolk- and egg-producing cells, eggs ectolecithal.

Order Seriata. Tubiform pharynx, serial arrangement of gonads and gut diverticula (Fig. 10.22).

Order Typhloplanoida. (No unique defining advanced features.) (Fig. 10.23A).

Order Dalyelloida. (No unique defining advanced features, although sharing with trematodes and cercomeromorphs a doliiform pharynx and reduction of double-gland adhesive system.) (Fig. 10.23B,C,D,E,F).

Order Trematoda. Vertebrate and mollusc parasites with complex life cycles, larval epidermal cells separated from those below by a syncytial "neodermis." Suborders Aspidobothrea (Fig. 10.23A) and Digenea (Figs. 10.18 and 10.17).

Order Cercomeromorpha. Larvae and at least younger postlarval stages with up to 16 marginal hooks. Suborders Monogenea (Fig. 10.25) and Cestoda (Figs. 10.26 and 10.27).

Selected Readings

Aria, H.P. (ed.). 1980. *Biology of the Tapeworm Hymenolepis diminuta.* Academic Press, New York.

Arme, C., and P. Pappas (eds.). 1983. *Biology of the Eucestoda,* vols. 1 and 2. Academic Press, New York.

Benazzi, M., and V. Gremigni. 1982. Developmental biology of triclad turbellarians. In *Developmental Biology of Freshwater Invertebrates* (F. W. Harrison and R. R. Cowden, eds.), pp. 151–211. Alan R. Liss, New York.

Bogitsh, B. J., and O. S. Carter. 1982. Developmental biology of *Schistosoma mansoni,* with emphasis on the ultrastructure and enzyme histochemistry of intramolluscan stages. In *Developmental Biology of Freshwater Invertebrates* (F. W. Harrison and R. R. Cowden, eds.), pp. 221–248. Alan R. Liss, New York.

Case, T. J., and R. K. Washino. 1979. Flatworm control of mosquito larvae in rice fields. *Science* 206:1412–1414.

Ehlers, U. 1985. Phylogenetic relationships within the Platyhelminthes. In *The Origins and Relationships of Lower Invertebrates* (S. Conway Morris et al., eds.), pp. 143–158. Clarendon Press, Oxford.

Erasmus, D. A. 1972. *The Biology of Trematoda.* Crane, Russak, New York.

Henley, C. 1974. Platyhelminthes (Turbellaria). In *Reproduction of Marine Invertebrates* (A. C. Giese and J. S. Pearse, eds.), pp. 267–343. Academic Press, New York.

Keenan, C. L., R. Coss, and H. Koopowitz. 1981. Cytoarchitecture of primitive brains: Golgi studies in flatworms. *J. Comp. Neurol.* 195:697–716.

Kozloff, E. N. 1972. Selection of food, feeding and physiological aspects of digestion in the acoel turbellarian *Otocelis luteola. Trans. Am. Micro. Soc.* 91:556–565.

Martin, G. G. 1978. A new function of rhabdites: Mucus production for ciliary gliding. *Zoomorph.* 91:235–248.

Moraczewski, J. 1977. Asexual reproduction and regeneration in *Catenula. Zoomorph.* 88:65–80.

Morita, M., and J. B. Best. 1984. Effects of photoperiods and melatonin on planarian asexual reproduction. *J. Exp. Zool.* 231:273–282.

Rieger, R. M. 1985. The phylogenetic status of the acoelomate organization within the Bilateria: a histological perspective. In *The Origins and Relationships of Lower Invertebrates* (S. Conway Morris et al., eds.), pp. 101–122. Clarendon Press, Oxford.

Riser, N. W., and M. P. Morse (eds.). 1974. *Biology of the Turbellaria.* McGraw-Hill, New York.

Ruppert, E. C., and P. R. Smith. 1985. A model to explain nephridial diversity in animals. *Am. Zool.* 25:40A.

Schockaert, E. R., and I. R. Ball (eds.). 1980. *The Biology of the Turbellaria.* Junk, The Hague.

Smyth, J. D. 1969. *The Physiology of Cestodes.* Freeman, San Francisco.

Smyth, J. D., and D. W. Halton. 1983. *The Physiology of Trematodes*, 2nd ed. Cambridge Univ. Press, Cambridge.

Wardle, R. A., J. A. McLeod, and S. Radovinsky. 1974. *Advances in the Zoology of Tapeworms, 1950–1970.* Univ. Minnesota Press, Minneapolis.

Wilson, R. A., and L. A. Webster. 1974. Protonephridia. *Biol. Rev.* 49:127–160.

11

Gastrotricha

Introduction to Pseudocoelomates

In the phyla of the preceding chapters, the area between the epidermal musculature and the gastrodermis is generally filled with a solid mesoglea or the mesodermally derived parenchyme. In all other animal phyla, the organs usually lie in a body cavity of some kind, making them independent of the body wall. A body cavity provides some real advantages. Organs can move more freely than if they are embedded in solid tissue where they otherwise are squeezed and stretched every time the body moves. Occasionally, even acoelomate animals lose or reduce the parenchymal tissues to accommodate body distortion. Moreover, a body cavity allows the exchange of materials between the gut and the other viscera as well as the outer body wall.

Two kinds of body cavities may be distinguished. A **coelom**, found in vertebrates and higher invertebrates, is a space that forms within the mesoderm during development, and its walls are lined with a cellular epithelium, the **peritoneum**. A **pseudocoel**, found in a number of invertebrates, is a space that appears between the endodermal gut and the body wall formed of mesoderm and ectoderm, and is not lined with true peritoneum.

Eight distinctive types of animals have a pseudocoel: Acanthocephala, Rotifera, Gastrotricha, Kinorhyncha, Priapulida, Nematomorpha, Nematoda, and Loricifera. (Sometimes Entoprocta are said to have a pseudocoel, although that issue is not clear—see Chapter 33.) These animals are sometimes included in a single phylum, Aschelminthes. However, because these groups do have different structural plans, each of the eight is now more commonly placed in a separate phylum.

These phyla have the pseudocoel as one of their unifying characters, and all have a body wall composed of epidermis, muscles, and a fairly extensive dermis within which the pseudocoel lies. The digestive tube and gonads are located in the pseudocoel, and are generally bathed in a perivisceral fluid. However, in a few instances, some gastrotrichs and nematomorphs, the body cavity is virtually filled with tissue to provide a type of hydrostatic skeleton.

Pseudocoelomates are generally somewhat wormlike and are not highly cephalized. The body is covered by a cuticle, often quite complex, that lies on either a cellular or syncytial epidermis. Generally the body wall does not contain tightly integrated circular and longitudinal muscle layers. The digestive tract is predominantly epithelial, but sometimes may have a delicate network of muscles, especially around the pharynx. Occasionally, membranes can be seen that suspend organs or drape the surfaces of the pseudocoel, for example, in priapulids and chaetognaths. However, these are either not true peritonea or are very peculiar in nature. There is no respiratory or circulatory system. All pseudocoelomates except the nematodes have protonephridia. Urinary, reproductive, and digestive systems often join terminally in a cloaca, a common chamber through which gametes, urine, and feces pass; but female nematodes have a separate gonopore. Most pseudocoelomates undergo determinate cleavage; many exhibit distinctive modes of spiral cleavage with monets or duets instead of quartets, and there is a strong tendency toward **eutely** or constancy in the number of cells in each organ. Pseudocoelomates lack primary larvae as seen among coelomates and many acoelomates. Such "larvae" as they have are really juvenile growth stages that through successive molts gradually lead to the adult form.

Each of the pseudocoelomate phyla are covered in a separate chapter, starting here with Gastrotricha.

Gastrotricha

Gastrotrichs are gregarious microscopic animals common as meiofauna in interstitial environments in both freshwater and marine habitats (Boaden, 1985). They attach themselves momentarily with posterior adhesive tubes, and then move freely among sediment grains or suddenly curl up like tiny hedgehogs. They usually glide along, retaining contact with the substrate. Superficially, they look somewhat like rotifers, but have no prominent anterior ciliary organ and generally have a characteristic scaly or spiny body. Two classes are recognized: Macrodasyida and Chaetonotida. The former are marine and live among vegetation or in sand. The latter include all of the freshwater and a few marine

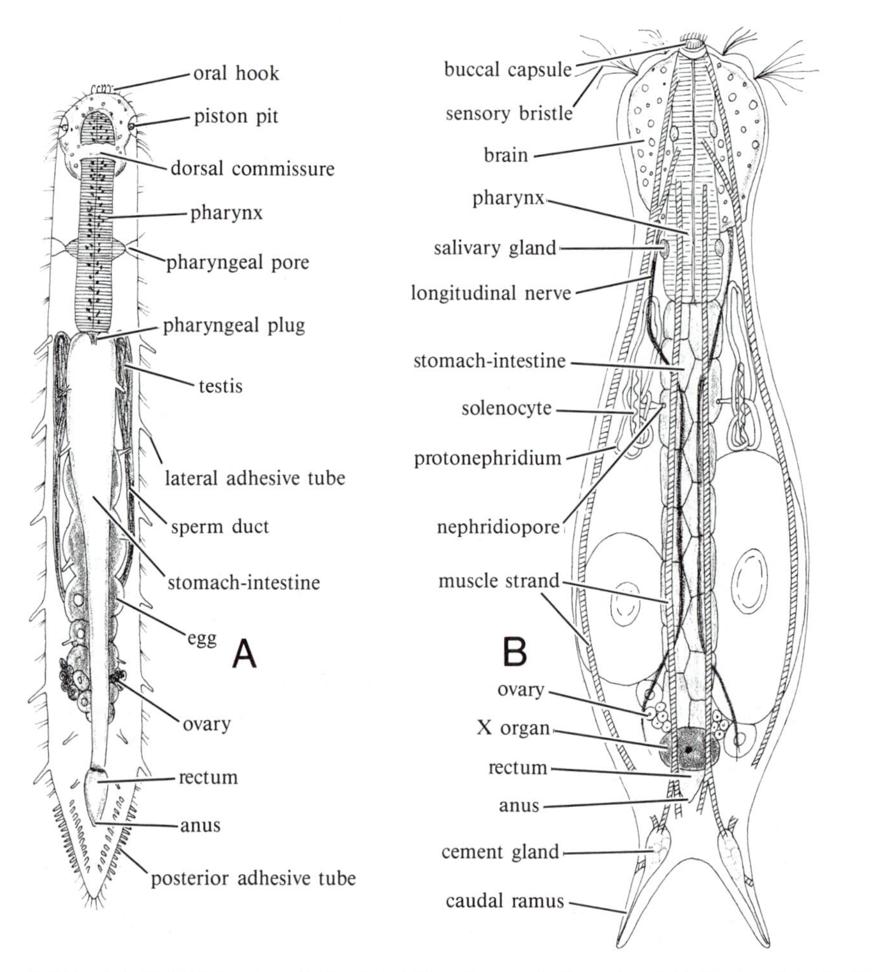

FIGURE 11.1. Gastrotrich anatomy. **A**. Structure of *Macrodasys*, marine Macrodasyida. **B**. Structure of chaetonotid gastrotrich. (**A** after Remane; **B** after Pennak.)

species. There are about 450 species of gastrotrichs currently recognized.

Definition

Gastrotricha are small (75–500 μm) meiofaunal forms. They are often marked with cuticle-encased monociliated cells on the ventral surface; the cuticle is specialized as spines or plates, or both. The body cavity is completely filled with organs. The gut has a mouth and anus, and the pharynx is well developed. The protonephridia have flame bulbs. Typically there are at least posterior pairs of adhesive glands.

Body Plan

Macrodasyids have a smooth body outline (Fig. 11.1A), while chaetonotids usually have a definite anterior region or head marked by a constricted neck, a convex trunk, and a forked caudal end (Fig. 11.1B). The gastrotrich head and ventral surface are ciliated in various patterns. Several tufts of sensory cilia and bristles usually project from the surface of the head. The dorsal body surface is sometimes covered with abutted or overlapping cuticular scales, usually ornamented with spines (Fig. 11.2A,B). A variety of surface adhesive organs as well as **caudal cement glands** are characteristic of gastrotrichs (Storch, 1984). The latter open through the **furca**, sometimes known as toes, in chaetonotids (Fig. 11.1B), and occur in marginal posterior rows or in clusters in the posterior end in macrodasyids (Fig. 11.1A).

The cuticle has a granular or fibrous basal layer, from which the various scales, spines, and hooks arise, and outer lamellar layers (Rieger and Rieger, 1977). Cuticle in chaetonotids is very reminiscent of that seen in nematodes. The epidermis is syncytial in chaetonotids, although cellular in macrodasyids, and secretes the cuticle and its derivatives. In macrodasyids, the ventral surface is undulating and possesses microvilli. Both multiciliated and monociliated cells occur (Ruppert, 1982), but the cilia are always enclosed in the cuticle. Circular muscle fibers lie just below the epidermis and are used to operate movable cuticular bristles. Dorsal, ventral, and ventrolateral longitudinal muscle strands retract the adhesive tubes and shorten or curl the body. The pseudocoel is completely filled with organs, chief of which is the all-encompassing **Y-organ** whose function is not fully understood, but seems to serve as a kind of hydrostatic skeleton.

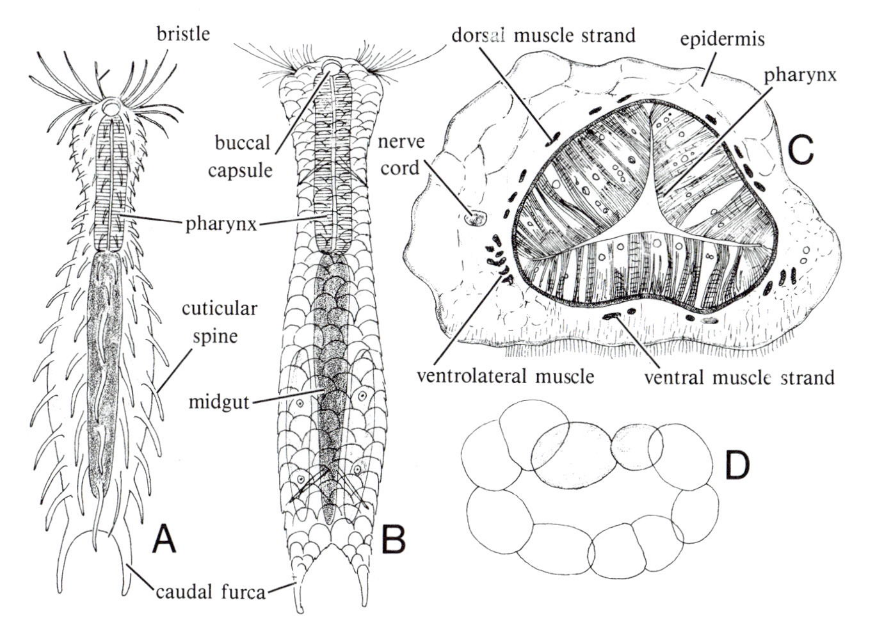

FIGURE 11.2. Gastrotrichs. **A**. Freshwater *Chaetonota,* with conspicuous surface spines. **B** *Lepidodermella,* chaetonotid with surface scales. **C**. Section through pharynx of *Macrodasys.* Note inverted triangular lumen and radial muscles. **D**. Sagittal section through gastrotrich embryo at time of gastrulation, with invaginating cells darkened. (**A**,**B** after Hyman; **C** after Remane; **D** after de Beauchamp.)

Digestive System A cuticular **buccal cavity**, often ridged or toothed, leads to the elongated, muscular **pharynx**. The pharynx (Fig. 11.2C), lined with smooth cuticle, is sometimes dilated to form a bulb and contains unicellular **salivary glands**. It has a triangular, Y-shaped lumen. However, the two groups are distinctly different in this regard since chaetonotids have a Y cross section, and macrodasyids have an inverted Y (Fig. 11.2C) as well as possessing pharyngeal pores that communicate with the surface of the head (Fig. 11.1A). The pharynx opens into the **midgut** composed of a single glandular epithelial layer of cells. The gut terminates in an **anus**, ventral in macrodasyids and dorsal in chaetonotids.

Excretory System Only chaetonotids have **protonephridia** with flagellated flame bulbs. A long **flame bulb** containing a one- or two-flagella flame opens into a coiled tube on each side of the body (Fig. 11.1B); each side empties through a separate **nephridiopore**.

Nervous System The central nervous system consists of two **lateral ganglia** (the "brain") connected by a narrow or broad **dorsal commissure**. A **lateral nerve trunk** issues from each ganglion.

Reproductive System Macrodasyids are hermaphroditic. The male system (Fig. 11.1A) consists of one or two **testes**, with **sperm ducts** opening through either a male gonopore, a common gonopore, or the anus. The female system is simple as well, with a single or paired **ovaries**, an **oviduct**, **seminal receptacle**, **bursa**, and **gonopore**. Paired glands also may be present, but no one species possesses all possible structures.

Most chaetonotids appear to be parthenogenetic females. However, a few vestiges of testes have been reported, and sperm has been found in a few species (W. D. Hummon, 1974). The female system (Fig. 11.1B) of chaetonotids is not very well understood. There are one or two ovaries, without a covering capsule. As ova mature, they secrete their own eggshells and move into a space often called the **uterus**, but there is little evidence of any oviduct and many authorities doubt its existence. A nutritive tissue is present and is called a **vitellarium**. Ripe eggs move into a sac, called the **X-body**, and eventually escape through a ventral gonopore, the encapsulating shell hardening after the eggs emerge.

Biology

Locomotion Movement is achieved typically by the beating of the ventral cilia. However, gastrotrichs can move in inchwormlike or leachlike fashion using the posterior adhesive discs and the anterior bristles. These muscular movements apparently operate in opposition to the hydrostatic skeletal support supplied by the Y-organ. A few forms appear to be facultatively sessile.

Feeding Gastrotrichs feed on detritus, algae, protozoa, and bacteria. These tidbits are collected either by action of the buccal cilia or by pumping actions of the pharynx. Food passes to the midgut, a straight, tubular organ generally without associated gland organs, but with gland cells in its wall thought to secrete enzymes. Digestion appears to be extracellular. Feces are voided through a short rectum that leads to the anus: ventral in macrodasyids, dorsal in chaetonotids.

Excretion Only chaetonotids are known to have protonephridia, and these are thought to be primarily osmoregulatory in function.

Sense Organs Cephalic **sensory bristles** *and* **ciliary tufts** appear to be sensitive to touch and water currents. Clusters of red pigment bodies in some brain cells are probably photosensitive. **Ciliated pits** and **sensory palps** on the sides of the chaetonotid head appear to be modifications of the piston pits of macrodasyids (Fig. 11.1A). They are largely chemoreceptors, although the palps appear to have terminal mechanoreceptors.

Reproductive System Some macrodasyids are so extremely protandric that they seem to be dioecious with separate sexes. Functioning males vibrate their posterior ends to attract a functioning female. Copulation occurs when tails are entwined to allow sperm transfer. Males can copulate several times before becoming females. Egg laying is achieved typically by rupturing the body wall (Hummon and Hummon, 1983).

The parthenogenetic chaetonotids produce thin-walled summer ova and thick-walled winter ova, both of which develop parthenogenetically. Summer ova are formed under more favorable environmental conditions by younger animals, and thick-walled ova by older females in less favorable environments. Most females produce only four or five eggs in a lifetime that at most lasts but a few weeks.

Development Gastrotrich embryonic development has been worked out in some detail (see Sacks, 1955; Teuchert, 1968; Hummon, 1974). Cleavage is holoblastic and determinate. The first division is transverse, that is, perpendicular to the axis of the embryo. The AB cell is anterior, and the CD cell posterior. The second divisions are parallel to the axis but at right angles to each other. The C cell then shifts anteriad in macrodasyids and will give rise to much of the primary ectoderm. The blastomeres shift during the course of early cleavage in patterns that are somewhat different between the two orders, such that the germ layers of macrodasyids arise from different blastomeres than do those of chaetonotids. A **coeloblastula** invaginates as two blastomeres move into the interior (Fig. 11.2D), derived from the D cell in macrodasyids and the A cell in chaetonotids, to produce a midgut. The blastopore stretches and forms a ventral groove that eventually closes over. In macrodasyids, this closing is completed with the formation of a **stomodeum**. In chaetonotids, the formation of the stomodeum and mouth is followed by the appearance of a second furrow posteriorly to form a **proctodeum**, no such event occurring in macrodasyids. There are no larval stages. Young, which look like diminutive adults at hatching, grow rapidly to mature in two to three days.

Phylogeny

Gastrotrichs would seem to be among the most primitive of bilaterians and often are compared to the acoelomate Gnathostomulida. The monociliary epidermis (Rieger, 1976), monoecious reproduction, and the acoelomatelike nature of the body are features shared with platyhelminths or gnathostomulids. However, some of these characters are generally thought to be plesiomorphic or primitive, and thus cannot be used to link taxa (Boaden, 1985), while others may be convergent. Gastrotrichs clearly have a complete gut with anus, and the absent pseudocoel is due to a structural adaptation of the Y-organ to aid locomotion in these tiny animals. Of particular, and still not fully assessed, interest is the close similarity in cuticle structure and early embryonic cleavage patterns with those seen in nematodes.

Taxonomy

The two gastrotrich groups are so decidedly different to Inglis (1985) that he suggested each should be placed in at least separate classes, if not separate phyla. This arrangement may have some merit, but it is not clear whether the consequent defining characters of each class are unique derived features. In addition, those characters that the orders share appear to be largely primitive or general bilaterian features. Thus, we retain the use of orders here.

Order Macrodasyida. Anterolateral and posterior adhesive tubes; gut with pharynx with pharyngeal pores and of inverted Y-shaped cross section, anus ventral; hermaphroditic; no protonephridia (Fig. 11.1A).

Order Chaetonotida. Distinct head lobe, with prominent sensory bristles or cilia; gut with pharynx of Y-shaped cross section, anus dorsal; protonephridia with flame bulbs; males generally absent (Fig. 11.1B).

Selected Readings

Boaden, P. J. S. 1985. Why is a gastrotrich? In *The Origin and Relationship of Lower Invertebrates* (S. Conway Morris, et al., eds.), pp. 248–260. Clarendon Press, Oxford.

Hummon, M. R. 1974. Reproduction and sexual development in a freshwater gastrotrich. 3. Postpartheongenic development of primary oocytes and the X-body. *Cell Tissue Res.* 236:629–636.

Hummon, W. D. 1974. Gastrotricha. In *Reproduction of Marine Invertebrates*, vol. 1 (A. C. Giese and J. S. Pearse, eds.), pp. 485–506. Academic Press, New York.

Hummon, W. D., and M. R. Hummon. 1983. Gastrotricha. In *Reproductive Biology of Invertebrates*, vol. 1 (K. G. Adiyodi and R. G. Adiyodi, eds.), pp. 211–221. John Wiley, New York.

Inglis, W. G. 1985. Evolutionary waves: patterns in the origins of animal phyla. *Aust. J. Zool.* 33:153–178.

Remane, A. 1936. Gastrotricha. In *Dr. H. G. Bronn's Klassen und Ordnungen des Tierreiches*, Band 4, Abt. 2, Buch 1, Tiel 2, Lief. 1–2, pp. 1–242. Akad. Verlagsges., Leipzig. [An old but classic work.]

Rieger, G. E., and R. M. Rieger. 1977. Comparative fine structure study of the gastrotrich cuticle and aspects of cuticle evolution within the Aschelminthes. *Z. zool. Syst. Evolut.-forsch.* 15:81–124.

Rieger, R. M. 1976. Monociliated epidermal cells in Gastrotricha: Significance for concepts of early metazoan evolution. *Z. zool. Syst. Evolut.-forsch.* 14:198–226.

Ruppert, E. E. 1982. Comparative ultrastructure of the gastrotrich pharynx and the evolution of myoepithelial foreguts in Aschelminthes. *Zoomorph.* 99:181–220.

Sacks, M. 1955. Observations on the embryology of an aquatic gastrotrich, *Lepidodermella squammata*. *J. Morph.* 96:473–484, pls. 1–5.

Storch, V. 1984. Minor pseudocoelomates. In *Biology of the Integument*, vol. 1 (J. Bereiter-Hahn et al., eds.), pp. 242–268. Springer-Verlag, Berlin.

Teuchert, G. 1968. Zur Fortplanzung und Entwicklung der Macrodasyoidea. *Z. Morph.* 63:343–418. [Good English abstract and summary.]

12

Rotifera

Rotifera (also known as Rotatoria) range in size from 100 to 2000 μm, and some 1800 species are known. They are found in fresh and marine water as well as in moist terrestrial habitats among mosses and lichens.

Definition

Rotifera are minute, unsegmented pseudocoelomates, with an anterior ciliated region or corona. The cuticle is perforated by radial tubules from a syncytial epidermis. There is typically a complete digestive tract with an anterior mouth, a highly differentiated pharynx containing movable pieces that act as jaws, and a posterior anus emptying into a cloaca along with the protonephridia and female reproductive tract. In some groups, females produce two kinds of eggs, one developing parthenogenetically, another requiring fertilization, but the class Bdelloidea lacks males altogether.

Body Plan

Rotifer anatomy is summarized in Figure 12.1. The body is divided into a head bearing the **corona**, a **trunk**, and a posterior region called the **foot**. A **neck** may separate the head and the trunk.

Often the central part of the head lacks cilia and is known as the **apical field**. The ciliated corona encircles the apical field (Fig. 12.1). Projecting sensory bristles and one or two papillae with pores of the **retrocerebral organs** lie on the apical field. Retrocerebral organs may be homologous to the frontal glands of turbellarians. Each consists of a pair of **subcerebral glands** near the brain, and a median **retrocerebral sac** (Fig. 12.2A). Either sac or glands may be missing, and both vary markedly in size and form in different species. The function of the retrocerebral organ is unknown, although the presence of the organ appears to be negatively correlated with the number of protonephridia.

The form of the corona determines the overall appearance of the head. The coronal cilia beat metachronally and give the impression of a revolving wheel, and from this the name rotifer, or wheelbearer, is derived. The corona of primitive, creeping species is composed of an evenly ciliated buccal field arranged around the mouth, and a circumapical band that encircles the apical field (Fig. 12.2B). All resemblance to the primitive form is lost in specialized species. Tufts of cilia may be present at the sides of the buccal field, and from these, lateral projections known as auricles may develop. Swimming rotifers usually have a reduced buccal field divided into suboral and supraoral fields. Stiff cirri, several cilia covered by a single membrane, form a **pseudotroch** in the supraoral field of some species. The apical field of some rotifers is bilobed (Fig. 12.2E,F), and in bdelloids it is separated into two fields (Fig. 12.2G,H). Cilia at the anterior and posterior margins of the ciliated field are enlarged, forming two special circlets, an anterior **trochus** and posterior **cingulum**. The apical field of those Flosculariacea who are sessile is lobed, with the trochus more prominent than the cingulum (Fig. 12.3B). The sessile Collothecacea have a funnel-shaped apical field, usually with marginal lobes bearing filamentous setae (Fig. 12.3A) and with coronal cilia missing or reduced. Protozoa and algae are guided into the funnel by the setae; the funnel then closes to bring the prey to the mouth. The Bdelloidea (Fig. 12.3C) have a **rostrum**, or proboscis, that is only extended when the wheel organ is retracted. The latter are withdrawn into an epidermal inpocketing, possibly a **lemniscus**, which extends down into the **pseudocoel** in the retracted state.

The trunk, cylindrical or flattened, is covered by a variously modified cuticle. The cuticle of swimming species sometimes forms an enveloping case, the **lorica**. The cuticle of creeping species is divided into telescoping rings that allow the body to extend and contract. Single or paired dorsal antennae occur at the anterior end of the trunk, and lateral antennae may be found on the posterior end of the trunk. The anus, middorsal in position, lies at the point of union of the trunk and the foot.

The foot of most rotifers is covered by cuticular rings. It ends in an adhesive organ in sessile species, but in motile rotifers often ends in one to four movable toes containing pedal cement glands. These glands secrete an adhesive used to attach the animal permanently, or temporarily for feeding or creeping.

The body wall or integument is composed of the cuticle, epidermis, and subepidermal muscles. The syncytial integument, with keratinlike lamina within the cytoplasm, covers the body surface and forms the lorica, spines, or other surface structures (Storch, 1984; Clement, 1985; Bender and Kleinow, 1988). The syncytium contains a constant number of nuclei, formed into regularly arranged epidermal cushions that contain one or several nuclei. The modified epidermis of the corona contains a ciliary root system and large epidermal cushions, and is termed the coronal matrix. Subcerebral glands, pedal glands, and other surface glands are epidermal in origin, although they extend below the epidermis.

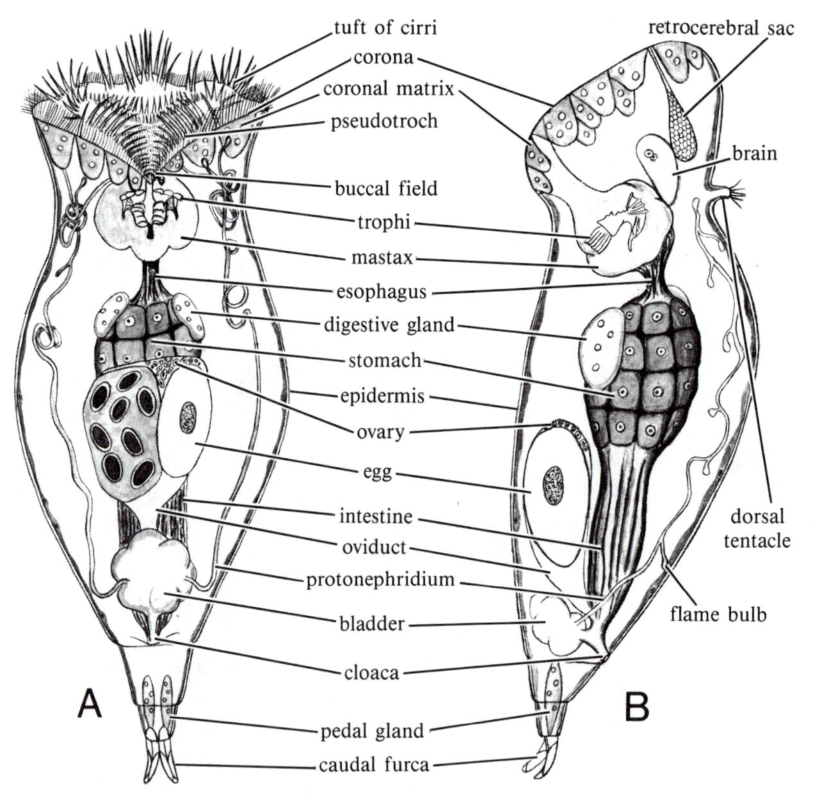

FIGURE 12.1. Rotifer body plan. **A**. Scheme of *Epiphanes senta,* as seen in ventral view; **B**. from lateral view.

The foot epidermis or pedal glands secrete the material used to make the tubes of sessile rotifers and the gelatinous envelopes of some pelagic rotifers.

Rotifer muscles are not arranged in sheets or layers. Visceral muscles occur in some organs, and cutaneovisceral muscles extend from the body wall to the viscera. Subepidermal muscles form the body wall musculature and consist of circular muscle bands, which when they constrict lengthen the body, and longitudinal bands, which when they constrict shorten the body or retract the corona and foot. The subepidermal circular muscles lack nuclei. They are arranged either in complete rings or as laterally placed transverse arcs, but are much reduced in loricate species. Longitudinal muscles pass through several cuticular rings in creeping forms, or are detached for most of their length and joined to the cuticle only near the head or the foot to serve as retractors. Longitudinal muscles are nucleate, and there are both smooth and striated types.

Digestive System The gut is typically well developed in rotifers, but the genera *Asplanchna* and *Asplanchnopus* lack a complete digestive system. The mouth usually lies in the buccal field or below a supraoral field. Sphincter and dilator muscles may open and close the mouth. Raptorial species usually have oral dilator muscles and a mouth opening directly into the pharynx. Other species have a ciliated buccal tube leading to the pharynx.

The pharynx, called the **mastax**, is a unique rotifer feature. It is a muscular chamber containing a set of jaws, formed of cuticularized trophi. Jaw structure is complex (Fig. 12.4A). A median fulcrum and pairs of rami, unci, and manubria make up the major parts. The fulcrum and rami together form the incus; the unci and manubria compose the malleus. The intricate jaw architecture is adapted along a variety of patterns, associated with the kind of food eaten and feeding habits, but the component parts usually remain recognizable. Jaw homologies have been worked out in detail (Fig. 12.4B,C,D, E,F,G,H,I), and jaw structure is very important in taxonomy.

The thin-walled, syncytial intestine is sometimes ciliated. A posterior cloaca receives the gut, protonephridia, and oviducts. The dorsal anus often has dilator muscles.

Excretory System Rotifers have two syncytial protonephridial tubules, sometimes joined by a transverse tube. These are coiled and usually fork anteriorly, terminating in flame bulbs. Flame bulbs are different from the typical flame cell in several respects, but most notably, instead of cilia for a flame, have 30 or more flagella. The protonephridia drain

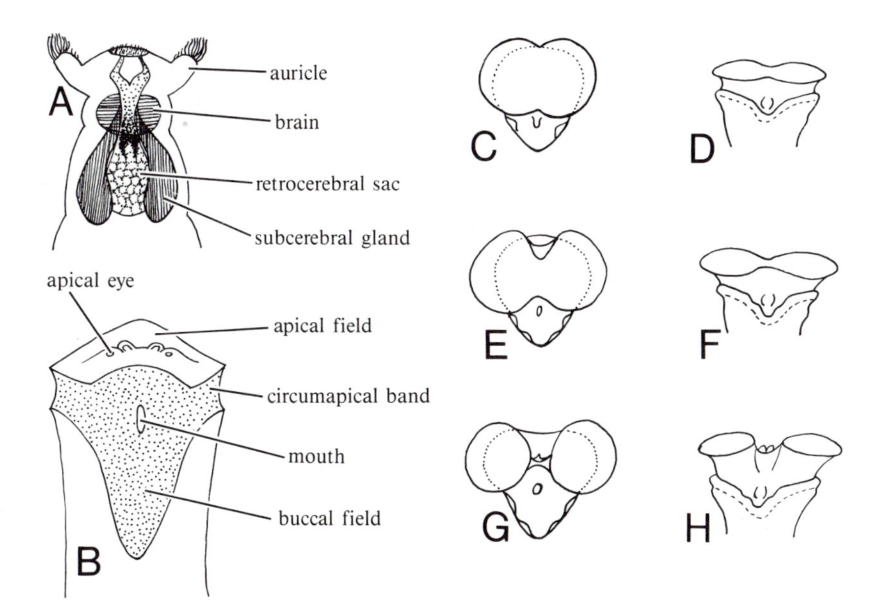

FIGURE 12.2. **A**. Unusually prominent retrocerebral organs of *Notommata*. **B**. Ventral view of primitive corona. **C**,**D**. Top and ventral views of basic corona. **E**,**F**. Top and ventral view of floscularian type of corona. **G**,**H**. Front and ventral views of bdelloid type of corona. (All after de Beauchamp.)

into the cloaca by way of a common duct or a bladder.

Nervous System The nervous system is bilaterally symmetrical. A bilobed brain gives rise to a pair of longitudinal, ventrolateral nerve trunks. Several ganglia are connected to the brain or nerve trunks (Fig. 12.5A). For example, the anterior end of each nerve trunk bears a small anterior ganglion and a large geniculate ganglion. Posteriorly, the nerve trunk ends in a pedal ganglion, associated with the foot, and a vesicular ganglion, associated with the urinary bladder, although these two ganglia may be united as a single caudovesicular ganglion. A mastax ganglion lies on the surface of the pharynx and is connected to the brain by pharyngeal nerves; visceral nerves to the intestine issue from the mastax ganglion.

Sensory and motor nerves are distinct, and most body ganglia are wholly motor or sensory. The cephalic sense organs and dorsal antennae are innervated from the brain. Lateral antennae receive nerves from the geniculate ganglia. Any caudal sense organs present are innervated by nerves from the pedal or caudovesicular ganglion. Motor nerves from the brain innervate the salivary glands and anterior retractor muscles. Posterior retractors receive motor nerves from the geniculate ganglia or the ven-

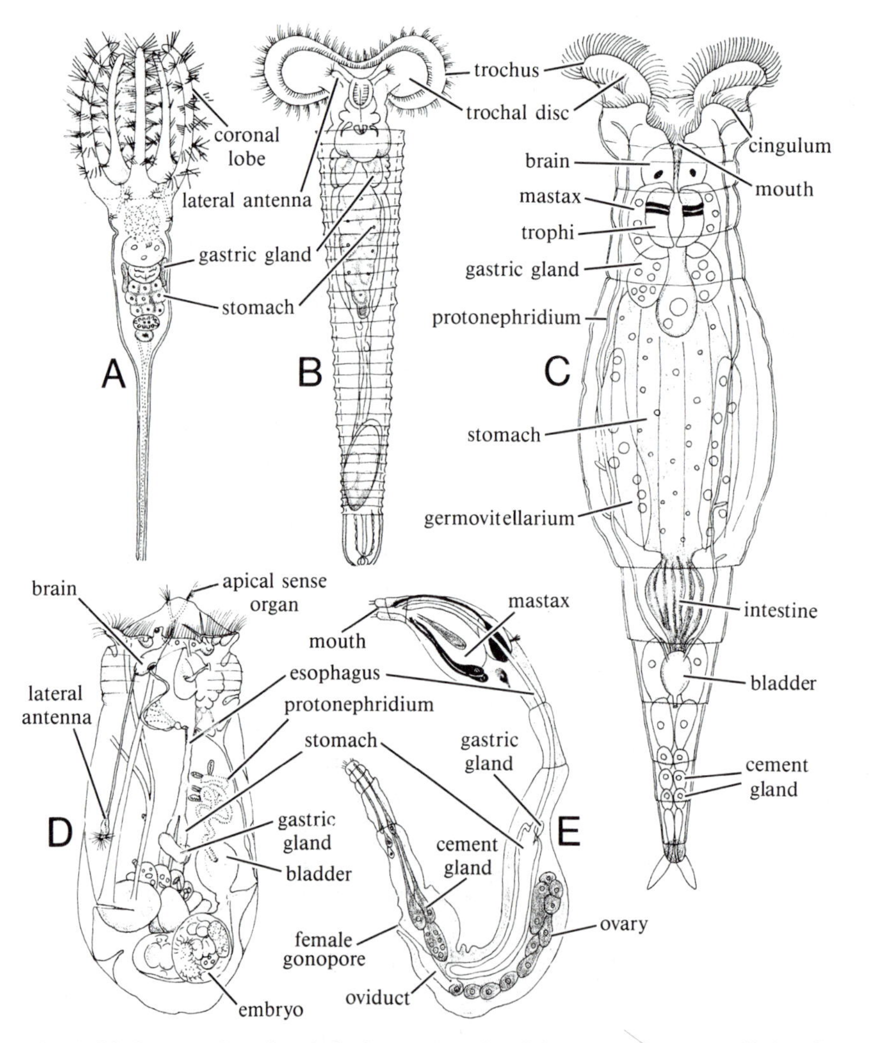

FIGURE 12.3. Representative rotifers. **A.** *Stephanoceros,* sessile collothecacean, with corona modified into funnel equipped with bristles but no cilia. **B.** *Limnias,* sessile flosculariacean, with corona of trochal and cingular circlets, and foot with no toes. **C.** *Philodina,* bdelloidean, with two germovitellaria and retractile anterior end. **D.** *Asplanchna,* ploiman, with normal corona and foot. **E.** *Seison,* characterized by epizoic habits and poorly developed corona. (**A,B,D** after Edmondson; **C** after Hickernell; **E** after Plate.)

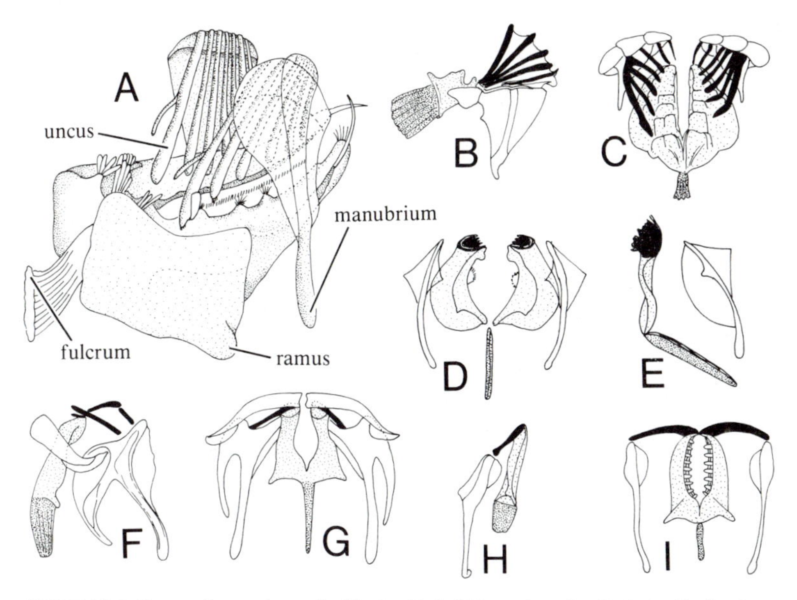

FIGURE 12.4. Comparative anatomy of rotifer trophi. **A**. Oblique view of malleate trophi, showing arrangement of component parts. **B**,**C**. Lateral and dorsal view of grinding trophi of *Epiphanes senta.* **D**,**E**. Dorsal and lateral views of piercing trophi of *Synchaeta.* **F**,**G**. Lateral and ventral views of suctorial trophi of *Lindia.* **H**,**I**. Lateral and ventral views of forcepslike trophi. Fulcrum (*dark stipple* in **B**,**C**,**D**,**E**,**F**,**G**,**H**,**I**); ramus (*light stipple* in **B**,**C**,**D**,**E**,**F**,**G**,**H**,**I**); uncus (*black* in **B**,**C**,**D**,**E**,**F**,**G**,**H**,**I**); manubrium (*unshaded* in **B**,**C**,**D**,**E**,F,G,H,I). Certain accessory pieces also unshaded. (**A** adapted from Stossberg by Edmondson; **B**,**C**,**D**,**E**,**F**,**G**,**H**,**I** adapted from various sources.)

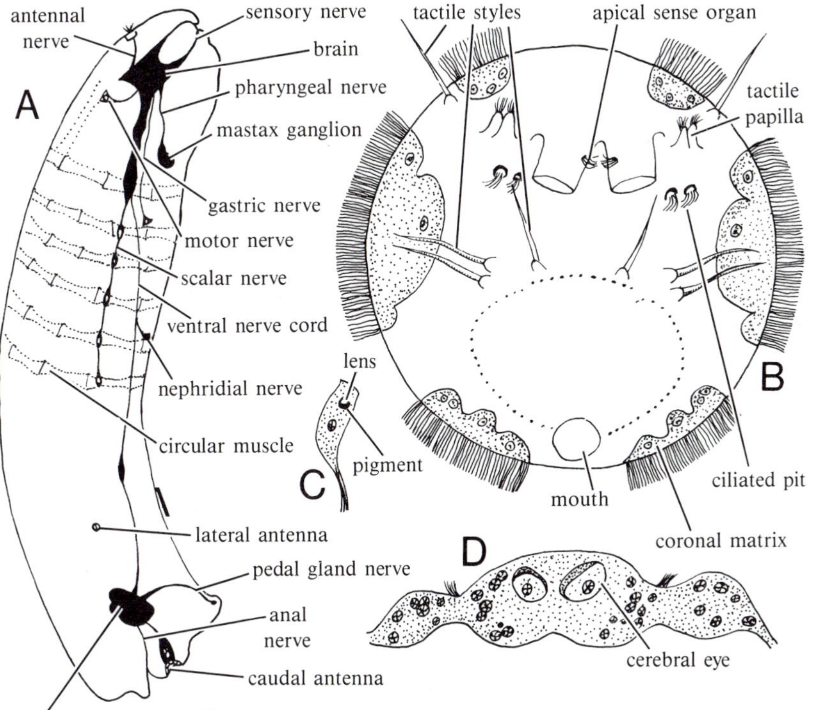

FIGURE 12.5. Rotifer nervous system and sense organs. **A**. Nervous system of *Lindia.* Note partially segmental character, related to partially segmental arrangement of musculature. **B**. Scheme of corona of *Euchlanis,* showing position of sense organs. **C**. Frontal eye of *Euchlanis,* containing intracellular lens and pigment cup. **D**. Section through brain of *Synchaeta,* passing through cerebral eyes. (**A** after Diehl; **B**,**C** after Stossberg; **D** after Peters.)

tral nerve trunks. A scalar nerve from the geniculate ganglion goes to the circular muscles of the body wall; it contains a ganglion cell in each muscle band. The coronal sphincter receives a motor nerve from the geniculate ganglion or nerve trunk on each side of the body.

The outstanding features of the nervous system of rotifers are (1) the clear separation of sensory and motor elements and (2) the phenomenon of cell constancy. Brain cells and nerve cells, like the epidermal syncytium, are constant in number and position, providing a remarkable consistency in the nervous system of each species.

Reproductive System The female reproductive organs are very simple. Monogonont females have a single syncytial **ovary** and a syncytial **vitellarium** that is enclosed in a membrane that continues as the **oviduct** and opens into the cloaca (see Fig. 12.1). Bdelloids have a pair of **germovitellaria** and a common oviduct (see Fig. 12.3C). Females of the monogeneric Seisonidea have paired ovaries, but no vitellaria.

Males can be as little as a quarter the size of females, or less in most cases. Monogonont males have a single **testis**, opening into a sperm duct ending in the male gonopore (Fig. 12.6A). The last part of the sperm duct is sometimes modified as an eversible **cirrus**, and some males have a part of the body wall specialized as a protrusible penis. Impregnation is usually hypodermic. Sperm mature during embryonic development and are of two kinds: typical, ciliated sperm, and rod-shaped sperm (Fig. 12.1B). The latter probably help to pierce the cuticle. In *Seison* the males equal the females in size, and the male system is quite different from that in other rotifers. Testes are paired, and the sperm duct forms chambers and tortuous coils as it passes through a syncytial mass. The sperm are cemented together into a packet, the spermatophore, in the syncytium.

Biology

Locomotion Locomotory efficiency is size dependent. Large *Asplanchna* move relatively more slowly than smaller individuals. Rotifer size appears to test the upper limits of the mechanical efficiency of cilia as locomotory organs. *Brachionus* expends 62 percent of its total energy output in swimming (Epp and Lewis, 1984). Rotifers can move about either by swimming or crawling. The beating of the coronal cilia move the animal about in swimming. Pelagic forms facilitate maintaining buoyancy with fat glob-

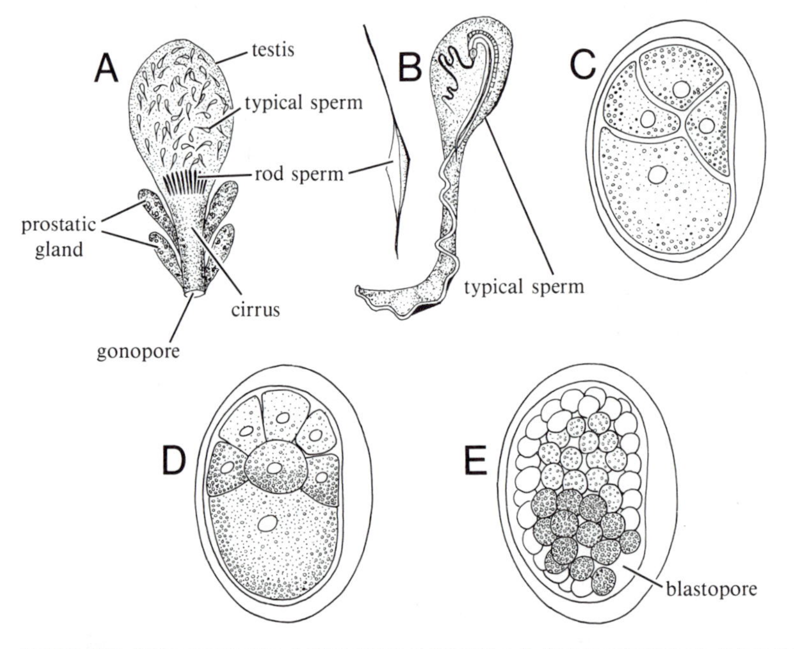

FIGURE 12.6. Rotifer reproduction. **A**. Male system of *Rhinoglena*. **B**. Typical and rod sperm of *Sinantherina*. **C,D,E**. Three stages in development of *Monostyla;* **C**. four-cell stage; **D**. eight-cell stage, with A, B, C, and d^1 filled with yolk and forming germ ring; **E**. gastrulation by epiboly and involution. Ectodermal cells have covered surface of embryo; derivatives of **D** with moderate amount of yolk have moved away from blastopore and will form ovary and vitellarium; cell derivatives of A, B, C, and d^1 filled with yolk lie near blastopore. (**A** after Wesenberg-Lund; **B** after Hamburger; **C,D,E** after Pray.)

ules. The foot and its associated glands in bdelloids help move the body around in a leechlike manner in crawling. Those rotifers with a rostrum or proboscis also employ this structure in alternation with the foot to creep about on the substrate. Some highly specialized forms are sessile (Wallace, 1980) and remain permanently affixed to a substrate once the juvenile attaches and undergoes metamorphosis. A few species form tight epizootic relationships with microcrustaceans, but most sessile rotifers attach to common water plants. Two families of the order Flosculariacea form colonies such as in *Conochilus unicornis* and *Floscularia conifera* (Wallace, 1987).

Feeding The jaws exhibit wide varieties of form related to dietary preferences. For example, some grind food (see Fig. 12.4B,C); others are used to pierce and/or immobilize food such as plankton or plant cells (see Fig. 12.4D,E); one types rolls the elements to facilitate sucking (see Fig. 12.4F,G); another type acts as a forceps to grab prey and pull it into the mastax (see Fig. 12.4H,I). Food leaving the pharynx is often fully macerated, but some families swallow their prey whole.

Food passes through an esophagus into the stomach, a thick-walled, ciliated sac or tube composed of a constant number of cells. A delicate mesh of circular and longitudinal muscles covers the surface, and a pair of syncytial gastric glands open into it. The gastric glands secrete enzymes for extracellular digestion; however, in the rotifers that digest food intracellularly, the gastric glands are reduced. Carbohydrate digestion is apparently limited. Digestion is very rapid; within 15 to 20 minutes protein spherules and fatty droplets begin to accumulate in the stomach wall where they may be stored for some time. The bdelloid stomach is a heavy, syncytial mass pierced by a narrow, ciliated tube. Food enters the syncytium to be digested intracellularly.

Excretion Flame bulb activity varies with the osmotic pressure of the environment, but typically the bladder or cloaca is flushed several times per minute. The protonephridial system is probably primarily a water-regulating device. Although the salt concentration of the pseudocoel is maintained in tandem with that of the environment, the expelled fluid is lower in salt content than the fluid of the pseudocoel (Epp and Winston, 1977). Bdelloids can remain dormant for years by forming a desiccated, noncyst, resistant stage if water becomes scarce.

Respiration Generally, swimming and creeping rotifers have high oxygen requirements. However, some live in habitats with little free oxygen, and presumably these have some ability to respire anaerobically. Extreme drops in salt concentration of the environment decreases rotifer oxygen consumption, but changes in pH seem to have little effect (Epp and Winston, 1978). On the other hand, changes in temperature have limited effect on rates of respiration. *Brachionus* experiences no change in the metabolic rate at temperatures of 20 to 28°C, although below and above that range, metabolism is much higher (Epp and Lewis, 1980).

Circulation The spacious pseudocoel is filled with a perivisceral fluid and a loose syncytial reticulum composed of amoeboid cells. Such small creatures with a high degree of body mobility, flexibility, and an open pseudocoel probably have little trouble in moving materials about the body. No special circulatory adaptations occur in the phylum.

Sense Organs A variety of sensory projections occur on the body surface, especially in the apical field (Fig. 12.5B). Stiff bristles (**styles**) along the anterior edge of the circumapical band have one or two sensory cells at their bases. Paired ciliated pits, probably chemoreceptive, occur on the apical field of some rotifers, and palps or other projections with sensory bristles are commonly found.

Rotifer eyes are unusual. A mass of red pigment granules forms an intracellular pigment cup with a refractive spherule (Fig. 12.5C). Rotifers may have cerebral eyes (Fig. 12.5D), coronal eyes, or lateral eyes, depending on whether the eye cell is on or in the brain, the coronal matrix, or endodermal cushions lateral to the corona. The eye cells apparently have some features in common with the retinal cells of acoelomates (Clément, 1980). Rotifers respond positively or are indifferent to light. When positively phototactic, general body sensitivity to light induces swimming toward the light at a speed inversely related to the wavelength, and stimuli received by the eyes control movements to maintain a predetermined angle toward the light rays.

Most rotifers have a dorsal antenna, and some have lateral antennae. **Antennae** are tactile organs and may be conspicuous tentacles or reduced, in some cases to a single style.

The retrocerebral organs had been long thought to be neurosecretory. However, investigation reveals that they are composed of two kinds of mucous cells and appear to be glands (Clément, 1977).

Reproduction Most rotifers are females. Bdelloids are exclusively so, producing only parthenogenetic ova. *Seison* is an exception, for males are always present and females cannot reproduce parthenogenetically. In Monogononta, males appear only rarely.

Reproduction is simple enough in bdelloids: the females all produce ova that develop parthenogenetically into parthenogenetic females. It is simple in *Seison* also: the females produce only ova that must be fertilized and develop into a male or a female. Female Monogononta, however, produce three kinds of ova. During most of the year the females are all diploid and produce diploid **amictic eggs** that develop without fertilization into more diploid, amictic females. Just before the species is to become quiescent, either in response to the effects of crowding or in response to striking environmental changes, **mictic females** appear, which, except in *Asplanchna*, are morphologically similar to amictic females. Mictic females produce haploid **mictic eggs** that, if unfertilized, may develop parthenogenetically into **haploid males** but, if fertilized, form a thick-walled case and remain dormant as diploid **resting eggs** until favorable environmental conditions return for new amictic females to hatch (Gilbert, 1963).

Rotifer phenotypes can be environmentally controlled. The form of amictic females gradually changes during the year in response to environmental conditions, a process referred to as **cyclomorphosis**. *Asplanchna brightwelli*, a carnivorous species, releases a factor into rotifer cultures that appears to cause *Brachionus calyciflorus* to develop an extra set of spines that neither the mother nor siblings in untreated controls possess. Such spines are also noted in natural populations of *B. calyciflorus* when *A. brightwelli* is present (Gilbert, 1966).

Development Cleavage is holoblastic. In the four-cell stage (Fig. 12.6C), there are three smaller blastomeres (A,B,C) and one large one (D). D divides unequally; the smaller daughter (d^1) comes to lie at about the same level as the A, B, and C cells. These also divide unequally, giving rise to three smaller, granular blastomeres that with d^1 form a germ ring, and three larger, clear blastomeres above them (Fig. 12.6D). The descendants of these clear blastomeres form the ectoderm and can be recognized easily in later stages. The descendants of the densely granular blastomere of the germ ring, and of the less granular D cell, can also be recognized easily in later stages (Fig. 12.6E).

The blastopore forms opposite the ectodermal cells, and epiboly and involution result in the clear ectodermal cells covering the surface, while the descendants of D and of the germ ring blastomeres move into the interior. Eventually the blastopore closes as the ectodermal cells meet. The progeny of the D cell form the germovitellarium, and the yolk-rich cells of the germ ring form the gut from esophagùs to intestine. The mastax arises from a part of the stomodeum, and the foot from ectoderm at the opposite end of the embryo.

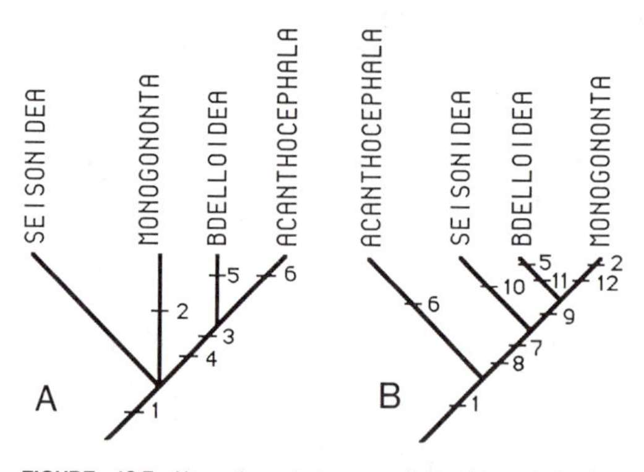

FIGURE 12.7. Alternative cladograms illustrating relationships among rotiferan groups and Acanthocephala. Derived characters are: (1) epidermal cuticle pierced by tubules; (2) one gonad; (3) lemnisci projecting into pseudocoel; (4) introvert proboscis; (5) lack of males; (6) acanthocephalan autapomorphies (see Chapter 13 for details); (7) retrocerebral organ; (8) epidermis with intrasyncytial lamina with several layers; (9) vitellarium; (10) spermatophores; (11) noncyst desiccated resting stage; and (12) mictic resting egg. (**A** modified from Lorenzen, 1985.)

Fossil Record and Phylogeny

Some fossilized rotifer cysts are known from the Eocene, but beyond this serendipitous occurrence the phylum has no fossil record.

Evolution in the group has been constrained by its elaboration of asexual modes of reproduction and by the phenomenon of **eutely**, or cell constancy. The latter has been most crucial in that it restricts the capacity to repair injury and poses embryological and physiological problems of great importance. What makes cells stop dividing, and how can a syncytial animal differentiate? On the other hand, rotifers have enjoyed great success because of their great reproductive capacities under optimal conditions.

Although rotifers are clearly characterized by the presence of the corona, mastax, foot with adhesive glands, and retrocerebral organ, Lorenzen (1985) suggests this is not entirely sufficient to establish Rotifera as a monophyletic group. This is because Acanthocephala could be considered as highly specialized rotifer types that have secondarily lost these features. Under such an interpretation, Rotifera would be a paraphyletic group (Fig. 12.7A). An alternative view held by many rotifer authorities (e.g., see Clément, 1985) interprets the lemnisci and introvert proboscis as convergent characters. In this case, only the nature of the epidermal tubules would serve to unite Acanthocephala and Rotifera, and both groups would be monophyletic. The inclusion of some additional characters of reproductive anatomy and epidermal biochemistry would further serve to resolve the rotifer/acanthocephalan relationships (Fig. 12.7B). Be that as it may, a consensus seems to be emerging that Rotifera and Acanthocephala at least are related in some way.

Taxonomy

Class Seisonidea. Marine rotifers, living on gill surfaces of Crustacea; body elongated, with a long neck; jaws (trophi) as forceps; small corona; paired gonads; without vitellaria; males present; sexual reproduction obligatory, females do not form parthenogenetic ova. Example: *Seison* (Fig. 12.3E).

Class Monogononta. Swimming or sessile; trophi not ramate; males more or less reduced and with one testis; females with one germovitellarium.

Order Ploima. Swimming forms with normal corona; foot reduced or with two toes and two pedal glands. Examples: *Epiphanes* (Fig. 12.1) and *Asplanchna* (Fig. 12.3D).

Order Flosculariacea. Sessile or swimming; corona often circular or lobed, with trochal and cingular circlets; foot without toes; males much reduced. Example: *Limnias* (Fig. 12.3B).

Order Collothecacea. Generally sessile rotifers with anterior end modified as funnel and without definite ciliary circlets; often with immobile bristles; foot without toes; males much reduced. Example: *Stephanoceros* (Fig. 12.3A).

Class Bdelloidea. Swimming or creeping forms, with retractable anterior end; corona with two trochal discs or some modification of this pattern; mastax ramate; foot often with more than two toes, and with spurs; with more than two pedal glands; males unknown, females with two germovitellaria. Example: *Philodina* (Fig. 12.3C).

Selected Readings

Bender, K., and W. Kleinow. 1988. Chemical properties of the lorica and related parts from the integument of *Branchionus plicatilis*. *Comp. Biochem. Physiol.* 89B:483–487.

Clément, P. 1977. Ultrastructural research in rotifers. *Arch. Hydrobiol. Beih. Engebn. Limnol.* 8:270–297.

Clément, P. 1980. Phylogenetic relationship of rotifers as derived from photoreceptor morphology and other ultrastructural analysis. *Hydrobiol.* 73:93–117.

Clément, P. 1985. The relationships of rotifers. In *The Origins and Relationships of Lower Invertebrates* (S. Conway Morris, et al., eds.), pp. 224–247. Clarendon Press, Oxford.

Dumont, H. J., and J. Green (eds.). 1980. *Rotatoria*. Junk, The Hague.

Epp, R. W., and W. M. Lewis. 1980. Metabolic uniformity over the environmental temperature range in *Brachionus plicatilis*. *Hydrobiol.* 73:145–147.

Epp, R. W., and W. M. Lewis. 1984. Cost and speed of locomotion for rotifers. *Oecologia* 61:289–292.

Epp, R. W., and P. M. Winston. 1977. Osmotic regulation in the brackish-water rotifer *Brachionus plicatilis*. *J. Exp. Biol.* 68:151–156.

Epp, R. W., and P. M. Winston. 1978. The effects of salinity and pH on the activity and oxygen consumption of *Brachionus plicatilis*. *Comp. Biochem. Physiol.* 59A:9–12.

Gilbert, J. J. 1963. Mictic female production in the rotifer *Brachionus calyciflorus*. *J. Exp. Zool.* 153:113–123.

Gilbert, J. J. 1966. Rotifer ecology and embryological induction. *Science* 151:1234–1237.

Koste, W. 1978. *Rotatoria*. Borntraeger, Berlin.

Lorenzen, S. 1985. Phylogenetic aspects of pseudocoelomate evolution. In *The Origins and Relationships of Lower Invertebrates* (S. Conway Morris, et al., eds.), pp. 210–223. Clarendon Press, Oxford.

Schramm, U. 1978. On the excretory system of the rotifer *Habrotrocha rosa*. *Cell Tiss. Res.* 189:515–524.

Storch, V. 1984. Minor pseudocoelomates. In *Biology of the Integument*, Vol. 1 (J. Bereiter-Hahn et al., eds.), pp. 242–268. Springer-Verlag, Berlin.

Wallace, R. L. 1980. Ecology of sessile rotifers. *Hydrobiol.* 73:181–193.

Wallace, R. L. 1987. Coloniality in the phylum Rotifera. *Hydrobiol.* 147:141–155.

13

Acanthocephala

There are almost 1000 species of acanthocephalans, or thorny-headed worms. They occur as intestinal parasites in marine, freshwater, and terrestrial hosts. Because of their highly derived anatomy (little more than gonad bags on an anchor), determination of the closest affinities of this phylum has been a problem. Acanthocephala come close to being the ultimate parasites, with evolution having focused their biology almost entirely on the business of reproduction.

Definition

Acanthocephala are parasitic pseudocoelomates without a digestive tract. A proboscis is armed with cephalic, recurved hooks that originate from the basement membrane of the epidermis. The proboscis is retractable into a proboscis receptacle. The cuticle is pierced by radial tubules extending from a syncytial epidermis, inner portions of the syncytial epidermis being lacunar. Females possess a uterus bell. Acanthella larvae parasitize arthropods, while adults parasitize vertebrates.

Body Plan

The acanthocephalan body is divided by a cuticular furrow into an anterior presoma, consisting of a **neck** and a **proboscis**, and a posterior trunk region (Fig. 13.1). The trunk and the proboscis vary in shape and size between species. The trunk composes the vast bulk of the body and often is armed with spines, and the proboscis armature is extremely variable but species specific.

The body wall (Fig. 13.2B) is composed of an outer cuticle, a thick, syncytial epidermis, a thin dermis, and a thin layer of circular and longitudinal muscle. The syncytial epidermis is made up of several fibrous layers, and contains a constant number of giant nuclei or many small nuclear fragments. The outer cuticle is pierced by tubules or canals extending from the epidermal syncytium below. The lower layers of the epidermis have an elaborate system of **lacunae** (Fig. 13.2A), filled with fluid high in glycogen and lipid. In addition, a separate system of canals, the **rete system**, is associated with the muscle layers (Fig. 13.2B). Two inpocketings of the body wall, the **lemnisci**, project into the pseudocoel in the neck region (Fig. 13.1A,D). The lemnisci may serve as reservoirs for the body fluid when the proboscis is retracted.

The proboscis is a formidable organ, often causing serious damage to the host's intestine. A retractor muscle attaches the proboscis tip to the proboscis receptacle wall. Since neither end is stationary, the proboscis retractor is as likely to pull the receptacle up as to pull the proboscis down. This former is prevented by the receptacle retractor muscle, actually a continuation of the proboscis retractor that attaches the receptacle to the body wall. Neck retractor muscles connect the neck to the body wall, and these muscles often surround the lemnisci, serving as lemnisci compressors.

Digestive and Execretory Systems A digestive system is lacking. However, the suspensory ligament (Fig. 13.1A,C,D), to which the gonads are attached, is thought to be a remnant of that system. The ligament bears some similarities to the reduced intestinal cords seen in some rotifers.

Protonephridia located near the gonads are present only in some archiacanthocephalans and contain multiciliated flame cells.

Nervous System The brain (Fig. 13.1A) is a simple ganglion situated in the proboscis receptacle, issuing nerves to the muscles and the proboscis sense organ. A pair of lateral nerve trunks course along the body wall, sending branches to the genital organs. A ring commissure in the base of the penis contains a pair of genital ganglia.

Reproductive System The two testes are attached to the ligament in a ligament sac (Fig. 13.1A). Muscles from the body wall join the **ligament sac** below the **testes**, forming a **genital sheath** that encloses the sperm ducts and associated structures. Sperm accumulate in **spermiducal vesicles** in the sperm ducts,

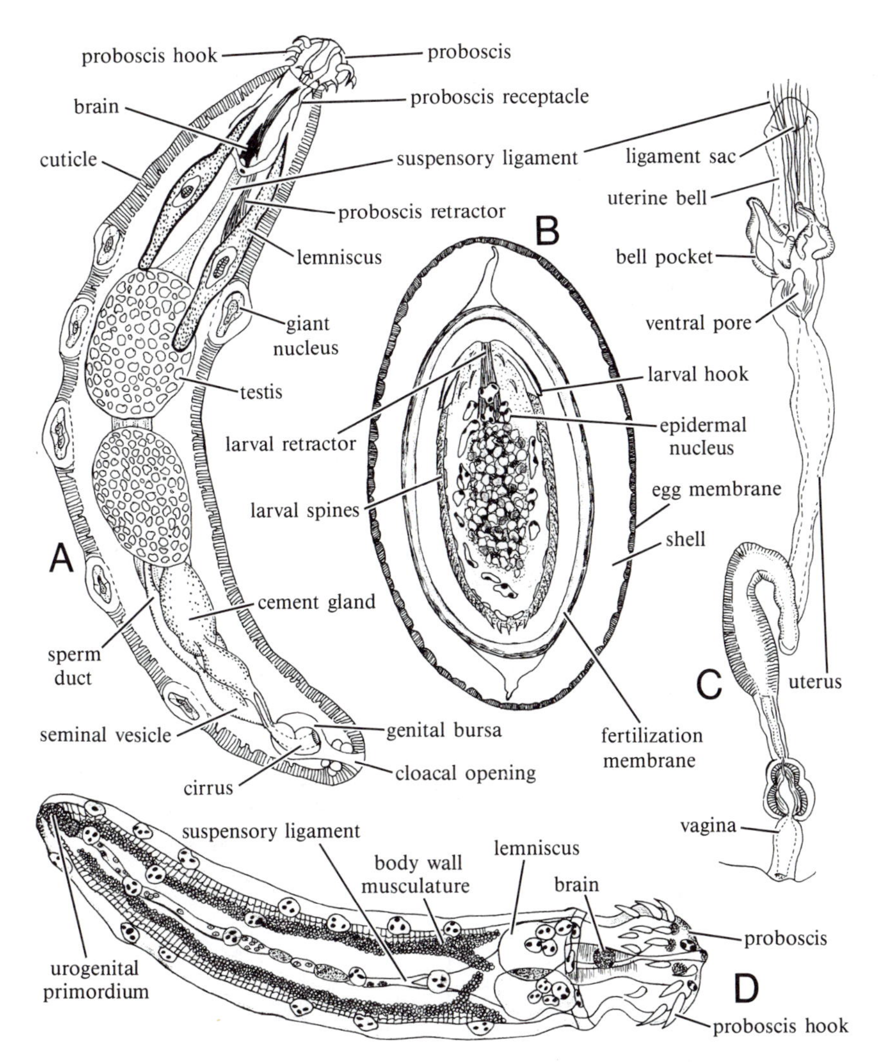

FIGURE 13.1. Acanthocephala. **A**. *Neoechinorhynchus rutili* male, showing main organs in position. **B**. Shelled acanthor embryo of *Macracanthorhynchus hirudinacea*. **C**. Uterine bell system of *Bolbosoma*. **D**. Acanthella of *M. hirudinacea* from beetle larva. (**A** after Biehler, from Goodchild, in Brown; **B** after Meyer; **C** after Yamaguti; **D** after Van Cleave.)

or a **seminal vesicle** at the base of the common **sperm duct**. **Cement glands**, usually six or eight, surround the sperm ducts. Cement passes through cement ducts, sometimes after storage in a cement reservoir, into the common sperm duct near the base of the **penis**. Urine also drains into the common sperm duct if protonephridia are present. The **urogenital duct** thus carries cement, urine, and sperm, and may be the vestige of an ancestral cloaca that may have included an opening of the gut. The endodermal origin of the ligament supports this point of view. The penis projects into a **bursal cavity** with a heavily muscled **bursal cap**, and sometimes connected with a fluid-filled sac, called **Saefftigen's pouch**.

The female system is relatively simple. The **ovary** is suspended in the ligament sac. This organ is quite separate from the uterus **bell**, an organ where eggs are either allowed to proceed to development in the **uterus** or are cycled back into the body to be fertilized.

Biology

Locomotion Movement of these parasites from host to host is passive as eggs or larvae are eaten. Once within the primary host and permanent attachment occurs, subsequent body movements are achieved by the body wall muscles and are largely directed toward groping about the gut to find a mate.

Feeding Digested food is taken from the host's gut into the acanthocephalan through the worm's porous cuticle. The adjacent tissues of the worms are very rich in glycogen, Golgi bodies, mitochondria, and digestive-type enzymes. Carbohydrate metabolism is the main source of energy in these parasites. Glucose and glycogen are metabolized anaerobically, although details as to specific chemical pathways used are not yet clear.

Excretion It is assumed that the body wall functions in waste elimination and osmoregulation. Recent studies indicate that all acanthocephalans are probably osmoconformers, that is, passively adjusting to prevailing conditions in the host gut.

Circulation Since the body of thorny-headed worms is relatively large, a means to move nutritive materials around by other than passive diffusion is a functional necessity. The lacunae (Fig. 13.2) serve such a function, and indeed the fluid contained within them is largely nutritive in nature, high in glycogen and lipid (Miller and Dunagen, in Cromp-

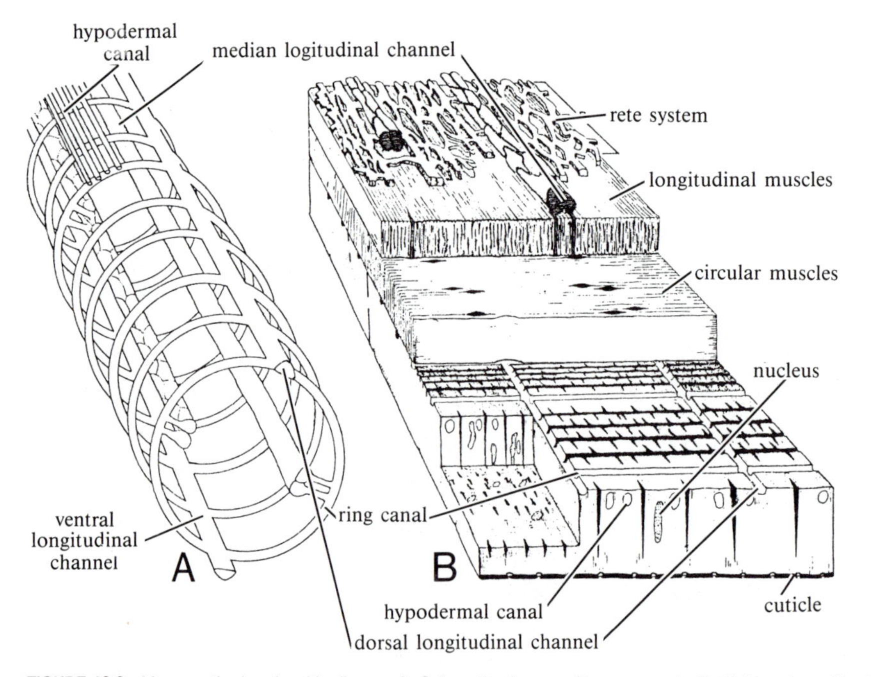

FIGURE 13.2. *Macracanthorhynchus hirudinacea*. **A**. Schematic diagram of lacunar canals. **B**. Oblique view of body wall from within looking out, innermost layers on top, outermost layers and cuticle on bottom. (After Miller and Dunagen.)

ton and Nichol, 1985). There is no pumping mechanism, however, and movement within the lacunae is facilitated only by muscular contractions of the body wall.

Sense Organs Sense organs have degenerated to a few bulbous endings in the reproductive organs and a single sensory pit at the proboscis tip; sometimes an additional pair of sensory pits lie in the neck.

Reproduction During mating, the muscles of the bursal cap in the male contract, and fluid is injected from Saefftigen's pouch, if such is present. The bursa everts and enters the female. Sperm are injected into the vagina, which is then in turn plugged with cement. The sperm travel up the genital duct and escape into the fluid of the pseudocoel.

In the female, the ovary breaks up into fragments that float as **ovarian balls** in the pseudocoel (Crompton, in Crompton and Nichol, 1985). As ova mature, they escape from the ovarian balls and are fertilized by the sperm in the pseudocoel. A fertilization membrane forms within the egg membrane, and development begins.

A unique organ, the uterine bell (Fig. 13.1C), somehow evaluates the condition of ova that pass through it. If fertilized, eggs pass into the uterus through the bell pockets. If unfertilized, the ova are cycled back into the pseudocoel through the ventral pore of the bell for reexposure to sperm. When the uterus is full, the cement plug is somehow popped, allowing the developing eggs to be passed into the host's gut for elimination with the feces.

Development Acanthocephalan cleavage is rather peculiar but is reminiscent of the spiral, determinate cleavage of flatworms. Cleavage is complicated by the formation of polar bodies at what seems to be the anterior end of the animal, just the opposite seen in other animals (Schmidt, in Crompton and Nichol, 1985). A stereoblastula forms, and the cell membranes break down to produce a syncytial embryo. The only vestige of gastrulation is the inward movement of a small group of nuclei that will become the primordia of gonads and the ligament sac. Differentiation does not proceed while the embryos reside in the mother's body. The embryo develops six hooks, and a shell forms between the inner fertilization membrane and the outer egg membrane. This hooked, shelled larva is known as an **acanthor** (Fig. 13.1B), and it is ready to be infective.

When the acanthor is eaten by a suitable arthropod secondary host, it emerges from the shell and works through the intestinal wall into the host hemocoel. Here, it becomes an **acanthella** (Fig. 13.1D), a growing embryo that gradually becomes a juvenile acanthocephalan, with all of the adult organs save the reproductive system. Development then stops and will not begin again until the juvenile and its host are eaten by a suitable primary host. In some cases the primary host does not ordinarily eat the larva-bearing host, and in such instances the juvenile worm passes through several transport hosts, reencysting each time it is eaten.

Fossil Record and Phylogeny

No fossil acanthocephalans are known, although much has been made of the similarities of some fossil priapulids, especially the Middle Cambrian form *Ancalagon*, to a hypothetical protoacanthocephalan (Conway Morris and Crompton, 1982). These comparisons ignore the important fact that the acanthocephalan proboscis, which lacks a mouth or a gut, is functionally and structurally different from the introvert "proboscis" of priapulids and their presumed relatives, which contain a mouth. These two types of evertable organs are apparently not homologous or analogous.

The nearest relative of acanthocephalans now appears to be rotifers (Whitfield, 1971), with whom they share a number of features (see Chapter 12). However, assessing relationships within the Acanthocephala does not appear to be so easily achieved at this time. As to which of the three classes is most primitive, it is unclear. Archiacanthocephalans possess protonephridia and cement glands, but palaeacanthocephalans have what appears to be a primitive ligament anatomy.

Taxonomy

Acanthocephala are currently divided into eight orders within three classes.

Class Archiacanthocephala. Main lacunar vessels dorsal and lateral, or only dorsal; eight uninucleate cement glands; trunk spines absent; proboscis receptacle with single muscle layer, often with ventral cleft;

ligaments dorsal (attached to uterine bell) and ventral; protonephridia possible; uniramian intermediate hosts, terrestrial primary hosts. Four orders (Fig. 13.1B,D).

Class Palaeacanthocephala. Main lacunar vessels generally lateral; two to eight multinucleate cement glands; trunk spines possible; proboscis receptacle as closed sac, usually two muscle layers; single ligament, posterior attachment inside uterine bell; no protonephridia; crustacean intermediate hosts, aquatic primary hosts. Two orders.

Class Eoacanthocephala. Main lacunar vessels dorsal and ventral, anteriorly developed; single syncytial cement gland with giant nuclei; trunk spines possible; proboscis receptacle closed, with single muscle layer; ligaments dorsal and ventral (attached to uterine bell); no protonephridia; crustacean intermediate hosts, aquatic primary hosts. Two orders (Fig. 13.1A).

Selected Readings

Conway Morris, S., and D. W. Crompton. 1982. The origins and evolution of the Acanthocephala. *Biol. Rev.* 57:85–115.

Crompton, D. W., and B. B. Nichol (eds.). 1985. *Biology of the Acanthocephala.* Cambridge Univ. Press, Cambridge.

Meyer, A. 1932–1933. Acanthocephala. In *Dr. H. G. Bronn's Klassen und Ordnungen des Tierreiches.* Band 4, Abt. 2, Buch 2, Lief 1–2, pp. 1–582. Akad. Verlagsges., Leipzig. [An old but classic work.]

Whitfield, P. J. 1971. Phylogenetic affinities of Acanthocephala: An assessment of ultrastructural evidence. *Parasitol.* 63:49–58.

14

Loricifera

This newest phylum was described in 1983. Members of the group have been collected from many places around the world. Loriciferans are about 250 μm in length and live on the upper layers of marine sediment at depths from 15 to almost 8300 m. They are associated with other meiofaunal invertebrates such as nematodes, kinorhynchs, priapulids, copepods, and tardigrades.

Definition

Loricifera are bilateral animals with an invertible, spiny head. The buccal canal is sometimes telescopic, and there are oral stylets surrounding the mouth cone with encircling rows of spines or scalids. The neck region possesses spines or trichoscalids with basal plates. The abdomen is armed with a plated or plicate lorica, and is equipped with posterior sensory flosculi. The head and the neck are retractable into the lorica. The gut and the gonads open terminally, and the body is apparently pseudocoelomate.

Body Plan

The mouth is surrounded by either eight stylets (Fig. 14.1A) or six to twelve ridges on an **mouth cone**. The **oral stylets** can be extended from the distal end of the outer membrane by stylet protractor muscles. The mouth is on a **buccal canal** (Fig. 14.1B), which is retractable in *Nanaloricus,* but not so in the pliciloricids. The base of the head, or **introvert**, has nine rows of spines, or **scalids**. The first row, the **clavoscalids**, exhibits sexual dimorphism in *Nanaloricus* and may be adhesive in pliciloricids. The next eight rows, the **spinoscalids**, are directed posteriorly and can be highly modified. When the head is retracted into the lorica, all these spines fold up and are directed anteriorly.

The **neck**, or collar, is a region between the head and the thorax. This area is either naked or equipped with platelike scales. The more posterior **thorax** bears spines, **trichoscalids**, anteriorly and sometimes **basal plates** posteriorly, as in the pliciloricids. Internally, the thorax and the abdomen are separated by a fibrous ligament.

The **abdomen** is covered by the **lorica**. In *Nanaloricus* the lorica is composed of six plates, but the Pliciloricidae have longitudinal folds or plicae. The lorica has sensory organs, **flosculi**, near the posterior margin. The anterior edges of the lorica are sometimes heavily spined.

The buccal canal is a thin, cuticularized tube curved up within the mouth cone (Fig. 14.1B). This is supported by accessory stylets near the base. Salivary glands enter the buccal canal. The buccal canal connects to a bulbous **pharynx**, followed by a short

esophagus and a straight **midgut**. These end in a **rectum** and a terminal **anus**.

The **protonephridium** has a monociliary terminal cell. The **brain** fills half of the head and innervates each row of introvert scalids separately. Additional ganglia are noted in the neck and the abdomen, but details are not clear at this time.

The female reproductive system has paired **ovaries**. Only one egg is developed at a time. A small **seminal receptacle** may be present. The male system has two large dorsal **testes**. Fertilization may be facilitated by a pair of **penile spines** on the posterolateral edges of the lorica plates, and is assumed to be internal.

Biology

Little is known about matters related to movement, feeding, excretion, reproductive biology, or early development of loriciferans (Kristensen, 1986). The posterior flosculi are sensory in function. One type consists of a cilium and five microvilli sitting in pits in the lorica. Another type consists of tiny papillae on the lorica surface.

Reproduction appears to be seasonal. A larval or juvenal form is known, called **Higgins larva** (Fig. 14.2). The head is armed with scalids, and the neck is covered with five rows of plates. These plates prevent the neck from being completely retracted into the lorica. The anterior edge of the abdomen is equipped with pairs of three locomotory spines. The juvenile lorica consists of either plicae or only four plates, and has a pair of dorsal sensory setae and a series of small plates around the anus. The most prominent feature of the larva is a pair of **caudal appendages** or **toes** that are either leaflike or bladelike and able to move in all directions. The leaflike toes serve to assist the larva in swimming. There are glandular adhesive organs at the base, whose secretions allow the animal to attach and release from a substrate.

Several juvenile or larval stages separated by molts are seen before a postlarval stage is achieved.

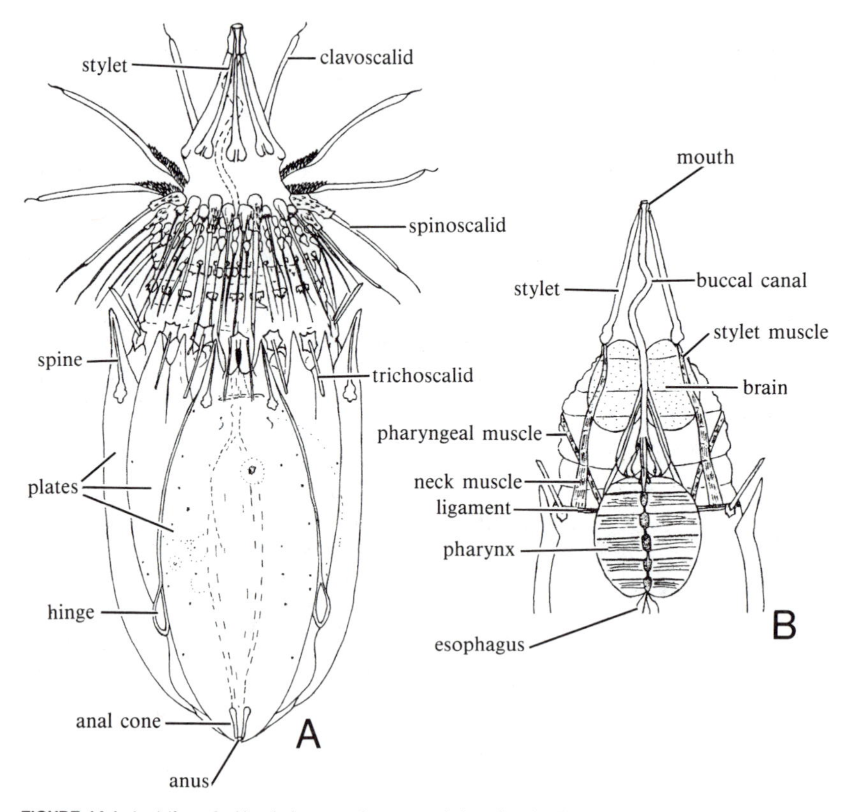

FIGURE 14.1. Loricifera. **A.** *Nanaloricus mysticus,* ventral view, female. **B.** *N. mysticus,* anterior end with buccal canal and mouth cone retracted. (After Kristensen)

Taxonomy and Phylogeny

A single order, Nanaloricida, is currently recognized that contains at least three families. New material is being described all the time, and our understanding of the group is in its infancy.

Kristensen (1983) thought that loriciferans may be the "missing link" that unites priapulids, kinorhynchs, and nematomorphs; however, the current consensus is that all these groups are monophyletic taxa. Some nematomorph larvae have a buccal canal similar to *Nanaloricus* as well as a bulbous pharynx. Nematomorph larvae also have a ligamentous diaphragm between neck and abdomen, and similar stylet types and scalids as loriciferans. Loricifera and Kinorhyncha share oral stylets, clavoscalids, and a similar ventral closing apparatus. The Higgins larva looks similar to those of priapulids and shares with priapulids flosculi (as do kinorhynchs) and caudal appendages.

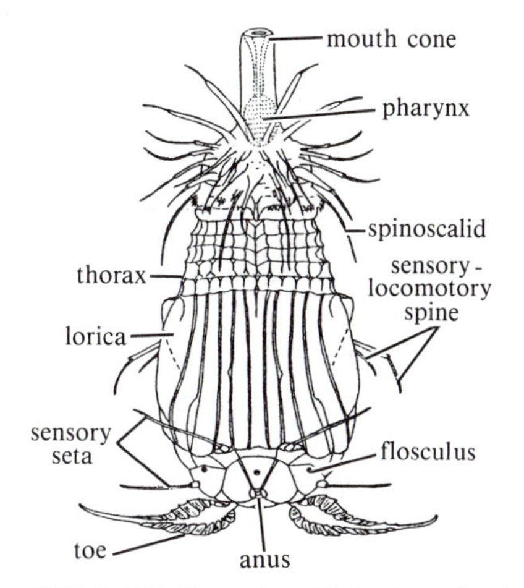

FIGURE 14.2. *N. mysticus,* Higgins larva, dorsal surface, with mouth cone everted. (After Kristensen.)

Selected Readings

Higgins, R. P., and R. M. Kristensen. 1986. New Loricifera from southeastern United States coastal waters. *Smith. Contr. Zool.* 438:1–70.

Kristensen, R. M. 1983. Loricifera, a new phylum with Aschelminthes characters from the meiobenthos. *Z. zool. Syst. Evolut.-forsch.* 21:163–180.

Kristensen, R. M. 1986. Loricifera. In *Stygofauna Mundi* (L. Botosaneanu, ed.), pp. 119–121. Brill, Leiden.

15

Kinorhyncha

The first kinorhynchs, also known as Echinodera or Echinoderida, were recognized in 1851. Since then only about 150 species have been described. There is still a great deal to learn about their basic biology because most research in the group has been taxonomic. They are important components of the meiofauna, preferring muddy or fine sand bottom sediments (Higgins, 1988).

Definition

Kinorhyncha are microscopic pseudocoelomates. Although they lack external cilia, the chitinous cuticle is well developed, with the epidermis and the cuticle divided into 13 segments. The head is invertible and is armed with circlets of spines. There is one pair of protonephridia, each with three terminal flame cells.

Body Plan

Kinorhynchs have a body subdivided into 13 segments. The first segment is the head, with up to seven rows of curved spines (**scalids**) and a **mouth cone** whose terminal mouth is surrounded by a whorl of **oral stylets** (Fig. 15.1B). Retractor muscles pull the head back into the trunk region. Major subgroups within the phylum can be distinguished on the basis of the arrangement of plates used to protect the withdrawn head. The withdrawn head of Cyclorhagae is closed over with **placids**, large plates on the second segment. The first trunk segment of Conchorhagae is bivalved; the laterally arranged plates are used to protect the retracted head and neck segments. The Cryptorhagae have no apparent apparatus to protect the retracted head segments. Homalorhagae close the trunk opening with one to four dorsal and ventral placids, each in combination with the "trapdoor"-like closure of three ventral plates on the third (first trunk) segment.

The 11 trunk segments generally have a single dorsal cuticular plate, the **tergite**, and a pair of ventral plates, the **sternites**. Peg and socket joints occur at the tergal–sternal junction of most homalorhagids. Body segmentation also involves the muscles (Fig. 15.1A). Cross-striated dorsoventral muscles, derived from the circular muscles, connect the tergite and sternites and serve to flatten the body and force the eversion of the head by increasing the internal pressure of the perivisceral fluid. The pseudocoel is filled with much cellular material. Circular muscles are prominent in the head and neck segments. Intersegmental, diagonal muscles may be found in the trunk. Paired intersegmental, longitudinal muscles are located dorsolaterally and ventrolaterally, and several head retractor muscles attach at various points of the trunk. Although many trunk segments can have dorsal and ventral spines, the last body segment, the thirteenth, is often modified with conspicuous movable spines, manipulated by means of special muscles, and sometimes an additional median movable spine, which may be as long as the entire trunk.

An epidermis, which lacks cilia, secretes a cuticle (Fig. 15.1C). There are adhesive tubes on the ventral surface of segments three and four, and there can be additional tubes located posteriorly.

Digestive System The anterior end of the digestive tract is highly specialized. The mouth cone contains the **buccal cavity** that ends posteriorly in a **pharyngeal crown** (Fig. 15.1B). The tapered, tubular **pharynx** has a triangular lumen in homalorhagids and circular in cyclorhagids. It resembles the nematode and macrodasyid gastrotrich pharynx, but has the radial muscles situated outside the epithelial lining (Fig. 15.1D).

The narrowed anterior part of the gut is the **esophagus**, and it receives secretions from small **salivary glands**. The wider **midgut** contains gland cells. A mesh of circular and longitudinal muscles covers the midgut. The hindgut is separated from the midgut by a sphincter and is lined with cuticle. The lining is probably derived from the **proctodeum**. The cuticular lining of the pharynx and esophagus indicates that they are derived from the **stomodeum**.

Excretory System Kinorhynchs have a pair of **protonephridia** on each side of the tenth and eleventh segments, with three terminal, biciliate **flame cells**. These cells are connected, by means of a nonciliated

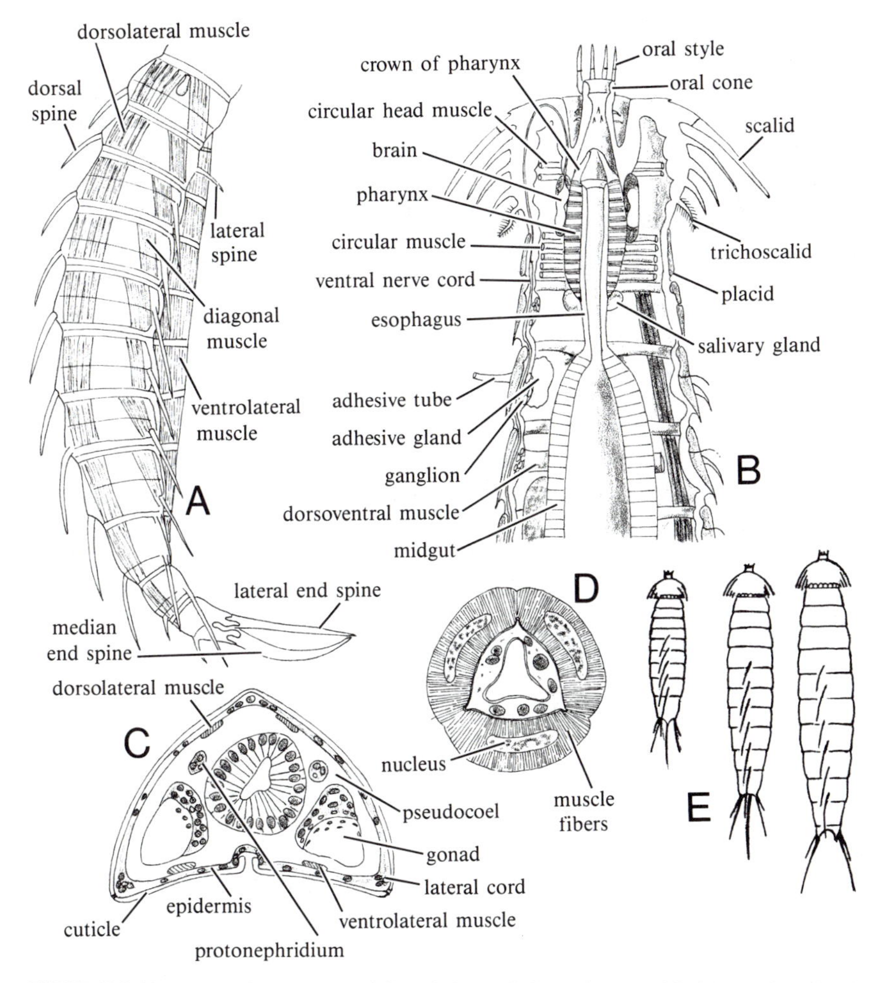

FIGURE 15.1. Kinorhyncha. **A.** Lateral view of *Campyloderes,* with introvert retracted. **B.** Scheme of kinorhynch anatomy, with introvert extended. **C.** Section through intestinal region of *Pycnophyes.* **D.** Section through pharyngeal region of *Pycnophyes.* **E.** Some stages in posthatch development of *Echinoderes bookhouti.* (**A,C,D** after Zelinka; **E** after Higgins; **B** after Remane, in Kükenthal and Krumbach.)

canal cell, to a nephridiopore cell with many microvilli and a cuticular sieve plate (Kristensen and Hay-Schmidt, 1989). This is an arrangement similar to that see in priapulids.

Nervous System The nervous system is epidermal and segmented. A **circumenteric nerve ring** surrounds the pharynx with ten lobes, each with a forebrain, midbrain, and hindbrain. Eight ganglionated **nerve cords** extend posteriorly, the most prominent being paired, ventral cords as well as dorsolateral, lateral, and ventrolateral cords.

The nerve ring around the pharynx resembles the circumenteric ring seen in many other pseudocoelomates. The segmentation of the cuticle is associated with segmentally arranged muscles, as well as the segmentation ganglia of the nervous system. This is similar to that seen in the rotifers with cuticular rings, where the ganglia that are associated with the diversions of the scalar nerve arise (see Fig. 12.5A).

Reproductive System Kinorhynchs are dioecious, although the separate males and females are typically similar. The paired, saccate gonads lie laterally along the posterior half of the gut, each opening by a gonoduct on the terminal body segment. The **ovary** contains **germinal** and **nutritive nuclei**. The short **oviduct** has a **diverticulum**, the **seminal receptacle** at the gonopore. The short **sperm duct** connects the **testis** to the gonopore or penile spines on the terminal body segment.

Biology

Locomotion Kinorhynchs move themselves forward by contracting the dorsoventral body muscles to laboriously thrust their spiny heads forward into the sediment to act as anchors. When the head retractor muscles are contracted, kinorhynchs then pull their bodies forward. The head scalids assist in gripping the substrate. When disturbed, the head is pulled back into the trunk, and the animals then lie quietly in the sediment (Higgins, 1987). They cannot swim.

Feeding The mouth cone is protruded during feeding. Most kinorhynchs live in ocean mud in shallow waters. They feed on bacteria or organic detritus in that mud. The exact feeding mechanism, including the function of the oral stylets, is poorly understood. Other kinorhynch species live among algae and may feed on diatoms.

Sense Organs Various sensory organs are recognized. These include eyespots, sensory setae, collar receptors (modified floculi), and scalids. Nearly all kinorhynchs are photosensitive, some having eyes similar to those seen in some polychaetes associated with the forebrain lobes.

Reproduction The penile spines may be used in copulation, although that act has never been observed (Higgins, 1974). Sperm are often "huge," up to one-fourth the body length. In at least some species, the male deposits a spermatophore on the female's terminal segment. Oöcytes are scattered about the ovaries, but only one, at most two, actually matures at any one time.

Eggs are apparently extruded into the sediment. *Echinoderes kozloffi* has been observed to lay an egg, and it contained an already developed juvenile.

Development Nothing is known about embryology. Postembryonic development results in an apparent 11-segment juvenile, although all 13 segments are really present (Fig. 15.1E). A series of molts then occurs in which posterior segments become better defined to achieve the adult number. At the same time, molts produce the adult pattern of spines. The development is essentially direct; there are no larval forms (Kozloff, 1972).

Taxonomy

Two orders of kinorhynchs are currently recognized, as outlined in Higgins (1968).

Order Cyclorhagida. First trunk segment entire or bivalved; round or oval body cross section; well-developed lateral, dorsal, and terminal spines (Fig. 15.1A).

Order Homalorhagida. First trunk segment with single arched dorsal plate, and with either single midventral plate and two ventrolateral plates or single ventral plate; trunk spines, if present, only as lateroterminal and, more rarely, midterminal spines (Fig. 15.1B).

Selected Readings

Higgins, R. P. 1968. Taxonomy and postembryonic development of the Cryptorhagae, a new suborder of the mesopsammic kinorhynch genus *Coteria*. *Trans. Amer. Microsc. Soc.* 85:21–39.

Higgins, R. P. 1971. A historical review of kinorhynch research. *Smith. Contr. Zool.* 76:25–31.

Higgins, R. P. 1974. Kinorhyncha. In *Reproduction of Marine Invertebrates*, vol. 1 (A. C. Giese and J. S. Pearse, eds.), pp. 507–518. Academic Press, New York.

Higgins, R. P. 1987. Kinorhyncha. In *The Encyclopedia of Science and Technology*, vol. 9 (S. Parker, ed.), pp. 523–524. McGraw-Hill, New York.

Higgins, R. P. 1988. Kinorhyncha. In *Introduction to the Study of Meiofauna* (R. P. Higgins and H. Thiel, eds.), pp. 328–331. Smithsonian Inst. Press, Washington.

Kozloff, E. N. 1972. Some aspects of development in *Echinoderes*. *Trans. Amer. Microsc. Soc.* 91:119–130.

Kristensen, R. M., and A. Hay-Schmidt. 1989. The protonephridia of the arctic kinorhynch *Echinoderes aquilonius*. *Acta Zool.* 70:13–27.

Zelinka, C. 1928. *Monographie der Echinodera.* Wilhelm Englemann, Leipzig.

16

Priapulida

Only 15 living species of priapulids are known, the largest reaching about 13 cm in length. Priapulids are mud dwellers, the smaller forms being virtually interstitial, and occur worldwide (van der Land, 1970) at all depths varying from littoral (Kirsteuer and van der Land, 1970) to abyssal (van der Land, 1972). An additional six very diverse fossil forms are recognized.

Priapulids were once thought to have a true coelom with mesenteries, but study with transmission electron microscopy has revealed no true peritoneum (McLean, 1984). Rather, the retractor muscles have nuclei that lie on the surface of the muscles outside the striated region (van der Land and Nørrevang, 1985). The muscle cells secrete extracellular membranes. These membranes and the adjacent muscle nuclei gave the false impression that a true mesentery was present. Priapulids, thus, have a rather distinctive pseudocoel (not unlike chaetognaths) that resembles a coelom only in that there are muscle layers associated with the walls of both the body and gut. Because of the lack of true mesenteries, as well as the overall body plan of the priapulids, the phylum is clearly allied with the pseudocoelomates.

Definition

Priapulida are marine pseudocoelomates. Their cuticle is warty and superficially segmented, and can be molted. The presoma is large, spiny, and retractable. Protonephridia and gonads share a common urogenital duct. The body often possesses a terminal caudal appendage.

Body Plan

The priapulid body is divided into a **presoma** and a **trunk** (Fig. 16.1A). The barrel-shaped, proboscislike presoma is usually extended, but retractor muscles can withdraw it into the trunk. The anteriormost end is a spiny, circumoral region, sharply set off from the rest of the presoma by a smoother collar (Fig. 16.1B). The spiny trunk is superficially segmented with rings and contains posteriorly an **anus** and a pair of **urogenital pores**. Several genera bear one or two branched **caudal appendages** (Fig. 16.1C) with hollow vesicular twigs that are covered by tissue histologically like the body wall. Their function is unknown, and animals seem undisturbed by their removal and merely grow a new set.

The thin, layered cuticle, produced by the single-layer epidermis, contains chitin and can be molted. The body wall has an outer layer of circular muscle and an inner layer of longitudinal muscle.

A toothed cuticle lines the muscular **pharynx**, which leads to the straight **midgut**. The midgut lining is deeply folded and is separated by a constriction from the cuticularized **hindgut**.

Priapulids have **protonephridia**. These are linked with the gonads to form a **urogenital organ**. Thousands of nucleated solenocytes, grouped in clusters of three cells, arise from one side of the urogenital duct. The bi-ciliated **flame bulbs** (Kristensen and Hay-Schmidt, 1989) take up injected carmine, and are apparently excretory in function. The gonads have an elaborate tubular form.

A circumenteric **nerve ring**, closely associated with the epidermis, encircles the body in the collar region. It gives rise to a midventral **nerve cord**, from which a system of peripheral nerves and segmental **ring commissures** arises (Fig. 16.1D). The nerve ring gives rise to subepidermal nerves that supply the trunk, and to four nerves that serve the pharynx.

Biology

Locomotion and Feeding Priapulids typically burrow in slime, usually lying quietly with their open mouths aflush with the surface ooze of the bottom mud. They explore their surroundings with the front part of the body when hungry, or plow about at random through the mud. They are carnivores, capturing and swallowing whole their slowly moving prey.

Respiration and Excretion The coelomic fluid contains many rounded cells that take up injected particles and convey them to the protonephridia; these cells functionally resemble athrocytes. In *Halicryptus* and *Priapulus*, coelomocytes are found that contain a respiratory pigment, hemerythrin.

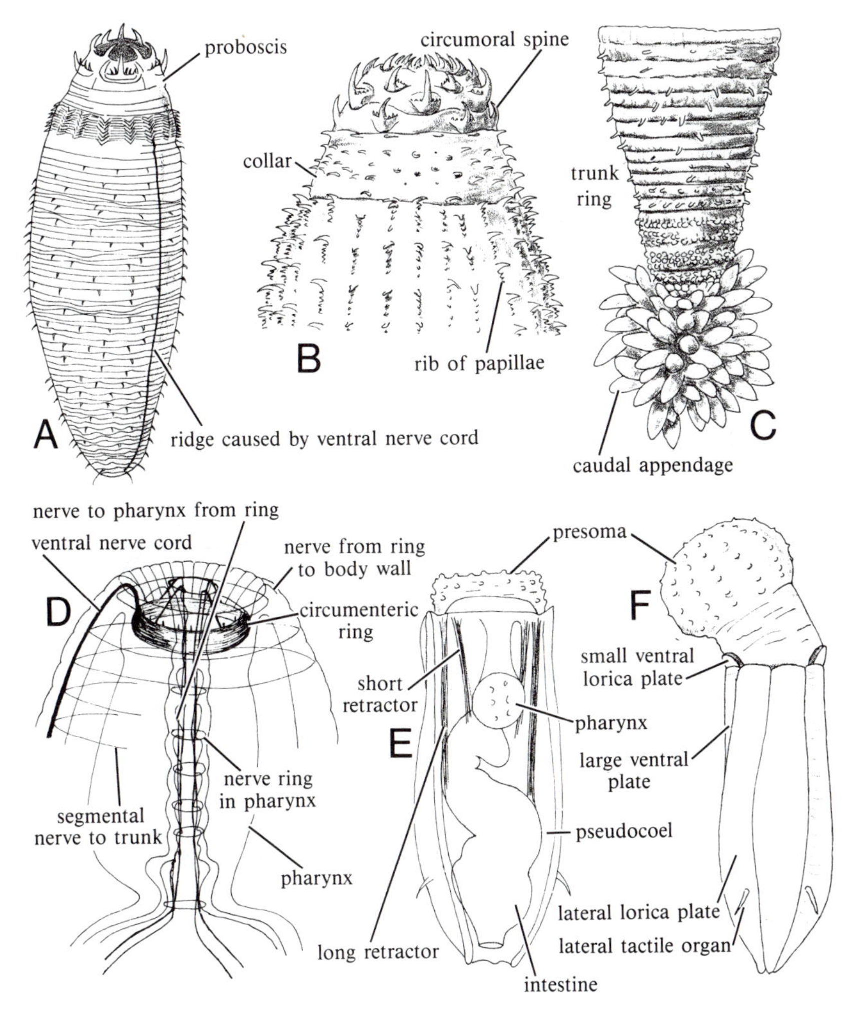

FIGURE 16.1. Priapulida. **A.** *Halicryptus spinulosus,* form that lacks caudal appendage. **B.** Everted anterior end of *Priapulus.* **C.** Posterior of *Priapulus caudatus,* with caudal appendage. **D.** Anterior nervous system of *Halicryptus.* **E,F.** Larva of *Priapulus caudatus,* with anterior end withdrawn **E** and extended **F.** (**A** after Shipley; **B,C** after Theel; **D** after Apel; **E,F** after Lang.)

Reproduction Fertilization is external. Males shed sperm into the water, and this triggers females to shed ova. Only the genus *Tubiluchus* may have internal fertilization, and exhibits, perhaps as a consequence, some slight sexual dimorphism (van der Land, 1975).

Development Little is known of early embryonic development, although cleavage is considered to be radial. A **stereoblastula** gastrulates by epiboly and forms a **stereogastrula** that emerges from the egg membranes. The **priapulus larva** has a **lorica**, composed of dorsal, ventral, and three pairs of lateral, cuticular plates (Fig. 16.1F). A terminal foot is found at the posterior end of the priapulus larva. Larvae live as juveniles for some time, eventually shed the lorica by molting, and thereafter gradually acquire adult characteristics through a series of molts.

Fossil Record and Phylogeny

Although Priapulida is considered a minor phylum in the modern fauna, the fossil record is surprisingly interesting. A Pennsylvanian form, *Priapulites konecniorum* is easily allied with the living Priapulidae (Schram, 1973). The Middle Cambrian forms (Fig.

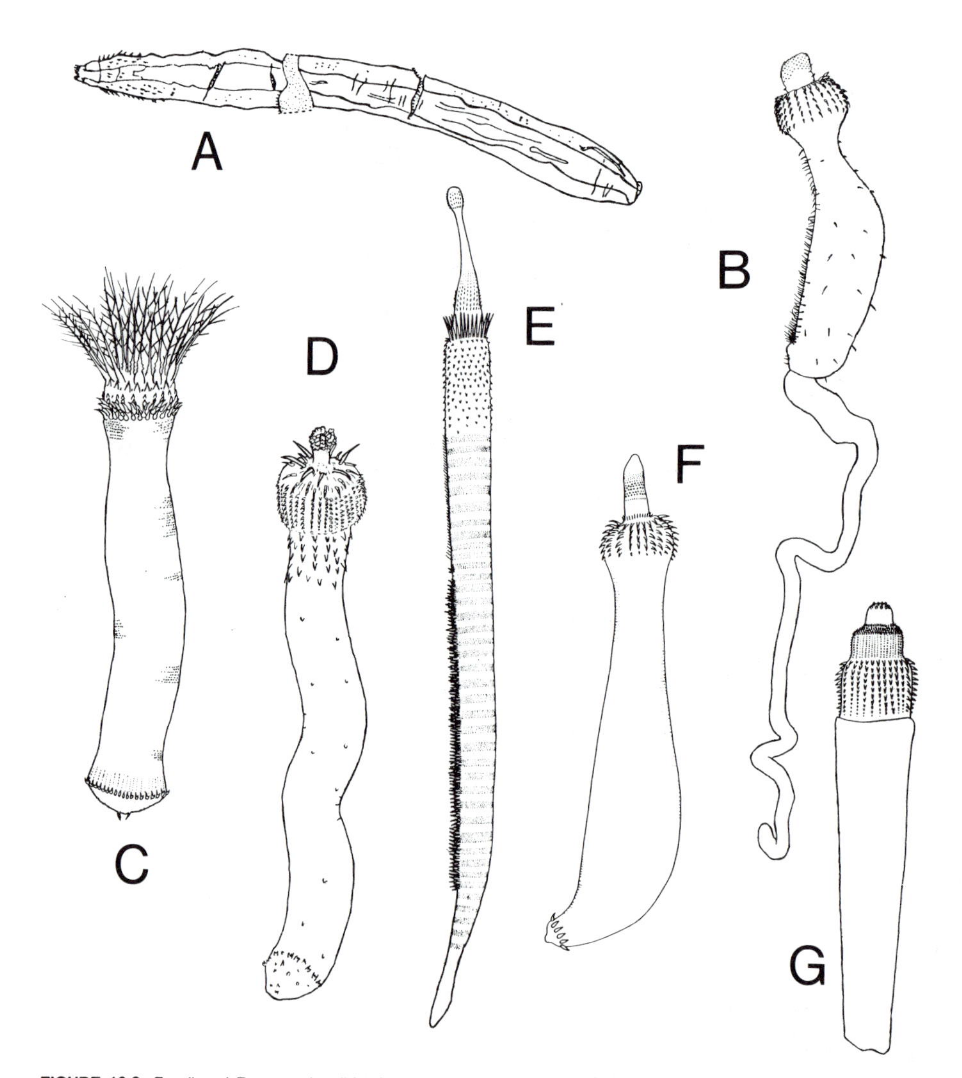

FIGURE 16.2. Fossil and Recent priapulids. **A**. *Ancalagon minor.* **B**. *Tubiluchus corallicola.* **C**. *Maccabeus tentaculatus.* **D**. *Meiopriapulus fijiensis.* **E**. *Louisella pedunculata.* **F**. *Ottoia prolifica.* **G**. *Selkirkia columbia.* (**A** after Conway Morris; **B**,**C**,**D**,**E**,**F**,**G** after van der Land and Nørrevang.)

16.2A,E,F,G), however, are so distinctly different from living priapulids and from each other that they are placed each in separate families. It appears that the living priapulids are but a mere vestige of the former diversity of the phylum (Conway Morris, 1977).

The similarity of the loricate priapulus larva to the Loricifera is significant (van der Land and Nørrevang, 1985). The possession of scalids on the introvert is characteristic of priapulids, kinorhynchs, and loriciferans; and the possession of the introvert itself is shared in these three phyla as well as the nematomorphs and primitive nematodes. The priapulids are thus clearly related to the other pseudocoelomates.

TAXONOMY

The 21 known living and fossil species of Priapulida are classified into eight families. Three of these are extant: Priapulidae (Fig. 16.1), Tubiluchidae (Fig. 16.2B,D), and Chaetostephanidae (Fig. 16.2C). Five are extinct: Ancalagonidae (Fig. 16.2A), Miskoiidae (Fig. 16.2E), Ottoiidae (Fig. 16.2F), Selkirkiidae (Fig. 16.2G), and Fieldiidae. No higher divisions above the family level are recognized at this time (van der Land, 1970).

Selected Readings

Conway Morris, S. 1977. Fossil priapulid worms. *Spec. Pap. Palaeontol.* 20:1–95.

Kirsteuer, E., and J. van der Land. 1970. Some notes on *Tubiluchus corallicola* from Barbados, West Indies. *Mar. Biol.* 7:230–238.

Kristensen, R. M., and A. Hay-Schmidt. 1989. The protonephridia of the arctic kinorhynch *Echinoderes aquilonius*. *Acta Zool.* 70:13–27.

McLean, N. 1984. Amoebocytes in the lining of body cavities and mesenteries of *Priapulus caudatus*. *Acta Zool.* 65:75–78.

Schram, F. R. 1973. Pseudocoelomates and a nemertine from the Illinois Pennsylvanian. *J. Paleo.* 47:985–989.

van der Land, J. 1970. Systematics, zoogeography, and ecology of the Priapulida. *Zool. Verhandl. Leiden* 112:1–118.

van der Land, J. 1972. *Priapulus* from the deep sea. *Zool. Mededelingen* 47:358–368. [2 pls.]

van der Land, J. 1975. Priapulida. In *Reproduction of Marine Invertebrates*, vol. 2 (A. C. Giese and J. S. Pearse, eds.), pp. 55–65. Academic Press, New York.

van der Land, J., and A. Nørrevang. 1985. Affinities and intraphyletic relationships of the Priapulida. In *The Origin and Relationships of Lower Invertebrates* (S. Conway Morris, et al., eds.), pp. 261–273. Clarendon Press, Oxford.

17

Nematomorpha

Previously grouped as the class Nematomorpha within the phylum Aschelminthes, the "horsehair" worms (or Gordiacea) are so designated based on their usual knotted and tight-coiling behavior as adults. The fabled "Gordian knot" of antiquity tied by Gordius, the mythical founder of the Kingdom of Phrygia (currently Turkey), on the yoke of his chariot was declared by an oracle to be impossible to untie, but he who did so would master all Asia. Alexander the Great averted the omen by severing it with his sword and thus extended his power and control over most of ancient Asia. Currently there are about 275 species placed in two orders under a now separated phylum Nematomorpha.

Definition

These pseudocoelomates are free-living (usually aquatic) as adults. Their extreme filiform or hairlike body is without a defined excretory system or patent digestive system. The nervous system is simple and nematodelike. The sexes are dioecious and dimorphic, with a cloaca present in both sexes. Larvae are parasitic in arthropod hemocoels and have a spinose presoma and annulate trunk.

Body Plan

Nematomorphs have a very thick cuticle (similar to that of mermithid nematodes) with an outer homogeneous layer often containing rounded or polygonal thickenings (areoles) that contain a mucoprotein secreted as a lubricant for unknotting of coiled states. Beneath this layer is a series of helically wound collagenous fibrous layers (up to 45) that alternate in a two-to-one packing order in males and a one-to-one order in females. In *Paragordius varius* the two-to-one packing order in males confers a preferred right-handed sense when tight-coiling occurs; this constraint is species specific and may have importance in recognition during reproduction where tight-coiling also occurs prior to semination. Some species have two or three **caudal lobes** at the posterior end (Fig. 17.1B), and all have a posterior **cloacal pore**. The head is not set off from the rest of the body, but a lighter **calotte** containing the mouth is usually present (Fig. 17.1A). The mouth is small and functionless, for adults do not feed, but rely solely on stored glycogen reserves for their brief, free-living existence during which reproduction occurs.

The adult nematomorph gut is greatly reduced. The pharynx is usually a solid cord of cells, and the intestine is a diminutive epithelial tube (Fig. 17.1C,D). The digestive tract is effectively of no more use to the larva than to the adult. Larval feeding appears to be by absorption directly across the cuticle from host hemocoel fluids.

The central nervous system consists of a circumenteric **nerve ring** in the head calotte and a **midventral cord** in the epidermis. In the more primitive genus, *Nectonema*, it remains in this position, but the cord migrates inward in the freshwater gordioids to become attached to the epidermis by a lamella (Fig. 17.1C,D). Neuroglial partitions divide the nerve trunk into three tracts. Giant neurons at the union of nerve ring and nerve trunk probably coordinate distant regions of the worm.

All muscle fibers are longitudinal, and in *Nectonema* they clearly resemble the coelomyarian muscles of nematodes. Gordioid muscle contains large paramyosin myofilaments all the way around the central cytoplasmic region, conferring an economical "catch" property similar to that initially found in the adductor muscles of lamellibranch molluscs. *Nectonema* has a spacious pseudocoel, and the gordioids have a pseudocoel largely filled with mesoglea.

Males have a pair of long, tubular **testes** that open independently into the cloaca through short **sperm ducts**. A **seminal vesicle** sometimes lies at the end of the testis. The cloacal wall of some species is armed with bristles and may facilitate its service as a **cirrus**. Young **ovaries** resemble the testes, but as they mature they form **lateral diverticula** in which the ova ripen. As many as 4000 diverticula may occur. The ovary is left vacant and serves as a uterus to store the ripened ova. The two **oviducts** combine to form a common **antrum**, from which a **seminal receptacle** arises. The antrum opens into the cloaca.

Biology

Locomotion Fully-formed adults emerge from pleural folds in the host arthropod integument often

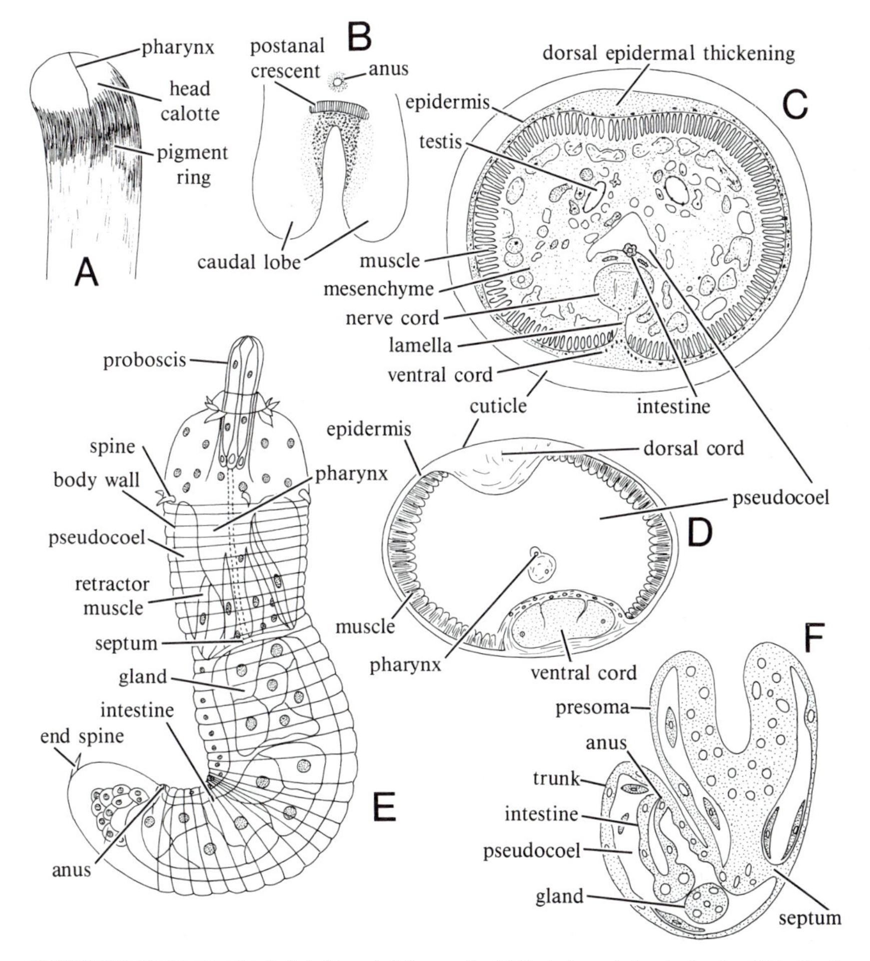

FIGURE 17.1. Nematomorpha. **A**. Anterior end of *Paragordius*. **B**. Posterior end of male *Gordius*. **C**. Section through *Paragordius*. **D**. Section through *Nectonema*. **E**. Older larva of *Paragordius*. **F**. Larva of *Gordius aquaticius*. (**A** after May; **B** after Heinze; **C** after Montgomery; **D** after Burger; **E** after Mühldorf; **F** after Montgomery.)

with an explosive force lethal to the host. Although adults are usually tightly coiled in masses or on other marginal aquatic debris during the daylight hours, they migrate into open water at night and move on the surface film with gyratory sine-wave-type motion by repeated flexures in the dorsoventral plane. In this they resemble typical nematode swimming. The diurnal variation in movement has been demonstrated in *Paragordius varius*, which also has been described to have a photosensitive "eye" anterior to the calotte. The synchrony of nightly swimming may be a mechanism to allow for passive increase in population density whereby worms swept downstream in lotic habitats settle at the margins of common eddy pools in high numbers for subsequent bouts of reproduction.

Feeding Food is apparently absorbed through the body wall of the larvae directly out of the arthropod host's hemocoel; adults rely solely on stored energy reserves (mainly glycogen) during their brief, free-living existence.

Excretion The cells of the degenerate gut resemble those of insect Malpighian tubules, an excretory organ, and as there is no other excretory organ, it may be that the gut functions as one.

Sensation Gordioids are highly thigomotactic, a sense derived by nerve endings penetrating the cuticle to varying extents. Presumably, some of the bristles and surface papillae are sensory, and nerve endings in the calotte look like touch receptors. The patterns of papillae in the cloacal region are species specific and, if innervated, may be involved in recognition prior to reproductive tight-coiling. *Paragordius* has a heavily innervated sac containing fusiform cells suspended in a coagulated fluid; it has a dark pigment ring behind it. If this is a photoreceptive organ, it may explain their photoperiodic changes in state from tight-coiling to surface swimming. This "eye" is probably not image-forming, but is rather merely a light detector. Other species without eyes also show some photoperiodic behavior that might be driven by a general dermal and/or central nervous system sensitivity to light.

Reproduction Nematomorphs usually mate soon after being freed from the host, but may overwinter if they emerge in the late fall. Males die soon after mating, and females die during or after egg laying. Mating usually involves a single male that wraps his body around the female and deposits sperm near the female cloacal opening. The sperm migrate into the seminal receptacle. When ova are mature, they emerge in long strings that may contain several million ova. Cloacal secretions provide the binding material to keep the strings intact. Once released, egg strings gradually disperse, and either the developing larvae encyst on marginal vegetation to be subsequently eaten by terrestrial arthropod hosts (usually crickets), or the eggs hatch immediately to infect other aquatic hosts (usually dytiscid or other beetles). The life cycle of *Paragordius varius* is fairly well known, but that of other species and the cycle for the marine genus *Nectonema* is less well documented.

Development Accounts of embryonic development differ and suggest generic differences in growth patterns. The egg undergoes holoblastic, equal cleavage and passes through nematodelike three- and four-celled stages. A coeloblastula undergoes atypical gastrulation, before or after the blastocoel is invaded by mesodermal cells. An ectodermal presoma invaginates to form a proboscis (Fig. 17.1F), which eventually develops several spines and three long stylets. The anterior part of the embryonic gut forms a gland, and the posterior part an intestine that is blind anteriorly but open posteriorly at the blastopore–anus.

A larva (Fig. 17.1E), unusual for pseudocoelomates, sometimes called an **echinoderid larva**, resembles a loriciferan, kinorhynch, or priapulid (Inoue, 1958). The larva may penetrate the cuticle of any common water animal soon after hatching, but more probably is consumed (as free-living or encysted late embryo) and subsequently penetrates the host gut epithelium into the hemocoel where continued development occurs. Once in the proper host the larva grows gradually, without a sharp metamorphosis, into a juvenile worm. It emerges when the host is in or near water, and molts to lose the last traces of larval structure.

Phylogeny

The degenerate body form of adult nematomorphs, and the marked development of a peculiar larval parasitism, makes it difficult to determine close relationships. The larvae are similar to loriciferans, priapulids, and kinorhynchs. The nature of the elaborate cuticle, presence of only longitudinal muscles in the body wall, and locomotion only in a dorsoventral plane are features shared with nematodes and suggest possible sister group status of these two phyla (Lorenzen, 1985).

Taxonomy

Two orders are recognized.

Order Nectonematoidea. All marine; larvae parasitic in decapod crustacean hosts; dorsal and ventral nerve cords, with single gonad in open, spacial pseudocoel (Fig. 17.1D).

Order Gordioidea. All freshwater; larvae parasitic in terrestrial or aquatic insect host; adults free-living in aquatic habitats with single ventral nerve cord, pseudocoel filled with mesenchyme, paired gonads, and ovaries with lateral diverticula (Fig. 17.1A,B,C,E,F).

Selected Readings

Coomans, A. 1981. Aspects of the phylogeny of nematodes. *Atti Acad. Naz. Lincei* 49:162–174.

Inoue, I. 1958. Studies on the life history of *Chordodes japonensis*, a species of Gordiacea. I. The development and structure of the larva. *Jap. J. Zool.* 12:203–218.

Lorenzen, S. 1985. Phylogenetic aspects of pseudocoelomate evolution. In *The Origins and Relationships of Lower Invertebrates* (S. Conway Morris et al., eds.), pp. 210–223. Clarendon Press, Oxford.

May, A. 1919. Contributions to the life histories of *Gordius* and *Paragordius*. *Ill. Biol. Monogr.* 5(2):1–118.

18

Nematoda

The Nematoda, or Nemata, may have as many as one million species (May, 1988), and are found everywhere: from marine and freshwater to terrestrial habitats; from wet to arid conditions; from tropical to alpine and polar habitats; and are all feeding types including herbivores, carnivores, saprophages, and parasites (Maggenti, 1982a). In aquatic habitats, nematodes are most abundant as individuals in muds, moderately so in fine sand, less so in coarse sand, but have the greater species diversity in fine sand rather than in mud (Platt and Warwick, 1980). They are almost unbelievably abundant: 90,000 in a single rotting apple; 236 species and 1074 individuals in 6.7 cc of coastal mud; 527,000 per acre in the top 1.5 cm of beach sand; and from 3 to 9 billion per acre in good farm land! They abound in soils containing fungi or bacterial material, they parasitize plants and animals, and they feed on the roots of plants (e.g., 13 species around the roots of a single 10-cm-long wheat seedling). They even live in the traps of the insectivorous pitcher plants. Cobb (1915) observed that there are more individual nematodes than any other phylum, and that if everything in the world were removed, there would still be a ghostly outline of what was there, composed entirely of nematodes.

Their abundance can be partly explained by their ability to resist factors toxic to other animals and by their general adaptability. For example, the vinegar eel, *Turbatrix aceti*, is often abundant in vinegar. It endures an acetic acid concentration of 13.5 percent, but can also tolerate a pH of 1.5 and can live for several hours in mercuric chloride concentrations instantly fatal to most animals. Nematodes range in size from the microscopic to the 8-m-long, 2.5-cm-wide *Placentonema gigantissima*, a parasite of sperm whales.

Definition

Nematodes are generally elongate, spindle-shaped pseudocoelomates. Their sensory system is well developed, with a radial array of anterior sensilla, anterior amphids, and sometimes posterior phasmids. The cuticle is complex. There are longitudinal body muscles only. The digestive system is complete, the anus subterminal. Renette cells, when present, exit through an anteroventromedian excretory pore. The circumesophageal nerve ring coordinates four longitudinal nerve cords. Nematodes are dioecious, the female with a separate vulvar opening, the male system opening into a cloaca equipped with penile spicules. There is a postembryonic development with four juvenile stages.

Body Plan

The success of nematodes does not depend on diverse anatomical specializations, but rather on minor modifications of an effective organizational plan characterized by simplicity rather than complexity. The body is composed of tubes. The body wall is a series of tubes composed of a resistant outer cuticle; an epidermis forms four longitudinal nerve cords, and within these are groups of peculiar longitudinal muscles. A pseudocoel containing the perivisceral fluid separates the body wall and viscera. The gut is a simple tube, with some regional specialization. The excretory system, when present, is unique; there are no flame cells or protonephridia, but is composed of a simple cell or a system of tubes developed from a renette cell. Even the characteristic sense organs, amphids and phasmids, are essentially tubular pits.

Nematodes are bilaterally symmetrical (Fig. 18.1), but in fitting into the narrow, elongated, tubular body, the internal organs are often coiled, or one member of a pair is often lost. Externally, nematodes have a strong tendency to have radially arranged parts. Radial tendencies are best seen in a front view of the head (Fig. 18.2A). Primitive marine nematodes have strong hexamerous tendencies. Six lips, three on each side, surround the mouth. Each **lip (labium)** bears a sensory bristle or papilla; together they form the inner ring of six **labial papillae**. An outer ring of six labial papillae surrounds the inner ring or attaches to the cephalic region around the lips. An outer circlet of four cephalic bristles or papillae completes the basic complement of the head sense organs. Species vary in that the circlets can be combined, the papillae lost or fused, and the lip structure modified, thus making the head structure useful in taxonomy. The sensory amphids are paired, and the excretory pore, when present, is midventral, so that a definite sagittal plane is defined in spite of the radial tendencies. The triangular lumen of the pharynx is an inner continuation of the hexamerous lip arrangement.

Body organization is best understood in cross section (Fig. 18.2B). The body wall consists of layers of cuticle, epidermis or hypodermis, and muscle. Ex-

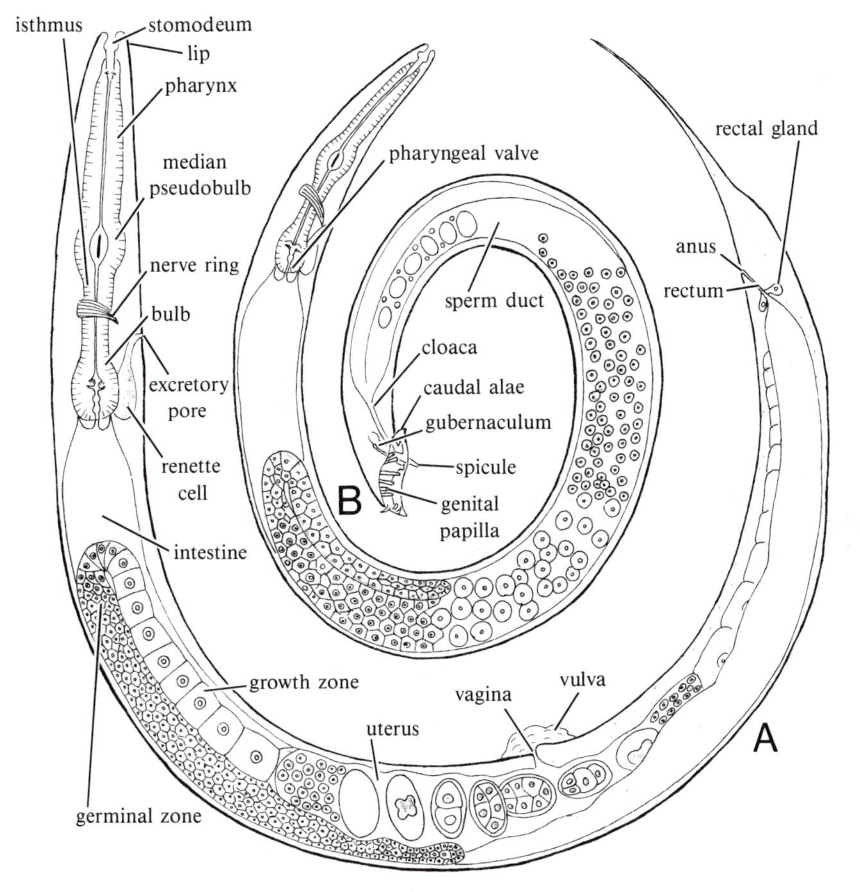

FIGURE 18.1. Typical nematode *Rhabditis*. **A**. Female. **B**. Male. (After Hirschmann, in Sasser and Jenkins.)

cept in the pharynx (= muscular esophagus), the gut wall is a single layer of epithelial cells. The reproductive organs lie in the pseudocoel.

The nematode **cuticle** is a remarkable product of the epidermal or hypodermal cells. It is elastic, but provides support; tough, yet permits increase or decrease of body volume without changes in the fluid pressure of the pseudocoel. It is readily permeable to water, yet protects the animal against many toxic compounds and extremes of habitat.

The cuticle is sometimes smooth and featureless, and sometimes ornamented with bristles or protuberances used as sense organs or in locomotion. In several groups, the cuticle is ringed, sometimes so deeply as to give the impression of segmentation (see Fig. 18.7A). The cuticle is structurally complex (Maggenti, 1979). At its most elaborate, it consists of an epicuticle; an exocuticle of two or more layers; a mesocuticle with a variable number of layers composed of organized rods, fibers, struts, and plates; and an endocuticle that is often poorly defined and penetrated by hypodermal extensions and muscle attachments (Fig. 18.2C). The fibers of each layer in the mesocuticle are parallel, but the direction differs between layers. Volume changes tolerated by the cuticle are sometimes remarkable. *Rhabditis*, a free-

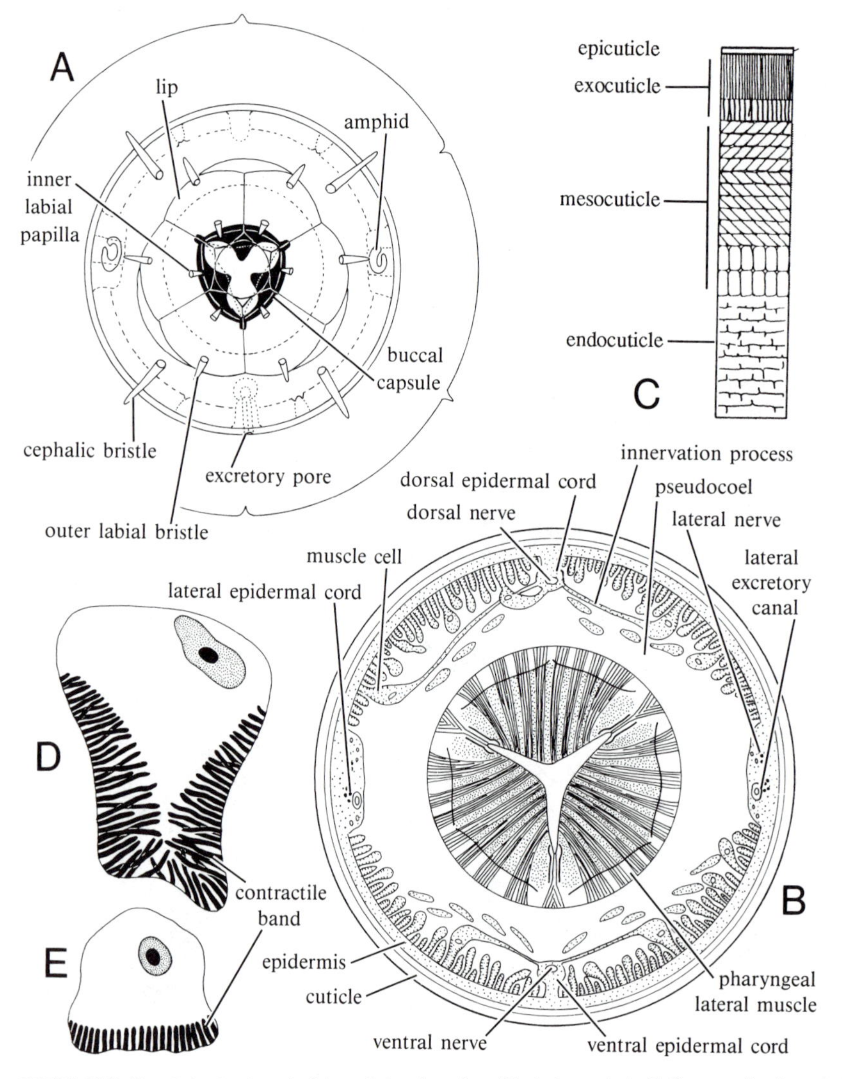

FIGURE 18.2. Nematode structure. **A**. Schematic head-on view of typical nematode. **B**. Cross section through *Ascaris* in region of pharynx. **C**. Diagram of *Deontostoma* cuticle. **D**. Schematic of coelomyarian muscle cell. **E**. Platymyarian muscle cell. (**A** after de Coninck; **B** after Hirschmann; **C** after Maggenti, in Sasser and Jenkins; **D,E** after Chitwood.)

living nematode, can lose 30 percent of its length in hypertonic solutions, and *Onchocerca*, a microfilarian, can increase in length up to 55 percent when removed from isotonic glucose. Water comes in through the cuticle, but is usually not discharged through it. *Ascaris* cuticle is permeable to respiratory gases and ions such as chloride or ammonium, but impervious to volatile organic compounds. *Ascaris* cuticle has an epicuticle composed of a thin, lipid layer and inner and outer cortical layers, perhaps composed of quinone-tanned proteins, for it resists digestion.

Flattened hypodermal cells lie just below the cuticle and project inward as four ridges, known as **cords** (Fig. 18.2B). The dorsal and ventral cords are less prominent and may disappear near the ends of the body. The cords enclose the nerve trunks and excretory canals, and divide the musculature into four fields. Nematodes have all of the hypodermal nuclei in the cords, and flattened extensions of the cells, without nuclei, connect the cords. Large nematodes, like *Ascaris*, have a syncytial hypodermis in which the original nuclei have fragmented to form many small nuclei; however, they remain within the cords.

A constant number of large longitudinal muscle cells lie grouped in fields between the hypodermal cords. Nematodes may have from as few as two to five muscles or many muscle cells in each field. The bulging, cytoplasmic portion of the muscle cells contains supporting fibrils and contractile fibers either next to the hypodermis (**platymyarian**) or that extend up the periphery of the cell (**coelomyarian**) (Fig. 18.2D,E). Nematode muscle cells possess innervation processes that extend *to the nerve cords* (Fig. 18.2B), rather than have nerve fibers extend to the muscles.

Digestive System The digestive tube consists of a stomodeal **buccal cavity** and **pharynx** (= esopha-

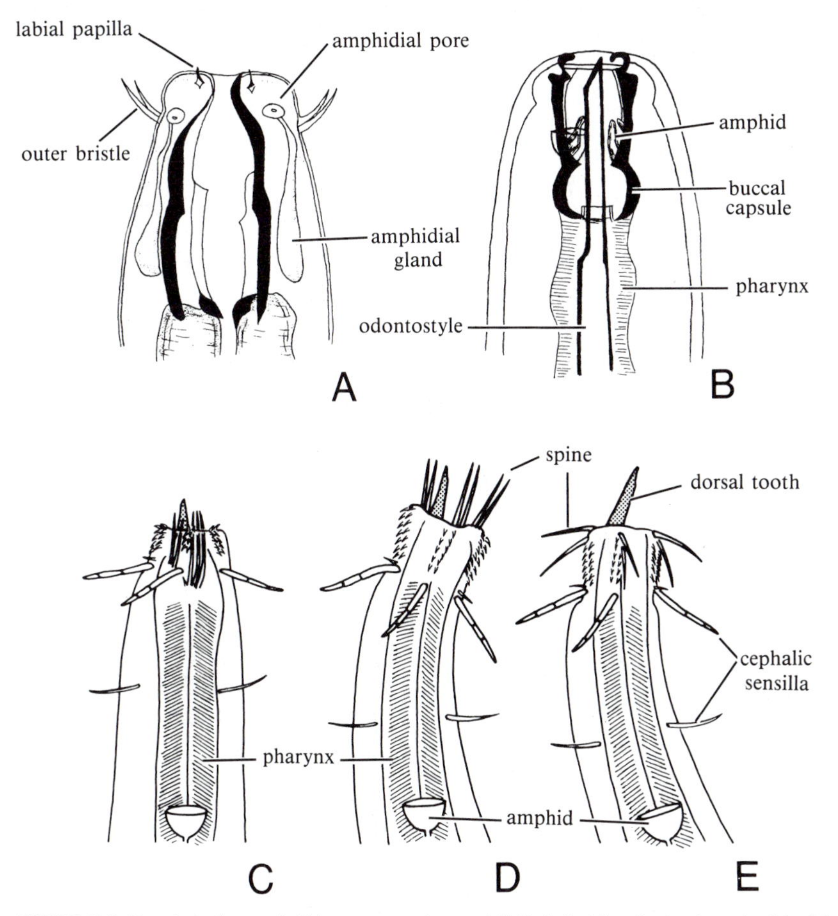

FIGURE 18.3. Nematode pharynx. **A**. *Pelagonema*, marine enoploid. **B**. *Actinolaimus* freshwater and soil dorylaimoid with piercing odontostyle. **C,D,E**. *Kinonchulus*, successive stages of eversion of buccal apparatus. (**A** after Kreis; **B** after Filipjeo; **C,D,E** after Reimann, in Lorenzen.)

gus), a midgut consisting of the intestine, and a proctodeal rectum or cloaca. The inconspicuous to large **buccal cavity** is often highly differentiated. It is lined with cuticle and often thickened to form ridges, plates, or teeth. When highly specialized, the buccal cavity may be divided into a long anterior chamber enclosed by the lips, the cheilostome, and a small posterior chamber in front of the pharynx, the esophastome (Figs. 18.3A,B and 18.4). Plant parasites have an **oral stylet**, used to puncture plant cells so the contents can be sucked out, formed either by the modified lining of the buccal cavity (**stomatostyle**) or sometimes from a highly modified tooth (**odontostyle**, Fig. 18.3B). In *Kinonchulus*, a primitive form, the buccal area is eversible, just as in priapulids, loriciferans, and kinorhynchs (Fig. 18.3C,D,E).

The **pharynx**, termed the **esophagus** by most nematologists, in its simplest form is an unspecialized tube, but in most nematodes it undergoes regional specialization (Fig. 18.4). Various distinctive patterns of muscular and glandular regions are recognized and are important in taxonomy. The pharynx has a characteristic anatomical structure, with a triangular lumen and strong radial muscle fibers (see Fig. 18.2B).

In most nematodes the **intestine** is a simple epithelial tube without any regional specialization. It is composed of a relatively small number of cells in small nematodes, for example, 18 to 64 in rhabditids and 16 to 24 in tylenchids. In some nematodes an anterior ventricular region, a midregion, and a posterior prerectal region can be distinguished on histological grounds.

The cuticle-lined proctodeal **rectum** leads from an intestinorectal sphincter muscle to the slit-shaped

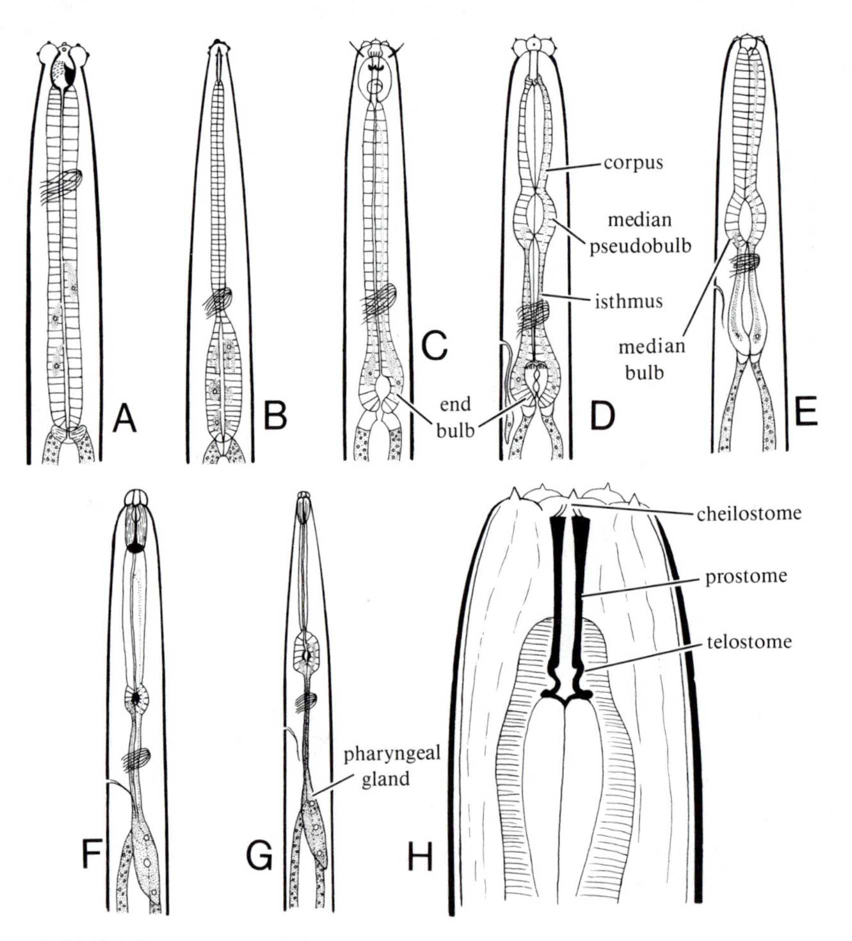

FIGURE 18.4. Pharynx structure. **A**. *Mononchus*, cylindrical type. **B**. *Dorylaimus*, dorylaimoid type. **C**. *Ethmolaimus*, bulboid type. **D**. *Rhabditis*, rhabditoid type. **E**. *Diplogaster*, diplogasteroid type. **F**. *Helicotylenchus*, tylenchoid type. **G**. *Aphelenchus*, aphelencoid type. **H**. Buccal capsule of *Rhabditis*. (**A,B,C,D,E,F,G** after Hirschmann, in Sasser and Jenkins; **H** after Stekhoven and Tenuissen.)

anus. A muscle lifts the lip of the anus and rectal wall to aid in defecation. Large, unicellular rectal glands are common in parasitic species. The male reproductive tract joins the rectum to form a **cloaca**, although it is not joined by the urinary system. The cloaca is highly modified to provide copulatory organs, discussed later with the male system.

Excretory System Primitive marine nematodes have a glandular excretory system, and other nematodes have a tubular system based on it. There are no flame cells or bulbs of any kind. The glandular cell (**renette cell**) making up the unicellular excretory system of *Enoplus* is on the ventral midline and opens through a midventral, anterior pore (Fig. 18.5A). Closely related parasitic and freshwater species also have a renette. Secernentea have a tubular excretory system (Fig. 18.5B). Rhabditids have an H-shaped system that in the adults is associated with two renette cells (Fig. 18.5C). Evolution in the excretory system has resulted in the reduction of the renette cells (Fig. 18.5D) and the loss of one arm of the H-shaped system of tubes (Fig. 18.5E).

The **pseudocoel** is derived from the blastocoel and contains the gut and reproductive organs. It is filled with the perivisceral fluid and a fibrous, lacy material that invests the gut and reproductive organs and lines the body wall. The fibrous material contains nuclei in some nematodes. In *Ascaris* and some other parasites, a single large cell above the pharynx produces fenestrated membranes covering the body wall and the digestive tube. Fixed cells, called **pseudocoelocytes**, are present in other nematodes. Among parasites, only two to six such cells are found, but in free-living species they may be more numerous. They are constant in position and are often highly branched. They do not take up particles or move, but may be oxidative centers.

Nervous System The central nervous system consists of a **circumenteric nerve ring**, several associated ganglia, and a system of longitudinal nerves (Fig. 18.6). The ring is predominantly formed of nerve fibers, and the nerve cell bodies are located for the most part in ganglia immediately posterior to the ring. Paired lateral and single or paired dorsal ganglia are attached to the ring. A variable group of smaller ganglia are associated with the nerve ring and nerves. Sensory nerves pass to the cephalic sensilla from the cephalic ganglion, and a pair of amphidial nerves arise from the amphidial ganglia associated with the lateral ganglia. The predominantly motor dorsal nerve cord runs in the dorsal hypodermal cord. The predominantly sensory lateral nerve trunks run in the lateral hypodermal cords and pass through lumbar ganglia to terminate at the anus. The main body nerve is the ventral nerve trunk in the ventral cord. It is formed by the union of two trunks at the retrovesicular ganglion behind the excretory pore and continues posteriorly to the single or paired anal ganglia. Commissural connectives are often present, and some nematodes have additional longitudinal trunks. Visceral nerves pass to the esophagus from the circumenteric ring and to the rectal region from the anal ring.

Reproductive System Nematodes are rarely hermaphroditic, with ova and sperm formed in the same gonad and with self-fertilization the rule. Most nematodes have separate sexes, with sex determined by environmental or hereditary mechanisms. Freshwater and soil nematodes tend to be predominantly female, and parthenogenetic development is common. An outstanding peculiarity is the continuity of gonads and gonoducts, with a continuous epithelial capsule enclosing both. Many nematode gonads have a terminal **stem cell**, from which all gametes arise.

Secernentean male nematodes have a single **testis**, but adenophoreans generally have anterior and posterior testes that join at the **sperm duct**. The testis consists of an apical germinal zone, a middle growth

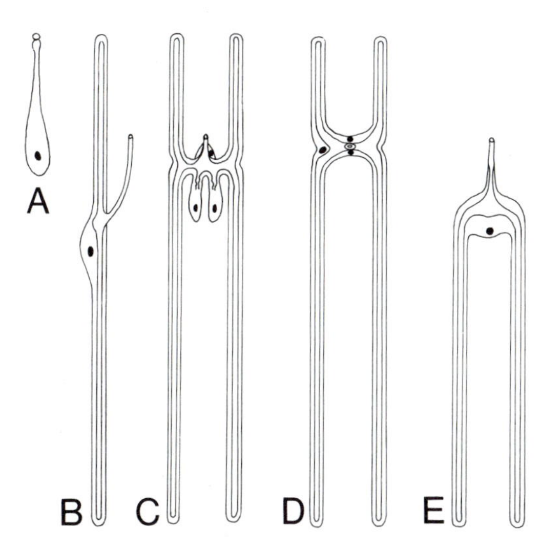

FIGURE 18.5. Forms of nematode excretory systems. **A**. Unicellular renette system. **B**. One-sided tylenchoid type, with renette cell and excretory tube, and median excretory pore. **C**. Rhabditoid system, with paired renette cells and H-shaped system of excretory tubules. **D**. Oxyuroid system, with H-shaped excretory tubules and no renette cells. **E**. Ascaroid type, without anterior excretory tubules. (All after Hirschmann, in Sasser and Jenkins.)

zone, and a terminal zone where gametes mature and are dissociated (see Fig. 18.1). The sperm duct usually dilates near the origin as a **seminal vesicle** with glandular walls, forming the cement used during mating. The seminal vesicle opens into the **vas deferens** that is generally subdivided into a tubular and a glandular region surrounded by muscle either terminally or over its entire length. The terminal portion is the **ejaculatory duct**. **Prostatic glands** may open into the end of the ejaculatory duct. Depressions in the cloacal wall, the **spicule pouches**, secrete cuticular rods or blades (**copulatory spicules**) used in mating. Most nematodes have two equal or unequal spicules, but a few have one or none. The spicule pouch sometimes secretes the **gubernaculum** that guides the spicule to the anus. Hook worms and some other nematodes have a prominent **bursa**, an umbrella-shaped affair formed of cuticular flanges (alae) and supported by radiating muscular bursal rays, arranged like the ribs of an umbrella (Fig. 18.7C). The bursa clasps the female during mating. The form of the spicules, gubernaculum, and bursa is important in species determination.

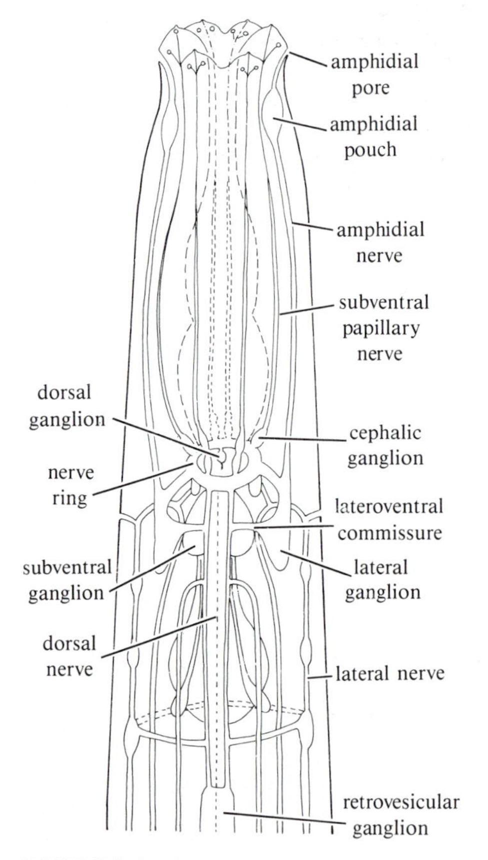

FIGURE 18.6. Anterior nervous system of *Rhabditis terricola*. Note nerve ring and relatively large number of ganglia and ring commissures. (After the Chitwoods, from Hirschmann, in Sasser and Jenkins.)

Most female nematodes have an anterior and a posterior ovary, but some species have one, and in one family there may be as many as 32. The **ovary**, like the testis, is divided into a germinal zone, growth zone, and terminal zone (see Fig. 18.1). The terminal zone leads quickly into the dilated **uterus**. The upper end of the uterus serves as a **seminal receptacle**, and eggs are fertilized as they pass through it. Shell formation and the early part of embryonic development take place in the uterus. In some nematodes, the eggs hatch into larvae in the uterus, but these ovoviviparous species are not common. The two uteri join at the **vagina**. Nematodes with one ovary usually have a posteriorly directed **diverticulum** at this position that serves as a seminal receptacle, probably the vestige of a second gonad. The heavily muscled vagina has a cuticular lining, and may be modified at the end to form an **ovijector** to expel the eggs. The female **gonopore** is a slit-shaped midventral **vulva**, often equipped with special dilator muscles.

Biology

Locomotion Nematodes have only longitudinal muscles. In the absence of circular muscle, the nematode body can be stretched out only as a result of the high turgor pressure of the perivisceral fluid and by the elongation of the crossed cuticular fibers, when present, as they change the angle of the lattice they form. Body length is determined by the fluid pressure in the pseudocoel and the tone of the longitudinal muscles, with the body diameter increasing to compensate for any decrease in body length.

Some nematodes with a rough cuticle move by alternately elongating and shortening the body, rather like earthworms. In this case the longitudinal muscles are arranged obliquely in an apparent chevron pattern, and those opposite one another contract together. Sinuous, dorsoventral, serpentine movements are more common and produce effective forward movement, especially when combined with roughened cuticle. This results from contraction of the longitudinal muscles in one region and relaxation of the longitudinal muscles directly below, and corresponding shifts in the angles of the cuticular lattice in the opposing surfaces. In a viscous medium or confined area, such sinuous movements are fairly effective, but result in aimless lashing in a less viscous habitat.

Draconematids (Fig. 18.7A) have long stilt bristles with adhesive glands, and inch along by attaching the head bristles and posterior stilt bristles alternately, or creep on the stilt bristles by alternately stretching and shortening the body.

Feeding Nematodes exhibit a wide array of feeding habits, including herbivory, saprophagy, carnivory, and parasitism of both plants and animals.

The secretions of the pharyngeal glands aid in penetration of host tissues and are important in food acquisition. Enzymatic secretions are forcefully ejected from the mouth to soften and predigest food before ingestion. The food taken in is often wholly or largely fluid, and many nematodes have developed a pharyngeal pump. A **pharyngointestinal valve** at the union of the pharynx and the intestine remains closed while the pharynx fills. It opens and the mouth closes while the pharyngeal muscles force food into the intestine.

Little if any intracellular digestion occurs, although details of the digestive processes are not well known. Microvilli thickly clothe the inner surface of the *Ascaris* intestine, and secretory bulbs can be seen on its surface. Feeding is usually very rapid and is achieved by a rapid pumping action of the pharynx. Digested food is stored within the intestinal wall. Undigested material is often violently ejected as turgor pressure builds; for example, *Ascaris* has been known to send a spray of feces flying two feet into the faces of investigators when animals are examined in the laboratory.

Glycogen and fat reserves in the intestinal cells are used during starvation and molting, and in some species food is accumulated in large amounts at specific periods of the life cycle in order to prepare for critical waiting periods or periods of intensive development.

Excretion Excretory physiology is poorly understood, but the predominant nitrogenous wastes are apparently ammonia and urea, most nematodes being ammonotelic. *Ascaris* can excrete more urea when water is scarce, up to 52 percent of its total nitrogen, but only 7 percent in more normal conditions. This ability to change excretory patterns with the availability of water may help to account for the adaptability of nematodes, especially to water deprivation in dry soils or in dehydrating organic matter. Urea is ordinarily formed from ammonia released during deamination by way of an ornithine cycle known to be centered in the intestine. Urease also is present and can convert urea to ammonia. The biochemical pathway needed to shift excretion products is present, but the way in which it is controlled is not understood. It is evident, however, that most, if not all, of nitrogen excretion occurs in the gut wall.

The nematode cuticle is permeable to water and other substances entering and leaving the animal, and many nematodes adjust to a wide range of salinities. Osmoregulation is brought about in part by the discharge of salts through the body surface, a process mediated by the epidermis. It also results from the discharge of water. Most water is eliminated by the intestinal wall, but the activity of the excretory system changes with the salinity of the surroundings, indicating that it has a hydrostatic function. The fine structure of the renette cells appears to be suitable for a filtration organ; it is possible that it also functions to some extent in the elimination of nitrogenous wastes. No clear functional understanding, however, has yet emerged for the renette cells. Although excretion and osmoregulation may occur there, the cells might also produce enzymes for digestion or anticoagulation, or they may be involved in regulating molting. The use of the intestine as the primary organ of nitrogen excretion and the use of the excretory system as primarily a hydrostatic de-

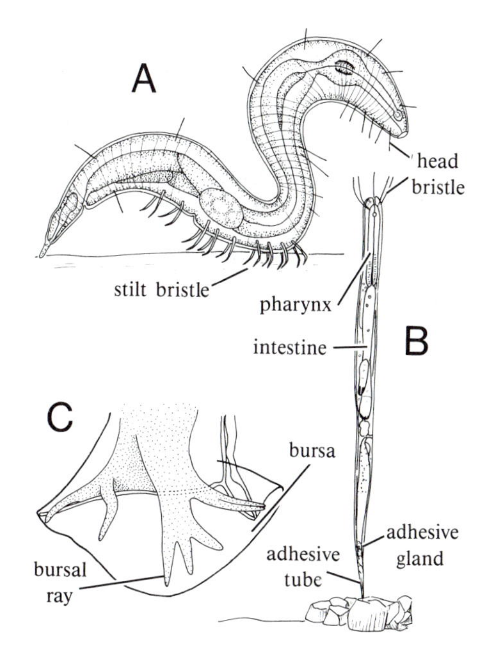

FIGURE 18.7. A. Draceonematid, characterized by enlarged anterior end, and stilt bristles containing adhesive tubes used in locomotion. **B**. Marine nematode, attached by posterior caudal glands. **C**. Bursa of a hook worm, *Ancylostoma duodenale*. (**A** after Steiner, from Rauther, in Kükenthal and Krumbach; **B** after Steiner; **C** after Looss, from Chandler and Read.)

vice are characteristic of the pseudocoelomates in general. However, nematodes typically have a very moderate capacity for osmoregulation, and in most cases their distribution depends on their ability to tolerate swelling or shrinking that results from the loss or entry of water.

Respiration Nematodes have no special respiratory organs, but a distinct nematode hemoglobin occurs in the perivisceral fluid of some parasitic nematodes. Hemoglobins of differing character sometimes occur in the body wall and the perivisceral fluid.

Most invertebrates are aerobic, but can survive periods of oxygen deprivation. Some capacity for anaerobic respiration is generally present, but this capacity usually is inadequate to permit normal activity levels and survival when oxygen is not available. Most aerobic animals take up larger quantities of oxygen immediately after oxygen deprivation, the so-called oxygen debt phenomenon. The debt must be paid to oxidize the end products of anaerobic respiration that have piled up in the tissues. An animal perfectly adapted to anaerobic life builds up no oxygen debt, but it is able to handle the lactic acid end products of its glycolytic energy release. In such organisms, oxygen actually may be toxic. For example, *Ascaris* is sensitive to high oxygen tensions, probably because it produces hydrogen peroxide, and lacks catalase for its disposal.

Complete anaerobes are extremely uncommon among nematodes. Trematodes, cestodes, and nematodes that live as intestinal parasites are predominantly anaerobic, but all can use oxygen when it is available. Adaptability to such changing respiratory situations is an important factor in survival. Nematodes live in a variety of environments, and many species are widely distributed, occurring in areas with both high and low oxygen tensions.

Selective pressures that favor respiratory adaptations in nematodes arise from low oxygen tensions in parasitic environments, benthic communities, or wet soils, as well as the result of increased size and activity. Many free-living species, because they lack a special respiratory system, remain small. They live on a narrow physiologic margin with little or no safety factor, and either adjust activity to levels that are suitable for a low oxygen uptake, or develop special methods that increase the efficiency of the oxygen uptake system or the oxygen transport system. The pseudocoel of nematodes undoubtedly improves respiratory transport by means of the respiratory pigments in the perivisceral fluid. Such pigments lower the critical oxygen tension at which aerobic respiration must be curtailed, increase the safety factor, and permit occupation of oxygen-poor habitats. Nematode hemoglobins retain oxygen at very low oxygen tensions, and although they are probably useless for oxygen transport, they may provide a small reserve of oxygen for times when the oxygen tension is very low. Host blood pumped through the gut of parasitic forms, such as hook worms and filarial species, provides oxygen to the worms. Thus, biochemical specializations are even more important in nematodes than anatomical specializations in adapting to a limiting respiratory situation.

Respiratory adaptation by way of aerobic enzyme systems is clearly shown in nematodes. The Krebs cycle has been identified in some nematodes, but is known to be absent in others such as *Ascaris*. Cytochrome–cytochrome oxidase is present in *Trichinella*, but absent in *Ascaris* where a similar system is found. The anaerobic respiration of carbohydrates releases a variety of end products, an amazing array of fermentative five- and six-carbon acids that is otherwise difficult to explain. The addition of oxygen to the environment reduces the output of organic acids in some nematodes, but has little effect on others, especially parasitic forms like *Ascaris* and *Trichinella*. In the absence of a Krebs cycle, or adequate substitute for it, these parasitic nematodes are dependent on glycolytic energy release. They depend on carbohydrates for food and ferments them in metabolic patterns resembling fermentations seen in plants. They are especially prone to store glycogen and to depend heavily on glycolysis. And above all, since a nematode extracts but a small part of the total available energy from its food, it must process a great deal of material, thus influencing its eating habits.

Circulation The perivisceral fluid is chemically complex in *Ascaris*, one of the few nematodes in which it has been carefully studied. It contains a variety of monovalent and divalent ions, with sodium the predominant cation, and chloride the predominant anion. A deficiency of anions is probably balanced by various organic acids and possibly bicarbonates. Proteins, fats, and carbohydrates are present, with proteins predominant. Urea, ammonia, and some free amino acids have been recovered, and, as mentioned earlier, a small amount of hemoglobin is present. The chemical complexity of this fluid reflects its function as an important transporting medium. Body movements agitate the fluid and increase its effectiveness. In addition, the perivisceral fluid is under high turgor pressure and is extremely important as a hydrostatic fluid in connection with body movements.

Sensation Although small and simple in form, nematodes have a number of sense organs and several muscles for moving specific bristles and papillae. Probably the most important development is the appearance of definitely specialized dendritic and axonic branches of the bipolar neurons.

So little is known of the functions of nematode sense organs that we have no good idea of their perception of the world they inhabit. Only a few marine nematodes have "eyes," pigment-cup ocelli with cuticular lenses. The cephalic bristles and papillae (see Fig. 18.2A) are evidently sensory and are complex histologically. Some must be tactile, but specific functions can only be guessed. **Amphids** are chemoreceptors. Although they vary with species, all have about the same components: an external pore, a duct, and an amphidial pouch (see Fig. 18.3A). An amphidial nerve forms a cluster of nerve endings, and the unicellular amphidial gland opens into the pouch. **Phasmids**, when present, occur on the posterior part of the body. When fully developed they consist of a gland, a pore, and nerve endings. They are evidently glandulosensory organs, but of unknown function.

Responses to single environmental stimuli have not been intensively studied. *Dorylaimus* and *Rhabditis* from sewage beds give no definite laboratory responses to temperature, light, or chemicals dispensed from capillary tubes. One species of *Rhabditis* is attracted by dilute acids. Many nematodes appear to be wholly insensitive to nearby prey, and must touch them with the lips before reacting. On the other hand, nematodes that feed on or enter plant roots locate roots from a distance and aggregate around them.

Knowledge of nematode nerve physiology is growing (Wood, 1988). Acetylcholine is more abundant anteriorly because of the numerous synaptic junctions associated with the ganglia, and also causes muscle contraction. Mammalian cholinesterase has not been identified; however, isozymes of acetocholinesterase have been demonstrated.

Reproduction Although nematodes do not reproduce by fission or polyembryony, soil forms and some parasites can reproduce by parthenogenesis. Different populations of the same species can use different ways to achieve this.

Most nematodes have separate sexes, although some species are demonstrably protandric hermaphrodites. In a few instances, intersexes have been observed: both females with some secondary male characters, as well as males with secondary female features. It is not clear whether the intersexes are examples of arrested serial hermaphroditism or whether they are induced by environmental factors.

Sexuality in dioecious forms is genetically determined: females have two X chromosomes while males have one. Nematodes do not seem prone to pass through successive cycles of reproductive fertility, but rather usually reach sexual maturity and then eventually engage in a single bout of copulation and egg laying before dying. Fertilization is internal, and mutual chemical attractants seem to bring males and females together. The males coil their tails around the female vulvar opening, inserting spicules guided by the gubernaculum, if present, to ensure that cloaca and vulva are aligned properly. The male bursa, when present, and secreted cements can help to hold the pair together. Sperm are injected quickly and move by amoeboid action to the seminal receptacle where they await use.

Development A membrane appears around the egg at fertilization, which thickens to form the shell, and a lipid membrane is deposited on its inner surface. Glands in the uterus wall add protein to the surface so the fully formed shell is three-layered. Some eggs have a break in the shell (**operculum**) through which the embryo escapes, and many parasitic and some free-living nematodes attach the ova with threads or branched filaments to submerged objects or the substratum.

The complete cell lineage of the free-living nematode, *Caenorhabditis elegans*, is known, and at present the locations of more than 700 genes have been mapped (e.g., see Davidson, 1986; Wood et al., 1988). The first cleavage is transverse and produces a germ cell (P1) and a somatic cell (S1) (Hope, 1974). S1 divides into blastomeres A and B, and P1 divides into EMST and P2. These are arranged into a very characteristic T-shape stage (Fig. 18.8A). The four cells arrange themselves into a rhomboid (Fig. 18.8B), with A anterior, B dorsal, EMST ventral, and P2 posterior. A and B give rise to the ectodermal covering of the body, and EMST is a stem cell for the endoderm, mesoderm, and stomodeum (Fig. 18.8C). It divides into blastomeres E and MST at its first division, and MST divides into M and ST at the next division, thus segregating as single blastomeres the future endoderm (E), mesoderm (M), and stomodeum (ST). The daughters of E move into the interior at gastrulation (Fig. 18.8D) and develop into a chain of cells that will form the midgut. The daughters of M form a strand of cells on each side of the endoderm cells and will form all of the mesodermal tissue. ST develops into a group of cells that becomes the stomodeum and establishes the foregut. The P line continues to produce for a time one P cell and one somatic cell at each division: P2 divides into P3 and C, P3 into P4 and D, etc. The P5 cell is the germinal cell, eventually producing the stem cell of ovaries or testes. The other cells are somatic, forming the posterior epidermis and proctodeum and, in some nematodes, some ectomesoderm. The nervous system and sense organs arise from the ectodermal cells, and the renette cells are also ectodermal. Cleavage is not spiral but is highly determinate and evocative of that seen in gastrotrichs.

A distinctive feature in the non-P-cell lineage is the phenomenon of chromosome diminution. After each successive division, only the middle part of the chromosomes is retained; the end pieces drift off into the cytoplasm. Only the germinal cells retain their chromatin intact.

Late in embryonic development, nuclear divisions cease so that at hatching all future growth results from changes in cell size. The only exceptions are the cells that produce the reproductive organs. The result is a great cell constancy and extends to all parts of the nematode body except the reproductive system. The nematode nervous system, for example, is formed of the same number of cells in each member of a species. Each ganglion consists of the same number of cells, and each cell is in a constant position, with its axon extending to a predetermined point.

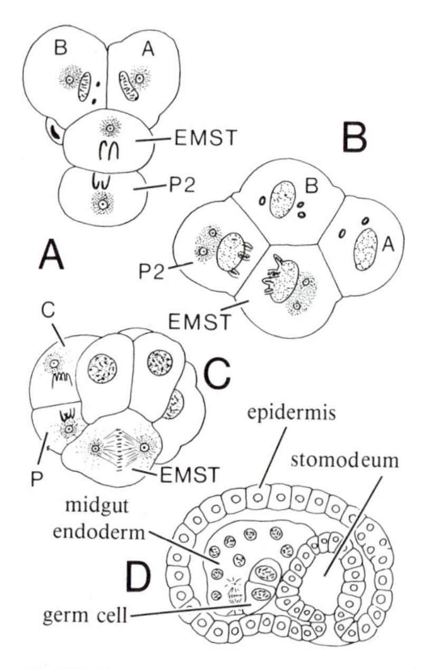

FIGURE 18.8 Nematode development, *Parascaris equorum*. **A**. T-shaped, four-cell stage. **B**. Four-cell stage, after rearrangement into rhomboid form. **C**. Seven-cell stage, showing stem cell for endoderm, mesoderm, and stomodeum (EMST). **D**. Sagittal section of embryo at time of stomodeum formation. (After Boveri.)

Life Cycles Fully developed, but juvenile, nematodes emerge from the eggs at hatching. Adenophoreans hatch as first-stage juveniles, whereas secernenteans emerge as second-stage juveniles having undergone one molt within the eggshell. Some changes in form occur during their growth, both at the anterior end where the pharynx may undergo progressive specialization, and in the reproductive system that develops gradually. Four molts occur during growth and maturation. When the mesocuticle and endocuticle are present, the entire cuticle is shed at molt (Maggenti, 1979), so the larva loses the lining of the buccal capsule, pharynx, and rectum. Each molt terminates one growth period and begins another, often separating distinctive phases of the life cycle. Parasites sometimes remain in a stage, awaiting introduction to an appropriate host, whereupon they will molt immediately and begin a new phase of growth. The cuticle shed at the second molt of some parasitic nematodes is not discarded but forms a sheath around the third-stage larva, increasing its resistance to unfavorable environmental conditions (Fig. 18.9B). After the fourth molt the worms are mature, but may grow somewhat in length.

Most nematodes are free living and develop directly into adults, but specialized behavior and growth produces specific and varied life cycles often centered on specialized feeding habits.

Phytophagous nematodes are small, and the plants on which they feed are relatively huge. The line between normal foraging on one or a few plant species and parasitism is indistinct, but eventually the association of nematode and food organism becomes so intimate that the nematode must be considered parasitic. Juveniles, adults, or both may be parasitic.

Some ability of adults and larvae to live freely in the soil is to be expected in the initial phases of parasitism, and some species are facultative parasites and may or may not enter plants. An early stage of parasitism is seen among species that have larvae that attach to, but do not enter, plant roots. Adults produce soil-inhabiting larvae that find their way to new roots. Some nematode larvae enter the plant tissues in which the adults feed and grow. Eventually parasitism becomes obligatory and the final stages of phytoparasitism are attained. The second-stage juveniles of *Meloidogyne*, for example, enter the host plant and pierce and suck out the contents of cells; their feeding causes hyperplasia and hypoplasia resulting in a gall, inducing giant nutritive cells that allow the adult worms to be sedentary. Juveniles mature in the gall. Male nematodes swell, as do females, but in the fourth stage metamorphose to a vermiform stage and embark from the home gall into the soil to seek other infected plants, where they fertilize the females and die. The females become more and more plump, losing all resemblance to the cylindrical larvae, and several hundred ova form within them (Fig. 18.9C). In *Heterodera*, the gravid females die, but their cuticle remains as a cyst membrane around the eggs (Fig. 18.9A), which can remain alive for years before germination.

Insects feeding on the host plant can transport larvae from plant to plant, and on this relationship complex and detailed life cycles may be based. Thus, adult females of *Tylenchinema oscinellae* live in the hemocoel of fruit flies. Their larvae enter the fly gut and are deposited on oat plants parasitized by the fly. Larval flies and larval nematodes develop independently on the oat plant, and as the nematodes mature, the females enter the fly larvae after mating, while the males die.

One avenue to zooparasitism is the habit of feeding on feces. Many soil forms have specific preferences for fecal material. Dung beetles and flies that visit feces become transporters of some *Rhabditis* species. This habit is fixed in the life cycle of some, whose third-stage larvae remain quiescent until they have been carried by insects to fresh dung. It is no large step from this kind of hitchhiking to endoparasitism. Diplogasterid larvae are transported to the galleries of bark beetles by attaching to the insects, and some enter the gut or hemocoel of the insect. Only the larvae of mermithids are parasitic. The eggs are eaten by grasshoppers and develop in its body. They emerge, complete the final molt, and grow to maturity in the soil. Scavenging adults may have larvae that enter the body of an animal and await its death to mature, as in *Rhabditis maupasi*. The young live in the soil; they enter and leave earthworms freely, but do not mature until the earthworm has died. They feed on the bacteria decomposing the carcass, producing larvae that live freely in the soil.

A number of nematodes parasitize humans and domestic animals, and a variety of fascinating life cycles can be recognized. *Ascaris* eggs leave the host with the feces and are infective in new hosts. They hatch when eaten. The larvae migrate into the bloodstream and then to the lung, where they pause and molt twice. After the last molt they pierce the lung wall, migrate up the respiratory tree, enter the pharynx, and are reswallowed, eventually becoming mature in the intestine. *Trichinella* females release up to 1500 live larvae that migrate through the lymphatics and bloodstream and eventually encyst in striated muscle tissue. Here they remain until the muscle is eaten by an acceptable host. While in the gut of this host, they emerge from the cysts and reach maturity in the intestine. Hook worm eggs

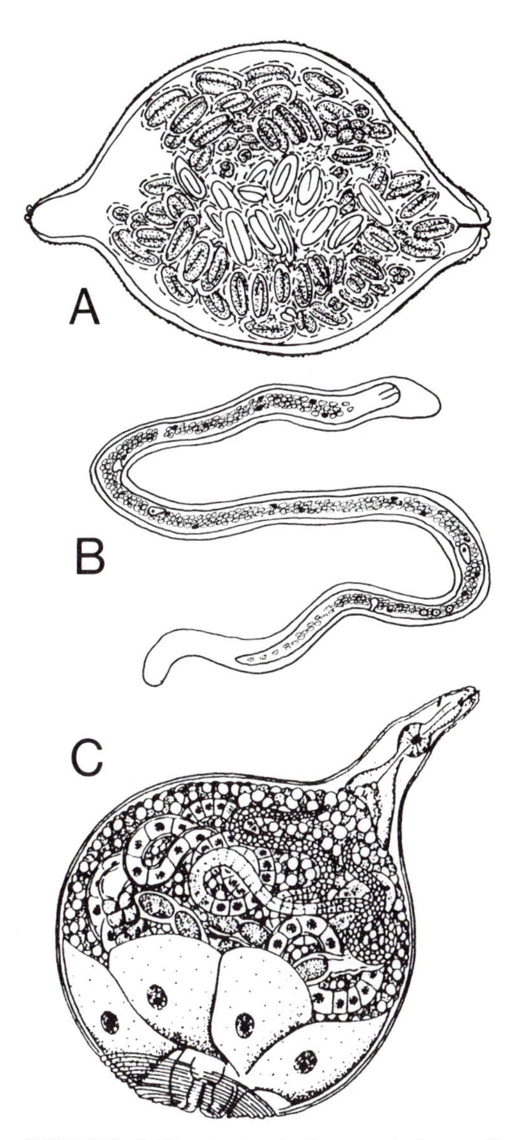

FIGURE 18.9. **A**. *Heterodera* female after conversion into cyst filled with inactive larvae. **B**. Sheathed larva of *Wuchereria;* cuticle from preceding molt has remained in place to form sheath. **C**. *Meloidogyne*, adult female with maturing ovaries. (**A** after Maggenti and Allen, from the Nobles; **C** after Maggenti.)

leave the body with the feces and hatch to form larvae that live in the soil. These larvae eventually become infective and penetrate the skin, migrate along pathways like those followed by *Ascaris*, and mature in the intestine where they attach to the intestinal lining and feed on blood from the wounds produced by their teeth. *Wuchereria* adults live in the lymphatics and give rise to larvae known as microfilariae. The larvae then enter the blood and are ingested by blood-sucking arthropods, in whose body they become infective larvae. When the arthropod takes another blood meal, the larvae reenter a vertebrate host and develop to maturity in the lymphatics.

Phylogeny

Some fossil nematodes are known, but these few fossils provide no information as to how the group evolved.

Shared features of cuticle structure, eversible buccal complex in primitive forms, body muscle form and function, and pharyngeal structure seem to clearly ally nematodes with nematomorphs (most closely) and kinorhynchs, loriciferans, priapulids, and possibly gastrotrichs, an arrangement formalized by Malakhov into a higher taxon, Cephalorhyncha.

The radial arrangement of nematode anatomy, especially at the head end, has suggested to some that nematodes were primitively sessile. Many sessile animals tend to develop radial symmetry, and it is thought that primitive nematodes were aquatic, attaching themselves to submerged objects by caudal cement glands and waving their bodies sinuously to obtain food (Fig. 18.7B,C). This is possible, but the tubular and, as a result, radial form of the nematode body plan could merely be a stereogeometric result of a wormlike animal with great body turgor. High internal pressure inflates the body into a cylinder with no clear dorsoventral or lateral surfaces, onto which external structures then tend to be arranged radially.

Discerning relationships within Nematoda, therefore, must rely on careful analyses of the anatomy of living forms. This is only just beginning. A further problem exists in that, at least among the Secernentea, there has been an oversplitting of species, genera, and families, and this is overlain by what appears to be an outmoded concept of host specificity among parasites (Maggenti, 1983). The suggested relationships in Figure 18.10, therefore, reflect the great imperfections of the current state of nematode taxonomy (Coomans, 1981) (discussion follows).

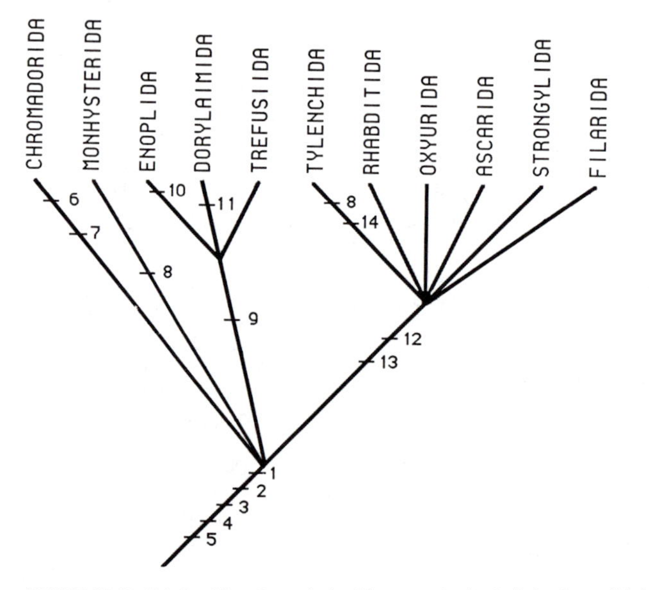

FIGURE 18.10. Relationships of nematodes. The unresolved polychotomies on this tree illustrate the marked uncertainty that still exists concerning nematode relationships. Characters are: (1) anterior sensilla in pattern of 6 + 6 + 4, (2) vulva, (3) male spicules, (4) renette cells, (5) four postembryonic juvenile stages, (6) elaborately ornamented cuticle, (7) some with 12-fold buccal vestibulum, (8) ovaries elongate, (9) amphids nonspiral, (10) metanemes, (11) pharyngeal glands open posterior to nerve ring, (12) only anterior testis present, (13) caudal glands absent, (14) anterior and posterior gonads on same side of gut. (Modified from Lorenzen, 1983.)

Taxonomy

Nematode classification is not stable. Serious attempts to bring all nematodes into a single classification are relatively recent, and disagreements grow out of the homogeneity of the group. A confusing array of different combinations of minor variations makes it difficult to define taxa. The system used here is based on that devised by Lorenzen (1981, 1983), focusing largely on free-living forms and to which have been added the parasitic groups. However, it must be pointed out that this system has not gained universal support among nematologists. The main problem is that not all the traditionally used taxa can be defined with unique sets of derived characters. For example, class Adenophorea (= Aphasmidia) is defined by primitive or generalized nematode features and so is paraphyletic. Though the system is imperfect, it serves a pragmatic function of providing a focus for continuing research. Alternative schemes can be found in Maggenti (1982a,b, 1983) and Inglis (1983).

Class Adenophorea. Postlabial amphids variable, cephalic sensilla labial or postlabial, setose or papillose; cuticle generally smooth with four layers; excretory duct not cuticularized; six or more pseudocoelomocytes; testes paired.

Order Enoplida. Amphids nonspiral and pocketlike, metanemes (Fig. 18.4A).

Order Dorylaimida. Amphids nonspiral and pocketlike, metanemes absent; all pharyngeal glands open behind nerve ring (Fig. 18.4B).

Order Trefusiida. Amphids nonspiral and pocketlike, metanemes absent.

Order Monhysterida. Ovaries elongate and outstretched.

Order Chromadorida. Cuticle elaborately ornamented; anterior part of buccal cavity or vestibulum with twelvefold symmetry in some (Fig. 18.7A).

Class Secernentea. Only anterior testis present; caudal glands absent.

Order Tylenchida. Stomatostyle; ovaries outstretched, on same side of gut (Fig. 18.4F).

Order Rhabditida. Pharynx with posterior and sometimes anterior bulb; amphids are small pockets; with or without buccal stylet. Example: *Rhabditis* (Figs. 18.1 and 18.4D,H).

Order Oxyurida. Without cervical papillae; pharynx with posterior bulb; terminal part of female reproductive tract heavily muscled, and females usually with slender tail; males with one or two copulatory spicules, and often with alae forming bursa.

Order Ascarida. Mouth surrounded by three, rarely six, lips; with cervical papillae; pharynx with or without bulb, but never valvulated; often with caeca from pharynx, intestine, or both; without buccal capsule; females with short, curved tail; males with two equal copulatory spicules and usually without alae. Example: *Ascaris.*

Order Strongylida. Simple, lipless mouth or with buccal capsule; pharynx without bulb; females usually with ovijector, and males with usually conspicuous copulatory bursa supported by 13 muscular rays.

Order Filarida. Simple, lipless mouth and usually without buccal capsule; without pharyngeal bulb; females larger, with anterior vulva and long, tubular vagina; males with two unequal spicules and with or without alae. Example: *Wuchereria* (Fig. 18.9B).

Selected Readings

Chitwood, B. G., and M. B. Chitwood. 1974. *Introduction to Nematology*, rev. ed. Univ. Park Press, Baltimore.

Cobb, N. A. 1915. *Nematodes and Their Relationships.* Dept. of Agriculture Yearbook 1914, Washington.

Coomans, A. 1981. Aspects of the phylogeny of nematodes. *Atti Acad. Naz. Lincei* 49:161–174.

Croll, N. A., and B. E. Matthews. 1977. *Biology of Nematodes.* Blackie, Glasgow.

Croton, H. D. 1966. *Nematodes.* Hutchinson, London.

Davidson, E. H. 1986. *Gene Activity in Early Development.* Academic Press, Orlando.

Hope, W. D. 1974. Nematoda. In *Reproduction of Marine*

Invertebrates, vol. 1 (A. C. Giese and J. S. Pearse, eds.), pp. 391–469. Academic Press, New York.

Inglis, W. G. 1983. An outline classification of the phylum Nematoda. *Aust. J. Zool.* 31:243–255.

Lee, D. L. 1965. *The Physiology of Nematodes*. Oliver and Boyd, Edinburgh.

Lorenzen, S. 1981. Entwurf eines phylogenetischen Systems der freilebenden Nematoden. *Veröff. Inst. Meeresforsch. Bremerh. Suppl.* 7:1–472. [An exhaustive treatment of free-living forms.]

Lorenzen, S. 1983. Phylogenetic systematics: Problems, achievements, and its application to the Nematoda. In *Concepts in Nematode Systematics* (A. R. Stone, et al., eds.), pp. 11–23. Academic Press, New York.

Lorenzen, S. 1985. Phylogenetic aspects of pseudocoelomate evolution. In *The Origins and Relationships of Lower Invertebrates* (S. Conway Morris, et al., eds.), pp. 210–223. Clarendon Press, Oxford.

Maggenti, A. R. 1979. The role of cuticular strata nomenclature in the systematics of Nemata. *J. Nematol.* 11:94–98.

Maggenti, A. R. 1982a. *General Nematology*. Springer-Verlag, New York.

Maggenti, A. R. 1982b. Nemata. In *Synopsis and Classification of Living Organisms* (S. P. Parker, ed.), pp. 879–923. McGraw-Hill, New York.

Maggenti, A. R. 1983. Nematode higher classification as influenced by species and family concepts. In *Concepts in Nematode Systematics* (A. R. Stone et al., eds.), pp. 25–40. Academic Press, New York.

May, R. M. 1988. How many species are there? *Science* 241:1441–1449.

Platt, H. H., and R. M. Warwick. 1980. The significance of free-living nematodes in the littoral ecosystem. In *The Shore Environment: Methods and Ecosystems* (J. H. Price, et al., eds.), pp. 729–759. Academic, London.

Wood, W. B. (ed.). 1988. *The Nematode Caenorhabditis elegans*. Cold Spring Harbor Lab., Cold Spring Harbor.

Zuckerman, B. M. 1980. *Nematodes in Biological Models*, vols. 1 and 2. Academic Press, New York.

19

Chaetognatha

The chaetognaths are a small group of about 115 species; but the arrow worms has never been securely allied with any phylum, since at one time or another they have been linked with almost every phylum. They are virtually worldwide in their distribution, but seem to prefer tropical and warm temperate waters. They are not uncommon; in some areas they comprise 12 to 15 percent of the total plankton. They are most effective swimmers, darting around in the water, and thus their common name of arrow worms. Although most live in the plankton, some chaetognaths are benthic at shallow and abyssal depths. Chaetognaths are all much alike. Differences in the position and number of lateral fins and in the shape of the caudal fin and detailed differences in head construction and in the seminal vesicles are the most important taxonomic characters.

Definition

Chaetognaths are generally planktonic, bilateral worms. The head is well delineated, with grasping spines, cephalic hood, retrocerebral organ, and ciliary loop. The trunk is fusiform, with one or two pairs of lateral fins and a horizontal tail fin. There are only longitudinal body wall muscles. There is no excretory system or distinctive larval phases.

Body Plan

The bilaterally symmetrical body is divided into head (**protosome**), trunk (**mesosome**), and tail (**metasome**). One or two pairs of **lateral fins** on the trunk and anterior tail region, and a **tail fin**, are lateral extensions of epidermis strengthened by fin rays (Fig. 19.1A).

A ventral depression on the head, the **vestibule**, leads to the mouth (Fig. 19.1B). Most species have anterior and posterior rows of **teeth**. Raptorial spines extend from the sides of the head. A conspicuous **hood** folds over the head between feeding periods to reduce drag while swimming and to protect the teeth and bristles. Each of the paired compound **eyes** is formed of a large **pigment-cup** ocellus and five smaller **ocelli** (Fig. 19.2). The **vestibular organs** are rows of papillae on each side of the head. Many chaetognaths also have glandular **vestibular pits**. Retrocerebral organs, or deep sacs, reach into the cerebral ganglia. They are enclosed by a connective tissue sheath, but receive fibers from ganglion cells. The organs open by way of paired ducts that unite before reaching a middorsal pore.

The **ciliary loop** or **corona ciliata** (Fig. 19.1C) is an oval or sinuous structure that lies at the anterior surface of the middorsal trunk. The *Spadella* loop has an outer zone of flagellated cells and an inner zone of glandular cells.

The body wall is composed of intracellular cuticle,

epidermis, basement membrane, and longitudinal muscles. There is a peculiar epithelial lining of the body cavities (Hyman, 1959, p. 14; Welsch and Storch, 1982). The basement membrane is thickened to form capsules for the eyes, and lateral and ventral head plates between the cuticle and the epidermis stiffen the head covering. The muscles have their origins and insertions on the thickened basement membrane.

A **head–trunk septum** divides a small head cavity from the trunk cavity. Muscles and other head structures encroach on the head cavity, reducing its volume significantly. A **trunk–tail septum** divides the trunk cavity from the tail cavity and defines the trunk and tail regions. The gut (Fig. 19.3A) is held in place by a dorsal and ventral membrane formed of double layers of basement membrane (Fig. 19.3A,B). A similar membrane divides the tail cavity into right and left compartments (Fig. 19.1A). The body fluid contains no cells, but small granules circulate slowly, passing through the lacy membranes, as a result of the beating of large cilia (flagella?) on the inner surface of the body wall.

Because chaetognaths are well cephalized, they have an intricate head musculature. However, the trunk and tail muscles are simple, and circular muscles are absent. Two dorsolateral and two ventrolateral longitudinal muscle bands bend the body and force the fins against the water to cause rapid darting movements. Histologically, the muscle is like coe-

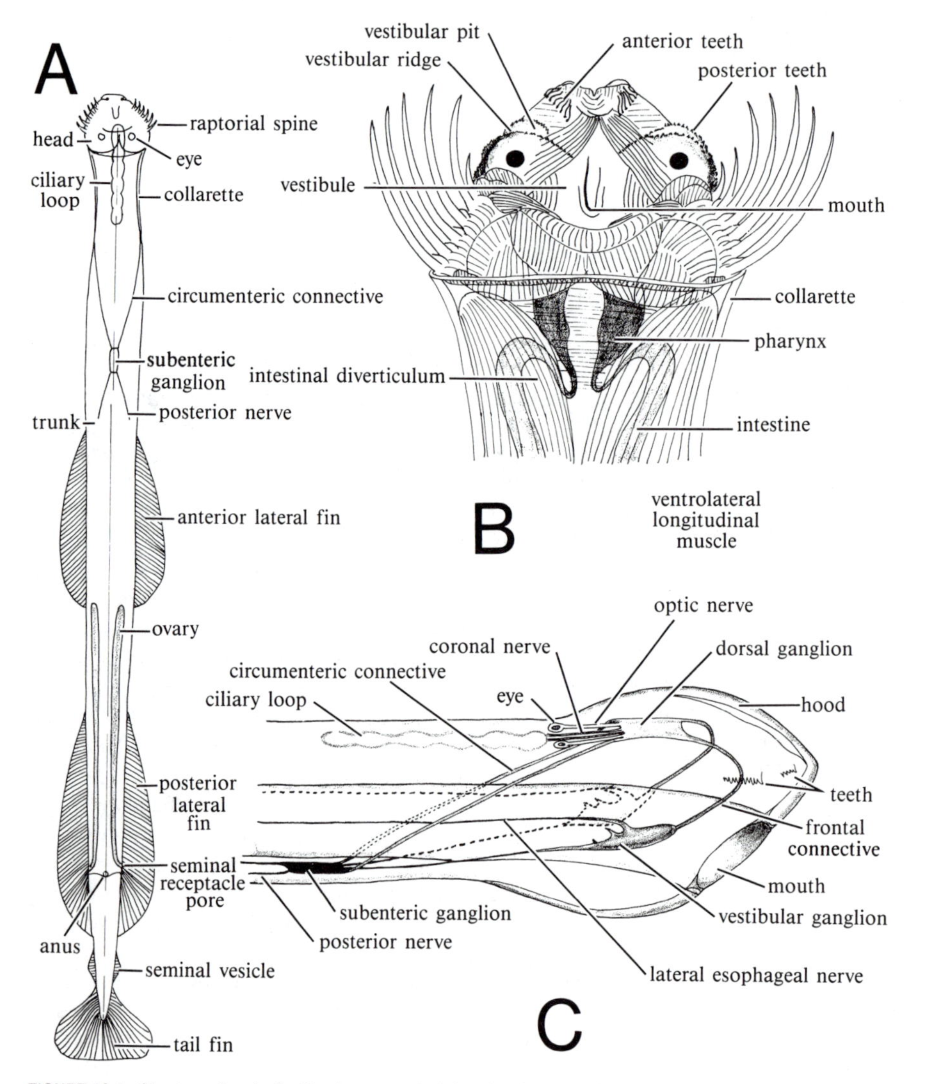

FIGURE 19.1. Chaetognatha. **A.** *Sagitta elegans,* ventral view showing general body form. **B.** Ventral view of everted chaetognath head. Note highly developed musculature. **C.** Lateral scheme of organization of chaetognath anterior end. (**A** after Ritter-Zahoney; **B,C** composite.)

lomyarian nematode muscle, with feathery bands of fibers arranged along the margins of cells.

Digestive System The circular or T-shaped mouth opens into a glandular **pharynx** that ends in a bulb at the level of the head–trunk septum. The lining of the straight **intestine** contains more gland cells anteriorly and more absorptive cells posteriorly, indicating some sequentialism in digestion. The outer wall of the intestine is marked with muscle filaments. An anteriorly directed set of gut **diverticula** can extend forward from the intestine to supply the complex cephalic tissues. The last part of the intestine is histologically differentiated as a **rectum**, strengthened with circular muscles and bearing a ciliated lining.

Excretory System No special excretory organs are recognized, although the ciliary loop has been suggested to possibly serve this function.

Nervous System Chaetognaths have a moderately complex nervous system, in keeping with their active life-style and complex musculature. A **frontal connective** joins the bilobed **dorsal ganglion** to a pair of ventral **vestibular ganglia** (Fig. 19.1C). Nerves to the eye and corona arise from the dorsal ganglion. Nerves to the pharynx, vestibular border, and head musculature arise from the vestibular ganglion. A pair of **circumenteric connectives** pass ventrally and posteriorly to a large **subenteric ganglion**. Pairs of nerves from the subenteric ganglion extend to the **subepidermal plexus** of the trunk and the tail.

Reproductive System Chaetognaths are hermaphrodites, though they are often protandric. Paired **testes** lie behind the trunk-tail septum. Masses of **spermatogonia** enter the tail cavity, and, when mature, the filiform sperm pass into the funnel-shaped openings (**gonostomes**) of the **sperm ducts**. Each of the sperm ducts ends in a **seminal vesicle** that bulges the body wall out near the base of the tailfin as they fill. Adhesive secretions mold the sperm into **spermatophores** in a sac at the anterior end of the seminal vesicle (Fig. 19.3B), and these are eventually freed by rupture of the vesicle and body wall.

Ovaries (Fig. 19.1A) lie at the posterior end of the trunk cavity, attached by a mesentery to the body wall (Fig. 19.3B). Chaetognath oviducts are unique. Each **oviduct** is double, formed of an outer tube, the oviduct proper, and an inner tube (Fig. 19.3B). The oviduct proper has no opening to the ovary or to the outside. The inner tube opens to the exterior by way of a short **vagina** and is expanded posteriorly to form a **seminal receptacle**.

Biology

Locomotion Chaetognaths are most effective swimmers. Their darting around in the water is achieved by the rapid dorsoventral flexure of their longitudinal muscles in combination with the lateral body fins. One form, the genus *Spadella*, is a bottom dweller and has large posterior adhesive papillae to help grasp the substrate.

Feeding Chaetognaths are voracious carnivores and use the venom tetrodotoxin, produced by commensal bacteria, to help capture prey (Thuesen and Kogure, 1989). *Sagitta bipunctata* has been seen to swallow baby herring as large as themselves. They usually swim about actively, striking at prey with a quick, flashing movement and a rapid snap of the bristles. However, the benthic *Spadella* attaches to the substrate and waits for prey to come by. During feeding, the hood is retracted, the mouthparts everted, and the teeth and spines spread. Little is known of digestive physiology. The pharynx secretes an adhesive substance that entangles, but does not stun, prey and probably prepares it for swallowing. *Spadella* gulps seawater after swallowing food, probably flushing enzymes into the rear part of the intes-

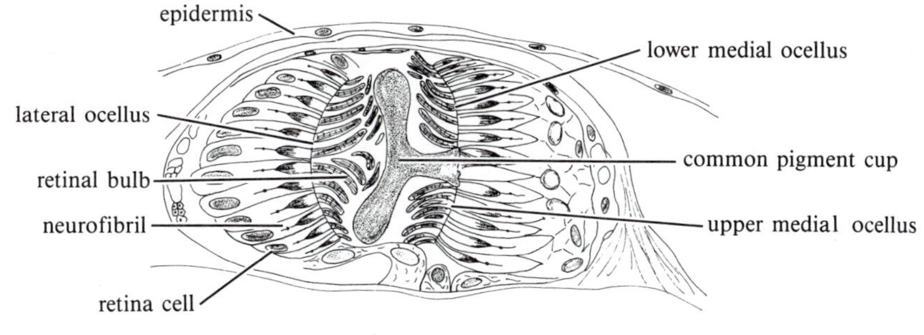

FIGURE 19.2. Section through eye of *Sagitta*. (after Hesse.)

tine where food usually lodges. Digestion seems to be wholly extracellular. Only glycogen- and starch-splitting enzymes have so far been reported. Digestion is rapid, taking from 40 minutes to four hours in various species.

Respiration and Excretion Respiration apparently occurs through the body surface. No true excretory organs are found, but the ciliary loop may have an excretory function. Its cells are known to take up and eject vital dyes, and some workers have suggested its ciliated cells are structural precursors of the protonephridial flame cells and are thus excretory in function. Soluble ammonia probably escapes through the body surface.

Sensation The ciliary loop of some species is a tactile receptor and may be used to detect water currents. In other instances it seems to have a chemoreceptive role as well. Mechanoreceptors are known to occur in rows along the body.

A pair of eyes occur on the posterior part of the head. These are composed of five units, although only three can be seen in any one section (Fig. 19.2). The pigment-cup ocelli are partially fused in the central region, and retinal cells face the pigment cup. This eye structure is very similar to that seen in some turbellarian flatworms.

Reproduction Some species are supposedly self-fertilizing, but most seem to outbreed (Reeve and

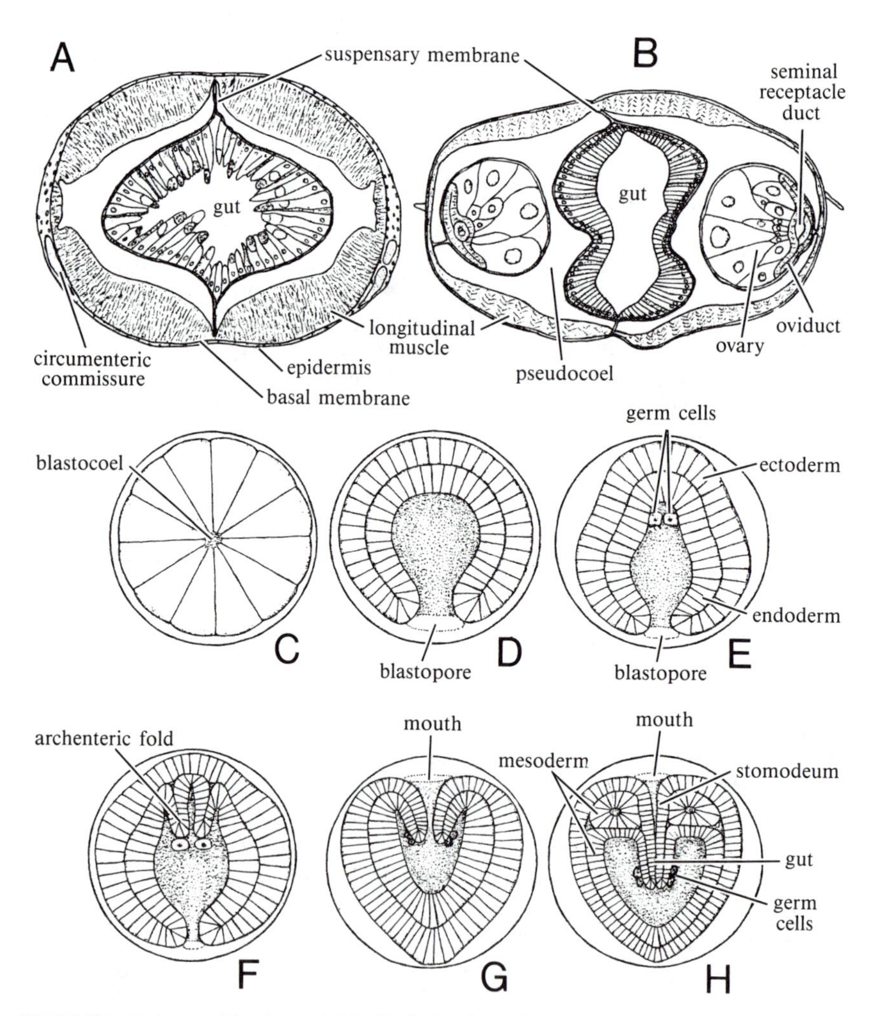

FIGURE 19.3. Anatomy and development of *Sagitta*. **A**. Anterior trunk, cross section. **B**. Trunk at level of ovaries, cross section. **C,D,E,F,G,H**. Early stages of development. (**A,B** after Hyman; **C,D,E,F,G,H** after Burfield.)

Casper, 1975). In the latter case, individuals align themselves head to tail, and spermatophores are attached to the partner's body surface after they break through the body wall. These are often placed on the dorsum of the animal, either on the neck or near the tail. The posterior part of the spermatophore dissolves, and sperm streams posteriad. It has been suggested that secretions and currents set up by the ciliary loop help get the sperm to the tail region where the stream bifurcates. Sperm eventually enter the vagina and move into the seminal receptacle. Fertilization occurs when sperm leave the receptacle and pass into the ovary through a duct formed by accessory nurse or fertilization cells of the female. The nurse cells form a hollow stalk that pierces the inner duct and conducts sperm to the ovum. The eggs then move into the outer oviduct. The eggs exit externally either by a temporary duct formed to permit their departure through the body wall, or by passing down the same duct the sperm came up. Both have been reported with supporting micrographs (Reeve and Lester, 1974).

Development Equal, holoblastic cleavage first produces blastomeres with staggered positions akin to that seen in spiral cleavage. These continue to cleave to form a blastula with a small blastocoel (Fig. 19.3C). The endoderm invaginates, coming in contact with the ectoderm to destroy all traces of the blastocoel (Fig. 19.3D). A pair of primordial germ cells arise early and remain unchanged during development (Fig. 19.3E). Lateral folds of the archenteron form anteriorly as the blastopore is closed off (Fig. 19.3F). A stomodeum is formed (Fig. 19.3G). The front end of the developing cavity is then cut off to form head mesoderm (Fig. 19.3H), and the trunk and tail cavities separate later.

It is noteworthy that the cavities flanking the embryonic gut disappear during the course of development as the mesodermal tissues consolidate. In fact, these cavities are reported to disappear before the mesodermal tissues seal off entirely from the gut (Burfield, 1927). The germ cells are embedded in mesoderm tissue. Although most people have termed these embryonic chambers "coeloms," they apparently do not in fact go on to form such cavities in the adult. This process of archenteric outpocketing merely seems to be a unique method to differentiate mesoderm. The body cavities that appear later in development have nothing to do with these early embryonic structures.

The newly hatched chaetognath is a diminutive adult and undergoes no metamorphosis. Each primordial germ cell divides once during early development, and one daughter cell comes to lie in each trunk cavity and tail cavity compartment. They later develop into the testes and the ovaries.

Phylogeny

Two fossil chaetognaths are known. One occurs in the famous Middle Cambrian Burgess Shale, and the other is from the well-known Mazon Creek fauna of Pennsylvanian age. They are recognizable as arrow worms largely from their body form, and tell us little about chaetognath history except that the group has been around for a long time.

Hyman (vol. 5) reviews the history of where researchers thought arrow worm affinities lie. Consensus of late has been on placing them among advanced coelomates as a degenerate deuterostome. But the early embryonic cavities appear to have nothing to do with the adult body cavities. Hyman (1959:64) considered the adult cavities to be closest to a pseudocoel, that is, a cavity that is not lined with true epithelia and that exists between the endodermal gut and the mesodermal tissues of the body wall. Welsch and Storch (1982) advanced evidence for true peritoneum in *Sagitta*. The visceral peritoneum has characteristic thick and thin myofilaments arranged around the intestine, which function to facilitate food movement in the gut. The parietal peritoneum is very thin, and only vesicles, granules, and mitochondria have been seen. The mesenteries are continuous with the basal membranes of the epidermis and gut epithelium, but are covered apparently on both sides with epithelium.

Despite the presence of a somewhat peculiar epithelial lining of the chaetognath body cavities, Hyman's original remarks remain germane (Hyman, 1959).

1. The intracellular cuticle and retrocerebral organs are similar to features present in rotifers.
2. The longitudinal muscles of the body wall have the gross arrangement, histologic structure, and function like those of nematodes.
3. The gut musculature is poorly developed.
4. The rupturing of gametes through the body wall commonly is seen in groups like gastrotrichs as well as in the acoelomate gnathostomulids.
5. The manner in which the mouth field unfolds and everts from the cephalic hood to feed is similar to that seen in several pseudocoelomate groups, but whether the vestibulum is structurally homologous to the proboscis seen in some rotifers and acanthocephalans, or the introvert type seen in other aschelminths, is not clear.

6. The very early ontogenetic delineation of germ cells is more characteristic of pseudocoelomates than coelomates.

7. The adhesive papillae of the bottom-dwelling *Spadella* may be homologues of the caudal adhesive glands seen in most pseudocoelomate phyla.

Taken altogether, it seems better at this time to think of the Chaetognatha as a very specialized pseudocoelomate adaptation to planktonic life, rather than as some kind of degenerate coelomate. Determination of exact affinities, however, must await further electron microscope and biochemical studies of chaetognaths.

Taxonomy

Two orders are recognized within a single class, Sagittoidea. One of these is based solely on a primitive character, and some future analysis of derived chaetognath features likely will rearrange the known families into some other pattern (Salvini-Plawen, 1986).

Order Phragmophora. Ventral transverse muscle present. (This supposedly primitive feature in the bottom-dwelling form *Spadella* suggests to some that chaetognaths were derived from benthic forms.)

Order Aphragmophora. Ventral transverse muscle lacking.

Suborder Ctenodontia. Cephalic spines gently curved; two pairs of comblike rows of teeth on head.

Suborder Flabellodontia. Cephalic spines sharply curved; teeth stout and fan-shaped.

Selected Readings

Burfield, S. 1927. *Sagitta. Proc. Trans. Liverpool Biol. Soc.* 41(appendix 2):1–104, pls. 1–12.

Ghirardelli, E. 1968. Some aspects of the biology of the chaetognaths. *Adv. Mar. Biol.* 6:271–375.

Hyman, L. H. 1959. *The Invertebrates*, vol. 5. McGraw-Hill, New York.

Michel, H. B. 1982. Chaetognatha. In *Synopsis and Classification of Living Organisms*, vol. 2 (S. P. Parker, ed.), pp. 781–783. McGraw-Hill, New York.

Nielsen, C. 1985. Animal phylogeny in light of the trochaea theory. *Biol. J. Linn. Soc. Lond.* 25:243–299.

Reeve, M. R., and T. C. Casper. 1975. Chaetognatha. In *Reproduction of Marine Invertebrates*, vol. 2 (A. C. Giese and J. S. Pearse, eds.), pp. 157–184. Academic Press, New York.

Reeve, M. R., and B. Lester. 1974. The process of egg-laying in the chaetognath *Sagitta hispida. Biol. Bull.* 147:247–256.

Salvini-Plawen, L. 1986. Systematic notes on *Spadella* and on the Chaetognatha in general. *Z. zool. Syst. Evolut.-forsch.* 24:122–128.

Welsch, U., and V. Storch. 1982. Fine structure of the coelomic epithelium of *Sagitta elegans. Zoomorph.* 100:217–222.

20

Coelomates

Although we seek to understand organisms by trying to organize them into lineages derived from single ancestors, nevertheless, certain broad structural features occur in phyla at different grades of organization. These phenomena are like themes that are repeated in remarkable variations within the context of different structural or body plans. Thus, we seek to understand organisms by recognizing grades of evolution. However, characteristic things happen within grades related to the limitations of the body plan.

Large and complexly structured bodies require more efficient ways of moving material within and between tissues than we have seen in previous chapters. Body cavities with contained fluids facilitate such transport requirements, and one body plan just studied, the pseudocoel, essentially uses a derivative of the embryonic blastocoel. Pseudocoelomates solve internal transport problems by transforming much or all of the parenchyme into open spaces filled with a perivisceral fluid. This mode of circulation allows protonephridia to become less diffuse and be associated with the pseudocoel, and they can service the whole body even when concentrated in a small region. Lateral diverticula for food distribution do not occur. The gut becomes a simple tube, with one-way traffic between the mouth and the anus. As a consequence, the body wall is independent of internal organs, and a type of pseudosegmentation of the cuticle and body wall musculature sometimes occurs. No special respiratory or circulatory systems appear, and although the pseudocoel suffices for transport, pseudocoelomates are limited at best to only moderate sizes.

All animals need some kind of structural support. Acoelomates use the solid parenchymal tissues of the mesoderm as a type of skeleton. Consequently, ramifying protonephridial and digestive diverticula are easy solutions to problems of transportation and metabolic efficiency in such a solid body. This body plan, however, is constrained by size limitations due to internal transport requirements.

Another solution to the problems of complex body plans involves the development of a coelom.

THE COELOM

The **coelom** is a secondary body cavity. It forms as a new space in the embryo within the mesoderm, completely surrounded by mesodermal tissue. As the coelom increases in size, the outer part of the mesoderm becomes intimately associated with the body wall as the **parietal** or **somatic mesoderm** and is covered by an epithelium, the **parietal peritoneum**. The inner mesoderm becomes intimately associated with the gut wall and other viscera as the **visceral** or **splanchnic mesoderm**, which is lined by another epithelium, the **visceral peritoneum**. The organs that lie in the coelom are connected to the body wall by suspensory epithelia, or mesenteries, actually continuations of the peritonea connecting parietal and visceral layers (Fig. 20.1). The coelomic fluid in the coelomic cavity does not come directly in contact with either the gut or the body wall tissues, but is separated from both by the peritoneal epithelia.

Metabolic Functions With the appearance of the coelom, important new potentials rise. The physical separation of parietal and visceral tissues favors their functional separation, leading to a more complex organ architecture.

The coelomic cavity itself becomes important. Organisms use the coelom and its fluid as a primary transport system. It is common for the coelomic lining to bear cilia or flagella that help to circulate the coelomic fluid and facilitate its use in picking up waste and delivering food materials. The coelom then is a functional analogue of the pseudocoel. The coelom comes to perform another important role. In most coelomates, the gonads are closely associated with the peritoneum, and gametes are discharged into the body cavity.

With wastes and gametes accumulating in the coelom, it is necessary to get them out again. Therefore, for both excretory and reproductive purposes, it is important for the coelom to be connected with the exterior so the wastes and gametes can be discharged. Rupturing body walls to do so, the solution of some coelomates and pseudocoelomates, is difficult in a structural plan in which the body wall is so complex.

The extent to which protonephridia enter into nitrogenous waste disposal is not clear, as most organisms with these structures seem to be ammonotelic. Strong evidence, however, links protonephridia with hydrostatic control. Although the first coelomates appear to have been equipped with protonephridia, most of them have **metanephridia**, a different kind of organ equipped with an open ciliated funnel or nephrostome in place of flame cells or bulbs. The appearance of a nephrostome means that coelomic fluid can enter directly into the nephridial tube. Reabsorption of substances present in the coelomic fluid is needed to avoid depletion of body fluids, and thus methods of urine concentration are closely linked to metanephridial function. Excretory controls are established, as internal pressures in the coelomic compartment would otherwise lead to a rapid depletion of the coelomic fluid.

A similar tube is the **coelomoduct**, which serves for the discharge of gametes in primitive coelomates. This gonoduct is also open to the coelom by way of a ciliated funnel. Thus, the typical coelomate situation is that shown in Figure 20.1, in which the coelom is open to the outside by way of a metanephridium and a coelomoduct (gonoduct).

Use of the coelom itself as both a hydrostatic skeleton as well as a transport system is not to be unexpected. Exploitation of it as a skeleton resulted in the coelom being subdivided into smaller compartments or segments. Under such circumstances the primary transport function of the body cavity is then hindered, so as a consequence a blood-vascular system serves in circulation (Ruppert and Carle, 1983).

Movement and the Coelom The body wall contains sheaths of longitudinal and circular muscles that can act as antagonists; as the muscles are contracted or relaxed, force applied to the coelomic

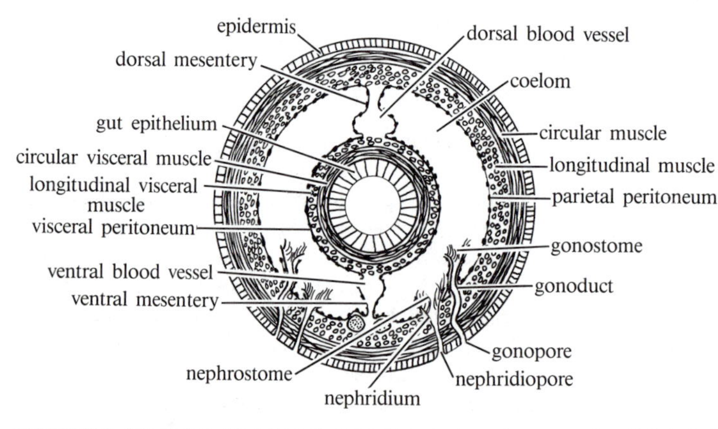

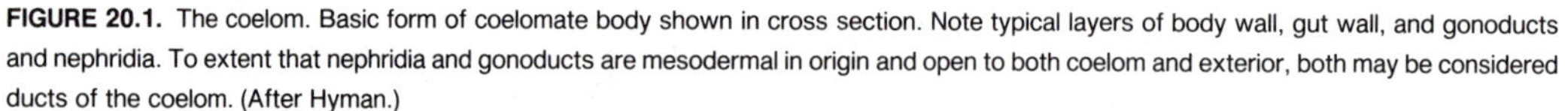
FIGURE 20.1. The coelom. Basic form of coelomate body shown in cross section. Note typical layers of body wall, gut wall, and gonoducts and nephridia. To extent that nephridia and gonoducts are mesodermal in origin and open to both coelom and exterior, both may be considered ducts of the coelom. (After Hyman.)

fluid facilitates in varied types of body movements (Fig. 20.2). The mechanics vary with the nature of individual coeloms (Clark, 1964). In annelids the coelom is divided into a series of separate compartments by septa, and each compartment acts as a separate unit. The fluid is not compressible, and the volume of the compartment remains constant regardless of muscle contractions. If the circular muscles are contracted, the compartment becomes narrower and longer; if the longitudinal muscles are

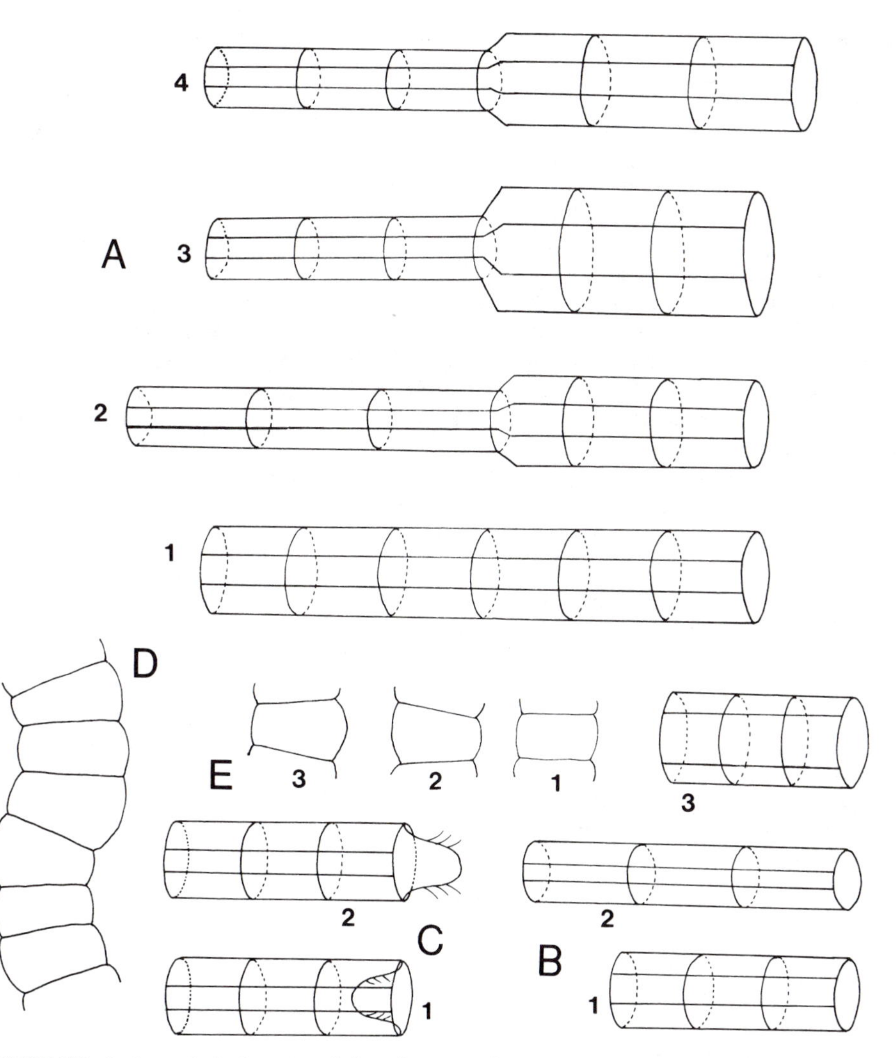

FIGURE 20.2. Coelom as hydraulic system. **A**. Four diagrams to illustrate possible changes of form from contraction of circular muscles. Bottom figure (**A1**) represents situation with all muscles relaxed. If circular muscles on left part of body contract, while circular and longitudinal muscles or connective tissues of body wall of rest of body do not permit it to change shape, segment with contracted circular muscles must elongate (**A2**). If circular muscles on left section contract, while longitudinal muscles on right do not permit body elongation and circular muscles on right relax, dilation of section on right results (**A3**). If right section longitudinal muscles relax while circular muscles retain original body diameter, elongation of right region without dilation results (**A4**). **B**. Three diagrams illustrating effects of common contraction of circular or longitudinal muscles. Bottom sketch (**B1**) represents body at rest. In next above (**B2**), circular muscles have contracted, with simultaneous relaxation of longitudinal muscles, and body elongates. In upper figure (**B3**), contraction of longitudinal and relaxation of circular muscles have permitted shortening and dilating of body. In organism with segmental coelomic compartments, like earthworm, each body segment behaves like this during movement. **C**. Extrusion of introverted part. When body muscles are relaxed (**C1**), introverted part withdrawn. Contraction of body musculature (**C2**) increases pressure in coelomic compartment and extrudes proboscis. **D**. Sinuous movements of body. Longitudinal muscles on one side of body act as antagonists to longitudinal muscles on other side, as suggested in **E**. Segment on right (**E1**) at rest and compressible on either side by contraction of longitudinal muscles on one side while muscles on other side relax (**E2**,**E3**). (Based on Chapman.)

contracted, the compartment becomes broader and shorter. To the extent that the antagonists of the contracting muscles relax, pressure in the compartment remains constant. If both sets of muscles contract equally, however, the compartment does not change shape. The internal pressure rises, and makes possible the extrusion of an introverted part.

When the coelom is not divided into compartments, as happens, for example, in echiurans and sipunculans, simultaneous contractions of all circular or longitudinal muscles provide the same whole body movements that a single coelomic compartment undergoes. If a region of the body is constricted by the contraction of circular muscles, however, several kinds of body movements are possible. If neither circular nor longitudinal muscles are relaxed elsewhere in the body, the contracting part of the body is elongated. Relaxation of the circular muscles or contraction of the longitudinal muscles elsewhere in the body will result in a shortening and expansion of the body. Thus, a portion of the body can be expanded to help hold an animal in its tube or to do other work in burrowing or feeding. Of course, the same potentials exist in organisms with individual coelomic compartments, for if some coelomic compartments are narrowed while others are broadened, then regional changes in body contour result. In creeping animals, alternate lengthening and shortening of body regions, coupled with some device to cause friction with the substrate, results in progressive body movements.

Circular and longitudinal muscles need not be used as antagonists. Contraction of longitudinal muscles on one side, with corresponding relaxation of longitudinals on the other, bends the body in one direction. When these contractions are reversed, the body is bent in the opposite direction. These sinuous movements are especially useful for swimming, but are also effective in some kinds of creeping movements. Mechanically, sinuous movements are more easily achieved in an organism containing many coelomic compartments, but are not impossible where a single body coelom is present.

Origin of the Coelom The appearance of any structure of such importance as the coelom raises two important questions. How did it originate? Where did it come from? Four major theories have been put forward. Two of them, the enterocoel and schizocoel theories, are actually related to how coeloms develop; the other two, the gonocoel and nephrocoel theories, are functional reasons for a coelom. None is probably entirely correct.

The **enterocoel** theory explains the coelom as a derivative of a radially partitioned enteric cavity, as seen in Anthozoa and Scyphozoa. It is inspired by the enterocoelous mode of development. The idea is that the tips of the gastric pouches were cut off to form coelomic cavities. As gonads are often found in the mesenteries of sea anemones or in the septa of scyphozoans, this supposedly explains the relationship of the gonads to the coelomic wall. This explanation, however, is not completely persuasive for two reasons. (1) This theory involves the derivation of the lower cnidarians from more highly organized cnidarian forms, and (2) the mechanics of enteric pocket formation differ from the mechanics of coelom formation. Enterocoelous coelom formation occurs as the result of outpocketings of the gut, while the mesenteries or septa of cnidarians grow inward to divide the coelenteron into gastric pockets. Hyman (1959:750) and others have severely criticized this idea as "fantastic nonsense." For a discussion of the theory and reasons for it, the writings of Marcus (1958) and Jägersten (1972) can be consulted.

The **schizocoel** theory assumes that spaces appeared independently in the mesoderm, rather than as derivatives of any preexisting part. The best features of this idea are that it does not demand that any major phylum be derived by degeneration from some more highly organized phylum, and that it is flexible enough to explain the appearance of a single large coelomic space or segmental coelomic spaces. Hyman (1951:27) suggests that the coelom may have appeared as a site for the accumulation of fluid, possibly dissolved wastes. Fluid does tend to accumulate in aquatic organisms, and it seems no more difficult to visualize the appearance of fluid-filled schizocoelous cavities than the appearance of a fluid-filled pseudocoel. It seems most likely, however, that the coelom first appeared, like the pseudocoel, as a space that facilitated internal transport of all kinds and then became important as a place to accumulate specific fluids and as a hydrostatic skeleton suited to the demands of locomotory efficiency in the absence of a hard skeleton.

The **gonocoel** theory derives the coelom from the lumen of the gonads. This theory also accounts for the relationship of gonads to the coelomic wall. A regular series of gonads occur in some flatworms and nemertines that alternate with the lateral diverticula of the gut. If the branches of the gut were withdrawn in such a scheme, presumably as an anus forms, and if the gonads expanded the cavities that contain the gametes to fill in the resulting space, a series of coelomic pouches surrounding the gut could appear. This proposal has some serious drawbacks, for it would link the origin of the coelom with metamerism and leave no explanation for unsegmented coelomates. This idea was strengthened by the discovery of a turbellarian with large testes in a regular series between the gut diverticula. It was common in the past to dwell heavily on similarities of living adult species to organizational plans of other phyla as supposed missing links. A peculiar rotifer, *Trochosphaera*, has a superficial resemblance to a trochophore larva but is probably nothing more than an unusual rotifer. The flattened, platyctene ctenophores are meaningless insofar as flatworm phylogeny is concerned, although the convergent similarities between them and polyclads have stimulated lots of argument. Today, it is generally agreed that for clues to relationships we must look to the most primitive rather than the most specialized members of phyla, and sometimes as well to larval forms and embryonic stages where they are not too highly modified to yield useful clues.

The **nephrocoel** theory sees the coelom as the derivative of the inner, expanded ends of the nephridia. This idea's principal disadvantages are the absence of any real supporting evidence and the total absence of nephridia in some coelomates. Moreover, the embryonic origin of the coelom does not reflect such a relationship with the nephridia. Its principal strength is the evident usefulness of the coelom to collect wastes from the body and gut wall.

However coelom cavities may have appeared in the first coelomates, they arise in three major ways in modern forms. In nemertines, echiurans, sipunculans, annelids, molluscs, arthropods, hemichordates, and some chordates, it appears as spaces in the mesoderm, that is, as a schizocoel. In echinoderms the coelom appears as outpocketings of the gut, that is, as an enterocoel. In phoronids and ectoprocts, the mesenchyme cells rearrange themselves to form the coelom and develop into a peritoneal lining. Whether this last type is degenerate or preserves traces of a method more primitive than either the schizocoelous or enterocoelous method of coelom formation remains uncertain.

SEGMENTATION VERSUS SERIAL REPETITION

A controversy has raged in invertebrate zoology over the entire question of segmentation or metamerism. What constitutes true segmentation? What animals are truly segmented? What is the origin of segmentation? Hyman (1951:28) offered a concise definition of what she considered to be true segmentation when she declared that segmentation is a primarily mesodermal phenomenon that extends its effects into the ectodermal and endodermal tissues. She recognized three independent evolutions of true segmentation in the animal kingdom: in the annelid/arthropod line, in the chordate line, and in the cestodes. Under the terms of this definition, all other segmentation in the animal kingdom is "pseudosegmentation," that is, not true segmentation. Beklemishev's (1969:188) definition of segmentation is much broader as he considers any serial repetition of structures as segmental. Thus, what Hyman terms the tricoelomic nature of echinoderms, Beklemeshev would designate as trisegmental.

Any attempts to resolve this debate should probably consider serial arrangement of structure as it occurs in *all* metazoans, and should be concerned with just what is the *functional* significance of, and/or the ontogenetic basis for, serial repetition.

Some triclad rhabditophoran flatworms, like *Procerodes lobata*, exhibit a striking serial repetition. The gut diverticula are very regularly arranged along the whole length of the tripartite gut. The gonads are placed in between these diverticula. The commissures between the nerve cords occur in the areas between the gut outpocketings. The excretory system, with paired dorsal and ventral longitudinal ducts, has connections between dorsoventral ducts at almost every diverticulum, and occasionally there are interdorsal connections between many of the gut outpocketings. This arrangement is very similar to that seen in many nemertines, where the gonads alternate with the gut diverticula. In the hoplonemerteans, which have paired lateral nerve cords and a dorsal cord, the lateral cords give off peripheral nerve branches into the body wall, and these come off between the diverticula. In *Neuronemertes* there are ganglionic swellings on the middorsal cord that correspond to the interdiverticula and the lateral nerve branches.

This kind of serial repetition might hint at some underlying developmental proclivity to proliferate identical structures in the metazoan genome. Indeed such seems to be indicated as a result of the analysis in Chapter 38.

Another type of repetition of structures occurs sporadically in various turbellarian orders: the formation of colonies by incomplete asexual fission. Here, such as in the macrostomidan *Microstomum* or the catenulidan *Stenostomum*, the budding of new zooids outstrips the actual separation so that chains of zooids of up to 16 individuals can result. A similar phenomenon occurs in cestodes, but whether there is "segmentation" in cestodes has been the subject of intense debate. The functional advantage for the serial repetition of the proglottids is reproductive. It mirrors the effect of strobilization seen earlier in some scyphozoan cnidarians. Replicating sex organs allows one individual to multiply its chances of reproductive success by several orders of magnitude.

Serial repetition of body wall structures is also observed in some pseudocoelomates, especially Kinorhyncha, Priapulida, and Nematoda, as noted in earlier chapters. The nematodes seem to be the least segmented of the three, with only the cuticle and the cuticular structures displaying repetition in some of the free-living marine forms. In the kinorhynchs, the cuticle is organized into typically 13 zonites or segments. Each zonite is externally delineated by cuticular plates, a dorsal tergum and two ventral plates. In addition, the body wall muscles are arranged in longitudinal fibers that match the cuticular zonites, and the ventral nerve cord has ganglia that occur in the middle of the zonites. Other organs partially reflect the zonation: the gonads are found in the twelfth segment and open on the thirteenth. The paired protonephridia are always in the tenth zonite and open by a nephridiopore on the eleventh. The priapulids have an externally annulated body, but this is reflected internally in only one of the muscle layers. The circular layer of priapulids is arranged in rings that seem to correspond to the cuticular annuli.

Serial repetition as a whole in the pseudocoelomates, when it occurs, seems to be related to locomotory factors with the correspondence of muscle fiber arrangements with the metamerism of the cuticle and the enhancement of wiggling movement by breaking up the cuticle into smaller units.

The annelid/arthropod serial repetition has been advanced traditionally as a pristine example of true segmentation in which the entire body is involved. Its functional role in locomotion has been outlined previously. Besides the locomotory advantages of this segmentation, there are developmental advantages. The replication of unit subdivisions of a system is one way of ultimately getting specialized, serially arranged organs in a system, for example, the digestive or nervous system.

The nature of mollusc serial repetition has been heatedly debated since the discovery of *Neopilina galatheae*. Not only are there eight sequentially arranged pedal muscles, but other organ systems, namely, nerve commissures, gonads, nephridia, ctenidia, and auricles, are sequentially arranged and coordinated with the muscle metamerism. One of the main arguments that has been employed against the monoplacophorans as being truly segmented is that the muscles are segmented while the shell is a single unit. This argument overlooks that the monoplacophoran muscles are not shell muscles but pedal muscles. They function in locomotion, not shell movement. There may be a good functional reason for a single-unit shell. One might argue that because the carapace of the crab is a single unit, the animal is not primarily segmented. However, the dorsal carapace in crabs has little to do with the ventral locomotory anatomy.

Finally, the Chordata need to be briefly considered. Here the serial arrangement of the nerves and chevronlike muscles is related to the sinuous locomotory habits of the group. Such an arrangement initially evolved in the Cephalochordata and, like that seen in the annelid/arthropod line, involves the whole body. This was elaborated in the endoskeletal system as it developed in vertebrates, the vertebral units being intersegmental for greater efficiency of muscle action. In fact, the development of the body musculature, combined with its effective neural coordination, is in large part responsible for the initial and ongoing success of the phylum.

A much broader perspective is gained when repetition of structure is looked at as a functional issue rather than as a strictly anatomical one of whether something is truly segmented. Serial repetition in some organ systems can be seen as a response to selective pressures that afford some specific functional advantage. In the chordates, the annelid/arthropod line, possibly the molluscs, and those pseudocoelomates that possess it, serial repetition is linked to locomotory efficiency. Seriality in cestodes has been shaped by the reproductive problems imposed by the vagaries of a parasitic life-style and the need to ensure enough offspring. The serial repetition seen in some turbellarians and the nemertines is harder to define, although it too appears to have a reproductive focus. In addition, serial repetition may have some substantial genetic foundation in the trochophorate/articulate phyla as well as the higher chordates, groups wherein whole bodies are involved in a special form of seriality, or true segmentation.

Serial repetition is a way of coping with problems, reproductive or locomotory, and develops when the selective pressure and ontogenetic channeling allows it. It is restricted only by the limits placed on it by the particular structural plan in which it appears. That many coelomates *appear* to use segmentation more effectively is merely a reflection of the more elaborate uses it can be put to, given a fundamentally more complex body plan and the developmental controls involved.

PROTOSTOMIA AND DEUTEROSTOMIA

Two main evolutionary lines of coelomates are traditionally recognized, the protostomes and deuterostomes. **Protostomes** includes molluscs, annelids, and arthropods, in addition to several minor phyla. **Deuterostomes** includes the echinoderms, chordates, and some minor related phyla. The concept seems to go back to Bateson in the last century and has achieved wide use in textbooks and in speculations on invertebrate phylogeny (Table 20.1). The idea is that coelomate phyla sort themselves on dichotomous differences in cleavage, blastopore fate, larval types, and coelom form. Løvtrup (1975) argued for the total abandonment of the concept and influenced the discussion that follows.

Spiral cleavage is said to be due to the oblique orientation of the third cleavage spindle apparatus. It appears, however, that the staggered arrangement of blastomeres is also in part due to the mechanics of linking a series of small cells with a series of large cells. Maximum stability is achieved when the cells, adhesive to each other, come to lie in furrows formed by sets of adjacent cells. There is no reason why a spiral arrangement could not occur in the so-called deuterostome phyla except that they are prevented from so arranging themselves. Echinoderm blastomeres are kept together in a radial format by a sticky hyaline membrane (Schroeder, 1986), and chordates have a syncytial surface coat of cell material to serve the same end.

The issue of determinate and indeterminate cleavage was based on classic experiments in which separated early blastomeres were allowed to develop. In echinoderms and amphibians the first two cleavage blastomeres go on to develop viable, although small, larvae and embryos. Løvtrup (1975), however, points out that no successful completions from eight-cell-stage blastomeres have been obtained from amphibians. It appears that once the animal-vegetal pole axis is established in deuterostomes, cell determination is present. Therfore, the difference between protostomes and deuterostomes in regard to cell determination is quantitative rather than qualitative.

The blastopore fate is another feature that is not nearly so clear as textbooks would have it. Many protostomes do develop a mouth from the blastopore. Others, however, such as some annelids, may develop both mouth and anus, or just the anus, or neither opening from the blastopore. In arthropods, the blastopore is the result of micromeres that grow inward toward the yolk mass, and it typically closes at the completion of gastrulation. The oral stomodeum and anal proctodeum later grow inward from the ectoderm to join the midgut that is developing from the yolk-laden endoderm.

Trochophore larvae indeed are characteristic of protostome phyla. The pluteus larva, however, is only characteristic of *some* echinoderms; other echinoderms use a more trochophorelike vitellarian larva. The hemichordate tornarian larva is somewhat pluteuslike, but chordates lack any such larvae.

Cavities in mesoderm can only form in certain ways, either by spontaneous appearance in a three-dimensional mesodermal mass, the schizocoel, or by deformation of an essentially two-dimensional surface, the enterocoel. The latter coelom results in the echinoderms in large measure due to the constraints placed on the blastomere surface by the adhering hyaline membrane. The sticky nature of that membrane forces gastrulation and mesoderm formation to occur by folding of the cell surface rather than delaminating a free-cell layer of mesoderm. Furthermore, in most chordates, coeloms form by splitting cell layers in the mesoderm.

Thus, the traditional dichotomy of protostome and deuterostome characters is not a consistent one, and in some instances these features can be seen as specializations to peculiar ontogenetic conditions unique to specific phyla. The use of this dichotomy probably should be moderated (see Chapter 38).

TROCHOPHORE LARVA

Trochophore larvae are characteristic of protostomes, especially annelids and molluscs. This larva is typically formed after spiral, determinate cleavage of eggs with little yolk. Blastomeres form a characteristic annelid rosette and cross during later cleavage stages (Fig. 20.3A), and a similar cross of blastomeres forms in molluscs (see Fig. 21.8B). The annelid and molluscan crosses are superficially similar but are derived from different blastomeres. The annelid rosette is formed by blastomeres $1a^{111}$, $1b^{111}$,

Table 20.1. Traditional characters of protostomes versus deuterostomes

Protostomes	Deuterostomes
Spiral cleavage	Radial cleavage
Determinate cleavage	Indeterminate cleavage
Blastopore → mouth	Blastopore → anus
Trochophore larvae	Pleuteus larvae
Schizocoel	Enterocoel

$1c^{111}$, $1d^{111}$, and the cross by the sister cells, $1a^{112}$, $1b^{112}$, $1c^{112}$, $1d^{112}$. The molluscan cross is formed of blastomeres $1a^{12}$, $1b^{12}$, $1c^{12}$, $1d^{12}$. The annelid cross is thus interradial in position, and the molluscan cross radial. (For review of notation, see Fig. 4.2.)

The four quartets of micromeres and their descendants give rise to remarkably consistent parts of the adult and the trochophore larva (also true of flatworms). Blastomeres 1a, 1b, 1c, 1d form the ectoderm near the animal pole of the blastula. They give rise to the apical part of the trochophore larva, including the apical sense organ, the cerebral ganglia that develop near it, and the apical ectoderm. They may also make some contributions to the head nephridia. The blastomeres $1a^2$, $1b^2$, $1c^2$, $1d^2$ are the primary trochoblasts and give rise to four ciliary tufts. Later, other blastomeres, known as secondary trochoblasts, develop cilia to complete a ciliary band around the trochophore larva known as the prototroch (Fig. 20.3E). The blastomeres of the second quartet, 2a, 2b, 2c, 2d, contribute to the surface ectoderm and to the stomodeum. Among annelids, 2d is known as the primary stomoblast and gives rise to the greater part of the trunk ectoderm, including the ventral nerve cord. Blastomeres 3a, 3b, 3c, 3d give rise to the circumanal ectoderm, and descendants of $3c^2$ and $3d^2$ give rise to most of the larval excretory organs (archinephridia). The endoderm arises from seven blastomeres, 4a, 4b, 4c and 4A, 4B, 4C, 4D, which eventually invaginate and form the endoderm

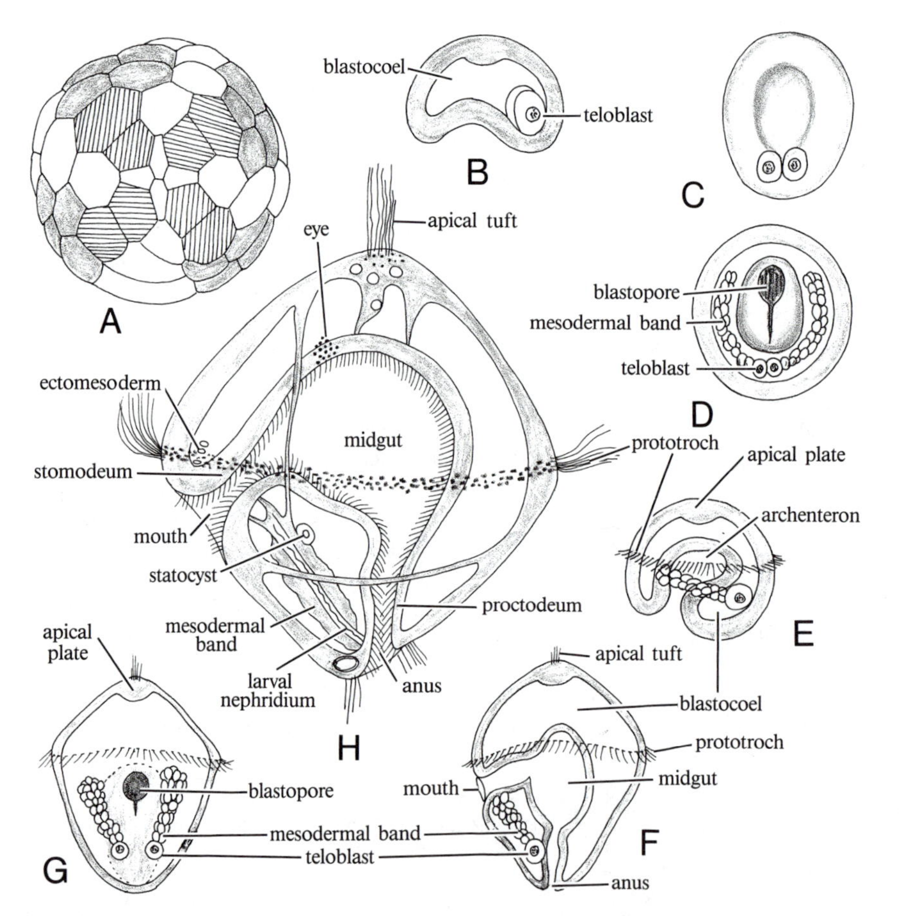

FIGURE 20.3. Trochophore development. **A.** Embryo of annelid, *Arcenicola,* in 64-cell stage, with annelid cross and rosette, *hatched,* composed of blastomeres $1a^{112}$, $1b^{112}$, $1c^{112}$, $1d^{112}$, and some cells from second quartet. Annelid rosette (*white cells*) at animal pole composed of blastomeres $1a^{111}$, $1b^{111}$, $1c^{111}$, $1d^{111}$. (Similar cross formed during molluscan development, but from different blastomeres.) **B,C,D,E,F,G.** Scheme of transformation of gastrula into trochophore larva, as it occurs in polychaetes. **B.** Gastrulation seen from left and (**C**) from animal pole. Note teloblasts. **D,E.** Teloblasts develop into mesodermal bands, and primitive gut differentiates. Locomotor band, prototroch, appears at surface. **F,G.** Apical tuft of cilia appears, and proctodeum develops to contact with primitive gut and break through to complete gut formation. Mouth arises directly from blastopore or from its anterior margin. **H.** Fully developed trochophore, showing its major differentiated parts. (**A** after Child; **B,C,D,E,F,G** after Korschelt, from Dawydoff, in Grassé; **H** after Shearer.)

of the stomach and the intestine. Blastomere 4d gives rise to a few endodermal cells, but its most important contribution to the embryo is two teloblastic cells that divide repeatedly to form the mesodermal bands on each side of the embryo in which the coelom forms (Fig. 20.3B,C,D,E,F,G).

The fully formed trochophore (Fig. 20.3H) is essentially biconical, with a band of cilia, the prototroch, passing just above the mouth. A second ciliated band, the paratroch, may encircle the anus. The U-shaped gut, lined with a ciliated endoderm, protrudes into the upper cone and is surrounded by remnants of the blastocoel, partially occluded by ectomesodermal cells and muscles derived from the ectomesoderm.

A main nerve ring encircles the larva below the prototroch and connects to a variable number of radial nerves that arise from the apical ganglion at the base of the apical sense organ. A pair of endomesodermal bands, derived from the teloblasts, parallel the digestive tube. The lower hemisphere contains a pair of nephridia, terminating in one or several flame cells.

The occurrence of a trochophore larva in annelids, molluscs, and other related phyla has made it the center of various phylogenetic schemes. The trochophore is thought to be at about the same level of organizational complexity as a rotifer, and one theory questionably derives the trochophore from ancestral rotifer stocks. This idea cannot be accepted for a variety of reasons, of which the most consequential are the gross difference in embryonic derivation of parts and the different rotifer cleavage pattern that is not at all similar to typical spiral cleavage. A ctenophore-trochophore theory has been advanced to explain the origin of bilateral animals from radial animals by deriving the trochophore from a ctenophore. The ctenophore embryo also develops a cross of blastomeres, superfically similar to the annelid cross, and the apical sense organ and the radial nerves of ctenophores and of the trochophore larvae have some resemblances. The greatest difficulty in this theory is its failure to account for flatworms other than as degenerate annelids, a possibility that seems too remote to be persuasive.

Selected Readings

Beklemishev, W. N. 1969. *Principles of Comparative Anatomy of Invertebrates.* Univ. Chicago Press, Chicago.

Clark, R. B. 1964. *The Dynamics of Metazoan Evolution.* Clarendon Press, Oxford.

Hadzi, J. 1963. *The Evolution of Metazoa.* Macmillan, New York.

Hanson, E. D. 1977. *The Origin and Early Evolution of Animals.* Wesleyan Univ. Press, Middletown.

Hyman, L. H. 1951. *The Invertebrates,* vol. 2. McGraw-Hill, New York.

Hyman, L. H. 1959. *The Invertebrates,* vol. 5. McGraw-Hill, New York.

Jägersten, G. 1972. *Evolution of the Metazoan Life Cycle.* Academic Press, New York.

Løvtrup, S. 1975. Validity of the Protostomia-Deuterostomia Theory. *Syst. Zool.* 24:96–108.

Marcus, E. 1958. On the evolution of the animal phyla. *Quart. Rev. Biol.* 33:24–58.

Ruppert, E. E., and K. J. Carle. 1983. Morphology of metazoan circulatory systems. *Zoomorph.* 103:193–208.

Schroeder, T. 1986. The egg cortex in early development of sea urchins and star fish. *Exper. Biol.* 2:59–100.

21

Mollusca

Mollusca are usually thought of as animals that have a calcareous shell, radula, and gills enclosed in a mantle cavity. Exceptions to almost all of these features, however, can be found in specific groups. There seems little doubt, though, that molluscs are all derived from a common ancestral stock in which these traits appeared (Morton, 1979). There are from 100,000 to 150,000 living species, and at least 45,000 fossil species have been described. Molluscs are found in the abyssal depths of the ocean and above high tide line, and are common in fresh water everywhere. Gastropods are reasonably common in moist, terrestrial habitats. Molluscan shells have always been economically important, having served as money and been fashioned into jewelry and buttons. Shells of molluscs are important components of kitchen middens of prehistoric man. Scallops, oysters, clams, squids, and octopi are still important food items, and considerable money is provided for their study and conservation.

Definition

Mollusca are strongly cephalized and equipped with special sense organs. The coelom is restricted to being pericardial. The ventral side of the body wall is developed as a muscular foot generally used for locomotion, while the dorsal body wall is extended as one or a pair of mantle folds that secrete the shell and enclose the mantle cavity. The mantle folds typically contain gills, anus, and excretory pores. The digestive tract is regionalized, with a sclerotized buccal cavity almost always containing a toothed radula, a stomodeal esophagus usually including specialized regions for food storage and fragmentation, a midgut stomach with a pair of large digestive glands or liver, and a hindgut or intestine that ends at the anus. The circulatory system possesses a well-developed heart that has one or more collecting chambers or auricles to convey blood to a median ventricle. Blood usually circulates through open spaces or perivisceral lacunae and sinuses that encroach on the coelom, and is supplied with a respiratory pigment of hemocyanin. There is a single pair of kidneys, sometimes reduced to one, closely associated with the pericardial cavity. The circumenteric nerve ring is typically associated with two pairs of nerve cords, one cord to the foot and one cord to the viscera and the mantle. Ganglia are usually associated with the nerve ring, the nerve cords, and the extensive subepidermal nerve plexus. Cleavage is generally spiral and determinate, and development leads to a trochophore larva.

Body Plan

There are a variety of molluscs (Fig. 21.1), adapted for extremely diversified habitats. Some of the general features that serve to link all of them are not always obvious.

The molluscan body wall serves several functions: it forms a protective barrier, is responsible for receiving sensory stimuli, and, far more than in most animal groups, is responsible for locomotion. It may be drawn out to form sensory or prehensile tentacles, or folded to form a mantle, or thickened to form a foot. A mantle may be a secretory device that produces a protective shell, serve as a protective cover itself, or in some cases act as a swimming organ.

The outer layer of the body wall is the epidermis, rich in gland cells with mucous glands predominant. Molluscs find many uses for slime, such as trapping food particles, preventing evaporation at the body surface in terrestrial species, and lubricating the foot surface. Special glandular regions derived from the epidermis provide secretions for specific uses, the most important of which are shell production, forming byssal threads to attach some mussels to rocks or other submerged objects, and producing cement for attaching ova to surfaces or for other reproductive activities. The epidermis has special cells for sensory reception, to be described later. Patches of epidermis are ciliated, generating currents that are important factors in feeding and respiration.

Beneath the epidermis is the dermis, consisting of connective and muscle tissue. In molluscs, these tissues are rarely arranged in definite layers, but are interwoven in complex masses. Rings of circular muscle and strands of longitudinal muscle are the exception rather than the rule. The evolutionary trend away from alternating layers of circular and longitudinal muscle is undoubtedly associated with the complex movements required of the body wall.

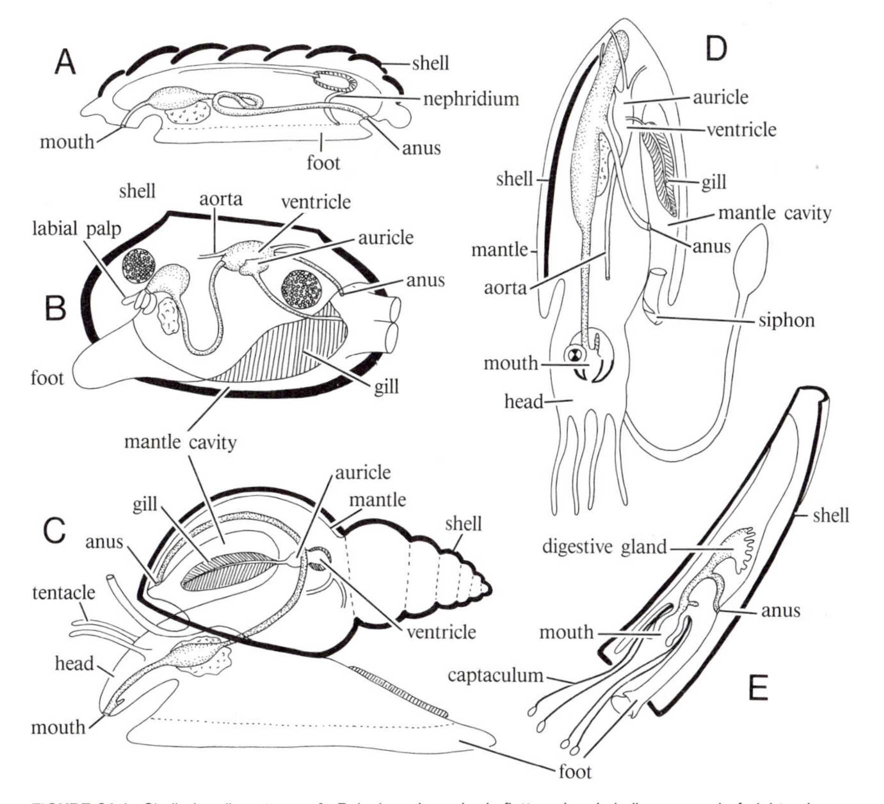

FIGURE 21.1. Shelled mollusc types. **A**. Polyplacophora, body flattened and shell composed of eight valves arranged in longitudinal series, head reduced, and ctenidia lie in pallial groove that encircles foot. **B**. Pelecypoda, bivalve shell with dorsal hinge, blade-shaped foot thrust between valves, reduced head. Typically, large gills on each side of body hang down from roof of huge mantle cavity. **C**. Gastropoda, visceral mass very tall and usually extending into spiral shell, ventral foot large, and head well developed. **D**. Cephalopoda, head well developed and encircled by tentacles, visceral mass very tall, typically with vestiges of shell enclosed within covering mantle. Head ventrally placed, and siphon, derived from foot, opens into mantle cavity. Original dorsoventral axis elongated and functions as anteroposterior axis. **E**. Scaphopoda, conical shell open at both ends, much-reduced head provided with captacula used in feeding, and cylindrical foot extrudes through anterior opening of shell.

Connective tissues are quite varied. Molluscs with reduced shells often have a very firm type of connective tissue known as chordoid tissue, built of vesicular cells that can become firm through turgor, somewhat like the notochord cells of chordates. Loose, spongy connective tissue is especially important in molluscs. For example, a snail tentacle or a foot can be withdrawn by the contraction of a retractor muscle, but how can they be extended once again? Because molluscs lack a spacious coelom to use as a hydraulic system for such purposes, the loose, spongy tissue filled with blood vessels or blood lacunae can expand and become firm and thus thrust anatomical parts outward. Such tissue is similar to the erectile penis tissues of vertebrates. Special valves are sometimes present in molluscs to hold the blood in the tissues throughout the period of activity.

Muscle tissue is irregularly distributed in the body wall of molluscs. It is abundant in the foot if the foot is used for locomotion, and also in the mantle of species with reduced shells, especially where the mantle is used in locomotion.

The shell is secreted by the outer, glandular surface of the mantle. Where the shell is internal, the mantle is turned back on itself so that the shell is completely surrounded by the glandular epithelium. Polyplacophoran shells (Fig. 21.2A) have a different structure than other molluscs and will be discussed later. Higher molluscs (sometimes referred to as Concifera) have shells that are essentially similar to each other, although they differ in mineralogical detail (Beedham and Trueman, 1967). The outer layer is the **periostracum** and is composed of a horny substance, **conchiolin** (Fig. 21.2C). It is secreted by a fold of the mantle edge, and so is laid down only at the shell margin where growth occurs, and older portions of shells are often badly worn (Saleuddin and Petit, in Wilbur, vol. 4, 1983). The rest of the shell is calcareous, known as the **ostracum**. In most molluscs, two layers can be recognized, an outer prismatic layer and an inner nacreous layer (Wilbur and Saleuddin, in Wilbur, vol. 4, 1983). The prismatic layer is built of densely packed, polygonal prisms of calcareous material, separated by delicate membranes of conchiolin. In areas where the mantle is reflected, secondary additions of prismatic substance may be added to the outer surface, forming such external structures as the callus in snail shells. Nacreous material is secreted constantly by the

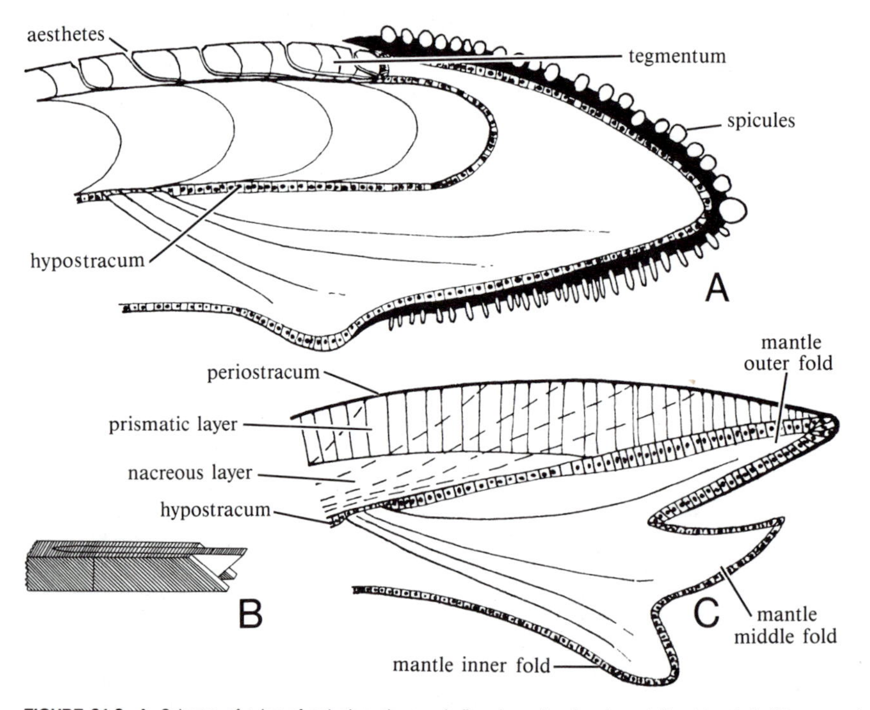

FIGURE 21.2. A. Scheme of edge of polyplacophoran shell and mantle showing relationship of shell layers and mantle edge. **B**. Crossed lamellar shell structure, showing three highly magnified primary lamellae, each composed of slanted secondary lamellae with direction of slant anternating. **C**. Scheme of edge of conciferan shell and mantle (clam) showing relationship of shell layers and mantle edge. (**A,C** after Beedham and Trueman; **B** after Newell.)

whole inner mantle surface and becomes thicker throughout the life of the animal. Nacreous material is laminated, parallel to the shell surface. When these lamina are straight and thick, the shell lining is dense and porcelaneous. But if they are thin and wavy, the lining of the shell is iridescent and pearly. Sometimes a crossed lamellar structure is seen. In this case the primary lamina are composed of secondary lamellae laid down at a characteristic angle (Fig. 21.2B). When a foreign particle comes to lie between the inner shell surface and the mantle, it is covered with nacre; when composed of pearly nacre and well formed, the resulting **pearl** is among the most treasured of precious stones. Some gastropods make pearls, but commercial pearls are obtained from pelecypods, especially from the pearl oyster, *Pinctada*. The inner and outer ostracum is calcite or aragonite, differing in various species and sometimes between the two layers of a single shell.

The shell first appears during the veliger larva stage, a stage that succeeds the trochophore, and grows throughout life. Growth at the margin of the shell produces regular growth lines that parallel the contour of the margin. Additional sculpturing is common and is sometimes very complex. The aperture of a shell grows with the animal, and the shell thus provides a growth history of the animal as a whole.

The molluscan coelom is a schizocoel. A primitive arrangement, shown in Figure 21.3A, consists of paired spaces, unconnected; each space is divided into a pericardial part, a gonocoel (the cavity of the gonads), and a nephrocoel that is connected to the exterior by way of ducts, and serves for gamete discharge and excretion. The molluscan excretory organ is a metanephridium, with a nephrostome opening into the pericardial cavity and a nephridiopore to the exterior. Two points of difference from the usual metanephridia seen in other phyla should be noted: (1) the glandular part of the excretory organ arises from the coelomic wall rather than from a part of the duct, as in most other coelomates; and (2) the duct is a coelomoduct, growing outward to establish a pore, rather than arising from the ectoderm and growing inward.

The reproductive and excretory systems tend to separate, with the excretory system retaining its connection to the pericardial cavity. The shell-less Aplacophora (see Fig. 21.9) essentially retain the primitive arrangement (Fig. 21.3D). In chitons the two systems are completely separated (Fig. 21.3C), with separate gonoducts, as in the most advanced clams (Fig. 21.3C). The Monoplacophora have the gonoducts join the excretory organs (Fig. 21.3B). Cords of cells connect some of the renal organs with the pericardial cavity, but no lumen has been convincingly demonstrated in them.

The molluscan coelom is very small and restricted to the vicinity of the heart. It has been generally thought that the coelom was reduced by the encroachment of blood sinuses and lacunae; therefore, only the pericardial region persists. As the mollusc body plan does not depend on coelomic fluid to act as a hydraulic skeleton for movement, it is quite possible that the coelom of the molluscs was *never* a large perivisceral space.

Digestive System Molluscs have a digestive tract divided into discrete, functionally specialized regions.

The foregut consists of the **buccal cavity**, the **esophagus**, and the parts derived from them. Its adaptations are closely correlated with the method used for food capture and with the nature of the food itself. Although some molluscs feed on large plants or animals and some are ciliary feeders, most molluscs gain their livelihood by scraping muck off of surfaces. The most characteristic feature of the molluscan buccal cavity is the radula apparatus, found in nearly all molluscs in one form or another, and peculiarly suited for scraping food particles from a surface, although it can assume other functions. The apparatus consists of the **radula**, a toothed belt, that passes over a supporting rod, the **odontophore**, and the **radula sac**, in which the radula typically lies (Fig. 21.4A). When the radula is used for scraping, the odontophore is thrust out of the mouth by the contraction of the protractor muscles, and the radula is applied to the surface that is to be scraped. The odontophore is composed of firm chordoid tissue, and hence suitable for pressing the radula against the detritus-covered rock. When in place, the radula may be moved back and forth across the odontophore by muscles attached to the belt, while the recurved teeth engage particles, pulling them away from the surface as the odontophore is retracted. Teeth used in this way have a short life expectancy; therefore, new teeth are constantly being formed by odontoblasts at the back of the radula. Teeth move forward as the front of the radula wears. In any given species a definite number of transverse rows of odontoblasts occur, so a definite number of radular teeth occur in each row. Teeth are valuable in taxonomy because the number, size, and shape of radular teeth are remarkably constant within a species. Typically, there is one central tooth in each transverse row, flanked by one or two lateral teeth on each side. One or several marginal teeth at each edge of the radula complete the row of teeth (Fig. 21.4B,C). At the two extremes, *Chaetoderma* has a single radular tooth, while *Umbrella* may have as many as 750,000 radular teeth.

Radular teeth exhibit many modifications. Some snails bore through clam or oyster shells with the aid of the radula, lapping the flesh from the helpless prey when the shell is pierced. Many herbivorous snails have cutting jaws, located at the aperture of the buccal cavity. They nip off pieces of vegetation, which are torn and fragmented as they pass over the radula. The radula of carnivorous snails is often borne at the tip of a proboscis that can be shot out with great speed and force. In this case the radular teeth serve as weapons. Cone shells (Conidae) have hollow radular teeth, packed with poison from poison glands. The poison is extremely powerful and can even cause human death. Although a radula is found in most molluscs, filter feeders have no use for one. In pelecypods, the radula is wholly lost, even in

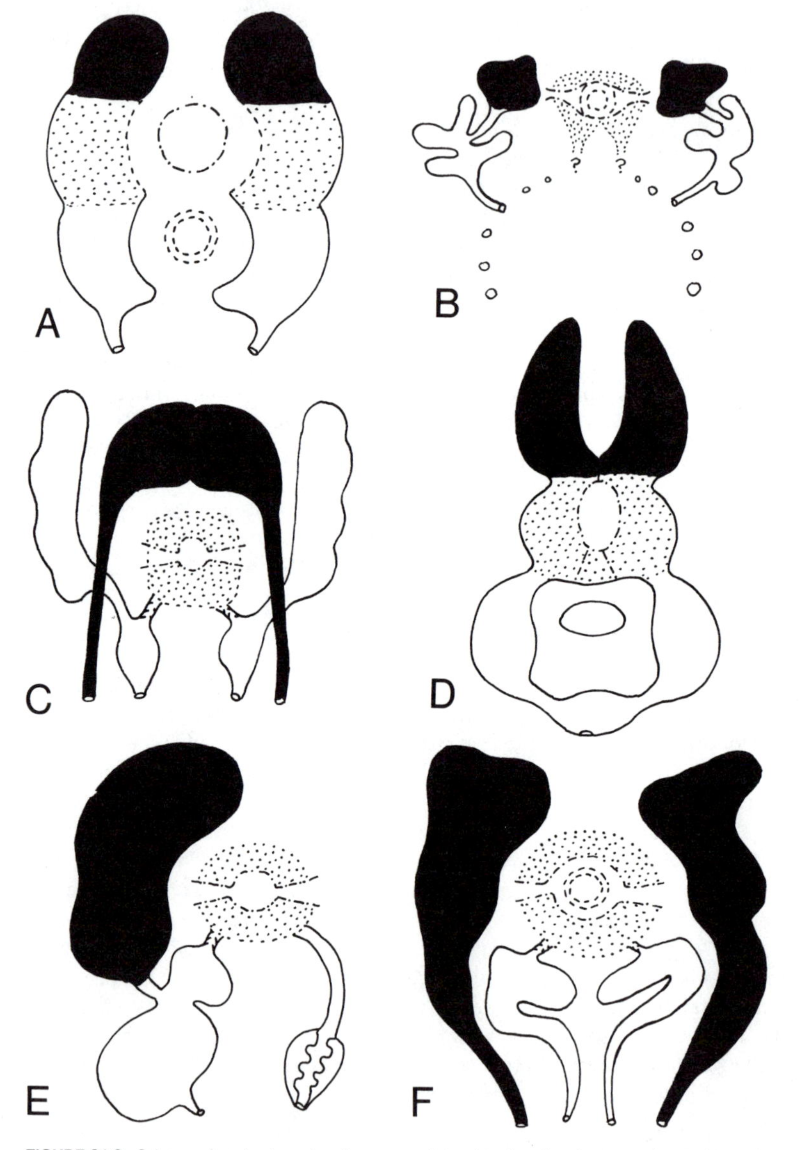

FIGURE 21.3. Scheme of coelomic regions in some molluscs. Region of coelom associated with nephridium *unshaded,* region associated with gonad *blackened,* and pericardial cavity *stippled.* **A.** Hypothetical primitive condition. **B.** *Neopilina.* **C.** Polyplacophora. **D.** Aplacophora. **E.** Primitive prosobranchs. **F.** Pelecypoda. (**A,C,D,E,F** based on Goodrich; **B** based on Wingstrand.)

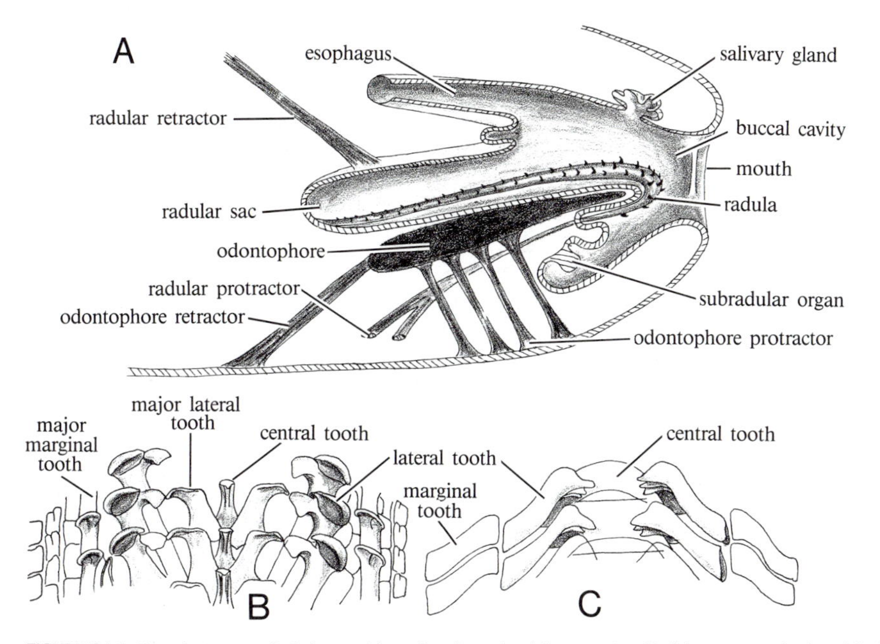

FIGURE 21.4. Buccal structure. **A**. Scheme of buccal cavity and radula apparatus. Radula can move back and forth over extruded odontophore. **B**. Part of *Chiton* radula, showing two rows of radular teeth. **C**. Part of *Dentalium* radula, showing two rows of radular teeth. (**B**,**C** after Cooke.)

species that, like the ship worms and boring clams, use a rasping surface. In these organisms, the modified shell valves serve as a rasp.

The esophagus is extremely variable. It is sometimes a simple tube and in other cases has a dilated region that serves as a crop or gizzard. When small particles are swallowed, they are usually conveyed to the stomach trapped in mucous particles, while muscles are used for swallowing in forms that feed on larger particles.

The complex stomach structure (Fig. 21.5) varies in detail with the feeding habits, but some features have a special molluscan quality. It is characteristically associated with a pair of large **digestive glands**. Although major changes in stomach form are associated with adaptation to various kinds of feeding, the **style sac** is usually retained. Clams, and some other molluscs, have a **crystalline style**, a transparent mass of material rich in amylase. In some cases, spirochaetes live in the crystalline style, providing more enzymes for digestion. The simplest stomachs are found in carnivorous species. Some have a simple crushing region and a digestive region, while others have a simple saccate stomach in which extracellular digestion occurs.

The hindgut is an **intestine**, often quite long and sometimes divided into definite regions. Although it sometimes participates in absorption, it is generally most important in feces formation. This mundane activity is of considerable importance because the

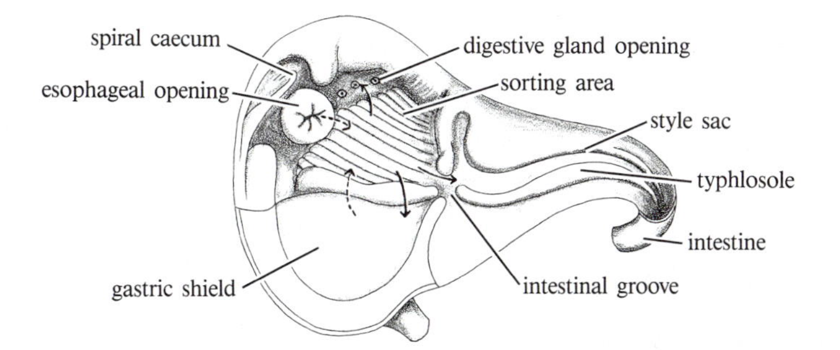

FIGURE 21.5. Scheme of dissection of *Diodora* stomach, thought to retain some primitive characteristics. (From Fretter and Graham.)

anus is near the gills in the mantle cavity (Fig. 21.6), and firm, well-formed feces are less likely to foul the respiratory surfaces. Feces formation generally involves the secretion of mucus, the compacting of the mucus and rejected material to form firm bodies, and some absorption of water.

Respiratory System Nearly all molluscs have a shell of some kind, and the shell was a factor in preventing molluscs from using simple external gills and body surfaces for respiration, since molluscs that have no shells or much-reduced shells have lost the gills and use the body surface or have simple external gills. The mollusc ctenidium is a kind of gill seen repeatedly in diverse types. The **ctenidium** has a simple form to permit easy passage of a quantity of water over a large respiratory surface. A long, usually flattened central axis contains incurrent and excurrent blood vessels (Fig. 21.6). A series of flattened gill filaments are attached to the axis, separated enough to permit water to flow freely between them, but close enough so that cilia on adjacent filaments can cooperate in generating water currents. Skeletal rods support the filaments and prevent their collapse. The ctenidia are located in the mantle cavity, attached by membranes above and below. Cilia in the mantle cavity generate inhalant currents to bring water in below the ctenidia, and exhalant currents to flush water out from above them. Special cleansing cilia sweep particles off the filaments, and hypobranchial glands in the roof of the mantle cavity secrete mucus in which the particles are trapped, then to be discharged through the opening of the mantle cavity. Respiratory exchange is most efficient if the blood and water flow in opposite directions. In molluscs, water is brought up to the ctenidium from below, while blood is brought in from above, making for the most effective use of the gill.

The common tendency for larger or more complex animals is to replace ciliary mechanisms with muscular ones. The clam *Yoldia*, for example, has muscle fibers in the membrane that attach the ctenidium to the roof of the mantle cavity. The muscles relax each time a pulse of blood enters the gills, and the gill moves downward. The gill is lifted each time the blood drains from it. These movements help ventilate the gill filaments. Some pelecypods have no gills, and in their place are muscles that pump water and draw it into the mantle cavity. The vascularized mantle cavity wall serves for respiratory exchange. Many swimming molluscs aerate the gill surface as a result of swimming movements that result in an automatic adjustment of the rate of gill ventilation with changes in the animal's activity.

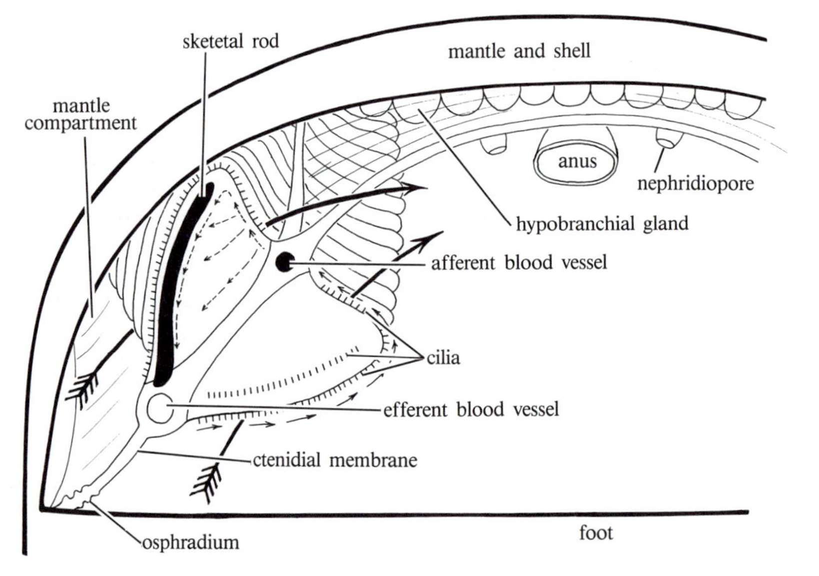

FIGURE 21.6. Scheme of arrangements in mantle cavity of generalized mollusc. Water enters, passes forward in dorsolateral mantle compartment and between filaments of ctenidium (*large arrows*.) Feces and kidney wastes discharged into mantle cavity and ejected with exhalant current in medial part of mantle cavity. Each ctenidial gill filament with cilia on its face, supported by skeletal rods. Lateral cilia bring water up between gill filaments, and frontal and abfrontal cilia cause cleansing currents (*small solid arrows*) that bring particles upward for discharge by exhalant current. Blood flows through filament (*dotted arrows* in left filament) in direction opposite to that of water flow. Hypobranchial glands in roof of mantle cavity secrete mucus to trap particles discharged with exhalant current. (Somewhat modified from Fretter and Graham.)

Pulmonate snails have no ctenidia, and the mantle cavity wall contains a rich supply of blood vessels that convert it to a **lung**. This change in snails is especially correlated with terrestrial habits and air breathing. Cilia cannot generate air currents; therefore, the mantle cavity has a muscular, dome-shaped floor that is raised and lowered by alternate contraction and relaxation of muscles, causing air to rush in or out through a narrow opening, the **pneumostome**. When air enters, the pneumostome is closed and the muscles relax. The elasticity of the tissues tends to compress the mantle cavity, increasing the partial pressure of oxygen and facilitating oxygen absorption. The pneumostome is then opened and air rushes out.

Excretory System Three parts of molluscan renal organs can be recognized (Martin, in Wilbur, vol. 5, 1983). An opening into the pericardial coelom more or less corresponds to a **nephrostome** that leads to a glandular region, usually much convoluted, that in turn empties through a tube at an external **nephridiopore** or, if gametes also leave by way of the renal organ, a **urogenital pore**.

Molluscs have very rarely increased excretory capacity by replication of parts. Only in the metameric Monoplacophora and *Nautilus* is more than a single pair of renal organs present, and in many cases one member of the original pair is lost.

Excretion occurs as a result of the filtration of fluid from the blood, reabsorption of compounds, and active secretion of substances into the tubule. The site of filtration has not been determined in all molluscan types but appears to be traditionally the pericardial cavity and the **pericardial glands**, patches of glandular and flattened cells. In aquatic forms, water is no problem unless too much is taken in with food, and blood pressure is high enough to provide a filtration pressure. The fluid that enters the renal organ from the pericardial cavity is essentially like the blood in terms of salinity, and is processed to form urine by active transport of compounds, usually through the walls of the nephridium or renopericardial canal, and reabsorption of glucose and other useful compounds, largely in the glandular body.

Adaptation to land living involves the reduction of the pericardial glands, and in the size of the renopericardial canal and its opening to the pericardial cavity, and thus cutting down the loss of fluid in the excretory process.

Circulatory System Considering the diversity seen in other organ systems, the molluscan circulatory system is remarkably uniform. The heart and major blood vessels retain about the same relationships with the main organs of the animal in all groups.

The blood circulates in part through large **hemocoel lacunae** and in part through blood vessels. The proportion of vessels and lacunae varies in different groups. Blood that has passed through the viscera and body wall is collected in one or two sinuses that convey it to the gills. It passes through an afferent vessel to the gill, through lacunae or capillaries in the gill filaments, and into an efferent branchial vessel. This vessel takes blood to the **auricle** associated with that gill (Fig. 21.7B,C). Except in Monoplaco-

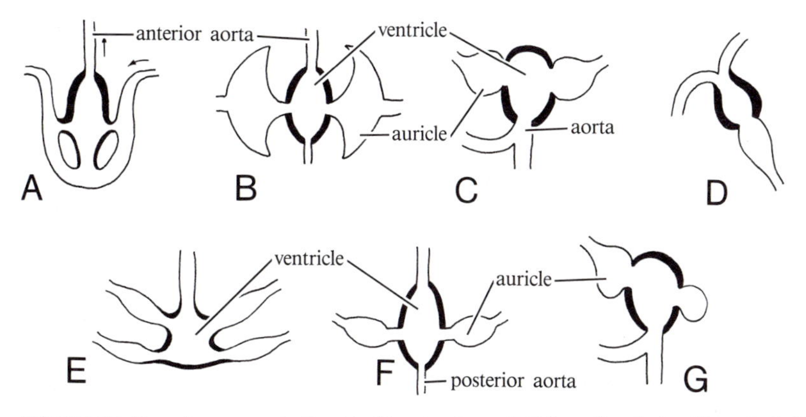

FIGURE 21.7. General arrangement of heart in different molluscs. **A.** *Chiton,* with pair of auricles collecting blood from paired rows of ctenidia, single anterior aorta. **B.** Clams, with large auricles collecting blood from large gills on each side, and anterior and posterior aortae. **C.** Prosobranch, with two gills, each drained by auricle, and auricles shifted to anterior position by torsion. Aorta emerges at "posterior" end of heart, but divides into anterior and posterior branches. **D.** Opisthobranch, with single auricle brought backward by detorsion. **E.** Nautiloid, with two pairs of gills and anterior aorta. **F.** General cephalopod with one pair of gills. **G.** Prosobranch with single pair of gills, showing partial reduction of one auricle.

phora and chitons (Fig. 21.7A), which have many gills served by a collecting vessel on each side of the body, each gill drains into its own auricle. The majority of molluscs have a single pair of gills and a single pair of auricles, but there are some modifications. Asymmetrical gastropods (Fig. 21.7D,G) tend to lose one gill and with it lose one of the auricles. *Nautilus* (Fig. 21.7E) has two pairs of gills and two pairs of auricles. The blood then passes into the **ventricle** and is distributed to the tissues by an **anterior aorta** and its branches, and in some cases a **posterior aorta** as well (Fig. 21.7F).

Nervous System In general, molluscs are well equipped with sense organs, but they are so variable that they are best discussed in connection with the various classes.

Molluscs tend to have very few ganglia. The system in the conciferan classes is built around four pairs of primary ganglia. A pair of **pedal ganglia** are assocated with the pair of **pedal nerves** that innervate the foot. A pair of **palliovisceral nerves** arise in a pair of **pleural ganglia** and end in a pair of **visceral ganglia**. Sometimes a pair of **parietal ganglia** lie between the pleural and visceral ganglia on the palliovisceral nerves. A pair of **cerebral ganglia** are associated with the **circumenteric nerve ring**. The cerebral, pleural, and pedal ganglia form a neural triangle on each side of the body, connected by cerebropedal, cerebropleural, and pleuropedal connectives. In the more highly developed molluscs like cephalopods, the cerebral, pleural, pedal, and parietal ganglia tend to crowd together and, to a greater or lesser extent, unite to form a complex brain.

Reproductive System The reproductive system is quite variable in each major group, and is discussed under the individual classes.

Biology

Locomotion and Muscle Movements Two general classes of muscles are recognized in molluscs: striated, phasic muscles that contract and relax rapidly, and unstriated, tonic muscles that contract slowly and maintain contraction for a relatively long time. A few examples may help to clarify the difference. It is important, and often critical, that a clam be able to clap its shell shut quickly in an emergency and hold it shut for a long time. On the other hand, scallops swim by opening and closing the shell valves rapidly. A tonic muscle that can hold a clam shell shut for a long time would be useless for swimming movements like those of a scallop, while a phasic muscle that can close a clam shell rapidly would not be suited for holding it shut (Chantler, in Wilbur, vol. 4, 1983).

Some idea of the functional constraints seen in molluscan muscles is shown by their time properties. The muscles used by a scallop for swimming are striated and have a contraction time of 0.046 sec and a half-relaxation time of 0.04 sec. A squid swims with the aid of mantle muscles that have a contraction time of 0.07 sec and a half-relaxation time of 0.11 sec. The unstriated muscle fibers in the tentacle retractor of a snail have a contraction time of 2.5 sec and a half-relaxation time of 25 sec.

Feeding Because feeding habits are highly specialized, specific comments are made for each group in the discussion that follows.

Excretion Aquatic species can eliminate ammonia through the body and gill surfaces as well as through the nephridia, thus compensating for lack of excretory surface. Ammonia is relatively toxic; few animals can tolerate much of it in their blood, although molluscs have a rather higher tolerance for ammonia than most animals. Cold-blooded aquatic vertebrates usually excrete a significant amount of ammonia, but normal blood ammonia levels are below 0.1 mg per 100 ml. Comparable figures for the cuttlefish, *Sepia*, are 2.8 to 4.8 mg, and snails vary from about 2.3 to 3.0 mg. The blood ammonia concentrations of molluscs run about 1000 times as high as mammals, about 30 times as high as fish, and twice as high as lobsters. *Sepia*, aquatic snails, and clams excrete about half of their nitrogenous wastes as ammonia.

A second line of biochemical accommodation is to convert nitrogenous wastes to relatively insoluble, nontoxic compounds. This is the common technique for terrestrial species that cannot excrete ammonia at the body surface effectively. Nitrogen is eliminated as amino acid or uric acid. Amino acids make up almost a fifth of the nitrogen excreted by the land slug, *Arion*; *Sepia* excretes less than ten percent of its nitrogen as amino acid. However, amino acid excretion tends to occur sporadically, and it is more often uric acid that is produced to deal effectively with the problem. The uric acid content of terrestrial snails, as shown by dry-weight measurements, is from 10 to 300 times as high as marine snails, and species that live in the intertidal zone and are sometimes exposed to air are intermediate. *Pila*, an amphibious snail, for example, has 1.68 mg of uric acid per gram of kidney when freshly collected, but comes to have up to 102 mg uric acid per gram of kidney when it is estivating under drying conditions.

Molluscs are peculiar in having few or no enzymes associated with the ornithine cycle, the pathway of urea production in other animals. Adjustments to water availability and osmotic stress are made without excreting urea. Clams, for example, depend heavily on ammonia excretion, although appreciable amounts of amino acids are excreted; and some clams release trimethylamine oxide, a soluble, nontoxic nitrogenous compound.

Respiration The demand for oxygen varies with environmental factors, such as fluctuating oxygen supply and temperature, and with the activity levels of the organism. Unless an animal has so much respiratory surface that it can meet the highest peaks of demand without regulatory mechanisms, it must depend on anaerobic energy release for the duration of the emergency or be forced into inactivity to balance its respiratory accounts.

An example of respiratory regulation is seen in lung-breathing pulmonate gastropods. The most important internal clues in animals that regulate breathing, generally, are low oxygen levels or high carbon dioxide levels. Pulmonates appear to be sensitive to both. Some of the pulmonates have returned to aquatic habitats, such as *Helisoma* and *Lymnea*, and come to the surface to breathe, presumably when the oxygen supply reaches a critical level. They remain submerged three times as long in water containing 6 ml of oxygen per liter as in water with 2 ml of oxygen per liter. They remain submerged 22 times as long at 11°C as at 21°C. They regulate respiration, at least partly, by the regulation of behavior. *Limax*, a slug, and *Helix*, a land snail, adjust the opening of the pneumostome in accordance with carbon dioxide levels. When the air contains three to five percent carbon dioxide, the pneumostome is held open constantly. At this point their regulatory system breaks down, for the permanently open pneumostome reduces the ability to absorb oxygen since lung pressure cannot be raised. Up to this point, the dilation of the pneumostome results in increased frequency of breathing movements and exchange of a higher percentage of lung air with each breath. A more significant change in lung ventilation results from changes in the muscle contractions of the lung wall. At reduced oxygen levels, muscles along the sides as well as the floor of the lung contract, increasing the amplitude of breathing movements. An oxygen debt system operates in many molluscs. Rather surprisingly, they often build up the debt during periods of inactivity, when respiration is low.

Molluscs use a variety of respiratory pigments (Table 21.1) including myoglobin and hemoglobin as well as hemocyanin (Bonaventura and Bonaventura, in Wilbur, vol. 2, 1983). The iron-containing globins occur in blood cells, hemolymph, and various body tissues, depending on species. Hemocyanin, a copper compound with a high molecular weight, is generally dissolved in the blood plasma. This last pigment is colorless when reduced, and bluish when loaded with oxygen. Mollusc blood is a very effective oxygen carrier by invertebrate standards, combining with about 1 to 7 mg oxygen per 100 ml. Most cold-blooded vertebrates have blood that will require between 6 and 15 mg per 100 ml. Oxygen-carrying capacity, however, does tend to be associated with activity. The copper content of the blood is distinctly greater in highly active cephalopods than in smaller and less active molluscs.

Circulation Because lacunae and hemocoel sinuses are so important in clams and snails (Voltzow, 1985), it is easy to overemphasize the importance of open molluscan circulatory systems and to underemphasize blood vessels. Some of the most active animals are cephalopods. Blood flow is necessarily rapid in large, active animals, and a reasonable blood pressure is maintained. An octopus has a blood pressure of 40 to 60 mm mercury at the heart and 5 to 6 mm mercury in the gill veins. This is a higher pressure than is found in many fishes, and far higher than the pressure in small molluscs. The snail *Helix* has a blood pressure of only 1.1 mm mercury at the heart and about 0.4 mm mercury in the gill veins, and a large clam has a blood pressure of about 4.4 mm mercury at the heart. The maintenance of high blood pressure depends in part on the strength of the heart and partly on the effectiveness of retain-

Table 21.1. Distribution of respiratory pigments in classes of Mollusca

Class	Pigment		
	Myoglobin	Hemoglobin	Hemocyanin
Polyplacophora	+		+
Gastropoda	+	+	+
Cephalopoda			+
Pelecypoda	+	+	

Modified from Bonaventura and Bonaventura, in Wilbur, vol. 2, 1983.

ing walls around the blood. Small arteries have an endothelial lining that can detach to form blood cells, and both large and small arteries have a thick basement membrane under the endothelium. The basement membrane is the primary source of strength in smaller arteries, but large arteries have smooth muscle fibers and a large amount of connective tissue. Although the arteries deliver blood into lacunar spaces rather quickly in some molluscs, the arteries of cephalopods divide and redivide, eventually emptying into true capillary beds in some organs, especially the gills.

A heart is essentially a muscular tube, whose contraction forces blood along. A single contraction is ineffective; it must contract repeatedly and rhythmically to maintain blood flow. Furthermore, blood must go in a definite direction, made possible by valves that prevent backflow. The rate of flow must be adjusted to the physiological needs of the organism at any given time, requiring feedback regulatory mechanisms. To do this, one can either deliver repeated stimuli to the heart, so that the rhythmicity of heartbeat depends on rhythmic firing of nerves (neurogenic control), or one can develop muscle that will spontaneously contract in a rhythmic way (myogenic control). Mollusc hearts are myogenic. They will beat if completely separated from the nervous system. Nerve cells in the heart wall of gastropods and cephalopods can modify the rate of beat. In most animals with myogenic hearts, some part of the heart wall is dominant and serves as a pacemaker. Pacemakers have been reported near the aorta and also near the auricles.

Various kinds of mechanisms can adjust the heartbeat to meet physiological demands (Jones, in Wilbur, vol. 5, 1983). The frequency and power of contraction increases with stretching of the muscle with increased venous return that results from greater activity. Control from without is no less important. Stimuli that inhibit and augment beat are provided by neurosecretions from special neurons to the heart. Molluscan heart tissue is inhibited by acetylcholine and excited by serotonin, in general, and there is good evidence that an acetylcholine–serotonin system of neuronal secretions regulates the heartbeat of some molluscs.

Sensation and Reproduction These processes are highly specialized to the needs of each group and are discussed later.

Development The course of development varies (Verdonk and van den Biggelaar, in Wilbur, vol. 3, 1983). Some molluscs produce large, yolky ova that undergo modified development. Where eggs are small, a typical spiral cleavage occurs (Fig. 21.8B), similar to the spiral cleavage of annelids (Fig. 21.8A) and polyclad flatworms.

Gastrulation (Fig. 21.8C) varies with the yolk content of the macromeres. Invagination and epiboly are the most common techniques used, but some species gastrulate by ingression. The embryo differentiates rapidly after gastrulation, forming a **trochophore larva**. The blastopore narrows and ectoderm cells turn in at the margin of the blastopore to form an ectodermally lined stomodeum (Fig. 21.8E). Two teloblasts arise from blastomere 4d. These lie on each side of the primitive gut and divide to form a band of mesodermal cells. A ciliated band, the prototroch, forms a girdle anterior to the stomodeum, and an apical plate bearing an apical tuft of cilia develops at the anterior end of the larva. The larva elongates and in most molluscs becomes a veliger. Veliger larvae differ in different classes. It may suffice here to note that the mollusc posttrochophore stage (Fig. 21.8G) shows no signs of the segmentation so prominent in the annelid posttrochophore larvae (Fig. 21.8F). During this period the fundamental differences in annelid and molluscan architecture are determined. The similarity of annelid and mollusc trochophores, however, would seem to attest to their relationship.

Review of Groups

CLASS APLACOPHORA

Traditionally, the Aplacophora, or solenogasters, have been placed in a class Amphineura along with the chitons, but embryological as well as other kinds of evidence have made it increasingly clear that they deserve a class designation of their own. They live in deeper waters (about 30 to 1800 m), feeding on detritus in the mud or living with corals or colonial hydroids. They are small creatures as a rule (about 2.5 cm long), but reach a maximal length of about 30 cm. They are of special interest because of the curious mixture of primitive and specialized features.

Solenogasters are so aberrant from the typical molluscan body plan. Thus, Hyman (1967:13) characterizes them as "without head, mantle, foot, shell, or nephridia." The class is characterized here by: the elongated, bilaterally symmetrical body; terminal mouth and anus; body surface covered with a spicule-studded cuticle; and straight digestive tract, usually with a radula.

The foot of the order Neomenioidea (Fig. 21.9A) is reduced to a ventral ridge. The **mantle** nearly encloses the animal, and there is little to choose between anterior and posterior ends. The cuticle is a thick, uniform layer of glycoprotein, although low in proteins and high in mucopolysaccharid, in which are embedded calcareous spicules (Beedham, 1968). The mouth opens at the anterior end, and at the posterior end a cloacal opening is found. The slitlike space on each side of the vestigial foot appears to represent a mantle cavity, which may or may not contain gills. Paired **nephridia** unite on the midline, and open through a single pore below the anus. There is a muscular **pharynx** with large **salivary glands**. The **radula** varies considerably in form, depending on feeding habits, and is sometimes missing. The intestinal digestive glands are reduced to patches of secretory epithelium in the stomach wall, and the **intestine** is a straight tube with small **diverticula** along its length.

The circulatory system is reduced. Blood containing hemoglobin in erythrocytes circulates through a system of sinuses and passes through the gills, if present, and through a pair of **auricles** to the **ventricle**. The neomenioids are monoecious. The gonads are **ovotestes**, although ova maturation is somewhat delayed over that of sperm. Glands in the walls of the nephridia beyond the point of union are thought to secrete eggshells, and no certain excretory role for them has been demonstrated. Fertilization is external.

Larval development has been followed (Fig. 21.9D,E). The young larvae are striking because a test of ciliated cells, derived from the velum, en-

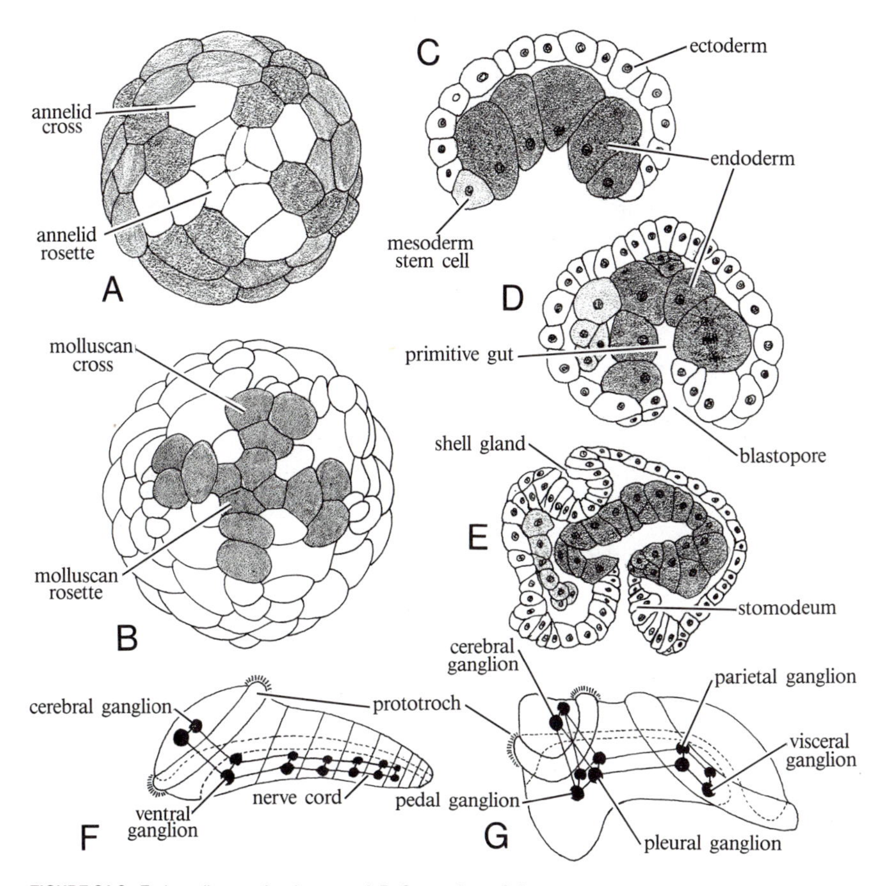

FIGURE 21.8. Early molluscan development. **A,B**. Comparison of cleavage patterns in annelids and molluscs. **A**. Cleavage stage of *Arenicola*, showing interradial annelid cross and rosette (*unshaded*). **B**. Cleavage stage of mollusc, *Crepidula*, showing molluscan radial cross (*shaded*) and rosette. Molluscan cross is radial in position. **C,D,E**. Stages in development of *Littorina*, showing gastrulation and early changes leading toward trochophore larva, including early proliferation of mesoderm stem cell (*lightly shaded*). **F,G**. Comparison of annelid and molluscan posttrochophore larvae. **F**. Annelid, in which paired ganglia are added with new segments. **G**. Mollusc, without evidence of segmentation and posterior part of body covered by mantle. (**A** after Child; **B** after Conklin; **C,D,E** after Delsman, from Korschelt; **F,G** after Naef.)

closes the larva. These test cells are shed suddenly, shortly before metamorphosis begins. During metamorphosis, supposedly a transitory set of dorsal plates, similar to those of chitons, appears; however, this apparently has never been verified since the original observation based on a single larva in the 1890s.

In the order Chaetodermatoidea the foot is missing entirely, and the mantle forms a cylindrical tube around the body. The anterior and posterior ends are marked by constrictions (Fig. 21.9B). The posterior end is expanded as a bell-shaped cloacal region containing a pair of large bipectinate gills (Fig. 21.9C) comparable to those seen in chitons. The gut differs from that of neomenioids in that a set of very large diverticula, the **hepatic glands**, arises from the midgut. Chaetodermatoids are dioecious. The single median gonad empties by way of the renal organs that are not modified for the purpose and lack any evidence of accessory glandular functions. The circulatory system is essentially like that of the neomenioids.

The aplacophoran nervous system closely resembles the nervous systems of polyplacophorans and monoplacophorans in basic pattern. A **circumoral nerve ring** gives rise to a pair of **pedal nerves** and a pair of **pallial nerves**. Right and left sides are connected by posterior commissures, and pedopallial connectives run between the main nerves. The principal difference between the aplacophoran and the

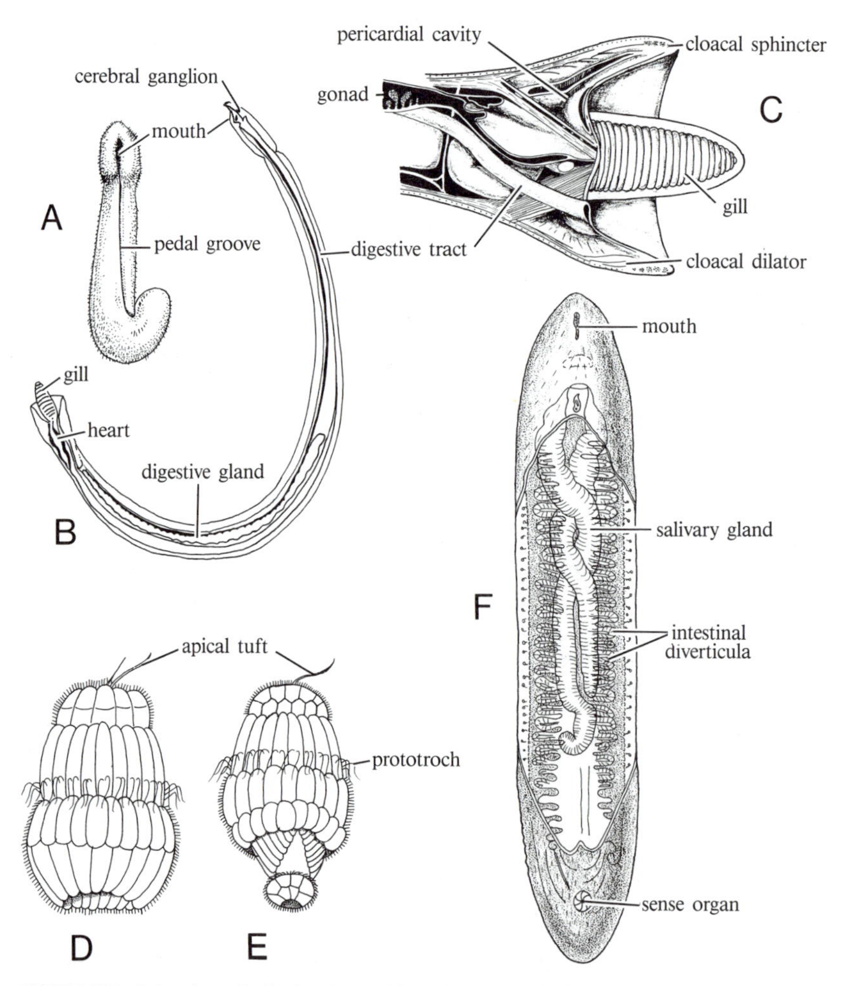

FIGURE 21.9. Aplacophora. **A**. *Paramenia cryophila,* neomenioid. **B,C**. *Chaetorderma nitidulum,* chaetodermatoid with no distinct ventral groove, and posterior cloacal chamber with gills. **D,E**. Stages in development of aplacophoran, younger and older trochophore larvae with enveloping test of large cells. **F**. Anatomy of *Proneomenia,* showing internal structures. (**A** after Pelseneer, in Lankester; **B,C** after Wirén; **D,E** after Pruvot; **F** after Kowalesky and Marion, in Grassé.)

polyplacophoran or monoplacophoran systems is the greater development in Aplacophora of the distinct **cerebral** and **pleural ganglia**.

It is generally agreed that the Aplacophora, although aberrant, are unmistakable molluscs. It is also evident that they have a number of very primitive characteristics. The wormlike form, the external spicules in place of a shell, possibly the posterior mantle cavity, the very primitive coelomic system, and the nervous system with distinct cerebral and pleural ganglia may all represent important evidence of the nature of early molluscan or premolluscan stocks.

CLASS POLYPLACOPHORA

Most chitons are inconspicuous creatures, only a few inches long at the most. Their dull shades of color are valuable camouflage, helping them to blend with the rocks on which they live. The largest species is *Cryptochiton stelleri*, found on the Pacific coast of North America. It is brick red and reaches up to about 0.3 m long. About 600 species are known. Chitons occur in the sea in all climates and at all depths up to about 4000 m. Although some live high in the intertidal zone, moistened only at full tide, they are especially abundant in shallow water in rocky areas (Kaas and Van Belle, 1985).

Polyplacophora are characterized by a shell composed of eight plates with aesthetes and apophyses.

Chitons are homebodies and move about very little unless disturbed. They usually stray away from a definite homesite only on short foraging expeditions, coming back to the original spot, to which they may have adapted by slight shell modifications. *Nuttallina* is even less mobile and lives permanently in depressions that fit its body almost perfectly, living on whatever food that fate happens to bring to their niche. These niches appear to have been carved in the rock by past generations of *Nuttallina*. Chitons tend to be photonegative, and many species will crawl slowly to the underside of their rock if it is turned over. They adhere to the rocks with the broad flat foot unless one tries to detach them, when the tough girdle is thrust against the rock and the foot is retracted. This forms a suction cup and enables them to hold tenaciously. If they are detached from their rock, they roll up like an armadillo, with their dorsal plates forming a continuous protective covering.

The shell is formed of eight valves arranged in a longitudinal row. The front margin of each of the last seven plates extends below the plate in front of it. Special articulating surfaces or **apophyses** mark the overlapping regions of the middle six plates (Fig. 21.10D). The anterior (cephalic) and the posterior (anal) plates are differentiated (Fig. 21.10C,E), as they do not articulate respectively with any valves in front or behind.

The shell is made of two layers (see Fig. 21.2A), different from other mollusc shells (see Fig. 21.2B). The upper layer (**tegmentum**) is composed of conchiolin impregnated with calcium carbonate. The lower layer (**articulamentum**) is wholly calcareous. The edge of the mantle turns back over the edges of the valves, and in some cases covers the valves entirely (Beedham and Trueman, 1967). Shell characteristics are valuable in species identification, and an extensive vocabulary has developed to describe the form and sculpturing of the valves (Fig. 21.10A).

The tegmental sculpturing is not mere ornamentation. It is caused by epidermal canals associated with unique sense organs known as **aesthetes** (see Fig. 21.2A). The development of these surface sense organs is possibly associated with the reduction of a head completely enclosed by the girdle and covered by the shell. Some chitons have only simple tactile aesthetes, but many also have large visual aesthetes. These shell eyes vary from simple ocelli to highly differentiated eyes with a calcareous cornea and a lens. Some species have thousands of eyes, but as the tegmentum gradually wears away, the visual sense declines. In this case the older animals are indifferent to light. Branches of the visceral nerve innervate the aesthetes.

The ventral edge of the mantle forms the tough girdle that completely encircles the body. The color and ornamentation of the girdle are important factors in taxonomy. On the ventral surface a pallial groove separates the girdle and the foot. This groove is the mantle cavity and contains a series of **gill pairs** attached to its roof (Fig. 21.11A). The primitive posterior mantle cavity has been reduced, and the mantle cavity has spread forward as a trough around the body. The number of gills expands into the available space. The edge of the mantle is lifted slightly in front, permitting water to enter. Drawn in by the ciliated mantle wall, water passes posteriorly between the gills and the mantle. Cilia on the gill filaments cause currents that sweep the water up over them and between the lamellae. As the water reaches the exhalant space between the gills and the foot, it continues to flow posteriorly, departing at one or two points at the rear, where the mantle again is lifted slightly.

Chitons feed by scraping algae and other material from the rocks. The foregut retains a primitive character. Seven jaws usually lie just inside the **buccal**

cavity, but the **radula** does most of the heavy work of food procurement. The radula is very long, often reaching back beyond the beginning of the **stomach**, and has many transverse rows of 17 teeth. Typically, each row is made up of three central teeth, three lateral teeth on each side, and four marginal teeth on each side (see Fig. 21.4B). Immediately below the opening of the radula sac is the **subradular sac** (see Fig. 21.4A), which contains a chemoreceptor, the **subradular organ**. A hungry chiton extends the subradular organ from time to time to survey the situation. It is withdrawn when food is located, and the radula is extruded. Mucus from the salivary gland lubricates the radula and catches small particles as they are loosened from the rocks. The mucus forms strings containing food particles that are swept toward the stomach by cilia in a ventral groove of the esophagus, while amylase is poured over the food from a pair of **sugar glands**.

The oddly shaped stomach (Fig. 21.11B) has no sorting areas. Proteolytic enzymes from the large **digestive gland** enter the stomach through two ducts. Food and enzymes are mixed by peristaltic movements of the stomach. The food eventually passes into the slightly expanded **anterior intestine** where it remains for a time, held back by the **intestinal valve**. It has been suggested that the anterior intestine is actually a style sac.

Absorption and digestion occur in the stomach and the anterior intestine. Both contain mucous glands, and by the time residues reach the **posterior intestine**, they are heavily charged with mucus. Each time the intestinal valve opens and closes, a fecal pellet is stamped out and passes into the posterior

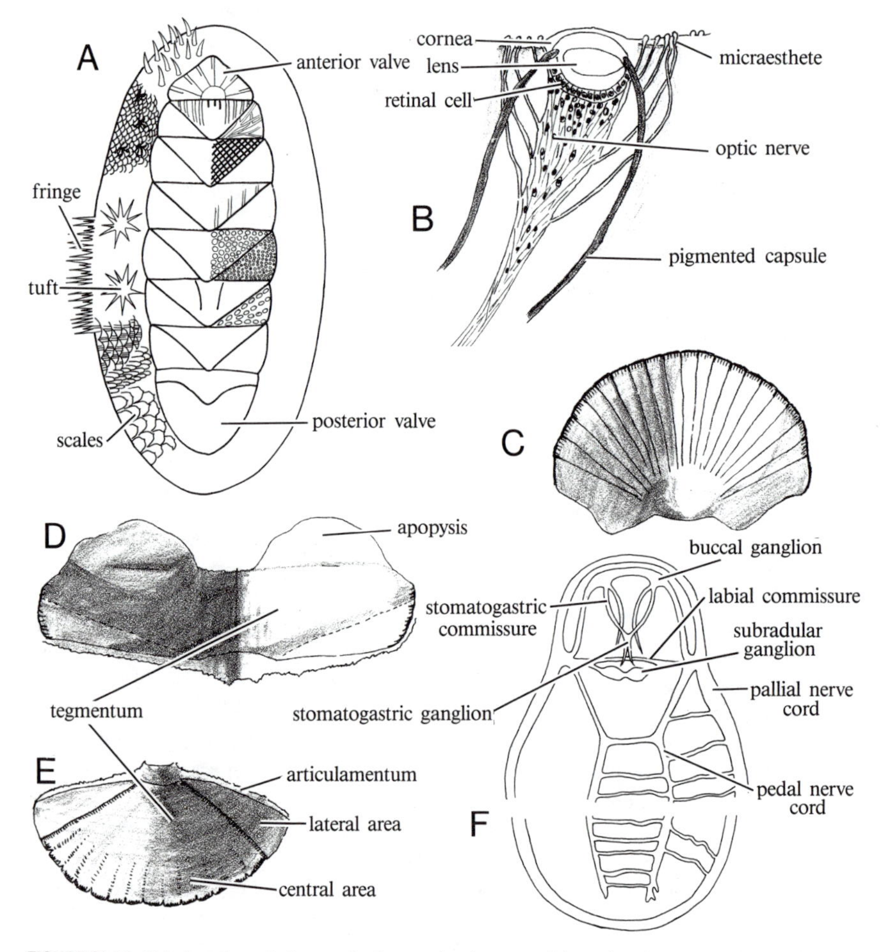

FIGURE 21.10. Polyplacophora. **A**. Composite diagram showing some of the sculpturings and girdle ornamentations seen in polyplacophorans. **B**. Section through aesthete of *Acanthopleura*. **C,D,E**. Anterior, middle, and posterior valves of shell of typical chiton. **F**. Dorsal view of nervous system of *Acanthochiton* (*middle portion omitted*). Note absence of ganglia, save for those associated with anterior stomatogastric region. (**A** after Turner; **B** after Mosley; **C,D,E** after Cooke; **F** after Pelseneer, in Lankester.)

intestine. The pellets are further condensed by reabsorption of water as they move toward the anus. Water currents in the mantle cavity waft the feces away as they are released.

Chiton renal organs have been carefully described, but little is known of excretory physiology. Urease has been reported from *Chiton*, so presumably urea is transformed into ammonia for excretion.

Chitons have a primitive arrangement, with a pair of kidneys open to the pericardial cavity (Fig. 21.11A). The **renopericardial canal** consists of an upper nonexcretory region followed by a second region that appears to be active in excretion. It opens into the long glandular part of the renal organ. This region gives rise to many blind **renal diverticula** and presumably is a site of secretory and perhaps reabsorptive activity. The glandular region runs forward and then runs back, emptying into a duct, expanded near the nephridiopore into a bladder region. Undoubtedly the intimate relationship of the diverticula of the glandular region and the hemocoel is an important factor in kidney function.

The circulatory system consists of two bilaterally symmetrical **auricles**, located at the back end of the body, that open independently into the ventricle through from one to four openings. The ventricle pumps blood into the aorta. Blood emerges into a system of sinuses that serve the head region and body wall and surround the internal organs. This blood eventually drains into two **pallial sinuses** that run back along the junction of the mantle and foot (Fig. 21.11A). Afferent branchial vessels extend to the medial side of the gill axis from the pallial sinuses. Aerated blood is returned by an efferent branchial vessel on the lateral side of the gill axis, and empties into the gill veins that pass back to the auricles.

The rudimentary head has no cephalic tentacles and no eyes, and there are relatively few special sense organs. The subradular organ is the principal chemoreceptor. Some chitons have sensory patches on the palps around the mouth. The mantle cavity has a number of sensory patches, thought to be homologous to the osphradia of other molluscs. Epidermal neurosensory cells provide sensitivity to touch and temperature changes, and the aesthetes are the principal tactile and, in some cases, photosensitive receptors of the dorsal surface.

The nervous system is relatively diffuse and decentralized for animals as large as the chitons. There is very little evidence of ganglion formation. The nervous system consists of a **subepidermal plexus**, a **circumenteric ring**, a pair of **palliovisceral nerves**, and a pair of **pedal nerves**. Commissures run between the two pedal nerves and connect the pedal

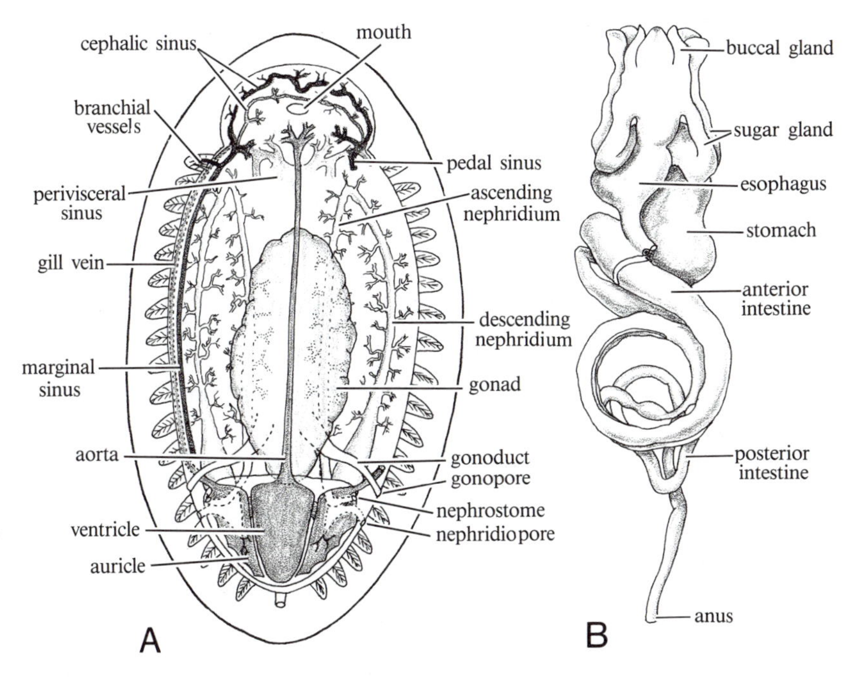

FIGURE 21.11. Polyplacophoran anatomy. **A.** Generalized scheme of internal organization of a polyplacophoran (*digestive tract omitted*). **B.** Digestive tract of *Lepidochiton.* (**B** after Fretter.)

and palliovisceral nerves (Fig. 21.10F). A small nerve ring arising from the circumenteric ring innervates the radula sac, and may contain a pair of tiny radular ganglia. A somewhat larger ring associated with the buccal cavity sometimes contains a pair of small buccal ganglia. The rings and ganglia are part of the sympathetic (stomatogastric) nervous system.

Chitons are dioecious but without dimorphism (Giese and Pearse, 1979). The two gonads unite to form a single large organ with paired gonoducts. The gonoducts open into the mantle cavity through gonopores that lie just in front of the nephridiopores. A few species, however, retain paired gonads. One species, *Cyanoplax dentiens*, has been reported as having hermaphroditism.

Sperm are shed into the mantle cavity. As chitons are gregarious, some sperm reach neighboring females. Males seem to predominate in populations of many species. Females do not shed ova until the males have shed sperm, presumably shedding in response to some chemical stimulus. Ova can be deposited singly, in masses, or in strings. In several species, the sperm enter the oviducts, fertilizaton is internal, and the young are brooded in the oviduct. These species have larger, yolkier eggs, and development is modified. For most chitons, however, fertilization is external, but in the pallial groove.

Development to the trochophore larva follows the usual molluscan pattern. Chitons do not pass through a veliger stage, and the trochophore metamorphoses directly into juveniles. The posterior end of the larva elongates as the shell starts to form, initially with seven plate primordia, then later with eight. The coelom appears as spaces in the mesodermal bands on each side of the body, derived from cell 4d. The buccal cavity and much of the esophagus arises from the stomodeum, but the proctodeum forms late and appears to give rise to only a small part of the intestine. Ctenidia appear late in development, with the most posterior pair appearing first. The apical cilia and prototroch decline during metamorphosis, and the young chiton sinks to the bottom and takes up life as a juvenile.

CLASS MONOPLACOPHORA

Monoplacophorans have been the focus of great controversy concerning their position in molluscan evolution. The first described species, *Neopilina galatheae*, and especially the more recent *Vema ewingi* provide models for understanding the group (Wingstrand, 1985).

Monoplacophorans are flat, with a dome-shaped shell, like the shell of a limpet. The protoconch, a remnant of the embryonic shell, is bilaterally symmetrical.

A shallow mantle cavity runs around the broad, flat foot, much as in chitons and limpets. There are five to six pairs of **ctenidia** in the mantle cavity. Their fine structure is somewhat different from that seen in other molluscs, and they may represent more primitive ctenidial structures. Respiratory currents are like those of chitons; water enters from the front and sides, and is discharged behind (Fig. 21.12A). Although superficially like chitons and limpets, monoplacophorans have distinctive structures. The **preoral tentacles** are small papillae, and the mouth is flanked by **lateral flaps**, considered the homologue of the larval velum and of the labial palps of clams. Conspicuous postoral tentacular tufts are also present.

The nervous system is amphineuranlike, but the other organ systems have unique characteristics, although all are evidently molluscan in character. The coelom (see Fig. 21.4B) is pericardial, with diverticula that extend forward adjacent to the aorta. Six to seven pairs of **nephridia** empty through **nephridiopores** near the gill bases. These are glandular organs whose cavities are not always separate from each other and are associated with five to six pairs of gills. Whether the kidneys communicate with the coelom through ciliated (?) funnels is not known, but cellular strands connect the pericardium and nephridia in the posterior part of the body.

The foregut contains a long, coiled **radula sac** and a **radula** with chitonlike features. The simple stomach is equipped with a **crystalline style sac**, and the **intestine** is coiled. Blood from the gills is drained into two pairs of **auricles** and enters paired **ventricles** discharging blood into paired aortae, which unite anteriorly and continue as a long **anterior aorta**.

The two pairs of gonads are associated with the fourth and fifth sectors, although *Vema* has a blind gonoduct in the third as well. The gonads empty by means of ducts into the nephridia, and gametes are voided then through the nephridiopores. Fertilization is external; however, no details of spawning are known. Currently, we have a lack of knowledge about early embryology and larval development.

The outstanding feature of the Monoplacophora is the serial repetition of parts (Table 21.2). To some workers, this repetition suggests that the primitive molluscs were segmented and that the monoplacophorans still exhibit this feature (Wingstrand, 1985; Yochelson, 1978). Soft parts show clearly a linear repetition of body parts, so characteristic of metameric animals, and many fossils clearly show the existence of various numbers of paired muscles to the shell in almost all groups of molluscs. No external demarcation between somites is present.

CLASS GASTROPODA

About 15,000 fossil and 40,000 modern species of gastropods are known. Their success need not be calculated on the basis of the number of species alone. Anyone who has visited a mudflat, where one can scarcely step without crushing some snails, or marveled at the many limpets attached to the rocks in the intertidal zone can testify to the success of individual species.

Gastropods are characterized by: a restriction of the pallial groove to the "posterior" of the body; body torsion; asymmetry of body organs; dorsoventral elongation of the body (frequently leading to a spiral body plan); and a well-developed head.

Body Plan Gastropods have two unique features. Most gastropods live with the body coiled up in a spiral. In addition, while they are embryonic, the

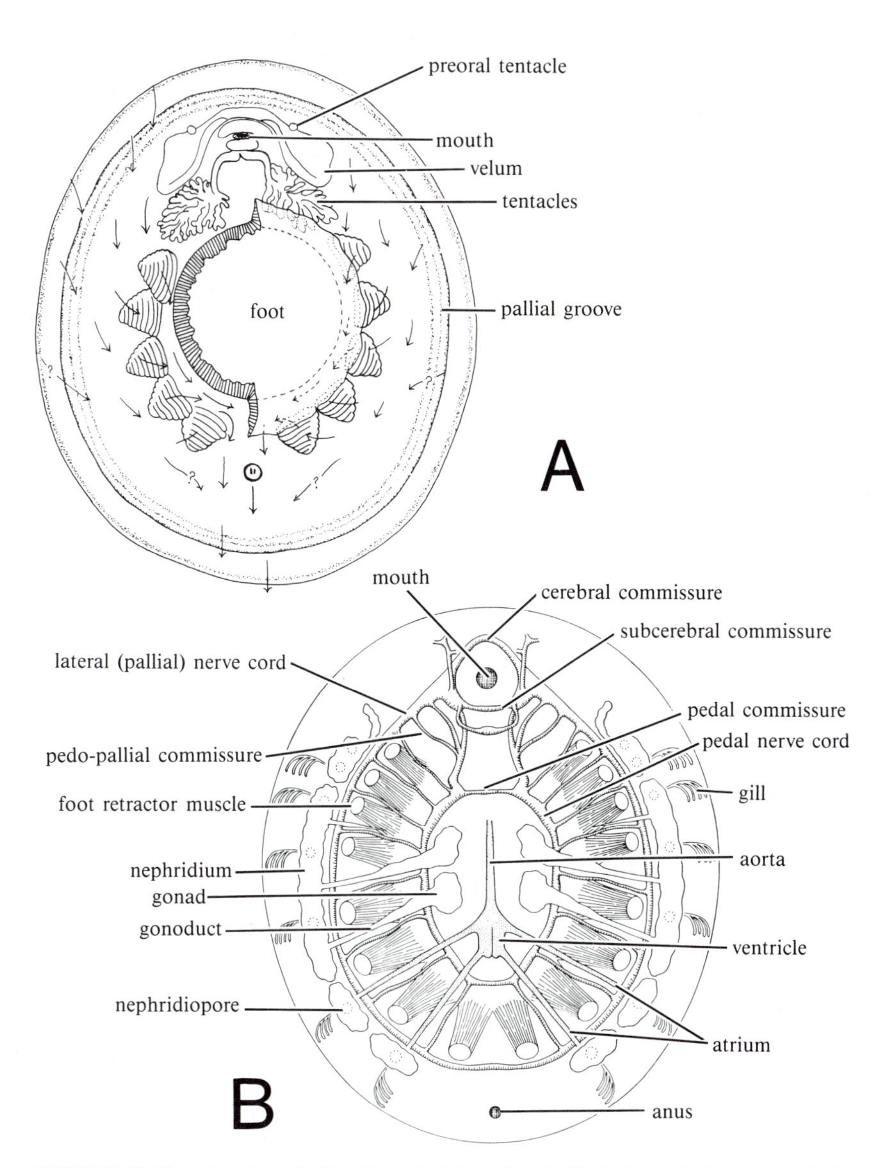

FIGURE 21.12. Monoplacophoran. **A.** *Neopilina,* ventral view, with part of foot cut away to show course of water currents indicated by *arrows.* **B.** Diagram of internal organization of *Vema.* (**A** after Lemche and Wingstrand, in Grassé.)

visceral mass is suddenly torted or twisted about, bringing the back end to the front over the head. Most gastropods remain permanently torted, and while a few partially untort during later growth, all show its results.

The gastropod shell grows with the body. New shell is laid down at the aperture and reflects the growing size of the body. If the body stops growing or grows more slowly, the shell also usually stops or slows its growth. If a snail hibernates or is stunted by a bad environment, shell growth slows down or stops. If the shell were unwound, it would often be a simple cone with a straight edge, or in species where the body grows more rapidly than the shell, the uncoiled shell would flare out like a trumpet. Shells of conical gastropods are of both types.

Modern gastropods do not have conical shells, but spiraled ones. **Spirals** are generated from a modification of circular movement about an apex or central point. Two general cases can be distinguished. (1) The moving point may move away from the center point at a constant rate. In this instance, each new whorl of the spiral is the same breadth as the preceding whorl. (2) The moving point may accelerate as it moves. In this case, each new whorl will be broader than the last if the acceleration is positive, or narrower if the acceleration is negative. The rate of movement of the moving point away from the center is, therefore, an important factor in determining the form of the spiral. Another factor is the direction of the movement of the moving point. The point may move away from the center in a straight line, remaining in the same plane. This simple outward movement results in a flat, plane spiral (Fig. 21.13A), with each whorl the same breadth if it moves at a constant rate, or each new whorl broader if it accelerates as it moves. Plane spirals with exactly equal whorls are never found, although some living and many fossil snail shells approach this form. Plane spirals with the whorls becoming larger with each revolution are common, but if the acceleration is great, the shell form is greatly modified, as in *Haliotis*. The point might move downward in a straight line, generating a tubular, helical spiral (Fig. 21.13L). If the moving point does not accelerate, a springlike form results, and if it accelerates, the whorls become farther and farther apart, so that they appear twisted rather than spiraled. This type of spiral is never perfectly achieved by gastropods, although some shells approximate it and some fossil cephalopods virtually achieved it. In most cases, the moving point moves both outward and downward and generates a conical spiral (Fig. 21.13D). If the rate of movement outward and down is constant, or if the rates remain proportionate throughout, a straight-sided spiral shell of conical form results. In many cases, however, movement outward and downward are not proportionate because of acceleration in one direction or unequal acceleration in the two directions. If the outward movement accelerates more rapidly than the downward movement, a conical spiral with concave sides results (Fig. 21.13J). If the downward movement accelerates more rapidly than the outward movement, a conical spiral with convex sides results (Fig. 21.13G). To achieve this kind of structure, right and left sides must grow unequally and in such a manner that a compact and manageable spiral results. An infinite variety of shell shapes results from slight changes in the rate and direction of growth. If the growth of the whorl diameter does not keep pace with the outward and downward coiling, the shell falls apart as a spiral and becomes so unwieldy that the animal must, perforce, adapt its habits to its house. A few species have survived, despite such abnormalities (Fig. 21.13M,N,O).

The conical spiral shell is more compact than a plane spiral shell, and mechanical advantages have apparently outweighed the disadvantages. At no time in geological history have plane spiral shells disappeared, but the conical spiral appeared early in gastropod history and is the predominant snail form today.

The other big event in gastropod history was the appearance of **torsion** (Linsley, 1978b). Torsion happens to modern gastropods when they reach the **veliger stage**. By this time the mantle has appeared, and a broad, ciliated region, the **velum**, forms the an-

Table 21.2. Metamerism of monoplacophorans *Neopilina galatheae* (x) and *Vema ewingi* (*) plotted with regard to the muscle sectors (A through H).

	Sectors							
Structure	**A**	**B**	**C**	**D**	**E**	**F**	**G**	**H**
Pedal retractors	x *	x *	x *	x *	x *	x *	x *	x *
Nerve connectives	x *	x *	x *	x *	x *	x *	x *	x *
Nephridiopores	*	x *	x *	x *	x *	x *	x *	
Gonoducts			*	x *	x *			
Gills		*	x *	x *	x *	x *	x *	
Atria						x *	x *	

Modified from Wingstrand, 1985.

terior end of the embryo. The body is curved, with both head and foot protruding from the mantle (Fig. 21.14D,E). The visceral hump has already begun to form a spiral. The whole visceral mass revolves 180 degrees (Fig. 21.14F,G). It may take a few minutes or far longer.

The results of torsion are drastic (Fig. 21.14A,B,C). Before torsion, the mantle cavity with its gills and anus is posterior, but after torsion it is anterior. Before torsion, the visceral nerves and ganglia form a simple loop, but after torsion they are twisted into a figure eight. As the twist is counterclockwise, the left visceral nerve passes dorsally to the right. As the twist occurs between the pleural and parietal ganglia, the pleural ganglia are not affected, but the left parietal is carried upward to become the **supraparietal ganglion**, while the right one is carried ventrally to become the **infraparietal ganglion**.

There are advantages to this peculiar arrangement. It may be that torsion is a mechanical side effect of forming conical spirals, but this idea is not

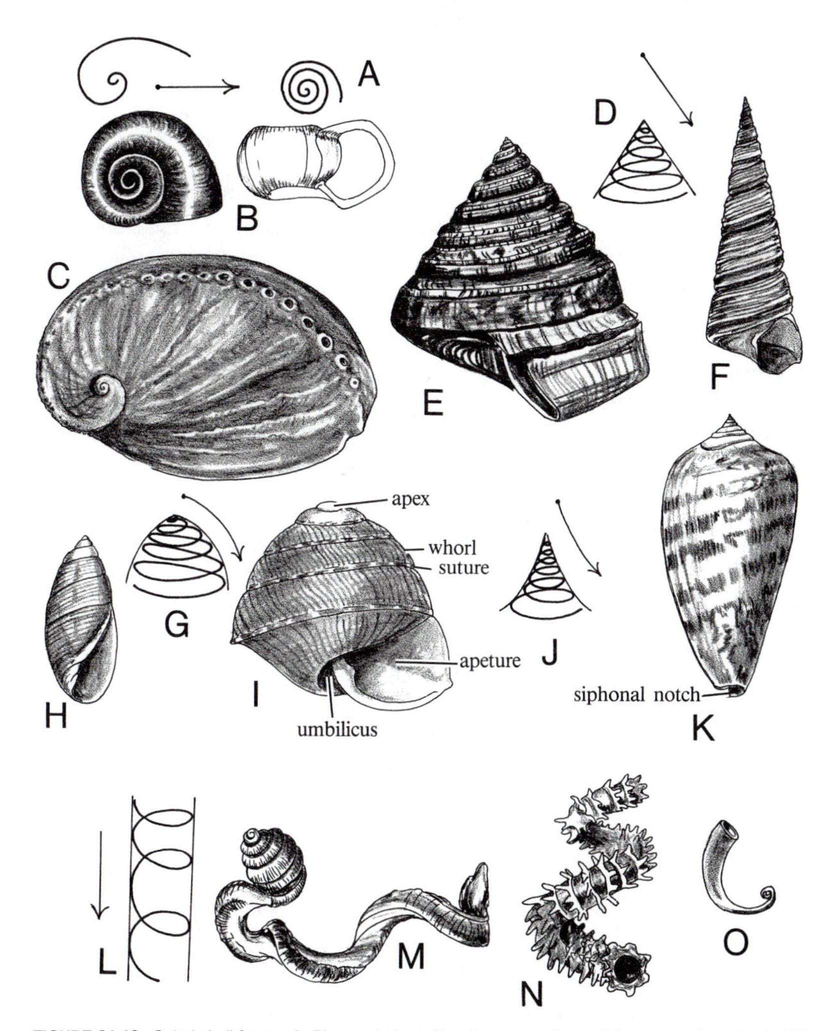

FIGURE 21.13. Spiral shell forms. **A.** Plane spiral results when generating point moves outward only. If it moves outward at nearly constant rate, each whorl is of nearly same diameter, as in *Heliosoma* **B**; if it accelerates rapidly, great increase in whorl breadth results, as in *Haliotis* **C**. **D.** If generating point moves downward as it moves outward, conical spiral results. Such straight-sided conical spirals relatively low and broad, as in *Pleurotomaria* **E**, or high and narrow, as in *Turitella* **F**, depending on ratio of outward to downward movement. **G.** If outward movement of generating point decelerates, a convex-sided conical spiral results, as in *Pupa* **H** and *Laoma* **I**. **J.** If outward movement accelerates, concave-sided conical spiral results, as in *Conus* **K**. **L.** Acceleration of downward movement of generating point results in whorls losing contact, as in *Siliquaria* **M** or *Cyclosaurus* **N**. Plane spirals may also fall apart if whorls do not grow in diameter as rapidly as they grow outward, as in *Cylindrella* **O**. (**B** after Walker, in Edmondson; **H,I** after Suter; **F,M,N,O** after Cooke.)

persuasive, since the first group of gastropods to undergo torsion were predominantly equipped with plane spiral shells. It has been suggested that the veliger larva gains by torsion, because after torsion it can pull the velum into the shell and fall to the bottom if pursued by enemies. Molluscan cilia, however, are generally under nerve control; therefore, the velum need not be retracted to stop swimming movements. Furthermore, torsion does not occur in other classes of molluscs whose larvae have no trouble surviving. The mantle cavity is quite plastic, and a far simpler solution would have been to expand the mantle cavity a little to accommodate the velum.

Torsion brings the mantle cavity to the front of the body. As a result, water currents generated by forward locomotion tend to bring water into the mantle cavity, while such locomotion works against water entering when the mantle cavity is behind. This would be especially important in freshwater gastropods with the habit of moving against the current in a stream. There are serious drawbacks to this idea also. Most gastropods move so slowly that currents set up by their own travels are not a very important factor. Besides, many snails have a narrow, tubular siphon for drawing water into the mantle cavity. In these snails, the opening of the mantle cavity is smaller, and any advantage coming from its anterior position is lost. As it stands, there is no entirely satisfactory explanation for torsion.

Be that as it may, torsion has created problems. The most obvious disadvantage of torsion is that the anus is brought directly over the head and threatens to foul the respiratory organs and the head sense organs. It is not surprising to find that the first adaptational trend evoked by torsion involved attempts to get the "sewer away from the front door." The anus is bent upward in a tube that fits into a special notch in the shell. As the shell grows, the anal notch becomes a slit band. The most ancient prosobranchs were often flat spirals with conspicuous slit bands (Fig. 21.15F), but then as now, there was considerable variation in shell form. In modern species with arrangements of this kind, respiratory currents enter the mantle cavity from in front, flow up between the gill filaments, and depart dorsally, past the anus, carrying away the fecal material (Fig. 21.15A). The site of the anus varies considerably in these gastropods. The anus may change position as the animal grows, fitting into a rather small notch at the margin of the shell (Fig. 21.15E), or it may remain in the original site, with the slit band of the shell eventually sealing up as the animal grows, leaving the anus centrally placed (Fig. 21.15B,C,D), as in the keyhole limpets. *Haliotis*, the abalone, has a series of openings in the selenizone (Voltzow, 1983).

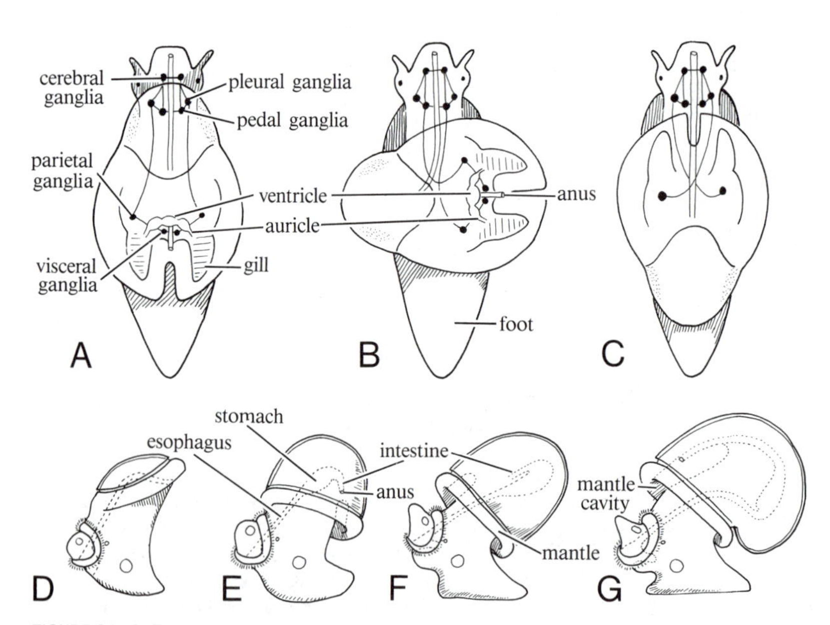

FIGURE 21.14. Torsion. **A,B,C.** Scheme of organization of gastropod before, during, and after torsion. Note, as mantle cavity rotates forward, visceral nerve trunks twist and position of heart reverses. General organization at **C** is that of two-gilled prosobranchs belonging to Archeogastropoda. **D,E,F,G.** Side view of torsion of *Paludina vivipara*. (All after Naef.)

Not all of the gastropods, however, have a perforate shell. Many modern species have a solid shell and exhibit another solution to the problem of feces disposal. The gill, the auricle, and the kidney on the larval left (adult right) side are reduced and disappear. This is linked with a new routing of water through the mantle cavity (Fig. 21.17D). Water enters on the left, first flows over the persisting gill, and then passes the anus and the nephridiopore on its way out of the mantle cavity to sweep wastes away.

Reduction and loss of the organs on the larval left side occur among some of the archeogastropods. For example, the limpet *Acmea* has a single gill, with water currents passing over the gill before reaching the anus and the nephridiopore. The shell is imperforate, with neither a slit nor a hole for the anus. Some limpets have gone even farther, developing pallial gills in a secondary mantle cavity like the mantle cavity of chitons. Respiratory currents in these limpets are like those of chitons. Others have two gills, but with the right gill smaller, as in *Haliotis*. All of the archeogastropods, however, have two auricles. The auricle associated with the reduced gill is smaller, and where the gill has disappeared completely, the auricle is rudimentary.

Most gastropods with a single gill have also lost the auricle associated with it. They belong to the Mesogastropoda, the largest gastropod order. Opisthobranchs and pulmonates have been derived from the mesogastropods. As a part of the remarkable adaptive radiation of mesogastropods and the establishment of the opisthobranch and pulmonate stems, several important tendencies have appeared.

1. Many gastropods have further reduced the gill surface. The remaining left ctenidium becomes one-sided, is fused into the mantle wall, and eventually may be reduced to a vascular area in the mantle. This vascularized wall of the mantle cavity then serves as the respiratory surface, acting as a lung.

2. Some gastropods undergo detorsion. This movement occurs during development, and although it does not wholly eliminate the results of the original torsion, it greatly reduces them. Gastropods that have followed this line have no shells or reduced shells, and some have wholly lost the mantle cavity. In this case, the body surface is the site of respiratory exchange and may bear simple or complex secondary gills.

3. A strong tendency to untwist the nervous system is evident. Untwisting grows naturally out of detorsion, but is also achieved by the centralization of the nervous system and the shortening of the visceral nerves.

The gastropod digestive system (Fig. 21.16) includes the foregut with the buccal cavity, the radula apparatus, and the esophagus. The salivary glands opening into the buccal cavity are somewhat variable, but by no means as much so as the esophageal glands. The esophageal glands are lacking in pulmonates and opisthobranchs, and in many prosobranchs as well.

The esophagus not uncommonly contains a crop, especially in opisthobranchs and pulmonates. Food is not only stored in the crop but is partially digested, and in some opisthobranchs two crops are present, separated by a muscular gizzard used to fragment the food mechanically.

Three distinctive stomach areas can be distinguished (see Fig. 21.5). The esophagus enters into a more or less globular region in which a sorting area

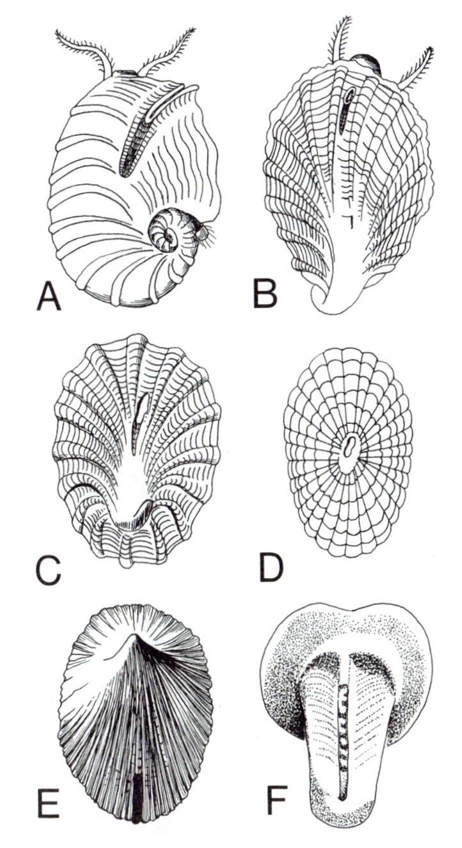

FIGURE 21.15. A,B,C,D. Four stages in metamorphosis of keyhold limpet *Fissurella.* Begins with spiral shell and slit-band **A**. New shell laid down at margin and slit-band encircled **B**, eventually coming to occupy central position **C**. Spire deteriorates, becoming callosity near aperture **D**. **E**. *Emarginula,* limpet with marginal slit. **F**. Internal cast of *Salpingostoma* shell, fossil form with partially filled-in slit-band. (**A,B,C,D** after Boutan; **E** after Cooke; **F** after Ulrich and Scofield.)

and a grinding area can be recognized. The sorting area consists of ciliated, grooved, and folded surfaces that in primitive forms extend into a long **caecum**. The grinding area is marked by hard, chitinous **gastric shields**. The chitinous lining was undoubtedly protective originally, but in many forms has been developed to reduce moderately sized particles or erode the surface of a crystalline style. The third region is a conical or tubular **style sac** that extends to the entrance of the intestine.

Torsion brings the mantle cavity forward, placing paired ctenida over the head, with the anus situated between the gills. This arrangement (Fig. 21.17A) is characteristic of the dibranchiate archeogastropods. In other archeogastropods, the right ctenidium is lost (Fig. 21.17C). These monobranchiate archeogastropods, however, retain some traces of the right auricle (Fig. 21.17B). The mesogastropods have a similar gill arrangement but have lost all evidence of the right auricle (Fig. 21.17D).

Next, the left ctenidium is reduced. It becomes attached along its whole axis to the mantle wall. In the primitive ctenidium, gill filaments are free of the mantle wall and occur on both sides of the axis. In gill reduction, the filaments, like the axis, become attached to the mantle wall, and the filaments on one side of the axis disappear entirely. The one-sided, adherent gill is termed **monopectinate**. The final stages of gill reduction involve the disappearance of the remaining gill filaments, with the mantle wall coming to contain the vascular elements previously contained in the gill filaments. The mantle wall functions as an air- or water-breathing lung. This arrangement is characteristic of the pulmonates, but also occurs in some prosobranchs.

Many mesogastropods living in the intertidal zone, alternately submerged in water and exposed to air, have greatly reduced left ctenidia, and in some the left ctenidium is wholly lost. In these forms a true lung, like the pulmonate lung, has developed. This has occurred independently in several prosobranch families. Prosobranchs with a lung can be recognized by the large opening into the mantle cavity, contrasting with the narrow opening, the pneumostome, characteristic of pulmonates.

As detorsion and shell reduction occur, opisthobranchs tend to lose their mantle cavity. Some have a single reduced ctenidium and a single auricle, demonstrating that they have arisen from prosobranch ancestral stocks (Fig. 21.17E,F). Many, how-

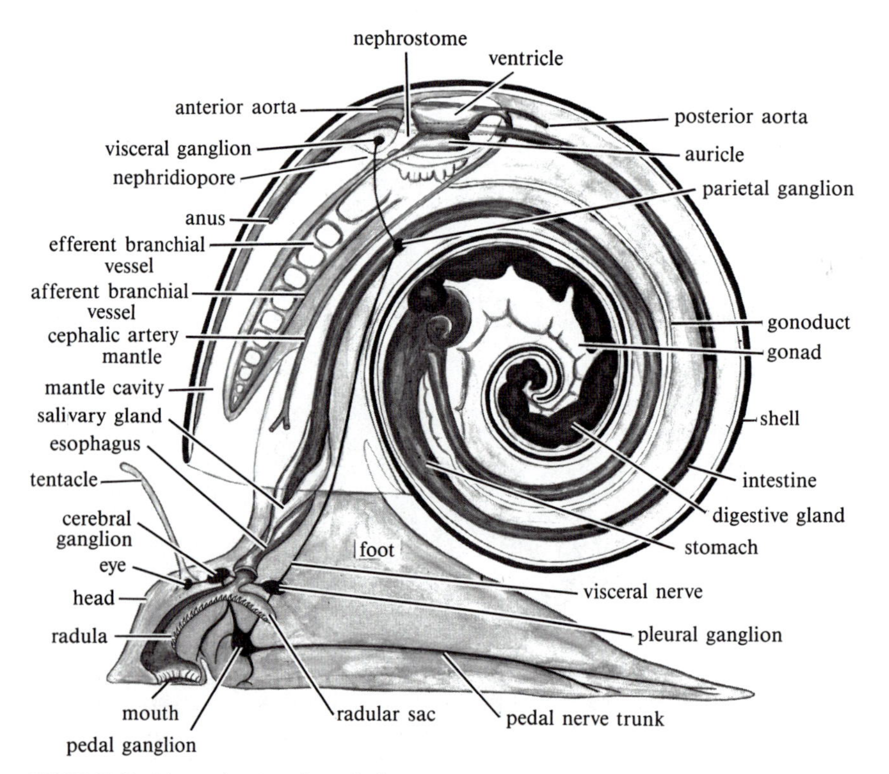

FIGURE 21.16. Scheme of gastropod organization.

ever, have no ctenidium, depending on the body surface, or secondary gills derived from the body surface for respiratory exchange.

Pulmonates have an anterior mantle cavity with highly vascular walls. Blood returns to a single auricle, giving evidence of their descent from monobranchiate prosobranchs (see Fig. 21.18). The mantle is closely pressed to the right side of the foot, leaving only a small opening, the pneumostome, through which air enters and leaves (see Fig. 21.19).

Primitive gastropods have a pair of U-shaped renal organs open that to the pericardial cavity. The right one is also used for the release of gametes. Most of the archeogastropods retain a pair of kidneys, but in all other gastropods the right kidney is sharply reduced, remaining only in the form of a **gonopericardial canal**. This canal connects the gonoduct and the pericardial cavity, or has become an integral part of the gonoduct itself.

The arrangement of the primitive circulatory system has been described. Paired ctenidia, located in a posterior mantle cavity, drain into a pair of auricles, and an anteriorly directed aorta leads from the ventricle. After torsion, however, the auricles lie in front of the ventricles, and the aorta extends posteriorly, as in dibranchiate archeogastropods (Fig. 21.17A,B). As the right ctenidium dwindles and is eventually lost, the right auricle dwindles with it, although the monobranchiate archeogastropods retain some vestiges of a right auricle (Fig. 21.17C). In mesogastropods, however, all traces of the right auricle have been lost. The left auricle drains blood from the left ctenidium, and the heart retains its position, with the ventricle discharging into a posteriorly di-

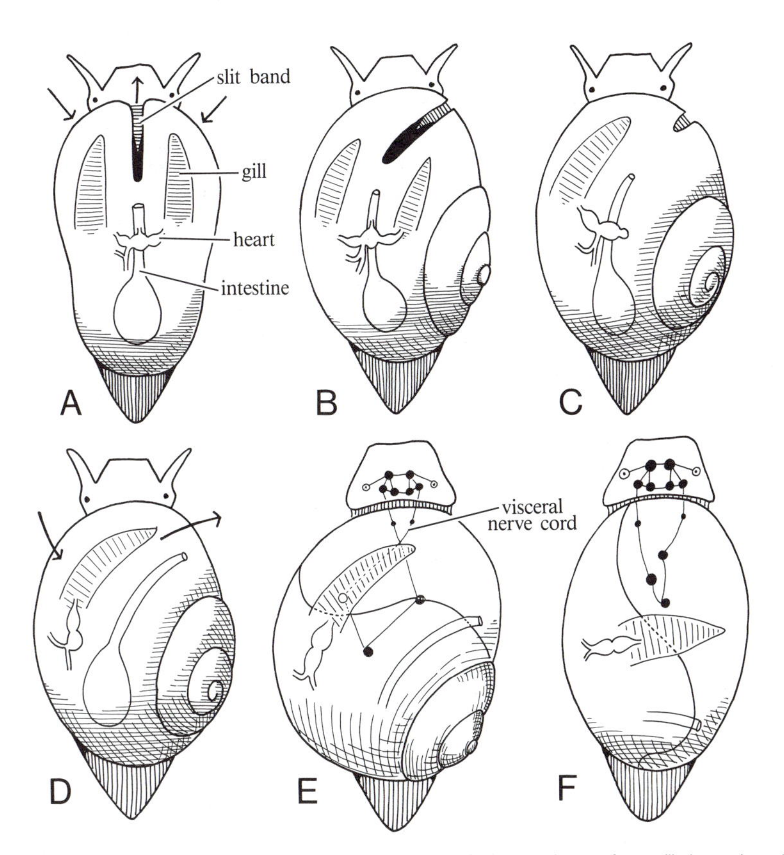

FIGURE 21.17. Gill reduction and mantle cavity currents. **A.** Ancestral type of two-gilled prosobranch, with slit-band and plane spiral shell. Water enters mantle cavity from both sides in front, and discharged upward and forward. **B.** Archeogastropod with slit-band and unequal gills. Right gill smaller, and shell conical. Water emerging from mantle cavity no longer passes directly over head. This type of organization seen in *Pleurotomaria*. **C.** Archeograstropod with reduced slit-band and single gill. Right gill completely missing and right auricle greatly reduced, although still present as in *Trochus*. **D.** Mesogastropod. Slit-band gone, and heart without trace of right auricle, more asymmetrically placed. Water enters only from left side and emerges on right, as in *Paludina*. **E.** Organism essentially like (**D**), showing advanced type of nervous system, as thought to have been present in ancestral forms leading to opisthobranchs. **F.** Opisthobranch, showing effects of detorsion. Shell opening now toward right, and nervous system detorted, as in *Bulla*. (All after Naef.)

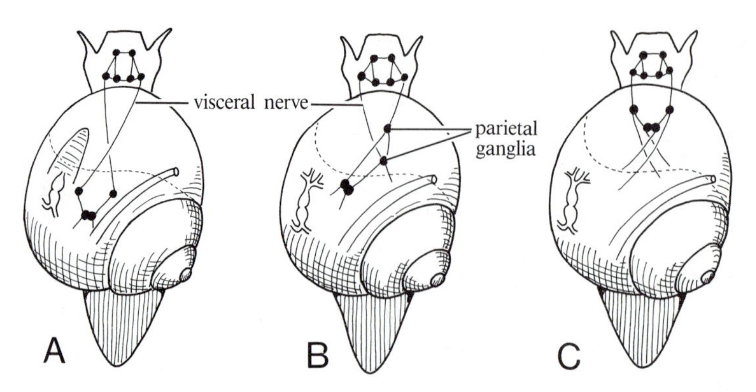

FIGURE 21.18. Modifications involved in pulmonate development. **A**. Pulmonates may have arisen from typical one-gilled prosobranchs. **B**. Loss of gill, coupled with vascularization of mantle cavity wall, converts mantle cavity into lung. Lung vessels have same relationship to heart as gill vessels. Meanwhile, visceral nerve somewhat shortened, partly untwisting visceral trunks. **C**. Further shortening of visceral nerve trunks results in straightening nervous system, resulting in short-looped condition characteristic of many pulmonates. (After Naef.)

rected aorta (Fig. 21.17D). This arrangement is retained in the pulmonates, although the auricle receives blood from the vascular mantle wall rather than a ctenidium (Fig. 21.18). Among opisthobranchs, detorsion returns the auricle to a posterior position, but they have a single auricle and have ev-

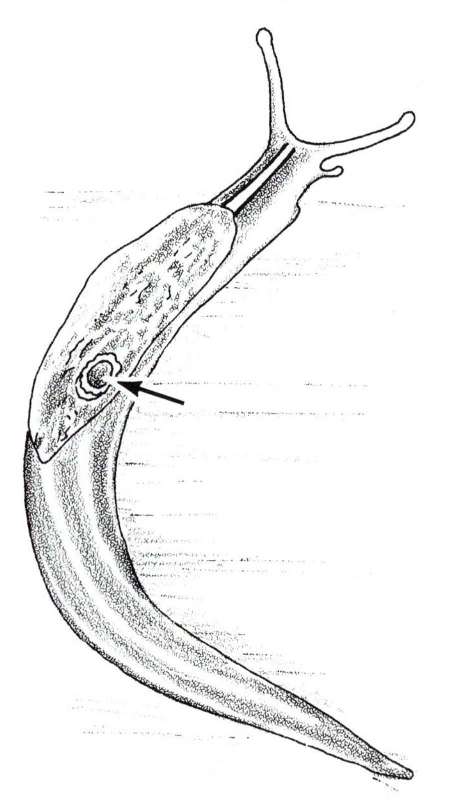

FIGURE 21.19. *Limax*, a typical slug. Mantle becomes more apparent with shell reduction. Note opening (*arrow*), pneumostome, permitting ventilation of mantle cavity, which serves as lung.

idently arisen from mesogastropod stocks (Fig. 21.17F).

It is easier to understand the form of the gastropod nervous system before torsion (Fig. 21.20C). Three pairs of ganglia lie near the esophagus. The **cerebral ganglia** lie above the esophagus, the **pedal ganglia** lie below it in the anterior midline of the foot, and the **pleural ganglia** are more laterally placed. Cerebropedal, cerebropleural, and pleuropedal connectives link the ganglia, forming a **neural triangle** on each side. A pair of small buccal ganglia are attached by buccal connectives to the cerebral ganglia. A pair of **pedal nerves** arise in the pedal ganglia and extend back in the foot. They are connected by cross-commissures and are ganglionated at the junction of commissures and nerves. A pair of **palliovisceral nerves** arise from the pleural ganglia and extend back into the visceral hump. Typically, a **parietal ganglion** is found on each of the palliovisceral nerves, and each nerve ends in a **visceral ganglion**. Cross-commissures connect the two cerebral ganglia, the two pedal ganglia, and the two visceral ganglia. The connectives of commissures can become so shortened as to bring the ganglia so close together that they effectively form a "brain" (Fig. 21.20B).

Torsion brings the visceral ganglia forward, twisting the palliovisceral (also known as the visceral) nerves between the pleural and parietal ganglia, but does not affect the rest of the nervous system (Fig. 21.20D). The visceral and parietal ganglia on the larval (pretorsion) right are brought to the left side of the adult. The adult left parietal ganglion is higher than the right and is the supraparietal ganglion; the right parietal is the infraparietal ganglion.

A nervous system twisted in this manner is characteristic of the more primitive of the modern gastropods. In many gastropods, however, a more or less complete return to bilateral symmetry of the nervous system is achieved by one of two methods: (1) detorsion, which untwists the visceral nerve and brings the parietal ganglia back to their original position and (2) shortening of the connectives between the pleural and parietal ganglia, which hauls the parietal ganglia back into a symmetrical position.

Most prosobranchs are dioecious, but sexual dimorphism is rarely very conspicuous. Males are sometimes smaller than females, and may have other peculiarities. All gastropods have a single gonad, usually located high in the visceral mass or at the apex (see Fig. 21.16). Gametes of archeogastropods are usually discharged into a short **gonoduct** that opens into the right nephridium by one or several pores. The gametes depart through the **right nephridiopore**, and fertilization is external. Two factors are important in further differentiation of the prosobranch genital system. The right nephridium is reduced, the vestiges becoming incorporated in the gonoduct system, and mating occurs, involving the differentiation of parts used for copulation. As the right nephridium is reduced, its renopericardial pore and the upper part of the tubule become a canal and pore opening into the pericardial cavity, the **gonopericardial canal**. A part of the gonoduct is derived from the rest of the nephridium, ending at the old nephridiopore. All of the organs used for copulation and some of the glandular parts of the gonoduct are beyond the nephridial part of the gonduct. This part of the gonoduct is termed the **pallial gonoduct** and is at first no more than a groove across the mantle, later closing over and becoming a tube.

The female system (see Fig. 21.21A,B,C) is derived in a similar manner, but the gonopore usually lies near the opening of the mantle cavity, and the last part of the oviduct is a pallial canal. The original oviduct is a short upper region, usually convoluted. Beyond this is a short nephridial region that may be connected to the pericardial cavity by a gonopericardial canal. The pallial oviduct contains the conspicuous glandular regions responsible for the secretion of the albumin and the capsules that cover the ovum. The albumin gland is smaller and more proximally located. A diverticulum, the seminal receptacle, usually lies near the albumin gland. In many cases, a copulatory bursa located just beyond the capsule gland first receives the sperm at mating. In this case, the sperm later migrate to the seminal receptacle for storage.

The male system is simple (Fig. 21.21D,E,F). Sperm may be stored before mating in the convoluted gonoduct, derived from the nephridial gonoduct. In many prosobranchs, the gonopore occurs at the end of this part of the gonoduct, although a **penis** is present, separated by some distance from the gonopore. In this case, a ciliated groove connects the gonopore and the penis. Sperm pass along the groove at the time of mating. More commonly,

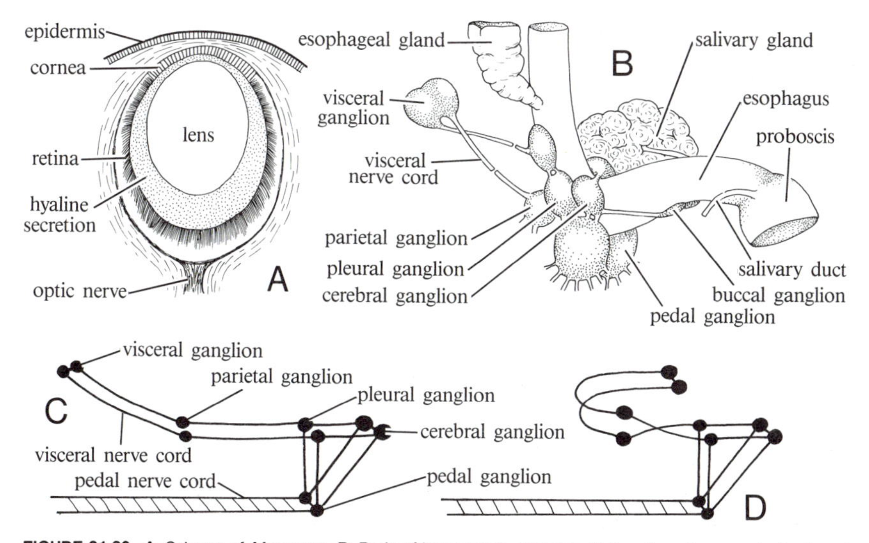

FIGURE 21.20. **A.** Scheme of *Murex* eye. **B.** Brain of large conch, *Busycon.* Note rather strong centralization of ganglia in brain region. **C,D.** Scheme of arrangement of nervous system before and after torsion. Twisting occurs between pleural and parietal ganglia, bringing right parietal above and left parietal below. Results in pretorsion right visceral nerve cord lying above and pretorsion left visceral nerve cord below digestive tract. (**A** after Hesse; **B** after Pierce, in Brown.)

however, the groove is roofed over to form a pallial gonoduct, and the penis is hollow instead of grooved.

Biology

Gastropods have undergone a remarkable adaptive radiation. Some are pelagic and others benthic; they have remarkably diversified food habits and have spread to all kinds of marine and freshwater habitats. They are one of the few invertebrate groups to successfully invade land. The group is so varied that only a few of the most important points can be mentioned here.

Shell and External Form. Shells present some interesting problems in biomechanics. Most gastropods have a dextral shell (coiled clockwise), but some have a sinistral shell (coiled counterclockwise) and a few species may have either kind of shell.

Limpets, ear shells, and slipper shells are flat; this form is characteristic of sedentary species that cling to rocks. Some flattened shells consist of a few very broad whorls, but others have lost their spiral character and are cup-shaped, resembling the cuplike protogastropods. All modern species with this kind of shell, however, begin life as definitely spiraled, snaillike juveniles, losing the spiral form as new shell

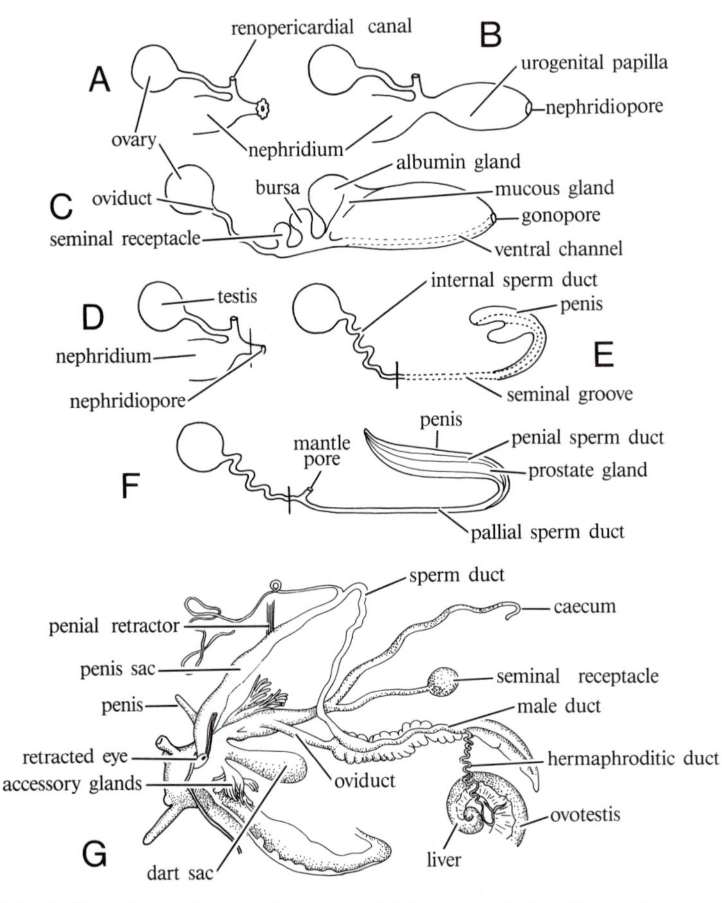

FIGURE 21.21. Gastropod reproductive systems. **A,B,C.** Female prosobranchs; system varies from simple, composed of gonoduct proper and parts of nephridium, to complex, which includes glands and other parts derived from wall of mantle cavity. **A.** Primitive system, as found in *Gibbula* and *Monodonta.* Oviduct joins renopericardial canal, and ova are discharged through nephriodiopore. **B.** Illustrating condition in *Calliostoma.* Urogenital papilla extends into mantle cavity with nephridiopore at tip. **C.** Mantle or pallial part of oviduct very complex in *Potamopyrgous,* with appearance of organs for sperm storage, and secretion of albumen and capsule. **D,E,F.** Male systems. **D.** Primitive relationships, as in *Calliostoma.* Only internal duct systems present, and sperm discharged through nephridiopore. **E.** *Calyptraea,* with penis developed on body surface. Sperm reach it by way of groove that passes across mantle cavity and along penile surface. **F.** Roofing-over of groove provides pallial portion of sperm duct system. **G.** Complex hermaphroditic system as in typical pulmonate, *Helix.* (**A,B,C,D,E,F** after Fretter and Graham; **G** somewhat modified, after Cooke.)

is added during growth (see Fig. 21.15B,C,D). When the shell is cup-shaped, the body is less firmly attached, for the attaching muscle must hold it without the aid of a spiral columella. Such shells may have an inner partition of variable shape (Fig. 21.22A,B), which holds the body firmly in the uncoiled shell. The attachment muscles are then somewhat reduced.

The shell provides protection, but at the expense of adding weight. A strong tendency toward the reduction of unnecessary shell weight is seen in many gastropods. Opisthobranchs show this most clearly. The shell is reduced in association with detorsion, and the amount of detorsion is usually correlated with the amount of shell reduction. In some cases, the shell is completely lost. The shell is a greater burden on land, and land gastropods usually have a very thin shell. The land slugs probably represent a final step in an adaptational line tending toward reduced shell weight. Even creeping aquatic snails sometimes have methods of reducing shell weight or making it less awkward to handle. The visceral hump sometimes withdraws from the uppermost whorls, which are broken off or drop off. Neritas, cowries, and olive shells reduce weight by reabsorption of the inner shell partitions. This has an added advantage, for the tight coils of the viscera relax as they conform to more open inner space.

Unless the shell is considerably reduced, the visceral hump is completely enclosed, and all that can be seen is the head and foot (see Fig. 21.16). The head is generally well developed and is distinctive. The mouth is more or less terminal, and nearly all gastropods have a pair of **tentacles**, primitively containing eyes near their bases. Land pulmonates have two pairs of tentacles, and some opisthobranchs have lost their tentacles; in some, the tentacles are fused together to form cephalic shields. The tentacles are important as chemoreceptors. In some opisthobranchs, a second pair of tentacles (**rhinophores**) have a large chemoreceptive region. The head is less developed in sedentary species.

The foot is a ventral creeping organ in its primitive form. In the foot, muscle strands run in all directions, permitting a variety of movements. A rich supply of mucous glands in the foot provide slime for lubrication. Prosobranchs and opisthobranchs usually have a large, anterior mucous gland, which empties into a median furrow. Mucus spreads over the foot surface from the furrow. Pulmonates have a dorsal mucous gland, which opens above the anterior border of the foot. Some terrestrial pulmonates also have mucous glands at the posterior end of the foot. The mucus produced is different in quality, hardening on contact with air or water. Some snails use this mucus to spin a filament from which the animal can suspend itself. The pelagic snail, *Janthina*, uses the secretions of the foot gland to make a bubble-filled float for itself and its eggs (Fig. 21.22C).

Foot movements can be highly organized in some species, and specific locomotory patterns are often related to shell form (Linsley, 1978a). In *Littorina* and some land operculates, a longitudinal furrow divides the foot into functional right and left sides that move forward alternately, so the animal skates over the surface. The foot is sometimes divided into an anterior propodium, a middle mesopodium, and a

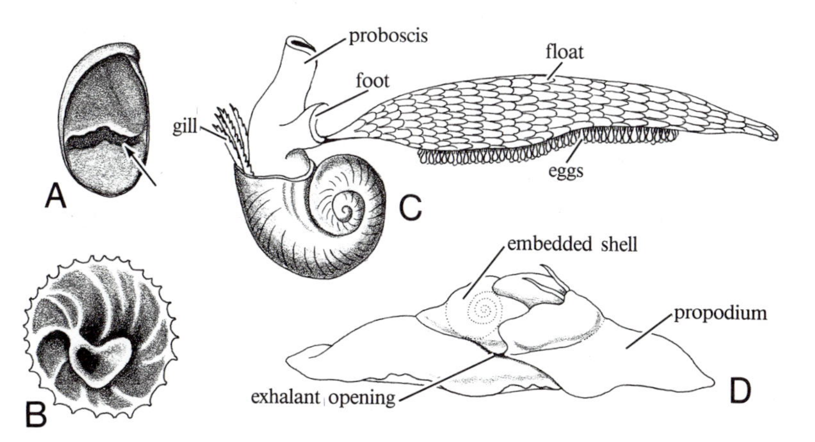

FIGURE 21.22. **A**. Inner shell of *Crepidula,* showing inner shelf (*arrow*) that aids in holding shell to body. **B**. Inner shell of *Crucibulum spinosum,* cup and saucer limpet, showing another arrangement for holding flattened shell to body. **C**. *Janthina,* with float that carries eggs, secreted by foot. **D**. *Natica* in movement, using very large foot with propodium that covers much of shell to serve as sand plow. (**C** after Quoy and Gaimard, from Cooke; **D** after Schiemenz, from Pelseneer, in Lankester.)

posterior metapodium (Fig. 21.22D). The propodium of *Melampus* is sharply set off by a deep transverse grove, and the animal moves by extending the propodium, attaching it, and pulling the rest of the foot after it. The propodium of species living in soft sand (e.g., *Natica*) is usually much enlarged and serves as a sand plow. The rest of the foot is very broad, and the animal slides over the sand without sinking in.

Usually, the mesopodium is not separated from the metapodium by external groves. The metapodium may contain special glands for the secretion of an operculum. The **operculum** is often horny but is sometimes composed of both calcareous material and conchiolin. It is a door that is brought into position in the aperture by the withdrawal of the head and foot into the shell. Prosobranchs with flattened shells, like the limpets, do not withdraw in this manner, and the operculum has lost its usefulness, often completely disappearing in adults although present in the young animal. The operculum is almost universally missing among opisthobranchs and pulmonates. The operculum grows with the organism by the addition of material at the margin, usually forming concentric or spiral growth lines. When the opercular growth lines are spiral, they invariably wind in a direction opposite to the coiling of the shell. Nevertheless, there is a very close correlation between the rate of opercular growth and the growth of the aperture, so that the operculum fits the aperture with remarkable exactitude.

It is not uncommon for the foot to develop lateral outgrowths of one kind or another. **Parapodia** are outgrowths of the ventral margins of the foot, and **epipodia** are outgrowths of the lateral surface of the foot (Fig. 21.23A,B,C). They are especially well de-

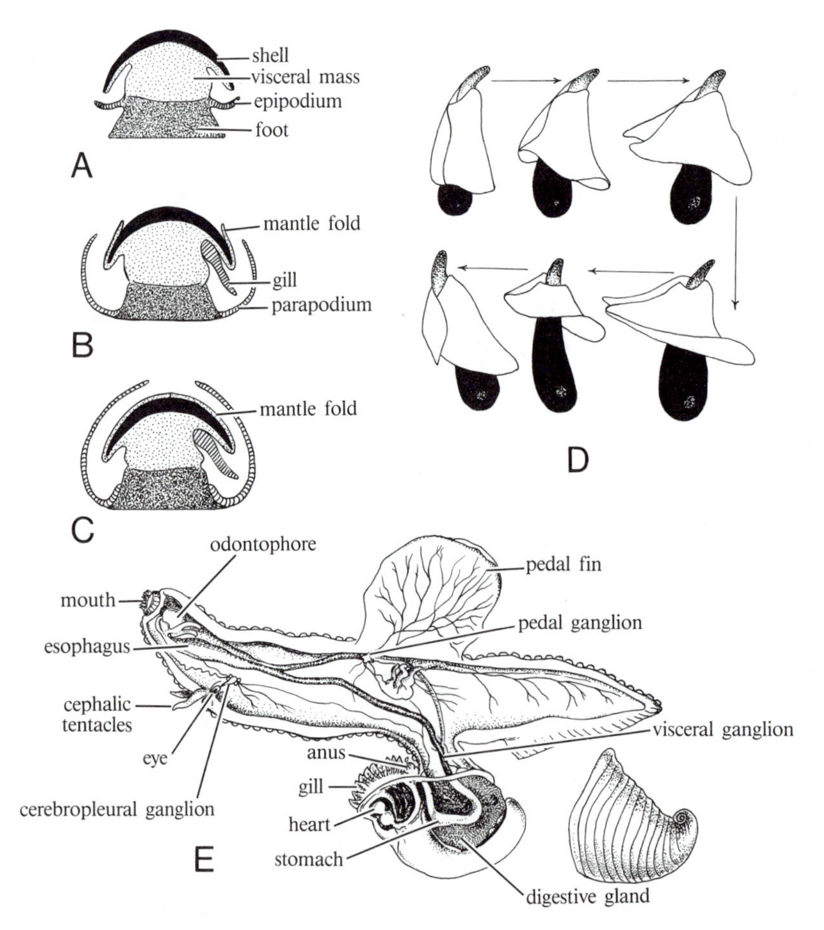

FIGURE 21.23. A,B,C. Schematic cross sections showing relation to parapodia and epipodia to foot, and enclosure of shell by mantle folds. **D.** Swimming movements of opisthobranch, *Akera bullata*. **E.** *Carinaria*, typical heteropod. Swims with foot upward, in position shown. Shell (*on the right*) removed from visceral mass. (**A,B,C** after Lang; **D** after Morton and Holme; **E** after Souleyet, from Pelseneer, in Lankester.)

veloped in opisthobranchs, where they serve as fins in swimming species (Fig. 21.23D). Heteropods are pelagic prosobranchs, whose evolutionary development has more or less paralleled that of opisthobranchs. The shell and the operculum are reduced, and the mantle is progressively reduced until in some species it is completely lost. Heteropods are characterized by varying degrees of detorsion and a tendency to return to bilateral symmetry (Fig. 21.23E). Some of the pelagic prosobranchs, like opisthobranchs, swim with the aid of epipodia.

Feeding. The great majority of gastropods are microphagous feeders, with particles being scraped or brushed from surfaces of rocks or seaweeds or sessile animals and similar objects. However, some gastropods feed on detritus, and some are scavengers (Kohn, in Wilbur, Vol. 5, 1983). The radula is important in feeding of this kind, as its form and function makes the most of abrasive or collecting properties of the teeth. In some cases, radular movements over the end of the odontophore erect the teeth to make them more effective as rasps or brushes. In other cases, radular movements cause the collapse of the teeth to grasp particles and bring them back into the mouth. In some detritus feeders, the radula serves to scoop up particles of mud and detritus for swallowing. Rasping and grasping action may be used to obtain algae, bits of food from larger plants, or dead or living animal flesh.

A number of carnivorous gastropods feeding on large prey are actually microphagous. They feed on slow-moving or sessile forms, boring through their shells or protective cover, and then thrusting the radula into the opening to feed on the unfortunate prey. Oyster drills consume enormous numbers of oysters and are a serious economic pest. Some boring gastropods apparently depend wholly on the abrasive action of the radula to penetrate the shells of their prey, as in *Natica*; however, boring gastropods are equipped with an accessory boring organ that is repeatedly applied to the region where a hole is being cut. In *Urosalpinx*, the accessory organ is repeatedly thrust into the hole, and boring is inhibited by the removal of the organ. The accessory organ may apply a shell-softening compound to the shell, preparing it for the action of the radula, but the nature of such a compound remains uncertain. An acid or chelating agent has been suggested to attack the calcareous bivalve shell. A substance that attacks conchiolin would also cause softening of the shell.

Some gastropods have developed ciliary–mucus–feeding mechanisms, and feed on suspended particles. In some cases, as in *Viviparus*, it is a relatively unimportant secondary method of food acquisition, but in *Crepidula* it is the primary one. These gastropods have modified gill structure and mantle cavity structure to develop suspension-feeding mechanisms. The lamellae of the ctenidia are elongated, more precisely defining inhalant and exhalant passageways; a mucus-secreting endostyle at the base of the gill is present; and a food groove conveys the trapped particles to the mouth region. Some gastropods have developed a special suction–feeding mechanism (Fig. 21.24A,B). Suction may be produced by the proboscis, or the buccal region may be converted into a pumping area. Although some particle feeders depend in part on suction, suction mechanisms are especially important in some opisthobranchs with radular teeth adapted for piercing plant cells, and in parasitic forms such as *Entoconcha* (Fig. 21.24G). In the latter, the radula is missing but a piercing stylet has been formed of the buccal jaws.

Many gastropods are macrophagous; they take in larger pieces of plant material or, in some cases, swallow prey organisms whole. Herbivores of this type use the radula to fragment pieces of food that are nipped off by horny jaws located on the walls of the buccal cavity. In the Conidae the radular teeth are greatly reduced to a single tooth. A poison gland, probably the homologue of esophageal glands, fills the hollow tooth with poison (Fig. 21.24D). Folding of the body wall forms a protrusible proboscis that can be shot out with great speed. A proboscis, however, is useful other than for carnivores; it can be useful to scavengers (Fig. 21.24C) as well.

Evolutionary trends are associated with the type of food and feeding, but involve a general trend toward extracellular digestion. Larger particles are taken in by many herbivores, particles that may be stored and partially digested in a crop; some herbivores, especially among the Mesogastropoda, develop a crystalline style. Reduction of the sorting area and caecum also tends to simplify the stomach area. A crystalline style is more typical of clams than of gastropods, and will be discussed in greater detail with pelecypods. As the crystalline style is proteinaceous, it is characteristic of herbivorous forms; proteases in the stomach of carnivores would tend to break down the crystalline style and make it inefficient. A few omnivorous gastropods, however, are known to have a crystalline style.

As a rule, carnivores tend to have far less complex stomachs than other gastropods. In the most advanced forms, the stomach has become a simple saccate organ, and digestion has become predominantly extracellular. It is doubtful, however, if intracellular digestion ever disappears wholly, although it is clear that it is far less consequential in the most advanced gastropods.

A remarkably wide range of carbohydrates are attacked by enzymes found in gastropods. In a number of gastropods, cellulolytic bacteria are found in the gut, and it seems evident that in at least many gastropods, cellulose digestion depends on these symbionts. In some, however (e.g., *Levantina*), elimination of the bacterial flora does not hinder digestion of cellulose. Chitinases are also found and may be produced by bacterial symbionts or by the gastropod itself.

A most remarkable adaptive line is found among the nudibranchs. Some nudibranchs have run-of-the-mill digestive glands, wholly contained within the general visceral mass. In others (Fig. 21.24E), the huge digestive glands have long diverticula that extend out into the surface gills (**cerata**). Glandular tips at the end of the diverticula secrete protease and diastase, and some intracellular digestion of glycogen may occur. These species are carnivorous, feeding in part on cnidarians, which when eaten, most of their nematocysts are digested. Some nematocysts, however, are not attacked and eventually reach the end of the diverticula. Here, they are passed to chambers at the ends of the cerata where they form a secondarily acquired protective organ for the mollusc (Fig. 21.24F). Different types of nematocysts occur and depend on prey and nudibranch species involved (Rudman, 1981).

Respiration. Gastropod lungs (Fretter and Peake, 1975) may function as diffusion lungs, with air reaching the lung surfaces by simple diffusion. It has

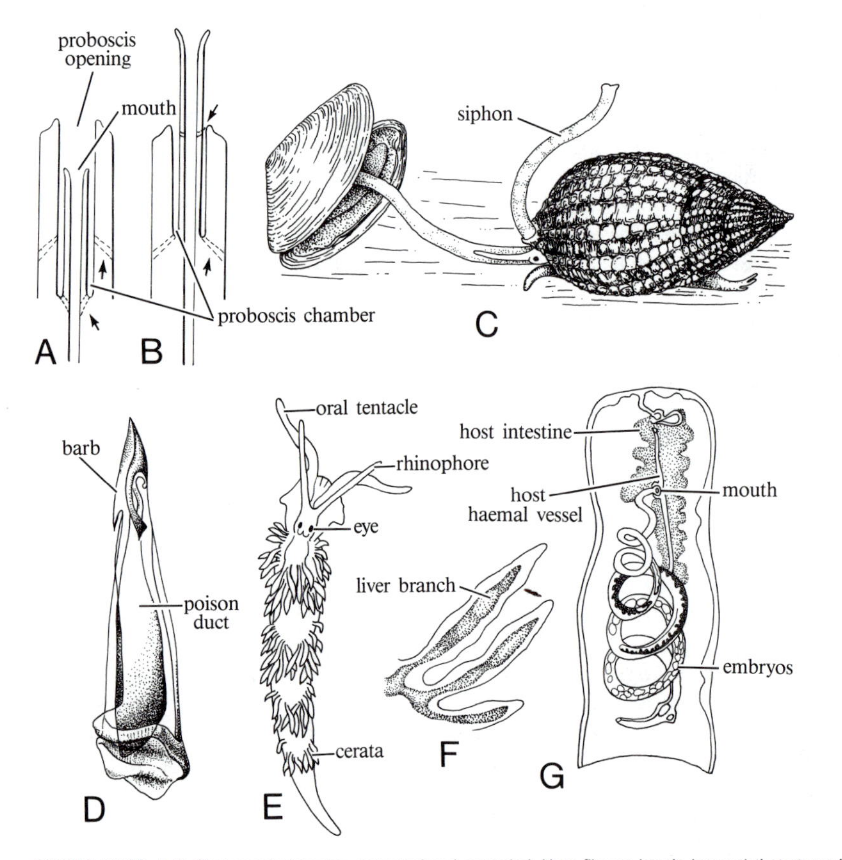

FIGURE 21.24. A,B. Gastropod proboscis, retracted and extended. Note fibrous bands (*arrows*) that stop withdrawal and extrusion at predetermined points. **C.** *Nassarius* feeding on dead bivalve, *Tellina*. Note long proboscis and nearly equally long siphon. **D.** Hollow tooth of poisonous cone shell. Poison, collected in tooth, enters prey when barb breaks off. **E.** *Eolis,* nudibranch, with cerata containing extensions of liver. **F.** Several cerata, showing liver branches extending into them. **G.** *Entoconcha mirabilis,* parasitic gastropod in position in host sea cucumber. Mouth embedded in hemal sinus on intestine of host. (**A,B** after Lankester, from Pelseneer, in Lankester; **C** after Fretter and Graham; **D** after Cooke; **E,F** after Alder and Hancock, from Pierce, in Brown; **G** after Bauer.)

been calculated that a differential of 2 mm mercury suffices to permit gaseous exchange in air-breathing pulmonates, as a large area is exposed. Some lungs, however, are ventilation lungs, with air being moved in and out in regular ventilation movements. Movements consist of opening and closing the pneumostome and flattening or raising the floor of the lung. A normal breathing cycle involves: (1) opening the pneumostome; (2) lowering the floor of the mantle cavity to draw air in; (3) closing the pneumostome; (4) relaxing muscles of the mantle cavity, reducing its volume, and putting the air under slight pressure; and (5) opening the pneumostome and permitting discharge of air. The increase in air pressure in the lung as the mantle walls relax facilitates gaseous exchange. Dilation of the pneumostome can serve to regulate respiration in a diffusion lung, but is inadequate in a ventilation lung.

The most remarkable lung development is seen in janellids. Tubular invaginations of the lung wall form trachea that extend *into* the body, penetrating the blood sinuses where oxygen absorption can occur with maximum efficiency. It is evident that the pulmonate lung developed as an adaptation to terrestrial life, but has not restricted pulmonates to land. Many have returned to fresh water, and a few to the sea. Some of these surface to breathe air; some have the pneumostome located at the end of a snorkel so air breathing can occur while the animal is submerged; still others pump water in and out of the lung. In water breathers, secondary gills have tended to develop.

Excretion. The **renopericardial canals** of archeogastropods are functional. Pericardial glands on the wall of the heart of the pericardium are usually present, and some structural differences between right and left kidney can be seen, with the right typically having a larger glandular region and close relationships between the glandular region and the blood sinuses. Mesogastropods and neogastropods have less conspicuous pericardial glands and a single kidney, the left, also with extensive glandular tissue. A study of *Haliotis* indicates that filtration occurs in the pericardial cavity, probably in the pericardial glands. The filtrate is concentrated by the secretion of materials into the tubule in the right kidney, although it appears that glucose is reabsorbed in the left. Thus, the two kidneys have different functions as well as different structures. A **nephridial gland** lies on the renopericardial canal between the kidney and the pericardial cavity, and appears to be the site of reabsorption of sugar and protein. The glandular region of the kidney is the site of active secretion into the tubule. The single pulmonate kidney is compact, with much-folded walls and a rather long tubular ureter, which apparently is important in water reabsorption. The terrestrial pulmonates have reduced openings for the renopericardial canals, perhaps to reduce the loss of water during urine formation.

When some marine gastropods are placed in dilute seawater, they promptly swell, showing very little control over water intake. *Doris*, a nudibranch, and *Onchidium*, a sluglike marine pulmonate, swell and show no sign of size regulation in 24 hours. *Aplysia* also swells in dilute seawater, but later excretes up to 37 percent of its body salt, apparently through the body surface, thus eliminating water to regain normal volume.

Water regulation by means of salt excretion is far from perfect, but may be a step toward adaptation to freshwater life. The blood concentration of freshwater snails and clams is distinctly lower than in marine species. Blood, for example, has an osmotic concentration equal to about 0.43% sodium chloride. The urine it excretes contains about 0.3% sodium chloride. Evidently salt is a precious material. Adjustment to salt lost in excretion is made possible by a remarkable capacity for selective salt absorption through the body surface.

Land snails have considerably more salts in their blood than do freshwater snails, but have poor mechanisms for controlling the concentration. A heavy rain may reduce the salt concentration of the blood by over half. Water loss resulting from excretion is held to a minimum by excretion of uric acid and by a few adaptations that reduce evaporation at the body surface. Hibernating or estivating snails, for example, secrete a partition over the shell aperture, thus reducing water loss during inactivity.

Circulatory System. Blood returning from the ctenidia, secondary gills, or lung is oxygenated, therefore, the heart is an arterial heart, pumping blood to tissues for the distribution of oxygen and pickup of carbon dioxide. The short aorta gives rise to a cephalic artery, conducting blood to the head and the foot, and a visceral artery, which empties into lacunae around the viscera. Blood collects in a system of sinuses that eventually return it to the gills or lung, generally by way of lacunae around the kidney. In some of the pulmonates, however, blood returns directly to the auricle from the kidney.

Sense Organs. Tactile cells are scattered over the body surface. They are concentrated in regions of high sensitivity, such as the head, the margin of the foot, and sometimes the edge of the mantle. The cephalic tentacles are well supplied with tactile cells, and other tactile projections appear sporadically in different groups. The most outstanding of these are the **epipodial tentacles**, found on the sides of the foot of many archeogastropods, and the **cerata**, which occur on the dorsal or lateral surfaces of many nudibranchs.

A patch of sensory cells, the **osphradium**, occurs at the base of each gill. The osphradium is often a strand of elevated sensory cells associated with the nerve to the ctenidium, but in some cases it is incorporated in the ctenidial axis. Sometimes lateral tracts of elevated cells extend on each side of a central strand, giving the osphradium a bipectinate appearance. Gastropods without ctenidia retain an osphradium if a mantle cavity and secondary gills are present. Aquatic pulmonates have an osphradium, but land pulmonates do not.

Osphradia lie in the path of inhalant respiratory currents. They are probably important in governing water flow into the mantle cavity. They may participate in chemoreception generally, but clear evidence is lacking.

Chemoreceptors are important mediators of gastropod behavior (Audesirk and Audesirk, in Wilbur, vol. 8, 1985). Some gastropods make frenzied flight responses to the tube feet of starfish. This response is sometimes initiated before contact with the tube feet is made, indicating both contact and distant chemoreception. Oriented responses to food are usually well developed in gastropods. In some species, removal of osphradia destroys the ability to detect food and to feed normally. **Rhinophores** are more sensitive to contact with acids and salts than are the oral tentacles, but there is no evidence that they are stimulated by substances not in contact with them. The oral tentacles of nudibranchs prove to be more important in contact chemoreception involved in feeding. The anterior margin of the foot is also an important site of chemoreception in many aquatic gastropods.

Hollow **statocysts**, innervated from the cerebral ganglia, containing calcareous concretions are almost universally present in gastropods. Generally, creeping gastropods have statocysts in the foot, while heteropods and most nudibranchs have the statocysts close to the cerebral ganglia.

Most gastropods have eyes at the base of the cephalic tentacles (Fig. 21.20A), converse pigment-cup ocelli. The simplest eyes are open epidermal cups, differentiated into sensory and pigment cells covered by a layer of hyaline, rod-shaped retinidia. Fibers from the sensory cells extend to the optic nerve. Further specialization of the eye, comparable to cephalopods, leads to: (1) the appearance of a crystalline lens, varying from a few hyaline cells to a complex structure covered by an inner cornea; (2) the secretion of a vitreous body in eyes containing a lens; (3) narrowing and, eventually, closing of the cup opening, with the concomitant development of transparency of the closing tissues; (4) in-sinking of the eye, accompanied by the formation of a transparent cornea from the superficial epidermis; and (5) the narrowing of the retina as the eye becomes more complex, with some histological specialization and the appearance of a limiting membrane between the retina and the vitreous body.

Some sluglike marine pulmonates have mantle eyes, located on tubercles on the dorsal surface. These eyes are independent developments. They are converse pigment-cup ocelli with a lens composed of a few hyaline cells.

No matter how the nervous system is organized, the connectives between the pedal and the cerebral ganglia are maintained whether or not they are visible from without. The cerebropedal connection is functionally important, particularly in control of locomotion. The immediate control of peristaltic creeping movements of the foot is vested in the pedal ganglia, working through fibers in the pedal nerve, and affected by stimuli from a well-developed subepidermal plexus in the foot. Stimulation of one side of the *Aplysia* foot evokes reactions on the other side. These cross-reactions are destroyed by disturbance of the connection between the two pedal ganglia, but not by destroying the commissure between the two cerebral ganglia. Evidently the reflexes that are important in operation of the foot are entirely within the pedal nervous system. If the cerebral ganglion is destroyed, the muscle tone of *Aplysia* rises, and the parapodial activity increases. Therefore, the cerebral ganglia evidently exercise an inhibitory effect on the pedal ganglia. Foot potentials of *Helix* also indicate an inhibitory effect of the cerebral ganglia on the pedal ganglion motor neurons. The cerebropedal connectives thus maintain relationships between the control centers of the cerebral ganglia and the motor centers in the pedal ganglia.

Stimuli from the main sense organs, tentacles, tentacular eyes, and statocysts go into the cerebral ganglia, which appear to be the overall coordinating centers. Motor innervation of the mantle is largely from the pleural ganglion, although the parietal is also involved. Sensory stimuli also appear to enter the pleural and parietal ganglia, although distinct osphradial ganglia, associated with the visceral nerve, are often present.

The sympathetic or stomatogastric nervous system centers in the buccal ganglia and serves the main visceral organs, heart, kidney, and reproductive organs.

Reproduction. Opisthobranchs and pulmonates are hermaphrodites (Giese and Pearse, 1977; Wilbur, vol. 7, 1984). The male and female systems are complex, with functional regions that are about the same as in the more highly differentiated prosobranch systems. The animals have only one gonad, an **ovotestis**. It opens into a **hermaphroditic duct**, usually divided by longitudinal folds into one channel for the sperm and another for the ova. The hermaphroditic duct ends in a **hermaphroditic pore** near the mantle cavity entrance on the right side of the body. The penis does not lie near the hermaphroditic pore, and a ciliated channel conveys the sperm over the body surface to the anteriorly placed penis. As in the prosobranchs, the ciliated channel tends to become roofed over, forming a definite pallial duct. As a result, there are two gonopores, a **female gonopore** at the old hermaphroditic pore, and a **male gonopore** associated with the penis. This arrangement is common in nudibranchs and aquatic pulmonates. The female gonoduct can be extended onto the pallial region. In some nudibranchs it lies near the penis. In terrestrial pulmonates, the secondary relationship of the male and female gonopores becomes more intimate. The end of the male and female ducts open into a common **genital antrum** that opens through a **common gonopore**. This common gonopore is at some distance from the site of the hermaphroditic pore of the more primitive opisthobranchs, for the lower part of both male and female tracts is pallial. In *Helix* (Fig. 21.21G), the upper part of the reproductive system is functionally, but not morphologically, divided into male and female channels. The lower part, beyond the old hermaphroditic pore, contains separate male and female duct systems, with associated glands.

A further complication is seen in some nudibranchs. A special **copulatory pore** provides an entrance into the copulatory bursa. The bursa is connected to the oviduct by an internal passageway. In such gastropods, there are three genital pores: a male gonopore associated with the penis, a female gonopore through which the ova emerge, and a copulatory pore through which sperm enter at mating.

In adapting to life in a variety of habitats, gastropods have developed many different reproductive habits and varied devices to protect the ova during early development under diverse conditions. The majority of these devices involve specializations of accessory organs associated with the female system; they are used to fabricate egg cases that will float or can be attached to objects. Mating behaviors can be quite bizarre. *Helix*, for example, perform a curious precopulatory ceremony. The two animals approach with their genital atria everted. Each fires a calcareous dart, produced in a specialized dart sac associated with the reproductive tract. The dart penetrates deeply into the internal organs. After this sadomasochistic performance, the orgasmic creatures mutually exchange sperm.

Development. Early development follows the general molluscan plan (see Fig. 21.8) but may be quite modified in ova with a considerable amount of yolk. The ova of archeogastropods are not enclosed in capsules, and develop in typical free-swimming trochophores. All other gastropod ova hatch at a later stage. Prosobranchs and opisthobranchs usually hatch as free-swimming larvae, somewhat more advanced than trochophores. This larval stage is known as a **veliger**. Pulmonates go a step further, retaining the veliger larva within the egg membranes, and hatching as diminutive snails. The trochophore larvae that are retained within the egg membranes are sometimes reduced, but are as a general rule well formed.

The transformation of a trochophore into a veliger involves considerable differentiation of parts and some changes in basic organization (Fig. 21.25). The blastopore closes; a stomodeum forms, makes connections with the archenteron, and begins to differentiate into buccal structures. The prototroch grows and gradually shifts to a more dorsal position, eventually forming a large, bilobed, funnellike **velum**, with prominently ciliated margins. Meanwhile, a **shell gland** appears on the posterodorsal surface and starts to secrete the **larval shell**. At first the shell gland is a posterodorsal cap, but it grows at the margins, eventually enclosing the whole visceral mass, and establishes the mantle. A larval mantle cavity, containing the anus, takes shape. The body projects ventrally from the mantle as a foot primordium, just ventral to the stomodeum. A pair of eyes develop above the mouth at the center of the velum, and eventually come to lie at the base of the primordia of the cephalic tentacles. Statocysts lateral to the foot are formed. While these changes are taking place, the nutritive material from the ovum is concentrated in a nutritive sac at the apex of the visceral mass. The digestive gland comes to occupy this position during later development. The body is enveloped by the growing mantle, and the internal organs grow rapidly on the dorsal side. This brings the anus forward, toward the aperture of the larval shell, and gives the digestive tract its characteristic, sharply U-shaped form. It is at about this time that torsion occurs.

As development continues, the foot grows larger and the velum becomes relatively smaller. In many species, the late veliger is able to swim with the velar cilia and to creep about with the foot. Eventually the velum is wholly reabsorbed, and the young snail becomes a juvenile, assuming an essentially adult way of life.

CLASS CEPHALOPODA

From the standpoint of complexity of structure and behavior, cephalopods stand at an apex of invertebrate evolution (Nixon and Messenger, 1977). Some fossil and modern cephalopods are tiny creatures, 2 to 3 cm long, but *Architeuthis*, the giant squid, may reach 18 m or more in length, and giant conical shells from the Ordovician were 4.5 m long and 30

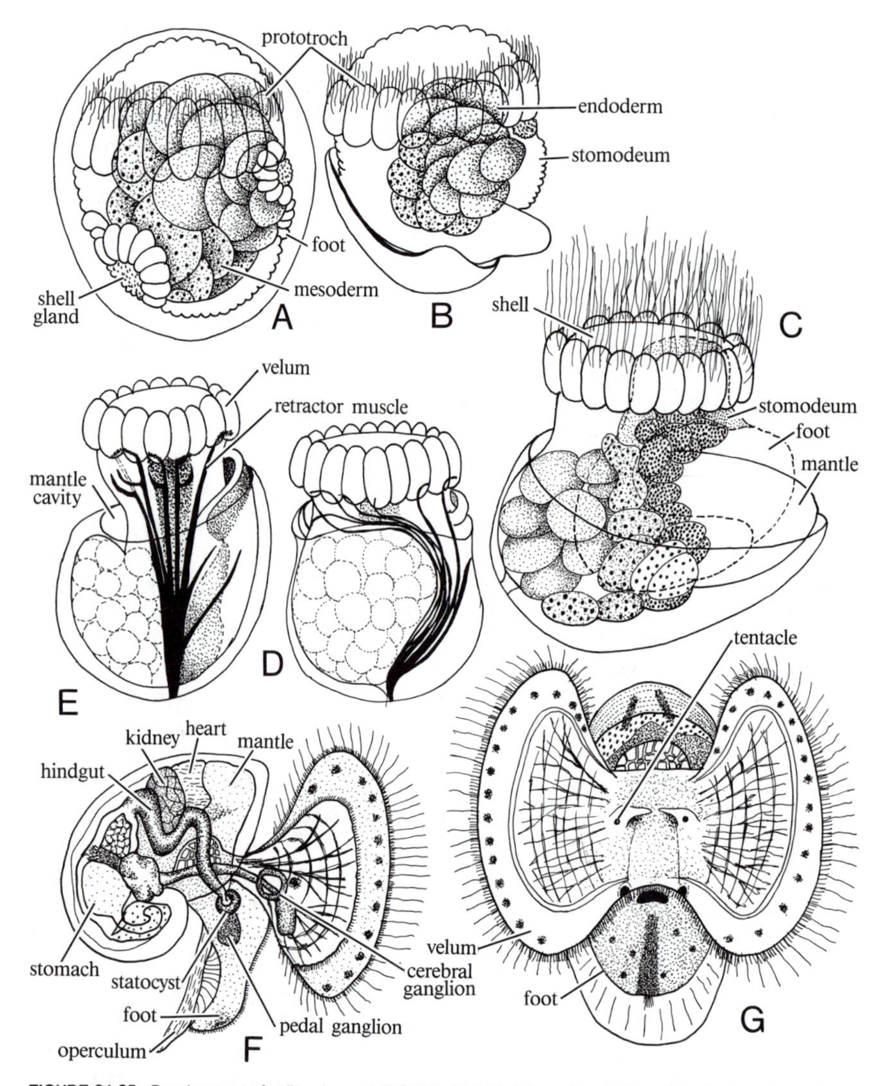

FIGURE 21.25. Development of veliger larva. **A,B,C,D,E**. *Haliotis tuberculata*. **A**. Young trochophore just before hatching. Blastopore closed and stomodeum present. Foot and shell gland have begun to form. **B,C**. Further development of prototroch into velum. Young shell begins to form, and mantle folds to enclose mantle cavity. **D,E**. Subsequently six larval retractor muscles develop. Contraction of larval retractor muscles responsible for torsion. At (**E**), 90 degrees of torsion has occurred. **F,G**. *Crepidula fornicata*, side and front views of veliger. Note bilobed form of velum in front view. (**A,B,C,D,E** after Crofts; **F,G** after Werner.)

cm wide at the aperture of the shell. The largest coiled shell, *Pachydiscus seppenradensis* is from the Cretaceous and reaches 2.5 m in diameter.

Cephalopoda are characterized by: pallial groove restricted to the posterior part of body; body developed in the dorsoventral axis; and head well developed with an array of arms or tentacles.

Cephalopods first appeared in the upper Cambrian. Three major groups can be recognized: the nautiloids, represented today by a single species, *Nautilus*; the ammonoids, now wholly extinct; and the coleoids, which include the modern squids, cuttlefish, and octopi. Although only *Nautilus* remains today, the nautiloids were very successful, reaching their peak during the Ordovician and Silurian periods, about 400 to 500 million years ago. Ammonoids reached their peak during the Mesozoic, about 200 million years ago. Coleoids did not appear until late in the Paleozoic. As they have reduced shells, their fossil record is incomplete. All modern cephalopods are marine animals, and the fossil record suggests that this has always been so. They are widely distributed today, occurring from the littoral zone to great depths.

Modern cephalopods are sometimes divided into two subclasses: Dibranchiata, with a single pair of gills and an internal shell, and Tetrabranchiata, with two pairs of gills and an external shell. *Nautilus* is the only living tetrabranch. We do not have enough information about the soft parts of extinct cephalopods to divide them on the basis of gill number, although it is possible that the extinct forms may have had two pairs of gills, like *Nautilus*. The other classification of cephalopods into three subclasses, Nautiloidea, Ammonoidea, and Coleoidea, is on the basis of shell characters. As this system separates the tetrabranch *Nautilus* from other forms, and as there are so many more extinct than modern cephalopods, it seems best to follow the fossil-based classification.

Body Plan Cephalopod evolution has resulted in the complete shifting of the functional axis of the body (Fig. 21.26). With the elongation of the visceral hump, and with the development of the funnellike hyponome from the posterior end of the foot, the primitive ventral surface was changed to the functional anterior end. The apex of the visceral hump, primitively the dorsal surface, has become the posterior end of the body. The funnel marks the primitively posterior end of the animal, but is functionally ventral. In discussing the cephalopods, functional rather than morphological axes are used; head and tentacles are considered anterior, and the funnel ventral.

During development, the head and the foot of cephalopods become indistinguishably merged. A circumoral ring, generally considered to be at least partly derivative of the foot, surrounds the head. From this ring the **arms** or **tentacles** develop. *Nautilus* has over 90 tentacles; sepioids and teuthoids, ten; and octopoids, eight. The **hyponome** is attached to the ventral head surface. It is apparently homologous to the epipodia of some gastropods. In *Nautilus* it is composed of two separate folds, but in other cephalopods has become a single structure. An unusual feature of the *Nautilus* head is the muscular **dorsal hood**, which serves as an operculum when the animal withdraws into the shell. One of the most prominent features of the head is the pair of large eyes, located on the lateral or dorsolateral surfaces.

The mantle invests the body. The *Nautilus* mantle lines the **living chamber** of the shell and extends through the septa as a **siphon**, reaching the last chamber at the apex. The tough, muscular coleoid mantle is external and contains the reduced shell. The mantle of squids projects forward as a collar around the head and the funnel, leaving a clear path for water to enter the mantle cavity. The mantle of octopi, however, adheres to the dorsal and lateral surfaces of the head, considerably reducing the aperture.

Nautiloids and ammonoids have an external, chambered shell. The **protoconch**, or embryonic shell, lies at the apex. It is present or absent in fossil species but is lacking in adult *Nautilus*. The upper part of the shell (phragmocone) is divided into **camera** or **chambers** by perforated partitions. The final large chamber is the living chamber, in which the body lies. A ventral notch, the **hyponomic sinus**, often marks the aperture, providing space for the funnel.

The first nautiloids (see Fig. 21.29) had a straight conical shell, like the first gastropods. The shells of many fossil nautiloids are slightly curved, whereas others are loosely or tightly coiled (Fig. 21.27A). Ammonoids appeared after the nautiloids, and originally had coiled shells (Fig. 21.27B). Secondarily straightened shells are common. The outstanding difference between the ammonoids and the nautiloids is the nature of the septa. Ammonoid septa are wrinkled and meet the shell in sinuous sutures, whereas the smooth nautiloid septa meet the shell in straight sutures.

Coleoids (Fig. 21.28) capitalized to a greater extent on the increased capacity for movement provided by the hyponone. Cephalopods are carnivores, using the tentacles for food capture (Figs. 21.26 and 21.27F). Although some carnivores are successful in ambushing or trapping prey, most are hunters. The earliest coleoids retained the maximal shell protection with a minimal weight. *Belemnoteuthis* (Fig. 21.27E) shows the primitive organization of the body and the shell. A thickened guard, the rostrum, forms the apex of the shell, and a platelike proostracum covers the dorsal surface. All modern coleoids can be derived from an ancestral stock of the *Belemnoteuthis* type (Fig. 21.29). Loss of the proostracum, followed by coiling and eventually loss of the rostrum, and coiling of the phragmocone leads to sepioids of the *Spirula* type (Fig. 21.27D). Conversion of the proostracum into a covering for the phragmocone, reduction of the rostrum, and slanting of the septa lead to forms like modern *Sepia*. Reduction of the rostrum, loss of the phragmocone, and conversion of the proostracum into a conchiolin pen (gladius) lead to such teuthoids as *Loligo* (Fig. 21.28A,B). Total loss of the shell would lead to octopods (Fig. 21.28C,D,E).

The buccal cavity contains a **radula**, typically with five teeth in each transverse row. Most cephalopods have two pairs of **salivary glands**. The anterior pair secrete mucus, possibly accompanied by some digestive enzymes. The posterior pair are the poison glands. *Nautilus*, however, has only the anterior pair.

The **esophagus** is muscular, and peristaltic contractions are used for swallowing. The esophagus may be a simple tube, but it is sometimes dilated to form a crop near the distal end, as in *Nautilus* and octopi.

The esophagus opens into a vestibular region from which **stomach**, **caecum**, and **intestine** arise. The large **digestive gland** or **hepatopancreas**, paired

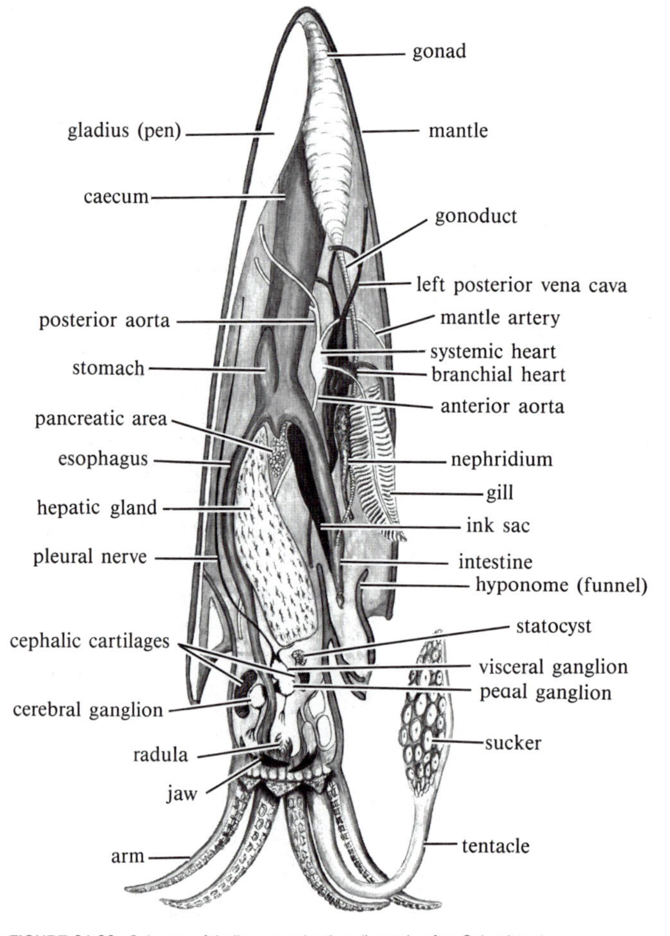

FIGURE 21.26. Scheme of *Loligo* organization. (Largely after Schechter.)

in *Sepia*, consists of a brownish "liver" region and a whitish "pancreas" region. The two regions occur as separate organs in squids (Fig. 21.26). Paired ducts from the digestive glands open into the upper part of the caecum. The stomach is lined with chitin and is presumably a derivative of the stomodeum. The caecum is typically coiled, but is straight in *Loligo* (Fig. 21.26).

The gills are evidently modified ctenidia, with a central axis containing the main afferent and efferent blood vessels, and flattened gill filaments containing capillary beds. Except for *Nautilus*, which has two pairs of gills, all modern cephalopods have a single pair.

Cephalopods have a relatively spacious coelom, but it retains a molluscan character. A flattened **pericardial cavity** is partly separated from the larger **perivisceral coelom** by a perforated or incomplete septum. Lateral extensions of the pericardial cavity of sepioids and teuthoids enclose the **accessory hearts** at the bases of the gills, but in octopi the pericardial cavity extends only to the base of the accessory hearts. The perivisceral coelom is generally restricted to the visceral hump, but in some squids it extends far forward over the pericardial cavity, becoming very large. The larger coelom is in part the result of a reduction of the hemocoel lacunae.

Nautilus differs from other cephalopods in having

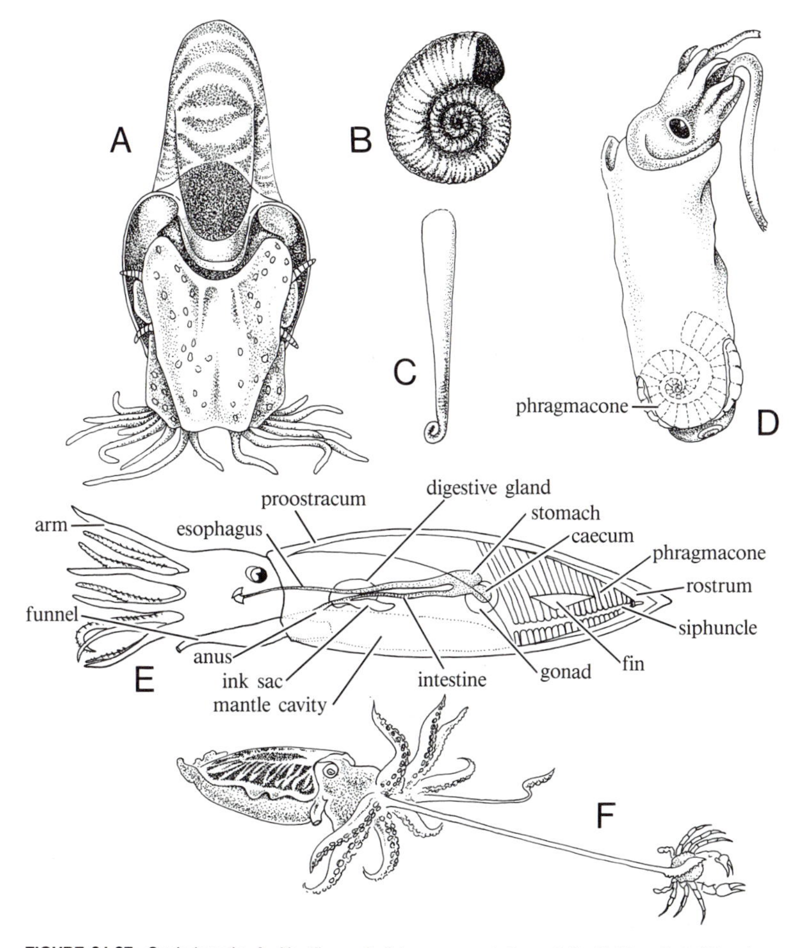

FIGURE 21.27. Cephalopods. **A.** *Nautilus,* only living representatives of Nautiloidea, that differs from ammonoids in having simple sutures where septa dividing shell into chambers meet outer shell. **B.** *Idoceras,* Jurassic coiled ammonoid. **C.** *Baculites,* Upper Cretaceous coleoid. Coleoids have single pair of gills and internal shell. Shell well developed in Belemnoidea, to which *Belemnoteuthis* (**E**) belongs. Fossil impressions make it possible, in some instances, to get fairly detailed ideas of internal organization. **F.** *Sepia,* cuttlefish, with ten arms and moderately reduced internal shell. **D.** *Spirula,* another sepioid, with small, coiled internal shell. (**A** after Willey, from Pelseneer, in Lankester; **B** after Imlay; **C** after Reeside; **E** after Roger; **F** after Boulenger; **D** after Owen and Adams, from Cooke.)

two pairs of kidneys. Is this duplication primitive or a secondary development? As with the duplication of gills, there is insufficient evidence to be sure.

Cephalopods have an advanced circulatory system, with a complex organization (Wells, in Wilbur, vol. 5, 1983). It differs from the circulatory system of other molluscs in having a much more extensive set of arteries and veins, and in having a number of capillary beds in place of the more open hemocoelic lacunae. Nevertheless, the basic pattern of blood flow follows the typical molluscan plan (Fig. 21.30). Each gill drains into an auricle; *Nautilus* has four auricles and other cephalopods two. An **aorta** carries blood from the ventricle, giving off branches to the viscera and continuing to the head, where branches to the cephalic organs and tentacles arise. Blood returns from the capillary beds through a complex venous system. **Abdominal veins** return blood from the viscera; **pallial veins**, from the mantle; and an **anterior vena cava**, from the head. All of these drain into the **efferent branchial vessels**, equipped with booster hearts that drive the blood into the gills. Blood returns to the auricles by way of **efferent branchial veins**.

The cephalopods have the most highly developed nervous systems to be found in invertebrates, and

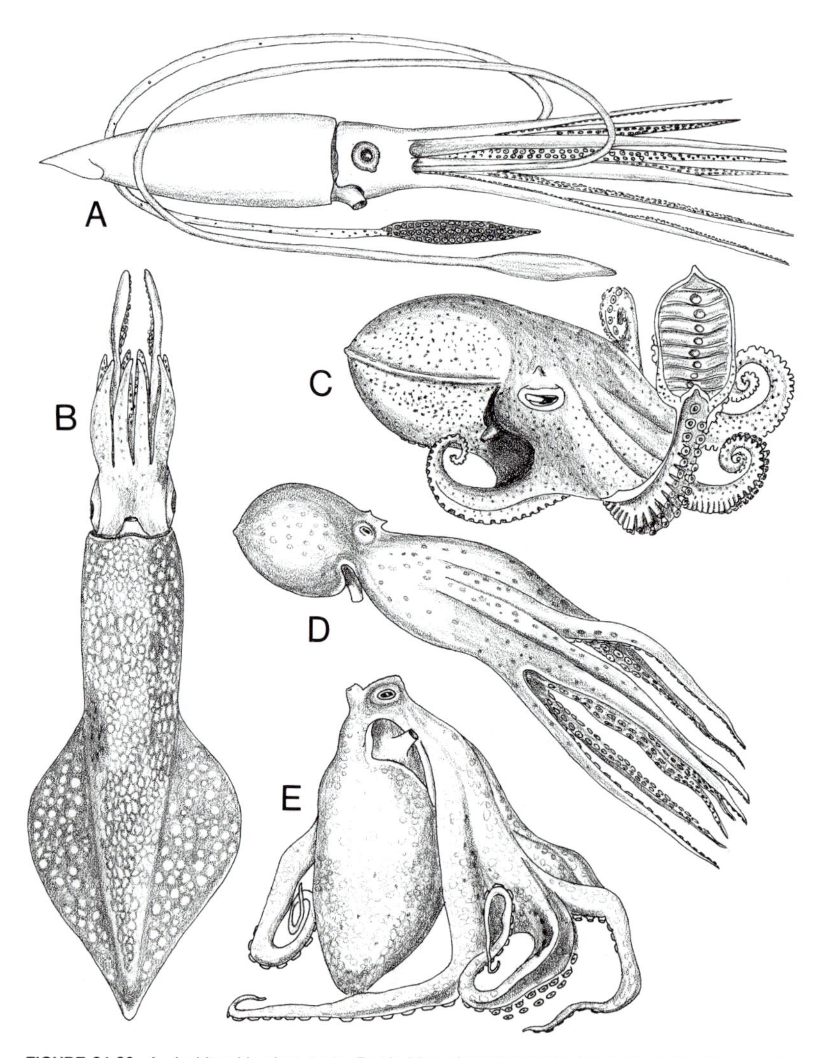

FIGURE 21.28. A. *Architeuthis,* giant squid Teuthoidea, characterized by ten arms, shell with no rostrum and either without or with much-reduced phragmocone, pen-shaped gladius formed from proostracum, without cornea, and without specialized fourth arms. **B**. *Loligo,* common squid, teuthoid with elongated fourth arms and cornea over eye. **C**. *Octopus lentus,* Octopoda, characterized by eight arms, no shell, and short, rounded body. Hectocotyl arm turned up, showing arrangement for carrying spermatophores. **D,E**. Resting and swimming positions of *Octopus vulgaris.* (**A,C** after Verrill; **D,E** after Merculiano, from Cooke.)

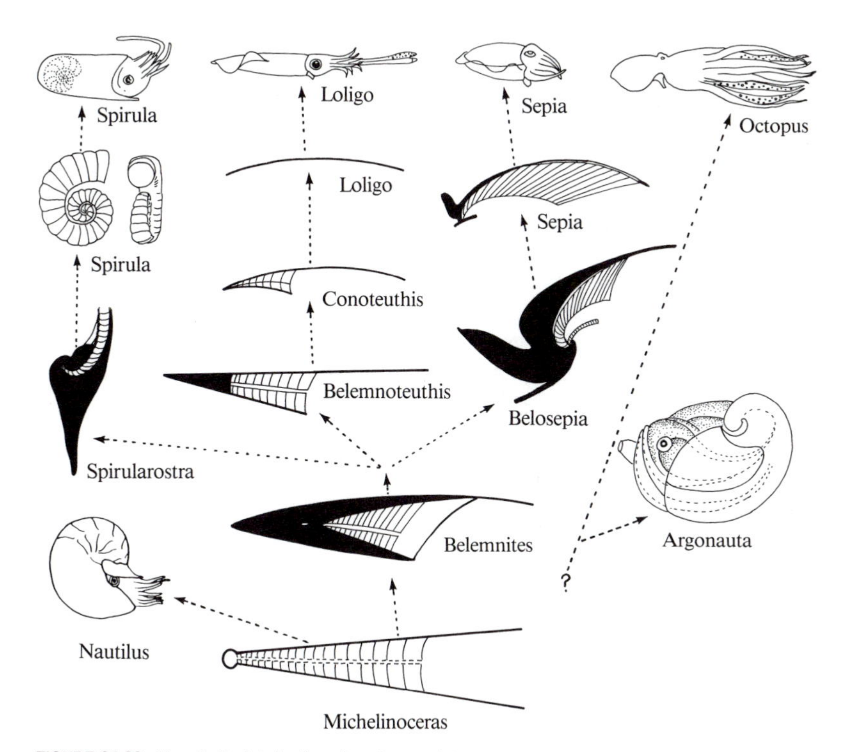

FIGURE 21.29. Hypothetical derivation of modern cephalopod types from fossil types. Presumably straight-shelled ancestral nautiloid (*Michelinoceras*) enclosed shell in mantle *Belemnites,* leading to differentiation of rostrum, phragmocone, and proostracum of shell, as in *Belemnoteuthis.* Total reduction of shell leads to Octopoda, although *Argonauta* has developed calcareous egg case that resembles shell. Different lines of shell reduction lead toward modern *Spirula,* with phragmocone retained; *Loligo,* with gladius and no phragmocone; and *Sepia,* with remnants of both rostrum and phragmocone. Modern nautiloids develop from straight-shelled forms by coiling shell and restricting body to living chamber. (After Shrock and Twenhofel.)

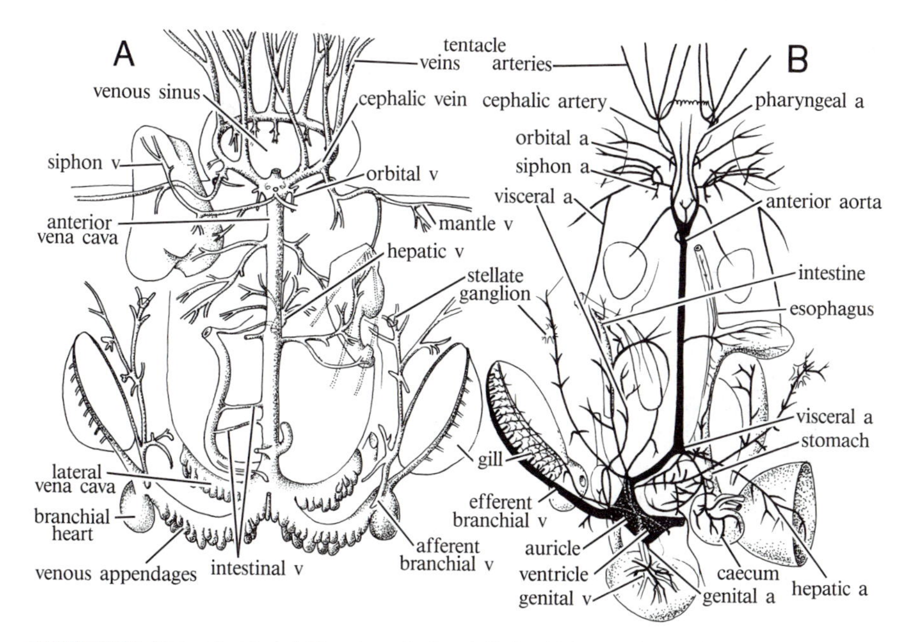

FIGURE 21.30. Venous **A** and arterial **B** systems of octopod, *Eledone.* Note extensive set of vessels that correlate with reduction of hemocoel. (**A**,**B** after Isgrove.)

correspondingly complex behavior patterns (Young, 1971). Masses comparable to the cerebral, pedal, pleural, and visceral ganglia can be recognized, but the ganglia have lost their integrity as individuals. They have been subdivided and assembled into a circumenteric nerve center, functionally a complex **brain** (Fig. 21.31A). A cortex of neurons covers a deeper neuropile, in which a number of organized tracts and pathways can be recognized.

Cephalopods are among the few invertebrates to have cartilage tissue composed of cartilage cells located in lacunae surrounded by matrix material. Cephalopod cartilage closely resembles the cartilage of vertebrates, but differs in having adjacent cells connected by intracellular bridges that pass through the matrix. The cartilage is used to support the mantle margin and the fins, and to form a skull-like capsule around the brain for protection and to provide sites for muscle attachment.

The unpaired gonads (see Fig. 21.26) lie in the wall of the coelom at the posterior tip of the body. Sexes are nearly always separate. Males are usually smaller than females and sometimes show other evidence of sexual dimorphism. Gametes break out of

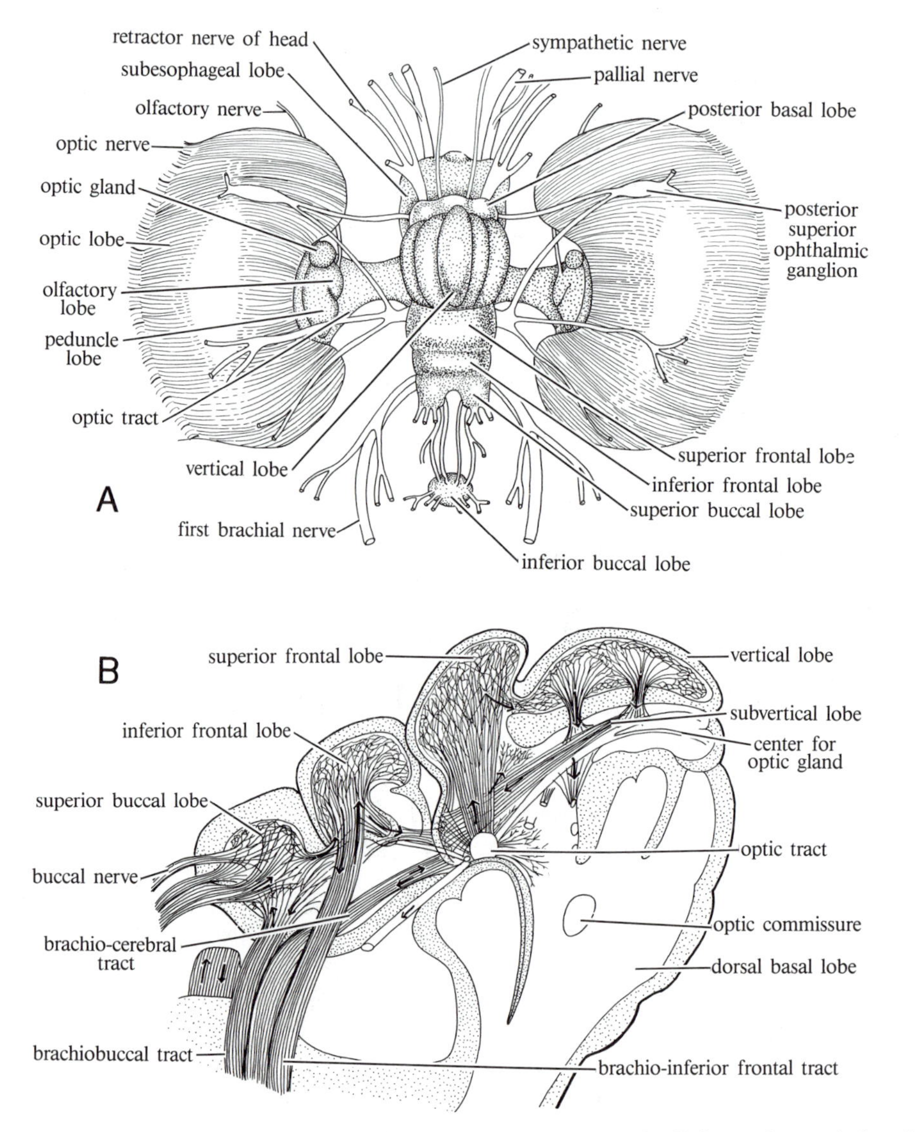

FIGURE 21.31. *Octopus* brain. **A**. Dorsal view. **B**. Section through vertical lobe region. Notice complex organization, with cortex, tracts, and subdivision of lobes. (After Boycott and Young, from Young.)

the gonads, entering the coelom, to be picked up by the gonoducts.

Female octopi and many teuthoids have symmetrical gonoducts that arise side by side and pass to the posterior end of the mantle cavity. *Nautilus* has a vestigial left and functional right oviduct. In other cephalopods the right oviduct is missing and the left is functional.

Several glands are associated with the oviduct. An **oviducal gland** secretes the inner albuminous or capsular covering of the egg. Where the oviduct courses over the mantle wall, **nidamental glands** may be found. These secrete gelatinous material around the eggs or, in some cases, material that hardens on contact with water and is used to stick the eggs to objects.

In *Nautilus* the left **sperm duct** is rudimentary, while in other cephalopods the right sperm duct is missing. Sperm are temporarily stockpiled in **seminal vesicles** near the upper end of the sperm duct. Here, they form masses, enclosed in a complex **spermatophore** (Fig. 21.32). Secretions from the **prostate gland** enter the seminal vesicle and are used as cement to hold the sperm together. The fully formed spermatophores pass into a distal, dilated part of the sperm duct (**Needham's sac**), where they are stored until the time of mating.

Biology

Locomotion. Cephalopods swim by ejecting a jet of water through the funnel. *Loligo* pumps water in and out of the mantle cavity by alternate contractions of antagonistic circular and radial muscles in the mantle. When the circular muscles relax and radial muscles contract, the collar flares out, and water enters through the space between the mantle and the head. Contraction of the circular muscles forces the collar against the head; therefore, the emerging water must pass through the funnel. The funnel can be aimed forward or backward, so squids can swim in either direction. Fins on the mantle undulate gently when the squid is hovering, but make powerful movements during rapid swimming. They are more important, however, as stabilizers than as propellors. Squids swim gracefully and are able to achieve speeds that compare favorably with the most active fishes.

The details of swimming movements differ in other cephalopod groups, but all are essentially similar. *Nautilus* forces water out of the mantle cavity by retracting the body and contracting the funnel musculature. Octopi can also swim by ejecting water through the funnel, but have assumed a benthic life and usually creep about, using the suction cups on the arms for attachment as they pull themselves along. They can move with considerable speed, even on land. Some octopi are especially modified for swimming and have webs between the tentacles.

Chromatophores. Special pigment cells, known as chromatophores, are found in the mantle epithelium of coleoids, each with a single pigment, yellow, red, brown, or blue, as the case may be. Different species have different combinations of colors, in different proportions. The chromatophores have an elastic cell membrane to which a circlet of smooth muscle cells is attached. Muscle contraction expands the cell, stretching the membrane; the cell membrane contracts as muscles relax (Fig. 21.33A,B). These unusual muscled chromatophores work very rapidly, blinking on and off like lights in an electric sign. Neural and humoral mechanisms cooperate to

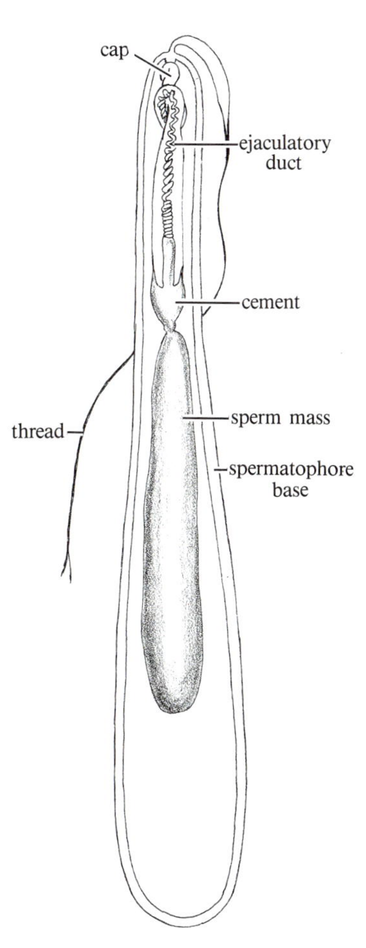

FIGURE 21.32. Spermatophore of *Loligo,* consisting of outer covering, ending in cap with attached thread. When cap detaches, coiled ejaculatory duct everts, pulling cement mass at base and large sperm mass with it, cementing sperm mass in place on mate.

control them. They are doubly innervated. A special color center in the central ganglion of the cerebrum is responsible for inhibition. Stimuli pass through motor centers in the subesophageal ganglion on their way to the chromatophores. Blood-borne tyramine acts like adrenalin, causing chromatophore expansion and a darker color. Betaine acts like acetylcholine, stimulating the inhibitory center and leading to a lighter color. The blood factors are not species specific, for blood transfusions between dark and light species cause color changes.

The presence of chromatophores permits color changes for background matching. Normal postural reflexes keep the ventral surface lighter than the dorsal surface, providing camouflage during swimming. Impairment of the optic nerves reduces, but does not stop, chromatophore changes and prevents background matching. If the suckers on the tentacles are removed, the animal becomes lighter. General stimulation and excitability increase chromatophore activity, producing responses that have an almost emotional quality.

Food Capture and Digestion. The arms are used for capturing prey. They are complexly muscled and under delicate control. *Nautilus* arms have no suckers but are made adhesive by secretions. Coleoids have muscular suction cups on the inner surfaces of the arms. Some are set on stalks, some have horny rims, and some have a hook that helps to attach the arms to the prey. They are quite powerful; even the tough skin of whales is marked by the suckers of giant squid. Other molluscs, crustaceans, and fish are the principal prey of cephalopods; the exact nature of the diet depends on the size and habits of the species.

Prey is brought to the mouth and attacked by the heavily muscled, horny jaws. Pelagic species typically kill the prey with a single bite and swallow small organisms whole or eat larger ones bite by bite, whereas *Sepia* and octopi use poisons secreted by the posterior salivary glands to immobilize prey. *Sepia* chews its food, but octopi feed more slowly, freeing soft parts from the skeleton by a preliminary digestion, and rejecting most hard parts.

Digestion begins in the stomach and continues in the caecum, depending largely, if not wholly, on enzymes from the digestive gland (Boucaud-Camou and Boucher-Rodoni, in Wilbur, vol. 5, 1983). Analysis of stomach and caecal fluids and of liver and pancreatic secretions has shown considerable variation in details and also in whether liver or pan-

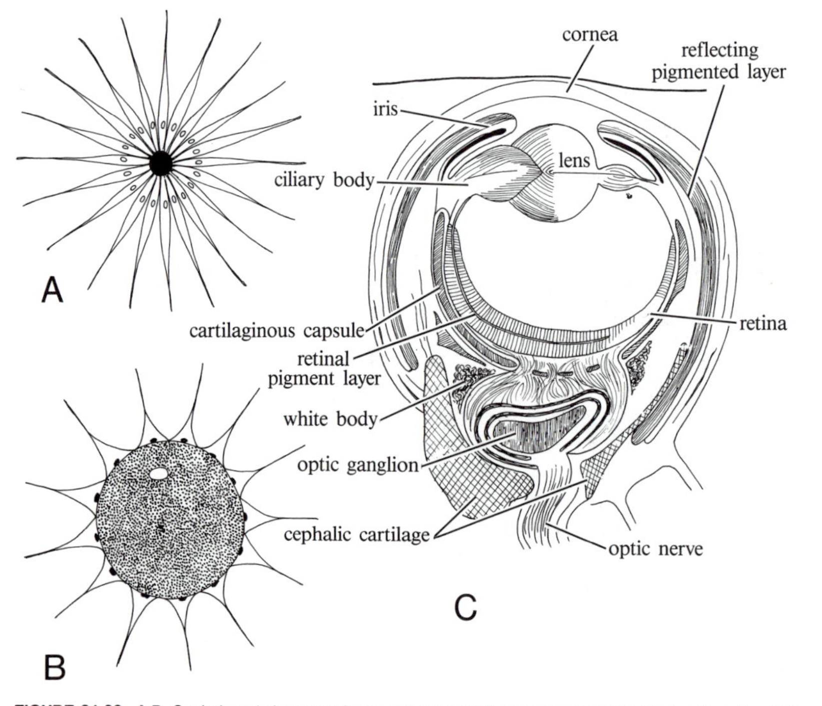

FIGURE 21.33. A,B. Cephaloped chromatophores contracted **(A)** and expanded **(B.)** Contraction of radially arranged muscle fibers expands chromatophore, causing expansion of pigmented area and darker color. **C.** Scheme of eye of *Sepia.* Note strong convergence between eye of cephalopods and vertebrates. (**A,B** after Bozler, from Prosser et al.; **C** after Henson, from Pelseneer, in Lankester.)

creatic secretions are predominant in stomach or caecal digestion. In any case, a variety of starch-, fat-, and protein-splitting enzymes have been recovered, and digestion is not completed in the stomach. Sphincters or valves isolate the stomach and the caecum, providing time for digestion to be completed. Ciliated leaflets in the caecum form a sorting region that directs rejected particles into the intestine, while rejected skeletal fragments pass directly from the stomach to the intestine. In *Nautilus*, food absorption appears to occur predominantly in the digestive gland, and distinctive liver and pancreas regions are lacking, and in other cephalopods it takes place in the caecum, or in the caecum and hepatopancreas, and perhaps in the intestine as well.

Gut motility and at least some aspects of secretion are under the control of the stomatogastric (sympathetic) nervous system. Liver and pancreas secretions may be delivered at different rates, depending on the presence of food in the stomach, for example. Details are not fully known.

Mucus-compacted feces are discharged from the anus by muscular action of the rectum. The intestine runs anteriorly, giving the whole digestive tract a U-shaped configuration.

A remarkable peculiarity of the coleoid digestive system is the specialization of the rectal gland into an **ink gland** and **ink sac**. The ink, known as sepia, has long been used as an artist's pigment. It usually consists primarily of melanin, but some abyssal cephalopods secrete a luminescent ink. The ink apparatus is a protective device. When the animal is disturbed, it can eject a cloud of ink from its anus, escaping behind its protective screen.

Respiration. Respiratory exchange occurs at the surface of the folded, highly vascular gills. Ciliary currents, so important in gill ventilation in other molluscs, is of no consequence in cephalopods, for the muscular movements used in filling and emptying the mantle cavity suffice.

The mantle musculature has a double innervation. Small nerve fibers are thought to be responsible for ordinary respiratory movements, while giant fibers are responsble for movements associated with locomotion. The strength and frequency of respiratory movements are affected by activity and carbon dioxide levels, and are also related to body size.

Coelom and Excretion. So little is known of excretory physiology except in the Octopoda that the following is intended to apply to them alone. As in other molluscs, the kidneys communicate by renopericardial canals with the pericardial cavity, and the filtrate from the pericardium enters the kidney. Unlike other molluscs, however, the **branchial hearts** and their heart appendages are primarily concerned. The latter apparently correspond to the pericardial glands of other molluscs. The kidneys consist of renal sacs, partially filled with urine, and renal appendages, which are intimately associated with the venous system. The renal appendages hang from the veins, and blood in them is separated from the contents of the renal sacs by a single layer of cuboid epithelium. Renal appendages beat at about the same time as the branchial hearts, providing a continuous flow of blood. The blood then enters the branchial hearts. The secretory aspects of excretion apparently are concentrated in the renal appendages. There is evidence of the reabsorption of glucose and animo acids at this point.

The principal excretory product is ammonia, and the actual output of nitrogenous wastes includes a considerable release from gill surfaces. In some of the cephalopods, the digestive glands are also important sites of nitrogenous waste disposal.

Sense Organs. Tactile cells are found in the epidermis and are concentrated in the tentacles and other points of high sensitivity. Tactile reception is particularly important in the suction cups of the coleoid tentacles. *Nautilus* has a pair of **osphradia** near the gills, and the tentacles immediately in front of and behind the eyes are specialized for chemoreception. A pair of **chemoreceptive pits** near the eyes are often found in coleoids, and some of the teuthoids have **rhinophores**. Cephalopods generally have a pair of statocysts, near the pedal ganglia in *Nautilus*, but otherwise embedded in the cephalic cartilages.

The eyes are the most remarkable sense organs. *Nautilus* has a relatively simple eye, with a large pigment cup and no cornea, lens, or other refractive parts. Other cephalopods, however, have eyes of complex structure; some of them are remarkably like vertebrate eyes, fully capable of forming a good image. The similarity of cephalopod eyes to vertebrate eyes is striking, but the cephalopod eye is derived wholly from the surface ectoderm, whereas much of the vertebrate eye arises from the embryonic neural tube. It is a remarkable instance of convergent evolution.

In *Sepia* and *Loligo* eyes (Fig. 21.33C), a transparent cornea covers the lens that is held in place by suspensors containing ciliary muscles. An adjustable pupil is centered in a pigmented iris. Pigmented and retinal layers line the optic cup. A capsule protecting the eye is formed from the cephalic cartilages. Special eye muscles attach to the cartilages and the external eye surface, making some eye movement possible. Unlike the vertebrate eye, the retinal rods face the open cup, and there is a large optic ganglion immediately below the retina.

Definite brain centers control specific activities. Responses are partially prepackaged by virtue of specific neuronal linkages, but associational centers provide for discrimination and learning. Some remarkable experiments with food placed in stoppered bottles have revealed cephalopods capable of learning and rudimentary problem solving (Boyle, in Wilbur, vol. 9, 1986).

The subesophageal part of the brain consists of several ganglionic subdivisions. The right and left sides have been partially fused to form a symmetrical lobed mass. Nerves to the arms, funnel, mantle, and viscera arise from the subesophageal region, and centers for the control of eye muscles, the iris and pupil, chromatophore activity, inhalant and exhalant respiratory movements, arm movements, movements of the mantle and fins, and the activity of various visceral organs are found here.

The supraesophageal region consists primarily of the cerebral ganglia and their subdivisions. Although the subesophageal region is predominantly a motor center, the supraesophageal region is predominantly composed of relay centers for the reception of sensory stimuli. Optic and olfactory nerves enter here, and sensory centers evoking specific responses to specific sensory stimuli are also built into this part of the brain. The cerebral region also contains an associative region. No responses are evoked when this part of the brain is stimulated, but its removal prevents the development of conditioned behavior in *Octopus*, and it must be considered the functional equivalent of the associative regions of the vertebrate cerebrum. Although many motor activities depend on subesophageal centers, motor centers that evoke the movement of large muscle groups are also found in the cerebral region.

The lobus magnocellularis contains giant neurons. It receives sensory stimuli that activate the giant neurons. Each giant cell crosses to the other side and synapses in the visceral ganglion with giant neurons of the second order. The second-order giant neurons pass along the pallial nerves to the stellate ganglia, where they synapse with third-order giant neurons, whose axons innervate the mantle muscles. The giant-neuron system transmits critical stimuli rapidly, evoking prepackaged startle reactions. More delicate controls are provided by smaller axons that run parallel to the giant fibers.

Reproduction. *Nautilus* males have four arms that are permanently modified for copulation, but most cephalopod males have a single specialized arm when sexually mature (Giese and Pearse, 1977; Arnold, in Wilbur, vol. 7, 1984). The specialized arm varies considerably in form and in the way in which it is used; the most highly specialized are seen in *Argonauta* and some octopi. The modified end of the arm is detached during mating and remains in the female mantle cavity, undoubtedly as a device to foil the attempts of other males to inseminate the female. The copulatory arm was correctly understood by Aristotle but was completely misinterpreted by zoologists of the last century. Cuvier described it as a parasite under the name *Hectocotylus octopodis*. Later, Kölliker believed it to be a complete male animal and described its heart, intestine, and reproductive system. It is still called a **hectocotylus arm**. The most highly specialized hectocotylus arms are flattened, containing a cup-shaped cavity in which the spermatophores are carried. At breeding seasons, a slender, threadlike filament grows from the tip of the arm; it enters the female mantle cavity and breaks off, carrying the spermatophores with it.

Spermatophores (Fig. 21.32) are deposited in the mantle cavity or elsewhere on the body surface depending on the species. *Nautilus* and squid spermatophores are attached to the head region or inserted into a seminal receptacle near the mouth. When the spermatophore cap is loosened, the ejaculatory organ is extruded, releasing the sperm. Eggs are fertilized as they emerge from the oviduct, before the secretions from the nidamental gland have hardened.

Egg masses or single eggs are attached to stones or other objects as a general rule, but some pelagic species have floating eggs. Octopi often tend the eggs, ventilating them with jets of water from the funnel, cleaning them as well as keeping them surrounded by fresh water. The most unusual system is employed by the female *Argonauta*, who has a pair of highly modified dorsal tentacles that secrete a thin, gently fluted capsule into which the eggs are placed for brooding. The female carries the nursery throughout the period of egg development, retiring into it when disturbed, and males sometimes occupy the shell with the female and the ova. An unfortunate consequence of mating and egg laying in cephalopods is death, since massive tissue degeneration is triggered by the sex act.

Development. Development is greatly modified. The ova contain a great deal of yolk, derived from follicle cells. A germinal disc free of yolk occurs at the animal pole. The germinal disc undergoes meroblastic cleavage, forming a discoid blastula. A layer of cells grows down from the margins of the germinal disc, turning under to form endoderm. Mesoderm forms from the posterior ends of the germinal disc, forming two crescentic strands of cells. The buccal cavity, radula, and associated parts rise from a stomodeum, and the end of the rectum from a proctodeum. Ectodermal fields corresponding to the ganglia give rise to the nervous system.

CLASS SCAPHOPODA

Scaphopods are distinctive creatures with a shell in the form of a tapered tube, usually slightly curved and open at both ends. About 200 species have been described. They burrow in sand or mud, with the larger, anterior end down and the other end thrust slightly above the bottom (Fig. 21.34A). Water enters and leaves through the posterior aperture, and the foot is thrust from the anterior aperture during burrowing. Some scaphopods live in shallow water, but most live at moderate depths. A few occur 5000 m or more below the surface.

Scaphopods are characterized by: reduced head and sense organs, mantle flaps forming a tube, tubular shell, and fused cerebral and pleural ganglia. The mantle unites along the ventral margin to form a tube around the body. Thereafter the shell is tubular, and the body lengthens as it grows.

The general body organization is rather like that of a clam, except the mantle and the shell are sealed dorsally and ventrally, although open at the ends. Scaphopods also resemble pelecypods during early development, and like pelecypods, the adults have reduced heads. A well-developed buccal mass containing a radula with five teeth in each transverse row sets them off from modern pelecypods. There are no gills in the mantle cavity and there is no heart; circulation occurs through a series of open sinuses.

Body Plan The long, slender foot ends in a discoid tip or in lateral processes that can be expanded to form an anchor. The strong foot retractors pull the shell up toward the anchored foot.

The head is shaped like a proboscis and is equipped with long, slender, capitate tentacles known as **captacula**. The captacula are prehensile and function as food gatherers, feeling about in the mud and being retracted to carry food to the mouth. The **esophagus** produces a pair of pouches. The rather small **stomach** gives off a **caecum**, into which the **digestive glands** open. The large digestive glands branch out into the mantle. A coiled **intestine** ends in a ventral **anus**; the feces are carried out of the mantle cavity with the respiratory currents.

Scaphopods do not have a well-organized flow of water through the mantle cavity. Cilia draw water in, and when fully engorged, sudden muscular contractions discharge the mantle contents.

It is generally held that scaphopods have no heart or pericardial cavity and that the kidneys have no connection with a pericardial coelom. Some morphologists have failed to find evidence of these structures, but others have reported a pulsating vessel in

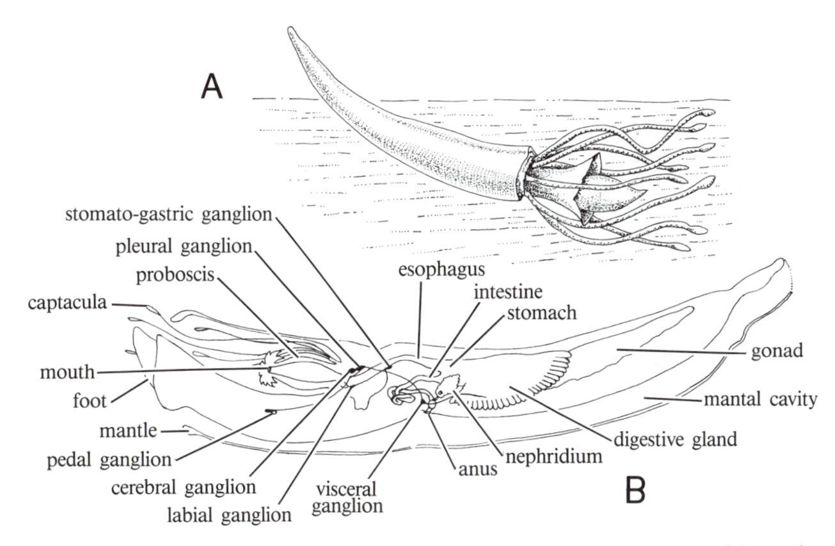

FIGURE 21.34. Scaphopods. **A**. Scaphopod in position, with captacula extended to feed. **B**. Scheme of organization of *Dentalium*. (**B** after Pelseneer, in Lankester.)

a closed pericardium, and a communication between the pericardial coelom and the left kidney has also been described. The saccate kidneys have much-folded walls and open near the anus on each side.

Sexes are separate. The single median gonad is long, extending from an anterior gonoduct to the posterior tip of the body. The gonoduct is connected to the right kidney, as in some gastropods.

Biology Little is known of scaphopod physiology. A subradular organ in the buccal cavity is thought to be chemoreceptive. The captacula are important tactile organs, and probably receive stimuli that mediate burrowing movements. Otherwise the only special sense organs are **statocysts**, located in the foot near the pedal ganglia. The nervous system is organized like that of protobranch pelecypods but with the addition of a buccal ganglion on each side of the buccal mass. The buccal ganglia are connected to the cerebral ganglia by connectives, and to each other by a commissure.

Fertilization is external. Eggs are sometimes released singly, others as a cloud of 1000 or more.

Early development leads quickly to a trochophore larva that lacks the usual excretory organ, anus, or cerebral plate. The veliger is pelecypodlike and forms early a mantle and a shell gland. The mantle grows ventrally as two lobes that unite to form a tubular structure.

CLASS PELECYPODA

Pelecypods are bivalved molluscs such as clams, mussels, oysters, and scallops. A number of them are commercially important shellfish, and all commercial pearls are obtained from pelecypods. Their shells are used to make buttons and other small pearly objects. Highly modified pelecypods, known as ship worms, are important pests, burrowing into and destroying wooden pilings and other structures in harbors throughout the world, causing millions of dollars of damage every year.

They occur widely in marine habitats (Vogel and Gutmann, 1981), from the upper part of the littoral zone to depths of almost 5000 m, and several families have established themselves in ponds and freshwater streams. Most of them plow through sand or mud with the aid of their wedge-shaped foot, feeding on organic detritus or plankton stirred up around them. A few burrow into wood or rocks or live permanently attached to rocks or other submerged objects. Nearly all of them are dependent on and specialized for ciliary feeding.

Pelecypods are characterized by: reduced head and sense organs, mantle as left and right flaps, shell bivalved and hinged dorsally, cerebral and pleural ganglia fused, highly muscular foot, and gills generally well developed and often specialized to assist in feeding. Pelecypods are often known as Bivalvia. This term should be avoided, however, because the

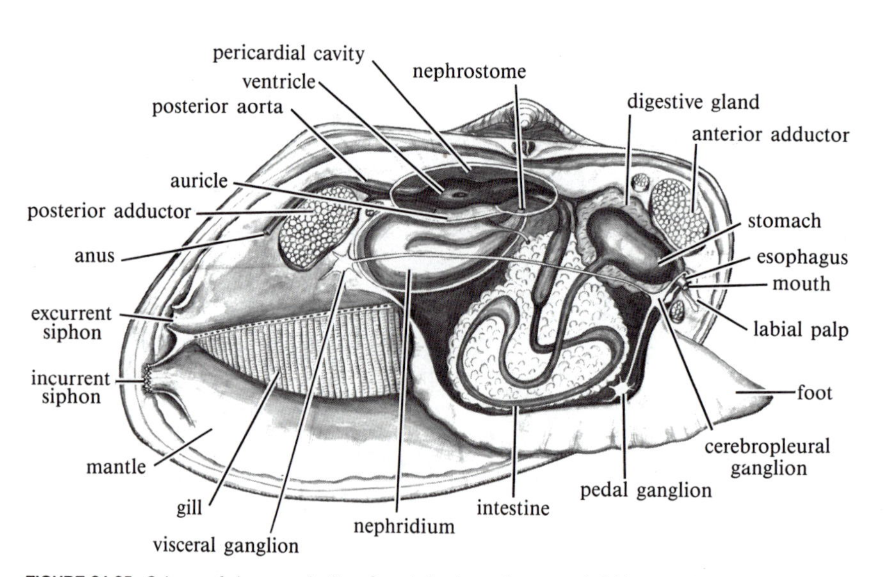

FIGURE 21.35. Scheme of clam organization. One shell valve and one mantle fold removed, and muscles clipped.

closely related, extinct class Rostroconchia are also "bivalved," but are clearly not pelecypods.

Body Plan Pelecypods are simply organized (Fig. 21.35). The body is typically symmetrical about a plane passing between the two valves, but the symmetry is sometimes spoiled by inequalities of the two valves.

A pair of **palps** enclose the mouth. Each palp is equipped with a long, slender palp appendage (Fig. 21.36A). When the animal is feeding, the appendages or **proboscides** are thrust out of the shell and feel about in the mud or sand. Particles are carried upward toward the mouth along a ciliated groove, eventually reaching the **palp lamellae** on each side of the mouth. A ridged and ciliated sorting field on each lamella grades out the large particles, which pass to the mantle and are ejected with sediment from the gills. Small particles are diverted to a deep, ciliated oral groove, along which they move to the mouth. Clams belonging to the orders Filibranchia and Eulamellibranchia have no palp proboscides, but use gills in feeding.

The mantle begins to form shell during the veliger larva stage. The two mantle folds secrete conchiolin to form two tiny valves connected by a narrow ligament. New material is added at the margins of the valves, producing a series of concentric growth rings. As the veliger settles to the bottom, shell growth continues, but more rapidly on the ventral side. As a result, the original larval shell is displaced dorsally, forming an elevation (**umbo**) directed anteriorly so that the right and left valves can be easily distinguished.

The mantle is pressed against the inner surface of the valves. It ends in three folds, the outer of which secretes the new shell (see Fig. 21.2C). **Periostracum** forms at the the inner margin, calcareous material of the **prismatic layer** at the outer border, and the **nacreous material** on the whole mantle surface, thickening the shell as it ages.

No calcareous material is added to the valve margin at the hinge line, and a **hinge ligament** composed of conchiolin is secreted instead. The ligament binds the two valves together. In most pelecypods, the articulation between the valves is ventral to the ligament, and the ligament is stretched when the adductor muscles close the valves. When the muscles relax, the ligament opens the valves (Fig. 21.37A). In other pelecypods, a different mechanical principle is used. The conchiolin of the hinge is laid down as an elastic **resilium** that extends between the margins of the valves. The resilium is compressed as the adductor muscles close the valves, and springs the valves open when the adductor muscles relax (Fig. 21.37B,C).

The hinge line, or cardinal margin, has various adaptations to strengthen the articulation between

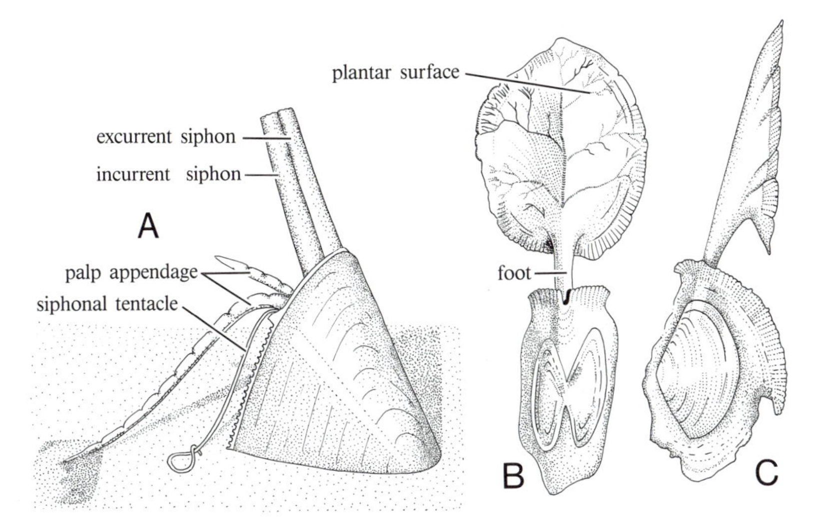

FIGURE 21.36. A. *Yoldia* feeding. Foot thrust deep into sand, and siphons elevated. One palp appendage in contact with surface so particles of detritus can move along groove in palp appendage to mouth. **B,C**. Dorsal and side views of *Entovalva semperi.* Note relatively small size of shell in comparison with mantle, and distal expansion of foot. Foot with plantar surface like this, thrust into sand or mud, expanded, and foot retractor muscles contracted to pull shell forward. Foot then contracts and thrusts forward again. (**A** after Drew; **B,C** after Ohshima, from Franc, in Grassé.)

the two valves. These take the form of interdigitating teeth and dental sockets that serve to prevent slippage (Fig. 21.37D). The articulating dentition is extremely varied and is an important factor in taxonomy.

Wherever muscles attach to the valves, the orderly secretion of the nacreous layer is disturbed and muscle scars are formed (Fig. 21.37F,G,H). The most prominent muscle scars are caused by the large adductor muscles that close the valves. The number and position of the adductor muscle scars are also useful in classification.

Other muscles also leave scars. The inner lobe of the mantle contains the orbicular muscle, which attaches to the shell in a scar known as the pallial line (Fig. 21.37D). The margin of the mantle is sealed to the shell along this line, preventing particles from entering the space between the mantle and inner shell surface. When particles do enter this space, they interfere with normal nacre depositon, and may serve as nuclei for pearl formation. In many clams, the right and left mantle folds are fused to define siphons. Special parts of the orbicularis muscle serve as siphon retractors, which cause an indentation of the pallial line, known as the pallial sinus. The size of the pallial sinus roughly corresponds to the size of the siphon, except in a few species that cannot pull the siphon into the shell.

Foot muscles also cause scars (Fig. 21.37G). A pair of anterior retractors attach to the shell just behind the anterior adductors, and a pair of posterior retractors attach in front of, and often a little above, the posterior adductors. A pair of anterior foot protractors attach near the anterior adductors, usually below the retractors. A median foot levator is also present in some pelecypods.

The flow of water through the mantle cavity is one of the most critical factors in pelecypod survival. As

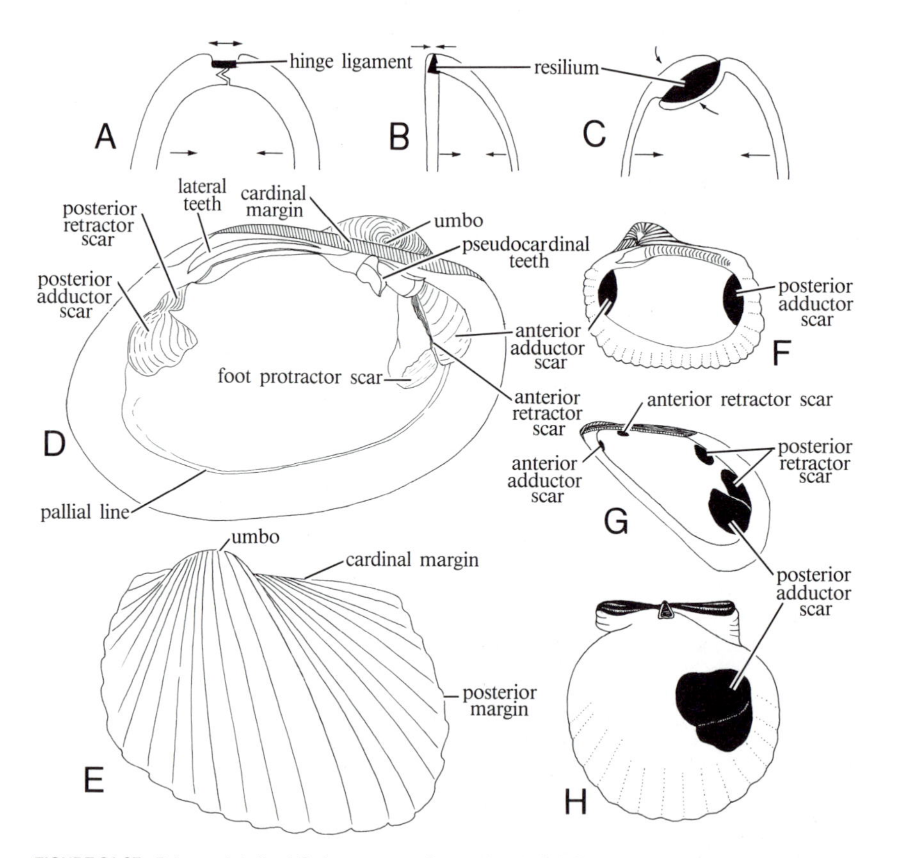

FIGURE 21.37. Pelecypod shells. **A,C**. Arrangements for opening shell. **A**. Elastic hinge ligament stretches when adductor muscles close shell. When muscles relax, hinge pulls shell open. **B,C**. Conchiolin built into elastic resilium, compressed by adductor muscles closing shell. Resilium springs back into shape when muscles relax, opening shell valves. **D,E**. General features of pelecypod shell. **F,G,H**. Muscle scar arrangements. **F**. Some clams have approximately equal anterior and posterior adductor muscles, termed isomyarian. **G**. Anisomyarian shell, with unequal anterior and posterior adductors. **H**. Monomyarian, with single adductor. (**A,B,C** after Shrock and Twenhofel; **D** after Turner, from Clench, in Edmondson.)

water enters and leaves, it brings oxygen and food, and wafts away wastes and carbon dioxide. It is not surprising that pelecypods have developed adaptations to make the entry and departure of water more orderly, or that the more primitive protobranchs and filibranchs include a number of species without any adaptations for this purpose (Martin, in Wilbur, vol. 5, 1983). The primitive condition, as seen in some protobranchs and filibranchs, is for the two mantle folds to be wholly free; however, in most pelecypods the right and left mantle folds have fused in one or several places, defining passageways through which water enters or leaves. The mantle folds of many eulamellibranchs are united posteriorly, defining a ventral opening for the foot and an exhalant opening for water. The exhalant opening of many freshwater unionids is subdivided into a more dorsal anal opening and a more ventral opening for water. Some protobranchs and septibranchs and many eulamellibranchs have mantle folds joined in a second place, near but in front of the first, leaving three openings: a dorsal and posterior exhalant opening, a ventral and posterior inhalant opening, and a more ventral pedal opening for the foot. A few pelecypods, especially those with the foot far forward or rudimentary, have a fourth opening for the entrance of water between the pedal and inhalant openings. The mantle folds may be feebly or tightly joined, depending on the species. In many cases, the united mantle folds extend out from the shell, forming inhalant and exhalant siphons. The siphons are extremely valuable to species that bury themselves deeply in the mud, and may be very long.

The buccal region, so important in the food gathering of most molluscs, has lost its function in pelecypods. The buccal cavity is usually much reduced, although a small one is found in *Nucula*, a relatively primitive pelecypod; *Nucula* also has a small pair of salivary glands. A radula is never present, and the mouth leads directly into a short esophagus that carries food to the stomach by ciliary currents.

As in more primitive gastropods, the proximal globular region of the stomach (Fig. 21.38A) contains a sorting area, prominent **gastric shields**, and pores for the digestive glands. The narrowed distal region contains a mucous **protostyle** and the **intestinal groove**.

Larval pelecypods, like gastropods, have asymmetrical **digestive glands**, the left one being the larger. The glands, however, are nearly symmetrical in adults. Originally, each digestive gland opened into the stomach by way of a single pore, but the pores are usually subdivided in modern species. *Mytilus*, for example, has a dozen openings into its digestive glands. In addition to the general function of phagocytosis and intracellular digestion, intestinal glands absorb food already digested, secrete enzymes, and aid in excretion; considerable diversity in histological structure is found. A two-way circulation of material is carried on in the main ducts (Fig. 21.38B).

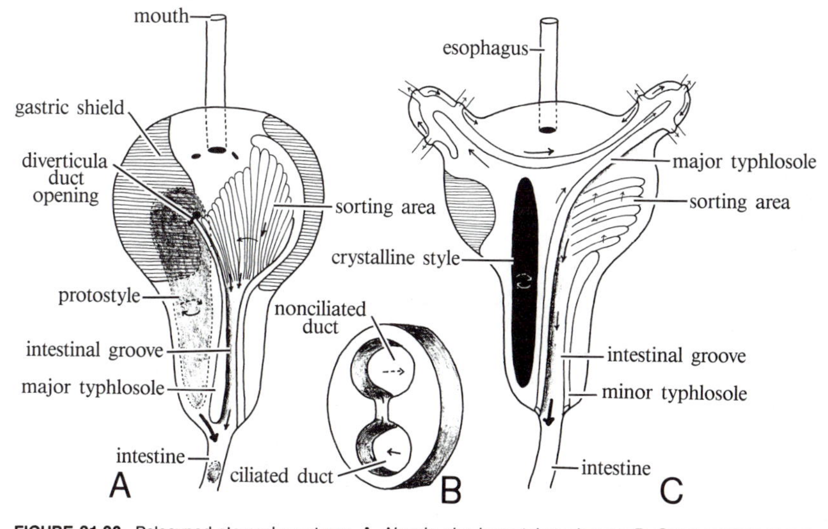

FIGURE 21.38. Pelecypod stomach anatomy. **A.** *Nucula,* simple protobranch type. **B.** Cross section of main duct of digestive diverticula in typical eulamellibranch. **C.** Stomach of typical eulamellibranch. *Small arrows* show protostyle and crystalline style rotation. In other parts of diagrams, *larger arrows* show direction of ciliary currents, whereas *smaller arrows* show sorting or inhalant countercurrents. (After Owen, in Wilber and Yonge.)

The **intestine** is unusually long. It passes through from one to many initial coils, then curves dorsally and eventually enters the pericardial cavity (Fig. 21.35). Here, it turns posteriorly, running immediately above or below the heart in some cases, but usually passing directly through the heart. It continues posteriorly, ending in an anus situated close to the exhalant opening. A dorsal ridge extends over the aperture, preventing the entry of particles except by way of the intestinal groove. The principal func-

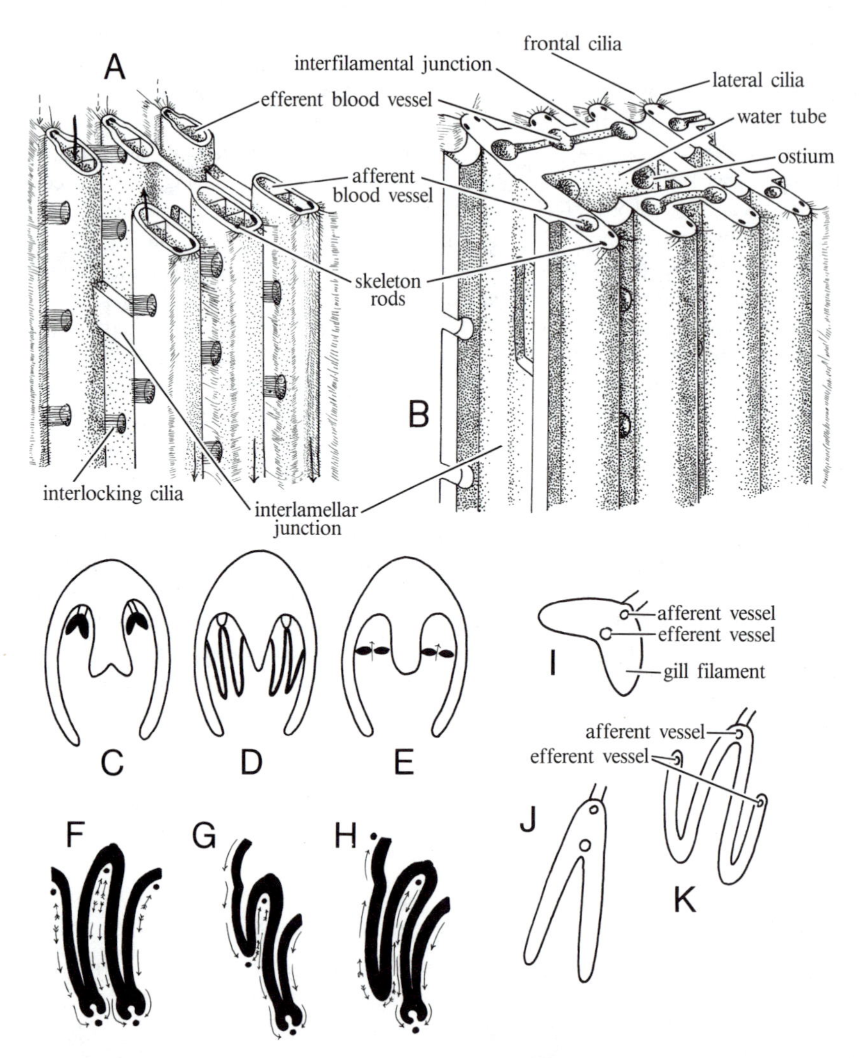

FIGURE 21.39. Pelecypod gills. **A.** Filibranchs have cilia connecting filaments and interlamellar junctions made of tissue to connect ascending and descending lamellae of demibranch. **B.** Eulamellibranchs have added interfilamental junctions composed of tissue that connect adjacent filaments. Interfilamental junctions partially close face of demibranch, leaving ostia for entry of water into water channels within. **C,D,E,I,J,K.** Protobranches have pair of typical ctenidia **I,C** with broad, flat filaments and axis containing afferent and efferent blood vessels. Conversion of filibranch and eulamellibranch gills into filter-feeding apparatus involved lengthening filaments **J,K** and reflecting them to permit further lengthening. As filaments lengthened, efferent blood vessel locates at tip of filament, forming pair of demibranchs on each side of body **D**, each demibranch composed of descending lamella on axis side and ascending lamella on outer side of gill. Septibranchs **E** convert axis of gill and its supporting membrane into muscular pump that pulls water through mantle chamber and ejects it from excurrent passageways above. Respiratory exchange occurs in vascular mantle wall. **F,G,H.** Powerful lateral cilia generate currents to move water between gill filaments, while delicate frontal cilia cleanse particles from surface of protobranch gills and move food particles along surface of gill filaments. At certain points, feeding currents convey particles longitudinally, from filament to filament, indicated by dark dots in **F,G,H**, eventually bringing them to palps. Position of feeding currents and direction of movement of particles over filaments varies considerably in different groups of clams. *Lutraria* **F** has two equal demibranchs and five feeding currents. In *Venus* **G**, very short outer demibranch has three feeding currents. Unionids **H**, outer demibranch slightly shorter with only three feeding currents. (**F,G,H**, after Atkins.)

tion of the intestine is feces formation, achieved by mucus secretion and water absorption.

In these orders, the gills have undergone a profound change, becoming organs for filter feeding as well as for respiration. The gills are very large, often extending the entire length of the mantle cavity, and each gill filament is greatly lengthened (Fig. 21.39I,J,K). Usually, the filaments bend back on themselves; therefore, the whole gill is W-shaped in cross section. Efferent blood vessels lie at the bottom of the central axis in protobranchs, but as the filaments lengthen they divide and come to lie at the tip of the ascending arm of the filament, on each side of the axis. The original ctenidium is thus changed into two **demibranchs**, each consisting of the descending and ascending arms of one row of filaments.

Although filibranchs and eulamellibranchs have gills with filaments similarly lengthened and folded, gill structure provides the major distinction between them. It is evident that gill filaments drawn out into long, recurved strands are mechanically weak. Strengthening is necessary if the gill is to retain any real integrity as an organ. Longitudinal strengthening is needed to hold the series of filaments together, and transverse strengthening to hold the ascending and descending arms together. In both filibranchs and eulamellibranchs, tissue connections between the ascending and descending arms of a demibranch strengthen the gill transversely. Filibranchs use interlocking cilia to hold the series of filaments together, in this resembling the protobranchs (Fig. 21.39A). Eulamellibranchs have tissue connections between filaments (Fig. 21.39B). Connections between the ascending and descending arms of a demibranch are known as **interlamellar junctions**. They do not interfere with the flow of water over the faces of the gill filaments, although they prevent the ascending and descending arms from falling apart. In some of the filibranchs these connections are relatively simple, consisting of connective tissue only. In others, and in eulamellibranchs, blood vessels may pass through the interlamellar junctions. Tissue connections between adjacent filaments, characteristic only of eulamellibranchs, are known as **interfilamental junctions**. These connections interfere, to some extent, with the passage of water over the gill filaments. In some eulamellibranchs, the interfilamental junctions are sparse, but in many, adjacent filaments are connected by large interfilamental junctions, leaving only small openings through which water can enter the space between filaments. These openings, known as **ostia**, open into **water tubes**, bounded by interfilamental and interlamellar junctions. Water enters the gills from the mantle cavity, passing through the ostia into the water tubes. The water passes upward in the water tubes to the exhalant space at the top of the gill. The large surface area exposed makes respiratory transfer a simple matter. The circulation of water through the gill is caused by lateral cilia, as in protobranchs.

Delicate frontal cilia cleanse the surface of protobranch gill filaments. These small cilia have been used to build a food collecting and sorting device in filibranchs and eulamellibranchs (Fig. 21.39F,G,H). They draw small particles over the gill surface to deliver them to ciliated food grooves. The particles move along the food grooves to the palps where they are sorted, and particles of acceptable size are conveyed to the mouth. Most eulamellibranchs have two food grooves, one just below the afferent blood vessel at the gill axis, and one at the point where ascending and descending arms of the inner gill filaments meet, but a number of other patterns occur. *Pinna* and *Mytilus* have five food grooves, one at each junction of ascending and descending arms, one at the gill axis, and one at each junction of ascending arms and the mantle cavity.

Septibranchs have evolved along an entirely independent line. They have no gill filaments, and respiratory exchange occurs at the surface of the mantle cavity (Fig. 21.39E). The gill base of protobranchs is muscular. In septibranchs, only the muscular gill base persists. It forms a perforated muscular septum that contracts rhythmically to pump water into the exhalant channel above it. Water is pulled in with considerable force; worms, small crustaceans, and other animals are drawn into the mantle cavity, where the large palps grab them and pop them into the mouth.

All pelecypods have a pair of **kidneys** open to the pericardial cavities by **renopericardial pores**. The kidneys vary considerably in structure and complexity. The most primitive ones are seen in protobranchs, where they are U-shaped, with two similar arms (Fig. 21.40A). In other pelecypods, and especially in larger species, one tubule is highly folded; therefore, a considerable amount of glandular epithelium is provided (Fig. 21.40B). The kidneys are intimately associated with the circulatory system. Details of circulatory arcs differ somewhat, but the main blood supply is from the venous sinus (Fig. 21.41). In septibranchs the kidneys hang down from the walls of the pallial sinus, completely bathed in blood.

A generalized circulatory scheme is presented in Figure 21.41. Protobranch gills are relatively short and lie toward the back of the mantle cavity. The two **auricles** arise at the end of the **efferent branchial vessels**. They empty into the **ventricle** that lies in a dorsal pericardial cavity and then forces blood into a single anteriorly directed **aorta**. The aorta gives rise to a mantle or **pallial artery**, which supplies blood to the mantle tissues, and a **visceral artery**, which supplies blood to the lacunae around the viscera. A **pedal artery** taking blood to the foot sometimes arises from the visceral artery. The development of strong posterior siphons in some forms favors the development of a posterior aorta, arising from the ventricle and carrying blood to the siphon and posterior part of the mantle cavity. A system of sinuses collect blood from the viscera, the mantle, and the foot. The blood passes by the nephridia, entering a pallial sinus, from which **afferent branchial vessels** arise.

The simplicity of the pelecypod nervous system is undoubtedly related to sedentary habits and poor sensory apparatus. The most primitive arrangements are seen in protobranchs. A neural triangle, composed of **cerebral**, **pleural**, and **pedal ganglia** with pleuropedal and cerebropedal connectives, lies on each side of the body. Cerebral and pleural ganglia are partly united; therefore, cerebropleural connectives are absent, although the ganglia are functionally connected. Commissures connect the two cerebral and two pedal ganglia. A pair of **visceral nerves** arise from the pleural ganglia, ending in **visceral ganglia** that are connected by a commissure. The arrangement is essentially like that of gastropods, but with no evidence of torsion and with the parietal ganglion omitted. In all other pelecypods, the cerebral and pleural ganglia have united to form a single ganglion, and a single pair of connectives pass to the pedal ganglia (Fig. 21.42). There are no buccal ganglia, although a **buccal connective** often passes between the cerebral ganglia on the two sides of the animal. The sympathetic nerves arise from the visceral nerves or the cerebral ganglion.

The reproductive system of pelecypods is very simple. Most of the pelecypods are dioecious, with a pair of gonads opening into a pair of gonoducts. In the more primitive pelecypods, the gonoducts join the nephridia on each side of the body. The point of junction of gonoduct and nephridium varies greatly. It is near the renopericardial pore in some protobranchs and near the nephridiopore in some filibranchs. The relationship of nephridium and gonoduct is lost in eulamellibranchs and some filibranchs. They have a complete gonoduct on each side of the body, ending in gonopores located near the nephridiopores. Hermaphroditism has apparently appeared independently in several groups of pelecypods, for the arrangements differ markedly. Some hermaphroditic species have a single gonad that produces both ova and sperm and opens by way of a single gonoduct. At the other extreme are species with testis and ovary completely separated, and

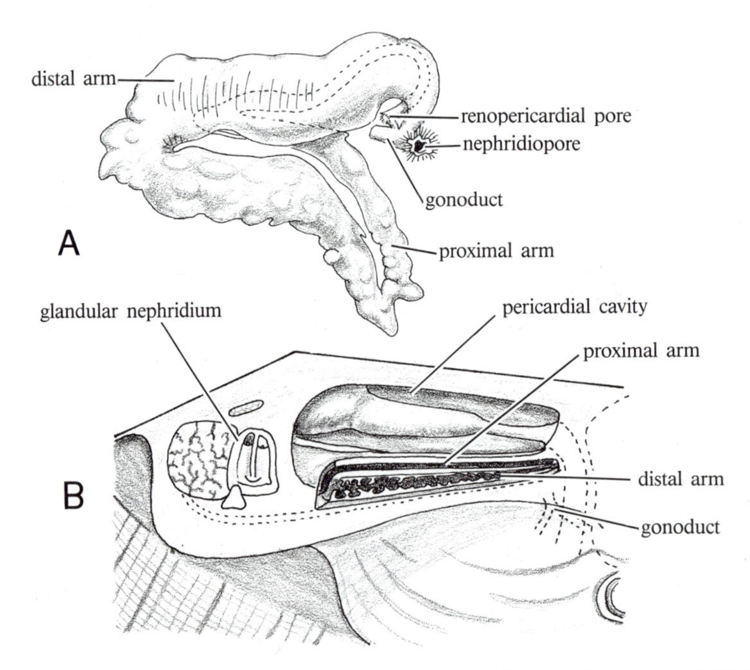

FIGURE 21.40. Pelecypod excretory organs. **A**. Left kidney of *Nuculana,* showing regional differentiation and relation to gonoduct. **B**. Detail of heart region and kidney of *Anadonta.* (**A** after Stempell; **B** after Fernau, from Franc, in Grassé.)

separate gonoducts and gonopores for each. A variety of intermediate arrangements is known.

Biology

Locomotion. The foot is used for locomotion in most pelecypods. In a number of protobranchs, the foot retains a plantar surface (see Fig. 21.36B,C). The plantar surface is withdrawn as the foot is thrust forward, but is expanded when the foot is fully extended. Contraction of the foot muscles pulls the animal forward, after which the anterior tip is constricted and thrust forward again. Most pelecypods have no true plantar surface, but the foot dilates distally when extended. Distal enlargements of the foot and expansion of the plantar surface depends on the engorgement of the foot with blood as well as on contraction of foot muscles. In some cases, pelecypods have developed specialized foot movements. *Cardium*, for example, achieves a sort of jumping movement by sudden foot contractions that push it backward forcibly.

A number of pelecypods are sessile. Oysters are permanently attached by one shell valve, and many pelecypods attach themselves to objects by tough byssal threads secreted by byssal glands in the foot. The secretion of the byssal thread by *Mytilus* occurs when the foot is extended and a thread is attached. As the foot is retracted, the thread is spun out, which hardens on contact with the water. Thread after thread is spun, the foot moving from place to place. Jingle shells are unusual in having byssal threads that pass through an opening in the right shell valve. Generally, the byssal threads extend through the pedal opening between mantle folds. As sessile pelecypods do not move about, it is not surprising to find that the foot is greatly reduced. Ship worms burrow in wood, and some pelecypods burrow into rocks. The foot is not used for this purpose; the shell valves are modified and serve as rasps. Burrowing pelecypods have a much reduced foot.

Scallops swim awkwardly, propelled by jets of water ejected by sudden clapping movements of the shell valves. Water is forced out at both ends of the hinge, pushing the animal toward the ventral margin. Scallops, however, can reverse the direction of movement by expelling water ventrally. *Lima* is a swimming clam that moves in much the same manner. Swimming pelecypods have reduced the foot markedly in adapting to their unusual method of locomotion.

Feeding. Except for the septibranchs, the pelecypods are all ciliary feeders. Fine particles are delivered to the mouth by ciliary currents passing along the palp proboscides or the gills, whereas coarse particles are rejected by sorting areas on the palps (Morton, in Wilbur, vol. 5, 1983).

Small particles enter the digestive glands to be digested intracellularly as they enter the stomach, and larger particles pass over the sorting area to be rejected and sent to the protostyle or sent to the digestive glands. Larger particles are reduced in the front of the protostyle by rotation against the gastric shields, and are resorted. Material from the back of the protostyle enters the intestine. Rejected fine par-

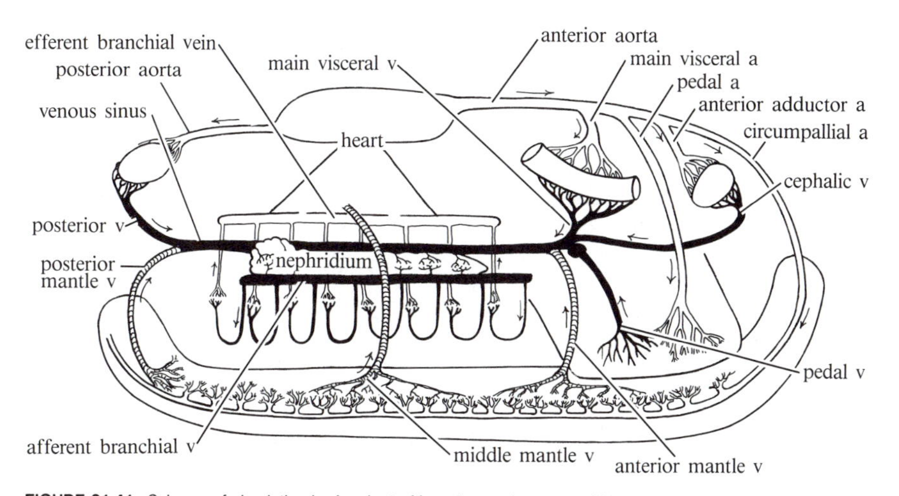

FIGURE 21.41. Scheme of circulation in *Anadonta.* Vessels carrying aerated blood *shown in outline,* and those carrying blood that has lost oxygen in tissues are *solid.* Notice blood passing through mantle partially aerated, and some returns directly to auricle, whereas other blood returns through kidney and gills. Extent to which mantle circulation separate from circulation through the kidney and the gills varies in different pelecypods.

ticles from the digestive glands pass to the intestine along the intestinal groove.

Gill feeders do not ingest coarse particles, and have a **crystalline style** in place of the protostyle (Fig. 21.38C) and reduced gastric shields. Amylases, lipases, and cellulases have been recovered from pelecypod crystalline styles. The rotation of the style, clockwise viewed from the esophageal end, wears away the style substance, releasing its enzymes and starting digestion of particles. Ciliary currents carry particles repeatedly past the entrance to the digestive glands. Digestion is completed in the phagocytic cells of the digestive glands.

Septibranchs have evolved along different lines, pumping larger food organisms into the mantle cavity and capturing them with the large palps. They have a very heavily chitinized stomach, which serves as an effective crushing mill. The style is greatly reduced and is thought to serve primarily to lubricate the food particles with slime, thus protecting the stomach lining. *Solemya reidi*, a gutless pelecypod, lives near sewage outfalls and derives its energy from the metabolism of sulfides directly in its tissues. Symbiotic bacteria in their gills provide carbon compounds for growth. Similar modes of feeding are seen in pelecypods from the deep-sea vents.

Respiration. The great enlargement of the gills of filibranchs and eulamellibranchs has provided a considerable respiratory reserve. Generally only about ten percent of the dissolved oxygen is removed from water passing over the gills. In intertidal clams, exposed to air during low tide, the reserve is somewhat less. *Mya*, for example, removes 25 percent of the oxygen from water passing over its gills for a time immediately after low tide. The low percentages of oxygen removed are related to the large volume of water passing through the mantle cavity, the great gill area, and, in all but the most active pelecypods, a relatively low oxygen consumption. Even scallops use less oxygen per gram of body weight than most snails. The critical oxygen tensions also give evidence of a great respiratory reserve. Oysters, for example, have a critical oxygen tension of 30 mm of mercury, considerably below critical oxygen tensions of *Helix* or squids.

Excretion. Most clams ingest a great deal of water with their food, and a considerable quantity of fluid

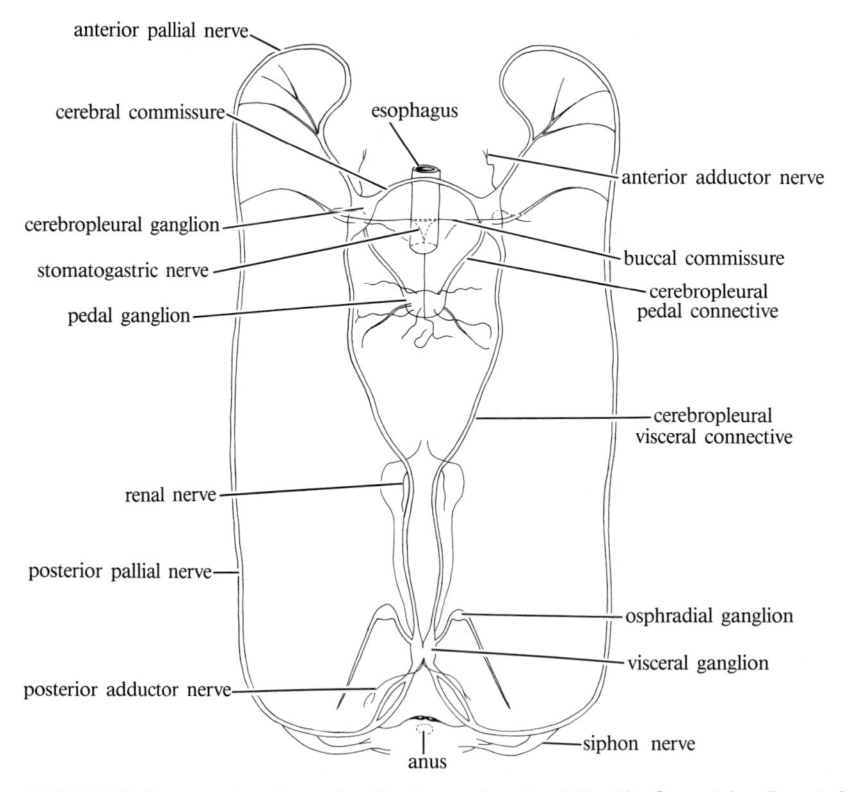

FIGURE 21.42. Nervous system of razor clam, *Tagelus*, seen from dorsal side. (After Stempell, from Franc, in Grassé.)

filters through the pericardial glands in the pericardial cavity. Athrocytic cells also passively take up carmine and are undoubtedly of importance in excretion. The filtrate, probably with additions from the athrocytic cells, passes into the nephridia proper for further treatment. Indigo sulfonate accumulates in the folded walls of the nephridia, indicating that the kidneys are the site of active transport. Evidence of the reabsorption of some ions has also been obtained.

Marine pelecypods are predominantly ammonotelic. Under conditions of osmotic stress, as in freshwater *Unio* and in intertidal species like *Mya* and *Mytilus*, ammonia excretion is sharply reduced. Unlike gastropods, pelecypods never form large amounts of uric acid. When ammonia excretion is cut down, nitrogen is excreted as amino acid or in other forms.

Circulatory System. The development of the glandular nephridium is accompanied by changes in the circulatory system that bring more blood to the nephridium and in some cases set up a more or less independent kidney arc. In *Anadonta*, for example, some of the blood returns to the heart from the kidney without having passed through the gills (Fig. 21.41). The mantle surface is very large, and blood circulated through the mantle is aerated. In a number of pelecypods, blood returning to the heart from the mantle is sent back to the heart, by way of the nephridium, without entering the gill vessels.

During active movement, the foot tissues are gorged with blood, producing a turgor pressure that is important in locomotion. The foot of sessile or swimming pelecypods is small, and pedal circulation is relatively insignificant, but the large foot of most pelecypods contains a considerable part of the blood. In these forms, a large pedal sinus at the tip of the foot drains into the median ventral sinus immediately above it. The opening between the two sinuses is guarded by a valve, which is closed during movement to hold blood in the foot. Various other devices are used to maintain blood pressure in the foot. These include aortic enlargement, booster hearts, and compensation chambers, and occur sporadically in different pelecypod groups. Where siphons are very large, engorgement with blood is a factor in their movement also. In some cases, aortic enlargements on the posterior aorta are found.

Neurosensory Apparatus. Pelecypods have very poor sense organs. The close relationship between the development of the sense organs and movement is shown clearly, for the more mobile species have somewhat better sense organs.

The sense of touch is important, even in sedentary animals. Tactile cells are abundant in exposed parts of the body and are concentrated on the mantle border, especially near the inhalant and exhalant openings. Swimming pelecypods have definite sensory papillae or sensory tentacles on the mantle border. Some of the pelecypods have sensory tentacles associated with the siphons also, as in the case of *Cardium.*

A sensory patch, sometimes pigmented, often lies near the gill attachment. It resembles an osphradium, but is innervated from a ganglion connected to the cerebral ganglion; therefore, the question of homology with gastropod osphradia is uncertain. Its function is in question also, for it is located in the exhalant channel instead of the inhalant channel.

Nearly all pelecypods have a pair of **statocysts** close to or embedded in the pedal ganglia. In protobranchs, the statocysts are deep pits, open to the outside by pores in the sides of the foot. Open statocysts are also found in some filibranchs (e.g., *Mytilus*). Eulamellibranchs and septibranchs have closed statocysts. A single large statolith is often present, while in other cases there are many small ones. In a few cases, a large one and small ones occur together. Although the statocysts are gravity orientors, organs of this kind can be used for detecting vibrations.

Larval pelecypods have **ocelli**, but the eyes are almost always lost during metamorphosis. They persist as cephalic eyes in the Mytilidae, but as they are inside the shell, they are of doubtful value. In other pelecypods, light reception has become a function of the mantle border equipped with simple pigmented cells in some cases, but some pelecypods have fairly complex ocelli. Photosensitive elements are often concentrated near the siphons. A simple type of compound eye is formed in the Arcacea by clusters of pigmented cells. The shallow-water scallops sometimes have complexly organized eyes along the mantle border. *Cardium* has equally complex eyes located on the sensory tentacles of the siphon.

Control over some specific functions is vested in specific ganglia. *Mytilus*, for example, loses its ability to open and close the shell in response to changes in water salinity when the visceral ganglia are removed, and the pedal ganglia are required for creeping movements and the spinning of byssal threads.

Reproduction. Methods of reproduction are fairly simple (Giese and Pearse, 1979; Mackie, in Wilbur, vol. 7, 1984). Both gametes may be shed freely. Typically, species that brood ova to a planktotrophic stage usually breed several times. Those that retain the ova until they are freed at a juvenile phase often die at the close of the season.

Freshwater sphaerids and unionids incubate the young in the gills. The velum is greatly reduced, and the shell valves develop quickly. Sphaerids develop directly into juveniles when they emerge from the parent. Unionids, however, are released as **glochidia** (Fig. 21.43C), small shelled larvae that are obligatory parasites. Some swim about by clapping the shell valves together, while others lie in wait on the bottom with the valves open wide. Within a few days they must attach to the gills, fins, or body surface of a fish. The shell valves are clamped together on the host tissue; in some cases, they are armed with glochidial hooks on the shell valves. Host epidermis grows around them, and they continue development in the resulting cyst. Phagocytic mantle cells feed on the host tissues. After two to six weeks, the young clams break out of the cyst and fall to the bottom. They attach themselves temporarily with a byssus and become juveniles. Most species require several years to reach maturity.

Development. Pelecypods give a good illustration of the principle that more primitive groups follow a more primitive course of development, and that larger, yolkier eggs, and eggs that are brooded, undergo modified development. In some of the protobranchs the **velum** is reflected, forming rows of ciliated **test cells** (Fig. 21.43A). The larva resembles an amphineuran larva or the larva of *Dentalium*. The test cells are suddenly shed. The abrupt dispersal of the test is reflected in the habit of casting off the velar cells as metamorphosis begins. Without the velum, the larva settles to the bottom, and sessile forms like the oyster attach to objects.

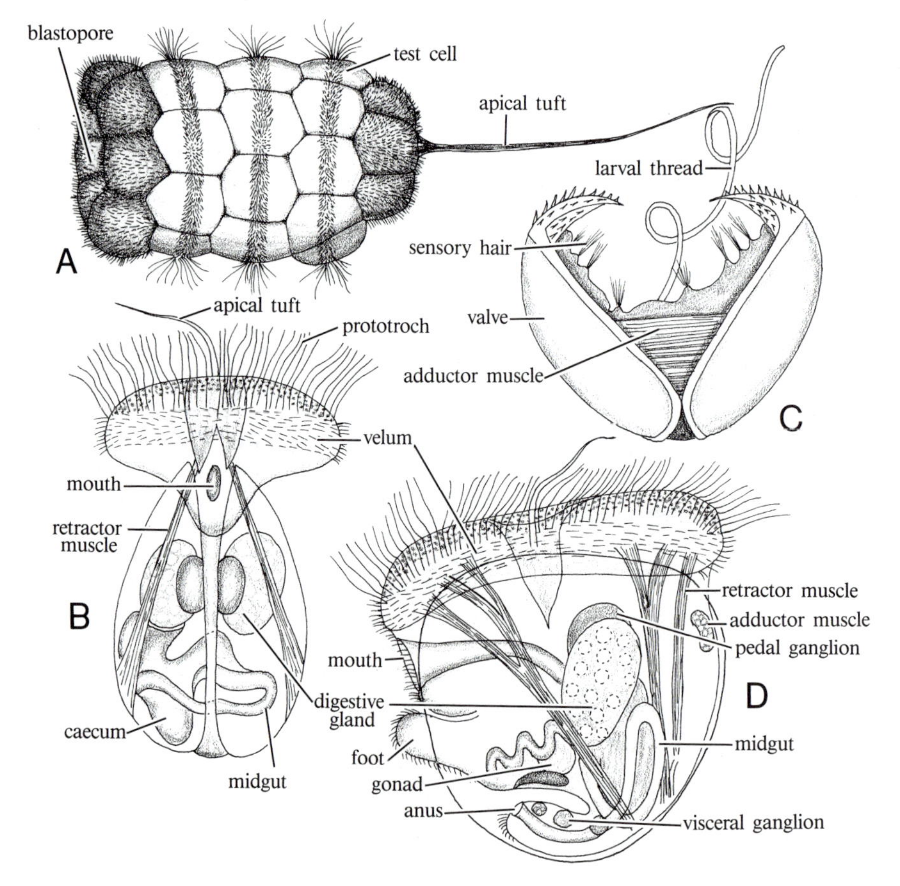

FIGURE 21.43. Pelecypod development. **A**. Trochophore of protobranch, *Yoldia*. Note five rows of ciliated test cells, which shed suddenly during development as in Aplacophora. **B,D**. Two views of pelecypod veliger. Veliger remains symmetrical, as torsion does not occur in pelecypods, and conspicuous bilobed form of velum, so common in gastropods, not apparent. **C**. Glochidium of freshwater clam, *Anadonta*. Glochidium passes through period of parasitism on fish gill. Larval thread protrudes from shell and probably aids in attaching to host. Larval shell bears hooks that clamp onto gill filaments of host fish. (**A** after Drew; **B,D** after Meisenheimer; **C** after Pennak.)

The pelecypod **veliger** has several peculiarities. It is organized like the gastropod veliger, but the velum is circular rather than bilobed. The apical tufts of cilia are often fused to form a conspicuous "**flagellum**." The larval foot is invaginated at the larval **byssal gland**, even in species that have no adult byssal threads. The larval byssal thread is used in a variety of ways. It sometimes attaches the young to the maternal gills and sometimes makes a temporary attachment to pebbles or sand. Gill formation follows a regular pattern. The most anterior gill filament appears first, and the others appear in sequence. The larval gills, unlike the adult gills, are never reflected. Folding occurs during the growth and metamorphosis of the juvenile clams. As torsion does not occur, the veliger retains its bilateral symmetry (Fig. 21.43B,D); the shell is laid down originally in a bivalved form.

Fossil Record and Phylogeny

Molluscs have been around since the Cambrian Period, and the interpretation of their earliest record has been hindered by a number of enigmatic forms of which we know little. In fact, there is some debate as to whether some of such groups like Stenothecoidea (Yochelson, 1969) and Hyolitha are molluscs, and to just which groups taxa like the polyplacophoranlike *Matthevia* or the clamlike *Fordilla* belong (Pojeta and Runnegar, 1974). Some of these fossil groups are major taxa and afford some insight into the diversity of the early mollusc radiation (Runnegar and Jell, 1976). The Paragastropoda are coiled but untorted Paleozoic forms that superficially resemble gastropods (Linsley and Kier, 1984). The Rostroconchia are bivalved molluscs that grew from an uncoiled, univalved larval shell; had no hinge teeth, ligaments, or adductor muscles; had a fused, inflexible hinge; and are related to pelecypods and scaphopods.

Although a discussion of evolution within molluscan classes is beyond the scope of a text such as this, current understanding of the major living and fossil classes does allow at least a reasonable assessment to be made of relationships among the classes. This outlook is due in large part to two factors: (1) the recognition of primitive forms of most classes in the fossil record that clearly show evidence of serial arrangement of at least some aspects of the soft anatomy (Runnegar and Pojeta, 1974, 1985), and (2) the verification of the serial nature of several organ systems in living monoplacophorans (Wingstrand, 1985). Three major groups of Mollusca are recognized (Fig. 21.44): the Aplacophora with their wormlike bodies lacking a shell, the Polyplacophora with their plated shells, and the conciferans sharing a host of features related to shell structure and nervous specializations. The Polyplacophora and conciferans are united by some authorities as the Testaria, denoted by the possession of larval velum, eight pedal retractors, and radular specializations. Conciferans are composed of three groups that cannot be sorted adequately yet: the Monoplacophora, retaining the primitive conciferan body plan; the cyrtosomes (gastropods and cephalopods) with their dorsoventral lengthening of the body; and the diasomes (pelecypods and scaphopods) with their degeneration of the head.

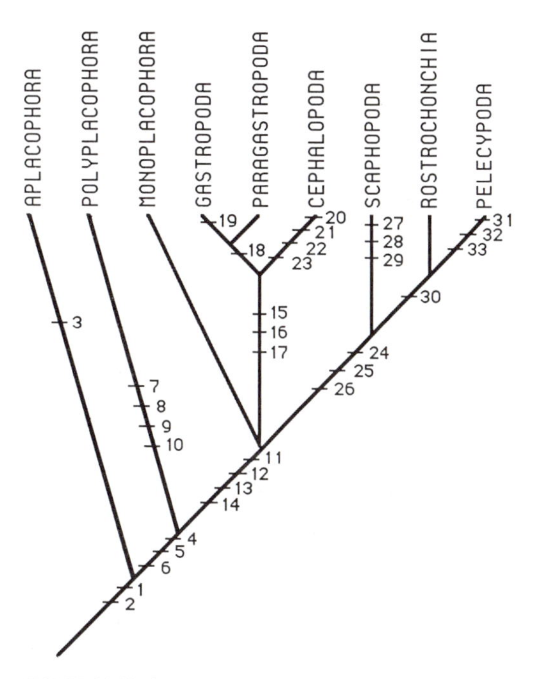

FIGURE 21.44. Cladogram depicting relationships of major molluscan classes. Derived characters are: (1) epidermis as mantle; (2) radula in buccal cavity; (3) reduction of foot; (4) velum on larvae; (5) radula elaborated and specialized; (6) eight pedal retractors; (7) sensory aesthetes in shell; (8) eight plated shell; (9) articulatory apophyses; (10) multiplicity of gills; (11) shell with surface periostracum and prismatic and nacreous layers; (12) margin of mantle as three folds; (13) larval shell; (14) statocysts; (15) dorsoventral extension of body; (16) "posterior" pallial groove; (17) well-developed head; (18) body coiled; (19) body torted; (20) foot developed as arms and tentacles; (21) spacious coelom; (22) highly developed nervous system and sense organs; (23) closed circulatory system; (24) degenerate head; (25) mantle developed as two folds; (26) cerebral and pallial ganglia fused; (27) lobes of mantle form a tube; (28) shell tubular; (29) captacula; (30) shell bivalved; (31) hinged shell; (32) tendency for gills to specialize for feeding; and (33) loss of radula. (Derived from Wingstrand, 1985.)

The central position of the monoplacophorans is seconded by the existence of transitional forms for the advanced classes with monoplacophoranlike serial features, namely, the sutured monoplacophoran *Knightoconus* for cephalopods (Yochelson et al., 1973), the bellerophontid gastropods (Runnegar, 1981), the rostroconch *Riberia* for scaphopods and advanced rostroconchs (Pojeta and Runnegar, 1976), and the lucinoid *Babinkia* for the pelecypods (Wingstrand, 1985). All these examples combine to yield a coherent view that throughout molluscan history there was a repeated tendency within various lines to deemphasize and lose the serial character of the basic body plan in favor of adaptive experiments that exploit relevant aspects of specific body forms, for example, cephalization in the cyrtosomes and decephalization in the diasomes.

Taxonomy

The following classification of the Mollusca focuses only on the living forms and does not include the numerous and often problematic fossil types.

Class Aplacophora. Vermiform molluscs, poorly cephalized, with terminal mouth and anus; body surface provided with spicules; gills in posterior mantle cavity, or missing; digestive tube straight, typically with radula; one pair of renal organs open to pericardial cavity, and used for discharge of gametes.

Order Neomenioidea (= Solenogastres). With midventral, longitudinal groove. Example: *Paramenia* (Fig. 21.9A).

Order Chaetodermatoidea (= Caudofoveata). Without foot or midventral groove; posterior cloacal chamber. Example: *Chaetoderma* (Fig. 21.9B).

Class Polyplacophora. Shell of eight plates with sensory aesthetes and articulatory apophyses (Fig. 21.11).

Order Lapidopleurida. Gills posterior and few in number; insertion plates lacking articulamentum layer on plates 2 through 8 as apophyses only.

Order Ischnochitonida. Girdle well developed but not covering plates; articulamentum with insertion plates and apophyses.

Order Acanthochitonida. Girdle covers plates; tegmentum reduced.

Class Monoplacophora.

Order Trybilidea. Univalved caplike shell; serial arrangement of soft anatomy (Fig. 21.12).

Class Gastropoda. Body coiled and torted; well-developed head; body dorsoventrally elongated; pallial groove restricted to posterior body.

Subclass Prosobranchia. Torted; primitively with pair of ctendia, but usually with single gill and auricle; primitively with pair of nephridia, but generally with only right nephridium retained; nervous system twisted into figure eight, which may be partially corrected by some concentration of nerves and ganglia near head.

Order Archeogastropoda. Prosobranchs with primitive mantle cavity without gills, or containing pair of ctenidia or single ctenidium; heart with pair of auricles, although one may be reduced in size; primitively with notch or slit in shell for anus and exhalant siphon, but with imperforate shell when ctenidia reduced. Examples: Ear shells, *Haliotis* (Fig. 21.13C); keyhole limpets, *Fissurella* (Fig. 21.15D); slit limpets, *Emarginula* (Fig. 21.15E); top shells, *Trochus*; neritas, *Nerita*; and cat's eyes, *Turbo.*

Order Mesogastropoda. Prosobranchs with single gill, auricle, and nephridium; with or without operculum; shell without slit band or hole for exhalant siphon and usually without groove for inhalant siphon; gill almost invariably one-sided and attached for its whole length to wall of mantle cavity (monopectinate); usually with well-developed, sometimes pectinate osphradium; sexes separate; male usually with penis. Examples: Pelagic snails, *Janthina* (Fig. 21.22C); moon shells, *Natica* (Fig. 21.22D);

Polynices; slipper shells, *Crepidula* (Fig. 21.22A); periwinkles, *Littorina*; worm shells, *Vermetus*; cowries, *Cypraea*; freshwater snails of many kinds; *Valvata*, an exception with a bipectinate gill; *Campeloma*; *Vivipara*; pelagic heteropods, *Carinaria* (Fig. 21.23E); and *Atlanta*.

Order Neogastropoda. Bipectinate osphradium; unpaired esophageal gland, becoming poison gland in some; strongly concentrated nervous system; with penis. Examples: Boring murexes, *Murex*, *Thais*; mud snails, *Nassarius* (Fig. 21.24C); whelks, *Buccineum*; conchs, *Busycon*; olive shells, *Olivo*; and cone shells, *Conus* (Fig. 21.13K).

Subclass Opisthobranchia. Marine gastropods with single auricle and nephridium and with one or no ctenidium; often with surface gills; nervous system untwisted by detorsion and often relatively centralized; with strong tendency toward shell reduction and return toward bilateral symmetry.

Order Cephalaspidea. With ctenidium and usually with shell; male genital groove open; mantle cavity well developed; with cephalic shield; usually with gastric shield for grinding. Example: *Bulla*, bubble shells.

Order Anaspidea. Shell poorly developed or missing; with well-developed natatory parapodia; mantle cavity small, on right side; male genital groove open; with several gastric shields and large radula with median tooth; ctenidium and jaws present; visceral hump attached along foot. Example: Sea hares, *Aplysia*.

Order Thecosomata. Variable shell development; with mantle cavity, but with gill usually reduced; with strongly developed, natatory epipodia; male genital groove open; nerve ganglia centralized and united.

Order Gymnosomata. Without shell or mantle cavity; with gill behind heart; foot with parapodia; male genital groove open; nerve ganglia free.

Order Acochlidiacea. Without shell, gill, or cephalic shield; with rhinophores; visceral mass distinct from foot; without jaws but with radula; without gastric shield; with nerve ring in front of buccal mass. Example: *Acochlidium*.

Order Saccoglossa. Jawless, herbivorous forms with rhinophores; with one series of teeth in sac; male genital groove closed over; never with cnidosacs containing nematocysts; nerve ganglia but little fused. Example: *Styliger*.

Order Acoela. Shell-less adults; with visceral mass attached along foot; without parapodia; with male genital groove closed over; with nerve ring behind buccal mass. Examples: *Doris* and *Eolis* (Fig. 21.24E).

Order Notaspidea. Shell usually internal; without parapodia or gastric shield, but some with chitinous spines; with one gill; male genital groove closed over; radula variable, but without radula sac. Example: *Pleurobranchia*.

Subclass Pulmonata. Predominantly terrestrial and freshwater snails without ctenidia, mantle cavity modified as respiratory sac, used for air or water breathing; with one auricle and nephridium; nervous system bilaterally symmetrical from concentration of nerve tissue to form complex brain; almost universally without operculum; shell sometimes lacking; hemaphroditic, producing yolky ova that develop without passing through larval stages.

Order Stylommatophora. Almost exclusively land pulmonates with anterior pair of tactile and probably chemoreceptive tentacles, and posterior pair of tentacles with eyes at tips; usually with common genital antrum into which both male and female organs open. Examples: land snails, such as *Helix*, *Polygyra*, *Succinea*, and *Zonitoides*; and slugs, such as *Arion* and *Limax* (Fig. 21.19).

Order Basommatophora. Predominantly aquatic pulmonates found in fresh water and rarely in marine habitats; with single pair of tentacles; usually with separate male and female gonopores. Examples:

Freshwater snails, such as *Lymnea*, *Helisoma* (Fig. 21.13B), and *Physa*; and a few saltwater forms, such as *Melampus*.

Class Cephalopoda. Head well developed with arms and tentacles modified from foot; closed circulatory system; highly advanced nervous system.

Subclass Nautiloidea. Cephalopods with coiled or straight external shells, divided into chambers by straight septa that meet shell in simple sutures. (At least 14 orders are recognized, of which only Nautilida is extant.) Example: *Nautilus* (Fig. 21.27A).

Subclass Ammonoidea. Cephalopods with coiled external shells, divided into chambers by variously shaped septa meeting shell in complex sutures; typically with external sculpturing of shell. (About 5000 extinct species have been described.) Example: *Idoceras* (Fig. 21.27B).

Subclass Coleoidea. Single pair of gills, and with internal shell considerably reduced or without shell.

Order Belemnoidea. Extinct coleoids with internal shell having well-developed apical part, rostrum; dorsal, platelike proostracum; and chambered phragmocone. Examples: *Baculites* (Fig. 21.27C) and *Belemnoteuthis* (Fig. 21.27E).

Order Sepioidea. Shell with reduced rostrum, modified proostracum, and rudimentry phragmocone; with ten tentacles and well-developed coelom; with fourth pair of tentacles specialized and equipped with special pit into which they can be retracted. Examples: *Sepia* (Fig. 21.27F) and *Spirula* (Fig. 21.27D).

Order Teuthoidea. Without rostrum; with phragmocone rudimentary or missing; proostracum modified to form pen-shaped gladius; with ten arms; fourth arms may be modified, never withdrawn into special pits. Examples: *Architeuthis*, the giant squid (Fig. 21.28A), and *Loligo*, the common squid (Fig. 21.28B).

Order Octopoda. Eight tentacles; reduced coelom; with shortened, rounded body, usually without vestige of shell. Example: *Octopus* (Fig. 21.25C,D,E).

Class Scaphopoda. Reduced head and sense organs; mantle flaps forming tube; shell tubular; captacula; cerebral and pleural ganglia fused (Fig. 21.34).

Order Dentaliida. Captacula thin and numerous; foot well developed; hepatopancreas bilobed.

Order Gadilida. Captacula thick and few in number; foot reduced; hepatopancreas as single lobe.

Class Pelecypoda. Reduced head and sense organs; mantle bilobed into right and left halves; shell bivalved, dorsally hinged; highly muscular foot; gills well developed and often specialized to assist in feeding; cerebral and pleural ganglia fused.

Order Protobranchia. Simple, ctenidalike gills, not reflected, and not used for feeding; hypobranchial glands at outer surface of each gill; foot with ventral plantar surface; large palps, modified for feeding, and typically with long palplike proboscides. Examples: *Solenomya*; *Nucula*; *Yoldia* (Fig. 21.36A); and *Malletia*.

Order Filibranchia. Pelecypods with reflected gills, used for feeding; adjacent gill filaments connected by tufts of interlocking cilia; without interfilamental junctions; without hypobranchial glands; foot usually equipped with byssal gland; palps small and lamellar. Examples: *Anomia*, jingle shells; *Arca*, ark shells; *Mytilus* and *Modiolus*, common mussels; and *Pecten*, scallops.

Order Eulamellibranchia. Gills composed of reflected filaments, with adjacent filaments joined by interfilamental junctions through which blood vessels pass; no hypobranchial glands; gonads with gonopores; usually with two adductor muscles and with mantle margins joined at one or two sutures. Examples: *Ostrea*, oysters; *Cardium*, cockles; *Unio*, *Sphaeridium*, *Quadrula*, freshwater clams; *Ensis*, *Tagelus*, razor clams; *Venus*, hard-shell clam; *Mya*, soft-shell clam; *Teredo*, ship worm; and *Petricola*, boring clam.

Order Septibranchia. Pelecypods with gills replaced by muscular septa used to pump water through mantle cavity; mantle with two sutures and two siphons; long slender foot without well-developed byssal gland. Examples: *Poromya* and *Cuspidaria.*

Selected Readings

Beedham, G. E. 1968. The cuticle of the Aplacophora and its evolutionary significance in the Mollusca. *J. Zool. Lond.* 154:443–451.

Beedham, G. E., and E. R. Trueman. 1967. The relationship of the mantle and shell of the Polyplacophora in comparison with that of other Mollusca. *J. Zool. Lond.* 151:215–231.

Fretter, V., and J. Peake (eds.). 1975. *Pulmonates*, vol. 1. Academic Press, London.

Giese, A. C., and J. S. Pearse. 1977 and 1979. *Reproduction of Marine Invertebrates*, vols. 4 and 5. Academic Press, New York.

Gotting, K.-J. 1980. Origin and relationships of Mollusca. *Z. zool. Syst. Evolut.-forsch.* 18:24–27.

Hyman, L. H. 1967. *The Invertebrates*, vol. 6. McGraw-Hill, New York.

Kaas, P., and R. A. Van Belle. 1985. *Monograph of Living Chitins*, vol. 1. Brill, Leiden.

Linsley, R. M. 1978a. Locomotion rates and shell form in the Gastropods. *Malacol.* 17:193–206.

Linsley, R. M. 1978b. Shell form and the evolution of gastropods. *Am. Scient.* 66:432–441.

Linsley, R. M., and W. M. Kier. 1984. The Paragastropoda: A proposal for a new class of Paleozoic Mollusca. *Malacol.* 23:241–254.

McAlester, A. L. 1965. Systematics, affinities, and life habits of *Babinka*, a transitional Ordovician lucinoid bivalve. *Palaeontol.* 8:231–246.

Morton, J. E. 1979. *Molluscs*, 5th ed. Hutchinson, London.

Nixon, M., and J. B. Messenger (eds.). 1977. *The Biology of Cephalopods.* Clarendon Press, Oxford.

Pojeta, J., and B. Runnegar. 1974. *Fordilla troyensis* and the early history of pelecypod mollusks. *Am. Scient.* 62:706–711.

Pojeta, J., and B. Runnegar. 1976. The paleontology of rostroconch mollusks and the early history of the phylum Mollusca. *Geol. Surv. Prof. Paper* 968:1–88.

Purchon, R. D. 1977. *The Biology of Mollusca*, 2nd ed. Pergamon Press, Oxford.

Rudman, W. B. 1981. The anatomy and biology of alcyonarian-feeding aeolid opisthobranch molluscs and their development of symbiosis with zooxanthellae. *Zool. J. Linn. Soc.* 72:219–262.

Runnegar, B. 1981. Muscle scars, shell form and torsion in Cambrian and Ordovician univalved molluscs. *Lethaia* 14:311–322.

Runnegar, B., and P. A. Jell. 1976. Australian Middle Cambrian molluscs and their bearing on early molluscan evolution. *Alcheringia* 1:108–138.

Runnegar, B., and J. Pojeta. 1974. Molluscan phylogeny: The paleontological viewpoint. *Science* 186:311–317.

Runnegar, B., and J. Pojeta. 1985. Origin and diversification of the Mollusca. In *The Mollusca*, vol. 10 (E. R. Trueman and M. R. Clarke, eds.), pp. 1–57. Academic Press, New York.

Runnegar, B., et al. 1975. Biology of the Hyolitha. *Lethaia* 8:181–191.

Salvini-Plawen, L. 1980. A reconsideration of systematics in the Mollusca. *Malacol.* 19:249–278.

Vogel, K., and W. F. Gutmann. 1981. The derivation of pelecypods: Role of biomechanics, physiology, and environment. *Lethaia* 13:269–275. [Nice functional morphologic treatment of bivalve history.]

Voltzow, J. 1983. Flow through and around the abalone *Haliotis kamtschatkana. Veliger* 26:18–21.

Voltzow, J. 1985. Morphology of the pedal circulatory system of the marine gastropod *Busycon contrarium* and its role in locomotion. *Zoomorph.* 105:385–400.

Wilbur, K. M. 1983–1986. *The Mollusca.* Academic Press, New York. [Ten-volume compendium on all aspects of biology.]

Wingstrand, K. G. 1985. On the anatomy and relationships of Recent Monoplacophora. *Galathea Repts.* 16:7–94, 12 pls.

Yochelson, E. L. 1969. Stenothecoida, a proposed new class of Cambrian Mollusca. *Lethaia* 2:49–62.

Yochelson, E. L. 1978. An alternative approach to the interpretation of phylogeny of ancient molluscs. *Malacol.* 17:165–191.

Yochelson, E. L., R. H. Flower, and G. F. Weber. 1973. The bearing of the new late Cambrian monoplacophoran *Knightoconus* upon the origin of Cephalolpods. *Lethaia* 6:275–309.

Young, J. A. 1971. *The Anatomy of the Nervous System of Octopus vulgaris.* Clarendon Press, Oxford.

22

Nemertinea

Nemertinea, or ribbon worms, also called Rhynchocoela or Nemertea, are abundant in the littoral zone of temperate oceans, where they live among rocks and algae or in mucous tubes embedded in mud or sand. A few have invaded fresh water, and some live in humid tropical and subtropical land habitats. The smallest are only a few millimeters long, but the longest, *Lineus longissimus*, is reported to reach 30 m. Most species are pallid or dull in color, but many are brightly pigmented. They are a relatively successful group, with about 800 known species and estimates of perhaps 2200 more species yet to be discovered.

In the past, nemertines have been compared to flatworms. Yet, their body plan is beyond anything seen at the acoelomate level. Discovery of the coelomic nature of their blood vessels, gonosacs, and the unique rhynchocoel cavity now places nemertines among more advanced phyla (Riser, in Bierne, 1985).

Definition

Nemertinea have a closed circulatory system, formed from epithelial-lined cavities in the mesoderm and containing a variety of pigments. The digestive system is complete, with mouth and anus. The coelom is developed as circulatory vessels, gonadal sacs, and the rhynchocoel. A unique ensnaring proboscis is carried in the rhynchocoel. There is a highly developed muscular and nervous system. Protonephridia are well developed, with flame bulbs.

Body Plan

The body plan of nemertines (Fig. 22.1) is wormlike, being either flattened or tubular. The histologic structure of the body wall is rather complex. Interstitial cells in the epidermis sometimes form a syncytium, and are crowded in between the narrowed bases of the ciliated cells (Fig. 22.2B,C). Nemertines employ mucus for many uses, and leave slime trails behind them. The epidermis has several kinds of unicellular glands and small clusters of gland cells (Fig. 22.2B) that exit through a common duct (packet glands). Similarities to the integument of tubellarian flatworms are superficial or based on primitive characters (Norenburg, in Bierne, 1985). In heteronemerteans, packet glands lie in the dermis (Fig. 22.2C).

A **dermis** lies below the epidermis. It is a thin, hyaline, gelatinous connective tissue in palaeonemerteans and hoplonemerteans (Fig. 22.2B); but it is differentiated into an outer hyaline layer and an inner fibrous region (Fig. 22.2C) and often contains some muscle fibers in heteronemerteans.

The complex body wall muscle layers differ markedly in the various nemertine groups. Palaeonemerteans and hoplonemerteans have an outer layer of circular and an inner layer of longitudinal muscles (Fig. 22.3A,C), and some have a second circular layer inside the longitudinal layer. That order is reversed in heteronemerteans, with outer longitudinal, middle circular, and inner longitudinal layers (Fig. 22.3B). The muscle layers of the everted proboscis correspond to those of the body wall and are reversed in the inverted proboscis.

The connective tissue of the body is highly specialized. It is represented by the dermis and by a gelatinous matrix that surrounds the blood vessels, nerves, and proboscis sheath and contains vesicular cells. The body wall musculature increases at the expense of the connective tissue. When it is very thick, little connective tissue is present.

The position of the nerve trunks also is characteristic of the orders. Some palaeonemerteans have epidermal nerve trunks, but they are located generally just below the epidermis (Fig. 22.3A). In other palaeonemerteans and in heteronemerteans, the nerve trunks are in muscle layers (Fig. 22.3B). In hoplonemerteans and bdellonemerteans, the trunks are in the connective tissue, completely within the muscle layers (Fig. 22.3C).

The proboscis is *not* a part of the digestive system. It arises separately and, although it is important in food capture, it is no more a part of the digestive system than eyes or other sense organs that also aid in capturing food. The **proboscis** apparatus consists of several parts (Fig. 22.2A). The **rhynchodeum** is a tubular cavity, open at and just inside the proboscis pore, continuous posteriorly with the cavity of the proboscis. The proboscis is a narrow, blind tube

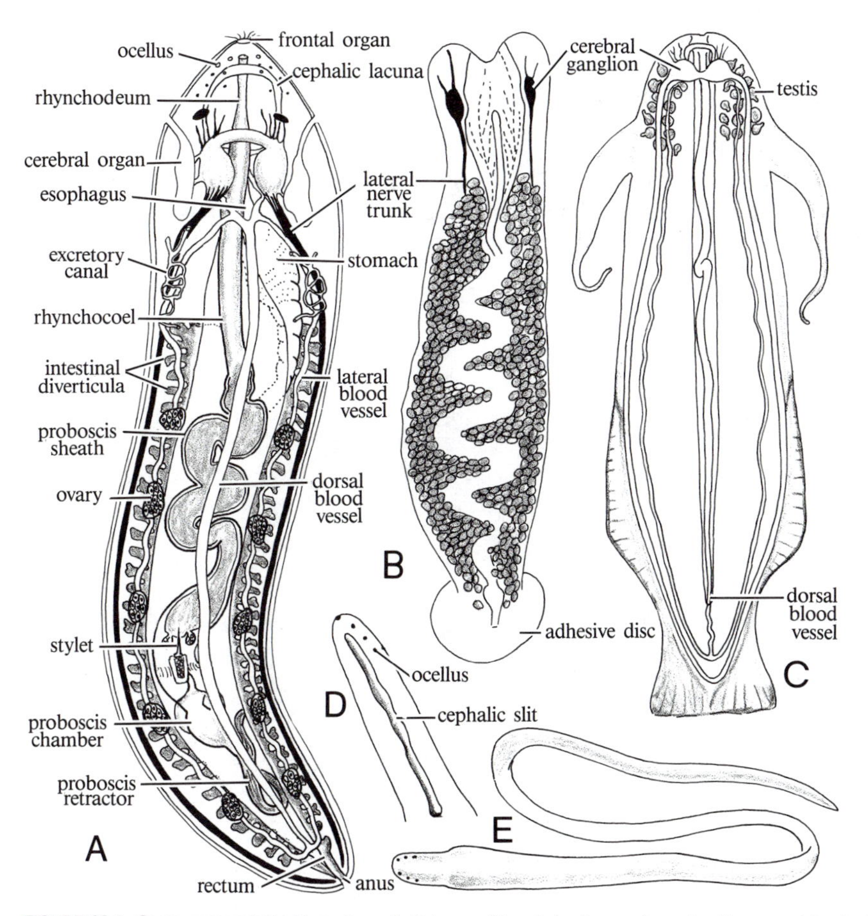

FIGURE 22.1. Some representative Nemertinea. **A.** Scheme of female hoplonemertean, *Amphiporus pulcher,* in dorsal view, with mouth in front of the brain and the proboscis armed with stylets. **B.** Bdellonemertean, *Malacobdella,* with brain behind mouth, unarmed proboscis, and posterior adhesive disc. **C.** Pelagic hoplonemertean, *Nectonemertes mirabilis,* with lateral and caudal fins used in swimming. **D,E.** Heteronemertean, *Lineus ruber,* showing head from side view and whole worm. (**A** after Bürger, from Goodchild, in Brown; **B** after Guberlet; **C** after Coe, from Pratt; **E**, after Hyman.)

with muscular walls and a glandular lining (Fig. 22.1A). It is often much longer than the body and forms coils in the **rhynchocoel**, the epithelial-lined, fluid-filled coelomic cavity into which the proboscis is withdrawn. Strong contractions of the muscles around the rhynchocoel raise the fluid pressure explosively to turn the proboscis inside out, evert it to the outside, and bring its sticky, glandular surface in contact with the prey. The stylet of the armed type of proboscis (Fig. 22.1A) wounds the prey and pours a toxic secretion into it. The proboscis then wraps around the prey and pulls it to the mouth. Prey is generally swallowed whole, but some nemertines can suck out the tissues of animals that are too large to swallow. A **retractor muscle** attaches to the posterior end of the proboscis and is used to haul in the everted proboscis.

Anopla have a simple, unarmed proboscis with rhabdites, and sometimes with nematocysts, in its lining. Hoplonemerteans have an armed proboscis, with one or more stylets. A **stylet** is a spine, sitting on a prominence, the **stylet base**. The armed proboscis is divided into a long anterior region, a bulbous middle region, and a posterior region containing the toxic secretion (Fig. 22.1A). The stylet sits on a glandular partition (**diaphragm**) that contains a pore to permit the escape of venom from the posterior region. Only the anterior part of the armed proboscis everts, bringing the stylet to the tip.

Digestive System Nemertines have a complete digestive tract, and in many species, regional specialization of the gut is evident. Thick, glandular lips lubricate the elastic mouth to enable large organisms to be swallowed. The esophagus of many hoplonemerteans opens into the rhynchodeum, the mouth having been lost.

More highly specialized nemertines have a foregut, composed of a **buccal cavity**, an **esophagus**, and a **stomach**; a midgut, composed of the **intestine**; and a hindgut, composed of a short **rectum**. The regions can sometimes be distinguished only by the histology of the lining cells, but are externally visible in hoplonemerteans. The stomach is variable; in some forms it is drawn out into a **pyloric tube** lying on top of the anterior part of the intestine. The intestine is often simple but usually has lateral **diverticula**, sometimes alternating regularly with the gonads (Fig. 22.1A). If the stomach opens into the intestine from above, an anterior **caecum** extends forward below the stomach. Diverticula disappear near the posterior end of the body, and the intestine continues to the anus as a rectum.

Circulatory System Blood corpuscles (Fig. 22.5B) can contain yellow, green, or orange pigments and also sometimes contain hemoglobin. Several kinds of amoeboid lymphocyte cells also occur in the blood of some nemertines.

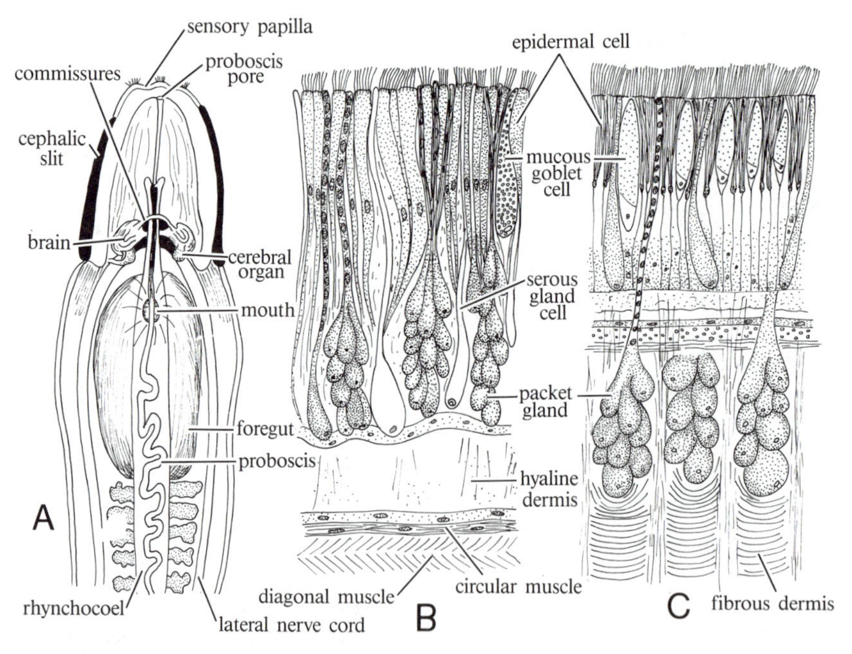

FIGURE 22.2. **A.** Dorsal view of front end of *Cerebratulus*. **B.** Epidermis of *Tubulanus*, palaeonemertean with gelatinous dermis. **C.** Epidermis of *Baseodiscus*, heteronemertean with fibrous dermis. (All after Bürger.)

Blood flow is not highly organized; blood may ebb and flow in a vessel, and transport is less efficient than in a system with one-way traffic following a definite circuit. The lack of valves and a central pump makes effective blood routing impossible. The circulatory system is closed, the fluid remaining in vessels with at least a double wall or in lacunar spaces with endothelial linings. Contractile blood vessels have a complex wall (Fig. 22.5C), with an inner endothelium, a layer of hyaline connective tissue, a layer of circular muscles, and an enucleated outer layer.

The simplest system consists of two lateral blood vessels, connected by a cephalic lacuna and an anal lacuna (Fig. 22.4A). Smaller, irregular branches arise from the main trunks. Subdivision of the primitive vessels and addition of new trunks result in more complex systems (Fig. 22.4B,C). A common development is a ventral connective passing between the two lateral vessels below the rhynchodeum. A middorsal vessel from the ventral connective runs beneath the rhynchocoel to the caudal lacuna. The system is made considerably more complex by vessels to the rhynchocoel and the gut.

The embryonic development of these vessels (Turbeville, 1986), similar to that in some polychaete annelids, reveals that they arise from solid bands of mesodermal tissues that split apart to form

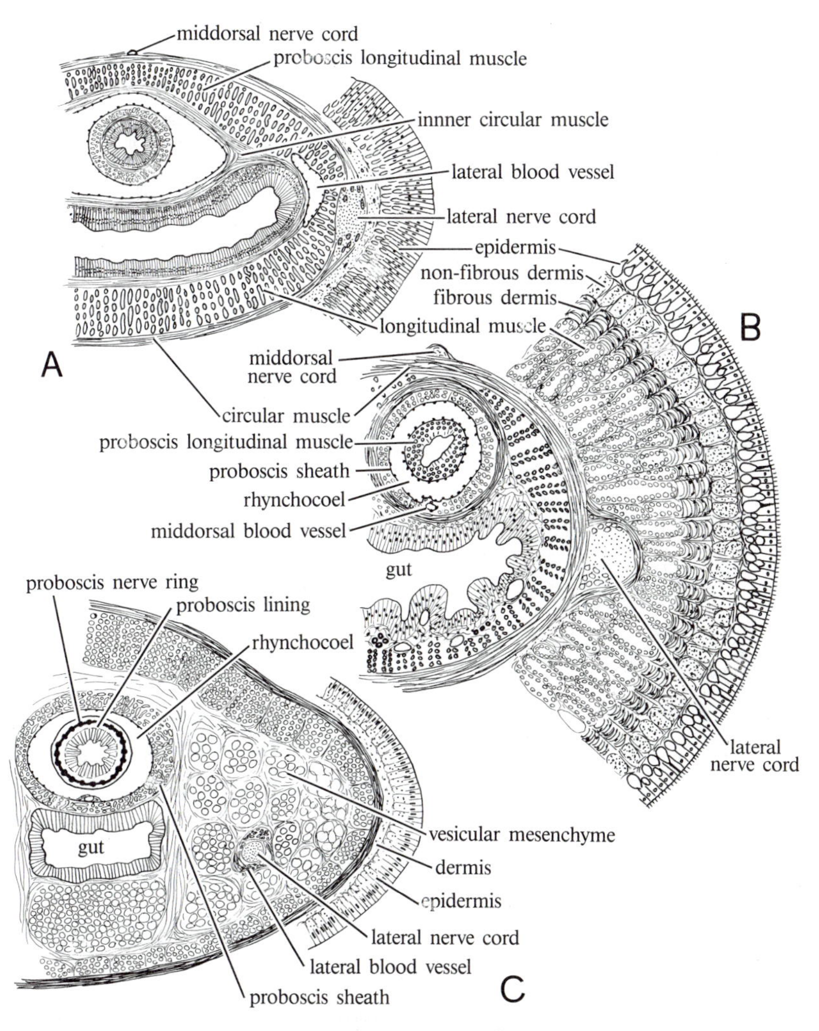

FIGURE 22.3. Comparison of cross sections of nemertines of different types. **A**. Section through palaeonemertean, *Tubulanus*, with three layers of body wall musculature, circular fibers forming outer layer. **B**. Section through heteronemertean, *Lineus*, with three layers of body wall muscles, outer layer of longitudinal fibers. **C**. Section through hoplonemertean, with nerve cords within internal layer of muscles. (All after Hyman.)

cell-lined body cavities (Fig. 22.4D,E). That is to say that the blood vessels are schizocoels, as is the case with the rhynchocoel and the gonad cavities.

Respiratory System Nemertines generally lack respiratory organs, but some species pump water in and out of a foregut richly supplied with lacunae. In this arrangement, the components of a respiratory system are seen: provisions for ventilating a respiratory surface, and vascularization of the surface to facilitate pickup and delivery of respiratory gases. It is not known how effectively this adds to respiratory exchange at the body surface.

Excretory System Nearly all nemertines have a protonephridial system. Protonephridial tubules of primitive nemertines are usually ciliated and end in multinucleate flame cells (Fig. 22.5D). Two important adaptive lines can be seen: (1) the development of a close association of **protonephridia** and the circulatory system (Fig. 22.5A) and (2) the tendency to break the protonephridial system into transverse segments. Some primitive nemertines have a pair of branched protonephridial tubules, with flame cells scattered in the tissues and with many **accessory nephridiopores**. The highly branched systems tend to break into separate nephridia, each associated with one nephridiopore. Each subdivision usually consists of a small cluster of capillaries that end in flame cells. These are connected with the nephridiopore by a regionally specialized tubule that has a thicker-walled, convoluted part and thinner-walled straight parts. In *Cephalothrix*, the flame cell has become highly modified to consist of a thin-walled capsule opening into the tubule by a pore passing through a ciliated section (Fig. 22.5E).

Some highly branched protonephridial systems have capillary branches closely applied to the walls of the lateral blood vessels (Fig. 22.5F). A more intimate association of flame cells and blood vessels is seen in palaeonemerteans, where the flame cells push into the wall of the blood vessel. The blood vessel wall may disappear, leaving the flame cells

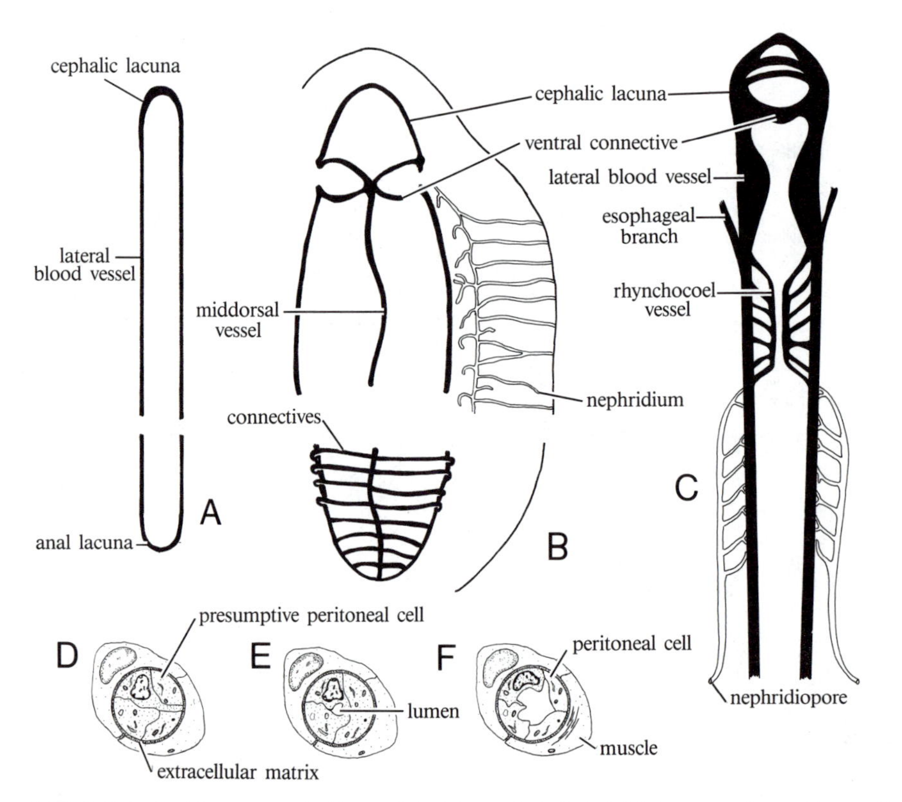

FIGURE 22.4. Circulatory system. **A**. Scheme of circulatory system of *Cephalothrix*, representing primitive condition. **B**. Somewhat more complex circulatory system of *Amphiporus*, with limited association of nephridial and circulatory systems. **C**. More advanced circulatory system of *Tubulanus*, with close association of nephridia and circulatory system. **D**,**E**. Schematic diagram of blood vessel formation in *Prosorhochmus americanus* sp. Note mesodermal cords split to form schizocoelous blood vessel. (**A**,**B** after Audemans; **C** after Bürger; **D**,**E**,**F** after Turbeville, 1986.)

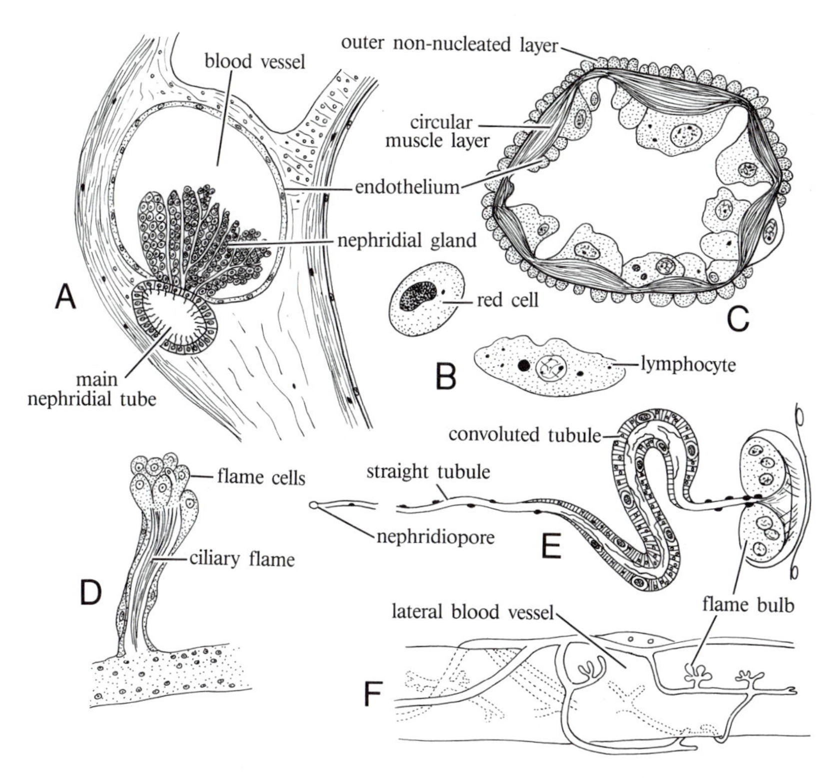

FIGURE 22.5. Excretory and circulatory system. **A**. Section through blood vessel and nephridial gland of *Procaranina*. **B**. Blood cells. **C**. Section of contractile blood vessel. **D**. Multicellular flame cell of *Drephanoporus*. **E**. Single nephridium of *Cephalothrix*. **F**. Nephridial tubules on surface of blood vessel, from *Amphiporus*. (**A** after Nawitzki; **B** in part after Prenant; **C** after Riepen; **D**,**F** after Bürger; **E** after Coe.)

bathed in blood. In some species, the capillary tubules extend into the blood vessel and the flame cells are missing (Fig. 22.5A), an arrangement known as a **nephridial gland**.

Nervous System The nervous system is highly centralized. The large brain is usually composed of four ganglia, a dorsal pair connected by a commissure above the rhynchodeum, and a ventral pair connected by a commissure under the rhynchodeum. A nerve ring thus encircles the rhynchodeum. A pair of lateral nerve cords, lying near the surface or deep, pass posteriorly, where they are connected by an anal commissure. A complex set of commissural and subsidiary longitudinal nerves may be present (Fig. 22.6).

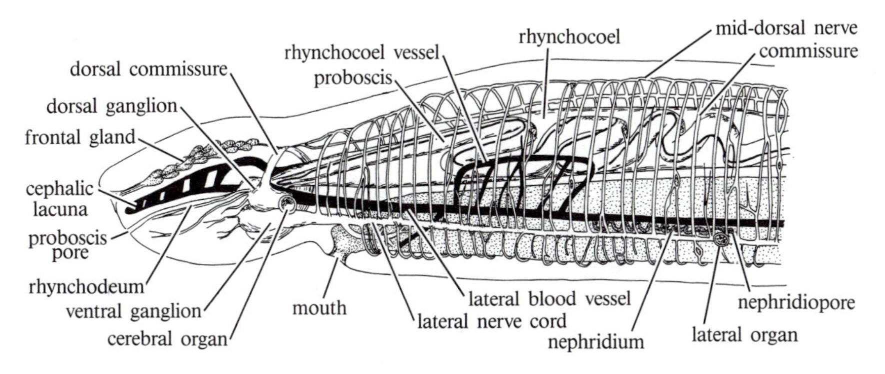

FIGURE 22.6. Anatomy of pelagic nemertine *Neuronemertes*. (After Coe.)

Reproductive System Considering the complexity of the rest of the organs, the reproductive system is surprisingly simple. The many **ovaries** or **testes** arise from aggregations of mesodermal cells that differentiate into coelomic sacs filled with developing gametes. As gametes ripen, each gonad develops a lateral **gonopore**, although males of *Carcinonemertes* form a common sperm duct. A few nemertines form body wall projections that may be copulatory organs. Sexes are almost always separate, and in the few hermaphroditic species, self-fertilization is prevented by early development of the male gonads (protandry).

Biology

Locomotion Although many nemertines depend heavily on cilia for locomotion, and such species cannot move unless in direct contact with the substrate, body wall muscles have undergone a dramatic development in nemertines, and in larger species muscular waves help to propel the animal in burrowing and surface gliding. Even the rhyncocoel has secondary uses as a fluid-filled coelomic cavity directly involved in locomotion by providing a hydrostatic skeleton for muscles to work against (Turbeville and Ruppert, 1983). Pelagic species have a caudal fin and sometimes lateral fins as well (Fig. 22.1C); they swim with undulatory movements of the body. Coordination of these activities is modulated through the peripheral nerve plexus, as transection of the nerve cords often has little or no effect on locomotion.

Feeding Most nemertines are carnivores, spearing or entangling their food with the everted proboscis. At least some nemertines may inject toxins from a gland at the base of the stylet. Both cytolytic proteins and neurotoxins have been noted (Kem, in Bierne, 1985). Some forms are scavengers, others are macrophagous feeders, some are suctorial, and one is a suspension feeder (McDermott and Roe, in Bierne, 1985).

Accounts of nemertine digestion vary markedly, probably reflecting diversity in the way in which the gut is divided into regions. Extracellular digestion is important, but intracellular digestion still occurs. Extracellular digestion occurs in the stomach of some nemertines and in the intestine of others. Proteins, fats, and carbohydrates are all used. Food is stored primarily as fat. Protein reserves and glycogen have also been reported.

Although the gut is tubular, its walls are embedded in solid tissue and are not free to move. Peristalsis cannot occur, and food is moved along the digestive tube by ciliary action.

Several species have been reported to take up dissolved organic material through the epidermis. These forms have digestive enzymes in the body wall and blood vessels.

Excretion and Respiration Little as yet is known of these functions in nemertines. Vital dyes collect in nephridial tubule walls but not in flame cells.

Sense Organs Nemertine sense organs are quite diverse, from two to several hundred. Many have pigment-cup ocelli in the head region (Fig. 22.1D). In some forms, the ocelli have lenses. Statocysts occur in one family in the ventral part of the brain. Ciliated **cephalic grooves** and **slits** on the head are thought to act as chemoreceptors (Fig. 22.1D). The heteronemerteans and hoplonemerteans also have an eversible chemoreceptive **frontal organ** (Fig. 22.1A), and some have **lateral organs** as well (Fig. 22.1C). The most characteristic sense organs are the **cerebral organs** (Fig. 22.1A), deep ectodermal pits that open into the ciliated grooves on the head. They occur in nearly all nemertines, and are closely associated with the cerebral ganglia and the vascular system. They may be chemoreceptors, but the close association of gland tissue, nerve tissue, and blood vessels suggests a possible endocrine function. In any case, staining reactions used to demonstrate neurosecretory activity are effective for some of the inclusions in cerebral organ cells. The cerebral ganglia control gonad maturation. The cephalic organs and glands also govern osmoregulation both directly, by effecting nephridial activity, and indirectly, by secreting mucopolysaccharides that in turn alter the osmotic tension of the blood and the extracellular fluid.

Reproduction Gonads exist as hollow cavities in muscle tissues. When the walls of these germinal sacs are formed, outpocketings of the gut grow between them (Riser, 1974).

Males and females shed gametes at the same time, probably in response to chemical stimulation. This is done only when worms are in contact, or two individuals sometimes enclose themselves in mucus before shedding. Fertilization is sometimes external, but in some species sperm enter the ovaries and fertilization is internal. Some of these are viviparous, the young developing in the ovaries.

More often ova are encapsulated and extruded in gelatinous masses or threads called **cocoons**. After spawning, the cells of the gut and gonadal sacs break down and pass out the gonoducts to mix with the matrix of the cocoon. It appears this mixture is used by the hatched juveniles until they are large enough to feed on their own.

One species, *Lineus sanguinensis*, reproduces only by asexual fission, and several species of *Lineus* have been reported to reproduce both by fission as well as by sexual means.

Development Cleavage is holoblastic, spiral, and determinate, but is noteworthy in that development produces "micromeres," that is, cells at the animal pole that are as large and sometimes larger than the "macromeres." Ectomesoderm generally arises from the micromeres, and endomesoderm from a mesendoblast, usually 4d. A coeloblastula is produced that gastrulates by invagination or by polar ingression to form either a hollow or a solid gastrula. In any event, the typical gastrula has an apical sense organ, endomesodermal cells in the old blastocoel, and an archenteron open through the blastopore (Fig. 22.7A).

Most nemertines undergo direct development, the embryo maturing to a wormlike juvenile. Heteronemerteans pass through a free-swimming larval stage, the **pilidium**, formed by the downward growth of lateral **oral lobes** (reminiscent of the molluscan velum) to enclose the blastopore region (Fig. 22.7B). Invagination continues, producing midgut and foregut regions.

Development at this point is interrupted by a unique metamorphosis. Discs, produced by ectodermal thickening and invagination, detach and move into the interior. They grow and unite around the pilidium gut to form a larva (Fig. 22.7B,C). There are three pairs of discs (cephalic, cerebral, and trunk) and sometimes unpaired dorsal and proboscis discs. The pilidium mesoderm and endoderm are retained, but the disc tissues produce the adult ectodermal derivatives. Eventually, the juvenile within the pilidium sheds the pilidial ectoderm to escape and develop into a young worm.

Fossil Record and Phylogeny

Some body-outline fossils from the Carboniferous period are evocative of the nemertine form. These tell us little about nemertine evolution except to suggest that the phylum has been around for a long time (Schram, 1979). The major groups of nemertines are rather distinctive from each other in histological arrangement; therefore, it is difficult, although not impossible (Norenburg, 1988), to envision at this time how they are related to each other.

Their possession of a blood-vascular coelom, prototrochal lobes in the larvae, and a tendency of some nemertines to develop serial structures in their soft anatomy is similar to what molluscs do. Consensus is now emerging that nemertines are coelomates and intermediate in organization between acoelomates and advanced coelomates (Turbeville and Ruppert, and Riser, in Bierne, 1985).

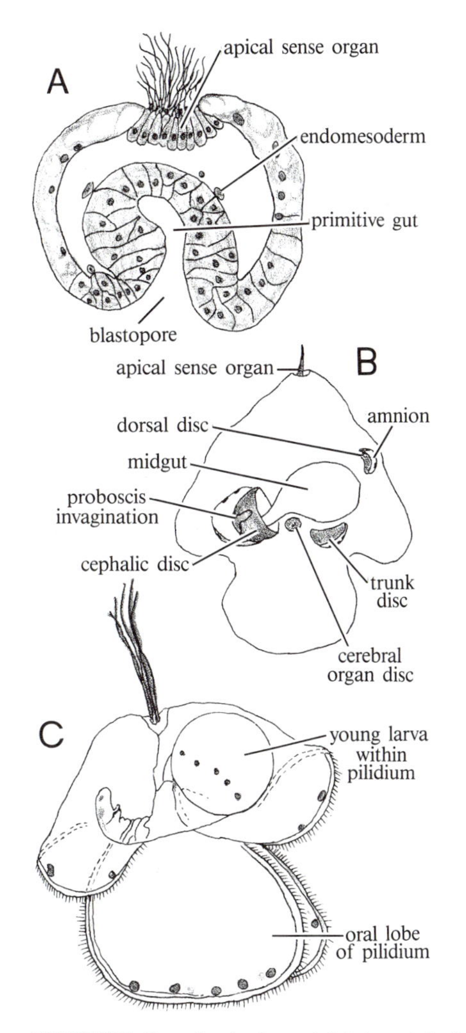

FIGURE 22.7. Nemertine development. **A.** Late gastrula of *Cerebratulus*. **B.** Pilidium larva, showing discs that develop into adult ectoderm. **C.** Philidium containing advanced larva. (**A** after Wilson; **B** after Salensky; **C** after Verrill.)

Taxonomy

Nemertinea is externally a relatively homogeneous group, far less variable in appearance than any phylum so far discussed. Internally, however, very distinctive arrangements of tissue layers are used to differentiate classes and orders. The phylum presently is divided into two classes.

Class Anopla. With mouth behind brain; central nervous system below epidermis or among body wall muscles; proboscis without stylets, consisting of simple tube, sometimes narrower anteriorly, but not divided into distinct regions.

Order Palaeonemertea. With outer circular and inner longitudinal layer of body wall muscles, sometimes with second, innermost circular layer; dermis of hyaline, gelatinous connective tissue. Example: *Tubulanus* (Fig. 22.3A).

Order Heteronemertea. Body wall musculature of outer longitudnal, middle circular, and inner longitudinal layers; dermis of fibrous material. Examples: *Cerebratulus* (Fig. 22.2A) and *Lineus* (Figs. 22.1E and 22.3B).

Class Enopla. With mouth in front of brain; central nervous system inside body wall muscles; sometimes with piercing organ or stylet associated with proboscis; proboscis divided into anterior, middle, and posterior regions.

Order Hoplonemertea. Proboscis armed with one or more stylets; straight intestine with paired, lateral pouches. Examples: *Nectonemertes* (Fig. 22.1C) and *Amphiporus* (Fig. 22.1A).

Order Bdellonemertea. Proboscis without stylet; intestine without diverticula and winding; with posterior adhesive disc; endocommensals in molluscs. Example: *Malacobdella* (Fig. 22.1B).

Selected Readings

Bierne, J., et al. 1985. *Comparative Biology of Nemertines. Am. Zool.* 25:1–151.

Norenburg, J. L. 1988. Phylogenesis of the phylum Nemertina. *Am. Zool.* 28(4):150A.

Riser, N. W. 1974. Nemertinea. In *Reproduction of Marine Invertebrates*, vol. 1 (A. C. Giese and J. S. Pearse, eds.), pp. 359–389. Academic Press, New York.

Schram, F. R. 1979. Worms of the Bear Gulch Limestone of central Montana. *Trans. San Diego Soc. Nat. Hist.* 19:107–120.

Turbeville, J. M. 1986. An ultrastructural analysis of coelomogenesis in the hoplonemertine *Prosophochmus americanus* and the polychaete *Magelona*. *J. Morph.* 18:51–60.

Turbeville, J. M., and E. E. Ruppert. 1983. Epidermal muscles and peristaltic burrowing in *Carinoma tremaphoros* (Nemertini): Correlates of effective burrowing without segmentation. *Zoomorph.* 103:103–120.

23

Sipuncula

There are about 238 species of peanut worms. Sipunculans are widespread marine animals that center in the littoral zone of warmer seas, but extend into polar waters and reach abyssal depths (Rice and Todorović, 1975). Sipunculans are sedentary creatures that burrow in the sand or mud, or occupy any protected den of appropriate size, whether a rocky crevice, a convenient niche among corals, an annelid tube, or a molluscan shell (Fig. 23.1B). A minute species lives in foraminiferan shells. They range from about 2 mm to over half a meter in length.

Definition

The body is divided into an anterior introvert containing the mouth and a posterior trunk. The mouth is surrounded by tentacles in one class, whereas there is a dorsal nuchal organ in others. The gut is spiral and U-shaped. The anus is located dorsally in the anterior end of the trunk. There is a single or one pair of metanephridia and no special circulatory system. The brain is dorsal and connects by means of a circumenteric nerve ring to an unsegmented ventral nerve cord. The sexes are separate, with simple gonads derived from coelomic peritoneum. Gametes mature in the coelom and are discharged through nephridia. There is spiral cleavage and a trochophore larva.

Body Plan

Sipunculans have a very simple external form. The long or short **introvert** ends in an anterior **oral disc**. Peripheral **tentacles** of variable number, size, and complexity surround the mouth of Sipunculidea, but in Phascolosomatidea there is only a crescent of simple tentacles around a dorsal nuchal organ. A smooth zone, the **collar**, may be found immediately behind the tentacles. Minute spines or hooks and sensory or glandular papillae often stud the surface of the introvert. A hardened **anterior shield** is found at the anterior end of the trunk in many species living in coral rocks, and in the genus *Aspidosiphon* a **posterior shield** also occurs (Fig. 23.1A). The trunk has no hooks or spines but usually has sensory and glandular papillae. The conspicuous **anus** is usually located at the anterior end of the trunk or, more rarely, on the introvert, marking the middorsal line. On the ventral side of the body, at about the same level, a pair of nephridiopores are usually found.

The body wall is composed of layers of cuticle, epidermis, dermis, circular muscle, longitudinal muscle, and peritoneum. The layers vary in thickness and complexity with the size and habits of the species. The circular and longitudinal muscles of *Sipunculus* can be seen through the body wall, giving it a gridded appearance. Introvert retractor muscles (Fig. 23.1D) merge with the longitudinal muscle layers of the body wall and insert at the anterior tip of the introvert, on the anterior end of the esophagus.

Coelom The **coelom** is spacious and undivided; only vestiges of mesenteries are present. The coelom and the body wall act together as a hydraulic system to operate the introvert. Contraction of the body wall musculature raises coelomic pressure to evert the introvert, to be later pulled back by the introvert retractor muscles.

A separate system of canals is associated with the tentacles. One or two **tentacular canals** extend into each tentacle; these open into a **ring canal** around the upper end of the esophagus. One or two **compensation sacs** or contractile vessels (Fig. 23.1D) arise from the ring canal; it is thought that contraction of vesicle muscular walls extends the tentacles. When the tentacular muscles contract, the fluid is forced back into the compensation sacs. In some cases, however, ciliated tracts on the lining of the tentacular canals causes some circulation of the fluid. Coelomocytes are able to enter the tentacular canal system and move about with the fluid, so there is some thought that this system may serve some respiratory role.

Digestive System The large mouth opens into a short muscular **pharynx** or leads directly into the slender esophagus. The posterior end of the **esophagus** of *Golfingia* has several longitudinal ridges and continues as the coiled descending gut. The ascend-

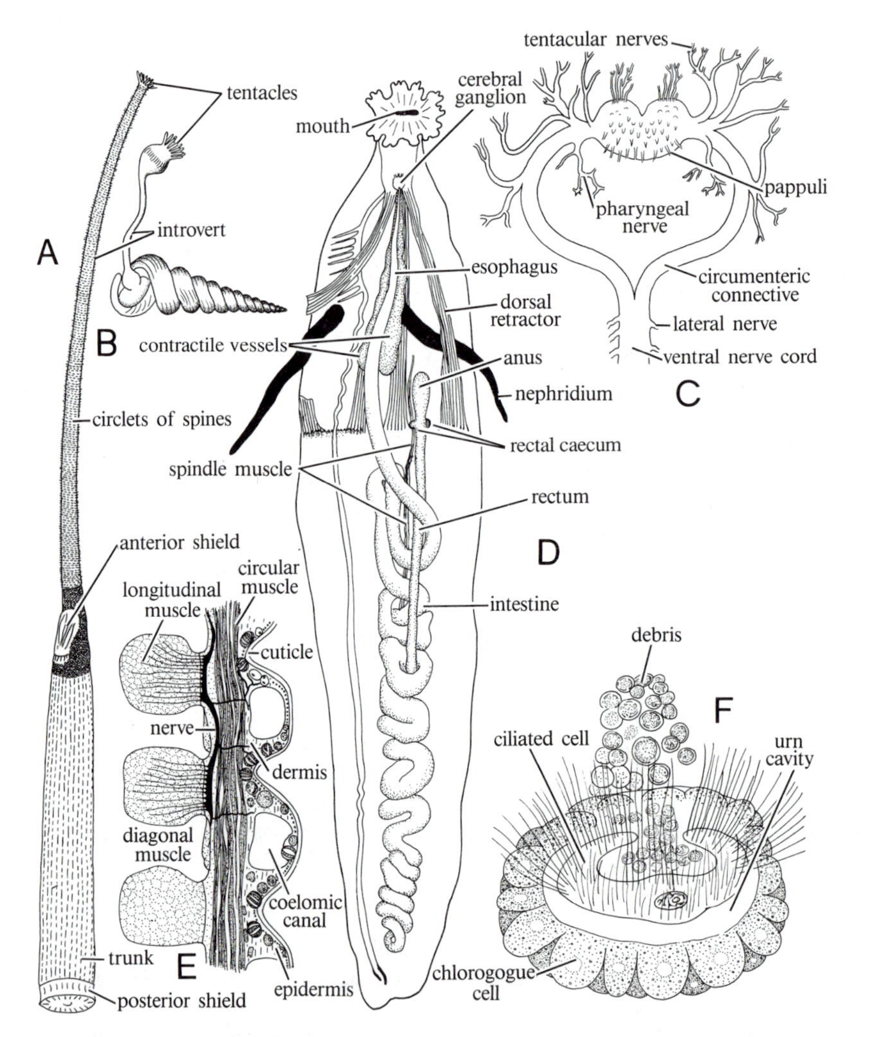

FIGURE 23.1. Sipunculans. **A**. *Aspidosiphon*, with anterior shield dorsal to base of introvert, and posterior shield. **B**. *Phascolion*, at home in snail shell. **C**. Central nervous system of *Phascolopsis gouldi*. Notice lack of segmentation in nerves arising from nerve cord. **D**. Dissection of *Sipunculus nudus*. **E**. Transverse section through wall of *Sipunculus*, with coelomic canals near body surface. **F**. Free coelomic urn of *Phascolosoma*, with attached debris. (**A** after Hyman; **B** after Théel; **C** after Andrews; **D,E** after Metschnikoff; **F** after Salinsky.)

ing arm of the gut is a coiled **intestine** (Fig. 23.1D) that straightens and continues anteriorly as the **rectum**. In some species blind sacs, the **rectal caeca**, extend from the rectum.

Excretory System Most species have a pair of large, saccate, brownish **metanephridia** (Fig. 23.1D) with open nephrostomes and nephridiopores. In two genera, *Phascolion* and *Onchnesoma*, one member of the pair has been lost. These organs act to drain the large coelomic body cavity.

Nervous System The nervous system is rather like that of annelids, but the only ganglion is the **cerebral ganglion** (Fig. 23.1C), and the lateral nerves show no sign of a metameric distribution. They may be opposite, alternate, or rise in a completely irregular manner. Nerves to the retractor muscles usually arise from the **circumenteric ring**. There is evidence that the sipunculan brain, like that of annelids, exercises an inhibitory effect on other parts of the motor nervous system. A single or a pair of dorsal **cephalic tubes** extend from the oral disc in toward the cerebral ganglion.

Reproductive System The gonads are peritoneal outgrowths, located at the base of the retractor muscles on the body wall. The germ cells are immature when they are released and ripen in the coelom. Sexes appear to be separate in all sipunculans except for one protandric and one hermaphroditic form.

Biology

Locomotion Burrowing sipunculans thrust the introvert forward into the mud, dilate the tip, and then contract the retractor muscles, effectively pulling the body forward.

Feeding The tentacles are sometimes used as a ciliary-mucus feeding system, but are too small to be adequate in most species. Most sipunculans eat sand and mud, and subsist on whatever digestible organic material that may yield. Diatoms, protozoa, larvae, and small particles of organic detritus provide most of their nourishment.

Little is known about digestion. The intestine has a conspicuous dorsal ciliated groove that aids in moving food along. Digestion and absorption appear to center in the descending arm of the gut, and the ascending arm is probably concerned primarily with feces formation. Food is stored in the intestinal wall as lipids, ribonucleids, and carbohydrates. What is known about the enzymes present indicates that they are similar to those of polychaetes with similar dietary habits.

Excretion The brownish glandular walls of the nephridia probably excrete nitrogenous wastes into the tubule. Sipunculans are ammonotelic, with up to 90 percent of the nitrogenous wastes being released as ammonia. When put in hypotonic media, they can expel salt through the intestine and nephridia, thus regaining normal body volume. There is some evidence that they also can expel water. Evidence of neurosecretory control of osmoregulation has been obtained.

Chlorogogue tissue, presumably functioning like annelid chlorogogue, arises from the visceral peritoneum on the intestine. Some sipunculans have curious peritoneal structures known as **coelomic urns** (Fig. 23.1F). They appear as elevations of the peritoneum, which contain a ciliated cell. In some species they remain attached to the peritoneum, but in others the urns break away and swim about actively in the coelom. Particles adhere to them and are eventually converted into brown bodies or discharged through the nephridia.

Respiration The coelomic fluid contains several kinds of amoebocytes and corpuscles that contain a respiratory pigment, hemerythrin. Hemerythrin also occurs in the fluid of the contractile vessels and tentacular canals, and in some species the properties of the hemerythrin in the coelomic and tentacular systems differ. This has a curious effect in *Themiste* and *Siphonosoma*. The coelomic fluid of *Themiste* has a higher affinity for oxygen than the tentacular fluid; in this case the tentacles pick up oxygen and transfer it to the coelomic fluid. *Siphonosoma* coelomic fluid has a lower affinity for oxygen than the tentacular fluid. It does not respire through the tentacles, and transfers oxygen from the coelomic to the tentacular fluid.

Circulation The coelomic fluid plays an important role in internal transport. Furthermore, coelomic channels can extend into the body wall of some larger species (Fig. 23.1E). The coelomic fluid has been noted to contain high levels of glucose.

Sense Organs Sipunculans have surprisingly varied sense organs. Scattered neurosensory cells, papillate sensory buds, and protrusible ciliated pits are found commonly on the body surface or tentacles. A ciliated region, the **nuchal organ**, a lobed pad on the dorsal edge of the oral disc, is probably a che-

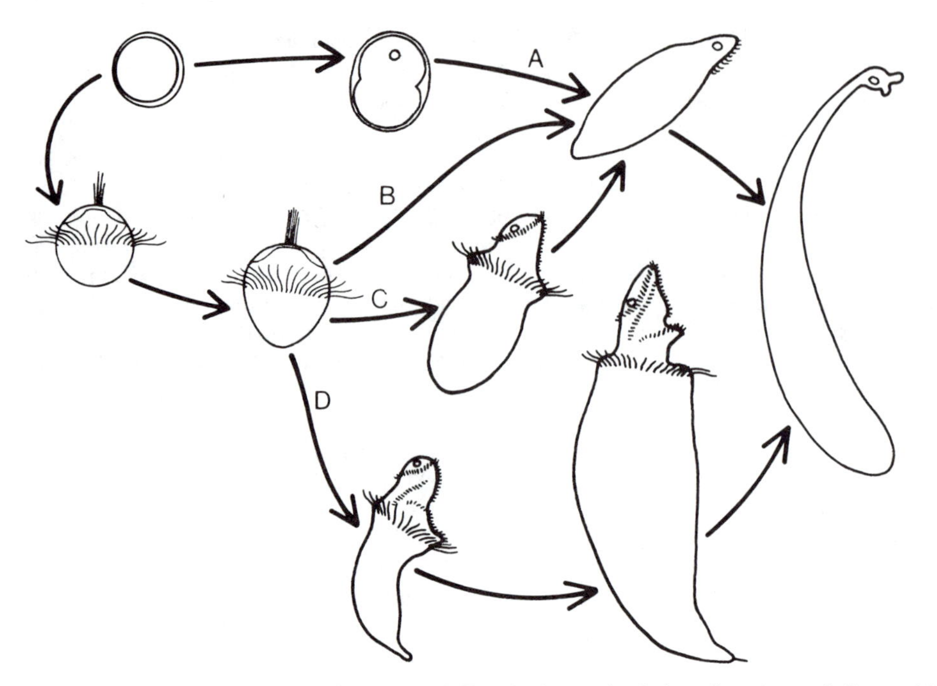

FIGURE 23.2. Developmental pathways of sipunculans. **A**. Direct development; hatched vermiform stage gradually grows into juvenile. **B,C,D**. Indirect development; **B**. trochophore larva metamorphosing to vermiform stage; **C**. lecithotrophic pelagosphera larva metamorphosing to vermiform stage; **D**. planktotrophic pelagosphera growing before metamorphosing to juvenile stage. (Modified from Rice.)

moreceptor. In some species, the deep cephalic tube or tubes extend back to the brain where it expands as a flattened sac, the **cerebral organ**. Although it may have a sensory function, some cells of the cerebral organ have been shown to contain neurosecretory granules in the larva. Nearly tubular pigment-cup ocelli may be embedded in the brain. Although sensory cilia are present, the eyes are considered to be rhabdomeric. In one genus, *Phascolopsis*, filamentous and leaflike processes or **pappuli** (Fig. 23.1C) from the brain surface extend into the coelom, and may provide information about coelomic conditions.

Reproduction Sexual reproduction is the general rule, although two species have been noted to reproduce asexually by fission and budding (Rice, 1975). These animals for the most part retain the ability to regenerate lost tissues. For example, about three percent of collected *Phascolion strombus* in Sweden had regenerating introverts, which apparently were autotomized when attacked by predators. Parthenogenesis has been observed.

Sipunculans have epidemic release of gametes into the water during the summers, and fertilization is external. *Golfingia* fills its nephridia with seawater and accumulates gametes in them when it is ready to spawn. Evidently some kind of sorting device excludes the immature reproductive cells and coelomocytes. The males tend to discharge sperm first, and males seem to spawn over longer periods than females. The presence of spermatic fluid induces the females to eject ova into seawater.

Development Cleavage follows the annelid pattern. In very yolky eggs the micromeres tend to be larger than the macromeres (not unlike nemertines). During cleavage, a cross of blastomeres, comparable to the molluscan cross, is formed. Gastrulation is by invagination and/or epiboly, depending on yolk content. The blastopore is elongate in form, and closes from posterior to anterior. Derivation of the various parts of the typical trochophore larva closely follows the line seen in annelids and molluscs. The larva elongates as it metamorphoses into a juvenile; at no time can evidence of metamerism be seen. It is interesting that the *Phascolopsis* metanephridia appear to arise from an ectodermal invagination and a mesodermal, peritoneal primordium that produces the nephrostome. In *Sipunculus*, on the other hand, the whole nephridium arises from mesodermal cells. (Where two such closely related members

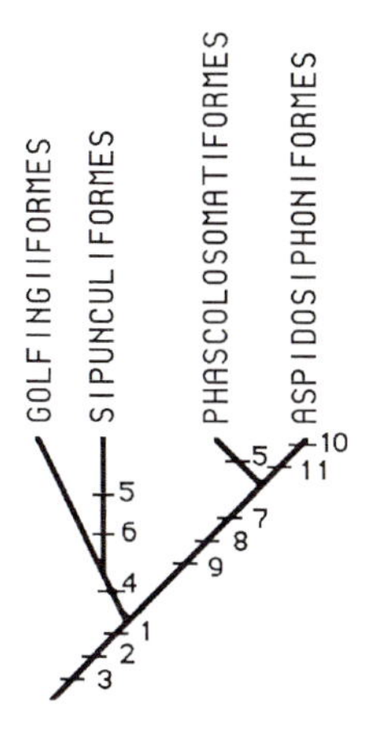

FIGURE 23.3. Cladogram depicting relationships of sipunculan orders. Derived characters are: (1) spiral gut with anterior middorsal anus; (2) introvert; (3) tentacular hydraulic system; (4) mouth surrounded with tentacles; (5) longitudinal muscle bands; (6) coelom extending into epidermis; (7) nuchal tentacles well developed, peripheral tentacles missing; (8) introvert hooks arranged in rings; (9) spindle muscle extended to posterior end of body; (10) hardened shield on anterior part of body; and (11) loss of dorsal retractor muscles. (Derived from Cutler and Gibbs.)

of a small phylum disagree, one wonders how much dependence can be placed on the embryonic origin of nephridia as a means of determining how the coelom may have arisen.)

Several distinct developmental pathways (Fig. 23.2) are used among sipunculans (Rice, 1981). These types cut across taxonomic boundaries and thus seem to have ecological rather than evolutionary import. Development can be direct (Fig. 23.2A), with the embryo hatching into a vermiform stage that gradually grows into a worm. Indirect development can follow three possible paths, all of which move through a trochophore larva stage. In one, the larva metamorphoses into a vermiform stage (Fig. 23.2B); in the other two paths, the trochophore produces a pelagosphaera larva. These latter can be of two types: either lecithotrophic (Fig. 23.2C), which live off of stored food reserves before metamorphosis to a vermiform stage, or planktotrophic (Fig. 23.2D), which feed and grow before metamorphosis into a juvenile worm.

PHYLOGENY

No fossil record exists for sipunculans; therefore, inferences about phylogeny must be based on living forms. The available data were used to propose a testable hypothesis about relationships (Cutler and Gibbs, 1985). Two main lines are indicated (classes Sipunculidea and Phascolosomatidea), and each of these has two orders (Gibbs and Cutler, 1987). The analysis, partially represented in Figure 23.3, used 12 characters to resolve 17 genera into four orders. Additional data (e.g., karyotypes, biochemistry, and ultrastructure) are needed to test this hypothesis.

Taxonomy

Class Sipunculidea. Tentacles encircling mouth; introvert hooks (when present) simple, hollow structures usually irregularly distributed.

Order Golfingiiformes. Longitudinal muscles continuous, never thickened into bands.

Order Sipunculiformes. Longitudinal muscle bands; generally coelom extended into epidermis as channels (Fig. 23.1D).

Class Phascolosomatidea. Tentacles confined to arc surrounding dorsal nuchal organ; introvert hooks recurved, solid, closely packed in regularly spaced rings.

Order Phascolosomatiformes. Four introvert retractor muscles (Fig. 23.1B).

Order Aspidosiphoniformes. Calcareous or horny anal shield on anterior part of trunk; two dorsal retractor muscles (Fig. 23.1A).

Selected Readings

Åkesson, B. 1958. *Undersökningar över Öresund. XXXVIII. A Study of the Nervous System of the Sipunculoideae.* C. W. K. Gleemp, Lund.

Clark, R. B. 1969. Systematics and phylogeny: Annelida, Echiura, Sipuncula. In *Chemical Zoology*, vol. 4 (M. Florkin and B. T. Scheer, eds.), pp. 1–68. Academic Press, New York.

Cutler, E. B., and P. E. Gibbs. 1985. A phylogenetic analysis of higher taxa in the phylum Sipuncula. *Syst. Zool.* 34:162–173.

Gibbs, P. E., and E. B. Cutler. 1987. A classification of the phylum Sipuncula. *Bull. Br. Mus. (Nat. Hist.), Zool.* 52:43–58.

Rice, M. E. 1975. Sipuncula. In *Reproduction of Marine Invertebrates*, vol. 2 (A. C. Giese and J. S. Pearse, eds.), pp. 67–127. Academic Press, New York.

Rice, M. E. 1981. Larvae adrift: patterns and problems in life histories of sipunculans. *Am. Zool.* 21:605–619.

Rice, M. E., and M. Todorović (eds.). 1975. *Proceedings of the International Symposium on the Biology of the Sipuncula and Echiura*, vols. 1 and 2. Inst. Biol. Research, 'Siniša Stanković,' Belgrade.

Stephen, A. C., and S. J. Edmonds. 1972. *The Phyla Sipuncula and Echiura.* Br. Mus. (Nat. Hist.), London.

24

Echiura

The Echiura, or spoon worms, are a small marine phylum of some 130 annelidlike beasts. Most Echiura have cryptic habits, dwelling in burrows and crevices and only thrusting out the remarkable proboscis as a food finder. They show modest diversity. *Urechis* and *Echiurus* build U-shaped mucus-lined tubes in soft bottoms. Others live among rocks in convenient crevices or live in the shells or tests of other animals. *Thalassema melitta*, when young, enters the empty tests of sand dollars and then grows too large to emerge, but lives well enough in its self-chosen cage with the help of its proboscis.

Definition

Echiura have the body divided into a permanently everted, grooved proboscis and a trunk. The convoluted digestive tube begins with a mouth situated at the base of the proboscis, and ends with a posterior anus and rectum, modified for respiration. There are one to many metanephridia. The circulatory system is simple, with dorsal and ventral vessels. The nervous system possesses no definite, well-developed brain or ventral ganglia, but there are circumesophageal connectives and a subepidermal ventral nerve cord. Sexes are separate, with reproductive cells arising from a modified peritoneum and maturing in the coelom to escape by way of the nephridia. Cleavage is spiral. The embryo has little trace of segmentation, and culminates in a trochophore larva.

Body Plan

The prominent, extremely sensitive **proboscis** is probably a center for chemoreception as well as a center for food procurement. Analogous to the nemertine proboscis, it could be a homologue of the annelid prostomium. The trunk is just a bag of guts and gonads.

The body wall consists of an epidermis, a dermis, and layers of longitudinal and circular muscle, followed by the coelomic peritoneum. Several types of glandular cells occur, and sensory papillae are found on the body surface. A pair of large, hooked, ventral **setae** project from setal sacs. They are secreted in the same manner as the setae of annelids; that is, they are formed by single chaetoblasts and are equipped with protractor and retractor muscles. Some Echiura also have one or two posterior circles of setae.

Digestive System The mouth opens into a short buccal tube, leading to an often large **pharynx** (Fig. 24.1B). The **esophagus** is a narrow, frequently coiled tube that leads to the midgut. A muscular **gizzard** and saccate **stomach** sometimes lie between the esophagus and the intestine. The long, irregularly coiled **intestine** is attached to the body wall by muscular, mesenteric strands. The intestinal musculature is delicate; therefore, food moves along largely by means of currents generated in a ciliated groove. A narrow ciliated tube, the **siphon**, arises from and parallels the intestine for some distance and then rejoins it. The intestine straightens and continues as a short **rectum**, from which a pair of diverticula, called **rectal sacs**, protrude. The rectal sacs are simple or branched, but in either case are open to the coelom through ciliated funnels. Their exact function is uncertain, but they may serve as accessory nephridial organs, provide control over coelomic pressure, or aid in respiratory exchange.

Excretory System The **metanephridia** resemble those of sipunculids. They vary greatly in number; for example, *Bonellia* has one, whereas *Ikeda* has hundreds. Most species, however, have from one to three pairs of nephridia. Typically, there are brownish inclusions in the glandular part of the nephridial tube, which may mean that this region is an active one for nitrogenous excretion.

Respiratory and Circulatory Systems The spacious **coelom** is undivided. The coelomic fluid contains amoebocytes and hemocytes bearing hemoglobin.

The circulatory system itself is simple, and the blood is without a respiratory pigment. In this the Echiura resemble annelids, which have a respiratory pigment in the coelomic fluid. A ventral vessel runs above the nerve cord. Blood passes to the dorsal vessel through contractile **circumintestinal** and **circumesophageal vessels**, varying in number with the species. The **dorsal vessel** arises at the junction of the

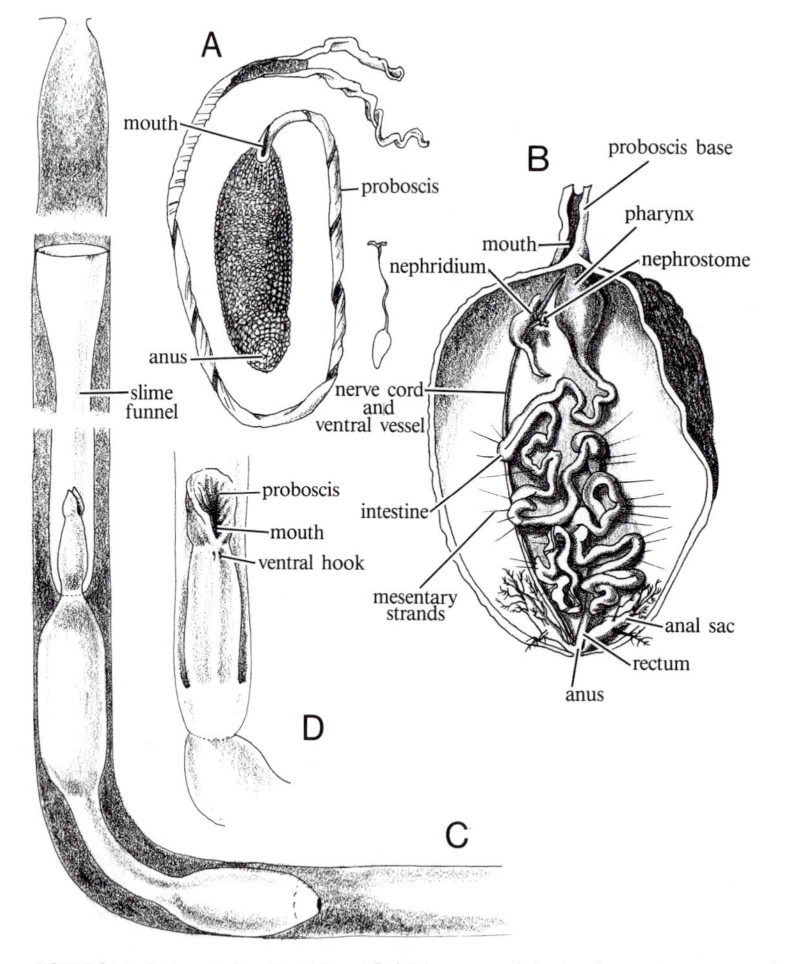

FIGURE 24.1. Echiura. **A.** *Bonellia viridis* and *B. fuliginosa*, natural size (smaller species more naturally arranged). **B.** Dissection of female *B. viridis*, opened laterally. **C,D.** *Urechis caupo*, in its tube, with mucous funnel in position. (**A,B** after Shipley; **C,D** after Fisher and MacGinitie, from the MacGinities.)

most posterior circumintestinal vessels and runs forward to the anterior tip of the proboscis. Here, it divides to form right and left branches that pass along the proboscis margins and join together behind the mouth to form the **ventral vessel**.

Nervous System The sedentary life of the echiurans is reflected in the simplicity of the neurosensory equipment. There are no special sense organs, unless the proboscis be considered one. The simple nervous system is built on the annelid pattern, but it contains no ganglia except for some quasimetameric clustering of nerve cell bodies in the cords. The circumenteric ring is drawn out into a long loop that follows the proboscis margin and connects with a ventral nerve cord. The lack of segmental ganglia and often irregular placement of lateral nerves emphasize the basic lack of organized metamerism in Echiura.

Reproductive System The character of a reproductive system loosely associated with the coelomic wall invites comparison of annelids, nemertines, and sipunculids with echiurans. Sexes are separate in echiurans, and the gonads are specialized regions of the ventral mesentery. Gametes are shed into the coelom before they reach maturity. After they have ripened, they escape through the nephridia.

Biology

Locomotion Most Echiura are sedentary in nature, but some move about considerably in captivity. *Bonellia minor*, when first put in an aquarium, probes about inquisitively with its sensitive proboscis. When an acceptable nook is found, it attaches and pulls the body with gyrations into the space, proboscis first. It turns about and extends the proboscis once more, exploring the neighborhood before it begins to feed.

Feeding The proboscis secretes a mucous cover on which food organisms and small particles stick, to be conveyed to the mouth by a ciliated groove and by muscular movements of the proboscis itself. The *Bonellia* proboscis is a remarkable organ (Jaccarini and Schembri, 1977). In *B. viridis* (Fig. 24.1A) the proboscis is only about 7 cm long when contracted, but can be extended fully to as much as a 1.5 m, almost transparent thread!

Urechis (Fig. 24.1C,D) has evolved a different method of feeding, capitalizing on its U-shaped tube in much the same manner as the annelid *Chaetopterus*. The proboscis secretes a slimy, mucous funnel that is attached to the wall of the tube and the proboscis base. Peristaltic movements of the body wall pump water through the tube, pulling it through the mucous funnel. When filled, the funnel is simply wrapped up and swallowed. The funnel is a remarkably effective strainer, for its openings measure only about 40 Å.

The precise function of the unique intestinal siphon is unknown, but it evidently serves as a shunt for some of the intestinal contents. It may keep the concentration of enzymes high in the main intestinal channel by shunting water to the lower intestine.

Excretion Little is known of waste product removal or osmoregulation in echiurans. Anal pumping occurs in *Urechis*, with rhythmic peristaltic movements passing coelomic fluid over the outer rectal wall while the inside is flushed with seawater, indicating that excretion might be taking place (Harris and Jaccarini, 1981).

Respiration Hemoglobin in the coelomocytes appears to serve primarily for oxygen storage during periods of low oxygen availability, such as when the tide is out. Enough oxygen is held in this way to maintain metabolism for some time at the reduced rates characteristic of *Urechis* at ebb tide. *Lissomyema* has been maintained for up to 18 days in anoxic conditions.

Reproduction Fertilization usually occurs externally in the ocean, but *Bonellia* and *Hamingia* have other arrangements. *Bonellia* females are very large, and the males are very small. Any larva that settles by itself develops into a female. A larva that settles near to or in contact with a female becomes a male (Jaccarini et al., 1983). The development of the male is retarded, and he remains a small, ciliated organism with many juvenile characteristics. He may cling to the proboscis of the female for a time, but eventually enters the female mouth and takes up residence in a fold of the nephridium, living the rest of his life as an endocommensal. In these genera, the ova are fertilized in the nephridia and remain there during early development.

Development The zygote undergoes spiral cleavage and goes through a sequence of events similar to that seen in the annelids, including the formation of an annelidan cross (Newby, 1940). The embryo then develops into a trochophore. Early studies of echiuran trochophores reported thickenings in the mesodermal bands and a transitory appearance of coelomic compartments, but these have yet to be confirmed. The ventral nerve cord is initially differentiated as two parallel rows of cells, connected by thin bands of ectoderm on the midline, with rudiments of serial ganglia. These rows eventually fuse into a single cord.

Fossil Record and Phylogeny

One fossil echiuran is known from the Pennsylvanian period of North America.

Similarities in early embryonic development, as well as in larval construction and many features of the adult organism, provide evidence of a link between the Echiura and Annelida. These similarities were the basis for maintaining echiurans as a taxon within the annelids until as recently as 1940. The lack of metamerism, however, in adult Sipuncula and Echiura and the true metamerism of Annelida indicate divergence of these stocks.

Taxonomy

As in nemertines, ordinal taxonomy within the single class Echiuroidea is subtle and is based in part on detailed histologic features (Clark, 1969; Stephen and Edmonds, 1972).

Order Echiuroinea. Longitudinal body wall muscles between outer circular and inner oblique muscles; seven or fewer pairs of nephridia; closed circulatory system. Example: *Bonellia* (Fig. 24.1A,B).

Order Xenopneusta. Longitudinal body wall muscles between outer circular and inner oblique muscles; two or three pairs of nephridia; proboscis greatly reduced; open circulatory system. Example: *Urechis* (Fig. 24.1C,D).

Order Heteromyota. Longitudinal muscles banded, outside circular and oblique layers; 200 to 400 unpaired nephridia. Example: *Ikeda.*

Selected Readings

Clark, R. B. 1969. Systematics and phylogeny: Annelida, Echiura, Sipuncula. In *Chemical Zoology*, vol. 4 (M. Florkin and B. T. Scheer, eds.), pp. 1–68. Academic Press, New York.

Harris, R. R., and V. Jaccarini. 1981. Structure and function of the anal sacs of *Bonellia viridis*. *J. Mar. Biol. Assc. U.K.* 61:413–430.

Jaccarini, V., and P. J. Schembri. 1977. Feeding and particle selection in the echiuran worm *Bonellia viridis*. *J. Exp. Mar. Biol. Ecol.* 28:163–181.

Jaccarini, V., et al. 1983. Sex determination and larval sexual interaction in *Bonellia viridis*. *J. Exp. Mar. Biol. Ecol.* 66:25–40.

Newby, W. W. 1940. The embryology of the echiurid worm *Urechis caupo*. *Mem. Am. Phil. Soc.* 16:1–219.

Rice, M. E., and M. Todorović (eds.). 1975. *Proceedings of the International Symposium on the Biology of the Sipuncula and Echiura*, vol. 2. Inst. Biol. Research, 'Sinilša Stanković,' Belgrade.

Stephen, A. C., and S. J. Edmonds. 1972. *The Phyla Sipuncula and Echiura.* Br. Mus. (Nat. Hist.), London.

25

Pogonophora

Pogonophora is among the most enigmatic of phyla and has been the subject of intense study to resolve its relationships to other groups. Long thought to be related to deuterostomes (Ivanov, 1963), current consensus has them more closely related to annelids. Pogonophorans, or beard worms, all live in tubes composed of 30 percent chitin and 50 percent protein. One group, the Perviata (traditional pogonophorans *sensu stricto*) lie buried upright in sediment; the other group, the Vestimentifera or Obturata, live in aggregations on the sediment surface. Pogonophorans are most characteristic of deep-ocean habitats, although some shallower-water forms are known. Perviates can reach 0.3 m in length, but are no more than 2 mm in diameter; vestimentiferans are up to 1.5 m long and 4 cm in diameter.

The terminology developed for the two classes, Perviata and Vestimentifera, arose independently, so we are plagued with different sets of terms for homologous regions of the body.

Definition

Pogonophora are tube-dwelling worms. The body is divided into three regions: an anterior forepart that includes a structure bearing tentacles or branchiae, an extremely elongate trunk or metasome, and a very small segmented terminus or opisthosoma. There is no gut, but rather nutrition is derived from a bacteria-filled trophosome. The circulatory system has a heart, but blood moves in portions of a diffuse coelom, whose coelomic chambers are often unlined. Excretion occurs with protonephridia, and the nerve system is epidermal.

Body Plan

The body is divided into three regions: an anterior tentacular region, or **forepart**; a very elongate **trunk**; and a tiny terminal **opisthosoma**. The forepart of perviates, called the **vestimentum** in vestimentiferans (Fig. 25.2B), anteriorly bears **tentacles** or **branchiae**, respectively. These are borne dorsally on a lobate, ventral **cephalic lobe** in perviates, or laterally on a bilobed cylindrical structure, called the **obturaculum**, in vestimentiferans. Species of the perviate genus *Siboglinum* have a single tentacle, very long and much coiled, extending forward from the forepart above the cephalic lobe. Other perviate genera have numerous tentacles (some species have over 200). Each tentacle contains an extension of the protocoel, and bears a double row of delicate **pinnules**, each an extension of one epidermal cell (Fig. 25.1C). The vestimentiferan branchiae in *Riftia* occur all along the length of the obturaculum, but in other genera they are generally restricted to near the obturacular base in front of the vestimentum (Fig. 25.2A). The **bridle** is a pair of oblique ridges of thickened cuticle on the forepart in perviates that is used to engage the tube rim when the body is extended. In vestimentiferans the forepart is developed into a folded vestimentum. The muscular folds double back on themselves to wedge the animal in its tube, and form a dorsally located **vestimental chamber** into which gametes are released.

The very long trunk is divided into preannular and postannular parts by a pair of setose ridges, the **girdle** or **annuli** (Fig. 25.1B). A deep midventral groove, bordered with prominent papillae, usually extends along the preannular region part of the way to the girdle, and a dorsal ciliated strip lies above it. The **papillae** may either secrete material for the tube or produce an adhesive substance, as the epidermis of this region is richly supplied with glandular cells (Southward, 1984). The undivided vestimentiferan trunk bears some scattered papillae, but displays no belt of setae. In both groups of pogonophorans, the body terminates in a small, segmented opisthosoma that bears pairs of setae on each segment (Fig. 25.3). Beard worms are enclosed in a tube. However, the tube is open at both ends in perviates to allow the opisthosoma to burrow downward as the animal grows in length.

Digestive System Internally, the most distinctive feature of the phylum is the total lack of a gut in adult worms. The larval gut quickly vanishes and is replaced by a peculiar organ, the **trophosome**, that forms throughout the vestimentiferan trunk and in the posterior of the perviate trunk. The trophosome is an irregular, lobate organ, heavily vascularized and often intimately associated with the gonads. The parenchymatous tissue is packed with endosymbiotic bacteria that are surrounded by membranes within the host cells.

Coelom The coelom can best be characterized as diffuse. It is composed of an irregular series of cavities in the mesodermal tissues, but often lacks any peritoneum. The coelom is almost totally filled by trophosome, gonads, and muscular tissues. Where the cavity is present, it is often filled with gametes or intimately involved with the circulatory system. This appears to be especially so in vestimentiferans (Fig. 25.3B), particularly the branchial circulation,

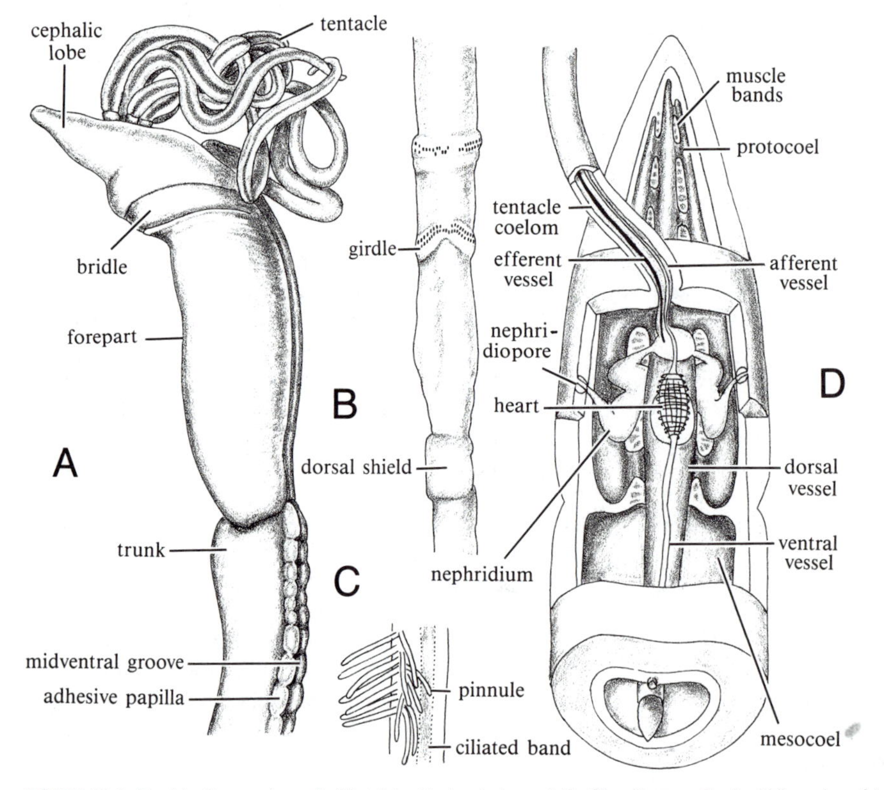

FIGURE 25.1. Perviate Pogonophora. **A.** *Birsteinia vitjasi,* anterior end. **B.** *Siboglinum caulleryi,* middle region of body showing annuli. **C.** *Galathealinum,* part of distal region of tentacle showing pinnules. **D.** *Siboglinum,* scheme of organization showing anterior end of forepart. (**A,B,C,D** after Ivanov.)

where the coelomic portion of the branchiae may function in conjunction with the efferent vessels in returning blood to the trunk. The coelomic cavities are often difficult to distinguish from the blood–vascular channels.

Perviates have a series of interconnected coelomic cavities in the tentacles, another around the heart (Fig. 25.3A). It has been suggested by some that these may represent right and left portions of a protocoel. Another chamber exists in the forepart, sometimes called the mesocoel, and is divided into right and left halves by a mesentery. A metacoel is found in the trunk and also is divided by mesenteries. The mesocoel and metacoel are separated from each other by a muscular diaphragm. A recent suggestion by Southward recognizes only one continuous coelom in the preopisthosomal region of perviates, but which is divided by muscles and connective tissues into apparently separate chambers.

Vestimentiferans, at least *Riftia*, have a similar arrangement (Fig. 25.3B). Coelomic cavities extend into the obturaculum; another is associated with the branchiae, the perivascular cavity of the dorsal vessel, and the anterior dorsal vessel itself. The vestimentum contains the posterior perivascular cavity. The trunk coelom is divided by a mesentery. It has been suggested that these four cavities represent the coeloms of four anterior "segments," although it is hard to see anything other than a correspondence to the four coelomic areas of the perviates (Fig. 25.3). In *Lamellibranchia*, indeed, the preopisthosomal region seems occupied by one continuous, all-penetrating cavity.

The segments of the perviate opisthosoma are septate, but with sheet muscles only on the posterior face of the wall. The vestimentiferan opisthosomal segments are not only septate but also divided by mesenteries. The septa have muscle sheets on both the anterior and posterior faces.

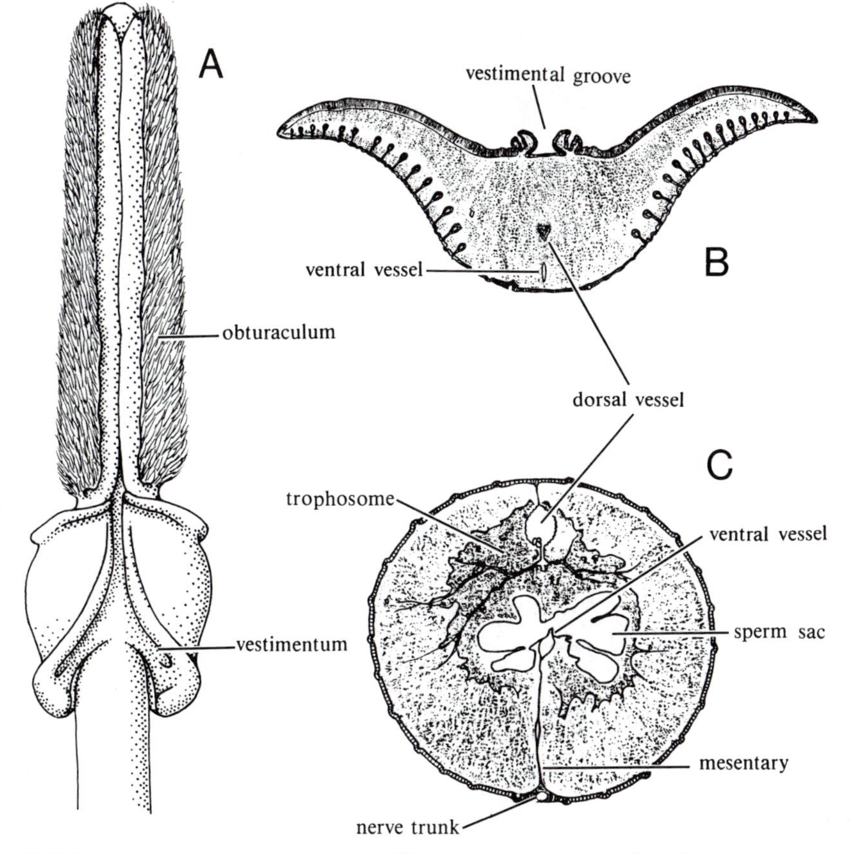

FIGURE 25.2. Vestimentiferan Pogonophora. **A.** *Riftia pachyptila,* anterior end with vestimentum and branchiae bearing obturaculum. **B,C.** *Lamellibranchia,* schematic cross section; **B** posterior vestimentum; **C** anterior of trunk. (**B,C** after van der Land and Nørrevang.)

Excretory System The excretory organs are located in the anterior portion of the forepart. In perviates they are a pair of **protonephridia**. A single dorsal excretory pore is found in the athecanephrian perviates, whereas in thecanephrians they are paired. In the vestimentiferans, an elaborate excretory organ is seen that seems to resemble a protonephridium. The ciliated excretory duct is subdivided again and again as it penetrates the tissue into a complex of nonciliated ductules, which at their end points do not appear to have any walls, but are merely open to intercellular spaces. The entire excretory ductule complex is enveloped in a sheath of musculoepithelial tissue. The excretory pores, as in perviates, can be either single or paired and are located dorsomedially near the base of the obturaculum. Excretory products are also accumulated in pigment cells of the trophosome, near the glands of the vestimentum, as well as in nephrocytelike cells in the body wall of the trunk.

Circulatory System The circulatory system is very similar in both pogonophoran groups (Fig. 25.4A). The median **dorsal vessel** drains the body and carries blood forward to a muscular **heart**. Blood is then pumped into the branchiae or tentacles as well as the

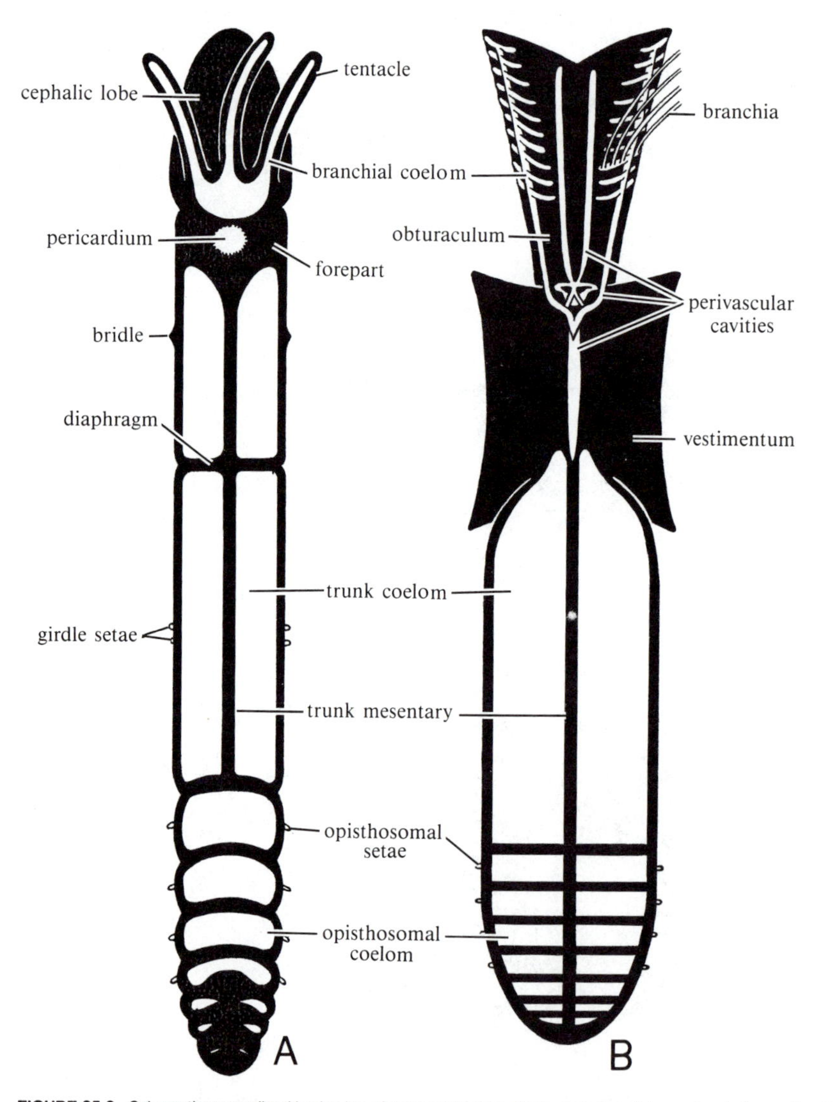

FIGURE 25.3. Schematic generalized body plan of pogonophorans. Body regions not drawn to relative scales; trunk region of both types very long, whereas opisthosome is very small. **A.** Perviate. **B.** Vestimentiferan. (Modified from Jones.)

obturaculum or the cephalic lobe. From these regions blood is then collected in a **ventral vessel.** In vestimentiferans a collecting chamber, the **sinus valvatus**, is located at the anterior end of the ventral vessel and also receives blood from the vestimentum. Branches of the ventral vessel along its length supply the various postcardiac regions of the body.

Nervous System The nervous system is epidermal. The dorsal **brain** is little more than a mass of tissue in the cephalic lobe or arranged around the base of the obturaculum. It extends laterally to give rise to tentacular and branchial nerves, and may also give off a middorsal nerve to the ciliated band of the trunk. There are ventral or lateral nerve cords, and the perviates have a nerve ring in the mesosome-metasome septum. The midventral cord in vestimentiferans is marked by a fluid-filled tube. There are two such tubes in the vestimentum, but posteriorly these join as one in the trunk and disappear completely in the posterior part of the body. In perviates, the preannular nerve cord is equipped with giant fibers whose cell bodies are located in the brain.

Reproductive System The sexes are separate. In perviates (Fig. 25.4B), males have a pair of **testes,**

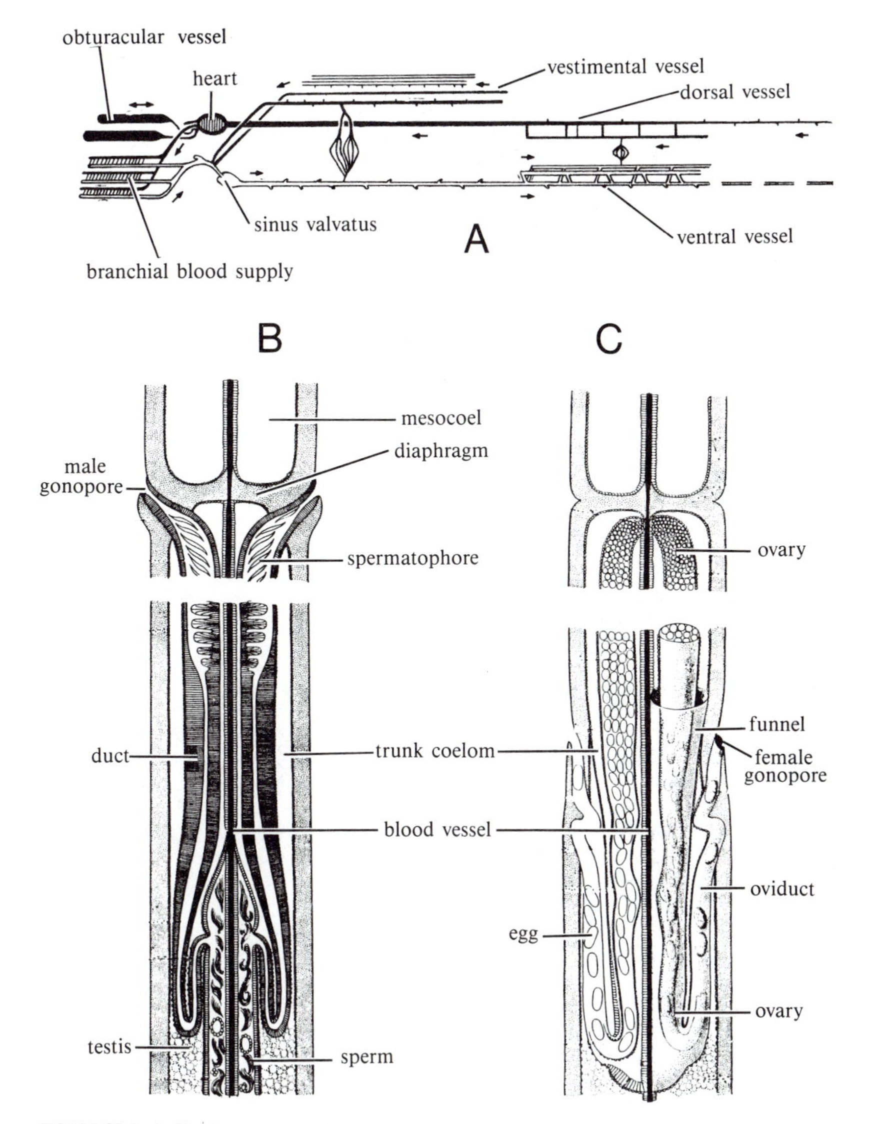

FIGURE 25.4. A. Circulatory pattern of vestimentiferans. **B,C**. Schematic perviate reproductive systems; (**B**) male; (**C**) female. (**A** after van der Land and Nørrevang, 1975; **B,C** after Ivanov.)

placed in the posterior half of the trunk. Long, ciliated, glandular **sperm ducts** extend forward to the forepart-trunk diaphragm, where they exit. **Spermatophores** crowd the sperm ducts, fusiform in Athecanephria and flattened in Thecanephria. **Ovaries** (Fig. 25.4C) lie in the anterior trunk and open through short **oviducts** to gonopores located near the middle of the trunk.

In vestimentiferans, the gonads are closely associated with the lobes of the trophosome. The male system is better understood. The testes are branched; each organ opens by a pore into the coelom that acts like a sperm sac (Fig. 25.2C). These sacs are drained by sperm ducts that pass forward and unite in the vestimentum. From this common sperm duct, two short ciliated ducts lead to the dorsal vestimental grooves from which sperm are shed, packed in spermatophores. The female system is apparently similar but developed on only one side. The oviduct opens to the outside in the posterior part of the vestimentum.

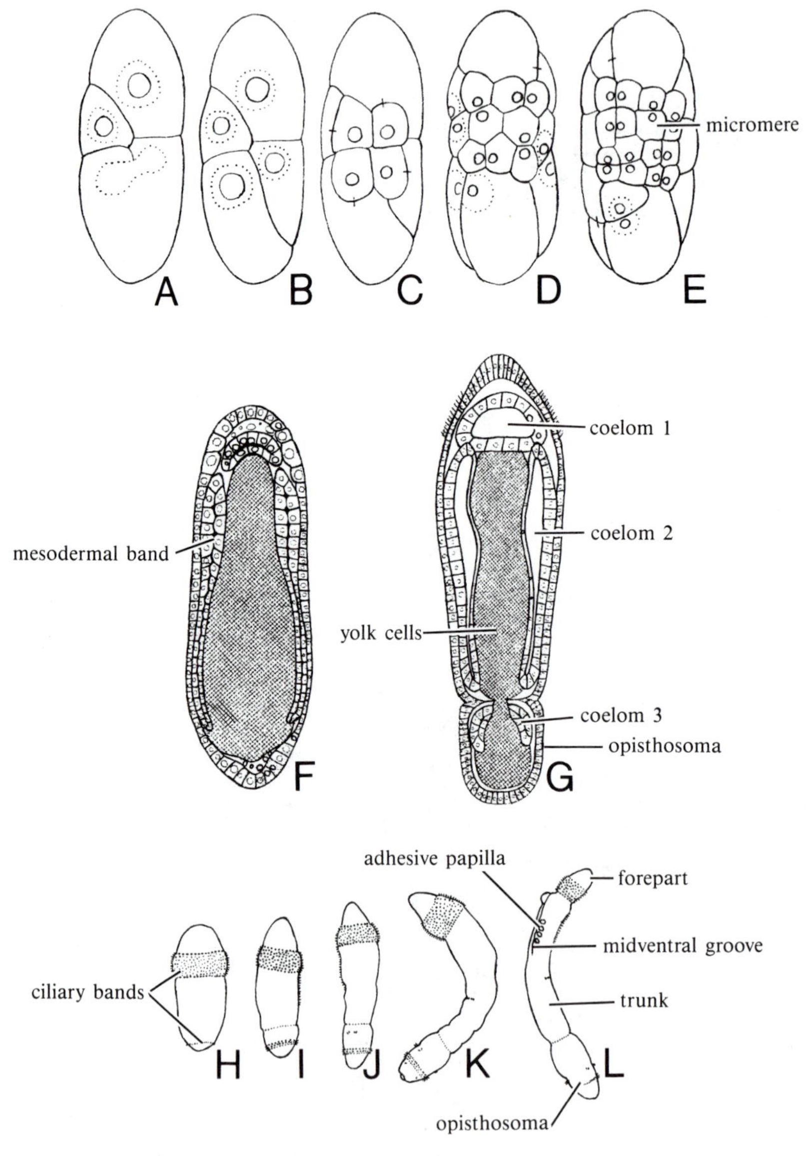

FIGURE 25.5. Pogonophoran development. **A,B,C,D,E.** *Siboglinum fiordicum,* early cleavage. **F,G.** *S. caulleryi,* mesoderm and coelom formation. **H,I,J,K,L.** *S. fiordicum,* ciliated larval development. (**A,B,C,D,E** after Bakke, 1980; **F,G** after Ivanov; **H,I,J,K,L** after Webb.)

Biology

Locomotion Pogonophorans are sessile. As larvae are expelled from the mother's tube, they have a very short and not too effective motile life before settling and building their own tube. Thereafter, motion is restricted to extending out from and pulling back into the tube. This latter is often very abrupt and facilitated by the giant axon fibers that fire the longitudinal body muscles. Movement up and down the tube is undoubtedly facilitated, especially in perviates, by the variety of setae, ciliated bands, papillae, and glandular and cuticular shields.

Feeding Because adults lack a gut, discerning the mode of food procurement in perviate pogonophorans has been a subject of considerable consternation for decades (Ivanov, 1963). At one time or another, several modes have been suggested. These have included external digestion of particles either within the tube or in a temporary cavity formed by the tentacles, surface pinocytosis of colloidal material, and absorption of dissolved nutrients through the body wall. Only the last has been experimentally verified, as *Siboglinum* concentrates radioactively labeled amino acids 21 to 28 times over that of the environment within 17 to 20 hours (Little and Gupta, 1969).

Although absorption is probably important in the perviate feeding economy, the trophosome, packed with symbiotic bacteria in both pogonophoran groups, is another, and probably the primary, source of nutrition (Southward et al., 1981). The worm provides oxygen, carbon dioxide, and hydrogen sulfide via the blood. The bacteria oxidize the sulfide and use the energy to fix carbon dioxide into organic carbon by chemosynthesis. The organic carbon builds the tissues of the host that provide protection for the bacteria (Southward et al., 1986). In the case of the vestimentiferans, this symbiosis has been crucial, allowing the worms to occupy the otherwise highly noxious thermal–vent habitats in the deep sea. The perviates have symbionts only in the posterior trunk, which extends into sulfide-rich sediments.

Larva and juveniles, however, have a gut, and subsist by means of ciliary feeding (Southward, 1988). The worms only gradually shift to dependence on bacterial endosymbionts as the juvenile gut is converted into a trophosome.

Respiration Pogonophoran blood contains hemoglobin. Little is known about any aspects of the respiratory physiology of any pogonophorans. Most of them live in the deep sea and are brought to the surface in such a dreadful state that, if not already dead, they soon expire. Some work on relatively shallow-water *Siboglinum* reveals ranges of oxygen consumption from 0.12 ml/hour/g wet weight at 15°C to 0.06 ml/hour/g at 5°C. *Riftia* blood has been shown to have a high oxygen affinity and carrying capacity that may allow the animal to rely on stored oxygen reserves when environmental oxygen is low (Arp and Childress, 1981). In addition, *Riftia* blood possesses a sulfide-binding protein that apparently serves to concentrate environmental sulfide and transport it to the trophosome for action by the bacterial endosymbionts (Arp and Childress, 1983).

Nerve Function The nervous system seems to have a basic structural peculiarity, at least in perviates. The system is entirely epidermal. There are no glial tissues, and fibers seem to be insulated only by their surrounding epidermal cells. Furthermore, there is no indication that axons ever penetrate the epidermal basal membrane to directly innervate the deeper muscle tissues. There are several areas in which the normal cell covering around the ends of the axons are gone, exposing the end of the fiber directly on the basal membrane. These could be the site of the release of transmitters that then diffuse to the muscles.

Reproduction One species, *Sclerolinum brattstromi*, fragments easily. Both halves are capable of growing the missing parts such that one can find two or more individuals of this species living in the same tube. No other pogonophorans are known to reproduce asexually (Southward, 1975).

Sperm in both pogonophoran groups grow and mature clustered about a **cytophore**. Sperm are distinctive, with filamentous mitochondria wound around the acrosome. In perviates, spermatophores are formed in the thick-walled glandular portions of the sperm duct. The females produce elongate, yolky eggs. Five to 100 eggs are produced at a time.

Modes of mating are uncertain. Spermatophores seem to be ejected one by one. *Siboglinum ekmani* has been observed in its tube to fold the tentacle back over the trunk and draw spermatophores toward the entrance of the tube. With successive actions, sperm packets are eventually pushed out. Because tentacles are short, sperm must float free and find their way to the female rather than be physically transferred there by the male. The spermatophore coat is soluble in seawater. Fertilization takes place within the female tube, but whether it is internal or external is not known. In vestimentiferans, spermatophores have been found on females, so fertilization may take place on the worm rather than in the water.

Development Knowledge of early embryonic development is currently based only on data from perviate species that brood their young. The elongate zygotes undergo holoboastic but unequal cleavage (Fig. 25.5A,B,C,D,E). The micromeres come to lie on a side of the embryo that will essentially define the dorsal surface of the animal (Bakke, 1980). No cross is formed, and the cleavage is probably a highly derived variation of a spiral type (Ivanov, 1988). Gastrulation is by epiboly, with the micromeres proliferating around the yolky macromeres (Fig. 25.5F). Considerable controversy has raged over just how the "mesoderm" forms (Ivanov, 1988). Only perviates have figured in this discussion. It appears that the mesodermal cells might arise at the posterior end of the embryo from the ectoderm and proliferate up the sides of the larva. The mode of cleavage and the presumed mode of mesoderm formation point to "protostome" affinities for the group. However, Gupta and Little (in Nørrevang, 1975) caution against trying to force a triploblastic concept of germ layer differentiation on a pattern of embryonic development that is essentially only "diploblastic."

A ciliated larva (Fig. 25.5H) is formed that is brooded in the mother's tube. The larva quickly delineates an anterior region from the opisthosoma. The anterior portion has a single coelomic compartment (Fig. 25.5F), and the delineation of the diaphragm, separating the trunk and forepart internally (Fig. 25.5G), occurs only after differentiation of the opisthosoma is under way. Subsequent growth of the larva is then confined to just the anterior, nonopisthosomal region (Fig. 25.5I,J,K,L). When liberated, the larvae are mobile for only a short time (Bakke, 1980).

Long thought to lack any gut or endodermally derived tissues, details of early ontogeny (Southward, 1988; Jones and Gardiner, 1988) reveal a transitory yet fully developed gut in juvenile forms. This gut later fills with endosymbiotic bacteria to form the trophosome, and the foreguts and hindguts disappear.

Fossil Record and Phylogeny

Fossil tubes, reminiscent of perviate types, are known from the Oligocene. The mineral deposition associated with hydrothermal vents is ideal for fossilization. Study of such ore deposits has produced some fairly convincing vestimentiferanlike tubes from the Mesozoic of Cyprus and Oman. More will undoubtedly be discovered as such work expands.

Consensus about evolution within the Pogonophora cannot come until two things occur. First, agreement must be reached on how many phyla are involved here. Three positions currently exist: one that places perviates (Hartman, 1954) and vestimentiferans (van der Land and Nørrevang, 1975) within the Annelida; a second that places the two groups in separate phyla (Jones, 1985a); and a third view, followed here, that treats both groups within a single phylum (Jones, 1981; Southward, 1988). Second, some progress must be made on understanding the basic biology and development of the vestimentiferans. This information will be difficult to come by since the animals typically live under extreme conditions in the deep sea.

Taxonomy

The phylum seems excessively split. Although we retain the separate classes Perviata and Vestimentifera, the current differences between orders are probably not equivalent to that seen in other phyla. Higher taxonomy of the group is not based on any rigid character analysis. For example, Jones (1985b) proposed an order of vestimentiferans, the Tevniida, based only on the possession of paired excretory pores, a primitive feature. Thus, classes of vestimentiferans are treated here only as orders (Southward, 1988). Rigorous character analysis is needed for all pogonophorans before a stable taxonomy will prevail.

Class Perviata. One or more tentacles arising from dorsal base of cephalic lobe; opisthosoma lacking mesenteries; males with spermatophores.

Order Athecanephria. Postannular region with ventral cuticular shields and glandular papillae regularly arranged; spermatophores spindle-shaped; single median nephridiopore (Fig. 25.1).

Order Thecanephria. Postannular region with transverse rows of adhesive cuticular plaques; spermatophores leaflike; tentacles can be fused at base.

Class Vestimentifera. Branchial filaments rising from sides of lobate obturaculum; winglike vestimentum located posterior to obturaculum.

Order Axonobranchia. Axially arranged branchial lamellae; dorsal and ventral blood vessels anteriorly branched (Fig. 25.2A).

Order Basibranchia. Basally arranged branchial lamellae; branches of dorsal and ventral blood vessels transversely arranged; some forms with single dorsomedian excretory pore (Fig. 25.2B,C).

Selected Readings

Arp, A. J., and J. J. Childress. 1981. Blood function in the hydrothermal vent vestimentiferan tube worm. *Science* 213:342–344.

Arp, A. J., and J. J. Childress. 1983. Sulfide binding in the blood of the hydrothermal vent tube worm *Riftia pachyptila*. *Science* 219:295–297.

Bakke, T. 1980. Embryonic and post-embryonic development in the Pogonophora. *Zool. Jahrb. Abt. Anat.* 103:276–284.

Hartman, O. 1954. Pogonophora Johansson, 1938. *Syst. Zool.* 3:183–185.

Ivanov, A. V. 1963. *Pogonophora*. Academic Press, London.

Ivanov, A. V. 1988. Analysis of the embryonic development of Pogonophora in connection with the problems of phylogenetics. *Z. zool. Syst. Evolut.-forsch.* 26:161–185.

Jones M. L. 1981. *Riftia pachyptila*, new genus, new species, the vestimentiferan worm from the Galapagos Rift geothermal vents. *Proc. Biol. Soc. Wash.* 93:1295–1313.

Jones, M. L. 1985a. Vestimentiferan pogonophores: Their biology and affinities. In *The Origins and Relationships of Lower Invertebrates* (S. Conway Morris, et al., eds.), pp. 327–342. Clarendon Press, Oxford.

Jones, M. L. 1985b. On the Vestimentifera, new phylum: Six new species, and other taxa from hydrothermal vents and elsewhere. *Bull. Biol. Soc. Wash.* 6:117–158.

Jones, M. L., and S. L. Gardiner. 1988. Evidence for a transient digestive tract in Vestimentifera. *Proc. Biol. Soc. Wash.* 101:423–433.

Little, C., and B. L. Gupta. 1969. Studies on Pogonophora. III. Uptake of nutrients. *J. Exp. Biol.* 51:759–773.

Nørrevang, A. (ed.). 1975. The phylogeny and systematic position of Pogonophora. *Z. zool. Syst. Evolut.-forsch. Sonderheft* 1:1–143.

Southward, A. J., and E. C. Southward. 1987. Pogonophora. *Animal Energetics* 2:201–228.

Southward, A. J., et al. 1981. Bacterial symbionts and low $^{13}C/^{12}C$ ratios in tissues of Pogonophora indicate unusual nutrition and metabolism. *Nature* 293:616–620.

Southward, A. J., et al. 1986. Chemoautotrophic function of bacterial symbionts in small Pogonophora. *J. Mar. Biol. Assc. U.K.* 66:415–437.

Southward, E. C. 1975. Pogonophora. In *Reproduction of Marine Invertebrates*, vol. 2 (A. C. Giese and J. S. Pearse, eds.), pp. 129–156. Academic Press, New York.

Southward, E. C. 1980. Regionation and metamerisation in Pogonophora. *Zool. Jahrb. Abt. Anat.* 103:264–275.

Southward, E. C. 1984. Pogonophora. In *Biology of the Integument*, vol. 1 (J. Bereiter-Hahn et al., eds.), pp. 376–388. Springer Verlag, Berlin.

Southward, E. C. 1988. Development of the gut and segmentation of newly settled stages of *Ridgeia*: Implications for relationships between Vestimentifera and Pogonophora. *J. Mar. Biol. Assc. U.K.* 68:465–487.

van der Land, J., and A. Nørrevang. 1975. The systematic position of Lamellibrachia. In *The Phylogeny and Systematic Position of Pogonophora* (A. Nørrevang, ed.). *Z. zool. Syst. Evolut.-forsch., Sonderheft* 1:86–101.

26

Annelida

There are about 12,000 species in this most successful phylum, exhibiting great diversity of body form (Fig. 26.1). The well-developed coelom and metamerism, or segmentation, of organs and tissues are hallmarks of the body plan of annelids. The coelom arises as paired spaces in the mesoderm of each body segment that enlarge as they develop to form a pair of large coelomic compartments in each somite. Coelomic compartments of adjacent segments meet to form the septa. Septa and mesenteries are formed of two layers of peritoneum. Septa are usually invaded by muscle tissue, but both septa and mesenteries may be partly open, permitting access between the various coelomic compartments in adult animals.

Definition

Annelida are worms with a body wall composed of an outer epidermis and with well-developed, segmentally arranged outer circular and inner longitudinal muscle layers. A cuticle secreted by the epidermis is usually present. The body is metamerically (segmentally) organized. Chitinous setae are usually present, secreted by single chaetoblasts. The head is variable, typically formed from a preoral prostomium segment and a peristomium containing the mouth. Typically the circulatory system is closed, and generally has a dorsal vessel in which blood flows anteriorly, and a ventral vessel in which blood flows posteriorly. The blood often has respiratory pigment dissolved in the plasma. A schizocoel, generally spacious, is usually divided into subequal segmental compartments. Primitively, one pair of nephridia are located in all but the first and last somites. The circumenteric nerve ring, containing a dorsal cerebral ganglion, a subpharyngeal ganglion, and connectives, and a ventral, fused, double nerve cord with paired segmental ganglia and commissural connections make up the central nervous system. Annelids have spiral, determinate cleavage and, in aquatic forms, a trochophore larva.

Body Plan

The annelid body wall (Fig. 26.2) consists of layers of **cuticle**, **epidermis**, a very thin to moderate layer of connective tissue forming a **dermis**, circular muscle, longitudinal muscle, and **parietal peritoneum**.

The cuticle is a nonchitinous, albuminoid material, somewhat variable in thickness but generally very thin and flexible (Fig. 26.3). Scale worms, however, have evolved a series of dorsal cuticular plates (see Fig. 26.23B). Slime and other secretions from the epidermis must pass through cuticular pores to reach the surface. Some oligochaetes and polychaetes have a cuticle with striae, which, when sufficiently delicate, confer iridescence to the body surface. Some polychaetes have brilliant, iridescent blue and green hues that result from such cuticular striations; for example, *Aphrodita*'s brilliant iridescence is due to a dorsal covering of felted filaments. Many of the polychaetes use surface epithelial cilia in some regions for food procurement or to generate respiratory currents. In these regions the cuticle is reduced.

The epidermis consists of a single layer of epithelium (Fig. 26.3A), usually columnar, but varying in thickness in different regions of the body (Fig. 26.3B). A scattered population of basal cells can be seen. There is some evidence that these cells arise from the mesodermal tissues below and migrate out as required to the epidermis to replace moribund epidermal cells. A variety of unicellular gland cells are interspersed among the columnar cells.

The dermis is more delicate in small species than in large, and in polychaetes and oligochaetes (Fig. 26.3B) than in leeches. In small polychaetes it is little more than a basement membrane. In large leeches it is an extensive mesenchyme layer (see Fig.

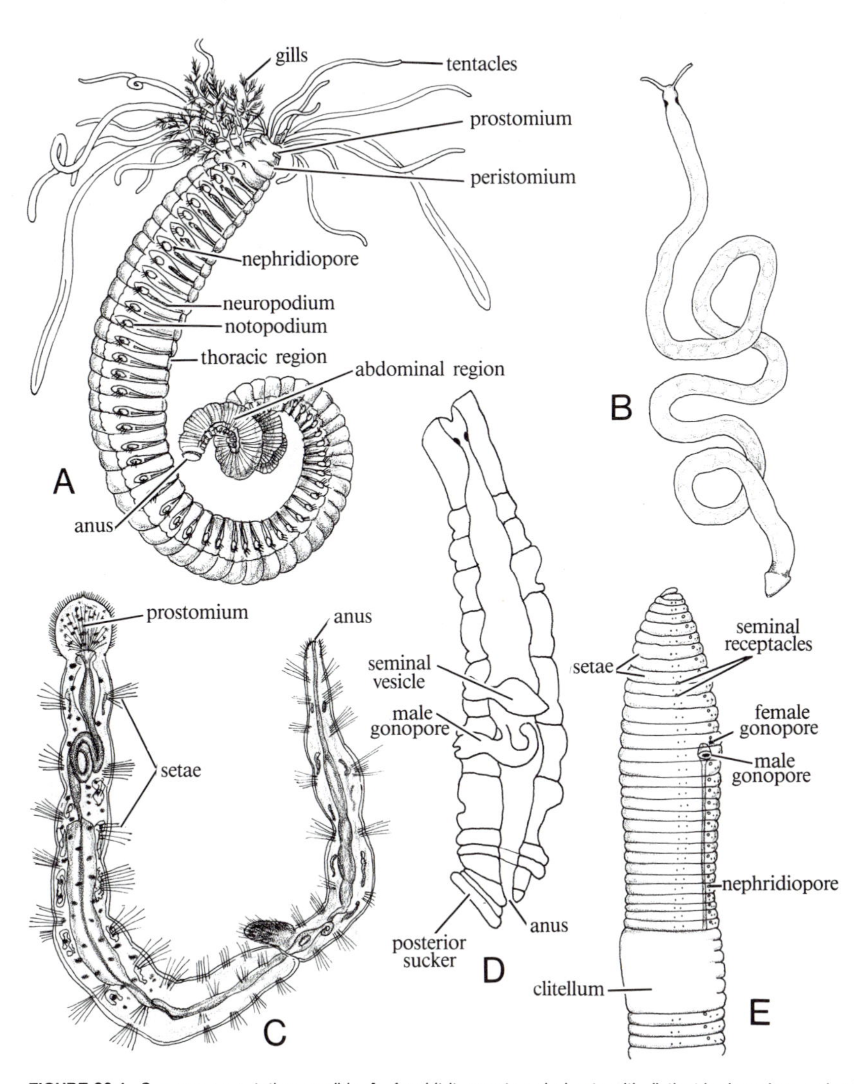

FIGURE 26.1. Some representative annelids. **A**. *Amphitrite ornata,* polychaete with distinct body regions and reduced parapodia. **B**. *Polygordius neapolitanus,* a meiofaunal form characterized by absence of parapodia, lack of septa, and general segment reduction. **C**. *Aeolosoma hemprichi,* oligochaete characterized by minute size, aquatic habits, and male gonopore in somite immediately behind that containing testes. **D**. *Cambarincola elevata,* branchiobdellid leech that parasitizes crayfish. **E**. Anterior end of *Lumbricus terrestris,* oligochaete characterized by long male gonoducts and male gonopores located well behind somite containing testes. (**A** after Brown; **B** after Fraipoint; **C** after Lankester, from Beddard; **D** after Goodnight, in Edmondson.)

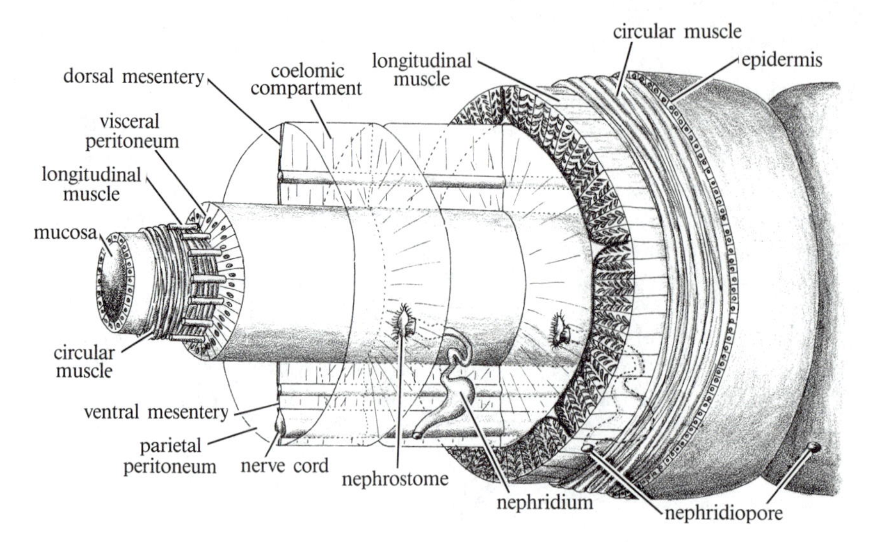

FIGURE 26.2. Scheme of basic organization of annelid somites. Successive layers of tissue removed to show their relationships, with cellular detail omited in peritoneum and mesenteries. Coelom arises as bilateral spaces, pair for each somite. Parietal peritoneum against body wall, visceral peritoneum against gut wall, and at midline two-leaved dorsal and ventral mesenteries. Peritoneum of adjacent somites forms double-layered septa. Paired excretory tubules, metanephridia, occur in coelomic compartments. Each opens into next anterior coelomic compartment of neophrostome, and to exterior by nephridiopore. Dorsal blood vessel in dorsal mesentery and ventral blood vessel in ventral mesentery. Nerve cord may be in body wall, or in ventral mesentery, as shown here. Mesenteries and septa tend toward reduction in many living groups.

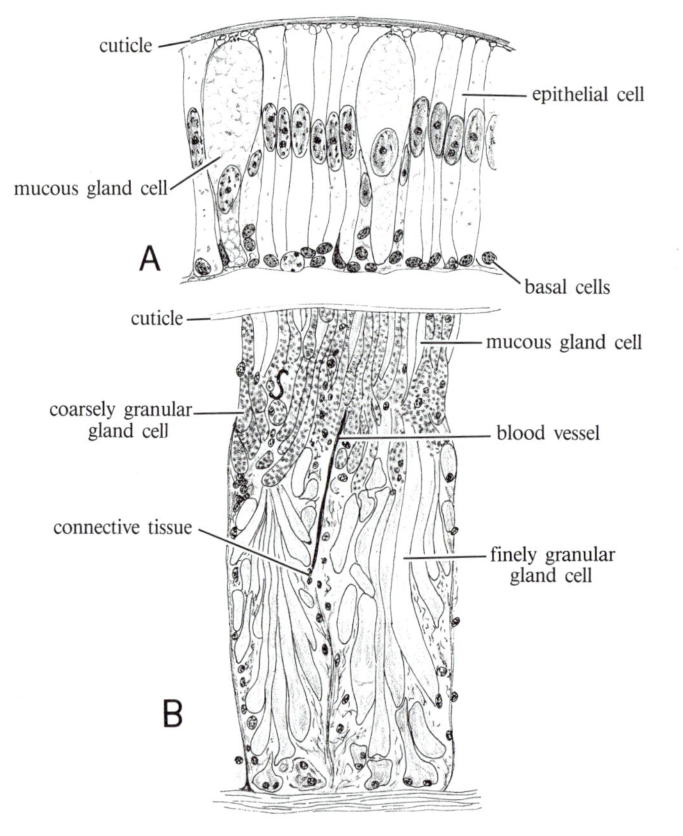

FIGURE 26.3. Epidermis of *Lumbricus terrestris.* **A**. Section through unspecialized somite. **B**. Section through clitellum. Notice specialized gland cells and increased thickness of epidermis. (**B** after Grove.)

26.28B) containing the tips of the dorsoventral muscle strands. In some of the oligochaetes, fibrils from the epidermal cells penetrate the basement membrane and pass through the dermis, attaching directly to the circular muscle. This arrangement is reminiscent of the epidermal fibers used to strengthen muscle attachments in many arthropods.

Annelids invariably have an outer layer of circular muscle and an inner layer of longitudinal muscle in the body wall (Fig. 26.2), but the two layers differ greatly in relative development in different annelids.

Coelom A compartmentalized, metameric coelom is characteristic of annelids (Fig. 26.2). Parietal peritoneum covers the body wall; visceral peritoneum covers the viscera, and septal peritoneum forms the septa. The right and left coelomic compartments begin as independent cavities and meet at the dorsal and ventral mesenteries in the midline. The ideal annelid plan is thus a series of discrete pairs of coelomic compartments, one for each somite, but the coelom is seldom quite like this in adults. The ventral and dorsal mesenteries are somewhat incomplete, and septa are often suppressed to produce large coelomic spaces that extend through several body segments.

A nearly primitive arrangement is seen in many oligochaetes. *Lumbricus*, for example, has a complete series of septa dividing the coelom into metameric compartments. The dorsal mesentery, however, persists only at the septum, and the ventral mesentery does not reach the body wall; therefore, right and left coelomic spaces are joined in each somite. Most leeches have no septa whatsoever, and the coelom functions in its original role as a circulatory system.

Placement and form of **septa** are often related to body movements or the movement of parts. Many oligochaetes and polychaetes have cupped or slanted septa that allow movement of the coelomic fluid when the septal muscles are contracted. The *Lumbricus* pharynx, for example, can be partly protruded by the contraction of strongly slanted septa attached to it. Septa are usually suppressed at the anterior end of polychaetes, converting the coelom into a hydraulic compartment used to evert the proboscis or extend the hollow tentacles. The septum immediately behind these hydraulic compartments is heavily muscled and is termed a **diaphragm**. In some polychaetes, *Arenicola* for example, several such diaphragms are present (see Fig. 26.13A).

Digestive System The digestive system is tubular, with an anterior mouth and a posterior anus. The gut is usually straight, although some part of it is spiraled or coiled in a few exceptional species. The relatively extensive foregut, consisting of **buccal cavity**, **pharynx**, and **esophagus**, is derived from the stomodeum and lined with epidermis, some of which secretes cuticle. The **midgut** is relatively long and consists of the **stomach** and the **intestine**. The **hindgut** is very short and lined with epidermis from the proctodeum.

The tubular esophagus is lined with cuticle or with ciliated epithelium. It carries food to the midgut, usually by peristalsis if there is no ciliary–mucus feeding system. Polychaetes with a very long **proboscis** often have an S-shaped esophagus, which straightens out only when the proboscis is extruded. Occasionally, lateral caeca occur for storage (Fig. 26.4A). Usually, however, the esophagus is a straight tube. Dilations of the esophagus form a thin-walled storage chamber, the **crop**, or a thick-walled, sclerotized grinding chamber, the **gizzard**. The crop and the gizzard are especially characteristic of the oligochaetes, although some polychaetes, the syllids for example, have a gizzard. Among the oligochaetes a crop may or may not be present, and some of the terrestrial oligochaetes have more than one. *Lumbricus* has a single crop and gizzard (see Fig. 26.25A). The esophageal wall is lined with a thin cuticle, continued without much change in the crop, but strengthened in the gizzard where it forms a protective lining for the muscular walls. The crop and the gizzard of terrestrial oligochaetes have no characteristic position (Fig. 26.4A). The *Lumbricus* crop and gizzard, for example, lie at the distal end of the esophagus, whereas they are proximal in megascolecids. The crop and the gizzard are usually more than one somite long, but where there are multiple gizzards, each occupies a single somite. Multiple gizzards are not uncommon; in eudrilids, for example, four to six gizzards are found. The so-called crop of leeches is typically a part of the midgut.

Esophageal glands occur sporadically among polychaetes, are absent in aquatic oligochaetes, and are present in terrestrial oligochaetes. The oligochaete esophageal glands are composed of epithelial plates or tubes in intimate contact with blood sinuses, and often contain tiny calcite crystals. Ingested soil contains a considerable quantity of calcium and other inorganic ions, some of which can unbalance the body fluids from a physiological point of view. Absorption of these ions appears to be controlled, in part, by the action of the esophageal glands, more commonly called **calciferous glands**. The calcite crystals formed there pass unchanged through the gut and emerge with the feces.

Carnivorous, herbivorous, and detritus-eating habits have relatively little to do with differences in the midgut of annelids. Leeches, however, do have a distinctive midgut, associated with their adaptation for bloodsucking. The simplest possible kind of midgut is a straight tube with no specialized regions. A midgut of this kind is seen in many polychaetes (see Fig. 26.13), in earthworms (see Fig. 26.25), and in most aquatic oligochaetes. It is termed the intestine or stomach-intestine and is the site of the greater part of digestion and absorption of foods.

The annelid intestinal wall is composed of a layer of visceral peritoneum on the outer surface, followed by layers of longitudinal muscle, circular muscle, and mucosa (Fig. 26.2). The **mucosa** consists of a single layer of epithelium resting on a delicate **submucosa** of loose connective tissue containing blood sinuses or capillaries. The mucosa is typically ciliated.

The capacity for digestion and absorption is partly determined by the area available for secretion and absorption. The shape of most annelids precludes extensive amplification of intestinal surface by lengthening and coiling of the gut, although the anterior intestine of *Amphitrite* is slightly coiled and some of the most highly modified sedentary forms have a coiled intestine (Fig. 26.4C). The most common surface amplification is provided by longitudinal ridges. Any part of the intestinal wall of polychaetes may be folded, and the same was probably true of primitive oligochaetes as well. Modern earthworms, however, have a single dorsal ridge (**typhlosole**), or, if several are present, have a larger dorsal ridge, through most of the intestinal length. Some of the earthworms have a simple typhlosole, with only the mucosa entering into the fold, but in *Lumbricus* (see Fig. 26.25B), the muscle and peritoneal layers are involved in the folding of the wall. Other methods of surface amplification are seen in the midgut of some oligochaetes and polychaetes. *Aphrodita* (Fig. 26.4B) has lateral diverticula in each somite that add absorptive surface and also facilitate food distribution. Only fluid and small, partly digested particles find their way into the diverticula, where the final stages of digestion and absorption occur.

Most annelids have no conspicuous hindgut. The intestine may change character slightly as it approaches the anus, but only leeches can be said to have a rectum. The anus is primitively situated in the last somite, but in most leeches is secondarily located outside the posterior sucker that contains the last somite.

Circulatory System The circulatory systems of oligochaetes and polychaetes have similar designs. There are two longitudinal vessels, a **dorsal vessel** above the gut and a **ventral vessel** below it (Fig. 26.5B). Blood flows anteriad in the dorsal vessel and posteriad in the ventral vessel. This longitudinal flow delivers food from the sinuses and capillaries of the midgut to the anterior parts of the body. Large size, greater complexity, and greater activity favor the development of additional lateral circulatory arcs. The most important of these is a subcutaneous plexus in the body wall, used for respiratory exchange and useful for the delivery of food and re-

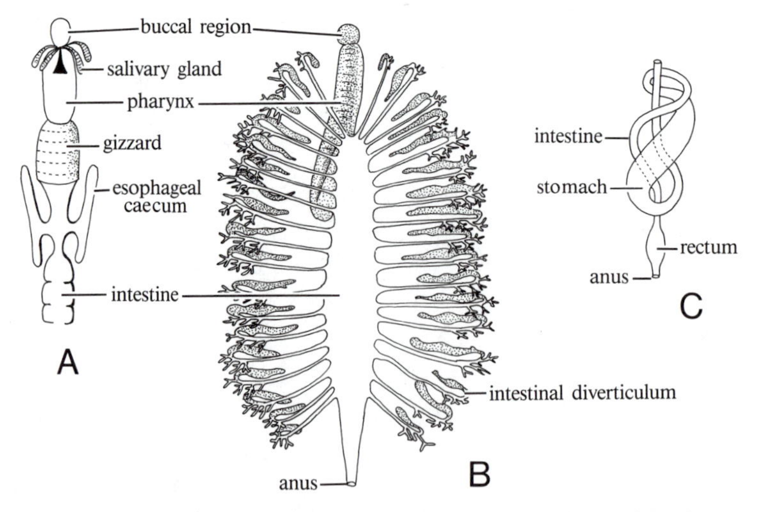

FIGURE 26.4. Surface amplification in polychaete digestive tract. **A.** Anterior end of digestive tract of syllid (schematic). **B.** Digestive tract of *Aphrodita*. **C.** Coiled digestive tract of *Petta*. (**A** after Benham; **B** after Gegenbauer; **C** after Wiren, from Benham.)

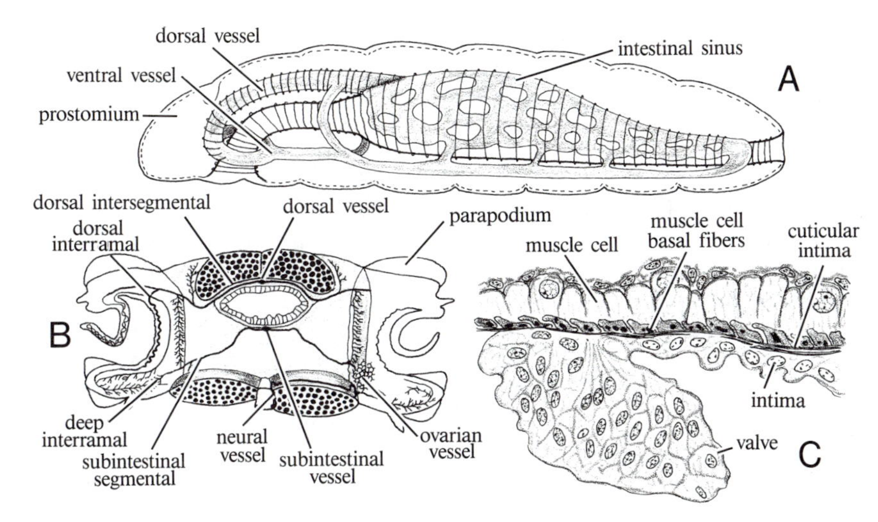

FIGURE 26.5. **A**. *Aeolosoma headleyi,* scheme of simple circulatory system. Dorsal vessel carries blood forward from intestinal sinus, and ventral vessel carries blood back and up to intestinal sinus. Vessels around foregut connect dorsal and ventral vessels. **B**. Schematic view of lateral circulation in middle of body of *Nephthys californiensis.* Blood passes to parapodium for aeration from subintestinal vessel and returns to dorsal vessel. Complete circulation can occur within single somites by way of circumintestinal connectives. **C**. Section through wall of heart of *Allobophora,* including part of valve that prevents backflow. *Arrow* indicates direction of flow. Circular muscles with fibrillar bases lie outside intima. (**A** after Marcus, from Avel, in Grassé; **B** after Clark; **C** after Dahlgren and Kepner.)

moval of wastes from the body wall tissues. Subsidiary circulatory arcs serve the parapodia of errant polychaetes, and smaller but functionally important arcs develop to serve the nephridia. Blood flows outward from the ventral vessel through paired afferent segmental vessels in each somite (Figs. 26.5B and 26.6A). They give rise to **nephridial vessels**, to branches reaching the body wall, and in polychaetes to branches for the parapodia. Efferent segmental vessels return the blood to the dorsal vessel (Figs. 26.5B and 26.6A). As blood flows through the intestinal wall from the dorsal to the ventral vessel, a complete lateral circulation is established, permitting the cycling of blood within a somite. The intes-

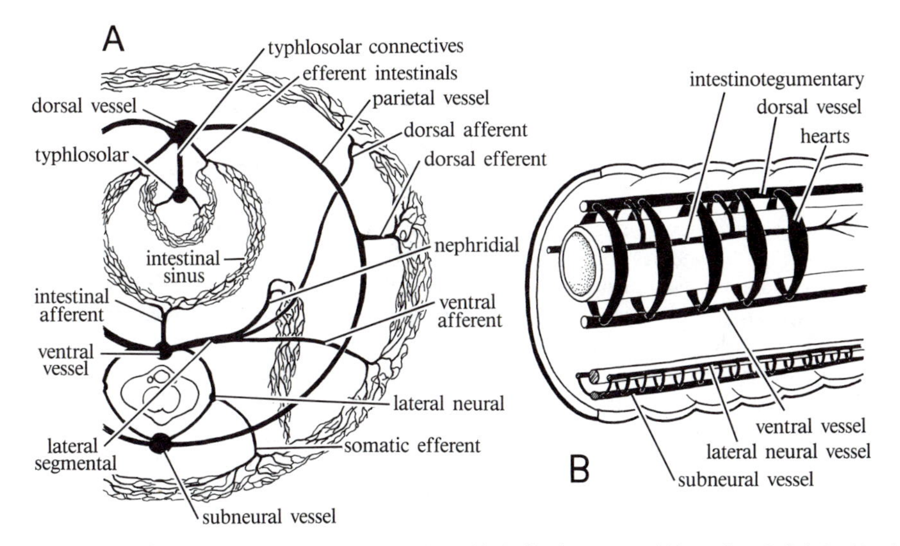

FIGURE 26.6. Circulation in *Lumbricus terrestris* (schematic). **A**. Circulatory arcs within somites. **B**. Relationship of vessels in esophageal region.

tinal circulation passes in the other direction in earthworms, entering the gut wall from the ventral vessel (Fig. 26.6A).

Additional longitudinal vessels are found associated with the nerve cord in many annelids, as in *Lumbricus* (Fig. 26.6A). In *Lumbricus* some freshly aerated blood is routed into vessels around the nerve cord, on its way to afferent segmentals (parietal vessels) that empty into the dorsal vessel. *Arenicola* has a supraneural vessel connected with the ventral vessel. Additional longitudinal vessels appear in the typhlosole of terrestrial oligochaetes, and special subintestinal and lateral intestinal vessels sometimes occur in polychaetes.

The most primitive method of blood propulsion is blood vessel contraction. Annelids generally retain a primitive, diffuse propulsion mechanism, but show some tendencies toward centralization in more highly specialized groups. The dorsal blood vessel is the most important contractile vessel, but other vessels may be contractile and contribute significantly to blood flow. Lateral vessels, esophageal vessels, and, to a lesser extent, nephridial vessels are sometimes contractile. None of the annelids have a definite "heart," although dilations of the dorsal vessel are sometimes called **heart bodies**. Some terebellids have a heart body (see Fig. 26.13B), but it is a spongy nonpulsating organ of unknown function. Pulsating chambers occur in a few species, as in *Chaetopterus* and *Arenicola* (see Fig. 26.13A), but these are never the sole pulsating parts and cannot be considered as true hearts. Oligochaetes have pulsating esophageal vessels connecting the dorsal and ventral vessels. These are sometimes termed hearts or **aortic arches**, but the dorsal vessel is also contractile and is at least as important in blood propulsion. The number of pairs of pulsating vessels in the esophagus region varies considerably in oligochaetes. Aquatic oligochaetes usually have a single pair, and in *Lumbricus* five pairs are found (Figs. 26.6B and 26.25A).

Annelids have a simply organized circulatory system, but the blood vessels have well-developed walls. Pulsating vessels have a thick endothelium (Fig. 26.5C). The wall is strengthened by a strong basement membrane around which a layer of circular muscles lies. In nonpulsating vessels, wall cells replace the circular muscles, although some retain a few muscle fibers, perhaps used for vasoconstriction. Capillaries are composed of endothelium only. Although annelids are said to have a closed circulatory system, the system of intestinal sinuses (Fig. 26.5A) does not have a complete endothelial lining. In larger species with well-developed circulatory systems, the dorsal vessel has a system of valves, derived from the endothelium, that prevent backflow (Fig. 26.5C). Where other pulsating vessels are present, they too may contain valves.

Respiratory System Respiratory exchange takes place at the body surface. Many annelids have a smooth body surface, and it is tempting to think of them as being without respiratory adaptations. The smallest oligochaetes and polychaetes have no circulatory system, or have one that does not serve the body wall. In larger forms, either blood or coelomic fluid is circulated through a subepidermal plexus where respiratory exchange can occur rapidly. A few of the polychaetes have coelomic subepidermal capillaries, but this is uncommon except in leeches. Respiratory exchange, however, is facilitated in several ways, such that all annelids exhibit some kind of respiratory adaptation.

First, gills for respiratory exchange can occur when a subepidermal plexus is present and respiratory stress is strong enough to favor it, but some of the largest annelids have no gills. Leeches are almost universally without gills, although gills are found in some piscicolids. This does not mean that other kinds of annelids suffer no respiratory stress, for there is evidence that leeches tolerate anaerobiosis rather better than most animals. Gills occur most commonly in polychaetes, but such is not clearly linked with body size. Polychaete gills are usually modifications of the dorsal or ventral cirri of the parapodia. Most polychaetes have dorsal gills (Fig. 26.7A), but some have gills on both notopodium and neuropodium, and a few have only ventral gills. The parapodial gills may be flat plates or filiform; filiform gills are often branched or plumose. They are always highly vascular and sometimes contrast with the body in color because of the hemoglobin or chlorocruorin in them. Prostomial and peristomial gills are common in sedentary forms, and are derived from various cephalic parts. The gills of chlorhaemids arise from the prostomial tentacles. *Amphitrite* gills are probably derived from the prostomial palps. The common occurrence of gills among sedentary worms is undoubtedly associated with the low oxygen tensions in the tube proper, which reduces the value of the general body surface for respiration. Very few oligochaetes have gills of any kind, even when very large. Some species of *Alma* have posterior gills, and a number of small aquatic oligochaetes have external gills, usually at the posterior tip of the body (Figs. 26.7B and 26.26H,I).

Second, ventilation of the respiratory surfaces by

ciliary currents or muscular movements can compensate for the lack of gills or make gills more efficient. Many polychaetes have dorsal cilia and cilia on the parapodial gill surfaces. Ciliary currents usually flow toward the posterior end of the body over the dorsal surface and to or from the gills to the main current. Ciliary tracts on the body of tube dwellers usually keep water flowing through the tube and aid somewhat in respiratory exchange. Nereids and glycerids have no body cilia but use undulatory movements of the body to ventilate the surface. Highly specialized respiratory currents are characteristic of the scale worms. Ciliary tracts bring water up between the parapodia and to a main stream that flows back under the elytra. As in molluscs, the principle of counterflow usually operates, as the respiratory currents flow back over the dorsal surface, in the direction opposite to blood flow. Aquatic oligochaetes usually thrust the posterior end of the body from their burrows (Fig. 26.7B), and have posterior gills when gills are present. Tubificids usually have no gills, but the posterior tip of the body is waved rhythmically to ventilate the surface. When oxygen tension falls, they lift the tail end of the body higher and increase the frequency of ventilating movements. Leeches also move the body rhythmically to ventilate the surface and change the rate of movements with increased temperatures or other factors producing respiratory stress.

Excretory System The primitive annelid somite contains a pair of coelomic compartments, each open to the exterior by way of two ducts (Fig. 26.8), a gonoduct used for gamete discharge and a nephridium used for excretion and osmoregulation.

Primitively, a pair of nephridia lie in each somite, and wastes reach the nephridia by way of the coelomic fluid. As the circulatory system becomes more highly organized, the nephridia come to be supplied with blood vessels. With the development of a vascular bed around the nephridium, the waste materials can arrive as well by way of the blood as by the coelomic fluid. The nephridia never lose their contact with the coelomic fluid, however, and a nephrostome remains as evidence of the importance of coelomic transport of wastes. However, with the development of nephridial blood circulation, longitudinal transportation of wastes is greatly improved, so that nephridia are reduced in some somites or assume functions other than elimination of wastes. Generally speaking, oligochaetes and errant polychaetes retain nephridia in nearly all somites. Tube-dwelling polychaetes tend to eliminate the posterior nephridia, this trend culminating in forms like the sabellids where a single pair of nephridia empty near the opening of the tube. Coelomic fluid in the posterior parts of the coelom is thus isolated from the nephridia, and this is linked with the increased importance of the circulatory system for waste transport.

A few annelids have **protonephridia** with flame cells (Fig. 26.8A,F). In *Vanadis* the protonephridium is entirely separate from the gonoduct, but as a rule the protonephridium is partially united with the gonoduct (Fig. 26.8B). Gametes enter through the funnel of the gonoduct, and wastes through the tu-

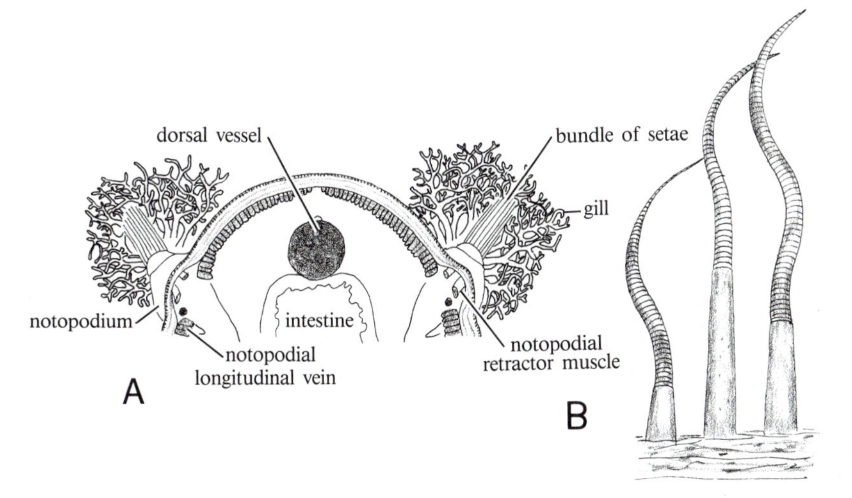

FIGURE 26.7. **A**. Gills of *Arenicola*, derived from notopodium. **B**. Tubificids extending posterior end of bodies from their tubes. Body surface aerated by spiral waving movements of tail, and increased oxygen uptake promoted by extending more of body from burrow. (**A** after Wells.)

bule walls, but both pass through the lower part of the compound tubule.

The capitellids have a metanephridium, open to the coelom by way of a ciliated nephrostome, and a completely distinct gonoduct used for the discharge of gametes (Fig. 26.8C). As in the case of the protonephridia, this primitive separation of gonoduct and nephridia disappears with the more or less intimate union of the gonoduct and nephridia (Fig. 26.8D). In all of these, two openings to the coelom are found, a nephrostome and a gonostome.

In still other polychaetes a single duct is present, showing no obvious evidence of a compound origin in the adult, as in *Arenicola* (Fig. 26.8E).

The matter is made more complex by deviations from the primitive state, in which a pair of gonads occur in each somite and the nephridia are active in all somites. In tube dwellers especially, regional specialization of the body leads to regional differences in segmental organs such as nephridia. In the sabellids, excretory functions are restricted to a single pair of enlarged and modified nephridia, while the nephridia of the fertile regions of the body are used as gonoducts.

All oligochaetes have metanephridia (Fig. 26.9A), and as the gonads lie in a few genital somites and have a distinct set of gonoducts, the nephridia are not used for the discharge of gametes. The primitive arrangement in oligochaetes has a pair of metanephridia in each somite. Each nephridium is made up of a preseptal nephrostome and a complex postseptal tubule (Fig. 26.9B,C,D,E) that ends in a nephridiopore, as in *Lumbricus*. The nephridial surface of some oligochaetes is amplified by subdivision of the pair of nephridia in each somite into many nephridia. In megascolecids the strand of embryonic tissue from which the nephridium arises divides into a series of loops, each of which becomes an independent nephridium. Some are equipped with a preseptal nephrostome; others are closed, without a nephrostome, and lie wholly within a single somite. Some open into the gut instead of the outside. Nephridia that open into the pharynx or esophagus are thought to have a digestive function, but those that open into the lower part of the intestine probably excrete wastes that are further concentrated in the gut by water reabsorption. *Pheretima*, with all of its nephridia behind the fifteenth somite draining into the gut, has a higher resistance to drying than most oligochaetes. In a number of genera, several kinds of nephridia are found in a single organism.

Polychaetes tend to have rather simple nephridia, consisting of a large, often frilled nephrostome, an excretory tube of varied length, and a terminal bladder. The tube and the bladder of shorter nephridia may be combined to form an excretory sac. Oligochaetes have more complex nephridia, probably in response to the greater osmotic stresses they encoun-

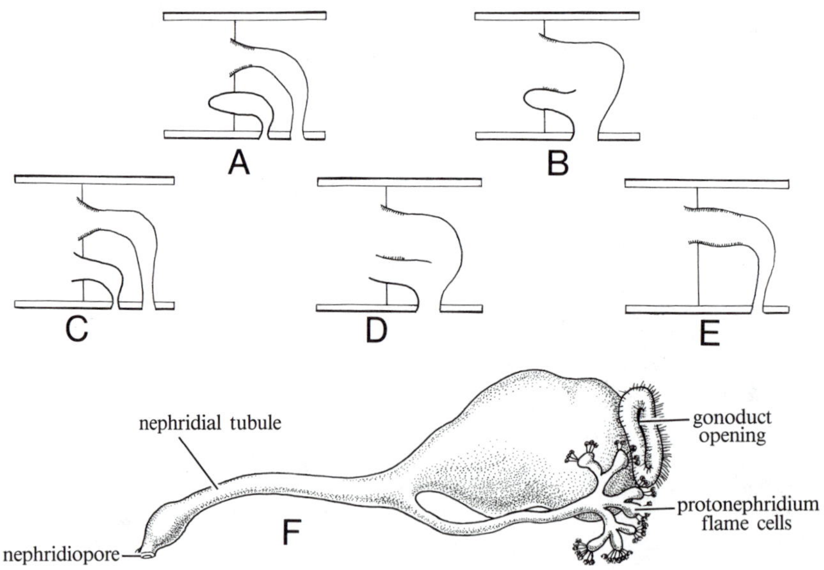

FIGURE 26.8. Annelid nephridia. **A,B,C,D,E.** Variations in union of gonoduct and nephridium. **A.** Primitive condition, with separate protonephridium and gonoduct with ciliated funnel, as in *Vanadis.* **B.** Protonephridium united with gonoduct, as in *Phyllodoce.* **C.** Completely separate metanephridium and gonoduct, as in *Notomastus.* **D.** United metanephridium and gonoduct, as in *Hesione.* **E.** Completely united nephridium and gonoduct, as in *Arenicola.* **F.** Nephridium of *Alciopa.* (After Goodrich.)

ter. The nephridium is divided into a series of histologically distinct regions in both aquatic and terrestrial species (Fig. 26.9B,C,D,E). The functional significance of histological specializations is not very well understood, and little is known of the urine of aquatic oligochaetes.

Some of the visceral peritoneum of oligochaetes and polychaetes is histologically differentiated as yellowish to brownish **chlorogogue cells**. They occur on the surface of the midgut, in patches on the nephridia and esophagus, or as small deposits on the major blood vessels. Deamination of protein compounds occurs here, some foods are stored, and some nitrogenous waste materials are formed. Chlorogogue cells release ammonia and urea. They become detached and float in the coelomic fluid as amoeboid cells. Bits of the chlorogogue cells may be budded off, to enter the nephrostome, and, in some cases at least, whole coelomic corpuscles may enter the nephridium, bearing a load of nitrogenous wastes. They may also be deposited as yellow bodies in the body wall, or pass completely out of the body. The chlorogogue cells are also important in food transport, collecting about the ova and nourishing them during maturation.

Nervous System The simplest annelid nervous system has a subepidermal plexus in contact with the epidermis. It is thickened to form a pair of **ventral nerve cords**, connected in each somite by **commis-**

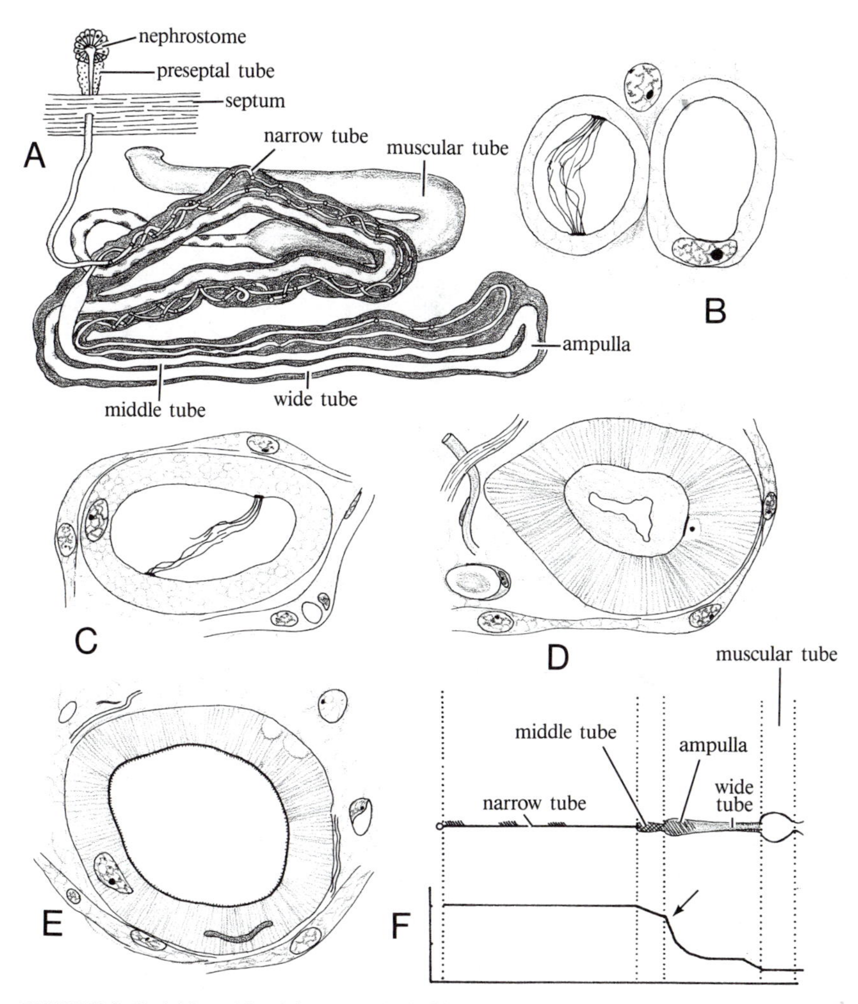

FIGURE 26.9. Nephridium of *Lumbricus terrestris*. **A**. Somewhat simplified scheme of whole nephridium. Note great length and extensive regionalization. **B,C,D,E**. Sections through parts of **A** at different regions; **B**. narrow tube; **C**. middle tube with ciliated tract; **D**. ampulla; **E**. muscular tube. **F**. Osmotic concentration (*arrow*) of urine in various parts of nephridium, in percent concentration of salt in Ringer's fluid. (**A** after Benham, from Michaelson, in Kükenthal and Krumbach; **B,C,D,E** after Dahlgren and Kepner; **F** after Ramsay.)

sures. At the junction of the nerve cords and commissures, **ganglia** occur. At the anterior end of the body a **circumenteric nerve ring** is formed of a pair of **dorsal cerebral ganglia**, or **brain**, connected by a **cerebral commissure** and a pair of connectives that join the first ganglia on the ventral nerve cords. The cerebral ganglia lie in the prostomium of polychaetes and give rise to nerves to the prostomial sense organs (Fig. 26.10A). This type of nervous system is seen in some small polychaetes and in *Aeolosoma*. The brain of polychaetes can be quite complex, with recognizable forebrain, midbrain, and hindbrain.

As in other phyla, the nervous system has tended to move inward to a more protected position (Fig. 26.10F,G,H). The nervous system retains its subepidermal position in some polychaetes, like *Clymenella*, but in the majority it lies in the muscle layer of the body wall. It is sometimes in the muscle layer, as in oligochaetes, but in larger species it has moved into the coelom, as in *Lumbricus*. Muscle cells, however, are incorporated in the sheath of the nerve cord, preserving a record of its past intramuscular position. The nerve cord of leeches lies within the coelomic sinuses (see Figs. 26.28C and 26.30A).

The originally paired ganglia in each somite tend to assume a more medial position and eventually to unite (Fig. 26.10A,B,C). The ganglia are still paired in some tube-dwelling polychaetes, notably the sabellids and the serpulids. In other polychaetes the ganglia have united to form a single median mass, which, however, preserves histological evidence of its double origin. *Aeolosoma* retains double ganglia and a double nerve cord, but most oligochaetes have united ganglia. The first ventral, or subpharyngeal, ganglion is distinctly larger than the others in nearly all cases. It is physiologically dominant over the rest of the ventral ganglia, although subsidiary to the cerebral ganglia in the chain of command.

The double nerve trunks also tend to unite as a

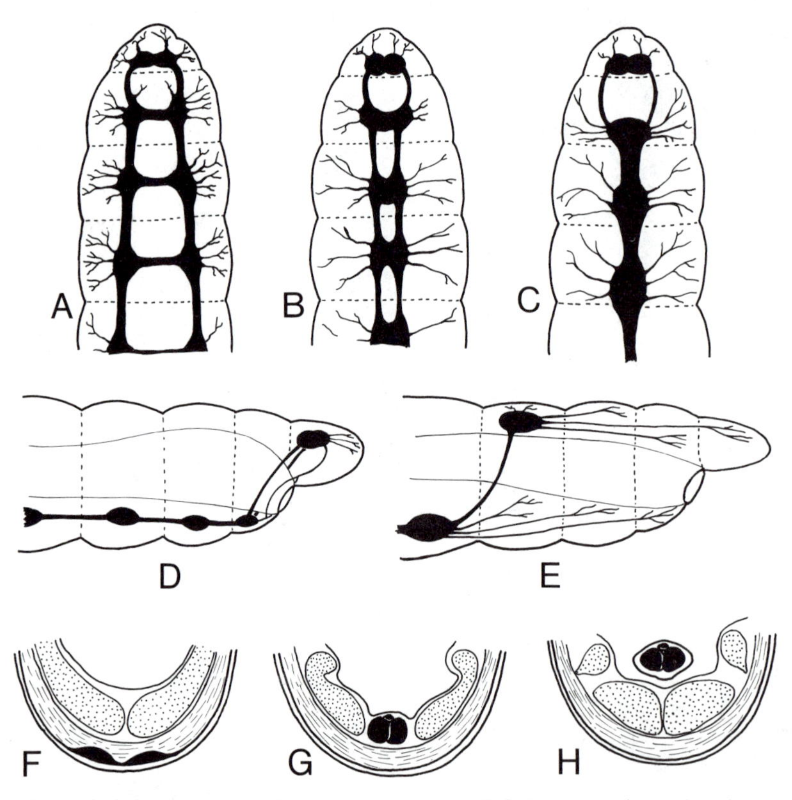

FIGURE 26.10. Variations in annelid central nervous systems. **A.** Brain in prostomium, with double ventral nerve cord. Ganglionic swelling located on each nerve trunk in each somite. Commissures connect ganglia in each somite. **B.** Paired nerve cords tend to assume more medial position, resulting in union of paired ganglia. **C.** Enclosure of two nerve cords in single sheath. **D.** In polychaetes, brain remains in prostomium. **E.** In oligochaetes and leeches, brain tends to move posteriorly, while nerves extend forward to sense organs and muscles of more anterior somites. **F.** Nerve cord in subepidermal position. **G,H.** Nerve cord tends to migrate inward, lying first in muscle layer of body wall (**G**), and eventually coming to lie in coelom (**H**).

simple **median nerve cord**, but not as quickly as the ganglia. Phyllodocid polychaetes have a double nerve cord between the united ganglia (Fig. 26.10B). The two nerve cords of *Amphitrite* lie side by side, enclosed in a single connective tissue sheath (Fig. 26.10C). In *Arenicola* the two cords have united but remain histologically double, with a double nerve tract. Nearly all of the oligochaetes, like *Arenicola*, have a single nerve cord with paired nerve tracts. Many of the leeches retain a double nerve cord between the median ganglia. *Hirudo* and the glossiphonids have a single nerve cord, but it too is histologically double.

The cerebral ganglia retain their primitive position in the prostomium in polychaetes (Fig. 26.10D), but tend to migrate backward in the oligochaetes and leeches, accompanied by the connectives and the subpharyngeal ganglion (Fig. 26.10E). Among the oligochaetes, only the aeolosomatids have the cerebral ganglia in the prostomium. In all others the cerebral ganglia are more posteriorly located, generally, as in *Lumbricus*, in the third somite. In leeches with an anterior sucker, the cerebral ganglia are still more posteriorly placed. As the brain moves back, the prostomial nerves are necessarily lengthened, as are the peristomial nerves from the subpharyngeal ganglion.

Superficial **follicles** lie on the surface of the ventral ganglia of leeches. Six such follicles, arranged in two rows of three, are associated with each ganglion (see Fig. 26.27A). As the cerebral ganglia move posteriorly, the anterior ganglia of the ventral chain unite with the subpharyngeal ganglion. The follicles remain discrete, however, and the number of fused ganglia can be determined by counting them. Typically, six ganglia unite to form the compound subpharyngeal ganglion of leeches (Fig. 26.11C). Leeches have some other peculiarities. The ganglia for the seven crowded somites in the posterior sucker also unite to form a compound ganglion. One of the annuli in each somite contains a subepidermal nerve ring associated with the sense organs. This is typically the sensory annulus. The nerve rings for adjacent somites are connected by longitudinal association neurons.

The cerebral ganglia are closely associated with the prostomial sense organs. The errant polychaetes have well-developed heads and prostomial sense organs, while oligochaetes have reduced heads, and leeches, although well cephalized, have a more diversified set of sense organs than the polychaetes. The errant polychaetes, primitive in many other respects, have the most complicated brains. The cerebral ganglia are differentiated into a number of functionally discrete lobes or centers, each associated with a special part or function (Fig. 26.11A,B). The development of specific centers and tracts is by no means as extensive as in the brain of cephalopods.

The ganglia on the nerve cords are not well marked in the more primitive nervous systems. Indeed, the ganglia of most polychaetes and oligochaetes have no sharp limits. In *Arenicola* the neurons have spread throughout the nerve cord, and ganglionic thickenings cannot be identified. In leeches, on the other hand, the ganglia are sharply defined and the nerve cords slender. Something of the same kind is seen in the circumenteric ring. In some annelids, notably the errant polychaetes, the brain is quite large and distinct, but in leeches the cerebral ganglia are small and the connectives large, so that the brain and the connectives together form a nerve collar.

Peripheral nerves arise from the brain, circumenteric connectives, and ventral ganglia. The number of nerves per somite varies. Polychaetes have from two to five pairs in each somite; aquatic oligochaetes usually have four pairs; terrestrial oligochaetes have three; and leeches have two or three. Primitively, the polychaetes had a series of ganglia associated with the peripheral nervous system. At the base of each parapodium, a pedal ganglion occurred on the lateral nerve. A longitudinal nerve connected the pedal ganglia on each side of the body. This primitive arrangement has been reduced in modern polychaetes. Some have retained the pedal ganglia but lost the lateral longitudinal nerves, while others have lost the pedal ganglia as well.

Polychaetes with an eversible pharynx have one or two pairs of motor nerves extending to the pharynx region, forming the sympathetic, or stomatogastric, nervous system (Fig. 26.11A,B). These nerves arise from the brain, the circumenteric connectives, or both, and often are associated with a series of small ganglia. This system provides motor control over the gut in *Lumbricus* and presumably in most other oligochaetes. In polychaetes the nerves are ganglionated and may be responsible for the control of secretion as well as muscular activity. Some kind of nervous control of secretion is indicated, for stimulation of the ventral nerve cord evokes secretion by the digestive glands.

Reproductive System Polychaetes have a very simple reproductive system. Sexes are nearly always separate. The reproductive cells arise from definite, small regions of the peritoneum, variously placed in different species. As cells proliferate, they become detached and float in the coelom, where both meiotic maturation divisions usually occur. When the animal is sexually ripe, the coelom is filled with gametes. Ova are brightly colored and often show through the thin body wall. Ova and sperm are shed into the water, the gametes emerging through separate gonoducts in a few, but generally passing out by way of the nephridium. Fertilization is external, and development usually occurs in the open sea. A few brood young in their tubes, however, and scale worms brood young in the respiratory channel beneath the elytra.

As a general rule, errant polychaetes produce gametes in most of the body somites. When the body becomes specialized into several regions, the number of genital somites is reduced. In most tube-dwelling species, gametes are produced only in the more posterior somites, and the posterior nephridia function as gonoducts. As a general rule, polychaetes have many genital somites, but the number is sharply reduced in some. *Arenicola* has only six genital somites, and *Trophonius* has but one.

Nearly all of the polychaetes have separate sexes, but a few sabellids and hesionids are hermaphroditic. The hermaphroditic species usually produce ova in the more anterior abdominal somites, and sperm in the more posterior ones, but some produce

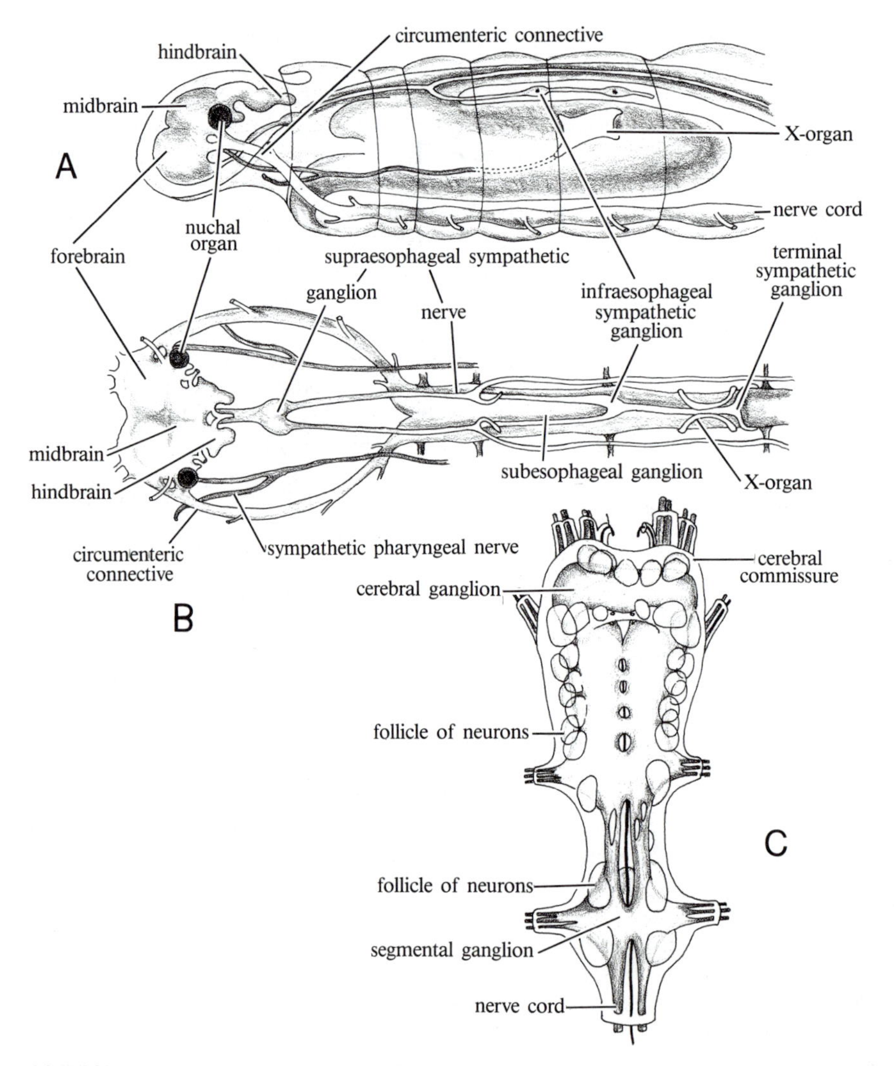

FIGURE 26.11. Annelid nervous systems. **A,B**. Scheme of central nervous system of *Eunice,* in lateral and dorsal views, showing relationship of sympathetic, stomatogastric, nervous system to brain. **C**. Brain region of leech, *Branchellion,* dorsal view. Note follicles of neurons that remain discrete on surface of compound subesophageal ganglion, three pairs of follicles for each original segmental ganglion; counting follicles can determine how many ganglia are united to form compound ganglion. (**A,B** after Heider; **C** after François, from Harant and Grassé.)

both sperm and ova in the same somite. It is not known what determines whether a young germ cell will become an ovum or a sperm.

The clitellates are characterized by hermaphroditism. Aeolosomatid oligochaetes are like polychaetes, forming gametes in many somites and using nephridia for the release of gametes. All other oligochaetes have a very few genital somites, served by a specialized set of gonoducts. The placement of the genital organs and arrangement of the gonoducts are variable and important in classification.

Aquatic oligochaetes typically have one pair of **testes** and one pair of **ovaries** located in the anterior half of the body. The testes and ovaries are hardly more discrete than in polychaetes, being essentially thickenings of the peritoneum. The young germ cells are released and mature in the coelom. The segment containing the testes contains a sperm sac or **seminal vesicle**, actually a portion of the coelom specialized for the reception of the developing sperm. In some of the aquatic oligochaetes the sperm sacs are elongated (Fig. 26.12F), occupying several somites, and in the tubificids a second sperm sac extends forward from the anterior septum of the male somite. The gonoduct ends in an open **seminal funnel**, located in the sperm sac, and the short male gonoduct opens through a male pore in the somite just behind the one containing the testes. **Prostatic**

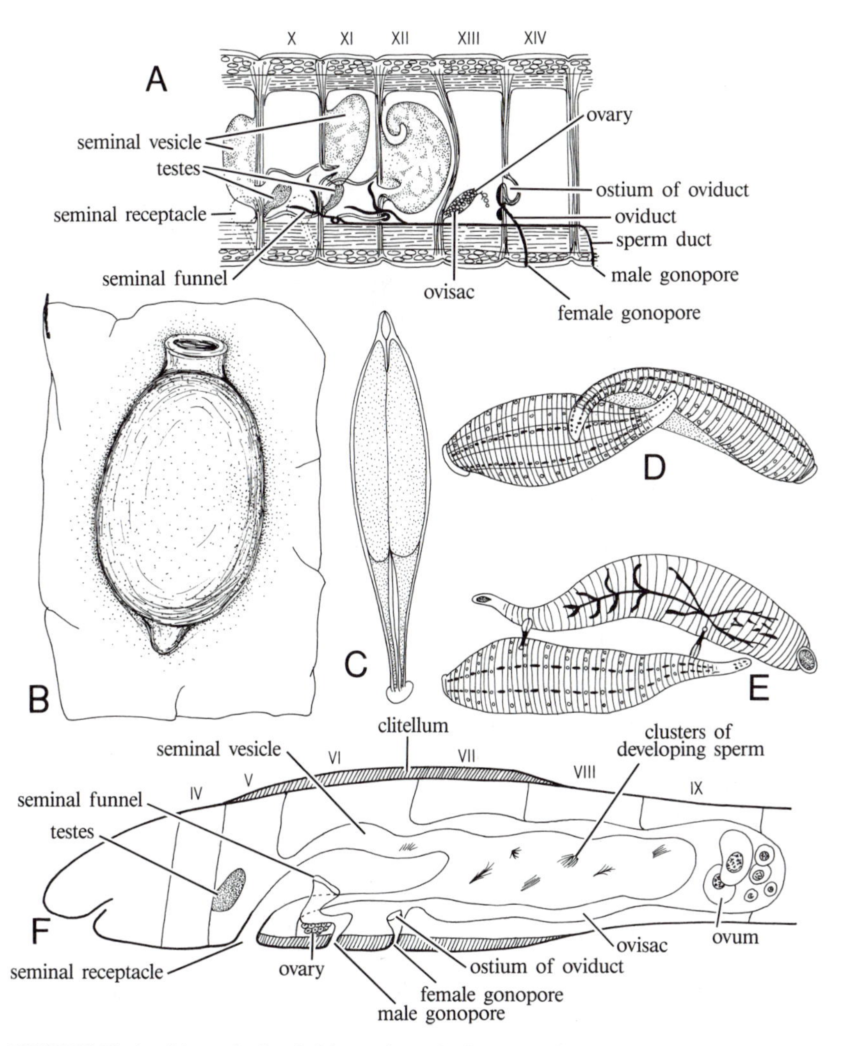

FIGURE 26.12. Annelid reproduction. **A**. Scheme of reproductive organs of *Lumbricus terrestris,* in longitudinal section. **B**. Cocoon of *Stylaria,* aquatic oligochaete. **C**. Spermatophore of glossiphonid leech. **D,E**. Copulation of *Glossiphonia*. Two leeches attach by means of ventral suckers. As they separate, each implants spermatophore on partner. **F**. Diagram of structure of reproductive organs of typical naid. (**A** after Hesse, from Stephenson; **B** after Pennak; **C,D,E** after Brumpt, from Harant and Grassé, in Grassé; **F** after Stephenson, from Pennak.)

gland cells usually surround a part of the gonoduct. The function of the prostatic secretion is unknown. The ovaries lie in the somite immediately behind the male genital somite. They release oöcytes that mature in an **ovisac**, comparable to the sperm sac. The short female gonoduct opens into the ovisac and leads to the genital pore in the somite immediately behind the ovaries.

The terrestrial oligochaetes differ only in details from the pattern seen in aquatic forms. They generally have two pairs of testes instead of one, with each pair in a different somite. The sperm ducts are long, opening several somites behind the two containing the testes. In *Lumbricus* (Fig. 26.12A), a third pair of sperm sacs is present, extending forward from the anterior septum of the anterior male somite, and the two sperm ducts unite to form a common sperm duct on each side. These open four somites behind the one with the posterior testes. The ovisac is very small and is missing in some earthworms. Two pairs of **seminal receptacles** are present instead of the single pair found in aquatic oligochaetes. Differences such as these are important to taxonomists.

The oligochaetes and hirudinidans have a unique specialization of the body wall, the **clitellum** (Fig. 26.1F). This is composed of glandular tissues (Fig. 26.3B) that produce the cocoon (Fig. 26.12C) into which eggs are deposited after mutual copulation.

The reproductive system is profoundly altered in leeches. Leeches have a single pair of ovaries, placed far forward between the anterior pair of testes (see Fig. 26.29B). The ovaries are not simple but include the specialized peritoneum that forms the reproductive cells, and a reduced ovisac, a vestige of the coelom. A short oviduct, possibly a modified coelomoduct, leads from each ovary. The two oviducts unite, and a short **vagina** leads to the median female gonopore.

The male system is more complex (see Fig. 26.29B). The testes occur in pairs, one pair to a somite. *Hirudo* has ten pairs, but some leeches have as few as four pairs. The testes are built like the ovaries, consisting of the proliferative peritoneum enclosed in a reduced sperm sac. Each testis opens into a **sperm ductule** that joins a longitudinal sperm duct on each side of the body. The sperm ducts are also vestiges of the coelom. In front of the most anterior testes, the sperm ducts are dilated to form seminal vesicles. The two seminal vesicles unite to form a median **antrum**, which is often complex, especially in the gnathobdellids, where a part of the antrum can be everted as a **penis**. The muscular antrum of the rhynchobdellids and pharyngobdellids cannot be everted, but serves as a chamber for the formation of the **spermatophores**. The single male gonopore lies just anterior to the female gonopore.

Biology

Movement and Locomotion Most polychaetes have a delicate layer of circular muscle and relatively powerful longitudinal muscles. The longitudinal muscles are especially strong in tube-dwelling species to facilitate withdrawal into their tubes. As a rule, polychaete longitudinal muscles are arranged in a set of four strong longitudinal muscle bands, two dorsolateral and two ventrolateral (see Fig. 26.20A). Both creeping and swimming movements depend on the antagonistic contractions of the longitudinal muscles of the two sides of the body. The circular muscles are not very important, for the use of the coelomic fluid and body wall as a hydraulic system is not important in polychaete locomotion. In burrowing forms, and especially in oligochaetes, a relatively even development of circular and longitudinal muscles is seen. The terrestrial oligochaetes have an almost continuous sheath of longitudinal muscle, separated into seven strips by narrow interruptions at the middorsal line, at the level of the setal groups, and at points about midway between the ventral setae and midventral line where the nephridiopores open (see Fig. 26.25B). In their movements, the circular and longitudinal muscles are antagonists, and the body wall and coelomic compartments serve in movement as a series of diminutive hydraulic systems. Although some leeches can swim by undulatory movements, the characteristic leech locomotion involves the whole body. All forward thrust depends on the use of the circular muscles of the body as a whole, and it is not surprising to find the circular muscle layer of leeches better developed than in oligochaetes of comparable size. When the whole body moves as a unit rather than in small sections, muscles more complicated than simple circular and longitudinal layers are needed to provide reasonable control. Thus, leeches have developed a layer of diagonal muscles between the circular and longitudinal layers, and slips of muscles pass through the leech body to form dorsoventral bundles or bundles that take a somewhat oblique course through the body.

Perhaps the most outstanding peculiarity of annelid muscle is the diversity of phosphagens that have been recovered from them. Several different phosphagens are sometimes found in the same species, and several phosphagens have been found only in annelids. Some of the polychaete muscles contain creatine, which is usually thought to be characteristic of vertebrates.

Body Wall Mucus is produced by unicellular goblets, which become the predominant cell in regions of high mucus production. Terrestrial oligochaetes have albuminous gland cells that contain finely dispersed secretions, closely resembling some of the gland cells of polychaetes, but the role played by the secretions is not certain in either type of annelid. Many of the polychaetes secrete mucus containing luminescent material (see Fig. 26.24B) produced in special epidermal gland cells. Some of the gland cells become very large and are deeply in-sunk. The clitellar epidermis of both leeches and oligochaetes is profoundly modified, with in-sunk gland cells extending into the deeper layers of the body wall (Fig. 26.3B).

The bright pigments of polychaetes and leeches are located mostly in tissues that are other than epidermal. The pigmented cells of polychaetes are often taller than the surrounding cells, forming low pigmented elevations. The pigments produced by annelids are quite diverse and are derived from diet (like caratenoids) or metabolic products (like hemoglobin breakdown). In some cases wastes appear to be used in pigment formation, providing an effective and unusual, although minor, adjunct to more ordinary methods of excretion. Most of the oligochaetes are unpigmented, but *Aeolosoma* has brightly colored oil droplets in its epidermal cells.

Feeding Annelids are quite diversified in their eating habits, but generally this has not had drastic effects on the basic design of the digestive tract. Among the polychaetes there are mud eaters like *Arenicola*, microphagous filter feeders like *Amphitrite*, microphagous plankton feeders like the sabellids, and predaceous species like *Neanthes*. The oligochaetes are predominantly detritus eaters when aquatic, and soil or decaying vegetation eaters when terrestrial. Leeches include herbivorous forms in addition to the more infamous bloodsuckers and carnivores.

The principal internal organs associated with food getting are the buccal cavity and pharynx. These parts clearly reflect the kind of feeding in the majority of cases.

Predaceous polychaetes have a proboscis (see Fig. 26.19B) equipped with cuticular jaws, denticles, or teeth. It is everted suddenly to grasp and tear food, and pulls the prey back into the mouth for swallowing. The sclerotized parts of the proboscis vary considerably in different groups of polychaetes. For example, *Glycera* has four jaws but no denticles, and *Neanthes* has two jaws and denticles. The most complex arrangements are found in the Eunicidae, which have upper and lower jaws formed of a series of plates as well as diversely shaped teeth held in a special sac. A few of the predaceous polychaetes have one or more sharp stylets used to stab the prey. *Autolytus* feeds on hydroids. It has a circlet of stylets used to pierce the unfortunate victim, while the main body of the pharynx remains within the worm and acts to pump out the soft tissues and the contents of the hydroid gastrovascular cavity.

The ciliary–mucus and plankton feeders usually have a small pharynx that cannot be everted. The pharyngeal walls are richly supplied with mucous glands. They are ciliated or have ciliated grooves used for swallowing mucous strings that contain food. *Amphitrite* is a good example of this type of polychaete (Fig. 26.13B).

Arenicola is a good example of a mud-eating polychaete. The pharynx is protrusible but unarmed, and picks up particles on an adhesive glandular surface. The much smaller aquatic oligochaetes have a pharynx that works on the same principle. The pharynx is attached to the body wall by diagonally placed muscle fibers used to extrude it for feeding. Some of the terrestrial oligochaetes feed in the same manner, but in most cases the pharynx is supplied with dilator muscles used to expand the pharynx and suck in particles of soil and bits of vegetation (Fig. 26.13A).

Some of the predaceous leeches have a proboscis used much like the proboscis of polychaetes for food capture. Others have a simple pharynx used for ordinary swallowing. Those that are adapted for taking blood meals use teeth on the walls of the buccal cavity to make an incision, and then pump in the flowing blood with the pharynx.

Polychaetes have simple mucous glands associated with the pharynx, but oligochaetes and leeches usually have conspicuous pharyngeal glands. The function of the pharyngeal glands of the aquatic oligochaetes is not clear. They undoubtedly secrete mucus but may also secrete some enzymes. Some species have very large pharyngeal glands, presumably of considerable importance. In the amphibious enchytraeids, for example, they extend up on the adjacent septa. The terrestrial oligochaetes take in dry food that cannot be swallowed until it is moistened.

The pharynx wall of *Lumbricus* is heavily infiltrated with dark-staining gland cells that secrete mucus containing a protease. In some earthworms, modified nephridia empty into the pharyngeal region to lubricate the food and to secrete enzymes.

The pharyngeal glands of leeches secrete saliva containing an anticoagulant that keeps the blood flowing during feeding. The form of the pharyngeal glands varies considerably. *Placobdella* has salivary glands composed of many scattered unicells with minute ducts that join the dorsal longitudinal muscles to form a strand on each side. The tracts enter the proboscis and continue through the proboscis muscle, opening through a pore at its tip.

The mucosal cilia stir up the intestinal contents, thus hastening absorption, and, in species with a delicate gut musculature, are important in conducting food through the tube. Generally, muscular peristaltic waves move along the intestine and are the main means of forcing food along, mixing digestive secretions with the food substances, and bringing new material in contact with the absorptive cells.

As organisms evolve into more complex creatures, sequentialism becomes a more significant feature of the digestive process. For example, three histologically differentiated regions of the midgut can be recognized in *Clymenella*. This trend culminates in the formation of a definite stomach. *Arenicola* (Fig. 26.13A) has a stomach with a highly glandular mucosa. The intestinal mucosa contains very few gland cells, so evidently there has been some separation of the secretory and absorptive functions. *Amphitrite* (Fig. 26.13B) has a stomach divided into two regions. The anterior stomach is glandular, and the posterior stomach has heavily muscled walls protected by a firm lining secreted by special cells. The posterior stomach is a gastric gizzard, where both chemical and mechanical trituration of the foods occur. Some of the aquatic oligochaetes also have a stomach between the esophagus and the intestine, but the details of histological differentiation in these regions are not known. Annelid digestion is at least predominantly extracellular. In *Arenicola*, however, particles are taken up by the mucosal cells,

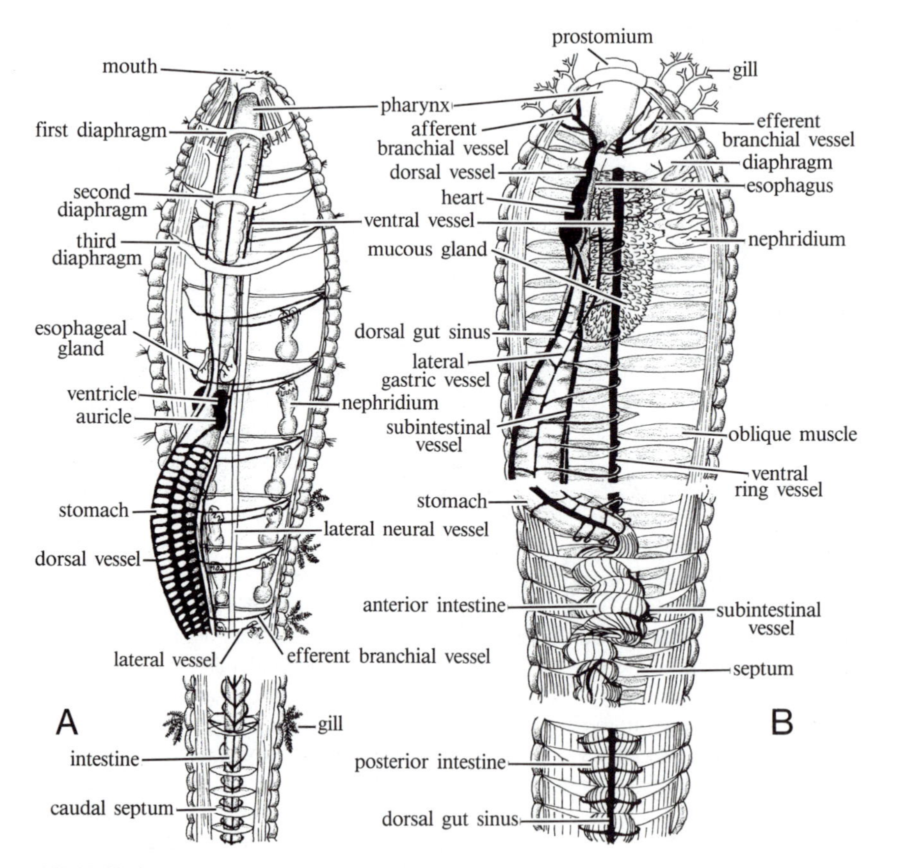

FIGURE 26.13. Internal anatomy of representative polychaetes. **A.** *Arenicola.* **B.** *Amphitrite.* (**A** after Ashworth; **B** after Brown, somewhat modified, from Brown.)

to be picked up by amoebocytes in which digestion is completed.

The leech midgut is highly specialized (see Fig. 26.27B). It is made up of two regions, here called the crop and the intestine. The crop is sometimes termed the stomach. Neither term is really apt, for it functions as a crop where food is stored, but it is derived from the midgut rather than the foregut. The crop may be a simple tube or be equipped with lateral diverticula or caeca. In predaceous species the crop is without caeca or has relatively few, whereas in bloodsucking species there are many large caeca. Some of the bloodsucking rhynchobdellids have as many as 14 pairs of caeca. Hirudids have fewer but can nevertheless store a large quantity of blood. A feeding leech may take as much as four times its weight in blood. Water and chlorides are removed from the blood as it enters the crop, and nephridia excrete copiously while feeding is going on. Condensation continues for a time, but the blood cells remain unchanged for long periods, up to 18 months in some cases. Where meals are taken at relatively long intervals (*Hirudo medicinalis*, for example, can survive from 18 to 24 months without feeding), some arrangement for slow digestion is needed. It appears that leeches may have adapted to this problem by greatly reducing their capacity for enzyme formation. Even the predaceous leeches appear to lack the enzymes needed to start protein digestion, and bloodsucking leeches produce even less in the way of enzymes than predaceous leeches. A bacterium, *Pseudomonas hirudinis*, isolated from the leech gut, causes a very slow breakdown of blood constituents and eliminates *Staphylococcus* when it is present. It may prove to be responsible for the breakdown of blood constituents. Food breaks down and is absorbed rapidly as it enters the intestine, which is generally short, but has caeca in some of the bloodsucking species (see Fig. 26.27B). The intestine opens into a short hindgut or rectum that discharges wastes through the anus.

Proteases, amylases, and lipases have been recovered from annelids. However, the ability to absorb amino and fatty acids and glucose through the body wall is widespread. The external secretion of protease by the tube dwellers may not only keep the tube clean of fouling organisms, but also may achieve extraorganismal digestion of proteins for epidermal absorption.

Defecation becomes a problem for an animal that lives in a burrow or tube, and habits or structures to prevent fouling the home are important. Most burrowers come to the mouth of the burrow to defecate, except when the burrow is U-shaped, in which case the feces are piled at the back door. Earthworms, *Arenicola*, and some of the tube dwellers behave in this fashion. Some species with tubes open at both ends maintain respiratory currents in the tube that also serve to waft away the feces. Sabellids and serpulids, however, have a special ciliated groove, used to convey the fecal pellets up to the tube mouth.

Excretion Annelids are ammonotelic, like other aquatic invertebrates. *Hirudo* excretes about three-fourths of its nitrogen as ammonia, and similar values have been obtained for *Aphrodita*. Significant amounts of other nitrogen compounds are also eliminated, such as urea, amino acid, and purines. It should be emphasized, however, that where a significant amount of nitrogenous material is permanently accumulated in the body tissues, for example in chlorogogue cells, the determination of the compounds released gives a very incomplete picture of excretory physiology. Not enough is known of the metabolic pathways used in excretory physiology, but the ornithine cycle is known to operate in some annelids. There is some evidence to indicate that leeches depend on bacteria for the production of ammonia from urea. When kept free of bacteria by antibiotics, the ammonia content of leech urine falls.

Adaptation to land life is accompanied by changes in excretory patterns, although earthworms excrete a considerable amount of ammonia. *Pheretima*, for example, excretes about half of its nitrogen as ammonia. Earthworms have the ability to change the excretory product with changes in the availability of water. When they are submerged in water, they are ammonotelic. When held in moist air, the urea content of the urine rises to 50 percent or more. Starvation also modifies the excretory physiology. Well-fed *Lumbricus* excretes large quantities of ammonia; starved ones switch to urea. *Eisenia*, on the other hand, works in the opposite direction, with urea output rising in well-fed organisms. It is probable that these differences are associated with the way in which acids are neutralized in the gut cavity. Ammonia is used for this purpose in *Lumbricus*, whereas *Eisenia* depends on carbonates and other compounds.

Coelomic fluid enters the nephrostome of open metanephridia and is processed as it passes through the tubule, eventually becoming urine. Differences between coelomic fluid and urine reveal something of what goes on in the tubule. Depending on the conditions to which the animal is adapted, the main changes of the fluid are: (1) changes in the water content, as urine becomes either hypotonic or hypertonic to the coelomic fluid, depending on whether water must be excreted or retained to preserve the osmotic balance; (2) changes in the salt concentration, as salts are excreted or retained; (3) changes in the concentration of the nitrogenous wastes; and (4) changes in the sugar content or concentrations of other organic compounds that may be useful to the organism.

Water regulation is less critical in marine than in freshwater and terrestrial species. Littoral marine species, however, may encounter considerable dilution of seawater in estuaries, and their distribution may be partly determined by their ability to cope with a more dilute medium. The properties of the body wall are also important in water regulation.

Some annelids are osmotic conformers, adjusting their salt content to that of the environment. *Arenicola*, for example, has blood with a freezing point identical to that of the surrounding water in a range of −0.29 to −1.72°C. It evidently has a high tolerance for the dilution of its body fluids.

The most detailed information about polychaetes is available for nereids. *Perinereis cultrifera* is an osmotic conformer, very much like *Arenicola*. Its saclike nephridia provide little surface. The body fluid of *Nereis diversicolor* becomes somewhat diluted, but remains hypertonic in diluted seawater. Its nephridia are long and convoluted. There is some evidence indicating that this species excretes water by forming urine more dilute than its body fluids, either by reabsorbing salts at the nephridial surface or by actively discharging water. In any case the body wall also helps in water regulation, for it is less permeable to water in dilute media and actively absorbs chlorine in a hypotonic medium. The ability of nephridia to regulate the osmotic properties of the urine probably depends in part on the length of the tubule, whether water or salts are being reabsorbed.

In earthworms the urine concentration changes as it moves down the tubule, indicating that materials are being absorbed or eliminated. The greatest changes in chloride content occur near the nephridiopore, where the wide part of the tubule is located (Fig. 26.9F). An extensive recovery of salts from the urine is indicated by comparisons of the freezing point of coelomic fluid and urine. Earthworms kept in tap water have a coelomic fluid with a freezing point of −0.31°C, while the urine has a freezing point of −0.06°C. The chloride content of the urine is only about a thirteenth that of the coelomic fluid. Similar values are quoted for *Pheretima*. Nevertheless, the low salt content of earthworm urine does not mean that water reabsorption is unimportant. Both water and salts are reabsorbed in the long, narrow tube, but in such proportions that the overall osmotic properties of the fluid remain about constant. The salt is probably reabsorbed as a means of recovering the water. In the wide tube, more salts than water are reabsorbed, and the osmotic concentration falls rapidly. Although more information is available about the movement of chlorides than of other substances, it is evident that useful compounds disappear from the fluid in the nephridium as it passes down the tubule. Glucose and amino acids are wholly reabsorbed from the nephridium of *Pheretima* during urine formation. Where the coelomic fluid participates in the transport of food—and this is certainly widespread among annelids—some such mechanism can be expected to develop.

Respiration and Circulation The compartmentalization of the coelom by the septa, and the resultant disruption of coelomic fluid flow, requires the development of a separate system to circulate fluids around the body (Ruppert and Carle, 1983). The coelomic fluid and the blood transport not only food and wastes but also respiratory gases, although to different degrees and in different ways in different annelids. Ancestral annelids were probably small forms, able to get along without respiratory pigments. Syllid, chaetopterid, and phyllodicid polychaetes have no respiratory pigments, and small aquatic annelids, like *Aeolosoma*, many naids, and even the amphibious enchytraeids have no respiratory pigment. The variety of respiratory pigments found in annelids also supports the idea that the primitive forms did not have any, and the pigments present in modern forms developed independently in different groups.

In both polychaetes and oligochaetes, the blood contains a respiratory pigment dissolved in the plasma, and is primarily responsible for respiratory transport, while the coelomic fluid remains important in the transport of food and wastes. Nevertheless, respiratory transport may be assumed by the blood, the coelomic fluid, or both. Some of the smaller annelids have no respiratory pigment in either the coelomic fluid or the blood. In the greater part of the annelids, the blood contains the respiratory pigment, and the coelomic fluid has none. In a few worms, *Terebella lapidarius* and *Travisia forbesii*, for example, the coelomic fluid contains hemoglobin in hemocytes, and the blood contains hemoglobin dissolved in the plasma. Generally, however, when the coelomic fluid contains a respiratory pigment, the circulatory system does not appear. This has occurred in the Glyceridae, Capitellidae, and some species of *Polycirrus*. In these forms the circulation of the coelomic fluid is enhanced by tracts of peritoneal cilia.

A variety of respiratory pigments occur in polychaetes. They often show through the body wall or tint the gills and the tentacles, contributing to a colorful appearance. **Hemoglobin** is the most common pigment, occurring in a variety of species. The green blood of sabellids, serpulids, and chlorhaemids is colored by **chlorocruorin**, an iron pigment with properties like hemoglobin. A few serpulids have both hemoglobin and chlorocruorin. *Magelona* blood is given a rose color by corpuscles containing **hemerythrin**. In all of these, the circulatory system is well developed. Oligochaetes and leeches are less variable; only hemoglobin has been reported from them.

A respiratory pigment, of course, must combine readily with oxygen and give off oxygen freely. Whether oxygen is to be given off or taken up depends most strongly on the oxygen tension. Exposure to carbon monoxide prevents oxygen transport by hemoglobin, but does not affect the transport of oxygen dissolved in blood plasma or coelomic fluid. *Lumbricus* oxygen consumption falls with exposure to carbon monoxide at high but not at low oxygen tensions. This unexpected result appears to indicate that the earthworm adjusts its metabolism downward when oxygen is in short supply, and uses its hemoglobin for transport primarily when oxygen tensions are high. *Tubifex* hemoglobin carries only a third of the oxygen used at high oxygen tensions, and the percentage falls at lower oxygen tensions, and in several species of leeches, the same kind of result has been obtained. Apparently the ability to reduce the dependency on hemoglobin transport at lower oxygen tensions is fairly widespread among annelids. Exactly opposed results are obtained with *Nereis diversicolor*. At high oxygen tensions, the hemoglobin carries about 50 percent of the oxygen used, while at low oxygen tensions, blockage of hemoglobin with carbon monoxide destroys all oxygen consumption. Evidently the hemoglobin of *Nereis* is especially important in transport when oxygen tensions are low, and depends more on dissolved oxygen with increased oxygen tensions. The polychaete *Euzonus mucronatus* is a facultative anaerobe, switching its physiology to depend only on glycolysis for as long as three weeks in the absence of oxygen (Ruby and Fox, 1976). The properties of chlorocruorin from sabellids and serpulids indicate that it can load up with oxygen at tensions normally present at the gills, and unload at tensions normally found in the body tissues. It is undoubtedly an important factor in normal internal transport of oxygen.

It is of interest to determine how pulsations are initiated in annelid circulatory systems. So far, the results have been problematic. "Heart" ganglion cells have been found in *Arenicola* and *Lumbricus*, but not in *Neanthes*. In some instances, the stretching of a vessel is an important factor in stimulating the contraction of myoepithelial cells in its wall. Generally, nerve-controlled hearts show fast oscillatory electrocardiograph waves, while muscle-controlled hearts show slow waves. In *Arenicola* both kinds of waves are seen.

However the pulsations originate, peristaltic waves sweep forward along the dorsal vessel every two to three seconds in *Lumbricus*, at a rate of about 25 mm/sec. These are in no way coordinated with the contraction of the "hearts." Hearts beat independently, with no common rhythm, although the two members of a pair usually contract together. *Arenicola* has a "heart" on each side of the esophagus. Blood passes forward in the dorsal vessel, through a plexus on the stomach wall to the lateral gastric vessels, and into the pulsating chambers. The lateral gastric vessels and the "hearts" beat at different rates. Evidently no overall coordination of pulsating parts is achieved in annelids as a general rule. Each part appears to be on its own in an essentially anarchic organization.

Little is known about annelid blood pressures. Pressures of 1.1 to 7.2 mm of mercury have been recorded in a resting *Neanthes*, and 17.6 mm Hg in active animals. Pressure during activity is enhanced by increased coelomic pressure, which reaches about 11 mm mercury.

Sense Organs Sense organs are well developed in errant polychaetes. Oligochaetes, however, adapting to a burrowing life, have greatly reduced sense organs. Leeches assume a more active life and have good sense organs, but as their ancestral oligochaete stocks had lost the types seen in polychaetes, a new set of organs developed.

Two kinds of light receptors occur in annelids: cephalic eyes, associated with the prostomium, and dermal eyes, which may appear at various points on the body surface. **Prostomial eyes** are characteristic of errant polychaetes and are rarely seen in any other group of annelids. They range from simple, converse pigment-cup **ocelli** to relatively advanced eyes. The simplest eyes have a cuticular cornea and characteristic neurosensory cells that extend as glassy rods above the pigment layer of the cup. As a general rule, however, the prostomial eyes have a cuticular lens (Fig. 26.14A,B). The fine structure of the neurosensory cells of *Nereis* shows that each contains a large number of light-sensitive microvilli. Microvilli of adjacent cells interdigitate to form **rhabdomes**, paralleling the **ommatidia** characteristic of the arthropod compound eye. In the sabellid, *Branchiomma*, the primitive ommatidia are clustered, evocative of compound eyes in the arthropods. Most of the errant polychaetes have one or two pairs of eyes, but some have as many as five pairs. Simple pigment-cup ocelli are also found in some naids, but otherwise this type of eye occurs only in polychaetes.

Polychaetes also have photoreceptor cells scattered in the epidermis, which can aggregate to form eyes at various points on the body surface. *Polyophthalmus* has eyespots on each somite, and many tubicolous species have eyespots on the tentacles that extend out of the tube. Even when conspicuous, these eyes are just clusters of photosensitive cells, each of which has some structural similarity to the dermal photoreceptors of oligochaetes in some way.

Terrestrial oligochaetes have no true eyes, but are equipped with scattered photosensitive cells. The visual cell contains a cavity filled with microvilli and several sensory cilia with a standard 9 + 2 structure. It is believed that light stimuli are received in this cavity.

Leeches have evolved cephalic and dermal eyes independently, using the same kind of epidermal photoreceptor for the purpose. Dermal eyes are clusters of photoreceptor cells (Fig. 26.14C). The sensillae are compound sense organs found on the sensory annuli, containing photoreceptors as well as other kinds of receptors. The large cephalic eyes are clusters of photoreceptors surrounded by a sheath of pigment (Fig. 26.14D). Leech eyes may be inverse or converse, and vary considerably in structural detail. In *Hirudo*, the cavity in the visual cell is continuous with the cell surface, a fact that has not been demonstrated in oligochaetes.

Very few annelids have specialized gravity receptors. A pair of open pits containing various foreign objects, such as diatoms and sand grains, are found in arenicolids. Sabellids and terebellids also have statocysts. These are burrowing forms, which probably depend more than most species on being able to distinguish between up and down, and which cannot depend on contact with the substrate for the purpose. *Arenicola* burrows downward, head first, and will slant its burrow if placed in a slanted tank. The sabellid *Branchiomma* orients itself to burrow backward instead of forward. The burrowing oligochaetes do not have statocysts, and it is not clear how they orient themselves, at least when in their burrows. All of the annelids will right themselves when overturned. These righting reflexes apparently depend on epidermal, tactile stimuli.

The sense of touch is important in the life of annelids, as might be expected in burrowing and tubicolous species and in forms that creep under rocks or other objects. Touch reception appears to center in the nerve endings and general neurosensory cells of the epidermis. These cells are concentrated in the tactile processes on the body of polychaetes, especially the palps, the parapodial cirri, and the prostomial and peristomial tentacles. As in other ani-

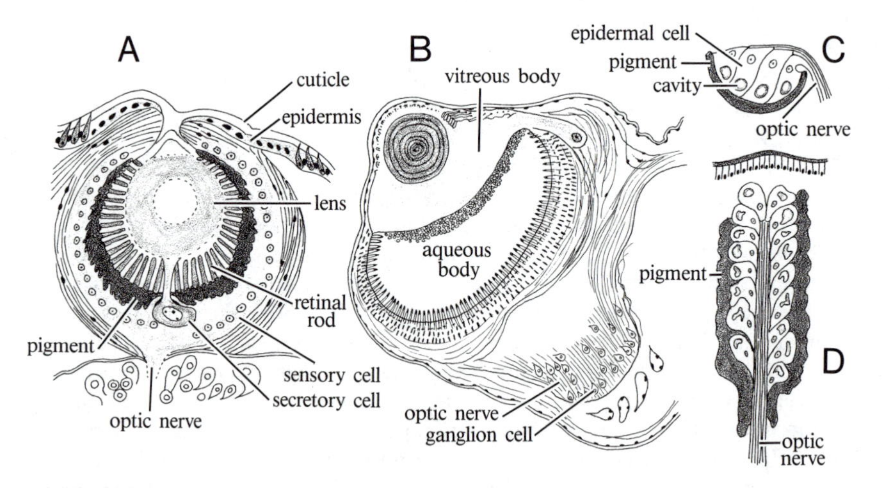

FIGURE 26.14. Annelid visual organs. **A**. Eye of *Phyllodoce,* polychaete, with fairly complex pigment-cup eye, with special cells for lens secretion and differentiated sensory rods. **B**. Eyes of *Vanadis,* polychaete, with considerably more complex organ; chambers containing vitreous and aqueous body present. **C**. Glossiphonid eye, showing inverse construction, and photosensory cells containing intracellular lenses. **D**. Converse pigment-cup eye from a hirudid leech. (**A,B,C,D** from Hesse.)

mals, most of the bristles are touch receptors. The setae of polychaetes and oligochaetes are important tactile organs. Some of the oligochaetes have special sensory hairs, especially at the anterior end of the body.

Ciliated nuchal organs are found on the heads of predaceous and burrowing polychaetes. They vary considerably in form, but all are associated with the posterior part of the cerebral ganglia. Nuchal organs are plaques of ciliated sensory epithelium, sometimes invaginated as pits and often interspersed with gland cells. Capitellids have unusual nuchal organs, set on protrusible stalks, somewhat resembling the rhinophores of molluscs. The nuchal organs are the only chemoreceptors to be positively identified in annelids. *Slavina*, an aquatic oligochaete, has sensory tubercles made up of a cluster of sensory cells that may be chemoreceptive. Other such cell clusters have been described, but their functions have not been demonstrated. There is evidence that the sensillae of leeches include chemoreceptive cells. Some of the terrestrial oligochaetes have a pair of water-sensitive receptors that permit orientation toward a more moist environment.

Although annelid chemoreceptors are not very well known, studies of behavior indicate a high sensitivity to chemical stimuli. Predaceous annelids usually recognize the presence and direction of prey. Leeches orient and move toward body fluids from prey animals, although *Piscicola*, which lives in turbulent water where chemoreception would be relatively inefficient, appears to lack this ability. Long-distance chemoreception is widespread among annelids, and contact chemoreception also occurs. Leeches respond strongly to surfaces coated with body fluids of normal prey, and other annelids avoid certain kinds of chemicals. Gustatory organs are also present, probably in the buccal cavity. Leeches stop feeding when sufficient quantities of quinine, salt, or sugar are added to the food, and earthworms can distinguish between and show preference for different foods.

The rather poorly developed oligochaete brain is not critically important. When the brain is removed, an earthworm is hyperactive but can still crawl, right itself, eat, copulate, and burrow slowly. Evidently most of the essentials of oligochaete life can go on without a brain. However, the more complex polychaete brain is important to behavior. Removal of the brain destroys the ability to feed and burrow, as well as most sensitivity to light and chemicals. The brainless polychaete is hyperactive, as are decerebrate leeches. Evidently the cerebral ganglia exercise an inhibitory influence on the motor centers of the subpharyngeal ganglion. Removal of the subpharyngeal ganglion has more drastic consequences than removal of the brain, for it effectively removes the influences of the cerebral ganglia as well. Nereids, for example, are almost motionless when it has been removed.

Removal of parts of the nervous system and observation of behavioral lesions show the extent to which the cerebral and subpharyngeal ganglia dominate activity. An earthworm normally turns away from light shone on it from the side. If the ventral nerve cord is cut, the segments anterior to the cut turn away from light, and those behind the cut turn toward the light.

In *Nereis* receptor cells are connected with a subepidermal nerve plexus, composed of multipolar associative neurons. Tracts leading to the nerve cord arise from the plexus, but these tracts join so that a very few (about 20) sensory fibers are present per segment. Likewise, only about eight to ten motor fibers emerge from the nerve cord in a segment. Stimuli are multiplied by the intervention of second-order and third-order motor neurons. It would appear, therefore, that only generalized information enters the nerve cord and relatively generalized commands can be given. The amount of sensory discrimination effected peripherally, and the peripheral development of detailed motor control, is not yet understood but evidently is of considerable importance.

A giant fiber system is used for rapid conduction, and evokes fast startle contractions. Most annelids have a giant fiber system, but the details differ greatly within the annelid groups. *Aphrodita* and the chaetopterids have no giant fiber system, but other polychaetes have giant fiber systems with from one to many fibers (Fig. 26.15A,B,C,D,E,F). The giant fiber system is more stable in terrestrial oligochaetes, with typically five giant fibers: a large middorsal fiber, a pair of dorsolateral fibers, and a pair of much smaller ventral fibers. The median fibers have anterior sensory connections and conduct impulses back to the muscles. The lateral fibers are connected transversely in each somite. They have sensory connections at the rear and conduct stimuli forward.

Two general types of giant fibers can be distinguished. Some are no more than grossly overgrown neurons, arising from a single cell. Others are formed by the end-to-end connection of large neurons in each somite that are converted into metameric conduction units of essentially noncellular form. In the earthworm, each unit meets the next in a slanting septum formed of vesiculated double membranes. Impulses move across the septa without synaptic delay, and although the giant fibers are polarized in the animal, they can conduct stimuli in either direction with equal rapidity under experimental conditions.

The rate at which impulses move along an axon is directly proportional to its cross-sectional area. The obvious advantage of giant fibers is rapid conduction. The fast neurons of vertebrates are myelinated, and giant fibers lack the conspicuous myelin sheath characteristic of fast neurons, although electron micrographs reveal a thin myelination in some. In any case, their large size makes up for whatever they lack in myelin. The large median giant fiber of *Lumbricus* conducts stimuli up to 45 m/sec, which compares favorably with the smaller fast fibers of cold-blooded vertebrates. The value of the giant fibers is emphasized by comparative conduction rates. The median giant fiber of *Lumbricus* carried impulses about 1600 times faster than the small neurons. Their length is a further advantage, as they can conduct an impulse the whole length of an animal without synaptic delays. In each somite, branches from the giant fibers extend out to the muscles (Fig. 26.15G).

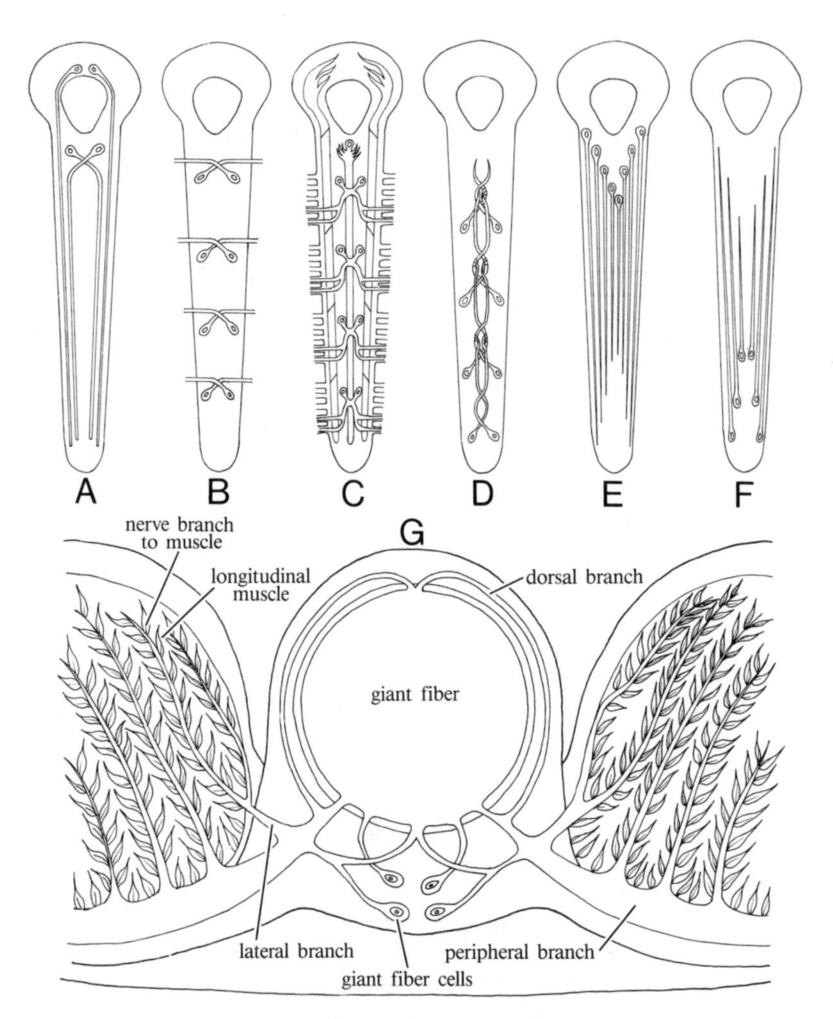

FIGURE 26.15. Diagrams of annelid giant fibers. **A**. Giant fibers of *Sigalion,* unicellular giant fibers from brain and anterior region of nerve cord. **B**. Intrasegmental giant fibers of *Lepidasthenia,* which decussate and pass to body wall in each segment. **C,D**. Giant fiber system of *Neanthes;* **C**. nerve cord with pair of lateral fibers and median giant fiber, latter associated with one or more cells in subesophageal ganglion that anastomose and arise from pair of giant fiber cells in each segmental ganglion, and former of multicellular origin; **D**. system of smaller giant fibers that decussate and extend through two somites, providing for side transmission as well as longitudinal transmission. **E,F**. Giant fiber system of *Halla,* consisting of distinct anteroposterior and posteroanterior fibers in nerve cord. **G**. Cross section of nerve cord, giant fiber, and branches of *Myxicola.* (**A** after Rhode; **B** after Haller, **C,D** based on Hamaker and on Stough; **E,F** based on Ashworth; **G** after Nicol, from Nicol.)

Reproduction Although the annelids are relatively highly organized, they retain the capacity for asexual reproduction, best developed in aquatic oligochaetes and certain of the polychaetes. Naids and aeolosomatids reproduce almost wholly by budding, and sexual reproduction is rarely seen. Budding occurs in a definite somite in any one worm, but the somite may differ among members of the same species. A budding zone appears, from which new posterior somites for the front animal and new anterior somites for the back one develop (Fig. 26.16D). The old segments behind the budding zone become the more anterior somites of the new worm.

Spontaneous fragmentation of the body, followed by regeneration of the missing parts, is a common method of asexual reproduction in some polychaetes. Syllids and sabellids have this habit (Fig. 26.16B). In many syllids, the points of breakage are predetermined and can be recognized by peculiarities in the septa between the somites where separation will occur. In such species, fragmentation and regeneration have become highly organized, and each fragment can develop a new anterior and posterior end. Two basic patterns of regeneration occur. In some cases, the original somites grow smaller as new somites arise, and, therefore, come to be normal parts of the new organism. The details of the use of the parent somites and the mechanisms that op-

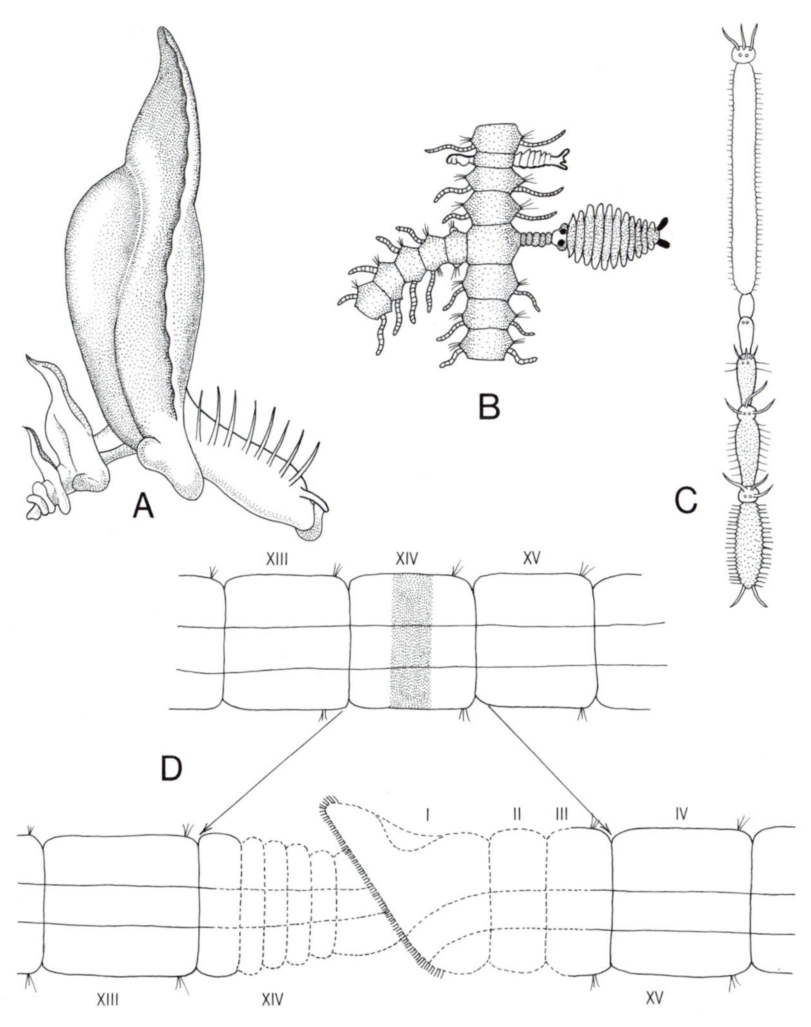

FIGURE 26.16. Asexual reproduction. **A.** Anterior and posterior regeneration from single fan segment of *Chaetopterus*. **B.** *Syllis,* at point of branch in body, and with chain of reproductive individuals budding off of parapodia. **C.** Chain of zooids of *Autolytus.* Note growth zone from which epitokes form. **D.** Budding zone of *Aeolosoma.* Definite somite serves as growth zone, where new posterior end for old, and new anterior end for young individuals forms. (**A** after Berrill; **B** after MacIntosh, from Hegner; **C** after Malaquin, from Benham; **D** after Marcus, from Pennak.)

erate as it dwindles to match the size of the new somites are not yet understood. In other cases the original somites remain large. A slender head and tail regenerate at each surface. These break off; the head end grows a new tail, and the tail end grows a new head. In *Dodececaria*, a second and sometimes a third set of anterior and posterior ends develop before the original somites are exhausted. The spontaneously formed fragments may be quite numerous; some species of *Dodececaria* and *Ctenodrilus* produce some single-somite fragments. Many species that do not reproduce by fragmentation have high powers of regeneration after experimental fragmentation. As a rule, polychaetes with specialized body regions do not regenerate as well as those without specialized regions, but there are exceptions. *Chaetopterus*, for example, can produce new individuals from any of several of the highly specialized anterior somites (Fig. 26.16A).

Annelids have evolved a variety of sexual reproductive methods and strategies, some of which use the asexual processes just discussed. Among polychaetes nereids, syllids, and eunicids live throughout most of the year as sexually unripe animals known as **atokes**. As the breeding season approaches, sexually ripe forms, known as **epitokes**, are produced by the atokes. At some sort of signal, the epitokes rise to the surface by the millions, swim about excitedly for a night, and, as the morning sun strikes, shed their gametes. Fish congregate to feed on the swarming worms, and birds gather to take their share of worms or fish, as the case may be. Humans, too, may await swarming time to capture the fish or the worms. *Eunice viridis* epitokes were highly prized as food, baked or eaten raw, by the Samoans.

Atokes produce epitokes by direct transformation or by budding. *Nereis* atokes become epitokes by transformation. The body becomes larger, the eyes enlarge, the parapodia change form, and the cephalic appendages are reduced (see Fig. 26.20B). The adult, sexually unripe atoke metamorphoses into a sexually ripe adult, the epitoke. The transformation is a profound one, changing the *Nereis* into a freely swimming, pelagic animal. The parapodial setae are adapted for swimming, and a number of the other traits resemble those seen in pelagic polychaetes.

Some polychaete atokes develop a budding zone, not unlike the budding zone of some aquatic oligochaetes. This method probably developed from the habit, still followed by some polychaetes, of transforming the posterior somites into an epitoke, which breaks away from the parent atoke as a new head develops. The parent atoke then develops a new growth zone at its posterior end, and adds new somites that in their turn will mature to form a new epitoke. This scheme of epitoke production is further developed in *Autolytus*. The first epitoke arises by transformation of the posterior somites and the development of a head. Before the epitoke is released, however, the growth zone, now a budding zone, proliferates to lay down a chain of epitokes (Fig. 26.16C); the most posterior epitoke is the eldest and is formed by direct transformation of the posterior somites. *Trypanosyllis* shows still further specialization for epitoke production. The budding zone protrudes as a stolon, from which a number of epitokes develop. An unusual method of epitoke production is seen in *Syllis ramosus*. Epitokes arise from multiple budding zones at the sides of the body. The young epitokes develop secondary buds of their own; therefore, a single atoke can produce a large number of epitokes (Fig. 26.16B). The varied methods used to produce epitokes suggest an evolutionary trend toward the production of numerous epitokes. This is probably related to the relatively low probability of completion of the life cycle. There can be no doubt that epitoky materially increases the reproductive potential of a species, especially when it is coupled with multiple budding.

Swarming is often under precise control, and the date or even the hour can be predicted with considerable accuracy (Caspers, 1984). As a rule, epitokes swarm at some definite lunar period and at some definite season. Swarming is not controlled by any single factor, for such environmental factors as light can sometimes suppress it. It is known that neurosecretions are involved in the production of at least some epitokes, and they are very probably involved in epitoke behavior. The removal of the brain of a young *Nereis* induces epitoky, and the grafting of a young brain into an adult *Nereis* suppresses epitoky. Hormonal secretions are triggered by such environmental factors as increased or decreased length of day.

When two oligochaetes copulate, sperm are exchanged mutually. The two worms often oppose the male gonopores and the pores of the seminal receptacles. Sperm from each worm move into the seminal receptacles, and the two animals disengage and move apart. In some species the pores are not opposed; in this case external grooves that lead to the seminal receptacles are usually present, as in *Lumbricus*. Rich mucous secretions help to hold the animals together during copulation and may also help to guide the sperm along the sperm grooves.

Egg laying need not occur immediately after mating. Although the oligochaetes mate, fertilization actually occurs externally. At the time of egg laying, the clitellum secretes a tough, membranous **girdle**. The worm backs out of the girdle, and gametes enter as it passes the female genital pores and the pores of the seminal receptacles. The two ends snap together as it slips over the head, enclosing the gametes in a relatively impervious **cocoon**. Aquatic species usually attach the cocoon to objects (Fig. 26.12C), while terrestrial species leave it free in the soil. Each cocoon contains a number of ova, but usually only one or two worms emerge.

Leech spermatophores (Fig. 26.12D) are double, consisting of a mass of sperm derived from each of two seminal vesicles that is enclosed by a tough membrane secreted by the antrum wall. During copulation the sharp point of the spermatophore is driven into the body of the mate, in some cases into a target area especially adapted for its reception. The spermatophore discharges the sperm into the underlying tissues. Sperm find their way to the ovaries. In some leeches a special vector tissue has appeared that conducts the sperm to the ovaries.

Gnathobdellid leeches have no spermatophores, and the penis is everted into the female gonopore. Sperm are discharged into the vagina, where they are stored.

Leeches have a clitellum and secrete cocoons, like oligochaetes. Glossiphonid cocoons, however, are very delicate and are attached to the ventral surface of the body. The young already have a posterior sucker when they emerge and attach to the ventral surface of the parent, where they continue their development. The parent leech remains rather quiet while brooding the young, cupping the body around them, but if the oxygen level falls, the leech makes rhythmic movements to aerate them.

Development Polychaetes undergo spiral determinate cleavage. Frequently, the egg membrane helps form the larval cuticle so, strictly speaking, the larvae do not "hatch." After gastrulation a trochophore larva develops that corresponds to the trochophore stage of molluscs (Fig. 26.16A and Chapter 20). The trochophore has a girdle of cilia, the prototroch; an apical sensory plate with an apical tuft of cilia; a posterior circlet of cilia, the telotroch; and in some cases a middle band of cilia, the metatroch. The curved gut opens at a mouth, located just below the prototroch, and at an anus, which is encircled by the telotroch. The mouth arises from the anterior margin of the blastopore, and the anus from a proctodeum that appears near the closed-over posterior margin of the blastopore. The part of the larva between the apical tuft and a point somewhat below the mouth will develop into the head of the young animal. When it is present, the line of the metatroch lies near the posterior margin of the future head. The telotroch and the anus will develop into the final somite of the body, the pygidium. All of the trunk arises from the short, tapered region between. During metamorphosis, this growth zone of the larva develops very rapidly.

As the growth zone lengthens, the anterior part is cut off as the first postcephalic somite. It continues to grow, and the second somite forms behind the first. At this point it can be termed a **metatrochophore**. The rhythmic production of new somites follows a regular pattern. Apparently, new tissue from the growth zone undergoes some maturation process, probably an early stage of differentiation, needed to trigger the division of the somite from the growth zone (Fig. 26.17B,C). Setal sacs and setae differentiate in the new somites. A girdle of cilia

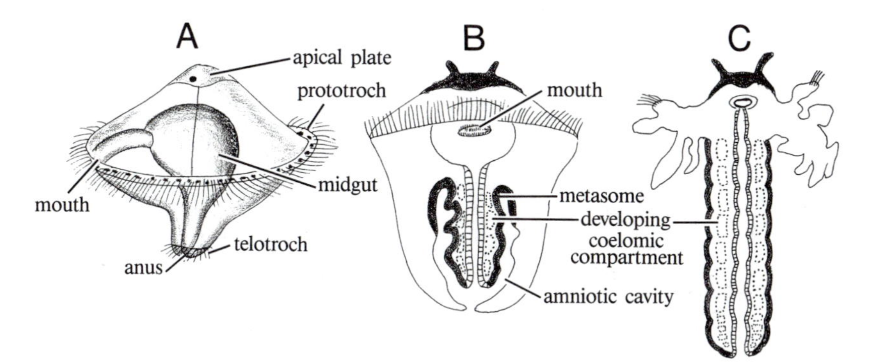

FIGURE 26.17. Polychaete larval development in *Polygordius*. **A.** Trochophore. **B,C.** Metamorphosis. Prostomium develops from region of apical plate. Body segments develop from growth zone that appears just in front of telotroch, with pygidium forming first.

around each somite often persists throughout the planktonic phase of larval life, and such a larva is called a **nectochaete**.

Blastomere 4d has, at an earlier stage, divided into two mesoblasts, the mesoderm mother cells. Each of the mesoblasts has divided repeatedly, establishing a pair of mesodermal bands that extend into the growth zone. As the somites form, paired spaces appear in the mesodermal rudiment of the somite, eventually enlarging to form the right and left coelomic compartments of the segment. All of the ectodermal covering of the trunk arises from the second quartet of micromeres. Blastomeres 2a, 2b, 2c give rise to the lateral dorsal ectoderm and to some other parts, including the stomodeum. Cell 2d proliferates to form a ventral plate of ectodermal cells, from which the ventral nerve cord and ganglia arise.

The brain develops from the apical sensory plate, and this is linked to an elaborate system of nerve rings throughout the larva. The part of the larva around the plate develops into the prostomium, and grows tentacles and palps on each side at this time. In *Polygordius*, the prostomial region is separated from the rest of the body by a large part of the trochophore. As growth continues, this part of the trochophore collapses, partly as a result of the contraction of larval muscle strands, bringing the prostomium back to the developing trunk. The peristomium develops from the collapsed part of the trochophore, which contains the mouth and through which the connectives to the subpharyngeal ganglion pass. In many polychaetes the subpharyngeal ganglion lies in the peristomium. In this case the adult peristomium is formed by the union of the first trunk somite and the peristomial region of the trochophore, as in *Nereis*, for example. As development continues, the larva settles to the bottom and takes up its normal way of life. Not all polychaetes pass through all of the stages mentioned. As in other animals, larval life may be shortened by delayed emergence from the protection of the egg membranes. Where this happens, the trochophore larva is often somewhat modified.

The primitive scheme of annelid development thus involves a ciliated larval stage. It is obvious that a drastic change is required to develop an embryo that can survive in a terrestrial habitat. Although clitellates are found in aquatic habitats, their main radiation is essentially terrestrial. All clitellates have the D blastomere giving rise to most, and sometimes

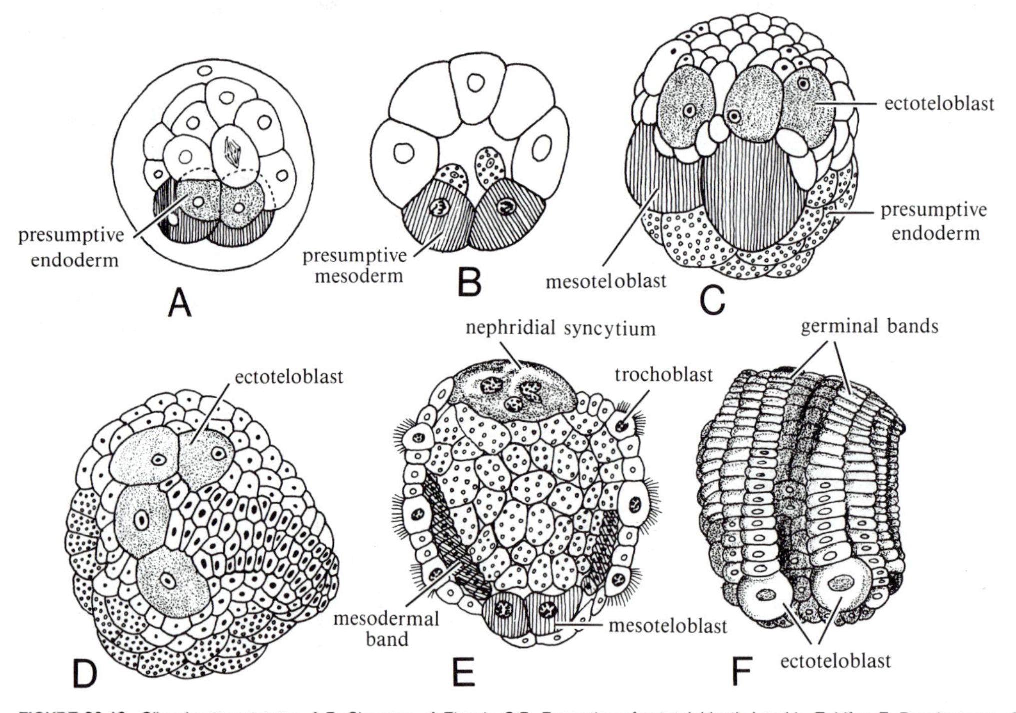

FIGURE 26.18. Oligochaete ontogeny. **A**,**B**. Cleavage of *Eisenia*. **C**,**D**. Formation of ectoteloblastic band in *Tubifex*. **E**. Development of mesodermal bands in *Bimastus*. Note trochoblasts on each side that become ciliated. **F**. Later stage in embryonic development of *Bdellodrilus*, showing fully developed germinal band. (**A**,**B**,**E** based on Swetlov; **C**,**D** based on Penners; **F** based on Tannreuther, from Grassé.)

all, of the embryo, and the teloblasts and their derivatives develop into paired germinal bands.

Early cleavage follows the typical annelid pattern, but differences soon appear that make the oligochaete development unique. Blastomere 2d gives rise to cells that will develop into ectoderm (Fig. 26.18A). Depending on the group, blastomere 3d or 4d gives rise to a pair of mesodermal stem cells. The endoderm arises from micromeres 4a, 4b, 4c, in part (Fig. 26.18A), but is joined by two enteroblasts arising from the mesodermal stem cells (Fig. 26.18B). The ectodermal teloblasts divide to form the four cells that give rise to a plate of ectodermal cells (Fig. 26.18C,D). A similar plate of ectoderm is seen in developing leeches (Fernandez and Olea, 1982). It gives rise to the ventral nerve trunk, among other things. Trochoblasts arise in the ectoderm and give rise to a homologue of the prototroch of the polychaetes (Fig. 26.18E), although the embryo does not emerge in this stage. Meanwhile, the mesodermal stem cells give rise to a chain of cells in the larva (Fig. 26.18E), within which the coelomic cavity will first appear. Postlarval coelomic spaces arise independently, thus completing the coelom. The late embryo develops from a ventral, germinal band, marked on the surface by a chain of ectoblasts derived from the ectoteloblasts (Fig. 26.18F).

Review of Groups

CLASS POLYCHAETA

Nearly all of the 8000 species of polychaetes are marine. They are especially abundant from low tide line to about 50 m, but many live in the intertidal zone, and some have been found at depths of over 5000 m. Some polychaetes burrow in the bottom; some crowd into crevices in or under rocks and shells, or live in the tubes or houses of other animals; some build tubes in the bottom material or on the surface of submerged objects. They are often extremely abundant; thousands of polychaetes may live in a single square meter of a mud flat, and they are important elements of the food chain. Some are primary or intermediate consumers, while others feed on organic debris, gathering small particles with a ciliary–mucus feeding apparatus, or swallowing quantities of mud. They are preyed upon by hydroids, flatworms, other annelids, crustaceans, echinoderms, or fishes. Different combinations of feeding habits and dwelling places have made for incredible diversification of form.

Body Plan. The primitive polychaete organization is best reflected in the simpler errant (free-moving) species; most features can be seen in *Neanthes* (Fig. 26.19). The head is composed of two pieces: a preoral **prostomium**, which projects forward over the mouth, and a **peristomium**, a modified somite containing the mouth. The rest of the segments are similar. All bear a **parapodium**, composed of a basal part, a dorsal part (**notopodium**), and a ventral part (**neuropodium**). Each part is supported by a skeletal rod, the **aciculum**, and bears depressions containing epidermal follicles with chaetoblasts that secrete setae. Muscles move the acicula (Fig. 26.20A), and as they move they extend or retract the setae.

The prostomium contains the **cerebral ganglia** and is equipped with peristomial **eyes**, a pair of ventral **palps**, and one or more pairs of dorsal **tentacles**.

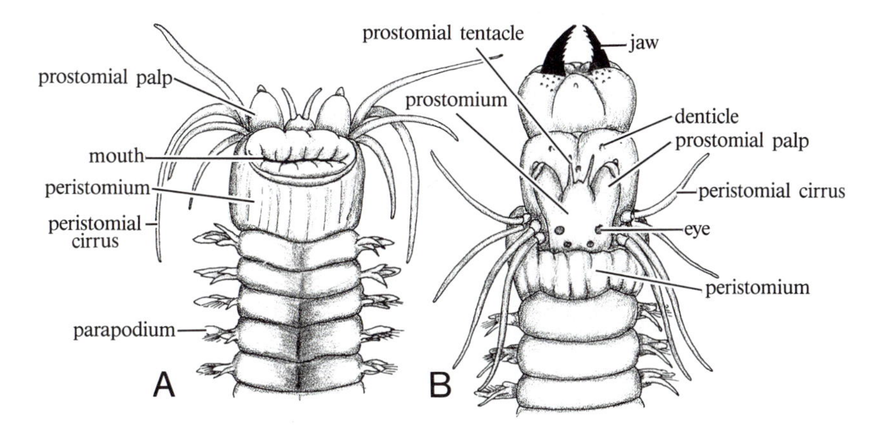

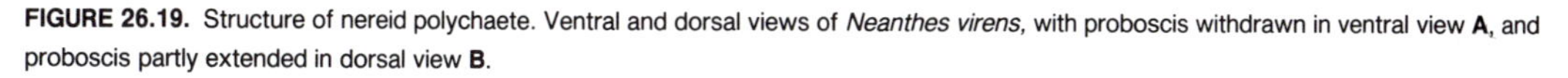
FIGURE 26.19. Structure of nereid polychaete. Ventral and dorsal views of *Neanthes virens*, with proboscis withdrawn in ventral view **A**, and proboscis partly extended in dorsal view **B**.

The peristomium loses its parapodia and may undergo other changes as it is incorporated into the head, but some modern polychaetes, like *Aphrodita* and *Nephthys* (Fig. 26.21C), have a peristomium with a pair of somewhat modified parapodia and setae. Most of the active errant polychaetes have a peristomium derived from a single modified somite that has neither parapodia nor setae, but a few, like *Phyllodoce* (Fig. 26.21E), have added a second somite, also without parapodia. Parts of the parapodia of somites behind the peristomium may be modified to serve as parts of a specialized front end, whether one wishes to consider them as a part of the head or not (Fig. 26.21A,B). The third segment of *Phyllodoce* has long peristomial cirri resembling the cirri of the peristomial region, and the hesionids have the first four body segments modified in this manner. The relative sizes and numbers of the sensory ap-

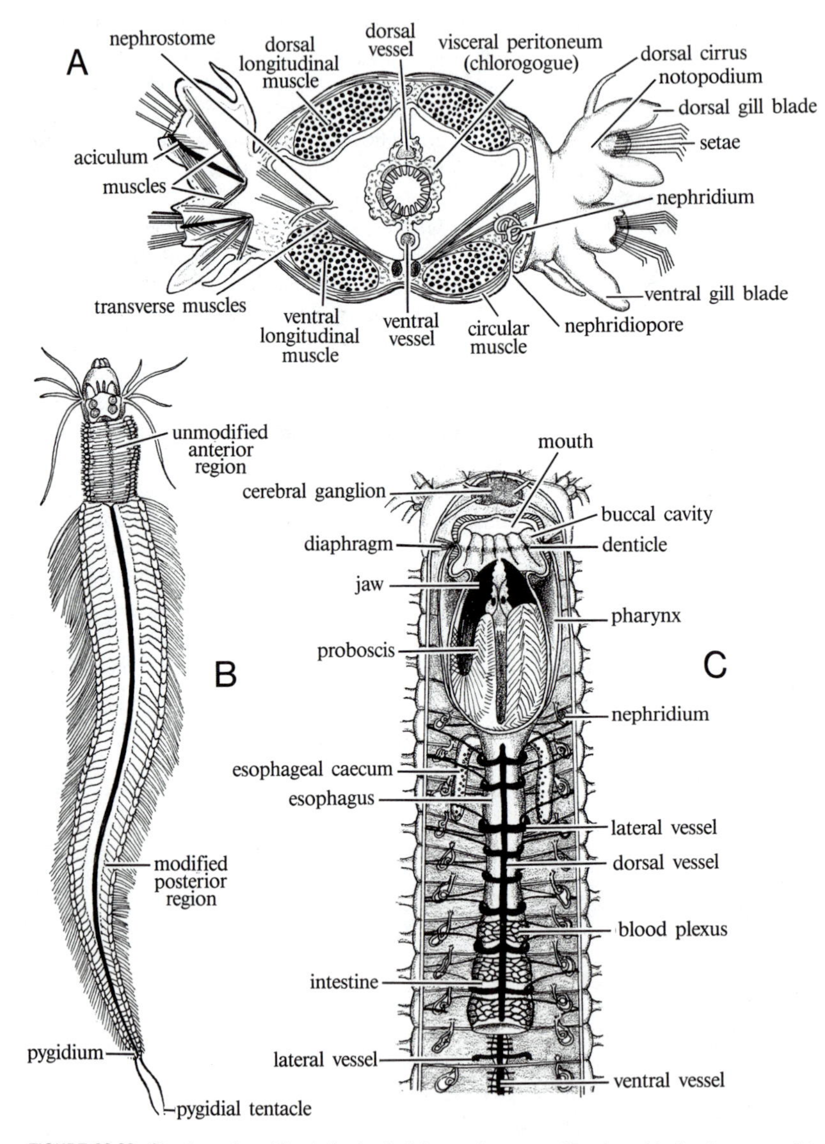

FIGURE 26.20. Structure of nereid polychaete. **A.** Scheme of cross section of nereid, showing parapodial structure on *left.* Parapodium on *right* shows form of normal parapodium of *Perinereis cultrifera,* before metamorphosis into heteronereid. **B.** Male heteronereid of *Nereis irrorata.* When sexually mature, posterior region develops into epitoke, with modified parapodia. **C.** Internal structure of nereid. (**B** after Rullier, from Clark; **A,C** composite.)

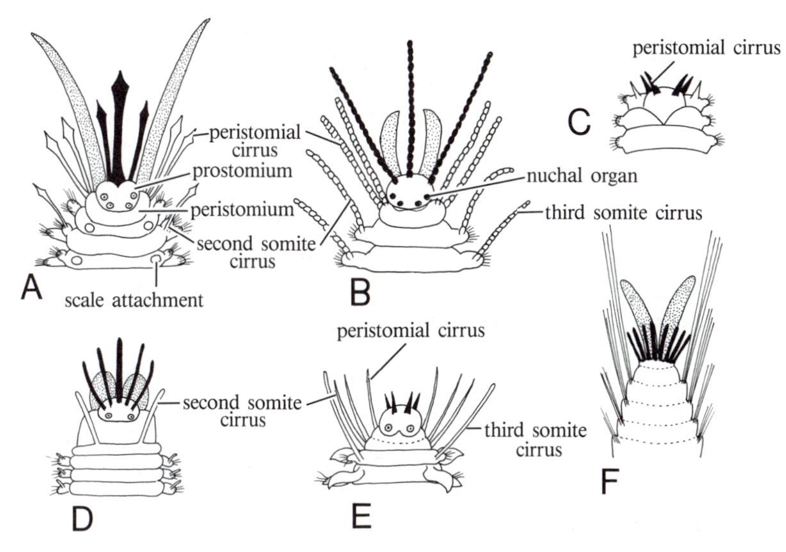

FIGURE 26.21. Polychaete heads (schematic). Notice differences in tactile organs associated with prostomium, peristomium, and, in some cases, second or third somites. **A**. Polynoid. **B**. Syllid. **C**. *Nephthys,* nephthyid. **D**. *Eunice,* eunicid. **E**. *Phyllodoce,* phyllodocid. **F**. *Trophonia,* chloraemid. For ease of comparison, prostomial palps are *stippled,* prostomial tentacles shown in *black,* and peristomial or notopodial cirri left *white.* (All after Benham.)

pendages of the head and first few somites (Fig. 26.21D,F) are characteristic of the various families of polychaetes.

Active errant polychaetes have strong, prominent parapodia, either elongated and leglike or flat and paddlelike. Modifications of parapodial form are important in taxonomy and reflect the habits of the various species (Fig. 26.22). The *Neanthes* parapodium shows the basic form (Figs. 26.19A and 26.22A). The dorsal lobe, the notopodium, bears a

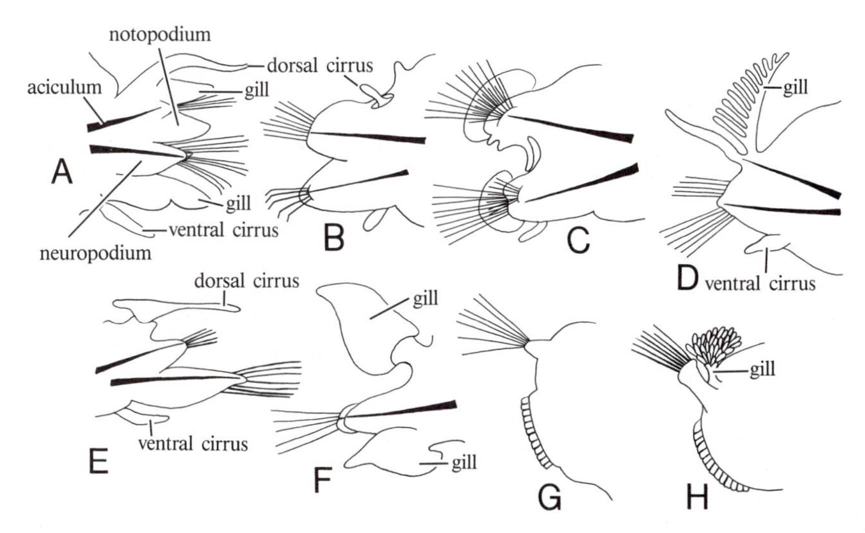

FIGURE 26.22. Schematic representation of polychaete parapodia, showing some variabilty found in different groups. **A**. Nereid, with about equal development of notopodium and neuropodium, and with dorsal and ventral cirri distinct. **B**. Glycerid, without conspicuous gill blades, and with relatively small cirri. **C**. Nephthyid, with about equally developed notopodium and neuropodium, although differently formed than in nereids. **D**. Eunicid, with notopodial setae absent and filamentous gill present. **E**. Polynoid, scale worm, with scale, or elytron, attached to flattened surface, developed from dorsal cirrus. **F**. Phyllodocid, with notopodium nearly absent, except for gill blade developed from it. **G**. Sabellid, lacking acicula, and neuropodium consisting of group of very short setae. **H**. Arenicolid, with notopodium rather like sabellid notopodium, but with filamentous gill. (From various sources.)

tentaclelike dorsal cirrus, and the ventral lobe, the neuropodium, bears a similar ventral cirrus. Vascular, flattened bladelike gills are attached to the notopodium and neuropodium. Modifications of parapodial structure (Fig. 26.22) involve changes in the number and type of setae, the form and position of the gills, the number of acicula, and the relative development of the notopodium and neuropodium, and sometimes by unusual modifications that set a group apart. A flat plate, the **elytron**, takes the place of the dorsal cirrus in *Aphrodita*, for example, and in the scale worms generally (Figs. 26.22E and 26.23B).

Locomotion Polychaetes swim, creep about over the surface of algae, rocks, and other objects or through the soft bottom material, and move about in their tubes. Errant forms are predominantly creeping organisms but are usually able to swim. Sedentary species are usually restricted to movements that extend the body or withdraw it into the tube.

Parapodia move alternately, with the right one making a forward, recovery stroke while the left makes a backward, effective stroke. It is not, however, a simple problem of alternate movements on the two sides of the body. When a parapodium completes its recovery stroke, muscles moving the aciculum are contracted, and it is forced outward, extending the setae. This increases the resistance of movement through the water and also tends to engage the bottom. As the effective stroke ends, the acicula are withdrawn. This withdraws the setae for the recovery stroke, and the parapodium is lifted slightly to disengage it from the bottom. These coordinated movements will produce a forward thrust with each parapodial "step," but the worm as a whole will get nowhere unless the parapodia of its many somites are working in a coordinated fashion.

Coordination of the various segments is achieved by activation waves that start at the posterior end of the body. The activation waves are initiated on the two sides of the body alternately, and move forward alternately on the two sides. Each activation wave affects about four to eight parapodia, and they are so timed that when the first group of parapodia complete the effective stroke, the next complete the recovery stroke. The alternating activation waves coordinate the body longitudinally and evoke the necessary stimuli to make the alternating parapodial movements. Forward impetus, however, comes mainly from body movements. Alternate contraction and relaxation of the longitudinal muscles on opposite sides of the body produce undulatory movements. These are so timed that the contraction wave affecting a group of segments coincides with the effective parapodial strokes. As a result, longitudinal muscles are contracting and pulling the body forward while the setae are engaged in the substrate.

Polychaetes can suddenly withdraw into their tubes or make sudden contractions of the whole body. Stimuli for such movements result from a giant fiber system in the nerve cord. These "startle" reactions are especially rapid in tube-dwelling species that have very strong longitudinal muscles. They are an important part of the behavioral equipment of errant, free-moving polychaetes as well.

Tubicolous polychaetes may use parapodia to produce water currents in the tube, but these are relatively unimportant and the parapodia are usually much reduced. The setae retain importance, for they engage the tube wall and help to withdraw or extend the body. They also make it more difficult to pull the worm out of its home. The setae often are hooked or have terminal structures that make them more effective in this regard.

Burrowing polychaetes also tend to have reduced parapodia (Fig. 26.23D), but some, like *Aphrodita*, have large parapodia and use them as shovels. Coordination of the movements for burrowing pose entirely different problems than for creeping. In creepers, the right and left longitudinal muscles are antagonists and cause an undulating movement. In burrowing forms, however, the circular and longitudinal muscles are antagonists. Relaxation of the longitudinal muscles of a segment and contraction of the circular muscles apply pressure to the coelomic fluid and cause the segment to elongate. When the circular muscles relax and the longitudinal muscles contract, the segment shortens. Each segment is a diminutive hydrostatic system; muscle contractions change its form but not its volume, and it increases or decreases in diameter to compensate for changes in length. Longitudinal muscles on both sides contract simultaneously rather than alternately. Forward progression depends on traction provided by the setae. The setae are brought against the substrate in short, stout, contracted segments, and lose contact in long, slender, elongated segments. The elongated segments thus thrust forward while the contracted segments serve as anchors, preventing backward movement.

Among the most highly modified polychaetes are the myzostomes (Fig. 26.23C) that live on crinoids, brittle stars, and asteroids. Some make limited movements; others are permanently attached. Each parapodium ends in two modified setae, one a hook and the other a guard, providing excellent anchorage. They have diverged so far from other polychaetes that some researchers would recognize them as a separate class of annelids.

Feeding Polychaetes have a diverse array of feeding habits (Fauchald and Jumars, 1979). Most of the active, predaceous errant polychaetes live under rocks, in crevices, or among seaweeds, corals, and other colonial animals, or emerge from burrows to feed. Swimming is not a very important part of their lives, and the parapodia are relatively short. The pelagic polychaetes, however, have very prominent parapodia, like *Tomopterus* (Fig. 26.23A), or may have natatory setae on the parapodia. An eversible pharynx is characteristic of the predaceous errant polychaetes (Fig. 26.19). It is operated by a hydraulic system formed of the body wall muscles and coelomic fluid. Contraction of the circular muscles compresses the coelomic fluid and shoots the proboscis out, sometimes so rapidly and powerfully that

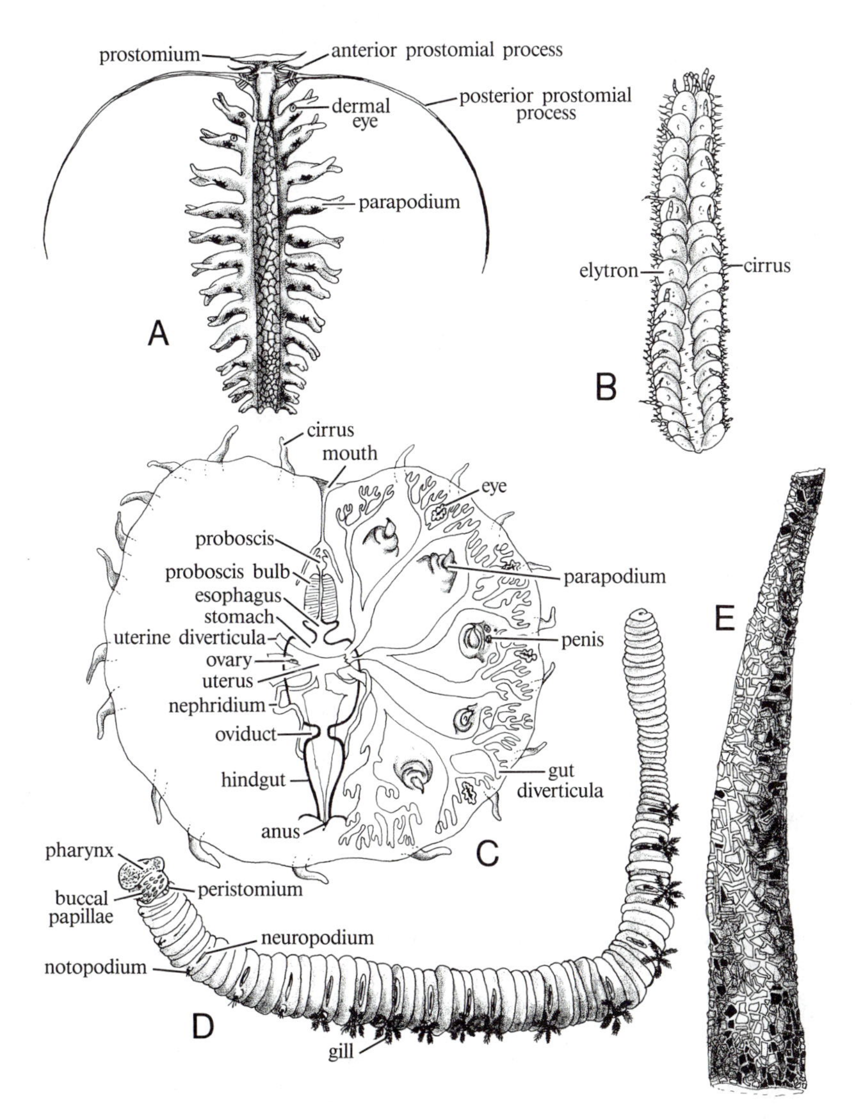

FIGURE 26.23. **A**. *Tomopterus rolasi,* pelagic polychaete with well-developed sense organs and long parapodia without setae. **B**. *Halosydna brevisetosa,* scale worm, with notopodial elytra. **C**. *Myzostomum antarcticum,* myzostomid, highly specialized ectocommensals or ectoparasites on crinoids and other echinoderms. Note gut diverticula, used for food distribution. **D**. *Arenicola,* lug worm, which burrows in sand, feeding on detritus. External rings do not correspond to internal somites. **E**. Tube of *Pectinaria,* sedentary polychaete that builds tube of small sand particles, glued together with mucoid secretions. (**A** after Greef; **B** after Ricketts and Calvin; **C** after Stummer-Traunfels, in Kükenthal and Krumbach; **E** after MacIntosh, from Benham.)

an audible pop is sounded. The proboscis jaws are inverted when the proboscis is withdrawn (Fig. 26.20C) and are positioned for use by proboscis extrusion (Fig. 26.19B). Most polychaetes have a single pair of jaws, but some have a single styletlike tooth, and others have one lower and several upper jaws. Small denticles sometimes stud the proboscis surface. The form of the jaws or other proboscis armament is important in taxonomy and varies considerably with the kind of food taken. When food is engaged by the extruded proboscis, retractor muscles invert it back and the food is swallowed. Nearly all errants with an eversible proboscis feed on living or dead animal tissue, but a few use the proboscis to crop tissues from seaweeds.

Some polychaetes are active burrowers, and these tend to have smaller parapodia, like *Glycera* (Fig. 26.22B). Some of the burrowing polychaetes are sedentary, but these, unlike *Glycera*, are mud eaters. *Arenicola*, the common lug worm, is a good example (Fig. 26.23D). *Arenicola* is one of the few sedentary worms with an eversible proboscis. It is thrust into the sand, and mucous secretions glue finer particles to the proboscis surface, while larger particles are thrust aside. The detritus-laden proboscis is then retracted, and adherent particles are swallowed. *Arenicola* pumps water out of its burrow, pulling water through the walls in the area in which it is working. The sand acts as a filter and, therefore, is highly charged with organic material when it is eaten. Even so, worms feeding on bottom deposits must swallow a great deal of material to obtain adequate nutrition. Large species of *Arenicola* reach a foot or more in length and burrow through the upper 50 to 60 cm of the ocean floor. Where they are abundant, they do a tremendous amount of work and keep the ocean floor in constant circulation. A burrowing *Arenicola* swallows every five seconds or so for a minute or two, and then rests for a few minutes. It works from five to eight hours a day, with occasional long rests of a hour or so. It is calculated that where *Arenicola* is abundant, some 4257 metric tons of sand are brought to the surface per hectare per year. All of the material in the upper 50 to 60 cm of the bottom of that hectare is worked through by *Arenicola* once in about two years.

Most of the sedentary polychaetes are permanent tube dwellers. Many feed on detritus, gathered by straight or branched filamentous processes, built of specialized prostomial palps or tentacles, or of peristomial cirri. The feeding apparatus is usually thrust into the upper parts of the bottom material. Particles are caught in mucus and move up ciliated grooves to the mouth. As the tentacles move about, and as new material is delivered by water currents, sedentary forms need not move about like the burrowing species. Those that feed farther below the surface do move around, as new material does not accumulate so rapidly. They either form no tubes or build unattached tubes that can be moved about, like *Pectinaria* (Fig. 26.23E). Annelids of this type usually have some system for sorting out food particles that are picked up, and some use the rejected particles for building the tube.

Not all of the permanent tube dwellers feed on detritus. Some lift their food-gathering structures above the bottom, trapping planktonic organisms as they move past. Among these are some of the most beautiful of the polychaetes. Serpulids and sabellids are sometimes called the feather duster worms for the crown of feathery tentacles extending above the tube. The ciliated, feathery processes are equipped with a mucous–ciliary apparatus to capture food and convey it to the mouth.

Polychaete tubes vary greatly in form, composition, and location. Some tubes are built of mucus and are little more than a consolidation of the particles of which the bottom is made. Others are built of onuphin, conchiolin, or other organic compounds that harden on contact with water to make tough, membranous tubes. The serpulids build calcareous tubes with a mucous base, and large aggregations of the tubes form soft rocky masses. Still other polychaetes build tubes of consolidated sand grains or other particles. They may be straight, irregularly curved, U-shaped, or spiraled, and may be branched or unbranched. The calcareous serpulid tubes are usually found on rocks or molluscan shells, and the tiny spiral tubes of *Spirorbis* are often seen on seaweeds. Soft membranous tubes are especially suited to fit into crevices or attach to seaweeds.

The relationship of the worm to its tube is a very intimate one, and the animals adapt in a variety of ways to the special self-made environment in which they live. *Chaetopterus* (Fig. 26.24) is a good example of the intensive adaptation to tubicolous life that is seen in so many polychaetes. The fourteenth to sixteenth parapodia are large fans that move back and forth rhythmically, sweeping a flow of water through the U-shaped tube. The twelfth notopodia are drawn out to form two wings. In position, they make a partial partition across the tube, leaving a small aperture through which all of the water must pass. The wings secrete abundant mucus, which moves down into the aperture. As water passes through, more mucus is secreted and moves into place. The sheet of mucus billows out to form a mucous sac that lengthens until it reaches a ciliated cupule midway between the wings and the fans. All of the water must pass through the mucous sac, leaving its burden of particles behind. When full, the sac is wrapped up to form a pellet in the cupule. Cilia in a special groove carry the pellet forward to the mouth for swallowing.

CLASS CLITELATA, SUBCLASS OLIGOCHAETA

Most of the approximately 3000 species of oligochaetes are terrestrial, but some live in freshwater environments (Brinkhurst and Jamieson, 1972), and a few are marine (Giere and Pfannkuche, 1982). Terrestrial earthworms live in moist soil almost everywhere, but are especially characteristic of good, rich soil in temperate regions. The aquatic species are also widespread.

Body Plan Oligochaetes are far less diverse than polychaetes externally (Fig. 26.1E). They have no parapodia and relatively few, inconspicuous setae. The greatly reduced head has no sensory appendages in the majority of species. As a result they are externally simplified, with relatively few features that can reflect adaptation. The terrestrial earthworms are larger than aquatic species. Some species become very large; *Megascolides australis*, an Australian species, reaches lengths of over 3 m. Some of the aquatic species are barely visible to the naked eye.

Polychaetes that burrow tend to have a smaller head and parapodia than those that do not. Oligochaetes have carried this even further. The head is scarcely a true head, for there are no sensory appendages and the prostomial eyes generally found in polychaetes have disappeared. With the reduction of the prostomial appendages and the migration of the brain to a more posterior position, the prostomium is greatly reduced (Fig. 26.25). The peristomium is very similar to the body somites, but contains the mouth, lacks setae, and has no nephridiopores. The setae are the only vestiges of the parapodia. Body somites may differ slightly in size because of tapering but, except for the clitellum, present few visible marks of specialization. Genital pores occur in specific somites and are useful in the identification of oligochaetes.

Terrestrial oligochaetes, the familiar earthworms, are generally larger than the aquatic forms. Setae are

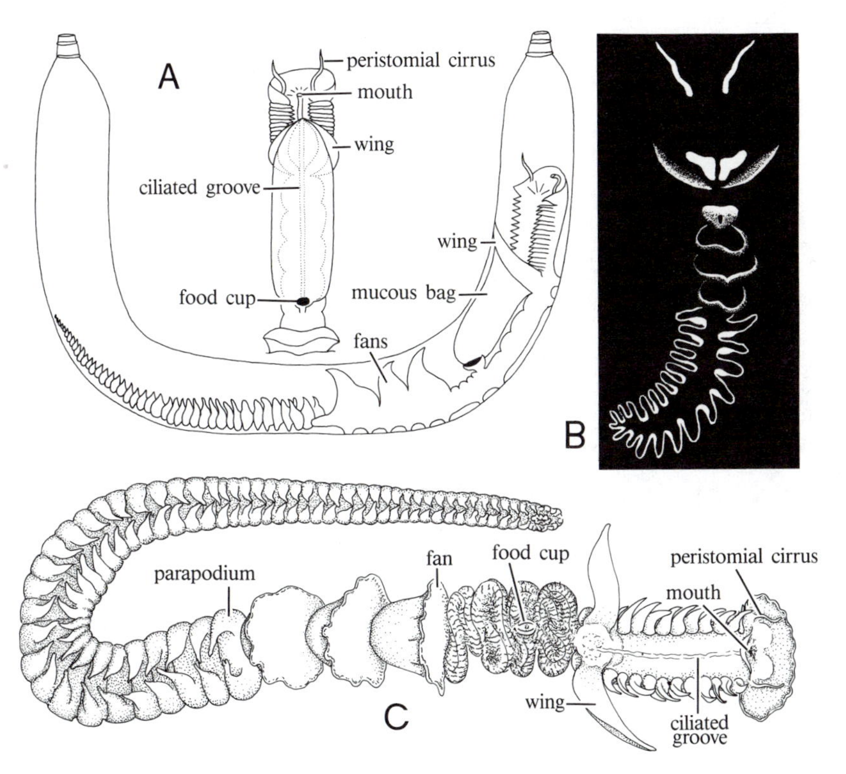

FIGURE 26.24. *Chaetopterus.* **A**. In feeding tube, with diagram of anterior end. **B**. Luminescence. **C**. Removed from tube. (**A** after MacGinitie from the MacGinities; **B** after Panceri, from Benham.)

usually short and stout, but quite variable. Primitively they occur as four pairs, two in the dorsolateral and two in the ventrolateral position, as in *Lumbricus* (Fig. 26.25B); however, the number is sometimes increased as in *Pheretima* where distinction between dorsolateral and ventrolateral positions is lost and an equatorial circle of as many as 200 or more setae is present. Setal shapes are characteristic of species, although often the setae are sigmoid (Fig. 26.26E) and frequently have a swelling (nodulus) in the middle (Fig. 26.26C,E,F). Slender, flexible hair setae increase surface resistance and are useful in swimming (Fig. 26.26A). Stiffer needle setae provide traction against the bottom or the walls of burrows or tubes (Fig. 26.26B,D). In many cases the dorsal and ventral setae of a somite are different. As a rule, the dorsal setae are hairlike and the ventral setae needlelike.

A system of dorsal pores connects each coelomic compartment with the exterior in most of the bur-

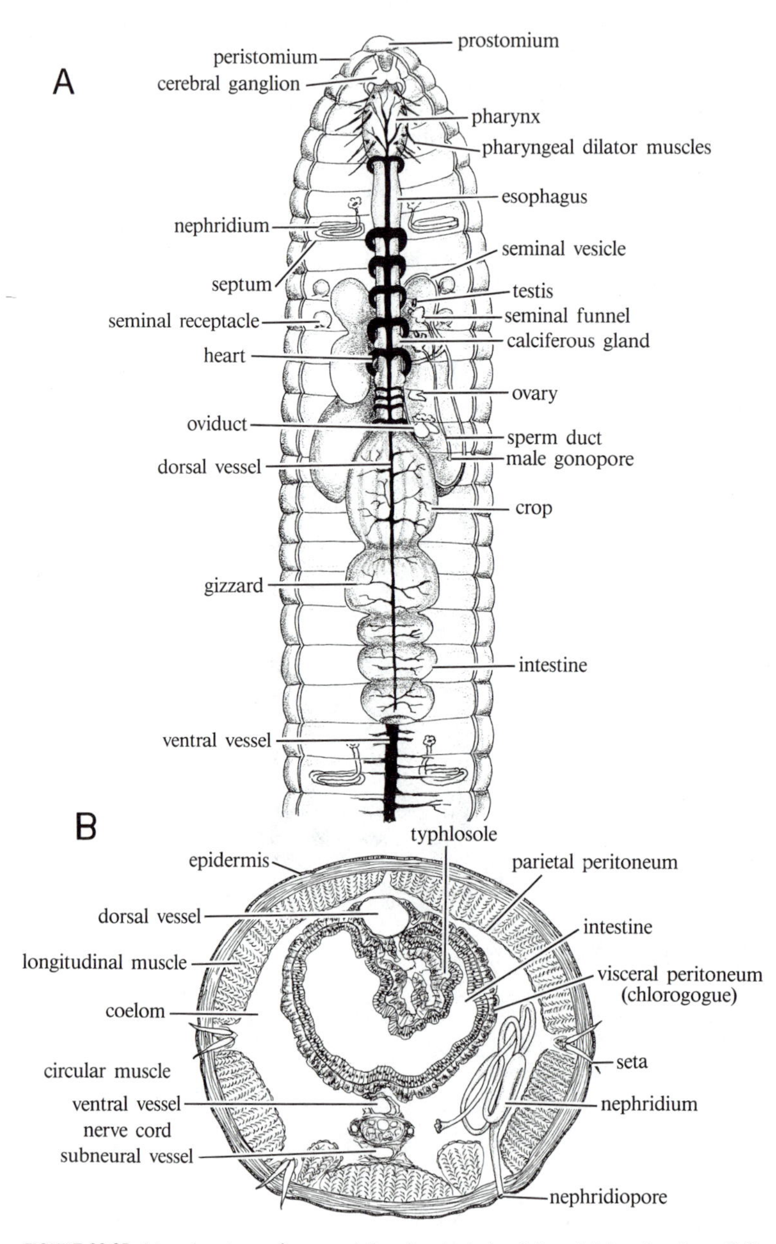

FIGURE 26.25. Internal anatomy of representative oligochaete *Lumbricus*. **A**. Internal anatomy. **B**. Transverse section.

rowing earthworms. Coelomic fluid can be extruded through the dorsal pores. Giant earthworms can squirt coelomic fluid several centimeters. The value of this talent is not very clear. The fluid may moisten the body surface, but numerous epidermal mucous glands are also present. Where it can be squirted out, it may startle predators. Neither function, however, seems adequate to account for the wide distribution of dorsal pores in terrestrial oligochaetes. One suggestion is that the extruded fluid may harden to consolidate the walls of chambers used for estivation or hibernation.

The glandular clitellum is characteristic of adult animals. It is a girdle that wholly or partly encircles several somites and secretes the wall of the cocoon as well as helping to form slime tubes during copulation. Modified setae in the clitellar region are not uncommon.

Locomotion Aquatic oligochaetes are somewhat more diversified in external appearance but are far less striking than the polychaetes. They are smaller than the terrestrial forms, and although they are benthic organisms, they are, as a rule, more mobile than the land species. They tend to have better-developed sense organs, and the setae are more highly developed. Aquatic oligochaetes creep about on the bottom or on submerged plants, and sometimes form soft mucous tubes on leaves or other objects. Some burrow in the soft bottom mud, and a few are ectoparasites. Terrestrial oligochaetes burrow and tunnel through soil, using their setae to anchor the body to faciliate pushing through the substrate.

None of the oligochaetes has a conspicuous head, undoubtedly related to their burrowing habits, but some have a large prostomium. *Stylaria* has an elongated prostomium, used as a delicately mobile tactile organ (Fig. 26.26G). *Stylaria*, *Nais*, and a few others have eyes. The peristomium has no setae, and sometimes the setae are modified or are missing in several anterior somites. These, however, are never incorporated into a definite head.

Feeding In the absence of external structures that can be specialized for food getting, food is taken in by the pharynx, sometimes with the aid of the prostomium. Pharyngeal pumping movements draw in humus or bits of dried vegetation, or the pharynx is extruded somewhat to pick up particles. All of the terrestrial oligochaetes are herbivorous detritus eaters. Charles Darwin pointed out the vast quantities of soil that earthworms move. In the area he studied, about 18 tons of soil per acre per year are brought to the surface by earthworms. A slow circulation of the upper layers of the soil results from the activities of earthworms and other burrowing animals, and is an important factor in determining the zonation and mixing of the soil strata.

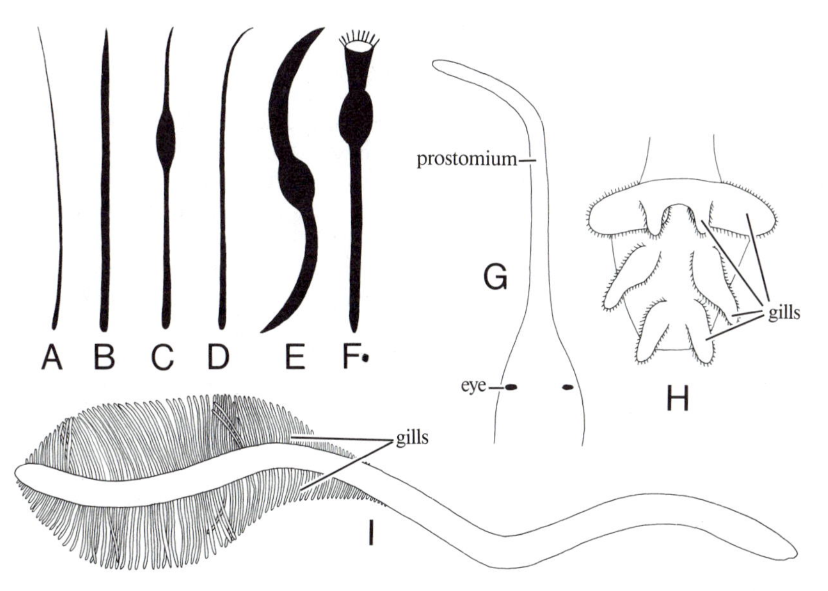

FIGURE 26.26. A,B,C,D,E,F. Some typical oligochaete setae; **A**. hair seta (capilliform); **B**. aciculate; **C**. aciculate with nodulus; **D**. uncinate; **E**. sigmoid, with nodulus; **F**. pectinate. **G**. Anterior end of *Stylaria fossularis,* with prostomium extended forward as proboscis, and with pair of eyes. **H**. Posterior end of *Dero digitata,* showing ciliated gills. **I**. Side view of *Branchiura sowerbyi,* showing dorsal and ventral gills. (**G,I** after Pennak; **H** after Bousefield, from Goodnight, in Edmondson.)

Detritus and algae, especially diatoms, are the principal foods of aquatic forms. Extrusion of the pharynx exposes an adhesive region coated with mucus. Particles adhere to it and are swallowed when the pharynx is retracted. The burrowing species swallow quantities of mud and digest out the organic material, casting the remainder at the end of the burrow. Like polychaetes and earthworms, they contribute to the circulation of the bottom material. Tiny *Chaetogaster*, less than a millimeter long, is an exception. It feeds on microcrustaceans, other worms, and insect larvae of appropriate size.

Respiration Some of the aquatic oligochaetes have gills. The most remarkable gills are seen in *Branchiura* (Fig. 26.26I), where long, digitate dorsal and ventral projections extend from the body surface. *Dero* (Fig. 26.26H) has four pairs of ciliated body projections at the posterior end that serve as gills, and *Aulophorus* has three. These worms extend the posterior end of the body from their tubes for respiratory exchange, and the surface cilia ventilate the gills. Body movements are also used to ventilate the respiratory surface.

Reproduction Most of the aquatic oligochaetes have a thin, inconspicuous clitellum. Many of the smaller species reproduce primarily by transverse fission, and in some cases chains of individuals are formed. The reproductive organs are often immature most of the year, and reproduction occurs only by fission.

CLASS CLITELATA, SUBCLASS HIRUDINIDA

Leeches make up a relatively homogeneous group of highly specialized annelids. About 500 species have been described, predominantly from freshwater habitats. A few, however, are marine, and some have adapted to terrestrial life in warm, moist regions. Leeches share with oligochaetes in the possession of a clitellum and a hermaphroditic reproductive system. They are thought by some to have arisen from an oligochaete stock that adapted to ectoparasitic life.

Leeches, however, are quite diversified in their habits. Some are free-living, eating detritus and plant material. Those leeches that engage in more aggressive feeding habits should not be thought of as true parasites, but rather as predators that take blood meals from vertebrates or invertebrates. Some live more or less permanently on a single animal and, therefore, verge on true ectoparasitism. Others make only infrequent visits to the prey for a blood meal (some subsist on one fill-up per 12 to 18 months), and still others are ordinary predators or scavengers.

Body Plan Leeches are usually flattened dorsoventrally and tapered fore and aft. Many are broad and leaflike; others look like plump worms. Most fall into a size range between 20 to 25 and 50 to 60 mm, but a few species are less than 10 mm long, and the largest leeches are said to reach nearly 0.5 m when creeping.

Leeches have a remarkably constant metameric organization, indicating both a close relationship and a high level of specialization. The body is formed of up to 34 somites. The major body divisions contain the same somites in all hirudineans: head (I–VI), preclitellar region (VII–IX), clitellum (X–XIII), postclitellar region or trunk (XIV–XXIV), anal region (XXV–XXVII), and posterior sucker (XXVIII–XXXIV). Superficially, they seem to have more somites (Fig. 26.27C), for each segment is subdivided into annuli by transverse grooves (Fig. 26.27A). The number of primary somites can be determined by counting the ganglia on the nerve cord or by counting the lateral nerves. Comparative studies indicate that three annuli occur in the primitive somite. However, some leech somites remain undivided and some have only two annuli, presumably as a result of suppression of one or two annuli. Many leeches have more numerous annuli formed by subdivision of the original three. The middle annulus is the sensory annulus, as it bears the sense organs and contains the main branch of the lateral nerve.

Piscicolid leeches (Fig. 26.27D) are divided differently, with a head, a trachelosome made up of the preclitellum and the clitellum, and a urosome extending to the posterior sucker. A definite collar at the posterior end of the clitellum may sharply define the body regions. Somites X–XIII still make up the clitellum, however, and the basic features of the metameric organization remain unchanged.

Branchiobdellids live on the gills or body surface of crayfish. They have a small but constant number of 15 somites. The first four somites are fused together to form a cylindrical head. Fingerlike projections and a sucker occur on the head. A pair of strong, chitinous jaws in the buccal cavity are used for feeding. The trunk consists of 11 segments. Fingerlike appendages are often found on the trunk somites, and in *Cambarincola* (Fig. 26.1D) a flangelike frill is found on the dorsal surface of each.

A single species, *Acanthobdella peledina*, has characteristics that are intermediate between the oligochaetes and the leeches (Fig. 26.28A), and is most interesting from an evolutionary standpoint. Its external anatomy reveals its distinctiveness from other leeches and shows some oligochaetelike traits. The body contains only 30 somites, and four of these are united to form the posterior sucker. There is no trace of an anterior sucker. The first five somites have setae, wholly lacking in all other leeches. The absence of an oral sucker, the simple posterior sucker, and the presence of setae are all primitive traits.

The posterior end of the typical leech body is formed of seven compacted somites, with no trace of external divisions. The nerve ganglia serving them are lumped together in a single mass. The last of these somites is the **pygidium**, which normally contains the anus in annelids. A few leeches have the anus located in the posterior sucker (Fig. 26.27B), no doubt associated with the pygidium, but in most cases the old anus is closed off and a new dorsal anus (Fig. 26.28A), just anterior to the posterior sucker, replaces it. The posterior sucker is a powerful suction cup, useful in locomotion as well as for attachment during feeding.

Locomotion Terrestrial leeches creep by attaching the posterior sucker, extending the body, attaching the anterior sucker, and pulling the posterior end forward. This inching along is essentially like the lo-

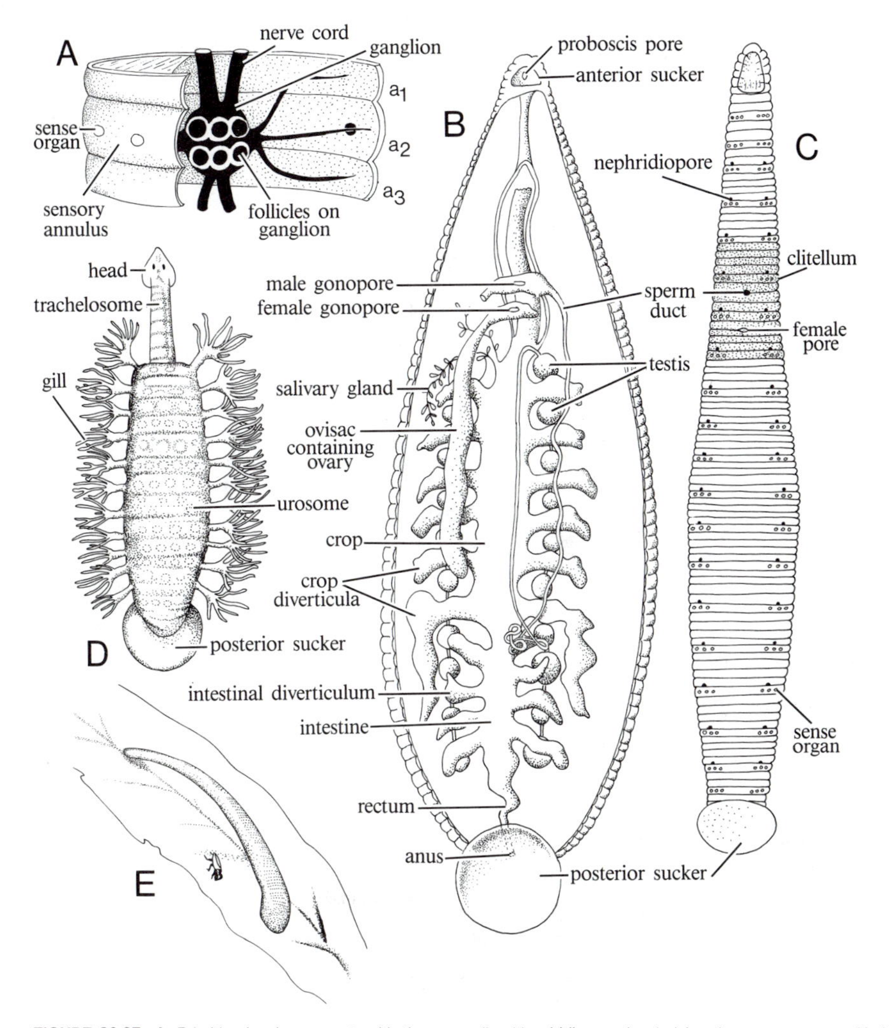

FIGURE 26.27. **A**. Primitive leech segment, with three annuli, with middle annulus (a_2) bearing sense organs. Notice six follicle masses of neurons associated with ganglion. **B**. Internal organization of *Glossiphonia complanata,* pharyngobdellid leech. **C**. Ventral view of *Hirudo medicinalis,* gnathobdellid leech. **D**. *Ozobranchus jantzeanus,* piscicolid leech with lateral gill filaments. Body of piscicolid leeches divided into head, trachelosome, and urosome. **E**. *Haemadipsa,* haemadipsid leech attached to leaf, waiting for appropriate host. (**A** after Mann, from Avel, in Grassé; **B** after Harding; **C** after Mann; **D** after Oka; **E** after Scriban and Autrum, from Mann.)

comotion of earthworms, except that the body as a whole extends and contracts alternately, behaving like a single segment of an oligochaete. Evidently the nature of the built-in intersegmental reflexes have been modified to bring about simultaneous contractions of the longitudinal or circular muscles of all somites together.

Aquatic leeches can creep in the same manner, but some can also swim gracefully, with undulatory movements of the body. The undulatory movements are similar to those of other annelids, involving the alternate contractions of the longitudnal muscles on the two sides of the body, so that they work as antagonists and with small groups of somites acting in unison. Movements of this type depend on activation waves of a type not required for creeping. Terrestrial leeches cannot swim, and drop to the bottom of a pond to creep out to the shore. Some of the aquatic leeches are also limited to creeping movements. As in other annelids, the activation waves responsible for undulatory movements are affected by, but are not dependent on, the brain. Removal of the brain and the compound ganglion at the base of the posterior sucker does not destroy the ability to swim.

Feeding Rhynchobdellid hurudineans have a proboscis and no jaws. These structural peculiarities are correlated with their habits, which differ somewhat from those of other leeches. Rhynchobdellids fall naturally into two groups, the glossiphonids and the piscicolids. Glossiphonids are flattened freshwater leeches with an inconspicuous anterior sucker. Most of them are predators or scavengers, moving about actively in search of worms or other small invertebrates. Prey is caught with the proboscis. Some feed on snails and live in or on snail shells, and some take blood meals from cold-blooded vertebrates. *Glossi-*

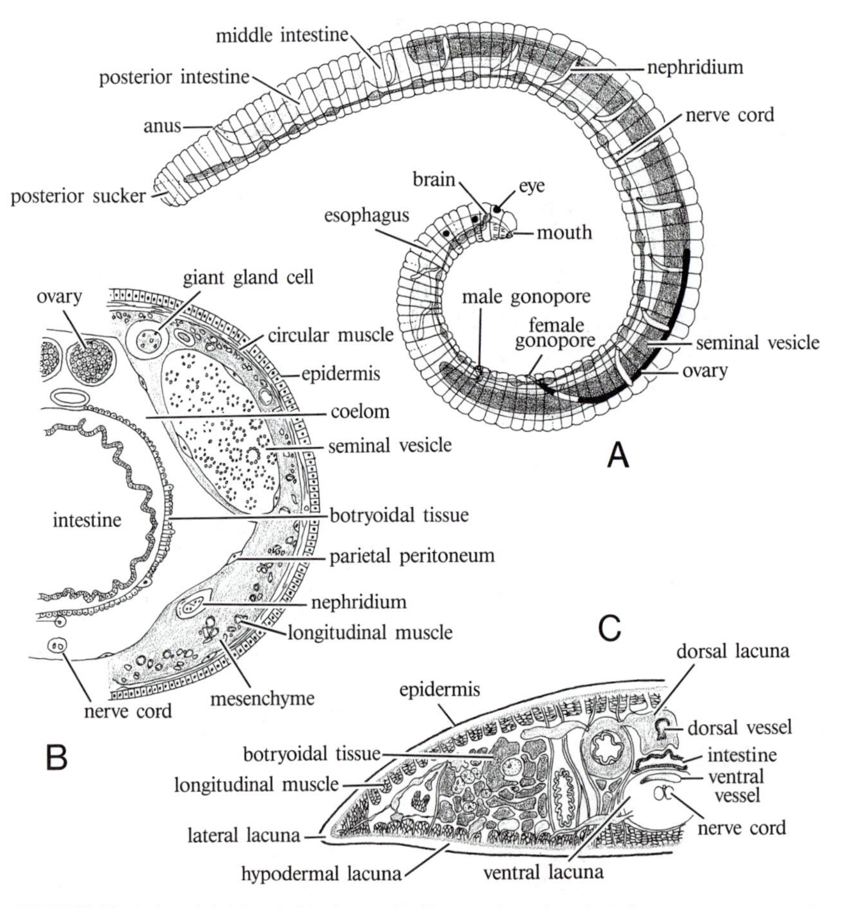

FIGURE 26.28. A. *Acanthobdella peledina,* characterized by posterior sucker of only four somites, presence of setae, and other primitive leech traits. **B**. Transverse section through *Acanthobdella.* Note open coelom, closed in part in other leeches by accumulation of botryoidal tissue. **C**. Transverse section through glossiphonid leech, *Placobdella,* with coelom reduced to narrow passages, lacunae, in botryoidal tissue. (**A,C** after Livanoff; **D** after Scriban, from Avel, in Grassé.)

phonia (Fig. 26.27B) is a good example. Piscicolids (Fig. 26.27D) are the fish leeches. Most of them live as ectoparasites on freshwater fishes and crustacea, but a few occur on marine elasmobranchs. Most of the piscicolids leave the host to breed and attach their egg capsules to submerged objects, but some attach the egg capsules directly to the host animal. They have no jaws, and the method used to penetrate the host skin remains a mystery. Piscicolids are usually long and cylindrical and have a conspicuous anterior sucker used for attachment.

Gnathobdellids have no proboscis, and three toothed jaws surround the mouth. The family Hirudidae includes the more familiar forms. *Hirudo medicinalis* (Fig. 26.27C), long used for bloodletting, is probably most often studied as an example of leech structure. *Hirudo*, like most of the members of its family, lives in relatively shallow water in temperate or subtropical regions. When it feeds, the jaws make a small, triangular incision. Saliva, containing an anticoagulant, hirudin, is applied to the wound. Some leeches promote blood flow by the injection of a second substance; its action resembles histamine and causes capillary paralysis. The muscular pharynx pumps blood into the stomach. As a rule, the hirudids visit animals to take blood meals infrequently. Some, however, enter the upper respiratory tract of cattle and horses as they drink, causing respiratory difficulties and secondary infections. The family Haemadipsidae includes the tropical land leeches with bloodsucking habits. They live on the ground in moist places, and some climb up on shrubs or other plants to wait for the arrival of a bird or mammal on which they can feed (Fig. 26.27E).

Some of the hirudids have quite small jaws and do not ordinarily depend on blood meals. They attack their prey, usually a small worm or other invertebrate, and swallow it whole. The pharyngobdellids have no jaws and feed in the same manner, but probably all hirudids and pharyngobdellids will take a blood meal when a bleeding wound is available. In some species this tendency is stronger than in others.

Excretion Leech nephridia have not been mentioned, because they are so unlike those of the oligochaetes and polychaetes. This is to be expected, of course, for the reduction of the coelom and its conversion into a circulatory system makes the metanephridial system with an open nephrostome inefficient or physiologically unsuitable. The rhynchobdellid nephridia are least modified (Fig. 26.29A). A nephrostome opens into the median ventral coelomic sinus and is connected by a short ciliated tube to a thin-walled sac, the capsule. The capsule is filled with phagocytic cells, often crowded together and with tiny ducts between them. In the hirudids the capsule is large and has multiple funnels, whereas in the pharyngobdellids, each nephridium is supplied with two capsules and two funnels (Fig. 26.29B,C). Openings to the capsule, however, are not entirely homologous with the old nephrostome, for in some cases the vestige of the nephrostome is found in ciliated bodies in the lining of the coelomic sinuses.

Unfortunately, there is little known about the details of leech excretion. They have no single cell comparable to the chlorogogue cells of other annelids, but the botryoidal cells take up injected particles in *Hirudo*, and probably partially correspond to the chlorogogue cells. Amoeboid corpuscles in the coelomic fluid are also phagocytic and play an important role in excretion, although the details are not understood. In any case, the phagocytic cells that fill the open capsule of the rhynchobdellid leeches appear to accumulate particles of material that are passed down into the nephridial tubule. It seems probable that the capsule was originally a chamber in which corpuscles could accumulate, for the capsule is not walled off from the tubule in the most primitive rhynchobdellids. However, it is closed off in others, and in these cases it is not easy to understand how the capsules function. In the hirudids, the cilia associated with the capsule beat outward, and it seems likely that as the coelom became a circulatory system, the capsule lost its original function and assumed a new one, the production of corpuscles. Evidence exists that the material in the corpuscles of the more advanced leeches never enters the nephridium, but is permanently stored in the botryoidal tissue. However the leech capsule works, it is evident that much of the waste material that emerges from the nephridium must pass through the tubule walls.

Beyond the capsule, the nephridium becomes a cord of cells containing a tortuous and usually branched, nonciliated intracellular duct. A short terminal section of the nephridium is formed of invaginated epidermal cells and serves as a bladder.

Respiration Most leeches respire through the body surface generally, but a few have definite gills. The gills are usually simple vascular vesicles on the sides of the body, but some have plumose gills (Fig. 26.27D).

Circulation *Acanthobdella* has a typical annelid coelom, divided into compartments by septa but somewhat reduced by encroachment of the parenchymal tissue and muscles on the cavity (Fig. 26.28B). The coelom of rhynchobdellids, however, is greatly reduced. Muscles pass through the coelomic region, and pigmented botryoidal tissue has completely filled the coelom in some regions (Fig. 26.28C). Narrow spaces in the botryoidal tissue form capillary lacunae, roughly analogous to the lymphatic system of vertebrates. The botryoidal tissue also divides the coelomic remnants into a system of larger passageways, together forming a coelomic lacunar system. Dorsal and ventral lacunae contain the dorsal blood vessel and ventral nerve and vessel, respectively (Fig. 26.30A). Intermediate lacunae connect the dorsal and ventral lacunae with the lateral coelomic lacunae at the body margin. Subepidermal lacunae connect with the lateral lacunae. The coelom is a significant factor in internal transport.

The rhynchobdellids also have a circulatory system that is not much different from that of other annelids (Fig. 26.30D). Blood flows forward in a dorsal vessel above the gut, and back in a ventral vessel below the gut. A system of looped vessels in the posterior sucker connect the dorsal and ventral vessels. Three commissural vessels, rather like those seen in oligochaetes but much elongated, connect the dorsal and ventral vessels in the esophageal region. Looped vessels for the anterior sucker arise from the anterior commissural vessel, and the most posterior one loops far back in the body.

Although some of the rhynchobdellids are quite active and relatively large, they usually have no respiratory pigment. The botryoidal capillaries, coelomic lacunae, and blood vessels do not constitute

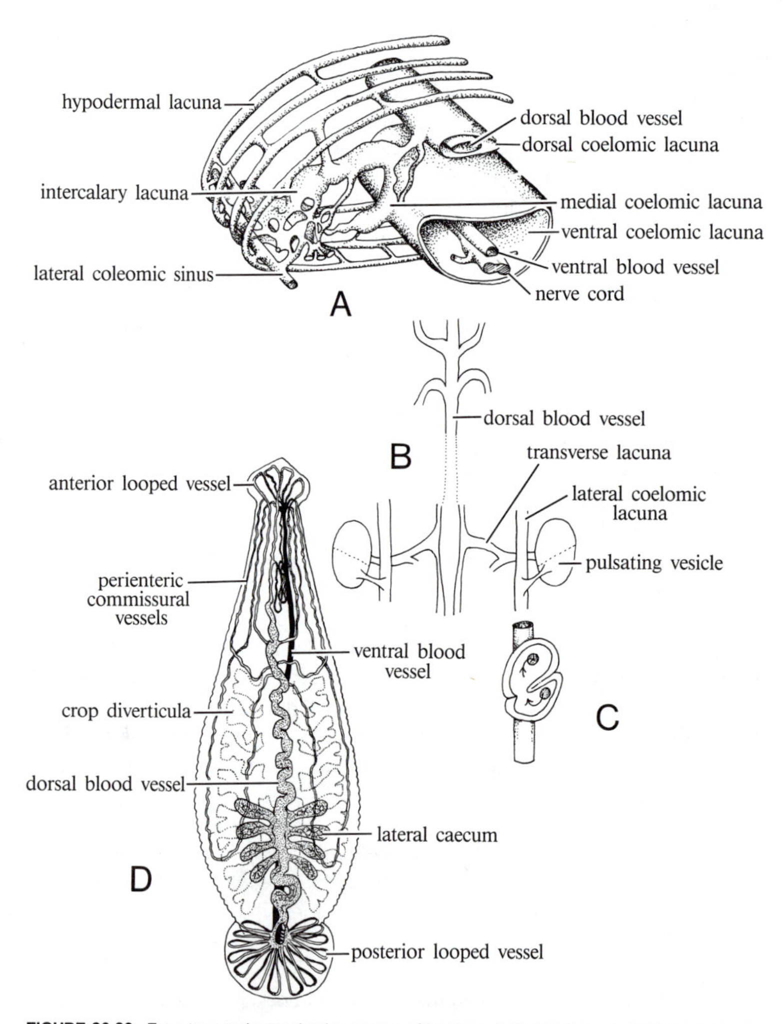

FIGURE 26.29. Excretory and reproductive organs of leeches. **A.** Nephridium of glossiphonid in place. Note ciliated nephrostome leading to coelomic capsule. **B.** Nephridial and reproductive system of *Haemopis*. **C.** Anastomosing nephridia of *Erpobdella*. Note disappearance of nephrostome with conversion of coelom into circulatory system. (**A** after Oka; **B** after Mann; **C** after Graf, from Harant and Grassé, in Grassé.)

three discrete systems. Injection experiments have shown that all three intercommunicate, forming to some extent a unified transport system.

Some hirudineans have lost the original circulatory system entirely. In adapting to assume the full responsibility for internal transport, the coelomic sinus system has lost all resemblance to a coelom and become a veritable circulatory system, and the coelomic fluid has become bloodlike. The coelomic fluid contains hemoglobin dissolved in the plasma, while other annelids carry any coelomic respiratory pigments in hemocytes. The simple walls of the rhynchobdellid coelomic sinuses are replaced by more complex walls resembling those of blood vessels, and the subepidermal lacunar system is replaced by an extensive capillary bed. *Hirudo* and some leeches retain dorsal and ventral coelomic sinuses; others have lost them entirely. In any event, the lateral sinuses are more important. They converge at the anterior and posterior ends of the body and give rise to segment connections with the subepidermal plexus and with the median sinuses, if they are present. Plexi extend to and around the nephridia, gonads, and other body parts, as in typical circulatory systems. The two lateral sinuses serve as tubular "coelomic hearts," pumping the coelomic

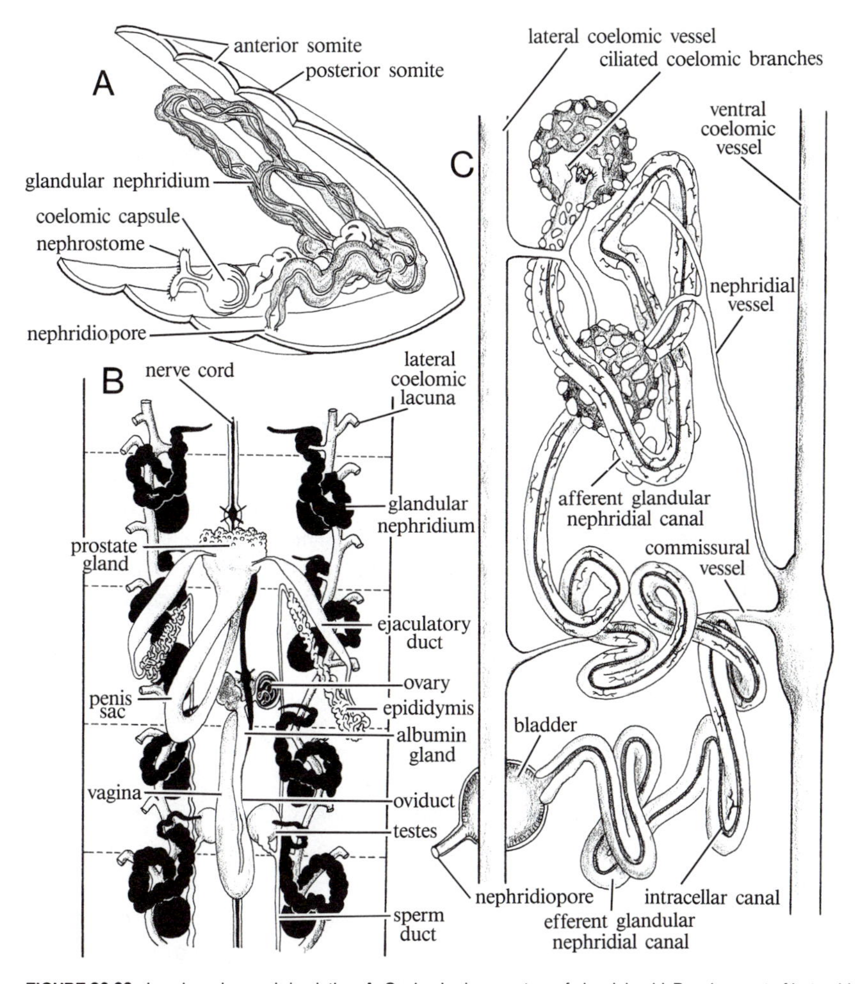

FIGURE 26.30. Leech coelom and circulation, **A**. Coelomic sinus system of glossiphonid. Development of botryoidal tissue has reduced coelom to canals around dorsal and ventral blood vessels, and system of lacunae responsible for much of internal transport. **B.** Pulsating vesicle system in piscicolids, associated with lateral coelomic sinuses. Coelomic fluid flows into them from dorsal sinus, and then into lateral sinuses. **C.** Pulsating vesicles of *Piscicola,* opened to show valvular openings that regulate direction of flow. **D.** Scheme of circulatory system of rhynchobdellid, *Hemiclepsis.* Original circulatory system retained in these leeches. Vascular caeca from dorsal vessel extend over digestive system diverticula, and looped blood vessels connect dorsal and ventral vessels. (**A,D** after Oka, from Beddard; **B** after Autrum; **C** after Selensky, from Scriban and Autrum, in Kükenthal and Krumbach.)

fluid around the circuit. Piscicolids are equipped with a series of special pumping chambers associated with the lateral lacunae (Fig. 26.30B,C), although they retain a circulatory system.

Sense Organs Leeches are well supplied with sense organs. Eyes of varying complexity, papillae, and tubercles are distributed over the body. Stripes of pigment or irregular patches of color are not uncommon, and in many cases, metamerically arranged pigment spots are placed in a definite relationship to the sense organs. The ocelli have been discussed already. The more complex sense organs, however, are associated with the sensory annulus.

Leeches are distinctly cephalized. The anterior sucker may be very prominent or may consist of little more than thickened lips around the mouth. Rhynchobdellid leeches have a protrusible proboscis, and the oral sucker lies behind the mouth, but in gnathobdellids and pharyngobdellids the mouth is centered in the oral sucker (Fig. 26.27B,C). The compound eyes also lie on the head. Their position, number, and form are important in the identification of species.

Fossil Record and Phylogeny

Annelids occur sporadically as fossils. Scolecodonts, annelid jaws, are well-known microfossils. Several ichnofossils (tracks, trails, and burrows) are attributed to the activity of annelids. Actual annelid body fossils are rare and are restricted largely to those few fossil assemblages with unusual preservation such as the Carboniferous Bear Gulch Limestone or Francis Creek Shale. In all cases, these fossils seem to be aligned with known living polychaete groups, and bear witness to Fauchald's view that polychaete divergence occurred very early, that is, in late Precambrian or early Cambrian time.

As a result, it seems almost impossible at present to develop an idea of how all the various annelids are related to each other. Two great groups are recognized: the polychaetes and the clitellates. The clitellates appear to be a monophyletic group, and have a unity in anatomical and developmental patterns (e.g., Dales, 1967). Beyond this, consensus does not exist. A principal problem in this regard is the nature of the polychaetes. Fauchald (1977) recognizes some 81 families in 17 orders within the Polychaeta. However, definitions of the families and orders are often characterized by qualifications and exceptions to a relatively few individual character states shared among them. Any attempt to sort 81 families should use hypothetically more than 81 characters. Under current conditions, clusterings of orders into clades is impossible. Nor is there agreement over facts and events in annelid history that might be used to polarize characters. Polychaetes of complex form have commonly been thought of as primitive, but Fauchald (1974) makes cogent arguments for an ancestral annelid being a simple oligochaete-like polychaete. It is also possible that annelids (at least the polychaetous forms), may have been the result of rapid radiation from several ancestral stocks. Brinkhurst (1982) indicated that he felt oligochaetes evolved independently of polychaetes. Thus, many characters that we interpret as similar, and which we are tempted to treat as shared derived features, might be parallelisms. Therefore, judgments about annelid interrelationships must await the conclusion of comparative cladistic studies actively being investigated.

Taxonomy

Class Polychaeta. Usually with paired appendages known as parapodia, from which chitinous setae project; usually strongly cephalized, with specialized head appendages; almost universally with separate sexes and with many peritoneal gonads segmentally arranged.

Order Phyllodocida. Segments similar; prostomium distinct, at least one pair of antennae; frequently with palps and eyes; one or more peristomial segments; pharynx eversible, one or two pairs of jaws; parapodia well developed (Figs. 26.19, 26.20, 26.21B,C,E, 26.22A,B,C,F and 26.23A).

Order Amphinomida. Prostomia and peristomia enclosed in anterior segments; pharynx eversible, without jaws.

Order Spintherida. Body oval and flattened; prostomium small; pharynx as eversible sac or funnel; parapodia modified for attachment.

Order Eunicida. Prostomium distinct; jaw apparatus complex; parapodia well developed (Figs. 26.21D and 26.22D).

Order Orbiniida. Prostomium conical; pharynx simple, lobed, or elaborately branched; body divided into anterior, flattened thorax and posterior, cylindrical abdomen.

Order Spionida. Prostomium with pair of long, grooved tentacular palps; pharynx eversible, ciliated; frequently tubicolous.

Order Chaetopterida. Prostomium tiny; peristomium as collarlike frill; pharynx noneversible; body divided into three highly specialized regions; tubicolous (Fig. 26.24).

Order Magelonida. Body threadlike, tapering, divided into thorax and abdomen; pharynx eversible.

Order Psammodulida. Larvalike meiofaunal form; segmentation poorly developed, body divided into thorax and abdomen.

Order Ciratulida. Prostomium distinct, fused more or less with peristomium; peristome with spioniform tentacles or numerous tentacular filaments; pharynx usually noneversible; parapodia without acicula or distinct setal lobes.

Order Ctenodrilida. Body minute, segments few; pharynx eversible; parapodia uniramous or biramous without distinct lobes.

Order Cossurida. Body small, threadlike; prostomium conical or rounded; pharynx eversible, with ciliated processes; single, very long, branchial filament located middorsally on anterior of body.

Order Flabelligerida. Prostomium reduced, fused to peristomium, both withdrawn into anterior segments; integument papillose; pharynx noneversible.

Order Opheliida. Segments divided into annuli; prostomia and peristomia small, bearing pair of nuchal organs; pharynx eversible.

Order Sternaspida. Body short, grublike; lacking parapodia lobes; pharynx small, eversible.

Order Capitellida. Pharynx eversible, papillose; burrowers, sometimes tubicolous (Figs. 26.13A, 26.22H, 26.23D).

Order Oweniida. Body spindlelike, segments not well developed and marked only by setae; parapodia reduced; pharynx noneversible.

Order Terebellida. Tubiculous; prostomia and peristomia fused; branchiae on anterior segments.

Order Sabellida. Tentacular, branchial crown of bipinnate filaments for filter feeding (Fig. 26.22G).

Order Nerillida. Body short and flat; prostomium with none or up to three filiform antennae, sometimes annulated, and spoonlike palps; mouth as ventral slit, pharynx eversible and usually armed with stylets.

Order Dinophilida. Small, meiofaunal forms; ventral surface with cilia, dorsal surface with cilia tufts or bands; pharynx eversible, though sometimes lacking.

Order Polygordiida. Parapodia and setae lacking; short, stiff frontal tentacles; pharynx eversible (Fig. 26.1B).

Order Protodrilida. Body elongate, flattened; long, mobile lateral tentacles; pharynx noneversible, muscular; parapodia small or absent.

Order Myzostomida. Body small, oval, flat, without external segments; parapodia uniramous (Fig. 26.23B).

Order Poebiida. Body elongate, oval, laterally compressed, without external segments or setae; internal segmentation weakly developed; body wall with thick gelatinous layer; prostomia and peristomia indistinct, with pair of long palps; mouth and buccal areas heavily ciliated. (May be a separate phylum.)

Class Clitellata. Clitellum used to secrete a cocoon for eggs; hermaphrodites; development direct without larvae.

Subclass Oligochaeta. With relatively few setae; small head without cephalic appendages; hermaphroditic, with testes anterior to ovaries and with only one or two pairs of male and female gonads located in a few genital somites; with complex gonoduct system and seminal receptacles.

Order Lumbriculida. Four pairs of setae per segment; male ducts in same segment as testes (Figs. 26.7B and 26.25).

Order Moniligastrida. One or two pairs of testes, with male pores in segments next posterior to testes (Figs. 26.1C and 26.5A).

Order Haplotaxida. Male pores at least one segment anterior to segments bearing testes (Fig. 26.26G,H,I).

Subclass Hirudinida. Body segment numbers fixed, segments divided into annuli, internal segmentation obscured, coelom often filled with botryoidal tissue.

Order Branchiobdellida. Fifteen body segments; anterior and posterior suckers; well-developed jaws (Fig. 26.1D).

Order Acanthobdellida. Thirty body segments; no anterior sucker; coelom with septa (Fig. 26.28A,B).

Order Hirudinea. Thirty-four body segments; anterior and posterior suckers (Figs. 26.12B,E, 26.27B,C,D,E, 26.28C and 26.30D).

Selected Readings

Brinkhurst, R. O. 1982. Evolution in the Annelida. *Can. J. Zool.* 60:1043–1059.

Brinkhurst, R. O., and D. G. Cook (eds.). 1980. *Aquatic Oligochaete Biology*. Plenum Press, New York.

Brinkhurst, R. O., and B. G. M. Jamieson. 1972. *Aquatic Oligochaeta of the World*. Toronto Univ. Press, Toronto.

Caspers, H. 1984. Spawning periodicity and habitat of the palolo worm *Eunice viridis* in the Samoan Islands. *Mar. Biol.* 79:229–236.

Dales, R. P. 1967. *Annelids*. Hutchinson, London.

Fauchald, K. 1974. Polychaete phylogeny: A problem in protostome evolution. *Syst. Zool.* 23:493–506.

Fauchald, K. 1977. The polychaete worms—definitions and keys to the orders, families, and genera. *Nat. Hist. Mus. Los Angeles Co. Sci. Ser.* 28:1–179.

Fauchald, K., and P. A. Jumars. 1979. The diet of worms: A study of polychaete feeding guilds. *Ann. Rev. Oceanogr. Mar. Biol.* 17:193–284.

Fernandez, J., and N. Olea. 1982. Embryonic development of glossiphonid leeches. In *Developmental Biology of Freshwater Invertebrates* (F. W. Harrison and R. R. Cowden, eds.), pp. 317–361. Alan R. Liss, New York.

Giere, O., and O. Pfannkuche. 1982. Biology and ecology of marine oligochaeta: A review. *Ann. Rev. Oceanogr. Mar. Biol.* 20:173–308.

Lasserre, P. 1975. Clitellata. In *Reproduction of Marine Invertebrates*, vol. 3 (A. C. Giese and J. S. Pearse, eds.), pp. 215–276. Academic Press, New York.

Laverack, M. S. 1963. *The Physiology of Earthworms*. Macmillan, New York.

Mill, P. J. 1978. *Physiology of Annelids*. Academic Press, London.

Ruby, E. G., and D. L. Fox. 1976. Anaerobic respiration in the polychaete *Euzonus mucronata*. *Mar. Biol.* 35:149–153.

Ruppert, E. E., and K. S. Carle. 1983. Morphology of metazoan circulatory systems. *Zoomorph.* 103:193–208.

Schroeder, P. C., and C. O. Hermans. 1975. Annelida: Polychaeta. In *Reproduction of Marine Invertebrates*, vol. 3 (A. C. Giese and J. S. Pearse, eds.), pp. 1–214. Academic Press, New York.

Shimizu, T. 1982. Development in the freshwater oligochaete *Tubifex*. In *Developmental Biology of Freshwater Invertebrates* (F. W. Harrison and R. R. Cowden, eds.), pp. 283–316. Alan R. Liss, New York.

Woodin, S. A. 1987. External morphology of the Polychaeta: Design constraints by life habit. *Bull. Biol. Soc. Wash.* 7:295–309.

27

Near Arthropods

A series of small groups with articulated bodies (pentastomids, tardigrades, and onychophorans) are a problem in terms of determining their affinities. These groups have been placed by various authors at different times in or near the arthropods. For example, Pentastomida are thought by some to be related to branchiuran crustaceans or fish lice (Wingstrand, 1972; Riley et al., 1978; Abele et al., 1989), but have also been allied with mites, tardigrades, and myriapods (Osche, 1963). Onychophora often have been treated as a separate phylum, but are also placed within the uniramian arthropods (Anderson, 1973; Manton, 1977). Tardigrades are sometimes allied to other arthropods within that phylum, and have been compared to onychophorans. All these groups are treated here as separate phyla.

Pentastomida

The Pentastomida, or Linguatulida, are a group of about 100 species of parasites of probable crustacean affinities. They live in respiratory tracts of higher vertebrates, especially tropical reptiles, but also occur in birds and mammals (including humans). Their range extends into temperate zones, but they are especially characteristic of tropical to subtropical areas.

DEFINITION

Pentastomida have a bilaterally symmetrical, wormlike body. They usually have two pairs of lobate, hooked legs posterior to the lobelike head.

BODY PLAN AND BIOLOGY

The wormlike body varies in length from about 2 to 15 cm. Two pairs of either hooked appendages or the **hooks** only are on the anterior end (Fig. 27.1C,D). The mouth usually lies on a medial prominence. The name Pentastomida, meaning five mouths, was mistakenly derived from a misinterpretation that these five anterior prominences found in the Class Cephalobaenida were mouths. The claws are used to attach to the nasal passages or lungs of the host.

The anterior end of the worm is the **cephalothorax** or **prosoma**, and the ringed posterior part is the **abdomen** or **opisthosoma**. The hooks inflict a wound on the soft tissues of the respiratory tract, and blood is sucked into the pentastomid gut. The hooks are secreted by **hook glands** in a manner reminiscent of tardigrades. Large **frontal glands**, which may secrete an anticoagulant, join the hook glands (Fig. 27.1A,B). Blood from the sucking **pharynx** moves into the large **midgut**. A short **hindgut** opens through a terminal **anus**.

There are no special circulatory, respiratory, or excretory organs. The nervous system is simplified. A large **ventral ganglion** is composed of three segmental ganglia. A **circumenteric ring** attaches to the ventral ganglion. It is a simple ring, without a dorsal ganglion. Nerves pass to the legs or hook musculature and to the body wall of the cephalothorax. A long pair of nerves pass back to the abdomen on the ventral side.

As is typical of parasites, the reproductive organs are large and relatively complex (Fig. 27.1A,B). The reproductive potential is enormous. A single female *Waddycephalus* may contain half a million fertilized ova in its uterus. The males are smaller than the females and more restless, moving about within the host in search of a mate. The sperm resemble those of maxillopodan crustaceans, especially those of the branchiurans, the fish lice (Wingstrand, 1972). The eggs are fertilized internally, and ova are stockpiled in the uterus for some time, probably to mature. They are discharged and then emerge with the secretions of the host respiratory tract.

When ova are eaten by an appropriate intermediate host, they hatch and release a migratory larva (Osche, 1963) that, when eaten by an intermediate host, pierces the stomach wall and encysts in the host tissues. Some of the larvae prefer vital organs, and cause illness or death of heavily infected hosts. The larva has two pairs of legs, and bears a superficial resemblance to a four-legged tardigrade or a mite larva (Fig. 27.1E). When the intermediate host is eaten by a suitable primary host, the larvae mature. They migrate to the respiratory tract by way of the esophagus and trachea if they live in the lung, or by way of the esophagus and pharynx if they live in the nasal passages (Esslinger, 1962). After several months they are mature and begin to produce gametes. The details of the life cycles are not very well

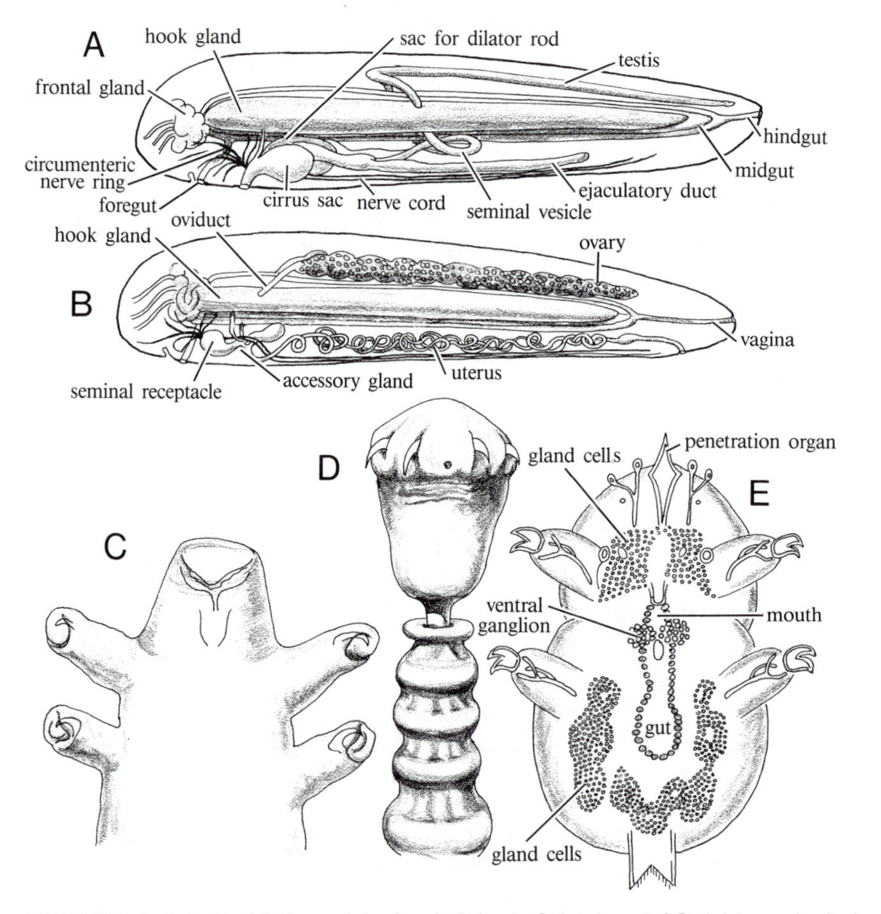

FIGURE 27.1. Pentastomida. **A**,**B**. *Porocephalus;* **A**. male; **B**. female. **C**. Anterior end of *Cephalobaena,* showing hooked legs and proboscis. **D**. Anterior end of *Armillifer,* which has lost all traces of legs except for hooks. **E**. Larva of *Porocephalus*. (**A**,**B** after Spencer; **E** after Stiles, from Shipley; **C** after Heymons, in Kükenthal and Krumbach.)

known, and intermediate hosts are unidentified for most species. All of the life cycles so far worked out involve intermediate hosts that are usually vertebrates in the food chain of the final hosts.

PHYLOGENY

The taxonomic affinities of pentastomids have been difficult to assess. Traditionally they have been treated as a separate phylum while being compared at one time or another with almost every group of annelids or arthropods. Consensus currently seems to be moving toward some crustacean affinity (e.g., Riley et al., 1978). However, the features used in this comparison present problems. The tailed sperm (Wingstrand, 1972) might be viewed as a generalized, primitive type, and the identification of a crustacean-type limb series in the larva (Osche, 1963) does not conform to what would be seen in a nauplius or egg-nauplius stage. Biochemical data are emerging that seem to indicate that pentastomids have some affinity with branchiuran crustaceans (Abele et al., 1989).

TAXONOMY

Class Cephalobaenida. Mouth on anterior prominence; cephalic hooks located on parapodial prominences. (Fig. 27.1C)

Class Porocephalida. Head flattened ventrally; cephalic hooks lateral to mouth. (Fig. 27.1D).

Tardigrada

Water bears are minute animals, ranging in length from about 50 μm to 1 mm. Only a few of the 600 species of tardigrades are marine, living among sand particles in the littoral zone as a rule, though a few have been found at depths of about 5000 m. Some of the rest live in freshwater habitats, in bottom detritus or attached to freshwater plants. The majority of tardigrades are semiaquatic, living on mosses, lichens, and liverworts.

Tardigrades have an odd combination of traits, shared by diverse groups. Limited cell constancy and a pseudocoelomlike hemocoel derived from the blastocoel, as well as cryptobiosis, give tardigrades some similarities to the pseudocoelomates. The mesoderm arises from outpocketings of the gut and then largely degenerates, as in chaetognaths, but the remnant coelomic compartments resemble those of "protostomes." The lobate appendages are located under the body, as in arthropods, but lack the jointed form characteristic of arthropods. The habit of molting is shared with arthropods, but the exoskeleton is not chitinous. All-in-all, this is an unusual combination of traits (Nelson, 1982b).

DEFINITION

Tardigrada possess a minute, bilaterally symmetrical body with four body somites, each somite bearing a pair of stumpy appendages ending in secreted claws. There is a cuticular exoskeleton, sometimes divided into segmentally arranged plates, that is molted during growth. They have a large dorsal brain, circumenteric connectives, and a subpharyngeal ganglion in the head, and paired nerve trunks containing metameric ganglia in the body. Sexes are separate, but females predominate, and some species lack males entirely. Development is direct, the embryo forming five pairs of coelomic spaces derived from the gut cavity that later deteriorate except for the posterior coelomic compartment, which persists as the lumen of gonad and gonoduct. Growth of postembryonic stages occurs without cell divisions, and there appears to be cell number constancy. The adult body cavity is derived from the blastocoel.

BODY PLAN

Sensory **cirri** (Fig. 27.2E) occur at the sides of the head of mesotardigrades and heterotardigrades, but are lacking in eutardigrades. A pair of red or black cup-shaped **ocelli** are usually embedded in the lateral lobes of the brain (Fig. 27.2A). The edge of the mouth is strengthened by one or more rings of thickened cuticle, and the dorsal cuticle is folded to form rings, typically two to a somite. However, scuteschniscids have dorsal, metameric plates of thickened cuticle (Fig. 27.2E). The short legs typically have single or double claws sometimes on long toes, formed anew at each molt.

Digestive System The **mouth** opens into a short **buccal cavity** (Fig. 27.2B), stiffened by cuticular rings. This is followed by a long, cuticularized **buccal tube** (**tubular pharynx**). The pair of stylets lie beside the buccal tube. At the junction of the buccal cavity and buccal tube, two slots, the **stylet sheaths**, introduce the stylets to the buccal cavity. Muscles protrude the stylets for feeding and retract them after use (Fig. 27.2B). The buccal tube opens into a heavily muscled **sucking pharynx** with four or six cuticular pieces, the **macroplacoids**, to which muscles attach. Food passes from the pharynx to the midgut, which is divided into an **esophagus** and a **stomach** (Fig. 27.2A). The short **hindgut** opens through a ventral **anus**, located between the third and fourth pairs of legs.

Three glands attach to the gut at the union of midgut and hindgut, the point in arthropods where the Malpighian tubules join the gut. The glands are often called **Malpighian tubules**, although they have not been shown to be excretory. The intestine itself also may participate in excretion and osmoregulation.

Body Cavity An extensive cavity occupies the space around the muscle fibers and between the internal organs and the epidermis. This cavity is sometimes termed the **hemocoel**. It contains a perivisceral fluid in which a large population of cells is suspended. The chemical nature of the perivisceral fluid is not known, and the functions of the cells have not been determined. The cells may be centers

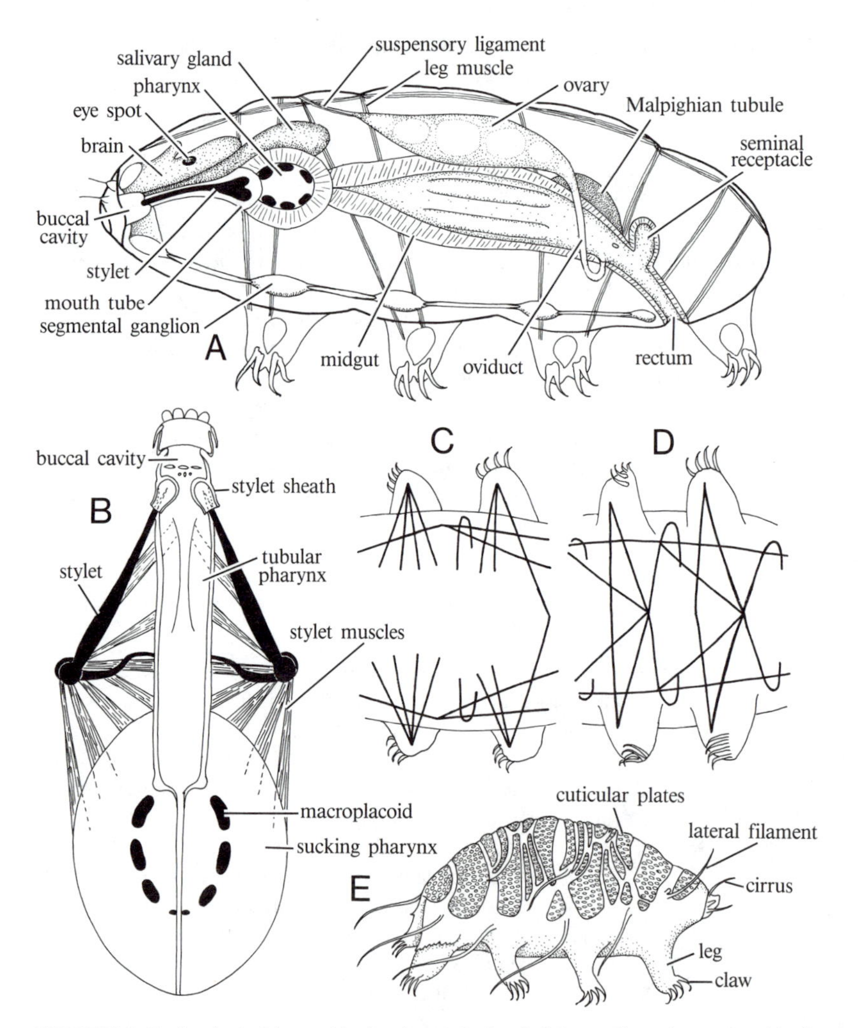

FIGURE 27.2. Tardigrada. **A**. Scheme of tardigrade organization. **B**. Scheme of buccal apparatus of tardigrade, with stylets withdrawn. **C,D**. Dorsal and ventral views of musculature of middle somites of tardigrade. Anterior end is left. Longitudinal muscle strands arch body, and strands extend at various angles into leg to give control over direction of leg movements. **E**. Tardigrade with metameric dorsal plates, *Echiniscus scrofa*. (**A** composite; **B,C,D** after Pennak; **E** after Richter, from Kükenthal and Krumbach.)

for the storage and transport of food. Undoubtedly the perivisceral fluid serves for internal transport of substances, but the small size of tardigrades makes internal transport a relatively minor problem.

Nervous System The nervous system has a complex dorsal brain composed of three median lobes and a pair of large lateral lobes (Fig. 27.2A). Connectives attach the brain to the double ventral nerve cords, along which median segmental ganglia occur.

Reproductive System A simple **testis** or **ovary** lies in the midline above the intestine (Fig. 27.2A). In males a pair of **sperm ducts** pass around the **rectum** and open through a median gonopore on the ventral surface, just in front of the anus. The single oviduct of eutardigrades leads to a similarly placed female gonopore, but in the heterotardigrades the oviduct opens into the rectum. In these tardigrades, a diverticulum of the rectum serves as a **seminal receptacle**.

BIOLOGY

Locomotion Tardigrades are clumsy animals. Even the aquatic tardigrades cannot swim, but creep awkwardly over plants or on the bottom, using the claws to cling to surfaces. The muscles are thin strands made up of only one or a few cells. Despite the simplicity of the muscular system, it shows some important features that are evocative of arthropods. The muscle fibers of tardigrades are attached to the exoskeleton and have specific origins, insertions, and actions. The wormlike invertebrates, on the other hand, generally depend on layers of muscles in the body wall to achieve locomotion. With the musculature of the tardigrade body wall dispersed to form distinct slips passing from one part of the exoskeleton to another, the body wall itself is greatly simplified. The tardigrade body is arched by the contraction of the longitudinal muscles. Five to six muscle fibers from the dorsal or ventral exoskeletal surfaces enter the legs from various angles, allowing five or six directions of movement (Fig. 27.2C,D). Tardigrade movements are somewhat clumsy because of the severe limitations imposed by the few muscle strands running into each leg, but the control of movement requires an independent stimulus for each muscle strand in each leg. Therefore, the nervous system is surprisingly complex for such a small animal.

Molting Animals encased in a confining, armorlike, cuticular exoskeleton find it difficult to grow. Soft tissues increase and crowd into the space constrained by the unyielding exoskeletal casing. The solution to the problem, independently evolved in several phyla, is that the cuticular covering is periodically shed and a new, larger covering is secreted. Little is known of the molting physiology of tardigrades. They prepare for a molt by shedding the hard parts of the buccal apparatus and growing new ones, and by secreting new claws for the legs. When ready to molt, the soft tissues draw away from the old cuticle, which then splits at the anterior end. The animal squirms out of its old exoskeleton and develops a new one very quickly. Most of the tardigrades molt from four to six times, becoming sexually mature after the second or third molt. Molting also is evoked by unfavorable conditions and, in some of the aquatic species, is associated with breeding.

Feeding and Digestion Tardigrades take no solid food. Many pierce plant cells and suck in the protoplasm, using a complex feeding apparatus with sharp stylets. A few tardigrades attack small animals with the stylets, and some may be true carnivores, but the majority appear to be herbivores sucking plant juices.

Little is known of digestive physiology, and the precise functions of the glands associated with the digestive tract have not been determined. Unicellular glands are found around the mouth. A pair of stylet glands open into the buccal cavity. They secrete new stylets before a molt, but whether they have other functions is unknown. Some zoologists believe stylet glands have an excretory function, but they are often termed salivary glands.

Excretion The methods of excretion are not understood. Some of the wastes are deposited in the cuticle and shed at molting. There is evidently some mechanism for controlling the salt and water content of the body, for tardigrades tolerate different salinities remarkably well.

Respiration Tardigrades have no respiratory organs and are quite sensitive to low oxygen tensions. When deprived of oxygen, they enter into a state of asphyxia, becoming motionless and swollen. They can survive several days in this state, recuperating promptly when given oxygen. They seem to have a very limited ability to carry on anaerobic respiration or to dispose of the end products that are formed during anaerobic oxidation.

Tardigrades live under remarkable conditions. Most of their lives are spent in a desiccated cryptobiotic state (Crowe and Cooper, 1971). In this condition they are shriveled, but are nonetheless slowly metabolizing. As soon as water is available, they swell up to become plump and active. The capacity of tardigrades to resist desiccation by entering a state of suspended activity is shared with bdelloid rotifers. A number of physiological problems must still be solved before the cryptobiotic state can be explained. In this state, tardigrades are quite resistant to low temperatures and to chemicals, they can pass through unfavorable periods of the year, and some may be dispersed by air currents while desiccated. Some of the aquatic tardigrades can enter cryptobiosis when desiccated, but others secrete a thick-walled cyst inside the cuticle.

Sense Organs Heterotardigrades have sensory cilia-bearing cells enveloped by a series of sheaths including a cell with microvilli as well as trichogen and tormogen cells. This arrangement is like that seen in arthropods (Kristensen, 1981). Cephalic cirri and leg spines appear to be contact chemo- and mechanoreceptors.

Reproduction As a rule, females of aquatic species molt just before depositing eggs. Males inject sperm through the anus or gonopore of the old cuticle before the female emerges, and the eggs are fertilized in the space between the old cuticle and the body surface of the female. The shed female cuticle serves as an egg case, within which development occurs.

Semiaquatic species must be ready to become cryptobiotic at any time, and do not usually combine breeding and molting. Males inject sperm into the anus or the female gonopore, and the eggs are fertilized internally. Some species produce a single egg and others as many as 30, but most species produce from two to six. Eggs not retained in the old cuticle are usually stuck to foreign objects by adhesive secretions.

Males are relatively scarce, and it is not improbable that many of the ova develop parthenogenetically. Some species produce two kinds of ova, one with thin shells and one with thick shells, rather like the summer and winter eggs of rotifers. There is more rapid development of thin-shelled eggs. The factors that determine whether thin- or thick-shelled eggs are produced, and the role of the two kinds of eggs in the tardigrade life cycle as a whole, are not presently clear. While some tardigrades produce eggs with a smooth covering, many species produce ova with prominent ornamentations. As a great many species have been named on the basis of egg characteristics alone, without demonstration of significant differences in the adults, it is very difficult to estimate how many species exist. However, since tardigrades have not been studied intensively, it is probable that only a small part of the total number of species are actually known.

Development The paucity of recent work on tardigrade development is unfortunate, since cleavage patterns and coelom formation do not follow the patterns so often described in other phyla (Nelson, 1982a) and includes the origin of the mesoderm from outpocketings of the gut.

PHYLOGENY

Despite their many peculiarities, few doubt that tardigrades are closely linked to the annelid–arthropod stem. However, because of those peculiarities, few would venture to be very specific about what position they have on that stem. The absence of evident homologues of antennae, and the lack of true appendages associated with the head or specialized as mouthparts, might argue against including them among the arthropods *sensu stricto*.

TAXONOMY

Three orders are recognized (Renaud-Morrant and Pollock, 1971), of which one, the mesotardigrades, is monospecific.

Order Heterotardigrada. Cephalic appendages and lateral cephalic cirri; no Malpighian tubules; gonopore preanal; legs digitate. (Fig. 27.2E)

Order Mesotardigrada. No cephalic appendages, lateral cephalic cirri, four peribuccal papillae; Malpighian tubules; legs terminating in simple claws.

Order Eutardigrada. No cephalic appendages or cirri; Malpighian tubules; cloaca; legs terminating in two double claws. (Fig. 27.2A).

Onychophora

About 75 species of onychophorans are alive today. They are an ancient group, as a Carboniferous fossil species indicates. They were probably widespread at one time and now have a relict distribution, for they occur sporadically and discontinuously in tropical to subtropical regions and in wet temperate regions of the southern hemisphere. They are specialized for life in leaf litter and plant debris and are ideally adapted for squeezing through narrow spaces and cavities.

DEFINITION

Onychophora have a wormlike body with homonomous segmentation. A pair of clawed, lobopodial limbs is found on each trunk segment. The cuticle is chitinous, very thin, and flexible. The head possesses antennae, jaws, and slim papillae. Nephridia are segmental. The coelom is reduced to gonadal cavities and nephridial sacs, with the principal body cavity being a hemocoel. Trachea are present.

BODY PLAN

The onychophoran head is composed of three limb-bearing somites. The first pair of appendages are the **antennae** (Fig. 27.3). They are mobile but not retractile, and each has an eye at its base. Small **peribuccal lobes** surround the mouth and hide the **jaws**. The jaws arise from small papillae on the embryonic second head somite. Flanking the mouth are the **oral papillae**, poorly developed, specialized appendages located on the third head somite. The head structure is unlike that of any arthropod, but nevertheless has an arthropod character for arthropod cephalization is advanced by the incorporation of somites with specialized appendages into the head.

The only real evidence of segmentation externally is provided by the paired appendages. The thin, flexible cuticle is not formed into external rings, like annelids, nor is it strong enough to form segmental plates, as in arthropods. Rather, the body surface is studded with minute scaled tubercles, some of which are equipped with sensory bristles. There are from 14 to 43 pairs of stumpy legs, each ending in two claws and a ventral pad. At the base of each leg are small **nephridiopores**, and nearby are prominences with the pores for the **crural glands** at their apex. The **anus** is posterior and ventral, and the **gonopore** lies directly in front of it.

Digestive System The chitin-lined foregut is divided into a dilated **pharynx** and a narrow **esophagus**. The somewhat dilated **midgut** fills much of the perivisceral cavity. The very short **hindgut** opens at the anus (Fig. 27.4A).

Respiratory System Many small openings, known as **spiracles**, occur on the body surface. Each leads to a short tracheal pit, or **atrium**, that penetrates the epidermis and ends in the muscular layer (Fig. 27.4D). It is lined with a delicate cuticle, and from it a tuft of tracheal tubes arises. The **tracheae** generally do not branch, but go directly to the organ they serve. In some species, spiral chitinous strands support the walls of the tracheae, as in insects.

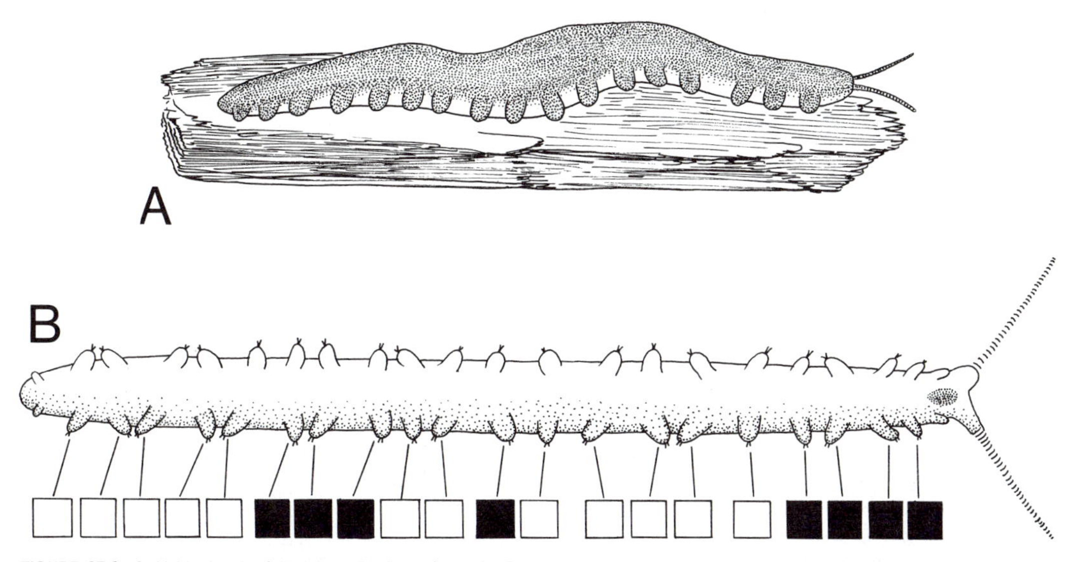

FIGURE 27.3. **A**. Habit sketch of *Peripatus*. **B**. View of moving *Peripatopsis* from below. In somites corresponding to *blackened squares*, paired legs move together, whereas in remaining segments, paired legs move in opposite directions. (**A** after Sedgewick; **B** after R. F. Lawrence.)

Excretory System Each somite except the first two has a pair of metanephridia, but the first pair serve as salivary glands (Fig. 27.4B,C). The coelom of Onychophora has been reduced and, as in the leeches, each nephridium opens into a saclike coelomic vestige (Fig. 27.4C). A ciliated tubule opens into this space and is followed by a coiled canal. The nephridium ends in an ectodermal bladder and nephridiopore.

Circulatory System The body wall is composed of layers of cuticle, epidermis, dermis, and circular, diagonal, and longitudinal muscle. The space between the body wall and the gut is partitioned to form the various hemocoel sinuses (Fig. 27.5C). A dorsal diaphragm separates the **pericardial sinus** in which the heart lies from the much larger **perivisceral sinus** in which the gut and other organs lie. A ventral diaphragm partially separates a ventral sinus from the perivisceral sinus. Muscles associated with the appendages cut off the lateral sinuses. None of the partitions are complete, and blood flows freely from one compartment of the **hemocoel** to another.

The **heart** is a contractile tube, open to the pericardial cavity by segmentally arranged openings called **ostia**. Heart pulsations force the blood forward and down into the perivisceral sinus. Body movements as well as heart movements help to cir-

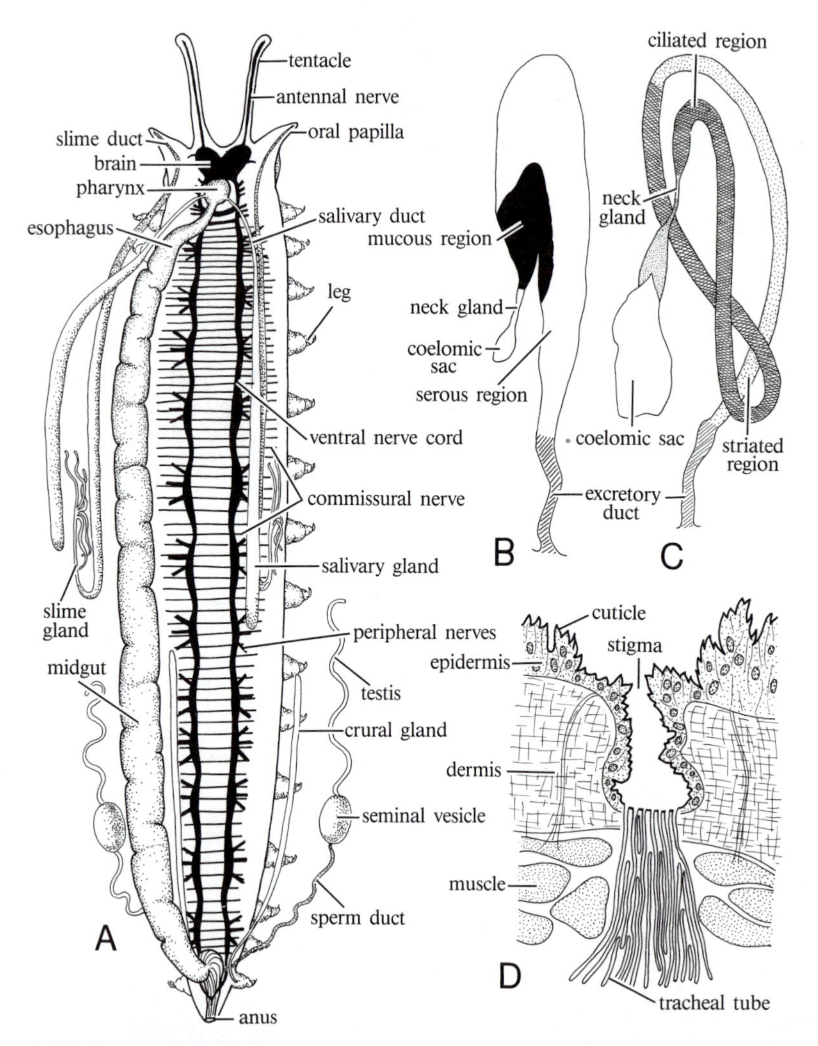

FIGURE 27.4. **A**. Scheme of internal organization of onychophoran. **B**,**C**. Scheme of slime gland (**B**), compared with that of nephridium from anterior body somites. **D**. Section through stigma, showing origin of tracheal tubes. Note that tubules occur in tufts and do not branch. (**A** composite; **B**,**C** after Gabe; **D** after Pflugpfelder.)

culate the blood. The heart, the ostia, and the system of hemocoelic sinuses closely resemble the similar parts of arthropods, although the system is less highly organized than in the majority of arthropod groups.

Nervous System The nervous system is well centralized (Fig. 27.4A). The dorsal brain is connected to the paired **ventral nerve cords** by **circumenteric connectives. Transverse commissures** connect the two nerve cords, but ganglionic enlargements of the nerve cords are inconspicuous. The jaws are innervated from the circumenteric connectives, and nerves to the digestive tube arise from the posterior end of the brain. Lateral nerves pass to the body wall and leg muscles.

Reproductive System Female onychophorans are somewhat larger than males and in some species have more somites. The female reproductive system consists of an irregular mass of peritoneal tissue, the **ovary**, served by a pair of recurved **oviducts**, derived from the coelom and perhaps to be considered highly modified nephridia (Fig. 27.6A). Near each ovary the **gonoduct** is expanded as a **seminal receptacle**.

The male system is also complex (Fig. 27.6B) and has a similar origin. The generally tubular **testes** open into a **seminal vesicle**, the first, dilated part of the sperm ducts. The right and left sperm ducts unite to form a common **sperm duct**, which may contain a compartment where **spermatophores** are formed.

BIOLOGY

Locomotion Onychophora live in inconspicuous places, under leaves, beneath logs, or along the margins of bodies of water. They creep about, with the body undulating as they walk. The legs swing back and forth with each step, coordinated with the regularity of such animals as millipedes or centipedes (Fig. 27.3B). Their intersegmental reflexes are highly organized. The extreme flexibility of their cuticle allows them to squeeze through cracks, crevices, and all manner of tight spaces that other animals would find to be barriers (Manton, 1977).

Feeding They feed on animal and plant material, with animal food probably predominating in most species. They capture food and defend themselves by ejecting a stream of nontoxic, sticky material from the oral papillae. They are probably the champion spitters of the animal kingdom, for they can project their secretions distances of up to 0.5 m. Strong contractions of the body wall propel the slime, formed by a pair of large adhesive glands in the lateral hemocoel spaces (Fig. 27.4A). The soft prebuccal lobes are pressed against the food. The jaws tear at it while the pharynx sucks in particles and juices. The space between the prebuccal lobes and the jaws is the vestibule or prebuccal cavity, comparable to the preoral cavity of insects and other arthropods. The true mouth is bounded by the jaws. The dorsal jaw is moistened by secretions from a pair of salivary glands (Manton, 1937).

Excretion Little is known of excretory physiology, but what is known indicates that they predominantly produce uric acid.

Respiration Onychophorans are adapted to air breathing, with a tracheal system resembling that of terrestrial arthropods. The tracheae are adequate for respiratory exchange, but cause one of the important physiological weaknesses of the Onychophora.

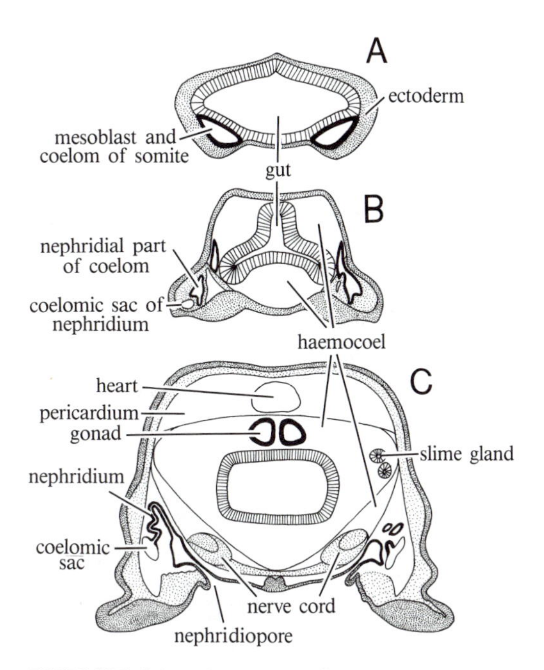

FIGURE 27.5. Schematic transverse sections, showing changes in coelom during development of onycophoran. **A**. Early stage of development, with small paired coelomic compartments on each side of primitive streak, between ectoderm (*stippled*) and endoderm (*hatched*). **B**. During later development, ectoderm and endoderm separate, leaving large hemocoelic spaces, and coelom divides into dorsal and ventral parts. **C**. At birth, hemocoel subdivided, with ventral part of coelom developed into nephridia, and dorsal as gonads. (After Sedgewick.)

Well-adapted tracheate arthropods can open and close the spiracles and so minimize the water loss that accompanies respiratory exchange. Onychophora lack this capacity. Although their cuticle is more impermeable to water than the cuticle of earthworms, onychophorans lose twice as much water through the body surface; for example, onychophorans lose 80 times as much as cockroaches. This deficiency has been an important factor in restricting onychophorans to humid habitats. Even there, they are predominantly nocturnal, forced to assume water-conserving habits in lieu of water-conserving structural features.

Sense Organs The principal sense organs are the antennae. They are richly supplied with tactile endings, and have direct, pigment-cup ocelli at their bases. A chitinous lens occupies the opening of the cup. Water is exceedingly critical for onychophorans, and hydroreceptors on both antennae and on the body surface permit orientation to water vapor.

Reproduction Apparently the male may insert his spermatophores in the female gonopore or place them on the body surface, depending on the species, but mating is not a common occurrence and has not been observed directly. The sperm make their way to the seminal receptacle for storage. Except for some oviparous Australian species, the eggs develop in the dilated main part of the oviduct, the **uterus**. The right and left uteri join near the gonopore.

Development Embryonic development follows the arthropod pattern (Manton, 1949; Anderson, 1973). The rather large ova cleave superficially, and a slit-shaped blastopore develops, below which the endoderm cells lie (Fig. 27.6C,D). Mesodermal bands develop on each side of the blastopore to form a prom-

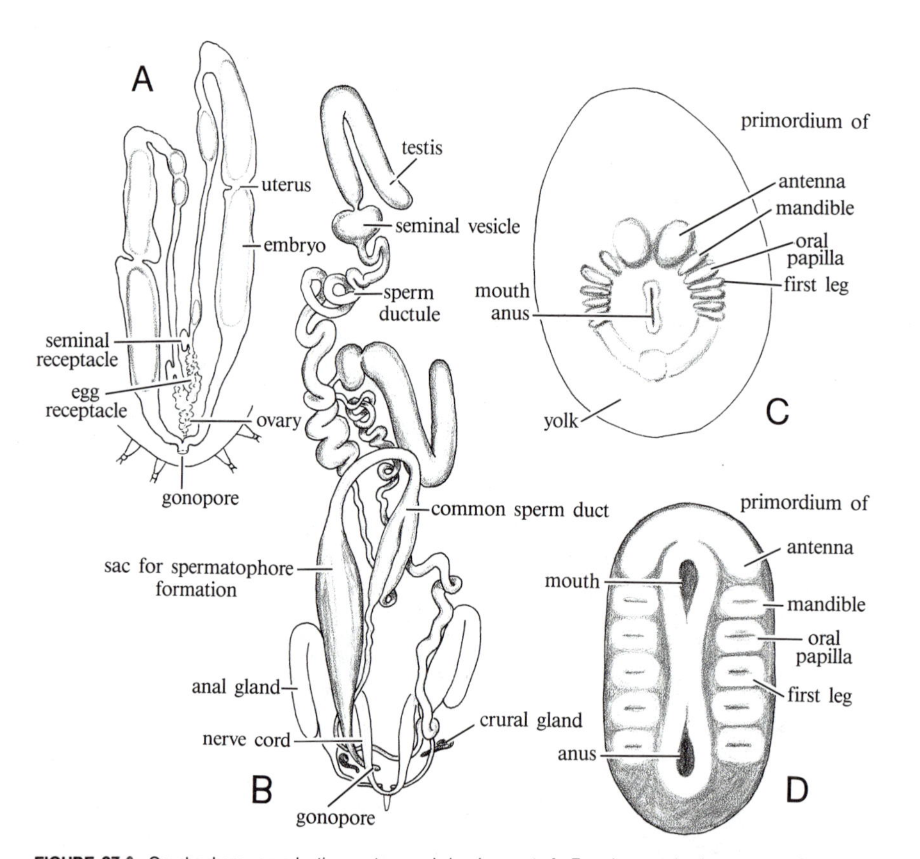

FIGURE 27.6. Onychophora reproductive system and development. **A**. Female reproductive system of *Eoperipatus*. **B**. Male reproductive system of *Peripatoides*. **C**. Young embryo of *Peripatopsis*. Blastopore elongating and somites developing from mesodermal bands on each side. **D**. Somewhat later stage of *Peripatopsis*, showing only germinal region. Blastopore elongated, and mouth and anus distinct. (**A** after Gravier and Fage; **B**,**C** after Bouvier; **D** after Sedgewick, from Zacher, in Kükenthal and Krumbach.)

inent primitive streak, from which the mesodermal somites develop. The two most anterior somites unite in front of the mouth. The blastopore elongates, and the anterior part forms the mouth while the posterior part forms the anus. Ectodermal cells turn in at the mouth and anus to form the stomodeum and proctodeum. A conspicuous coelom is formed during embryonic development, for each mesodermal somite contains a coelomic compartment. These divide into dorsal and ventral parts (Fig. 27.5A,B,C). The ventral compartments extend out into the base of the appendage, giving rise to the coelomic sacs associated with the nephridia. The dorsal compartments migrate up to the middorsal line, the anterior ones will eventually disappear, and the posterior ones contribute to the formation of the gonad.

PHYLOGENY

Onychophorans have been the subject of controversy as to their phylogenetic position. Traditionally they have been viewed as a separate phylum transitional between annelids and arthropods. However, Anderson (1973) and Manton (1977) argued that onychophorans were specialized uniramians, derived from the same stock as the myriapods and insects, as evidenced by their early ontogeny and locomotory functional morphology.

TAXONOMY

Only two families of onychophorans are recognized: the Peripatidae, with a circum-tropical-subtropical distribution, and the Peripatopsidae, with a circum-south-temperate distribution.

Selected Readings

Abele, L. G., W. Kim, and B. E. Felgenhauer. 1989. Molecular evidence for inclusion of the phylum Pentastomida in the Crustacea. *Mol. Biol. Evol.* 6:685–691.

Anderson, D. T. 1973. *Embryology and Phylogeny in Annelids and Arthropods.* Pergamon Press, Oxford.

Crowe, J. H., and A. F. Cooper. 1971. Cryptobiosis. *Sci. Am.* 225(12):30–36.

Esslinger, J. H. 1962. Morphology of the egg and larva of *Porocephalus crotalis. J. Parasitol.* 48:457–462.

Kristensen, R. M. 1981. Sense organs of two marine arthrotardigrades. *Acta Zool.* 62:27–41.

Manton, S. M. 1937. Studies on the Onychophora II. The feeding, digestion, excretion, and food storage of *Peripatopsis. Phil. Trans. Roy. Soc. Lond.* (B)227:411.64.

Manton, S. M. 1949. Studies on the Onychophora VII. The early embryonic stages of *Peripatopsis* and some general considerations concerning the morphology and phylogeny of Arthropoda. *Phil. Trans. Roy. Soc. Lond.* (B)233:483–580.

Manton, S. M. 1977. *The Arthropoda.* Clarendon Press, Oxford.

Nelson, D. R. 1982a. Developmental biology of the Tardigrada. In *Developmental Biology of Freshwater Invertebrates* (F. W. Harrison and R. R. Cowden, eds.), pp. 363–398. Alan R. Liss, New York.

Nelson, D. R. (ed.). 1982b. *Proceedings of the Third International Symposium on the Tardigrada.* East Tennessee State Univ. Press, Johnson City.

Osche, G. 1963. Die systematische Stellung und Phylogenie des Pentastomida. Embryologische und Vergleichend anatomische Studien an *Reighardia sternae. Z. Morph. Okol. Tiere.* 52:487–596.

Renaud-Morrant, J., and L. W. Pollock. 1971. Review of the systematics and ecology of marine Tardigrada. *Smith. Contr. Zool.* 76:109–117.

Riley, J., A. A. Banaja, and J. L. James. 1978. The phylogenetic relationships of the Pentastomida: the case for their inclusion within the Crustacea. *Int. J. Parasitol.* 8:245–254.

Wingstrand, K. G. 1972. Comparative spermatology of a pentastomid, *Raillietiella hemidactyli*, and a branchiuran crustacean, *Argulus foliaceus*, with a discussion of pentastomid relationships. *Korg. Danske Vidensk. Selsk. Biol. Skrift.* 19(4):1–72, 23 pls.

28

Arthropoda

No other phylum of animals rivals the arthropods in success: the arthropods contain the most speciose group, the insects, and the most variable, the crustaceans, of any other animal phyla. Much of their success can be traced to a remarkable flexibility of body plan. Even some of the smaller groups of arthropods have become adapted to remarkably diversified habitats, evolving fundamentally varied patterns of living, including parasitic habits. A number of the aquatic groups have invaded land independently with varying degrees of success, while a number of the terrestrial groups have successfully adapted to living in the water for all or a part of their lives. Arthropods are everywhere. There are four principal divisions of arthropods (Schram, 1978): the extinct Trilobitomorpha, the Cheliceriformes (chelicerates and pycnogonids), the Crustacea, and the Uniramia (myriapods and insects). Today, the living arthropods make up at least half to three-fourths of all living animal species! This is probably a conservative estimate if we were to consider yet-to-be-described species as well.

Definition

Arhropods are bilaterally symmetrical, segmented animals that, at least primitively, are equipped with jointed appendages attached to each body segment (though these can often be reduced or lost on some of the body segments). Appendages are typically specialized for specific functions. The surface of the body and appendages is covered by a continuous cuticle of complex structure. In most cases, the cuticle forms a series of heavy skeletal plates or rings, connected by thinner, flexible articular membranes; and the cuticle is turned inward at the stomodeum and proctodeum to form chitinous linings of the foregut and hindgut respectively. There is a well-developed cephalon. The paired, segmental coelomic compartments are greatly reduced with the development of a hemocoel. A contractile heart is present, derived from the dorsal blood vessel, and lies in a dorsal pericardial sinus. Blood enters the heart through pairs of apertures known as ostia. The nervous system is developed with double, ventral, nerve trunks, which primitively had segmental ganglia, and centers on a highly differentiated brain. Cilia are rarely found, and all movements of arthropod organisms or their body parts are dependent upon muscles. Eggs are richly supplied with yolk, and at best indicate only traces of spiral cleavage and a mesoderm stem cell. Mesodermal bands, in which coelomic compartments form, arise during early development from teloblasts, and the body typically forms from a germinal disc that serves as the growth center of the embryo.

Body Plan and Biology

Exoskeleton A single glance is usually enough to see that an animal is an arthropod; the jointed exoskeleton covering the body and legs is unique. The cuticle affords protection without causing a loss of mobility, allows appendages to specialize for specific tasks, and affords a foundation for the anchoring of muscles and the development of sense organs. However, there are some disadvantages too. The firm exoskeleton imposes limitations on growth, restricts the passage of materials in or out at the body surface, and makes it difficult to receive stimuli from the environment. Nevertheless, these limitations have been circumvented. The development of molting habits has overcome the restrictions imposed on growth. The reduced capacity for direct surface respiration and excretion has been overcome by the development of effective respiratory and excretory organs. In addition, the presence of a firm outer casing has favored the evolution of rather elaborate sense organs in all of the arthropods.

With the evolution of the exoskeleton, the body wall is profoundly changed (Fig. 28.1C). It contains only the cuticle, the epidermis that secretes it, and a basement membrane, sometimes somewhat amplified by a discontinuous dermis composed of connective tissue. The circular, diagonal, and longitudinal muscle layers found in the body wall of wormlike coelomates have been dispersed to form the specific musculature that manipulates the various joints of the body and legs. The peritoneum does not form an inner layer, for the coelom is reduced and a large hemocoel fills the space within the body wall in its place.

Three distinct layers can be recognized in the cu-

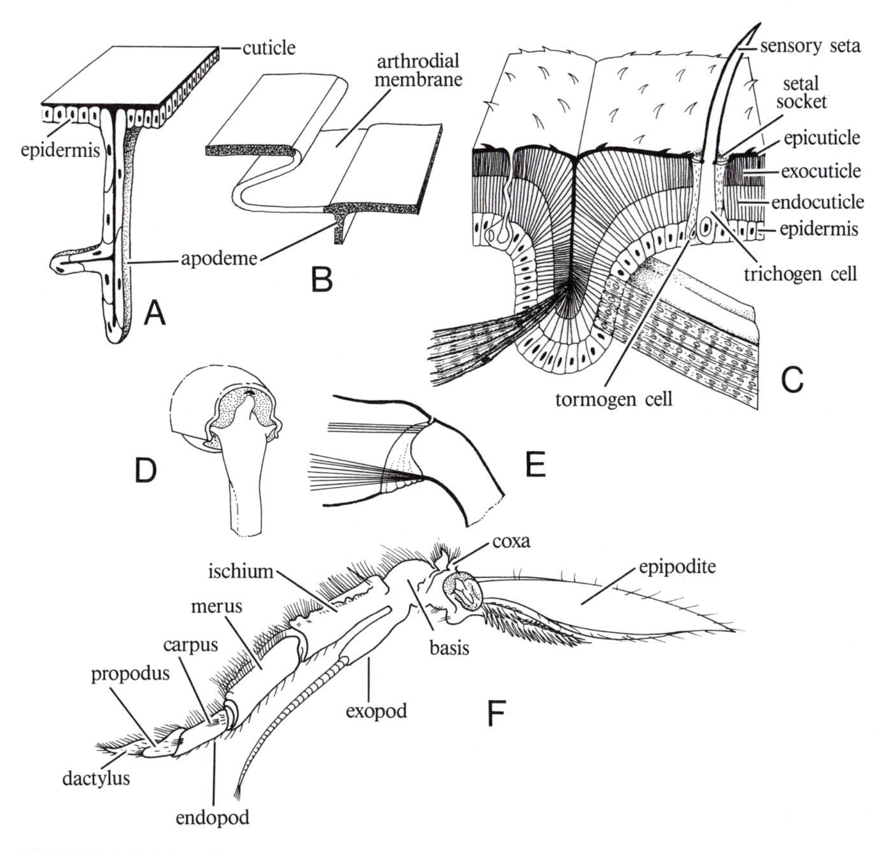

FIGURE 28.1. Cuticle, epidermis, apodemes, and limbs. **A.** Heavy cuticle extends inward to provide apodemes as projections or shelves for muscle attachment. **B.** Surface cuticle often indented between body segments, forming flexible articulating arthrodial membrane, permitting movement. Similar articulating membranes occur at joints of appendages and at base of movable spines or bristles. **C.** Relatively impervious cuticle interrupted at glands and at sense organs. **D.** Dicondylic joints; two condyles fit into corresponding depressions in adjacent segment, permitting flexing of joint. **E.** Monocondylic joints; single condyle located opposite wide, flexible articulating membrane, permitting free movement. **F.** Crustacean limb from thorax of malacostracan. (**D,E** after Snodgrass.)

ticle (Fig. 28.1C). A thin outermost layer of tanned proteins, waxes, and other fatty compounds makes up the **epicuticle**. Beneath it are two layers, an outer **exocuticle**, often pigmented and impregnated with the substances found in the epicuticle, and an inner **endocuticle**. The exocuticle is thickened to form skeletal plates or sclerites. The endocuticle is a continuous layer, sometimes thinner in some areas, and particularly important in the flexible membranes between the skeletal plates (Fig. 28.1B). Both these layers are laminated and contain **chitin**, an aminopolysaccharide. In some groups, especially among the Crustacea, the skeleton is impregnated with calcium salts. The thickened sclerites need not be richer in chitin than the other parts of the cuticle. Insect sclerites, for example, contain more carbohydrate and less chitin than the rest of the cuticle. The waxy compounds of the epicuticle and exocuticle are very important, for they reduce water loss in terrestrial forms. If the waxes are removed, water loss increases rapidly. It also increases at temperatures above the melting point of the exoskeletal waxes. Because of a symmetry of water movement, a sheet of tanned gelatin that is waxed on one side will absorb water from the atmosphere. This principle explains the way some insects and arachnids use layers of wax and tanned protein in the exoskeleton to absorb water from the air. It also explains the mechanism by which tick egg cases are prevented from drying. A waxy compound is secreted and used to treat the surface of the egg when it is laid. Without the waxy layer, the eggs quickly shrivel and the embryos die.

The capacity of the epidermis to secrete cuticle is important to functions other than protection. Wherever the epidermis is folded inward during development, it carries with it the capacity for cuticle formation. The epidermal lining of the proctodeum and stomodeum secretes the cuticular lining of the foregut and hindgut. The tracheal system of the terrestrial mandibulates is lined with an epidermis that secretes a lining cuticle, thickened at various points for support. Spiral cuticular ridges are found in the walls of the trachea, preventing their collapse. At the point where two skeletal plates meet, and especially in the vicinity of flexible joints, the epidermis is invaginated. Here it secretes internal ridges, plates, or spikes of cuticle, known as **apodemes** (Fig. 28.1A,B,C), forming an **endoskeleton** of varying degrees of complexity to which muscles attach.

Mobility of the body and its parts is maintained by specializations of the exoskeleton at the points where two sclerites or cuticular rings covering adjacent body segments, or somites, and leg podites meet. Immovable sutures occur in the head region and at other points. Here the epidermis often invaginates during development and forms apodemes. Movable sutures in the body wall and limbs usually involve two adaptations. The sclerites overlap, thus providing protection, and the cuticle between the sclerites is folded under (Fig. 28.1B). The folded-in cuticle, the arthrodial membrane, is thin and flexible, allowing a good deal of movement. At joints of the appendages, a somewhat different problem is encountered. If the joint were a simple infolding of more delicate, flexible cuticle, the weak cuticle would necessarily bear the full weight of the body or the full thrust of the leg. The joint is strengthened by gliding surfaces or condyles built of the hardened exoskeletal material. Opposing the condyle or surrounding it, the more flexible articular membrane connects the remaining surfaces of the two skeletal parts. This is especially important in legs. There may be either one or two condyles at the joints (Fig. 28.1D,E). The form and positioning of the condyles and the amount of articular membrane on the various sides of the joint determine the freedom and direction of movement.

Musculature There are a large number of joints in the body and its appendages, and each joint calls for its set of specialized muscle fibers. The body musculature, in particular, can become very complex (Fig. 28.2B). Most arthropods are small animals, and a muscle contains relatively few individual muscle fibers. The nervous system contains a limited number of neurons. In arthropods, a single axon may innervate several muscle fibers, and a single muscle fiber may receive stimuli from several different kinds of axons. In at least some cases, the same motor end plate serves more than one kind of axon (Fig. 28.2A). As a result, a single arthropod muscle fiber may contract more or less rapidly and more or less powerfully, depending on the kind of stimulus it receives.

The primitive arrangement seems to have consisted of a fast and a slow axon to each muscle fiber, but triple and even quintuple innervations of fibers have been described. A large axon delivers a stimulus that evokes a rapid, high-tension contraction, nearly as powerful as tetanic contraction. An intermediate axon evokes a slower response that builds up rapidly on repetition and follows a different fatigue pattern. A third, and usually the smallest, axon suppresses contraction, but does not necessarily prevent facilitation. As a result, an extra impulse from other fibers can evoke a sudden large response when the small axon is operating. The inhibitory stimuli have rather more effect on the slow than on the fast contractions. The large and the intermediate axons do not evoke contractions independently. The two kinds of axons cooperate. A series of slow impulses augmented by a single fast stimulus can sustain a high-tension contraction, for example, permitting a holding response.

It is not uncommon for arthropods to have some highly specialized muscles with unusual structural or physiological qualities. Muscles used in flight and in sound production are good examples. The physiological problems associated with flight muscles are very interesting. Generally speaking, a muscle responds to a single stimulus by a single contraction. When stimuli are delivered repetitively, a point is reached when the individual contractions fuse into a sustained contraction. As a general rule, the faster a muscle fiber contracts, the higher the rate of stimulus repetition it can tolerate before contractions become fused. Many insects must move their wings very rapidly in flight, and the problem centers around developing a set of muscles that are capable of making a rapid series of discrete contractions instead of going into a sustained contraction. Some insects have very fast muscle fibers that can make a distinct contraction to a very rapid set of nerve stimuli. They have flight muscles that contract with each wing movement and receive a stimulus for each contraction. Wing movements up to 20 to 40 per second can be maintained by this direct system. However, some insect flight, and all kinds of sound production, involves muscle contractions at frequencies far higher than this, frequencies at which only fused contractions are possible when separate stimuli are delivered to the muscle. It is clear that in this case the fiber must contract several times for each nerve stimulus, a conclusion that has been experimentally confirmed. Multiple contractions are achieved by opposing skeletal and muscular arrangements that stretch the muscle by recoil after the muscle has exerted its pull. The muscle contracts a second time, not because it has received a new nerve impulse, but because it has been stretched by the opposing recoil mechanism. In systems of this kind, the frequency of muscle contraction, and of wing movements, is determined by the nature of the recoil system, and nerves are used only to keep the muscle in an active state. In some other cases, repeated oscillatory contractions result from a single stimulus.

Jointed Limbs The cuticular exoskeleton and elaborate muscle specializations combine in *the* definitive arthropod feature, from which they get their name Arthropoda, "jointed limbs." Limbs exhibit an incredible array of variations and specializations, and this undoubtedly is linked, in large part, with the phylum's success (Manton, 1977). Limbs may be uniramous, as they are in uniramians and for the most part cheliceriforms, or multiramous, as is often the case in crustaceans. In the latter, the legs are composed of a basal piece, the protopod (often composed of two segments), and two branches, an outer exopod and an inner endopod. A dorsally placed epipodite, often bearing gill filaments, may also be present, attached to the protopod. Unfortunately, the segments of the appendages have been given two sets of names, one that is most commonly employed for the crustacean leg (Fig. 28.1F) and one that is used for the legs of spiders, insects, and other terres-

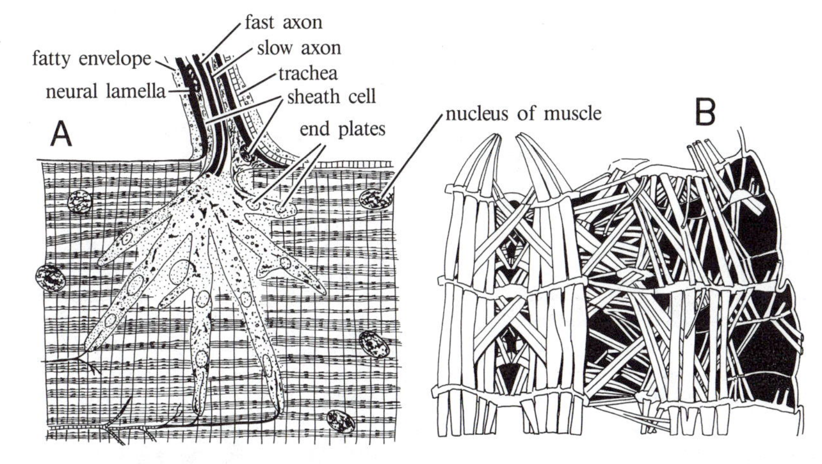

FIGURE 28.2. Arthropod muscles. **A.** End plate of muscle in locust leg, served by both fast and slow axons. Specific activity of muscle results from combined stimuli delivered by two kinds of neurons. **B.** Ventral muscles and muscles of right half of mesothorax and metathorax of caterpillar. Example of muscular complexity. (**A** after Hoyle, in Scheer; **B** after Snodgrass.)

trial arthropods (see, e.g., Fig. 29.6). Attempts have been made to homologize the segments, but the literature on arthropods does not always agree on these identifications.

Brain As part of the characteristic cephalization of the anterior region of the arthropod body, the brain becomes highly differentiated into distinct regions (Gupta, 1987). There is a clear continuity with what is seen in Annelida. The archicerebrum of polychaetes contains centers associated with the eyes and prostomial palps, and the first nerves to a body appendage arise from the first ganglion on the nerve cord (Fig. 28.3A). In uniramians and crustaceans (Fig. 28.3B), the archicerebrum is somewhat more complex, having divided into a **protocerebrum** associated with the optic nerves, and a **deutocerebrum** associated with the antennal nerves. A third part of the brain, the **tritocerebrum**, is associated with the nerves to the second antennae of Crustacea. Cheliceriforms (Fig. 28.3C) have no deutocerebrum. Apparently the parts that correspond to the first antennae are missing, and the portion of the brain associated with these appendages has been reduced. The chelicerae are innervated from the tritocerebrum, and so appear to correspond to the second antennae of crustaceans.

Molting Thinner, more flexible parts of the cuticle are folded in at each joint. These folds can be extended somewhat by the growth of tissues beneath the exoskeleton, but at most this permits only a very moderate amount of growth. Only by shedding the exoskeleton can the epidermis be relieved of external confinement so that the growing tissues can expand. After the remnants of the old cuticle are shed, an underlying new exoskeleton is hardened, larger than the last and providing some room for further growth. The young arthropod grows to fit its new case, as a small child grows to fill the clothes that were bought deliberately a little too large. The shedding of the cuticle is known as **molting** or **ecdysis**. The periods between molts are known as **instars**.

Embryonic growth of arthropods, as of other animals, is a precisely controlled affair, following a predetermined course. The more gradual change of the juvenile organism into an adult is no less controlled. In arthropods, this phase of growth is complicated by the problem of molting, but tends, nonetheless, to become organized by the development of a pattern of definite molts and instars, with each instar representing a definite stage with a characteristic form and size. Irregularities may result from unusual environmental conditions, but within the extremes normally encountered by the developing young, the number of molts that occur between hatching and the attainment of adult form only varies slightly within a species and sometimes is relatively constant for all members of a family or order. In this case, each instar may be considered as a stage of development, and the gradual development of the adult form can be followed from stage to stage. Among the common changes that occur in the form of young organisms at molts is the addition of body segments or pairs of appendages. In some cases, a remarkable transformation of the body structure oc-

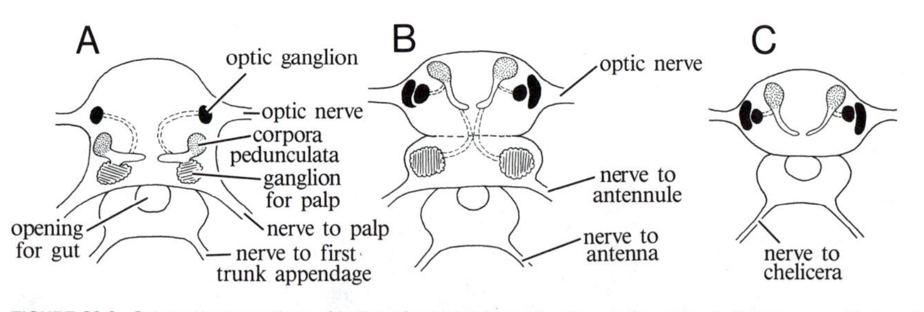

FIGURE 28.3. Schematic comparison of brains of polychaetes and various arthropods. **A.** Polychaete, with brain located in prostomium, and first trunk appendages innervated from subesophageal ganglion. Brain gives rise to optic nerve, associated with optic ganglion (*black*) and corpora pedunculata (*stippled*), and to palp nerve, associated with palp ganglion (*striated*). **B.** Crustacea and Uniramia have brain composed of protocerebrum, deutocerebrum, and tritocerebrum. Protocerebrum contains corpora pedunculata and optic centers, and gives off optic nerve. Deutocerebrum in Crustacea contains nuclei associated with antennules; in Insecta it is associated with antennae. Antennae of Uniramia correspond to antennules of Crustacea. It seems clear that protocerebrum and deutocerebrum correspond to polychaete brain. In Insecta tritocerebrum is small, whereas in Crustacea tritocerebrum is large and associated with crustacean antennae. Tritocerebrum derived from subesophageal ganglion of polychaete-like ancestors. **C.** Cheliceriformes have protocerebrum with optic nerves. No appendage corresponds to crustacean antennules and insectan antennae, and deutocerebrum is wholly absent. Tritocerebrum corresponds to tritocerebrum of other arthropods, and gives off nerves to chelicerae, which must, therefore, correspond to antennae of crustaceans and reduced somite in insect head. (After Hanström).

curs during a single metamorphic molt. A good example of this latter is the molt at the end of the pupa stage of some insects, in which a wingless larva is changed to a winged adult.

Molts are recurrent crises in the lives of arthropods. Until the new skeleton hardens, the protective case is weak, and without the hardened skeletal plates, movement is often restricted. Many arthropods have special behavior patterns that tend to minimize the dangers of the molting period. They may seek a secluded spot for molting, build a protective silken retreat, retire into burrows, or have special mechanisms to speed skeletal hardening. In any event, molting remains an especially dangerous time in the arthropod life cycle.

Ecdysis is far more than a simple splitting of the cuticle and the secretion of a new one. Arthropod life consists of alternating periods of premolt, molt, postmolt, and intermolt. Some arthropods escape into a permanent intermolt when they are adult; this period is known as anecdysis. However, others never achieve stability, molting at intervals throughout adult life.

The physiology of most of the body parts is affected by premolt. Glycogen reserves are built up. Some crustaceans reabsorb minerals from the old exoskeleton and store them, thus hastening skeletal hardening after molt. The epidermis secretes an enzyme that softens the cuticle at its base (Fig. 28.4B). The epidermal cells pull away, stimulating the formation of a new epicuticle (Fig. 28.4C). The epicuticle is impervious to the molting enzyme. Beneath its protective wall, new cuticle develops (Fig. 28.4D). At the end of premolt the old cuticle splits. The soft body within swells and emerges. Various adaptations are used to keep the body at a maximum volume while the new skeleton hardens. Insects inflate the crop with air, and sometimes other body parts are inflated with air or blood. The Crustacea imbibe water. The tissues grow rapidly, and the new cuticle hardens during the short postmolt period. At this time the stored mineral resources of Crustacea are rapidly exhausted, for they are deposited in the hardening cuticle. After postmolt, the animal may enter intermolt (Fig. 28.4A), but full intermolt probably occurs only in adults. At best, the rapidly molting juveniles must have a very short period of stability.

Molting is a complex event from the physiological point of view, and it is clear that physiological controls are required. In insects and decapod crustaceans, molting is under the control of hormonal mechanisms. Presumably, similar kinds of controls operate in other arthropods. For example, neurosecretory cells in the eyestalks of decapod crustaceans occur in the **X-organs** of the medulla terminalis and in the sensory papillae (Fig. 28.5A). These end in the **sinus glands**, made up of the hormone-filled terminations of the neurosecretory cells. A pair of glands in the maxillary or second antennal somites, the **Y-organs**, are controlled by secretions from the X-organs. The Y-organs also secrete hormones that stimulate molting. Removal of the Y-organs prevents molting, and implantation of new Y-organs restores the ability to molt. Y-organ secretions trigger the premolt changes, but once premolt begins the process goes onto completion even though the Y-organs are removed. The X-organ and sinus gland, on the other hand, inhibit molting. When the eyestalks, which contain the X-organs and sinus glands, are removed, molting is accelerated. Implantation of sinus glands into animals whose eyestalks have been removed retards molting. The sinus gland liberates molt-inhibiting hormones during postmolt and intermolt. Under appropriate internal or external conditions, the sinus gland secretion disappears, and the Y-organ, freed from restraint, secretes the molt-initiating hormone.

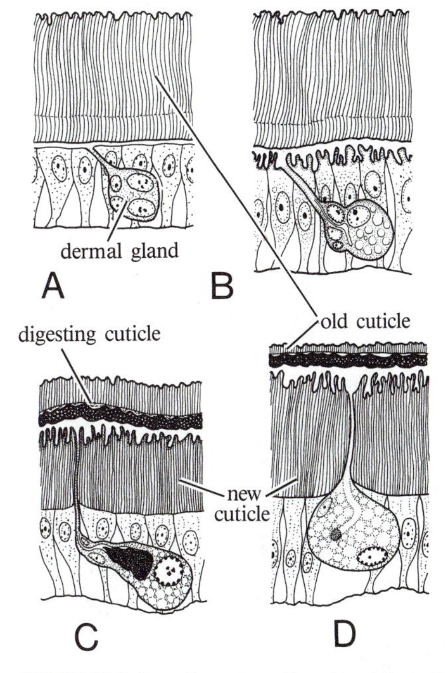

FIGURE 28.4. Four stages in molting of hemipteran insect, *Rhodnius*. **A**. Old cuticle beginning to separate from epidermis, and new dermal glands appearing. **B**. Day 3, dermal gland enlarged and first evidence of digestion of old cuticle seen. **C**. Day 5, dermal cells full of secretion, new cuticle being laid down. **D**. Day 6, old cuticle largely digested, new cuticle almost fully formed, and dermal gland cell very large. (After Wigglesworth.)

In insects, molting and metamorphosis are integrated by an endocrine mechanism different from that seen in decapods. Neurosecretory cells of the brain end in the **corpora cardiaca** (Fig. 28.5B). When a molt-initiating stimulus affects the organism, the corpora cardiaca release the brain hormone. The brain hormone activates the **prothoracic glands** (Fig. 28.5C) to liberate a hormone known as **ecdysone**. Ecdysone evokes an increase in the epidermal mitochondria and endoplasmic reticulum, and an elevation of protein and RNA content, and initiates premolt. A **juvenile hormone** is produced by small glands associated with the stomodeal nervous system, the **corpora allata** (Fig. 28.5B). The juvenile hormone prevents metamorphic molts. When corpora allata are surgically introduced into immature insects, metamorphosis is delayed, while experimental removal of the corpora allata results in the development of adultlike insects at early instars. Insects that are deprived of the brain hormone do not molt. However, they will molt under the influence of injected ecdysone. These hormones are not species specific but, like vertebrate hormones, act generally on all members of the group. It is evident that there are other controlling factors, but these are not yet understood.

The sensitivity of body tissues to the hormones changes during differentiation. The different patterns of development seen in various insect orders probably depend, at least in part, on the ways that the molting and juvenile hormones are released. Where the juvenile hormone is produced in large quantities during early instars, or where the tissues are highly sensitive to it, metamorphic changes are largely postponed until a metamorphic molt associated with pupation occurs. Where the amounts of juvenile hormone gradually change during development, or where the sensitivity of the tissue to the ju-

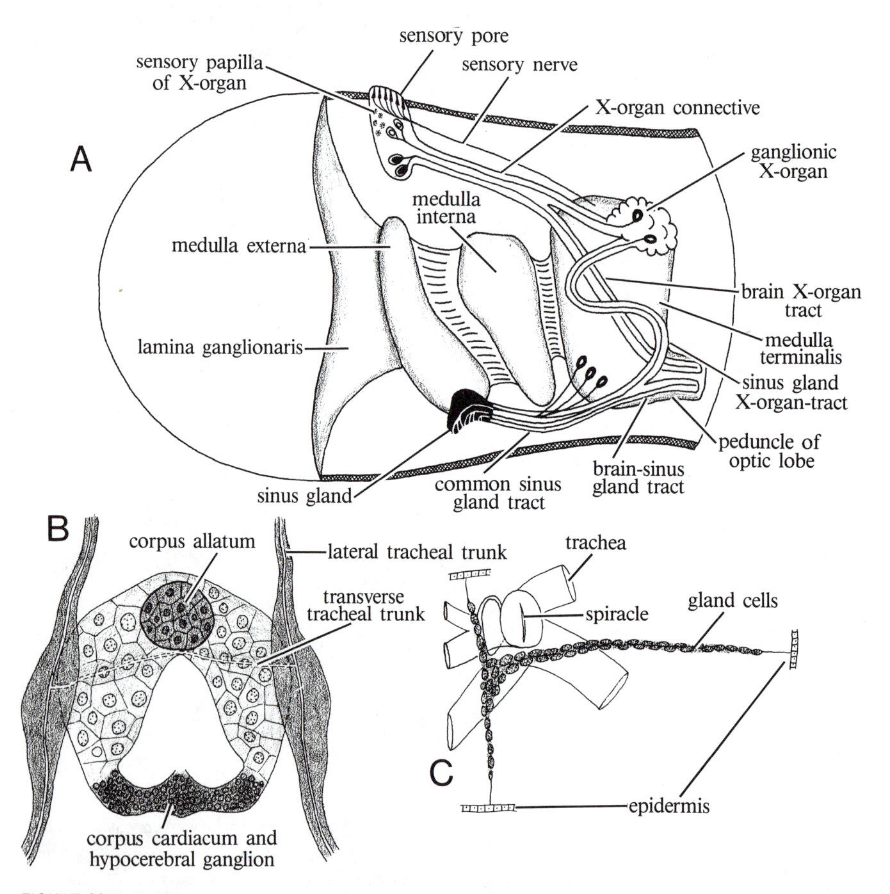

FIGURE 28.5. A. Neurosecretory system of crustacean, *Lysmata.* Sinus gland/X-organ system inhibits molting cycle. **B.** Semidiagrammatic figure of ring gland of mature *Drosophila* larva, showing corpus allatum and corpus cardiacum. **C.** Prothoracic gland of *Prodenia* late caterpillar. In insects, control of molt and metamorphosis linked with corpora allata (which produce juvenile hormone) and corpora cardiaca (which evoke ecdysone secretion from prothoracic glands and so stimulate molting). (**A** after Carlisle and Knowles; **B,C** after Bodenstein.)

venile hormone gradually changes, a gradual metamorphosis takes place.

Development Arthropod development (Anderson, 1973) is highly modified, partly because of the large, yolky eggs. Very few arthropods have small eggs that undergo holoblastic cleavage. In some of these, traces of spiral cleavage have been reported. However, most arthropods produce ova that are rich in yolk and in which cleavage patterns are greatly modified. Insects provide a good example of arthropod development in its more modified form. The zygote nucleus divides repeatedly to produce a number of cleavage cells in the yolk mass (Fig. 28.6A,B). These

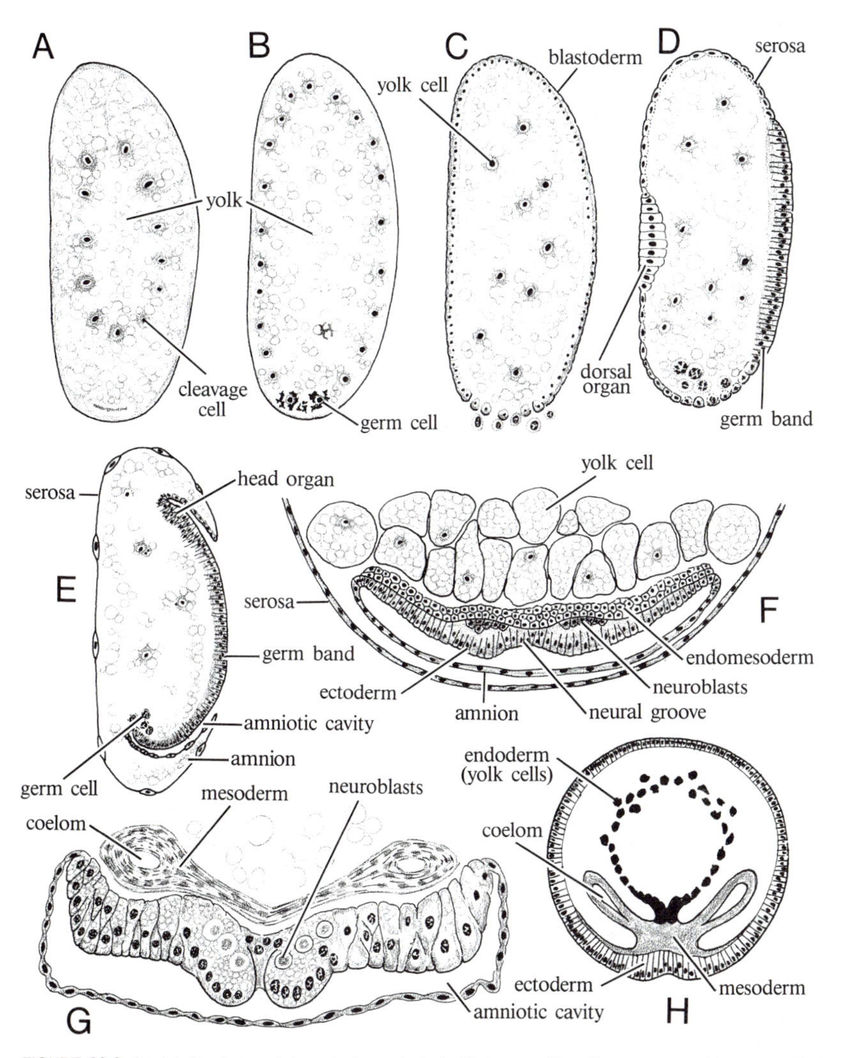

FIGURE 28.6. Insect development. Insects have centrolecithal egg, with yolk concentrated in center. **A.** First cleavage divisions produce cleavage nuclei. **B.** Nuclei migrate to more superficial position. **C.** Germ cells differentiate early, at one pole of developing embryo. Some cleavage nuclei remain in yolk, but most incorporate into continuous blastoderm that surrounds yolk mass. **D.** Blastoderm thickens on dorsal side to form dorsal organ, subsequently absorbed by yolk, and on ventral side to form germ band. **E.** Remaining blastoderm cells flatten out to form thin membrane, serosa. In many insects, head and tail ends of germ band invaginate, and amniotic folds of serosa grow up over germ band, forming amniotic cavity. **F.** When amniotic cavity formation ends, double membrane (outer serosa and inner amnion) covers germ band. Germ band forms outer layer of ectoderm marked by medial neural groove, and inner layer that arises from invagination along midline at point where neural groove eventually appears. **G.** Inner layer produces medial strip of mesoderm and two mesodermal bands in which cavities appear. **H.** Invaginating inner layer represents mesoderm. With blastopore "closure" (invaginating ceases) primary endodermal lining of primitive gut disintegrates. Yolk cells just below mesoderm serve as sources for secondary and permanent endoderm formed later in development. (**A,B,C,D,E,F,G** after Johannsen and Butt; **H** after Snodgrass.)

arrange themselves in an external layer, forming a **blastoderm** (Fig. 28.6C). A few yolk cells are usually left within the yolk mass, and in some cases germ cells can be recognized at a region near the posterior end of the embryo. At this stage, the embryo is a **coeloblastula** with the central cavity completely filled with yolk. The yolk is not divided into separate yolk cells, but is rather an undivided mass of material containing some nuclei (Fig. 28.6C,D).

The delineation of presumptive areas on the surface of the cleaving embryo is distinctly different in each of the major arthropod groups (Fig. 28.7). A clear similarity can be seen in this pattern between clitellate annelids (Fig. 28.7A), onychophorans (Fig. 28.7B), and uniramians, where the presumptive stomodeum lies in front of the presumptive endoderm, which in turn is in front of the presumptive mesoderm. In crustaceans (Fig. 28.7C), the position of the presumptive meso- and endoderm is reversed, and the mesoderm does not arise from the 4d lineage typical of almost all other protostomes, but rather arises from the 3A, 3B, and 3C cell lineages. In cheliceriforms (Fig. 28.7D), the endoderm arises from the yolky interior of the egg, and only the presumptive stomodeum and mesoderm are on the surface of the germinal disc.

Details of germ layer formation vary considerably in different kinds of arthropods. An example of the process is seen in insects. Germ layer formation is preceded by the appearance of a **germ band** on the ventral side of the embryo (Fig. 28.6D). In many insects, the embryo comes to lie in an amniotic cavity, formed by the growth of amniotic folds toward the midline on the ventral side of the egg (Fig. 28.6E). The amniotic folds gradually enclose the embryo while it gastrulates. In such cases, the remaining blastoderm becomes a squamous area immediately within the egg membranes. Gastrulation involves the involution or in-turning of some of the cells of the germ band (Fig. 28.6F). The cells that are turned under are endomesoderm, since they cannot at first be distinguished. The blastopore is never open. It corresponds to the long median part of the germ band where involution occurs.

The mesoderm forms a strip of cells beneath the germ band ectoderm, on each side of the midline. Paired coelomic cavities appear in the mesodermal bands (Fig. 28.6G). Meanwhile, the neural ectoderm

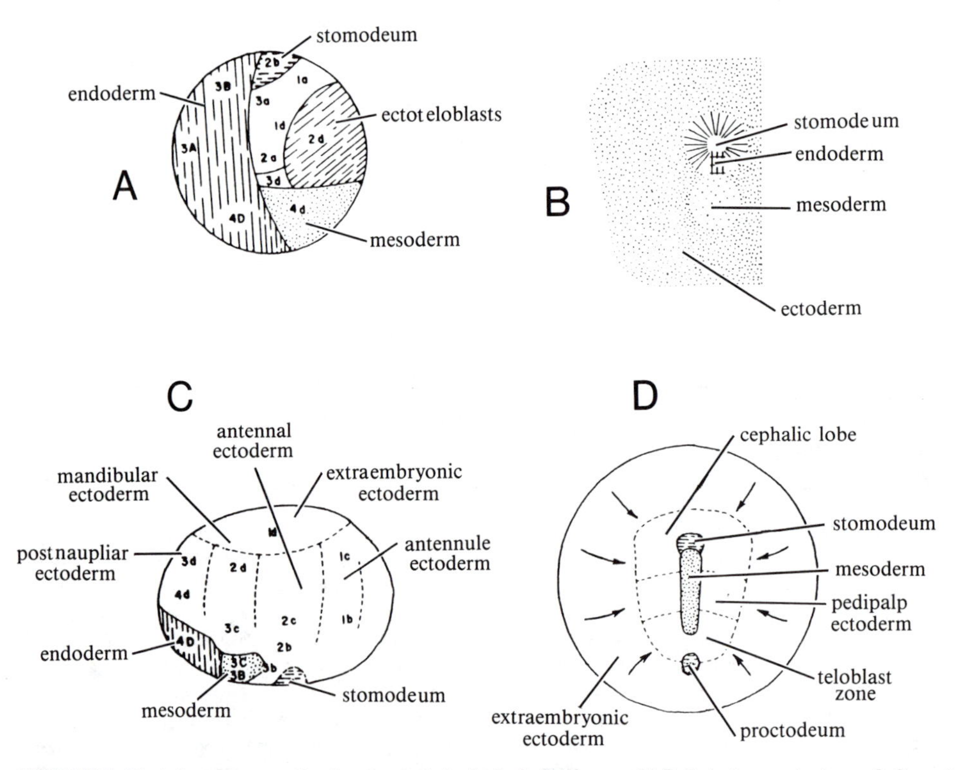

FIGURE 28.7. Blastoderm fate maps of various segmented animals. **A.** *Tubifex*, annelid. **B.** *Peripatus*, onychophoran. **C.** Generalized cirripede crustacean. **D.** Generalized xiphosuran chelicerate. (Derived from Anderson.)

separates off as a double primordium, from which the nerve cord and ganglia develop. The first endodermal cells are scattered through the yolk, devouring and digesting it. Eventually new endoderm is developed from three endodermal rudiments, one at the site of the future proctodeum, one at the future stomodeum, and one below the germ band mesoderm. These primordia eventually surround the remaining yolk, forming the midgut (Fig. 28.6H).

The generalized relationship of arthropod coelomic compartments to the embryonic segments is shown in Fig. 28.8A. The eyes as well as the first antennae and preantennal structures develop in the broad anterior part of the embryo, often termed the **acron**. New somites are added at the posterior end of the embryo, immediately in front of the last segment, the **pygidium** or **telson**, as in annelids.

Trilobite larvae also appear to have followed this general course of development. The youngest larvae were composed of the acron and the first four postoral segments. The pygidium (telson) formed very early. Once it was present, new segments were added between the pygidium and the next most anterior segment (Fig. 28.8B,C,D). The first four postoral segments were incorporated into the cephalon of adults.

After hatching, the young arthropod follows the general rules seen in embryos. If all of the body segments are not present at hatching (Fig. 28.8E), new segments are added in front of the last segment, at molts. When new somites are added to the body during molts, it is called **anamorphic development**. It is common in many groups of arthropods. When young hatch with essentially the adult complement of segments, it is called **epimorphic development**.

At the time of hatching, the young animal emerges as a larva in the great majority of cases. Larval forms vary greatly in different arthropod groups. Horseshoe crabs have a curious Euproops larva (Fig. 28.8F), named for its clear resemblance to the Carboniferous limulid, *Euproops*. A number of somites and appendages must be added before the larva becomes an adult. The sea spiders, or pycnogonids, have only three pairs of appendages when they hatch, and add others during molts (Fig. 30.16C). Some of the arachnids are anamorphic. Mites and

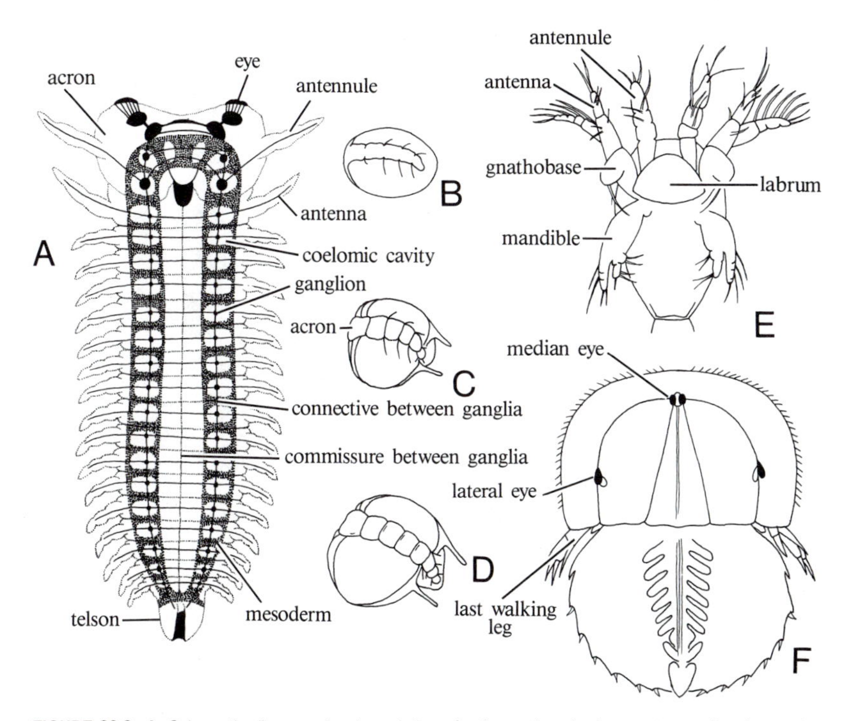

FIGURE 28.8. **A**. Schematic diagram showing relation of arthropod coelomic sacs to ganglia of central nervous system and to associated appendages. **B,C,D**. Trilobite protaspis larvae, showing stages in early development. Youngest larvae contain only unsegmented part of head and somites that will be incorporated in cephalon. In (**C**), terminal bud serving as growth center seen at posterior end of cephalon. **E**. Nauplius larva, typical of crustaceans, with three pairs of appendages: antennules, antennae, and mandibles. **F**. Euproops larva of *Limulus,* resembling Coal Age fossil genus *Euproops*. (**A** after Snodgrass, from Johanssen and Butts; **B,C,D** after Størmer, from Whittington; **E** after Calman; **F** after Kingsley and Takano, from Shipley.)

ticks hatch with three pairs of legs and add a fourth pair at the first molt.

The details of insect development vary greatly. A few of the most primitive insects have anamorphic development, adding segments during molts after hatching. Generally, however, the epimorphic insect has a full set of body segments at the time of hatching, although it does not have an adult form. If it changes gradually to resemble the adult, the juvenile stages are called **nymphs**. The aquatic nymphs are usually referred to as naiads, and must undergo a somewhat more extensive metamorphosis at the last molt, when they leave the water and become land animals. When the young insect is wingless and must pass through a **pupa** stage to assume the adult form, the juvenile stages are called **larvae**. In general, the most primitive hexapods are **hemimetabolic**, having no metamorphosis, and resemble the parents most at hatching, while the most highly evolved insects are **holometabolic** and metamorphose or pass through distinct larval, pupal, and adult stages.

Among Crustacea, the characteristic first larva is the **nauplius**. It has only three pairs of appendages, the first two being the two pairs of antennae, and the third the mandibles (Fig. 28.8E). New somites and appendages are added with each molt. The crustacean that has passed beyond the nauplius stage, but is not yet adult, is usually called a **metanauplius** or **zoea**, although a diverse array of special larval stages occur in a number of crustacean groups.

Fossil Record

An abundance of fascinating fossil arthropods are known, and contribute significantly to our understanding of the phylum's history. Unfortunately, extensive treatment of these is not germane to a book of this kind, and space limitations just do not allow consideration of these animals.

Chief among these fossils are the trilobitomorphs, of which the trilobites are the most common. The trilobite exoskeleton was predominantly chitinous,

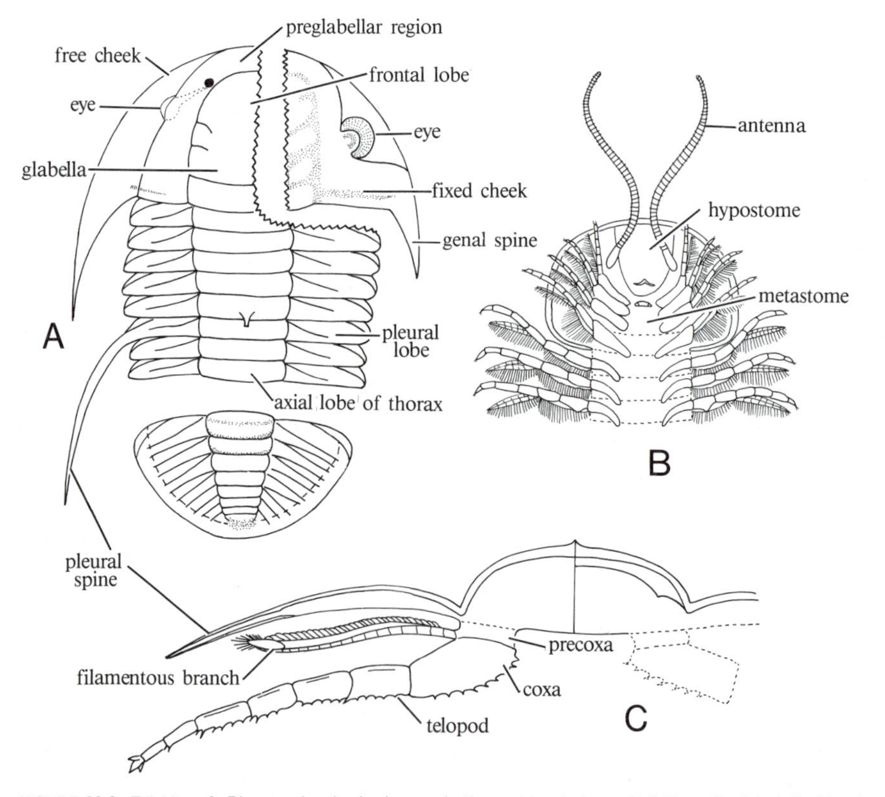

FIGURE 28.9. Trilobites. **A**. Diagram showing basic organization and terminology of trilobites. On right half of head, typical form shown, with facial suture cutting lateral margin of head, and fixed cheek carrying genal spine. On left half of head, variant form shown, with facial suture cutting posterior margin, and genal spine on free cheek. **B**. Ventral surface of trilobite, showing similarity of head appendages, except for antennae, to trunk appendages. **C**. Schematic cross section of trilobite, showing relationship of body and appendage. (**A**,**C** after Whittington; **B** after Beecher, from Woods.)

but was strengthened in some regions by calcium carbonate. Trilobites are so named because the body is divided into three lobes by a pair of longitudinal grooves (Fig. 28.9A). It is also divided into three tagma or regions: an anterior **cephalon**, a middle **trunk**, and a posterior **pygidium**. The cephalon is covered by a single, roughly semicircular, dorsal shield. The thorax is segmented, with separate skeletal plates for the various somites. The pygidium is also segmented in the more primitive species, but in more advanced forms is covered by a single dorsal plate, formed by the union of the dorsal plates of the component somites.

The trilobite head has complex markings of furrows and grooves. Most of the trilobites had a pair of large compound eyes (Fig. 28.9A). The appendages can be seen in ventral view (Fig. 28.9B). A pair of slender, annulate **antennae** occur on the head with three or four pairs of legs, whose structure will be considered later. A median lobe, the **labrum** or **hypostome**, lies over the mouth, and a smaller plate, the **metasome**, lies behind the mouth.

The thorax is covered by a series of overlapping dorsal plates. Fossils are often enrolled or curled up, suggesting that trilobites were quite flexible in life, probably to protect the relatively soft undersurface. Each dorsal plate is a single piece, but is divided into regions by furrows. The axial furrows define a median axial lobe and a pair of lateral pleural lobes.

The posterior pygidium contains from 2 to 27 somites, united in most cases by a solid dorsal shield. Axial and pleural furrows may continue into the pygidium, defining regions corresponding to the axis and pleura. The somites and appendages of the pygidium are usually somewhat reduced, but the basic form of the appendages remains unaltered.

A great deal of phylogenetic discussion has centered on the trilobite legs, and considerable difference of opinion still exists. The legs are attached to the ventral surface below the axial furrow by a basal piece, to which slender dorsal and heavier ventral branches are attached. The ringed dorsal branch bears filaments. The robust jointed ventral branch or telopod ends in several claws or hooks.

Little is known of the internal anatomy of trilobites. There is good evidence that the organs lay in the axis, while the pleura were little more than lateral hoods over the appendages. Evidence has been found of a dorsal circulatory system, resembling that of modern arthropods, as well as details of gut and body musculature.

Phylogeny

A mind-boggling array of hypotheses and scenarios exist related to arthropod evolution. Every authority seems to have his own version. There is not even agreement as to whether there is one phylum, Arthropoda (see, e.g., Snodgrass, 1952), or many. Of the latter school, there are warring factions between those (Cisne, 1975) that hold to two groups (the Uniramia, with a uniramous leg, and the Schizoramia, with a branched leg) and those (Anderson, 1973; Manton, 1977; Schram, 1978) that hold to four groups (Trilobitomorpha, Uniramia, Crustacea, and Cheliceriformes). Then there are those who would still unite the Crustacea and the Uniramia into the Mandibulata (Boudreaux, 1979), and those who would join Cheliceriformes and Trilobitomorpha into the Arachnomorpha (Stormer, 1944). Just about every permutation and combination has been proposed and argued for at some time in the last century. All of these speculations (including that of the junior author—FRS) are largely wasted verbiage. Despite the wealth of material, fossil and Recent, there has been no concerted cladistic analysis made of all groups using as many characters as possible. Until that is done—and several people are working on it—there will never be any hope for a consensus. At present, people agree that the three major living groups are each monophyletic (uniramians *sans* onychophorans, crustaceans, and cheliceriforms), but no one agrees as to how these might be interrelated (or whether they even are). The situation is in flux. For example, the junior author (FRS) once was convinced that the arthropods were polyphyletic and comprised four separate phyla. Now he is not so sure and feels a case can be made for arthropod monophyly, but not for *any* of the reasons that have been published in the literature to date (Gupta, 1979).

For the time being, in this edition, we prefer to treat each major group separately and leave issues of arthropod phylogeny for some future resolution. The matter can be explored by the student in some of the Selected Readings.

Selected Readings

Anderson, D. T. 1973. *Embryology and Phylogeny in Annelids and Arthropods.* Pergamon Press, Oxford.

Boudreaux, H. B. 1979. *Arthropod Phylogeny with Special Reference to Insects.* Wiley-Interscience, New York.

Bullock. T. H., and G. A. Horridge. 1965. *Structure and Function in the Nervous System of Invertebrates,* vols. 1 and 2. Freeman, San Francisco.

Cisne, J. S. 1975. The anatomy of *Triarthrus* and the relationships of the Trilobita. *Fossils and Strata* 4:45–64.

Gupta, A. P. (ed.). 1979. *Arthropod Phylogeny.* Van Nostrand Reinhold, New York.

Gupta, A. P. (ed.). 1987. *Arthropod Brain.* Wiley-Interscience, New York.

Manton, S. M. 1977. *The Arthropoda.* Clarendon Press, Oxford.

Schram, F. R. 1978. Arthropods, a convergent phenomenon. *Fieldiana: Geology* 39:61–108.

Schram, F. R. 1986. *Crustacea.* Oxford Univ. Press, New York.

Snodgrass, R. E. 1952. *Textbook of Arthropod Anatomy.* Cornell Univ. Press, Ithaca.

Størmer, L. 1944. On the relationships and phylogeny of fossil and Recent Arachnomorpha. *Skrift. Norske Videns. Akad. Oslo Mat. Nat. Kl.* 1944(5):1–158.

Tiegs, O. W., and S. M. Manton. 1958. The evolution of the arthropods. *Biol. Rev.* 33:255–337.

29

Uniramia

Uniramians have followed two main evolutionary pathways. One line is characterized by the division of the body into two regions or tagmata, the head and the trunk, with pairs of appendages on all of the trunk somites except the last. These are the myriapods. The other is characterized by a body of three tagmata, head, thorax, and abdomen, with appendages on the head and the thorax, but with the abdominal appendages greatly reduced or missing. This line includes the insects.

The head of uniramians somewhat resembles the head of crustaceans, but it is much more consolidated and has no equivalent of the crustacean antennae. The uniramian appendages are quite unlike the appendages of Crustacea. They are uniramous, the leg corresponding to the endopod of a crustacean appendage. At no time during development are the legs truly biramous. This is one of the arguments against the view that the mandibulates (uniramians and crustaceans) are more closely related to each other than to other arthropod stems, and one of the arguments used to support the view of some authorities that arthropods have arisen from several different protoarthropod stems.

The uniramians are the most successful of animal groups in terms of numbers of species. This is due solely to the insects. It is estimated that probably 750,000 insect species have been described. Estimates of total insect species range anywhere from one to ten million!

Definition

Uniramia are arthropods with uniramous trunk appendages. The head is composed of five sets of "limbs": antennae, labrum, mandibles, and first and second maxillae, with the latter forming a labium. Tracheae are used for respiration; Malpighian tubules are employed as excretory organs. In the brain, the tritocerebrum is small.

Body Plan

The uniramian head is divided into an anterior **procephalon** (also known as the **protocephalon**) and a posterior **gnathocephalon**. The **antennae** attach to the procephalon. These are sensory appendages, important as tactile organs and for chemoreception. The **compound eyes** and **ocelli**, when present, are also found on the procephalon.

The gnathocephalon bears the mouthparts, and includes the mandibular and two maxillary somites. The first somite of the gnathocephalon bears no appendages in the adult, but in some cases develops a pair of embryonic appendages, the **postantennae**, thought by some to be homologous to the second antennae of Crustacea.

The surface of the head is made up of a cranium and a part extending forward or down toward the mouth, the **epistome** (Fig. 29.1A,B,C) or **clypeus** (Fig. 29.2A). A partially free upper lip, the **labrum**, is attached to the epistome and hangs down over the top of the mouth. Although inconspicuous in myriapods, the labrum is equipped with setose sense organs thought to be chemoreceptive. The external form of the head in myriapods is simpler than in insects, where conspicuous head sutures associated with endoskeletal apodemes are important features that delineate a complex of plates (Fig. 29.2A,B).

The actual ventral surface of the head may be completely concealed by the mouthparts, which can extend forward from the gnathocephalon. A preoral cavity is formed that leads to the mouth. The upper surface of the preoral cavity is lined by the **epipharynx** (Fig. 29.1C), and its lower surface by a ventral fold, the **hypopharynx**, both derived from the floor of the head anterior to the mouth. The lateral walls of this preoral cavity vary a great deal in different groups, sometimes containing several sclerites and sometimes being partially composed of the mandible bases. The hypopharynx is concealed by the lower lip, the **labium** in insects (Fig. 29.2C) and the **gnathochilarium** in myriapods, formed by the union of the second maxillary elements and the sterna of the postoral head somites.

The mouthparts articulate with the membranous

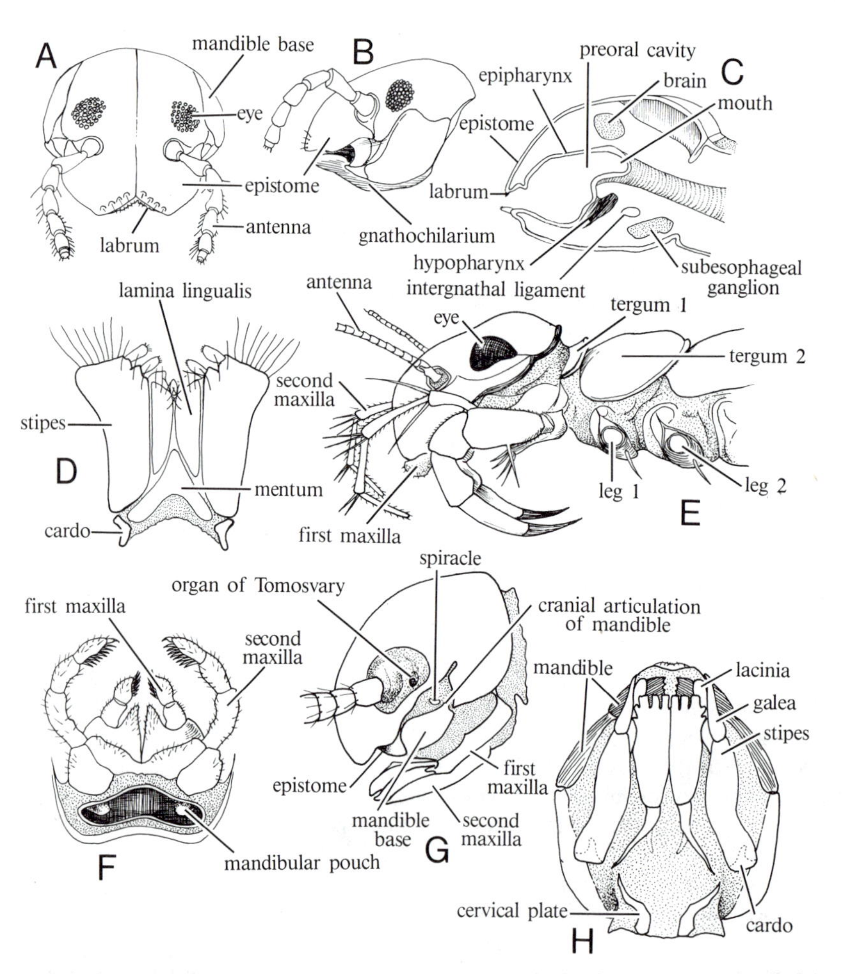

FIGURE 29.1. Myriapod head structure. **A,B**. Front and side views of head of millipede, *Arctobolus*. **C**. Schematic section through head of millipede, showing preoral cavity and surrounding mouth parts. **D**. Millipede gnathochilarium. **E**. Lateral view of head of centipede *Scutigera*. **F**. Maxillae of centipede *Lithobius* close together and extending forward to serve as lower lip. **G,H**. Lateral and ventral view of head of symphylan, with insectlike maxillae with second maxillae united to form underlip. (After Snodgrass.)

lower part of the head surface. The **mandibles** are so reduced that they have lost all resemblance to ordinary appendages (Fig. 29.2A,B). All segments are fused to form a single element with the distal tip developed for biting. As a general rule, insect jaws project below the head (hypognathous), but some jaws extend in front of the head (prognathous). Several types of mouths can be distinguished, with modifications of each type. The mouthparts that are described here, however, are those found in the generalized, chewing type of mouth.

The **first maxillae** are usually limblike. They are made up of a basal segment, the cardo, by which they are hinged; a main piece, the stipes, to which the other pieces attach; a lateral and usually sensory lobe, the galea; a medial lobe that is often equipped with spines or teeth along its border, the lacinia; and a sensory palp usually composed of several segments (Figs. 29.1H, 29.2D). The **second maxillae** are fused medially (Fig. 29.2C), forming the lower lip, or labium, and have parts corresponding to those of the first maxillae (Figs. 29.1H and 29.2E). The mentum

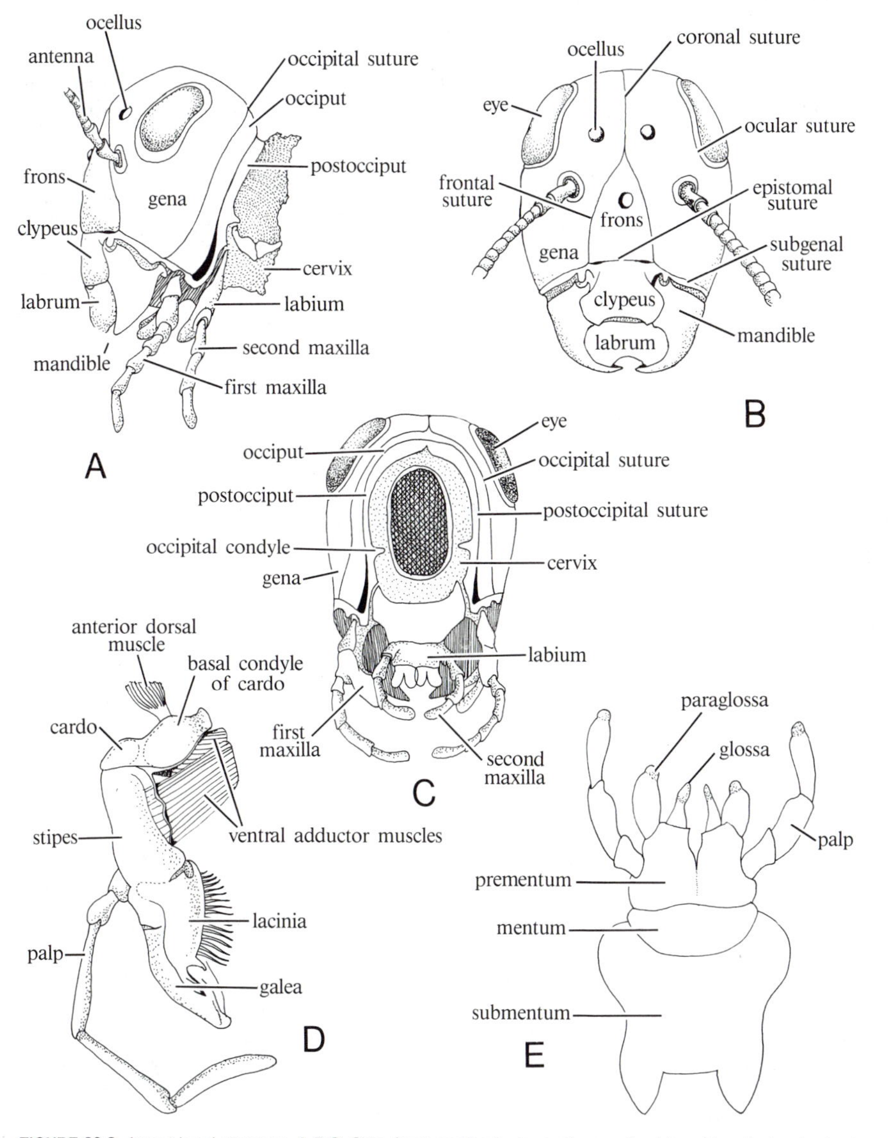

FIGURE 29.2. Insect head structure. **A,B,C.** Side, front, and back views of generalized insect head, showing basic regions and sutures, and position of mouth parts. **D.** Insect maxilla showing parts of base and palp. **E.** Generalized type of insect second maxillae, which unite to form labium. (**A,B,C,D** after Snodgrass; **E** after Imms, from Ross.)

and the submentum are medial plates, formed by the union of the cardo. The prementum corresponds to the stipes, and the median glossa and lateral paraglossa correspond to the medial lacinia and the lateral galea. The labial palps are usually shorter than the maxillary palps (Fig. 29.2C).

In myriapods, the nature and development of the second maxillae are related to the degree of formation of the lower lip. The second maxillae of some myriapods are leglike, whereas in others the second maxillae have been reduced and form a part of the lower lip. Most diplopods and all pauropods have a legless somite, the **collum** (Fig. 29.3E), about which there has been considerable difference of opinion. The incorporation of reduced appendages from the collum into the lower lip of some diplopods has been described, leaving little room for doubt that it is the second maxillary somite. The pselaphognathan diplopods have a leglike second maxillae on the collum. Presumably the head of pauropods is similarly built. It appears that the second maxillary somite, the most recent addition to the gnathocephalon in insects, has not fully incorporated into the head in many myriapods.

In pauropods, the parts of the gnathochilarium are not fully fused, and in pselaphognathans also, the parts are partially separate. In the majority of diplopods the gnathochilarium (Fig. 29.1D) is composed of the united gnathal lobes of the mandibles, the first maxillae, the sternum of the first maxillary somite, and the vestiges of the second maxillae, and is hinged on the sternum of the collum. The two first maxillae are attached at the base, but are free distally and retain some freedom of movement (Fig. 29.1F).

Symphylans and chilopods, as a rule, have more conspicuous second maxillae (Fig. 29.1F). The second maxillae are leglike in the scutigeromorph centipedes, but in the other groups are incorporated into the gnathochilarium to a greater or lesser extent. As a rule in centipedes, with the first thoracic legs modified to form maxillipedes, the large bases extend forward over the maxillae, whereas the two

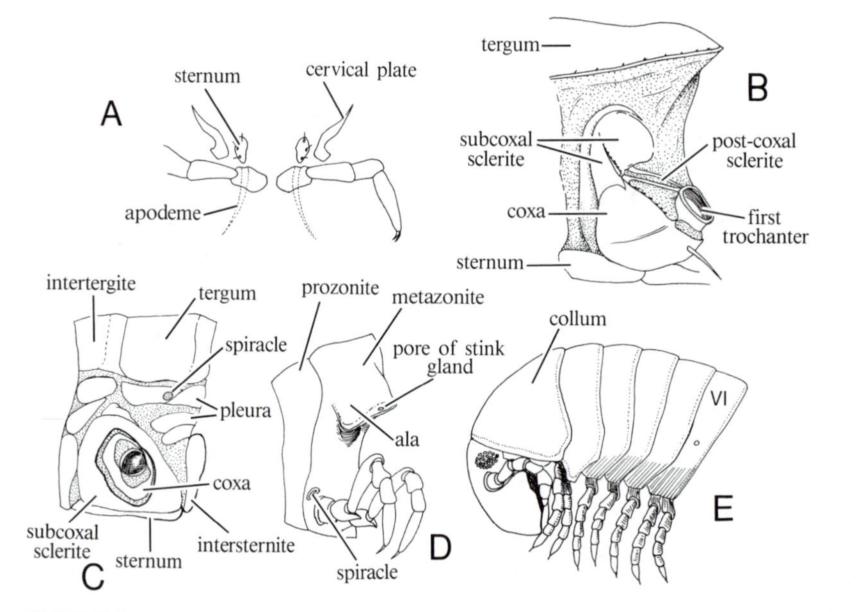

FIGURE 29.3. **A**. Ventral sclerites associated with first pair of legs of symphylan. **B,C**. Comparison of skeletal plates of somites of scutigerid and geophilid centipede. **B**. Scutigerids have single tergum over dorsal surface, sternum below, and soft pleural region, strengthened by few sclerites associated with appendage base. **C**. Geophilids have two dorsal plates, tergum and intertergite, and two ventral plates, sternum and intersternite. Pleural plates and sclerites associated with appendage bases strengthen lateral wall. **D**. Somite of polydesmoid millipede. Diplosomite usually covered by continuous skeletal ring in which tergum, pleura, and sternum cannot be distinguished. Polydesmoids have two such rings for each diplosomite, anterior prozonite and posterior metazonite. Lateral flares, which give polydesmoids their flattened appearance, located on metazonite. **E**. Head and anterior trunk of juliform millipede, *Arctobolus,* each diplosomite covered by single skeletal ring that overlaps next posterior, permitting animal to roll up in spiral. Collum does not belong to somite on which first pair of legs insert. Small sternum, without corresponding tergum, lies immediately behind sternal plate below collum, and first pair of legs insert on this intercalated sternum. First five pairs of legs single, attached to different somites. Somite VI carries sixth and seventh legs, and succeeding diplosomites have two pairs of legs. (After Snodgrass.)

maxillary appendages and the sterna of their somites project forward toward the labrum (Fig. 29.1E,F).

On the other hand, the symphylan head is more like the head of insects with the first maxillae more laterally placed and the second maxillae partially united to form a labiumlike ventral plate (Fig. 29.1G,H). The symphylan mandibles, however, have movable gnathal lobes and, therefore, are myriapodlike. The constituent parts of the two maxillae appear to correspond to the parts of the maxillae that are seen in the labium of insects in some detail, and are given similar names (Fig. 29.1H). The first maxillae are long and are formed of a basal stipes and two apical segments, a somewhat ventrally placed galea, and more dorsally oriented lacinia.

The uniramian trunk, as a rule, has the dorsal surface more heavily armored than the ventral surface, although smaller forms have generally softer bodies. In some cases this appears to increase the protective qualities of the dorsal exoskeleton, whereas in others, particularly where two or more segments are covered by a single, fused tergum, they may help to prevent sinuous bending of the body during rapid movement.

The membranous body side walls are termed the pleural regions (Figs. 29.3B,C and 29.4) in which subcoxal sclerites associated with the leg bases are placed, and which are further strengthened in many species with additional pleural plates. In the insects, abdominal somites have no articulations for legs, and the thoracic somites are greatly modified by the formation of wings and provisions for their attachment to the body (Fig. 29.4)

In insects, the term **notum** is applied to the dorsal thoracic plates. The least modified thoracic somite is the anterior **prothorax**, although it is somewhat more strongly consolidated in winged than in wingless species. The wing-bearing **mesothorax** and **metathorax** are considerably modified. The notum is divided into an anterior and a more posterior plate (Fig. 29.4), the latter of which is fused to the lateral skeleton to form a solid bridge behind the wings. At each end of the somite a deep, platelike apodeme, the **phragma**, provides a surface for the attachment of the flight muscles.

The wings are attached at a membranous region of the pleural wall. Two free sclerites, the **basalare** and the **subalare**, are found in the pleural wall near the wing bases (Fig. 29.4), and small axillary sclerites in the base of the wings articulate with the edges of the notal plates.

The insect wings (Fig. 29.5) are lateral flaps of the body wall and, therefore, are a double fold, consisting of upper and lower membranes tightly adherent to each other and supported by **veins**. The main veins run obliquely from the wing base to the wing margin, and are connected by small **crossveins**. The veins contain a tracheal tube, bathed in blood within a hemocoelic diverticulum, and may contain nerves. So besides serving as support struts, the veins serve for transport of materials as well. Wing venation is remarkably varied between major groups and is important in insect identification. The basic pattern of wing venation is shown schematically in Figure 29.5, with the major vein systems labeled.

The body legs are uniramous. The structure of

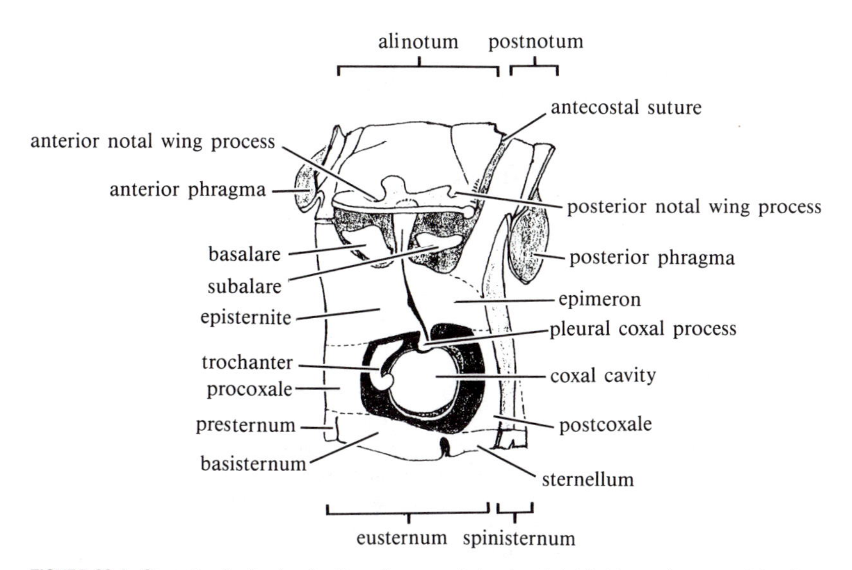

FIGURE 29.4. Generalized, wing-bearing thoracic segment showing skeletal plates and sutures. (After Snodgrass.)

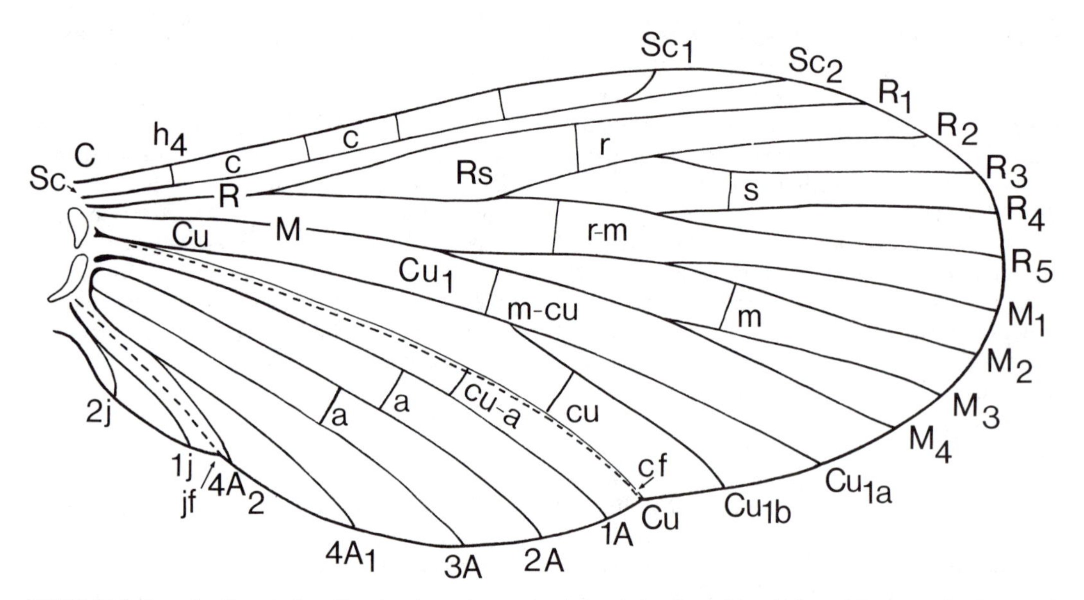

FIGURE 29.5. Generalized insect wing with main veins and crossveins. Letter designations of the widely used Comstock–Needham terminology. Veins shown in *capital letters,* and crossveins in *lowercase letters.* Numbered branches of veins sequentially numbered, from front to back of wing. **Veins:** Costa (C); Subcosta (Sc); Radius (R); Media (M); Cubitus (Cu); Anal (A); Jugal (J). Posterior main branch of radius is radial sector (Rs). **Crossveins:** humeral (h), between costa and subcosta near their base; costa; (c); radial (r); sectoral (s), between branches of radial sector; radiomedial (r–m); medial (m); mediocubital (m–cu); cubital (cu); cubitoanal (cu–a); anal (a). (After Ross.)

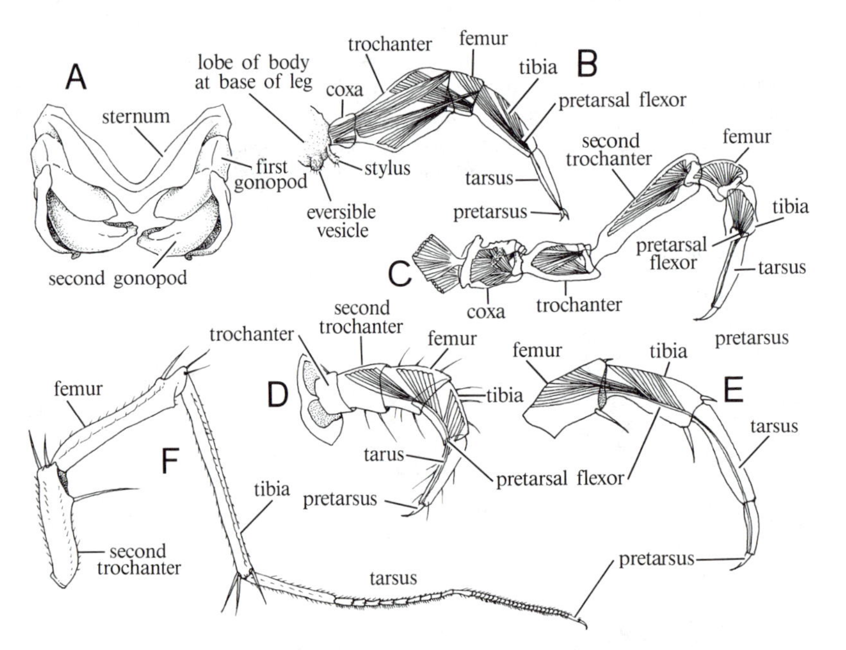

FIGURE 29.6. **A**. Sternum and gonopods found on seventh somite of juliform millipede, *Arctobolus.* **B**,**C**,**D**,**E**,**F**. Comparison of some myriapod legs. **B**. Symphylan leg (*Scutigerella*), arising from more or less conical lobe of body wall, from which eversible vesicle and stylus extend. Styli resemble similar processes found in some insects. **C**. *Euryurus,* millipede, with two trochanters. **D**. Geophilid centipede, with short, powerful legs, with two trochanters characteristic of chilopods and single tarsus. **E**. End of leg of lithobiomorph centipede, fast funner, with legs lengthened by addition of second tarsal piece. **F**. End of leg of scutigerid centipede. Faster than the lithobiomorphs, they have remarkably long legs, with individual podites prolonged and many tarsal pieces. (After Snodgrass.)

walking legs remains quite stable within groups, but gonopods (Fig. 29.6A) differ markedly from typical legs. Pauropods and symphylans have legs with six podites (Fig. 29.6B). They are quite similar, but the longer legs of the pauropods have an extra tarsal piece. Diplopods (Fig. 29.6C) and chilopods (Fig. 29.6D,E,F) have legs with seven segments, with a double trochanter in place of the single trochanter of the pauropods and the symphylans. Diplopod legs are short and resemble the legs of geophilid centipedes (Fig. 29.6C,D). Long legs are associated with fast running; the herbivorous millipedes are slower than the carnivorous centipedes, on the whole. Geophilids, however, live in the soil and surface litter. Many of them are burrowers, and none are fast runners. Legs can be elongated by subdivision of the tarsus. Most of the chilopods have at least two tarsal segments, and the very fast scutigeromorphs have a larger number (Fig. 29.6E,F). In all of these cases, the long tendon of the pretarsus flexor muscle passes through the tarsal pieces without inserting on them, as is typical of the terrestrial mandibulates.

Other external structures are worth noting. Sensory setae and hairs are found on the trunk somites and also on the legs. The tracheal system opens through spiracles with characteristic positions in centipedes and millipedes. Millipedes usually have two spiracular openings on each diplosomite, whereas centipedes have one on each somite. In nearly all centipedes and millipedes, the spiracles are openings in the membranous pleura near the leg bases, but the scutigeromorph centipedes are unique in having a series of middorsal spiracles in the middle of the terga. Symphylans have a stylus and eversible vesicle located near the leg bases. The vesicle arises from an embryonic rudiment known as the ventral organ. Its function is unknown. The styli resemble styli found in the abdominal somites of some insects. Symphylans also have a pair of spinnerets on the preanal somite. The spinnerets resemble the cerci of insects, but are probably not homologous to them.

The abdomen of insects is relatively simple. The exoskeleton consists of a dorsal tergum and a ventral sternum, connected by a membranous pleural region (Fig. 29.7). In some cases pleural sclerites are present, probably vestiges of the subcoxal sclerites of the lost abdominal appendages. In some of the primitive insects, however, remnants of the coxae of abdominal somites persist and bear rudimentary appendages, known as **styli**. A pair of **cerci** extend back from the posterior end of the abdomen in a number of insects. These are posterior, usually tactile organs derived from the appendages of the eleventh abdominal somite (Fig. 29.7A,D).

Highly specialized appendages on the last few somites are used for copulation and egg laying (Fig. 29.7). These parts are lost completely in the primitively wingless insects, but are extremely diverse in pterygote insects. In general, the female insect gono-

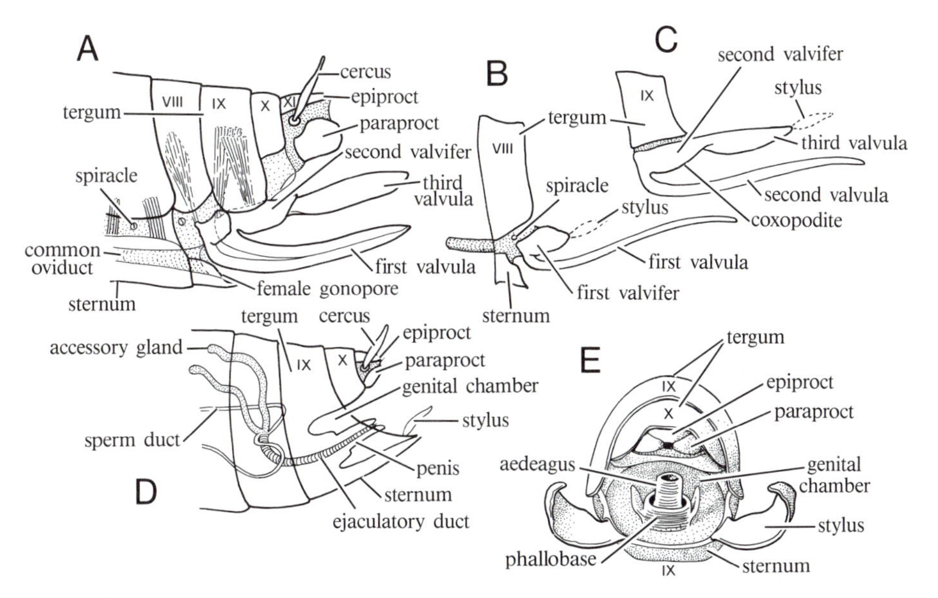

FIGURE 29.7. A. Generalized structure of posterior end of female winged insect. B,C. Lateral view of detached eighth and ninth abdominal segments showing parts of ovipositor associated with each. D. Generalized structure of genitalia of male winged insect (*lateral view*). E. End view of same, with claspers developed from styli. (After Snodgrass.)

pore is found on the eighth somite, partially covered by a projection of the sternum, the subgenital plate. Highly modified appendages (Fig. 29.7A) are attached to the ventral side of the eighth somite, composed of a basal **valvifer**, derived from the coxa, and a **valvula**, derived from the main branch of the leg. The ninth somite also bears a plate, the valvifer, and two valvulae, a dorsal one behind and a ventral one in front. The valvulae form the ovipositor, variously modified in different insect groups. The sting of a worker bee is the most remarkable modification of the ovipositor. In some insects the abdomen itself serves as an ovipositor, and the highly modified genital appendages are missing. The eleventh somite contains the anus and is covered by an often triangular tergum.

The male genital appendages are found on the ninth somite but are often partly fused with the tenth somite (Fig. 29.1D,E). A midvental penis is located on the ventral articular membrane between the ninth and tenth somites. Penile structure varies greatly in different insect groups and is sometimes important in the identification of species. During mating, the penis serves for sperm transfer and the genital appendages act as claspers.

Digestive System The digestive system is fairly uniform among uniramians. In myriapods a slender **esophagus** opens into a wide **midgut**, separated by a valvular constriction from the short, muscular rectum (Fig. 29.8A). Although the intestine of midgut is usually straight, it is coiled in oniscomorphs. The esophagus arises from the stomodeum and is lined with a thin cuticle, whereas the **rectum** arises from the proctodeum and is also lined with cuticle. A delicate cuticular lining is noted in the midgut of some species, but it is derived from special secretory cells in the stomodeum and closely resembles a similar membrane found in many insects. From one to three pairs of **salivary glands** open into the esopha-

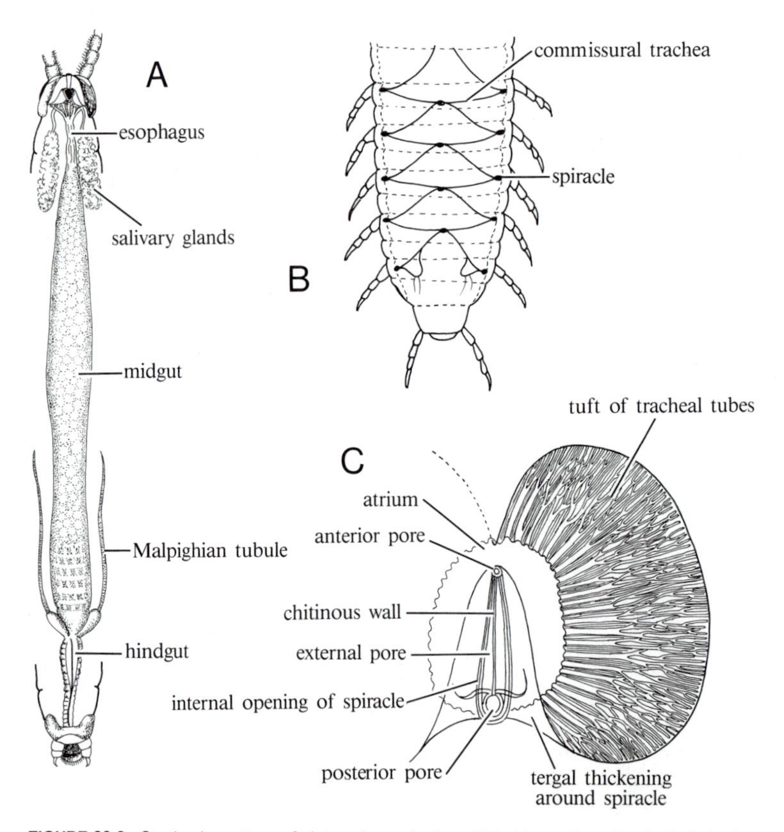

FIGURE 29.8. Centipede anatomy. **A**. Internal organization of lithobiomorph centipede. **B**. Spiracles and main tracheae of geophilid centipede with pair of spiracles on all but first and last body somites. Some connections between tracheae of adjacent somites, but no really effective longitudinal trunks. **C**. Tracheal lung of scutigerid centipede, unique respiratory system with middorsal spiracles opening into tracheal lungs that hang down into hemocoel. (**A** after Plateau, from Attems, in Kükenthal and Krumbach; **B** after Kaufmann; **C** after Hasse.)

gus, but myriapods lack the extensive system of diverticula so often found in other arthropods.

The insect gut is much like that of myriapods, although in more advanced forms the gut tends to be divided into a number of special regions (Fig. 29.9). The stomodeum consists of a **buccal cavity**, **pharynx**, **esophagus**, **crop**, and **proventriculus**. A constriction marks the position of the stomodeal valve, which guards the entrance into the midgut. The **midgut** is short and consists of the **stomach** and the **gastric caeca**. A deep constriction at the junction of the midgut and the hindgut contains the proctodeal valve. Malpighian tubules are attached to the gut at this point. The hindgut is relatively long and is divided into an anterior **intestine** made up of a somewhat dilated **pylorus**, a more slender **ileum**, and a slightly dilated **colon**, and a posterior intestine composed of a short **rectum**, dilated anteriorly as a **rectal sac**.

Respiratory System Myriapods are all air breathers and often have **tracheae** (Fig. 29.8B,C). These open to the surface by means of **spiracles**. Respiratory transfer occurs across the body surface of pauropods, as their minute size and soft exoskeleton evidently make this method adequate. One cannot safely conclude, as pauropods have no tracheae and show a number of primitive characteristics in other ways, that the myriapod groups developed tracheae independently. Minute insects, also, are sometimes without a tracheal system. Apparently the tracheal system can be reduced and lost in minute organisms. Nevertheless, the tracheal systems of different types of myriapods differ markedly.

Although gills are used in some aquatic insect larvae and nymphs, insect respiration is predominantly by means of a tracheal system that ramifies throughout the body (Figs. 29.9 and 29.10). The basic system can be supplemented with special **air sacs** (Fig. 29.10B) and specializations of the spiracles. Surface respiratory exchange is most important in small insects with a large surface area and in species with a thinner, more permeable cuticle, and is especially important in primitive, wingless insects that are both soft and small. Collembola, for example, have no tracheae and depend entirely on surface exchange. The first instar larvae of many other insects respire only across the body surface.

Juvenile insects that live beneath the surface of the water do not normally come in contact with air, and are necessarily restricted to respiratory exchange at the surface of the body. Two courses are open to such insects. They may depend on the blood for respiratory transport, or they may make an exchange between the water and a tracheal system. Insect nymphs or larvae that depend on the blood for respiratory transport tend to develop a rich distribution of blood to the body surface, and may also develop special amplified surfaces that serve as gills for respiratory exchange. A number of aquatic insect larvae or nymphs have well-developed tracheal systems that are completely closed, but nevertheless are responsible for delivering oxygen to the tissues. Tracheae extending to the surface are finely branched, and the tracheal capillaries function in much the same manner as blood capillary beds in other animals (Fig. 29.11A). The tracheal capillaries may be found over the general body surface, as in midge larvae, or may be concentrated in specialized gills. Many of the larger aquatic insect larvae have filamentous, plumose, or lamellate gills and contain repetitively branched tracheal trunks (Fig. 29.11B).

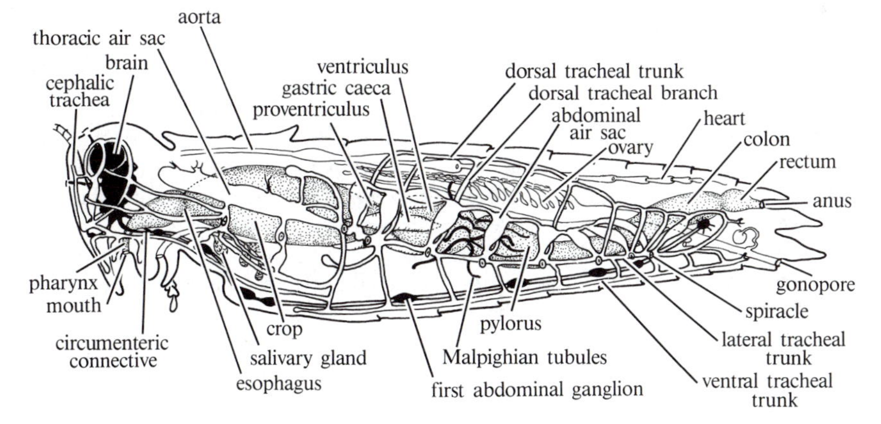

FIGURE 29.9. Scheme of internal organization of locust.

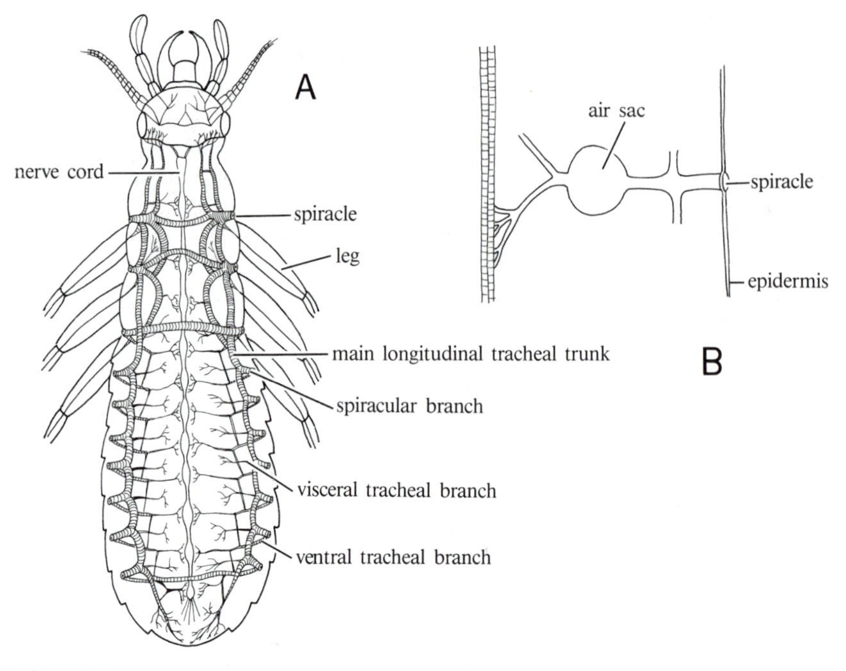

FIGURE 29.10. Generalized insect showing arrangement of tracheae.

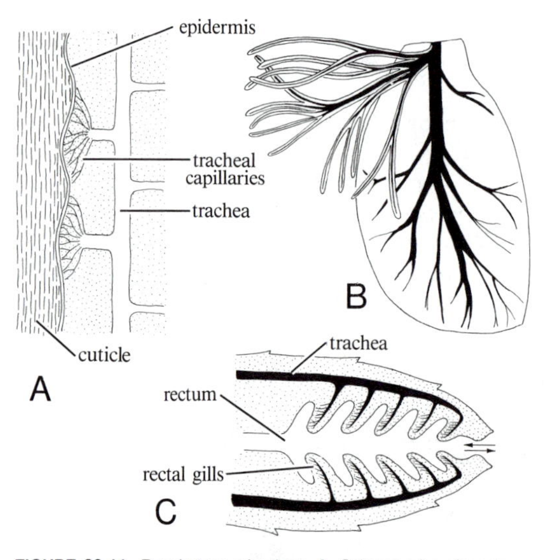

FIGURE 29.11. Respiratory adaptions. **A**. Scheme of surface trachea in aquatic insects with closed tracheal systems. Oxygen absorbed in tracheal capillaries moves into main tracheal trunks that branch through body tissues. **B**. Tracheal gill from mayfly nymph. Surface amplification with broad, flat surfaces or in filamentous tufts; branching produces bed of capillary tracheae more or less resembling blood capillary beds in animals with respiratory organs served by blood vessels. **C**. Rectal gills of dragonfly nymph (schematic). Anal pumping over folded rectal wall provides ample surface for respiratory exchange with tracheal capillary system extended into gills.

Gills may be found on or in any part of the body (Fig. 29.11C), but are most common on the abdomen.

Excretory System Uniramians excrete through **Malpighian tubules** attached at the junction of the midgut and the hindgut (Figs. 29.8A and 29.9). Two or four Malpighian tubules are present in myriapods. Amplification of the tubule surface is achieved by elongation, and the tubules often extend forward to the esophagus and turn back, ending near the posterior end of the body. In large-bodied myriapods, like *Spirobolus*, they are greatly coiled. Each tubule ends blindly and is composed of a single layer of tall, columnar cells.

In the insects, the amount of excretory surface is increased by increasing the number of Malpighian tubules, as well as by the elongation of the tubules. Where a small number of very long Malpighian tubules are present, they are often greatly coiled. Wastes formed in the Malpighian tubules are discharged through the rectum. This arrangement permits a maximum reabsorption of water.

Circulatory System Myriapods have a circulatory system in which the very long **heart** is more like a modified dorsal blood vessel. It lies in a long pericardial sinus that extends the whole length of the trunk. Segmental alary muscles attach the heart to the dorsal wall of the sinus, and contract to dilate the heart. Blood flows into the heart through paired **ostia**, with one pair in each chilopod somite and two pairs in each diplopod segment. Between the ostia, lateral blood vessels arise. These pass through the adjacent fat bodies and open into the hemocoelic lacunae. As a rule, the centipedes have a stronger heart than the less active millipedes.

In the circulatory system of insects, the body cavity serves as a hemocoel and is divided into several sinuses by **diaphragms**. An incomplete plate of muscles, the dorsal diaphragm, cuts off the most dorsal part of the hemocoel as the **pericardial sinus**. The viscera lie below the dorsal diaphragm in a large **perivisceral sinus** (Fig. 29.12A). A ventral diaphragm below the perivisceral sinus usually cuts off the **ventral sinus**, containing the ventral nerve cord. A dorsal blood vessel lies in the pericardial sinus. It is differentiated into a heart in the abdominal region, and an aorta in the thoracic region.

Blood flows forward in the **aorta** and is discharged into the tissue spaces of the head region (Fig. 29.12B). It flows back in the perivisceral and ventral sinuses, entering the legs from the ventral sinus and returning into the perivisceral sinus. In the abdominal region the blood flows upward, entering the pericardial sinus and passing through the ostia into the heart. Booster hearts in the mesothorax and the metathorax maintain circulation in the wing veins.

The tubular **heart** is usually dilated in each somite (Fig. 29.12C). Blood enters the heart through the paired ostia, and ostial valves prevent backflow during heart contraction. A few insects have chambered hearts, with valves to prevent backflow, but the heart is usually a continuous tube, closed at the posterior end. In a few insects, especially the orthopterids, a series of paired lateral arteries arise from the heart,

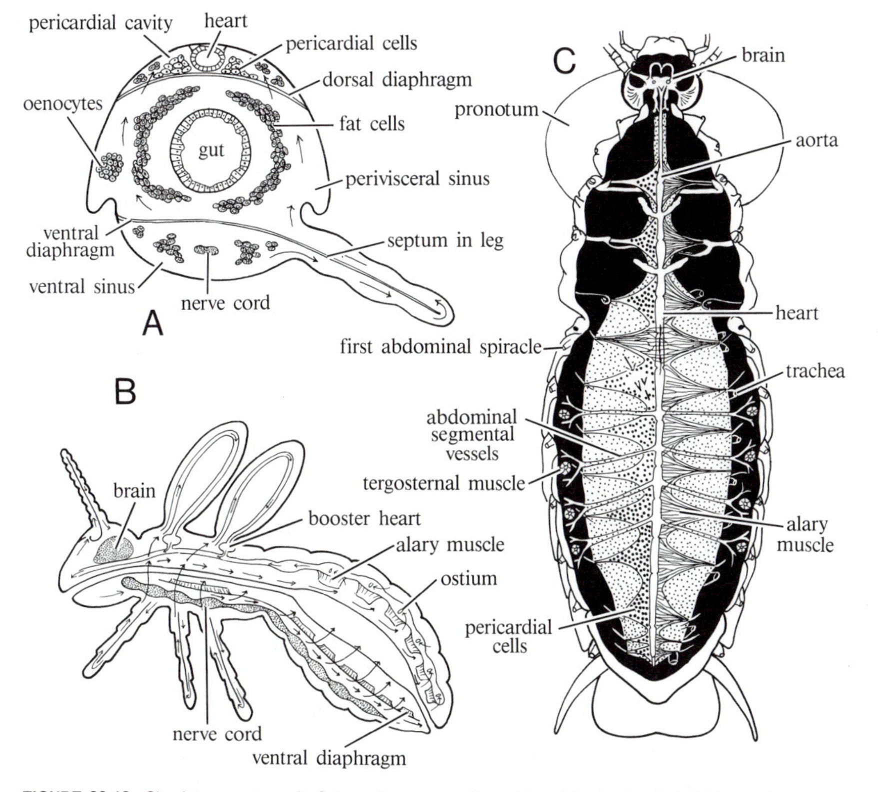

FIGURE 29.12. Circulatory system. **A.** Schematic cross section of insect body showing divisions of body cavity and course of circulation. Dorsal diaphragm formed by dorsal transverse muscles and connective tissue permits blood to pass through into pericardial sinus above. Ventral diaphragm, when present, formed of ventral transverse muscles, permits blood to enter perivisceral sinus. Septa in legs divide internal space into afferent and efferent channels. Oenocytes and fat cells in body cavity and pericardial cells in pericardial sinus provide for food storage, enter into excretory activities, and probably secrete substances. **B.** Scheme of circulation in insect. *Arrows* show direction of flow. **C.** Ventral dissection of cockroach, showing heart and associated structures. Alary muscle dilates heart. Valves close ostia when heart wall contracts. In some insects, segmental vessels arise from heart, with valves at junction with heart to prevent backflow into heart.

passing to the fat body. In the majority of insects, however, the aorta is the only blood vessel found in adult stages.

Nervous System The uniramian central nervous system centers on a dorsal **brain** connected to a double **nerve cord** by a **circumenteric ring**. Segmental ganglia occur on the nerve cord. They are actually double, although in many cases they appear single to the naked eye. Commissures connect the ganglionic pairs, and segmentally arranged lateral nerves issue from each ganglion to the sense organs and muscles of the somite.

The union of ganglia on the ventral nerve cord leads to further centralization of the nervous system. However, this has not progressed very far in any myriapod. The ganglia of the first three trunk somites of millipedes are close together and sometimes have partly fused. Although the brain is rather highly differentiated, the nerve cord and metameric ganglia are relatively primitive.

A close relationship between sense organ development and brain structure can be seen in myriapods (Joly and Descamps, in Gupta, 1987). In eyeless species the optic centers in the corpora pedunculata are dwarfed, and olfactory centers are correspondingly increased. As a general rule, the olfactory centers constitute a relatively large part of the brain of myriapods, although there are only two instead of the three such ganglia seen in other arthropods. The scutigeromorph centipede brain seems to be closely related to that seen in insects.

Insects have been studied more intensively than other arthropods, and more is known of the details of their nervous anatomy (Gupta, 1987). The outstanding developments are (1) continued progressive specialization of the brain, with the formation and development of specific nuclei and relay centers; (2) centralization of the nervous system by anterior migration and coalescence of ganglia on the ventral nerve chain; and (3) the lengthening of the peripheral nerves to compensate for the new position of ganglia.

The insect brain (Fig. 29.13A) is divided into a protocerebrum, associated with the eyes and containing the preantennal brain centers; a deutocerebrum, associated with the antennal nerves; and a tritocerebrum, possibly associated with the dwarfed, appendageless postantennal segment.

The **protocerebrum** is composed of the large protocerebral lobes, an intercerebral region, and sometimes accessory lobes. It contains a number of important centers (Fig. 29.13B). The pons cerebralis is an anterior, medial center that appears to have an associational function. The corpus centrale (central body) lies immediately behind, and is divided into several definite masses. It is also an associational center, receiving fibers from all other parts of the brain. The pedunculate bodies, also found in other arthropods, are the most important associative centers. They contain a number of discrete groups of neurons that form definite divisions in the stalked part of the pedunculate bodies. The size and complexity of the pedunculate bodies correlate well with the complexity of the behavior patterns of different species, and tend to vary with the size and the development of the compound eyes. Another associational center is found in a pair of ventral bodies, connected by a commissure. They are especially well developed in Lepidoptera, where they form a pair of accessory lobes. As a general rule, insects with larger ventral bodies tend to have smaller pedunculate bodies, suggesting that they have somewhat similar general functions. The most complex nerve centers are associated with the compound eyes.

The **deutocerebrum** contains the antennal relay centers, connected by a commissural tract. The **tritocerebrum** is small in insects, as there is no appendage in the somite to which it belongs. A commissural tract connects the two sides of the tritocerebrum. It is sometimes embedded in the circumesophageal connectives, and sometimes free. It is most closely associated with the labral nerves and the frontal connectives, which are important parts of the stomodeal nervous system.

The **subesophageal ganglion** is a compound ganglion and innervates all three of the somites associated with the mouthparts, the salivary glands, and some of the neck muscles. It is generally composed of a cortex containing the nerve cells and an inner neuropile. It is more than a center for the gnathocephalon, however, and exercises an important influence on motor activities in the body somites.

The trunk ganglia are complex. At the surface there are dorsal motor and ventral sensory tracts that are continuous with the ventral nerve trunks. Peripheral nerves from the ganglion extend to the musculature and the sense organs of the somite. As in other arthropods, the ganglia on the ventral nerve trunks tend to aggregate by moving forward and coalescing (Fig. 29.13C,D,E,F). As the thoracic ganglia aggregate, the abdominal ganglia tend to diminish in the Diptera, and something of the same tendency is seen in other orders.

The stomodeal nervous system corresponds functionally to the autonomic nervous system of vertebrates. It centers in the **frontal ganglion** and **recurrent nerve** (Fig. 29.13A), which passes along the floor of the brain and extends to the stomodeal region. Subsidiary ganglia and nerves are associated with some of the mouthparts, the salivary glands, the upper part of the digestive tract, and the aorta. It serves in feedback mechanisms that affect the physiological functioning of the viscera.

Reproductive System Myriapods have a relatively simple reproductive system. The **gonads** are dorsal and unpaired, and open into paired gonoducts that pass to the anterior genital somite behind the head. In some cases the males form **spermatophores**; in others the sperm is stored in large **seminal vesicles** associated with the **sperm ducts**. One or two pairs of accessory glands are present (Fig. 29.14A). The female system is essentially similar in its construction, and includes **seminal receptacles**, which store the sperm obtained during mating.

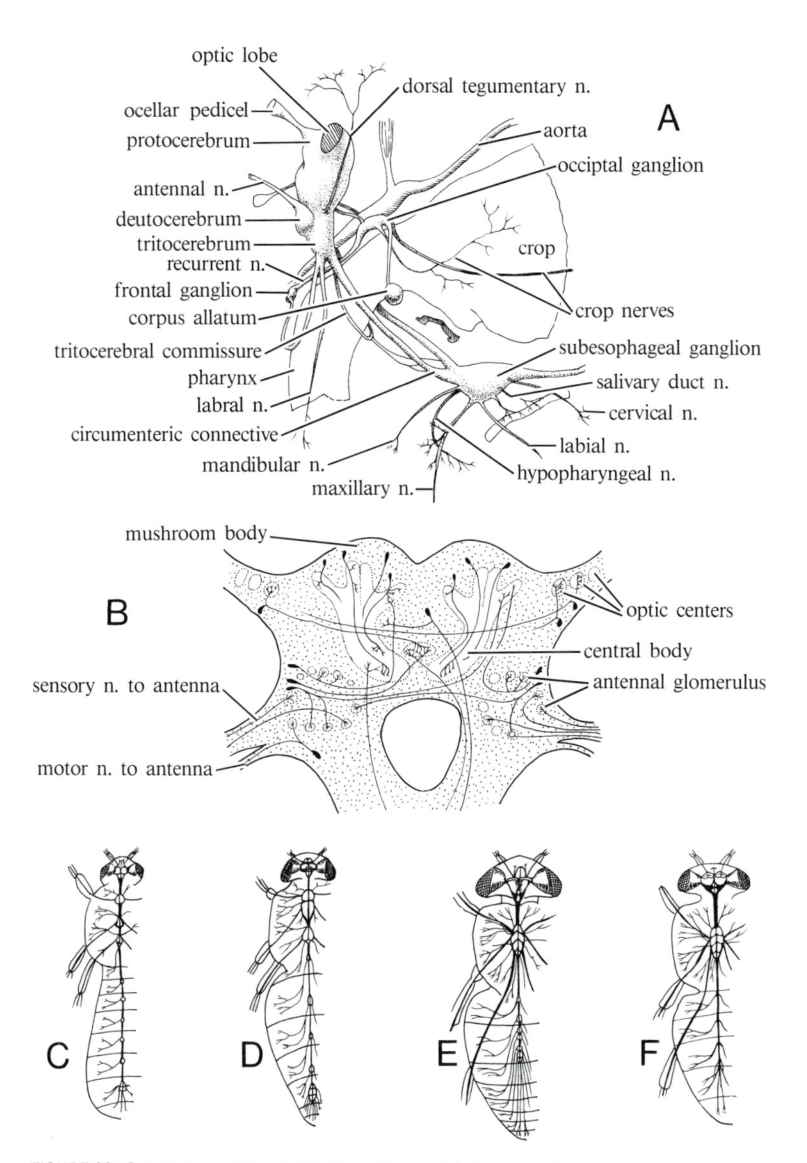

FIGURE 29.13. Nervous system. **A**. Side view of brain and head nerves in grasshopper. **B**. Chief nuclei and tracts in the brain of the cockroach. **C,D,E,F**. Evolutionary trends in central nervous system of Diptera with concentration of nervous system by union of ganglia and predominance of more anterior centers. Most primitive condition seen in *Chironomus* **C** where separate ganglia occur in divisions of thorax and in each abdominal somite. In *Empis* **D**, first two thoracic ganglia lie together. In *Tabanus* **E**, all thoracic ganglia assembled in compound thoracic ganglion, abdominal ganglia close together but discrete. In Sarcophaga **F**, thoracic ganglion more predominant, with abdominal ganglia scarcely seen. (**A** after Snodgrass; **B** after Hanström; **C,D,E,F** after Brandt, from Folson and Wardle.)

Insects are almost universally dioecious. Details of the reproductive system vary greatly from group to group, although the essentials remain quite stable.

The elements of the male system are the **testes**, **sperm ducts**, **ejaculatory duct**, and **intromittent organ**. The testes are composed of groups of tubular follicles bound together by a sheath of epithelial tissue (Fig. 29.14B). Each **follicle** contains an apical cell at the tip, thought to be a specialized spermatogonium that has assumed a nutritive function. A series of clusters of cells fill the follicle. The cells of each cluster are in the same stage of development. The youngest stages are found at the apical end of the follicle, and in a testis actively producing sperm, spermatogonical clusters are found at the apex of the follicle, followed by clusters of primary spermatocytes, secondary spermatocytes, spermatids, and mature sperm in regular sequence. Each follicle ends in a sperm ductule that unites with the sperm ductules from other follicles to form a sperm duct on each side of the body. The two sperm ducts pass posteriorly and join the ejaculatory duct. The ejaculatory duct arises from an invagination of the body surface and is lined with epidermis and cuticle, has muscular walls, and receives the secretions of a pair of tubular accessory glands. The **male accessory gland** secretion serves as a vehicle for suspending the sperm, or else hardens to mold the sperm into a spermatophore. The ejaculatory duct typically ends on the ninth abdominal somite or in the articular membrane between the ninth and tenth somites. An evagination of the body wall forms a **penis** around the opening of the ejaculatory duct. Various kinds of periphallic organs are found lateral to the penis

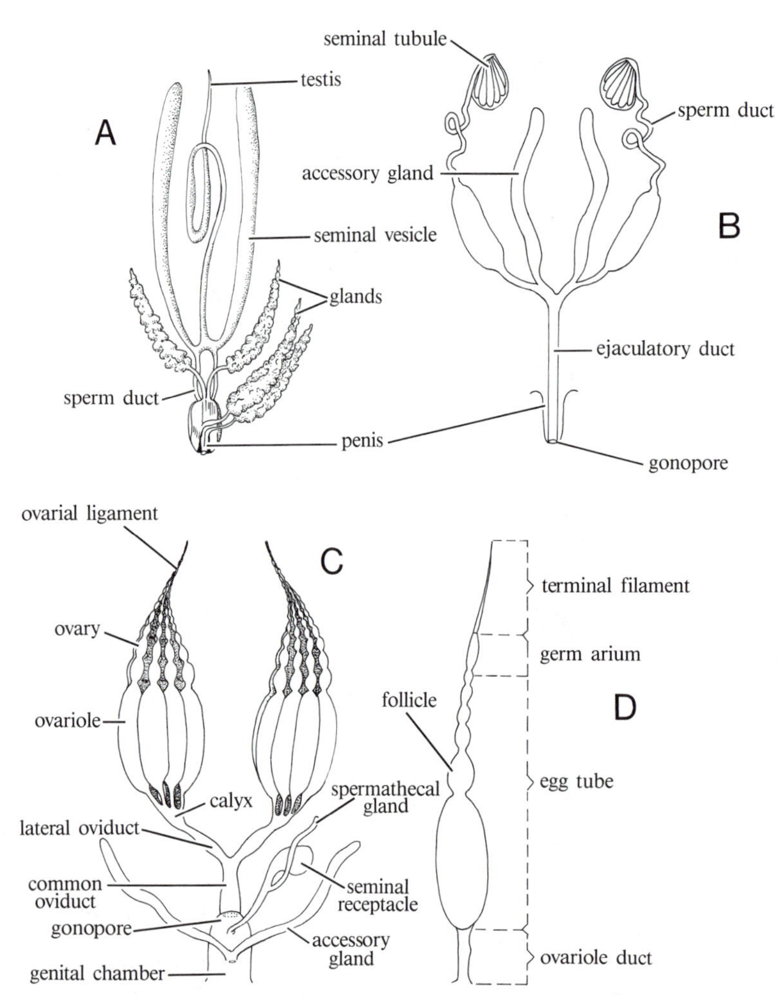

FIGURE 29.14. Reproductive system. **A**. Organs of male lithobiomorph. **B,C,D**. Insect organs: **B**. scheme of male reproductive system; **C**. scheme of female reproductive system; **D**. scheme of single ovariole. (**A** after Schaufler, from Snodgrass; **B,C,D** After Snodgrass.)

(Fig. 29.7D,E). These consist of modified appendages of the ninth somite, and various movable or immovable parts derived from the ninth or adjacent somites.

The female system closely parallels the male system in its basic organization. Each **ovary** consists of a group of ovarian tubules known as ovarioles, bound together by an epithelial sheath. The ovarioles open into oviductules that unite to form an oviduct on each side of the body. A typical ovariole consists of a terminal filament, often used to suspend the ovary from the body wall, a germarium in which the oögonia form, and a follicle or **egg chamber** where the ova mature (Fig. 29.14C,D). The paired oviducts unite to form a common **oviduct** that opens into the **genital chamber**. The genital chamber is formed by an invagination of the body surface on the eighth somite. It functions as a copulatory bursa, receiving the sperm at the time of mating. Its external orifice is the **vulva**, located on the eighth somite. In some insects the genital chamber is drawn out into a tubular part and is then called the **vagina**. In more generalized insects, a diverticulum arises from the genital chamber. This is the **seminal receptacle**, used to store the sperm received during mating.

Typically, a second invagination occurs on the female ninth somite. This gives rise to the **female accessory glands** and their external pore. In many insects the accessory glands have been joined to the genital chamber. The accessory glands commonly produce an adhesive used to glue the eggs to a surface, or a substance that hardens to form an egg case, but unusual secretions are formed in a number of insects. For example, it is the accessory glands that secrete the toxic materials that transform the ovipositor of a worker bee into an effective weapon.

Biology

Locomotion The movement of uniramians has been the object of much study (Manton, 1977). Millipedes, or diplopods, feed on plant material, as a rule, and have no particular need for speed. They move slowly, burrowing in the leaf mold or forcing themselves into crevices under bark or in rotting wood. When disturbed, many roll up, presenting the heavily protected dorsal surface, and eject toxic or nauseating substances from the stink glands. They have short, stout legs, suitable for thrusting the body forward against a resistance, and a battering-ram head particularly useful in burrowing (Fig. 29.3E). Each leg makes a step consisting of a quick recovery movement and a slow power stroke. An animal with as many legs as a millipede cannot be disorganized in its stepping. It is important that the many legs work together, each stepping in advance of the leg behind. Harmonious cooperation of the appendages is achieved by the passage of activation waves along the body. At a fast gait, wave follows wave rapidly, and each wave involves only a few pairs of legs. During the slow gait of burrowing movements, on the other hand, the waves are very slow and involve a large number of legs. In a long millipede, 50 or more legs may step together in a single powerful shove. The double somites, the head, the consolidated tergites, and the many legs with their slow activation waves are adaptations for burrowing or forcing the body into tight places. The overlapping of the dorsal borders of the diplosomites also aids in holding the body rigid when burrowing.

Centipedes are predominantly predaceous animals. Many live in the surface litter or under objects, but these can often run very rapidly in pursuit of prey. The longer legs are important for high speed, but create problems as well. As many as four pairs of legs may have overlapping strides, and if the steps were not very precisely coordinated, centipedes would be too tangle-footed to run. In millipedes, fast movement is achieved by the more rapid movement of activation waves along the body and by the reduction in the number of legs involved in a wave. The same principles apply in centipede movement, but the application is a good deal more complex. Centipedes have far fewer legs than millipedes, on the whole, and as running speed increases and the waves pass along the body more rapidly, a single metachronal wave does not suffice. Alternating waves pass along the bodies of the speediest forms. The step as well as the pattern of activation waves is different in centipedes. The longer legs make very quick backward effective movements and somewhat slower forward recovery movements. The fast centipedes make some of the fastest effective leg thrusts to be found among arthropods. As a result of the quick thrust and the fast activation waves, only two or three pairs of legs at a time may touch the ground together, and in fast species the rigidity of the body is as important as in the burrowing millipedes, although for a different reason. The extremely long-legged and fast scutigeromorphs have terga covering two or three somites, increasing the rigidity of the body. The somewhat slower lithobiomorphs have alternating long and short terga, which also appear to be an antiundulatory adaptation. Millipedes may move the legs slowly enough to feel for good footing, but centipedes step far too rapidly for such niceties.

Gripping hairs and other compensatory structures are found at the tips of the legs. The fast centipedes, *Scutigera*, for example, have legs that become longer toward the rear, helping to avoid tangling, and all of the more rapid groups have fewer body somites and fewer legs.

The geophilid centipedes, however, have followed an entirely different adaptational line. They live in cramped quarters beneath objects and can burrow like millipedes. However, they achieve their power in a different manner from millipedes. The very short legs do not provide the propulsive power. The body elongates and contracts during burrowing in much the same manner as in earthworms, and the short legs play about the same role as the setae of earthworms. When the geophilids walk, the legs move relatively slowly and feel about for a foothold. Each leg is on its own, in a sense, for metachronal stepping waves are not characteristic of the group. It is likely that the very short legs, with little overlap in step, have de-emphasized the importance of the close coordination of the legs during walking.

Walking movements are different in adult insects and in caterpillars, where peristaltic movements of the body accompany leg movements, and also differ in different kinds of insects. Jumping, digging, and running all present different problems and require different kinds of movements. The problems associated with walking are the same in principle, if not in detail, as those involved in jumping, digging, or other kinds of locomotor activity. As all cannot be discussed, walking has been chosen for discussion as the most basic type of leg locomotion.

Walking involves three kinds of problems. Each leg must make the movements of a step, the two members of a pair of legs must be suitably coordinated, and the three pairs of legs must work in cooperation. When a leg makes a step, the tarsal segments must be alternately flexed and extended to grip the substrate and free the foot, and the leg must be alternately flexed and extended to provide propulsion. If the insect is not to fall flat on its face, it must maintain a tripod at all times, and the legs tend to move in a tripod fashion. The first and third left legs and middle right leg work together as one tripod, and the other three legs act as the other tripod. Coordination for walking, thus, involves the control of individual legs to alternate properly the leg and the tarsal movements in a step, and the coordination of the six legs to produce organized forward movements, as well as an appropriate adjustment of the walking movements to the information about the environment available at the time.

The legs of isolated thoracic segments will make stepping movements. These are controlled by the thoracic ganglion and are triggered by impulses received through the proprioceptors of the somite when the somite is isolated. In the intact animal, however, stepping can be evoked by impulses moving through the nerve cord from other somites. Coordination of the several legs is achieved by intersegmental reflexes in the nerve cord. It is by no means a simple mechanical coordinating system, for an insect can compensate for the loss of a leg by modifying its method of stepping. In establishing a stepping rhythm, the prothoracic ganglion acts as the pacemaker, and several different pathways may be used for the movement of the stimulus to the more posterior thoracic somites. A second, and important, factor is the proprioceptive or positional information situation resulting from the stepping movements in the somite immediately in front of the middle and hind legs. The coordinating system is thus double failsafe.

An insect wing may be thought of as a lever, pivoting on the pleural process of the lateral exoskeleton (Fig. 29.15). The wing base projects inward from the pivot and is moved as the notum of its thoracic segment moves, but in the opposite direction. The notum is depressed when the dorsoventral flight muscles contract and the wings are lifted. The notum is bowed upward by the longitudinal muscles that extend from the anterior and posterior phragma at the ends of the notum; this depresses the wings.

It is evident that a simple flailing of the wings up and down would never result in flight. Forward thrust must be generated. Three factors enter into the forward drive. First, the costal vein is the most rigid vein in the majority of insect wings, and the wing does not remain flat as it moves up and down. Air is spilled from the buckled back margin, providing some forward impulse. The great variations in wing venation have been mentioned. It is very probable that these variations in veins, and the minor modifications in the relative rigidity of parts of the wing, are important factors in flight. Certainly, the wing movements of various insects are different, and they are very probably associated with differences in the qualities of the wings as airfoils, as well as with differences in the center of gravity and wing placement. Second, the wings are tilted or feathered as they move up and down. Tilting the wing aids in spilling air at the trailing edge, and helps markedly in providing forward thrust. Third, wings do not move straight up and down, but make a figure eight movement (Fig. 29.15F,G), intensifying the tendency of air to spill from the trailing edges of the wings. Narrow, fast-moving wings with a strong tilt and a strong figure eight component are found in the fastest flyers. Insects with flat wings with little give, and with wings that tilt little during flight and move more or less up and down with a minor tendency toward a figure eight motion, are slow and hesitant flyers. Dragonflies are good examples of strong flyers, and most butterflies are relatively poor flyers.

Where all of the factors are favorable, insects can achieve quite good speeds, although there has been a tendency to overestimate them. Sphinx moths and horse flies are the best that have been tested, achieving speeds of up to 30 miles per hour. Dragonflies reach speeds of about 25 miles per hour. It should be noted that the good flyers are heavy insects. The air resistance or viscosity of air becomes greater in proportion to the muscle weight and body size in smaller insects. In much the same way as the smaller planktonic crustaceans must cope with the viscous water as they swim about, the tiny insects are "air plankton," carried about by air currents even when wind velocities are not unusually high.

The frequency of wing beat varies with age, sex, species, and the composition, density, and temperature of the air, as well as the speed of air currents and the physiological condition of an insect. The wing beat of very light-bodied insects with large wings is very slow. Butterflies, for example, may move the wings as slowly as four times per second. Where the body is heavier and the wings are smaller, the frequency of beats rises. Bees and flies vibrate the wings far more rapidly, at speeds of 100 times per second or more. The frequency of wing beats becomes audible in insects of this kind, and serves as a very reliable guide. The heavy drone of a bumblebee corresponds to measured wing beat frequencies of 130 to 240 beats per second, whereas the high-pitched whine of the mosquito indicates wing beats of nearly 1000 per second. It is clear that the demands placed on the flight muscles vary enormously with the species as well as the conditions under which it flies.

Dragonflies, moths, butterflies, and orthopterans,

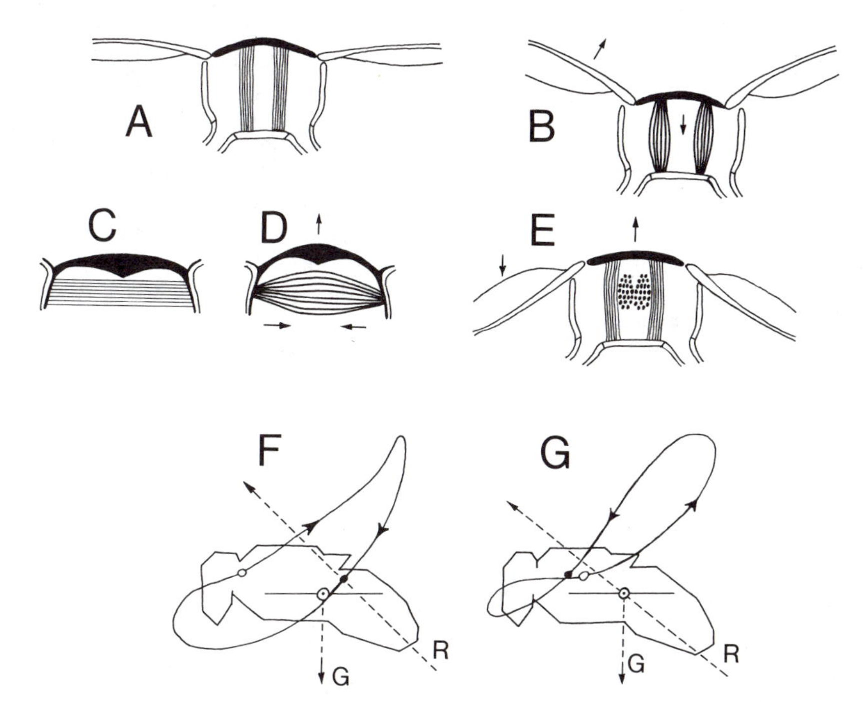

FIGURE 29.15. Wing movements. Wing bases articulate with tergal plates and pass over pleural plates. Wings raise when dorsoventral muscles contract, as in **B**, and rest when muscles relax, as at **A**. Further downward stroke requries more marked upward movement of tergal plate, made possible by longitudinal muscles that pass from anterior to posterior phragma in somite, shown at **C**. Contraction of these muscles bows tergum, lifting it in middle, as at **D**. Resulting movement permits depression of wings, seen at **E**. Wing movement also includes forward and backward components, which add to speed and stability of flight. Suspended in still air, fly moves its wings in path, shown in **F**. Resultant force passes along *dotted line,* marked R, inclined at angle of 48 degrees and sufficient to suspend fly. This line passes behind center of gravity **G**, therefore, suddenly released fly would dive downward to ground. Light air currents stimulate mechanoreceptors, and wing beat changes to assume figure eight shape, as shown at **G**. Resultant wing movements shift forward, passing through or in front of center of gravity. When released, fly moves forward. (**A,B,C,D,E** after Ross; **F,G** after Hollicks, from Wigglesworth.)

among others, have direct flight muscles. These contract with each wing movement and are not much faster than the leg muscles. The movements of the wing fall into the range of about 20 to 40 per second. In these insects the muscles are not arranged in the system shown in Figure 29.15, but are attached directly to the bases of the wings. In most of the more highly specialized orders, the flight muscles are attached to the cuticle of the somite rather than the wing bases, and the muscles function to move the wing indirectly.

Wing development follows a different line in minute insects than in larger ones. A tiny insect has an enormous surface in proportion to its weight; therefore, it encounters a very high air resistance. Such creatures do not fall rapidly through the air and can scarcely be said to fly; they very nearly swim or row through what to them is viscous air. The wings are not so extensively veined, and in some cases the wings are nearly devoid of veins. The minute thrips have very slender wings, equipped with a fringe of hairs, not unlike the natatory fringes on the swimming appendages of crustaceans, and the feather-winged beetles have featherlike tufts that "row" them through the air.

Food Getting and Digestion Except for the colobognaths, which have suctorial mouthparts and pierce plants to suck the sap, millipedes and symphylans generally feed on dead plant tissues. Some will attack living plants and, where they become abundant, can be quite destructive. Pauropod food habits are not very well described. Some feed on fungi, whereas others are thought to feed on dead animal or plant tissues.

Centipedes, as carnivores that lack appendages appropriate for easy manipulation of food, use a mechanism that quickly stuns their victims. The maxillipedes, with their sharp fangs and poison glands, are formidable weapons that immobilize animals that are fair prey.

Not much is known about myriapod digestive physiology. Presumably it is similar to that of insects, where much information exists on food procurement and processing.

The vast majority of insects are herbivores. The generalized mouthparts described above are used for cutting and chewing. Those insects that are carnivorous use, for the most part, the same kinds of mouthparts. The rolling action of the mandibles serve to bite and chew (Manton, 1964). A few insects have rather specialized eating habits and exhibit some mouthpart specializations. A few examples will suffice. Blood-eating flies have a cutting-sponging mouth. The mandibles are modified as sharp blades, and the maxillae as styles. The labium is spongy and is used to mop up blood from the wound made by the mandibles. Blood is sucked in through a tube formed by the modified epipharynx and hypopharynx. Many other flies have a sponging mouth. The mandibles and the maxillae are reduced, and a large, spongy labellum is found on the tip of the labium. Bees have a chewing-lapping mouth. The mandibles are not much modified and are of the biting type. The maxillae and the labium are modified to form a proboscis that can be thrust into flowers to gather nectar. Bugs, fleas, sucking lice, and mosquitoes have a piercing-sucking mouth. Details of construction vary, but the general principle is that the labrum, maxillae, and mandibles are united to form a beak containing an inner, sclerotized, sharp tube, formed of the epipharynx and the hypopharynx. When the plant or animal surface has been pierced, juices are sucked in through the tube.

Salivary secretions are extremely diverse. They lubricate the mouthparts and moisten the food in preparation for swallowing. In blood-sucking insects, the saliva contains an anticoagulant that keeps blood flowing while the blood meal is taken. Some insects live within the food mass, like the larvae so often seen in apples and other fruits. In these, the saliva often contains enzymes that are drooled over the food, partially digesting it before it is eaten. Occasionally salivary glands are modified for purposes not connected with digestion. In the Lepidoptera, the salivary glands are modified to form silk glands, and the mouthparts and buccal cavity contain specialized parts used in spinning the cocoon.

The midgut may be expanded or it may be a rather narrow tube, but it is generally equipped with prominent gastric caeca. In a large number of insects the midgut is lined with a chitinous tube, the peritrophic membrane (Fig. 29.16). The **peritrophic membrane** is not a product of the midgut cells, but is secreted by a ring of cells, apparently derived from the stomodeum, located near the stomodeal valve. In some species the peritrophic membrane is secreted continuously. It forms at the rate of 6 mm/hour in *Eristalis* larvae, and the nymphs of *Aeschna* discard two peritrophic membranes per day when unfed. The disposition of the peritrophic membrane varies in different insects. As a general rule, it is destroyed in the hindgut. In some cases special spines used for this purpose are found, and the musculature of the proctodeal valve sometimes helps to break the peritrophic membrane into small pieces. In other cases, however, it passes through the hindgut unharmed and encloses the fecal pellets. The part that the peritrophic membrane plays in digestion and absorption is uncertain. Many of the insects that feed on fluid foods have no peritrophic membrane, whereas insects that eat solid food usually have good peritrophic membranes, although exceptions occur on both sides. It is generally thought that it might protect the midgut cells from mechanical injury by hard food particles. The membrane is known to be permeable to enzymes as well as to the end products of digestion.

The high cost of enzyme production, as measured by cell regeneration, would make it advantageous to secrete enzymes only during active digestion. Controls of this kind, however, appear to be only moderately developed. Undoubtedly this is in part a normal result of the feeding habits of some insects. Many of them live within or upon food plants and feed more or less continuously. It may also be in part a reflection of the techniques used to estimate enzyme production, which are usually dependent on the histological condition of the midgut cells and may not be wholly dependable. Nevertheless, it seems fairly certain that in some carnivorous insects, enzyme secretion does change markedly when food is present in the gut. As all of the nerves reaching the intestine appear to be motor nerves to the intestinal musculature, the mechanism used for the control of secretion remains a mystery. Regulation by means of hormones has been suggested, and recent experiments have shown that, in a few cases at least, extirpation of the neurosecretory cells results in reduced ability to ingest and digest food.

As might be expected, a variety of enzymes have been recovered from the intestine of various kinds of insects. Lipases, carbohydrases, and proteases have been reported. A few produce cellulase, and it seems certain that special enzymes must be present in wool-eating and silk-eating insects and in others that have very specialized diets. Most insects are so small that knowledge of specific enzymes has been accumulating rather slowly. It is known that at least some enzymes are formed by microorganisms present in the insect gut. The most famous examples of this kind are the flagellates found in the intestine of wood-eating roaches and termites, some of which produce cellulase. Deprived of their protozoan fauna, these insects starve to death in the laboratory, even though they have a digestive tract full of food. In recent years, considerably more interest has been expressed in the contributions that microorganisms make to digestion in insects. It appears that the hemolysis of blood in the stomach of mosquitoes may depend on enzymes from bacteria, and the carbohydrases found in the blow fly larvae appear to be formed entirely by the action of bacteria. If this system of digestion is very widespread, it may also be a factor in reducing the control of enzyme formation in insects.

As a general rule, most absorption occurs in the midgut and in part of the hindgut. The sites of the absorption of substances may be very sharply defined in some species. In some insects, for example, a very short section of the midgut absorbs the greater part of copper and iron salts present in the gut. Fat absorption seems to center in the anterior part of the midgut. Carbohydrates are absorbed in diverse regions in different kinds of insects, but as a rule, the stomodeum appears to be impermeable to carbohydrates. In some cases the deposition of glycogen during the absorption of food indicates that the posterior half of the midgut is the primary site of carbohydrate absorption. Where sugars are not fully absorbed in the midgut, they may be absorbed in the hindgut. It appears that some proteins are absorbed in the midgut, but in some insects protein absorption occurs in the hindgut.

Although the hindgut participates in digestion and absorption, it is important in feces formation and especially in water conservation. No small part of the success of insects can be credited to the remarkable ability of insects to conserve water that passes into the hindgut. Cushions of tall epithelial cells located in the rectal lining, the rectal pads, are the most active sites of water absorption. They can be extremely efficient. Insects that live in dry places and feed on dry foods, for example, produce dry, powdery feces containing practically no water.

Desiccation is not a problem for some insects. Aquatic insects often produce relatively liquid feces. Some insects face the problem of too much fluid. Aphids and other Homoptera take in plant juices containing very little protein and must process a

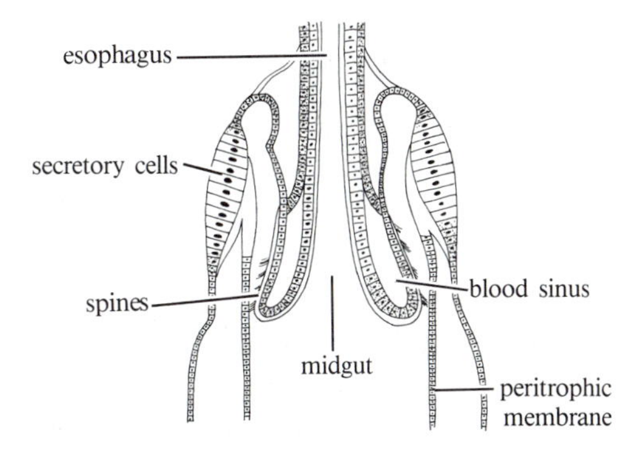

FIGURE 29.16. Scheme of posterior esophagus and anterior midgut showing site of peritrophic membrane secretion in simuliid larva.

very large amount of material to obtain an adequate protein supply. They have a unique filter chamber. As a rule, water and some carbohydrates pass from the filter chamber into a tube that leads to the hindgut, or pass directly into the hindgut. The rest of the fluid food passes through the long midgut, where it is processed and where protein is absorbed. The feces are rich in sugar and are voided as a sweetish liquid known as honeydew. The relationship between certain ants and aphids is well known. Some species of ants regularly tend aphids, placing them in position on plants and visiting them to collect the honeydew they produce. There is evidence that the rate of passage of fluids through aphids is a factor in growth rates, for aphids attended by ants pass more honeydew and reproduce at a higher rate than unattended aphids.

Respiration In primitive tracheal systems, the tracheal tubes from each spiracle form an independent system, and each tracheal tube in the tuft arising at a spiracle passes directly to the tissues without branching. Improvements of this primitive arrangement follow several lines. (1) Individual tracheal tubes may branch on their way to the tissues, and later main trunks appear from which branching tributaries arise. This increases the capacity of the tracheal system for holding air and increases the potential number of tracheae, permitting more nearly equal distribution to the tissues. (2) Tracheae from the various spiracles can be integrated into a continuous system by the development of commissural tracheae within the somite and by the anastomosis of tracheae from adjacent somites to form longitudinal tracheal trunks. As a result, the plugging of a spiracle or some local injury does not necessarily lead to respiratory failure in any of the body tissues. An even more important advantage is the fact that body movements may partially compress the longitudinal trunks, resulting in movement of air in the tracheal tubes, and in more rapid delivery of oxygen to the tissues. This may be capitalized on by the coupling of dilated air sacs with the longitudinal trunks. Pressure on the air sacs may result in an appreciable flow of air in the outer part of the tracheal system throughout the body. (3) The development of spiracle valves that can close the tracheae off when respiratory needs are low, and open when the animal is under respiratory stress, is an important factor to reduce water loss. Control of the spiracles, when accompanied by the development of longitudinal trunks and especially when air sacs are present, can result in the passage of air along a definite course, with some spiracles serving as inhalant and other spiracles as exhalant openings.

Where respiratory exchange occurs at the body surface, it is evident that blood is responsible for distributing the oxygen to the tissues. Respiratory pigments are extremely rare in insects, and the blood has about the same oxygen-carrying capacity as the blood plasma of vertebrates. Since insects are small, this suffices to make a significant contribution even in the larger species. About 25 to 50 percent of the respiratory exchange of various kinds of insect larvae occurs at the body surface, and even adult roaches and grasshoppers carry on ten percent or more of their exchange at the body surface. Under conditions of stress, surface respiration increases in importance and may reach as high as 50 percent in adult insects with blocked tracheal systems.

Although structural modifications may amplify the respiratory surface and provide for more efficient drain-off of oxygen, the ventilation of the respiratory surface is a very important factor in determining the rate of respiratory exchange. Gill ventilation is relatively unimportant for insect larvae that live in habitats with high oxygen content or with rapid currents, but is more consequential in still waters with considerable bacterial activity where oxygen tensions may be relatively low. A number of insect larvae have gills that move rhythmically, or make body movements that ventilate the gills at times of respiratory stress. Dragonfly nymphs fall back on a system used in a number of other invertebrates. Water is pumped in and out of the anus, and rectal folds contain a rich bed of tracheal capillaries, used for respiratory exchange (Fig. 29.11C).

It would be easy to overevaluate the importance of gills in aquatic insect nymphs and larvae. Where species of about equal size and activity live side by side, the survival of some without gills raises the question of the value of the gills in branchiate species. In some cases the gills can be removed without appreciably changing the rate of oxygen consumption. Presumably the gills are important only at times of respiratory stress in these forms. On the other hand, the removal of the gills of some mayfly nymphs reduces oxygen consumption to about one-fourth of the normal rate.

Calculations show that the diffusion of gases in a tracheal system is adequate to account for the delivery of oxygen and the elimination of carbon dioxide in most insects. The system of branching tubes provides a continuous column of air that extends to the smallest tracheae. Gaseous diffusion is quite rapid, and the tracheal system is surprisingly efficient. The least understood aspect of tracheal respiration by diffusion is the way that the tracheole end cells function. The end cells persist after the tracheoles have formed. A fluid extends partway up the tracheoles.

In fresh tissue the fluid extends further than in exhausted tissue. Presumably osmotic changes in the tissue cells and blood cause different levels of tracheolar filling. In any case, the mechanism tends to increase the delivery of oxygen at times of maximal needs. In fireflies, it appears that the tracheole end cell may pump this fluid while flashing, thus actively increasing the rate of oxygen delivery.

The margin of safety for oxygen diffusion is increased and higher levels of activity are maintained if the tracheal system is actively ventilated, made possible by the development of tracheal air sacs (Fig. 29.10B). Small, somewhat isolated air sacs may be found in legs or other parts of the body, where muscular movements will help fill and empty them.

As a general rule, complex ventilation mechanisms tend to appear in insects with a complex tracheal system. Grasshoppers, for example, relax the abdominal muscles for about 0.25 sec. During the last 0.2 sec of this period, the anterior four spiracles are opened and air enters through them. These spiracles close just before the abdominal muscles contract, thus retaining the air that has entered. With all ports closed, the abdominal muscles are contracted. This compresses the air sacs, and air is driven into the smaller, deeper air sacs, beyond which diffusion is the main mechanism for delivery of oxygen. The contraction of the abdominal muscles is maintained for about 0.4 to 1.5 sec. During the last half to third of this period, the spiracles that were not used for inspiration are opened, and air escapes.

Excretion Little is known of excretion in myriapods. In insects, uric acid is the most important waste, but small amounts of ammonia, urea, and allantoin are also excreted. Some wastes are stored as nontoxic compounds in the cuticle, and some are used for the synthesis of pigments. Nitrogenous wastes are sometimes excreted by the salivary glands or the intestine, and athrocytic cells known as nephrocytes are often found around the gut wall and the Malpighian tubules. However, the Malpighian tubules are the principal excretory organs, and the other methods of excretion are minor factors.

Two kinds of Malpighian tubules can be distinguished. One picks up water and waste materials throughout its length and pours fluid urine into the gut. In tubules that work in this way, all of the reabsorption of water occurs in the rectum. Other insects have Malpighian tubules that are divided into a proximal tubule filled with fluid urine, and a distal tubule where water is reabsorbed and crystals of uric acid appear. The crystals, suspended in water, pass to the rectum where further water reabsorption completes the drying of the wastes. Water reabsorbed in the rectum is returned to the blood. This recycling of water makes the most of the modest water resources of the insect body. Both types of Malpighian tubules may be either free at the tip or bound by connective tissue to the rectal wall.

The Malpighian tubules have muscular walls and can move about in the hemocoel vigorously. As insects have no cilia, these movements are important to agitate the contents of the tubules and help to keep the diffusion gradients as favorable as possible at the cell surfaces. The muscular movements also convey the contents of the Malpighian tubules to the hindgut. Tubule movements appear to be uncoordinated and independent of nerve stimuli.

Circulation Heartbeat physiology is imperfectly understood, and the data so far available suggest considerable diversity. Triangular alary muscles are arranged segmentally along the heart wall. When they contract, they dilate the heart, allowing it to fill. In some insects the rhythmic contraction of the alary muscles has been shown to alternate with the heart contractions, but in some insects transection of the alary muscles does not interfere with heartbeat. Some insects have myogenic hearts, with the beat originating in the striated cardiac muscle itself. Others have neurogenic hearts, with definite ganglion cells in the wall. Some tendency for insect hearts to change from myogenic to neurogenic during development has been noted, and there are other indications that myogenic hearts are more primitive and have tended to be improved by the addition of neurogenic mechanisms.

Although the insect circulatory system is open, the large hemocoelic spaces are enclosed by the body wall, and appreciable pressures may be present. For example, from 18 to 87 mm water pressure has been reported in dragonfly nymphs. However, there is surprisingly little information about the blood pressure of insects. In many instances no hemorrhage occurs at a wound, suggesting that the pressure is very low. It is probable that the pressure gradients involved in blood flow are quite low. Certainly the flow of the blood is not dependent only on heartbeat. Contraction of the abdominal muscles during respiration and of the thoracic muscles during flight plays an important part in propelling the blood. Blood flow increases with temperature and is affected by activity, carbon dioxide content, and oxygen availability.

Insect blood transports foods, wastes, and hormones. It undoubtedly aids somewhat in respiratory transport, especially of carbon dioxide, but this is a less important function than in organisms without a tracheal system. Only scutigeromorph blood contains the respiratory pigment hemocyanin, absent in other myriapods and insects (Magnum et al., 1985). The blood is also an important hydraulic fluid, in this way contributing to respiratory movements of the air sacs and tracheae, to stretching the new cuticle after a molt, and to inflating the wings before they harden as the insect emerges from the pupal stage.

A number of morphological types of blood cells can be distinguished (Gupta, in Gupta, 1979). Some hemocytes contain fats or carbohydrates, and probably function in food storage and transport. Marked fluctuations of some of the blood cells are noted in premolt and postmolt stages and in prepupation and postpupation stages, and some of these probably reflect changes in phagocytic cells as well as cells used for food storage. Some of the phagocytic cells have the usual functions of protecting the body from bacterial invasion and removing foreign particles, but others appear to be extremely active during metamorphosis. Hemocytes also migrate into injured areas and appear to be important in wound healing, where they agglutinate to prevent the loss of blood.

Sense Organs There is very little specific information on the function of the mechanoreceptors and chemoreceptors of myriapods. Sensory setae, bristles, and pits, essentially like those of other arthropods, presumably function in much the same way. However, the structure of some of the special sense organs is of interest.

A pair of peculiar sense organs of pauropods, known as pseudoculi, occur on the head (see Fig. 29.21A), about in the position ordinarily occupied by eyes. At first they were thought to be unusual photoreceptors, but their structure seems to preclude any possibility of their being light-sensitive. Each pseudoculus consists of a convex field of cuticle that roofs a large, fluid-filled space. Beneath the fluid there is a sensory epithelium made up of large cells innervated from the protocerebrum. No definite function has been ascribed to them, but it has been suggested that they might be sensitive to vibrations.

Symphylans, most millipedes, and some centipedes have a pair of sensory pits on the head, immediately behind the antennae. These are the organs of Tömösvary. They vary considerably in different species. They are sometimes open pits or grooves. At the bottom of the depression are sensory cells innervated from the optic lobe of the brain. Their function is uncertain.

Many of the myriapods are blind, but others have eyes of one kind or another. Their eyes are ocelli, innervated like the lateral compound eyes found in other arthropods. Details of ocellar structure are variable, but the general scheme remains about the same. A corneal lens covers the optic cup. The cup is lined with peripheral and central sensory cells equipped with striated sensory borders, somewhat like the rhabdomeres found in the retinula cells of ommatidia in compound eyes. Ocelli are sometimes isolated, but are usually clustered in smaller or larger groups. In many cases the members of a cluster are closely packed, and in this case the corneal lenses unite to form a single lens structure. The ocelli beneath, however, do not unite; each retains its individuality. The scutigeromorph centipedes are exceptions. They have large compound eyes, with each element rather like the ommatidia of insects and crustaceans. Although the eye of *Scutigera* has a far more advanced structure than the eyes of other centipedes, it does not approach the complexity of most insect or decapod eyes. At most, only a couple of hundred ommatidia are present. It resembles the eyes of some larval insects more than the eyes of adults.

Strong tactic responses are common among myriapods. They are usually negatively phototactic, as is suitable for organisms not well protected against desiccation. Even the eyeless species are usually photonegative, indicating a considerable dermoptic sensitivity.

A great deal more is known about the sense organs of insects than of other groups of arthropods.

Other than the eyes, the principal sense organs may be classified as (1) hair organs, long or short hairs or bristles, (2) campaniform organs, dome-shaped organs with a sensory ending, (3) plate organs, covered by a flat plate beneath which the sensory elements lie, and (4) compound scolopophorous organs, usually a bundle or cluster of associated sensillae.

Hair organ mechanoreceptors function when the sensory ending is mechanically deformed. They are used primarily for touch, orientation in space, and vibration reception, and are characteristic mechanoreceptors in arthropods generally. A hair sensillum (Fig. 29.17A) is set in a socket covered by a flexible membrane. It is formed by the cooperation of two cells, a tormogen cell and a trichogen cell. The tormogen cell is a modified epidermal cell that secretes the delicate membrane on which the hair rests. The trichogen cell projects through the tormogen cell and secretes the cuticle of the hollow hair. A sensory cell is associated with the tormogen and trichogen cells, typically extending through them and ending in a sensory projection that extends into the hair. In some cases, however, the sensory tip ends at the base of the hair. Movement of the hair excites the sensory ending to evoke an impulse in the sensory cell.

How the cell receives the energy of stimulation is not known, but extremely high sensitivities are attained, especially in vibration receptors. A cockroach vibration receptor can react to vibrations of less than 10^{-9} cm, actually less than the diameter of a hydrogen molecule.

The nerve endings may discharge only when moved or may discharge continuously while under pressure, depending on whether they adapt rapidly or slowly. Rapidly adapting endings are ready to discharge in small fractions of a second and are especially valuable for vibration reception, whereas the slow-adapting endings are better suited for tactile reception. Vibration reception is important in providing information about the movements of the air or water in which the animal lives. Vibration receptors are especially well developed in swimming and flying insects.

Hair sensilla are sometimes used to form compound sense organs. Hair plates, for example, are associated with the joints of cockroaches. They are deformed by movements of the joints and adapt slowly. Their characteristics would make them especially useful in providing information about the position of the body and the attitude of its parts, and are thought to be important in proprioception.

The **campaniform organs** are also used for proprioception. Each sensillum is formed by a thin dome of two-layered cuticle. The tip of a sensory cell extends to the inner layer of the dome (Fig. 29.17B). A single epidermal cell is associated with the campaniform organ, presumably laying down the modified cuticle in the dome. The campaniform organs are sensitive to cuticular stress and are important proprioceptors. Most of the proprioceptors are stretch receptors, providing information about the tension of a particular muscle. The campaniform organ, however, is responsive to the stress affecting the cuticle at a given point, and therefore provides information on the forces being applied to the exo-

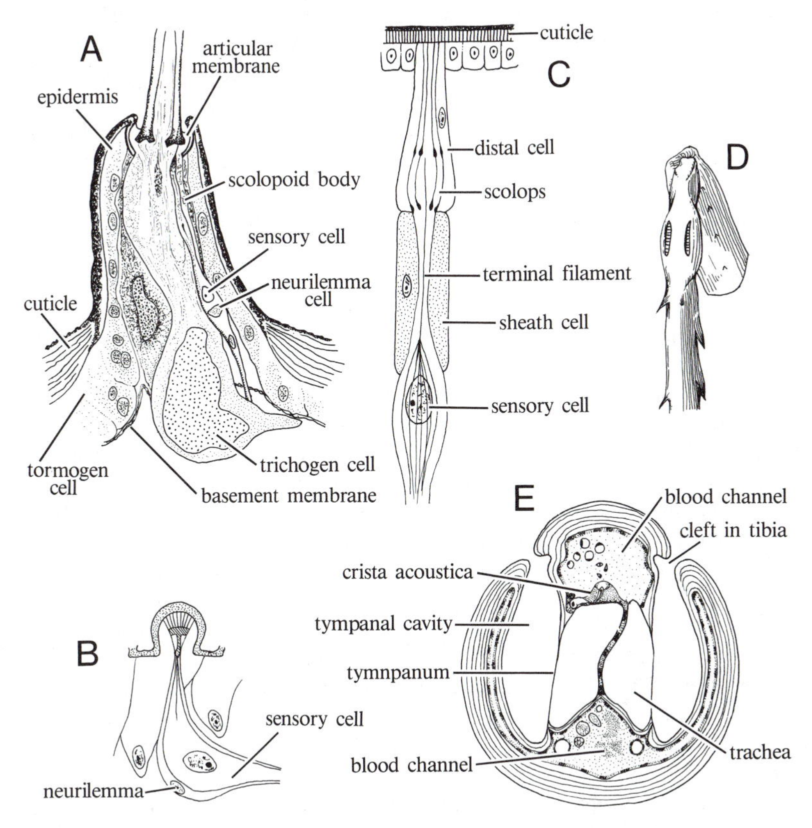

FIGURE 29.17. Sense organs. **A**. Section through base of hair sensillum. **B**. Campaniform organ. **C**. Scolopophorous organ. **D**. Tibia of grasshopper showing openings of proprioceptive tympanal organ. **E**. Section through tibia showing structure of tympanal organ. (**A** after Hsü; **B** after Pflugstaedt; **C** after Snodgrass; **D** after Weber; **E** after Schwabe, from Wigglesworth.)

skeleton. Proprioception has replaced gravity reception insofar as righting reflexes are concerned. Righting reflexes are initiated when no stimuli are received from the leg proprioceptors. The same mechanism is used in takeoff and landing in flying insects where the tensions recorded by the leg proprioceptors initiate predetermined movements of the body and the wings.

Scolopophorous organs are groups of sensilla that consist of a cap cell that touches the cuticle and an enveloping cell that encloses the sensitive tip of a sensory cell (Fig. 29.17C). The sensory ending is characterized by a scolops, or sensory peg, that contains a central axial filament to the cuticle, to which it is usually attached. The scolopophorous organs are widely distributed in insects and are found in various positions in the body. Some of them are definitely used for vibration reception, but these are found in connection with other parts that are suitable for sound reception. Very little is known of the function of the other scolopophorous organs. They may be sensitive to inner pressures placed on them by the surrounding tissues, for the scolops is often tethered to the cuticle and would remain in place even though the tissues around were to be pushed to one side. If this idea proves to be correct, the scolopophorous organs would be ideal for the reception of stimuli associated with acceleration. There is evidence to show that an acceleration sense may be of importance in the flight of dragonflies. These organs would also be useful in orientation to gravitational pull, which deforms the soft tissues around them. Even in flight, however, orientation to gravity appears to be relatively unimportant. A dragonfly flies with its dorsal surface toward the light and, if illuminated from below, flies upside down, and light has a very strong influence on the flight behavior of all insects.

Scolopophorus sensilla are used for sound reception. In this case the cuticle associated with the scolophorous organs forms a tympanic membrane, as in the phonoreceptors in the legs of some grasshoppers (Fig. 29.17D,E). Phonoreceptors are found on the thorax, abdomen, legs, and antennae of various kinds of insects.

Chemoreception is especially associated with modified sensory hairs and **plate organs**. They are concentrated in the antennae and mouthparts and in the vicinity of the ovipositor. One form of chemoreceptor is a peglike sensory hair with a thin cuticle. Cap and enveloping cells like those of mechanoreceptors may or may not be present. Arguments about whether sensory endings of chemoreceptors are exposed or covered by cuticle are partially solved. At least some have tiny pores through which exceedingly delicate microvillilike extensions of the neurosensory cell extend. It is possible that very thin cuticular coverings of some sensillae are permeable to chemicals. At the present time, a great many sensillae are of unknown function. There is evidence to indicate that the several different endings in chemoreceptors produce different firing patterns, and that mechanoreceptors accompany the chemoreceptive cells.

Insects are capable of receiving chemical stimuli from distant sources as well as from those in contact with the receptor. Presumably, the principal difference between these receptors is the threshold at which impulses are initiated. The most delicate chemoreceptors are known to be located on the antennae. Another point of high sensitivity is the maxillary palp. In the larvae of cabbage butterflies and in blow flies, removal of the antennae raises the concentration of substances required to cause response by the animal. Removal of the palps when the antennae are intact does not alter the threshold for response, but removal of the palps when the antennae have been destroyed causes a second increase in the threshold. Chemoreception is now known to be the most important element in the foraging patterns of honey bees. Foraging bees return to the hive and stimulate other workers to go out and seek a nectar source. The entire process is linked to a well-developed chemoreceptive sense rather than any purported language abilities (Wenner and Wells, 1987).

Humidity reception is a special case of chemoreception and appears to involve special chemoreceptors, built on similar lines but separated from the distance chemoreceptors. Humidity reception does not involve simple acceptance or rejection. As a rule, some humidity level is preferred, and negative reactions are evoked by drier or moister surroundings. Undoubtedly the humidity receptors play an important role to help insects choose a home area, sites for egg deposition, and the like. Well-fed and starved insects of the same species may differ in their humidity choices, as may different stages in the life cycle. It is likely that humidity reception is an important factor in some of the characteristic movements that many insect species make, passing from one part of the habitat to another as they mature, or after feeding, resting, or carrying out other activities. In some insects all humidity reception occurs in antennal organs, and removal of the antennae completely destroys the ability to orient to water vapor. In others the humidity receptors are more widely distributed.

Photoreception centers in the compound eyes and

ocelli, although some insects also have dermoptic sense organs. The compound eyes are remarkably like the eyes of the higher Crustacea, and their structure need not be discussed here. As in the Crustacea, pigment migrates distally and proximally about the ommatidia, converting the eye from an apposition to a superposition eye and back again, and diurnal rhythms of pigment movement follow patterns essentially like those of Crustacea. It appears, however, that nerve stimuli are responsible for pigment migration in insect eyes, instead of hormones as in crustaceans.

A great deal of work has gone into studies of the sensitivity of insects to light waves of differing amplitudes and frequencies, and of the ability of insects to distinguish between colors and shapes of objects. In many respects the compound eye compares very unfavorably to the human eye as it contains far fewer ommatidia than the number of rods and cones per unit area in a human retina. Experiments have shown that bees can be taught to distinguish some shapes by providing food at sites marked by symbols of a specific shape. Similar experiments have been used to determine the ability to distinguish colors. Only the simplest kinds of shapes are effective for training, and the kinds of errors that bees make when faced by what we consider very simple shapes are remarkable. The physical analysis of the insect eye indicates that it could form a far better image than experiments with bees would indicate. It is not clear whether bees cannot distinguish between shapes very well, do not respond adequately to the system of rewards that is used, cannot use information of this kind in relation to the search for food, have a central nervous system that is unable to record patterns of the kind we naturally think of, or have eyes with receptors that do not record images. It is the problem of analyzing sensory reception in nonhuman subjects.

On the basis of the available information, it would appear that the visual acuity of insect eyes is very low in comparison with human eyes, in the range of 0.01 to 0.001 where the human eye is 1. The ability to discriminate between different intensities is also far below that of humans, but this differs markedly with the level of illumination. At relatively high levels of illumination the insects are considerably better, although still well below humans. The difference between humans and the insects tested falls in the range of a factor of about 50 to 250. Insects fare far better on a flicker-fusion test. At a frequency of about 45 to 55 per second, flickers fuse in the human eye, but fusion does not occur until frequencies of from 200 to 300 per second are used in bees and blow flies. Flicker-fusion frequencies, however, differ markedly in different species and are considerably lower in slow-moving forms. Evidently insects with eyes having high flicker-fusion frequencies would be very well suited for the analysis of a fast-changing landscape as while flying. A considerable portion of the brain is occupied by visual centers.

Larval insects commonly have lateral eyes, composed of a single ocellus. The ocellus is not constructed like most ocelli; it has a double lens, and the retinula cells form a rhabdome, rather like the rhabdome in an ommatidium. The lens system is capable of projecting an image, and in some cases projects a triple image, but it is generally conceded that the neuronal connections are inadequate to convey an image.

Most adult insects have two or three ocelli located in the center of the head. These are typical ocelli, with a corneal lens, and are not adapted to image formation. Their function is uncertain, but there is evidence that when they are not illuminated, the ocellar nucleus in the brain initiates a characteristic pattern of firing in the circumenteric connectives. The firing pattern may indicate a change in the excitation of the brain that may well initiate or augment a group of reflex activities. In this case, darkening or illumination of the ocelli may change behavior patterns.

Reproduction Males have specialized legs, located in different areas depending on the myriapod group. They are used as claspers during mating or for sperm transfer. Some of the millipedes have penes associated with the sperm ducts at the gonopore. Mating habits are quite different in the various myriapods. Gonopores may be opposed, penes may inject sperm into the seminal receptacles, gonopods may be charged with sperm and brought in contact with the female gonopores, or spermatophores may be transferred with the gonopods. In any case, the seminal receptacles are charged with sperm, in some cases long before the eggs are mature.

The females produce shelled ova and exhibit a variety of kinds of maternal care. In some cases the eggs are produced more or less continuously and left in the soil or surface litter. Many millipedes construct nests for their eggs, using secretions from the anus to consolidate particles of soil or other materials. They tend the nests for a few days before leaving the young to develop on their own. Centipedes may produce nests or leave the eggs in small clusters in the soil or litter. It is not uncommon for centipedes to guard the young as well as the eggs for a time (Lewis, 1981). Apparently the nests and the guard duty are important to prevent the growth of fungi, which is one of the principal hazards of early development.

The female insect has arrangements for storing sperm and usually mates once or a few times in her lifetime, even though an extensive period of egg laying follows. The stimuli that initiate mating vary widely, ranging from odor signals, sound signals in insects with sound-producing or stridulating organs (Fig. 29.18), color signals, and various kinds of courtship behavior (Thornhill and Alcock, 1983). Mating occurs very quickly in some insects, and in others may take several hours or several days. As a general rule, the penis is inserted into the bursa or seminal receptacle to inject fluid sperm suspensions, or sperm are introduced in a saclike spermatophore. In any case, the sperm find their way to the seminal receptacle where they are stored.

The course of early development (Anderson, 1973) has been described (see Chapter 28). At the end of embryonic development, the young insect faces its first crisis. If it is to continue its development, it must escape from the chorion. In some insects the way has been prepared. The chorion contains an opening, covered by an easily detachable caplike operculum, and escape is not difficult. Otherwise, the embryo inflates itself with fluid, beats its head against the front face of the chorion, or splits the chorion by pulling at it. In spite of its efforts, the chorion of the egg sometimes becomes a shroud for the young insect. Some insects have reduced the dangers of hatching by the development of a special chitinized part of the exoskeleton used during escape from the egg.

Once out, the young insect faces a long period of growth and development before maturity. Covered by a firm cuticle, it can only grow and change form at molt. Development consists of a series of molts and instars. As a general rule, the number of molts is predetermined, and the form of the young insect at each instar follows a precise pattern characteristic of the species. Development is a regular sequence of instars that have an inherent form in each species.

When eggs are large and well protected, there is a tendency for embryonic development to be longer, with a corresponding reduction in the developmental period of the juvenile. This tendency has been very marked in arthropods generally in that they do not hatch in such early stages as do polychaetes. A number of the myriapods, and some of the most primitive insects, hatch before all of the body somites have formed. In these species, new somites are added during molts, and development is said to be anamorphic. In all but a few insects, however, the period of development in the egg is prolonged, so that all of the somites are present when it hatches.

Insects that hatch with all of the somites characteristic of the adult do not always resemble the parent, as do myriapods. No insect hatches with its wings fully developed, and even the wingless insects may have early instars that do not resemble the adult. Exceptions to this general rule are seen in insects that are primitively wingless. Insects belonging to the division Apterygota, the Thysanura and the

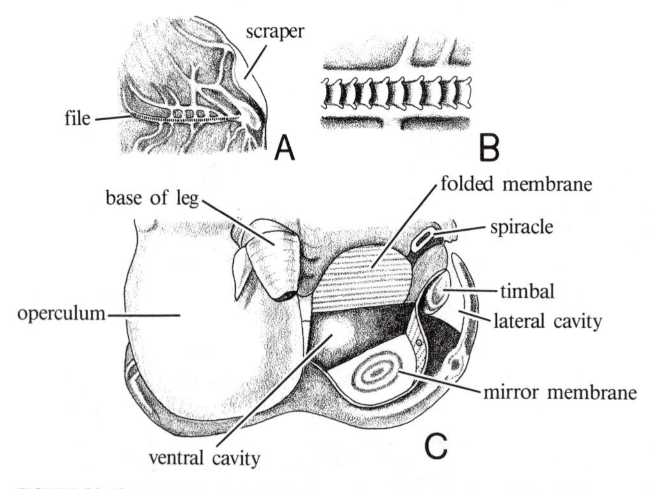

FIGURE 29.18. Musical organs. Crickets produce their chirp by rubbing filelike rasp **B** on wing over special wing surface, scraper (**A**), that vibrates to create sound. Cicadas have complex stridulating organs **C**. Muscle attached to dome-shaped membrane contracts and relaxes in quick succession causing membrane to vibrate. Cavity serves as sounding chamber to reflect sound off other membranes to reach outside. (After Comstock.)

Diplura, rather closely resemble the adult, although they are diminutive. Their development is essentially like the epimorphic development of some centipedes. As insects generally must undergo some metamorphosis to become adults, these orders are unusual. They are ametabolous, or undergo inconspicuous metamorphosis.

All of the winged or secondarily wingless insects undergo marked changes in appearance during the posthatching period of development. A large number of the insect orders are hemimetabolous, changing gradually as they molt, the body slowly assuming adult shape as well as size. In this large group of insects, the wings are present as wing buds at the time of hatching unless the adult is wingless. A juvenile insect that undergoes this gradual metamorphosis is called a **nymph**. Each developmental molt is accompanied by some change in form, and no one molt can be considered more metamorphic than another (Fig. 29.19). Development involving gradual metamorphosis is said to be **paurometabolous**.

Some of the hemimetabolous insects undergo early development in ponds and streams. The term **naiad** is used to designate the nymphal stages of aquatic insects. Dragonflies, stoneflies, and mayflies are common examples of insects with this type of development. The juvenile lives in and is adapted to an aquatic habitat. It usually has external gills or other specialized respiratory specializations, and may be adapted for swimming or other activities that are suitable for its aquatic home. Wings would be a nuisance in the water, and the wing buds of these young insects do not grow as rapidly as in land insects that undergo gradual metamorphosis. Although each molt is accompanied by some metamorphic changes, the last aquatic instar by no means resembles the adult. When the time for the last molt comes, the naiad creeps up on emergent vegetation, deserting its watery home for the air. The last molt then strips the body of its aquatic adaptations and provides the animal with terrestrial adaptations, including a fully grown set of wings. This molt is not far removed from the metamorphic molt that occurs in insects passing through a pupal stage. This type of development is sometimes termed **hemimetabolous**, in contrast to paurometabolous development. Some entomologists, however, use the term hemimetabolous for all cases of gradual meta-

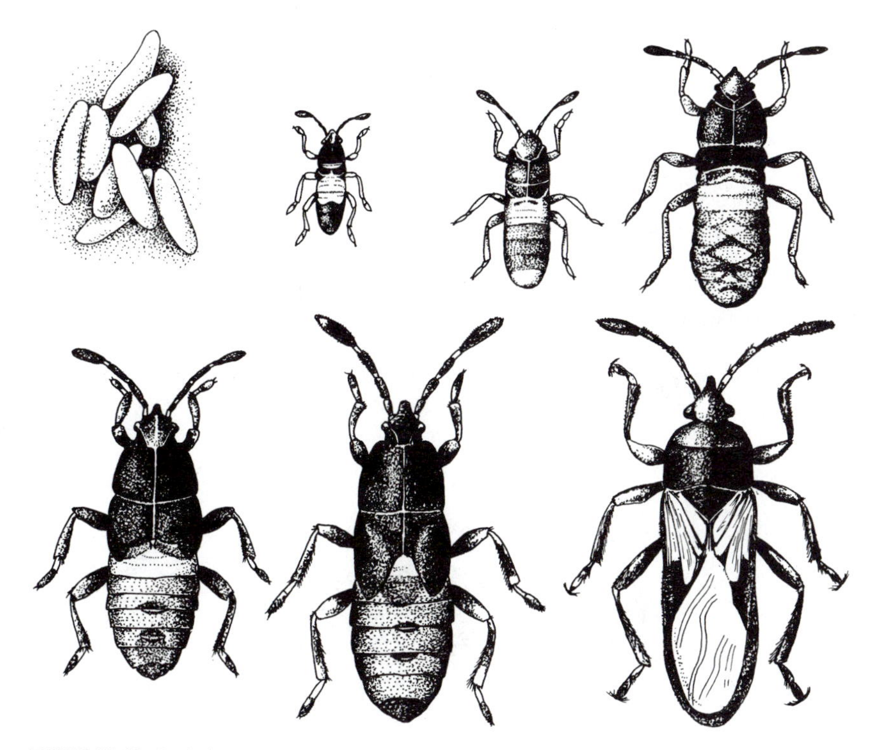

FIGURE 29.19. Gradual metamorphosis in hemipteran, *Anas*. Hemipterans, or true bugs, characterized by wings with leathery bases and membranous tips that fold over back, with membranous part of one wing lying over membranous part of other wing. Wing buds, seen in youngest nymphs, increase in size as insect grows with each molt. (From Ross.)

morphosis, and when reading about insect development, it is necessary to determine how an author uses the term hemimetabolous.

Holometabolous insects undergo a different kind of development. They bear no resemblance to the adult when they hatch from eggs (Fig. 29.20A), and are called **larvae** (Fig. 29.20B). Insect larvae are wormlike creatures, with no traces of the wings that will appear later. Most insect larvae gorge themselves like pigs, growing rapidly and achieving a weight and volume considerably greater than that of the adult. They build up food reserves that will be required later. Eventually they enter a quiescent, nonfeeding, and metamorphic stage known as the **pupa** (Fig. 29.20C). During pupation the insect is helpless, and most species have developed burrowing or cocoon-spinning habits that protect them during this critical period. Pupation usually requires a considerable period of time, and many insects overwinter in the pupal stage. Eventually the fully formed adult emerges from the pupa (Fig. 29.20D), having undergone a complete metamorphosis. Development of this kind is termed **holometabolous**. Pupation is in actuality a metamorphic molt. It is a complex event involving the complete destruction of some tissues and the formation of new organs. These profound changes are partly under the control of hormones.

Insects can live a long time as juveniles, and probably in most species the insect lives longer as a juvenile than as an adult. There are exceptions, of course, but it is far more usual for crop destruction to be caused by juveniles than by adults. In some insects, this has been carried to extremes. A mayfly naiad may live for a long time in a pond, sometimes for as long as three years, gradually changing with its naiadal molts. The cicada spends most of its 13- or 17-year life underground as a juvenile feeding on plant roots, to emerge for a short-lived, yet vocal, copulatory bout as an adult. Species such as these emerge for but a few hours of winged life as an adult. The adult stage is no more than a migratory set of reproductive organs, useful for distributing eggs but for little else. Such adults cannot feed, for they lack mouthparts, and are suitably adapted to their ephemeral reproductive life.

Review of Groups

SUBCLASS PAUROPODA

Pauropods include about 60 species of small (0.5 to 2.0 mm), soft-bodied, blind myriapods living under objects in moist surface litter, or in the upper soil. They are found in temperate to tropical regions. A small, conical head bears large, uniquely branched antennae; solid, deep-set mandibles with a comb of delicate blades at the tip; and a pair of maxillae that are united with the sternum of the maxillary somite to form a floor for the mouth cavity. Eyes are missing, but characteristic head sense organs, the pseudoculi, are present near the position ordinarily occupied by eyes. The last head segment has no appendages, and is separated from the remaining head segments, forming the collum, a separate ring (Fig. 29.21A).

The trunk is made up of 11 segments, counting the pygidium, with six-segmented legs on the first nine somites (ten in one species). Each tergal plate covers two somites, as in millipedes. All of the cuticle is membranous except for the terga.

Pauropods have no heart, no arteries, and no respiratory organs. Blood circulates in a system of hemocoelic lacunae, and respiratory exchange occurs at the body surface. The gonads are primitively ventral, but the testes become secondarily dorsal. The epidermal gonoducts open through paired gonopores located between the coxae of the second legs.

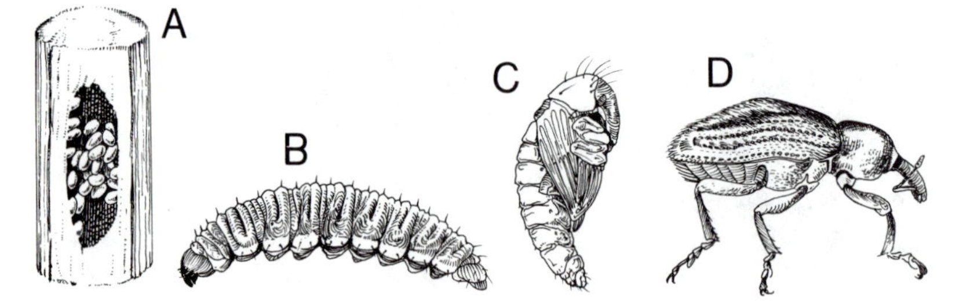

FIGURE 29.20. Holometabolous development of beetle *Hypera*. **A.** Eggs. **B.** Larva, one of larval series. **C.** Pupa, nonfeeding metamorphic instar. **D.** Adult or imago. Wings and other structures typical of adult develop within pupa, and fully formed adult emerges. Insects that undergo pupation undergo no molts after emerging from pupa. (After Michelbacher, from Essig.)

Male gonoducts end in ejaculatory ducts that contain penes. Female gonopores and a seminal receptacle open into a genital depression. Development is anamorphic, as the juvenile hatches with three pairs of legs. Example: *Pauropus* (Fig. 29.21B).

SUBCLASS DIPLOPODA

The millipedes (Fig. 29.21C,D,E,F,G) are herbivorous myriapods, with the trunk composed of double somites (diplosomites) of which all but the first three bear two pairs of legs. The first three trunk somites are modified, having only one pair of legs and the gonopores on the second somite. The head has antennae, mandibles, and typically a gnathochilarium formed of sternal elements, first maxillae, and the vestiges of the appendages of the last head somite. The last head somite typically appears legless as the result of gnathochilarium formation, and forms a separate ring, the collum. The body is typically circular in cross section.

The millipedes have the most thoroughly consolidated and strongest of exoskeletons. Each of the diplosomites is enclosed in a continuous skeletal ring, with no articular membranes between the tergal, pleural, and sternal elements. The polydesmoid millipedes (Figs. 29.3D and 29.21F) have two complete skeletal rings for each diplosomite, an anterior prozonite, and a posterior metazonite. More typically, the single or double rings of millipedes overlap

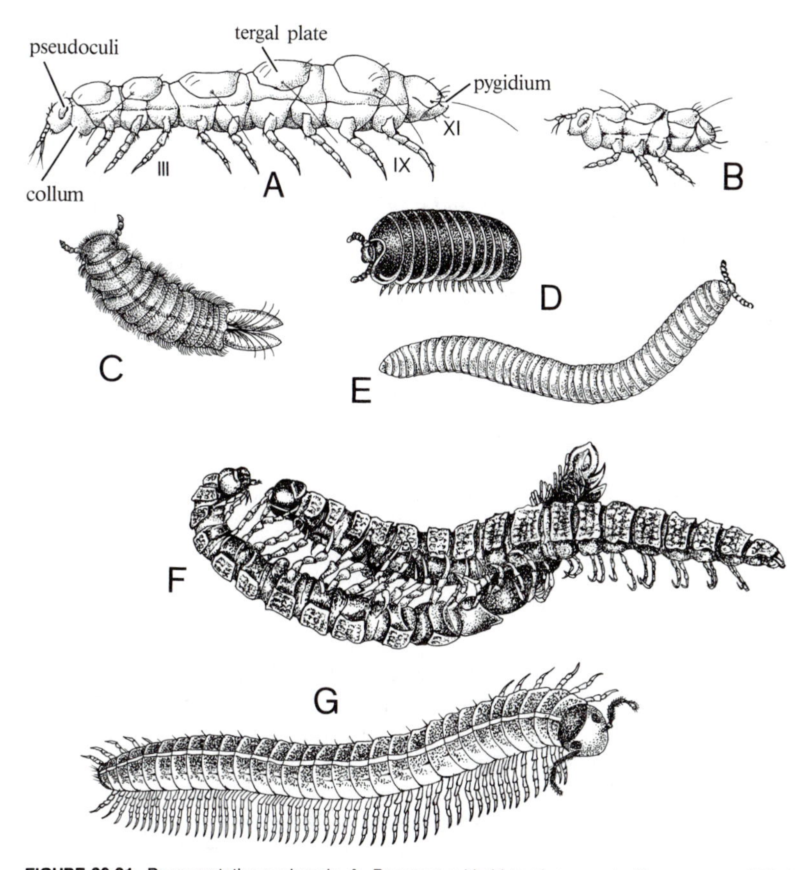

FIGURE 29.21. Representative myriapods. **A**. *Pauropus,* with 11 trunk segments (Roman numerals) but only six dorsal plates or terga, thus resembling millipedes. Last head segment, appendageless somite known as collum, bears no maxillae. Characteristic sense organs, pseudoculi, found on head. **B**. First instar of *Pauropus* with only three pairs of legs. **C**. *Polyxenus,* diplopod belonging to Pselaphognatha. These tiny, blind millipedes have no gonopods or stink glands, and maxillae do not form gnathochilarium. Characteristic short trunk of 10 to 12 segments and tufts of bristles. **D**. *Glomeris,* pill millipede, Oniscomorpha, with gnathochilarium and only 11 to 13 ventrally flattened trunk somites. Last two pairs of appendages modified in males. **E**. *Polyzonium,* colobognath, with elongated body, gonopods on sixth diplosomite of males, and suctorial mouth parts. **F**. Copulation in *Polydesmus,* flat-backed millipede, Polydesmoidea, characterized by dorsolateral flares of exoskeleton and double skeletal rings for each diplosomite. **G**. *Chordeuma,* ascospermomorphan, resemble polydesmoids but have more diplosomites, silk glands, and lack stink glands. (**A**,**B**, after Tiegs; **C**,**D**,**E**,**G** after Koch; **F** after Seifert, from Cloudsley-Thompson.)

on the dorsal side, so that none of the more membranous articular membranes are exposed. The rings slide past each other at the point of the dorsal overlap, and permit the body to be curled without any loss of protection (Fig. 29.3D,E). The polydesmoid millipedes, however, have a different kind of dorsal specialization. Lateral alae on the metazonites give the dorsal surface a flattened appearance.

Nearly all millipedes have a single pair of legs on the first three trunk diplosomites, counting the legless collum as the last head somite. This pattern is broken in the pselaphognathans (Fig. 29.21C), which have a pair of leglike maxillae on the collum, and in the juliform millipedes (Fig. 29.3E), which have no appendages on the collum but rather a single pair of legs on the first five trunk diplosomites. Legs are never found on the pygidium, and in many myriapods the somite immediately in front of the pygidium is also legless.

Diplopods are adapted for plowing their way through the restricted confines of leaf litter, humus, and the debris of rotting wood. The body is designed to be held rigid and the legs to develop great motive force. Experiments have been done that harnessed millipedes to small carts and forced them to tow many times their own weight in mass.

Millipedes have many spiracles and a simple tracheal system with a number of primitive characteristics. No spiracles are found on the first three trunk somites. From the fourth somite on, each diplosomite is equipped with two pairs of spiracles, located near the coxae of the legs. Air enters the spiracles and passes to air reservoirs in the transverse apodemes to which the leg muscles attach. Tufts of tracheae, differing in nature with the millipede order, arise from the air reservoirs. The majority of millipedes have long, slender tracheae that pass without branching to the tissues they serve. In some cases, however, the tracheae branch on their way to the tissues. Anastomosis of the tracheae does not occur, and thus the tracheal system consists of a series of small independent systems within each somite, except at the front of the body where the fourth somite takes in air for the more anterior part of the animal.

The slow-moving millipedes have no offensive weapons, as is often the case in herbivorous animals. As a general rule, the herbivore prefers a quiet, undisturbed life of munching. But as this is rarely possible in a world filled with carnivores, herbivores develop a variety of protective devices. When disturbed, many millipedes roll the body up into a ball or coil, presenting an unbroken expanse of tough exoskeleton to the unfriendly world. This habit has another advantage. Many predators are better at sensing movement than perceiving form, in some cases they lose sight of the motionless prey. Still another advantage comes from coiling. It reduces water loss at the surface and is helpful when the surroundings become less humid. A large number of millipedes, however, have another defensive adaptation in the form of stink glands that open through pores on the body surface (Fig. 29.3D). The duct from the stink glands and the soft tissues around the pore are equipped with muscles, and small quantities of the secretion are liberated when the duct and the pore are opened. Once such a device appears, it tends to be improved by further adaptation, and a few species have evolved powerful defensive weapons. Where the secretion is copious and the pores and the ducts are held open, contractions of the body wall muscles can spray the surrounding vicinity with the forbidding material. A few species have developed this habit, and in some it is extremely effective. A Haitian *Rhinocrisis* is said to spray its foul secretions almost a meter away. The secretions of the stink glands generally contain such noxious substances as hydrocyanic acid, iodine, quinine, and small amounts of chlorine. The fluid can be extremely irritating, even to humans, and causes blindness in chickens and other small animals unwary enough to attack millipedes that have strong stink glands. Here again is a weapon powerful enough to provide protection without use if properly signaled, and it is probable that the more brightly colored tropical millipedes are protected by their bright color.

Diplopods have anamorphic development and generally add four somites per instar molt.

SUBCLASS SYMPHYLA

This is a widespread but small group of centipedelike myriapods that have 12 pairs of legs and a pair of spinnerets on the next to last somite (Fig. 29.22A). The head bears long antennae, mandibles, maxillae, and a lower lip formed by the union of the second maxillae. Symphylans, along with most diplopods and some chilopods, have sensory pits, known as organs of Tömösvary, at the bases of the antennae. The gonopores are on the fourth trunk segment; development is anamorphic, with juveniles hatching with six to seven pairs of legs.

Symphyla are somewhat larger than pauropods and have a slightly more solid exoskeleton. The dorsal surface is covered by terga, and paired skeletal plates (sclerites) are found at the bases of the legs (Fig. 29.2A).

The Symphyla have a tracheal system that differs

markedly from that of millipedes and centipedes. They have a single pair of spiracles, uniquely located on the head. The branching tracheae extend to the tissues of the head and the first three trunk somites. All other parts of the body must receive oxygen from the blood, and probably a considerable part of respiratory exchange occurs at the body surface.

SUBCLASS CHILOPODA

The centipedes (Figs. 29.22B and 29.23) are carnivorous myriapods with relatively long, flattened bodies and heads with long antennae of 12 or more segments. Some of the chilopods have extremely long antennae to which segments are added at each molt throughout adult life, whereas others have a constant number of segments.

The trunk has one pair of seven-segmented legs on each somite. The first pair of legs are modified as maxillipedes with poison glands, and the bases of maxillipedes fuse to form a lower lip. Many centipedes have maxillipedes powerful enough to penetrate the skin of large animals like humans. Most of the smallest species cannot harm humans, but species as small as *Scutigera* can give a painful bite. Centipede bites cause severe local pain, swelling, and, in at least one authenticated case, human death. The maxillipedes may have developed primarily as offensive weapons, but they are no less effective for defense. With their rapid locomotion and fangs, centipedes are very well protected. A potential predator, however, may injure its centipede prey even though forced to drop it because of a painful bite. The brilliantly colored tropical centipedes are probably protected by their color, for a predator large enough to survive the bite of a centipede may be trained not to attack another member of the same species.

Other than the maxillipedes, the legs are all much alike, but close examination shows some small differences in the proportions of the leg segments, and in some cases the first and last legs have one less segment. The back legs of *Scutigera* (Fig. 29.23C) are held extended and pointing to the rear, and are somewhat modified to serve as posterior "feelers." Among chilopods, the outstanding leg specializations are seen in interruptions of the general rule of two pairs of legs for each diplosomite. Some of the diplosomites are legless or have a single pair of legs. The last two somites in front of the pygidium are the pregenital and genital somites, and are legless. The gonads are dorsal.

Chilopods have a strong sternum on each somite, and the dorsal surface is covered by strong tergal plates. The sides of the body are largely membranous in the majority of centipedes, although several subcoxal sclerites, located at the bases of the legs, provide points for muscle attachment. The geophilids are burrowing forms, and drive the head forward as a battering ram in some cases. The exoskeleton is strengthened by distinct pleura (Fig. 29.2B). The scutigeromorphs (Fig. 29.23C) also have some tergal plates that cover two or several somites. The lithobiomorphs (Fig. 29.23A) have alternating broad and narrow terga, and geophilids have intertergal and intersternal plates between the main tergites and sternites.

Centipedes have a highly developed tracheal system, characterized by branching and anastomosing of the tubes. Geophilids have spiracles in each somite except the first and last, and a relatively simple tracheal system (Fig. 29.8B) without conspicuous longitudinal trunks. The tracheae of other centi-

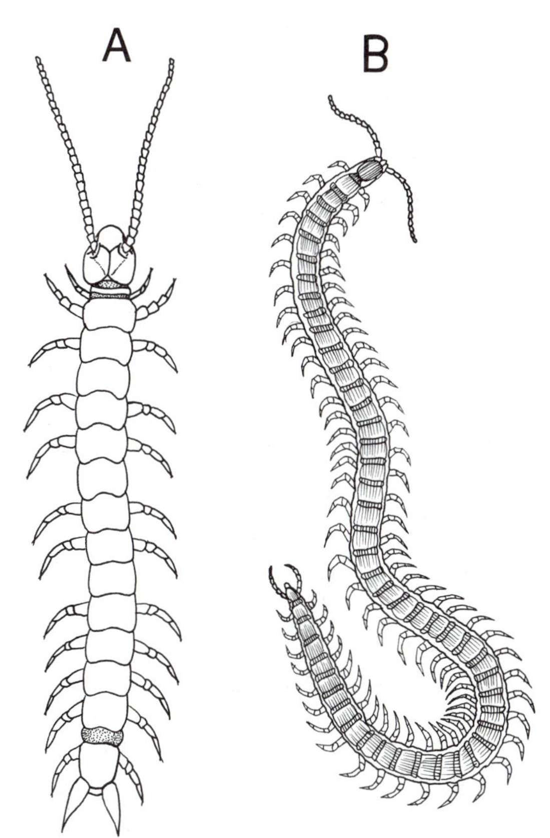

FIGURE 29.22. **A**. *Scutigerella*, symphylan with 12 pairs of legs and one pair of spinnerets on thirteenth trunk somite. **B**. Geophilid centipede, burrower with short legs and antennae. Somites with two dorsal and two ventral plates. All but first and last somites have spiracles. (After Snodgrass.)

pedes anastomose more freely and result in an integrated system with longitudinal trunks like those of insects. Many of the somites have no spiracles, the pattern differing in different groups. In lithobiomorphs, for example, there are usually no spiracles in the somites with narrow tergal plates. Little is known of the ways in which the tracheal system is ventilated. Presumably they use the same techniques as wingless insects.

The respiratory system of the scutigeromorphs is so different that it is probably an independent development. Each of the terga except the last contains a middorsal spiracle (Fig. 29.8C). The spiracle opens into an air sac, from which innumerable short tracheae arise. These are immersed in the blood of the pericardial sinus, aerating the blood before it is circulated to the tissues. This tracheal lung, similar in function, although so different in structure, to the tracheal lung of some spiders, is a unique development. Aeration of blood by tracheae occurs in some other mandibulates, but in no case is this lung structure similar to that of scutigeromorphs.

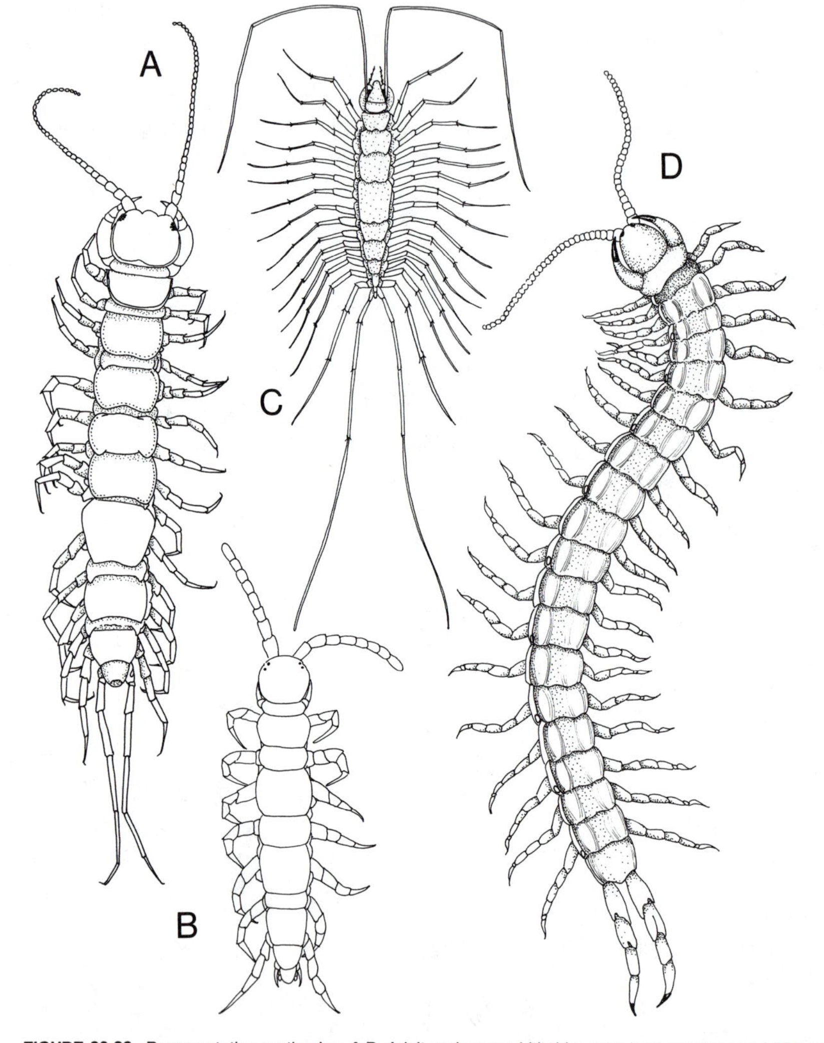

FIGURE 29.23. Representative centipedes. **A,B.** Adult and young *Lithobius,* with long antennae and 15 pairs of legs. Alternating short and long somites serve as stabilizing adaptation to rapid running. They hatch with only seven pairs of legs. **C.** *Scutigera,* house centipede, with 15 pairs of very long legs, last pair serving as anal feelers. Large, faceted eyes and dorsal spiracles, opening into tracheal lungs, are unique. **D.** *Scolopendra,* with never more than 23 pairs of legs. Scolopendromorphs become quite large; one species nearly 0.3 m long. Bite of some of larger species may be dangerous. Scolopendromorphs have 9 to 11 pairs of spiracles and all somites present at hatching. (**A,B** after Eason; **D** after Koch.)

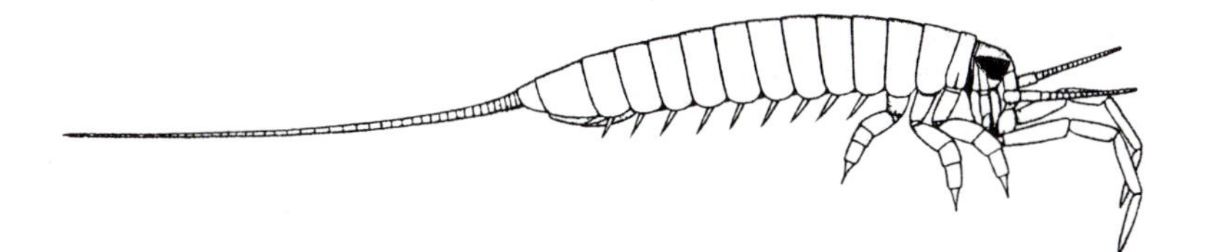

FIGURE 29.24. *Dasyleptus brongiarti,* Permian monuran from Soviet Union, intermediate between myriapods and insects. Note vestigial abdominal limbs. (After Sharov.)

CLASS INSECTA

The insects, or hexapods (= six legs), were covered in great detail previously, under the sections on Body Plan and Biology. The group is extensive but generally separated into two types, the apterygotes and the pterygotes. The former are the wingless insects. These are generally thought to be primitive forms, but are characterized by a lack of copulatory appendages on the terminus of the abdomen. The lack of these gonopods and the wings could be viewed as derived features related to the adaptation of apterygotes to cryptic habitats of leaf litter and rotting vegetation.

Pterygotes comprise the overwhelming bulk of "winged" insects, although in fact many of them (e.g., many of the social hymenopterans) secondarily lack wings. They have incredible diversity, but all variants on the same body plan (already covered), an idea of which can be gotten from perusing the taxonomic review.

Fossil Record and Phylogeny

The fossil record is uneven. Several interesting forms for all uniramian groups are known from the Paleozoic, the earliest occurring in the Devonian period. These either fall easily into known groups, or form interesting problematica in their own right, for example, the gigantic myriapod *Arthropleura*, which plodded around the Coal Swamps of North America and Europe in the Carboniferous. The insects of the Paleozoic, although easily recognized as hexapods, are in a number of orders that are now extinct, and so the fauna of that time had a distinctly different cast to it than the modern one. Relatively few ancient insects are body fossils; most are just wings. One interesting exception is the Permian monuran *Dasyleptus* (Fig. 29.24), which makes an excellent transition type from myriapods to insects. Debate still rages over whether wings were developed from parts of thoracic limbs or from folds of the cuticle and body wall (Kukalova-Peck, 1983).

Little consensus is evident among authorities concerning phylogenetic relationships within uniramians. The scheme presented here (Fig. 29.25) is only tentative. The two major classes, Insecta and Myriapoda, are evident. Within myriapods the transition is focused on the movement of gonopores anteriorly. Chilopods and insects have their gonopores located posteriorly on the body, whereas other myria-

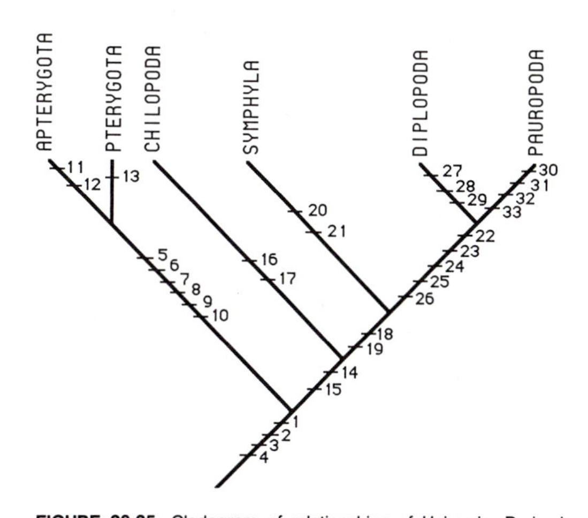

FIGURE 29.25. Cladogram of relationships of Uniramia. Derived characters are: (1) uniramous limbs; (2) excretory Malpighian tubules developed from proctodeum; (3) whole limb mandible; (4) tritocerebral segment suppressed; (5) tendency to develop protowings (either as epimeral folds of body wall or from limb parts); (6) mandible with rolling promoter–remotor action; (7) first maxilla coxa divided into plates; (8) abdomen of 11 segments; (9) abdominal limbs suppressed except posteriorly; (10) trunk limbs reduced to three pairs in thorax; (11) wingless; (12) no gonopodal limbs on abdomen; (13) wings greatly elaborated; (14) gnathic lobe on mandibles; (15) coxa articulated midventrally with body; (16) maxillipedes; (17) mandibles retracted into head pouches; (18) first maxilla with reduced or absent palp; (19) tendency toward anterior gonopores; (20) head tracheae; (21) gonopod on fourth trunk segment; (22) collum free; (23) gonopod on second trunk segment; (24) antennae generally short; (25) diplosomites; (26) hatchlings with four postcephalic somites; (27) stink glands; (28) rigid diplosomites; (29) solid tergoscutal body rings; (30) antennae with several rami; (31) eyeless; (32) pseudoculi; and (33) body reduced to 11 to 12 segments.

pods have them associated with either the second or fourth trunk limbs, symphylans on the fourth, and diplopods and pauropods on the second. Myriapods then are not seen so much as ancestors to insects, but rather are a distinct radiation in their own right.

TAXONOMY

Class Myriapoda. Gnathal lobes on mandibles; coxa articulated midventrally with body, functions with rotating action; tendency to locate gonopore anteriorly on body.

Subclass Chilopoda. First trunk limb as maxillipede, with poison glands; mandible retractable into head pouch.

Order Lithobiomorpha. Predominantly temperate and subtropical forms, with antennae of about 19 to 70 segments; nine large and six small somites; coxal glands associated with last four pairs of legs; eyes usually with many ocelli; development anamorphic. Example: *Lithobius* (Fig. 29.23A).

Order Scutigeromorpha. Fifteen sternal plates, but with only eight tergal plates; long slender antennae; large faceted eyes; seven middorsal spiracles; development anamorphic. Example: *Scutigera* (Fig. 29.23C).

Order Geophilomorpha. Elongated, blind forms with from 31 to over 180 pairs of legs; each trunk somite with dorsal tergite and intertergite, and ventral sternite and intersternite; antennae with 14 segments; pair of spiracles in all but first and last somites; development epimorphic. Example: *Geophilus* (Fig. 29.22B).

Order Scolopendromorpha. Predominantly tropical and subtropical creatures with 21 or 23 pairs of legs, and antennae with 17 to 31 segments; with 9 to 11 pairs of spiracles; eyes absent or composed of four ocelli; tergal plates correspond to sternal plates and are equally large; development epimorphic. Example: *Scolopendra* (Fig. 29.23D).

Subclass (and Order) Symphyla. First maxilla with no palp; head tracheae; gonopore on fourth trunk segment; 12 trunk segments (Fig. 29.22A).

Subclass Diplopoda. Antennae generally short; collum free; well-developed diplosomites; gonopores on second trunk segment; terga and scuta formed into solid rings; repugnant; hatch with only four postcephalic segments.

Order Pselaphognatha. Broad, soft-bodied, with trunk of 10 to 12 somites, and dorsal and lateral tufts of hairy bristles; maxillae on head and collum leglike; not fused to form gnathochilarium; with eyes; without stink glands or legs modified for copulation (gonopods); anus in next to last trunk segment. Example: *Polyxenus* (Fig. 29.21C).

Order Oniscomorpha. (Pill millipedes.) Mostly tropical, with no limbs on collum, gnathochilarium; 11 to 13 trunk somites, flattened on ventral surface; with last two pairs of legs modified in males, and last pair used for sperm transfer; body can be rolled into ball. Example: *Blomeris* (Fig. 29.21D).

Order Limacomorpha. Blind, with no limbs on collum, gnathochilarium; 19 to 20 trunk somites and last pair of legs modified as gonopods in males; without stink glands; cannot roll into ball.

Order Colobognatha. Suctorial, largely tropical, elongated, with no limbs on collum, gnathochilarium; 30 or more diplosomites; exoskeleton hardened, but sternal plates not fused with rest of exoskeleton; small head with mouthparts usually forming suctorial proboscis; row of stink glands on each side of body; both pairs of legs on sixth diplosomite modified as gonopods in male. Example: *Polyzonium* (Fig. 29.21E).

Order Polydesmoida. (Flat-backed millipedes.) Relatively short body, with no limbs on collum, gnathochilarium; 18 to 19 trunk diplosomites, each covered by two skeletal rings, anterior prozonite and posterior metazonite; flattened lateral carinae on exoskeleton give back a flat shape; stink glands on some trunk somites; no eyes; first pair of legs on sixth diplosomite modified as gonopods in males. Example: *Polydesmus* (Fig. 29.21F).

Order Ascospermomorpha. (Resembling the Polydesmoida.) No limbs on collum, gnathochilarium, but with 25 to 31 diplosomites; usually with cluster of small ocelli on each side of head; sometimes with stink glands; silk glands open on papillae in last tergal plate. Example: *Chordeuma* (Fig. 29.21G).

Order Juliformia. No limbs on collum, gnathochilarium; many diplosomites; second genital somite usually without legs; both pairs of legs on sixth diplosomite modified as male gonopods; stink glands on each trunk diplosomite (Fig. 29.3E).

Subclass (and Order) Pauropoda. Antennae multibranched; diplosomites; gonopores on third trunk segment; 11 trunk segments. Example: *Pauropus* (Fig. 29.21A, B).

Class Insecta (= Hexapoda). Mandible with rolling promotor-remotor action; first maxilla coxa divided into plates; functional legs reduced to three pairs on thorax; abdomen reduced to 11 segments, abdominal limbs generally suppressed except for modified posterior gonopods.

Subclass Apterygota. Wingless; lacking all abdominal limbs.

Order Protura. No antennae (perhaps vestigial on sides of head) and with forelegs used as tactile appendages; without compound eyes or ocelli; pseudoculi; Malpighian tubules reduced to papillae; nonpulsating pericardial cord in place of heart; simple or no tracheal system; 12-segmented abdomen without cerci; hatching before full complement of abdominal somites present, so development anamorphic. Example: *Eosentomon* (Fig. 29.26A).

Order Aptera. (Campodeids and japygids.) Small, soft, wingless insects with slender, flattened bodies, seeking dark and humid environments; long antennae; without eyes; 11-segmented abdomen with styli and with conspicuous cerci; papilliform or no Malpighian tubules. Example: *Campodea* (Fig. 29.26C).

Order Collembola. (Springtails.) Minute, soft body with short antennae; compound eyes very simple or missing; simple, chewing mouthparts with sharp mandibles; usually with springing mechanism composed of spring (furcula) on fifth abdominal somite and clasp (tenaculum) on third abdominal somite, used to hold spring for sudden release; ventral tube (colophore) for producing adhesive secretions on first abdominal somite; without abdominal cerci; without Malpighian tubules; cleavage holoblastic but young have all somites present at hatching and mature by epimorphic molts. Example: *Entomobrya* (Fig. 29.26D).

Order Microcoryphia. (Bristletails.) Small to medium, soft, wingless insects with elongated bodies, living in surface litter or under stones; long antennae; long cerci; long ten-segmented abdomen with eleventh segment a long caudal filament; large eyes; chewing mouthparts. Example: *Machilis* (Fig. 29.26B).

Order Thysanura. (Silverfish.) Small insects with long, often scaly bodies; long antennae; with compound eyes and with or without ocelli; biting mouthparts; short cursorial legs, sometimes with "exopodites" on middle or hind coxae; ten abdominal segments, some with styli; eleventh abdominal somite drawn out as caudal filament; long abdominal cerci; Malpighian tubules and tracheae well developed. *Thermobius* (Fig. 29.26E).

Subclass Pterygota. Insects with wings, or secondarily wingless.

Order Ephemeroptera. (Mayflies.) Slender body with short hairlike antennae; large compound eyes and three ocelli; chewing mouthparts in aquatic nymphs and vestigial mouthparts in adults; forewings larger than hindwings and both pairs held vertically over back when at rest; two very long abdominal cerci and caudal filament; ten abdominal somites and vestigial eleventh somite; tracheal gills in aquatic nymphs. Example: *Hexagenia* (Fig. 29.27A).

Order Odonata. (Dragonflies and damselflies.) Slender, predaceous forms with strong chewing mouthparts; small, hairlike antennae; huge compound eyes and three ocelli; stout legs; two pairs of similar, narrow, net-veined wings held straight out from body when at rest (dragonflies) or held vertically (damselflies); long abdomen with ten distinct somites and two rudimentary anal somites; very short, inconspicuous cerci; nymphs aquatic and with or without caudal gills. Example: *Macromia* (Fig. 29.27B).

Order Dictyoptera. (Cockroaches, mantids, and walking sticks.) Variable, moderate to large body, wingless or with tough, leathery forewings (tegmina) covering larger, folded, membranous hindwings; with antennae, eyes, and generalized chewing mouthparts. [Divided into three suborders: Blattaria (cockroaches), Mantodea (mantids), and Phasmida (walking sticks and leaf insects).] Examples: *Blatta* (Fig. 29.27C) and *Stagomantis* (Fig. 29.27D).

Order Isoptera. (Termites.) Social forms with soft exoskeleton; members of society differentiated as reproductive members, workers, and soldiers; with biting mouthparts, modified in soldiers; eyes and

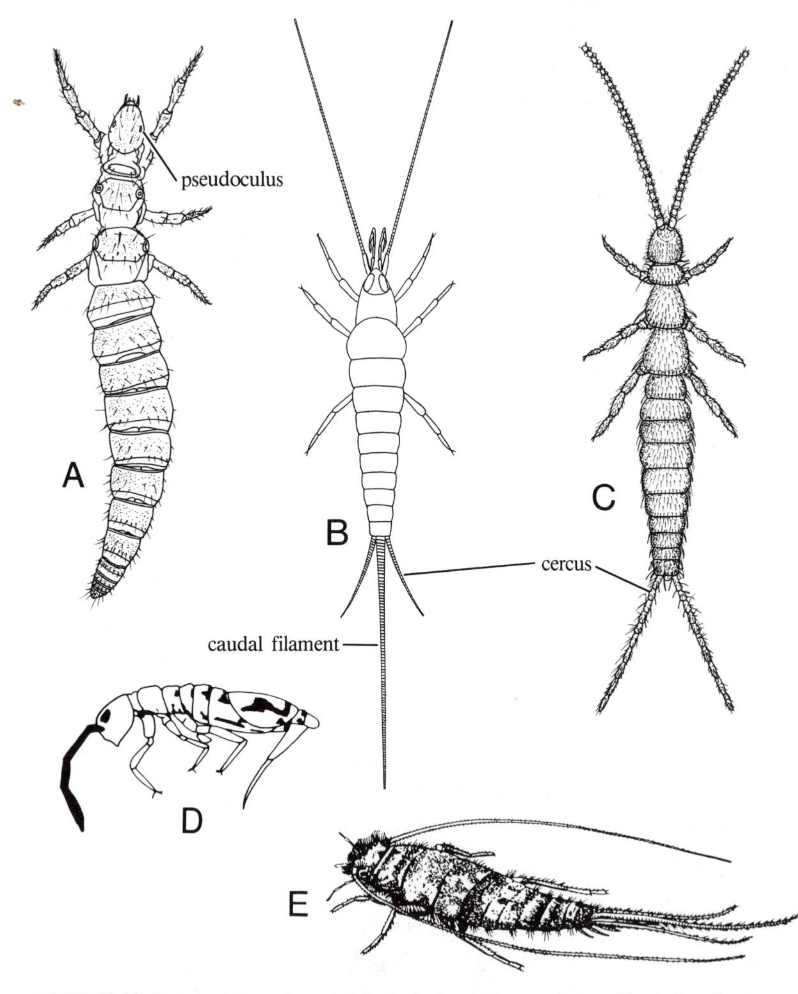

FIGURE 29.26. Representative apterygote insects. **A**. *Eosentomon,* proturan, with primitive chewing mouth parts and pseudoculi on head. **B**. *Machilis,* bristletail or microcoryphian, with caudal filament formed by prolonged last abdominal somite and long cerci. **C**. *Campodea,* apteran, with cerci, but without caudal filament. Japygids have cerci in form of forceps, whereas campodeids have filamentous cerci. **D**. *Entomobrya,* collembolan. Springtails jump by means of forked spring, furcula, held in place by tenaculum. Its sudden release sends them off in great hops. **E**. *Thermobius,* thysanuran, silverfish, common household pest, with characteristic cerci and abdominal filament. (**A** after Berlase; **B** after Lubbock; **C** after Essig; **D** after Pruvost; **E** after Ross.)

ocelli variable; antennae slender; short, stout legs; wings, when present, similar, with basal fractures at point where they break when shed. Example: *Zootermopsis* (Fig. 29.27E).

Order Orthoptera. (Grasshoppers and crickets.) Typically with long, saltatorial (jumping) hindlegs; forewings typically with tegmina covering folded, membranous hindwings; long antennae; large eyes; chewing mouthparts; abdomen with short cerci. Example: *Melanoplus* (Fig. 29.9).

Order Dermaptera. (Earwigs.) Elongated body with tough exoskeleton; long antennae; usually without

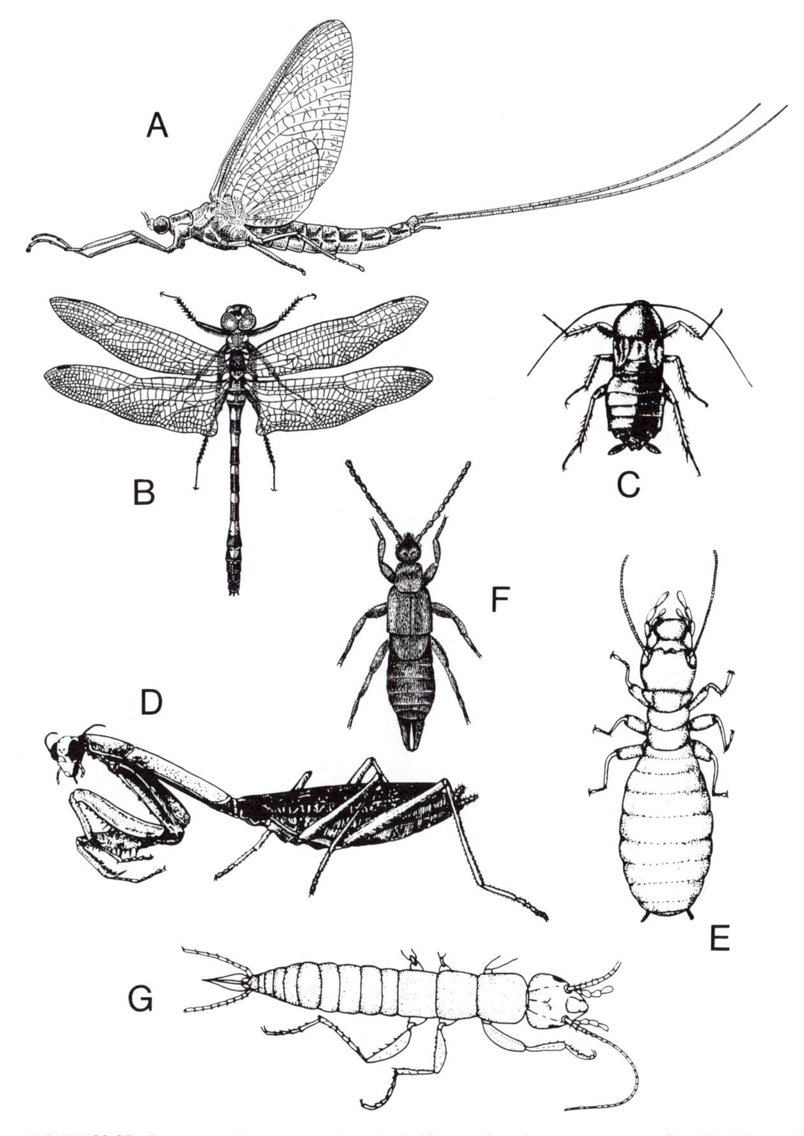

FIGURE 29.27. Representative pterygote insects. **A**. *Hexagenia,* ephemeropteran, mayfly, with delicate, dissimilar, net-veined wings, forewing being larger. Characteristic two long abdominal cerci and caudal filament. **B**. *Macromia* (Odonata), dragonfly, with huge compound eyes, narrow net-veined wings approximately equal in size, long slender abdomen with inconspicuous cerci, and strong biting mouth parts. **C**. *Blatta,* typical cockroach (Blattaria), with leathery forewings and membranous hindwings. **D**. *Stagomantis,* typical mantid, with raptorial forelegs. **E**. Worker termite, *Zootermopsis* (Isoptera). Termites, or white ants, are not true ants, distinguished by light color, broad junction of thorax and abdomen, and social habits. **F**. *Labia,* earwig (Dermaptera), characterized by conspicuous abdominal forceps formed by cerci and short leathery forewings that cover only part of membranous hindwings. **G**. *Grylloblatta,* grylloblattoidean. (**A** after Ross; **B,E** after Kennedy, from Ross; **C** after Essig; **D** after Ill. Nat. Hist. Surv.; **F,G** after Bebard.)

ocelli but with compound eyes; wingless or with short forewings forming truncate tegmina that cover all but tip of much-folded, membranous, radially veined hindwings; abdomen ending in prominent cerci modified to form forceps. Example: *Labia* (Fig. 29.27F).

Order Grylloblatodea. (Grylloblatids.) Wingless, living in soil, surface litter, and under rocks where snow-covered most of year; with long antennae, small eyes, and chewing mouthparts; abdomen with long cerci. Example: *Grylloblatta* (Fig. 29.27G).

Order Embioptera. (Webspinners.) Elongated body, forming large social aggregations without division of labor; without ocelli but with compound eyes; slender antennae; biting mouthparts; short, stout legs; foreleg tarsus enlarged and containing silk glands and spinnerets used to spin common web for colony;

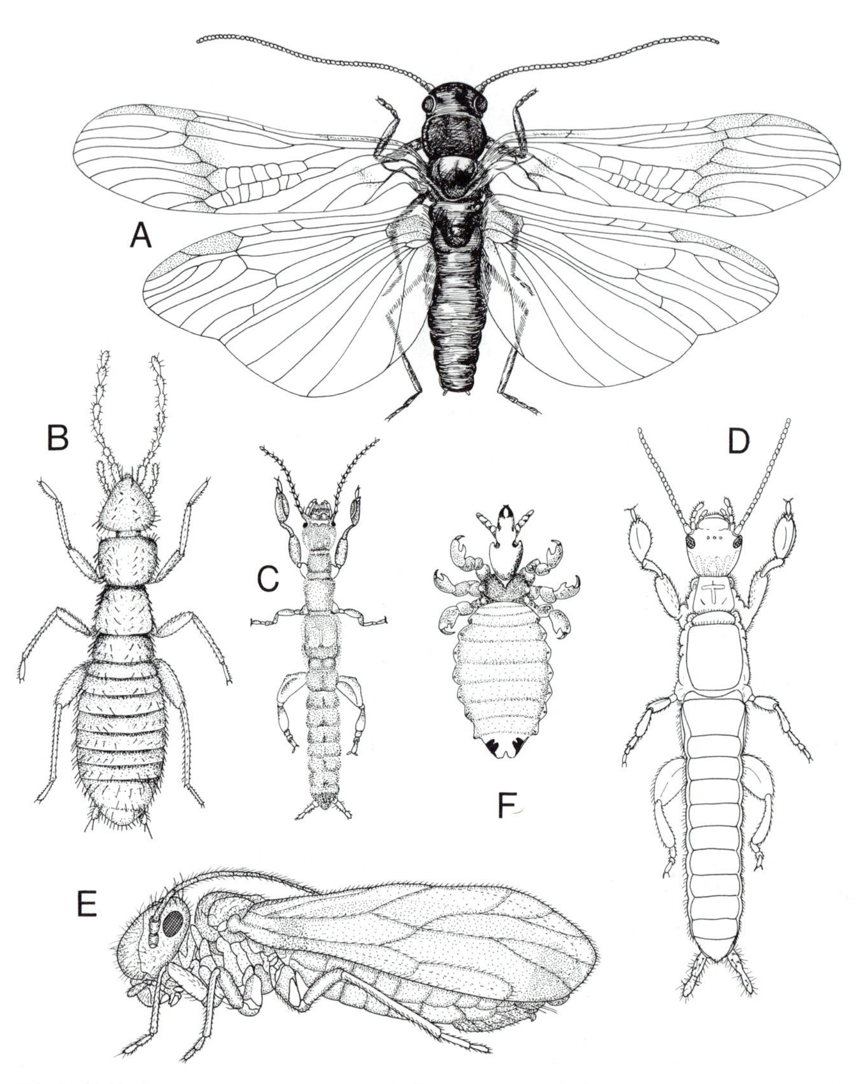

FIGURE 29.28. Representative pterygote insects. **A**. *Taeniopteryx,* stonefly (Plecoptera). Long membranous wings folded over back when at rest. Forewings generally smaller. **B**. *Zorotypus,* zorapteran, may be winged or wingless. **C**. *Embia* and **D**. *Oligotoma,* webspinners, with silk glands and spinnerets in foreleg tarsi. Winged and wingless forms occur. **E**. *Ectopsocus,* booklouse (Psocoptera), living in houses and warehouses where they attack packaged goods, books, and other paper goods. Labial spinnerets and silk glands used to form webs for brooding young and sometimes for residences. **F**. *Haematopinus,* sucking louse. (**A** after Newcomer; **B** after Caudell; **C** after Essig, from Essig; **E** after Sommerman; **F** after Ill. Nat. Hist. Surv.)

males sometimes with two similar, membranous wings, held flat over body when at rest; abdomen with ten segments and short cerci. Example: *Embia* (Fig. 29.28C,D).

Order Plecoptera. (Stoneflies.) With long antennae; modest compound eyes and variable ocelli; chewing mouthparts; two pairs of membranous wings folded over back when at rest; forewings usually smaller; abdomen with 11 segments, last segment reduced; long abdominal cerci; nymphs aquatic, with or without gills. Example: *Taeniopteryx* (Fig. 29.28A).

Order Zoraptera. (Zorapterans.) Minute forms, forming small social aggregations without division of labor but with winged and wingless members; winged members with compound eyes and ocelli, and wingless members without eyes; all with slender antennae; biting mouthparts; short, stout legs; wings long, slender, with few veins and with basal fractures; wings shed after mating; abdomen with ten full segments and with short cerci. Example: *Zorotypus* (Fig. 29.28B).

Order Psocoptera. (Booklice.) Chewing mouthparts; usually long antennae; large compound eyes and variable ocelli; slender legs; sometimes with two pairs of wings, folded rooflike over back when at rest; forewings larger; spinnerets on labial palps used to produce webs; abdomen with ten full segments and without cerci. Example: *Ectopsocus* (Fig. 29.20E).

Order Phthiraptera. (Lice.) Flattened, wingless forms, living as ectoparasites on birds or mammals. Antennae short; eyes reduced or absent; without ocelli; abdomen with five to eight somites; chewing or sucking mouthparts. [Two suborders recognized: Mallophaga (chewing lice) and Anoplura (sucking lice).] Examples: *Cuclotogaster* (Fig. 29.29A) and *Haematopinus* (Fig. 29.28F).

Order Thysanoptera. (Thrips.) Rasping, sucking mouthparts; short antennae; conspicuous compound eyes and three ocelli in winged species; when winged, forewings larger and both pairs narrow, nearly veinless, and with fringe of fine hairs on posterior margin; short, stout legs; abdomen with 10 to 11 segments and without cerci. Example: *Taeniothrips* (Fig. 29.29B).

Order Hemiptera. (Bugs.) Diversified group; piercing and sucking mouthparts; large compound eyes and variable ocelli; legs absent or adapted for running, jumping, digging, grasping prey, or swimming; wings absent in some; typically with similar, membranous wings in *suborder Homoptera* (cicadas, scale insects, and leafhoppers) and forewings modified as hemielytrae, with membranous apex and leathery base, used to cover membranous hindwings, in *suborder Heteroptera* (true bugs); abdomen with few to ten segments and without cerci. Example: *Anas* (Fig. 29.19).

Order Neuroptera. (Lacewings, dobson flies, snakeflies.) Carnivorous, with chewing mouthparts; antennae usually long and slender; large compound eyes; ocelli variable; forelegs sometimes raptorial; two pairs of similar, net-veined wings held rooflike over back at rest; abdomen with ten segments and without cerci. Example: *Chrysopa* (Fig. 29.29C,D).

Order Coleoptera. (Beetles.) With tough or hard exoskeleton, chewing mouthparts, antennae variable; large compound eyes and usually no ocelli; forewings modified as hardened elytra, meeting in straight line at middle of back; hindwings membranous and usually completely hidden by elytra when at rest; usually ten-segmented abdomen without cerci; largest and most varied insect order. Example: *Hippodamia* (Fig. 29.30A).

Order Mecoptera. (Scorpion flies.) Carnivorous, chewing mouthparts borne on snout; long, slender antennae; large compound eyes; ocelli variable; long, slender legs; wings absent, reduced, or two similar pairs held flat or rooflike over head at rest; abdomen with ten full segments and with cerci; male with tip of abdomen swollen by bulbous genital capsule, superficially resembling scorpion stinger. Example: *Panorpa* (Fig. 29.29E).

Order Trichoptera. (Caddisflies.) Weak, chewing mouthparts; long, slender antennae; large compound eyes; ocelli variable; long, slender legs; two pairs of large, hairy, sometimes scaled wings folded over back at rest; hindwings somewhat larger than forewings; abdomen with nine to ten segments and with short cerci; aquatic larvae, usually living in cases. Example: *Rhyacophila* (Fig. 29.30B).

Order Lepidoptera. (Moths and butterflies.) Mouthparts reduced to form coiled tube for sucking liquid food; antennae long and often feathery; large compound eyes; ocelli variable; two pairs of wings, often large and showy; forewings often larger than hindwings; wings with overlapping scales and often hairy; abdomen with ten segments; without cerci. Example: *Papilio* (Fig. 29.30C).

Order Siphonaptera. (Fleas.) Ectoparasitic on birds and mammals; with piercing and sucking mouthparts; short antennae; simple compound eyes; no ocelli; legs long, stout, adapted for jumping; wingless; abdomen with ten segments and no cerci. Example: *Pulex* (Fig. 29.30E).

Order Diptera. (Flies and mosquitoes.) Lapping, sucking, piercing, or vestigial mouthparts; large

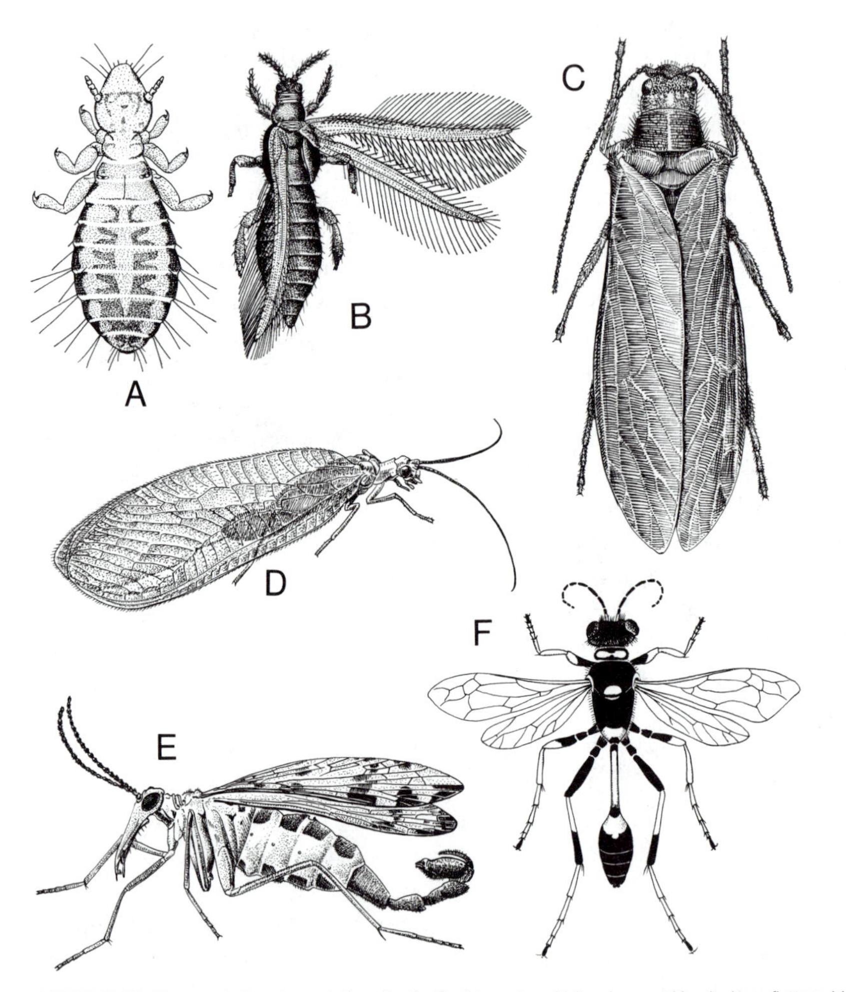

FIGURE 29.29. Representative pterygote insects. **A**. *Cuclotogaster,* chicken louse, with wingless, flattened body and modifications of appendages suitable for their parasitic life. **B**. *Taeniothrips inconsequens,* pear thrips, typical thysanopteran, with fringed wings. **C**. *Sialis,* alder fly or dobson fly, with two pairs of large, many-veined wings, folded over back at rest. **D**. *Chrysopa,* lacewing (Neuroptera), with two pairs of similar, net-veined wings held over back at rest. Typical large eyes and slender legs, sometimes with one raptorial pair. **E**. *Panorpa,* scorpion fly (Mecoptera), with swollen tip of male abdomen resembling scorpion sting. **F**. Mud dauber, *Sceliphron* (Hymenoptera), with slender pedicel at front of abdomen, characteristic of group. (**A**,**E** after Ill. Nat. Hist. Surv.; **B** after Moulton; **C** after Ross; **F** after Essig.)

compound eyes and usually three ocelli; antennae short or long; legs short or long; forewings membranous and used for flying, and hindwings reduced to balancers (halteres); abdomen with four to ten visible somites and without cerci. Example: *Drosophila* (Fig. 29.30D).

Order Hymenoptera. (Ants, bees, wasps, and sawflies.) Diverse structure and habits; with strong exoskeleton; chewing, lapping, or sucking mouthparts; slender antennae; large compound eyes and usually three ocelli; slender legs; wings absent, reduced, or two pairs of stiff membranous wings, coupled by hooks on hind pair; forewings larger; abdomen with slender pedicel; often living in nests in large societies with division of labor among several specialized castes. Example: *Sceliphron* (Fig. 29.28F).

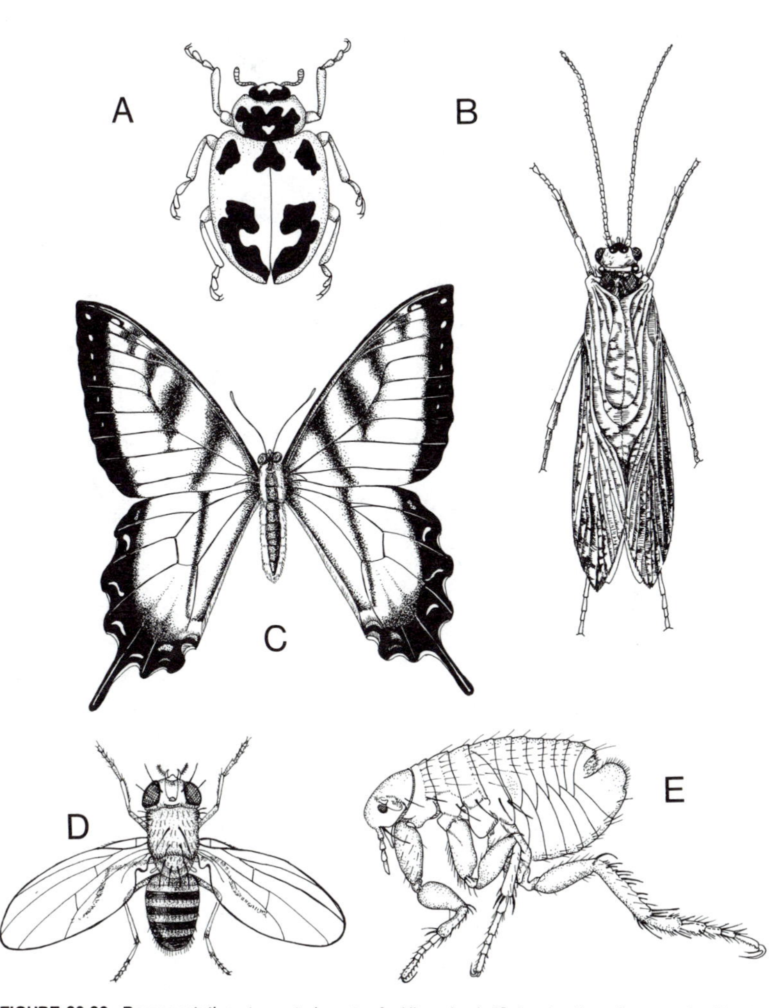

FIGURE 29.30. Representative pterygote insects. **A.** *Hippodamia* (Coleoptera), easily recognized by heavy forewings that completely cover larger, membranous hindwings. Forewings meet in straight line. **B.** *Rhyacophila,* caddisfly (Trichoptera), with large hairy or scaly wings that fold over back. Hindwings larger than forewings. Adults have large eyes and weak chewing mouth parts. **C.** *Papilio,* swallowtail butterfly (Lepidoptera), characterized by overlapping scales, often beautifully colored on wings, and sucking proboscis. **D.** *Drosophila,* fruit fly (Diptera), with one pair of wings; hindwings reduced to small balancers known as halteres. **E.** *Pulex,* flea (Siphonaptera), wingless and with greatly developed hindlegs. (**A** after Essig; **B** after Ross; **D** from Ross.)

Selected Readings

Anderson, D. T. 1973. *Embryology and Phylogeny in Annelids and Arthropods.* Pergamon Press, New York.

Borror, D. J., D. M. deLong, and C. A. Triplehorn. 1981. *An Introduction to the Study of Insects*, 5th ed. Saunders, Philadelphia.

Camatini, M. (ed.). 1979. *Myriapod Biology.* Academic Press, London.

Gupta, A. P. (ed.). 1979. *Arthropod Phylogeny.* Van Nostrand Reinhold, New York.

Gupta, A. P. (ed.). 1987. *Arthropod Brain.* John Wiley and Sons, New York.

Kukalova-Peck, J. 1983. Origin of the insect wings and wing articulation from the arthropodan leg. *Can. J. Zool.* 61:1618–1669.

Lewis, J. G. E. 1981. *The Biology of Centipedes.* Cambridge Univ. Press, London.

Magnum, C. P., et al. 1985. Centipedal hemocyanin: Its structure and its implications for arthropod phylogeny. *Proc. Nat. Acad. Sci.* 82:3721–3725.

Manton, S. M. 1964. Mandibular mechanisms and the evolution of arthropods. *Phil. Trans. Roy. Soc. Lond.* (B)247:1–183.

Manton, S. M. 1977. *The Arthropoda.* Clarendon Press, Oxford.

Snodgrass, R. E. 1935. *Principles of Insect Morphology.* McGraw-Hill, New York.

Snodgrass, R. E. 1952. *Textbook of Arthropod Anatomy.* Cornell Univ. Press, Ithica.

Thornhill, R., and J. Alcock. 1983. *Evolution of Insect Mating Systems.* Harvard Univ. Press, Cambridge.

Wenner, A. M., and P. M. Wells. 1987. The honey bee dance language controversy: The search for "truth" vs. the search for useful information. *Am. Bee J.* 127:130–131.

30

Cheliceriformes

Although not having as many described species as Uniramia or Crustacea, Cheliceriformes is nevertheless a large group of about 65,000 forms, many of which are economically important. Two major groups of cheliceriforms are known: the pycnogonids, or sea spiders, and the chelicerates, or merostomes, scorpions, and arachnids.

Everyone is familiar with chelicerates. Spiders, scorpions, and ticks are perhaps the most easily recognized forms, although some of the mites are familiar enough, if generally unseen. The chigger bites that irritate the skin so severely are caused by mites, and the "red spider" that destroys gardens and crops is also a mite. A number of the other mites are destructive to crops before or after harvest. Those who live in coastal areas may be familiar with the horseshoe crabs, which are not crabs at all but belong to the Merostomata, one of the chelicerate classes.

Definition

These arthropods have two regions: the prosoma and the opisthosoma. The prosoma is never divided into a separate head and thorax. Antennae and a distinct deutocerebrum are missing. The first pair of appendages or chelicerae on the prosoma are formed on a postoral somite, and move anteriad during development. The second pair of prosomal appendages, or pedipalps, are sometimes leglike and sensory, but at other times chelate and prehensile. The next four prosomal somites each bears a pair of walking legs. The prosoma has an internal endosternite skeleton used to anchor muscles and channel blood flow. The opisthosoma (vestigial in pycnogonids) contains up to a maximum of 13 somites and a tail spine, and bears appendages that are always reduced or highly modified. The first opisthosomal somite is a pregenital segment reduced or modified in some manner, while the second opisthosomal somite is the genital segment that contains the gonopores. Several succeeding opisthosomal somites may bear appendages modified for respiration by supporting book gills or book lungs (replaced by "fan" or "sieve" tracheae in some arachnids). Excretory organs may be highly modified nephridia known as coxal glands, or may be Malpighian tubules attached to the digestive tract at the junction of the midgut and the hindgut.

Body Plan

External Anatomy Two outstanding external traits of the cheliceriforms set them apart from other arthropods: they lack antennae, and the first pair of appendages are pincerlike **chelicerae**. The body of chelicerates is divided into two regions, or tagmata. The anterior is the six-segment **prosoma**, consisting of the head, mouthparts, and somites bearing the walking legs. The posterior region is the **opistho-**

soma, which, if not vestigial, has reduced or modified appendages. The prosoma and the opisthosoma are sometimes called the cephalothorax and the abdomen, respectively, but they do not correspond exactly with the parts of other arthropods that are termed head, thorax, and abdomen. The exoskeleton does not have discrete pleural, or lateral, sclerites.

The pycnogonid body (King, 1973) is greatly reduced, almost wholly prosoma (Figs. 30.1, 30.17E and 30.18A). The opisthosoma is reduced to a tiny, unsegmented protuberance with the anus at its tip. The prosoma is unconventionally formed. The mouth is at the end of a proboscis, sometimes nearly as long as the body. The remainder of the body typically has seven pairs of appendages: chelicerae, pedipalps, ovigers, and four pairs of legs.

The chelicerae have three segments and are fairly standard. The palps are reminiscent of the pedipalps of arachnids; they are sometimes missing and have a variable number of segments when present. The ovigers differ with the species, and appear to be reduced walking legs with four tarsal pieces. The rest of the prosoma is segmented and usually contains three segments, each with a pair of walking legs, but in some species one or two additional segments and pairs of legs are present. The legs are unusual in having an extra podite. The normal walking legs have nine segments, the tarsal segment being divided into two pieces. The most unusual feature of the leg is the independent musculature of the tarsal segments of the ovigerous legs; this is found nowhere else among arthropods.

In the living xiphosuran chelicerates (Bonaventura et al., 1982), the limulids (Fig. 30.2), the prosoma is circular in front and covered by a solid **prosomal shield** (see Fig. 30.17A). The opisthosoma is also covered by another dorsal shield. It is hinged to the prosoma and to the long, spikelike tail. The prosoma bears six pairs of appendages on its ventral side. The first pair are the chelicerae, pincerlike appendages used to partially macerate the food. The lack of any significant difference between the second pair of appendages, which in the majority of chelicerates are modified appendages known as pedipalps, and the third pair of appendages, which in most chelicerates are the first walking legs, is a distinctive characteristic of the females. Males, however, have the second pair of appendages modified for grasping the female while mating. The next four pairs are walking legs, bear **gnathobases** along the midline, and end in small pincers or chelae (see Fig.

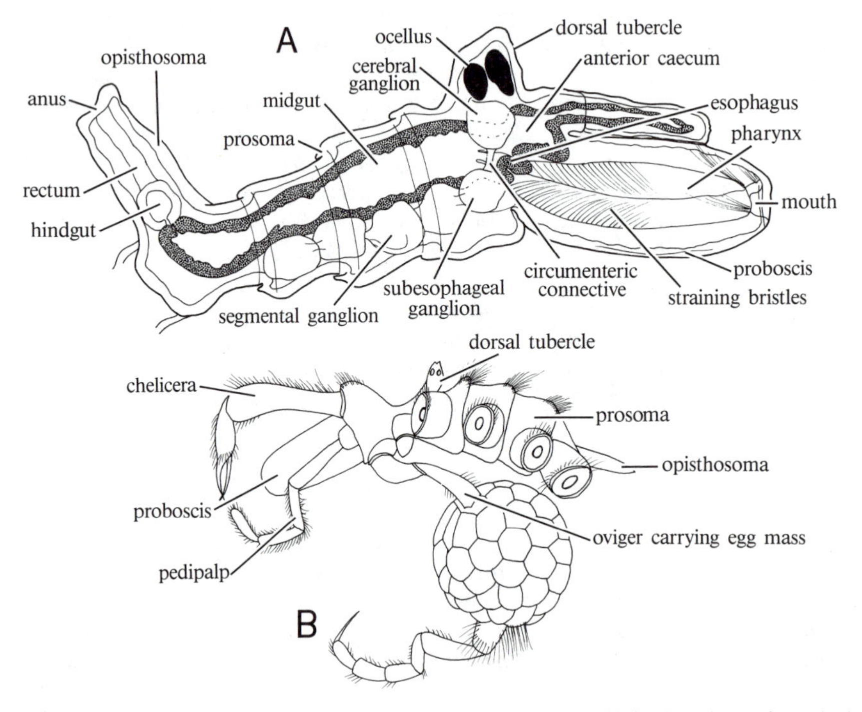

FIGURE 30.1 Pycnogonids. **A**. Longitudinal section through mature pycnogonid showing scheme of organization. **B**. Lateral view of male pycnogonid, with all but ovigerous legs removed. (**A** after Dohrn; **B** after Sars, from Helfer, in Kükenthal and Krumbach.)

30.10C,D). The next pair, also leglike, end in curious leaflike tips, called flabella, used to sweep away mud that clings to the body surface during burrowing. The seventh pair of appendages are attached to the reduced **pregenital somite** and are very small, highly modified **chilaria**.

Limulus has an endoskeleton independent of the apodemes that develop at skeletal sutures. An endosternum, composed of cartilagelike tissue, is suspended from the shield by muscles and provides sites for the attachment of body musculature and muscles reaching the leg bases. The leg musculature of *Limulus* is typical of that of arthropods in general (see Fig. 30.10D), although details differ considerably in different groups. The coxal podite is attached to a membranous, flexible cuticle, and is moved by nine muscles attached to its margin, five from the shield and four from the endosternum (see Fig. 30.10C). The large number of muscle strands at the leg base provide relatively delicate control of the leg movements. The musculature of the leg itself is relatively simple. Each joint of the leg is typically a hinge joint, permitting movement in one plane only. In a mechanism of this kind, a set of flexors and extensors for each of the various joints between podites is to be expected. Legs that terminate in chelae are equipped with tarsi that project forward as immovable fingers, and are opposed to movable fingers formed of the pretarsi. At these joints the flexor and extensor muscles become levators and depressors of the pretarsal finger of the pincer. Leg muscles are believed to be relatively stable and are sometimes useful in determining the homologies of leg podites when two legs with differing numbers of podites are being compared.

Six pairs of flaplike appendages are found on the limulid opisthosoma. Although they have lost all resemblance to ordinary legs, they are undoubtedly highly modified appendages. The first pair forms a flattened **operculum**, used to cover most of the posterior appendages. The gonopores are found on the operculum. The rest of the opisthosomal appendages have gill plates on their posterior surface. The gill plates are fitted together, rather like the leaves of a book, and are sometimes termed **book gills**. The **tail spine**, or "telson," starts just behind the anus. The spine is used to right the body when it is overturned, and pushes the body forward during burrowing. A strong hinge attaches the spine to the opisthosoma.

The scorpion body (Figs. 30.3A and 30.19B) is composed of a prosoma covered by a solid prosomal shield, and an opisthosoma with separate somites. The opisthosoma is made up of a broader **mesosoma** (**preabdomen**) of seven somites and a narrow **metasoma** (**postabdomen**) of five somites that ends in a sting. Three pairs of lateral eyes are found on the shield margin, and a pair of median eyes on the dorsal surface.

The fused terga that form the **prosomal shield** make up most of the prosomal exoskeleton (Fig. 30.3B,C). The prosomal cuticle folds down at the shield margin as a narrow pleural membrane, and the appendage bases occupy the whole lower surface of the prosoma except for a narrow sternum. Each mesosomal somite is covered by a dorsal tergum and a ventral sternum, connected by the membranous pleural membrane on each side. The tergal and sternal sclerites are united to form complete skeletal rings in the metasoma.

At the anterior end of the body, the prosomal shield is turned under as a doublure, covering the bases of the chelicerae. The membranous undersurface is thickened to form a large, median **labrum** between the bases of the pedipalps and a horizontal plate, the **epistome**, immediately behind (see Fig. 30.7A). The first pair of appendages are the short, powerful chelicerae, made up of three segments. The next pair are pedipalps, which end in strong chelae (Fig. 30.3F) and have the pretarsus either reduced to

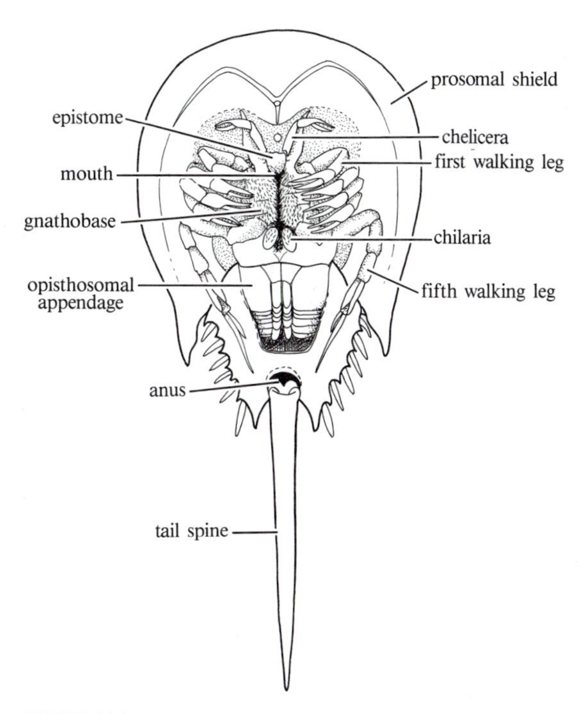

FIGURE 30.2. Ventral view of *Limulus polyphemus,* horseshoe crab, showing prosomal and opisthosomal appendages. (After Snodgrass.)

a terminal claw or entirely missing. The pedipalps are large, with well-developed claws (Fig. 30.3E). The walking legs have eight podites, two of which are tarsal segments.

The first opisthosomal somite is the pregenital somite, and as this is lost during development, the mesosoma begins with somite VIII. A **genital operculum** covering the gonopore is found on the genital somite, and it is a remnant of a pair of appendages. The sternum of somite IX is narrow and is flanked by a pair of **pectines**, comblike sensory appendages. The next four sternites are broad, flat plates, perforated by pairs of **spiracles** opening into the book lungs. Book lungs are also considered to be vestiges of appendages.

The rest of the somites have no appendages. The last metasomal somite contains the anus. The sting (Fig. 30.3F) may be a somite also, for it arises from an embryonic primordium with a ganglion. A pair of poison glands make the sting a formidable weapon.

The spiders (Gertsch, 1979) have a prosoma with segments generally fused (Fig. 30.4B), typically bearing two rows of four eyes (see Fig. 30.22B). The number and arrangement of the eyes are taxonomically important. The ventral surface of the prosoma (Fig. 30.4A) is covered by a large sternum, formed by the fused sclerites of somites III and IV, and a small sternal plate of somite III, the **labium**, which serves as an underlip. The spider labium is not ho-

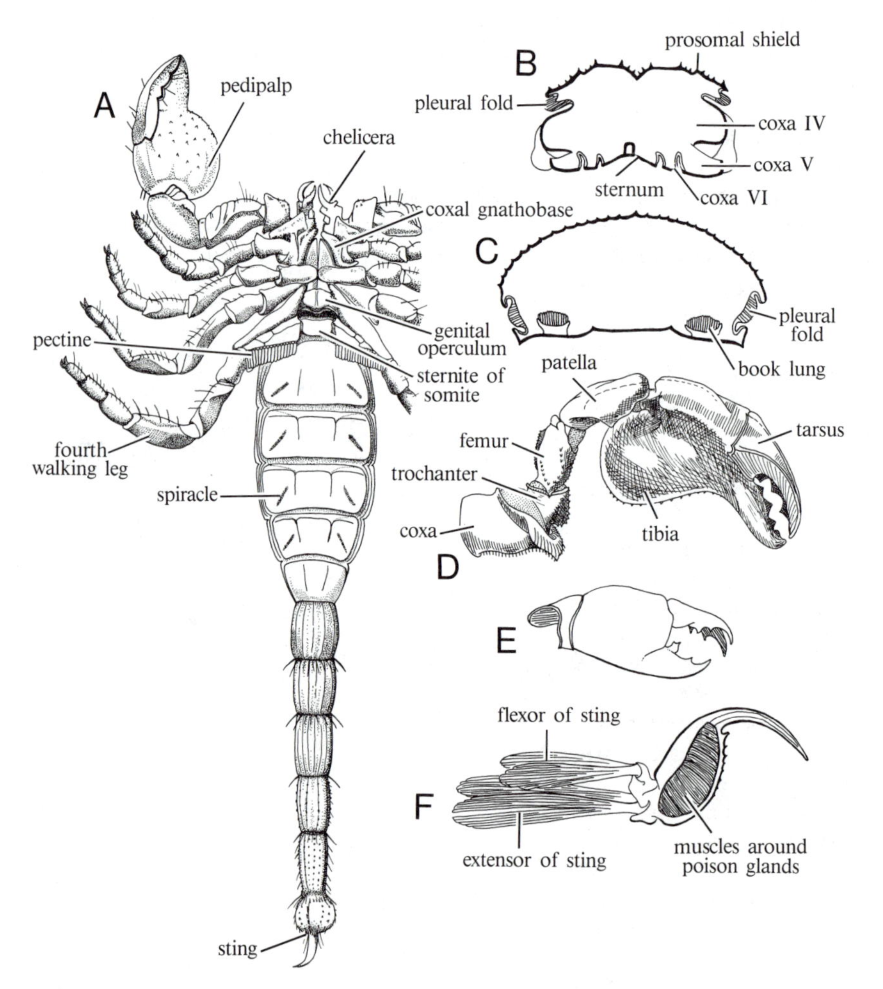

FIGURE 30.3. Scorpion anatomy. **A**. Ventral view of scorpion, *Pandinus*. **B**. Outline cross section through scorpion prosoma showing relationship of exoskeletal regions. **C**. Similar cross section through mesosomal region. **D**. Chelicera of *Centruroides*. **E**. Pedipalp of *Pandinus*. **F**. Scheme of section through scorpion sting, showing musculature. (**A** after Demoli and Versluys, from Kaestner, in Kükenthal and Krumbach; **B,C,E** after Snodgrass; **D** composite based on Snodgrass.)

mologous with the labium of insects, although similarly named. The rest of the prosomal wall is membranous, except in spiders, which have a narrow lateral plate, probably formed by fused pleura, immediately above the bases of the legs. A hard plate, the **epistome**, lies behind the pedipalp coxae in a few spiders. A flaplike fold of cuticle in front of the mouth forms a large **labrum** or **rostrum**. It contains muscles and a pair of glands of uncertain function.

The prosoma and the opisthosoma are connected by a very slender **pedicel**, through which the gut, aorta, tracheae, and nerve trunks pass. As a general rule, the opisthosoma is oval to globose, but in some cases it is bizarrely shaped. In some of the more primitive spiders (Fig. 30.4C), discrete opisthosomal terga can be recognized, but the exoskeleton is usually fused into a continuous dorsal covering. The pedicel is the first opisthosomal somite. Behind the pedicel, a transverse **epigastric furrow** crosses the ventral surface of the opisthosoma (Fig. 30.4A). The gonopore lies immediately in front of the furrow, and in most females a heavy plate, the **epigynum**, lies in front of the gonopore. Ducts to the seminal receptacles open on the epigynum, and the **spiracles** that lead to the anterior pair of **book lungs** lie at the edge of the epigastric furrow. In primitive spiders a second pair of spiracles that open into a second pair of book lungs lie immediately behind the first pair. In many spiders, however, the second pair of spiracles unite to form a single opening in front of the spinnerets, leading into a tracheal system. The anterior spiracles lie in somite VIII and the posterior pair in somite IX, but as somite IX includes the whole ventral surface from the epigastric furrow to the spinnerets, the spiracles can move over a large area without losing connection with their somite (Fig. 30.4C). Six **spinnerets** are usually associated with somites X and XI. A small anal lobe represents somite XVIII.

As in other arachnids, spiders have six pairs of prosomal appendages, a pair of chelicerae, a pair of pedipalps, and four pairs of walking legs (Fig. 30.4A). The two-segmented **chelicerae** arise from a postoral somite during development, but move forward and come to lie above the mouth in adults. The second cheliceral segment is the **fang**, which

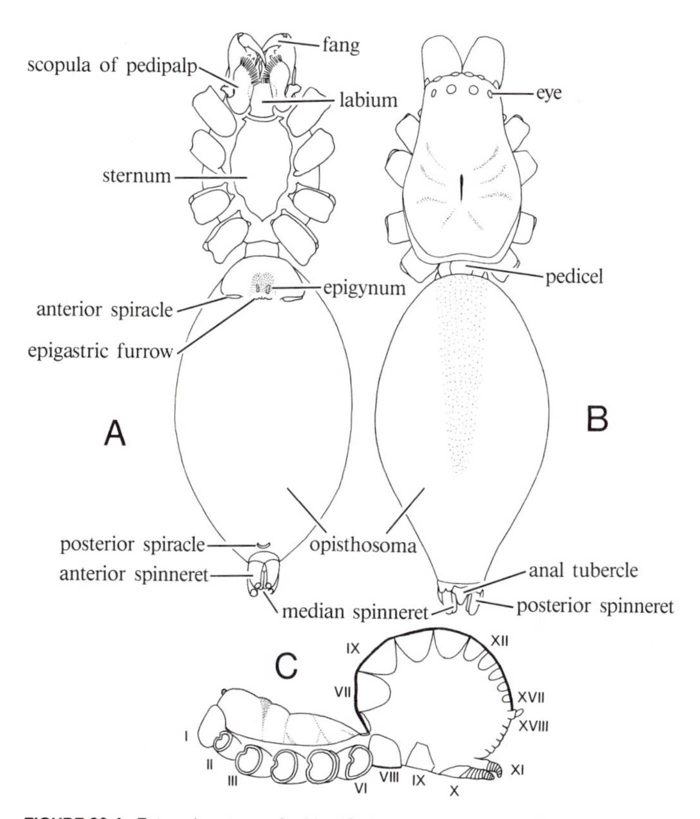

FIGURE 30.4. External anatomy of spider *(Clubiona)*. **A.** Ventral view. **B.** Dorsal view. **C.** Scheme of longitudinal section showing relationships of segments. Numbers of segments shown in Roman numerals. (**A,B** after the Kastons; **C** after Bristowe and Millot.)

contains the pore for the **poison gland**. The large poison glands often extend deep into the prosoma. They are homologous to the silk glands of pseudoscorpions, and sometimes the glands can secrete both poison and silk.

The **pedipalps** are leglike and are often carried forward as sensory appendages. A few species use the pedipalps as legs and carry the first walking legs forward as a kind of "antennae," as in amblypigids. The base of the pedipalp is an accessory mouthpart. The modified coxae or endites carry dense brushes of bristles, scopula, used to strain out particles too large to swallow. The so-called **salivary glands** are also found in the pedipalps. Despite the name, their function is unknown. The pedipalps of males are enlarged at the tip to form copulatory organs. Some of the pedipalps are equipped with extremely complex copulatory sclerites, and the structure of the male pedipalp is valuable for species identification. The pretarsus, and sometimes the tarsus, is enlarged (Fig. 30.5A). The pretarsus is modified to form a **basal bulb** and a spirally coiled **embolus** in which the seminal vesicle is located. The articular membrane between the pretarsus and the tarsus may be swollen to form a chamber, the **hematodocha**, that erects the embolus when it is gorged with blood, preparing it to be thrust into the female gonopore.

The walking legs contain seven true podites, but the last one, the pretarsus, is usually obscured by the hairy tarsus, and the tarsus is divided into two pieces. Muscles attach to the coxal border and to the shield and endosternum, giving considerable delicacy of control over the coxal movements. Intrinsic leg muscles move the leg at the various joints. The leg muscles are relatively complex; for example, each tarantula leg has 31 muscles, but most of the muscles are flexors. Extension of the legs is caused by increasing blood pressure.

Embryonic development shows that two pairs of **spinnerets** (Fig. 30.5C) are derived from the appendages of somites X and XI, whereas the remainder of the spinning apparatus arises from two pairs of outgrowths on the ventral body wall. The course of development of the spinnerets varies with the species; therefore, the number of spinnerets is variable, although three pairs is the most common number. The spinnerets derived from the appendages are more laterally placed; the pair associated with somite X are the anterior or first pair of spinnerets, and the pair on somite XI are the posterior or third pair of spinnerets. As a general rule, the ventral parts of these somites are so crowded together that the spinnerets lie very close to the anus. A pair of median or second pair of spinnerets derived from the abdominal body wall of somite XI usually lie between the bases of the posterior spinnerets, and a small **colulus**, the vestige of the body wall primordia on the tenth somite, lies between the bases of the an-

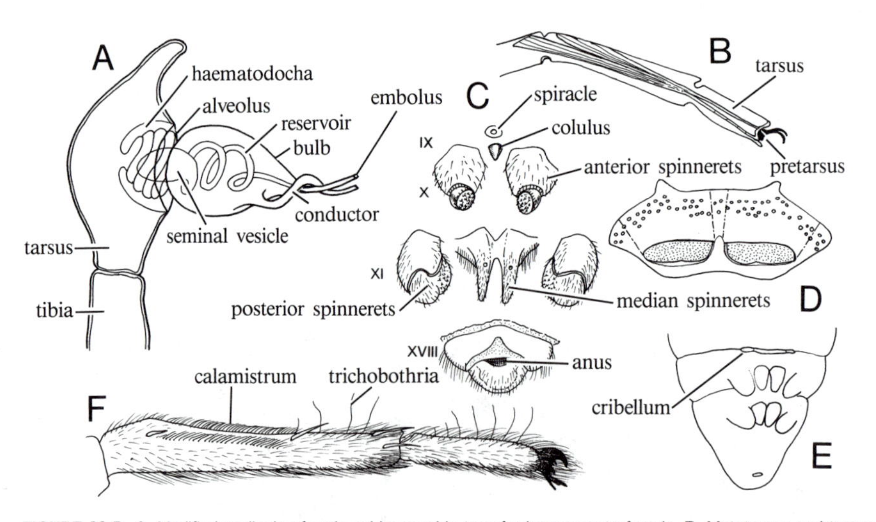

FIGURE 30.5. A. Modified pedipalp of male spider used in transferring sperm to female. **B.** Metatarsus and tarsus of spider leg. Note that muscles and tendons to pretarsus pass through two podites. **C.** Spinnerets of *Argiope*. Roman numerals indicate somite numbers. Most spiders with three pairs of spinnerets have median colulus, remnant of cribellum or anterior median pair of spinnerets. **D.** Cribellum of *Filistata*. In cribellate spiders, cribellum homologous to colulus. **E.** End of abdomen of embryo of *Filistata*, showing primordia of spinnerets. Cribellum arises from median, anterior spinnerets. **F.** Calamistrum of *Amaurobius*, used in web spinning. (**A** after Warburton; **B,C** after Snodgrass; **D,E** after Montgomery; **F** after the Kastons.)

terior spinnerets. In some spiders, rather than the colulus, a porous plate, the **cribellum**, is present (Fig. 30.5D,E). Sticky silk is spun from the cribellar pores and is combed out by one or two rows of spines on the metatarsal segment of the most posterior pair of legs. The comb is known as the **calamistrum**, and is found only in spiders with a cribellum (Fig. 30.5F).

Each spinneret is a compound spinning organ, made up of a number of small spinning tubes and large spinning tubes that vary considerably depending on the species (Kovoor, 1977). There has been considerable independent modification of the silk glands in the different spider groups, and attempts to classify the types of silk glands have not been wholly satisfactory. Threads of silk differ markedly in their properties. Some are relatively thick; others are quite thin. Sticky strands are used especially for the capture of prey. Spider silk is remarkable for its strength and elasticity. A thread 0.1 mm in diameter will support a weight of 80 g and will stretch up to 20 percent before breaking.

Digestive System The digestive system of cheliceriforms is quite variable. The pycnogonids have a muscular, suctorial **pharynx** in the proboscis. The **midgut** or intestine is equipped with **caeca** that extend into each limb. A short **hindgut** opens at the **anus** (Fig. 30.1A). The mouth in *Limulus* opens into the foregut, lined with cuticle continuous with exoskeleton. A narrow, tubular **esophagus** leads forward and upward, expanding into a thin-walled, distensible **crop** that opens into a heavy-walled, sclerotized **gizzard**. The opening of the gizzard into the midgut is guarded by a conical valve. The cuticular lining of the valve is folded, and strains out large particles of food. The midgut consists of a small **stomach** and a relatively wide **intestine**. Large digestive glands open into the stomach through pores on each side of the stomach. The hindgut is a short **rectum** (Fig. 30.6A).

Scorpions have a midgut with six pairs of **diverticula** (Fig. 30.7B). One pair of diverticula are located in the prosoma and serve as **salivary glands**. The remaining diverticula are located in the mesosoma. They are greatly folded and branched, and are gathered together into a general mesosomal mass by connective tissue. The midgut, whose epithelial cells are tall and columnar, continues through the metasoma to the short hindgut without any conspicuous dilations or other modifications.

Nearly all of the arachnids (see Fig. 30.9) depend on a pumping system to handle the fluid food. The pumping chamber is usually a part of the foregut and lies between the nerve ring and the mouth. The pharyngeal pump of a scorpion illustrates the basic design. The mouth leads into a chamber, infolded dorsally and here strengthened by an elastic rod. Compressor muscles cover the wall of the pharynx, and dilator muscles attach the body wall to the dorsal and lateral walls of the pharynx. Compressor and dilator muscles work in antagonism and pull the food into the pharynx and force it down into the midgut. Pseudoscorpions have a valve and straining device at the entrance to the esophagus, beyond the pharyngeal pump. The back part of the pharynx is lined with stiff bristles, and a conical, chitinous valve is located at the esophagus entrance.

Spiders are exceptional in having the pump located behind the nerve ring instead of in front of it (see Fig. 30.9). The mouth opens into a slender esophagus that extends to the middle of the prosoma. Here it expands suddenly to form a chitinized sucking stomach, located immediately above the endosternum. Dilator muscles extend from the endosternum to the ventral wall of the sucking stomach, and from the dorsal wall to the prosomal shield. The dorsal muscles attach to a heavy chitinous plate. Intrinsic muscles of the stomach wall are constrictors. A chitinous lining or chitinous plates in the sucking stomach are, of course, important factors in improving its performance. Although it is usually termed a sucking stomach, the spider pumping apparatus is a part of the foregut.

Solpugids have a pedicel, but an internal diaphragm separates the prosoma and the opisthosoma. The midgut is tubular, without conspicuous dilations, but is constricted at the diaphragm. At the front of the opisthosomal midgut, two long, tubular diverticula arise. Innumerable small branches cover the surface of the main diverticula. Near the union of the midgut and the hindgut, a stercoral sac arises. The stercoral sac is used for the concentration of nitrogenous wastes, and is discussed in conjunction with excretion.

Despite their small size, the pseudoscorpions have a very complex midgut. There are two five-lobed lateral diverticula, two median ventral diverticula, and a pair of large dorsal digestive glands. The functions of the various kinds of diverticula are as yet uncertain. Behind the fourth opisthosomal somite, the midgut narrows sharply. At the junction of the midgut and the hindgut is a stercoral sac.

The broad midgut of the harvestmen is completely surrounded by diverticula. Clusters of small caeca cover the narrower anterior part of the midgut, and two large caeca parallel the midgut, giving rise to secondary branches that extend throughout the opisthosoma. Branched pairs of lateral diverticula add to the confusion. Seven ventral diverticula, different in appearance, are thought to correspond to the digestive glands of other arachnids.

As might be expected from their diverse feeding habits, the acarines vary enormously in the detailed structure of the digestive tube. The most complicated arrangements are seen in ticks (Fig. 30.8). The esophagus is S-shaped and opens into a four-lobed stomach with thin, extensible walls. When gorged with blood, the walls dilate greatly. The lining cells put out pseudopodia and ingest blood cells that are then digested intracellularly. A large rectal sac apparently corresponds to the stercoral sac of other arachnids.

The spider midgut is composed of a dilated prosomal stomach, a constricted tube that passes through the pedicel, and an opisthosomal stomach (Fig. 30.9). A narrow intestine continues to the stercoral sac at the junction of the midgut and the hindgut. The sucking stomach opens immediately into the prosomal stomach, from which a pair of prosomal diverticula arise. These extend forward, around the dorsal dilator muscles, and may join anteriorly. From each of the prosomal diverticula, four branches arise, passing to the bases of the legs. The branches of the prosomal diverticula vary greatly in

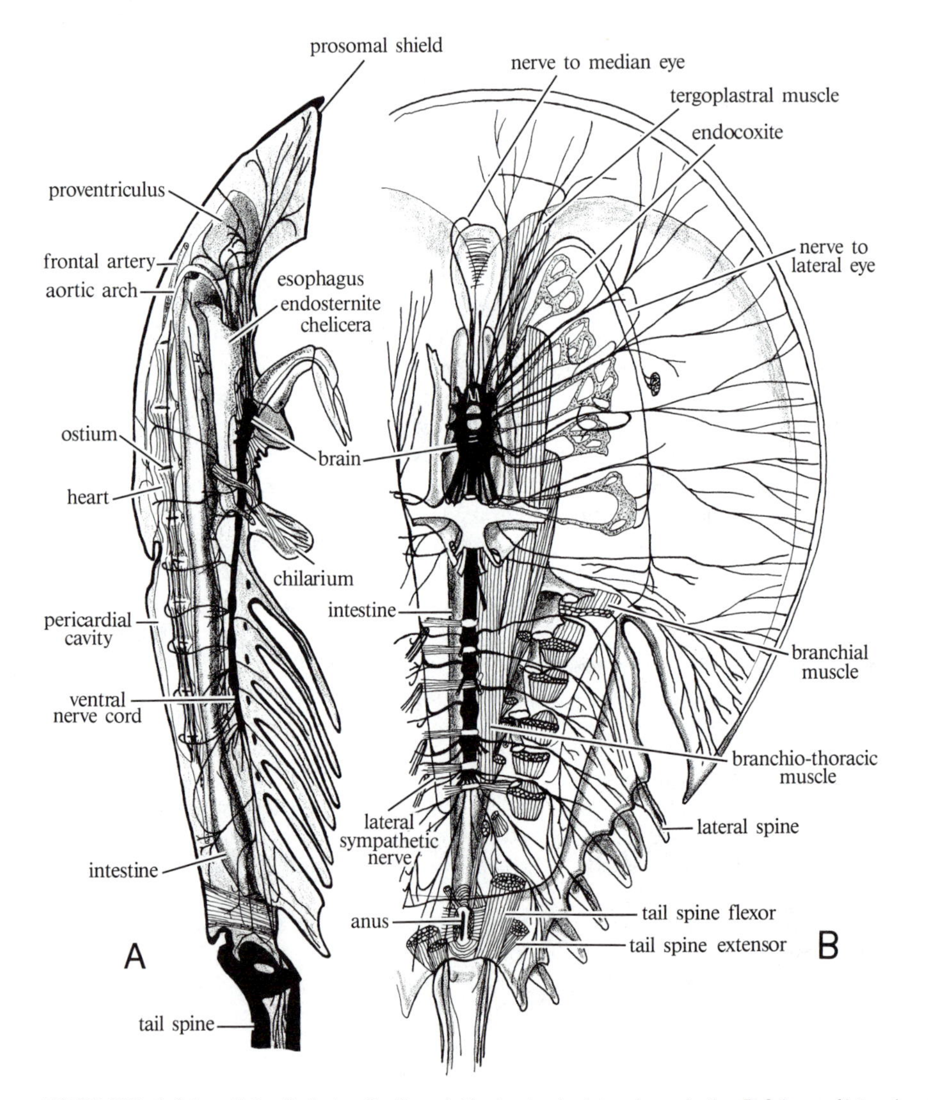

FIGURE 30.6. A. Schematic longitudinal section through *Limulus* showing internal organization. **B**. Scheme of internal organization of *Limulus* as seen in ventral view. (After Patten and Redenbaugh, from Shipley.)

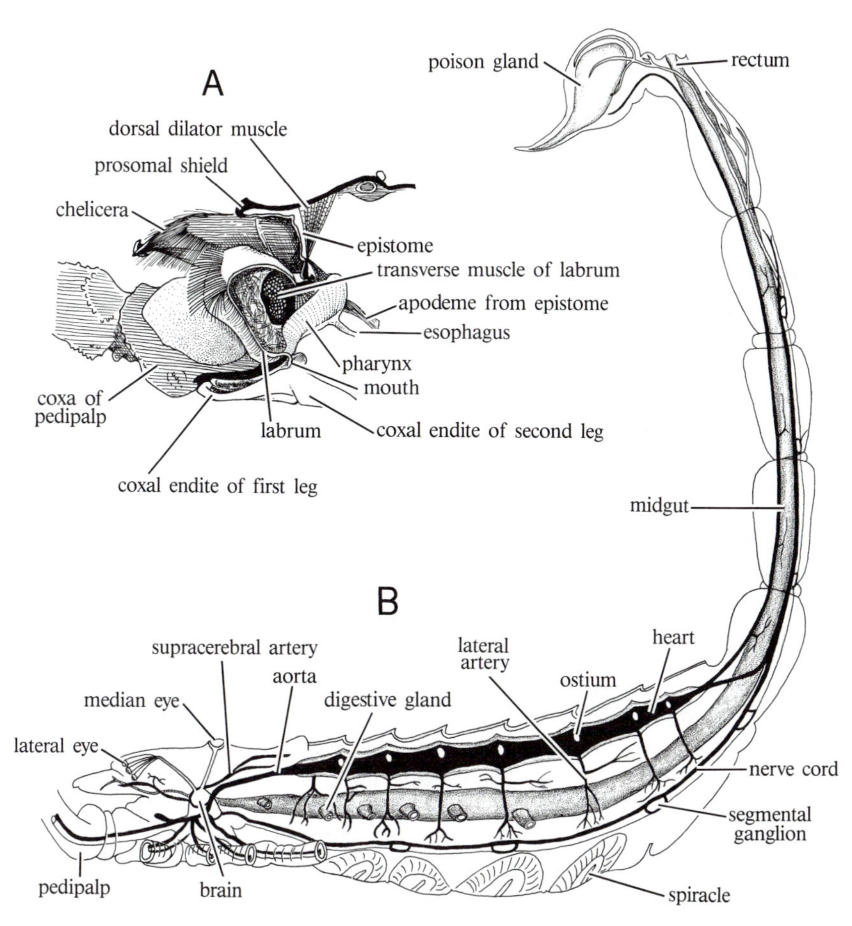

FIGURE 30.7. Arachnid anatomy. **A**. Oral area of scorpion. Note narrow pharynx and esophagus. **B**. Anatomy of scorpion. (**A** after Snodgrass; **B** after Kaestner, in Kükenthal and Krumbach.)

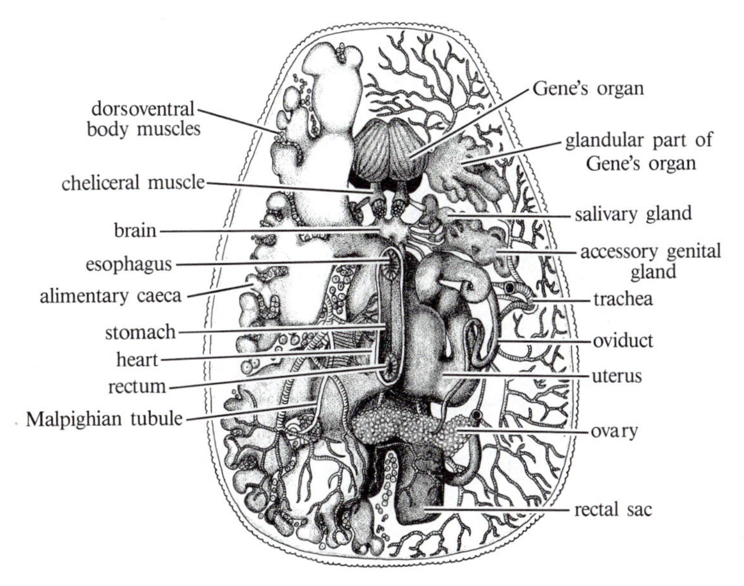

FIGURE 30.8. Anatomy of generalized tick. (After Hegner, Root, and Augustine.)

complexity. In some instances they branch out into the appendages or into the head. Beyond the pedicel the midgut expands again as an opisthosomal stomach. The opisthosomal diverticula branch complexly, forming the digestive glands. In all probability, however, the spider digestive glands are not true glands. The lining cells appear to be simple continuations of the lining of the opisthosomal stomach, and are probably absorptive.

Excretory System Pycnogonids have no excretory or circulatory systems. *Limulus* has a system of **coxal glands** (Fig. 30.10B). Four pairs of red-colored glands lie at the bases of the walking legs. The glands contain a tangle of fine ducts that merge to form a single duct. These ducts converge on each side in an excretory sac derived from the coelom. A coiled tubule arises from the excretory sac. The tubule is dilated terminally as a bladder. Urine is discharged from an excretory pore at the base of each fifth leg. The coxal glands may be highly modified derivatives of the metanephridia (Bonaventura et al., 1982).

Two entirely different kinds of excretory organs are found in scorpions and arachnids: coxal glands and Malpighian tubules. The **coxal glands** resemble similar glands of *Limulus* and are, in general, more suitable for aquatic excretion than excretion under terrestrial conditions. These coxal glands (Fig. 30.11A,B) are not united longitudinally. They begin as a **sacculus**, a derivative of the coelom. It is connected to a tortuous tubule that forms the **labyrinth**, and where nitrogenous material appears to be added to the tubule contents. The labyrinth is followed by a straight tube that extends to the excretory pore. Additional regions are not uncommonly seen. A dilated part of the tubule between the labyrinth and the straight tube is often present, serving as a bladder. In some cases a labyrinth sac lies between the sacculus and the labyrinth.

Malpighian tubules (Fig. 30.9) are blind tubules attached to the gut near the union of the midgut and the hindgut in terrestrial arachnids. As a general rule, these tubules are branched and become more or less tangled with the diverticula of the midgut. In a number of arachnids a **stercoral sac** arises from the gut near the point where the Malpighian tubules enter, or the Malpighian tubules are attached directly to the stercoral sac.

Respiratory System The respiratory systems of chelicerates are quite specialized. In *Limulus* each **book gill** is made up of a hundred or so thin "leaves" with stiffened outer margins and a delicate cuticle (Fig. 30.10A). Most arthropods lack cilia, and ventilation of the respiratory surfaces must be achieved without dependence on ciliary currents. Rhythmic movements of the opisthosomal appendages circulate water over gill surfaces and also drive blood into the leaves of the gill with each forward movement and drain blood from the leaves with each backward movement.

Living scorpions and other arachnids use book lungs and tracheae for respiration. Each **book lung** is a packet of hollow, flat plates through which air circulates (Fig. 30.12A,B,C). Blood flows over the inner surface of the leaves delivering carbon dioxide

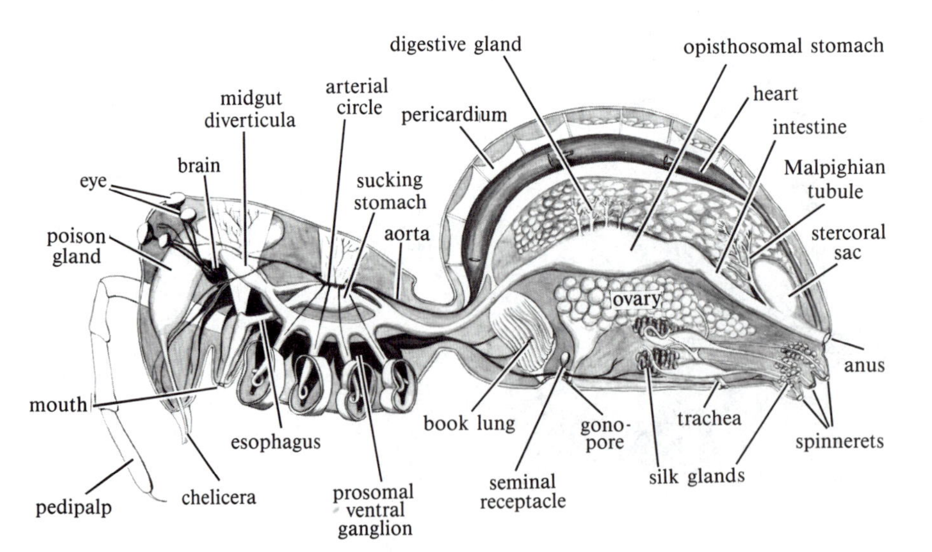

FIGURE 30.9. Internal anatomy of generalized spider.

and absorbing oxygen. The book lungs arise as ventral ectodermal invaginations, and have a remarkably similar organization in divergent arachnid groups. The external opening, the **spiracle**, leads to a chamber, the **atrium**, in which the leaves occur. The atrial walls are flexible except at the dorsal surface, where folding to form the leaves occurs. Firm cuticular bars cross the dorsal wall between adjacent leaves, preventing their collapse and bounding the hemocoelic sinus around them. The delicate cuticular covering of the leaves provides little resistance to the respiratory exchange. Book lungs are found in scorpions, in the more primitive arachnids such as uropygids and amblypygids, and in most spiders. Primitive spiders have two pairs of book lungs, in the second and third opisthosomal segments.

A number of arachnids use **tracheae** for respiratory exchange. Arachnid tracheal systems are considerably more diversified than book lungs, and have probably developed independently in several lines.

Tracheal systems are often formed by ectodermal

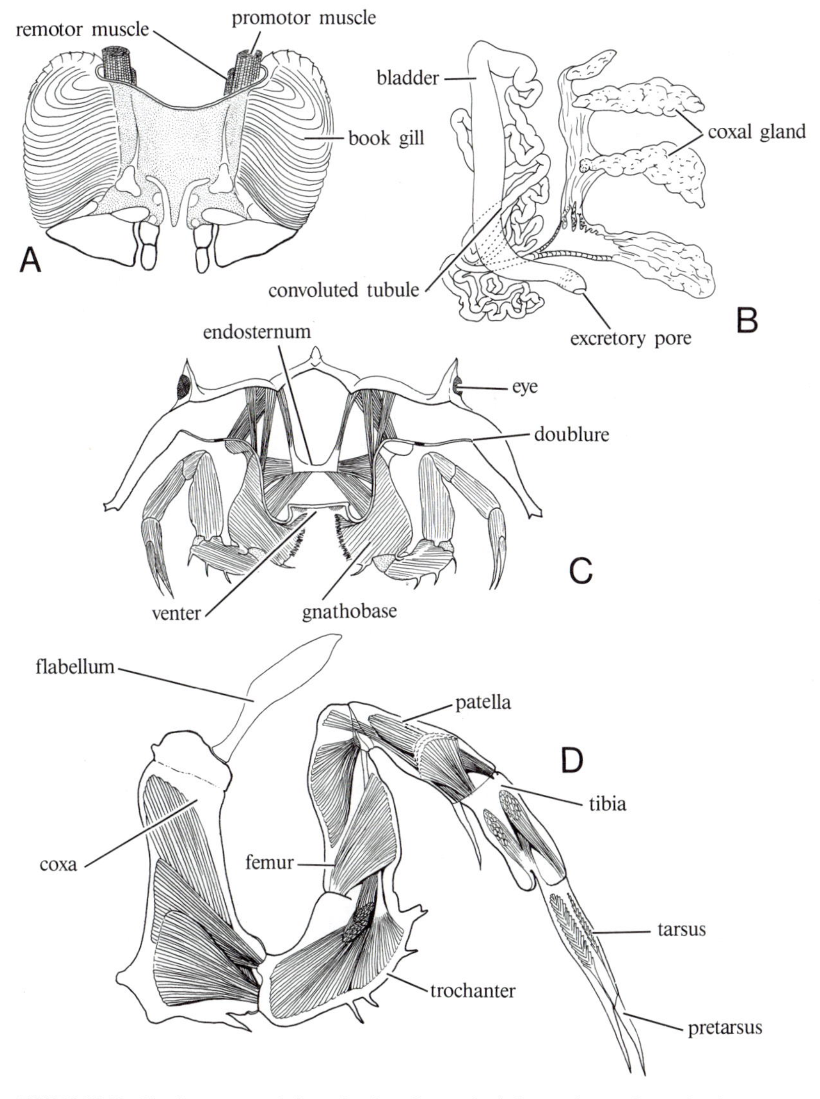

FIGURE 30.10. *Limulus* anatomy. **A.** Posterior view of second opisthosomal appendages showing arrangement of gill plates. **B.** Coxal glands formed from organs in four somites all opening through single duct and pore. **C.** Scheme of cross section of *Limulus* showing relationship of skeletal regions and appendages. **D.** Last leg of *Limulus* showing podites and muscular arrangement. (**A,C,D** after Snodgrass; **B** after Patten and Hazen, from Gehardt, in Kükenthal and Krumbach.)

invaginations developed from the second pair of book lungs, and typically open through spiracles. The spiracle opens into a vestibule or an atrium with a relatively thick cuticular lining. A system of internal tracheal tubes arises from the vestibule (Fig. 30.12D). Two types of tracheal systems can be distinguished.

Tracheae used to oxygenate blood hang down in a dense cluster into a hemocoel sinus. They are sometimes referred to as **tracheal lungs**, an apt term, as they are very like book lungs with tubular rather than lamellar respiratory surfaces. Tracheal lungs are found only in spiders and especially in spiders with book lungs.

Tracheae that carry oxygen directly to the tissues are much more common. They are found in spiders, pseudoscorpions, solpugids, harvestmen, and acarines. The simplest tracheal systems consist of simple tufts of unbranched tracheae associated with each spiracle, and differ from tracheal lungs only in that the tracheal tubules extend to the tissues instead of hanging in the hemocoel.

As a rule, the simpler tracheal systems are found in smaller arachnids. Pseudoscorpions have two pairs of tracheal spiracles. The tracheae from the anterior pair extend forward, and those from the posterior pair extend backward. They appear to be independent. Mites have very simple tracheal systems or none at all, but the somewhat larger ticks have tracheal tubes that branch extensively, passing to all parts of the body (Fig. 30.8).

The respiratory organs of spiders are particularly interesting. Primitive spiders have two pairs of book lungs, others have one pair of book lungs and one pair of tracheal tubes, and still others have two pairs of tracheal tubes. Some spiders have tracheal lungs, whereas others have extensively branched tracheal systems. In all cases the spiracles are associated with somites VIII and IX. It appears that the tracheae are of independent developments. In the embryo the rudiments of the book lungs do not appear to be identical with those of the tracheal spiracles.

The most highly developed tracheal systems among arachnids are found in harvestmen and solpugids. Opiliones have large longitudinal tracheae that pass forward into the prosoma from spiracles at the base of the last legs, sending branches to the legs and prosomal parts (Fig. 30.13B). Tracheae in the opisthosomal region are much less well developed. In solpugids one pair of prosomal and two pairs of opisthosomal spiracles are connected by a main longitudinal tracheal tube on each side of the body. Small branches from the tracheal trunks extend to all parts of the body.

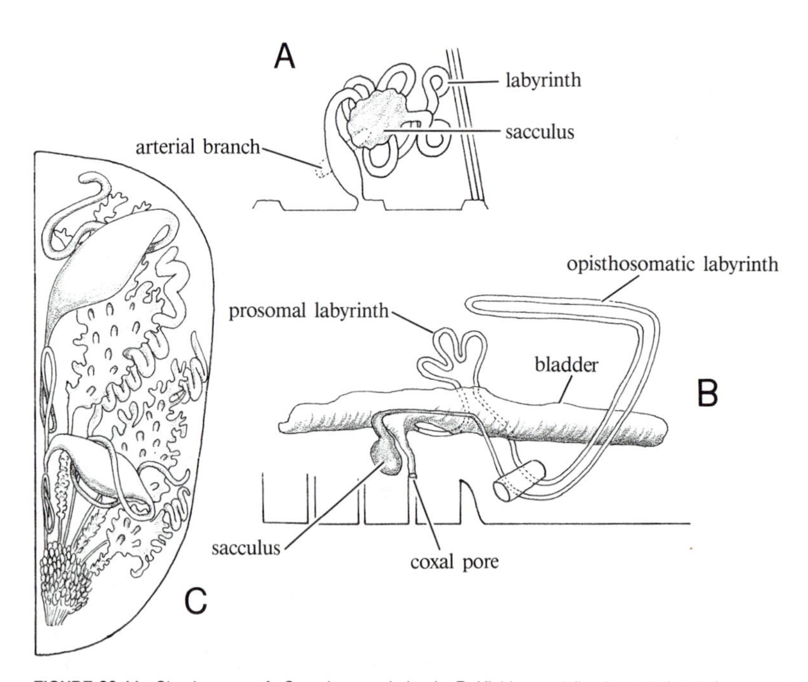

FIGURE 30.11. Gland organs. **A.** Scorpion coxal glands. **B.** Highly specialized coxal gland of harvestman. **C.** Silk glands of Araneae, typical in having several kinds of morphologically distinct silk glands and producing silk of varying properties. (**A** after Buxton; **B** after Kaestner, from Kaestner, in Kükenthal and Krumbach; **C** after Apstein, from Gerhardt and Kaestner, in Kükenthal and Krumbach.)

The circulatory system of cheliceriforms exhibits some unique features. In pycnogonids a transverse membrane, the **endosternite**, separates the dorsal pericardial sinus from the rest of the hemocoel. The **heart** lies in the pericardial sinus. Blood enters the heart through paired ostia and is driven forward to enter the perivisceral cavity anteriorly. It flows back in the perivisceral sinus and into the appendages. Blood returns to the pericardial sinus through openings in the endosternite that separates the two sinuses (Firstman, 1973).

In *Limulus* the heart is a long, tubular structure, held in the pericardial sinus by a system of nine pairs of suspensory ligaments. No veins return blood to the heart. Blood returns into the pericardial sinus above the endosternite (Fig. 30.10C), and the heart is bathed by blood. Blood enters the heart by way of **ostia**, a series of paired openings through the heart wall. Each ostium is guarded by a valve, preventing the exit of blood when the heart is contracting. The long, tubular form of the heart and the large number (eight pairs) of ostia are primitive characteristics. Blood returning from the gills enters the pericardial sinus during diastole and flows through the ostia into the heart lumen. A long pacemaker ganglion on the dorsal side of the heart initiates the beat. Inhibitory nerves from the brain pass to the pacemaker ganglion. During systole the ostial valves close, and blood is forced into the arteries. At the end of systole, the backflow of blood is prevented by the closure of semilunar valves at the arterial bases.

The limulid arterial system is complex. Four pairs of lateral arteries pass from the heart to a pair of collateral arteries that tend to run parallel to the heart (Fig. 30.6B). The collateral arteries serve as feeder vessels for several circulatory arcs. From the front of

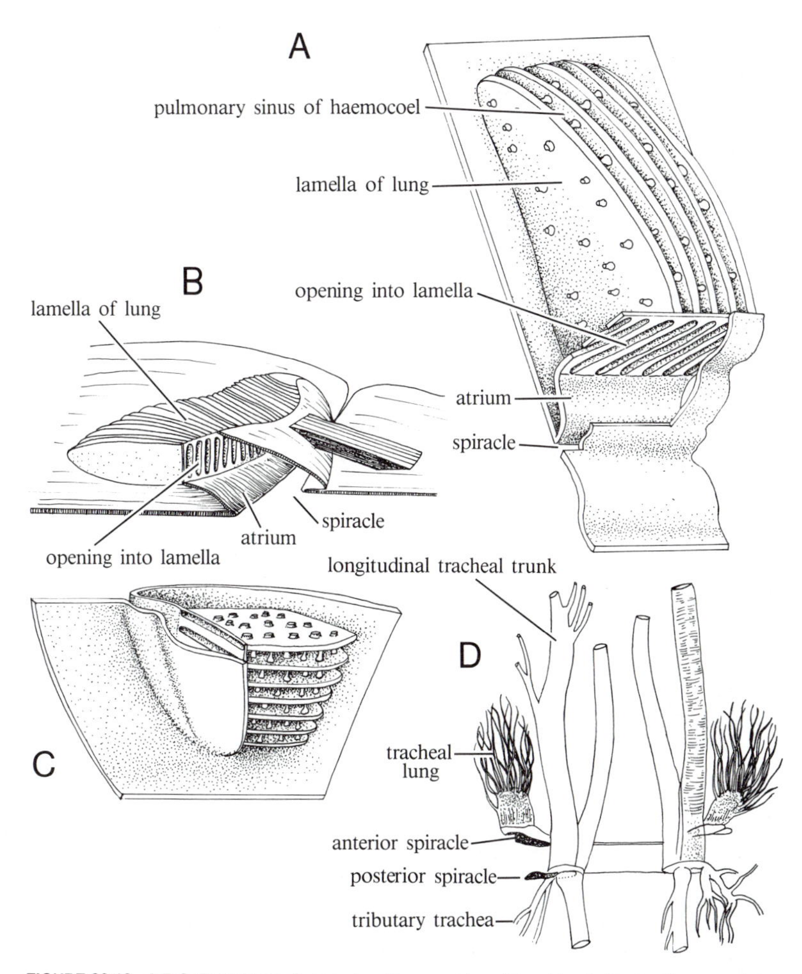

FIGURE 30.12. A,B,C. Book lungs: **A**. scorpion; **B**. uropygid; and **C**. spider. **D**. Tracheal lung consisting of tuft of tracheal tubules surrounded by blood sinus. (**A,B** after Kaestner; **C** after Gerhardt and Kaestner; **D** after Bertkau, from Gerhardt and Kaestner, in Kükenthal and Krumbach.)

the heart a pair of aortae and a median frontal artery arise. These are feeder vessels for cephalic arteries and, as the aortae combine and turn backward below the gut, for the circulatory arcs arising from the ventral vessel.

Four pairs of intestinal arteries arise from the collateral arteries and convey blood to the midgut. The collateral arteries also give rise to 14 pairs of lateral arteries that branch to the tissues in their region and extend to a system of marginal arteries at the edge of the prosoma and the opisthosoma. The two collateral arteries unite behind the heart to form the superior abdominal artery that extends to the telson and gives rise to a pair of lateral arteries that extend to the posterior marginal arteries of the opisthosoma. The anterior marginal artery runs along the prosomal margin and supplies blood to the eye and the digestive gland. The system of arteries arising from the collateral arteries is a predominantly dorsal arterial network and is strongly metameric in its organization.

The frontal artery supplies blood to the foregut, and branches each way to anastomose with the anterior marginals.

The paired aortae curve down and back, giving off branches to the foregut. They join near the mouth,

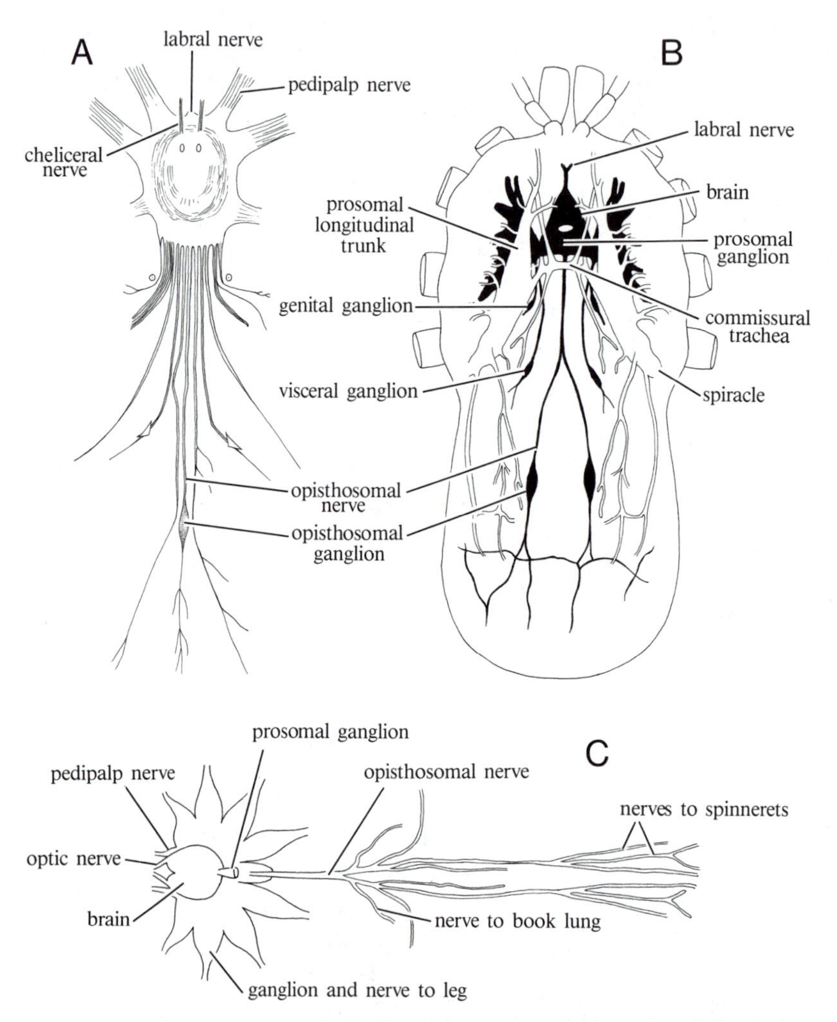

FIGURE 30.13. Nervous system in arachnids. Brain, connectives, and subesophageal ganglia fused to form complex nerve ring. **A**. Solpugid. **B**. Harvestman, nerves and tracheae. **C**. Spider. Prosomal nerve tissue forms massive prosomal ganglion with extensions, in many cases, into bases of legs. Abdominal nervous system with no segmental ganglia in most arachnids, and in many instances no ganglia whatsoever. Nerves to various body parts extend from ganglionic mass in prosoma. Harvestmen have complex tracheal system, with longitudinal trunks expanded to function in part like air sacs of insects. (**A** after Bernard; **B** after Warburton, from Warburton; **C** based on Kaestner, in Kükenthal and Krumbach.)

forming a vascular ring around the base of the esophagus. Most of the ventral arteries arise from this ring. The relationship of the nervous system and ventral arterial system is interesting. The brain lies within a vascular ring, and most of the major nerves run within arteries, a similar arrangement seen in many arachnids. A large ventral artery encloses the nerve cord and runs caudad, sending a branch up to the superior abdominal artery and ending near the origin of the nerves to the telson. The ventral artery supplies blood to the midgut and the ventral tissues. A number of vessels to ventral prosomal parts arise from the vascular ring. Nerves lie along these arteries.

At the ends of the finest arterial twigs, the blood enters hemocoelic lacunae, where most of the exchange between blood and tissues occurs. All of these empty into lateral and dorsal collecting vessels that unite with a main ventral collecting vessel on each side of the body. An afferent branchial vessel to each gill arises from the ventral collecting vessel. Efferent branchial vessels return aerated blood to the pericardial sinus.

The scorpion circulatory system is relatively simple (Fig. 30.7B). The heart is long and tubular and has seven pairs of ostia, each associated with a segment of the heart from which a pair of lateral arteries arise. A posterior aorta runs back into the metasoma, and an anterior aorta runs forward into the prosoma. As in *Limulus*, an intimate relationship exists between the prosomal branches of the anterior aorta and the nerves, and some of the nerves lie within arteries. A supraneural artery, also known as a ventral artery, runs immediately above the nerve cord and extends far back into the metasoma. As in the merostomes, the ventral circulation is largely established by branches from the anterior aorta, whereas lateral branches are associated with the more dorsal body parts.

Other arachnid circulatory systems have a dorsal, tubular heart that lies in a pericardial sinus, into which blood returns on its way to the heart. Blood enters the heart through paired ostia, and is driven through arteries into a hemocoelic sinus. Blood from the hemocoelic sinuses drains into an efferent sinus system that returns it to the pericardial sinus. Where the blood is responsible for transporting oxygen, the returning blood is routed past the book lungs or the tracheal lungs.

Spider hearts are shorter tubes, usually with three pairs of ostia and three pairs of lateral arteries that ramify to the tissues of the opisthosoma (Fig. 30.9). The anterior aorta passes through the pedicel, curves downward near the pumping stomach, and gives rise to a number of arteries to the legs and anterior appendages. Spiders have no ventral vessel, and the nervous and arterial systems are not as intimately associated as in scorpions. A system of ventral sinuses collect the blood from the tissues and return it to the lungs, from which it passes through a pair of pulmonary veins to the pericardial sinus.

The circulatory system varies considerably in the other arachnid orders. The primitive uropygids have hearts with nine pairs of ostia. Solpugids, although not otherwise as primitive, have a long heart with eight pairs of ostia. Most of the tracheate arachnids are small, and it is difficult to determine whether small size or efficiency of tracheal oxygen transport is responsible for the reduction of the circulatory system. Pseudoscorpions have quite small hearts with a single pair of ostia. In mites circulatory reduction is more extensive, and a system of irregular hemocoelic sinuses without a heart suffices.

Nervous System The nervous system in the proboscis of sea spiders contains a dorsal, ganglionated nerve from the **brain**, and a pair of ventrolateral nerves from the **subesophageal ganglion** (Fig. 30.1A). An anterior ring ganglion passes around the proboscis, and additional nerve rings occur at intervals. The nerve cord is double. It arises at the subesophageal ganglion and has four segmental ganglia on it. Nerves to the opisthosoma arise from the end of the nerve cord.

In *Limulus* all of the prosomal ganglia are united with the brain and **circumenteric connectives** to form a complex **nerve collar** (Fig. 30.6B). The collar lies in the vascular ring and gives rise to a double nerve cord located in the ventral vessel. Five opisthosomal ganglia lie on the ventral nerve cord, indicating some centralization of the opisthosomal nervous system. A sympathetic (stomodeal or stomatogastric) nervous system provides nervous control of the heart and digestive tract. Lateral longitudinal nerves parallel the nerve cord and give rise to nerves to the heart ganglion to stimulate or depress heartbeat. Nerves arise from the nerve collar to the sense organs, prosomal appendages, and segmental musculature, and from the opisthosomal nerve cord to the musculature of the opisthosoma.

Scorpions have a relatively simple nervous system. The nerve cord is a recognizable nerve trunk, and bears seven distinct ganglia (Fig. 30.7B). As there are fewer ganglia than body segments, it is evident that some centralization has occurred.

As a general rule, most of the nerve tissue of arachnids is gathered together to form a nerve collar around the esophagus. Nerves to various parts of the body arise from the nerve collar (Fig. 30.13). Some of the arachnids retain a recognizable nerve cord, but in others no definite nerve cord is found (Fig. 30.13A). Where a nerve cord can be distinguished, one or several ganglia may be found associated with it. Uropygids, for example, have a posterior ganglion at the end of the nerve cord associated with the operation of the tail filament. Solpugids have an abdominal ganglion that gives rise to nerves serving the alimentary tract, but as a general rule, all peripheral nerves arise from the brain or the nerve collar.

The brain proper is dorsal or anterior to the esophagus. It is made up of a **protocerebrum** closely associated with the visual organs and a **tritocerebrum** associated with the chelicerae. The complex mass of the rest of the nerve collar lies below the esophagus and contains the subesophageal ganglion united with all of the other ganglionic material that

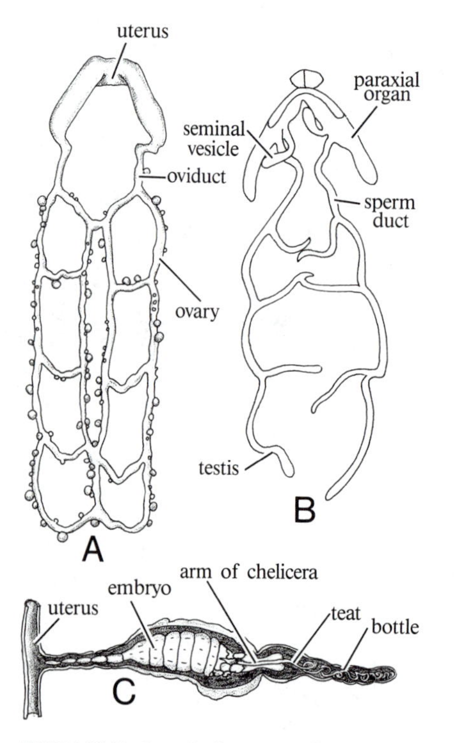

FIGURE 30.14. Reproductive system of scorpions: **A**. female and **B**. male. Systems similar, with branching gonads and paired gonoducts. **C**. In ovoviviparous scorpions, uterus gives rise to diverticula in which embryos develop. Placental extension of diverticulum extends to branches of digestive gland. Food processed in coiled, tubular bottle, that ends in teat held by embryonic chelicerae. Embryo uses its sucking pharynx to nurse. (**A,B** after Pawlowsky, from Kaestner, in Kükenthal and Krumbach; **C** after Vachon, from Cloudsley-Thompson.)

is retained. Cheliceriforms are without a deutocerebrum.

Reproductive System In Pycnogonida the sexes are separate, and females have reduced, ovigerous legs. The gonads are U-shaped and branch out into the legs and open through **gonopores** on the coxal leg podites. Some males have gonopores on the last pair of legs only, but most males and all females have gonopores on several or all pairs of legs. In females the eggs are stored in the femurs of legs with gonopores.

In merostomes the reproductive organs are quite similar in males and females. A median gonad is formed of loosely organized masses of follicles, in which the sperm or ova are produced. The gonad extends forward into the prosoma, where it tends to become tangled with the digestive gland, and back into the opisthosoma, where it lies below the intestine. In each sex, paired gonoducts arise from the gonad and lead to paired gonopores located on the operculum, formed by the first pair of opisthosomal appendages.

Scorpion **ovaries** are composed of a series of tubules with cross connections between the main lateral tubes on each side (Fig. 30.14A). A short **oviduct** leads from each ovary to the dilated **seminal receptacles**, located just within the gonoducts. A short genital atrium separates the gonopores and the seminal receptacles. The **testes** (Fig. 30.14B) are built like the ovaries and open into a long **sperm duct**, with several accessory glands.

Spiders have a pair of large **ovaries**, opening into **oviducts** that unite a short distance in front of the gonopore (Fig. 30.9). A pair of copulatory pores open into ducts that lead into one or more pairs of **seminal receptacles**. These are usually connected to the median oviduct, sometimes called the vagina. The male system is even more simple. The paired **testes** open into **sperm ducts**, which are coiled or convoluted and contain space for sperm storage. These unite before they open through the gonopore.

Biology

Locomotion Movement in cheliceriforms is almost exclusively by means of the walking legs. The system is based generally on a coxa that is hinged to the body wall by a single ventral condyle, and thus the leg achieves a characteristic rocking motion. In pycnogonids the system is slightly different, where opposing pairs of depressor and levator muscles in the coxa work independently to achieve a buckling of the arthrodial membrane and thus a rocking motion of the distal limb (Fry, 1978).

Limulus also uses the opisthosomal limbs to facilitate locomotion. Terrestrial chelicerates use the prosomal limbs in walking and running (Manton, 1977). Cheliceriforms, however, are not restricted to movement along surfaces. A unique form of flight known as ballooning is seen in small spiders. A line of silk is let out that is eventually long enough to be caught by the wind and loft the arachnid into the air. Such aeronauts can be encountered several thousand feet up.

Feeding and Digestion Pycnogonids are suctorial carnivores, inserting their probosces into their victims and sucking away. Pycnogonids generally creep about over the surface of colonial hydroids and bryozoa, and live on or in clams. A part of the prey is grasped by the chelicerae and sucked into the mouth. Sometimes the chelicerae clip off fragments, but this is usually unnecessary. The muscular proboscis contains the pharynx with a hardened lining. The sucking movements of the proboscis fragment soft food particles as they are swallowed, and bristles at the end of the pharynx strain out all but the finest particles. Food passes through a narrow esophagus into the midgut. Digestion and absorption of food occur in the midgut. Branches of the midgut extend out into the appendages, reaching almost to the tips. Intracellular digestion occurs in the mucosal lining of the midgut and its branches. Pycnogonids are unique in many ways, but in no point are they more so than in the method of intracellular digestion. Certain of the mucosal cells take up food material particles until they are gorged with food. They strip away from the mucosa and float freely in the cavity of the midgut. Other mucosal cells absorb food from them until, with the food exhausted, the floating cells are eliminated from the anus.

The limulines have a specialized mode of food processing, using the gnathobasic coxae (Fig. 30.10C) of the prosomal limbs to shred and pound food before swallowing (Manton, 1977). Even so, large pieces of food material are often swallowed. These pieces are mashed in the gizzard. The hardened gizzard walls protect the lining cells and aid in the breakup of food. Any material that cannot be ground up in the gizzard cannot pass the valve to the midgut and must be regurgitated.

Food entering the stomach undergoes a preliminary digestion caused by trypsin and alkaline lipase. The partly digested food enters the digestive glands, where the final phases of protein digestion and much absorption take place. Food residues returned to the stomach pass into the intestine. In the last part of the intestine, mucus-bound fecal pellets are formed.

The scorpions and most other arachnids are carnivores that capture the prey with chelicerae or pedipalps. The slow ingestion process may have favored the development of venom glands to immobilize the prey. Venom glands appear at various points: in the posterior sting of scorpions, in the chelicerae of spiders, and in the pedipalps of pseudoscorpions. It is evident that venom glands have developed independently in different arachnid groups. Food is held in front of the mouth and is broken into small fragments by the chelicerae, sometimes with the aid of endites on the pedipalps or leg coxae. Generally, salivary secretions are poured on the food, predigesting and softening it for swallowing. These secretions are supplied by glands from various parts of the body as well. Salivary glands are found in the chelicerae or pedipalps of some arachnids, but others regurgitate digestive juices secreted by glandular diverticula of the midgut. Arachnids have no jaws and the esoph-

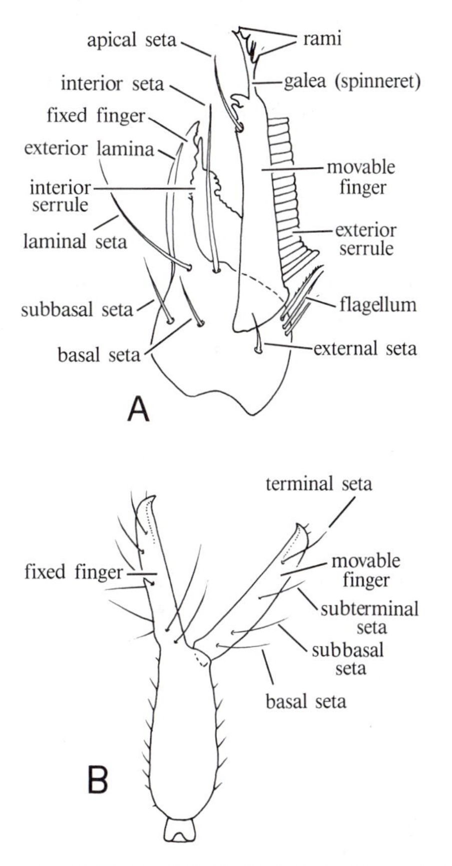

FIGURE 30.15. A. Chelicera of *Lamprochernes*, pseudoscorpion. **B.** End of pedipalp of pseudoscorpion, *Chelifer cancroides*. (After Hoff.)

agus is very narrow; therefore, only very small food particles can be swallowed. They solve this problem in different ways. For example, the chelicerae of pseudoscorpions are equipped with comblike serrules and a field of bristles that are used to clean the mouth after eating and to strain out particles too large to swallow (Fig. 30.15A).

The development of glands in different parts of the body, which are used for the same function, such as enzymes, silk, or venom, is an example of parallel evolution. In addition, homologous glands are used for different functions in different groups of animals; for example, cheliceral and pedipalpal glands may be used to secrete venom, silk, or enzymes, as the case may be.

Opiliones are voracious creatures, feeding on arthropods of suitable size and in some instances attacking terrestrial snails. They will feed on dead organisms, and some feed on bits of fruit or vegetation. Food is grasped by the pedipalp chelae and is broken up by the chelicerae with the aid of gnathobases on the coxae of the pedipalps and first two pairs of walking legs. Opiliones take a more varied diet and swallow larger particles than most arachnids.

The feeding habits of pseudoscorpions are rather like those of true scorpions, but on a diminutive scale. The prey are usually minute arthropods, such as psocids, collembola, and small dipterans, although species that live with ants attack the ants, often in small groups. Prey is caught in the pedipalps (Fig. 30.15B) and stung into submission if necessary. The chelicerae are thrust into the body of the prey, at the articular membranes between joints if the surface exoskeleton is too strong to pierce. Body fluids are sucked in, and enzymes are added to the tissues to soften them before they are ingested.

The outstanding feature of the solpugids is the huge, three-segmented chelicerae, used to kill prey and tear it apart. The huge pincers can bite viciously, and in areas where they are common, solpugids are much feared. No poison glands have been found in the chelicerae, although bites are sometimes slow to heal. Some human deaths have been attributed to solpugid bites, but the deaths are generally thought to be the result of secondary infections. The chelicerae are formidable weapons and can kill scorpions so rapidly that they have no chance to sting. The prey is held crosswise in the chelicerae, and pulped by lateral and vertical movements of the pincers. The macerated prey is held against the mouth, and the softened parts are sucked in.

The usually partly digested, liquid food enters the midgut where digestion is completed and absorption occurs. As a general rule, the sucking pharynx opens into an esophagus that passes through the nerve ring and opens into the midgut, but in spiders the sucking stomach opens directly into the midgut. The arachnid midgut is always equipped with some kind of diverticula, and in a number of arachnids large digestive glands, sometimes termed the liver or hepatopancreas, are found. In the majority of arachnids the greatly amplified secretory surface of the midgut diverticula is important, in part toward producing enzymes for preliminary digestion, and in part associated with the absorption of food. The lining of the midgut may participate in intracellular digestion, but the extent to which arachnids depend on intracellular methods of digestion and absorption is as yet uncertain. They may be used more extensively than has generally been thought. In one group or another, the diverticula arising from the midgut are differentiated into two or more types. In very few cases is there real information about the functional significance of these differentiated parts. Even the enzymes used by arachnids and scorpions in digestion are not very well known. Proteinases, lipases, and amylases have been recovered from the scorpion, but there is no significant information about where they are formed and where they act during digestion.

Excretion Details of excretory physiology in *Limulus* are not clear. Digestive glands excrete excess calcium into the gut, and it is probable that the gut wall also has a secondary excretory function.

Unfortunately, the details of arachnid excretion are only partly known. At least some of the arachnids excrete most of their nitrogenous wastes in the form of guanin. Guanin excretion is advantageous for terrestrial animals, as its solubility changes markedly with the pH of the medium. In an acid medium, guanins precipitate as insoluble crystals. Water can be reabsorbed and the crystalline guanin can then be excreted in a solid form. Malpighian tubules and the stercoral sac appear to be organs that are especially suited for guanin excretion. For example, in spiders the midgut walls contain a number of guanin cells that empty guanin into the lumen of the gut. These crystals may or may not go into solution in the midgut, but in any case they are crystallized out by the acid contents of the stercoral sac, preparing the way for water reabsorption before the guanin is eliminated. Guanin compounds are also picked up by the Malpighian tubules, and emptied into the stercoral sac or into the gut near the stercoral sac. The contents of the Malpighian tubules are acid at the time of their discharge into the digestive tube, and the stercoral sac undoubtedly aids in keeping the gut contents acid enough to prevent the guanin compounds from going into solution. The movement of nitrogenous wastes to the Malpighian tubules is not yet described. Certain cells in the prosoma of spiders and some other arachnids are basically athrocytes, but it is not clear how materials that are taken up by these cells are delivered to the Malpighian tubules. Presumably the nitrogen compounds are transported in a more soluble form, or the body fluids are alkaline enough to permit the movement of guanin in solution. It is probable that the cells of the Malpighian tubules can convert nitrogenous wastes of other kinds into guanin compounds.

Respiration The blood contains a respiratory pigment, hemocyanin, dissolved in the plasma and has an oxygen-carrying capacity about a fourth as high as most fishes. It may, like the hemocyanin of some other arthropods, circulate at less than full saturation with oxygen.

Two lines of improvement are seen in organisms with book lungs, one leading to control of the spiracle and the other to muscular movements to ventilate the lung. Neither appears to be a critical problem, for the body generally is small and a relatively large surface area is exposed in the book lung. Apparently, ventilating movements are not usually required, and diffusion suffices for respiratory exchange at an adequate rate. The effectiveness of book lungs is shown by removing them from scorpions. This reduces respiratory exchange to zero, indicating that the general body surface plays no part in respiratory exchange. Some scorpions have muscles that can expand the spiracles and the atrium, and some spiders have similar muscles. These appear to be brought into play only at periods of great activity, and may be more closely associated with water economy than with respiratory efficiency. It is clear that the smaller the spiracles, the less the loss of water through the book lung. Mechanisms that permit dilation of spiracles and expansion of the atria also permit the reduction of spiracles to a size that is normally adequate for moderate activity.

Sense Organs Little is known of pycnogonid sense organs. Four inverse pigment-cup ocelli are found on the posterior end of the cephalon directly over the brain. The brain is generally distinct from the circumesophageal connectives.

Limulus has three pairs of eyes: a pair of larval eyes that degenerate during development; a pair of median eyes in the form of pigment-cup ocelli with cuticular lenses and corneas; and a pair of compound lateral eyes. The structure of the lateral eyes is particularly interesting (Fig. 30.16A). Each visual component of the eye is an **ommatidium**, formed by 10 to 15 photosensitive cells surrounding a central rhabdome. The **rhabdome** is formed by the union of the photoreceptive rhabdomeres present in the component retinula cells. When light strikes the rhabdome it evokes an impulse in a ganglion cell associated with the rhabdome. This is an interesting development for, in general, the neurosensory cells of invertebrates are receptors and transmitters of stimuli. The sensory neuron associated with the ommatidium is structurally independent of the receptor. Each of the ommatidia is an independent unit and serves as the source of an independent nerve impulse. The compound eye of *Limulus* is made up of a number of ommatidia, each with an independent lens covered by a transparent cuticular cornea.

The larval eyes originally lie at a point immediately in front of the bases of the chelicerae. At this point, peculiar papillate sense organs are found in the adult. They are innervated from centers in the brain associated with the compound eyes. Spiny taste receptors are found on the gnathobases of the legs, and sensory bristles can be seen on the gill blades. At various points on the body, tactile bristles and pores, thought to have a sensory function, have been described, but a full analysis of sensory reception has not yet been made.

Arachnids use a variety of mechanoreceptors to analyze aspects of their environment (Seyferth and Barth, 1972). The hairs and the bristles used to convey a stimulus to receptor cells vary greatly. Some are short and stubby, whereas others are long, flexible filaments. Depending on the mechanical properties of the hair or bristle, and the frequency with which the nerve fiber from the receptor cell can accept new stimuli, the mechanoreceptors can provide surprisingly diverse bits of information about the environment. Simple tactile stimuli are important, of course, and provide information that helps in escaping enemies or capturing food, as well as helping to set up reflexes that are important in walking, running, fighting, or mating. The sagging of a long, slender hair can provide information about orientation to gravity or, as it blows in the wind, give clues to the direction and strength of air currents. Filamentous receptors can also record vibrations and, therefore, are sources of information about the texture of surfaces with which the receptor is in contact, or the frequency and amplitude of vibrations in the sound range. They can describe the changing tensions in a spider's web and give clues to the arrival of food, a mate, or other things that affect the web. Every arachnid has not one but several of these filamentous mechanoreceptors, but there is a dearth of information about their specific functions.

Just as every bristle is a potential point of sensory reception, so is every pit or thin spot in the exoskeleton. A number of sense organs are small, depressed pits or slits with attenuated cuticle at the floor. The tarsal organs of spiders and the lyriform organs (Fig. 30.16B) of spiders and some other arachnids are examples. Pit-shaped receptors, like filamentous receptors, may be used for a variety of purposes. They are particularly useful for recording exoskeletal tensions and for vibrations, but are also commonly used for chemoreception. The lyriform organs of spiders serve as good examples (Barth, in Merrett, 1978). Each receptor is a narrow pit, covered by a delicate floor against which sensory endings project. Several functions have been ascribed to them. There is some evidence that they function in chemoreception, and some have considered them as hydroreceptors. On the other hand, there is evidence that they are used to recognize vibrations in the web and evoke attack reactions when stimulated. It is possible that several kinds of lyriform organs exist, each sensitive to different stimuli.

Photoreception is better understood. Arachnid eyes are simple ocelli, with a circular corneal lens filling much of the cavity of the pigment cup. The lens is formed by corneagenous cells differentiated from the epidermis. The corneagenous cells of scorpion eyes do not extend over the light-receptive retinula cells (Fig. 30.16C), but do in spider eyes (Fig. 30.16D). As in *Limulus*, the retinula cells form clusters around a rhabdome, formed by the combined microvillar rhabdomeres from each retinula cell of the cluster. Each individual rhabdome serves as a source of light reception and initiates a separate nerve stimulus. The quality of the image depends on the number of rhabdomes present in the ocellus. This number varies greatly in the arachnids, even within the same group. Lycosid eyes, for example, may have from 100 to 4500 rhabdomes, depending on the position of the eye and the species. At the very best, arachnid eyes have very few rhabdomes compared with the number of rods or cones in a vertebrate eye, and the image is sketchy (Fig. 30.16E). As a rule, the medial eyes are inverse, with the rhabdomes directed away from the lens, whereas lateral

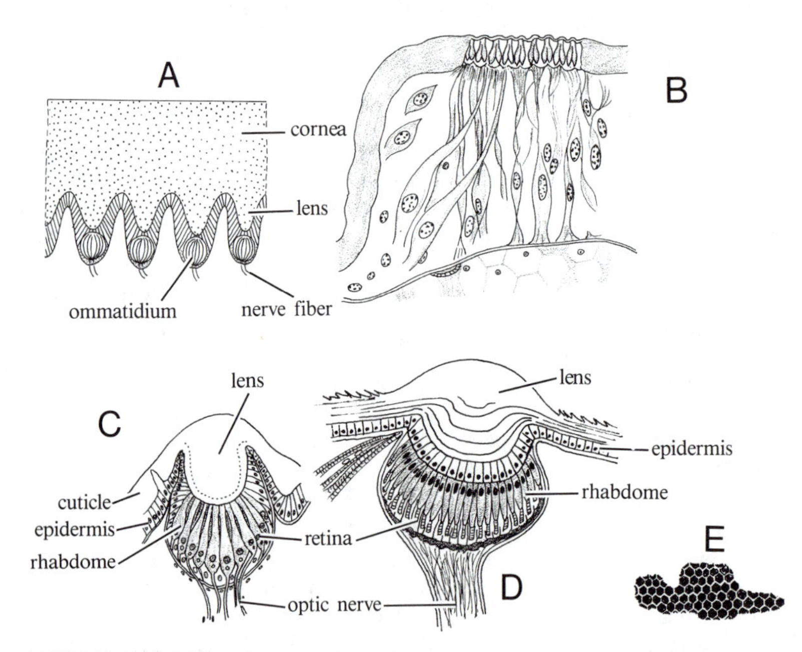

FIGURE 30.16. A. Scheme of structure of part of compound eye of horseshoe crab. Note continuous cuticle and separate ommatidia, each with its own nerve. **B**. Lyriform organ of spider, whose function may be chemoreception. **C**. Direct eye found in scorpion, with rods and rhabdome extending toward lens. **D**. Indirect eye from spider *Araneus,* with rods and rhabdome extending away from lens. In either type of eye, each rhabdome serves as distinct receptor to form mosaic image. **E**. Image of another spider, as seen when 8 cm away from spider eye, shown at (**D**). (**A** after Lankester and Bourne, from Snodgrass; **B** after Kaston, from Bristowe; **C** after Lankester and Bourne; **D** after Grenacher, from Dahlgren and Kepner; **E** after Homann, from Carter.)

eyes are direct, with the rhabdomes directed toward the lens. An unusual feature of inverse spider eyes is the presence of intrinsic eye musculature that can tilt the eye, thus directing it to different points without changing the position of the body, and change the focal depth by compressing the corneagen cells below the lens. The only other invertebrates that have definite eye muscles are the cephalopods. In arachnid eyes, a layer of cells forms behind the eye capsule. This postretinal layer forms the back of the eye, and is sometimes converted into a tapetum, a layer containing crystals that reflect light. The eyes of some spiders, for example, gleam when a beam of light strikes them, as a result of the reflection of light by the tapetum.

The relationship between brain regions and behavior has been partially described. The protocerebrum is associated with the optic nerves, but although arachnids usually have several pairs of eyes, the visual sense is not the most important in determining behavior, and no more than three percent of the brain is occupied by visual centers. In this, arachnids contast sharply with insects, in which a much larger part of the brain volume is turned over to visual centers. Two centers, the corpora pedunculata and the central body, are primarily responsible for integrating behavior. The corpora pedunculata are visual centers as well as integrating centers in arthropods, and they tend to become relatively larger in the arachnids that are larger or more active. The central body appears to be most closely associated with behavior patterns generally considered to be instinctive and also related to visual coordination.

Reproduction and Development In pycnogonids, the males shed sperm on the eggs as they leave the female. The eggs are deposited directly on the ovigerous legs of the male, or the male gathers them together. The eggs are cemented to bristles on the ovigers (Fig. 30.1B). The young may remain attached to the male parent for a time after hatching, or they may scatter at once to become internal or external parasites on hydroids or clams.

Development follows the general arthropod pattern. The zygote undergoes holoblastic cleavage, either equal or grossly unequal depending on the amount of yolk in the egg. Bands of ectodermal and mesodermal tissue form on the surface of the solid gastrula. The appendages develop from the bands, the chelicerae first and the remainder in regular sequence, from front to back. The young (Fig. 30.17A) usually have three pairs of appendages at hatching and acquire additional appendages with succeeding molts. The larvae develop cheliceral glands that secrete an adhesive used for attachment to the male parent or to hydroids.

In *Limulus*, the female builds nests in the sand for the eggs or attaches them to the opisthosomal appendages (Cohen and Brockman, 1983). The male clings to the female and releases sperm over the eggs as they are put into place. The eggs are then enclosed in a leathery capsule that eventually breaks free.

The egg undergoes holoblastic cleavage and eventually becomes a stereoblastula. The gastrula stage is also solid. The mesoderm is arranged in two growth centers, one forming the first four prosomal somites and the other the last three prosomal and the opis-

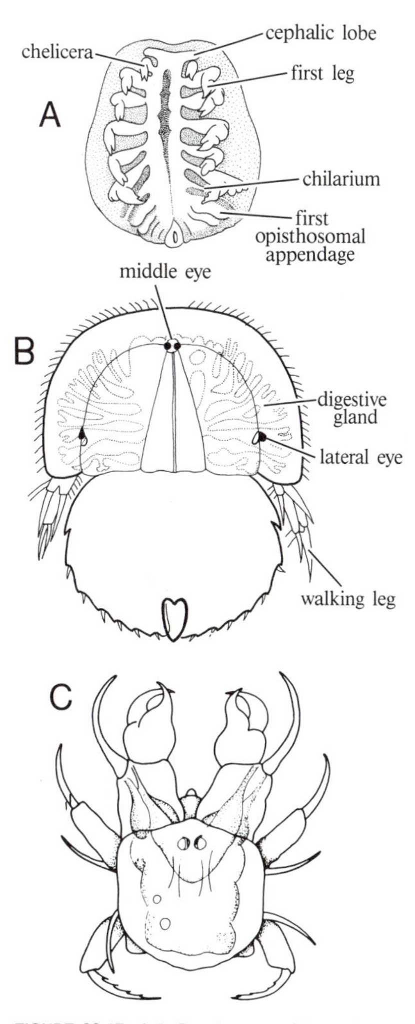

FIGURE 30.17. **A**,**B**. Development of horseshoe crab: **A**. embryo with developing appendages; **B**. Euproops larva. **C**. Pycnogonid larva. (**A** after Ivanoff; **B** after Kingsley and Takano, from Shipley; **C** from Kaestner.)

thosomal somites (Fig. 30.17B). The young eventually emerge as the Euproops (often inaccurately referred to as the "trilobite") larvae (Fig. 30.17C). The larvae are active creatures, clambering about on the bottom and sometimes burrowing in the mud. The telson gradually develops during the early molts.

In the scorpions and other arachnids, the simplicity of the reproductive system contrasts with the complexity of reproductive behavior. The structure of the modified male pedipalp has been described. When sexually ripe, the male charges the pedipalp with sperm, dipping it into a drop of semen secreted on a tiny sperm web.

Male spiders make a number of different kinds of approaches to females (Robinson and Robinson, 1980). The female has formidable fangs and in some cases is perfectly willing to kill and eat the male. The male makes an approach that will give him the maximum chance of sexual success and personal survival.

The male *Tetragnatha* makes a direct approach. He has larger chelicerae than the female. The male wedges the female's fangs open with his own so that she cannot bite, maintaining this hold throughout mating. When finished, he rapidly disengages and lowers himself to the ground on a thread of silk. Some of the other spider males are also forceful. The male *Xysticus* ties the female to the ground with silk before mating. This is counterbalanced by the conduct of the female *Argiope*, which tries to wrap the male in silk during mating and, if the male is not strong and fast enough to escape, feeds on him afterward. But the battle of the sexes is not always so fiercely fought. Some lycosid males present the female with a fly, wrapped in silk, and mate with her if she accepts. Others perform courtship dances, involving vibrations of the body or appendages. The female may retreat, attack, or, if in the mood, signal her acceptance by similar vibrations. Vibration of the body or its parts plays an important role in courtship affairs generally, and reinforces the observations about the importance of mechanoreception in arachnids and scorpions. The male often vibrates the web of the female in a characteristic fashion, as one might rap on a lover's window, and in some cases the male has a stridulating organ that is brought into play during courtship, perhaps soothing the "savage beast" in the female.

However the preliminaries may be effected, the male gorges the hematodocha of the pedipalp with blood, erecting the embolus in readiness for mating. The embolus is inserted into the copulatory pores, and the seminal receptacles are filled with sperm.

The female does not lay eggs immediately. When the eggs are ready, they are deposited in masses on a silken film. Sperm from the seminal receptacles is poured over them, and a covering sheet is added. The whole mass is wrapped up in layers of silk to form a cocoon. Cocoons may be carried about, attached to leaves or under objects, hung on threads from leaves or twigs, or deposited in a nest or burrow. The silken cocoon protects the eggs from mechanical injury and desiccation, serving the same function as the heavy egg membranes secreted by the accessory glands in some animals, or as the special brood chambers found in others.

Scorpions carry on a long courtship dance. Eventually the male persuades the female to retire with him into a burrow, or digs a special hole into which they retire. A true copulation occurs, with terminal parts of the male system serving as a penis. Not all scorpions, however, work in this manner. In some cases the sperm are cemented together to form a spermatophore in a special chamber in the male system. During the courtship dance the male sticks the spermatophore into the ground and maneuvers the female over it. He presses her down on it so that the tip enters the genital atrium. However sperm transfer is effected, the sperm move up the female ducts, eventually reaching the seminal receptacles.

Development is internal. Some scorpions produce large, yolky eggs that develop in the oviducts. Others have small ova with insufficient nourishment to carry the young creature through development. In these species, the young animal gets nourishment through a placentalike arrangement (Fig. 30.14C). The oviduct develops diverticula that extend to the nearest diverticula of the midgut. The embryo is contained therein, and nourishment is passed from the gut to the growing scorpion baby.

A variety of different solutions to the problems of mating and development are seen in other kinds of arachnids. The pseudoscorpion male, like some scorpion males, thrusts a spermatophore into the ground and presses the female down upon it. The male harvestman is equipped with a very long penis that conducts sperm to the female, whereas the female has a long ovipositor that is used to scoop out a burrow for the eggs.

Most arachnids provide a certain amount of postnatal care for the young. This may involve little more than building a burrow or nest in which the young develop and guarding it for a time. A number of spiders guard the cocoon zealously throughout development, and many others carry it until the young hatch (Fig. 30.18A). Pseudoscorpions usually retire to a silken nest for brooding the eggs. An incubation chamber attached to the gonopore is formed by the mother and the eggs are laid in the sac. The eggs hatch in a few days, and the larvae develop a sucking beak used to ingest a secretion produced by the female from the transformed ovaries. They grow enormously, and before they emerge from the incubation chamber, the mass of young in the brood pouch may be larger than the mother.

As a general rule, the mother's duties are over when the young hatch, but in some cases the period of postnatal care continues for a time. Scorpions, wolf spiders (Fig. 30.18B), and some of the pseudoscorpions carry the young about for a time, and in some cases the young organisms have special structures to help them hang on. Some of the arachnids feed the young for a time (Fig. 30.18C), either with special secretions or with regurgitated food.

Review of Groups

SUBPHYLUM PYCNOGONIDA

The sea spiders (Figs. 30.19E and 30.20A) are common among the bryozoan and hydroid colonies upon which they feed. They are found in tropical and polar seas, and in both shallow and deep waters. Most of the 500 species are only a few millimeters long, but the largest have a legspread of over 0.4 m.

Pycnogonids are queer-looking animals, often with strange habits. Their movements are very deliberate, generally uncoordinated, and lumbering. They are certainly cheliceriforms, and might appear superficially to be more like the arachnids than the merostomes; however, they are so highly modified that it is not difficult to define their sister group relationship to members of the other subphylum. The larva (Fig. 30.17A) is peculiar and is, in a sense, highly specialized because of its habit of attachment to the hydroids or male parent. Yet in many ways the larva, with so few limbs, is one of the most primitive of arthropod larvae. Some have felt that the larva relates pycnogonids to the protoarthropod stem, and thus that they are derivatives of the cheliceriform stem near its point of origin. Be that as it may, the pycnogonids are very interesting creatures and deserve more attention than they have had from both the physiological and evolutionary points of view (King, 1973; Fry, 1978).

SUBPHYLUM CHELICERATA

Class Merostomata The five existing species of horseshoe crabs are the last remnants of the merostome tree. Many of the ancient merostomes were large and had heavy exoskeletons, and thus fossilized well (Fig. 30.19B). Although they probably appear to have been more common than they actually were, there is no reason to doubt that they were relatively abundant throughout the Late Ordovician

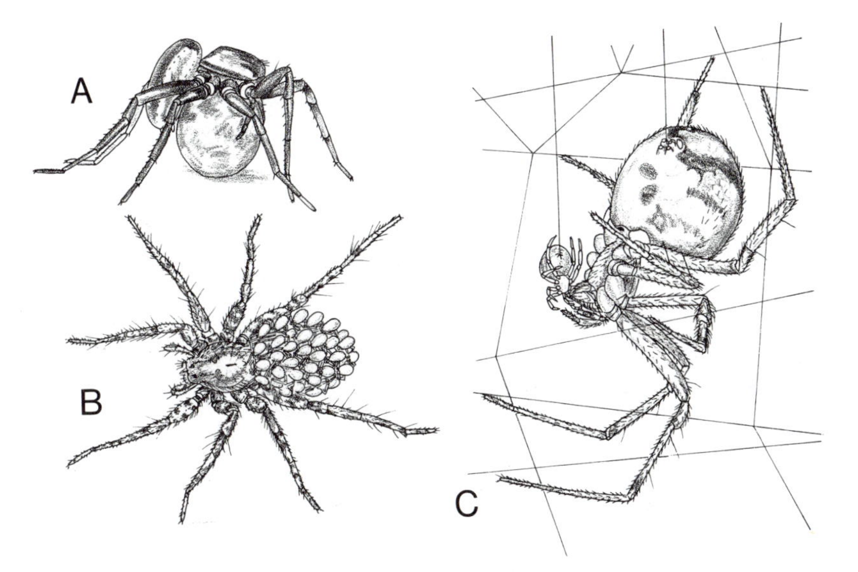

FIGURE 30.18. Maternal care. **A**. Female *Dolomedes fimbriatus* carrying egg sac until young emerge. **B**. Female *Lycosa,* covered with her young. Wolf spiders carry young for first week after they hatch, after which they fend for themselves. **C**. Spider madonna and child. Female *Theridion* regurgitates food for her young for first several days after they hatch. (After Bristowe.)

and Silurian. Forms resembling *Limulus* first appeared about 200 million years ago during the Triassic. In a sense, the few remaining horseshoe crabs are living fossils, thus deserving of the extensive study that they have received.

A horseshoe crab looks ungainly (Fig. 30.19A). It creeps about slowly and awkwardly and swims very poorly, pushing off with the prosomal limbs and using the flattened opisthosomal appendages as paddles. One might expect that any creatures constructed on such an ancient pattern must be beautifully adapted to some kind of environment. For horseshoe crabs, this environment is soft mud or sand. The sharpened margin of the shield makes an excellent shovel edge, and as the legs and telson can thrust the body forward against relatively high resistance, *Limulus* gets about very handily in the soft-bottom material.

The prosomal appendages function to some extent in locomotion but are particularly important in burrowing and capturing prey. Horseshoe crabs feed on worms, especially nereids, in the soft bottoms that they most commonly prefer. The prey is caught by the chelicerae while *Limulus* burrows in the mud. The first four pairs of walking legs have projections known as gnathobases on the inner side of the basal coxae. They are used to fragment the prey into particles that can be swallowed (Bonaventura et al., 1982).

Class Arachnida Arachnids are the most successful chelicerates today (Savory, 1977). Well over 60,000 species have been described. Arachnids compensate for their small size by large numbers, and are ecologically important in most terrestrial environments. Arachnids were found in one study to

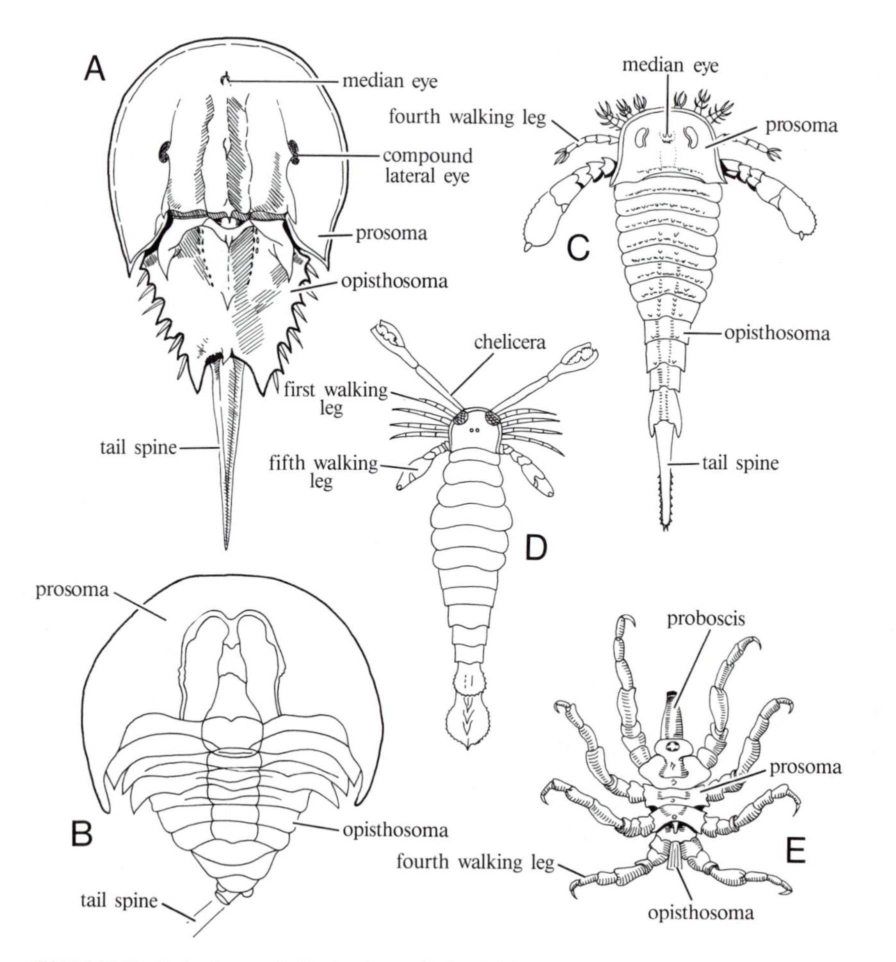

FIGURE 30.19. Cheliceriformes. **A.** *Limulus,* dorsum. **B.** *Koenigiella,* extinct synziphosuran. **C,D.** Eurpterids: **C.** *Pterygotus,* and **D.** *Eurypterus.* **E.** *Pycnogonum.* (**A,C,D,E** after Snodgrass; **B** after Eller.)

make up 18 percent of the arthropod macrofauna of Illinois deciduous and Utah coniferous forests. They are predominant in the mesofauna of the soil. About 20,000 mites per cubic meter are reported from soil 10 to 20 cm below the surface in tropical and temperate forests; over 160,000 mites per cubic meter have been found in disturbed grassland soil in England. The large number of orders reveals an extensive adaptive radiation of arachnids.

The scorpions (Fig. 30.21B) and the extinct eurypterids (Fig. 30.19C,D) are similar in many ways, and may yet prove to be more closely allied than previously thought (Kjelleswig-Waering, 1986). Scorpions prefer tropical and subtropical climates, but some species occur in temperate regions. They live at the ground level, hiding in crevices or under things, and emerge at night to feed. Some species require a humid environment, but others have adapted to arid habitats.

A muscular sheath covers the poison glands, forcing the poison through the paired poison ducts to a terminal pore in the barb. Two pairs of muscles from the last somite extend to the base of the sting. Scorpions carry the sting aloft and ready, curving the postabdomen up and forward. The ventral pair of muscles thrust the sting upward and into the prey, whereas the dorsal pair retract the sting. The scorpion poison is a neurotoxin, highly effective against the organisms on which it feeds, killing them immediately. The sting of most scorpions is painful but

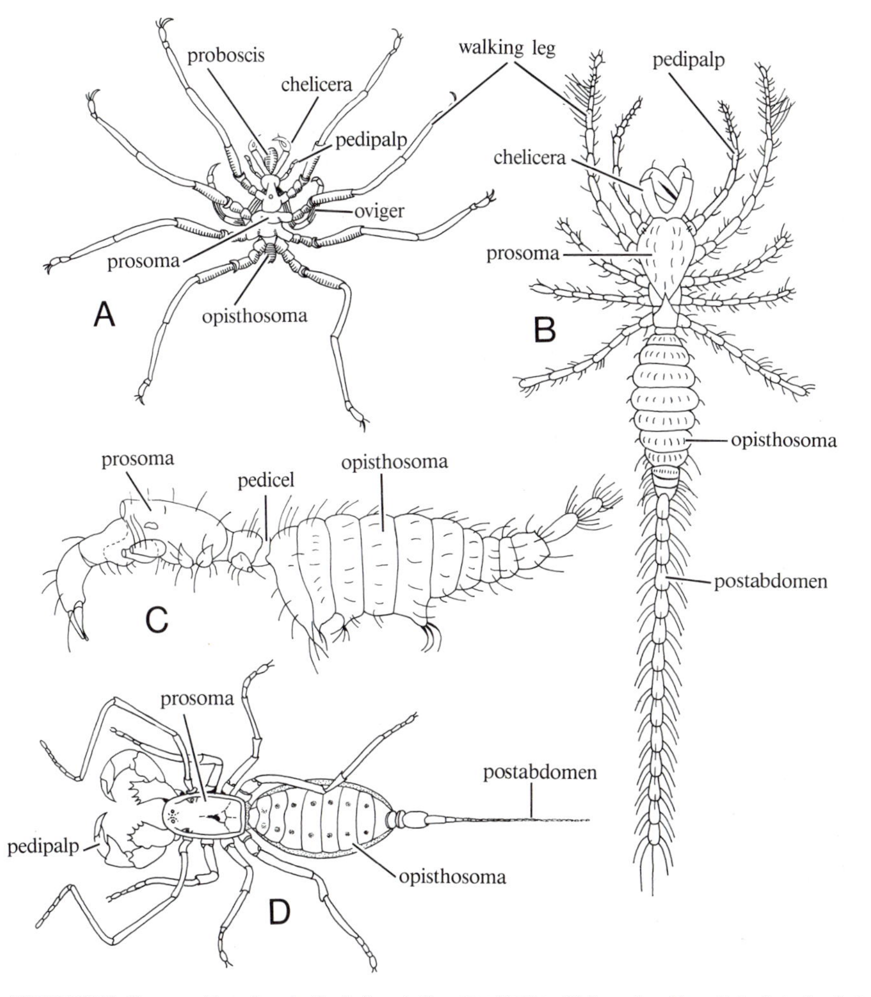

FIGURE 30.20. Pycnogonids and arachnids. **A.** Female *Nymphon hirsites* with five pairs of legs. Note diminutive first pair of ovigers on which males would carry eggs. **B,C.** *Koeneia*, palpigrade: **B.** dorsal view, note long, jointed postabdomen; **C.** lateral view with legs and distal postabdomen removed. **D.** *Mastigoproctus*, uropygid, with opisthosoma with eight broader somites of mesosoma and three narrower ones forming metasoma. Long postabdomen attached. (**A,D** after Snodgrass; **B,D** after Kraepelin, from Savory, 1977.)

not dangerous to humans, but there are some exceptions. The most dangerous is *Androctonus australis* from the Sahara, which causes death in about seven hours if the victim is not treated with antitoxin. A number of deaths in Mexico have been attributed to scorpion stings. The primary cause of death is respiratory and cardiac failure.

Most scorpions prey primarily on other arthropods, especially on the larger insects. Prey is caught by the powerful pedipalp chelae and torn up by the chelicerae, sometimes with the aid of the coxal endites on the first legs. Relatively quiescent prey may not be stung, but struggling prey is quickly overpowered by stinging. Scorpions are nocturnal animals and cannot use the eyes for recognition of prey. So little is known of sense organ physiology that the way food organisms are identified and located remains unknown.

The less abundant arachnid orders, Palpigradida, Uropygida, Amblypygida, and Ricinuleida, are predominantly inhabitants of the upper soil and surface litter in warm temperate to tropical lands. Small terrestrial animals have a relatively great body surface, at which loss of water can occur, and the majority of the more minute arachnids are limited to humid habitats. Larger arachnids are more common in cooler and drier places.

The Palpigradida (Fig. 30.20B,C) is a small group

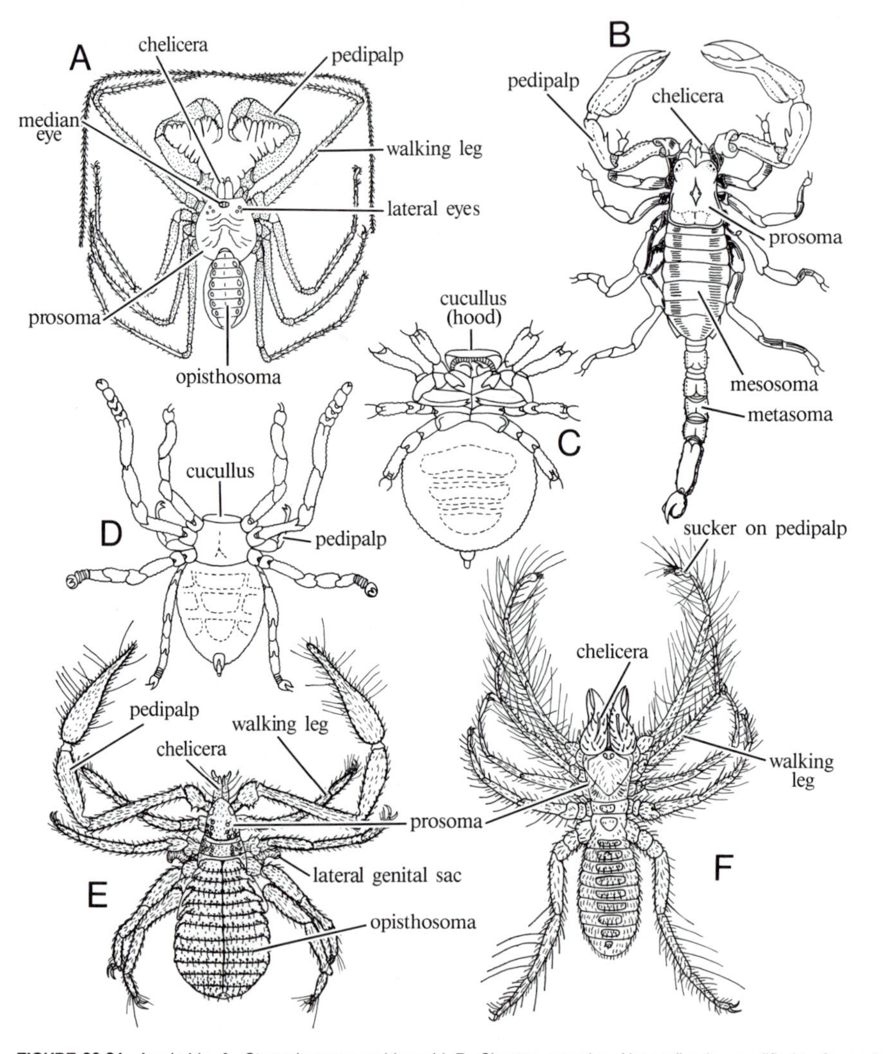

FIGURE 30.21. Arachnids. **A.** *Stygopharynus,* amblypygid. **B.** *Chactas,* scorpion. Note tail spine modified to form sting and large pedipalps with conspicuous chelae. **C,D.** *Ricinoides,* ricinuleid; **C.** ventral view; **D.** dorsal view. Note hood at front of shield to cover mouth and chelicerae. **E.** *Chelifer,* pseudoscorpion. **F.** *Galeodes,* solpugid. (**A,E,F** after Savory; **B** after Snodgrass; **C,D** after Hansen and Sorensen, from Savory, 1977.)

of about three dozen species. They lack eyes and their bodies are unique in that the anus is not terminal but followed by a many-segmented postabdomen or flagellum. The pedipalps are not specialized and, therefore, are used as another set of legs. Rather, the first walking legs are elongate, equipped with copious distal setae, directed forward, and used as sensory appendages. They live in soil, interstitially in sand, and under rocks, and seem to prefer tropical and warm temperate climates.

The Uropygida (Fig. 30.20D), or whip scorpions, possess the well-developed, clawlike pedipalps of scorpions and the long, jointed postabdomen of palpigrades. They have four sets of eyes, a pair on the anterior edge of the prosomal shield and three sets lateral and somewhat posterior to these. The chelae and the pedipalps are formidable but are not equipped with poison glands; however, they secrete either formic or acetic acid in a set of large anal glands.

The Schizomida is a small order, sometimes placed within the uropygids but having many (although possibly superficial) similarities to palpigrades. The pedipalps are large but leglike, and the first walking leg is adapted as an antennalike structure. The opisthosoma is regionalized into a large anterior portion and a constricted postabdomen of three segments. The body then terminates in a short flagellum of one to four segments.

The Amblypygida (Fig. 30.21A) are spiderlike creatures. They bear a set of formidable, raptorial, spiny pedipalps. Lacking poison glands, they grab their prey with the pedipalps, then proceed to dismember the still living, hapless victim by slowly shredding each tidbit with the chelae. The first legs are developed as a remarkable set of antennalike sensory limbs.

The Pseudoscorpionida (Fig. 30.21E) are relatively common at the ground level in all but the most arid climatic regions. They live quiet lives in the surface litter, among grasses or mosses, or under stones. As they are only 1 to 8 mm long, they are often overlooked. Some live in ant or termite nests, and a few live in trees. *Garypus* is found on sandy beaches in the intertidal zone hiding under stones. *Chelifer cancroides* lives with humans in houses and outbuildings, feeding on mites or other small arthropods. Pseudoscorpions superficially resemble scorpions, for they have large pedipalps and a broad opisthosoma, but they lack the narrower metasoma and sting. One or two pairs of laterally placed, indirect eyes may be present on the prosoma.

In pseudoscorpions the prosomal silk glands open through pores on the movable finger of the chelicera or on a spinneret or galea located on it. Silk is spun by attaching the spinneret to a surface and drawing the chelicera away. Pseudoscorpions use the silk to build nests for hibernation or into which they can retreat during molting. Although they are not given to long treks, they are not homebodies, and a new nest is built for each molt. The females also build a nest in which the young are brooded. The chelate pedipalps (Fig. 30.15B) resemble the pedipalps of scorpions superficially. They are especially important as tactile organs, but often contain poison glands with ducts opening on one or both jaws of the chelae.

The Solpugida (Fig. 30.21F), the wind or sun spiders, are characteristic of hot deserts, where they hide in burrows or crevices until night, when they come out to feed. They are formidable-looking creatures that reach up to about 7 cm long. The prosoma is covered by several sclerites, with a pair of eyes on the anterior piece. The opisthosoma is composed of ten segments and is broadly united to the prosoma, with a slight pedicel. The four pairs of legs are long, and most solpugids are very fast runners. A pair of spiracles open behind the coxae of the second legs, and two or three pairs open on the ventral side of the opisthosoma.

The Ricinuleida (Fig. 30.21C,D) are a small, little-known group of tropical litter inhabitants. The chelae and the pedipalps are greatly reduced. The prosoma has an anterior fold or hood that covers the chelicerae, and the opisthosoma is a wide oval structure.

The Araneae (Fig. 30.22), or true spiders, are the best known of the arachnids and are common nearly everywhere (Foelix, 1982). They are extremely successful, small carnivores, and play an important role in the ecological balance of their communities, serving as checks against some insect species. Unable to fly, the spiders are somewhat more restricted than insects, but they have nevertheless adapted to a wide range of habitats, living in the surface litter, in trees, and in herbaceous plants. Some trap their prey; others pounce on the prey, or wait for food organisms in ambush.

The fangs are flexed and extended by powerful muscles. As a rule, they are not strong enough to penetrate human skin, but some species can and do bite humans. The black widow (*Latrodectus*) produces a powerful neurotoxin (Fig. 30.22D) that can cause great pain, nausea, convulsive muscular contractions, and death by respiratory failure. In recent years, the brown recluse spider (*Loxosceles reclusa*) has emerged as another species with an extremely dangerous bite. The bites of some spiders cause a local necrosis of tissues, apparently as a result of a necrotoxin or a hemolysin, and are often very slow to heal.

Spiders are extremely diverse in their habits, but in almost all, the ability to spin silk is an important factor in their lives (Foelix, 1982; Shear, 1986). Spider silk is a scleroprotein that hardens as it is pulled out and comes in contact with air. It may be tough and elastic or sticky, and strands of different sizes and qualities are produced by different spinnerets (Fig. 30.11C). Silk is used for a variety of purposes. Some spiders spin tiny sperm webs, on which sperm is discharged. The seminal vesicles of the pedipalps are charged by dipping the tips into the suspended sperm on the sperm webs. Silk is used to build a cocoon for the eggs and to suspend the cocoons or attach them tightly to surfaces. One of the most remarkable uses of silk is for the construction of gossamer threads. Many young spiderlings clamber up on grasses or other plants when they emerge from the cocoon. When they reach the uppermost point, they spin strands of silk so light that they float the young animal on gentle air currents. Aerial ballooning is common in several spider families and plays an important role in dissemination. It is particularly important for carnivorous animals to spread out, avoiding high population densities in small localities. Spiders cannot fly and the ability to float about on silk is a remarkably effective method of getting about. Spiders were certainly among the first to reach Krakatoa after the great volcanic eruption had destroyed animal life there. The most common and best-known use of silk is for the spinning of webs.

Mygalomorphae are medium to large spiders that burrow in the ground or produce sheetlike or funnel-shaped webs. Among them are the tarantulas and the trapdoor spiders, which form a hinged door for the burrow. In many families of spiders, a rather flattened or irregular web is built, but in some cases the web has a characteristic and precise form. The irregular or flattened webs of the Dictynidae and Aglenidae contain a funnel-shaped hole into which the spider retreats. Araneidae are orb weavers, producing a beautifully regular web of radiating strands and a spiraled strand woven among them. The web of the orb weavers makes an effective trap. There are many ground spiders. Lycosidae (wolf spiders) are large, active species (Fig. 30.22A) that spin no web but sometimes line a nest with silk. Salticidae are jumping spiders that live on the ground or on plants. They too make no web, but make silk-lined nests. Thomisidae (crab spiders) make no web (Fig. 30.22C). They are commonly found on flowers or on leaves of herbaceous plants, beautifully camouflaged and lying in wait for unsuspecting insects that visit the plants. In general, spiders that do not spin a web set a dragline that protects them if they slip.

The Opiliones (Fig. 30.23A,B), harvestmen or daddy-longlegs, are familiar to all, but few laymen distinguish them from spiders. They are most com-

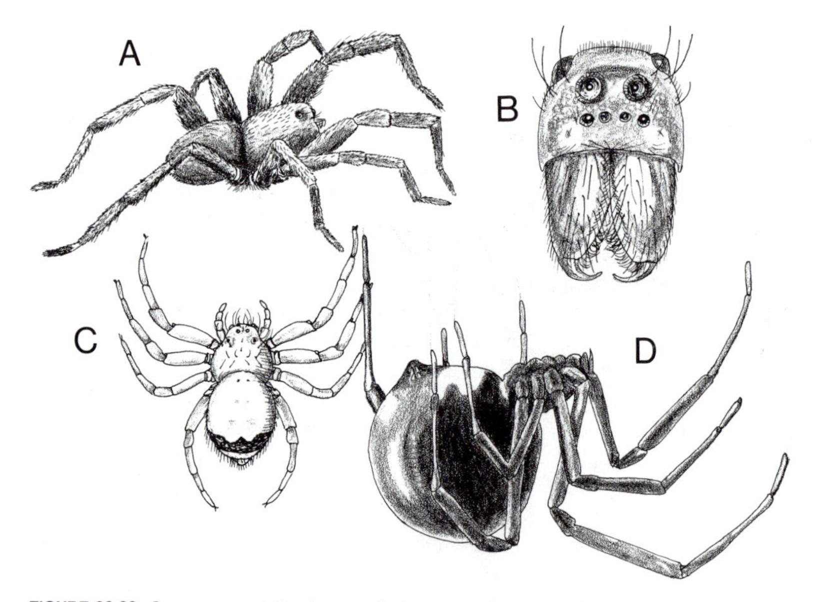

FIGURE 30.22. Some representative Araneae. **A.** *Lycosa carolinensis*, wolf spider. **B.** Face of *Lycosa*. **C.** One of the crab spiders, *Synema*. **D.** *Latrodectus mactans*, black widow spider, (**A,D** after the Kastons; **B** after Bristowe.)

mon in reasonably humid, temperate to tropical forests, where they live among litter on the forest floor and on or around fallen logs. Most species are from 4 to 10 mm long, discounting the legs. The common *Leiobunum vittatum* has a maximal legspread of about 9 cm, but some tropical forms are about twice as large. Harvestmen are the only arachnids that ingest solid food, not the liquified soup other arachnids take in.

A single glance suffices to distinguish between a long-legged spider and a harvestman, for a harvestman has no pedicel. The prosoma is usually covered by a single shield and joins the opisthosoma in a broad union. A dorsal tubercle with an eye on each side lies on the middle of the dorsal prosoma (Fig. 30.21C). Near the base of each pedipalp is a small pore that leads to the **stink glands**. These give off an offensive secretion when the animal is disturbed. In some species the secretion is odorless but repels attackers because of its unpleasant taste, whereas in other species the secretion has a nauseating odor. The opisthosoma contains nine or ten somites, usually so completely fused as to be indistinct. The sternites of the first and second opisthosomal and probably the sixth prosomal somites are joined to form a **genital operculum** in which the gonopores are located. The third opisthosomal sternum contains a pair of spiracles opening into the tracheal system.

Opiliones have the usual set of prosomal appendages. The chelicerae are small and the long, leglike pedipalps end in small chelae. The four pairs of legs are striking, primarily because of their great length. As is generally true of arthropods, the leg is lengthened by the elongation of the podites.

The Acarina (Fig. 30.24A,B) are a remarkably successful group of mostly minute arachnids that have adapted along a number of independent lines. The acarines may be polyphyletic, possibly including several independent groups of arachnids that might have undergone convergent evolution. They occupy extremely diverse habitats. Some live in plant galls, some live in or on birds or mammals, some contaminate grain or other products, and many live in mosses or surface litter or the upper levels of the soil. Some have become aquatic and live in salt water or fresh water.

All acarines have a capitulum, but the details vary considerably in different groups. The **capitulum** is a sclerotized, compound part formed of the mouthparts and the first two pairs of prosomal appendages (Fig. 30.24E,F,G). The base of the capitulum is attached at a notch in the body. In ticks the rostrum is formed of cylindrical extensions of the basis, containing the slender chelicerae, and of a ventral, concave hypostome. The basis is formed of the cheliceral sheaths. In most ticks the hypostome is studded with teeth that aid in attachment during feeding. These parts surround the prebuccal cavity. The mouth lies at the base of the hypostome and is equipped with a slender stylet, derived from the labrum immediately above it. The palps are modified pedipalps and are attached to the base of the capitulum. A wound is made with the chelicerae, and the rostrum is thrust into it. The palps spread to the side as the rostrum enters. During feeding, the hypostome armature engages with the soft tissues, holding the capitulum in place.

Ticks are the largest acarines, but share the same body plan as mites, with poorly defined prosoma and opisthosoma for the whole dorsal surface is united. Some female ticks have a dorsal shield, but this is not a prosomal shield. Three arbitrary divisions of the acarine body can be recognized (Fig. 30.24A,C). The specialization of the tick body is best seen from below (Fig. 30.24D). The gonopore is far forward, lying between the second and third pairs of legs, and the anus lies in the middle of the ventral surface. The primitive ventral surface has been crowded together and moved anteriorly, especially at the midline. Ticks have legs with eight podites. The tarsus is divided into two pieces, and there is a second trochanter between the femur and the first trochanter. Mites generally have six-segmented legs.

Mites live in so many different habitats and have such varied habits that they have become very di-

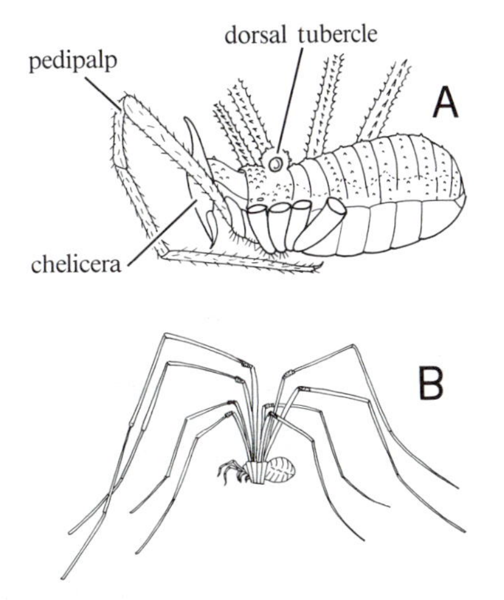

FIGURE 30.23. Body **A** and overall view **B** of harvestman. Opiliones recognized by lack of pedicel, long slender legs, and dorsal tubercle containing eyes. (**A** after Roewer, from Savory, 1977.)

versified in appearance. Many of them show few resemblances to ticks. Among the most highly modified mites is the wormlike *Demodex* (Fig. 30.25A), the mange mite, which lives in the hair follicles and sebaceous glands, and *Eriophyes* (Fig. 30.25B), which causes galls on various plants. Feather mites (Fig. 30.25C) live on the surface of birds and are modified for clinging; some are fantastically shaped. The Tetranychidae (Fig. 30.25E) include the spider mites, or red spiders, that are so destructive to many crops. They use the needlelike chelicerae to pierce plant tissues, sucking in the cell contents. They spin silk from oral silk glands, and use a tarsal comb to walk over and manipulate the silken strands. Freshwater mites (Fig. 30.25F) are often brightly colored and have spherical to oval bodies. They usually have prominent swimming hairs on the legs, which increase surface resistance and make the legs effective paddles. Some of these are huge, by mite standards, reaching lengths of up to 7 mm or so. A few have become parasitic on clams or other aquatic molluscs. *Trombicula* (Fig. 30.25D) is a good example

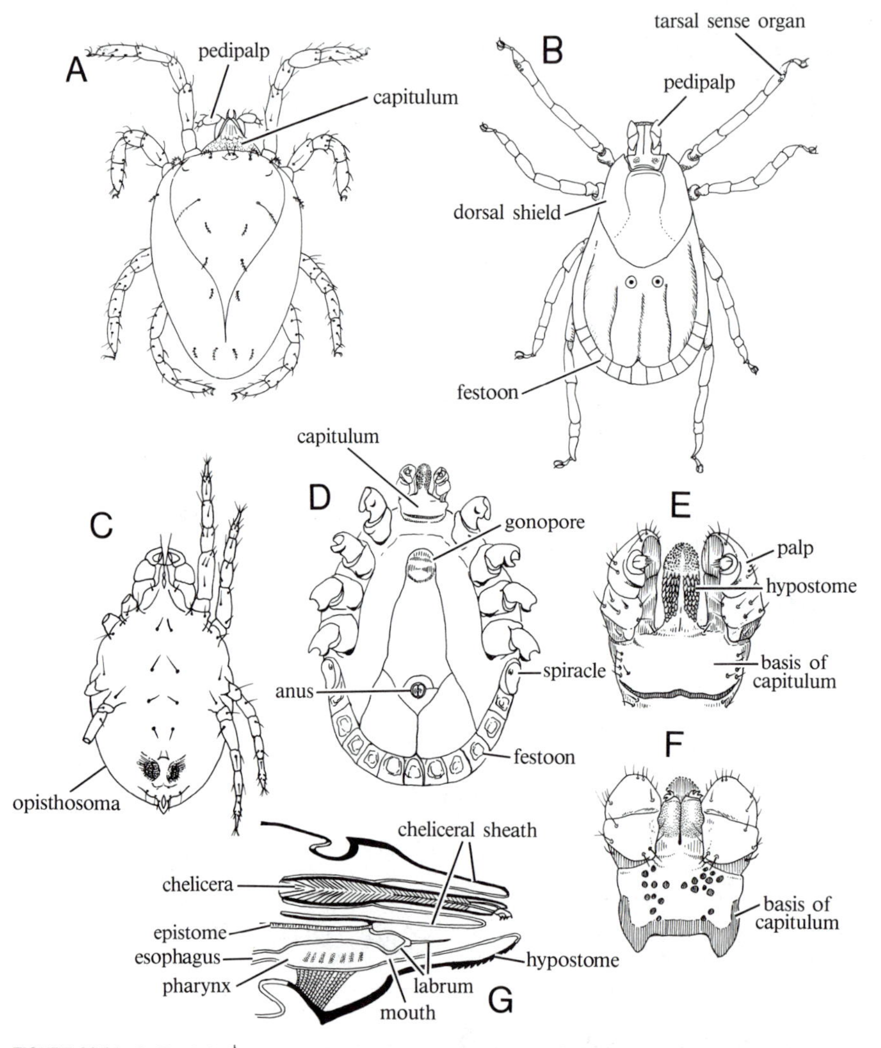

FIGURE 30.24. **A**. *Penthalodes*, mite. **B**. *Dermacentor*, tick. Note capitulum formed by head and mouth parts, characteristic of Acarina. **C**. Ventral view of mite. **D**. Ventral view of tick, *Haemaphysalis aponommoides*. **E**,**F**. Ventral and dorsal view of capitulum of tick. **G**. Scheme of longitudinal section through capitulum of tick. (**A** after Baker and Wharton; **B** after Snodgrass; **C** after Vizthum, from Baker and Wharton; **D**,**E**,**F** after Hoogstraal; **G** after Snodgrass.)

of a mite with a parasitic larval stage. Eggs are laid on the ground, and the young larvae, or chiggers, creep onto the surface of a convenient vertebrate host. They secrete saliva into a minute wound on the skin. Enzymes digest away a minute crater in the skin. The crater is filled with partially digested tissues on which the chigger feeds. When the animal has completed its larval development, it leaves the host and takes up life as an adult. Mites often live in improbable and unpromising habitats, such as grains, flour, and other milled organic substances, and are the cause of serious economic loss. One species is a common occupant of house dust!

Fossil Record

Cheliceriformes do not have an extensive fossil record, but it is nonetheless a rather interesting one as it clearly reveals a group that was much more diverse than the living representatives would lead us to believe. Just a few of the more important groups are mentioned here.

Pycnogonids are known from the Devonian and Jurassic (Bergström et al., 1980). These fossils are important in that they reveal that modern sea spiders evolved from forms with an opisthosoma of several segments.

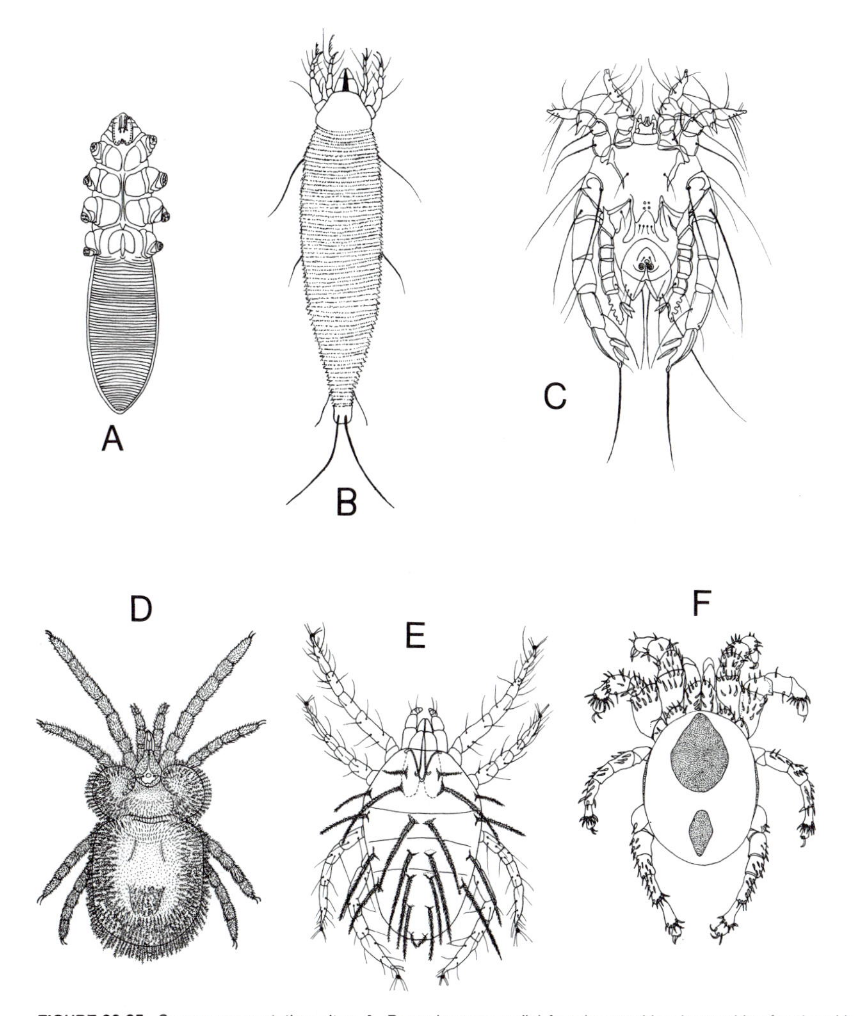

FIGURE 30.25. Some representative mites. **A**. *Demodex muscardini,* female parasitic mite on skin of various kinds of mammals. **B**. *Eriophyes pyri,* pear leaf blister mite causing galls, blisters, and other disturbances in many important domestic plants. **C**. *Megninia columbae,* feather mite. **D**. *Trombicula alfreddugési,* chigger, common pest in many parts of United States. **E**. *Metatetranychus ulmi,* spider mite, pest affecting many species of domestic plants. **F**. *Tyrrelia,* water mite. (**A,C** after Hirst; **E** after Baker and Wharton, from Baker and Wharton; **B** after Essig; **F** after Newell, in Edmonston.)

Among the merostomes, the aglaspids are Cambrian and xiphosurids (Fig. 30.19B) appeared in the upper Silurian. However, doubt has been raised whether all these are chelicerates. At least some aglaspids are now known to lack chelicerae. The synziphosurans, however, still are considered to be merostomes. The limulines have been around since Carboniferous time. A considerable number of other kinds of fossil Limulida, with fused opisthosomal tergites, have been described.

The eurypterids (Fig. 30.19D,C) were once very prominent (Clark and Ruedemann, 1912). They appeared first in marine habitats but evolved into freshwater and perhaps amphibious environments. They are among the biggest arthropods that have ever lived and include the largest arthropod known. *Pterygotus* reached almost 3 m, and some other eurypterids were nearly as large. The first eurypterids appeared in the lower Ordovician, and the last ones were found in the Permian.

Eurypterids resemble scorpions. The opisthosoma was divided into a broader mesosoma and narrower metasoma. The telson was sometimes sharp and recurved, bearing a superficial resemblance to the sting of a scorpion to which it is homologous, and sometimes flattened, as if for swimming. The prosoma bore a pair of chelicerae and five pairs of legs, modified for creeping or swimming.

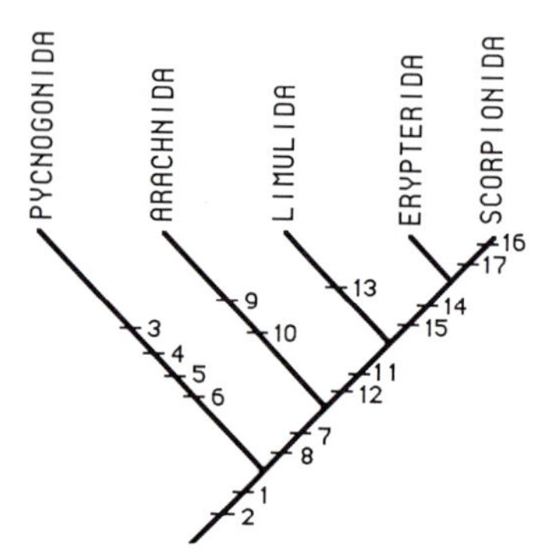

FIGURE 30.26. Cladogram depicting possible relationships of, except for eurypterids, living Cheliceriformes. Derived characters are: (1) loss of deutocerebrum; (2) first limb as chelae; (3) proboscis; (4) vestigial opisthosoma; (5) third limbs as ovigers; (6) gonads and gonopores segmentally arranged in legs; (7) first opisthosomal segment reduced to small pregenital somite; (8) second opisthosomal segment as genital somite with gonopores; (9) opisthosoma highly modified or reduced; (10) opisthosomal legs lost; (11) prosoma dorsally modified into broad shield; (12) opisthosomal limbs modified for respiration; (13) prosomal leg bases gnathobasic; (14) opisthosoma differentiated into anterior mesosoma and posterior metasoma; (15) specialized pectine or medial organ on second opisthosomal somite; (16) terminal sting; and (17) well-developed and chelate pedipalps.

An interesting eurypterid feature was the peculiar median organ that may be homologous to the scorpion pectines (Kjelleswig-Waering, 1986). Also, there were flat, gill-like appendages on the ventral surface of the opisthosoma, protected by extensions of the skeletal plates. These concealed gills have been suggested as a step toward the formation of the book lungs of terrestrial scorpions. It is possible that eurypterids are the ancestral stock from which scorpions arose. The earliest eurypterids were considerably less scorpionlike than the later ones, and by the time they came to resemble scorpions, true scorpions had already appeared.

Arachnids were among the first arthropods to invade terrestrial habitats. The oldest arachnid fossils are Silurian (scorpions), but by the end of the Paleozoic Era, a number of arachnid orders had appeared, including the Opiliones, Acarina, Araneae, Uropygida, Ricinuleida (Fig. 30.21C,D), and Solpugida. The arachnid fossil record is scanty as most arachnids lived under conditions that were not conducive to fossilization.

Phylogeny

There is a long-standing debate about how closely the xiphosurids and trilobites are related. The two growth centers of the xiphosurid embryo play the same role as the acron and pygidial growth center of trilobites, and the Eurypterida had similar growth centers that appeared during development. It is certainly possible that the cheliceriforms are related to the trilobitomorphs more closely than either were related to other arthropods (Schram, 1978).

The cheliceriforms are the most distinctive of creatures, very unlike other arthropods. The group has diverged into two basic lines (Fig. 30.26): the pycnogonids, in which the opisthosoma is almost completely lost and unique reproductive and feeding behaviors were exploited, and the chelicerates sensu stricto, with various opisthosomal specializations. Most arachnids exhibit a trend to reduce the opisthosoma, whereas the merostomes and scorpions develop a large prosomal shield and specialize the opisthosoma with unique respiratory limbs. The limulids have specialized the prosomal limbs with gnathobases. Scorpions share with the extinct eurypterids a division of the opisthosoma into an anterior mesosoma and posterior metasoma, and the development on somite VIII or IX of a special structure ("pectines" in scorpions and "median organs" in eurypterids). Whether these features are sufficient to unite scorpions and eurypterids as sister groups is not clear at this time.

Taxonomy

Several fossil groups are of great importance; however, the taxonomy here largely concentrates on living forms.

Superclass Pycnogonida. Mouth on proboscis; opisthosoma vestigial or reduced; ovigers, segmental gonopores.

Order Pantopoda. Opisthosoma as a single segment (Figs. 30.19E and 30.20A),

Superclass Chelicerata. Anterior opisthosomal segment reduced and/or specialized as pregenital somite, second opisthosomal segment bearing gonopores.

Class Merostomata. Prosoma covered by large, continuous dorsal shield; five or six opisthosomal somites bear flattened appendages associated with respiration; well-developed spine or telson at posterior end; pair of compound and pair of simple eyes.

Subclass Xiphosura. Semicircular prosoma; four- to six-segmented walking legs; six pairs of book gills on opisthosoma.

Order Limulida. Gnathobasic prosomal limbs (Fig. 30.19A,B).

Class Arachnida. Opisthosoma greatly modified, often reduced, and lacking limbs.

Order Scorpionida. Prosoma covered by single shield and with opisthosoma divided into mesosoma of seven broad somites and metasoma of five narrower somites, ending in telson modified as poison gland and sting; prosoma with pair of powerful three-jointed chelicerae and large, chelate pedipalps; four pairs of walking legs with nine podites, two of which are tarsi; pair of modified, comblike appendages (pectines) on second opisthosomal somite; medial triad and two lateral groups of eyes; four pairs of opisthosomal book lungs; one pair of coxal glands and Malpighian tubules (Fig. 30.21B).

Order Palpigradida. Prosoma divided into two or three parts with separate shields; chelicerae well developed and with sternal plate; pedipalps used as walking legs, and first pair of legs carried forward and used as sensory appendages; opisthosoma of 11 segments, ending in long, jointed postabdomen or "flagellum"; with pair of coxal glands; without respiratory organs (Fig. 30.20B,C).

Order Uropygida. Prosoma covered by single or double shield. Two-jointed chelicerae end in hook or fang. Pedipalps jawed and raptorial. Four pairs of eyes usually present. Elongated first pair of legs as sensory appendages, with many terminal segments. Opisthosoma contains a narrow first somite, followed by eight broad anterior somites and three posterior, narrower somites, ending in long or short, jointed postabdomen. Two pairs of book lungs open on third and fourth opisthosomal somites. Pair of coxal glands open behind first pair of legs, and Malpighian tubules present. Pair of large anal glands discharging formic or acetic acid for defense (Fig. 30.20D).

Order Schizomida. Pedipalps leglike. First walking leg as antennalike sensory limb. Opisthosoma as large anterior portion, and tiny postabdomen or "pygidium" of three segments with terminal flagellum.

Order Amblypygida. Prosoma bears three pairs of reduced lateral eyes, usually with pair of median ocelli. Two-segmented chelicerae subchelate, and powerful pedipalps used as raptorial legs. First legs extremely long and slender, used as tactile appendages; remaining legs cursorial. Flat, oval opisthosoma ends in simple, reduced telson, attached to prosoma by pedicel. Two pairs of book lungs open on second and third opisthosomal sternites. Between ninth and tenth opisthosomal sternites, pair of thin-walled chitinous sacs, everted by blood pressure and retracted by muscles. No spinnerets (Fig. 30.21A).

Order Ricinuleida. Chelicerae with two segments, and pedipalps end in small pincers. Third legs of males modified as copulatory organs. Prosoma with anterior, jointed hood, or cucullus. Prosoma and opisthosoma separated by narrow pedicel, opisthosoma formed of 12 united somites and ending in anus on short tubercle. Spiracles at sides of prosoma open into tufts of tiny, unbranched tracheal tubules. Malpighian tubules and pair of coxal glands present (Fig. 30.21C,D).

Order Pseudoscorpionida. Broad, flattened opisthosoma; prosoma with two-jointed chelicerae containing silk glands, and powerful pedipalps containing poison glands; walking legs with five to seven podites; eyeless or with one to two pairs of eyes; two pairs of spiracles leading to tracheae; pair of coxal glands opening on third pair of legs (Fig. 30.21E).

Order Solpugida. Prosoma covered by shield of a single large anterior and three posterior plates, ten-segmented opisthosoma; prosoma with two-jointed, powerful chelicerae, pair of leglike adhesive pedipalps, and four pairs of walking legs with five to seven podites; pair of eyes on anterior shield; tracheal system open through one pair of prosomal and two pairs of opisthosomal spiracles; one pair of coxal glands and Malpighian tubules (Fig. 30.21F).

Order Araneae. Prosomal segments fused; opisthosoma of 12 united segments, attached to prosoma by narrow pedicel; prosoma with two-jointed chelicerae containing poison glands, leglike pedipalps used for food handling, modified as copulatory organs in males, four pairs of walking legs with seven podites; up to eight eyes; with book lungs, tracheae, or both; with one to two pairs of coxal glands and Malpighian tubules (Fig. 30.22).

Order Opiliones. Undivided prosoma broadly attached to nine-segmented opisthosoma; chelicerae three-jointed; pedipalps long, leglike, and sometimes chelate with chewing plate at base; walking legs very long, with multiple tarsal segments; base of first walking leg modified to form chewing plate; pair of eyes on prosoma; usually with single pair of spiracles opening to tracheal system in first opisthosomal somite; pair of coxal glands opening between third and fourth legs (Fig. 30.23A,B).

Order Acarina. Single prosomal and opisthosomal shield and broad junction between prosoma and opisthsoma; opisthosomal somites fused; prosoma with chelicerae typically modified for piercing, leglike or chelate pedipalps, and four pairs of legs with variable number of podites; spiracles and tracheae sometimes present; one to four pairs of coxal glands, or Malpighian tubules, or both (Figs. 30.24 and 30.25).

Selected Readings

Bergström, J., W. Stürmer, and G. Winter. 1980. *Palaeoisopus, Palaeopantopus,* and *Palaeothea,* pycnogonid arthropods from the Lower Devonian Hunsrück State, West Germany. *Paläont. Z.* 54:7–54.

Bonaventura, J., C. Bonaventura, and S. Tesh (eds.). 1982. *Physiology and Biology of Horseshoe Crabs.* A. R. Liss, New York.

Clark, J. M., and R. Ruedeman. 1912. The Eupypterida of New York. *New York State Mus. Mem.* 14:1–628.

Cohen, J. A., and H. J. Brockmann. 1983. Breeding activity and mate selection in the horseshoe crab, *Limulus polyphemus. Bull. Mar. Sci.* 33:274–281.

Firstman, B. L. 1973. The relationship of the chelicerate arterial system to the evolution of the endosternite. *J. Arachnol.* 1:1–54.

Foelix, R. F. 1982. Biology of Spiders. Harvard Univ. Press, Cambridge.

Fry, W. (ed.). 1978. Sea spiders. *Zool. J. Linn. Soc. Lond.* 63:1–238.

Gertsch, W. J. 1979. *American Spiders,* 2nd ed. Van Nostrand Reinhold, New York.

Kaestner, A. 1968. *Invertebrate Zoology,* vol. 2. Wiley-Interscience, New York.

King, P. E. 1973. *Pycnogonids.* St. Martin's Press, New York.

Kjelleswig-Waering, E. 1986. A restudy of the fossil Scorpionida of the world. *Palaeontgr. Am.* 55:1–287.

Kovoor, J. 1977. Silk and silk glands of Arachnida. *Année Biol.* 16:97–172.

Manton, S. M. 1977. *The Arthropoda.* Clarendon Press, Oxford.

Merrett, P. (ed.). 1978. *Arachnology. Symp. Zool. Soc. Lond.* 42. Academic Press, London.

Robinson, M. H., and B. Robinson. 1980. *Comparative Studies of the Courtship and Mating Behavior of Tropical Araneid Spiders. Pac. Insect Monogr.* 36. Bishop Museum, Honolulu.

Savory, T. H. 1977. *Arachnida*, 2nd ed. Academic Press, London.

Schram, F. R. 1978. Arthropods: A convergent phenomenon. *Fieldiana: Geology* 39:61–108.

Seyferth, E. A., and F. G. Barth. 1972. Compound slit sense organs on the spider legs: Mechanoreceptors involved in kinesthetic orientation. *J. Comp. Physiol.* 78:176–191.

Shear, W. A. (ed.). 1986. *Spiders: Webs, Behavior, and Evolution*. Stanford Univ. Press, Stanford.

31

Crustacea

Crustacea are the most diverse of arthropods, although they have not differentiated into as many species as insects, for only about 75,000 species are known. They are extremely variable and abundant individually, however, and a few are relatively large. They are as important in freshwater and marine habitats as insects are on land, and their disappearance would greatly disturb the natural balance of aquatic ecosystems. Only a few have ventured onto land, however, one exception being the region just above high tide line where a number of species live and are among the most important animal inhabitants.

Definition

Crustacea are arthropods with typically five pairs of head appendages: antennules, antennae, gnathobasic mandibles, maxillules, and maxillae. The body is often divided into different tagmata in subsidiary groups, but typically there is a recognizable head, thorax, and abdomen. The posterior anal somite or telson contains the anus and often bears caudal rami. Often the excretory system consists of antennal glands or maxillary glands. A median naupliar eye and usually a pair of lateral compound eyes are present. Sexes are usually separate. Crustacea generally exhibit indirect development and typically possess a nauplius larva or egg–nauplius stage.

Body Plan

The primitive body plan of crustaceans consists of a head and a long, uniform or homonomous trunk. This arrangement, however, is seen in only one class, the Remipedia. Almost all crustaceans have the trunk divided into an anterior **thorax** and a posterior **abdomen** with its terminal **urosome** (Fig. 31.1A,B). There is a clear trend among crustaceans to shorten the trunk and to increase the heteronomy of the body regions by aborting the development of limbs on the abdomen and the posterior thorax.

The segments of the head are, with one exception, fused to form a dorsal head shield. The head appendages are of two types: the anterior sensory antennules and antennae, and the posterior feeding mandibles, maxillules, and maxillae. The **antennules** are uniquely biramous limbs (Fig. 31.1A), although there is a tendency to form triramous and uniramous types. There is some evidence, however, that indicates that these latter are derived from biramous precursors. The **antennae** (Fig. 31.1B) are a characteristic limb that evolved from the biramous limb associated with the tritocerebral part of the brain, and are thus apparently homologous to the cheliceriform chelae and the uniramian labrum (Schram, 1986). The **mandibles** (Fig. 31.1C) are developed from the gnathobases of a limb, the distal parts of which are either lost or reduced to a small uniramous or biramous palp. The **maxillules** (Fig. 31.1D) are small food-processing limbs behind the mandibles. Sometimes these can take on a major role in processing food, for example, as in nectiopodan remipedes where the mandibles are specialized as an internalized mill and the base of the maxillules function as jaws. The **maxillae** (Fig. 31.1E) are generally mouthparts, but in some groups, especially phyllopods, they are developed as thoracopodlike limbs. In additon to these head limbs, often the anterior thoracopods are developed into a **maxillipede** (Fig. 31.1F) to help process food. Maxillipedal segments are often incorporated into the head, but an increase in the number of head somites and appendages, from a functional point of view, is considered evidence of more advanced structure.

The appendages of crustaceans (McLaughlin, in Bliss, vol. 2, 1982) are basically biramous, consisting of a basal part, the **protopod**, an inner ramus, the **endopod**, and an outer ramus, the **exopod**. The appendages can be multiramous or uniramous, but often in the latter with evidence during early development of an initial endopod and exopod. The development of uniramy means the loss of the exopod; multiramy results from the development of accessory **exites** and **endites** on the limb base. The protopod tends to split into two segments, a proximal **coxa** and a distal **basis**. The endopod typically has five segments or less (Fig. 31.1B).

In Crustacea, trunk regionalization often involves development of a carapace. Apparently this has hap-

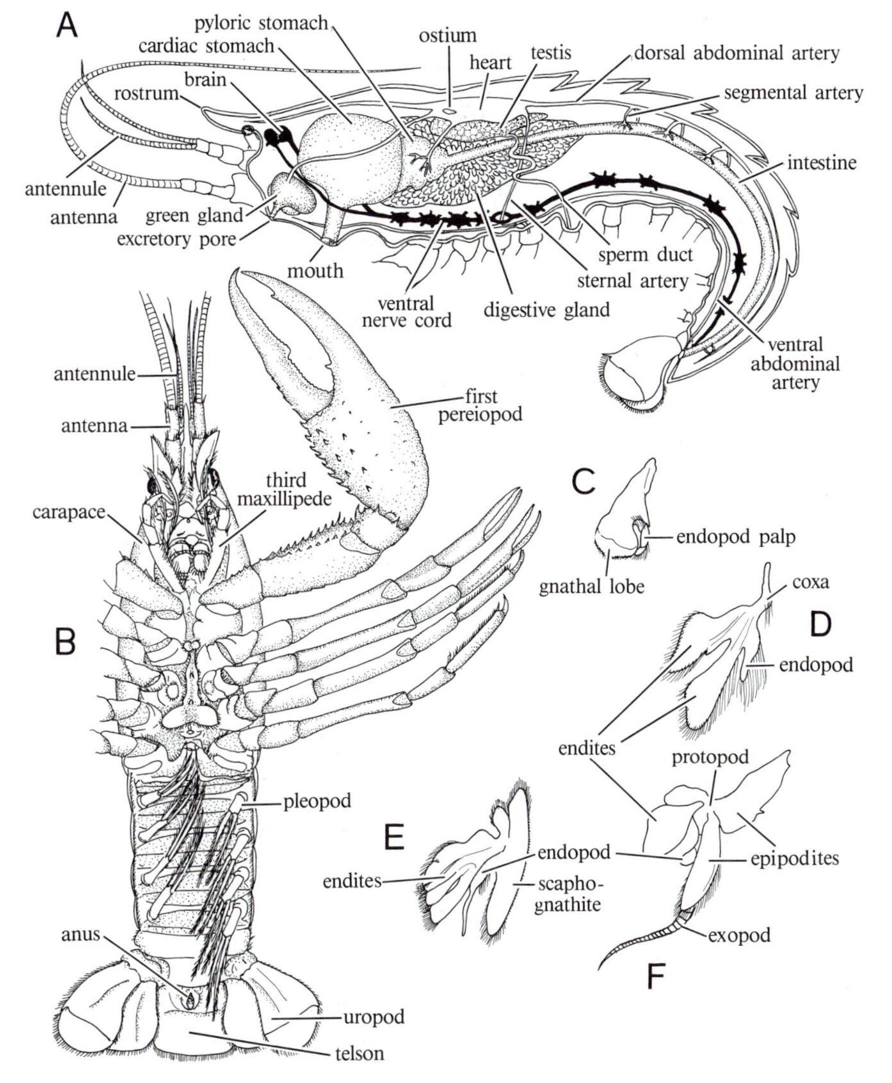

FIGURE 31.1. Crustacean anatomy based on crayfish. **A.** Scheme of internal organization. **B.** Ventral view. **C.** Mandible. **D.** Maxillule. **E.** Maxilla. **F.** First maxillipede.

pened several times independently within the group. The crustacean **carapace** is a fold of the integument that develops typically from the back part of the head and comes to partly or wholly enclose the body. The exoskeleton of the carapace forms an important protective cover where it is well developed, and provides an enclosure into which water may be channeled over gills for respiration or toward the mouth for feeding. As the carapace encloses the thoracic somites, the head and the thorax may fuse into a **cephalothorax**. The most primitive crustaceans have trunk appendages on all body segments that, as in remipedes, are all alike, whereas in the more advanced crustaceans the appendages are specialized for specific uses or may be reduced or wholly lost in some parts of the body.

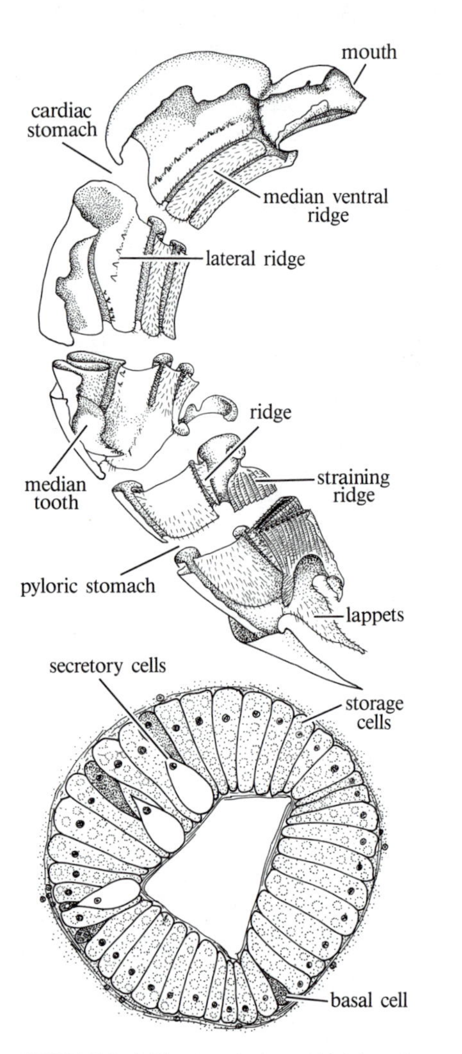

FIGURE 31.2. A. The stomach of a peneid shrimp, *Cerataspis*. Median and longitudinal ridges of cardiac stomach develop hard, grinding lateral teeth with large median tooth. Much-folded walls of pyloric stomach provide filtering system of some delicacy, and terminates in lappets that project into midgut. **B.** Cross section through branch of digestive caecum of crab, *Callinectes*. Small basal cells appear to replace secretory and "hepatic," or absorptive cells, as they become exhausted. (**A** after Bonnier, from Calman, in Lankester; **B** after Regan, from Andrew.)

Digestive System The crustacean **foregut** is quite variable, as might be expected considering the differences in size and dietary habits found among Crustacea. On its way into the foregut, food passes the mandibles, which vary greatly in form and effectiveness. As a general rule, the mandibles fragment food for swallowing but do not adequately reduce it to particles that can be effectively digested. This is especially true of the larger Crustacea that feed on larger plants or animals. Here, special grinding chambers are an almost constant feature of the foregut. Devices of this kind cannot be interpreted as providing evidence of relationship as they appear to have developed independently in different groups.

The simplest kind of foregut is seen in forms that feed on tiny food particles. Copepods, for example, have a short, tubular foregut, equipped only with extrinsic dilator muscles used for swallowing. Filtering setae in the foregut are used to strain out larger particles, preventing their entry into the midgut.

Crustacean foregut grinding chambers are usually called **gastric mills**. Generally they consist of one to several plates, equipped with bristles or teeth, that are manipulated by the intrinsic musculature of the gut wall and by extrinsic muscles attached to the body wall. Gastric mills occur sporadically in some groups. Boring barnacles, for example, have a well-developed gastric mill, whereas most other barnacles do not.

Malacostraca have the most elaborate gastric mills. They may have one or two dilated regions in the foregut. Where two are present, the first is called the **cardiac** and the second the **pyloric stomach** (Fig. 31.1A). The simplest system is found in *Anaspides*, consisting of a single stomach with longitudinal ridges bearing setae. The gastric mills of decapods (Felgenhauer and Abele, 1989) have complex cardiac and pyloric regions (Fig. 31.2A). A constriction separates the cardiac and the pyloric stomachs. A large median tooth is on the dorsal wall at the point of constriction. The ventral part of the cardiac region contains a median ventral ridge and a pair of toothed, lateral longitudinal ridges. Food is ground in the cardiac region and passes to the pyloric region, which serves primarily as a strainer. The pyloric stomach narrows sharply near the junction with the midgut where it is heavily setose. A groove in the floor delivers particles to a wedge-shaped straining chamber on the ventral side that projects into the midgut as a valvular apparatus. With increasing specialization of hard parts and musculature, gastric mills of considerable complexity appear. Sclerotized plates that serve as a platform for grinding or chewing teeth are a common development. Extrinsic muscles attach to the plates, making for more powerful movements. The sorting apparatus also becomes more complex. As a rule, the sorting device consists of grooves equipped with setae that divert larger or smaller particles along appropriate channels. Smaller particles are conveyed to the pyloric stomach, and larger ones are held in the cardiac stomach for further treatment or regurgitation. Intermediate particles follow a more dorsal course, and enter a setose muscular press in the pyloric stomach, where they are further compressed in size. A ventral groove carries the smallest particles to the openings of the digestive caeca, where they must pass through special filters to gain entrance.

Most of the Crustacea have a short **midgut**, but in forms with a simple, tubular esophagus, the midgut is usually long. In at least some of the Malacostraca, the midgut forms the posterior end of the pyloric stomach. **Diverticula** or **caeca** are most important regions of the midgut from a functional point of view (Fig. 31.2B). Enzymes are secreted and food is absorbed in the diverticula. A variety of names have been applied to the branches of the midgut, such as caeca, digestive glands, and hepatopancreas.

The midgut proper of Malacostraca is usually a tubular **intestine**. The intestine is coiled in some cumaceans, and here there are no caeca. Generally, however, the intestine is straight, and all surface amplification depends on diverticula. As a rule, there are from one to four pairs of diverticula, although each diverticulum may be very extensively ramified. The larger decapods have a single pore to the diverticula on each side of the midgut, leading to a very extensively branched duct system extending to tubular tips. Remipedes have a pair of diverticula in each trunk segment. An enormous amount of surface is provided by the diverticula. Most of the midgut diverticula are laterally placed, but in some crustaceans median caeca are found. Some groups, for example, mysids, amphipods, and some decapods, have a single or a pair of dorsal caeca that extend forward at the junction of the midgut with the foregut and/or hindgut.

Even when the chitin-lined **hindgut** is long, it remains relatively simple in form. A narrow tube, it is typically equipped with intrinsic muscles used in defecation. In some instances the hindgut has extrinsic dilator muscles used for anal pumping, thought to aid in intestinal peristalsis. Crustacean feces are somewhat compacted and often mixed with mucus. Presumably, at least the terrestrial and amphibious Crustacea absorb considerable quantities of water from the hindgut.

Respiratory System Crustaceans are generally small animals; therefore, most have a delicate, transparent exoskeleton. A considerable respiratory exchange occurs at the body surface. Nevertheless, many crustaceans have **gills** formed from the epipodites of some or most of the appendages. The gills are flat plates through which blood circulates (see Fig. 31.8C,D,E). Where the gills are a part of the appendages, normal swimming movements tend to ventilate them, but especially in the larger species or in species with heavier exoskeletons, the gills tend to be ventilated by definite respiratory currents, produced either by the appendages that bear the gills or adjacent appendages. In filter feeders, the feeding currents may be identical with the respiratory current. It is probable that the strong tendency of gills to be placed on the thoracic appendages is related to the advantages of being able to combine feeding and respiratory currents.

Development of the respiratory capacity of the body wall may be associated with gill reduction. Ostracodes, for example, have a relatively heavy shell, and in some cases a network of blood vessels converts it into a respiratory surface. Barnacles have developed their **mantle**, a special saccular kind of carapace, as a respiratory organ and have no need of special appendage gills. Some of the Malacostraca have followed the same line; for example, tanaids have a branchial chamber formed by the carapace and supplied with a rich plexus of blood vessels.

Amphibious or terrestrial habits have appeared in several malacostracan groups. One infraorder of isopods has been most successful, but some amphipods and decapods have also spread to amphibious or semiterrestrial habitats with considerable success. Gills are not usually suitable for land respiration, and it is interesting to observe how terrestrial crustaceans have modified structure and altered behavior to manage their respiratory problems.

Amphibious amphipods and decapods tend to reduce the importance of gills and to develop a vascularized branchial cavity, in this case paralleling the changes seen in the mantle cavity of terrestrial snails. The reduction of gill surface is important because the gill surface is a potent site of water loss.

Amphipods are restricted to humid surroundings, as their small size makes water loss an important factor. The much larger crabs have a heavier exoskeleton and are more resistant to water loss. Furthermore, decapod gills are protected by a branchial chamber, and in crabs the entrance to the cavity is small (Fig. 31.3A). Nevertheless, amphibious crabs generally do not circulate air through the branchial chamber. Many fill the chamber with water, using an opercular flap on the maxillipede to retain the water while on land. The water is circulated in the gill chamber by the **scaphognathite**, and in some species, water circulates through special canals on the carapace surface for aeration. Many crabs must return to water often to charge the gill chamber, but some can live for many days without moisture. The final step is the circulation of air in the gill chamber, converting it to a lung, as in *Birgus*, the coconut crab, and grapsid crabs. Movements of the carapace and fanning movements of the scaphognathite circulate the air, and most of the respiratory exchange occurs in the vascularized lining of the branchial chamber.

Isopods have followed a different tack. The platelike pleopods are divided into a respiratory endopod and a protective exopod that can be folded over the endopod. The arrangement cuts down water loss and favors easy ventilation by appendage movements. Immediately below the endopod cuticle, there is a system of thin, branching tubules or **pseudotracheae** that are filled with air and participate in respiratory exchange (Fig. 31.3B).

The importance of the blood in respiratory transport is evidenced by relationships between the gills and the circulatory system. The heart takes a more posterior position in crustaceans that have abdominal gills, as in isopods and stomatopods. In crabs (Fig. 31.3A), blood returns to the pericardial cavity directly from the gills, and passes to the gills from the hemocoel after circulating about the internal organs.

Excretory System Crustacea, like other arthropods, have a small coelom. As a consequence, the excretory organs are highly modified. It is unclear whether these are really metanephridia. **Antennal** or **maxillary glands** occur in adults, and there is some evidence that these might be the remnants of a segmental series of secretory glands in the head (Schram and Lewis, 1989). Although both may be found in larvae, one or the other predominates in adults. Well-developed antennal glands and maxillary glands are rarely found in the same adult animal. Both kinds of glands have the same basic struc-

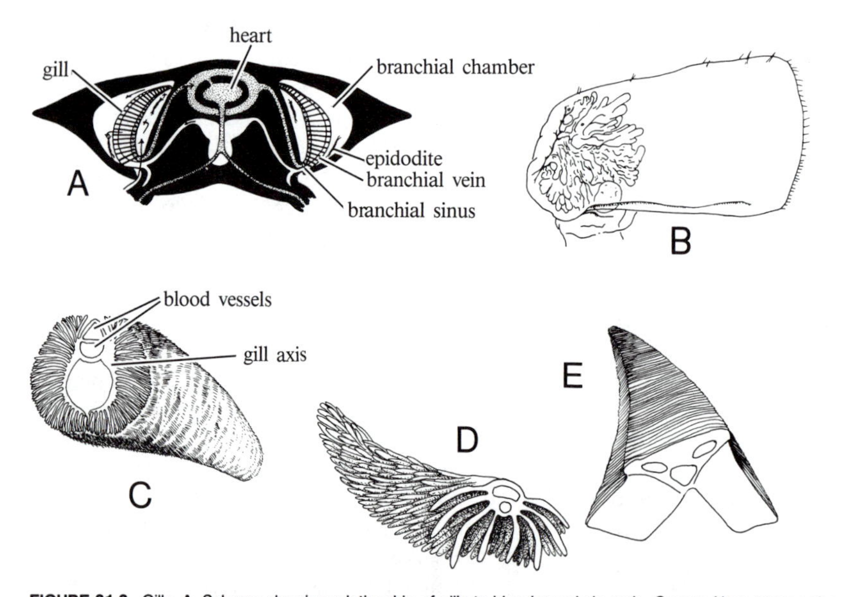

FIGURE 31.3. Gills. **A**. Scheme showing relationship of gills to blood supply in crab, *Cancer*. Note *arrows* showing direction of flow of blood and water. **B**. Diagram of tracheal system in gill of terrestrial isopod, *Porcellio*. Air enters through a definite stigma and circulates through tracheal tubes. **C,D,E**. Types of gills seen in decapods: **C**. dendrobranchiate gills, with two series of primary filaments, and each primary filament complexly branched to form gill of many tiny filaments; **D**. trichobranchiate gills, with series of unbranched filaments arranged in various patterns around axis; **E**. phyllobranchiate gills composed of platelike filaments, usually arranged in two series. (**A** after Wolvekamp and Waterman, in Waterman; **B** after Herold, in Kükenthal and Krumbach.)

ture: an end sac from which a tubule arises that continues to an excretory pore in the base of the antenna or maxilla. A second type of excretory organ also occurs in some Crustacea. Clusters of athrocytic cells can appear at various points in the body. In some cases, these form definite masses with a special blood supply located at the base of the thoracic legs or, more rarely, the abdominal appendages. Where the athrocytic organs are best developed, they appear to form a definitely organized excretory system. Where neither antennal glands nor maxillary glands are well developed, the athrocytic organs are most highly developed.

In most crustaceans, the antennal or maxillary glands are relatively simple. Others have an organ of some complexity. For example, crayfish live in fresh water and may excrete a considerable quantity of hypotonic urine. The end sac is greatly folded and lies at the anterior and dorsal end of the large antennal or **green gland** (Fig. 31.4A). It opens into the upper part of the excretory tubule, essentially a large sac with greatly folded walls dividing the lumen into many channels and providing a very large excretory surface. This portion of the tubule is the labyrinth and is greenish in color. In crayfish, the labyrinth opens into a long, white, elaborately convoluted tubule. When the antennal gland is in place in the animal, the labyrinth makes up the outer, cortical part of the antennal gland, and the convoluted tubule fills the central, medullary part of the gland. The convoluted tubule opens into the bladder, the expanded end of the tubule in simpler kinds of antennal glands. It is a large, membranous region that extends over the intestinal caeca and in some decapods is extensively ramified throughout the thoracic region. The bladder opens to the exterior through the excretory pore. As is typical of the more complex excretory organs, the green gland is intimately associated with the circulatory system. Arterial branches from the antennal and ventral thoracic arteries supply the substance of the glandular part of the green gland, ending in microscopic hemocoelic sinuses.

Circulatory System The demands placed on the circulatory system in a small crustacean are quite dif-

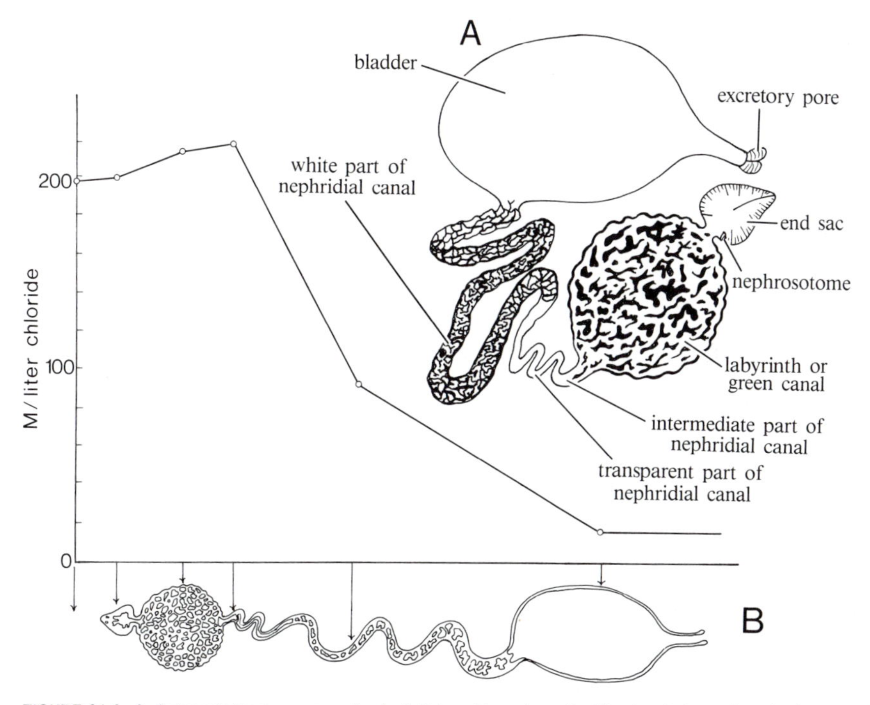

FIGURE 31.4. A. Antennal gland, or green gland, of *Potamobius* schematic. Gland ends in small coelomic remnant, end sac, that opens by vestige of nephrostome to a much-coiled labyrinth. Nephridial canal leading from labyrinth has short intermediate part, short transparent part, and longer convoluted white part to end in dilated bladder region. **B**. Diagram to show correspondence of chloride content with regions of green gland of *Astacus*. Chloride content remains high but falls in convoluted part of nephridial canal. At minimum in bladder. Water probably reabsorbed with chloride ion. (**A** after Marchal, from Balss, in Kükenthal and Krumbach; **B** after Marchal.)

ferent from a large crustacean. It is not surprising to find that the circulatory system varies markedly in different groups of Crustacea.

Wherever a circulatory system is found, it is patterned on a basic design (Fig. 31.5A,B). A dorsal **heart** lies in a **pericardial sinus**, into which blood returns from the gills. Blood enters the heart through paired valves or **ostia**, and leaves the heart through an **anterior aorta**; usually a **posterior aorta** is also found. As a rule, other arterial channels are also present. These usually include some lateral arteries that arise from the heart or aortae, a descending artery that passes ventrally from the heart or the base of the posterior aorta, and a system of median ventral arteries to the viscera. However the arteries are arranged, the blood eventually reaches open hemocoel lacunae where exchange of substances with the tissues occurs. Occasionally, as in mantis shrimps and barnacles, the system is virtually closed in vessels. Blood from the lacunae collects in a system of ventral venous sinuses. These converge on a median ventral sinus from which blood flows into the gills. Efferent branchial vessels lead to branchiopericardial vessels that deliver the aerated blood to the peri-

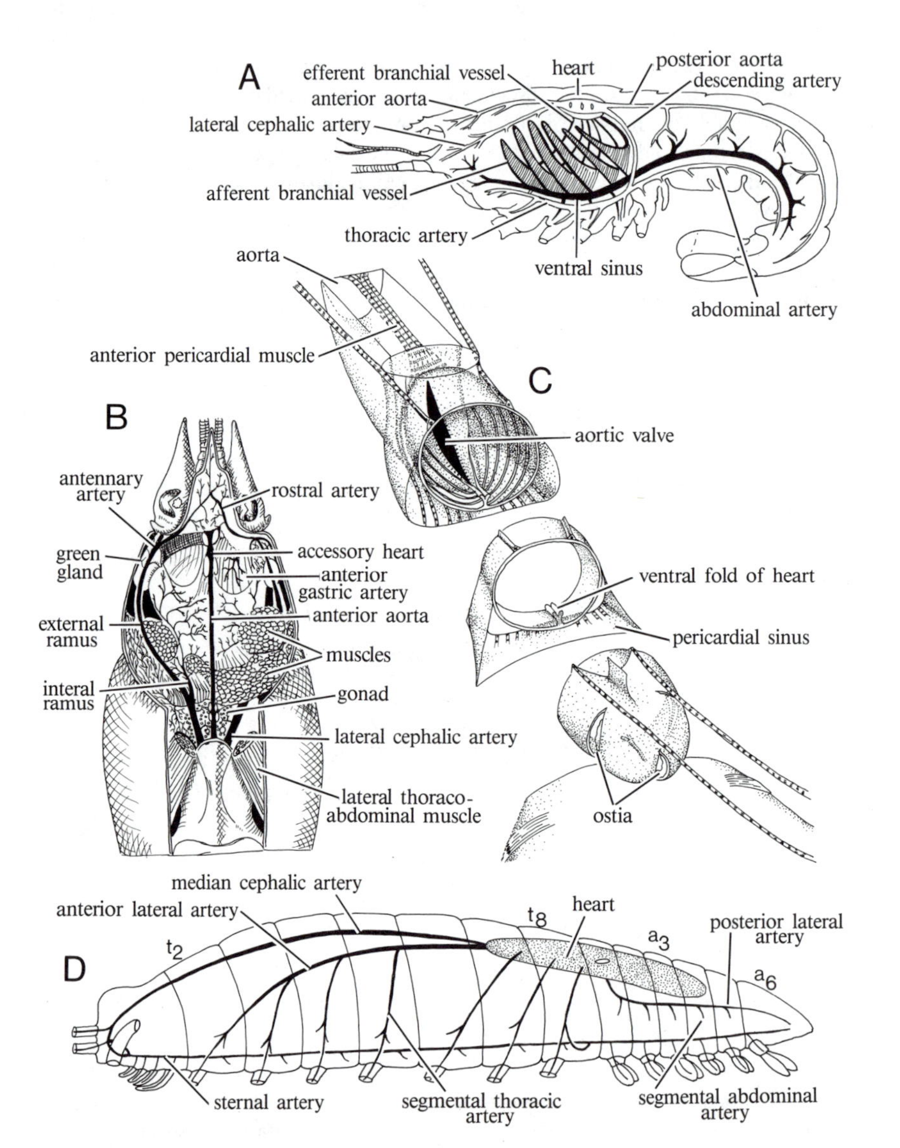

FIGURE 31.5. Circulation. **A**. Scheme of circulatory system of lobster with arteries *clear* and venous channels shown in *black*. **B**. Anterior end of macruran, *Potamobius,* showing anterior arteries. **C**. Heart of copepod, cut into three sections, **D**. Circulatory system of isopod, schematic. (**A** after Gegenbauer; **B** after Baumann; **C** after Marshall and Orr; **D** after Delage, from Balss, in Kükenthal and Krumbach.)

cardial sinus. Therefore, crustacea have an arterial heart pumping freshly aerated blood to the body tissues. In some cases accessory booster hearts are found in the head, but as a general rule, the movements of the appendages, the body musculature, and peristalsis in the gut keep blood flowing in the venous channels. Valves are often found at key points, such as the entrance and exit of the gill vessels, and play an important part in maintaining the direction of blood flow.

The heart has developed from a dorsal blood vessel and, in its primitive form, is a long, slender tube with many pairs of ostia. Anostracans have a heart that runs the full length of the body with ostia in all but the first and last somites, and the more primitive Malacostraca also have a long heart with many ostia. *Nebalia*, for example, has seven pairs of ostia. As the body regions become more distinct and the tendency for the body to shorten is expressed, the heart becomes shorter and has fewer ostia. For obvious reasons, it tends to lie above the appendages that bear the gills, and so is more posterior in isopods (Fig. 31.5D) with abdominal gills, but is in the thoracic region of most other eumalacostracans.

Some smaller Crustacea have a bulbous heart (Fig. 31.5C), and body movements help drive blood through a system of **hemocoel lacunae** (as in many maxillopodans such as copepods and ostracodes). In these forms, valves are sometimes found that prevent backflow in the lacunae. Barnacles, although considerably larger, lack a heart and depend on a muscular vessel to pump fluid through a variable and formal system of vessels and lined lacunae.

A primitive arrangement of the nervous system is

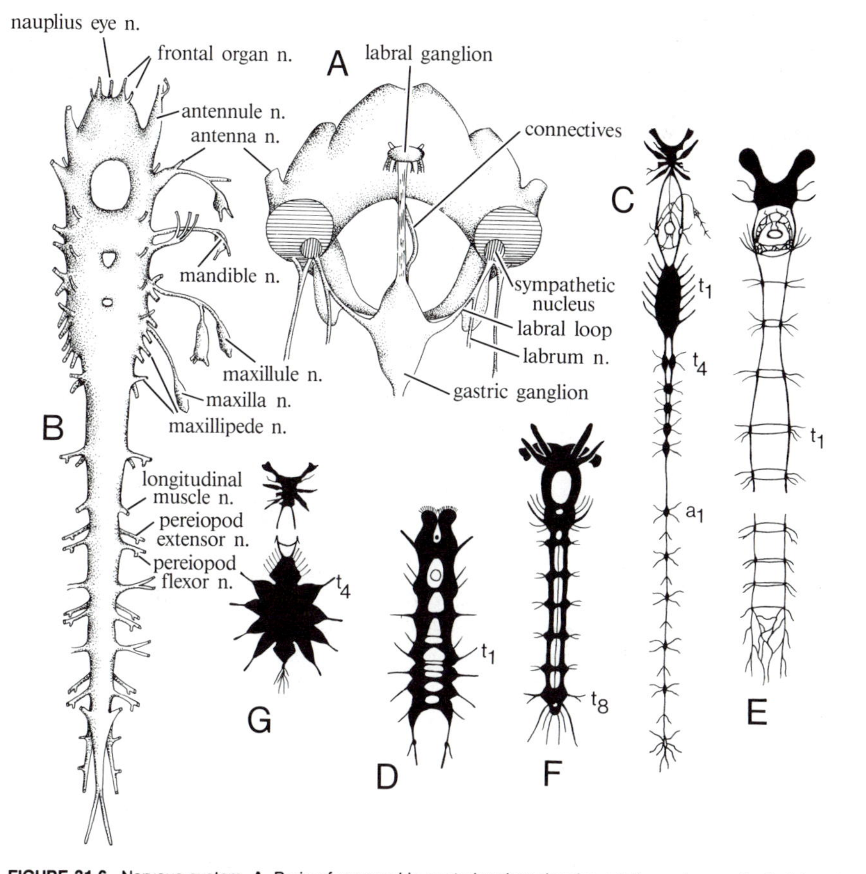

FIGURE 31.6. Nervous system. **A**. Brain of copepod in posterior view showing relations of sympathetic (stomatogastric) system. **B**. Dorsal view of copepod brain and nerve cord. **C**. Scheme of nervous system of macruran decapod with several of anterior thoracic ganglia united with subesophageal ganglion, but remaining thoracic and abdominal ganglia distinct. **D**. Scheme of nervous system of cladoceran. Nerve cord double and ganglia discrete in thoracic region. **E**. Scheme of anostracan nervous system, primitive in having nerve cord paired, thoracic ganglia distinct, and commissures relatively long. **F**. Scheme of nervous system of isopod. Thoracic ganglia distinct, although some evidence of concentration of ganglia at anterior end. **G**. Scheme of nervous system of brachyuran decapod with thoracic ganglia concentrated in large, compound thoracic ganglion. (**A**,**B** after Marshall and Orr; **C**,**D**,**E**,**F**,**G** after Giesbracht, from Hanström.)

seen in Branchiopoda. Fairy shrimps have a simple, segmental nervous system, with ganglia and the commissures in all thoracic somites (Fig. 31.6C). In abdominal somites that have no appendages, the ganglia and commissures are missing. The antennules are innervated from the deutocerebrum in front of the esophagus, and are evidently preoral. Nerves to the antennae arise from the connectives, although the nerve fibers pass to the brain. All other appendages are evidently postoral, as indicated by their innervation.

Nervous System As a general rule, the nervous systems of Malacostraca are often less centralized than those of higher forms (Fig. 31.6E,F), for at least some of the abdominal ganglia are distinct, although there may be fewer ganglia than somites. The thoracic ganglia are often crowded together and in many cases are partly or wholly united to form a compound thoracic ganglion. With the reduction of the abdomen, the nervous system becomes more highly centralized. Brachyuran crabs (Fig. 31.6G) have a single thoracic ganglion from which nerves pass to the various parts of the body. Giant fiber systems are common in Crustacea. They conduct impulses rapidly and evoke escape movements, prepackaged responses that are sometimes of considerable complexity.

The nervous system of maxillopodans and some phyllopodans is far more highly centralized. As a general rule, there is a ventral ganglionic mass in which some of the component ganglia can sometimes be distinguished (Fig. 31.6D). Some ostracodes have two or three such ganglionic masses, and the ganglia are partly separated in some of the copepods (Fig. 31.6B).

Table 31.1. Body length and gonopore location of various crustaceans

	Total body somites (including 5 head segments)	Location of gonopore on trunk segment	
		Female	Male
Nectiopoda	up to 37	8	15
Malacostraca	19	6	8
Brachypoda	25	6	6
Most branchiopods	up to 42	11	11
Anostraca	25–32	12	12
Copepoda	16	7	7
Cirripedia	16	1	7
Mystacocarida	16	4	4
Branchiura	9(+?)	4	4
Ostracoda	9	3	3

Modified from Schram, 1986.

Reproductive System Sexes are generally separate, although some notable examples of hermaphroditism occur, such as in barnacles, brachypodans (cephalocarids), and nectiopodan remipedes. Sex reversal is also not uncommon (Ginsburger-Vogel and Charniaux-Cotton, in Bliss, vol. 2, 1982).

Reproductive organs are fairly simple, and gonads may be either compact or tubular, in the abdomen or in the thorax. In males, **testes** are drained by **sperm ducts** that may have special **seminal vesicles** to store sperm. Females have **ovaries** drained by **oviducts** and may or may not have **seminal receptacles** to store sperm. The location of gonopores varies with groups, but the general trend seems to be to move gonopores forward in more specialized groups as the trunk is progressively shortened (Table 31.1).

Biology

Locomotion Aquatic crustaceans are generally very efficient swimmers. This is achieved by metachronal beats of entire trunk limb series (e.g., nectiopods, conchostracans, anostracans) or just the abdominal limbs (malacostracans). The development of a urosomal tailfan, as in malacostracans, helps in maintaining body orientation while swimming, and in the eumalacostracans is used for a backward, rapid escape reaction achieved with an abrupt flexion of the tailfan under the abdomen. Specialization of the limb endopod in malacostracans and some phyllopodans is useful in both aquatic and terrestrial forms for walking or in benthic aquatic species for pushing off the bottom.

Feeding The majority of crustacean orders use only cephalic limbs in feeding. A substantial number of the rest integrate anterior thoracic limbs as maxillipedes into a posteriorly extended cephalon. Some orders, like barnacles or fairy shrimps, have specialized the entire thorax for food getting.

At one time it was thought that filter feeding was the principal form of food procurement—and the most primitive. Studies of the physics of feeding, especially in the more common small forms, have revealed that true filter feeding is often impossible. In such tiny crustaceans, water is too viscous to allow filtration (Koehl and Strickler, 1981). Rather, it appears that almost all cephalic feeders use some kind of grappling action of mouthparts to forcefully bring food into the mouth field (Fig. 31.7), either by waving particles into the oral areas, grabbing prey that swim by, scraping organic films off sediment particles, or grabbing and holding onto a host (Schram, 1986, Chap. 44). Some current work is revealing that many thoracic filterers are also probably grappling with food rather than engaging in true filtration (Fryer, 1988).

In one crustacean class, the Maxillopoda, repeated adaptations to parasitic life-styles have occurred. Almost every subclass is, in whole or in part, parasitic. Some of these adaptations are extreme, such as the rhizocephalans, who are little more than a periodic external eruption of a gonad bag on the abdomen of their host.

The digestive caeca are the main centers of secretory and absorptive activity (Dall and Moriarty, in Bliss, vol. 5, 1983). Tall cells serve for the absorption and storage of food, and pyriform enzyme-secreting cells are scattered among them (Fig. 31.2B). At the base of the mucosal epithelium and at the tips of tubules are small basal cells that are thought to replace the absorptive and secretory cells as they deteriorate. While particles enter the digestive glands, as well as the caeca, digestion appears to be wholly extracellular in most Crustacea. It should be noted that despite histological similarities, experimental evidence of functional differences between the main gut and the hepatopancreas has been obtained.

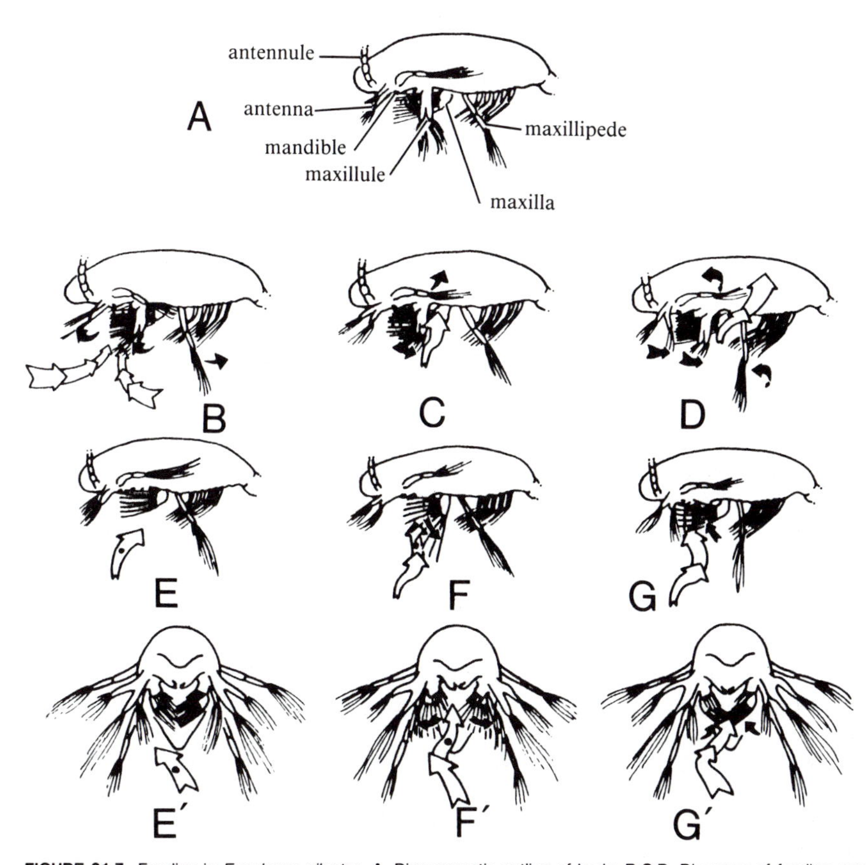

FIGURE 31.7. Feeding in *Eucalanus pileatus*. **A**. Diagrammatic outline of body. **B,C,D**. Diagrams of feeding appendage movements (*black arrows*) and surrounding water currents they produce (*white arrows*) (arrow with narrow shaft and wide head indicates lateral movement out of plane of page toward reader; arrow with wide shaft and narrow head indicates medial movement away from reader). **B**. Outward movements of second antennae and maxillipedes suck water toward copepod's maxillae. **C**. Posteromedial movement of maxillules and dorsolateral movement of mandibular palps suck water laterally. **D**. Inward movements of antennae and maxillipedes coupled with dorsolateral movement of mandibular palps shove water posterolaterally. **EE′,FF′,GG′**. Particle capture (*white arrows* are water currents; *black arrows* indicate movements of maxillae and of a maxillule in G′). **E,F,G**. Copepod viewed from its left side, and maxillule has been left off for clarity. **E′,F′,G′**. Animal viewed from its anterior end. Feeding currents bypass maxillae (**EE′**) until alga nears them. Alga captured by outward fling (**FF′**) and inward sweep (**GG′**) of maxillae. (Modified from Koehl and Strickler.)

Most of the information about enzymes is based on studies of Malacostraca. Hydrolytic enzymes for carbohydrates, fats, and proteins have been found, and the properties of some described. Except for the isopod *Porcellio*, no evidence of symbiotic cellulose-digesting bacteria has been found, even in wood-eating forms. Bile acids have been demonstrated in several decapods, and may be important in the aborption of fats. In any case, the foregut diverticulum, or hepatopancreas, appears to have a number of properties that justify its name, including the storage of fat, glycogen, and calcium, and an active role in purine metabolism.

Excretion The physiology of antennal, green gland excretion is not very well understood. Fluid certainly enters the green gland, and there is some reason to think that the end sac and labyrinth take up fluid from the blood by filtration, as in the vertebrate glomerulus. Filtration pressure would be very low, however, and filtration may not be an important factor. It appears that the fluid at the upper end of the tubule is rather more like blood than urine. Changes in the composition of the urine as it passes along the tubule involve the reabsorption of some materials. Glucose is present in considerable quantities in the blood. It is normally absent from the urine, but if a substance that antagonizes the reabsorption of glucose is administered, glucose appears in the urine. It is also clear that chlorides are reabsorbed from the lumen of the green gland, especially in the region of the convoluted tubule and, perhaps, the bladder (Fig. 31.4B). Droplets can be seen forming in the epithelial lining of the tubules in the glandular part of the green gland, indicating that some substances are actively secreted into the tubules by the lining cells.

Some of the differences in the form of the excretory organs in Decapoda appear to be related to the kind of urine formed. Lobsters produce a relatively small amount of isotonic urine. In lobsters, the convoluted tubule is missing. Brachyuran crabs that live in estuaries and are exposed to both brackish and salt water excrete a relatively large amount of isotonic urine. They also lack a convoluted tubule, but are equipped with a greatly expanded bladder. It is tempting to conclude that the convoluted tubule is primarily responsible for osmotic control of the urine, on purely morphological grounds, but until a good deal more experimental evidence is available, no firm generalizations are warranted. It is clear that osmotic control of the body fluids is by no means entirely dependent on the function of the antennal glands. Stoppage of the excretory pores has but a limited effect on the excretion of ammonia or urea. Apparently much of the soluble nitrogen wastes are or can be eliminated at the gill surfaces, and the digestive tube is also important in the excretion of some nitrogenous compounds, especially of uric acid. The gill surfaces are also important in eliminating salt when the regulation of the osmotic pressure of the body fluids is achieved by salt excretion.

Crustacea are predominantly ammonotelic, with most of the marine species excreting about 70 to 90 percent of their nitrogenous wastes in the form of ammonia. Urea and uric acid make up but a small part of the nitrogen excreted. In some cases amino acids and other nitrogen compounds make up a significant part of the excreted nitrogen. The successful terrestrial invertebrates have developed methods of reducing ammonia excretion, and something of the same kind is sometimes seen in freshwater forms. The Crustacea have never achieved this solution to the excretory problem. Freshwater Crustacea do excrete somewhat more urea, and terrestrial Crustacea excrete somewhat more uric acid, but the differences are small at the best, and in no case does a significant shift in excretory patterns occur. The failure of the land isopods to develop an excretory pattern suitable for land animals (they excrete much ammonia) may have been one of the factors that has prevented them from competing favorably with chilopods, diplopods, arachnids, and insects.

Crustacea occur in a wide range of aquatic habitats and differ markedly in the details of water and salt regulatory mechanisms (Mantel and Farmer, in Bliss, vol. 5, 1983). Some of the marine Crustacea have very little control over water intake, absorbing water and swelling when placed in a hypotonic medium. They soon excrete enough salt to make the blood isotonic and bring body volume back to normal. The spread of such forms into brackish water depends on the tolerance of the tissues to lower osmotic tensions in the body fluids. Some can tolerate a wide range of salinities, whereas others can live only in very nearly full-strength sea water. As a general rule, crustaceans lose water when placed in a hypertonic medium, but absorb salt until the blood can be isotonic with the medium and normal volume of the tissues restored. Most of the marine crabs, lobsters, and barnacles work in this way. They are osmotic conformers and secrete isotonic urine in larger or smaller quantities, as the situation demands. Curious osmotic relationships sometimes develop in osmotic conformers. *Lernaeocera*, a parasitic copepod, remains hypotonic in relation to the aquatic environment while feeding on the relatively hypotonic blood of its host, but quickly becomes isotonic with the medium when detached from the host.

Shore and estuarine crabs are usually osmoconformers in more saline habitats and become hypertonic in brackish water. The body surface and gills are less permeable to water than in marine forms, and the animals swell less and adjust more quickly when placed in hypotonic media. Under these conditions, urine production rises. The urine may not be hypotonic, but it accounts for enough water discharge to regulate volume while enough salt regulation occurs at the gills and the intestinal lining to keep the animal close to its normal salinity. Some of the Crustacea can absorb salt in very dilute salt concentrations. *Eriocheir*, the wool-handed crab, normally lives in freshwater streams and returns to the sea to breed. It thus tolerates both extremes. The body surface is very impermeable, permitting no change of volume in differing osmotic environments. It forms very little urine, but absorbs salts actively at the gill surfaces when in freshwater environments.

Artemia, the brine shrimp, lives in very strongly saline habitats and can be hypotonic. This is arranged by absorbing water and sodium chloride as water passes through the gut. Other salts are eliminated at the anus, and salt is actively eliminated at the gills.

Respiration The respiratory efficiency of Crustacea varies markedly. Some crustaceans can withstand a considerable reduction of oxygen tension before oxygen consumption falls. Others are oxygen conformers, whose oxygen consumption falls with reduced oxygen tensions in the ranges normally encountered in their habitats. Closely related species may fall into different categories, and a great deal more information is needed before generalizations are justified. For example, lobsters are oxygen conformers, whereas the closely related crayfish are independent of oxygen tensions to about 40 mm mercury. One group, the nectiopodan remipedes, appears to live under permanent anoxic conditions in the range of 0.1 parts per million of oxygen.

Some Crustacea have no respiratory pigment, whereas others have hemoglobin or hemocyanin. As a rule, crustacean blood has only moderate oxygen-carrying capacities (Mangum, in Bliss, vol. 5, 1983). Crab blood and lobster blood hold about 1.2 to 2.3 ml oxygen per 100 ml. This compares unfavorably with the blood of larger molluscs (*Sepia*, 7 ml; *Octopus*, 3.9 to 5.0 ml) or with that of the lower vertebrates (skate, 4.2 to 5.7 ml; carp, 12.5 ml). Under normal physiological conditions, the hemocyanin is not saturated with oxygen, but it has been shown that hemocyanin picks up oxygen at the gills of some Crustacea and loses oxygen before returning. Although only a small part of its potential is used, it contributes about 90 percent of oxygen transport in *Panulirus*.

The ability to adjust ventilation movements when under respiratory stress varies markedly, and may be one of the more important factors in determining whether organisms are oxygen conformers or oxygen regulators. A crayfish, for example, increases its ventilation rate in response to reduced oxygen or increased carbon dioxide, and is an oxygen regulator. A lobster reduces the ventilation rate with reduced oxygen or increased carbon dioxide, and is an oxygen conformer, adjusting its metabolism rather than its respiration. Crustacea are generally more sensitive to a fall in oxygen than to an increase in carbon dioxide. The Malacostraca, however, that are able to regulate the ventilation of the gills usually respond to changes in either oxygen or carbon dioxide tensions. Nevertheless, aquatic isopods and some amphipods increase gill ventilation only with a reduction in oxygen tension. No conclusions are as yet justified except the obvious ones that some Crustacea have developed a feedback system that makes it possible to compensate for respiratory stress by increased respiratory movements, whereas others live very well as oxygen conformers without the ability to regulate respiratory movements.

Circulation An open circulatory system is dependent on heartbeat and shows greater fluctuation in pressure than a closed system. Movements of internal organs and other body parts move blood in open lacunae more effectively than in capillary beds, and it is not impossible for small arthropods to get along very well without a heart. Pressure rises and falls dramatically with each heartbeat, as there are few elastic blood vessel walls to absorb part of the pressure. Blood pressure in a resting lobster is 13 cm H_2O at systole and only 1 cm at diastole. Blood pressures rise markedly with body activity. Systolic pressure is doubled in an active lobster, and diastolic pressure is brought to about the level of systolic pressure in a resting animal. Pressure varies considerably in different parts of the animal, and may differ in adjacent appendages, depending on their activity. Although pressures are low, blood flow is not sluggish and approximates the rates seen in cold-blooded vertebrates. Apparently the small resistance to flow compensates for small heart volume and low pressures.

The more advanced crustaceans seem to have

myogenic hearts, the stimulus for the beat originating in the heart muscle itself, but most crustacean hearts have a pacemaker ganglion made up of relatively few neurons. The heart ganglion of a lobster contains five large and four small neurons. The small neurons are pacemakers and fire spontaneously. The large neurons are motor neurons that follow the pacemakers and do not fire spontaneously unless they are separated from the small neurons. Apparently the pacemaker and follower neurons interact, but the details are not yet clear. Accelerator and inhibitor fibers come to the heart ganglion from the central nervous system. They do not necessarily have an equal effect on the pacemaker and follower neurons. It is probable that the heart was originally myogenic, without innervation, and the direction of adaptation has been toward heart innervation and the development of a neurogenic beat. With the appearance of inhibitory and accelerator fibers, the machinery needed to regulate heartbeat effectively by feedback controls is achieved.

Nervous Control and Sense Organs Crustacea have a sympathetic (stomatogastric) nervous system associated with the gut (Fig. 31.6A). Details differ considerably in the groups (Sandeman, in Bliss, vol. 3, 1983), but as a general rule, the nervous system centers in swellings located on the connectives, a postesophageal commissure, and a system of ganglia associated with the anterior end of the gut. In copepods, a small gastric ganglion is tucked into the junction of the esophagus and the midgut, and a larger labral ganglion lies below the cerebral lobes. Nerves to the midgut, and eventually to the heart, arise from this system. The effect of these nerves is modulated through a wide array of neurohormones, such as serotonin and dopamine, that either stimulate or inhibit target organs.

Neurosecretions also play an important role in controlling some physiological processes, like molting, and also control some immediate aspects of behavior (Fingerman, 1987). Color changes are more fully analyzed than most kinds of behavior assocated with such secretions. The mechanism differs in detail in various crustaceans, but centers in the **X-organ/sinus glands** of the eyestalks and the postesophageal commissures. Substances that evoke the expansion or contraction of the pigment cells in the cuticle of prawns and crabs have been extracted, and there is evidence that a complex of secretions affects the color-control system of many crustaceans (Rao, in Bliss, vol. 9, 1985).

Crustacea have a varied set of sense organs, similar in many respects to those found in other arthropods. Many of the sense organs are mechanoreceptors, associated with movable bristles or pits. Eyes are varied in structure and function, as well as in origin. The most outstanding development in eye structure is the appearance of a compound eye, which rather parallels the development of the compound eye found in many other fossil and living articulates. Chemoreceptors also occur and are important for the recognition of food and mates, as well as general orientation to the environment.

Stretch receptors are mechanoreceptors that are primarily concerned with internal conditions. They consist of small muscles surrounded by many sensory endings. The sensory endings are stimulated by passive stretch and contraction of the muscles. Fast and slow receptors occur in each abdominal somite. The fast receptors respond to stimuli of short duration, whereas slow receptors fire repeatedly for several hours when continuously stretched. The presence of both types of receptors enables the animal to distinguish between temporary short changes resulting from movements and long-lasting posture stimuli. Rather similar proprioceptors are found in muscles, tendons, or the articular membranes of joints. They are stimulated by the stretching of tissues during movements and by vibrations of the substrate. Stretch receptors and general proprioceptors are important in establishing postural reflexes and in reflexes associated with body control during movement (Page, in Bliss, vol. 4, 1982).

Many of the bristles and hairs at the body surface contain sensory neurons and are mechanoreceptors of one kind or another (Bush and Laverack, in Bliss, vol. 3, 1982). Sensory bristles located at a joint may provide information about the bending of the joint, whereas similar bristles at the tip of a leg may be stimulated by contact with the substrate. Some of the mechanoreceptors adapt rapidly and respond less to stimuli that are repeated. Others continue to discharge for long periods when the stimulus is sustained. Endings that adapt rapidly are especially suitable for tactile sensations. The sensory bristles at the margins of a crayfish uropod, for example, discharge once when the bristle is bent. The rigidity of the bristle determines the ease with which a stimulus is evoked, and thus determines the kind of event to which it is sensitive. When rapid-fire stimulation of a hair results in a matching pattern of rapid-fire neuronal discharge, the sense organ is suitable for vibration reception. Delicate hairs of this kind are useful in detecting water currents and, in land forms, air currents.

Statocysts are specialized mechanoreceptors found in the base of the antennules or, in some cases, in the abdomen (Fig. 31.8F). Structurally and functionally they resemble the georeceptors of other animals. Each is a cavity lined with sensory epithelium and containing one or more heavy bodies. Some secrete statoliths, whereas others use sand or other particles present in the immediate surroundings. The sensory epithelium contains several kinds of fibers. In a lobster some nerve stimuli denote position, others direction of movement, and still others acceleration or vibration. The frequency of nerve impulses varies with the strength and nature of the stimulation of the sensory hairs. Removal of one of the statocysts causes the animal to circle toward the defective side, indicating that the signal evoking behavioral responses depends on a summing of the stimuli from both statocysts.

A median **naupliar eye** is found in crustacean larvae. It is composed of several **ocelli**, united as a single structure (Fig. 31.8A). This kind of eye persists in some adults and exhibits a variety of forms useful in defining crustacean classes. In the majority of Crustacea a pair of **compound eyes** are found (Cronin, 1986). Each compound eye is composed of a number of discrete optical units, the **ommatidia**. Compound eyes vary greatly in size and structure. Some contain only a few dozen ommatidia, whereas others have well over 10,000. Each ommatidium sends a single stimulus to the central nervous system, and the precision of the image depends on the number of ommatidia present. Some compound eyes are sessile, that is, flush with the head shield surface, whereas others are on short or long stalks. The large stalked eyes of decapods or stomatopods can have a considerable advantage over sessile eyes, for they have a very wide visual field, reaching up to 200 degrees.

Each ommatidium (Fig. 31.8B) is covered by a **cornea**, produced by the corneagen cells that lie immediately below it. Below the cornea a **crystalline cone** is found. This accessory lens is formed, typically, of four cells. Below is a circlet of **retinula cells**. Each retinula cell contains a differentiated border, directed toward the center of the cluster to form the **rhabdome**. The plane of interdigitation varies in different organisms and may be important in some aspects of light reception. In simpler types of crustacean eyes the retinula cells contain pigment, but in more complex eyes the pigment around the crystalline cone is held in the special pigment cells. Processes from the neurosensory retinula cells extend

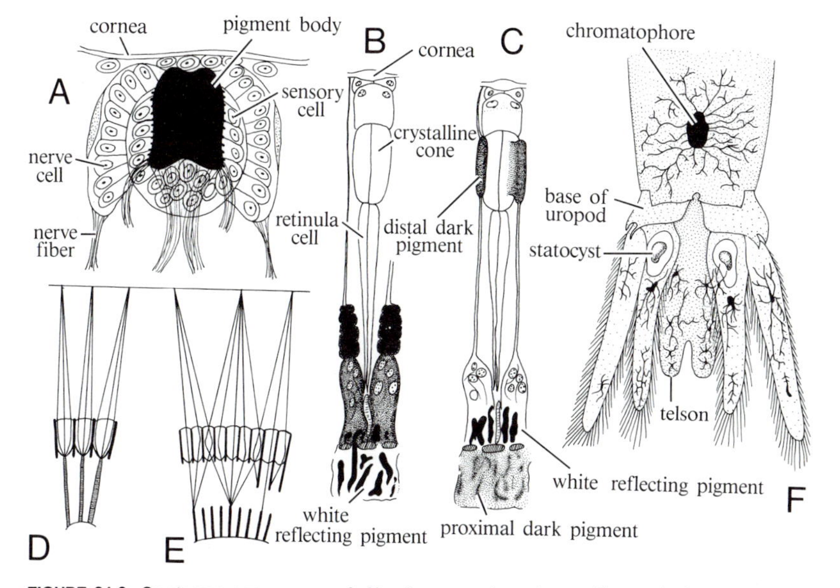

FIGURE 31.8. Crustacean sense organs. **A.** Nauplius eyue of anostracan. Pigment body provides complex pattern of shading. Bilaterally arranged nerve fibers carry symmetrical or asymmetrical stimuli depending on direction from which light comes. **B.** Light-adapted ommatidium of *Palaemonetes.* Distal dark pigment and proximal dark pigment cluster about light-sensitive elements, whereas white, reflecting pigment lies below basement membrane. Under these conditions, eye functions as apposition eye **D**, with each ommatidium functioning independently. **C.** Dark-adapted ommatidium of *Palaemonetes.* Distal pigment has migrated up to surround crystalline cone, and proximal dark pigment has migrated below basement membrane. White reflecting pigment surrounds rhabdome. Eye now functions as superposition eye **E**, with light from several ommatidia reaching light-sensitive elements. **F.** Posterior end of mysid, *Macromysis,* showing statocysts in uropods. (**A** after Claus, from Broch, in Kükenthal and Krumbach; **B,C** after Kleinholz; **D,E** after Kühn, from Wigglesworth.)

through the basement membrane at the base of the eye.

The ommatidia contain three pigments, a black distal pigment and two proximal pigments, one black and the other white. The black pigments are melanin or ommochrome, and the white reflecting pigments are purines or pteridines. The pigments migrate in response to changes in illumination, under the influence of neurosecretions from the eyestalk sinus glands and probably also under the influence of nervous stimuli. In a light-adapted eye the distal and proximal black pigments are clustered about the retinula cells, and the white reflecting pigment is below the basement membrane (Fig. 31.8B). Under these conditions, only light passing down directly from above can reach the rhabdome. In a dark-adapted eye, the distal pigment migrates around the crystalline cone, whereas the proximal black pigment moves below the basement membrane. Meanwhile, the white proximal pigment moves up around the rhabdome forming a reflecting layer, the tapetum. Light from several crystalline cones can now strike a rhabdome. Injections of eyestalk extract causes the distal pigment and the white proximal pigment to assume the position normally held in light-adapted eyes, but leaves the proximal black pigment below the basement membrane (Fig. 31.8C). Evidently the eyestalk secretions are an important, but not the only, factor in charge of pigment migration.

Eye function changes with pigment migration. When each ommatidium responds only to light from its own crystalline cone, as in a light-adapted eye, an apposition image is formed (Fig. 31.8D). When each ommatidium responds to light from several crystalline cones, a superposition image is formed (Fig. 31.8E). The superposition image is better suited to the analysis of a rapidly changing visual field and, of course, makes better use of the light available for vision. The apposition image is better adapted to a clear, mosaic picture. Superposition eyes occur in animals from deep-water or turbid habitats.

Chemoreception is still not well understood in crustaceans (Ache, in Bliss, vol. 3, 1982). As a general rule, the gnathobases and the mouthparts are responsive to food substances, and sensory bristles or pads located on the antennae and antennules are also chemoreceptive. No very full description of chemical sensitivity is available, but work so far clearly shows that Crustacea can receive chemical stimuli from a distance as well as on contact. Contact chemoreception, however, appears to predominate. A few special chemoreceptors are known, such as organs sensitive to salinity in some of the brackish-water species.

Many species undergo cycles of diurnal activity, seeking deeper water during the day and surfacing at night, or otherwise modifying behavior during day and night hours. Terrestrial species are usually nocturnal to minimize desiccation. Shore species often have cyclical behavior patterns, associated with the tides, which continue when placed under constant laboratory conditions. Even when kept in constant darkness, the eye pigments of crabs and crayfish maintain the diurnal rhythm of light and dark adaptation.

Reproduction and Development As discussed earlier, crustacean reproductive organs are basically very simple, basically gonads, gonoducts, and occasionally seminal storage sacs. However, location of gonopores and reproductive behaviors are extremely variable, the latter so much so that they are discussed individually in the Review of Groups section. Some generalizations are offered here.

Fertilization can range from internal, through external with spermatophores, to external with free broadcast of gametes. Likewise, sexuality ranges from simultaneous hermaphrodites (remipedes, cephalocarids, and barnacles), through sequential hermaphrodites, to separate sexes. It is this second alternative that is most intriguing, as a growing number of crustaceans are proving to have protandrous or protogynous life-styles. **Protandry**, functional male first, is now known to occur in several groups of isopods, some amphipods, two genera of caridean shrimps, and several ascothoracidans, and has been suggested for sand crabs. **Protogyny**, functional female first, occurs in anthuridean isopods and tanaidaceans. Whether an individual is a male or a female in at least some of these cases seems governed by the secretions of a small endocrine organ, the androgenic gland. Some of the variant morphologies of precopulatory and postcopulatory stages of some of these are extreme, and, in the case of a few amphipods, variant male phases have been mistakenly placed in different genera and families. It suggests that developmental shifts in life cycles might be one possible source for the wide diversity noted in crustacean species.

Cleavage is of a modified spiral type with a distinctive cell lineage for the mesoderm (Anderson, 1973). The wealth of larval types among the crustaceans is often confusing, even for the expert. The terminology is resolvable (Table 31.2), however, in the recognition of three distinct stages: **nauplius**, **metanauplius**, and **juvenile** or **postlarva**. Although the original development pattern appears to have been one in which the series of metanauplii and juveniles represented essentially a sequence of gradual changes, the disruption of that sequence into discretely different larval types was an important one for crustacean evolution. The larvae are adaptively significant in that they occupy distinctly different niches from their parents, and thus offer and suffer no competition to or from the adults. The disruption of developmental sequences is yet another way variant morphologies were produced in crustacean evolution. **Progenetic paedomorphosis**, the development of functional gonads in a larval body, is common. Many orders and suborders, like bathynellacean syncarids, thermosbaenaceans, or cladocerans, and one whole class, the Maxillopoda, undoubtedly arose through progenesis (Schram, 1986, p. 547).

After hatching, the young crustacean launches into a series of developmental molts, changing in appearance from instar to instar. Decapods offer a nice example (Wenner, 1985). The characteristic crustacean larva is the nauplius (Fig. 31.9A). The body of the nauplius is not segmented, and it has three pairs of appendages: antennules, antennae, and mandibles. All aid in swimming. During molting, additional appendages are added, and the thoracic and abdominal regions begin to form, resulting in the appearance of the **protozoea** larva, a kind of metanauplius (Fig. 31.9B). The regions are defined, all of the head appendages are present, and two pairs of maxillipedes are also formed. A small carapace is developed. The protozoea swims with the head appendages like the nauplius, and depends heavily on the natatory setae of the antennae; the thoracic appendages may make some contribution to locomotion. Subsequent molts lay down the rudiments of the thoracic appendages and the uropods. The dependence on head appendages for locomotion and the lack of connection between the thorax and the carapace are important attributes of the protozoea. Most decapods pass through both nauplius and protozoea stages in the egg. Eukyphid and euzygid shrimps, however, hatch as protozoea.

Table 31.2. Variant terminology used among groups of crustaceans for various stages in larval development

	Nauplius	Metanauplius	Juvenile or postlarva
Lysiosquilloidea		Antizoea	Erichthus
Gonodactyloidea			Erichthus
Squilloidea			Erichthus
Bathynellacea		Parazoea + bathynellid	
Euphausiacea	+	Calyptopis + furcilia	Cyrtopia
Amphionidaceae		Amphion	Megalopa
Penaeoidea	+	Protozoea + mysis	Postlarva
Sergestoidea	+	Elephocaris + acanthosoma	Mastigopus, decapodid
Eukyphida		Zoea	Megalopa, decapodid
Euzygida		Zoea	Megalopa, decapodid
Eryonoidea		Eryoneicus	Eryoneicus, decapodid
Palinuroidea		Phyllosoma	Megalopa, decapodid
Astacidea		Zoea	Puerulus, decapodid
Anomura		Zoea	Megalopa, decapodid
Brachyura		Zoea	Megalopa, decapodid
peracarids			Manca
Cephalocarida		"Nauplius"	Cephalopodid
Branchiopoda	+	Metanauplius	
Mystacocarida	?	Metanauplius	
Ostracoda	+	Instars	
Copepoda	+	Copepodid	
Cirripedia	+	Cyprid	

The protozoeal stage is followed by the **zoea** larva (Fig. 31.9C,E). Three shifts are important in defining the zoeal stage. With the addition of thoracic limbs and the development of exopods on them, locomotion comes to depend on the thoracic appendages. As the carapace grows, it begins to fuse to the thoracic tergites, and the stalked eyes appear. The pleopods also appear, but remain functionless. As the young decapod emerges from the zoea stage, it becomes a postlarva of one kind or another (Fig. 31.9D), usually rather closely resembling the adult. All of the appendages are now present, and the pleopods become functional and responsible for swimming movements. No further metamorphoses are required, as a rule. The exception is seen in the brachyuran crabs. They pass through a **megalopa** stage (Fig. 31.9F) in which the abdomen is extended, very like a macruran, although the body has a number of crablike features. At this stage the young crab swims with the aid of the pleopods, and eventually settles to the bottom to undergo a last metamorphosis and assume the adult shape.

Review of Groups

CLASS REMIPEDIA

This most recently discovered crustacean class includes a single living order, the Nectiopoda (Fig. 31.10). They live in anoxic, brackish waters in caves

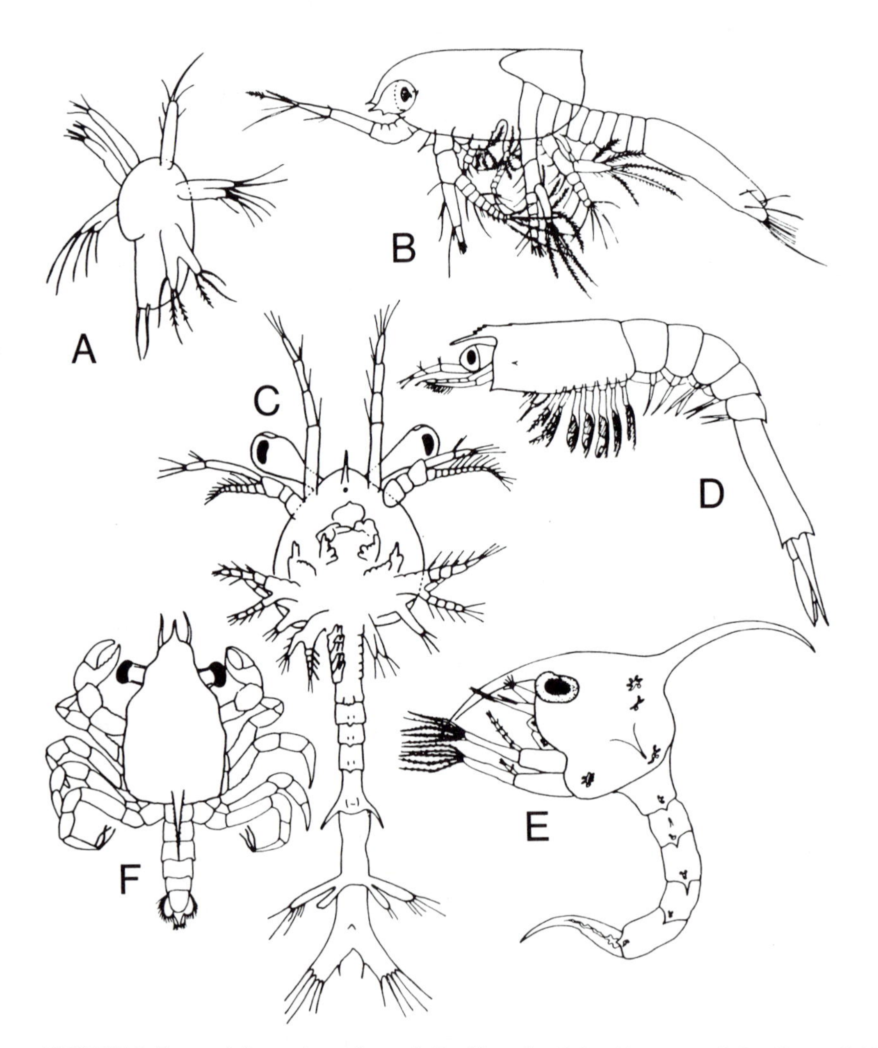

FIGURE 31.9. Representative crustacean larvae. **A**. Peneid nauplius. **B**. Peneid protozoea. **C**. Peneid zoea. **D**. Mysis larva. **E**. Crab zoea. **F**. Crab megalops. (**A**,**B**,**D** after Gurney; **C** after Claus; **E**,**F** after Poole.)

and drowned lava tubes found in the Bahamas, Turks and Caicos, and Canary Islands. The animals are noteworthy for the primitive body plan they display: biramous antennules and antennae, feeding exclusively with cephalic limbs (augmented with a set of maxillipedes); no trunk tagmosis on a body 20 to 30 segments long, each segment with a pair of biramous paddlelike limbs; segmentation of internal organs such as nerve cord, digestive diverticula, and possibly reproductive organs.

The specific mode of feeding is quite specialized (Schram and Lewis, 1989). Nectiopods are active carnivores using a method not unlike that seen in arachnids. The uniramous mouthparts are well developed, and prehensile or subchelate. The first trunk segment is fused to the head, and bears a set of maxillipedes. Prey crustaceans (typically shrimps) are grabbed and pulled into the mouth field by the maxillae and maxillipedes. The maxillules, bearing a terminal fang, close down and, in piercing the victim, inject a secretion. The fang connects to a large duct that drains a large maxillulary gland in the head. The maxillules are armed with large endites at their base that cut a hole in the victim's cuticle. The anterior foregut is equipped with powerful dilator muscles, and as material is sucked out of the prey, it passes through a chamber just outside the true mouth. This chamber is formed by the enlarged labrum to enclose the mandibles developed as a mandibular mill. Thoroughly macerated and liquified food material is then passed to the midgut for absorption in the segment diverticula.

Nectiopods are hermaphrodites (Ito and Schram, 1988). The female system extends through the anterior trunk. An ovary is in the first trunk segment, and two oviducts extend back to and exit on the eighth trunk limb. The testes extend from the eighth to tenth segments, and sperm ducts exit at the base of the fifteenth trunk limb.

Nothing else is yet known about nectiopod biology, although what appear to be juvenile forms have been collected. These generally resemble the adults except they are smaller and have fewer segments.

CLASS MALACOSTRACA

About two-thirds of crustacean species belong to the Malacostraca. The differences between the malacostracan groups are subtle and are rather bewildering at first. The head and the thorax are covered by a **carapace** that ends in an anterior **rostrum**. Stalked eyes, biramous **antennules**, and **antennae** with a scalelike exopod are prominent features of the head. The thoracic legs are called pereiopods and can have exopods that may help in swimming. The female gonopores are on the sixth thoracic appendages, and the male gonopores are on the eighth thoracic appendages. There is a total of eight thoracic somites and six abdominal somites. The first five abdominal somites have **pleopods**, and the last bears a pair of **uropods**. The terminal unit is the **telson**, which has no appendages and, with the uropods, makes up the **tailfan**. In classifying a malacostracan, the main questions that must be asked are related to the alterations of basic form.

1. What is the carapace form, and how intimate is the union of the head and the thoracic regions of the body? The carapace may not cover the whole thoracic region. There is a strong tendency for the thoracic segments to unite with the head. One or two of the thoracic somites may be united with the head even when a carapace is missing. As a part of this general tendency, the first thoracic appendages may be modified for food handling, serving as maxillipedes.

2. How have the thoracic appendages been adapted for specific functions? In addition to the tendency to alter the anterior thoracic appendages into maxillipedes, there is a tendency for one or two of the legs behind them to be raptorial legs, used for food capture. Raptorial legs are longer and more

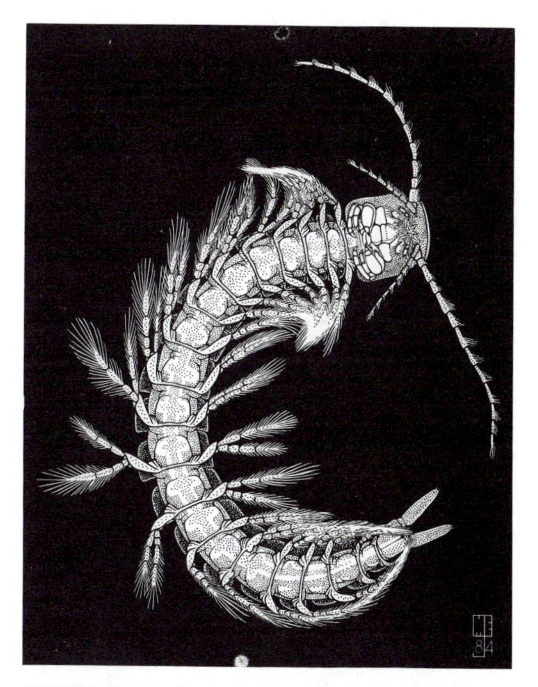

FIGURE 31.10. Remipede, *Speleonectes ondinae*, among the most primitive of living crustaceans, with a long unregionalized trunk with each segment bearing simple biramous limbs. (From Schram.)

powerful than other thoracic legs and are modified for grasping. The most common type ends in powerful pincers, or chelae, as in crabs and lobsters. In this case the appendage is a chelipede. Some raptorial legs are subchelate, closing as a jackknife closes, ending in a toothed or ridged podite that can be forcibly bent back against the preceding podites. Stomatopods have powerful subchelate raptorial legs that can cut a shrimp in two so effectively that the two halves fall in their relative positions as if chopped by a guillotine. A few raptorial legs end in other ways with harpoonlike spines or stabbing blades. Thoracic legs can also be reduced or absent, and in some groups the somites as well as the legs may be reduced.

3. What kind of gills are present, and where are they located? Gills are a characteristic feature of Crustacea generally, but the position and form of gills vary greatly (Fig. 31.3C,D,E). In some groups the gills are bladelike processes on the appendages, whereas in others the gills are filamentous or branched. The gills are formed of epipodites, or of filaments of uncertain origin attached at the base or joints of legs. In the great majority of Malacostraca, the gills are attached to the thoracic legs, but in a few groups the gills are abdominal. As they occur in only a few groups, the presence of abdominal gills is a rather critical diagnostic feature.

4. What provisions are made for brooding eggs? In some of the Malacostraca, medial plates are attached to the thoracic appendages. These are the **oöstegites**, and form a ventral brood chamber. They are usually found only in females, but in some cases the males also have rudimentary oöstegites. Some Malacostraca brood eggs on the pleopods instead of the thoracic appendages, but this does not usually involve structural specialization of the pleopods.

5. What kinds of appendages occur on the abdominal somites? The pleopods may be relatively large or small. The first abdominal appendages are sometimes modified as gonopods, used for the transfer of sperm or spermatophores to the female. Pleopods may be used for walking or crawling, but they are typically swimming appendages and are equipped with natatory setae. The appendages on the somite next to the telson are particularly valuable. Most of the Malacostraca have a tailfan, composed of the flattened telson and the flattened uropods. A few, however, have elongated and multiple uropods, unusual enough to be a relatively critical characteristic.

The Malacostraca include the largest Crustacea and also the smallest ones. They are extremely abundant and very diversified in habits. Some of them are economically important. Lobsters, crabs, and shrimps are among the most valuable commercial shellfish.

Subclass Hoplocarida Subclass Hoplocarida includes a single living order, Stomatopoda. They are rather specialized in many ways and have followed quite different evolutionary lines than the Eumalacostraca. Among their outstanding features are a short carapace, fused to the anterior thoracic somites, a head with an anterior movable piece, and abdominal gills.

Stomatopods resemble preying mantids and are sometimes called mantis shrimps. The 350 species include some interesting organisms from structural and behavioral points of view. They range from 4 to 34 cm long, with a number of relatively large species. They are especially characteristic of tropical seas, but extend into temperate zones, where they may be abundant locally. They live in littoral regions or in relatively shallow waters, and are very rarely found below 500 m. *Squilla* (Fig. 31.11A) is the best-known genus.

Stomatopods like to lie in wait in crevices or burrows until a likely looking bit of food swims or creeps by, when the huge raptorial legs are shot out to drag the prey in. They accept any food of appropriate size, including fish. Some species leave the burrow to hunt, swimming about with the aid of the pleopods and with the huge scales on the antennae serving as rudders. The chopping movement of the raptorial leg is fast and effective, and large species inflict lethal wounds on victims. They are often brightly colored, with greens and browns predominating, but reds and blues are occasionally seen. Their distinctive color patterns are intimately associated with elaborate display and agonistic behaviors (Reaka and Manning, 1981).

The anterior head is movable, containing the stalked eyes and the antennules. The remaining head somites and the first three thoracic somites are covered by the carapace. The thoracic somites are structurally joined to the head by the carapace, and the anterior thoracic appendages are converted into mouthparts or limbs for food handling. The last three thoracic and six abdominal somites are free.

Stomatopod appendages have some unusual features. The antennules have three flagella in place of the usual two. The antennae have an unusually large two-segmented scale on the exopod. The maxillules are unusual in having only a vestigial palp. The maxillae (Fig. 31.11B), composed of four indistinct segments, are unique with little resemblance to other crustacean maxillae. The maxillary glands open near the bases of the maxillae. The first five thoracic legs are subchelate; the second pair are huge, powerful raptorial legs and are the most characteristic feature of the stomatopods (Fig. 31.11A). The last three thoracic legs are biramous, have a three-segmented protopod, and are quite unlike the more anterior ones. The flattened abdominal appendages are pleopods with one unusual feature, the presence of filamentous gills that branch from an axial stem, actually a process from the exopod (Fig. 31.11C), and are quite distinctive. The broad, conspicuous tailfan has an unusual feature also. The uropod has a spiny, median plate on the protopod (Fig. 31.11D), sometimes used to plug up the entrance to the burrow.

In the male reproductive system, the testes are delicate tubes, located in the abdomen and extending back to a median part in the telson. The sperm ductules pass to paired sperm ducts that open in the last thoracic segment. Each gonoduct is equipped with a long, jointed, chitinous penis.

The large ovaries lie close to the midline and join in the telson. The oviducts pass to gonopores on the sixth pair of thoracic legs. A small seminal receptacle lies between them. Glands in the last three thoracic somites are used to cement the eggs together into a mass, which may be kept in the burrow or attached to the thoracic legs.

Adult stomatopods are benthic organisms common in the tropics but rarely taken on temperate zone collecting trips. The larvae sometimes occur in large numbers in tropical plankton and are seen more often than the adults. They are large, trans-

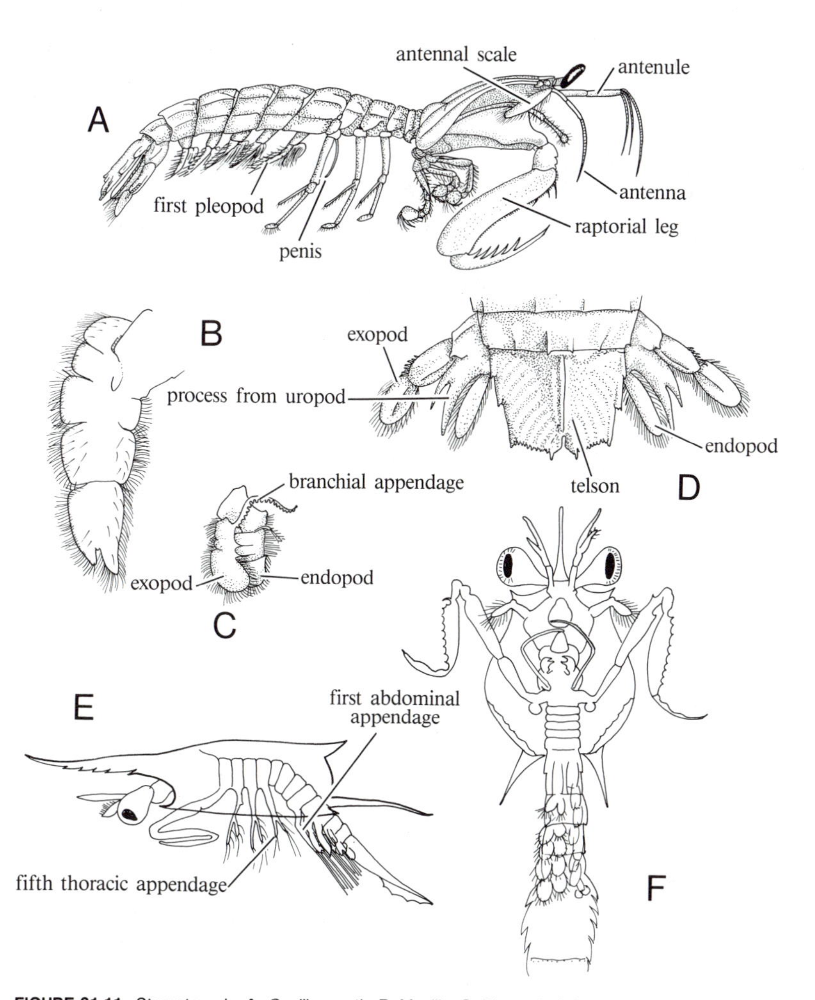

FIGURE 31.11. Stomatopods. **A.** *Squilla mantis.* **B.** Maxilla. **C.** Pleopod of *Squilla,* with gill filaments removed form axis. **D.** Tailfan of *Squilla* with unique process extending from protopod of uropod. **E.** One kind of erichthus larva, seen from side. **F.** Second type of erichthus larva from ventral view. Note suppression of posterior thoracic appendages, also observed in zoea larvae of decapods. (**A,B,C** after Calman, in Lankester; **D,E** after Claus, from Korsheldt and Heider.)

parent creatures with a characteristic form (Fig. 31.11E,F). Not all of the larvae have been matched with the adult stomatopods into which they develop. Two larval types have been recognized, an **erichthus** and an **alima**. An interesting feature of stomatopod larvae is that they pass through a long period when the last three thoracic somites have no appendages, although somites in front and behind have appendages. This is surprising, because arthropods generally develop appendages in regular sequence, with the front ones appearing first. A number of Eumalacostraca pass through a zoea stage in which these three somites and their appendages are suppressed.

Subclass Eumalacostraca This is the larger of the two malacostracan subclasses. It is characterized by biramous or uniramous antennules, gills (when present) generally restricted to the thoracic limbs, and thoracic protopods divided into a coxa and basis. The group is divided into several orders.

Order Syncarida. Syncarids are generally recognized to have many primitive features. Among these are the presence of exopods on most of the thoracic appendages, the relatively unspecialized thoracic legs, the absence of chelate or subchelate legs, and the similarity of function of thoracic and abdominal appendages. They are distinctive in lacking a carapace. Modern species occur only in restricted habitats, in widely remote places like Tasmania and Chile, typical of relict populations, or fixed in groundwater habitats worldwide.

Anaspidacea is a small suborder of syncarids living in caves and mountain pools, known best for *Anaspides* (Fig. 31.12B). The head is covered by a head shield (Fig. 31.12A). The first thoracic somite is joined to the head, and the first thoracic legs are modified as maxillipedes with distinct gnathobases. The five pairs of pleopods are powerful walking and swimming organs. The first two abdominal legs of the male are modified as copulatory organs. Sperm are transferred to a seminal receptacle on the ventral side of the last thoracic somite of the female. The eggs are not brooded, and development does not include free larval stages.

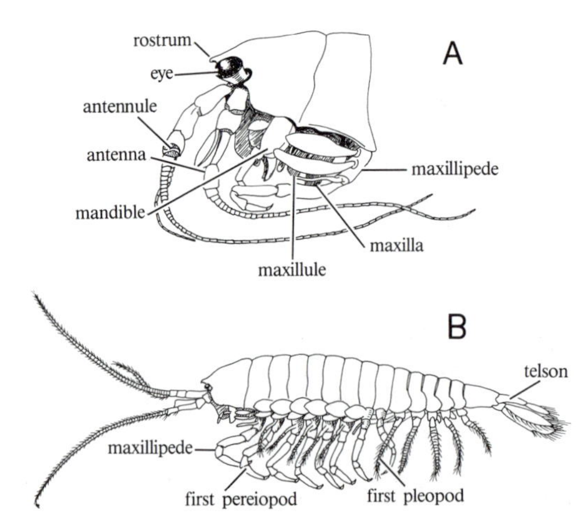

FIGURE 31.12. Syncarids. **A.** Head and appendages. **B.** Lateral view of *Anaspides tasmaniae.* (After Snodgrass.)

Bathynellacea is a syncarid suborder of small, mostly groundwater- and cave-dwelling syncarids. One species of *Bathynella* is only about 2 mm long. The body form is similar to that of *Anaspides*, but the first thoracic somite is not united to the head. They are worldwide in distribution.

Mysidaceans. These are a series of orders of almost exclusively marine, shrimplike Crustacea. The group is currently divided into two living orders: the Mysida and the Lophogastrida. There are a few freshwater species. Mysids creep about on the bottom or live among littoral vegetaton; some are pelagic. *Mysis oculata* lives only in deep lakes and undergoes a remarkable diurnal movement. During the daylight hours they stay within a meter of the bottom, but during the night they rise to the surface. Most of the mysids are about 2 to 3 cm long, but some reach 15 cm and some are no more than 3 mm long. In favorable habitats, they can become very abundant and are important food organisms for economically important fish. Mysids sometimes account for 80 percent of the diet of lake trout in the Great Lakes, and marine species are important food organisms for flounders.

Most of the thoracic somites are covered by the carapace, but only up to the first three somites may be united with it (Fig. 31.13A). The first thoracic appendages are maxillipedes and the rest are pereiopods, with a slender setose flagellum on the exopod, used in swimming or to circulate water over the gills. Many of the mysids have branched gills on the pereiopods. Oöstegites on from two to all of the thoracic legs form a marsupium, or brood pouch, justifying the common name "opossum shrimps." Sexual dimorphism is common in the pleopods, which are always present in males but may be missing in females. When the pleopods are reduced, the animals swim with the pereiopods or creep over the bottom mud or vegetation.

Most of the mysids appear to be filter feeders, although some deep-sea forms are scavengers. The filter is formed by the thoracic limbs and is cleaned by the two maxillae and the maxillipede. Direct feeding involves the manipulation of food by endopods of the pereiopods.

Eggs are brooded in the marsupium, and development is direct, without larval stages. At the time of hatching, the juveniles have all of their appendages.

Order Hemicaridea. Hemicarideans contain forms with a short carapace covering only the anterior thoracic segments, and the epipodite of the maxillipede developed as a special movable flap or bailer to move water through the carapace chamber for respiration and sometimes filter feeding. There are three suborders.

Cumacea (Fig. 31.13B) are exclusively marine, living in burrows or mucous tubes in bottom mud or sand. Although most numerous in the littoral region, some occur in deep waters. They vary from a few millimeters long to about 3.5 cm.

The greatly inflated carapace covers the first three to four thoracic somites. The slender abdomen is usually without pleopods in females, and bears two to five pleopods in males. The uropods are slender, filiform structures, and are equipped with rows of bristles used to clean off the anterior appendages.

The burrowing habit is associated with modifications in respiratory and feeding habits. The fifth thoracic legs are often reduced. The sixth to eighth legs are modified for burrowing. As the back end of the carapace may be buried, water is pulled in from the front, propelled largely by the beating epipodite of the first maxillipede, but aided by the maxillae and

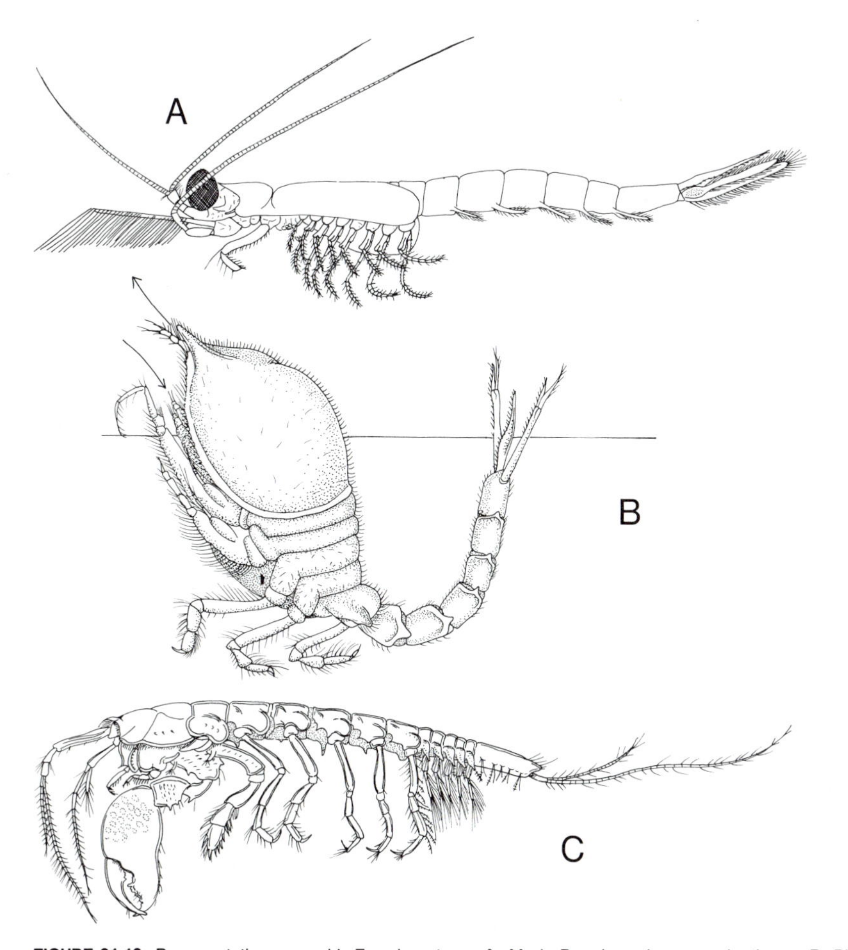

FIGURE 31.13. Representative peracarids Eumalacostraca. **A.** *Mysis.* Brood pouch seen under thorax. **B.** *Diastylis,* Cumacea. These mud dwellers characterized by inflated carapace, three pairs of maxillipedes, reduced fifth thoracic legs, and modified appendages for burrowing on next three somites. *Arrows* show combined feeding and respiratory currents. **C.** *Apseudes,* Tanaidacea. Short carapace covers only first two thoracic somites. First thoracic legs are maxillipedes, and second are chelate raptorial legs. Small abdominal somites sometimes without appendages. (**A** after Pennak; **B** composite; **C** after Sara, from Smith and Weldon.)

first maxillipede endites. Water passes over the mouthparts and over the gills, located on the epipodites on the first maxillipedes. It is diverted forward, to be discharged dorsally on each side of the head.

Filter feeding involves straining with the aid of setae on the maxillae. Some cumaceans also feed by picking up sand grains with the maxillipedes, and rubbing off organic material for swallowing (Dennell, 1934).

Swarming habits are common among Cumacea, especially among males. It is thought that this behavior may be related to mating. The eggs are retained in the brood chamber and hatch in the manca stage, a postlarval juvenile with the last thoracic legs incomplete.

Tanaidacea (Fig. 31.13C), like cumaceans, burrow in the mud, or live in mucous tubes. Tanaids prefer shallow waters, but extend to depths of 4000 m. The largest reach about 12 mm in length, but most are very small, ranging from about 3 mm. They have adapted for a life in the soft bottom material like Cumacea. Thoracic exopods are missing or vestigial, reducing the ability to swim.

The greatly reduced carapace covers and is fused with the first two thoracic somites. The remaining thoracic somites are free; the abdomen is short. The last abdominal somite is fused with the telson and often bears slender, filiform uropods.

The eyes are lobed, on lateral immovable processes. The first thoracic appendages are maxillipedes equipped with flattened epipodites. The second thoracic appendages are large, chelate raptorial legs. The third thoracic appendages are sometimes specialized for burrowing, but the third to eighth thoracic appendages are usually similar pereiopods. Females have a marsupium formed by oöstegites on the sixth, or second to sixth, thoracic legs.

The carapace forms a small gill chamber on each side. Respiratory currents are generated by the epipodites and the reduced exopods of the maxillipedes. At least a part of the respiratory exchange probably occurs in the vascular wall of the carapace.

Filter feeding is not common (Dennell, 1937), and tanaids are mostly raptorial feeders. Feeding currents are generated, as are the respiratory currents, and in the Tanaidae the epipodites and exopods of the first thoracic legs are reduced, as are the gill chambers and the maxillae. Setae on the maxillipedes and the maxillae serve to screen particles in filter-feeding forms. Excretion is by maxillary glands, sometimes with the aid of vestigial antennal glands.

Each tanaid begins life as a female. After having served in this role at least once, it molts to a male and serves in that capacity at least once. Eggs are brooded in the marsupium and hatch at the postlarval manca stage, as in Cumacea.

Order Thermosbaenacea and hemicaridean Suborder Spelaeogriphacea. These are small groundwater forms that would seldom be encountered by students. Spelaeogriphaceans are known from one locality in South Africa and another in Brazil. Thermosbaenaceans occur around the Mediterranean Sea and on the islands of the Caribbean.

Order Edriophthalma. Edriophthalmans contain forms that are united by sharing such derived features as lack of a carapace, uniramous thoracic limbs, sessile compound eyes, and a tendency to form coxal plates (Schram, 1986). It is the most diverse of eumalacostracan orders and includes as suborders the isopods and the amphipods.

Isopoda, although noteworthy for wide differences in body shape, the extent of their fusion of abdominal somites, and the nature of the thoracic coxae, are characterized by having abdominal gills.

Recognition of the various infraorders is not particularly difficult. Most of them have one or two outstanding traits that make them distinctive. The members of the Oniscoidea are distinguished by being amphibious or terrestrial. The ventral surface of the body is partially consolidated by expansions of the thoracic coxae that form ventral coxal plates fused with the body. They also have vestigial antennae (example: *Ligidium*, Fig. 31.14A). Most of the isopods have a thorax and an abdomen of about the same breadth, but Gnathiidea (Fig. 31.14B) is characterized by having an abdomen much narrower than the thorax. The first and seventh thoracic somites are reduced, and the seventh thoracic legs have been lost. Larval gnathiids are also distinctive, living as ectoparasites on marine fishes. Two of the infraorders of isopods are characterized by elongated, cylindrical bodies, contrasting sharply with the flattened bodies of most isopods. Of these, the Anthuridea are marine, and have the first pleopods modified to form a protective operculum that folds over the gill-bearing pleopods behind (example: *Cyathura*, Fig. 31.14C). The other elongated isopods, the Phreatoicidea (Fig. 31.14G), are freshwater forms found in the southern hemisphere. They are characterized by a circular to bilaterally compressed body with styliform uropods. As a general rule, isopods have a full complement of mouthparts, but members of the parasitic suborder Epicaridea (Fig. 31.14D) have suctorial mouths with both of the maxillae reduced or missing. The epicarids are parasites of other Crustacea. The females are greatly modified, often bizarrely shaped, and sometimes lose all traces of segmentation and appendages. The remaining infraorders of isopods are all aquatic, and all have some evidence of fusion of the abdominal somites. They can be distinguished by the form of the uropod and the telson. The predominantly marine Flabellifera (Fig. 31.14E) have a tailfan formed by the uropods and the telson, and also have ventral coxal plates formed by the widened thoracic coxae. The Valvifera (Fig. 31.14F) have highly modified uropods. The exopod has been lost, and the endopods fold over the pleopods to form a protective operculum. The Asellota are found in marine and freshwater habitats, and are characterized by styliform uropods that do not form a tailfan. The telson is fused with the abdominal plates to form a single continuous dorsal plate.

Many isopods can roll up, armadillolike, when disturbed. Land isopods, often called pill bugs, sow bugs, or wood lice, are frequently found under logs in wooded areas. Isopods have uniramous walking legs without exopods. They creep like insects, and many can run rapidly when out of water. Undoubtedly, these have been factors that favored the spread of isopods into amphibious and terrestrial habitats. Some of the aquatic isopods can also swim. They have flattened pereiopods or, more often, swimming pleopods. In some cases the anterior three pairs of pleopods are modified for swimming and have natatory setae, whereas the last three pairs are modified for respiration.

Isopods are not filter feeders. The mouthparts are somewhat consolidated, with a superficial resemblance to those of insects. The labrum hangs over the mouth, covering the strong mandibles. The maxillules have no palp and the maxillae have overlapping endites. The maxillipedes can be coupled temporarily through the use of special coupling spines.

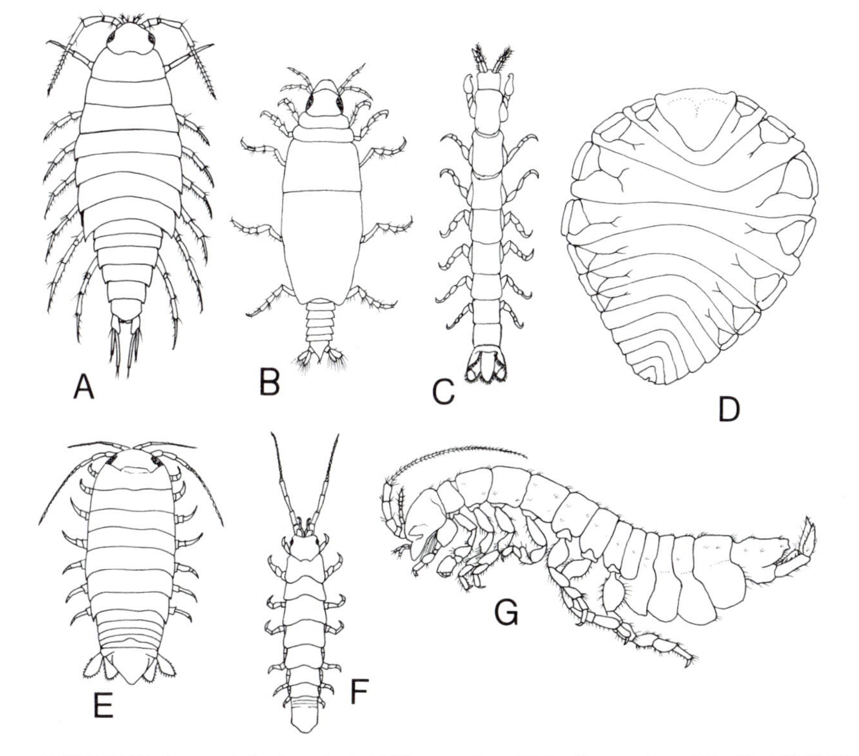

FIGURE 31.14. Representative isopods. **A.** *Ligidium,* member of Oniscoidea, are terrestrial with vestigial antennae. **B.** *Gnathia,* member of Gnathiidea, ectoparasites on fishes, as larvae, and unusual in having broad thorax and narrow abdomen. **C.** *Cyathura,* member of Anthuridea, with cylindrical rather than dorsoventrally flattened body. First pleopod forms operculum that covers more posterior pleopods. **D.** *Probopyrus,* member of Epicaridea, with suctorial mouth parts for life as parasite. **E.** *Cirolana,* member of Flabellifera, characterized by well-developed tailfan and ventral coxal plates. **F.** *Idothea,* member of Valvifera, with uropodal endopods as operculum that covers pleopods. **G.** *Phreatoicus,* member of Phreatoicidea, freshwater, southern hemisphere form with round to bilaterally compressed body. (**A** after Walker; **B** after Sars from Calman, in Lankester; **C** after Harger, from Pratt; **D** after Van Name, from Van Name; **E,F** after Johnson and Snook.)

The different methods used to protect the abdominal gills provide an interesting example of adaptation. The first pair of pleopods are sometimes modified to form an **operculum** that fits over the branchial appendages. In land isopods the exopods of the pleopods have been modified to serve as protective covers, and are folded over the respiratory endopods. In the Valvifera still another solution has appeared; the uropod has been modified as an operculum. The development of protective coverings for respiratory appendages was undoubtedly an important preparatory step for the invasion of terrestrial habitats. Protection of the abdominal gills reduces the loss of water at the respiratory surface.

The reproductive system is simple, and the gonoducts open on the sixth or eighth thoracic somites, as in other Malacostraca. In most species, the first one or two pleopods are specialized for sperm transfer. As a general rule, oöstegites on at least some of the thoracic legs form a marsupium in which the ova develop, but in a few species the sternal plates of the thorax are modified to form a brood pouch. The eggs are carried until they hatch, when the young usually leave the female. They are in a late stage of development but with the last pair of thoracic legs still incomplete.

Amphipoda are edriophthalmids that are somewhat less diversified than the isopods. Although they are predominantly marine, there are many freshwater species of amphipods. A few are semiterrestrial, living above the high tide lines of temperate shores and sometimes extending inland in humid, tropical areas. Although some amphipods are pelagic, most of them creep about on the bottom or among algae or other aquatic plants. They are more common in shallow waters, but an extensive deepwater fauna is also known.

The first, and sometimes the second, thoracic somite is united with the head, and some of the abdominal somites are fused together and typically bear several sets of uropods. The body is often laterally compressed, and the gills are thoracic.

The main infraorders are the Gammaridea, Hyperiidea, and Caprellidea. The Gammaridea (Fig. 31.15D) includes the many, essentially ordinary amphipods, without unusually large heads or short abdomens. Most of them are laterally compressed, but some are dorsoventrally flattened and can be mistaken for isopods if one does not look at the position of the gills. Gammarids are characterized by having only the first thoracic somite united with the head; an abdomen, usually of separate somites, with a full complement of appendages; a thorax with a full set of legs having coxal plates; and maxillipedes with palps.

The marine Hyperiidea (Fig. 31.15B) includes amphipods with unusually large heads and eyes. The body is sometimes elongated or oddly shaped in other ways. The head is united with the first thoracic somite, the mandibles have no palps, the abdominal somites are distinct and have well-developed appendages, and the thoracic region has a full set of legs with small coxal segments. Hyperiids are transparent pelagic forms, and many live in jellyfish or transparent tunicates.

There are two major types of Caprellidea, both greatly modified insofar as body shape is concerned. Caprellids are marine and have several unique features. The second, as well as the first, thoracic somite is joined to the head. The fourth and fifth thoracic legs are usually vestigial or missing. The small abdomen is made of fused somites and bears vestigial appendages. The Caprellida (Fig. 31.15A), one of the adaptive lines, are known as skeleton shrimps because they are narrow and elongated. They creep about among plants and are equipped with large, conspicuous raptorial legs. The other line is the Cyamida (Fig. 31.15C), known as whale lice. They live on the body surface or external orifices of whales, and are modified for clinging to the host. The body is strongly flattened, and the appendages are converted into anchoring devices.

Amphipod appendages have some interesting features. The antennules are usually biramous, thus differing from uniramous isopod antennules, but the inner branch is sometimes small and occasionally wholly missing. The antennae never have a scale, and the female Hyperiidea have both pairs of antennae greatly reduced. The maxillules are usually larger than the maxillae, reversing the general trend in most Crustacea. The maxillipedes usually are fused at the base to form a median, ventral plate, but are otherwise quite variable. They are well developed in the gammarids but far less so in the other amphipods. The first two pairs of thoracic legs tend to be modified as grasping gnathopods with chelate or subchelate tips, and in caprellids are sometimes very large. Gills may be found on the last six pairs of thoracic legs, but in many cases some of the legs are without gills. Caprellids have gills on the fourth and fifth pairs of thoracic legs only. The pereiopods behind the gnathopods are essentially alike and have a variety of terminal specializations fitting them for burrowing, clinging to plants, clinging to a host, and the like. Two to four pairs of oöstegites form a marsupium under the thorax. The three anterior pairs of abdominal appendages are usually directed forward and are used in swimming, while the three posterior pairs, usually termed uropods, are directed posteriorly.

Amphipods usually use the pereiopods to scramble over plants or objects on the bottom, but sometimes rely on their pleopods for swimming. The body is normally curved, but powerful muscles can straighten it suddenly, giving a sudden shove. In aquatic species this results in sudden, darting movements, and enables beach fleas to hop vigorously. Amphipods that live on the bottom often have cement glands on the fourth or fifth thoracic legs, used to construct tubes of slime or to cement particles together to form a tube. Others do not build tubes of their own but usurp the tubes or shells of other animals.

A number of amphipods are filter feeders. The pleopods or thoracic appendages are used to generate feeding currents, and setose portions of the gnathopods, maxillipedes, or maxillae filter out particles. Other amphipods live on the surface of algae or colonial animals and feed on them, or capture other animals that visit the plants for food with the gnathopods. Some amphipods burrow in tunicates or sponges, living in a shadowland between predaceous and parasitic life. Most of them feed on masses of plant or animal material as it is available, handling food with the gnathopods and fragmenting it with the strong mandibles.

The male and female reproductive systems are simple, with paired gonads on the sterna of the usual genital somites. The male pleopods are not specialized for copulation. In some cases the sperm are liberated in the respiratory current of the female and, therefore, are carried past the eggs in the marsupium. The ova develop in the brood pouch and hatch after all of the appendages have been developed. The young generally remain in the marsupium until it is cast off by the female at the next molt.

Order Mictacea. Mictaceans is a small group found in Bermudan caves and the Atlantic deep sea. They lack a carapace, like isopods and amphipods, but bear biramous thoracic limbs. They appear to be a primitive sister group of the isopods and amphipods.

Order Euphausiacea. Euphausiaceans, or krill (Fig. 31.16), contains the most important species. These eucarids, ranging from about 1 to 6 cm long, are pelagic creatures, characteristic of the open ocean, and are found from the surface to depths of 3500 m and more. They often swarm in almost unbelievable numbers, and are a crucial part of the ocean food chain. Krill are important to baleen whales, and play a central role in the oceanic energy flow. Some are transparent, some are red, nearly all are luminescent. Luminescence depends on special photo-

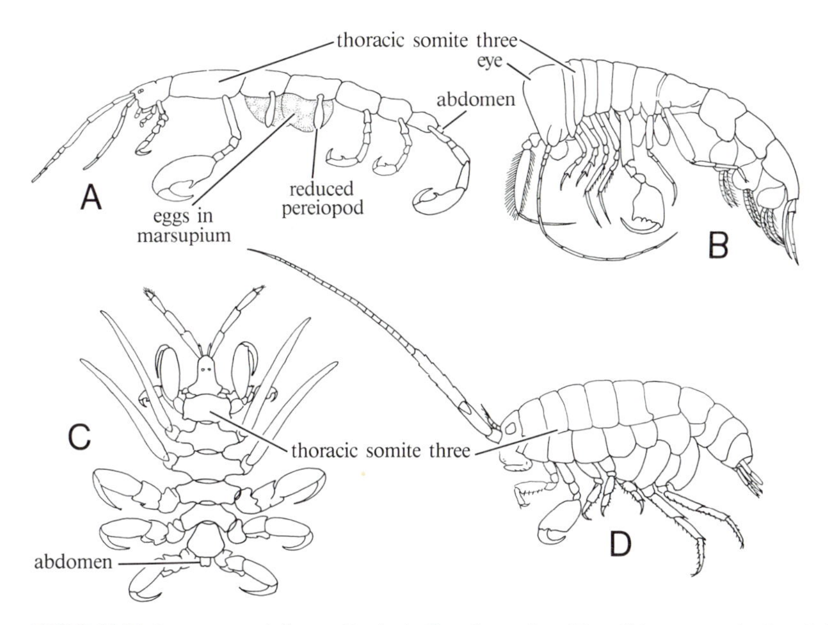

FIGURE 31.15. Some representative amphipods. **A.** *Caprella,* member of Caprellidea, representing line of skeleton shrimps with small fused abdomen and elongated thorax, and large raptorial leg. **B.** *Phronima,* member of Hyperiidea, parasites of jellyfish or tunicates. Extremely large heads and eyes characteristic. **C.** *Paracyamus,* Caprellidea, whale louse. Ectoparasite with fused abdominal somites forming diminutive abdomen and large thorax dorsoventrally flattened. Appendages modified for clinging. **D.** *Orchestoidea,* member of Gammaridea, with thoracic legs with coxal plates and first thoracic somite fused with head. (**A** after Ricketts and Calvin; **B** after Claus; **C** after Sars, from Calman, in Lankester; **D** after Johnson and Snook.)

phores, typically located on the penultimate thoracic appendage, the ventral side of some abdominal somites, or the optic stalks. No function is known for the luminescent habit, but it may be significant that the ones that are not luminescent are also eyeless.

The carapace is not developed ventrally to form a gill chamber. The first thoracic appendages are not modified as maxillipedes. All of the thoracic legs are similar and carry a gill formed from a typically branched epipodite, as well as natatory setae. The last one or two pairs of thoracic legs, however, are often reduced or vestigial. The pleopods are large and have natatory setae. The first two pairs are modified as copulatory parts in males. Uropods are flattened and biramous, and form a tailfan with the telson.

The majority of krill are exclusively filter feeders. The setose endopods of the first six thoracic appendages form an elongated, tapering filter chamber. Water enters it as the animal swims, and food is shoved forward to the mouth. Macroplankton is taken in this manner, but variations in the setae may be used to filter out particles of different sizes in different regions.

Order Amphionidacea. Amphionidaceans contain only one species of planktonic shrimp of worldwide distribution. It is notable for a unique brood pouch formed by enlarged first pleopods that extend forward under the ventral side of the female thorax.

Order Decapoda. Decapodans got their name because generally only five pairs of the thoracic appendages are leglike pereiopods. The remaining three pairs are maxillipedes. Decapoda is the largest and most successful order of Crustacea, with about 10,000 species. They have invaded nearly every kind of aquatic habitat, and some are semiterrestrial to nearly terrestrial.

Dendrobranchiata, or peneids, is a relatively homogeneous decapod suborder (Fig. 31.17A). Unless reduced, the third pereiopods are chelate, but not much heavier than the others. The pleura of the second thoracic somite do not overlap the pleura of the first somite, and the body is nearly straight. Peneids occur in both deep and shallow waters, but they are more abundant in the littoral region, where they may swarm in huge numbers. Like euphausiaceans, dendrobranchs hatch at a nauplius stage and then pass through protozoea, zoea (mysis), and postlarva (mastigopus) stages.

Eukyphida is a large and varied group. Eukyphids lack a chela on the third pereiopod, have a second abdominal somite with pleura overlapping the first somite, and have a lobe on the base of the exopod of the first maxillipede. The principal group of eukyphidans is the Caridea. They are common, small, shrimplike forms found worldwide in marine, brackish, and freshwater habitats (Fig. 31.18B,C).

Euzygida, or Stenopodidea, is characterized by having the pleura of the second abdominal somite overlapping the first, and chelae on the first three pereiopods. At least one of the third pair of pereiopods is heavier and stronger than the first two pairs. These shrimps are found in the warmer seas; one genus, *Spongicola*, lives as a commensal in hexactinellid sponges.

Reptantia are decapods characterized by strong, uniramous, thoracic legs, and often with the first pereiopod modified into a strong chelipede; a reduced first abdominal somite; and pleopods often reduced and not used for swimming. The reptant traits are evidently more suited to crawling than to swimming. An often reduced antennal scale, a generally more cylindrical body shape, and pleopod reduction suggest a trend away from swimming, whereas the more substantial pereiopods indicate more dependence on walking. Reptantia include

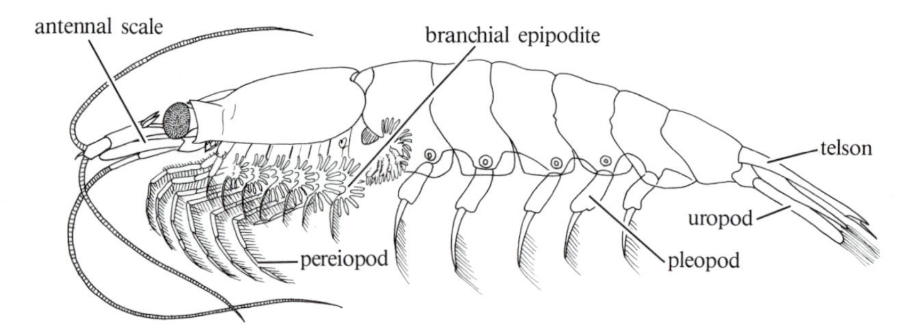

FIGURE 31.16. *Meganycitiphanes* (Euphausiacea) with no maxillipedes and carapace not covering branchial processes on thoracic legs. (After Holt and Tattersall, from Zimmer, in Kükenthal and Krumbach.)

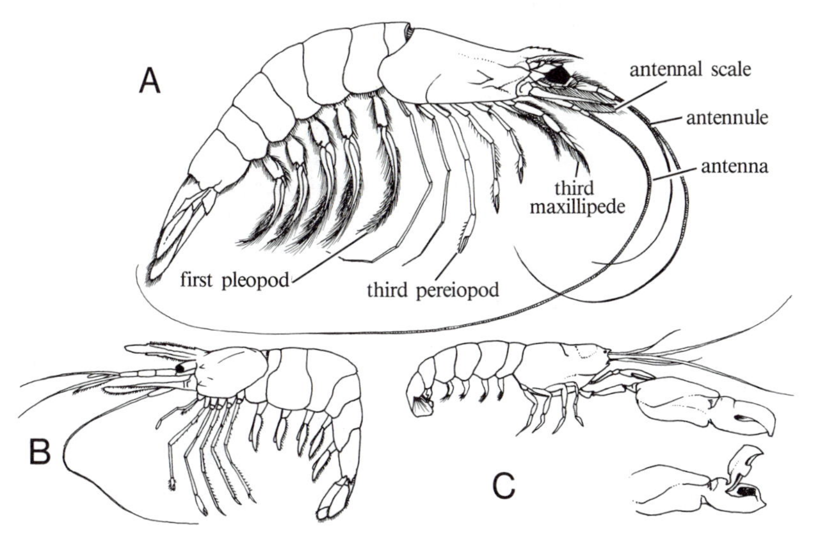

FIGURE 31.17. Some representative swimming Decapoda. **A**. *Parapeneopsis* (Dendrobranchiata) with well-developed natatory appendages and large antennal scale. Most peneids have chela on third pereiopod and pleuron of second abdominal somite does not anteriorly overlap adjacent abdominal pleura. **B**. *Palaemonetes,* freshwater shrimp, belonging to Caridea, that have no chelae on third pereiopods and have second abdominal pleura that overlap adjacent plates. **C**. *Crangon,* pistol shrimp. Modified chela snapped shut with enough force to kill or stun nearby prey. (**A** after Balss, in Kükenthal and Krumbach; **B** after Creaser, from Pennak; **C** after the MacGinities.)

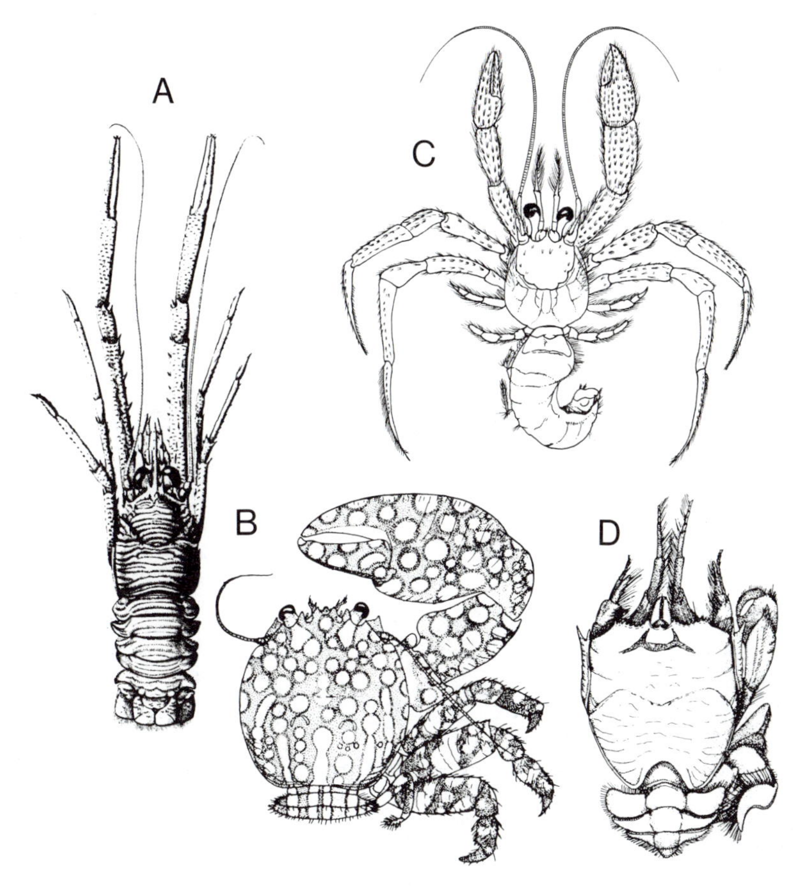

FIGURE 31.18. Anomuran decapods. **A**. Galathea crab, *Munida.* **B**. Porcelain crab, *Porcellana.* **C**. Hermit crab, *Catapagurus,* with borrowed shell. Body considerably adapted to permit entrance into shell. **D**. Sand crab, *Albunea.* Sand crabs occur in vast numbers in surf zone, where waves churn up sand as they break. (**A** after Edwards and Bouvier; **B**,**D** after Williams, from Williams; **C** after Balss.)

such diverse forms as lobsters and crayfish (Fig. 31.1), porcelain crabs (Fig. 31.18B), sand crabs (Fig. 31.18D), galathea crabs (Fig. 31.18A), hermit crabs (Fig. 31.18C), and true or brachyuran crabs (Fig. 31.19).

A central feature in the success of the decapod radiation has been the level of gill elaboration that in turn allowed for a degree of respiratory efficiency that enabled these species to occupy the wide range of habitats. Respiratory surface can be increased by adding to the number of gills and by increasing the surface area of individual gills. The gills are attached to the thoracic appendages and lie in a cavity formed by lateral extensions of the carapace, the branchiostegites. As many as four gills may occur on each side of a somite: one **pleurobranch**, attached to the body wall above the base of the appendage; two **arthrobranchs**, attached to the articular membrane of the coxa; and one **podobranch**, derived from the epipodite. Each gill consists of a central axis containing the afferent and efferent vessels and variously arranged filaments. Gills made up of branched filaments are **dendrobranchiate gills**, those composed of unbranched filaments are **trichobranchiate gills**, and those made up of flat plates are **phyllobranchiate gills** (see Fig. 31.22E). The position and form of decapod gills are valuable in classification and taxonomy. As decapods adapt to a semiterrestrial life, the respiratory surface is reduced, and the same principle applies to those that live in the intertidal zone and are exposed to air for short periods of time. Crabs living above high tide line have less gill surface and fewer gills than crabs living in the intertidal zone, and crabs living in the intertidal zone have fewer gills and less gill surface than those living below low water line.

CLASS PHYLLOPODA

The phyllopods are a class of crustaceans characterized by the possession of distinctive multiramous, leaflike limbs on the thorax at least, a tendency to reduce or eliminate abdominal limbs, and generally uniramous antennules.

Subclass Phyllocarida Phyllocarids have eight segments in the thorax and an abdomen with seven somites. Phyllocarids are thought by many to be the most primitive of the Malacostraca; however, their structural and functional affinities seem to be with branchiopods and cephalocarids. They appeared in the Cambrian.

A single modern phyllocarid order is known, the Leptostraca. Most of them live in the bottom mud among seaweeds in the marine littoral zone, but *Nebaliopsis* lives as a planktonic organism in the open ocean. Other leptostracans, or nebaliaceans, are widely distributed in shallow waters worldwide.

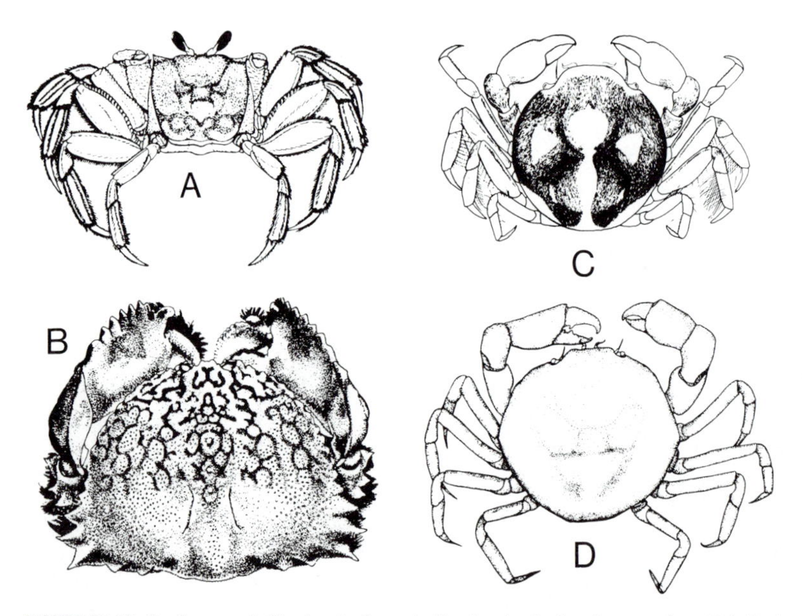

FIGURE 31.19. Brachyurans. **A**. Ghost crab, *Ocypode*. Ghost crabs dig deep burrows above high tide line and are remarkably active on land. **B**. Box crab, *Calappa*. **C,D**. Male and female pea crabs. Pea crabs live as symbionts or relatively innocuous parasites in mantle cavity of molluscs. (**A** after Crane; **B** after Holthuis, from Williams; **C,D** after Williams, from Williams.)

Nebalia (Fig. 31.20A) has a bivalved carapace that can be closed by an adductor muscle. It encloses the head, the thorax, most of the appendages, and the first abdominal somites. The head is covered by a hinged rostral plate and bears a pair of stalked eyes, sometimes rudimentry. The slender abdomen ends in an anal somite with prominent caudal rami.

The antennules are unique in being equipped with a conspicuous movable scale, and the antennae have very long flagella, longer in the male. The strong mandibles have large palps, and the maxillules typically have long slender palps that extend up under the carapace. The maxillae are not adapted as mouthparts, but resemble the thoracic legs. The eight pairs of similar thoracic legs are leaflike and phyllopodous (Fig. 31.20B). The exopod is flattened, and both exopods and endopods have long setae. A flattened epipodite is present, serving as gill blades. The first four abdominal segments bear pleopods (Fig. 31.20C) used in swimming; the next two have reduced appendages; the last lacks limbs.

The flattened thoracic legs generate a feeding current along a midventral food groove that passes forward toward the mouth and aids in respiratory exchange. The inner surface of the expanded carapace is also used in respiration. The long heart has seven pairs of ostia, the last above the sixth thoracic legs. Small antennal and maxillary glands are present, and pairs of glands at the bases of the thoracic legs are also excretory.

Eggs are brooded in a chamber between the thoracic legs of the female. They pass nauplius and metanauplius stages before hatching. The juveniles resemble adults at hatching, but have a small carapace.

Subclasses Sarsostraca and Calmanostraca Branchiopods include two distinctive types of organisms

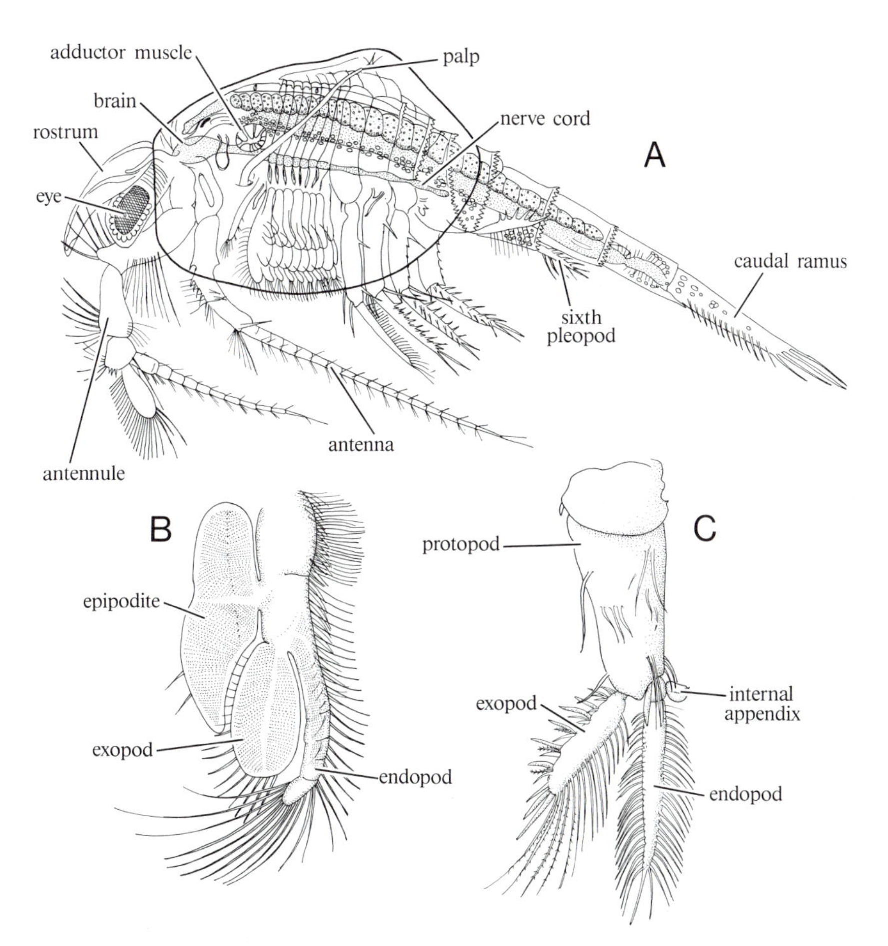

FIGURE 31.20. Phyllocarida. **A.** *Nebalia bipes.* **B.** Phyllopodous thoracic limb. **C.** Pleopod. (After Claus.)

based on the form of the carapace. The fairy shrimps (Fig. 31.21A), the Anostraca, or Sarsostraca, have no carapace. The Calmanostraca possess a carapace. Of these latter, Notostraca, the tadpole shrimps (Fig. 31.21B), have a shieldlike dorsal carapace. The bivalved shells of the order Diplostraca (Fig. 31.21C,D), are equally distinctive. Conchostraca, the clam shrimps (Fig. 31.21C), are considerably larger than the Cladocera and have the head enclosed by the shell. Growth lines on the shells aid in making them distinctive. Cladocera (Fig. 31.21D) are characterized by the failure of the carapace to enclose the head, and the presence of a single compound eye. A few marine Cladocera are known, but nearly all of the branchiopods are freshwater animals. Except for the Cladocera, they are especially characteristic of small temporary ponds so common over most of the temperate zone.

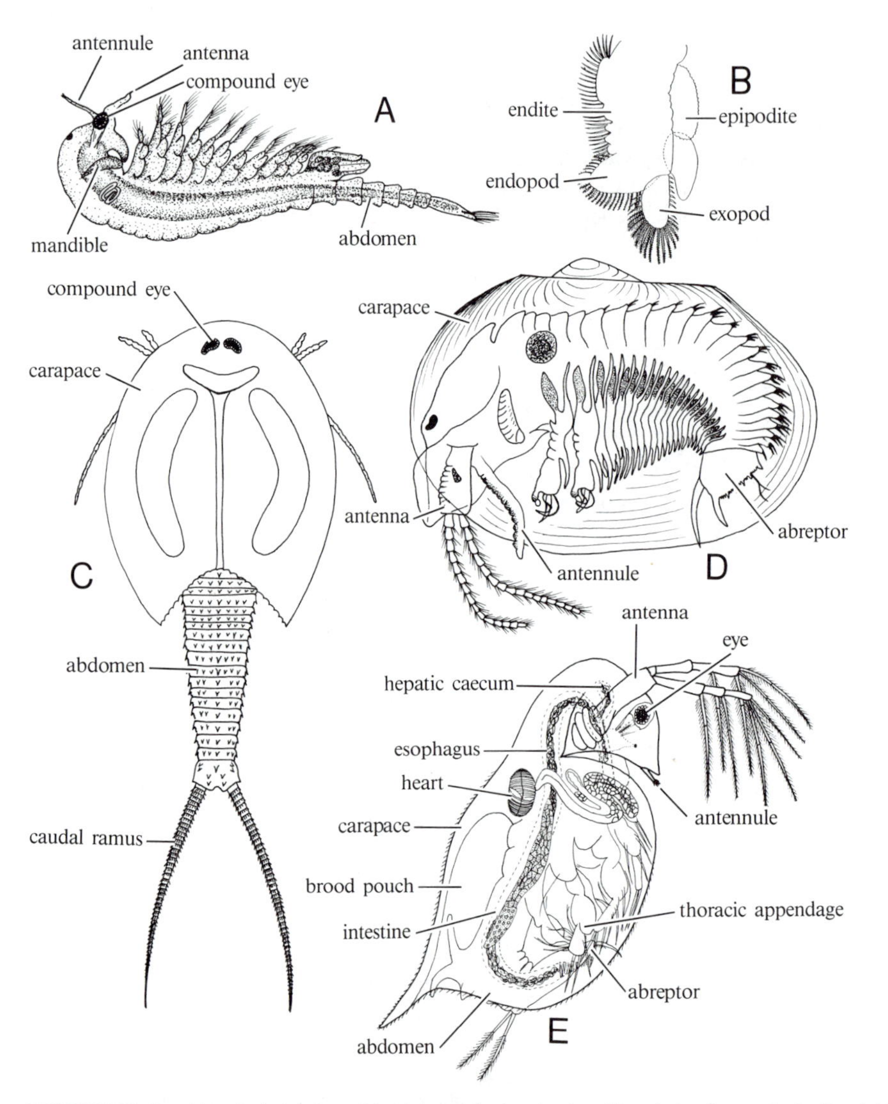

FIGURE 31.21. Branchiopods. **A**. Anostraca, fairy shrimp, *Artemia* swimming with ventral surface up. Lack of head shield as well as carapace characteristic of Anostraca. **B**. Phyllopodous trunk appendage of fairy shrimp, *Branchinecta*, soft and flattened, without joints so commonly seen in arthropod legs. Prominent exite serves as gill lobe. **C**. Notostraca, *Triops*, with low, broad carapace not divided into valves or laterally compressed. **D**. *Cyzicus*, clam shrimp. Carapace as bivalved shell, closed by adductor muscle, and encloses whole body, including head. **E**. *Daphnia*, typical water flea. Carapace not hinged and head extends forward. Large antennae used for swimming. (**A** after Vehsteat from Kükenthal and Krumbach; **B** after Sars, from Calman, in Lankester; **C** after Pennak; **D** after Mattox, in Edmondson; **E** after Storch, from Pennak.)

The branchiopods generally appear to be filter feeders. Although details of the feeding currents may differ, all employ essentially similar tactics. Movements of the thoracic appendages create water currents that ventilate the limb surfaces and deliver food, predominantly algae, to a food groove between the appendages. Although the main flow of water is backward to propel the animal forward, eddying currents at the bases of the appendages generate a small stream that flows forward in the food groove. Glands on the labrum may secrete slime in which particles are trapped. The gnathobases push the food forward, delivering it to the maxillules. The maxillules are used to shovel the food between the mandibles, which grind it before it is swallowed.

Branchiopods generally have simple reproductive systems and complex reproductive habits. Sexes are separate, and paired gonads are found on each side of the abdomen. The relatively simple gonoducts usually open on the genital somites, but the sperm ducts of Cladocera open near the end of the postabdomen, whereas the oviducts open into the brood chamber.

Males clasp the females for mating, using special clasping appendages. Genital pores are opposed in Notostraca and Conchostraca, but the male fairy shrimps have a pair of penes, and the postabdomen is modified for copulation. All of the branchiopods brood the ova, but they use somewhat different techniques. Anostraca generally have an elongated brood sac attached to the ventral surface of the body. In the Notostraca the flabella of the eleventh trunk appendages are enlarged and, in conjunction with the gills, form a brood chamber. The Conchostraca use the space inside the shell as a brood chamber, secreting gelatinous material to hold the ova in place. The Cladocera, of course, have a built-in brood chamber between the carapace and the dorsal body surface.

In some species males have never been found, and parthenogenetic development is common in all. In a number of species the males appear capriciously, and the factors that favor male production are not described. A number of branchiopods form summer and winter eggs. The more resistant winter eggs are covered by a heavy, brownish membrane and are slower to develop. In some species the thickness of the egg membrane varies at different times of the year, and the rate of development varies with the membrane; the thicker-walled ovum is undoubtedly associated with the temporary pond habitat for it permits branchiopods to withstand winter cold and desiccation. As a rule, eggs with heavy membranes are highly resistant and appear only when conditions are unfavorable. The relationship of parthenogenesis and winter egg production is not clear. In some cases the heavy ova can be produced by either fertilized or parthenogenetic females.

Anostraca have 11 to 19 thoracic somites to which phyllopodia are attached, followed by two modified somites containing the reproductive organs. Seven limbless abdominal somites follow the genital somite. The anal somite is equipped with a pair of caudal rami. Typical trunk appendages have a protopod, an exopod known as the flabellum, an endopod, one or two epipodites on the lateral margin of the protopod, and five medial endites on the inner margin of the protopod, used for the manipulation of food. Setae and filaments on the endites, endopod, and exopod are used to strain out food particles as water flows over the appendages.

As a general rule, fairy shrimps are sensitive to high temperatures, and usually disappear during late spring or summer. They are easy prey for carnivorous fish and insects, and occur predominately in habitats without carnivores. They tolerate a wide range of pH and salinity, important for animals living in temporary ponds that change rapidly from week to week. *Artemia salina*, the brine shrimp, can excrete a hypertonic urine and tolerate extreme salinities. It lives in Great Salt Lake and the Dead Sea, for example.

The appendages of the fairy shrimps and the notostracan tadpole shrimps beat metachronously as they swim gracefully through the water, with obvious activation waves passing along the body in regular sequence. The speed of movement is adjusted by changes in the angle at which the exopods are held, and to some extent by independent movements of the exopods. Although they can swim in any orientation, the fairy shrimps normally swim with the ventral side uppermost, and the tadpole shrimps with the dorsal side up. Tadpole shrimps often scramble about awkwardly on the bottom or burrow in the mud.

The Notostraca have unique external cuticular rings that do not correspond to the somites. Eleven rings encircle the thorax, and a variable number are found on the abdomen. The rings have no constant relationship to the placement of the appendages. The first two thoracic appendages have long rami and setae, and are probably used as tactile sensory appendages. The appendages become progressively smaller at the rear, eventually becoming difficult to recognize. There are no appendages in the abdominal region. Although there are minor differences in lobulation, the thoracic appendages are essentially like the appendages of the anostracans. The telson bears a pair of annulate caudal rami.

The Notostraca are also found in small freshwater

pools in the western United States and arctic regions. They prefer alkaline water and are not disturbed by high turbidity.

Diplostraca are covered by a carapace, and the enlarged second antennae serve as paddles, propelling them through the water. Each stroke of the antennae produces a forward thrust, and they move rather jerkily and erratically, although some of the Cladocera achieve respectable speeds. The larger, heavier Conchostraca have relatively smaller antennae and, therefore, are weak swimmers. When disturbed, they close the shell by contracting the adductor muscle, and sink to the bottom, looking like small clams with the foot retracted. When Cladocera are clinging to plants or in the mud, they may use the postabdomen to push the body along. Some of the benthic species depend primarily on this type of locomotion.

The Conchostraca live in small, temporary ponds, but prefer warmer water and tend to be abundant in late spring and summer. The conchostracan body is curved to fit the contour of the shell (Fig. 31.21C). The 10 to 28 pairs of trunk appendages diminish in size posteriorly, and one or two anterior pairs are modified as claspers in males. Although differing somewhat in detail, the appendages are phyllopodia. The anal somite has a pair of dorsal ridges, between which a biramous filament arises, to end in a pair of spiny rami.

Cladocera is the most diverse of branchiopod groups. Haplopods (e.g., *Leptodora*) are bizarre, elongated animals, up to 18 mm long, with the body and legs outside the carapace (Fig. 31.22B). The small carapace is used only as a brood chamber. The legs of haplopods are cylindrical instead of flattened and have no gill blades. Eucladocera have a shortened body, and trunk appendages have gill blades. One family, the Polyphemidae, resemble the haplopods somewhat, in not being enclosed by the carapace (Fig. 31.22A), but differ from haplopods in having a larger carapace and gills on the thoracic appendages.

The body, but not the head, of Cladocera lies within the carapace, and a space, the brood chamber, lies between the dorsal body surface and the carapace (Fig. 31.21D). They have five or six pairs of thoracic phyllopodia and a short abdominal region without appendages. The body ends in a recurved postabdomen or telson with two terminal claws and a pair of dorsal abdominal setae. Unlike most of the branchiopods, Cladocera usually have thoracic appendages that are markedly different from each other (Fig. 31.22C,D,E,F,G). Because of the shortening of the body and the differentiation of the thoracic appendages, the Cladocera appear to be quite specialized.

Cladocera are widespread, occurring in small, temporary ponds as well as in large lakes. A number of species are common among the emerging vegetation of the littoral zone and others abound in the limnetic region. A few are adapted to bottom life, but most of them swim about with sporadic movements in open water or among vegetation. They are less sensitive to temperature than most branchiopods, and occur throughout most of the year, although many species are characteristically abundant in one season. A few genera, such as *Evadne* and *Podon*, are marine.

The reproductive habits of the Cladocera are highly organized. Thin-walled ova are parthenogenetic, and under as yet unspecified conditions, males appear. Females able to copulate appear with them and produce winter eggs in the heavy-coated ephippium (Fig. 31.22H). A cladoceran population may contain parthenogenetic females producing thin-walled ova, sexual females producing ephippia, and males, at the same time. Evidently the parthenogenetic females produce ova of three kinds, but the factors that control development and the production of the different kinds of eggs are not understood. Parthenogenetic ova are diploid; eggs that are to be fertilized undergo meiotic divisions. Male-producing ova have not been adequately characterized. In some cases only females are produced by the fertilized ova.

Cladocera tend to fall into three groups: species with a double population peak, one in the spring and one in the fall (dicyclic); species with a single population peak (monocyclic); and species that fluctuate little during the year (acyclic). Similar patterns of seasonal populations are sometimes seen in other branchiopods. The cycles are probably associated with temperature. Acyclic species are especially characteristic of cold waters with little seasonal temperature change. It is not yet clear to what extent the monocyclic and dicyclic peaks are dependent on food and other kinds of external factors, and to what extent they are related to internal changes in reproductive potential and activity. A number of the Cladocera, like some of the rotifers and protozoa, gradually change in form as the season progresses (cyclomorphosis).

Subclass Cephalocarida Cephalocarids (see Fig. 31.23A) are marine and generally found in soft bottom sediments. They are tiny, eyeless creatures who seem to feed on organic residues in bottom deposits.

The head is broadly rounded. The trunk is tapered

very little and is divided into a thoracic region of eight segments, each bearing a pair of appendages, and an abdominal region without appendages. A pair of large caudal rami arise from the anal somite.

The appendages are especially interesting. The two pairs of antennae are small and the mandibles are simple. The maxillae are almost identical in form to the thoracic appendages. They are flat phyllopods, like the appendages of the branchiopods. An interesting feature is the presence of endites on all of the trunk appendages; they resemble gnathobases, and help to form a feeding groove along which particles of detritus apparently move up to the mouth region. Cephalocarids are hermaphrodites with a common gonoduct opening on the sixth trunk segment. They produce eggs that hatch at the metanauplius stage.

CLASS MAXILLOPODA

The class Maxillopoda includes crustaceans that have no more than six thoracic segments, five limbless abdominal segments, uniramous antennules, and a distinctive naupliar eye. There are a number of constituent groups, of which only the major ones are mentioned here.

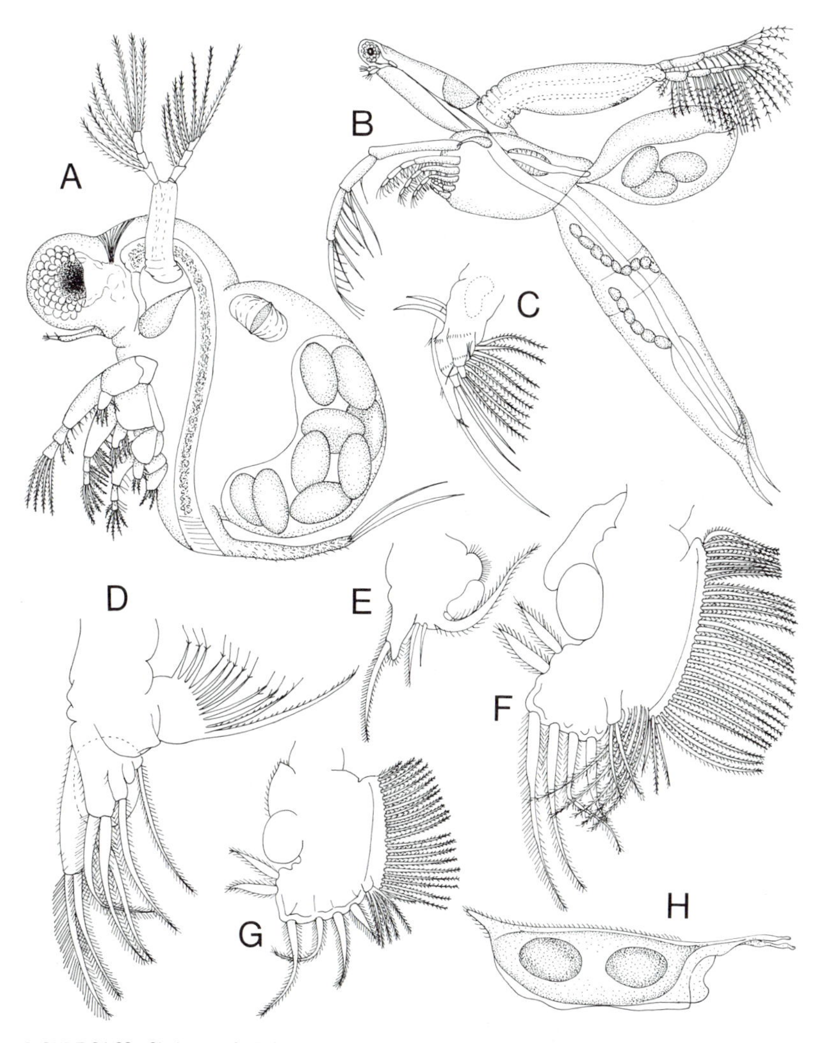

FIGURE 31.22. Cladocera. **A**. *Polyphemus pediculus,* eucladoceran that somewhat resembles haplopods. Unlike haplopods, they have phyllopods as thoracic appendages, and large carapace. **B**. *Leptodora kindtii,* haplopod cladocera, with carapace not covering body and legs. Instead of flattened phyllopodia, they have special cylindrical stenopodia. **C,D,E,F,G**. Trunk appendages of *Daphnia;* **C**. first leg; **D**. second leg; **F**. third leg; **G**. fourth leg; **E**. fifth leg. **H**. Recently shed ephippium of *Daphnia,* containing two winter eggs. (**A** composite; **B,C** after Lilljeborg, from Calman, in Lankester; **D,E,F,G** after Uéno, from Pennak; **H** after Smith and Weldon.)

Subclass Mystacocarida Mystacocarids (Fig. 31.23B) have been found in the Mediterranean and from widely scattered places around or adjacent to the Atlantic Ocean.

The tiny (about 0.5 mm) body is long and cylindrical. The anterior part of the head, bearing the antennules, is divided from the rest of the head by a dorsal groove, and only a nauplius eye is present. The antennae and the mandibles are biramous. Like a nauplius larva, mystacocarids swim with the aid of these head appendages. The most anterior thoracic appendages are maxillipedes, and cooperate with the two pairs of maxillae and the gnathobases of the mandibles in grazing. The four pairs of thoracic appendages are simple.

Subclass Branchiura The fish lice (Fig. 31.23C) are highly modified Crustacea living on the gills or body surface of freshwater and marine fishes and a few Amphibia. Some species are relatively common,

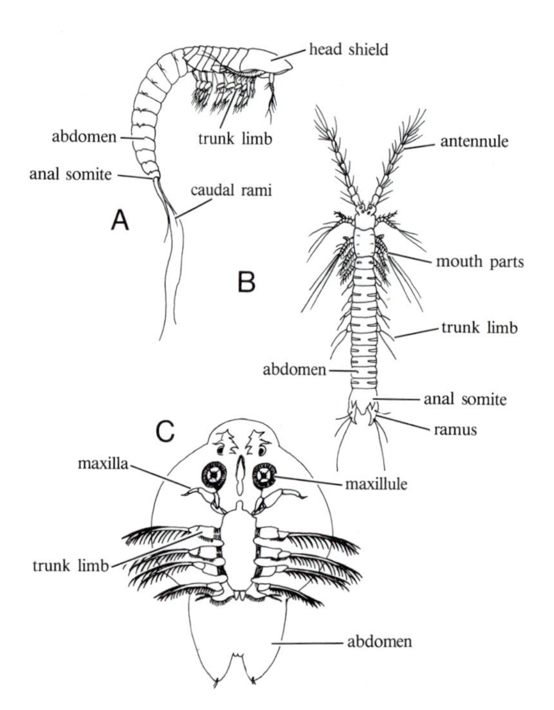

FIGURE 31.23. Some representative Crustacea. **A.** *Hutchinsoniella,* a cephalocarid, with maxillae resembling trunk appendages and abdominal somites without appendages. **B.** *Derocheilocaris,* mystacocarid, with head with usual set of appendages, four thoracic segments with legs, and prominent rami on last abdominal segment. **C.** *Argulus,* branchiuran, with flattened body suitable for clinging. Head appendages reduced and adapted for clinging, and labium and labrum specialized to form sucking proboscis. (**A** after Waterman and Chace, from Waterman; **B** after Noodt, from Green; **C** after Yamaguti.)

and the adaptations for an epizoic life seen in the Branchiura are most interesting.

The flattened body has a short abdomen, and the head and most of the thorax are covered by a posteriorly notched carapace. The two pairs of antennae are small; the antennules have a large claw used for attachment to the host. The large compound eyes are sessile. The labrum and labium are modified to form a sucking oral cone, sometimes with a hollow spine in front. The maxillules and maxillae are also highly modified, forming suckers or bearing spines or claws used for attachment to the host. The four pairs of thoracic appendages are biramous, have natatory setae, and are used for swimming from host to host. The bilobed abdomen is without appendages.

Relatively little is known of branchiuran physiology. The circulatory system is simple; in at least one species, hemoglobin is present. Branchiura have separate sexes; females have a single ovary and a single oviduct, opening between the last pairs of thoracic legs. The paired testes open into sperm ducts that unite to form a common sperm duct equipped with a seminal vesicle, and the male gonopore is located at the same point as the female gonopore.

There have been some suggestions that pentastomids may be related to branchiurans (see Chapter 27).

Subclass Ostracoda Ostracodes are a group of small animals, with most of the species not far from a millimeter in length. There are some fairly large species, and the largest ostracodes attain a length of over 10 mm. Enclosed in a bivalved carapace similar to Conchostraca, ostracodes are typically calcareous, smaller, and have a much more highly reduced body. The trunk of the body is very short, and has lost its external segmentation. Some ostracodes are excellent swimmers, but most of them live on or about the bottom and creep in the bottom mud or over the surface of plants. Some of them have become parasitic on aquatic animals. Some are predaceous and others pick up particles of food from the bottom mud. They are widely distributed in fresh water and the seas. Although most species prefer shallow waters where vegetation is present, some species have been taken at depths of 2000 m.

The two valves of the carapace are attached by a dorsal hinge of elastic, noncalcified cuticle. The adductor muscles stretch the hinge as they close the shell, and the hinge springs the valves apart when the muscles relax, somewhat as in clams. A notch in the carapace of one group permits the extrusion of the antennae while the carapace is closed. Bristles, pits, and other external markings are found on the carapace, but growth lines like those seen in many Conchostraca are absent. The carapace is often pigmented, but is rarely brightly colored.

The large ostracode head is completely enclosed by the shell (Fig. 31.24A) and bears four pairs of appendages. There are two pairs of antennae, both well developed. A pair of mandibles flank the mouth. The mandibles usually have a toothed cutting edge and a prominent three-segmented palp (Fig. 31.24B). Only a pair of maxillules occur; each of the pair has a conspicuous gill blade and four basal processes, the largest serving as a palp. Ostracode eyes are extremely variable. Some ostracodes are eyeless. Some have a median ocellus, derived from the eye of the nauplius larva. Others have a pair of ocelli, sometimes intimately united. Still others have a pair of compound eyes, somewhat like cladoceran eyes.

The trunk is short and not segmented externally. It bears no more than three pairs of appendages. The first pair of trunk appendages are thought by some to be maxillae and, therefore, to belong to the head. In the male it has a palp and is used as a clasper. The first trunk appendage is leglike, usually ending in a claw. The second trunk appendage is often turned back and can be highly specialized to serve in clearing mud away from the shell cavity (Fig. 31.24C). The abdomen is greatly reduced. It consists of paired caudal rami, sometimes very long, forming a furca that usually ends in setae or claws. It is extruded

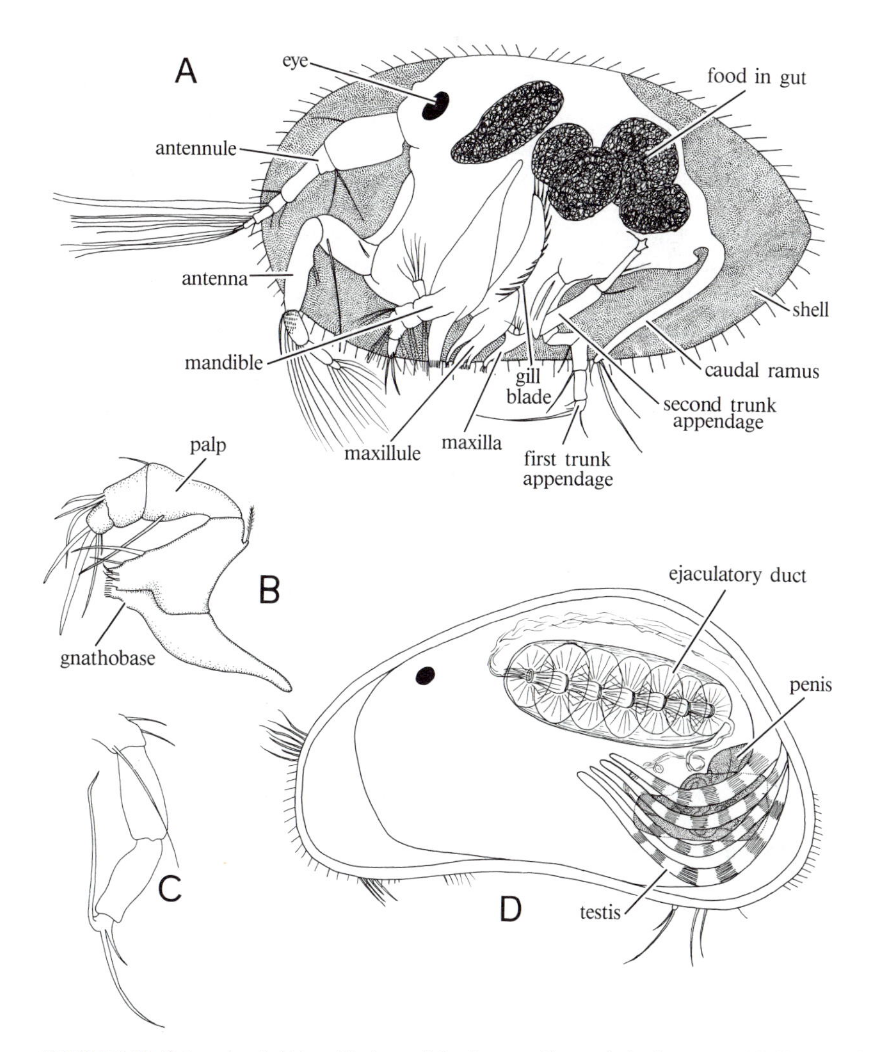

FIGURE 31.24. Ostracodes. **A**. External features of *Cypricercus,* with one shell valve removed. **B**. Mandible. **C**. Third leg of ostracode, *Cypridopsis.* This appendage is used to clean out material from inside shell. **D**. Male reproductive system of *Candona.* (**A**,**C** after Hoff; **B** after Muller, from Calman, in Lankester; **D** after Pennak.)

from the shell to push the body forward through the mud.

Antennal modifications seen in the various orders are correlated with differences in habits. Species with long swimming setae are good swimmers. Species with a few setae or bristles on the antennae are usually bottom dwellers, using the abdomen to push themselves along. Cytherids have leglike antennae, used to capture food and cling to vegetation. A few African and New Zealand ostracodes live in wet forests in the upper layer of humus, burrowing with the aid of antennules specialized to form stubby shovels.

The majority of ostracodes are said to be filter feeders. However, the issue needs careful restudy as the minute size of ostracodes ensures very low Reynolds Numbers that would indicate that they live in a viscous medium in which it might be physically impossible to "filter" in the classic sense of that term.

Ostracodes are dioecious. Females have a pair of ovaries, which extend into the shell valves in Cypridae. The oviducts are usually paired and are dilated at some point to form a seminal receptacle. The female gonopores are located behind the last pair of trunk appendages. Males have a pair of testes, each consisting of four long tubules that extend into the shell valves. The testicular tubules unite in the body to form a sperm duct. A complex ejaculatory duct leads to a penis on each side of the body (Fig. 31.24D). The spermatozoa are very peculiar in form and are sometimes longer than the body. Details of the male system are quite varied and are useful in taxonomy.

Some ostracodes reproduce only by parthenogenesis; males do not occur. In some species the males and the females are about equally abundant. All gradations between these extremes are seen.

Subclass Copepoda Copepods are abundant in marine and freshwater habitats, and many are successful and highly modified parasites. Free-living species are from about one to a few millimeters long. They make an important link in aquatic food chains because they feed on microscopic organisms and are eaten by macroscopic carnivores. In many kinds of freshwater habitats, especially, they are the most important group of organisms that do this.

Copepoda can fuse one or two thorax segments with the head somites. The last thoracic somite is the genital somite, which in many females is united with the first abdominal somite.

Free-living copepods have tended to develop along lines that suit them to a pelagic or a benthic life. The harpacticoids (Fig. 31.25C), with short antennae, are obviously less suited for active swimming than the calanoids (Fig. 31.25A), with very long antennae. The rather straight sides of the harpacticoids are suitable for creatures living in the shifting material of soft bottoms, whereas the highly developed swimming setae of calanoids help them to float when at rest and to swim when active. The cyclopoids (see Fig. 31.29B) have not committed themselves strongly to either way of life, and remain to some extent intermediate in character as well as in habits.

The antennules are conspicuous uniramous appendages of variable length (Fig. 31.25A,C). They are usually important for swimming and are also equipped with sensory hairs. The antennules of males are modified as claspers. Male calanoids have only the right antennule modified, whereas the male harpacticoids and cyclopoids have both antennules altered. The male antennules are bent more or less sharply, and in extreme cases have a hinge joint.

The antennae are shorter than the antennules and may be uniramous or biramous. In cyclopoids the exopod is missing. As a rule the antennae are sensory, but they are prehensile in male harpacticoids.

The first thoracic somite is fused to the head. Its appendages are accessory mouthparts, the maxillipedes, usually uniramous and to some extent intermediate in structure between the maxillae and the other thoracic legs.

The next four pairs of thoracic appendages are essentially alike in structure. They are flattened and have marginal bristles, making them more effective paddles for swimming (Fig. 31.25A). In most copepods the fifth pair of thoracic legs are modified, but in female calanoids they are like the other thoracic legs. The fifth thoracic legs of male calanoids are modified for the transfer of spermatophores to the female and are asymmetrical. In cyclopoids and harpacticoids the fifth thoracic legs are modified in both males and females but are symmetrical in both sexes. Modification of the appendages of the pregenital somite is a common feature of many Crustacea. The seventh thoracic somite is the genital somite; by some it is considered the first abdominal somite. At the most it has rudimentary appendages, and in most females it is wholly legless.

The abdomen has up to six legless somites. At the posterior end of the body is a pair of caudal rami. Pelagic copepods often have long plumose bristles on the rami and thoracic appendages that help them to swim. Some of the calanoids are fantastically ornamented with such natatory setae (Fig. 31.26A).

The planktonic copepods, especially the calanoids, are excellent swimmers. At times they move smoothly, and at other times they make sudden jerky movements. Movements of the antennae and the mouthparts are responsible for smooth swimming movements during feeding, and the sudden, jerky spurts result from powerful thrusts of the thoracic legs alternating with antennal movements. The mouthparts of cyclopoids contribute little to swimming, and they depend primarily on the antennae and the thoracic legs. The harpacticoids are almost entirely benthic and clamber about on the bottom or creep over the surface of aquatic plants.

Female copepods have a relatively simple reproductive system. The one or two ovaries open into paired oviducts. When ova are released they pass into paired lateral diverticula from the oviduct, the uteri, where they grow and accumulate yolk (Fig. 31.25B). The oviducts continue to the gonopore, known as the vulva, on the genital somite. At the last molt a sac is formed on the genital somite. A short duct connects the sac to the oviduct. At mating time the sperm from the spermatophores move into the sac that serves as a seminal receptacle. The ova are fertilized as they pass down the distal part of the oviduct after leaving the uteri. As they emerge they are accumulated in one or two ovisacs, attached below or at the sides of the genital somite.

Free-living male copepods usually have a single testis (Fig. 31.26B). The end of the sperm duct is glandular and secretes the vehicle used to cement

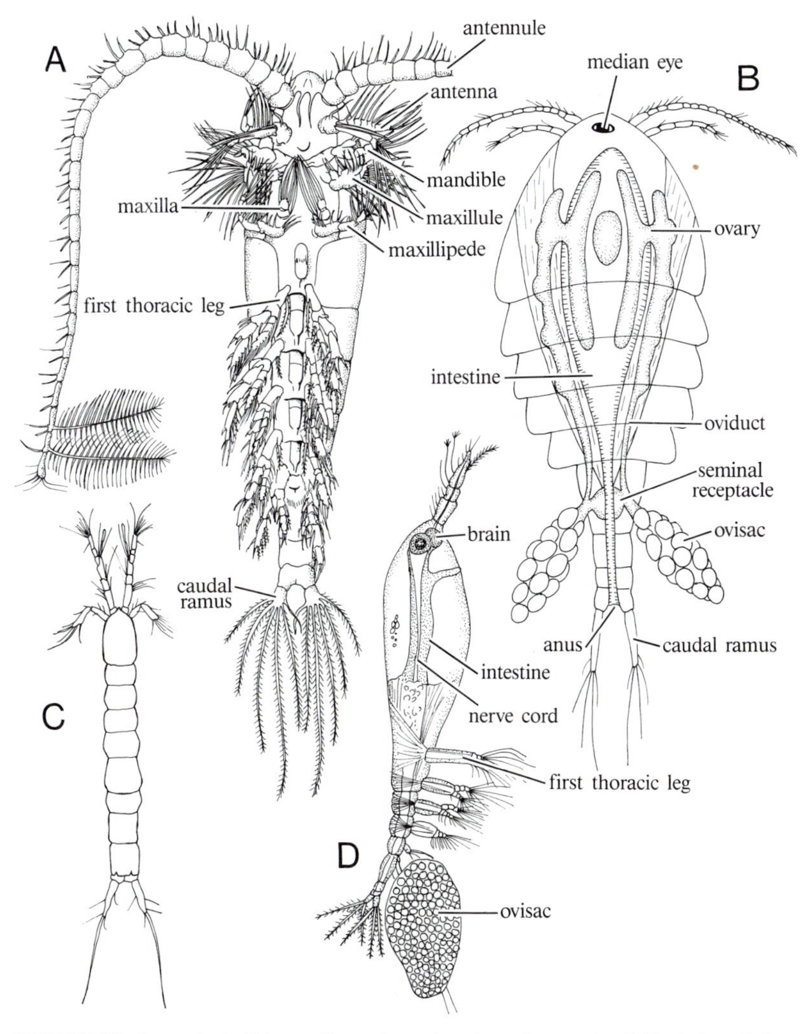

FIGURE 31.25. Copepods. **A**. *Calanus*, with very long antennules and many setae. **B**. Semischematic figure of *Cyclops*, cyclopoid, with broad anterior and narrow posterior, and antennae relatively short. **C**. *Parastenocaris*, harpacticoid, with very short antennae and with little difference in breadth along body length. **D**. *Haemocera*, adult monstrilloid, parasitic as larva in polychaetes, but adults free-swimming but with no antennae and mouth parts. (**A** after Giesbrecht; **B** after Matthes from Kaestner; **C** after Pennak, in Pennak; **D** after Malaquin.)

the sperm together to form a spermatophore. In the genital somite the sperm duct expands as a seminal vesicle, where the secretions harden around the sperm and the spermatophore assumes a form characteristic of the species.

At mating the male uses the modified antennae, and in some cases the fifth thoracic leg, to clasp the female. The spermatophore is transferred to the female by the thoracic appendages, where it is attached by cement, probably derived from the seminal vesicle wall. The sperm move into the seminal receptacles, where they may be stored for some time before the ova mature.

Copepods hatch as nauplii and go through several metanaupliar stages. Eventually the first copepodid larval stage is reached. The copepodid resembles the adult in general form, but may have only three pairs of thoracic legs and an unsegmented abdomen. After several copepodid instars the adult form is finally attained.

A few cyclopoids and harpacticoids live as parasites, but most of the parasitic or commensal forms belong to other orders.

Many copepod groups have adapted to a parasitic life-style. Often these show little differences from free-living forms (Fig. 31.27A) except for the mouthparts. Others, like monstrilloids, have parasitic larvae and adults who do not feed and exist only to reproduce (Fig. 31.25D). Some forms exhibit highly aberrant body forms (Fig. 31.27B,C), hardly recognizable as arthropods, let alone crustaceans, except for their larvae.

Subclass Thecostraca Thecostracans include the familiar barnacles, or cirripedes, found all over the seas of the world, attached by the millions to rocks, pilings, boats, and other objects in shallow waters and the intertidal zone (Southward, 1987). They do not limit themselves to inanimate substrates, but many are ectocommensal on crustaceans, echinoderms, whales, turtles, and other marine animals. Many are found in the deep sea. Adaptations for a sessile life and extensive exoskeletal modification make barnacles the most highly derived Crustacea. Barnacles have been extremely successful, and in favorable habitats they are remarkably abundant, cov-

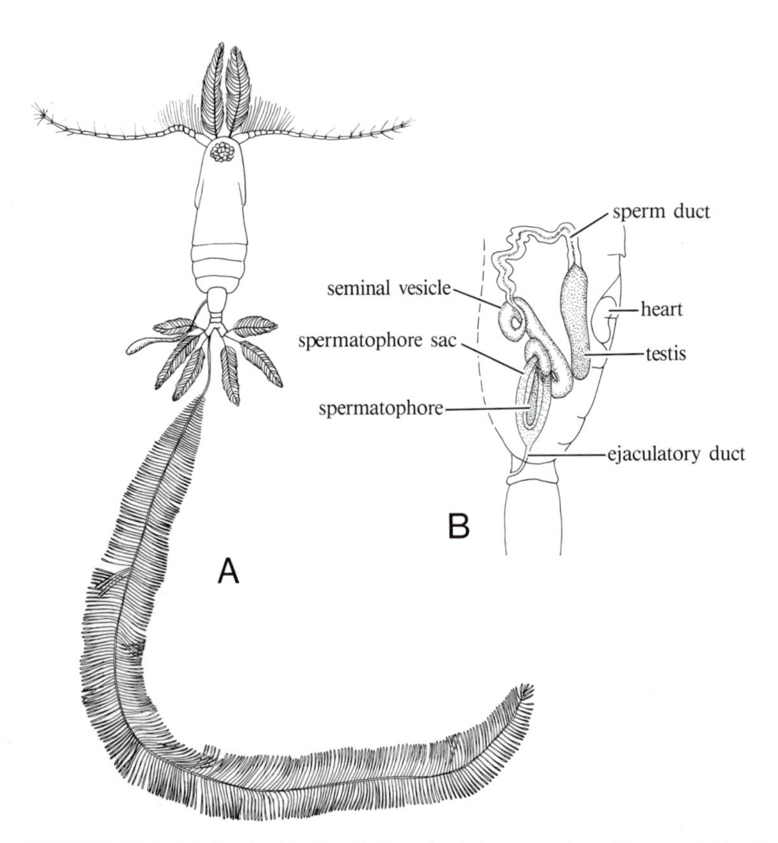

FIGURE 31.26. **A**. Pelagic calanoid with projecting setae to increase water resistance and aid flotation. **B**. Genital system of male. (**A** after Giesbrecht; **B** after Marshall and Orr.)

ering every available surface. They are economically important as fouling organisms, slowing boats down and encrusting harbor installations. The skeletal plates that make up the barnacle shell tend to scatter, but a well-documented fossil record is known. The first identified species lived in the Cambrian.

Barnacles have followed two main lines of evolutionary development. Some of them are pedunculate, with a long stalk, the **peduncle**, and a **capitulum**, consisting of the main mass of the body enclosed in a shelled "carapace" (Fig. 31.28A,B). The peduncle is actually a stalk formed by the drawn-out front part of the head focused on the antennules. Other barnacles are sessile and have no stalk (Fig. 31.28C). In these forms the front of the head is modified to form a flattened, membranous or calcareous basis attached to the surface upon which the barnacle lives.

Barnacles have become so highly modified that the terms anterior and posterior or dorsal and ventral have lost their significance. The barnacle carapace is a fleshy mantle, separated from the body by a mantle cavity except at the attachment point. The side where the body attaches to the mantle is the rostral, and the opposite is the carinal side.

The forepart of the head is the most highly modified part of a greatly altered body, and the anterior head appendages have been involved. The antennules are degenerate in adults, and the antennae are wholly missing. Barnacles attach themselves with cement formed in glands that open near the site of the antennules. The middle part of the head attaches the body to the **mantle**, a modified carapace, and the mouthparts lie near this junction. The mouth lies behind a large labrum and is flanked by a pair of mandibles with setose palps. The maxillules are small, flaplike parts, and the maxillae are fused medially to form a lower lip.

The thoracic appendages, or **cirri**, are used only to kick food into the mouth, and are quite different from the appendages of most Crustacea. The basal part is the protopod, made up of the usual two podites. The exopod and the endopod, however, are long, slender, and with many setae (Fig. 31.28B,C). The cirri are thrust through the shell aperture when the barnacle is feeding (Fig. 31.28A,C). The cirri work busily, making an effective stroke downwards and toward the midline. Food organisms are caught on the cirral setae and shoved into the mouth by the maxillae. The anterior cirri of some barnacles have been altered to serve as filters, and the remaining cirri, in this case, generate water currents. Most barnacles feed on bits of organic matter and small plankton organisms, but large barnacles can eat siz-

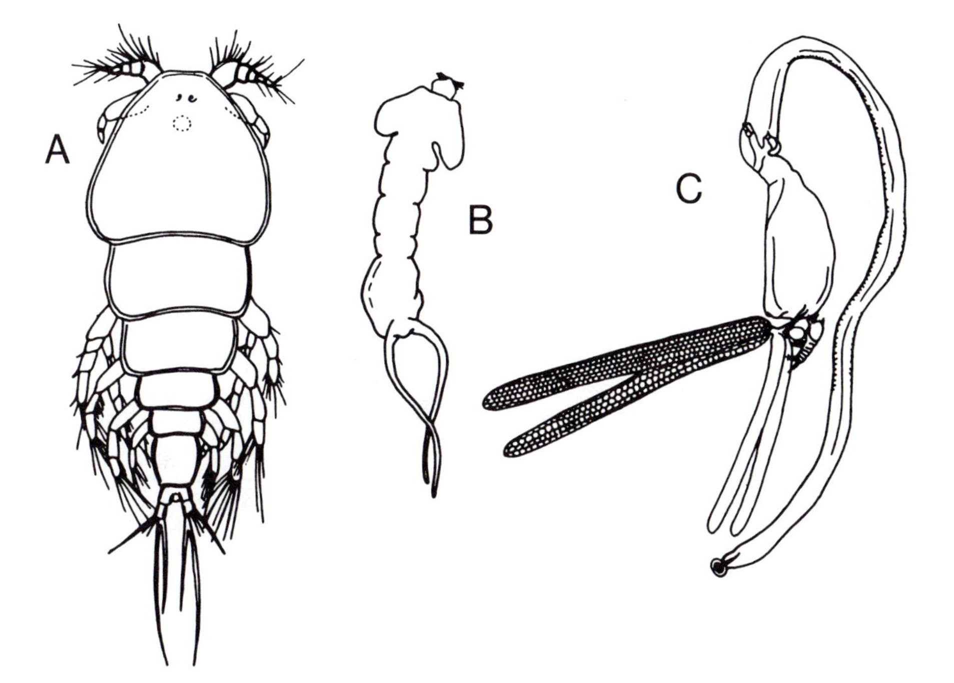

FIGURE 31.27. Parasitic copepods. **A.** Female *Ergasiloides*, Cyclopoidea. **B.** Male *Sarcotaces*. **C.** Female *Lerneopoda*, with pigmy male attached. (After Yamaguti.)

able organisms that are wrapped up in the cirri and brought individually into the range of the maxillae. The thoracic appendages of the boring barnacles and the parasitic barnacles are reduced to a greater or lesser degree.

Ectocommensal barnacles have developed parts that are especially suited to their way of life. *Conchoderma aurium* is an interesting pedunculate barnacle that may attach to the acorn whale barnacle, *Coronula diadema* (Fig. 31.29A). The carapace plates of *Conchoderma* are small, and the mantle forms a hoodlike covering around the operculum, drawn out into two funnels. *Conchoderma* attaches with the operculum toward the front end of its whale host. As is often the case with epizoic forms, the shape and the orientation of the body are means of avoiding hard work. As the whale swims, water currents flow into the mantle cavity of *Conchoderma*, who thus expends little energy. In modern times a new, mechanical type of host has appeared for *Conchoderma*, for it often fastens to the hulls of boats, which are equally good at providing the water currents it needs.

Barnacles are hermaphrodites. Isolated barnacles can presumably reproduce after self-fertilization. The ovaries are found in the peduncle and sometimes extend into the mantle. The oviducts are paired and extend to the gonopores, situated near the first thoracic appendages (Fig. 31.28B). The testes are also in the head region, but often extend into the thorax or the thoracic appendages. Sperm ductules unite to form a pair of sperm ducts that join at the base of a long, slender penis attached in front of the anus on the ventral side. Wherever barnacles are attached in dense aggregations, they outbreed, the long, groping penis slithering out to inseminate neighboring animals.

Some of the pedunculate barnacles have devel-

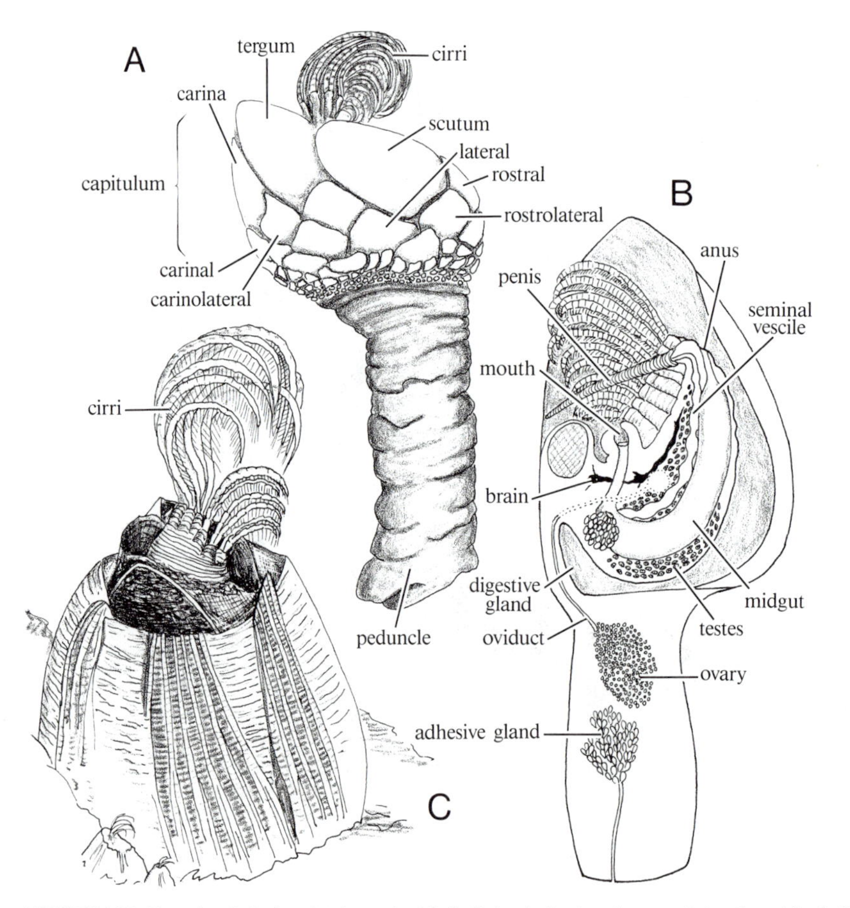

FIGURE 31.28. Barnacles. **A**. Pedunculate barnacle, *Mitella*. Peduncle develops from greatly lengthened front of head, and all important parts of body are in capitulum enclosed by scales and plates of carapace. **B**. Internal anatomy of pedunculate barnacle. **C**. Sessile barnacle, *Balanus tintinnabulum*. (**B** composite; **C** after Ricketts and Calvin.)

oped another system to achieve cross-fertilization while living a sessile life. The males are dwarfed and live in or near the mantle cavity of the females (Fig. 31.29B). The dwarf males may be like diminutive females or in some cases are extensively simplified, retaining a number of larval characteristics. In some of these species, the large barnacles have lost all traces of male organs and, therefore, are structural as well as functional females.

Eggs are usually brooded in the mantle cavity and hatch in the nauplius stage. The barnacle nauplius has a characteristic triangular carapace (Fig. 31.30A) and is equipped with the usual three pairs of appendages: antennules, antennae, and mandibles. After several molts it becomes a bivalved cypris larva (Fig. 31.30B). The cypris attaches with its antennules, and metamorphoses into an adult (Fig. 31.30C,D,E).

The most remarkable life cycles are seen in the rhizocephalan thecostracans (Fig. 31.29E). The body of the adult rhizocephalan is more extensively affected by parasitic habits than in other crustaceans. Only the larvae betray their crustacean affinities. The *Sacculina* cypris larva swims about for a time, eventually attaching to a suitable crab. It undergoes a dramatic metamorphosis in which the whole trunk is discarded and a cuticular tube is formed through which the remains of the larva gain entrance to the host body. The parasite is little more than a mass of undifferentiated cells at this stage. It migrates through the host hemocoel and attaches to the intestine. Rootlike processes grow out, eventually extending to all parts of the host body. At the next host molt, a central mass pops out of the host body and hangs down from the external surface of the host abdomen. This externa is the female par-

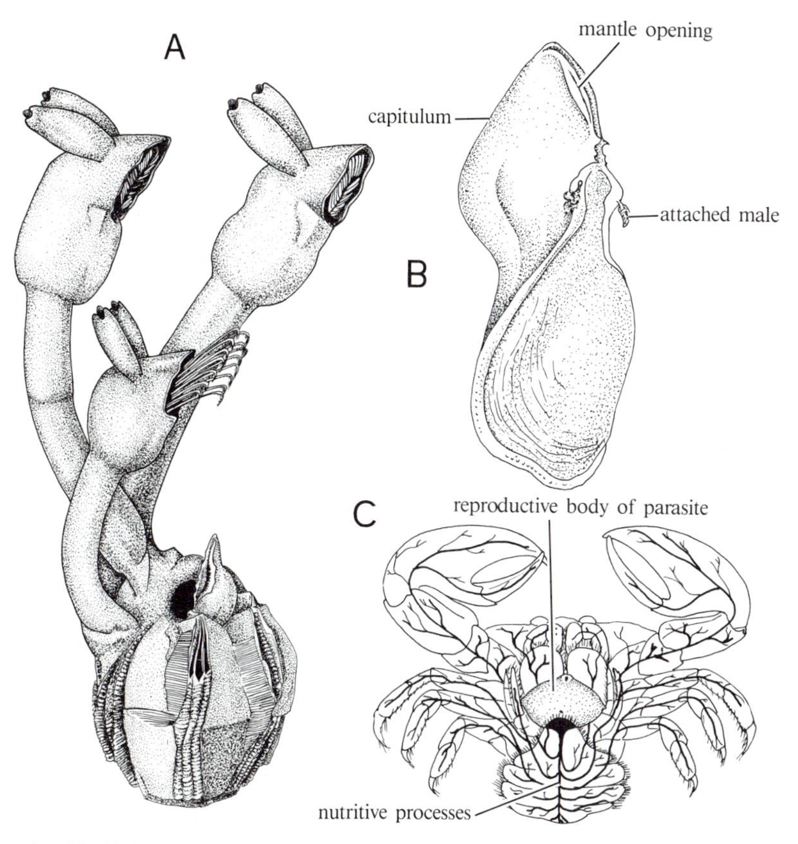

FIGURE 31.29. Some representative barnacles. **A.** Two whale barnacles. At bottom, note cask-shaped acorn barnacle, *Coronula*, that attaches to whale surface. Above are three rabbit-eared barnacles, *Conchoderma*, attached to *Coronula*. **B.** *Trypetesa*, one of acrothoracic or boring barnacles that bore into corals and shells, and have lost calcareous plates so characteristic of free-living and many parasitic barnacles. **C.** Crab infected by *Sacculina*, rhizocephalan. Adults lose all resemblance to normal barnacles, becoming more or less saccate gonad bags with branching, absorptive processes that extend into host tissues. (**B** after Berndt, from Broch, in Kükenthal and Krumbach; **C** after the MacGinities,)

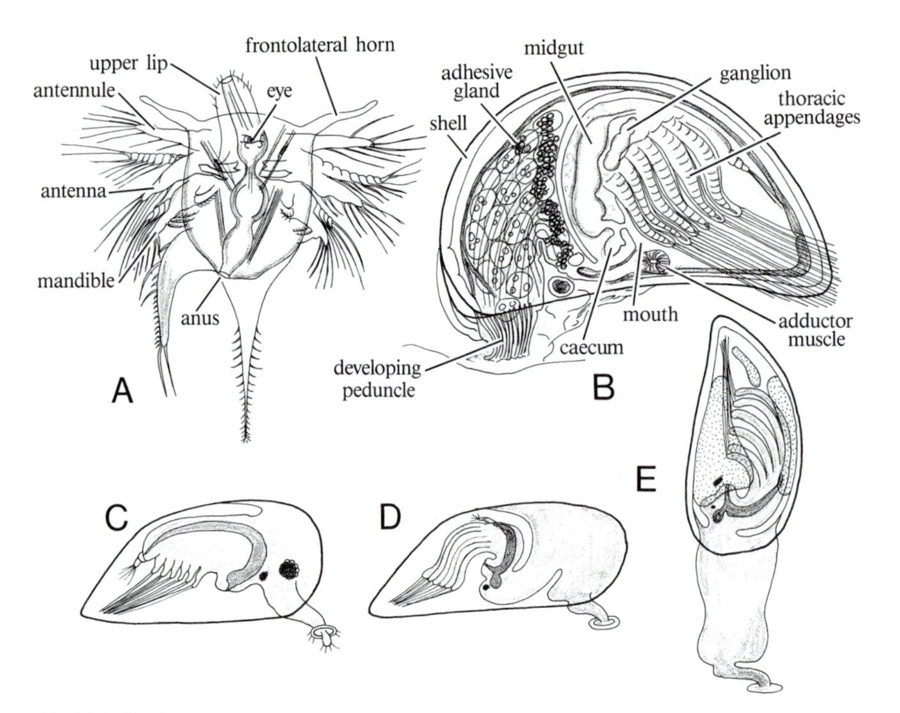

FIGURE 31.30. Barnacle development. A. Nauplius. Note frontolateral horns. B. Cypris larva, which develops adhesive glands and grows peduncle from portion of head adjacent to antennules, eventually becoming adult barnacle. C,D,E. Settlement and metamorphosis of cypris larva. (A after Claus, from Calman, in Lankester; B composite, based especially on Claus; C,D,E after Korsheldt.)

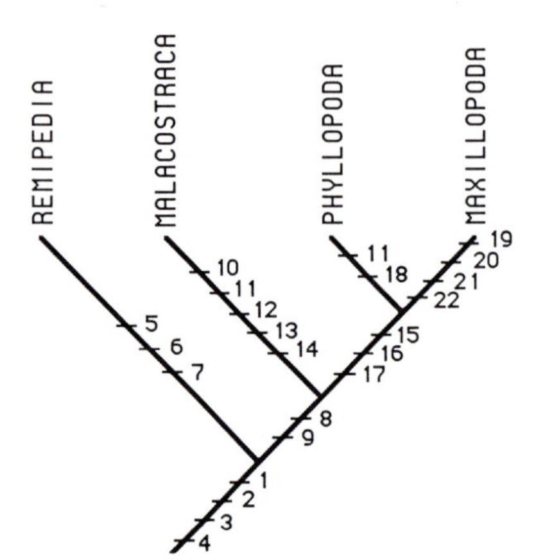

FIGURE 31.31. Cladogram of relationships of crustacean classes. Derived characters are: (1) biramous antennules; (2) second set of head limbs as biramous antennae; (3) two pairs of maxillae; (4) nauplius larva or egg–nauplius stage; (5) raptorial mouth parts; (6) maxillulary gland; (7) maxillipede segment fused to head; (8) trunk regionalized; (9) typically no more than eight thoracic segments; (10) malacostracan naupliar eye; (11) polyramous limbs, (12) thoracic endopod as stenopod; (13) uropods; (14) carapace that covers only, or only parts of thorax, and does not envelop limbs or abdomen; (15) abdomen with reduced number of entirely lacking limbs; (16) uniramous antennules; (17) tendency to lose mandibular palps; (18) leaflike, foliaceous thoracic limbs; (19) at most 11 trunk segments; (20) no more than six thoracic segments; (21) heart, when present, short and bulbous; and (22) maxillipodan naupliar eye.

asite's gonads. The male is a tiny parasite inside the female externa. *Sacculina* inhibits the host reproductive system, causing the phenomenon of parasitic castration. The female crab takes on a more juvenile appearance when it molts, and the male becomes more femalelike. If the *Sacculina* dies, a male crab host may develop a hermaphroditic gonad, producing both sperm and ova. Evidently the hormonal secretions of the parasite are enough like those of the host to scramble its normal physiology.

Fossil Record and Phylogeny

The fossil record of the Crustacea is often stated in textbooks to be poor, but the truth is that the record is good. All four classes of crustaceans have fossils; some of them, like those of Malacostraca and cirripede Maxillopoda, are remarkably informative (Schram, 1983) and provide us many "missing links." The record is so extensive, in fact, that limitations of space do not allow much discussion, although the student can consult Schram (1986) if details are desired. As a result of integrating information from fossil and living crustaceans, a cladogram of relationships among the major classes is possible, something that was not available until relatively recently (Fig. 31.31).

Crustaceans are now thought to be derived from a form that was a cephalic feeding organism, somehow grappling its food, and with a homonomous trunk lacking any regions and bearing biramous paddlelike limbs. From such a form the highly carnivorous Remipedia evolved, as well as all other crustaceans. The latter line was characterized by a regionalization of the trunk into an anterior thorax, posterior abdomen, and terminal urosome. Two great branches of this clade evolved. The Malacostraca retained limbs on every body segment, but specialized individual limb types around a polyramous thoracic appendage on which the endopod was a stenopod, and elaborated the urosome with the addition of uropods to form a tailfan. The other clade (sometimes referred to in old textbooks as the entomostracans) is characterized by several tendencies among which are uniramous antennules, a reduction or loss of abdominal limbs, mandibles often without palps, and a tendency to shorten the body in both the thoracic and abdominal fields. Entomostracans evolved along two lines. The Phyllopoda developed and exploited a polyramous leaflike thoracic appendage and used this limb in forms of thoracic filter feeding. The Maxillopoda retained the primitive biramous limb, but reduced the thorax to no more than six segments and the abdomen to no more than five. This last class also exhibits clear tendencies to exploit parasitic life-styles.

Taxonomy

The classification given here focuses only on living forms. For a more complete version that includes fossils, Schram (1986) can be consulted.

Class Remipedia. Cephalic limbs large and raptorial, subchelate or prehensile mouthparts; maxillipede segment fused to head; trunk unregionalized, trunk limbs as biramous paddles.

Order Nectiopoda. Maxillule terminating in large, clawlike fang; maxillulary glands to facilitate arachnidlike feeding; mandibles enclosed within atrium oris under the bullate labrum to function as mandibular mill (Fig. 31.10).

Class Malacostraca. Thorax of eight, abdomen of no more than six segments plus a telson; carapace typically covers thorax; female gonopore on the sixth and male gonopore on the eighth thoracic limbs; polyramous thorocopods with stenopodous endopods; telson and uropods forming tailfan.

Subclass Hoplocarida. Triflagellate antennules; thoracic protopods composed of three segments, endopods composed of four segments; pleopods with dendrobranchiate gills.

Order Stomatopoda. Anterior thoracic limbs as subchelate raptorial claws (Fig. 31.11).

Subclass Eumalacostraca. Antennal scale as single segment; thoracic limb protopod composed of two segments, endopod composed of five segments; gills when present as special structures restricted to thoracic limbs.

Order Syncarida. No carapace (Fig. 31.12).

Order Mysida. Gills on thoracic appendages; oöstegite brood pouch (Fig. 31.13A).

Order Lophogastrida. Maxillipede segment incorporated into head; oöstegite brood pouch.

Order Hemicaridea. Carapace reduced to covering anterior thoracic segments; maxillipede segment fused to head, maxillipede epipod as modified bailer; oöstegite brood pouch. Includes suborders Cumacea (Fig. 31.13B), Tanaidacea (Fig. 31.13C), and Spelaeogriphacea.

Order Thermosbaenacea. Loss of oöstegite brood pouch; carapace on female as a brood pouch; pleopods reduced or absent.

Order Mictacea. Head shield with lateral folds extending around sides of head; maxillipede segment fused to head, maxillipede without epipods; oöstegite brood pouch.

Order Ediophthalma. No carapace; sessile compound eyes; uniramous thoracic limbs; tendency to fuse coxa to thoracic pleura to form coxal plates; oöstegite brood pouch. Includes suborders Isopoda (Fig. 31.14) and Amphipoda (Fig. 31.15).

Order Euphausiacea. Gills on base of thoracic limbs outside carapace (Fig. 31.16).

Order Amphionidacea. Two sets of maxillipedes; brood pouch formed by enlarged first pleopod under thorax.

Order Decapoda. Three pairs of maxillipedes; specialized gills under carapace formed from epipodites or extensions of body wall. Includes suborders Dendrobranchiata (Fig. 31.17A), Eukyphida (Fig. 31.17B,C), Euzygida, and Reptantia (Figs. 31.18 and 31.19).

Class Phyllopoda. Polyramous leaflike limbs, typically thoracic filter feeding; generally uniramous antennules; abdominal limbs reduced in number or lacking.

Subclass Phyllocarida. Carapace bivalved, generally enveloping thorax, thoracic limbs, and anterior abdomen; abdomen of seven segments plus anal somite with caudal rami well developed.

Order Leptostraca. Antennule with small, scalelike outer branch; seventh pleopod lacking, fifth and sixth pleopods reduced (Fig. 31.20).

Subclass Cephalocarida. Maxilla pediform; large labrum in adults; egg-brooding appendages on females.

Order Brachypoda. Ambulatory thoracic endopods (Fig. 31.23A).

Subclass Sarsostraca. No head shield.

Order Anostraca (Fig. 31.21A).

Subclass Calmanostraca. Bilobed carapace.

Order Notostraca. Caudal rami as cerci; posterior trunk segmentation obscure (Fig. 31.21B).

Order Conchostraca. Caudal rami as clawlike abreptors; limbs on all trunk segments, trunk unregionalized (Fig. 31.21C).

Order Cladocera. Caudal rami as clawlike abreptors; body reduced to head and anterior thorax; cephalon exposed, separate from carapace (Figs. 31.21D and 31.22).

Class Maxillopoda. Thorax not more than six, abdomen not more than five segments; uniramous antennules; short bulbous heart when present.

Subclass Tantulocarida. No cephalic limbs, oral disc.

Order Tantulocaridida.

Subclass Branchiura. Less than six thoracic segments; mandibular palp lost; flat, shieldlike carapace; cirriform thoracic limbs.

Order Arguloida (Fig. 31.23C).

Subclass Mystacocarida. Less than six thoracic segments; mandibular palp; naupliar eye lacking; maxillipede present; caudal rami as pincers.

Order Mystacocaridida (Fig. 31.23B).

Subclass Ostracoda. Less than six thoracic segments, mandibular palp; body reduced to head and part of thorax; naupliar carapace; vertically oriented caudal rami. Several orders (Fig. 31.24).

Subclass Copepoda. First thoracomere fused to cephalon, maxillipede; mandibular palp; six naupliar stages. Several orders (Figs. 31.25, 31.26, and 31.27).

Subclass Thecostraca. Cypris larva; antennules as attachment organ.

Order Facetotecta. Certain distinctive naupliar and cypris larvae called "Y-forms."

Order Rhizocephala. Frontolateral horns on nauplii; parasitic, with body reduced to ramifying stolon and gonad sac; four naupliar stages (Fig. 31.29C).

Order Ascothoracida. Saccular carapace; parasitic; female gonopore on first thoracic segment; mandible as blade; mandibular palp; two naupliar stages; antennae lacking.

Order Cirripedia. Saccular carapace; female gonopore on first thoracic segment; mandible as blade, with palp; thoracic limbs as cirri; frontolateral horns on nauplii; carapace divided into plates; abdomen lacking. Two suborders, Acrothoracica (Fig. 31.29B), and Thoracica, the latter with two infraorders, Pedunculata (Figs. 31.28A,B, and 31.33D) and Sessilia (Fig. 31.28C).

Selected Readings

Anderson, D. T. 1973. *Embryology and Phylogeny in Annelids and Arthropods.* Pergamon Press, Oxford.

Bliss, D. (gen. ed.). 1982–1986. *The Biology of Crustacea,* vols. 1–10. Academic Press, New York.

Cronin, T. W. 1986. Optical design and evolutionary adaptation in crustacean compound eyes. *J. Crust. Biol.* 6:1–23.

Darwin, C. 1851–1854. *A Monograph on the Subclass Cirripedia.* Ray Society, London.

Dennell, R. 1934. The feeding mechanism of the cumacean crustacean *Diastylis bradyi. Trans. Roy. Soc. Edinb.* 58:125–142.

Felgenhauer, B. E., and L. G. Abele. 1989. The evolution of the foregut in the lower Decapoda. *Crust. Issues* 6:205–219.

Fingerman, M. 1987. The endocrine mechanisms of crustaceans. *J. Crust. Biol.* 7:1–24.

Fryer, G. 1988. Studies in the functional morphology and biology of the Notostraca. *Phil. Trans. Roy. Soc. Lond.* (B)321:27–124.

Gurney, R. 1942. *Larvae of Decapod Crustacea.* Wheldon, London.

Ito, T., and F. R. Schram. 1988. Gonopores and the reproductive system of nectiopodan Remipedia. *J. Crust. Biol.* 8:250–253.

Koel, M. A. R., and J. R. Strickler. 1981. Copepod feeding currents: Food capture at low Reynolds number. *Limnol. Oceanogr.* 26:1062–1073.

McLaughlin, P. A. 1980. *Comparative Morphology of Recent Crustacea.* Freeman, San Francisco.

Reaka, M. L., and R. B. Manning. 1981. The behavior of stomatopod Crustacea, and its relationship to rates of evolution. *J. Crust. Biol.* 1:309–327.

Schram, F. R. (gen. ed.). 1983-present. *Crustacean Issues.* Balkema, Rotterdam. [Volumes with in-depth coverage of fields, e.g., phylogeny, larval and adult growth, biogeography, barnacles, feeding and grooming, egg production.]

Schram, F. R. (ed.). 1983. *Crustacean Phylogeny. Crustacean Issues* 1. A. A. Balkema, Rotterdam.

Schram, F. R. 1986. *Crustacea.* Oxford Univ. Press, New York.

Schram, F. R., and C. A. Lewis. 1989. Functional morphology of feeding and grooming in Nectiopoda. *Crust. Issues* 6:115–122.

Southward, A. J. (ed.). 1987. *Barnacle Biology. Crustacean Issues* 5. A. A. Balkema, Rotterdam.

Wenner, A. M. (ed.). 1985. *Larval Growth Crustacean Issues* 2. A. A. Balkema, Rotterdam.

32

Phoronida

A small phylum, only two genera with ten species, these wormlike animals form aggregations that attach to pilings, build tubes in soft bottoms, or burrow in calcareous rocks and shells in shallow coastal waters, especially in temperate regions (Emig, 1982). Similar tubes are found in great numbers in early Paleozoic sandstones; therefore, phoronids may have been extremely abundant in past ages. Phoronids live quiet lives, never emerging from their tubes. The tentacles of the lophophore are thrust out for feeding, and when disturbed, the animal merely retreats into its abode. One species forms aggregations by asexual reproduction, but most phoronids cannot reproduce asexually.

Definition

Phoronida are bilateral animals with a spirally coiled lophophore at the anterior end. The coelom is divided by a septum into mesocoel and metacoel, with some evidence of a protocoel in a larval epistomal flap that covers the mouth. The circulatory system has contractile vessels and blood corpuscles containing hemoglobin. There is a metanephridial excretory system, with nephrostomes draining the coelom. A complete U-shaped digestive tract has the mouth and the anus located at the lophophore. A simple intradermal nervous system exists, with an anterior nerve ring. Gamete production is associated with the peritoneum, gametes escaping through the nephridia. Biradial cleavage results in an actinotroch larva.

Body Plan

The slender, wormlike body expands in an **end bulb** that anchors it in the tube (Fig. 32.1A). A constriction or **collar** sets the trunk off from the upper anterior end on which the lophophore and all conspicuous external features are placed. The body wall is relatively thin; as in many other animals, it is protected by a tube. The circular muscles are delicate, but the longitudinal muscles are strong and complexly folded, crowding many fibrils into a single muscle strand (Fig. 32.1C). A delicate peritoneum invests the muscle layers and lines the coelom (Fig. 32.2).

The **lophophore**, a tentacular ring, is the most conspicuous feature of the body. New tentacles are formed at a median gap, where several short, growing tentacles may be found. Each arm of the lophophore is double, with inner and outer ridges, bearing inner and outer rows of tentacles (Fig. 32.1B). Space for tentacles is increased by spiral coiling; a great many tentacles are present in some species. A ciliated buccal groove between the lophophore ridges leads to the corners of the crescentic mouth.

The slender tentacles are hollow; their walls are a

continuation of the body wall and contain weak longitudinal muscles permitting limited movement. The thickened basement membrane supports the tentacles and makes them rather stiff. Each tentacle contains a branch of the coelomic compartment found in the lophophoral ridge. A flaplike extension of the body wall, the **epistome**, covers the mouth and contains a coelomic cavity continuous with the coelom of the lophophore. This **mesocoel** is divided from the trunk or **metacoel** by a slanting transverse septum that extends from the body wall to the esophagus. The lophophore with the mesocoel may be called the **mesosome**. The trunk with the metacoel may be termed the **metasome**.

The coelomic fluid acts as a hydraulic skeleton. Undoubtedly important in internal transport, it contains albuminous material and a variety of coelomocytes.

A ventral mesentery is formed during early growth of the coelom, and a dorsal and two lateral mesenteries are usually added later. The middle part of the trunk coelom is thus divided into four longitudinal compartments, usually confluent at the proximal end of the trunk (Fig. 32.2).

Digestive System Food passes from the buccal tube into the **esophagus**, which is attached to the body wall by radial muscle fibers and has heavily folded, glandular walls. The **prestomach**, which makes up the rest of the descending arm of the digestive tube, has thinner walls (Fig. 32.1D). The **stomach** lies in the end bulb. A band of robust cilia begins in the prestomach and passes through the stomach. The ascending arm of the digestive tract is made up of the three regions of the **intestine**: a wide intestine in the end bulb, a long, ascending, narrow intestine, and a short rectum ending in the anus.

Excretory System Phoronids have a pair of **metanephridia** that open to the coelom through at least

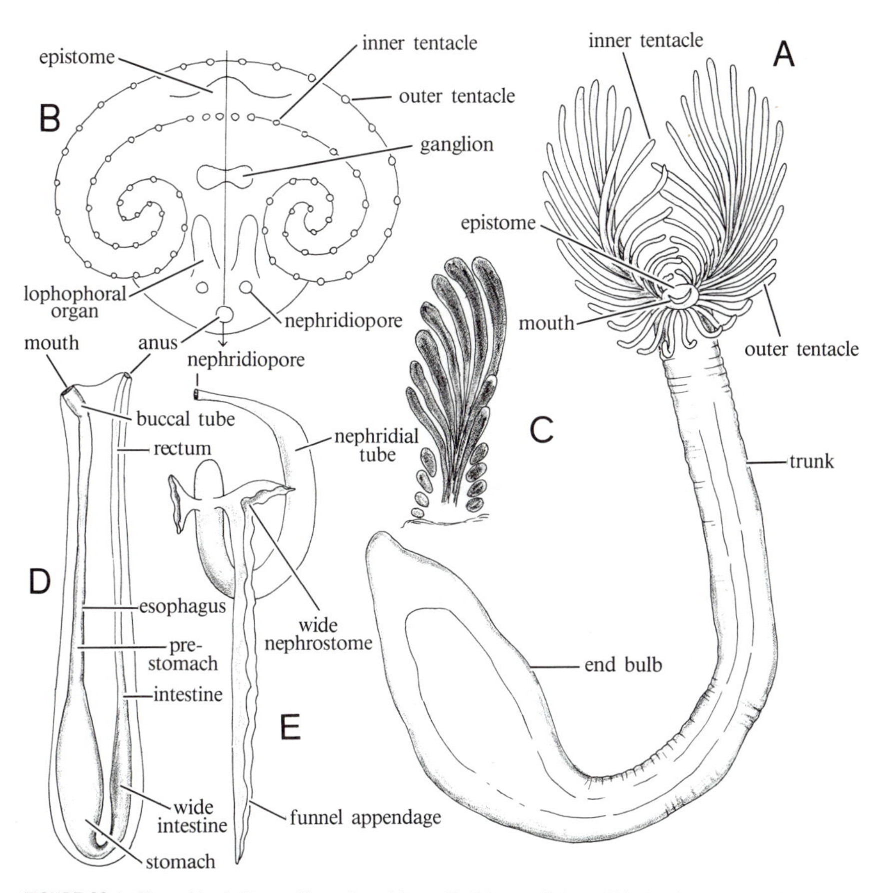

FIGURE 32.1. Phoronids. **A**. Young *Phoronis architecta*. **B**. Scheme of phoronid lophophore organization. **C**. Muscle bundle of *Phoronopsis pallida* showing infolding with associated multiplication of muscle fibers. **D**. Scheme of digestive tract of phoronid. **E**. Metanephridium showing two funnels. (**A** after Wilson; **B** after Sélys-Longchamps, from Dawydoff and Grassé, in Grassé; **C** after Silen; **D** after Hyman; E after Benham.)

one, but sometimes two, ciliated funnels, called **nephrostomes** (Fig. 32.1E). The metanephridial tubules open to the exterior through a pair of **nephridiopores** located near the anus. Gametes as well as wastes escape by way of the nephridia.

Circulatory System The coelom is divided into distinct compartments. The circulatory system is simple and contains blood with red corpuscles holding hemoglobin (Emig, 1982). Each tentacle contains a single blood vessel (Fig. 32.3A), in which blood ebbs and flows. Blood is drained from the **tentacle vessels** by a **lophophore ring vessel** that lies in the lophophore ridge. Two vessels from the lophophore ring vessels join to form the **lateral vessel** after passing through the septum. The lateral vessel runs to the stomach, where it branches to form a **hemal plexus** on the stomach wall. Blood from the hemal plexus drains into the **median vessel**, which carries blood straight to the lophophore. A number of **capillary caeca** branch from the lateral vessel. Gonads cluster about the caeca during their development. There is no heart; the major vessels contract and force the blood along.

Nervous System The nervous system has many primitive characteristics. It is an intraepidermal nerve plexus, conducting impulses in all directions. The nervous layer is thickened to form a dorsal nervous field, from which arises a **nerve ring** paralleling the outer ridge of the lophophore (Fig. 32.3B). Nerve fibers pass from the ring to the outer circle of tentacles, and the inner tentacles are innervated from the preoral field. Motor fibers pass through the septum to innervate the anterior end of the body wall muscle tracts. One or two lateral nerves may be present; if single, it typically lies on the right and is a giant fiber, originating on the right, crossing to the left side of the nerve ring, and then turning to course along the body wall. The giant fiber coordinates the longitudinal muscles for rapid withdrawal into the tube; when it is cut, the muscles below the cut fail to contract.

Reproductive System Discrete gonadal organs are not known. In appropriate seasons, peritoneal cells adhering to the capillary caeca are transformed into gametes, thus forming temporary gonads. Distended fat body cells associated with the capillary caeca seasonally contribute to the nourishment of gametes that are shed into the coelom. Both hermaphroditic and dioecious species are recognized. Gametes escape the body through the nephridia.

Biology

Feeding and Digestion Phoronids are ciliary suspension feeders. The cilia on the lophophore reverse their beat to capture particles (Strathman, 1973; Gilmour, 1978). The resultant water current moves toward and along the lophophoral ridges to the epistome and the mouth, and then flow out over the anus and the nephridiopores. Plankton in the water currents are entrapped and conveyed to the mouth by the beating cilia. Mucus plays no role in actual food capture, but is a factor in particle rejection (Emig, 1982).

There is some ultrastructural evidence of enzyme production in the adult stomach, but there appears to be no evidence of the syncytial region that, in the larva, is the site of intracellular digestion. It seems probable that intracellular digestion is wholly abandoned by the adult.

Phoronis psammophila has been shown to absorb amino acids from seawater (Emig, 1982). Whether this is a widespread phenomenon among phoronids has not been determined.

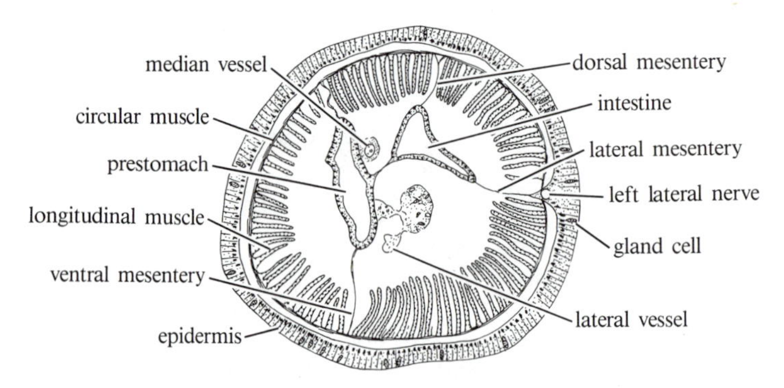

FIGURE 32.2. Transverse section through trunk of *Phoronopsis harmeri*. Note deep muscular ridges. (After Pixell.)

Excretion and Respiration Excretory physiology is unknown, but dark inclusions in the nephridial wall, believed to be wastes, have been seen emerging from the nephridiopore. In addition, white crystals, which pierce the body wall to appear on the surface, are apparently guanin.

There are no special respiratory organs, but the circulation of blood through the tentacles undoubtedly oxygenates the hemoglobin.

Sense Organs Reflecting the reduced form of the central nervous system, the sense organs are also little developed. Clusters of neurosensory cells occur over the body surface, especially on the tentacles and the upper part of the trunk. They are long, thin cells with a surface bristle.

Reproduction Sperm are shed through the nephridiopores. Little is known about the details of spawning. Fertilization is poorly understood, but apparently sperm from the spermatophores can eventually reach the female metacoel and come in contact with the ova (Emig, 1982). Fertilization can be either internal or external, depending on species. The depression between the arms of the lophophore serves as a brood chamber in some species. Lophophoral organs important in reproduction include nidamental glands, which secrete mucus used to attach eggs and young embryos, and accessory spermatophoral organs. A ciliated groove passes from the nephridiopore to the accessory spermatophoral organs; masses of sperm move along these to be compacted into spermatophores.

Development Cleavage is a radial or biradial type, and lacks complete cell determinancy (Emig, 1977). A coeloblastula is formed. Gastrulation is by invagination and leads to the development of an apical sensory tuft. The blastopore forms the mouth. Mesoderm cells move into the blastocoel from the archenteric cells during gastrulation. A specialized kind of trochophore, the **actinotroch** larva, is formed (Emig, 1982). A preoral lobe grows over the originally large blastopore, enclosing a vestibule that opens into the esophagus. The intestine ends in an anus, formed without a proctodeal invagination. A ciliated band appears, along which hollow projections, the larval tentacles, develop (Fig. 32.4A). A posterior ciliated band, the telotroch, forms. Larval tentacles continue to develop, and the larva be-

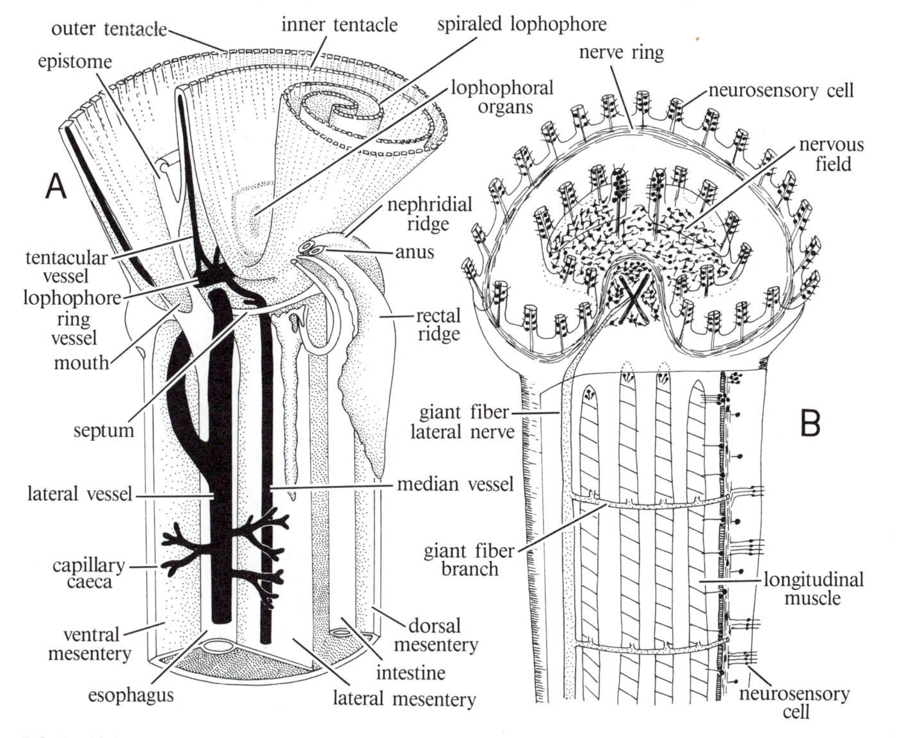

FIGURE 32.3. Phoronid structure. **A.** Anterior end of *Phoronis australis* showing circulatory system and major organs. **B.** Scheme of phoronid nervous system. (**A** after Benham; **B** after Silén.)

comes a fully developed, free-swimming, plankton-feeding actinotroch (Fig. 32.4B).

Coelom formation is complex; patterns differ depending on species (Zimmer, 1980). Three coelomic compartments are formed. The most anterior, preoral coelom forms early and typically disappears entirely, but occasionally may persist. This occupies the apical area of the larva and the epistome, and its remnant tissues develop as muscles and later may contribute to the epistomal coelom of the adult. The posterior trunk coelom arises somewhat later, as a schizocoel; the two ventral arms of the U-shaped space meet to give rise to atypical midventral mesentery. The anterior mesocoelomic compartment arises much later in development, but details of its formation are not clear. The origins of the cells that form the coelomic lining and mesodermal tissues are from the endodermal archenteric cells. These mesoblasts migrate into the blastocoel and come to line it. Mesoderm and coelom formation is thus very unique (Hyman called it a mesenchymal coelom).

The actinotroch undergoes a complex metamorphosis (Fig. 32.4C,D) during which a 90-degree reorientation of the functional body axis occurs (Emig, 1982).

Phylogeny

The fossil record of phoronids is virtually nonexistent except for the suspicious Paleozoic tubes mentioned previously. The aberrant structural plan makes it difficult to assess relationships, and that which has occurred focuses on speculations regarding the nature of the coeloms.

Taxonomy

Only two genera, *Phoronis* and *Phoronopsis*, within a single family are known. Conclusive species identifications require microscopic examination of internal anatomy, especially the nephridia, as well as the form of the lophophore.

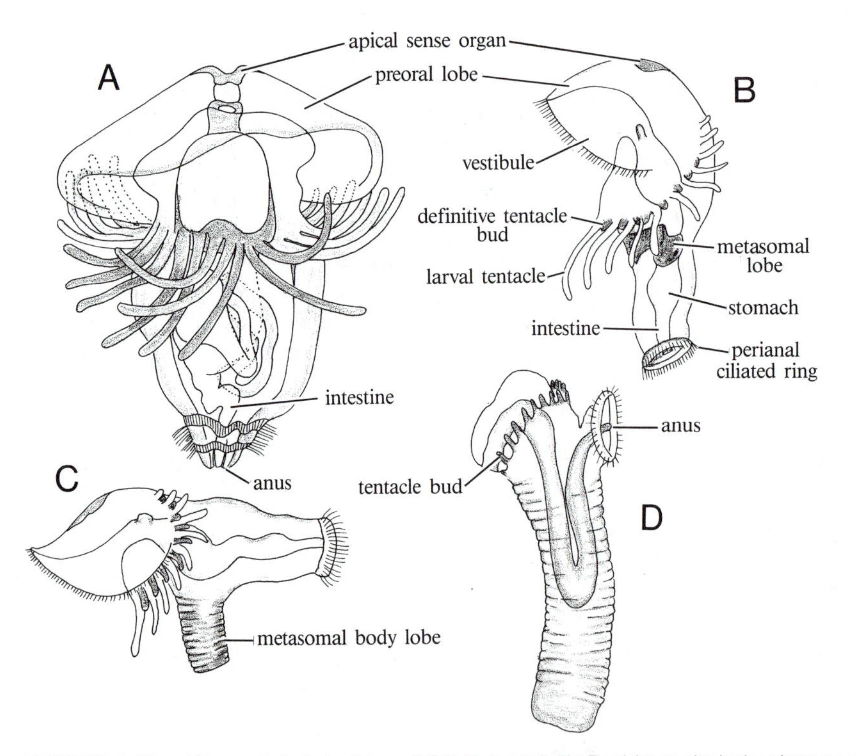

FIGURE 32.4. Phoronid larvae. **A**. Actinotroch larva. **B,C,D**. Metamorphosis: **B**. adult tentacles bud out between gradually degenerating larval tentacles; **C**. trunk region develops metasome lobe that grows down into stalklike extension of larval trunk; **D**. lobe becoming greater part of adult animal. (**A** after Sélys-Longchamps; **B,C,D** after Meeck, from Dawydoff and Grassé, in Grassé.)

Selected Readings

Emig, C. C. 1974. The systematics and evolution of the phylum Phoronida. *Z. zool. Syst. Evolut.-forsch.* 12:128–151.

Emig, C. C. 1977. Embryology of Phoronida. *Am. Zool.* 17:21–37.

Emig, C. C. 1982. The biology of the Phoronida. *Adv. Mar. Biol.* 19:1–89.

Gilmour, T. H. J. 1978. Ciliation and function of the food-collection and waste-rejection organs of lophophorates. *Can. J. Zool.* 56:2142–2155.

Strathman, R. R. 1973. Function of lateral cilia in supension feeding of lophophorates. *Mar. Biol.* 23:129–136.

Zimmer, R. L. 1973. Morphological and developmental affinities of the lophophorates. In *Living and Fossil Bryozoa* (G. P. Larwood ed.), pp. 593–599. Academic Press, London.

Zimmer, R. L. 1980. Mesoderm proliferation and formation of the protocoel and metacoel of early embryos of *Phoronis vancouverensis*. *Zool. Jahrb. Abt. Anat.* 103:219–233.

33

Bryozoans

These animals are perhaps among the most problematic of all in the animal kingdom. Originally they were treated as a single group, the Bryozoa or Polyzoa. Subsequently, they were split by Hyman into two phyla, Entoprocta and Ectoprocta, the former allied with pseudocoelomates, the latter with lophophorate coelomates. Recently, they have been rejoined into a single phylum by Nielsen (1971). Just how closely they might be allied with each other is still debated, however, although our analysis (see Chapter 38) seems to indicate lophophorate affinities for both groups. We treat them here together, but leave them as separate phyla for now.

Entoprocta

Entoprocta (= Kamptozoa, Endoprocta) are curious, sessile creatures, superficially resembling hydroids. They are predominantly marine, but one genus, *Urnatella*, occurs in fresh water in the United States and India. About 120 species are known and they are widely distributed geographically. They are so named because their *anus lies within* the lophophore.

DEFINITION

Entoprocta are bilateral, sessile animals, either colonial or solitary. They are perhaps pseudocoelomate, but with a tentacle ring or lophophore. The U-shaped gut has the mouth and the anus located near the lophophore, but the anus is on an anal cone within the tentacular crown. Entoprocts have protonephridia. Asexual reproduction is used to form colonies. They have spiral, determinate cleavage, and a 4d mesoderm. There is a metamorphosis of the larva to the adult condition, and the larval apical organ degenerates.

BODY PLAN

The adult body consists of a **calyx** and a **stalk** (Fig. 33.1B). During metamorphosis the ventral side of the larva comes to form the upper surface of the calyx. The stalk usually either ends in a padlike basal plate or sends out lateral **stolons** from which other stalks arise to form a colony. Some stalks, however, attach by means of secretions from pedal glands.

A **lophophore**, that is, a crown of ciliated **tentacles**, encircles the upper surface of the calyx (Fig. 33.1D), which contains four midsagittal openings: **mouth**, **nephridiopore**, **gonopore**, and **anus**, in that order. The mouth and the anus lie near the tentacles, whereas the gonopore and the nephridiopore usually lie in a depression (**vestibule**) that serves as a **brood chamber** for the developing young (Fig. 33.1C). A marginal flange, the **velum**, lies just external to the tentacles; it encloses the tentacles when they are folded into the vestibule.

The calyx wall consists of a cuticle, a cellular epidermis, and a few longitudinal muscle strands. Transverse muscles below the stomach compress the body, and muscles from the base of the calyx to the vestibule retract the calyx. The tentacles are simple extensions of the body wall, without a cuticle, and are moved by subepidermal, longitudinal muscles.

Entoprocts are said to have a pseudocoel. The cavity, however, is generally filled with mesenchyme and muscle such that the exact nature of the interior is not clear (Fig. 33.1C).

The U-shaped gut nearly fills the calyx. It is heavily ciliated; food slowly revolves in the ciliary currents. Food accumulates in the **buccal cavity** and is swallowed by contractions of the **esophagus**. A glandular region of the stomach secretes enzymes and mucus, and food particles are trapped in clumps or strands.

A pair of **flame bulbs** lie just above the stomach, opening into protonephridial tubules equipped with large amoeboid athrocytes. The two tubules join just before reaching the nephridiopore.

The tentacles and the velum have many tactile cells (Fig. 33.1A). On each side of the oral end of some entoprocts is a tuft of bristles. The **brain** is a ganglion above the stomach. Pairs of nerves pass to the body wall, gonads, and stalk, and three pairs of nerves branch and connect with ganglia at the base of each tentacle.

Entoprocts may all be hermaphroditic, but most

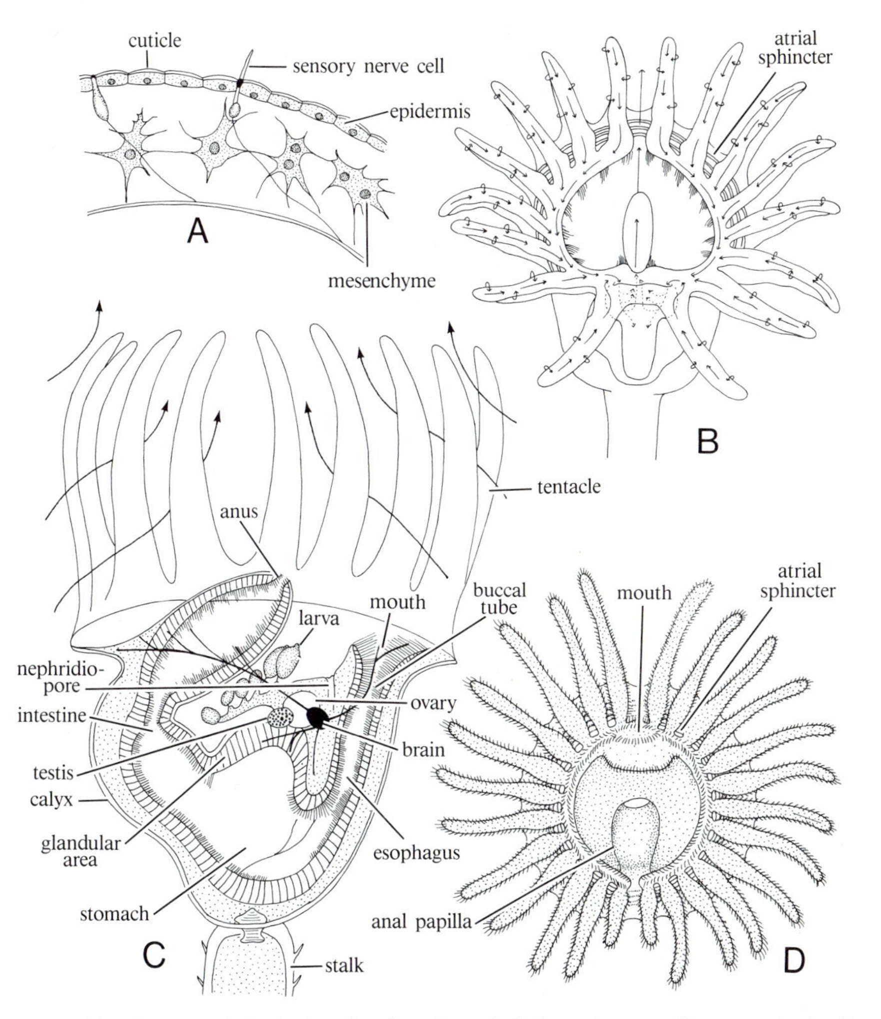

FIGURE 33.1. Entoprocta. **A**. Section through surface of tentacle. **B**. Tentacular crown of *Loxosoma* showing ciliary currents on lophophore associated with feeding. **C**. Scheme of organization of entoprocts, based largely on *Pedicellina*, with currents in surrounding medium generated by lophophore. **D**. Scheme of upper surface of *Pedicellina*. (**A** after Harmer; **B** after Atkins; **C** composite; **D** after Marcus, in Grassé.)

species appear to be highly protandric, and thus in the past the phylum was described as dioecious. They have a single pair of saccate **gonads** with short gonoducts that unite near the gonopore. Monoecious species have an additional pair of gonads and gonoducts, but all gonoducts unite near the common gonopore.

BIOLOGY

Movement The stalk of primitive entoprocts has a sheath of longitudinal muscles, but the muscles are usually concentrated in swollen sections of the stalk (Fig. 33.2A). This limits movement to a quick flicking of the calyx when the otherwise straight stalk is bent at the muscular region. Large or small colonies are formed by stolonic branching, and in many cases the stalk also branches at the muscular enlargements. The stalk is a continuation of the body wall. It is filled with gelatinous material through which amoebocytes wander. A constriction at the calyx base is filled with a plug of cells that separates stalk and calyx "pseudocoels."

Feeding Entoprocts are ciliary-mucus feeders. Long lateral cilia waft diatoms, protozoa, and bits of detritus over the calyx surface, and short cilia on the inner face of the tentacles capture them, sweeping them downward toward the ciliated vestibular groove (Fig. 33.1B) and eventually into the mouth.

Ciliated tracts revolve the mucous food strings and move them into the narrowed intestine. Digestion is extracellular, and absorption occurs in stomach and the intestine. A sphincter at the junction of the intestine and the rectum and at the anus isolates the rectal region. Waste is held here for some time

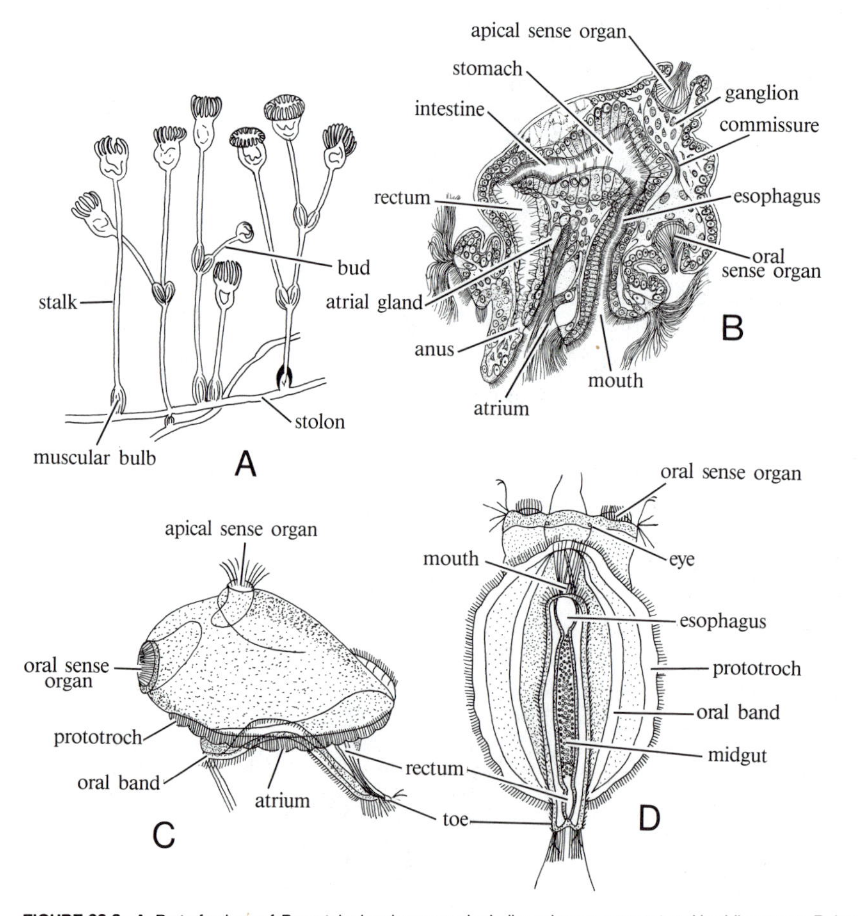

FIGURE 33.2. A. Part of colony of *Barentsia* showing muscular bulbs, where movement and budding occur. B. Larva of *Barentsia*, somewhat like cyphonautes of ectoprocts, although with extensive peculiarities. C,D. Larva of *Loxosomella* seen from side and below. (A after Hyman; B after Marischal; C,D after Cori, from Marcus, in Grassé.)

when the animal is not gorged with food; probably some absorption occurs here also.

Excretion In addition to the osmoregulatory protonephridia, the gut wall contains brownish inclusions of uric acid and guanin that are discharged into the stomach lumen.

Reproduction Asexual reproduction occurs by budding. Solitary species form buds in the calyx, whereas colonial species produce buds only on the stalks or stolon. Buds contain only epidermis and mesoderm, and the new digestive tube is formed from an epidermal vesicle. Stalks regenerate new calyces, and in unfavorable conditions a large number of headless stalks are sometimes found; new calyces form when conditions improve.

Eggs are fertilized in the ovaries or oviducts. As they are shed, the prominent eosinophilous glands cover the eggs with secretion and form a stalk by which the eggs are attached (Mariscal, 1975). The stalk adheres to the depressed vestibular wall in a region known as the **embryophore**. The thickened embryophore wall of *Pedicellina* is filled with food inclusions used by the embyros. Each embryo pushes the next in line further along, so a regular series, decreasing in age, is formed (Fig. 33.1C).

Development Spiral determinate cleavage occurs. Five quartets of micromeres and one set of macromeres are formed. Cell 4d gives rise to endoderm and mesoderm, and the first quartet of micromeres forms the ciliated girdle of the larva. The mouth forms at a stomodeal invagination near the sealed-over blastopore, and a proctodeal invagination also occurs. Mesenchyme cells arise from the stomodeum.

The larvae are curious creatures (Mariscal, 1965), with some resemblance to trochophores and the ectoproct cyphonautes (compare Figs. 33.2B,C and 33.8D,E). It swims or creeps about for a time, and then attaches by means of the ciliated girdle. A complex metamorphosis then occurs. Either the settled juvenile buds off adult tissues, or only portions of the swimming or settled larval tissues actually give rise to lateral or internal buds while the rest of the larva degenerates.

FOSSIL RECORD AND TAXONOMY

No fossil entoprocts are known. Four families are recognized, and these are not clustered into any higher taxa.

Ectoprocta

About 4000 living species of bryozoans are recognized. Despite their small size, ectoprocts are also common fossils; about 4000 fossil species have been described. Ectoprocts occur in freshwater and marine habitats everywhere in the world. They live in shallow waters and to depths of about 6000 m, and are abundant in both polar and tropical seas.

Like phoronids, ectoprocts secrete a covering for themselves, but whereas the phoronid and its tube remain separate, the ectoproct case is an integral part of the body wall. Ectoprocts are colonial. Colonies either form gelatinous or firm encrusting masses or are aborescent. They are so named because their *anus lies outside* the lophophore.

DEFINITION

Ectoprocta are bilateral, sessile, colonial coelomates, with a surface skeleton or zoecium. They have a lophophore. The U-shaped gut has a mouth and an anus near the lophophore, but the anus lies outside the tentacular crown. Simple gonads are derived from the peritoneum. Ectoprocts have a high capacity for asexual reproduction to form colonies. They lack nephridia, have a funiculus, and possess a nervous system with a central ganglion in a ring coelom. Ectoprocts have radial, determinate cleavage, with a marked metamorphosis of the larva to the adult condition.

BODY PLAN

Each individual of the ectoproct colony is termed a **zooid**. The zooid contains a spacious coelomic cavity (Fig. 33.3). The lophophore contains a separate coelom, and in phylactolaemates the epistome also contains yet another coelom. Once thought to be separate, it appears that these three coelomic areas are commonly continuous with each other. Zooid form is extremely variable. The **zoecium**, or colony skeletal framework, may be gelatinous, membranous, or strongly calcified. The mechanism for withdrawing and extruding the lophophore varies with the flexibility of the zoecium and involves extensive modification of the form of the wall and the details of the musculature. An important source of variability is the extensive polymorphism of some marine ectoprocts. The following description relates only to the unmodified, feeding polyp (autozooid).

Skeleton Zoecium is the term applied to the specialized skeletal framework formed from the cuticle.

A formidable vocabulary is used to describe detailed differences in the zoecium of different ectoproct orders, and much of the classification depends on zoecial traits. The zoecium can only be discussed meaningfully for each group separately.

Phylactolaemata are freshwater species with either a delicate or a tough membranous zoecium, although some form massive colonies with gelatinous zoecia. Membranous zoecia are built of chitin. Each zoecium is tubular, with a circular to subcircular orifice through which the lophophore protrudes. The zoecium wall is never complete; therefore, the coelom of the whole colony is continuous, but partial divisions are usually present.

Gymnolaemata zoecia are more complete. The zooids are still connected by rosettes of tiny interzooidal pores in the zoecial wall, but these openings are plugged by epidermal cells. The main body wall (Fig. 33.4A) consists only of the zoecial wall, the epidermis, and the peritoneum; the musculature has been reduced to just the parietal muscles used to manipulate the lophophore.

The simplest zoecia are seen in some Ctenostomata. They are cylindrical or vase-shaped cases with a thin, chitinous wall and no calcareous layer (Fig. 33.3J). Ctenostomata are considered the most primitive gymnolaemates, whereas Cheilostomata are more advanced.

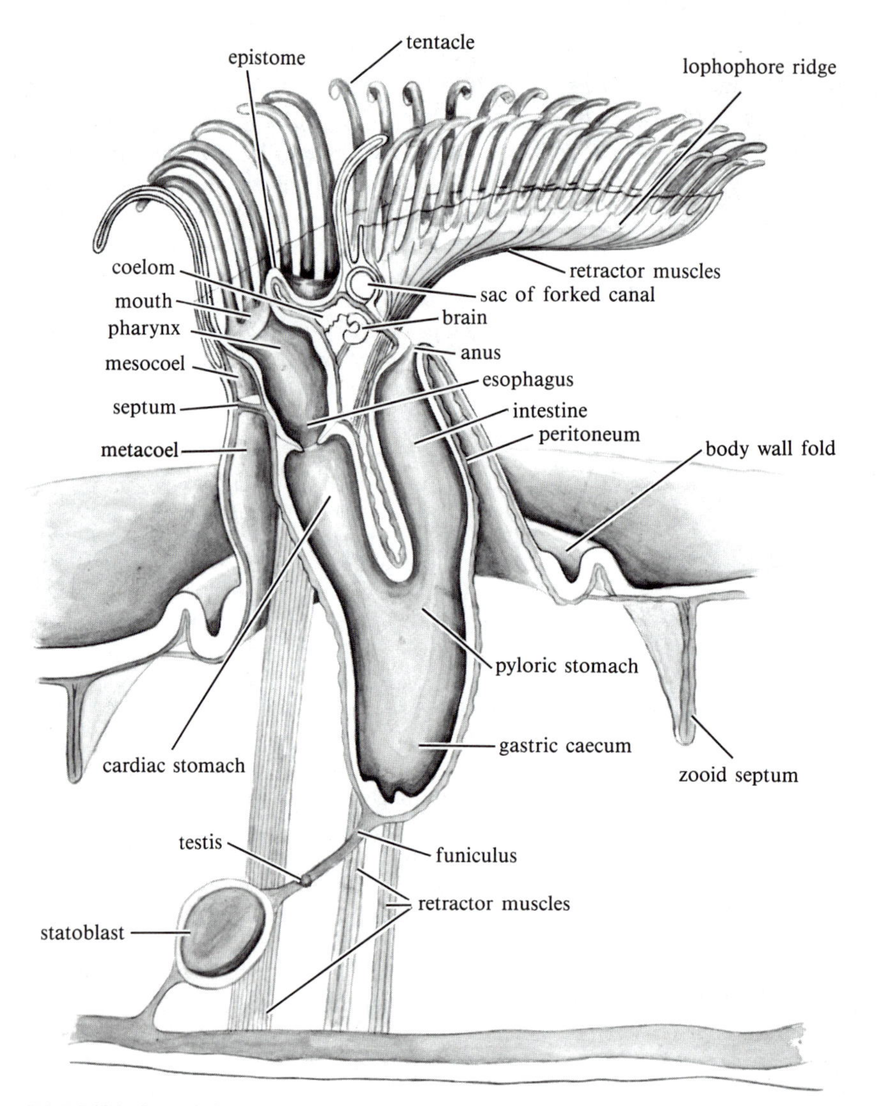

FIGURE 33.3. Scheme of phylactolaemate *Cristatella* anatomy. (Adapted from Cori.)

The zoecium provides protection and support; both are increased by thickening the zoecial wall. A common cheilostome trait is a hinged lid (**operculum**) that can cover the orifice when the lophophore is withdrawn. The major adaptational trend in this group has been toward strengthening the zoecial wall by means of calcification. This interferes with the mechanism for extruding the lophophore that is always achieved by a hydraulic mechanism. Pressure is exerted on the coelomic fluid to force the lophophore out, like a jack-in-the-box. Phylactolaemates and ctenostomes generally have an elastic zoecial wall that can be easily compressed. Where the zoecial wall is inelastic, a special mechanism is required.

In many cheilostomes, all but the frontal zoecial

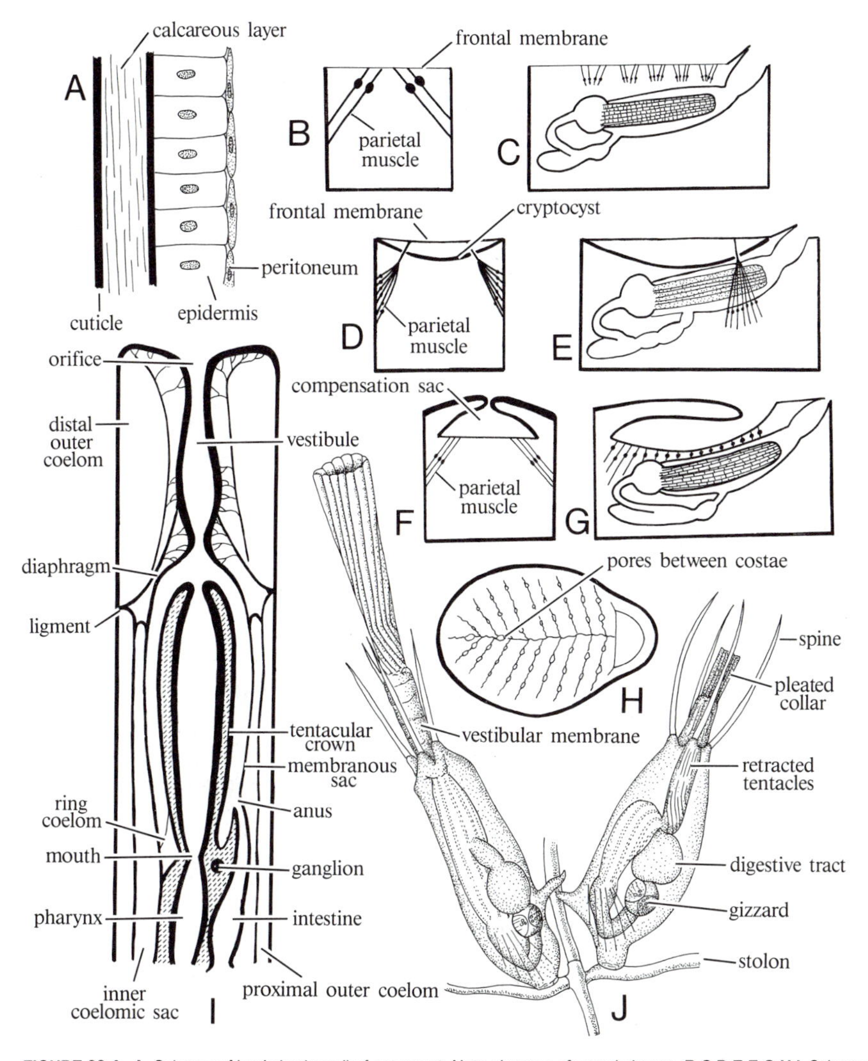

FIGURE 33.4. A. Scheme of basic body wall of ectoproct. Note absence of muscle layers. **B,C,D,E,F,G,H,I**. Scheme of extrusion mechanisms of various types of ectoprocts. **B,C**. End and side views of simple zoecium with frontal membrane. When parietal muscles depress frontal membrane, zooid extrudes. **D,E**. End and side views of zoecium with cryptocyst wall located beneath frontal membrane. In such ectoprocts, extrusion mechanism resembles forms with simple frontal membrane, although parietal muscles must pass through concave cryptocyst wall. **F,G**. End and side views of zoecium with compensation sac. In this case, parietal muscles depress lower membrane of compensation sac. Water flows into compensation sac and coelomic pressure rises, extruding zooid. **H**. Top view of zoecium with compensation sac showing pores between partially consolidated costae that protect sac. **I**. Scheme of cyclostome construction showing extrusion mechanism. Contraction of vestibule dilators forces fluid from distal, outer coelomic space into proximal, outer coelomic space. This raises pressure in inner coelomic space into proximal, outer coelomic space. This raises pressure in inner coelomic space and extrudes zooid. **J**. Two *Averrillia* zooids showing pleated membranes often seen in Ctenostomata. (**A,B,C,D,E,F,G,H,I** modified, from various sources; **J** after Marcus.)

wall is calcified. The frontal wall remains an elastic, chitinized, frontal membrane (Figs. 33.4B,C and 33.10D) that covers all or part of the exposed surface. Parietal muscles attach to the firm zoecial wall and the frontal membrane; they depress the frontal membrane to extend the lophophore (Fig. 33.4B). The frontal membrane is the unprotected part of the cystid wall; thus many species partially protect it by spiny outgrowths of the hard parts of the zoecium. Some cheilostomes have a calcareous shield, the cryptocyst, below the frontal membrane (Fig. 33.4D,E). Parietal muscles pass through openings and attach to the frontal membrane (Fig. 33.4D,E), permitting the same mechanism of lophophore extrusion as in species without a cryptocyst. Some cheilostomes have a compensation sac in place of the frontal membrane; in these the whole frontal wall is calcified except for the orifice into the compensation sac (Fig. 33.4F,G,H). When parietal muscles depress the floor of the compensation sac, the lophophore is forced out while water rushes into the compensation sac from without.

The Stenolaemata has only one living order, the Cyclostomata. The tubular zoecium of cyclostomes is completely calcified and has no frontal membrane or compensation sac. An entirely different system of polypide extrusion is used. The orifice is covered by a terminal membrane, turned in to line a saclike vestibule (Fig. 33.4I). The coelom is divided into inner and outer parts by a membrane. When the dilator muscles extend the vestibule, coelomic fluid around the vestibule in the outer coelomic chamber is forced down around the inner coelomic sac. This increases pressure in the inner coelomic sac and forces the lophophore out.

The ectoproct body wall is everywhere firmly attached to the zoecium, just as epidermis clings to the cuticle in other kinds of animals. The zoecium is, in a sense, a highly modified cuticular covering.

Zooids Zooids refer to the ectoproct animal. The literature distinguishes between the body wall tissues, or **cystid**, and the business part of the animal, or **polypide**, that is, the lophophore and the internal organs. The crescentic to horseshoe-shaped phylactolaemate **lophophore** carries more tentacles than the circular cyclostome and gymnolaemate lophophore, and an **epistome** is found only in phylactolaemates. Ciliated currents caused by the lophophore have been described in detail. The cilia on the lateral surfaces of the tentacles pull water down and out between the tentacles, directing particles toward the mouth (Fig. 33.5A). Large particles are rejected by retraction or muscular movements of the tentacles, and small particles are delivered to the mouth by ciliary currents at the lophophore base.

Ectoprocta are colonial organisms and have tremendous powers of asexual reproduction and regeneration. The individual zooids are exceptionally small, and in some cases, specialization of zooids for particular functions has occurred. Feeding zooids are called **autozooids**, but other zooids specialized for specific functions are known.

Avicularia are zooids found only in cheilostomes. They are essentially monstrous opercula attached to greatly reduced zooids. *Bugula* avicularia consist of a peduncle of variable length, attaching the zooid to the colony, a head that is actually the modified zoecium, and a mandible that is a modified operculum (Fig. 33.6A,B). Avicularian form is variable and sometimes several kinds occur in a single species. The reduced parts of the polypide and cystid can be

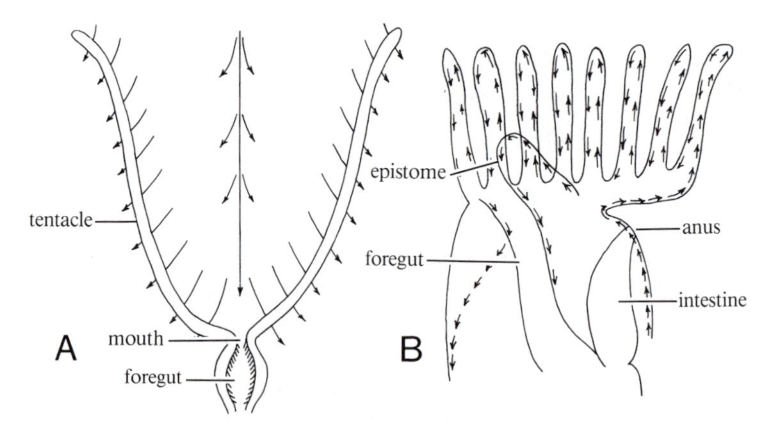

FIGURE 33.5. Ciliary currents. **A**. Currents in gymnolaemate lophophore sweeping food in and toward mouth. **B**. Coelomic ciliary currents in lophophore of phylactolaemate, *Pectinatella*. (**A** after Borg; **B** after Oka.)

seen in the interior (Fig. 33.6A). The most conspicuous internal elements are the opening (abductor) and closing (adductor) muscles. Avicularia snap viciously when disturbed; they grasp and hold small organisms that visit the colony, and probably prevent fouling of the colony surface by other sessile organisms.

Vibracula (Fig. 33.6C) are also highly modified zooids with the operculum altered into a long, movable bristle. Muscles similar to those of avicularia manipulate the bristle, sweeping the colony surface free of particles or organisms that cling to it.

Kenozooids are extremely variable. All are modified zooids that serve as stolons, rhizoids, or other hold-fast parts.

Gonozooids are zooids specialized for reproductive uses, and are important colony components in cyclostomes.

Digestive System In the autozoid digestive system, the short, ciliated **pharynx** leads to a long, delicately muscled **esophagus** (Fig. 33.3). A **cardiac valve** usually separates the esophagus and the **stomach**. Some ctenostomes have a **gizzard**, containing grinding denticles, at the base of the esophagus (Fig. 33.4J). Other ectoprocts have three stomach regions: cardia, caecum, and pylorus. The caecum forms the bottom of the U, and the **rectum** is sharply set off from the pylorus.

Respiratory and Excretory Systems There are neither special respiratory nor excretory organs in ectoprocts.

Coelom Ectoprocts are small, and the coelomic fluid is, for the most part, adequate for internal transport within zooids. The peritoneum is partially ciliated, and a constant, definite circulation of coelomic fluid is maintained (Fig. 33.5B). Examination of the **funiculus**, the tubelike extension of the body wall with muscle fibers and that is attached to the stomach, reveals fluid-filled cylinders with walls of peritoneal cells, which in phylactolaemates and cyclostomes also have bundles of longitudinal muscle fibers within the tubes. The funicular system is associated with statoblast formation in phylactolaemates. The funiculus is apparently a homologue of the circulatory system of brachiopods and phoronids (Carle and Ruppert, 1983).

Nervous System There are no special sense organs, but sensory cells provide for reception of tactile, chemical, and water current stimuli. The nervous system is simple (see Lutaud, in Woollacott and Zimmer, 1977). Gymnolaemates have a **nerve ring** around the pharynx. A small **cephalic ganglion**, with peripheral neurons and central neuropile, is attached to the ring and lies within the lophophoral coelom. Two motor and two sensory fibers pass to each tentacle from the ring. Pairs of sensory and motor fibers pass to the tentacle sheath from the ganglion, probably connecting with the subepidermal nerve plexus. Motor nerves to the various muscles and to the gut arise from the tentacle sheath nerve or cephalic ganglion. The cystid wall contains a nerve net, which has been reported to pass through the pores to connect with that of adjacent zooids.

The nervous system of phylactolaemates is somewhat different. The hollow **cerebral ganglion** lies in a peritoneal vesicle open to the coelom. A pair of ganglionated tracts parallel the lophophore, issuing sensory and motor nerves to the tentacles (Fig. 33.7B). The epistome on the one side and the mouth and the pharynx on the other interfere with the innervation of the tentacles in the middle of the lophophore. Two nerve rings, a **circumoral ring** and an **epistomial ring**, surmount this difficulty. Nerves to the tentacle sheath, the pharynx, and the body muscles arise from the ganglion.

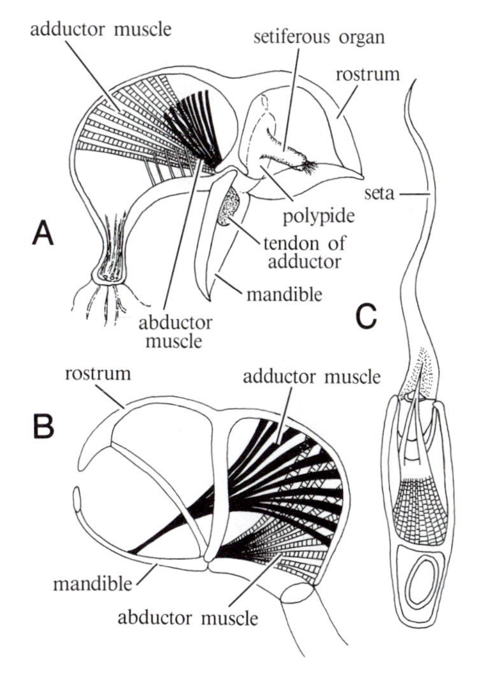

FIGURE 33.6. Some specialized zooids. **A**. Avicularium of *Bugula*, with mandible open, showing reduced polypide within. **B**. *Bugula* avicularium with mandible closed. Antagonistic muscles *shaded.* **C**. Vibraculum of *Scrupocelaria*. (**A** after Calvert and Marcus, from Cori in Kükenthal and Krumbach; **B,C** after Hincks.)

Most marine ectoprocts are protandric hermaphrodites; uncommonly, different members of a colony are of different sexes. The **gonads** develop in the peritoneum of the coelom, on the body wall, gut, or funiculus. All phylactolaemates are hermaphroditic with **ovaries** on the body wall and **testes** on the funiculus. Ova may be released through a supraneural pore or an intertentacular organ, and in some cases sperm escape through open tips of the two dorsomedial tentacles. In other cases, embryos are brooded within the coelom and escape with the degeneration of the adult polypide, whereas in cheilostomes a special brood chamber is common.

BIOLOGY

Locomotion and Movement Ectoprocts have few muscles whose function consists largely of withdrawing or extruding the lophophore and manipulating the tentacles. Special muscles close the operculum, and move the epistome when it is present. The most conspicuous muscles, other than the parietal muscles previously mentioned, are the lophophore retractors, large strands of muscle on each side of the body that connect the lophophore base to the bottom or sides of the skeleton.

An unusual ability to move is seen in some phylactolaemates. The circular and longitudinal muscles on the bottom of the body wall are strongly developed, forming a creeping sole (Fig. 33.7A). Colonies creep about, apparently by alternate contraction and relaxation of these muscles, sometimes with the aid of currents produced by the lophophores.

The operation of the coelomic hydraulic system to extrude the polypide lophophore demonstrates some of the advantages of a hydraulic system. Muscles can only shorten and pull on an object; they can never push something directly. This is a severe limitation when something must be lifted from the body surface. Pressure in the lophophore coelom inflates the tentacles as the lophophore is extruded. To achieve the same result by other means, a very complex arrangement of levers would be required.

Feeding and Digestion Bryozoans filter food particles from the surrounding water by altering the direction of beat of the impinging cilia, effectively reversing it to bring the particles into position for movement to the mouth (Strathman, 1973, 1982). Food accumulated in the pharynx is swallowed by the esophagus. The stomach and the caecum are very active, and food moves back and forth between the caecum and the pylorus. Food is entangled in an elongated mucous cord that is rotated by ciliary currents up to 150 times per minute in the caecum and the pylorus. Digestion is very rapid in phylactolaemates; food, when plentiful, may be no more than an hour in the digestive tract. Digestion is correspondingly inefficient, however, for an occasional rotifer or nematode will simply shake itself and swim away after having been through the whole process and voided through the anus. Digestion is wholly extracellular in phylactolaemates, but food vacuoles are formed in the cardiac and the caecal epithelia of some gymnolaemates. Mucus is used to consolidate the residues into fecal balls, an important progressive step and useful, as the anus is so

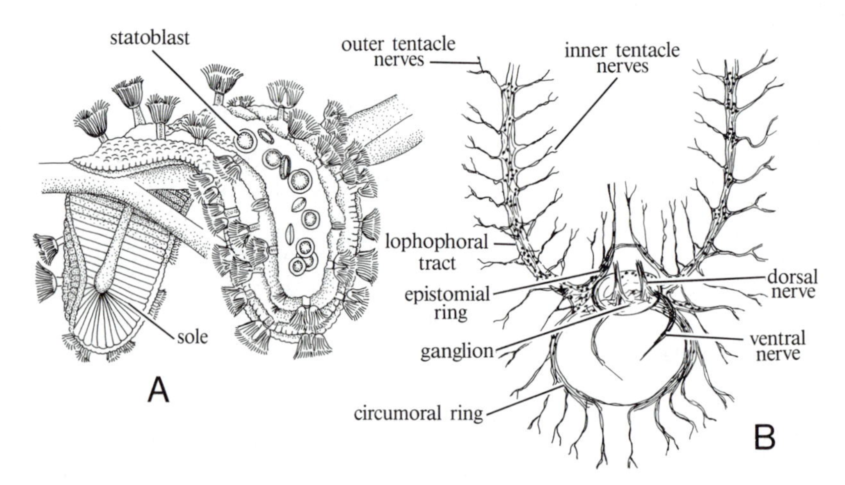

FIGURE 33.7. **A**. *Cristatella mucedo* creeping over twig. Note muscular creeping sole of colon. **B**. Nervous system of *Cristatella*. (**A** based on Allmann; **B** after Gewerzhagen.)

close to the lophophore. The lophophore is bent away at the time of defecation.

Starch is digested slowly and deposited as glycogen in the gymnolaemate *Zoobotryon*. Proteins are hydrolyzed in the stomach and are probably absorbed there. Fat globules are phagocytized by stomach epithelial cells. Glycogen and fat are eventually distributed throughout the body, and food reserves are laid down in the stolon as fat, glycogen, and protein masses.

Excretion There are no excretory organs, although it has been suggested that the forked canals connecting the bases of the inner tentacles to the coelom may be vestiges of nephridia. Not many have accepted this view, but there is evidence that nitrogenous wastes build up there and are ejected. Yellowish inclusions in the stomach wall are also generally thought to be wastes, and coelomocytes appear to pick up wastes for storage excretion.

Polypides do not live long and are replaced by regeneration. An exhausted polypide contracts powerfully; its tissues break down and some are phagocytized, but the stomach tissue balls up to form a brownish mass, the **brown body**. During life the stomach wall builds up dark inclusions, perhaps of an excretory nature, which give the brown body its color. A bud from the cystid wall develops into a new polypide. In some species the brown body remains in the coelom; sometimes several are present, attesting to several cycles of degeneration and regeneration. In others, the brown body is surrounded by the forming gut and is eventually discharged through the anus. Senility and physiological depression resulting from poor environmental conditions cause degeneration of the polypide, but some believe that it is also caused by accumulation of excretory products. Whatever causes degeneration, brown body formation may be an unusual kind of storage excretion, for waste materials appear to be incorporated into a body part that will eventually be shed or be retained in an inert condition. Regeneration in some ectoprocts, however, results in a breakdown of the brown body, and for some species the brown body may be a source of stored reserves.

Respiration The exchange of respiratory gases occurs at the body surface, no doubt largely through the lophophore.

Reproduction A few ectoprocts shed ova into the sea, but most marine and all freshwater species brood the young. Phylactolaemates breed in late spring and early summer. The gametes unite in the coelom, and the young develop in an embryo sac that forms on the body wall near the ovary. Sperm are shed through the tips of the tentacles.

Most cheilostomes have ovicells, essentially a hoodlike device into which ova are deposited (Fig. 33.9E). Ova removed from ovicells do not develop normally; evidently a special environment is maintained for the young. Cyclostome ova develop in gonozooids, forming a germinal ball that develops surface lobes, pinched off to form secondary embryos. The formation of several embryos from a single ovum is called **polyembryony**.

Freshwater ectoprocts reproduce, of course, asexually during colony formation, but also reproduce asexually by means of **statoblasts**. Buds appear on the funiculus. They grow and differentiate into a dark, resistant covering and a mass of yolky, undifferentiated germinal cells. Three kinds of statoblasts are recognized: floatoblasts with air-filled cells making up the capsule wall, sessoblasts attached to the zoecium wall by cement, and spinoblasts with spiny projections. Statoblast form is characteristic of the species and is used in species identification. Statoblasts are liberated when polypides degenerate. They are resistant and protect the species during the winter or when low water levels make conditions unfavorable. They are also important in the dissemination of the species. Sexually produced larvae do not swim far and in any case cannot reach another pond or lake. Statoblasts can be carried by air or water currents for long distances, can pass unharmed through the digestive tract of some aquatic animals, and can cling to the surface of animals, thus spreading the species widely. When conditions are right, cells of the statoblast begin to grow, protrude from the capsule wall, and produce a young polypide.

Development In phylactolaemates, holoblastic cleavage produces inside the embryo sac a coeloblastula. No endoderm seems to be formed. Mesoderm arises from the ectoderm, and arranges itself to form a coelom. The early embryo is a cystid, which buds off two or more polypides, develops cilia, and emerges as a free-swimming larva (Fig. 33.8A). It settles down after a brief migratory phase and develops into a new colony.

Gymnolaemate and cyclostome cleavage is apparently radial or biradial, and produces a coeloblastula (Fig. 33.9B). Four cells elongate and move into the interior, giving rise to endoderm and endomesoderm. Large ciliated coronal cells form a girdle around the embryo, an apical organ begins to form, and several ectodermal invaginations appear. These become an adhesive sac and a pyriform organ. A vibratile plume develops in connection with a ciliary cleft. Ova developing externally sometimes form a plankton-feeding **cyphonautes larva** (Fig. 33.8D,E). Modified coronate larvae that subsist on yolk are formed by some brooding species of cheilostomes. After swimming about for a time, the larva uses the vibratile plume as a sensory tuft in selecting a suitable site for attachment. The adhesive sac is everted and sticks the larva to the spot selected. The pyriform organ may also secrete an adhesive substance.

After attachment, the larva undergoes metamorphosis to produce the first zooid of the new colony. In cyphonautes larvae and the coronate *Bugula neritina*, the first zoecium arises from the adhesive sac, and the polypide develops from two primordia originally associated with the apical complex of the larva. In *Bowerbankia* the adhesive sac eventually degenerates after serving to attach the larva, and various apical elements metamorphose into the adult. Daughter zooids are budded off in whatever pattern is characteristic of the species. Budding patterns are complex and varied; therefore, a great variety of colony forms are known.

FOSSIL RECORD

The fossil record of ectoprocts is as extensive as the modern forms. Equal numbers of species are known in each, but as more sophisticated techniques for studying fossilized anatomy are utilized, more species will be recognized (Ross, 1987). Most fossils are assigned to the Stenolaemata, which contains three

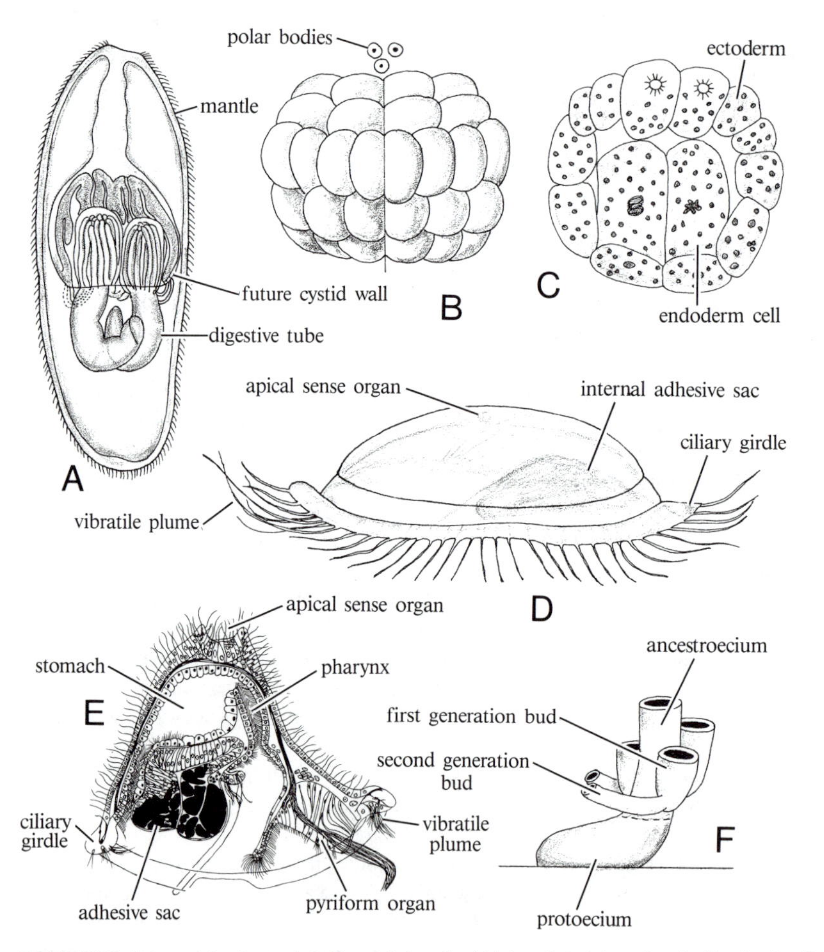

FIGURE 33.8. Ectoproct development. **A**. *Plumatella* larva in which two phylactolaemate polypides develop. **B**. *Bugula* embryo in 64-cell stage. **C**. Section through *Frustrellidra* embryo during gastrulation. **D**. Cyphonautes larva of *Frustrella*. **E**. Sagittal section through cyphonautes larva. **F**. Scheme of early development of *Prasopora* colony. (**A** after Brien; **B** after Correa; **C** after Pace; **D** after Barrois; **E** after Kupelwieser; **F** after Cumings.)

extinct orders in addition to the living cyclostomes. The great post-Cretaceous radiation of great numbers of cheilostome species is apparently related to the evolution of the coronate larva that is brooded in the parent for some time before shedding (Taylor, 1988). The limited dispersal capabilities of these larvae appears to be linked to limited gene flow among populations, and thus fosters increased levels of speciation. Brooding structures, such as ovicells, first appeared in cheilostome fossils in the middle of the Cretaceous just prior to the major radiation of this order.

PHYLOGENY

Although entoprocts and ectoprocts have been considered as separate groups for several decades, it does appear that they have some striking similarities and that their differences are not irreconcilable (see Nielsen, 1971, and Chapter 38). Entoprocts have spiral cleavage, whereas ectoprocts appear to be radial, although some species of *Loxosomella* apparently digress from a strict spiral pattern. The entoproct larva is similar to that of the ectoproct cyphonautes. Both larvae are bell-shaped, with an

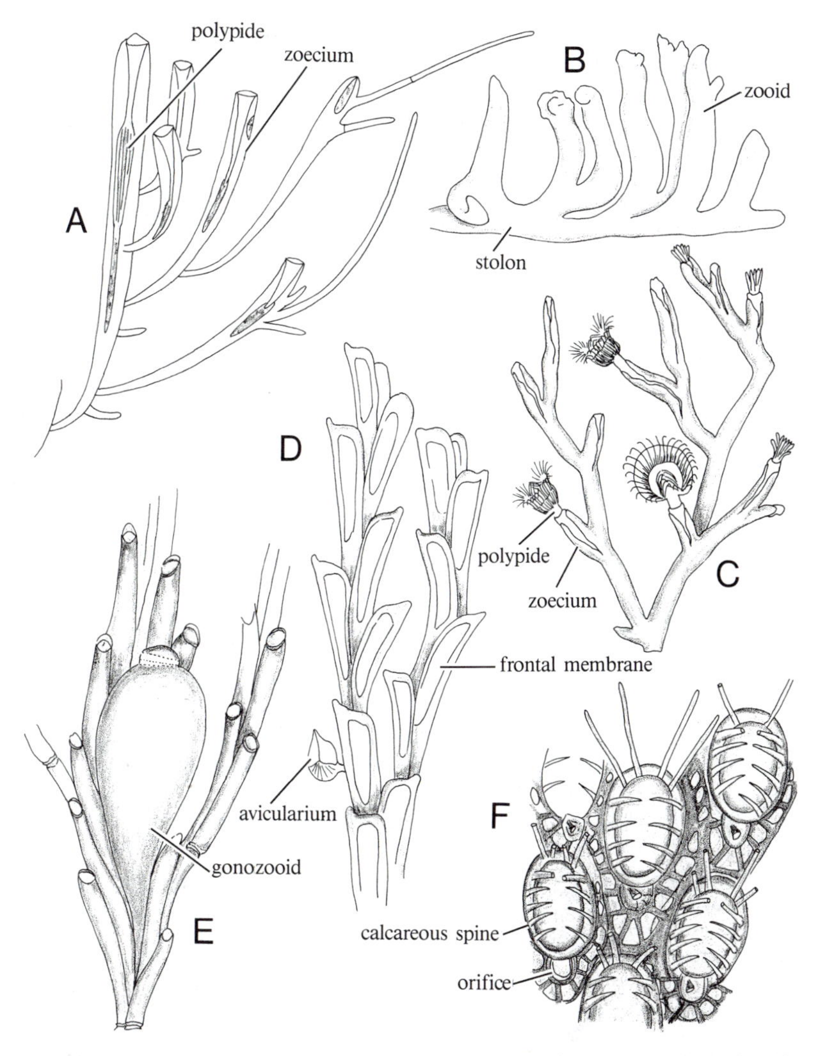

FIGURE 33.9. Representative ectoprocts. **A.** *Victoriella,* gymnolaemate, Ctenostomata, with simple, chitinous zoecium. **B.** *Paludicella,* another ctenostome. **C.** *Fredricella,* arborescent phylactolaemate. **D.** *Bugula,* cheilostome. **E.** *Crisia,* cyclostome. **F.** *Membranipora,* encrusting cheilostome. (**A** after Braem; **B,C** after Rogick, in Edmondson; **D** after Hyman; **E,F** after Robertson.)

encircling girdle of cilia on the bell margin, both have preoral and pyriform organs in front of the mouth, both possess well-developed apical sense organs, both utilize a ventral atrium or adhesive sac, and both undergo an abrupt larval metamorphosis into the adult.

Entoprocts are said to be pseudocoelomate, although the "cavity" is so small and packed with mesenchyme that it might just as well be viewed as acoelomate. The lack of a well-defined cavity in entoprocts is undoubtedly related to the economy of space in the body plan and the overall small size of the animals, and in this respect they are reminiscent of the coelomate nemertines and pseudocoelomate gastrotrichs, which also fill their body cavities with mesenchymal tissues. Ectoprocts are coelomate. The "coelomic" nature of the ectoprocts is problematic, however, as body cavities are lacking in the larvae, and the massive metamorphosis to the adult condition sees a complete de novo generation of adult tissues. Therefore, the ectoproct adult coelom bears little in common with that of other coelomates in terms of its genesis.

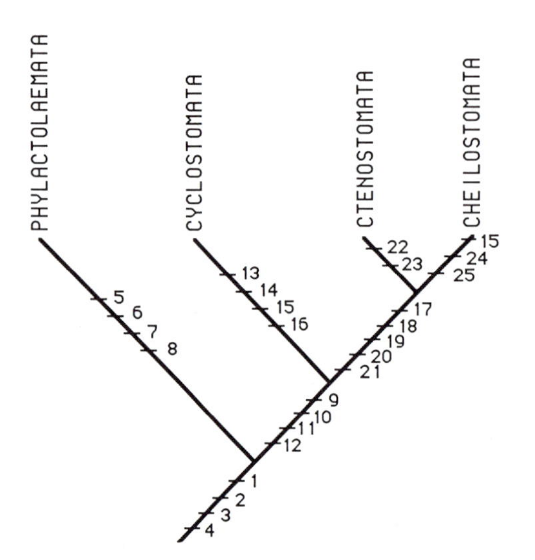

FIGURE 33.10. Cladogram illustrating relationships among ectoprocts. Derived characters are: (1) colonial; (2) funiculus, blood vessels, with longitudinal muscle cells and noncontractile epithelium; (3) nephridia lost; (4) cuticle developed as zoecium; (5) statoblasts; (6) larva as juvenile colony; (7) brooding in embryo sac; (8) polypides bud first; (9) two-stage metamorphosis; (10) funiculus without basal lamina; (11) loss of epistome; (12) cystids bud first; (13) membranous sac; (14) brooding occurs in gonozoid; (15) calcareous body wall; (16) polyembryony; (17) loss of body wall muscle; (18) funiculus without longitudinal muscle cells; (19) multiple funiculi with each autozooid; (20) septula; (21) cyphonautes larva; (22) pleated collar orifice; (23) brooding in introvert; (24) boxlike zoecium; and (25) operculum. (Modified from Carle and Ruppert, 1983.)

The body parts of the two groups are similar. Entoprocts, however, are ciliary-mucus feeders, whereas ectoprocts are ciliary-filter feeders. The nervous systems are similar, and both arise as ectodermal invaginations of the mouth region. The sperm of entoprocts and that of the gymnolaemate ectoprocts is also similar, but also of a general type like those sperm generally found among "protostomes." Both groups produce resting bodies or cysts or hibernaculae.

Taken together, there do seem to be some compelling reasons to reunite the groups. Some workers would derive the ectoprocts from the entoprocts; others would see the entoprocts as degenerate ectoprocts. Problems nevertheless remain. The general spiral determinate cleavage of entoprocts is difficult to reconcile with the radial but apparently determinate cleavage of ectoprocts. The former would seem to ally with protostome types, whereas the latter, along with their lophophore, would seem to fit in with phoronids and brachiopods. Furthermore, ectoprocts have a ciliary feeding mechanism using upstream capture of particles, identical to that seen in phoronids, brachiopods, pterobranchs, enteropneusts, and echinoderm larvae (Strathman, 1973). Nevertheless, both groups are so peculiar in terms of their structure and development, and need to have so much more known about their body plans and biology, that the issues of their interrelationships and position with regard to other groups should probably not be forced at this time.

The ectoprocts alone, however, seem to clearly bear connections with other lophophorates. Some of the characters uniting lophophorates, such as the lophophore itself and the U-shaped gut, are shared with other groups, such as pterobranch hemichordates (see Chapter 38). Nonetheless, the phoronids, brachiopods, and ectoproct bryozoans seem related in some way. The brachiopods and phoronids may in fact have a common ancestry based on their sharing monociliary epidermal cells and an intraepidermal nervous system, but both of these characters are throwbacks to primitive conditions seen among acoelomates and pseudocoelomates, and it is unclear whether these characters might not be independently developed homoplasies in these phyla. The ectoprocts form a monophyletic group (Fig. 33.10) within the lophophorate line. Their autapomorphies, unique advanced characters, are focused on achieving a communal, sessile life-style.

Taxonomy

Ectoprocts have an extensive fossil record with many extinct higher taxa, but only modern forms are given here.

Class Phylactolaemata. Zooids with horseshoe-shaped lophophore; with epistome; body wall with musculature; zoecium incomplete and not calcified; coelom of adjacent zooids continuous; without polymorphic zooids; exclusively freshwater species. Four families generally not grouped into higher taxa (Fig. 33.9C).

Class Stenolaemata. One living order.

Order Cyclostomata. Tubular, fully calcified zoecium having terminal, typically circular orifice, with operculum but orifice closed by perforated membrane; coelom divided by membranous sac into inner and outer regions used in lophophore extrusion; polymorphism common, with formation of gonozooids, kenozooids, and nannozooids; many young formed by polyembryony from single embryo in brood chamber (Fig. 33.9E).

Class Gymnolaemata. Lophophore circular, without epistome; body wall musculature reduced to parietal muscles; funiculus lacking longitudinal muscle cells, multiple funiculi in autozooid; without direct connections between coeloms of adjacent zooids; cyphonautes larva.

Order Ctenostomata. With simple, flexible zoecium of chitinous material, having terminal or subterminal orifice; without closing plate, but often with pleated, collarlike membrane on diaphragm that occludes vestibule; polymorphism restricted to ordinary feeding zooids and specialized zooids for stolonic attachment; mostly marine, but including some freshwater forms (Figs. 33.4J and 33.9A,B).

Order Cheilostomata. With membranous or calcareous, boxlike zoecium, variously arranged to form colonies, but with separate wall between zooids; zoecium with subterminal opening and typically with hinged lid, operculum; sometimes with collar containing secondary, permanently open orifice; commonly polymorphic, with specialized zooids for grasping (avicularia), for sweeping colony surface (vibracula), or for attachment (kenozooids); many with coronate larva; marine (Fig. 33.9D,F).

Selected Readings

Carle, K. J., and E. E. Ruppert. 1983. Comparative ultrastructure of the bryozoan funiculus: A blood vessel homologue. *Z. zool. Syst. Evolut.-forsch.* 21:181–193.

Mariscal, R. N. 1965. The adult and larval morphology and life history of the entoproct *Barentsia gracilis*. *J. Morph.* 116:311–338.

Mariscal, R. N. 1975. Entoprocta. In *Reproduction of Marine Invertebrates*, vol. 2 (A. C. Giese and J. S. Pearse, eds.), pp. 1–41. Academic Press, New York.

McCammon, H. M., and W. A. Reynolds (eds.). 1977. Biology of Lophophorates. *Am. Zool.* 17:3–150.

Nielsen, C. 1971. Entoproct life cycles and the entoproct/ectoproct relationship. *Ophelia* 9:209–341.

Ross, J. R. P. (ed.). 1987. *Bryozoa: Present and Past*. Western Washington Univ., Bellingham.

Ryland, J. S. 1970. *Bryozoans*. Hutchinson, London.

Ryland, J. S. 1976. Physiology and ecology of marine Bryozoa. *Adv. Mar. Biol.* 14:285–443.

Strathman, R. R. 1973. Function of lateral cilia in suspension feeding of lophophorates. *Mar. Biol.* 23:129–136.

Strathman, R. R. 1982. Cinefilms of particle capture by a induced local change of beat of lateral cilia of a bryozoan. *J. Exp. Mar. Biol.* 62:225–236.

Taylor, P. D. 1988. Major radiation of cheilostome bryozoans: Triggered by the evolution of a new larval type? *Hist. Biol.* 1:45–64.

Woollacott, R. M., and R. L. Zimmer. 1977. *Biology of Bryozoans*. Academic Press, New York.

34

Brachiopoda

At the beginning of the Cambrian period, 600 million years ago, brachiopods or lamp shells were already abundant and the two main divisions of what was then a phylum, inarticulates and articulates, were well established. Their shells fossilize easily and have kept paleontologists busy, and some 45,000 extinct species have been described. Brachiopods, however, are not abundant today; only about 350 species are still alive (Rudwick, 1970).

Like other lophophorates, brachiopods lead quiet lives. Protected in their secreted shells, brachiopods are superficially like clam shells, and indeed until about a hundred years ago they were classified as molluscs. However, the symmetry of brachiopod and clam shells differs, in that pelecypod shells are right and left valves, and brachiopod shells are dorsal and ventral valves.

Definition

Brachiopoda are bilateral coelomates, with a bivalved shell with dorsal and ventral valves that are generally attached by a pedicle or stalk. They have a lophophore with a single row of alternating tentacles leading to a food groove bound by an epistomal brachial fold. The lophophore arms come equipped with two mesocoelic channels. An archimeric coelom is divided into mesocoel and metacoel compartments. These have one or two pairs of metanephridia, and the circulatory system is open with a pulsating dorsal vessel. The gonads are formed from peritoneal cells.

Body Plan

Brachiopods are divided into two classes, Articulata and Inarticulata, the former with a hinge connecting the two shell valves, and the latter without such a hinge. Most inarticulates have a long posterior **pedicle**, an anchorlike stalk, and the valves are held together by muscles only; they burrow in sand or mud flats and pull the body out of danger by contracting the pedicle (Fig. 34.1D). Other inarticulates are permanently attached to objects by the ventral valve, much like oysters (Fig. 34.1C). In articulates, interlocking processes form a hinge. The pedicle passes through a posterior notch in the valves at the hinge line, or through an opening in the ventral or **pedicle valve**, which curves up as a beak. The dorsal or **brachial valve** is usually smaller (see Fig. 34.1A,B). The main mass of the body occupies only the posterior third of the space between the valves. The trunk metacoel extends into the mantle, but the upper and lower layers of the mantle unite as it matures, leaving a number of tubular coelomic channels known as mantle canals. The edge of the mantle bears prominent, stiff setae, formed by single chaetoblasts, supposedly protective and sensory in function.

The formidable vocabulary, useful primarily to

paleontologists, that has been developed around the details of shell structure is avoided here. The **shell** is tightly adherent to the two **mantle** folds, extensions of the body wall. Shell and mantle are intimately associated; tubular mantle caeca penetrate the shell, reaching the uppermost horny layer, the periostracum (Fig. 34.2B). There are alternating layers of chitin and phosphate in the inarticulate *Lingula* shell. The calcareous shell of most other species has an outer, fibrous or laminated and an inner, prismatic layer of calcium carbonate. The shell begins as a small plate, which, as additonal shell is laid down, comes to lie at the growth center of the shell (beak). Once formed, additional material cannot be added to the periostracum or outer calcareous layer, but the inner prismatic layer is thickened throughout the life of the animal.

Hinge **teeth** on the pedicle valve of articulate shells fit into **dental sockets** on the brachial valve that extend out as a **cardinal process** between the lateral dental sockets. The hinge teeth serve as articulating surfaces when the shell opens or closes (Fig. 34.2A). **Adductor muscles** attached to both valves close the shell, and **diductor muscles** attached to the pedicle valve and the cardinal process of the brachial valve open the shell (Fig. 34.2C).

The inner surface of the brachial valve is often sculptured to accommodate the lophophore, and articulates bear the **brachidium** or lophophoral skeleton, often of complex form (Fig. 34.2C). The simplest brachiopod lophophores are discoid. The number of tentacles is increased by folding and lobation of the lophophore, as indicated in Figure 34.8D,E,F,G,H. Additional space is obtained by adding lobes or by extending the lateral arms of the lophophore. The plectolophous type (see Fig. 34.8H), with two simple lateral arms and a median arm, is most common in modern terebratulid articulates, and the spirolophous type (see Fig. 34.8G), without a median arm but with spirally coiled lateral arms, is seen in modern rhynchonellid articulates and inarticulates. A **brachial fold** extends over the mouth and is thought to represent a modified epistome; in inarticulates the fold contains a coelom that is open to the lophophoral mesocoel. The articulate brachial fold is solid tissue. Details of mesocoel construction vary with the lophophore type and need not be discussed here. As in other lophophorates, the tentacles contain extensions of the mesocoel (see Fig. 34.6B).

Digestive System Food passes into the dorsally arched **esophagus** and is pushed by peristalsis to the dilated **stomach** (Fig. 34.3). Conspicuous paired **di-**

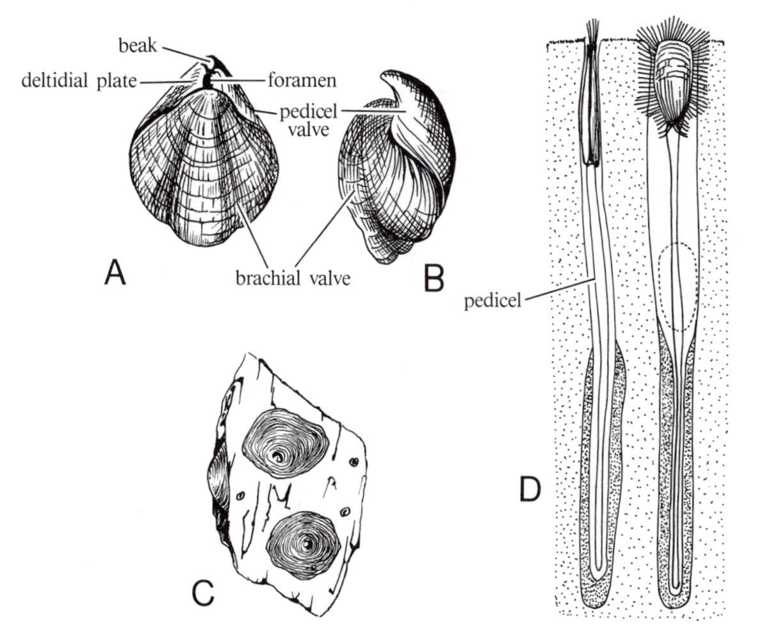

FIGURE 34.1. Brachiopods. **A,B.** Dorsal and lateral views of *Hemithyris psittacea*, an articulate. Note prolonged pedicle valve and opening for pedicle. **C.** Three *Crania* attached to rock. Inarticulates with no pedicle, but like other acrotretids they attach with pedicle valve downward using calcareous secretions. **D.** Two views of lingulid inarticulate *Lingula* embedded in mud. Contractile pedicle secretes mucus; when contracted it pulls body down into position indicated by *dotted lines* on right. (**A,B** after Blockmann; **C** after Shipley; **D** after François.)

verticula arise from the stomach wall. They are called digestive glands or the liver, but little is known of their function. They are probably sites of intracellular digestion and absorption. Articulates have a narrow, blind **intestine**, but the inarticulate intestine curves or loops in the metacoel and ends in an **anus**.

Circulatory System Brachiopods have two circulating body fluids. Coelomic fluid flows throughout the coelomic system. The **mesocoel** and the **metacoel** are only partially separated by a diaphragm at the level of the esophagus. Branches of the mesocoel extend into the tentacles, and branches of the metacoel extend into the mantle. Peritoneal cilia or flagella circulate the coelomic fluid. The fluid moves rapidly in the narrow passageways of the mantle and tentacles, and was once thought to be blood.

Blood, containing hemerythrin in cells, circulates in an open circulatory system. The middorsal **blood vessel** contains a pulsating vesicle, or **heart** (Fig. 34.3). The middorsal vessel divides into two anterior branches, each communicating with an extensive sinus around the gut and continuing to the lophophore, where branches reach each tentacle. The middorsal vessel bifurcates posteriorly to form dorsal and ventral mantle vessels that branch extensively in the mantle. The ventral mantle vessel also communicates with sinuses in the nephridia and gonad. The heart beats slowly, and circulation is sluggish. Nothing is known of details of circulation; presumably the blood ebbs and flows. The blood and coelomic channels communicate extensively through tissue spaces, and blood and coelomic fluid must exchange components everywhere.

Excretory System The **metanephridia** lie in the metacoel (Fig. 34.3). The ruffled, flagellated **nephro-**

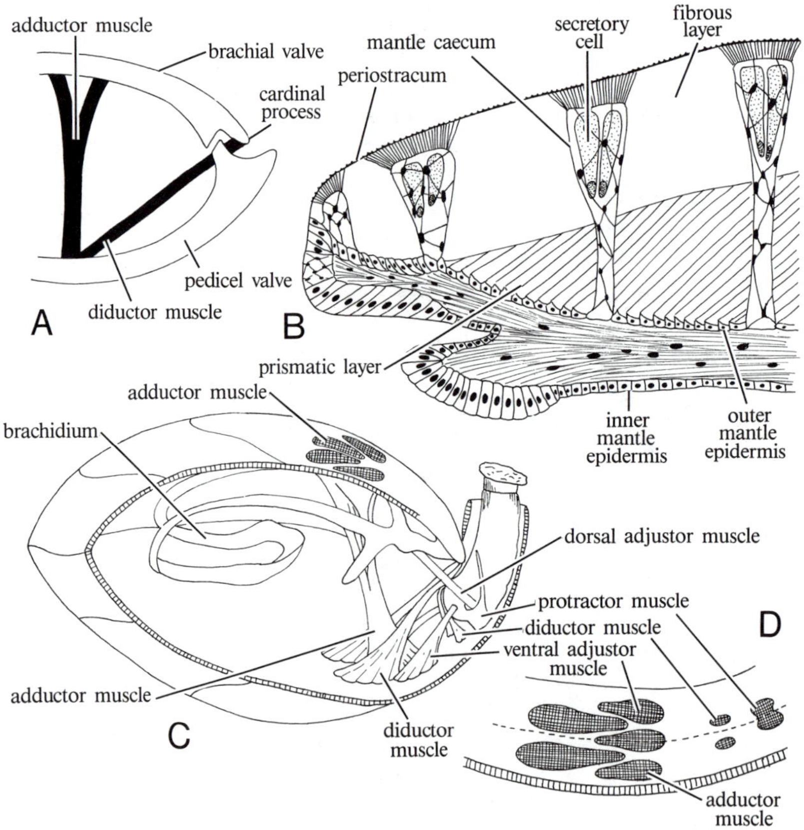

FIGURE 34.2. Structure of articulate brachiopod shells. **A.** Hinge arrangement. Rocking surface of hinge leads to opening of valves when diductor muscle contracts, and closing of valve when adductors contract. **B.** Scheme of longitudinal section through mantle and valve. Note caeca that extend into calcareous and fibrous parts of shell, conferring characteristic porous texture. **C,D.** Muscles and muscle scars of *Magellania*. Muscle scars of brachial valve shown in **C**, and of pedicle valve in **D**. (**B** after Williams; **C,D** after Hancock.)

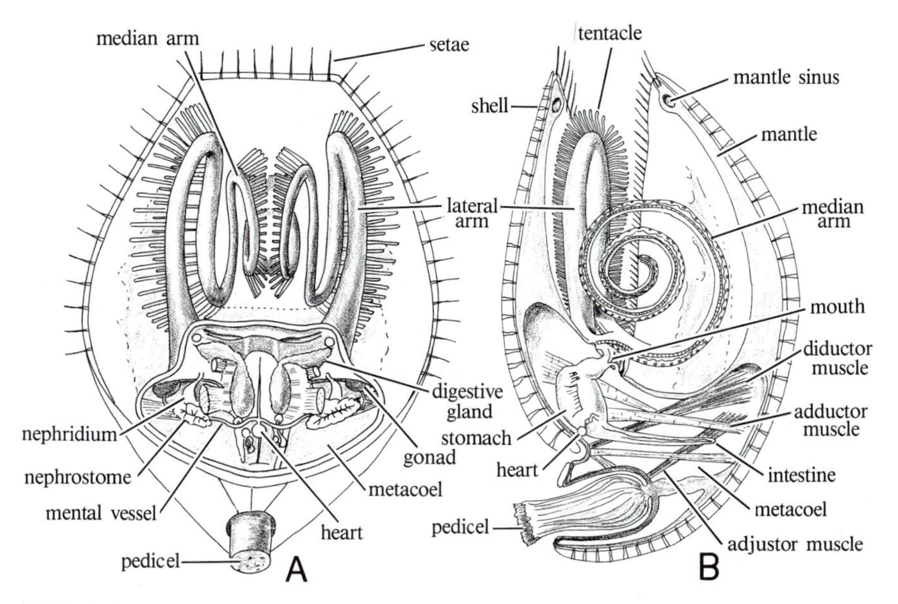

FIGURE 34.3. Brachiopod anatomy, based on *Magellania,* as seen in dorsal and lateral views. (After Delage and Hérouard, from de Beauchamp, in Grassé.)

stomes are supported by the lateral mesenteries that also support the gut. The tubular part of the nephridium narrows as it approaches the **nephridiopore**, which opens into the mantle cavity. Fingerlike glandular processes occur on the nephridia of *Terebratulina* (Fig. 34.4); their function is unknown.

Nervous System The sensory cells and nervous system are diffuse (Fig. 34.5). The inarticulate nervous system appears to be in the epidermis, and the articulate nervous system just below the epidermis. Central, small supraesophageal ganglia send nerves to the lophophore, and the large subesophageal ganglion supplies nerves to the mantle, pedicle, and shell muscles.

Reproductive System Gametes arise from the peritoneum, forming indistinct though sometimes extensive gonadal masses. When present, four pairs of gonads usually are noted.

Biology

Locomotion Movements are limited in brachiopods. Inarticulates, like *Lingula*, can contract the pedicle and, therefore, achieve some evasive movements by moving deeper into sediments. All brachiopods operate a shell opening and closing mechanism. Inarticulates use extensive adductor muscles to close the shell, while a single longitudinal muscle contracts the body, causing it to bulge, and in doing so opens the shell indirectly by hydrostatic pressure (Gutman et al., 1978). Articulates use a modified adductor muscle to open the shell as a diductor. Diductor muscles work at a mechanical disadvantage and are, therefore, larger than the adductors. The shell cannot be opened wide, but sufficient space is provided to permit water to enter and leave. Additional muscles, known as adjustors, pass from the pedicle valve to the pedicle (Fig. 34.2C,D,E). They tilt the shell with respect to the pedicle. Inarticulates can make more complex shell movements and have correspondingly more complex muscles arrangements.

Feeding The brachiopod lophophore can capture food as in other lophophorates. The shell valves are

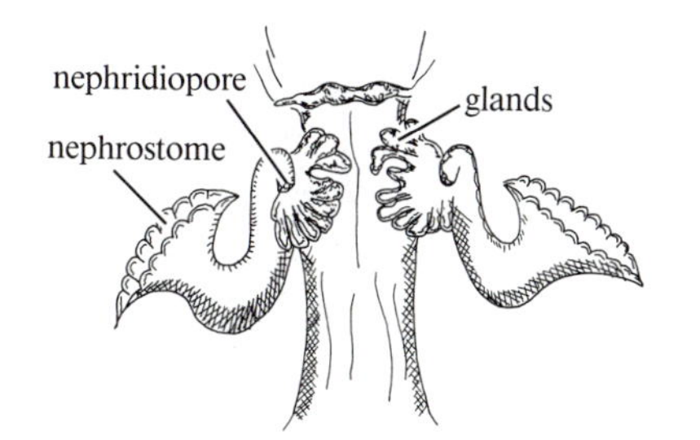

FIGURE 34.4. Nephridia of *Terebratulina retusa.* (After Heller.)

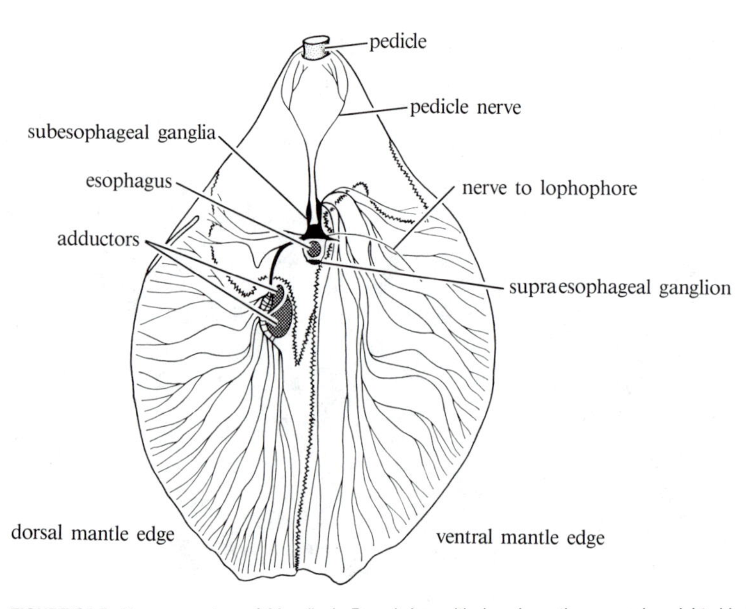

FIGURE 34.5. Nervous system of *Magellania.* Dorsal view with dorsal mantle removed on right side to show ventral mantle structures. (After Hancock.)

held apart during feeding. Lateral tracts of cilia produce lateral intake currents and a median exhaust current (Fig. 34.6A). The tentacles are held high, and large particles drop to the mantle to be ejected by the mantle cilia or the sudden clapping of the valves (Thayer, 1986). A brachial groove at the base of the tentacles collects the food particles brought to it by frontal cilia on the tentacles (Fig. 34.6B). The brachial groove cilia convey the food particles, trapped in mucus, to the mouth. The brachial fold projects over the mouth, helping to direct traffic inward (Gilmour, 1981).

Inarticulates subsist on particulate matter and are especially adept at sorting edible and inedible particles as they enter the tentacle field. To this end an anus is necessary to void wastes. Articulates, on the

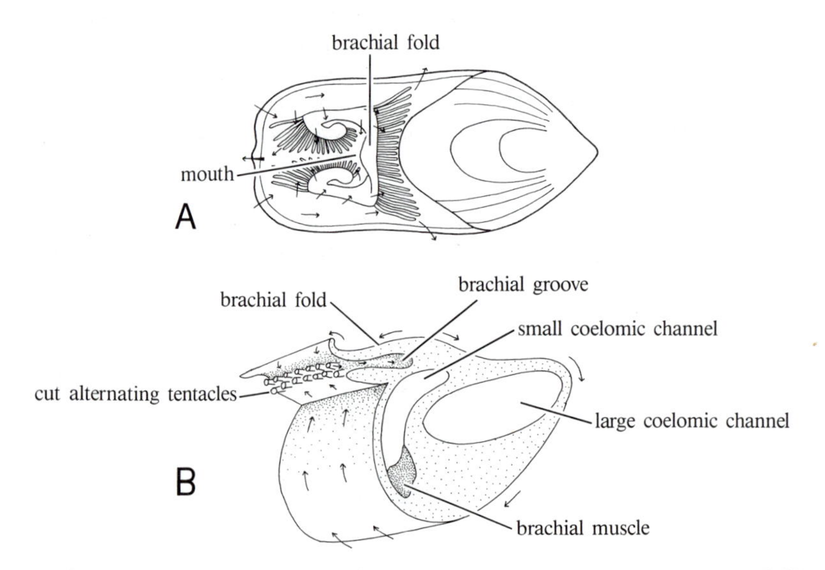

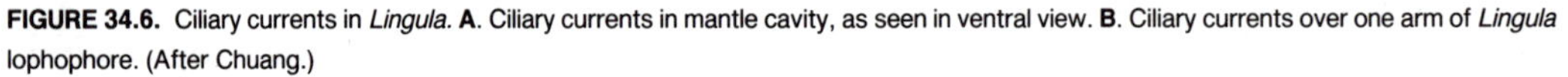
FIGURE 34.6. Ciliary currents in *Lingula.* **A.** Ciliary currents in mantle cavity, as seen in ventral view. **B.** Ciliary currents over one arm of *Lingula* lophophore. (After Chuang.)

other hand, derive most of their nutrition from the absorption of dissolved and colloidal organic nutrients in seawater (McCammon, 1969; McCammon and Reynolds, 1976). The gut, dealing as it does with a low-waste food source, lacks an anus. Tellingly, modern articulates thrive in polar and subpolar areas where dissolved organics are in high concentrations, and such localized populations rival in density Paleozoic occurrences of brachiopods.

Excretion *Lingula* is largely ammonotelic, although the enzymes to produce urea are present (Hammen, 1977). Particulate material is taken up by coelomic phagocytes and by peritoneal cells that detach and disintegrate. Streams of particles move to the nephridium and are taken up by the epithelial lining, eventually to be discharged, suspended in mucous strands, from the nephridiopore. Particles of nitrogenous waste may be expelled in this way.

Respiration There is no evidence that respiratory deficiencies limit brachiopod size. Indeed, brachiopods normally have low rates of oxygen consumption, 2.3 to 4.4 μmol oxygen/hour/g tissue, or can survive for long periods with no oxygen consumption at all (Hammen, 1977). Very large species once existed, fossil shells over 0.3 m in diameter having been recovered, and some modern species are quite bulky. A respiratory pigment known as hemerythrin

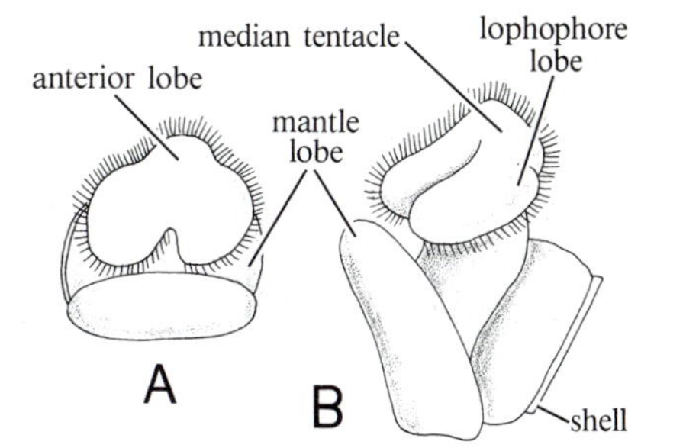

FIGURE 34.7. Two stages in development of inarticulate *Lingula*. **A**. Development of mantle lobes shown. **B**. Developing body extends upward from mantle lobes, and developing median tentacle and lophophore lobes evident. (After Yatsu.)

occurs in blood and coelomic fluid cells. It does not take up oxygen as readily as hemoglobin, but gives it up very readily, and is no doubt involved in respiratory transport.

A prime source of respiratory oxygen intake is the shell punctae with their mantle caeca. When the shell is experimentally sealed, brachiopods still can maintain 30 percent of their oxygen uptake and do not have to shift to glycolysis, as do clams under similar manipulations (Thayer, 1986).

Reproduction Sexes are separate, although some hermaphrodites are known. Gametes usually depart

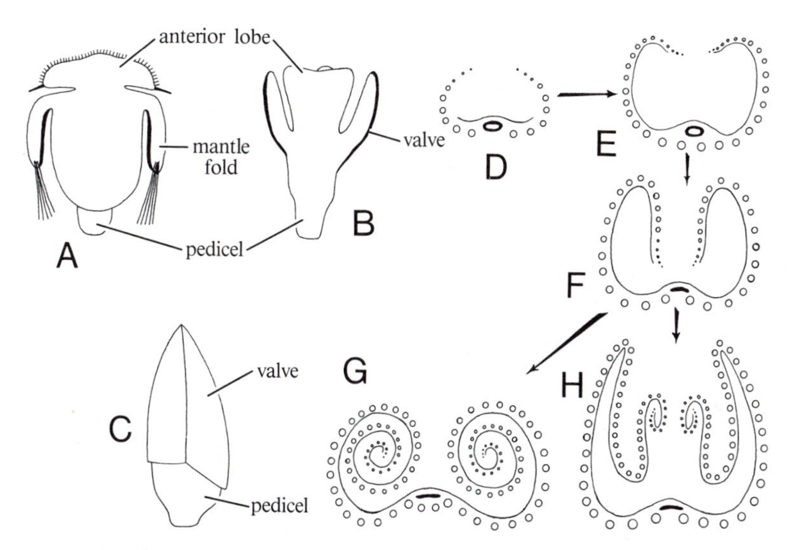

FIGURE 34.8. A,B,C. Three stages in articulate development, showing lifting of mantle lobes to enclose anterior lobe of larva, which then develops into greater part of body. **D,E,F,G,H**. Scheme of lophophore evolution in brachiopods. Tentacles shown as circles, with decreasing age indicated by decreasing size. Primitive type **D** gains tentacles by becoming bilobed (trocholophous) as shown at **E**. Further changes in this direction produce schizolophous type **F**, with two recurved arms. From this type, either highly spiraled spirolophous type **G** or horseshoe-shaped and spiraled plectolophous type **H** may develop. (**A** after Shipley; **B,C** after Kowalevsky; **D,E,F,G,H** after de Beauchamp, in Grassé.)

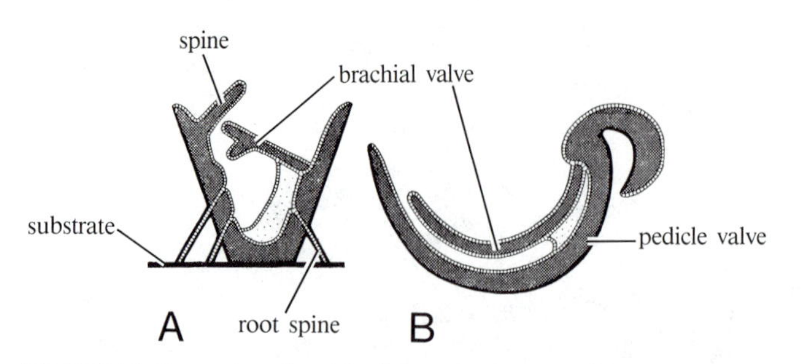

FIGURE 34.9. Some schematics of fossil forms showing some aberrant adaptations of shell and body chambers seen among extinct strophomenidan brachiopods. **A.** Richthofeniid. **B.** Lyttoniacean. (After Rudwick, 1970.)

through the nephridia and develop in the sea, but a few species brood young in the mantle cavity, nephridia, or lophophore arms.

Breeding patterns vary. Some species breed constantly; other species are seasonal. Lingulids breed all year round in the tropics, but are seasonal in temperate water. *Glottidia* releases about 10,000 eggs per spawning and may achieve close to 130,000 in a lifetime.

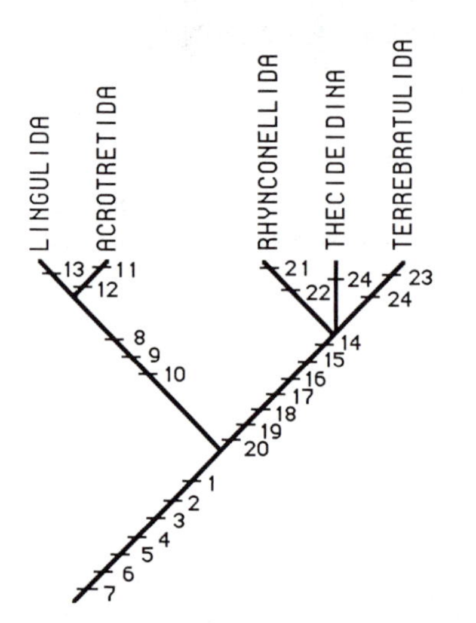

FIGURE 34.10. Cladogram of relationships of living orders of Brachiopoda. Derived characteris are: (1) tentacle filaments in a single palisade of lophophore axis; (2) tentacle filaments in a double row; (3) food groove bounded by brachial fold; (4) lophophore arms with two mesocoelic channels; (5) dorsal and ventral mantle lobes; (6) coelomic chambers in mantle lobes; (7) sensory setae on mantle margin; (8) hydraulic opening of shell; (9) shell muscles peripheral in body cavity; (10) larval shell; (11) holoperipheral growth (i.e., valves circular) of both valves; (12) closely comparable lateral and internal oblique muscle paths; (13) $CaPO_4$ in shell; (14) hinged shell; (15) mantle lobes fused posteriorly; (16) fibrous or prismatic secondary shell; (17) pedicle as larval rudiment; (18) mantle folds back at settlement; (19) open delthyrium; (20) lophophore with skeletal support; (21) lophophore support reduced or lost; (22) lophophore spiralophous; (23) lophophore plectolophous; and (24) single pair of nephridia. (Modified from Rowell, 1982.)

Development Embryonic development has been followed in articulates and inarticulates (Hammond, 1982). Early development involves holoblastic, radial cleavage, somewhat modified in the articulates *Terebratulina* and *Terebratella.* A coeloblastula forms that gastrulates by invagination. Accounts of mesoderm and coelom formation vary with species differences. *Lingula* and some *Terebratulina* mesoderm arises as two masses from the sides of the archenteron that later hollow out to form two coelomic pouches by schizocoely. In other terebratulids, the mesoderm arises as lateral, anterior, or posterior sheets or pouches of the archenteron, a type of modified enterocoely. The blastopore closes, and its relationship to the mouth remains uncertain. What is still unclear are the details of formation of the various coelom compartments. For example, disagreement occurs over whether or not a protocoel occurs.

Later development of inarticulates and articulates differs markedly. This may be due in part to differences between the inarticulate plankton-feeding larva developing from an egg with little yolk, and the articulate yolk-subsisting larva growing from a very yolky egg. In inarticulates, a mantle fold differentiates from the body mass, divides into dorsal and ventral lobes, and partly encloses the body (Fig. 34.7A). The body develops lobations for the future lophophore, and as the larval tentacles form, an epistome appears that grows into the brachial fold. The larval tentacles never surround the mouth, for a median tentacle appears above the mouth and new tentacles form at this point, gradually moving laterally as new ones form. The mantle folds eventually start to secrete a shell, and as the pedicle appears as a posterior bud, the typical bivalved free-swimming larva results.

Articulate embryos develop three body regions: apical, mantle, and pedicle lobes. The apical lobe becomes the lophophore and most of the body. The mantle lobe grows posteriorly, at first as a ring, but later bifurcating and bending forward to enclose the body, as mentioned earlier (Fig. 34.8A,C).

Fossil Record

Brachiopod fossils are known from the earliest Cambrian deposits onward. Their greatest radiation was in the Paleozoic (Rudwick, 1970). The greatest diversity occurred among the articulates, and several extinct orders thrived in the past. Besides standard brachiopod body forms, many strange adaptations are evidenced in the fossil record (Fig. 34.9), informing us that what is now a minor phylum was not so in the past, and warning us that phylogenetic relationships based solely on analysis of living brachiopods must be qualified by the realization that such is not nearly the whole picture.

Phylogeny

Some question has arisen as to whether brachiopods are monophyletic (Rowell, 1982) or polyphyletic (Wright, 1979). The weight of evidence seems to be on the side of monophyly, with a series of autapomorphies related to shell form and lophophore specializations. Essentially, two great divergences occurred from the ancestral stock (Fig. 34.10). On one hand are the inarticulate forms with their distinctive hydrostatic method of shell opening, peripheral location of shell adductor muscles in the body cavity, and development of a larval shell. The articulates exhibit several specializations related to shell closing and in holding the dorsal and ventral shells in position relative to each other. The two living orders in each class sort on the basis of one of each class being less derived (Lingulida and Rhynchonellida) and the other being more advanced (Acrotretida and Terebratulida). This simplistic scheme, however, is misleading given the great number of fossil orders left out (see, e.g., Rowell, 1982).

Taxonomy

Class Inarticulata. Shell held together by muscles only; lophophore without skeletal support; digestive tract complete with anus.

Order Lingulida. With pedicle valve ventral, but both valves modified to permit pedicle passage; shell mineral calcium phosphate. Example: *Lingula* (Fig. 34.1D).

Order Acrotretida. Without pedicle, or with pedicle emerging from notch or foramen in pedicle valve (some attached directly to objects by pedicle valve); shell minerals are calcium carbonate and/or calcium phosphate. Example: *Crania* (Fig. 34.1C).

Class Articulata. Valves hinged by interlocking processes; pedicle emerging through pedicle valve; shell mineral calcium carbonate; lophophore generally with skeletal support; digestive tract incomplete without anus.

Order Rhynchonellida. Lophophore spirolophous, skeletal support reduced or lacking; two pairs of nephridia.

Order Terebratulida. Lophophore plectolophus, with supporting loops; one pair of nephridia. Example: *Hemithyris* (Fig. 34.1A,B).

?Order Thecideidina. Lophophore schizolophous, with lobes fitting into hollow in dorsal valve; one pair of nephridia.

Selected Readings

Gilmour, T. J. M. 1981. Food collecting and waste rejecting mechanisms in *Glottidia* and the persistence of lingulacean inarticulate brachiopods in the fossil record. *Can. J. Zool.* 59:1539–1547.

Gutman, W. F., K. Vogel, and H. Zorn. 1978. Brachiopods: Biomechanical interdependences governing their origin and phylogeny. *Science* 199:890–893.

Hammen, C. S. 1977. Brachiopod metabolism and enzymes. *Am. Zool.* 17:141–147.

Hammond, L. S. 1982. Breeding season, larval development, and dispersal of *Lingula anateria* from Townsville, Australia. *J. Zool. Lond.* 198:183–196.

McCammon, H. M. 1969. The food of articulate brachiopods. *J. Paleo.* 43:976–985.

McCammon, H. M., and W. A. Reynolds. 1976. Experimental evidence for direct nutrient accumulation by the lophophore of articulate brachiopods. *Mar. Biol.* 34:41–51.

Rowell, A. J. 1982. The monophyletic origin of the Brachiopoda. *Lethaia* 15:299–307.

Rudwick, M. J. S. 1970. *Living and Fossil Brachiopods.* Hutchinson, London.

Thayer, C. W. 1986. Respiration and the function of brachiopod punctae. *Lethaia* 19:23–31.

Thayer, C. W. 1986. Are brachiopods better than bivalves? Mechanisms of turbidity tolerance and their interaction with feeding in articulates. *Paleobiol.* 12:161–174.

Wright, A. D. 1979. Brachiopod radiation. In *The Origin of Major Invertebrate Groups* (M. R. House, ed.), pp. 235–252. Academic Press, London.

35

Echinodermata

Nothing more certainly symbolizes the sea than starfish, sea urchins, and sand dollars. There is good reason for this, for echinoderms cannot adequately control the water content of their bodies and so have never been able to invade freshwater or land habitats.

Echinoderms, although exclusively marine, are still very successful. They are found at all depths and in all climates, and have been well established since the start of the Cambrian, 600 million years ago. Their heavily armored bodies, composed of calcareous plates and spines, and their localized great abundance make them the panzer divisions of the marine realm. About 8000 modern species and over 30,000 fossil species have been described.

Definition

The uncephalized body displays biradial symmetry with an oral-aboral axis. Echinoderms are composed of five fields with podia (ambulacra) alternating with five fields without podia (interambulacra). The body is unsegmented and enterocoelous, and with endomesoderm derived from the primitive gut or archenteron. An endoskeleton of calcareous ossicles is covered by a thin epidermis that is usually ciliated. The coelom is complexly partitioned during development, giving rise to hemal and visceral systems and a unique hydraulic ambulacral or water-vascular system equipped with tube feet or podia used for respiration, movement, and sensory reception. The digestive tract is usually complete (except in Concentricycloidea and Ophiuroidea). There is a diffuse, uncentralized nervous system, typically composed of three nerve rings, arranged around the digestive tube, and from which radiating peripheral nerves arise. There are no excretory organs. Echinoderms are typically dioecious with simple reproductive organs and gonoducts, and development begins with holoblastic, radial cleavage followed by a series of ciliated, bilateral larval stages that drastically metamorphose to become the radial adults.

Body Plan

Echinoderms are predominantly five-sided or pentamerous (Fig. 35.1), as the typical five arms of a starfish or the five petals on the surface of a sand dollar testify. There is nothing mythical about pentamerism; it merely seems to be a structural response to provide some mechanical rigidity to an otherwise basically weak globose or spherical shell (Nichols, 1969). However, the pentamerous radial form is imposed on a bilateral body, since essentially a plane can be drawn through the mouth and one of several off-center structures, either anus, gonopore, or madreporite. These structures are found between ambulacral fields; therefore, the plane cuts through the ambulacrum opposite the anus, gonopore, or madreporite. This transected arm or groove is then termed the A ambulacrum, and all the rest are named B, C, D, and E in a clockwise manner. Only a few fossil echinoderms, whose origins apparently are older than biradial symmetry, show no evidence of radial tendencies. Several modern groups of echinoderms, however, have taken up a free-moving life and show tendencies to replace the ancestral secondary biradial symmetry with a tertiary bilateral symmetry, for example, the sea cucumbers (Figs. 35.1E and 35.18).

The six types of modern echinoderms (Fig. 35.1) are known. Each is easily recognized by the normal orientation of the body and by the distribution of the alternating **ambulacral fields** containing the **podia** of the water-vascular system, and **interambulacra**, fields without podia. The mouth typically lies at the center of the oral surface, and the anus, if present, is usually placed on the opposite surface; thus typical echinoderms have an oral–aboral axis like cnidarians. Echinoids (sea urchins, sand dollars, and heart urchins) normally lie with the mouth down and have the ambulacra and interambulacra on the sides of the body (Figs. 35.1D and 35.19). Holothurians (sea cucumbers) also have the ambulacra and the interambulacra on the general body surface, but usually lie on their side, with the oral-aboral axis parallel to the substrate (Figs. 35.1E, 35.18). Concentricycloidea (sea daisies) are minute forms that lie with "mouth" down, although they lack a gut and have concentric rings of tube feet (Figs. 35.1F and 35.26). In all other echinoderms, the ambulacra extend along arms or rays, which attach to a central discoid body mass. Crinoids (sea lilies, feather stars) are attached at the aboral surface by a stalk (Fig. 35.1A), and so are normally oriented with the mouth up. Asteroids (starfish) normally lie with the mouth down and have open ambulacra extending along the rays (Fig. 35.1B). Ophiuroids (brittle stars)

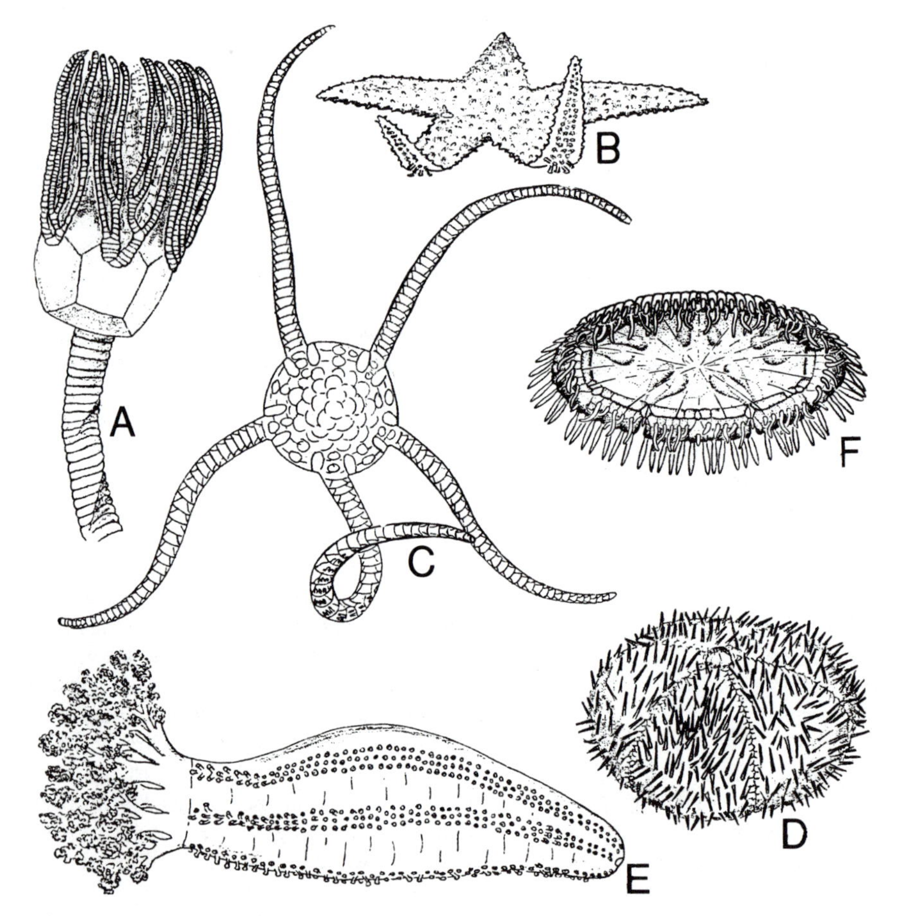

FIGURE 35.1. Living echinoderm types. **A**. Crinoidea, single surviving pelmatozoan. **B**. Asteroidea. **C**. Ophiuroidea. **D**. Echinoidea, regular type. **E**. Holothuroidea. **F**. Concentricycloidea.

also lie with the mouth down, but have narrow, flexible rays or arms, with the ambulacra enclosed and not visible from without (Fig. 35.1C).

Coeloms There are basically three coeloms in echinoderms: visceral, hemal, and water-vascular. Although the unique form of the coeloms varies in the different echinoderm groups, certain consistencies emphasize the unity of the phylum (Table 35.1). The details of the coelomic structures reveal a veritable maze of interconnected canals and tubes, seemingly designed to plague the minds of students.

The main coelomic space is the **perivisceral coelom**. Sometimes it can be filled with tissue, as in crinoids, or encroached upon by bursal sacs, as in ophiuroids (Fig. 35.2). The main perivisceral coelom (Fig. 35.3C) typically extends into the rays and is partly partitioned by mesenteries supporting the organs (see Fig. 35.25C). The rest of the visceral coelom consists of a complex of small spaces associated with the nervous and hemal systems, and with the gonads. Three coelomic ring sinuses are generally present. Their basic arrangement is seen most clearly in asteroids (Fig. 35.3A). A hyponeural ring sinus is separated from an epineural ring sinus by a thin membrane in which the nerve ring lies. The hyponeural ring sinus is connected with an aboral coelomic ring sinus (Figs. 35.3A and 35.4C) by an axial sinus that contains the axial gland and stone canal.

Radial canals or sinuses arise from all the ring sinuses (Figs. 35.3A and 35.4C). The canals from the aboral coelomic ring enclose the gonads (Figs. 35.3A and 35.4C), and the canals from the hyponeural ring sinus lie just internal to the radial nerves. Because asteroids have open ambulacral grooves, they have no epineural radial sinuses, but echinoids and holothurians have epineural radial sinuses located above the radial nerves (Fig. 35.5C).

The basic arrangement of the visceral coelom just outlined is sometimes modified or reduced. For example, the aboral ring sinus of crinoids is the chambered organ, associated with the cirri and the stalk. The hyponeural radial sinuses of asteroids are divided by an incomplete partition into two canals.

Table 35.1. Derivatives of echinoderm coelomic pouches, excepting concentricycloids

	Crinoids	Holothurians	Echinoids	Asteroids	Ophiuroids
Right					
Axocoel	Not formed	Missing	Lumen of axial gland and aboral coelom into which axial gland extends	Dorsal sac?	Right side of axial sinus
Hydrocoel	Not formed	Lost	Lost	Lost	Lost
Somatocoel	Aboral part of the perivisceral coelom; chambered organ; aboral or genital coelomic canals	Aboral part of the perivisceral coelom; perianal sinus	Aboral part of the perivisceral coelom; periproctal, perianal, and aboral sinuses	Aboral part of the perivisceral coelom; mesenteries of pyloric caeca; aboral coelom	Aboral part of the perivisceral coelom; aboral sinus
Left					
Axocoel	Axial sinus; hydropore	Stone canal?	Hydroporic canal; ampulla	Axial sinus; inner hyponeural sinus	Left side of axial sinus?; hydroporic canal
Hydrocoel	Water-vascular system; hydroporic canal	Water-vascular system; hydroporic canal	Water-vascular hydroporic canal, in part	Water-vascular system; hydroporic canal	Water-vascular system; hydroporic canal
Somatocoel	Oral part of perivisceral coelom; subtentacular coelomic canals	Oral part of perivisceral coelom; peribuccal and peripharyngeal sinuses	Oral part of perivisceral coelom; peripharyngeal cavity and gill cavities	Oral part of perivisceral coelom; outer hyponeural sinus; radial hyponeural sinus	Oral part of perivisceral coelom; periesophageal ring sinus; arm coeloms; peristomial ring sinus

Holothurians have a small perianal sinus at the aboral end, but the gonads are not radially arranged, so there are no radial genital sinuses. Regular echinoids have a complex set of three aboral ring sinuses that consist of a small perianal sinus, a periproctal sinus (separated from the main coelom by a ring mesentery supporting the rectum), and a genital ring sinus system. These three ring sinuses, however, tend to be united into a single aboral coelomic ring in irregular echinoids (Fig. 35.4C). Ophiuroids have an aboral sinus system like that of asteroids.

In echinoids and holothurians a unique arrangement exists. The epineural sinus lies in the buccal membrane and is termed the peribuccal sinus. Small epineural radial sinuses lie above the radial nerves, and although there is no hyponeural ring sinus, hyponeural radial sinuses lie just below the radial nerves (Fig. 35.5C). A part of the perivisceral coelom of regular echinoids and holothurians is cut off as a peripharyngeal coelomic sinus, which lies in the calcareous pharyngeal bulb of holothurians (Fig. 35.5B) and in Aristotle's lantern in regular echinoids. The chamber so formed helps in manipulating the oral tentacles and the peristomial gills of regular echinoids. Irregular echinoids have no peristomial gills and, therefore, do not have a peripharyngeal coelomic sinus.

The echinoderm **hemal system** is closely associated with the visceral coelom and the **axial gland**. The asteroid system (Fig. 35.3A) illustrates the basic pattern. A hemal ring inside of the hyponeural visceral coelomic sinus has radial branches that pass outward and inside of the hyponeural radial canals or sinuses. There is also an aboral hemal ring inside the aboral coelomic ring sinus, and radial genital lacunae that extend inside the genital sinuses. These oral and aboral systems are connected by a large axial gland that contains hemal passageways from which hemal tufts to the digestive tract arise. The axial gland lies inside the axial sinus and has an extension lying in a special coelomic sac, the dorsal sac (Fig. 35.3A).

Several modifications of this arrangement are seen in echinoids. The axial gland (Fig. 35.4C) in regular echinoids is a long, dark body that projects aborally as a fingerlike process enclosed in a small coelomic compartment. In irregular echinoids, the axial gland is much shorter and contains a plexus of

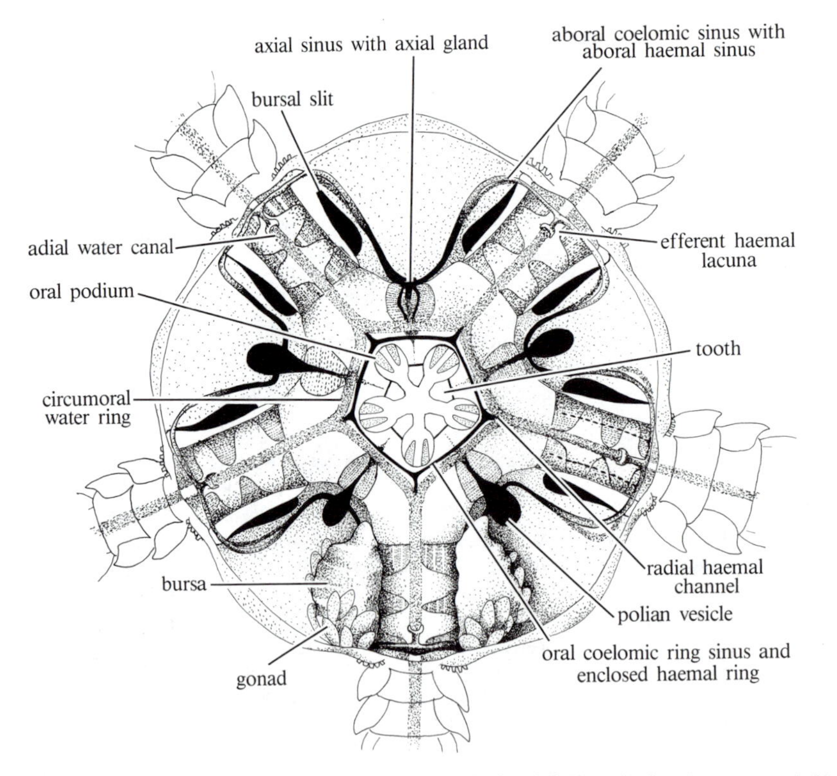

FIGURE 35.2. Ophiuroid anatomy. Scheme of internal organization of *Ophiura* with digestive system and all but one pair of bursae removed. (After Cuénot, in Grassé.)

water canals as well as a plexus of hemal vessels. A system of genital hemal lacunae arise from the aboral hemal ring (Fig. 35.4C), associated with the aboral coelomic ring sinus, as in asteroids. An oral hemal ring passes around the pharynx and gives rise to radial hemal lacunae that run between the water canals and the hyponeural coelomic sinuses along the body wall giving off branches to the podia. A series of short hemal sinuses extend to spongy bodies of unknown function on the interradii between the radial lacunae. Two main hemal vessels arise from the oral ring and pass along the opposite sides of the gut. These are the inner and outer marginal sinuses. In some sea urchins, a contractile, collateral hemal sinus, which is connected to the outer marginal sinus by many branches, parallels the marginal sinuses for a considerable part of their length (Fig. 35.4C).

Holothurians are unusual in having, at the most, only vestiges of an axial gland. Sea cucumbers have no radially arranged aboral gonads, and the aboral hemal ring and system of genital hemal vessels are missing. Otherwise, the system is much like that of echinoids. A hemal ring encircles the pharynx (Fig. 35.5A) and gives rise to radial hemal vessels that lie beneath the radial nerves and give rise to podial branches. Two hemal vessels associated with the gut arise from the oral ring (Fig. 35.5A). One, the dorsal hemal sinus on the mesentery side of the gut, is the only contractile hemal vessel. It is not attached to the intestinal wall but gives rise to many tufty branches, making up the rete mirabile, an important

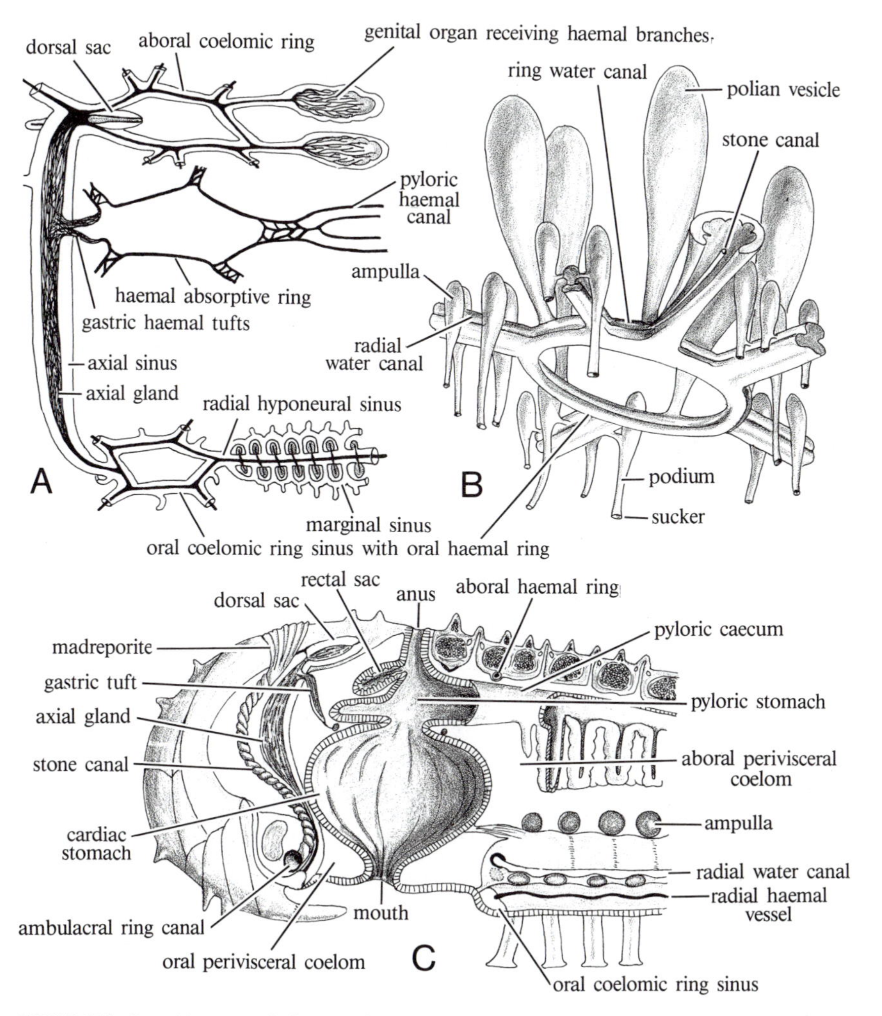

FIGURE 35.3. Asteroid anatomy. **A.** Scheme of hemal system and perihemal sinus system. **B.** Ambulacral system in mouth region. **C.** Schematized section through central disc and longitudinal cut through ray. (**A** after Cuénot, in Grassé.)

organ in digestion. A collecting vessel parallels the rete mirabile and gives rise to many small branches that attach to a hemal plexus in the intestinal wall. A ventral hemal sinus lies on the opposite side of the intestine.

The ophiuroid hemal system presents some interesting variations of the basic plan seen in asteroids. The oral hemal ring and the radial hemal vessels are essentially like those of asteroids. The aboral hemal ring (Fig. 35.2) is long and lies inside the aboral coelomic sinus, and on the radii it lies on the aboral side of the central disc just below the lateral shields. On each interradius, however, it turns orally to loop down near the oral shields. The aboral hemal ring vessel gives rise to branches to the bursal sacs and to the gonads, and receives a branch from the hemal plexus of the stomach of each radius. Because of the unusual sinuous course of the aboral ring, the axial gland inside an axial sinus turns orally from the oral hemal ring to reach the aboral ring (Fig. 35.2).

The ambulacral or **water-vascular system** is one of the most stable and characteristic elements of echinoderm anatomy. The radial water canals extend into the arms of asteroids, orphiuroids, and crinoids, and lie on the inner surface of the body wall in echinoids and holothurians. In any case, podia are attached to the **radial water canals** by **lateral canals** (Fig. 35.3B). Although crinoid and ophiuroid podia have no ampullae, the mechanics of the system's operation remain essentially the same throughout the phylum. Variations do occur in the number and position of **stone canals**, the **madreporite**, the external opening of the system, and the number and kinds of **polian vesicles**, saccate diverticula associated with the radial water canal around the stomodeum (Fig. 35.3B).

The oral water ring canal of sea cucumbers encircles the pharynx behind a calcareous ring (Fig. 35.5B). One or more saclike polian vesicles hang from the ring canal. Some holothurians have a single stone canal attached to the ring canal that sometimes reaches the body surface. In other holothurians the stone canal is branched, and several or many porous, ciliated madreporic bodies hang inside the coelom (Fig. 35.5A). Apodacea are unique among modern echinoderms in having no radial canals. All other holothurians have five radial canals that arch up over the calcareous pharyngeal bulb, enter the bulb near the oral end, and give off branches to the oral tentacles (Fig. 35.5A,B). They continue, much diminished, over the surface of the body wall beneath the ambulacra. Each radial water canal ends in a terminal podium, often associated with the system of anal papillae. Echinoids have a very similar system.

Ophiuroids may have a single hydropore in the

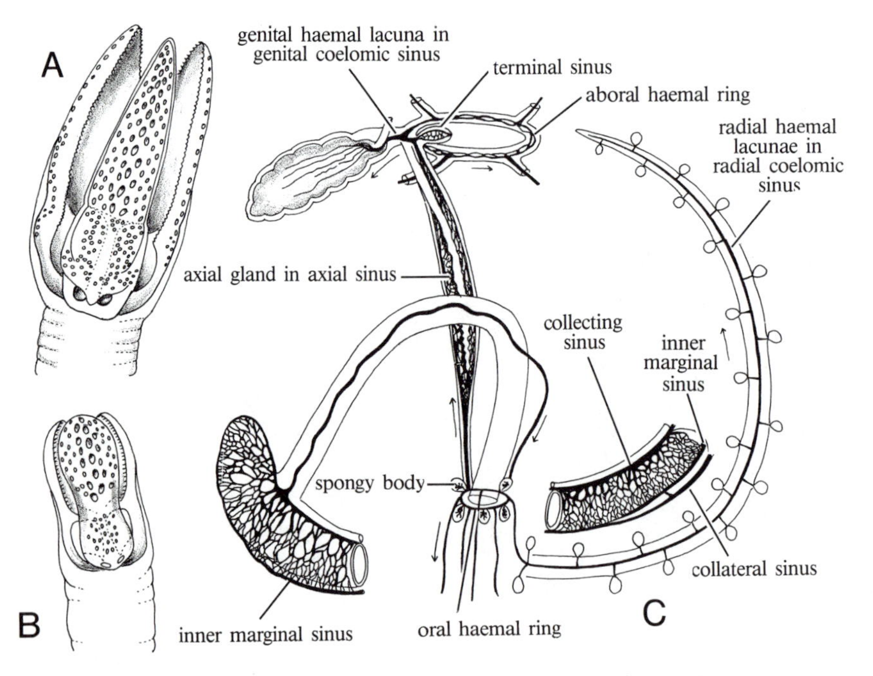

FIGURE 35.4. Echinoid pedicellariae and hemal system. **A.** Tridentate pedicellaria of *Arbacia*. **B.** Triphyllous pedicellaria of *Arbacia*. **C.** Scheme of hemal system of *Echinus*. (**A,B** after Reid, in Brown; **C** after Cuénot, in Grassé.)

oral shield, or may have a madreporite in each of the interradii. Typical ophiuroids have a pore canal between the hydropore and an ampulla. The ampulla and the ring canal are connected by a stone canal that passes through the axial sinus. The ring canal gives rise to polian vesicles on all of the interradii except the one occupied by the stone canal (Fig. 35.2). Special branches from the ring canal pass to the ten oral podia. The radial water canals pass along the aboral surface of the vertebrae in a notch. The water canal is expanded in each ossicle, and paired podial canals arise from these regions to the podia.

The locomotor podia of echinoids, holothuroids, and asteroids end in a plate of thickened epithelium. Marginal extensions of the epidermis and the connective tissue usually form **terminal disc** suckers (Fig. 35.5C). A podial nerve from the oral part of the radial nerve connects with a general epidermal nerve plexus in the podial wall and ends in a terminal nerve pad at the podial tip. One or more nerve rings may pass around the podium, most commonly near the tip.

Sensory podia are without terminal suckers and are often set on papillae in sea cucumbers (see Fig. 35.7A). They are especially abundant on the dorsal surface of holothurians, and in ophiuroids and echinoids. Irregular echinoids have flattened, saccate branchial podia used in respiration and located on the dorsal surface. Special larger and more powerful

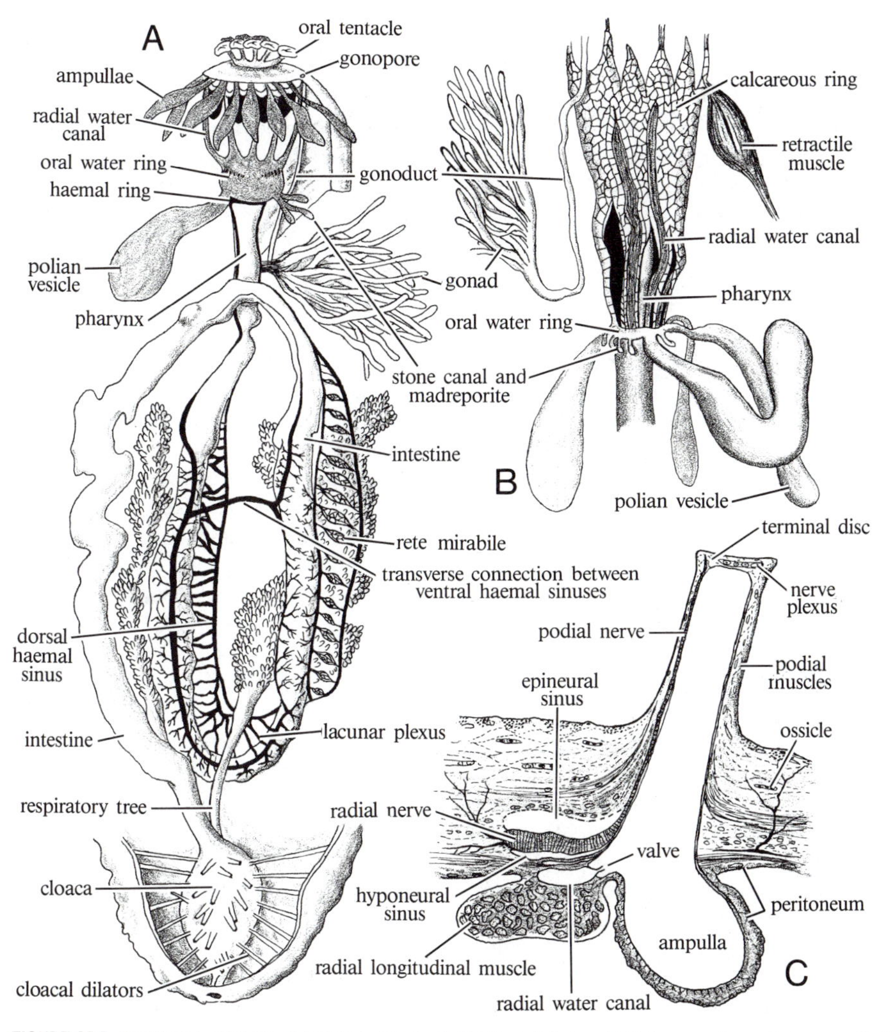

FIGURE 35.5. Holothurian anatomy. **A**. Internal anatomy of *Holothuria tubulosa*. **B**. Calcareous ring and pharyngeal bulb of *Thyone sacellus*. **C**. Section of podium and ambulacrum of *Holothuria*. (**A** after Cuénot; **B** after Selenka; **C** after Delage and Hérouard.)

oral podia occur in ophiuroids (Fig. 35.2) and echinoids. The oral tentacles of sea cucumbers (Fig. 35.5A) are modified podia.

Digestive System Although the echinoderm coelomic systems are hopelessly complicated, the other organ systems are quite simple as if to make up for it. Ophiuroids have the simplest of the echinoderm digestive tracts. The **mouth** lies in the middle of the **peristomial membrane**, covered by the modified arm skeleton forming the jaw apparatus. The short **esophagus** opens into a saccate **stomach**. The stomach bulges out between the bursae, nearly filling the cavity of the central disc, but there is no intestine, anus, or other digestive organs. The guts of asteroids and crinoids are complete, but very simple.

Echinoids and holothurians have an elaborate and elongated digestive tract. The **pharynx** passes through the pharyngeal bulb and emerges in the main body cavity (Fig. 35.5A). Echinoids have a large jaw apparatus, **Aristotle's lantern**. The main part of the gut is the **intestine**, but in some species a slender esophagus and stomach lie between the pharynx and the intestine. A **caecum** usually occurs at the union of the esophagus and the intestine. The intestine is far too long to fit in the body without coiling in most sea cucumbers. It is arranged in a clockwise spiral (see Fig. 35.23C), viewed from the oral end, descending to the aboral end of the coelom, and it ascends to a sharp aboral bend, the beginning of the rectum. A curious feature of the echinoid gut is the **siphon**, a narrow, ciliated tube branching from the first part of the intestine and running parallel to the intestine to rejoin it near the point of its recurvature (see Fig. 35.23B). The siphon is thought to shunt water past the first part of the gut and thus allow for the concentration of food and enzymes in the anterior intestine. The **rectum** is often swollen with a load of sand or mud. It leads directly to the **cloaca**, which is attached to the body wall by dilator muscles. **Respiratory trees**, highly branched diverticula of the cloaca (Fig. 35.5A), are used for respiration. The cloaca empties through an aboral **anus**.

Excretory and Respiratory Systems No special organ systems exist in echinoderms to address excretory and respiratory function. Rather, other parts of the body plan are adapted to these ends (see Biology section).

Nervous System The nervous system is extremely diffuse in echinoderms. A subepidermal nerve plexus extends everywhere on the body surface, and every projection from the body surface is richly supplied with sensory cells. Podia, papulae, pedicellariae, and spines react when stimulated. Special sense organs, however, are not abundant.

Most echinoderms have some remnants of three major nerve rings: the **oral** (ectoneural), the **deep oral** (hyponeural), and the **aboral** (endoneural). Except for the crinoids, the oral system predominates. The aboral nervous system has disappeared entirely in holothurians, but is much reduced in echinoids and ophiuroids, although some have a definite aboral nerve ring that gives rise to nerves to the gonads. Asteroids have a well-developed aboral nerve ring associated with a subperitoneal plexus that operates the muscles of the body wall appendages on the aboral surface.

The oral and deep oral nervous systems are closely coordinated. The oral system gives rise to sensory and motor fibers, whereas the deep oral system is at least predominantly motor. In ophiuroids the oral and deep oral systems are about equally well developed. The main nerve ring lies between the epineural and hyponeural ring sinuses. In addition, the radial nerves as well as the nerve ring are divided by a median septum into oral and deep oral parts. The deep oral nerve band produces only motor nerves to the intervertebral muscles, whereas the oral nerve band is larger and gives rise to motor and sensory nerves to podia and to the body surface.

Holothurians have a greatly reduced deep oral system. The oral system centers in a nerve ring in contact with the peribuccal coelomic sinus and gives rise to radial nerves that lie just external to the water canals. The deep oral system consists only of delicate motor nerves lying on the inner surface of the oral radial nerves that innervate the body wall muscles and make contact with the subepidermal plexus. The deep oral fibers do not reach the nerve ring.

In echinoids, the oral system centers in a nerve ring around the pharynx, and gives off nerves to the peristome and mouth region and five radial nerves that lie between the epineural and the hyponeural coelomic sinuses. The radial nerves give off branches to the podia and make contact with the rich subepidermal plexus. The plexus is strongly developed, forming rings around the spines and the pedicellariae. Five nerve masses on the pharyngeal wall just internal to the oral ring appear to be the vestiges of the hyponeural system. They innervate the lantern musculature.

Reproductive System The structure of the echinoderm reproductive system is relatively simple. Sexes are generally separate. **Ovaries** and **testes** are simple globose or diffuse organs with simple **sperm ducts** or **oviducts** leading to the exterior.

Biology

Locomotion Echinoderms will never earn any speed records among members of the animal kingdom. The basic form (Fig. 35.6) and mechanics of podia movement remain much the same between groups, regardless of the specific function of the podia or the presence or absence of ampullae. The lateral canals that connect the podia and the water canals usually have a valve that prevents backflow into the main water canal (Fig. 35.5C). Ampullae project into the coelom. The podia and the ampullae are attached to the water-canal system and are lined with the coelomic epithelium of the water-vascular system. Each podium contains the normal elements of the body wall: epidermis, dermis, muscle, and a lining of ambulacral peritoneum. Ampullae have perivisceral peritoneum, connective tissue, muscle, and the lining peritoneum forming their walls (Fig. 35.5C). When the valve of the lateral canal is closed, contraction of the ampullar muscles forces water into the podium (Fig. 35.6). The podium is prevented from dilating by connective tissue in the wall and so must lengthen. When the longitudinal muscles of the podial wall contract, the podium shortens and fluid is forced back into the ampulla. The degree of podial expansion is actually determined by the tone of ampullar and podial muscles and the amount of fluid on the podial side of the valve. Evidently, if the longitudinal muscle of one side of the podium contracts, the podium will bend (Fig. 35.6). If its terminal sucker is attached, a certain amount of leverage is applied to the whole animal or to the object to which the animal is attached. It is a common misconception that podia pull the animal forward. Most starfish and echinoid podia *push* the animal forward in a step much like that of a land vertebrate; pulling movements are most common in heavy-bodied echinoderms.

Despite the plodding movements and the diffuse nature of the nervous systems, there is effective locomotory coordination. In echinoids, movement calls for coordinating podia and spines, asteroid movement involves hundreds of podia and sometimes the body wall muscles, and holothurian movement depends on body wall muscles and podia. Podia can respond to stimuli as individual parts. In asteroids, the podia contract when exposed to strong illumination and may bend if touched lightly. Podia step toward the oral ring in isolated arms, but step toward the tip in arms containing parts of the nerve ring. Whichever happens to be the leading arm gains dominance over the others, which then step in its direction.

The podia of an isolated arm, like a well-trained army squad, march in unison. Local coordination is controlled by the radial nerves; destroying a radial nerve destroys coordination in that arm. The nerve ring governs general coordination; cutting the ring can isolate arms functionally. Some of the most interesting studies of coordination concern righting movements. Touching the aboral surface of a starfish or brittle star evokes an aboral reflex consisting of an aboral curvature of the arms caused by contractions of aboral muscle strands. Brittle stars right themselves by extending two adjacent arms in opposite directions, whereas the other three tilt the disc up and flip it over. Echinoids right themselves by attaching suitable podia and pulling themselves over. Sea cucumbers twist the body, bringing ventral podia in touch with the surface, and move off with the twist progressing along the body as it advances.

Asteroid righting has been most intensively studied and presents several enigmas. An overturned

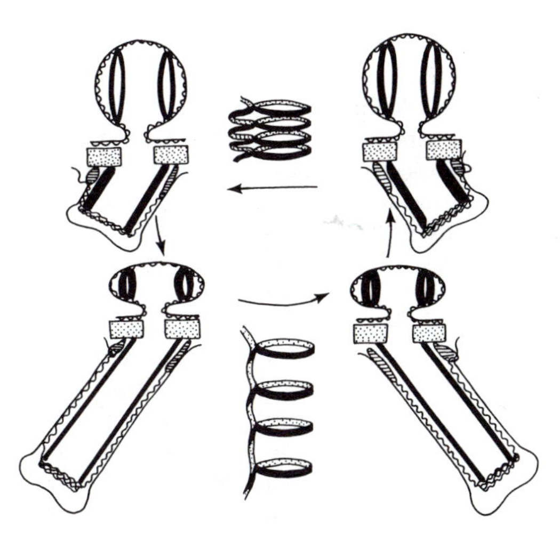

FIGURE 35.6. Mechanics of starfish step. Starting at upper left, blackened retractor muscles of podium contract. Podium shortens, with appropriate folding of longitudinal connective tissues, but does not dilate because of relatively inelastic circular connective tissue fibers. As result water flows into ampulla above. When protractor muscles of ampulla contract, water flows into podium, which elongates, straightening out longitudinal connective tissue fibers. Podium does not dilate, again because of inelastic circular connective tissue fibers. Stretched postural muscle (*shaded with lines*) on right side of podium contracts while its antagonist relaxes, and podium swings to right. Contraction of podial retractor muscles shortens podium, and as left postural muscle contracts, podium swings back, ready to make another step. (Based on Smith, 1984.)

starfish begins to right itself with a strong dorsal reflex, curving the body so that the podia at the tips of the arms eventually make contact with the substrate. Then the animal moves off on only two or three arms, the body flopping over when it has progressed far enough. According to some studies, righting cannot depend on podial stimulation, for starfish deprived of podia can still right themselves. Righting cannot depend on the nerve ring, for isolated arms, and even pieces of arms, if of suitable shape, can right themselves successfully. It cannot be dependent on contact of the aboral surface with the substrate, for it will take place despite removal of the aboral surface; nor can it be evoked by the sag of the stomach, as internal organs can be removed without interfering with it.

Brittle stars and feather stars use the arms, or arms and cirri, in movement. Arm movements are mediated by the radial nerves, while cirral movements of comatulid crinoids depend on the aboral nerve center.

Digestion Much attention has been directed toward the study of feeding and nutrition in echinoderms (Jangoux and Lawrence, 1982). The stomodeum is associated with the reception of food and its conduction to the midgut, and is sometimes adapted for food storage and the mechanical trituration of food. Its development varies considerably in different echinoderm groups. In ophiuroids, for example, modifications of the arm skeleton and spines enter into jaw structure, whereas in echinoids, the lantern is associated with the stomodeum.

Food is moved along the gut by muscular action as well as by cilia or flagella. Some ciliation of the gastrodermis occurs in all echinoderms, but it is sparse in holothurians. Holothurians with short tentacles do not capture food, but eat the mud or sand around themselves to digest out whatever suitable material it may contain. A great deal of mud must be processed to avoid starvation. A *Paracaudina* handles up to 64 kg of mud every year, and in one Bermuda bay the sea cucumbers pass from about 315 to 450 metric tons of sand through their digestive tracts per square mile per year. The dendritic tentacles of other holothurians are covered with mucus and are used to catch algae, diatoms, and small plankton organisms. The tentacles are licked off in a regular series, one by one, as a child might lick sugar from the fingers. Sea cucumbers have muscular intestinal walls. Peristaltic waves move along the intestine of *Synaptula* at intervals of about two seconds to push the food along.

Crinoids, on the other hand, depend more than most echinoderms on currents generated by a monociliary epidermis, in association with the habit of particle filtration. Usually cilia and muscles cooperate in moving food materials and digestive juices within the gut. A similar mode of filter feeding is also seen in some ophiuroids and asteroids under certain conditions.

Strong musculature develops in forms in which the stomach is everted for feeding. Starfish mouth muscles are relaxed by acetylcholine and contracted by adrenalin. The relaxation of the mouth is accompanied by contraction of the body wall muscles; the increased internal pressure forces the stomach out. The stomach is withdrawn by active contraction of the gastric musculature and relaxation of the body wall, presumably also as a result of muscles that respond antagonistically to neurohumoral substances.

Enzymes have been specified for several echinoderms; no great consistency is observed, as might be expected in view of dietary differences among them. Generally, protein digestion seems most active, carbohydrate digestion intermediate, and fat digestion least active.

Echinoderm glands have little surface amplification to form compound glands. Only asteroids have very large gut diverticula, and, although the gastric caeca secrete enzymes, they are also used in absorption and intracellular digestion. The intestinal caeca of *Asterias* pulsate rhythmically, the muscular movements helping cilia to propel food particles and digestive juices. Evidently species that evert the stomach must secrete large quantities of enzymes for preliminary digestion of food. In other echinoderms, unicellular glands are intermingled with ciliated epithelial cells. These glands occur abundantly in the pharyngeal region of holothurians, in the esophagus of echinoids, and in the esophagus and stomach wall of asteroids, but they are sparse in the holothurian intestine. Gland cells have not been described in the digestive tract of ophiuroids. Compound glands are occasionally seen, as in the esophagus of echinoids, but are rare.

Extracellular digestion is evidently important in asteroids that evert the stomach. The relative scarcity of gland cells in the gut wall of the other echinoderms suggests that some other source of digestive enzymes may be important. The transparent mucous covering of food balls in the echinoid intestine is known to include bacteria, and an agar-dissolving bacterium has been recovered from the intestine of the purple urchin, *Strongylocentrotus*. The hemal plexus of sea cucumbers contains the same enzymes as the intestine. It is thought that coelomocytes from the hemal plexus enter the gut to bring in enzymes and in turn to load up with food for the journey out. Intracellular digestion may occur in the coelomocytes. It is probable, in view of the dearth of gland cells, that the microbial flora and coelomocytes play an important role in digestion. A pH gradient is evident in the echinoderm digestive tube, with the acidic conditions in the stomach or the beginning of the intestine, and basic conditions in the rectum. As enzymes are active in specific pH ranges, some sequentialism of extracellular digestive processes probably occurs.

Modifications of the digestive tract are sometimes clearly related to function. The reduction of the intestine in asteroids that feed by stomach eversion, and in asteroids and ophiuroids that egest shells or hard fragments through the mouth, is an evident case. The pyloric caeca of asteroids amplify surface for secretion, absorption, and intracellular digestion, but the function of other diverticula is unknown. Probably the festoons of the ophiuroid stomach are associated with absorption rather than secretion, and the lengthening of the digestive tube of crinoids, echinoids, and holothurians may also be associated with something other than secretion. Peculiarities of the absorptive process make it uncertain to what extent echinoderm absorptive surface is increased by lengthening the digestive tube. Possibly some of the modifications in gut shape are important in delaying the expulsion of food, thus giving more time for the digestive processes to be completed.

Excretion Echinoderms have no excretory organs, but excretion is carried out by a variety of processes. Echinoderms are largely ammonotelic, but some excrete significant percentages of other compounds. They release nitrogenous wastes, ranging from 10 percent to more than 50 percent. Uric acid and allantoin are common excretory products of purine metabolism; both have been recovered from echinoderms. When the enzyme uricase is present, uric acid is converted to urea. All urea does not stem necessarily from uric acid, however, as ammonia may be formed from urea, or urea from ammonia. In *Asterias*, the ratio of ammonia to urea nitrogen released is approximately 4 to 1. Some echinoderms apparently liberate significant amounts of amino acids. Creatine and creatinine have been reported, and as they are also liberated by vertebrates, these substances have been considered to be phylogenetically significant. These compounds, however, are relatively unimportant in echinoderms. The overall impression is that echinoderms are for the most part ammonotelic, with urea the next most important excretory product. Uric acid is not excreted in large amounts, and amino acids are often more significant.

Soluble wastes leave at the body surface as well as the intestinal surface. Intracoelomic injection of the dye chlorophenal red is eventually excreted through the anus. Insoluble wastes are picked up by amoebocytes for disposal in various ways. Many amoebocytes contain brownish inclusions that may be urates. Special devices capitalizing on the phagocytic capacities of coelomic amoebocytes are sometimes found. Apodan holothurians have ciliated urns in the peritoneal lining. Here, coelomocytes accumulate to form masses that pass into the body wall or fall into the coelom as brown bodies. Loaded amoebocytes leave asteroids through the papulae or podia, and echinoid amoebocytes also leave the body through the podia. The axial gland seems to be an important way station, because loaded amoebocytes are abundant there and may make their way from there to the exterior through the madreporite.

Osmoregulation Echinoderms have little ability to control the entry or the exit of water or salts. The sea cucumber, *Paracaudina*, swells in diluted seawater and loses weight if returned to normal seawater. Pieces of the body wall are fully permeable to water and to inorganic cations and anions. Various asteroids (*Asterias*, *Solaster*, and *Astropecten*) have demonstrated the same inability to control water intake, but have some control over ions as more potassium and hydrogen ions occur in the coelomic fluid than in seawater. They can survive some dilution of seawater, but only by tolerating a more dilute coelomic fluid, a swollen body, a soft body wall, or other morphological changes. Echinoids imbibe water when placed in a dilute medium but cannot swell as they are enclosed within a solid shell. The coelomic fluid of some echinoids has a higher potassium, sodium, and chloride content than the external medium, as well as a lower pH.

Echinoderms evidently live in "glass," or in this case calcitic, houses. Their metabolism is carried on without the protective physical barrier that the body wall of most animals provides. Therefore, it is imperative that echinoderm cells be able to retain ions. Many ions are more concentrated in tissues than in coelomic fluid in contrast with the usual arrangement in animals where the body fluid provides a controlled isotonic environment for tissue cells. Consequently, echinoderm cells must use more energy for the control of inorganic cations and anions than is the case in most other animals.

The inability to control the internal ionic and water content, except at the level of individual cells, has probably been a factor in the restriction of echinoderms to marine habitats. Most echinoderms do not venture into brackish water, and they avoid estuaries.

Respiration Echinoderms are aerobic, but most echinoderms are not very effective in controlling respiratory exchange. Oxygen uptake of starfish and echinoids is dependent on environmental oxygen tension; under ordinary conditions they have no safety factor. This is not always the case. For example, although the sea cucumber *Paracaudina* takes up oxygen at rates dependent on the oxygen tension, the oxygen consumption in *Thyone* remains constant until oxygen tension has fallen to about one-seventh of normal figures. Evidently some echinoderms can have some physiologic controls of respiration, but the activities of most are limited by the availability of oxygen. Considering their relatively active life, it is clear that the respiratory arrangements are poor.

Respiratory adaptations differ in each class and sometimes in smaller subgroups. Most respiratory modifications seem to be expedient solutions to respiratory problems that may have appeared with increased size or activity. They involve surface amplifications bringing the coelomic fluid in close proximity to the sea, or developing a pumping device that "breathes" seawater, such as at the end of the hindgut.

Crinoids have no special respiratory structures. Seawater enters the coelom through pores in the tegmen, but this exchange is small and serves primarily to maintain coelomic pressure. Respiratory exchange occurs at the podia, but the rest of the body surface must also play a minor role. Podial respiration is important in all echinoderms, even when other respiratory structures are present. In starfish, for example, the covering of one ambulacral groove reduces respiratory exchange by about ten percent. Undoubtedly, body surface and podial respiration are the principal modes of respiration in the phylum.

Currents in the ambulacral grooves or over the general body surface ventilate the podia and the surface epithelium and help the exchange of oxygen. The thin podial walls bring the fluid of the water-vascular system close to the exterior, and therefore, podia have an excellent adaptation for gaseous exchange. They are not primarily for respiration, however, and the aerated fluid in the water canals does not reach many tissues, although oxygen in it can pass into the general coelomic fluid. If no other respiratory structures are found, podia may be specially modified for respiration; for example, irregular echinoids have flattened or ruffled branchial podia that extend from the petaloid, aboral ambulacral regions.

Commonly, echinoderms resort to nonpodial respiratory adaptations. Special body wall extensions, with podialike walls, occur in some. Asteroids have papulae, and regular echinoids have peristomial gills. Both are thin-walled evaginations of the body surface, contain coelomic fluid agitated by peritoneal flagella, and are equipped with external flagella that ventilate the outer surface. Starfish papulae occur over most of the body surface or on the major part of the aboral surface. The peristomial gills of echinoids occur in a circlet on the peristome.

The elevation and depression of the compasses of Aristotle's lantern by the lantern musculature raises and lowers the aboral membrane that bounds the peripharyngeal coelomic sinus, thus pumping coelomic fluid in and out of the gills. The substitution of muscular for ciliary or flagellar ventilation of surface membranes is a general tendency of larger and more complex invertebrates. A somewhat similar mechanism has developed in some starfish where the upper surface of these starfish is a false front, so to speak. A membrane is supported by the outer ends of the aboral paxillae. A spacious nidamental chamber lies beneath this membrane, with a floor that consists of the body wall covered by many papulae. The upper wall of the chamber is pierced by a large central osculum and many minute pores (spiracles). Openings also occur at the ray margins. Water is pumped in and out of this nidamental chamber, entering through the marginal openings and spiracles, and departing through the osculum. With each pulsation, the papulae expand and contract.

Most ophiuroids have a different respiratory mechanism, consisting of ten bursal sacs, open to the exterior through bursal slits (Fig. 35.2). Ciliary currents carry water into and out of these bursae. In some ophiuroids, disc movements pump water in and out of the bursae once or twice a minute. The bursae are sometimes very large and partially united, and encroach on the perivisceral coelom.

Many sea cucumbers have extensively branched diverticula of the hindgut known as respiratory trees (Fig. 35.5A). These are filled and emptied by means of regular pumping movements of the cloaca. Although gross, this method is effective. The anus is opened, and contractions of dilator muscles between the body wall and the cloacal wall fill the cloaca. The anus is closed, the cloacal wall contracts as a valve at the base of the respiratory trees opens, and water is forced into the respiratory trees. To expel the water, the respiratory trees contract, often simultaneously with body wall muscles. When the respiratory trees are lost by evisceration, cloacal pumping continues to move water into and out of the coelom. Cloacal pumping can be disturbed by stimulation of the oral tentacles, but the main regulatory mechanisms lie in the aboral end. The beat of the cloaca ceases if it is separated from the body wall; evidently the body wall nerve plexus exerts control. In isolated strips of cloaca a reduction in oxygen tension increases the vigor of pumping, but if it falls too low pumping rapidly ceases.

If a sea cucumber anus is covered by a rubber membrane, body oxygen consumption falls immediately by half or more. The animal extends its oral tentacles fully and stretches its body out in a reflex that favors oxygen absorption at the body surface. Anal pumping is not unique in sea cucumbers. Some crinoids pump water in and out their anus, and this activity also occurs in some other animal phyla, like the Echiuroidea and aquatic uniramians.

The fact that most echinoderms are oxygen conformers—and most are quite able to increase their metabolic rate and oxygen consumption if placed in a milieu providing higher than normal environmental oxygen tensions—indicates that respiration is a limiting factor in echinoderm biology.

Internal Transport Despite the diversity in respiratory adaptations for gaseous exchange, special internal transporting mechanisms have not been effectively developed by most echinoderms. In a situation such as exists in echinoderms, where water and salts pass freely through the body wall, how effective can coelomic fluid alone be in internal transport? Peritoneal cilia or flagella agitate the coelomic fluid of all echinoderms, and there can be no doubt that this agitation is useful in the transport of materials. The use of coelomic components as a hydraulic skeleton, and as a hydraulic system for moving gills or other parts, changes but does not wholly destroy definite currents in the perivisceral coelom and its extensions into the arms. In sea cucumbers, the coelomic fluid flows aborally along the inner surface of the body wall, and returns toward the mouth along the central axis of the body; in the tentacles the current passes toward the tip at the oral surface and back toward the body on the outer side. The coelomic fluid of starfish tends to flow distally along the aboral surface and the center of the central disc and arms, and returns along the margins of the rays.

The flowing coelomic fluid moves coelomocytes and aids in the distribution of oxygen. But what of food distribution? There is little organic material in the coelomic fluid, except in the form of coelomocytes. Amino acids injected into the digestive tract of a sea cucumber, *Arbacia*, appear in the coelomic fluid. Sugars injected into the gut may or may not appear in the coelomic fluid, but sugars injected in the coelom disappear as soluble sugar as they are rapidly converted into glycogen. Amino acids have been recovered from the coelomic fluid of feeding starfish. The organic content of sea cucumber coelomic fluid is also low.

The high permeability of the body wall to water and salts, so characteristic of most echinoderms, suggests that the coelomic fluid has no more than a limited role in food distribution, but evidence for this is at best equivocal. The chemical constitution of the coelom, however, is too close to that of the ocean and there are too many thin-walled podia, gills, and other structures to have faith that the coelomic fluid is the sole agent in nutrient transport. Another possibility is the use of coelomocytes in transport. These cells occur in the gut lumen, in the coelom, in the hemal system, and in the dermis of the body wall. They are often filled with food inclusions, supporting the idea that they are important in food transport. The clearest evidence for food transport by coelomocytes is found in experiments on sea cucumbers, where these cells appear to convey enzymes into the intestine and food out of the gut. Good evidence for similar coelomocyte activities in other echinoderms is lacking, although such has been suggested by many investigators. Food may also be stored in the gut wall. This is most clearly demonstrated in asteroids, where specially differentiated storage cells occur in the walls of the pyloric caeca.

The close relationship of the hemal system and the gut, as evidenced by the extensive hemal sinus system of holothurians and echinoids and to a lesser extent in other echinoderms, suggests that the hemal system is important in food distribution. In holothurians the hemal fluid contains hemocytes and, thus, plays a role in oxygen distribution, but the echinoderm hemal system has never become an effective circulatory system. Coelomocytes in the water canals of one species of brittle star have been reported to contain hemoglobin, but this has not been confirmed. Other pigments found in coelomocytes, however, have proved not to be respiratory.

Perhaps extensive currents in the perivisceral and water-vascular coeloms tend to make further development of the hemal system redundant. Movement of fluids in the hemal system is caused by flagella on the peritoneal lining, to some extent aided by contractility of parts of the hemal walls. The head of the axial gland contracts in some asteroids, the dorsal hemal sinus of holothurians is contractile, and the collateral sinus of some echinoids is also contractile. In some instances a local circulation is established in the hemal vessels and plexus associated with the gut. However, the walls of the hemal system are incomplete; all of the coelomic spaces communicate more or less directly with each other, varying in degree with the group. Therefore, internal transport occurs to some extent in all coelomic systems: the perivisceral coelom, the water-vascular system, and the hemal system. This illustrates again the diffusely spread functions typical of echinoderms.

Sensory Organs The neurosensory cells of echinoderms are like those of other invertebrates; that is, they terminate in sensory tips and have basal fibers that make contact with the nerve plexus. In some cases they form sensory buds (Fig. 35.7B), especially common in the jaws of globiferous echinoid pedicellariae and in the apodan holothurians in the oral and anal regions, or in papillate projections of the body surface. Generally speaking, neurosensory cells are concentrated on the peristomial membrane or around the mouth opening, and are also abundant near the anus.

The light-sensitive cells of most echinoderms are scattered over the body surface. Asteroids often have an optic cushion at the tip of each ray, consisting of a cluster of small pigment-cup ocelli. Some sea urchins have bright blue bodies on the genital plates, and in some species, similar blue bodies are found on the peristome. In *Astropyga radiata*, they form stalked bodies that histologically resemble a compound eye. Although there is good evidence for a light-receptive function of asteroid optic cushions, the echinoid "eyes" are not definitely shown to be light receptors.

Responses to light vary; some echinoderms are photonegative and others photopositive. Strong light usually evokes contractions of gills and podia, and affects the individual reactions of most body wall appendages. Two kinds of light reception occur in starfish: reception by ocelli and at the body surface. Oriented responses depend on the ocelli, whereas choice of a suitable illuminated background is dependent on general body surface. An isolated starfish arm exhibits a beautiful orientation to light, curving its tip up toward the cut end, thus pointing the optic cushion at the light source and stepping off toward it. Generally, sensitivity to light declines in species that live in deeper water.

Some echinoderms have balance organs. For example, some sea cucumbers have interesting statocysts set on stalks and hanging like hammers over the nerve ring or the base of the radial nerves (Fig. 35.7C). Many echinoids have somewhat similar organs. Geotactic responses, however, can occur in echinoderms that lack balance organs, presumably as a result of tactile stimulation of the upper or lower body surface.

Local responses vary with the part stimulated. Most pedicellariae respond positively to moderate contact, but bend away from strong mechanical stimuli. Pedicellariae with poison glands are not repelled by strong mechanical force. They attack fearlessly and break off, if necessary, still clamped to the intruder. Echinoid spines have a double muscula-

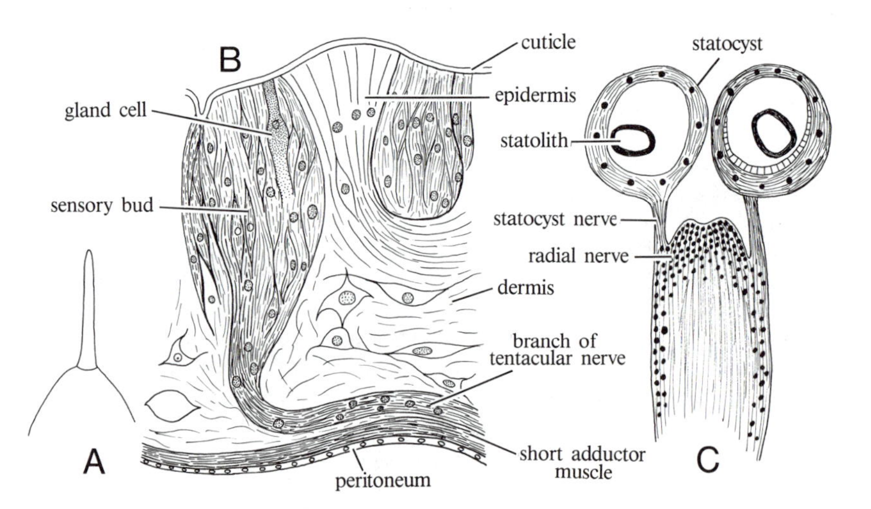

FIGURE 35.7. **A**. Papillate sea cucumber podium with sensory filament extended. **B**. Section showing sensory buds in oral tentacle of sea cucumber *Caudina*. **C**. Statocysts of sea cucumber *Caudina*.

ture, one to set the spine in position and one to manipulate it. Touching a spine causes contractions that set it, but the holding muscles relax, and manipulating muscles come into play when the body surface at the base of the spine is touched. Reactions of this kind are wholly controlled by local reflexes in the subepidermal nerve net, and are not disturbed by removal of all traces of the radial nerves. The echinoderm subepidermal nerve plexus, however, is in no way comparable to the nerve net of cnidarians. True, it transmits impulses in all directions; but it is characterized by definite, often ring-shaped ganglia located at the bases of spines and pedicellariae and in the podia.

The oral (ectoneural) nervous system is both sensory and motor, and is closely associated with the subepidermal nerve net. The deep oral (hyponeural) system is motor, innervating the podia and muscles of the body wall. Except in crinoids, the aboral (entoneural) nervous system, when it occurs, is largely concerned with the innervation of the gonads and the gonoducts. There is nothing in echinoderms comparable to a "brain." The nerve rings are essentially condensations of a peripheral nervous system.

Despite the "mindless" state of echinoderms, some studies have been directed toward the ability to modify behavior through "learning" experience. Starfish and brittle stars can be "trained" to right with arms not normally used, and animals so programmed retain the modified behavior for several days. They have also been trained to draw back from a boundary they would normally cross, but instruction is difficult and retention is short. These reactions occur despite the absence of a central brain.

Reproduction The specifics of echinoderm reproduction are described with each particular group. Sexes are separate and fertilization is most often external. Some specific cases of brooding do occur, for example, in ophiuroids, the asteroid *Leptasterias*, and concentricycloids. Brooders usually form larger, yolkier eggs that undergo modified types of cleavage, and development is shortened by the omission of some larval stages. These seem to be exceptions to the rule among echinoderms of little or no parental care of offspring. Asexual reproduction of echinoderm larvae has been reported (Bosch et al., 1989), as well as in adults (Emson and Wilkie, 1980).

Development Echinoderms undergo radial or biradial, holoblastic cleavage (Fig. 35.8A,B). The first cleavage divisions are indeterminate; the blastomeres of two-cell or four-cell stages, if separated, can each give rise to complete, although diminutive, larvae. Cleavage ends in the formation of a coeloblastula with a spacious blastocoel. A sticky, hyaline coat often combines with microvilli of the cell surface cortexes to maintain the radial, sheetlike arrangement of cells and thus predetermines the stereogeometry of cell layers in succeeding stages of development (Schroeder, 1986).

Invagination at the vegetal pole produces a small archenteron during gastrulation. Sometimes invagination is preceded by the unipolar ingression of some cells, and mesenchyme cells can move into the blastocoel during gastrulation (Fig. 35.8C). These cells arise from the differentiating primitive gut or its margins, and give rise to the skeletal spicules of the larvae.

The blastopore becomes the larval anus. It will be convenient to call this the posterior end. The anterior archenteron eventually forms the coelom, and the middle portion develops into the gut cavity. Larva formation involves (1) the origin of coelomic primordia, (2) the completion of the larval gut, and (3) the development of one or more locomotor bands of cilia on the surface. The timing of these events differs in various groups and is different between species in the same group.

Coelom development is essentially archimeric, similar to that seen in lophophorates and hemichordates. The coelom initially arises from the anterior portion of the archenteron (Fig. 35.8D,E,F) and then divides into right and left compartments. Each compartment lengthens and is subdivided into anterior and posterior parts, although the single, paired compartments may remain for a time (Fig. 35.8G). Eventually the posterior compartments form the right and left somatocoels, from which the perivisceral coelom develops. The anterior compartments are the axohydrocoels and become partially separated to form an axocoel and hydrocoel on both sides of the embryo. The right and left axohydrocoels are not symmetrically developed (Fig. 35.8I,J); in two groups of modern echinoderms, the crinoids and the holothurians, the right axocoel and hydrocoel never appear. Only four effective coelomic compartments develop further: right and left somatocoels, left axocoel, and left hydrocoel (see Fig. 35.11A,B). The left hydrocoel grows into an arc that becomes scalloped as five evaginations develop (Fig. 35.8I), which are the beginning of the ring canal and the radial canals. A stone canal connects the left hydrocoel with the axocoel, and the axocoel opens to the exterior by way of a hydropore and pore canal (Fig. 35.8H,I,J). The right and left axocoel and right hydrocoel are much reduced. Typically the madreporic vesicle arises from the right axocoel. The relationship of these first compartments of the coelom to the adult coelomic divisions is summarized in Table 35.1.

The development of the gut is completed while the coelomic compartments are being partitioned off. A stomodeum appears on the surface and grows inward to make contact with the enteric sac (Fig. 35.8G), that part of the expanded archenteron left behind when the coelomic primordia form. The larval anus develops from the blastopore, and the larval mouth from the stomodeum. The foregut contains the stomodeum and part of the enteric sac. The midgut or stomach is formed by the rest of the enteric sac, and the hindgut or intestine by the posterior end of the primitive gut. The larval gut is fully functional in free-swimming larvae; only in highly modified embryos that develop from yolky eggs does the gut remain nonfunctioning until metamorphosis.

The embryo is initially entirely covered by cilia or flagella. While the early steps in coelom formation take place, the surface cilia disappear except in one or a few bands. The development of a locomotor band transforms the embryo into a larva. Ciliated larvae of different types appear in different echinoderm groups. The simplest larval form is the auricularia of holothurians (Fig. 35.9A). A single locomotor band curves forward as a preoral loop, and forms a similar anal loop at the posterior end. Parts of the band degenerate, and the remaining parts are rearranged into three to five transverse locomotor

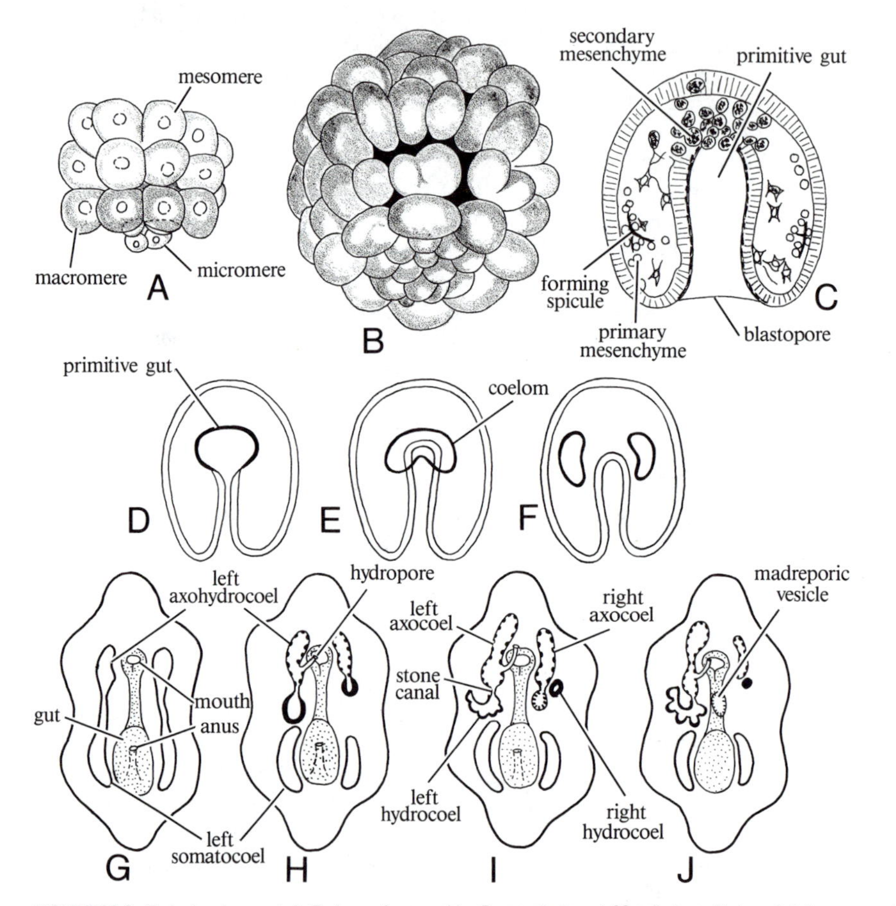

FIGURE 35.8. Early development. **A**. Embryo of sea urchin, *Paracentrotus,* at 32-cell stage. Note radial cleavage and micromeres at vegetal pole. **B**. Later cleavage stage of sea urchin, *Psammechinus,* showing field of micromeres at vegetal pole. **C**. *Paracentrotus* embryo at end of gastrulation. Primary mesenchyme from borders of blastopore have begun to fill blastocoel and form skeletal spicules, and secondary mesenchyme forming from inner end of primitive gut. **D,E**. Coelom cut off from inner end of primitive gut. **F**. Coelom divides into right and left compartments. **G,H**. Gut forms, right and left coelomic compartments elongate, and eventually divide into posterior somatocoel on each side and anterior axohydrocoel. **I,J**. Small right axohydrocoel, sometimes not formed, further reduced during development. Left axohydrocoel divides into axocoel that opens to outside through hydropore, and hydrocoel connected to axocoel by stone canal. Hydrocoel on left develops into water-vascular system. (**A,C** after Boveri; **B** after Selenka; **D,E,F,G,H,I,J** after Dawydoff, in Grassé.)

bands, resulting in a doliolaria larva (Fig. 35.9B,C). The initial crinoid larva is a doliolaria, similar to the holothurian doliolaria (Fig. 35.9D). The first starfish larva is a bipinnaria, which resembles an auricularia when young, but develops a number of larval arms and loses its auricularian form (Fig. 35.9E). The bipinnaria swims about for several weeks (Fig. 35.9F) and eventually develops three brachiolar arms to become a brachiolaria (Fig. 35.9G).

Ophiuroids and echinoids have a pluteus larva, termed ophioplutei and echinoplutei, respectively. The echinopluteus forms as the gastrula (Fig. 35.9I) becomes flattened on the future oral surface. The locomotor band follows the margin of an oral lobe that bends over the oral surface, and the young pluteus (Fig. 35.9H) grows two pairs of arms, a postoral pair and an anterolateral pair. More arms develop during later growth; most echinoplutei have five pairs, but spatangoid echinoids have six pairs of well-developed arms (Fig. 35.10A). The ophiuroid gastrula develops a ciliated girdle (Fig. 35.10B), and arms form more slowly. Only four pairs appear, and

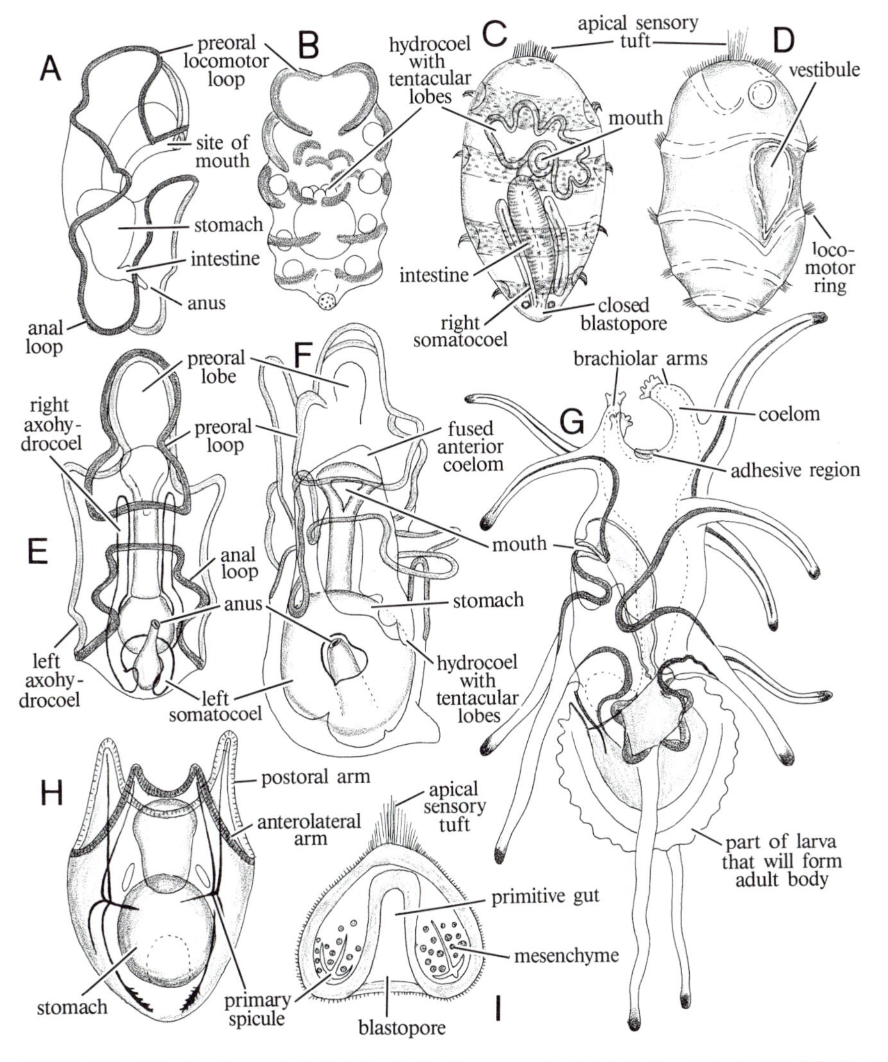

FIGURE 35.9. Echinoderm larvae. **A**. Auricularia, simplest larva, occurring in holothurians and asteroids. **B,C**. By partial degeneration of locomotor bands and their rearrangement, auricularia of holothurians becomes doliolaria. **D**. Crinoids do not form auricularia, but have doliolaria larva much like that of holothurians. **E,F**. Young asteroid larva develops arms to become bipinnaria; and then later **G**, brachiolar arms and sucker develop, transforming bipinnaria into brachiolaria that attaches and metamorphoses into starfish. **H,I**. Echinoids do not go through auricularia stage; gastrula flattens and develops into young pluteus larva. (**A,B,H,I** after Mortensen; **C** after Runnström; **E,F** after Hörstadius; **G** after Mead.)

nearly three weeks are required to complete their formation (Fig. 35.10C). The arms of echinoplutei and ophioplutei contain skeletal supports, and many are pigmented.

All echinoderm larvae undergo metamorphosis. Starfish and crinoids attach in order to metamorphose, but holothurians, echinoids, and ophiuroids continue to swim about, eventually sinking as they become too heavy to swim. It is not possible to discuss here the details of metamorphosis in all echinoderm groups; it will suffice to indicate the main changes in a starfish as an example (Fig. 35.11).

A starfish larva ready for metamorphosis feels about for a suitable site with the brachiolar arms. It eventually attaches with the aid of an attachment organ between them. The adult starfish forms from the rounded, posterior part of the brachiolaria; the arms and anterior part of the larva deteriorate during metamorphosis (Fig. 35.11A). The left side becomes the oral surface, and the right side the aboral surface. As the old anterior part of the larva degenerates, the crescentic hydrocoel rounds out to make a five-lobed disc around the esophagus (Fig. 35.11C,D,E). It is attached to the left axocoel by the stone canal. The axocoel opens to the exterior through the hydropore. A small remnant of the right axocoel (Fig. 35.11B) becomes the dorsal sac in which the head of the axial gland lies, and the axial gland develops in the coelomic compartment of the left axocoel, paralleling the stone canal. As the left hydrocoel completes the water ring, the first buds grow out to form the first podia (Fig. 35.11D,E). Shortly afterward, the young asteroid moves away from the sucker, which had attached it during metamorphosis, and takes up an independent existence.

Review of Groups

CLASS CRINOIDEA

Crinoids (Figs. 35.1A and 35.12) are the only living Pelmatozoa, stalked echinoderms. At their peak in the Middle Mississippian period, about 350 million years ago, they were diverse and abundant (Fig. 35.12B). Many limestones of that period are virtually pure crinoid fossils. Four of the five crinoid sub-

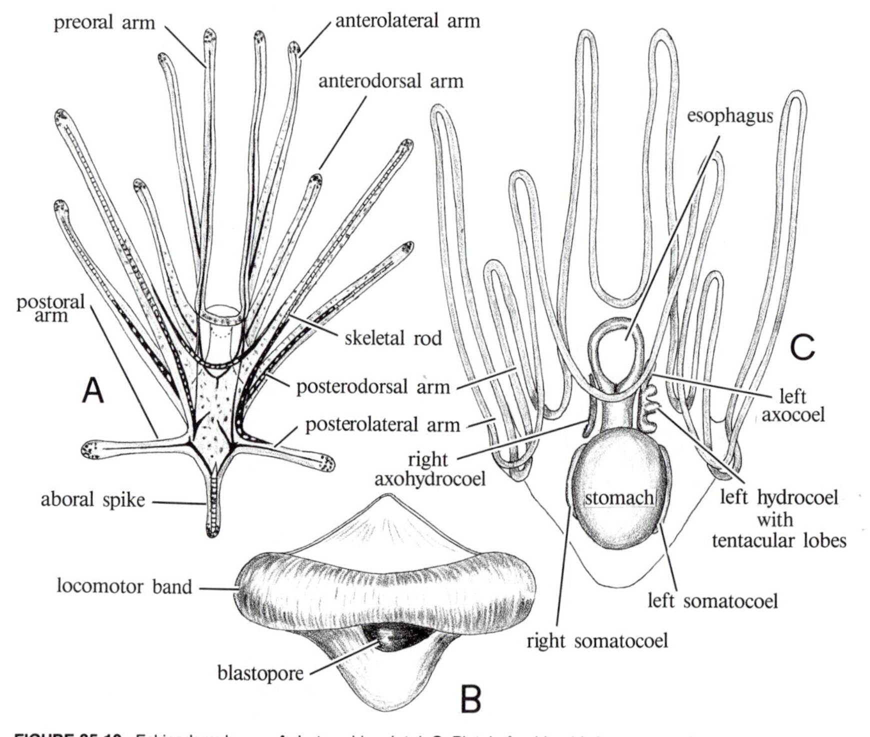

FIGURE 35.10. Echinoderm larvae. A. Late echinoplutei. C. Plutei of ophiuroids known as ophioplutei, developing directly from motile gastrula B, and structurally similar to echinoplutei. Skeletal spicules omitted from ophiopluteus shown here. (A after Mortensen; B after MacBride; C after Narasimhamurti.)

classes are extinct, and about 6000 extinct species have been described; therefore, one can think of crinoids as nearly extinct as there are only about 600 modern species. Nevertheless, they can be locally abundant. As many as 10,000 feather stars were taken once in a single haul of the *R. V. Albatross*'s nets.

All modern crinoids belong to the Articulata. Modern crinoids are stalked sea lilies belonging to either the order Isocrinida, with cirri on the stalk (Fig. 35.12A), or the order Millericrinida, without cirri. Feather stars belonging to the order Comatulida are stalked as juveniles but break free and move about on the bottom as adults (Fig. 35.12C). Feather stars are among the most beautifully colored animals. They creep about awkwardly on the aboral cirri or swim weakly but gracefully by undulating movements of the arms. Sea lilies are usually found in deep waters, and are pale or colorless. Feather stars from deep water are also pallid.

Crinoid stalks can have a number of holdfast adaptations: rootlike extensions, basal discs, or anchoring hooks. The stalk looks jointed, for it is supported by a series of hollow skeletal columnals. The longer nodal columnals have whorls of movable cirri attached to facets. Between them are many internodal columnals without cirri or facets. Cirri look jointed too, for they also are supported by a series of discrete ossicles known as cirrals. Cirri were primitively located only at the base of the stalk, as in millericrinids, but have spread upward, eventually reaching the aboral surface of the calyx. Comatulid cirri attach directly to facets on the aboral body surface; they vary greatly in form, depending on whether the feather star perches on rocks, creeps about on the bottom, or clambers about on algae.

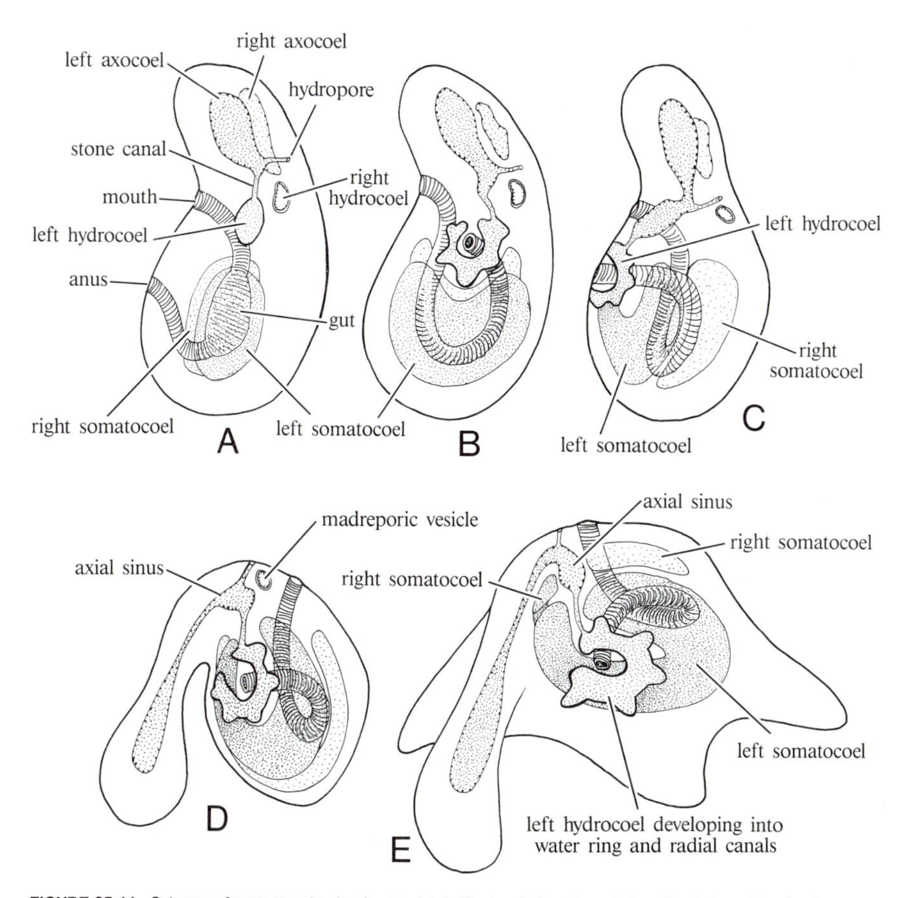

FIGURE 35.11. Scheme of metamorphosis of asteroid. **A.** Basic relationships of digestive tube and coelomic compartments as seen from left side. **B.** As result of flexion, larva becomes asymmetrical and anus comes to lie on aboral surface. Left hydrocoel encircles mouth and develops, eventually, into water-vascular system. **C,D,E.** Left somatocoel carried to oral surface and forms oral part of perivisceral coelom. Right somatocoel forms aboral part of perivisceral coelom. (After Heider, from Dawydoff, in Grassé.)

The main part of the crinoid body is the corona. It is covered by an aboral calyx formed of loose-fitting skeletal ossicles, and a membranous oral tegmen often containing small skeletal ossicles. The arms arise at the junction of the calyx and the tegmen.

Primitive crinoids had only five arms, but most modern species have more. The primary arms fork to form ten, or fork repeatedly to form as many as 200 arms. Diminutive branches (pinnules) attached to the arm margins give them a feathery appearance. The skeleton of arms and pinnules is jointed; brachial ossicles support the arms, and pinnulars support the pinnules. Details of skeletal form and articulation are characteristic of the species. Less is known of the soft parts.

Food Capture and Digestion The crinoid ambulacral groove system located on arms corresponds functionally to a lophophore. Many extinct Pelmatozoa had no arms, but ambulacral grooves converged at the mouth and were no doubt used for ciliary feeding, as in modern species. Arms are body extensions that increase the length of the crinoid feeding surface; the oral faces of the arms and most pinnules bear the ambulacral grooves. A hungry feather star clings to a rock with its arms spread wide and its pinnules at attention, forming an interlock-

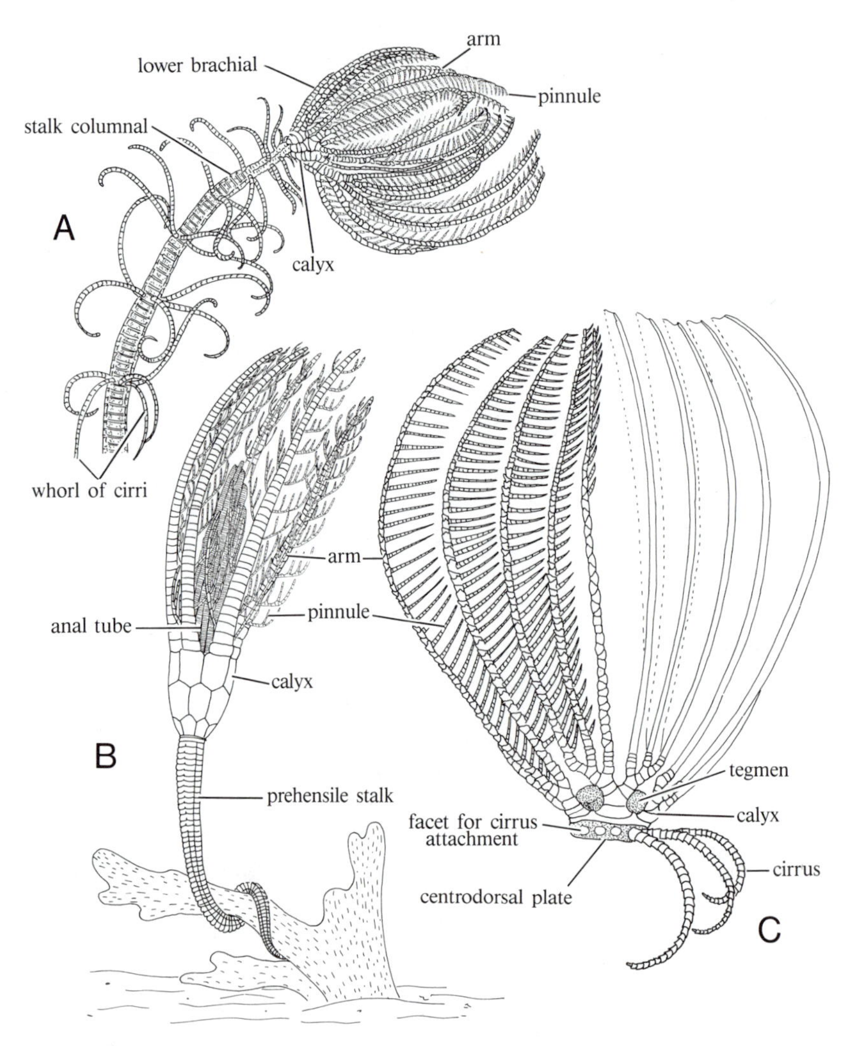

FIGURE 35.12. Some representative crinoids. **A**. Modern isocrinoid with stalk cirri, *Cenocrinus asteria*. **B**. Extinct Devonian crinoid without stalk cirri, *Eifelocrinus dohmi*. **C**. Comatulid, *Neometra acanthaster*, with cirri attached to calyx. (**A** after Carpenter; **B** after Haarmann, from Cuénot, in Grassé; **C** after Clark.)

ing net to filter out small food particles carried in passing water currents. If food is added to the water, arms and pinnules thrash about, and the three-branched podia at the margins of the ambulacra (see Fig. 35.14) push food into the groove. Diatoms, algae, protozoa, and small plankters are staple items of the diet. The epidermis is thin over most of the body surface, but its cells are tall and ciliated in the ambulacral grooves. Food is trapped in mucus, and the mucous strings move along the ambulacral groove like conveyer belts (Fig. 35.13B). Interambulacral cilia on the tegmen surface cause cleansing currents that sweep toward the grooves, turn, and run peripherally just outside the grooves. Epizoic polychaetes (myzostomes) live almost exclusively on crinoids, feeding on the moving mucous strands in the ambulacral grooves.

The digestive tract is simple, consisting of an esophagus, a midgut or intestine, and a hindgut or rectum. The whole gut lining is flagellated except for the rectum; food is propelled and stirred by currents. Circular muscles form a sphincter at the start of the midgut. The intestine may be somewhat dilated, may have several diverticula, or may be a simple tube externally indistinguishable from the esophagus (Fig. 35.13D,E). When the mouth is central, the

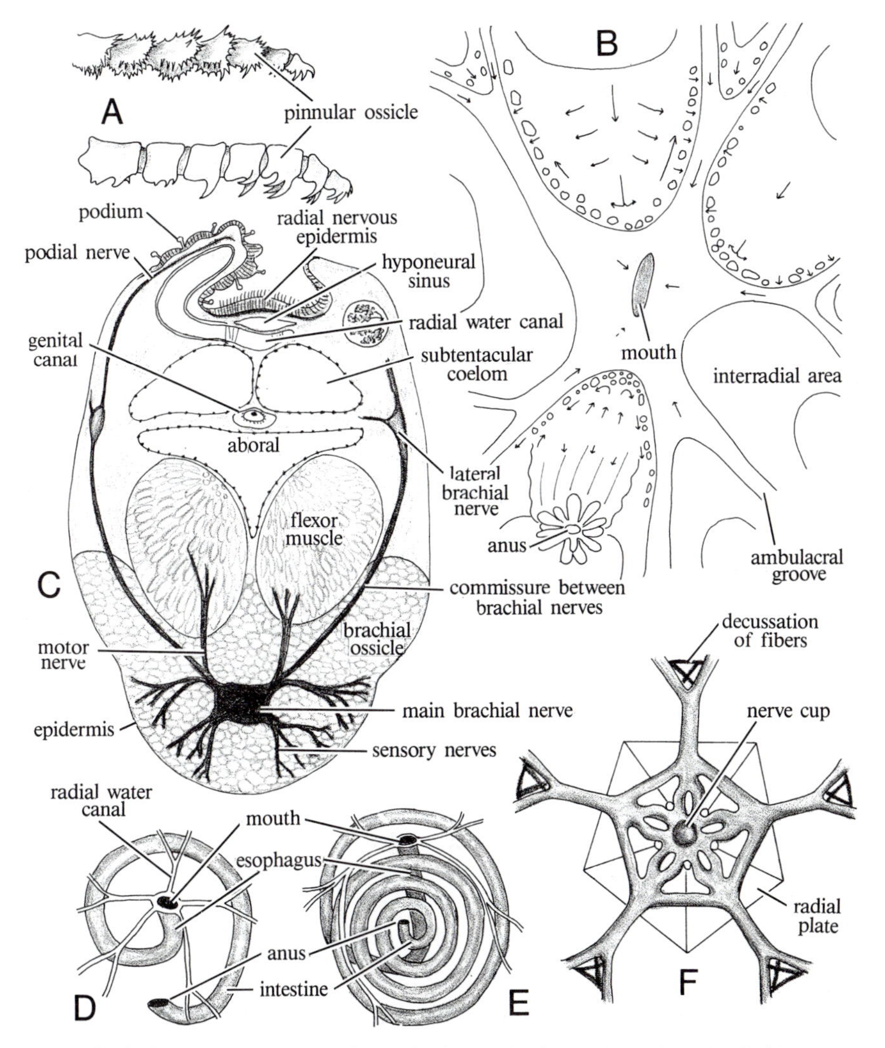

FIGURE 35.13. Crinoid structure. **A**. Ends of comatulid pinnules showing pinnulars and spination. **B**. Ciliary currents on oral surface of feather star, *Antedon*. Note cleansing currents and food-collecting currents in food grooves. **C**. Scheme of crinoid arm in cross section. **D**. Scheme of endocyclic digestive system characteristic of crinoids with central mouth. **E**. Scheme of exocyclic digestive system, characteristic of crinoids with eccentric mouth. **F**. Aboral nerve center of *Antedon*. (**A** after Clark; **B** after Gislén; **C**,**F** composite; **D**,**E** after Carpenter.)

gut arches aborally in a single clockwise turn (Fig. 35.13D). When the mouth is eccentric, the intestine is longer, making four coils (Fig. 35.13E). An anal cone lifts the anus above the tegmen surface. Fecal balls, bound together with mucus, are ejected and drop away from the oral surface or fall when the animal moves.

Perivisceral Coelom The crinoid coelom is reduced by webs of connective tissue, often containing calcareous spicules, and it is complexly compartmented. All coelomic compartments are lined with cuboidal peritoneum, and all parts intercommunicate to some degree. A space, the axial sinus, surrounds the esophagus and opens aborally into the perivisceral coelom that fills most of the corona. Each arm contains four intercommunicating but distinct coelomic spaces (Fig. 35.13C). The large aboral coelomic canal arises from the perivisceral coelom. Paired subtentacular coelomic canals branch from a common stem rising from the axial sinus. A tiny genital coelomic canal houses the genital cord, along which primordial germ cells migrate to the arms and pinnules where gonads form. The coelomic canals branch with the arms and continue into the pinnules. Here heavily ciliated pits of unknown function occur; similar pits may occur in the arm canal. The rest of the coelom is cut off as the chambered organ, embedded in the cup-shaped center of the aboral nervous system. Branches from its five separate compartments extend into the cirri, and in stalked crinoids extend as a five-chambered canal throughout the stalk, giving off branches to any cirri that may be present.

Coelomocytes occur everywhere in the coelom and wander through the connective tissue of the corona. Some are phagocytic, and probably participate in food transport, waste elimination, and other processes, as in other echinoderms.

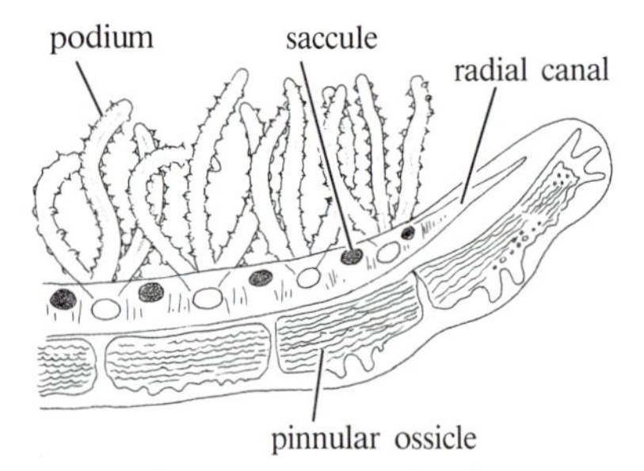

FIGURE 35.14. Podia and saccules on pinnule of crinoid. (After Chadwick.)

Hemal System and Axial Gland Crinoid connective tissue contains many spaces, and the hemal lacunae can be recognized only because they contain the coagulated hemal fluid. As a result, their distribution is imperfectly known. A plexus of hemal tubes occurs around the esophagus, and another just below the tegmen. The subtegminal plexus communicates with the genital tubes of the arms. At the aboral end of the periesophageal plexus, the lacunar walls thicken and merge with a type of spongy tissue of unknown function. Many lacunae connect the spongy organ to the axial gland, a group of ductless glandular tubules. The function of this gland is unknown in crinoids.

Water-Vascular System Crinoid podia have no ampullae. The crinoid ring canal is hexagonal, running around the margin of the oral depression. Branches to the large oral podia arise from the ring canal, and radial canals to the arms branch with the ambulacral grooves. Crinoids have treble podia (Fig. 35.14) containing sensory endings; they are used to secrete mucus and to push food into the ambulacral groove. Many small stone canals arise from the ring canal and hang free in the coelom. Small ciliated funnels pierce the tegmen, drawing seawater into the coelom.

Neurosensory Apparatus Little is known of crinoid sense organs. Crinoids react to touch and right themselves when overturned. Many are sensitive to light, choosing dark or moderately lighted areas. Comatulids are well coordinated, creeping about with the cirri, or clutching rocks or other objects. Some, like *Antedon*, can swim for short distances by means of undulating movements of the arms.

The oral nervous system is located superficially on the oral surface, centered in a nerve sheath around the esophagus, and gives rise to radial nerve bands located in the epidermis of the ambulacral grooves (Fig. 35.13C). The deep oral system is centered in a hexagonal ring in the tegmen, external to the ring canal, and gives rise both to nerves that supply the tegmen and the oral podia and to the lateral brachial nerves of the arms. The aboral nervous system is the predominant one in crinoids. It forms a nervous cup that contains the chambered organ, and gives rise to a nerve cylinder that runs through the center of the stalk (Fig. 35.13F). The nerve center lies in the radial plates of the calyx and sends branches to the cirri. Stalk cirri are innervated from the nerve cylinder. Motor nerves to the podia and water canals rise from the lateral brachial nerves, and a commissure connects them with the main brachial nerves that rise from the aboral nerve center.

Reproduction Crinoid gonads are indefinite masses of gametes located in the arms or pinnules. Genital pinnules usually lack an ambulacral groove and in some species have a cement gland whose secretions attach the ova to the body surface. Primordial germ cells arise elsewhere in the body and reach the arms through the genital cords.

Gametes are usually shed epidemically into the sea. As it is a large ocean, it is imperative that ova and sperm be released at the same time and in the same neighborhood. Generally, males shed sperm first; this then stimulates the females to shed eggs. *Comanthus japonicus* spawn between 3 and 5 PM on the same day during the last half of October, with the moon in the first or last quarter. Even when they are held in the laboratory, spawning occurs at this time. Consequently, the control of spawning must be independent of temperature or chemical factors in the water.

CLASS ASTEROIDEA

Sea stars (Fig. 35.1B), or starfish, perhaps *the* quintessential echinoderms in the public mind, have a worldwide distribution, occur abundantly in cold as well as tropical seas, prefer a rugged and rockbound coast, but also live in deep waters. Some species are widely distributed, whereas others are quite localized. For example, nearly all of New Zealand's 30 littoral species are endemic, being restricted to those waters. Starfish first appeared in the Ordovician, and although they do not fossilize as well as some echinoderms, a number of fossil species are known. About 2000 modern species have been described.

Starfish are usually pentamerous, with the five rays or arms not sharply set off from the central disc. Variations in the number of rays and their distinctness from the disc lead to considerable diversity in body form. Starfish also vary in surface texture. Most are roughened by conspicuous spines, but others are nearly smooth, and some are outlined by definite marginal plates. Many are drab, but some are brightly colored, with characteristic banding or mottling.

The body wall (see Fig. 35.17A) consists of cuticle, flagellated epidermis, a nerve plexus at the base of the epidermis, a delicate basement membrane, a dermis containing the skeletal ossicles, a delicate layer of circular muscle and a thicker layer of longitudinal muscle, and a cuboidal, flagellated peritoneum. Asteroids have a generalized body wall with all of the elements represented in equal quantities, whereas most other echinoderms have a preponderance of either skeletal or soft elements.

The body wall bears many minute structures. **Papulae** (see Fig. 35.17A,C) are inconspicuous, thin-walled, saclike extensions of the coelom, used for respiration. They occur on the aboral or on both body surfaces. Muscles permit extension or retraction. **Pedicellariae** (Fig. 35.15) are jawlike parts on

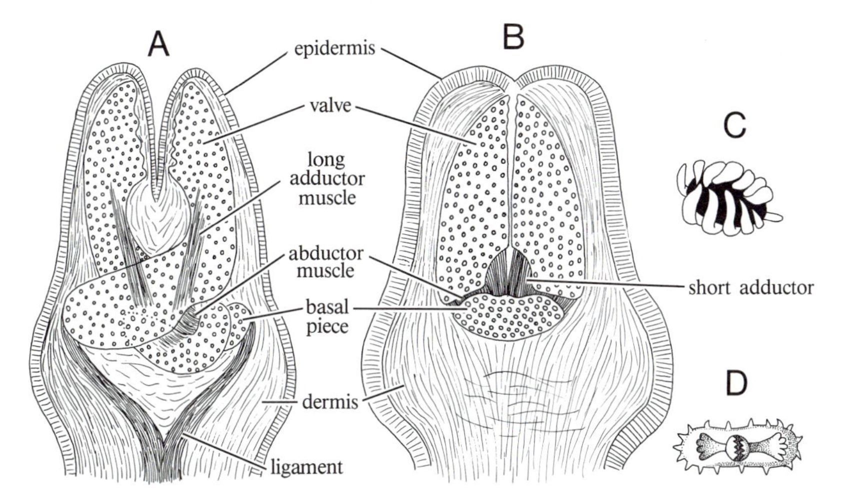

FIGURE 35.15. A. Cross-jawed pedicellaria of *Asterias.* **B.** Straight-jawed pedicellaria of *Asterias.* **C.** Sessile pedicellaria of *Astropecten.* **D.** Alveolar pedicellaria of *Hacelia.* (**A,B** after Hyman; **C** after Fisher; **D** after Cuénot, in Grassé.)

the surface that snap at intruders and protect the delicate papulae. There are three kinds: pedunculate (stalked), sessile (attached directly to skeletal ossicles and often composed of several spines) (Fig. 35.15C), and alveolar (somewhat in–sunk, sitting in small depressions) (Fig. 35.15D). Pedunculate pedicellariae may be cross-jawed (Fig. 35.15A) or straight-jawed (Fig. 35.15B), and have special musculature. **Spines** are covered with epidermis and surrounded by dermis at the base. Slips of body wall muscle insert on movable spines.

Asteroid podia have ampullae within the body (see Fig. 35.17C), above the ambulacral plates. The podial tube projects outward between the ambulacral plates. Where there are two rows of podia, the lateral canals are equal in length, but where there are four rows of podia, longer and shorter lateral canals alternate (see Fig. 35.17B). The podial and ampullar walls are continuous with the body wall and contain the same tissues. Each podium is, in a sense, an oddly shaped blister in the body wall, connected to a canal system. One of the most conspicuous features of the asteroid aboral surface is the water-vascular system.

The mouth is centered on the oral surface (Fig. 35.16D), surrounded by a membranous peristome and encircled by protective spines. A conspicuous ambulacral groove extends from the border of the peristome to the tip of each ray. It contains two or four rows of podia and a single terminal podium at the tip. A red spot, the **optic cushion**, lies at the tip of the ray.

Endoskeletal plates support the rays and the central disc. The two sides of the ambulacral groove are formed of ambulacral ossicles (Fig. 35.16A,B) that rest on adambulacral ossicles at the edge of the groove. Muscles between the ossicles flatten or deepen the groove (Fig. 35.16B). Inframarginal and supramarginal ossicles support the sides of the rays, and specialized carinate ossicles sometimes lie at the middle of the aboral surface of the rays (Fig. 35.16B). Some starfish have specialized inframarginal and supramarginal plates that outline the body margin. One group of starfish has skeletal ossicles

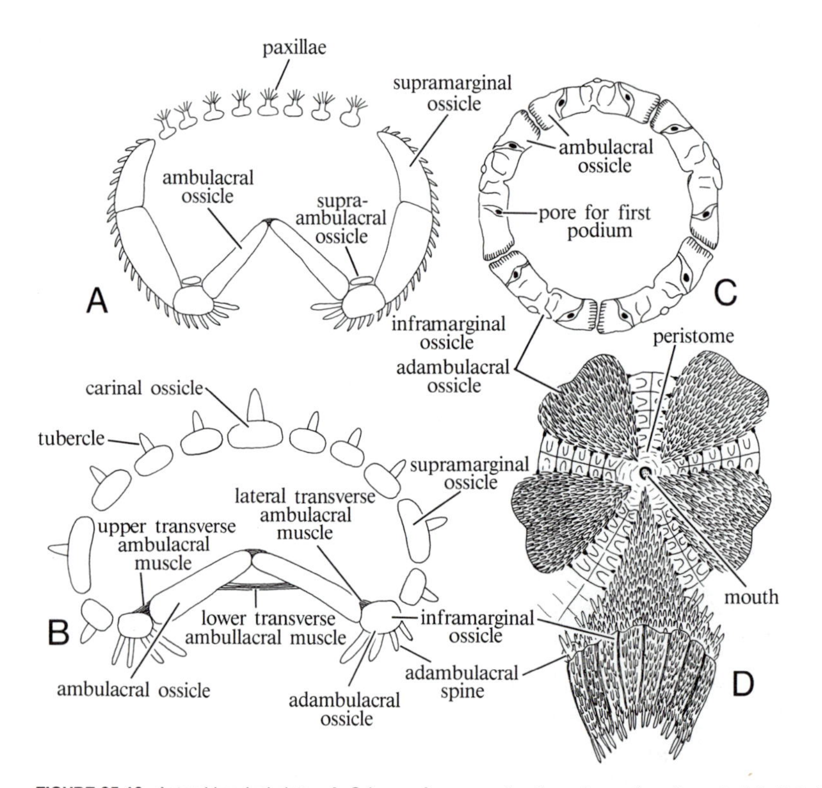

FIGURE 35.16. Asteroid endoskeleton. **A**. Scheme of cross section through ray of paxillose starfish. **B**. Scheme of cross section through ray of forcipulate starfish. **C**. Peristomial ossicle ring of ambulacral type of *Asterias* from inside. **D**. Peristomial ossicles of adambulacral mouth frame of *Astropecten*, seen from without. (**A**,**B**,**D** after Hyman; **C** after Chadwick.)

known as paxillae, with many small, movable spines on the expanded external face (Fig. 35.16A). The spines are also endoskeletal ossicles, and have special muscles at the base that serve to move them slightly.

A skeletal ring, the mouth frame, is formed by the ambulacral and adambulacral plates at the margin of the peristome. An ambulacral mouth frame is composed predominantly of ambulacral ossicles; in an adambulacral mouth frame, the adambulacral plates predominate (Fig. 35.16D).

Coelom The asteroid coelom is complexly partitioned, and all of its derivatives are lined with flagellated peritoneum. The parts of the starfish coelom have been discussed previously in the Body Plan section, and it is quite pervasive (Fig. 35.17C). The young gonads lie in the genital sinuses of the arm rays (Fig. 35.17C).

Hemal System The brownish or purplish axial gland lies in the axial sinus, with its head in the dorsal sac (Fig. 35.3A). One or more hemal tufts connect the axial gland with the hemal sinus system of the digestive tract. The axial gland tissue is spongy and contains many coelomocytelike cells. The hemal lacunae have incomplete walls and contain amoeboid cells indistinguishable from amoeboid coelomocytes.

The hemal system has no heart. Some circulation

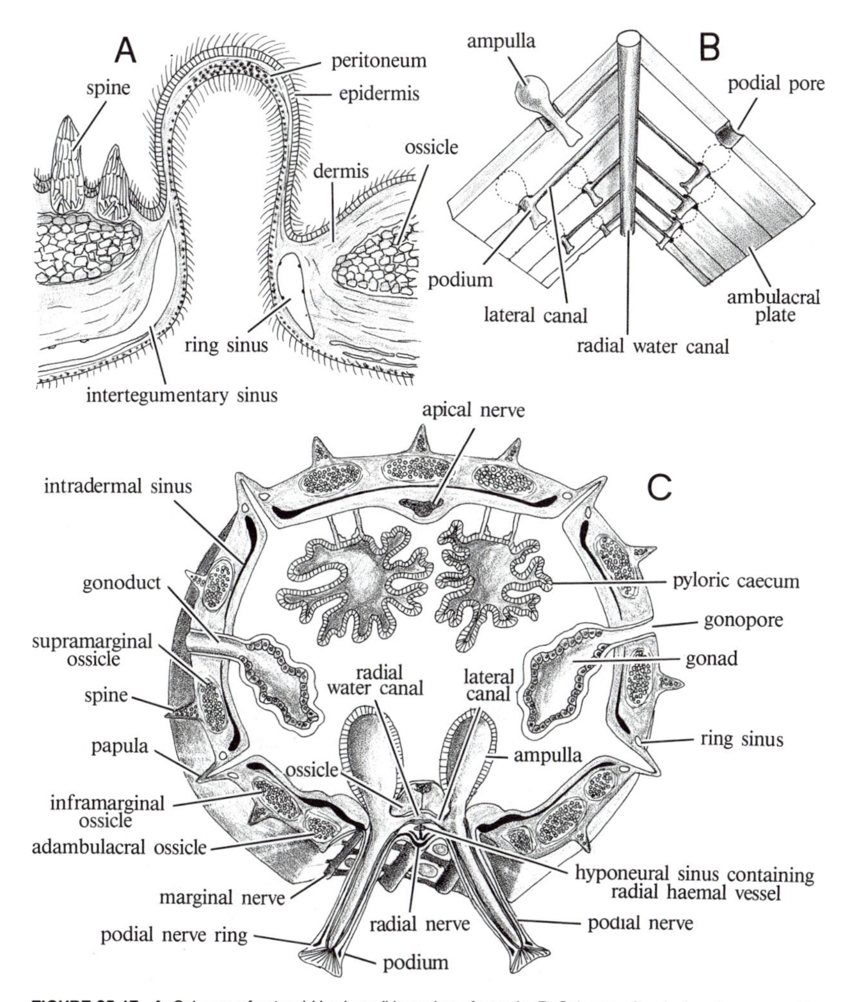

FIGURE 35.17. A. Scheme of asteroid body wall in region of papula. **B.** Scheme of ambulacral groove, with one row of podia on right and two rows on left. **C.** Schematized section through asteroid ray. (**A** after Cuénot, in Grassé.)

occurs, however, and contractions of the head process, the aboral hemal ring, and the gastric tufts have been described.

Food Capture and Digestion Starfish are carnivorous, and although some species have definite food preferences, most will take any kind of animal food they can get. Clams, oysters, snails, and other shelled molluscs are staple items, but brachiopods and hermit crabs are also eaten, and inactive or dying fish are not ignored.

Some starfish swallow food whole, but the anus is too small to accommodate empty shells, and such remains are necessarily ejected through the mouth. Other starfish evert the stomach through the mouth. It is pressed against the food organism, enzymes are secreted to soften the food, and the predigested tissues are swallowed. In this case there is little residue to egest, and the anus becomes unimportant and is lacking in some deep-sea forms. Everting the stomach increases the variety of animals that can be eaten, for the stomach can be pressed against sessile animals, like corals, digesting them in place. The attacks made by a starfish on a clam or oyster have often been described. The starfish grasps the shell with its podia and exerts a steady, outwardly directed pressure, while the mollusc attempts to hold the shell closed. The problem is for the starfish to get its stomach inside the shell. It was once incorrectly thought that the hydraulic system of the starfish exerted a long, steady pull that the pelecypod could not match. Now it is realized that the stomach is inserted through very tiny gapes or irregularities in the shell margin, and the clam opens wide only as it is digested. A lobe of an *Asterias* stomach can enter a gape of no more than 0.1 mm, according to one study, but this small gape was produced by muscular pull. In at least some cases, muscular pull by the starfish does not appear to be constant, and the unwary clam may open and close its shell normally during the first part of an attack. Not all clams show the striking escape reactions when near starfish, but it is clear that where a clam does make such movements it will not be unwary.

Some starfish, especially those living on muddy bottoms, like *Porania* or *Ctenodiscus*, are ciliary feeders. Starfish known to feed in this way stay alive and healthy in aquaria when denied large food organisms, whereas other species starve under identical conditions. This method of feeding, however, is at least generally no more than supplementary and does not appear to be very common.

The starfish digestive system consists of an esophagus, a stomach divided into cardiac and pyloric parts, large diverticula known as digestive glands or pyloric caeca, and a short intestine giving off intestinal caeca and leading to the anus (Fig. 35.3C). Some starfish also have ten esophageal pouches, and in some a Tiedemann's pouch, of unknown function, rises from the pyloric caeca. The gut wall is built of an inner gastrodermis (mucosa) with a nerve plexus at its base, a layer of connective tissue, circular and longitudinal muscle layers, and a peritoneum continuous with the peritoneum of the gastric ligaments. Regional specialization of various areas is accomplished by modifying the thickness or nature of the wall tissues.

Sense Organs The epidermis is supplied everywhere with neurosensory cells sensitive to touch, chemicals, and light. Mechanical stimuli cause contraction of papulae and podia. Acids cause contraction of papulae and podia when applied locally, but evoke a flight response when applied to the aboral surface. The presence of food arouses positive reactions. An impinging light ray causes contraction of podia and papulae, and may evoke coordinated responses of the whole animal as well.

Asteroids have the oral nervous system centered in a pentagonal nerve ring in the peristomial membrane, and give off a radial nerve at each arm. The radial nerve extends to the tip of the ray in the floor of the ambulacral groove (Fig. 35.17C) and connects with the subepidermal nerve plexus. The plexus is thickened in sensitive areas and at the margins of the ambulacral grooves, where it forms marginal nerves. Motor nerves to the ambulacral muscles and to the nerve plexus of the coelomic lining, which controls the body wall muscles, arise from the marginal nerves, which in this sense correspond to the adoral endoneural nervous system of other echinoderms. The only remnant of the hyponeural nervous system is a plate of nerve tissue known as Lang's nerve, which lies on the oral wall, separated from the oral nerve ring by a delicate septum. Predominantly motor branches from Lang's nerve run above the hyponeural coelomic sinuses.

Reproduction The common starfish has ten gonads, two in each ray, lying free in the coelom except at the gonoduct. Gonopores open near the interradius on the ray margin. Some starfish have many gonads in each arm, with a series of marginal gonopores or a collecting duct leading to a normally placed gonopore. Each gonad is enclosed within the delicate walls of a genital sinus that arises as a branch from the genital coelomic ring.

Sexes are usually separate, but an occasional hermaphroditic asteroid occurs. Some hermaphroditic species appear to be dioecious because of marked protandry. Gametes are usually released into the sea and fertilization is external, but some species have developed brooding habits. The females of *Leptasterias* cup their body over the egg mass and retain that position without feeding while the young develop. The young of some species implant themselves among the paxillae of the aboral surface. Some forcipulates, for example, the Heliasteridae, have a nidamental chamber, floored with many papulae, used for brooding the young as well as for respiration. Brooding starfish produce large ova with lots of yolk, and these undergo direct development. Evidently the chances of successful development are increased by brooding, because brooders produce only a few dozen to several hundred ova, whereas nonbrooding females may release millions of ova.

The sea star, *Luidia*, has been noted to form asexual buds from the arms of bipinnarian larvae. As much as 30 percent of larval populations (Bosch et al., 1989) in the mid-Atlantic Sargasso Sea were in such states. Such asexual cloning is viewed as a way to prolong the larval stage to increase potential for recruitment into pelagic trans-Atlantic benthic populations of adults.

CLASS HOLOTHUROIDEA

Pliny mentioned the *cucumis marinus*, and it is customary to use this as the beginning of our knowledge of sea cucumbers, but the Chinese were undoubtedly gathering them as delicacies for succulent dishes long before Pliny's day. Sea cucumbers (Figs. 35.1E and 35.18) are familiar littoral animals throughout the world, but they are not restricted to shallow waters. A rich deep-water fauna extends to depths of about 6000 m. Most of the 800 species of sea cucumbers are of moderate size, well named as cucumbers, but the group as a whole varies from the smallest gherkins to a truly noble flesh bags. *Stichopus variegatus*, from the Philippines, reaches a meter in length and 21 cm in diameter. Most species are dull, but an occasional one is brightly colored.

The soft leathery body is elongated, with an anus at or near the aboral end, and the mouth at or near the oral end. The standard five ambulacral regions usually mark the sides of the body, but sea cucumbers lie on one side, and dorsal CD and ventral ABE surfaces are often differentiated. They lie with ambulacrum A on the midventral line. The ventral surface is sometimes flattened, and the dorsal surface arched. Some glide on a creeping sole, propelled by muscular waves and locomotor podia (Fig. 35.18E,F). The mouth tends to migrate to the ventral, and the anus to the dorsal surface. The bilateral symmetry seen in some elasipods is emphasized by saillike extensions of the body, used in floating. These creatures have an amazing ambiguity concerning symmetry. Beginning as bilateral larvae, they become secondarily biradially symmetrical, and finally return to a tertiary bilateral symmetry.

The mouth is surrounded by modified podia known as oral tentacles. They may be digitate (fingerlike) (Fig. 35.18C,F), dendritic (arboreally branched) (Fig. 35.18D), pinnate (with side branches on a central axis) (Fig. 35.18B), or peltate (flattened, with a short stalk and terminal disc, often with horizontal branches) (Fig. 35.18A,E). The tentacles can be withdrawn and the body wall closed over them for protection. Some sea cucumbers have a definite collar around the oral end, forming an introvert that can be pulled in by retractor muscles. In others, a simple tentacular collar folds over the withdrawn tentacles.

Usually the body wall contains microscopic skeletal ossicles, but a few species have a protective layer of small, close-fitting plates. Various warty prominences, papillae, and other projections may mark the surface that are often modified podia connected to the water-vascular system. The anus is often surrounded by special anal papillae.

The body wall consists of a cuticle, an epidermis, a dermis, a sheath of circular muscle, and bands of longitudinal muscle in the ambulacral regions. The tiny, bizarrely shaped ossicles lie in the dermis. All echinoderm ossicles are fenestrated with holes, but this is seen most easily in the lacy form of holothurian ossicles. The outer part of the dermis is loose, but the deeper parts are strengthened by collagenous fibers that form a strong, protective layer. Strands of longitudinal muscle interrupt the sheath of circular muscles at each ambulacrum. Circular muscles are strengthened as sphincters at the mouth and in the collar or behind the introvert. The oral region is supported by a calcareous ring, the calcareous pharyngeal bulb (Fig. 35.5B), to which the retractor muscles attach. The unusually strong body wall musculature of sea cucumbers is associated with the absence of a skeletal sheath.

The sea cucumber body wall structures are much reduced. No spines, papulae, or pedicellariae occur. Podia may occur in definite ambulacral fields, as in *Cucumaria*, but in many cases they are more or less randomly distributed, and in some sea cucumbers, the podia are completely missing or restricted to a whorl of anal papillae. A number of podia are associated with papillae, and serve as sense organs (Fig. 35.7A). A variety of papillae and other prominences occur, and a number of these are known to have sensory functions. Clusters of sensory cells, supplied with sensory nerves, have been described in oral tentacles (Fig. 35.7B) and elsewhere on the body surface.

A peculiar process of evisceration has been observed in laboratory animals kept in crowded conditions, as well as in certain natural populations. The cloaca or parts of the body wall rupture and allow the expulsion of internal organs. It is suggested (Byrne, 1985) that this may be a peculiar way to expel waste-laden tissues.

CLASS ECHINOIDEA

Sea urchins and sand dollars (Figs. 35.1D and 35.19) are familiar animals of sea coasts, and are separated into the regular and irregular echinoids. Sea urchins

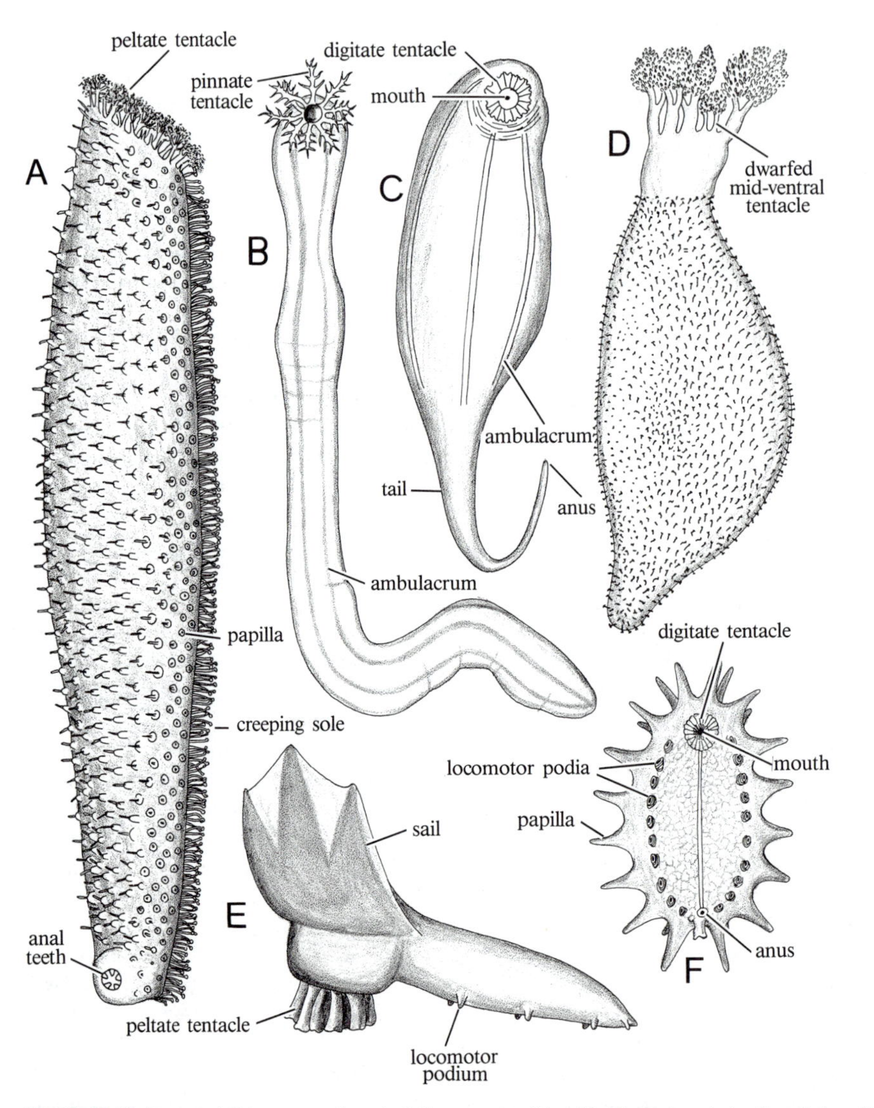

FIGURE 35.18. Representative sea cucumbers. **A.** *Actinopyga aggasizi,* aspidochirote, characterized by peltate oral tentacles and many podia. These cannot withdraw oral region. **B.** *Leptosynapta,* apodan, characterized by absence of podia and reduction of water-vascular system. **C.** *Molpadia musculus,* molpadonian, characterized by reduction of podia to few anal papillae and digitate oral tentacles. Posterior end tapers to caudal region. **D.** *Thyone,* common dendrochirate, characterized by dendroid oral tentacles and oral region modified as introvert. **E.** *Peniagone,* elasipod, one of deep-sea forms tending toward bilateral symmetry, with peltate oral tentacles and unable to retract oral region. **F.** *Deima,* in ventral view, another elasipod. (**A,D** after Hyman; **C** after Clark; **B** after Coe, **E,F** after Théel.)

are regular echinoids, with a radially symmetrical, globose test and forbidding spines. Sand dollars are irregular echinoids, with a flattened test, less prominent spines, and fairly marked bilateral symmetry. The firm shell, with its projecting spines, pedicellariae, podia, and gills, is characteristic of echinoids.

Echinoids occur in all climatic regions, and from the littoral zone, where they are abundant, to depths of 5000 m or more. Fossil echinoids date back to the Middle Ordovician, 470 million years ago. Deep-sea forms live on slimy bottoms, but regular echinoids usually prefer hard bottoms and rocky areas, whereas irregular echinoids prefer muddy or sandy bottoms. The regular sea urchins often lie in protected, rocky crannies; boring urchins enlarge these spaces with the teeth and spines when they are too tight for comfort. There are two major types of irregular echinoids: sand dollars and heart urchins (Fig. 35.19). The greatly flattened sand dollars are round to oval in outline, whereas heart urchins are indented at one ambulacrum and are deeper, with more evident bilateral symmetry. Sand dollars live

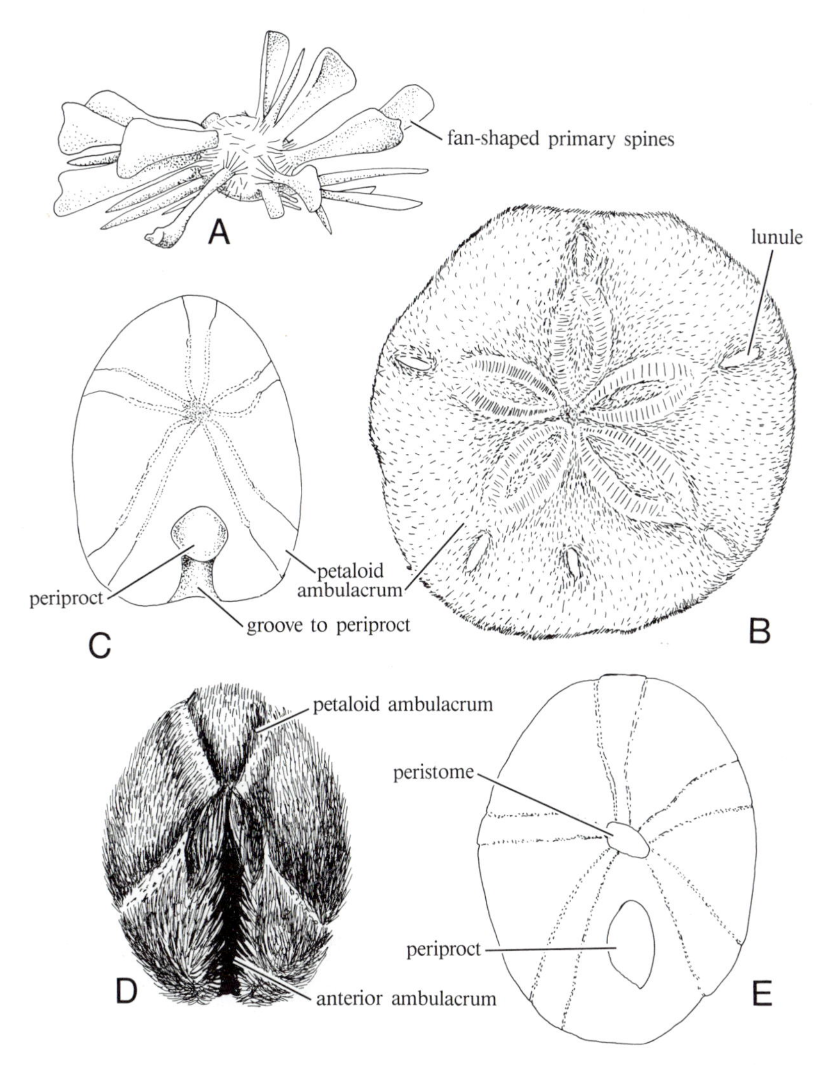

FIGURE 35.19. Representative echinoids. **A**. Cidaroidea with single large spine encircled by smaller spines on each ambulacral plate as in *Cidaris blakei.* **B**. *Mellita,* true sand dollars, clypeastroids, with small spines, oval to circular test, and petaloid ambulacra. **C**. *Apatopygous,* aboral view, example of Cassiduloida with mouth central or somewhat anterior, guarded by modified plates forming floscelle. **D**. *Echinocardium* with oral ambulacrum insunk and tending toward bilateral symmetry, and more flattened forms commonly known as sand dollars. **E**. *Echinoneus,* holectypoid, with regular test and central peristome, but without petaloid ambulacra. (**A** after Mortenson; **C,E** after Hyman; **D** after MacBride.)

just below the surface of the sand. When exposed, they bury themselves by piling sand in front of themselves and then moving into it. Heart urchins burrow in the mud; the burrows are strengthened with mucus to prevent their collapse.

Radial water canals lie just beneath the test. Pores for the podia make distinctive ambulacral regions, not always evident until the spines are removed. The ambulacra are often petal-shaped (petaloid) in irregular urchins (Fig. 35.19B,C,D), forming a flowerlike design on the test.

Aristotle's lantern is a unique, characteristic feature of echinoids associated with the mouth. It is a complex structure (Fig. 35.20A), with some 40 ossicles in addition to the actual teeth. The main external framework is dominated by five large pyramids, each consisting of two united half-pyramids (Fig. 35.20B). At the aboral end, ossicles termed the epiphyses, rotules, and compasses form complete or incomplete bars across the top of the pyramids. The long teeth are supported by the pyramids. At the aboral end, they become membranous and extend into the dental capsule of the peripharyngeal sinus. The musculature is as complex as the ossicle arrangement. Muscles rock the pyramids back and forth and transfer the pyramid movements to the teeth. Lantern protractors move the whole lantern downward, extruding the teeth, and lantern retractors withdraw the lantern.

The rear part of the lantern of sea urchins is used for respiration. The cavity of the peristomial gills is continuous with the peripharyngeal coelomic sinus. A flat sheet of muscle forms a pentagonal ring, connecting the compasses; its contraction elevates the compasses, increasing the volume of the sinus and collapsing the gills. Narrow compass depressors pass along the margins of the protractor muscles. When they are contracted, coelomic fluid is forced back into the gills. Only sea urchins have a fully developed lantern. Clypeasteroids also have a lantern but have no peristomial gills, compasses, or compass musculature, and the lantern is broader, lower, and star-shaped (see Fig. 35.23B).

The Test The echinoid body is completely enclosed in an armorlike shell or test. The test is best examined with all spines and other surface structures removed (Fig. 35.21B). A denuded sea urchin test is a beautifully symmetrical object, as the tubercles for spine attachment occur in meridional rows, adding symmetrical ornamentation to an otherwise symmetrically arranged set of plates. The test is typically built of 20 rows of plates; each radius centers on a double row of ambulacral plates that contain paired pores for the podia, and each interradius centers on a double row of interambulacral plates with conspicuous tubercles on which spines attach (Fig. 35.21A).

The skeletal plates are more or less straight on three sides and V-shaped on the fourth. The plates

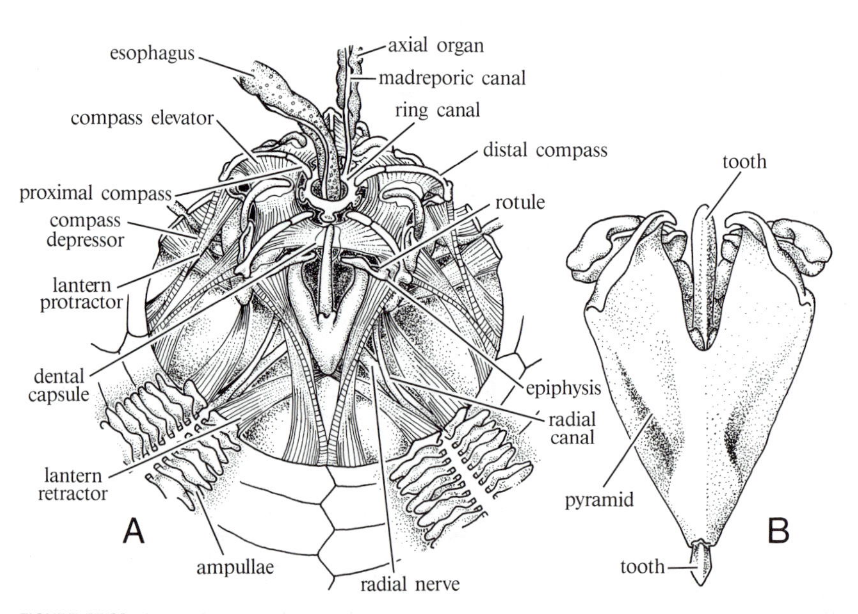

FIGURE 35.20. Aristotle's lantern. **A**. Lantern and musculature of *Arbacia punctulata*. **B**. Lantern ossicles of *Arbacia punctulata* showing relationship of tooth and pyramid. (After Reid, in Brown.)

of a row meet at two of the flat sides; two adjacent rows of ambulacrals or interambulacrals meet at the V-shaped ends and so are set alternately, as a mason lays bricks, increasing the strength of the test (Fig. 35.21B). The margins of the double rows of interambulacrals and ambulacrals meet in a straight line.

Echinoid podia have an inner canal leading to the water canal, and an outer canal leading to the ampulla; thus each ambulacral plate is pierced by two pores for each podium. Simple ambulacral plates are very narrow, with a single pore pair (Fig. 35.21A). Compound plates, formed by the fusion of simple plates, are common, and several patterns of fusion are recognized. Interambulacral plates are larger than ambulacral plates and have larger tubercles to accommodate the spines.

The test is modified at oral and aboral ends. The membranous periproct with the anus is surrounded by five larger genital plates on the interradii and five smaller terminals on the radii (Fig. 35.21B). The madreporite lies in a modified genital plate. Genital plates are pierced by gonopores, and terminal podia pass through the terminal plates. Both the periproct and peristome may contain small but flexible plates. The ambulacral and interambulacral plates of most echinoids generally end at the margin of the peristome (Fig. 35.22) but continue across the peristome in cidaroids; in lepidocentroids only the ambulacrals cross the peristome. Most echinoids have a thickening (perignathic girdle) around the peristome where jaw muscles attach.

The flat test of irregular echinoids is not so impressively designed as that of regular echinoids, but is stronger because internal struts extend from oral to aboral surfaces. The spines are small and the regularly symmetrical pattern of tubercles is missing. Clypeasteroids have wider ambulacrals, and spatangoids have wider interambulacrals. The petaloid ambulacra of many irregular echinoids are formed by very narrow ambulacral plates and their podial pores (Fig. 35.19B). Some irregular echinoids have expanded ambulacrals forming petallike phyllodes around the mouth, and prominences on the interambulacrals (bourrelets) may produce a flowerlike floscelle around the mouth (Fig. 35.22). The lunules or slots noted in many irregular echinoids are a de-

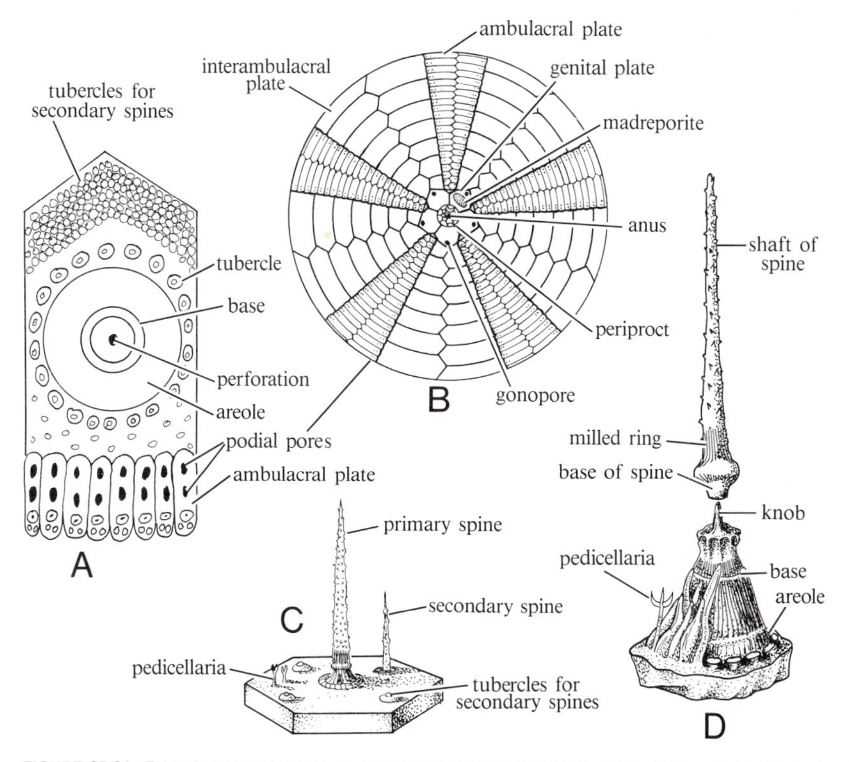

FIGURE 35.21. Echinoid test and spines. **A**. Ambulacral and interambulacral plates of cidaroid, *Eucidaris,* showing tubercles. **B**. Scheme of test of regular echinoid. **C**. Scheme of coronal plate showing pedicellariae, tubercles, and spines. **D**. Tubercle and spine detail. (**A** after Hyman; **C** after Jackson; **D** largely after Prouho.)

vice to reduce lift on an otherwise flat shell lying on the bottom and subject to the vagaries of swift currents (Telford, 1983).

The tubercles of regular echinoids consist of a basal tubercle, the base, raised to form a ring around a central knob (Fig. 35.21D). The base of the spine ends in a flange-shaped condyle that pivots on the knob. A circular area (areole) around the base provides a site for muscle attachments. Tubercle size varies with spine size. Primary tubercles are usually much larger than secondaries, but some species have intergrading sizes. Cidaroids have very large tubercles for the huge primary spines, surrounded by a ring of smaller tubercles for secondary spines (Fig. 35.21C).

Echinoids have a simple body wall composed of an external ciliated epidermis, a dermis composed almost entirely of skeletal plates, and a flattened peritoneal lining. As in other encased animals, body wall muscles have disappeared except for specialized slips; in echinoids these are associated with the body wall structures: spines, pedicellariae, podia, and gills. Sphaeridia are small, transparent, stalked bodies, thought to be balance organs and probably derived from spines. They lie along the ambulacra of regular echinoids and in depressions near the peristome or on the aboral surface of irregular echinoids. A circle of gills occur on the peristomial border of all regular echinoids. They are thin-walled extensions of the peristome. Irregular echinoids are without peristomial gills.

Spines are surprisingly dexterous and used as levers for locomotion or digging and for protection. An exposed *Evechinus choroticus* can cover itself with small pebbles in a few minutes, using spines and pedicellariae. Spine shape, size, and form vary with spine function. Longer spines are more effective than small ones for movement and protection, and spadelike ones are best for digging. The clavules of spatangoids are racket-shaped and heavily ciliated; they generate currents that cleanse the body surface. Some echinoids like *Diadema* have spines that can cause painful puncture wounds and may be made even more effective by venoms.

Echinoid podia are like locomotor podia of other echinoderms, but attach to the ampullae and water canals by a double canal system, as described previously. Diadematoids have ten specialized buccal podia used in food manipulation. Podia are more highly specialized in heart urchins. Digitate or discoid sensory podia lie in the anterior ambulacrum. Leaflike branchial podia from the petaloids serve as gills. Penicillate podia arise from the phyllodes; they correspond to buccal podia of regular echinoids and are used for food manipulation and chemoreception. Sand dollars have only branchial podia and simple, suckered podia used for locomotion and food gathering.

Four major types of pedicellariae are found among echinoids: dentate, foliate, ophiocephalous, and globiferous. Dentate pedicellariae are the most common; they usually have three jaws (tridentate), but sand dollars have two-jawed dentates, and some echinoids have dentates with four or five jaws (Fig. 35.4A). Ophiocephalous pedicellariae occur on the peristome (Fig. 35.23D). They are clamped shut by a special valve when closed. Dentate and ophiocephalous pedicellariae drive off or grasp intruders at the

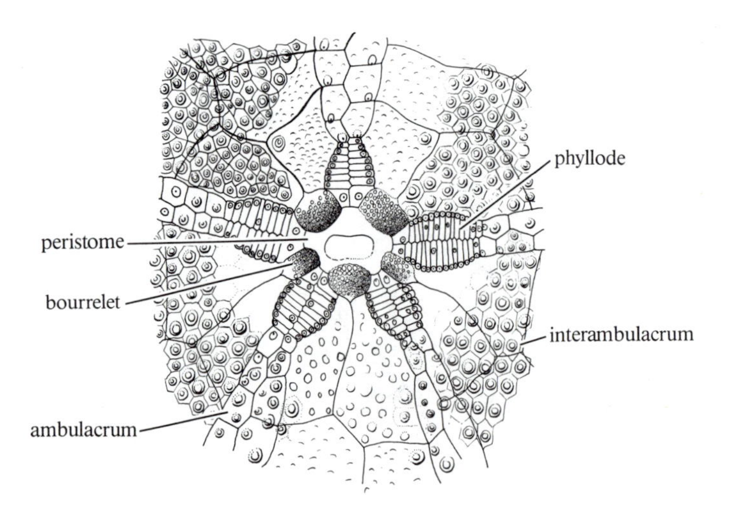

FIGURE 35.22. Peristome of *Cassidula* showing floscelle. (After Lovén.)

surface and may be used in food capture. Foliate pedicellariae usually have three jaws, although some sand dollars have two-jawed foliates. They are used to keep the test clean. Globiferous pedicellariae (Fig. 35.23A) have jaws and ducts specially formed to deliver poison from the poison sacs. They snap at intruders, viciously paralyzing small organisms, and are toxic enough to make starfish withdraw from combat. Some can cause severe pain and even systemic symptoms in humans.

CLASS OPHIUROIDEA

Brittle stars (Figs. 35.1C and 35.24A,G) or serpent stars are smaller, more agile, and on the whole are at least as successful as asteroids, if not more. About 2000 species have been described, but they are inconspicuous and, although as numerous, are considerably less well known than asteroids. Although abundant in the littoral zone, they are often overlooked, for they hide in the sand or under objects, and some live in sponges or in other colonial animals. They occur from tropical to polar regions and have been collected from depths of 6000 m.

Brittle stars are not as diverse in appearance as most other echinoderm groups. The arms can be simple or branched, longer or shorter in relation to the disc size, and smooth or scaly. Most are dull, but there is some variation in color and pigmentation patterns. The most striking forms are basket stars, with arms that branch repeatedly.

Three recent orders are generally recognized. Ophiuridans have simple arms, supported by skeletal ossicles with zygospondylous articulations (Fig. 35.24B,C), permitting only horizontal arm movements. The arms and disc are usually scaly. Oego-

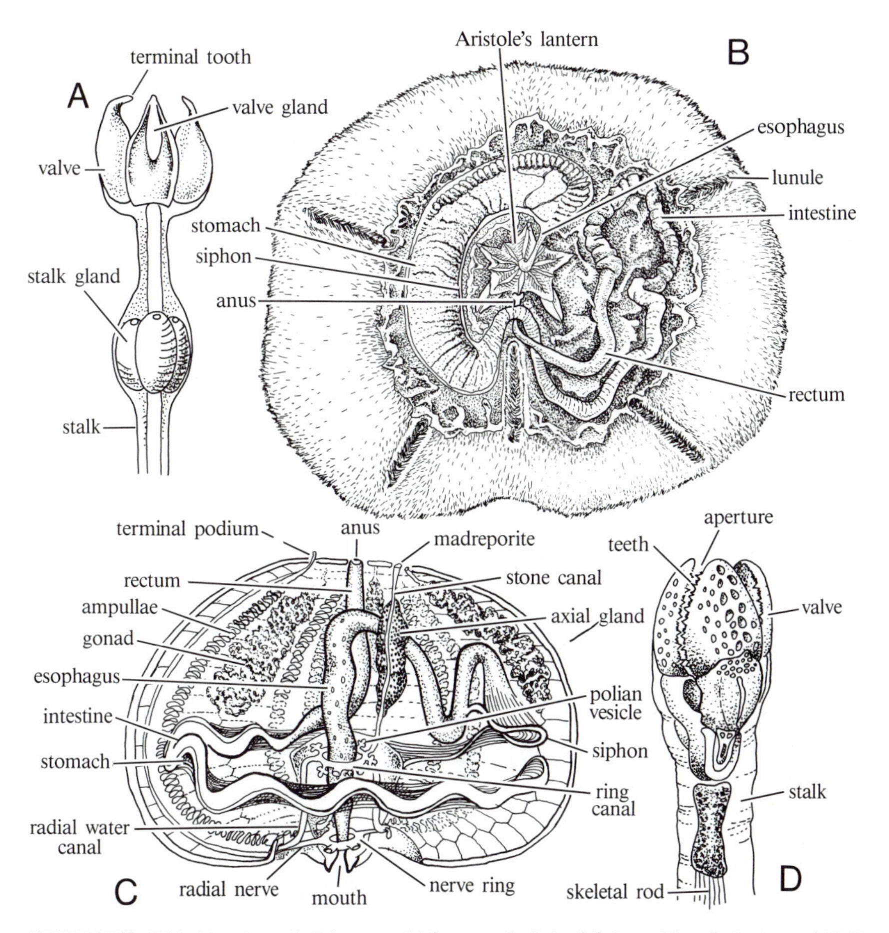

FIGURE 35.23. Echinoid anatomy. **A**. Poisonous globiferous pedicellaria of *Sphaerechinus*. **B**. Anatomy of *Mellita pentapora*. **C**. Scheme of internal organization of *Arbacia*. **D**. Ophiocephalous pedicellaria of *Arbacia*. (**A** after Cuénot, in Grassé; **B** after Coe; **C** after Petrunkewitch; **D** after Reid, from Reid, in Brown.)

phiuridans and phrynophiuridans have simple or branched arms, with streptospondylous articulations (Fig. 35.24D,E), permitting the arms to twine about objects. They do not have conspicuous scales on arms or disc.

During ophiuroid late development, the epidermis is invaded and largely replaced by mesenchyme. Gland cells are thus scant except on the podia. Cilia or flagella occur only in small tracts or patches, usually on the oral surface of the disc or arms. Epidermal cells do cover the sense organs and form a normal covering on the podia.

The dermis and the superficial elements of the skeleton form most of the brittle star surface. In many species the spines and granules of the superficial skeleton are obscured by dermis. Spines are relatively sparse, occurring only at the arm margins or on the central disc. Larger spines attach to tubercles and are movable.

The deeper skeleton of the aboral disc of a young ophiuroid consists of a central plate, surrounded by a series of concentric rings of plates with five (or multiples of five) plates to a ring (Fig. 35.24F). This primitive arrangement is so modified by secondary plates that the pattern is wholly obscured in many species; in others, granules of the superficial skeleton

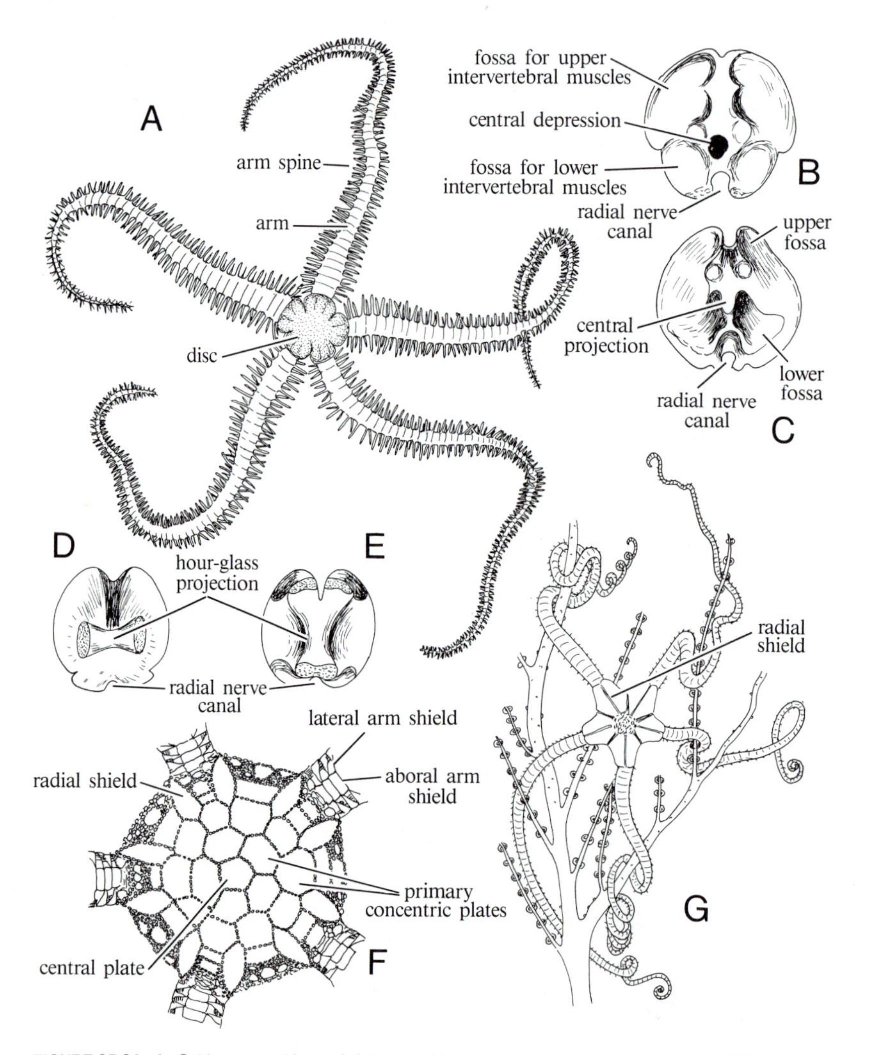

FIGURE 35.24. **A**. *Ophiocoma,* ophiurous brittle star with scaly arms, moving only in one plane. **B**,**C**. Proximal and distal surfaces of ophiurous brittle star vertebrae showing zygospondylous articulations that permit movements in horizontal plane only. **D**,**E**. Proximal and distal surfaces of vertebrae of phrynophiurid brittle star showing hourglass articulations of streptospondylous type, permitting arm movements in all directions. **F**. Aboral surface of disc of *Ophiolepus* with primitive concentric arrangement of plates. **G**. *Asteronyx excavata,* brittle star with arms that can twine about objects. (**A**,**F**,**G** after Hyman; **B**,**C**,**D**,**E** after Lyman.)

conceal it. The most conspicuous and constant of the secondary plates are the radial shields on each side of the arm bases.

The arms appear jointed, each joint corresponding to one of the arm vertebrae. The margins of the vertebrae are thinner, providing space for muscles (Fig. 35.25B), and meet at special articulating surfaces (Fig. 35.24B,C,D,E). Oral and aboral pairs of muscles lie between adjacent vertebrae. The mechanics of arm movement are determined by the articular surfaces; the hourglass form of the streptospondylous articulation permits movement in all directions. More combinations of directional muscle pulls are possible than skeletal flexibility will actually permit in the ophiuridans, which are limited to lateral arm movements by the vertebral articulation shape. Conspicuous arm shields are associated with each vertebra: one oral, one aboral, and two lateral shields. The two lateral shields are the most important and are often enlarged at the expense of the oral and aboral shields. They are homologous to the adambulacral ossicles of asteroid arms, and bear the arm spines. The extremely variable arm spines are of value in species identification. Little is known

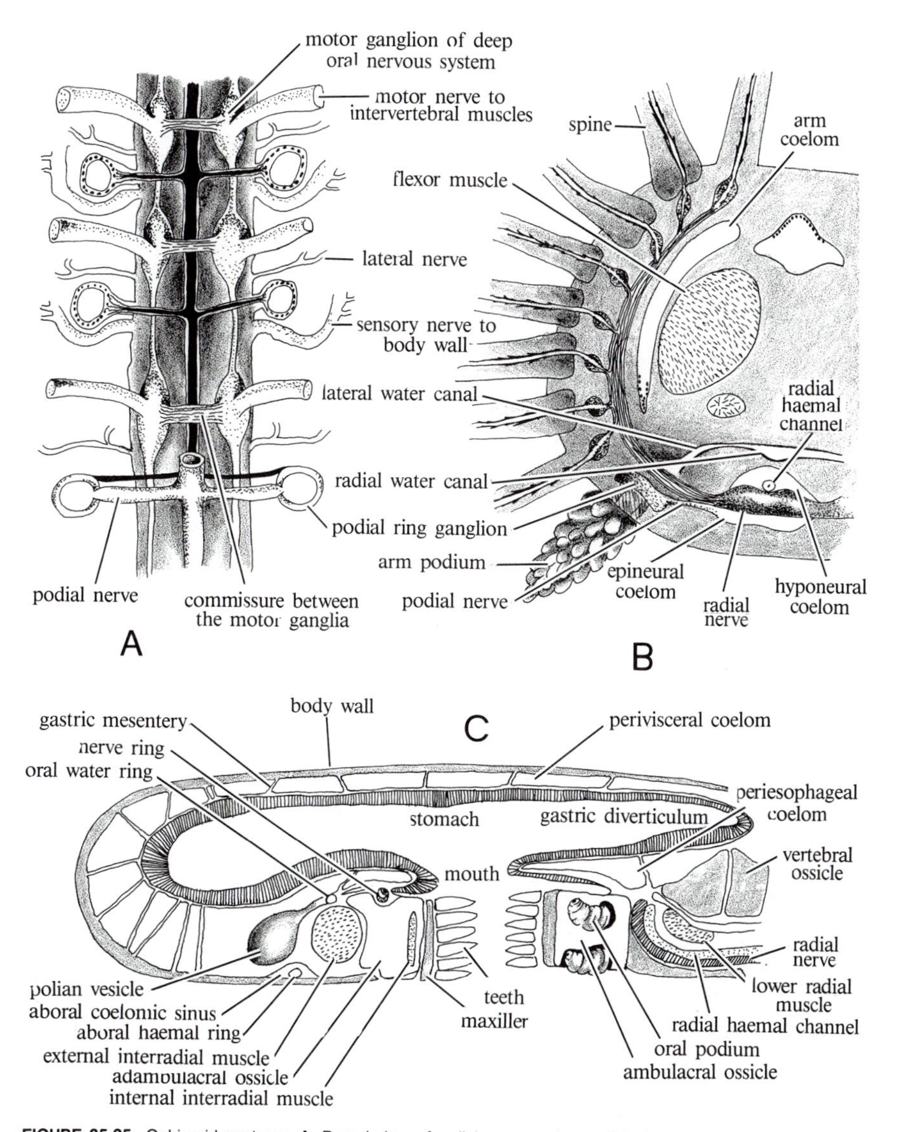

FIGURE 35.25. Ophiuroid anatomy. **A**. Dorsal view of radial nerve and associated structures in *Ophiocomina*. **B**. Section through arm of *Ophiothrix*. **C**. Scheme of section through central disc of ophiuroid passing through interradius on left and radius on right. (**A** largely after Ludwig; **B**,**C** after Cuénot, in Grassé.)

of the functional significance of variations in arm spines; presumably they roughen the surface and, therefore, aid both in clinging to surfaces and in locomotion.

The arm continues across the oral surface to the mouth. The oral and lateral shields are modified at the mouth border. The mouth frame is composed of five wedge-shaped interradial jaws, built of two half-jaws composed of the first two ambulacral plates and lateral arm shields. They bear spines modified to form teeth (Fig. 35.25C). A maxiller plate lies across the tips of the half-jaws; through this plate the muscles operating the teeth pass. The oral and two aboral shields conceal a considerable part of the jaw apparatus of most species. One of the oral shields is the madreporite. Bursal slits at the bases of the arms open to the genitorespiratory bursal pits. Two genital plates lies beside each bursal pit.

CLASS CONCENTRICYCLOIDEA

These tiny echinoderms (Fig. 35.26A), 3 mm in diameter, are collected from sunken logs at 1000 to 1200 m off New Zealand (Baker et al., 1986). The body is disclike, with the dorsal surface equipped with plates and the ventral surface developed centrally as a membranous **velum**. The margin of the disc is developed with a series of spines, interior to which on the ventral surface are the podia or tube feet arranged in a ring.

Preliminary knowledge of the internal anatomy reveals that there is no mouth or gut. This might indicate that the single known species, *Xyloplax medusiformis*, may subsist by absorbing nutrients in solution from its environment. A series of five pairs of gonads (Fig. 35.26B) contain embryos in all stages of development up to plated juveniles. The water-

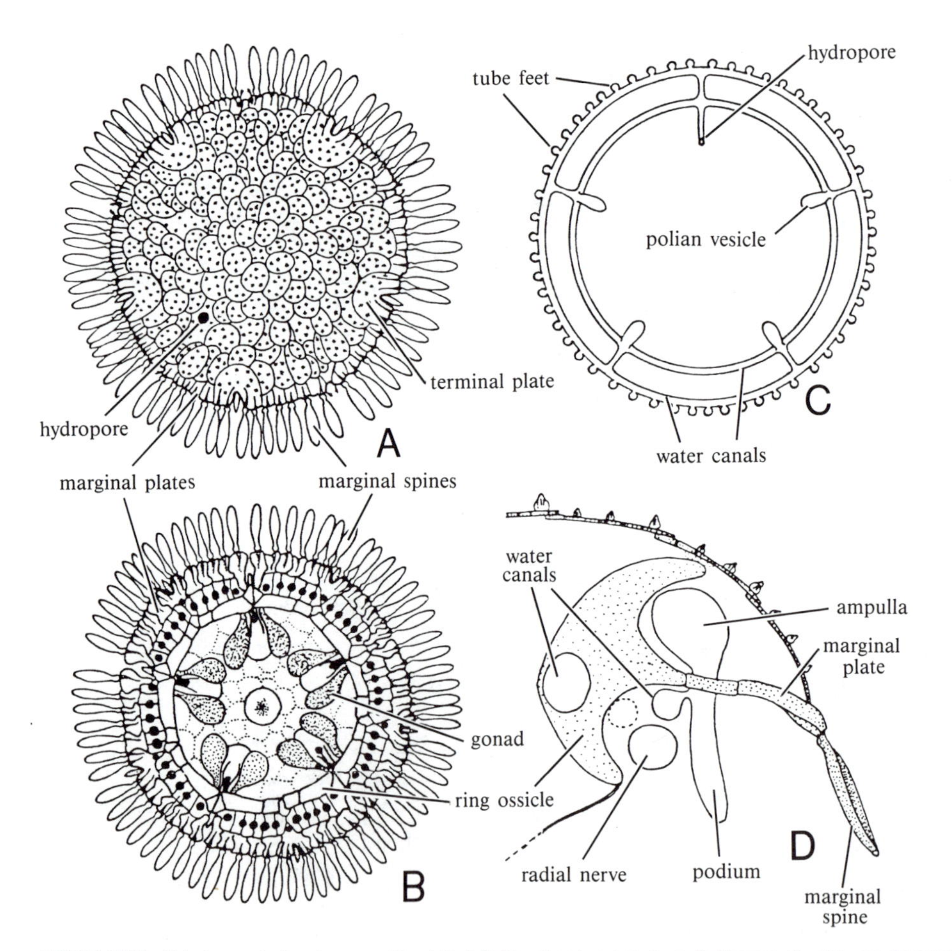

FIGURE 35.26. *Xyloplax medusiformis*, concentricycloid. **A**,**B**. Dorsal and ventral views. **C**. Diagram of water-vascular canals. **D**. Schematic section of margin. (After Baker et al., 1986.)

vascular system is composed of two rings (Fig. 35.26C). The inner ring canal is situated in a duct within the ring ossicles; the outer radial canal is connected to each tube foot and its ampulla (Fig. 35.26D). The system is connected to the outside by a hydropore on the dorsal surface.

The preliminary description of *Xyloplax* makes comparisons to the extinct class (Ordovician to Devonian) Cyclocystoidea. Many features, however, resemble those of the caymanostellid asteroids, especially the clublike marginal spines and their inhabiting sunken pieces of wood from the deep sea. More exact judgments must await detailed descriptions of adult anatomy and development (Nichols, 1986), in not only *Xyloplax* but also *Caymanostella*.

Fossil Record

The presence of calcareous skeletal plates, the relatively large size of many echinoderms, and the high population densities that echinoderms have achieved in the past have combined to make them among the most abundant invertebrate fossils. The richness of this record has allowed us to assemble some reasonable hypotheses about echinoderm relationships (see below). There are over two dozen extinct echinoderm classes, and treatment of all of them here is impossible. Because these classes are interesting and are important for an understanding of modern forms, a brief description of only some of the more prominent is included.

Class Heterostelea (Fig. 35.27B) includes unusual species that show no traces of radial symmetry. They are Paleozoic forms, which are known from Cambrian to early Devonian times. They perhaps represent the most primitive of all echinoderms. It is generally agreed that the flattened theca was horizontal on the sediment, whereas the surface containing the mouth was upward. In some, one or two small arms (brachioles) equipped with food grooves led to the

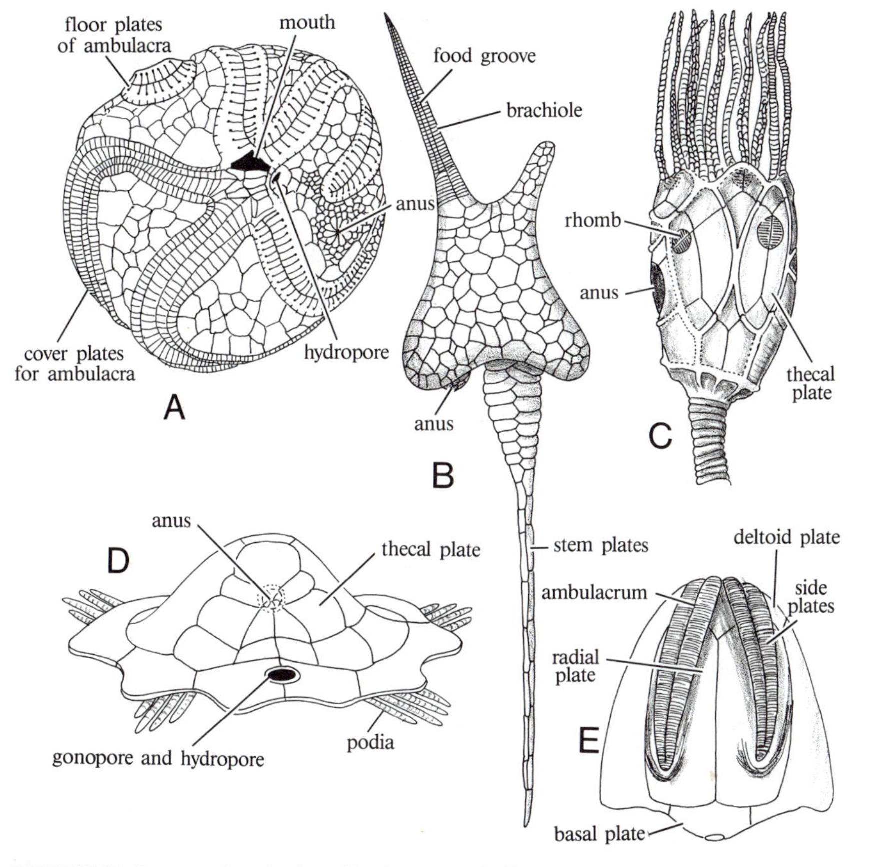

FIGURE 35.27. Representatives of extinct echinoderm groups. **A.** *Edrioaster bigsbyi*, Edrioasteroidea. **B.** *Dendrocystes scotica*, Heterostelea. **C.** *Cheirocrinus walcotti*, Cystidea. **D.** *Volchovia*, restored, Ophiocistoidea. **E.** *Pentremites*, Blastoidea, with brachioles removed. (**A,B** after Bather; **C** after Jaeckel, from Cuénot, in Grassé; **D** after Haecker; **E** after Hyman.)

mouth, and a style or stem was also present. They may have used a ciliary–mucus feeding system. At least a part of the stem contains two rows of ossicles, unlike stalked echinoderms, and the only openings in the theca are the mouth and the anus. There is no evidence of an ambulacral water-vascular system.

Class Cystoidea (Fig. 35.27C) had a somewhat elongated theca, sometimes attached directly at the aboral end and sometimes attached with a hollow stalk. The mouth was centered on the oral surface of the rigid theca, and primitively three food grooves (ambulacra) led to the mouth. Later, division of the two lateral grooves led to a pentamerous radial arrangement. Unbranched brachioles surrounded the mouth or flanked the food grooves and carried branches of the grooves. Unique perforations of the thecal plates, scattered or aggregated in fields (rhombs), were perhaps associated with a special respiratory system. The cystoids appeared in the middle Ordovician and persisted to the Permian.

Class Blastoidea (Fig. 35.27E), like cystoids, were either attached directly or by an aboral stalk. The bud-shaped, pentamerous theca was formed of 13 primary plates arranged in three tiers. The stalk was attached to the three basals. Five deltoid plates formed the oral surface and interdigitated with five radial plates notched for the prominent ambulacra. The five ambulacra were floored by a lancet plate and flanked by lateral plates on which the brachioles attached. The sometimes pinnate brachioles contained branches of the food grooves. Like the cystoids, the blastoids had characteristic structures thought to be respiratory in function. A system of folds or folds and pores (hydrospires) lay beside or beneath the ambulacra and are thought to have permitted water to circulate in a special coelom for respiratory exchange. Certainly the rigid theca of both cystoids and blastoids would make respiratory exchange difficult without some special system to provide ingress for water. Blastoids appeared in the Ordovician period and apparently became extinct in the Permian.

Class Edrioasteroidea (Fig. 35.27A) consists of a rather uniform group of fossil forms that appeared during the middle of the Cambrian and disappeared during the Carboniferous period. The theca was composed of small, irregular plates and was probably somewhat flexible. Some edrioasteroids were attached aborally, but others apparently could move freely. Most were at least somewhat flattened. Three to five ambulacra radiated from the mouth. In more primitive species, the ambulacra were straight, but later they became curved and twisted over the thecal surface. As edrioasteroids lacked brachioles and did not develop rays, the only method available for amplifying the food grooves was to increase the length of the ambulacra.

Although a relatively small class, edrioasteroids have attracted considerable attention, as they have some progressive features. Tubercles on the plates suggest that they may have borne spines and perhaps were able to move about with their aid. Furthermore, the ambulacral plates were pierced by pores for water-vascular podia, and a double series of cover plates made it possible to protect the food grooves and podia.

Class Ophiocystoidea (Fig. 35.27D) fossils are so few and so fragmentary that not much is known about them, and some conflicting interpretations leave further room for confusion. The flattened oral surface was directed downward, and the dome-shaped aboral surface was composed of plates not falling into a pentamerous pattern. On the oral surface, a central flexible peristome surrounded the mouth, which was equipped with five plates serving as jaws. The five ambulacra had a median row of plates and two lateral rows, and contained pores for giant podia. The podia were covered by scales that had pores between them, possibly for respiratory projections. The anus lay on a prominence on the aboral surface, and a water pore, located interradially, apparently supplied the ambulacral system with water. The few known specimens range from the Ordovician to the Devonian. They are an example of how strange some fossil echinoderms could be, and may be part of a group of which the living concentricycloids are a remnant.

Phylogeny

There is probably no greater snake pit in the field of invertebrate phylogenetic speculation, with the possible exception of arthropods, than that of echinoderms. Mental gyrations on the subject have been rampant since the last century. However, three recent systems deserve some comment.

First is the arrangement employed (Ubaughs, 1967) in the *Treatise on Invertebrate Paleontology*. Four subphyla were based on general external form. Homalozoa contain the bilateral and asymmetric forms, Asterozoa the armed types, Echinozoa the globose groups, and Crinozoa the stalked species. Although this system has achieved widespread use, it had the disadvantage of neglecting developmental and internal anatomical information, and ignored judgments on the role of environment in determining superficial external appearance.

The most fascinating system is that proposed by Haugh and Bell (1980). They interpret evidence of soft anatomy in several extinct groups that apparently indicates the existence of another coelomic system, what they termed the subdermal respiratory coelom, not present in living forms. Their Subphylum I has both water-vascular and subdermal respiratory coeloms (except living crinoids that seem highly modified from more typical types seen among the fossils). Subphylum II, consisting of only extinct classes, is distinguished by a lack of a water-vascular coelom and the link of the subdermal respiratory coelom with various thecal respiratory organs. Subphylum III is characterized only by a water-vascular coelom and includes most of the living classes. This system of classification has the advantage of perhaps offering some insight into why certain echinoderms survive and why so many others went extinct. The extinct members of Subphylum I and all of Subphylum II had a body plan in which the subdermal respiratory coelom may have been somehow less effective than the body plan of Subphylum III which uses the water-vascular coelom.

The final approach, by Smith (1984), is a consideration of relationships using cladistic techniques. This analysis (Fig. 35.28) lends some credence to the old classic system whereby stalked forms, subphylum Pelmatozoa, are separated from the free-living forms, subphylum Eleutherozoa. Outgroup comparisons of echinoderms to their immediate sister groups leave the issue open as to whether echinoderms were primitively sessile or free living. The characters used in Figure 35.28 do not resolve this issue, and sort the five major classes on the basis of other features.

One final theory deserves some passing comment. A student can hardly open any book or symposium volume without encountering papers by R. P. S. Jefferies, whose view it is that the extinct homalozoan echinoderms (Fig. 35.27B), what he calls "calcichordates," are direct ancestors to vertebrates. An excellent paper by Jollie (1982) effectively refutes Jefferies' contentions.

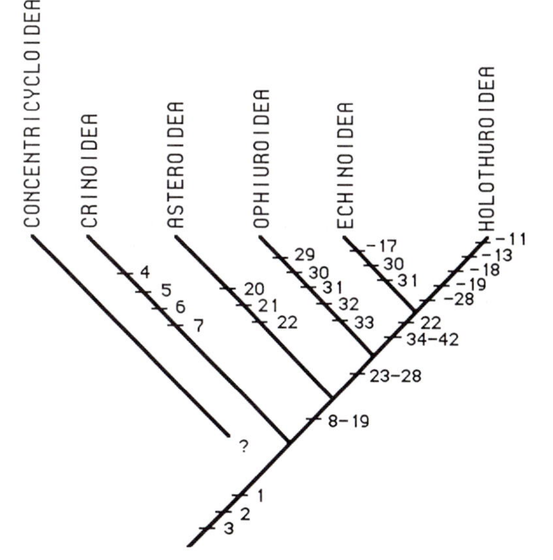

FIGURE 35.28. Cladogram of relationships of living echinoderm classes. Derived characters are: (1) calcareous ossicles; (2) left side of larval coeloms developed, right side suppressed; (3) pentamerous, radial symmetry; (4) right hydrocoel not formed; (5) gonads on arms; (6) ambulacra free from central theca, developed as arms; (7) blastopore seals after archenteron formed; (8) larvae with auricularia; (9) madreporite; (10) no anus, larval anus sealed; (11) ambulacra added adorally to terminal plate; (12) tube feet from lateral branches of radial water vessel; (13) spinea; (14) ectoneural oral plexus for motor control; (15) hyponeural sinuses; (16) larvae with mouth; (17) polian vesicles; (18) Tiedmann's bodies; (19) multiple internal gonads from genital rachis; (20) new adult anus laterodorsal in BC interradius; (21) no larval vestibule; (22) tube feet with internal ampullae; (23) entoneural adoral nerve complex absent; (24) larvae unattached; (25) radial water vessel and nerve enclosed by epineural fold; (26) epineural sinus present; (27) endomesoderm forms before invagination from one side of blastula wall; (28) adoral ambulacral ossicles as jaw apparatus; (29) vestibule remains open; (30) larval processes supported by skeletal rods; (31) larval mouth lost, new mouth forms to left of larval mouth; (32) vertebral ossicles; (33) ambulacral furrows covered; (34) tube feet with skeletal disc plates; (35) peripharyngeal coelom; (36) perianal coelom; (37) hemal system well developed; (38) adoral ossicles internal, around esophagus; (39) radial ambulacral muscles internal and unsegmented; (40) adult anus dorsal to site of larval anus; (41) radial water vessel meridional; and (42) tube feet wall with spicules. (Modified from Smith, 1984.)

Taxonomy

The classification used here presents only living forms, but in many respects reflects the influence of the extinct taxa that are not included.

Subphylum Pelmatozoa. Stalked forms, right hydrocoel not formed, ambulacra on arms free from theca, blastopore closes after archenteron forms.

Class Crinoidea. Calyx equipped with arms bearing pinnules.

Subclass Articulata. Single living subclass of crinoids (Figs. 35.1A and 35.12).

Order Isocrinida. With cirri on stalk.

Order Millericrinida. Without cirri on stalk.

Order Comatulida. Without stalk.

Subphylum Eleutherozoa. Free-moving, with body globose or discoid, with or without arms; with madreporite; tube feet on lateral branches of radial water canal; larvae with auricularia or lobes.

Class Asteroidea. Arms radiate, arranged horizontally, open ambulacral groove, double row of ambulacral ossicles, arms not offset from body disc; original larval anus lost and replaced by new anus located in BC interradius (Fig. 35.1B).

Order Platyasterida. Ambulacral furrow permanent, interpinnular grooves with cilia, no suctorial tube feet.

Order Paxillosida. With marginal plates, aboral paxillae, granule type pedicellariae; no suctorial tube feet.

Order Spinulosa (= *Notomyotida*). Usually without conspicuous marginal plates; pedicellariae usually wanting; aboral skeleton commonly with paxillae; mouth frame adambulacral; podia with suckers and single or bifurcated ampullae; aboral longitudinal muscles in arms.

Order Valvatida. Valvate pedicellariae, consisting of two robust contiguous plates or valves; suctorial tube feet, brachiolarian larva; Tiedmann's bodies.

Order Forcipulata. Without conspicuous marginal plates; spines not grouped and aboral skeleton usually reticulate; papulae on oral and aboral surfaces; pedunculate pedicellariae; suctorial podia usually arranged in four rows; mouth frame ambulacral; brachiolarian larva.

Class Ophiuroidea. Ambulacral furrows covered; anus lost; paired ambulacral plates forming arm segments and fused into single vertebrae; larval processes supported by skeletal rods; ossicular jaws; Tiedmann's bodies; larval mouth lost and replaced by new mouth.

Order Ophiurida. Zygospondylous articulations allowing only horizontal movements (Fig. 35.24A).

Order Oegophiurida. Lateral plates well separated, exposing oral surface of radial ossicles; madreporite marginal; vertebrae with streptospondylous (hourglass) articulation allowing horizontal and vertical movements.

Order Phrynophiurida. Lateral plates small; madreporite oval; streptospondylous articulation (Fig. 35.24G).

Class Echinoidea. Body globose or discoid; mouth generally with elaborate jaw apparatus (Aristotle's lantern); peripharyngeal coelom; test spinose; Tiedmann's bodies; larval processes supported by skeletal rods; larval mouth lost; larval anus replaced by one just dorsal to it.

Subclass Perischoechinoidea. One tube foot per plate.

Order Cidaroidea. Test rigid, with two rows of narrow, simple ambulacral plates and two rows of interambulacral plates, continuing to mouth rim; perignathic girdle marked by interradial muscle attachments; each ambulacral plate with single large spine encircled by small spines; without gills or balance organs; tridentate and globiferous pedicellariae (Fig. 35.19A).

Subclass Euechinoidea. Two columns of ambulacral and interambulacral plates.

Superorder Diadematacea. Anus aboral; tubercles perforate; gills usually present.

Order Echinothuroida. Flexible test of separated plates, ambulacral plates with peristome extending to mouth rim.

Order Diadematoida. Test globose, with two rows of intrambulacral plates and two rows of compound ambulacral plates, both ending at peristome margin; muscle attachments on perignathic girdle; radial gills prominent; small bodies (sphaeridia) on ambulacra thought to be balance organs; all types of pedicellariae or without globiferous pedicellariae; spines hollow.

Order Pedinoida. Ambulacral plates ending at peristome margin; spines solid; gill notches slight.

Superorder Echinacea. Regular urchins; gills present; well-developed dentition (Fig. 35.1D).

Order Salenoida. Stirodont with epiphyses separate and teeth crested on inner surface; interambulacral plates with single spine.

Order Pymosomatoida. Stirodont; tubercles nonperforate.

Order Arbacioida. Stirodont; periproct covered by four or five anal plates.

Order Temnopleuroida. Camerodont with epiphyses joined across pyramid tops; ambulacral plates compound; test sculptured.

Order Echinoida. Camerodont; three to sixteen pores per plate; gill notches shallow.

Superorder Gnathostomata. Irregular urchins; lantern present; ambulacral plates simple or weakly compound.

Order Clypeasteroida. Test flattened and oval or circular, usually densely covered by small spines; test often supported by internal strutting; peristome and apical system usually central; petaloid ambulacra with single sphaeridium near peristome; without phyllodes or bourrelets; without gills. Example: *Mellita* (Fig. 35.19B).

Order Holectypoida. Test regular, with apical plates centered at aboral pole; anus somewhat separate from apical plates; peristome central; ambulacra not petaloid; mostly extinct. Example: *Echinoneus* (Fig. 35.19E).

Superorder Atelostomata. Irregular urchin; lantern absent; ambulacral plates simple.

Order Cassiduloida. Test round to oval; periproct variable in position; peristome central or slightly anterior; ambulacral plates modified at peristome as petallike phyllodes and interambulacral plates with bourrelets; ambulacra petaloid, but open at margin 35.19C).

Order Spatanguloida. Test oval or cordiform, bilaterally symmetrical about long axis; peristome displaced anteriorly, with oral ambulacrum in-sunk and not petaloid, and remaining four ambulacra petaloid; phyllodes present but bourrelets wanting; with posterior interambulacrum often modified on aboral surface, with short spines; without gills (Fig. 35.19D).

Class Holothuroidea. Body warty, biradial or bilateral; peripharyngeal coelom; ambulacra reduced; gonads not radially arranged; no ossicles as jaw apparatus; no Tiedmann's bodies; larval anus lost and replaced by one just dorsal to it.

Subclass Dendrochirotacea. Retractile oral area with tentacles; internal madreporite.

Order Dendrochirotida. Ten to thirty tentacles, much branched (Fig. 35.18D).

Order Dactylochirotida. Eight to thirty digitiform tentacles; body test of imbricating plates.

Subclass Aspidochirotacea. Tube feet; no introvent; ten to thirty tentacles; bilateral.

Order Aspidochirota. Many podia, peltate oral tentacles, respiratory trees (Fig. 35.18A).

Order Elasipoda. Deep-sea holothurians with many podia, peltate oral tentacles, and without oral retractor muscles or respiratory trees (Fig. 35.18E,F).

Subclass Apodacea. Tube feet reduced or absent; ossicles with anchors or anchor plates.

Order Molpadinida. Digitate oral tentacles and respiratory trees; podia reduced to anal papillae, and posterior region tapering to caudal region (Fig. 35.18C).

Order Apodida. Oral tentacles, reduced water-vascular system, without respiratory trees (Fig. 35.18B).

Class Concentricycloidea. Discoid body; without mouth, anus, or radiating arms; water-vascular system concentrically arranged.

Order Peripodida. Single order known (Figs. 35.1F and 35.26).

Selected Readings

Baker, A. N., F. W. E. Rowe, and H. E. S. Clark. 1986. A new class of Echinodermata from New Zealand. *Nature* 321:862–864.

Binyon, J. 1972. *Physiology of Echinoderms.* Pergamon Press, Oxford.

Boolootian, R. A. (ed.). 1966. *Physiology of Echinodermata.* Wiley, New York.

Bosch, I., R. B. Rivkin, and S. P. Alexander. 1989. Asexual reproduction by oceanic planktotropic echinoderm larvae. *Nature* 337:169–170.

Byrne, M. 1985. Evisceration behavior and the seasonal incidence of evisceration in the holothurian *Eupentacta quinquisemita. Ophelia* 24:75–90.

Emson, R. H., and I. C. Wilkie. 1980. Fission and autotomy in echinoderms. *Oceanogr. Mar. Biol. Ann. Rev.* 18:155–250.

Gale, A. S. 1987. Phylogeny and classification of the Asteroidea. *Zool. J. Linn. Soc. Lond.* 89:107–132.

Haugh, B. N., and B. M. Bell. 1980. Fossilized viscera in primitive echinoderms. *Science* 209:653–657.

Jangoux, M., and J. M. Lawrence. 1982. *Echinoderm Nutrition.* Balkema, Rotterdam.

Jollie, M. 1982. What are the "Calcichordata" and the larger question of the origin of chordates. *Zool. J. Linn. Soc. Lond.* 75:167–188.

Nichols, D. 1969. *Echinoderms.* Hutchinson, London.

Nichols, D. 1986. A new class of echinoderms. *Nature* 321:808.

Schroeder, T. E. 1986. The egg cortex in early development of sea urchins and starfish. *Exper. Biol.* 2:59–100.

Smith, A. B. 1984. Classification of the Echinodermata. *Palaeontol.* 27:431–459.

Telford, M. 1983. An experimental analysis of lunule function in the sand dollar *Mellita quinquiesperforata. Mar. Biol.* 76:125–134.

Ubaughs, G. 1967. General characters of echinodermata. *Treatise on Invertebrate Paleontology*, Part S(1):53–560.

36

Hemichordata

Hemichordates are more or less wormlike animals, living in temporary and permanent tubes in the bottom mud or among rocks or masses of plant material. The more familiar acorn worms are solitary, but the pterobranchs are social and some form definite colonies. The two groups are so dissimilar, however, that it was not until the end of the nineteenth century that they were allied together, and even today some feel that their affinities are weak and that they are separate phyla (see Chapter 38). The only derived feature they seem to consistently share is the presence of a middorsal diverticulum of the buccal chamber. We treat them separately here within a phylum Hemichordata.

Enteropneusta

Most acorn worms are moderate in size, up to about 30 cm or so long, but *Balanoglossus gigas* may reach a length of 2.5 m. They are slimy, beige to pinkish worms, very sluggish, and so soft that they break easily when handled. Many smell like iodoform. Some live in tubes and eat sand or mud; others are ciliary-mucus feeders on the bottom surface that do not secrete tubes.

DEFINITION

Enteropneusts are free-living, and divided into a proboscis or protosome containing an unpaired protocoel, a collarlike mesosome containing paired mesocoels, and a trunk metasome containing paired metacoels. The protocoel and mesocoels open to the outside by pores. They possess a buccal middorsal diverticulum, a hollow middorsal nerve center in the mesosome, and typically one or more pairs of pharyngeal slits that connect the pharynx lumen with the exterior. They have a well-developed circulatory system, usually with a contractile heart, but have no nephridia.

BODY PLAN

Two constrictions divide the body into an anterior, conical, or elongated **proboscis** (**protosome**), a middle **collar** (**mesosome**), and a long, worm-shaped **trunk** (**metasome**) (Fig. 36.1).

The ciliated epidermis is richly glandular, composed of tall, slender cells that vary considerably between different body regions. Gland cells are mucus-secreting goblets or coarsely granular or reticulated mulberrylike cells. The collar is often marked by five transverse bands of alternating glandular and nonglandular epithelium. A thick nerve layer at the base of the epidermis lies immediately above the tough basement membrane. The outer circular and inner longitudinal muscle layers are well developed in the proboscis and collar, but the trunk musculature is thin and equipped almost wholly with longitudinal muscles.

The proboscis contains remnants of the protocoel (Fig. 36.2A). The embryonic peritoneum differentiates into muscle and connective tissue, and the coelom is partly filled with tissue. The proboscis has rather delicate circular muscles. Prominent strands of longitudinal muscles, sometimes arranged in a fan and often mixed with diagonal fibers, nearly fill the anterior part of the protocoel. The protocoel widens posteriorly, but is invaded by the anteriorly directed **buccal diverticulum**, which is covered by peritoneum and attached to the proboscis floor by a midventral mesentery that partly divides the protocoel into right and left chambers. A canal extends from the protocoel into the proboscis stalk on each side of the buccal diverticulum. Here the peritoneum is sometimes frilly and forms the "cauliflower organ." A **proboscis canal** connects the protocoel with a **proboscis pore**. Portions of the mesocoel extend anteriorly to meet the posterior canals of the protocoel. Where they meet, the coelomic peritoneum disintegrates and disappears and forms chondroid tissue, reminiscent of vertebrate cartilage. The basement membrane in the proboscis is thickened, and muscles insert onto it. At the proboscis base, it forms the proboscis skeleton consisting of a median plate and two anterior horns.

Adults have no definite proboscis–collar septum, but the protocoel and the mesocoel are not openly confluent. The mouth lies on the ventral side, between the proboscis base and the collar margin, where the latter is flared out as the **collarette** (Fig. 36.2A). A buccal tube extends through the collar. The rest of the space in the collar is occupied by the mesocoel, considerably reduced by connective tissue and muscle fibers. *Protoglossus* is considered the most primitive enteropneust genus; it has right and left mesocoel chambers divided by a median septum. In other forms, the mesentery, especially in the ventral side, is incomplete. The mesocoel opens to the exterior on each side through a collar canal that leads to a collar pore located in the first gill slit. The collar musculature is variable. Radial muscles, attached to the body wall and buccal tube, and longitudinal muscles are usually present, but some species have only a feeble collar musculature. A deep external constriction divides collar from trunk; here the internal collar–trunk septum lies, which, although simple in *Protoglossus*, is complicated in many forms by one or two pairs of metacoel pouches that extend through the septum. Perihemal spaces extend forward over the top of the buccal tube, and the dorsal blood vessel lies in the septum between them. In some species, a pair of flattened peribuccal folds contain that portion of the metacoel

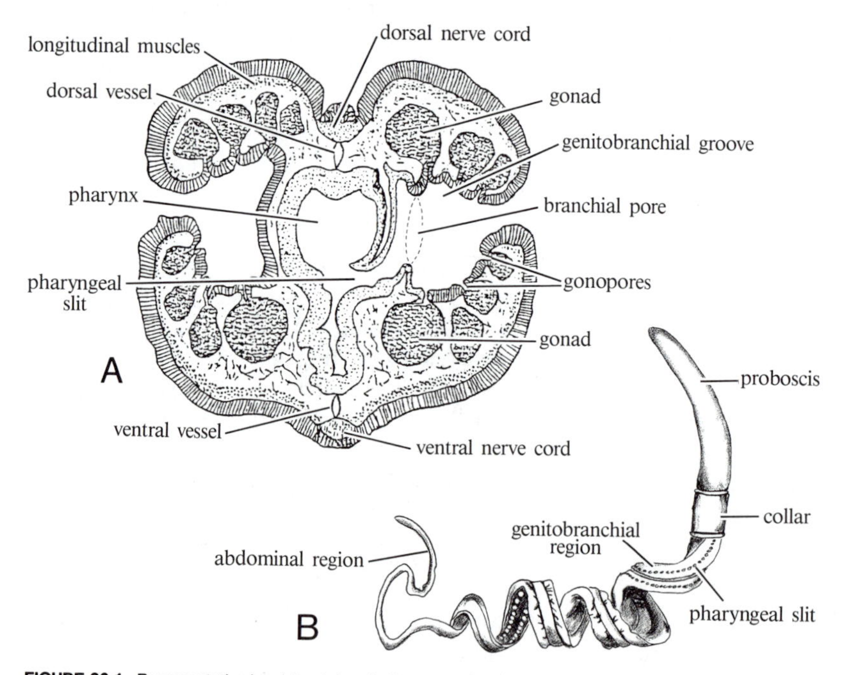

FIGURE 36.1. Representative hemichordates. **A.** Cross section through pharynx of *Stereobalanus canadensis;* note pharyngeal slit and relation of gonads. **B.** Typical enteropneust, *Saccoglossus kowalewski.* (**A** after Dawydoff, 1948; **B** after Bateson.)

that can extend forward on each side of the buccal tube.

The long trunk is usually divided into regions. A series of pairs of branchial pores mark the dorsolateral surface of the genitobranchial region of the trunk (Fig. 36.1). The number of pores is variable, for new ones are added as the animal grows. The gonads lie in this part of the trunk, and a genitobranchial ridge can often be seen on the surface (Fig. 36.1A). The anterior part of the postbranchial region is sometimes darkened by the hepatic part of the gut located internally; in this case, it is then termed the hepatic region. The metacoel is partly divided into right and left compartments by a mesentery, but the dorsal mesentery is incomplete. The metacoel does not communicate with the exterior by a canal or a pore. The coelomic fluid of the protocoel and mesocoel must approach the composi-

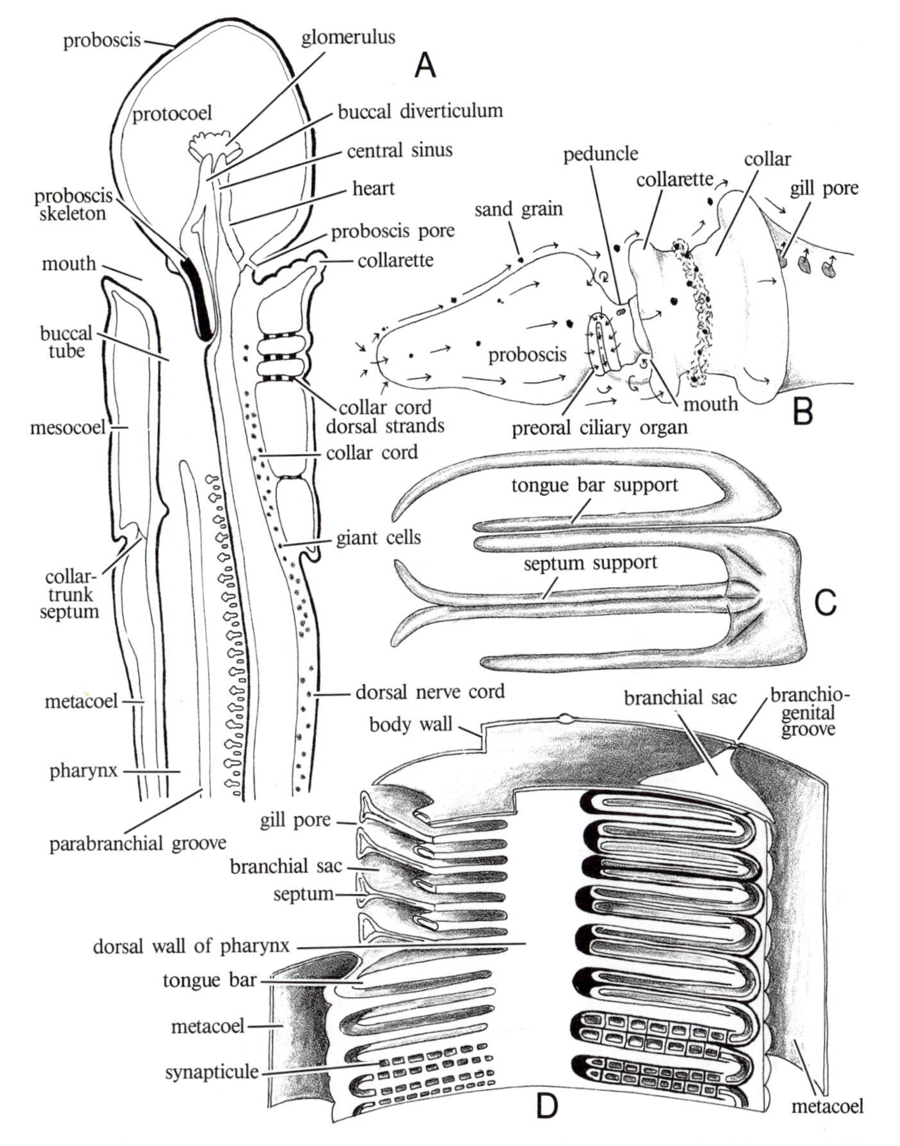

FIGURE 36.2. Enteropneust structure. **A**. Scheme of sagittal section of *Saccoglossus pusilus*. **B**. Anterior end of *Protoglossus köhleri* showing ciliary currents used in feeding and burrowing. **C**. Skeletal rods supporting branchial region, from *Saccoglossus kowalewski*. **D**. Scheme of gill region. (**A** after Bullock; **B** after Burdon-Jones, in Kükenthal and Krumbach; **C** after Spengel; **D** after Delage and Herouard, from van der Horst, in Kükenthal and Krumbach.)

tion of sea water as the chambers are open to the outside, but the metacoel contains a distinct coelomic fluid with many coelomocytes.

Digestive System The straight gut has no conspicuous dilations and usually is without intrinsic muscles, but it is connected to the body wall by radial muscles that span the metacoel. The **mouth** and **buccal tube** lie within the collar region, and the **pharynx** and **intestine** are in the trunk.

The buccal diverticulum is associated with the central sinus of the circulatory system, and with a mass of blind peritoneal tubes known as the **glomerulus**. The buccal diverticulum, central sinus, and glomerulus together form the **proboscis complex**.

The digestive tube widens to form a pharynx in the trunk. Here **pharyngeal slits** open to the exterior (Fig. 36.1A); their beating cilia help to pull water through the anterior part of the gut. The pharyngeal slits are primitively a food-gathering device; no gill structures are associated with enteropneust slits.

An epibranchial ridge usually attaches the dorsal ends of the pharyngeal slits. The paired slit pores open into a series of outer branchial cavities (Fig. 36.2D), which in some species are merged as a single long chamber on each side. Ciliary currents force water from the pharynx through the slits and branchial chambers to the outside. The slit pores are usually guarded by sphincter muscles and open in the genitobranchial groove on each side of the body (Fig. 36.1A). Pharyngeal slits are oval at first appearance, but a tongue bar pushes down from the dorsal side during development, giving them a chevron shape (Gilmour, 1982). Each tongue bar contains an extension of the coelom. A series of septa divide the slits (Fig. 36.2D). Each septum is supported by a skeleton, originally composed of two U-shaped pieces, with one prong in the septum and one in the tongue bar (Fig. 36.2C). Fusion of the two pieces in each septum converts them into three-pronged supports. In some species, small septa known as synapticules (Fig. 36.2D) attach the tongue bars and adjacent septa, holding the tongue bars firmly in place.

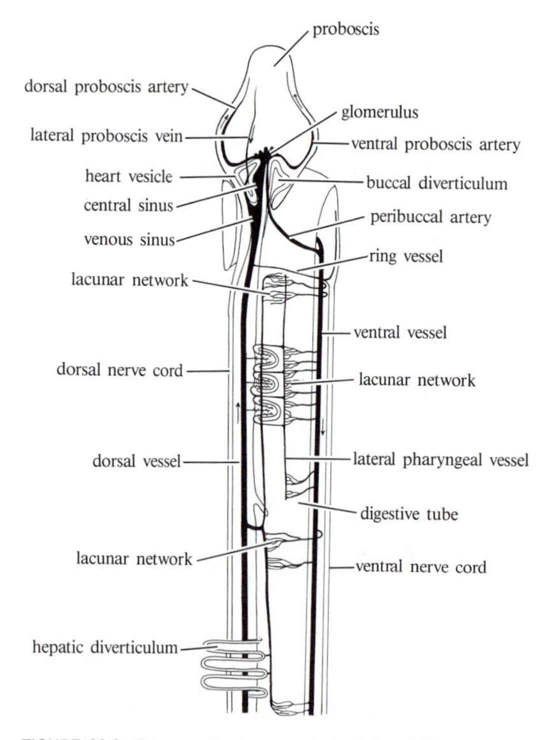

FIGURE 36.3. Scheme of enteropneust circulation. (After van der Horst, in Kükenthal and Krumbach.)

The intestine is regionalized. The **esophagus** has a relatively large lumen, and in some species is histologically differentiated into several subregions. In two families, esophageal canals lead to esophageal pores at the body surface. These are probably abortive slits. The anterior part of the intestine proper is the hepatic region. The dorsal epithelium is filled with brownish or greenish inclusions, and in some species is thrown into sacciform projections. This part of the intestine is richly vascular (Fig. 36.3), and one species has a siphon reminiscent of the echinoid intestinal siphon. The next part of the intestine has a wide lumen and a low epithelial lining. Dorsoventral ciliated bands are sometimes found. In some species a midventral band of cells extends between the two leaves of the ventral mesentery. This band of cells, sometimes hollow and often interrupted, has been called the pygochord because its vacuolated cells bear some resemblance to notochord cells. The intestine ends in a dilated or histologically differentiated **rectum** that opens through an **anus**, which is sometimes guarded by a sphincter.

Circulatory System The circulatory system (Fig. 36.3) consists of closed vessels, lacunar spaces, and a definite pulsating organ, generally known as the **heart**. The blood contains no or very few cells and no respiratory pigment. Most of the circulatory system is located between the lamellae of the basement membrane and the leaves of the mesentery.

The **dorsal vessel** conducts blood forward, and the **ventral vessel** backward (Fig. 36.3). The dorsal vessel is dilated at the front of the collar, forming a **venous sinus** that empties into a **central sinus** above the buccal diverticulum. Blood from the central sinus enters the glomerular sinuses of the proboscis complex. A contractile epidermal sac, the **heart vesicle**, lies immediately above the central sinus. No blood enters it; it is not comparable to a true heart, but its contractions help to force the blood from the central sinus. Dorsal and ventral blood vessels also contract to force blood along.

All of the blood flows through the glomerulus, where it is possibly cleared of nitrogenous wastes. Paired efferent glomerular vessels run back along the sides of the proboscis complex and then turn ventrally, as peribuccal arteries, meeting below to form the ventral blood vessel. The ventral vessel continues into the tail region.

A series of circulatory arcs arise from the major vessels. The proboscis arc is formed by a middorsal proboscis artery and a midventral proboscis artery arising from the glomerulus, and paired lateral proboscis veins that empty into the central sinus. A body wall arc receives blood from the ventral vessel, never far from the body wall and sometimes in contact with it. Blood enters a lacunar plexus in the body wall and returns to the dorsal vessel through many small branches. Short branches from the ventral vessel bring blood to a similar lacunar plexus in the gut wall; the blood enters the dorsal vessel through similar short branches.

The lacunar plexus of the collar body wall is continuous with the buccal lacunar plexus at the mouth and at the constriction between the collar and the trunk, where definite ring vessels occur. Blood from the ventral vessel reaches this double plexus by means of branches in the collar-trunk septum and then drains into the dorsal vessel through small branches. A ventral collar vessel also supplies blood to the collar plexus; it extends forward from the point where the peribuccal arteries unite.

The pharyngeal lacunar plexus is modified as a branchial circulatory arc. The pharynx is composed of a dorsal branchial region and a ventral digestive region. A lateral pharyngeal vessel at the junction of the two pharyngeal regions appears in the lacunar plexus. It gives rise to an afferent branchial vessel into each gill septum, which in turn gives off a branch that curves ventrally to enter the tongue bar. The lacunae of the branchial region empty into efferent branchial vessels, formed by branches from the septum and tongue bar (Fig. 36.3).

Nervous System The nervous system is composed primarily of an epidermal plexus, just external to the basement membrane. Thickened strands of the plexus form a **middorsal nerve cord**, extending the length of the body, and a **midventral nerve cord** that starts at the collar-trunk constriction and continues posteriorly. The nerve plexus is also thickened at the base of the proboscis, forming an **anterior nerve ring**, and again at the junction of the collar and the trunk where a **circumenteric nerve ring** connects the dorsal nerve cord with the start of the ventral nerve cord.

The dorsal nerve cord in the collar, sometimes called the **collar cord** (Fig. 36.2A) is especially interesting, for it lies in the mesocoel instead of outside the basement membrane and is hollow. In some species it opens by anterior and posterior neuropores to the exterior. In other species the cavity is discontinuous, consisting of scattered spaces. The collar cord arises from the epidermis, but has sunk inward to take a deeper position. Its similarity to the chordate nerve cord, formed by invagination of the neural ectoderm at the dorsal midline, is evident.

The collar cord, and sometimes the anterior part of the dorsal cord in the trunk or nerve ring, contains giant neurons. Each gives off a single large nerve fiber that crosses to the other side of the body and runs through the circumenteric ring to the ventral nerve cord. In some species giant neurons at the front of the collar cord cross over and enter the proboscis. The number of giant neurons varies from about 10 to 160. They are responsible for rapid conduction of stimuli leading to quick retraction of body parts.

A nerve plexus below the gut epithelium is most highly developed in the buccal region, forming a diffuse oral ring. It becomes less conspicuous posteriorly. It tends to be thickened into a lesser middorsal tract.

Reproductive System Sexes are separate, and male and female reproductive systems are simple. Each of the many gonads open through a gonopore, typically located beside the gill pores in the genitobranchial groove (Fig. 36.1A). The gonads sometimes extend into the genital ridges or genital wings on the sides of the body.

BIOLOGY

Locomotion and Food Getting The conical to elongate proboscis is the principal active part of the body (Barrington, 1965). Covered with cilia, it probes inquisitively into the mud or sand and collects food in mucous strands, which are carried by cilia to the margin of the collarette for swallowing. The proboscis is used for excavation by tube-dwelling species. It is thrust into the sand; cilia move sand and mucus back over the surface to the collar (Fig. 36.2B) where the sand and mucus either may be swallowed or may slow down and push back over the collar surface to form a ring around the body. In these cases, body cilia slowly move the sand ring back as sand is added at the front. The acorn worm tubes are U-shaped, with one or more "front entrances" and a "back door." The proboscis is thrust from the tube entrance for feeding, and the anus dangles from the back for defecation. A great deal of sand is eaten, and the telltale spiral fecal mounds are a clue to the presence of enteropneusts in a mudflat.

Food collects in the collarette and is diverted into the mouth. Strong cilia in the buccal tube cause a current that flows toward the pharyngeal slits. The horns of the proboscis skeleton extend along the upper surface of the buccal tube, which is often evaginated as a groove at this point.

Digestion Little is known of digestive physiology. Cilia are primarily responsible for food conduction within the gut, although some peristalsis is known to occur in the esophagus of *Saccoglossus*. An amylase is found in the proboscis secretions and is swallowed with the mucous strings. Weak lipase and protease have been recovered from extracts of body pieces. It is possible that the brownish inclusions of the hepatic region are enzymatic, but there is no direct evidence to support this idea.

Respiration Respiratory exchange occurs at the body surface and to some extent within the branchial apparatus and the foregut epithelium. No gills as such are attached to the slits, although the slits are often considered a part of the respiratory system.

The dorsal portion of the pharynx is supplied with a series of blood vessels. Undoubtedly this vascularization of the pharynx wall is associated with the development of some respiratory functions by the pharyngeal region, perhaps as a simple matter of efficiency in utilizing the water for respiratory exchange that streams through the pharynx. Unfortunately, experimental data that compare the amount of respiratory exchange at the body surface versus the pharynx are not available.

Excretion Nothing is known of excretory function. The assumption that the glomerulus is excretory is purely circumstantial. Brownish crystals occur in glomerular cells, and the peritoneum adjacent to the proboscis complex is also brownish and seems to have an athrocytic function in some species.

Sense Organs and Nerve Function The enteropneust body surface is supplied with neurosensory cells of the usual invertebrate type. The only special sense organ is thought to be a U-shaped preoral ciliary organ on the posterior face of the proboscis, which may be chemoreceptive.

Although the collar cord is the most complex part of the nervous system, it is no more than a conduction area and the site of giant neuron formation. The anterior proboscis ring is dominant, and activity is greatly reduced when it is removed. Conduction in the plexus is diffuse, as is typical of nerve nets. Digging movements are hampered but not totally inhibited when the dorsal proboscis cord is transected. Sudden startle contractions, much more rapid than ordinary peristaltic movements, are not disturbed when the dorsal nerve cord of the trunk is transected, but are interrupted when the collar cord or ventral cord is cut, where the giant fibers are found. Giant neurons conduct impulses rapidly and are responsible for rapid, coordinated reactions of the body as a whole, as in a shortening reflex. The normal peristaltic body movements are disturbed but not totally destroyed when the dorsal and ventral cords are severed. The epidermal plexus can control peristaltic movement, but is less effective in doing so than the nerve cords.

Reproduction Spawning is epidemic; one individual triggers a whole population (Hadfield, 1975). Females shed gametes first, releasing 2000 to 3000 eggs about half an hour after the tide has ebbed. Males release sperm about half an hour later. The eggs and sperm are embedded in mucus, but are scattered by the returning tide.

Enteropneusts fragment easily and regenerate reasonably well. Trunk pieces regenerate a new proboscis and collar, but oddly, the isolated proboscis and collar cannot regenerate a new trunk. At least one species reproduces asexually by transverse constrictions that cut off pieces of the trunk to form new individuals.

Development *Saccoglossus* produces large, yolky eggs that undergo direct development (Burdon-Jones, 1952), but most enteropneusts produce small ova with little yolk. These small ova undergo indirect development through a ciliated **tornaria larva** stage that strongly resembles some echinoderm larvae. (One large, transparent larva is known of the tornaria type but with highly branched, ciliated bands on the surface. It has been collected several times, but cannot be related to any known hemichordate, and is referred to as *Planctosphaera*.)

Cleavage is equal, holoblastic, and usually radial. A coeloblastula gastrulates by invagination. The blastopore soon closes, but marks the posterior end of the embryo. The embryo elongates, develops a covering of cilia, and escapes from the fertilization membrane after a day or so of development. At hatching, a long apical tuft of cilia is present (Fig. 36.4A). The tuft and surface ciliation degenerates as a locomotor ciliary band develops. This sinuous band transforms the embryo into a young tornaria larva, with an apical plate at the site of the former ciliary tuft (Fig. 36.4B).

The locomotor band forms a preoral and ventral ciliary loop. Later a circumanal telotroch loop is formed, completely separated from the first band. Tentacles appear at the margins of the locomotor band in some species. Tornaria larvae are not uncommon in plankton samples, and are interesting from a structural and functional point of view. They filter feed using an upstream form of particle capture, and some workers have observed a reverse beat at the affected cilia. The tornarian is capable of sorting particles. Excess water taken into the larval esophagus with the food is passed back out again along sparsely ciliated lateral grooves in the esophagus wall. The pharyngeal slits develop from these larval esophageal grooves.

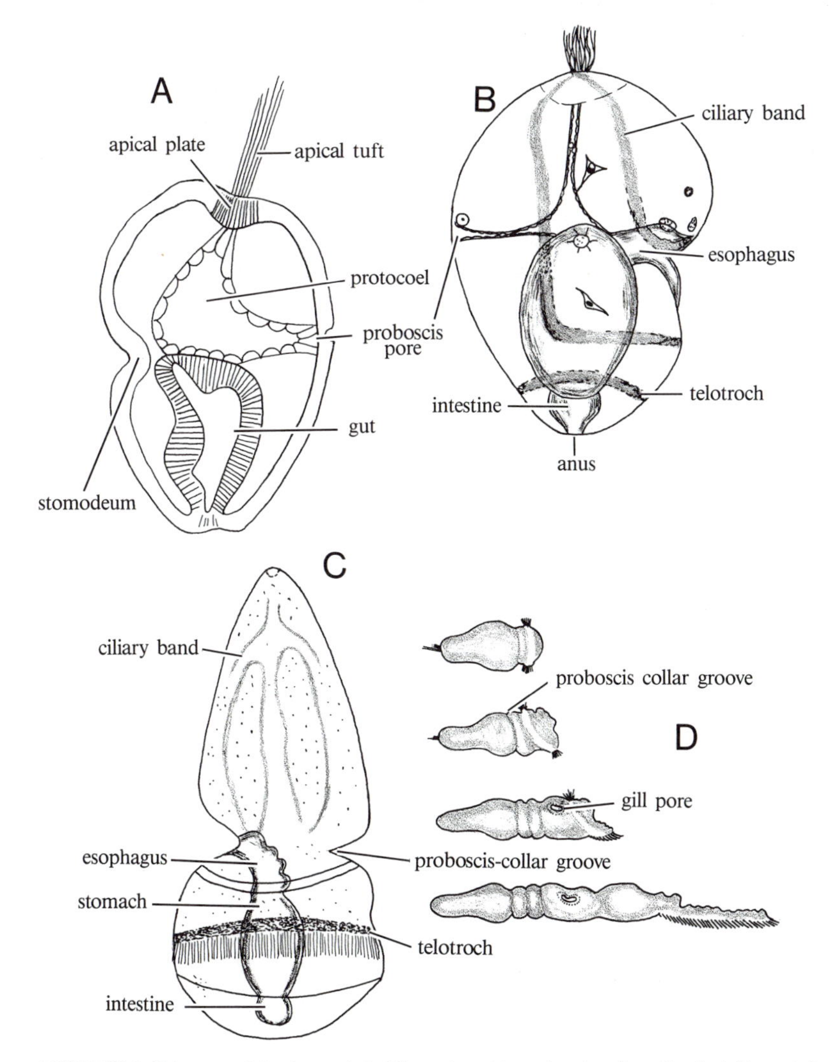

FIGURE 36.4. Enteropneust development. **A**. After protocoel has arisen from inner tip of primitive gut, it comes to open to exterior through proboscis pore. **B**. Mouth and anus form, ciliary bands develop to convert larva into tornaria. **C**. During later development, anterior end of larva loses its ciliary bands, separates from posterior part by proboscis–collar groove, and develops into proboscis, posterior part of larva forming collar and trunk. **D**. Some enteropneusts develop directly without passing through tornaria stage, undergoing similar metamorphosis to develop three body regions characteristic of group. (**A** after Heider; **B** after Stiasny; **C** after Morgan; **D** after Burdon-Jones, 1952.)

More distinct kinds of tornaria have been described than there are known species of enteropneusts, so it may be assumed that several undescribed taxa must exist. The fully grown tornaria has a pair of pigment-cup ocelli at the sides of the apical plate, and a newly generated, short tuft of cilia on the plate (Gilmour, 1982). The tornaria eventually regresses, becomes constricted, foreshadowing the proboscis–collar constriction, and becomes wormlike as the locomotor band gradually declines (Fig. 36.4C,D).

The protocoel arises from the anterior tip of the archenteron. A pore canal and pore form during development; these become the adult proboscis canal and proboscis pore. The mesocoel and metacoel arise in a variety of ways, depending on the species. The principal methods are: (1) the protocoel gives rise to two lateral extensions that grow backward and are cut off as right and left mesocoels and metacoels; (2) solid mesodermal cords on each side of the gut are divided into collar and trunk regions, hollow out, and become mesocoels and metacoels; (3) the mesocoels and metacoels arise independently as either solid or hollow outgrowths of the gut wall; or (4) they appear independently as spaces in groups of mesenchyme cells. The mesocoels, like the protocoel, eventually become open to the exterior by pore canals and pores.

The enteric sac, left after coelom formation, becomes the gut. A stomodeum makes connection with it, and then the foregut, midgut, and hindgut differentiate. Enteropneusts derive the adult mouth and anus from the larval mouth and anus.

Pterobranchia

Pterobranchs are lophophoratelike creatures, largely confined in their distribution to antarctic and subantarctic regions, but extending into the tropics in the Indo-Pacific region. Pterobranchs are poorly known, for they are rarely taken, and then only on collecting expeditions, and are extremely tiny (1 to 5 mm in length). Only three genera have been described: *Atubaria*, *Cephalodiscus*, and *Rhabdopleura*. *Atubaria* is known from a single dredge sam-

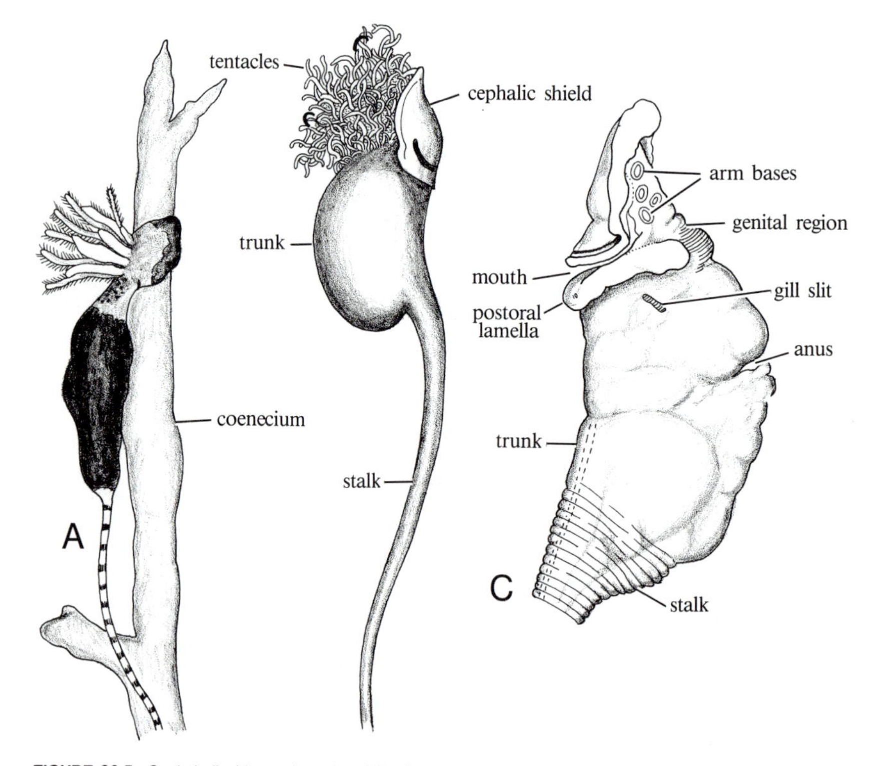

FIGURE 36.5. Cephalodiscid pterobranchs. **A,B.** *Cephalodiscus hodgsoni* **A.** Individual zooid outside its compartment, clinging to zoecium with its stalk. **C.** *Atubaria heterolopha* with arms removed to show body form. *Atubaria* produces no coenecium, individuals cling to colonial animals with their stalks. (**A** after Andersson; **B** after Ridewood, from van der Horst and Helmeke, in Kükenthal and Krumbach; **C** after Sato.)

ple taken near Japan, and is unique in having no tube or case, attaching by its stalk to colonial hydroids (Fig. 36.5C). *Cephalodiscus* zooids (Fig. 36.5A,B) are not connected; they live in tubes that are as independent as separate apartments in an apartment house. *Rhabdopleura* are permanently attached to the common stolon from which they arise as buds (Fig. 36.6A).

DEFINITION

Pterobranchs are sessile and divided into a cephalic shield bearing the lophophore arms with tentacles, a collar with a mesocoel, and a trunk with a metacoel and a posterior stalk typically enclosed in a secreted coenecium. The protocoel and the mesocoel open to the outside by pores. There is a small buccal diverticulum and, at most, one pharyngeal slit but often none.

BODY PLAN

A glandular region on the discoid cephalic shield secretes the **coenecium**. The coenecium of *Cephalodiscus* is a soft, gelatinous mass. It may be stiffened by firmer spines or have filamentous projections, giving it a plantlike appearance. Each zooid inhabits a firmer, darker tube embedded in the coenecium. The *Rhabdopleura* coenecium is a branching tube, giving off upright tubes for the zooids. The creeping tube is somewhat irregular, and contains the black stolon, formed of a firm tube containing a cellular column from which new buds arise. The upright tubes are built of regular, successive rings (Fig. 36.6A).

Each zooid is composed of a **cephalic shield (protosome)**, a **collar (mesosome)**, and a **trunk (metasome)**, and contains the same five coelomic compartments as in enteropneusts (Fig. 36.6B). The protocoel opens to the exterior through paired **pore canals** and paired pores at the base of the arms. The mesocoel also opens through paired pore canals and collar pores.

The flattened cephalic shield of *Cephalodiscus* always is crossed by a red band of unknown function (Fig. 36.5B). The shield is highly muscular, with radiating subepidermal muscles and radial muscle fibers attached to the shield-collar septum and that cross the protocoel. A buccal diverticulum complex, essentially like that of enteropneusts, extends into the protocoel from the shield-collar septum (Fig. 36.6B).

The mouth is on the ventral side of the collar and can be closed by the cephalic shield. Flanges of the body wall (oral lamellae) guide food into the mouth. One to nine pairs of lophophorelike arms are marginally fringed with 25 to 50 tentacles and ventral ciliated grooves or gutters. Mucus glands on the dorsal sides of the arms help to trap food. The collar coelom (mesocoel) extends into the oral lamellae and arms.

The trunk narrows posteriorly into a long, muscular stalk. A single pair of gill slits, without skeletal supports, can open just behind the collar-trunk septum. The trunk coelom (metacoel) is nearly filled anteriorly by gonads and the digestive tube, and in the stalk by muscles and connective tissue. The stalk terminates in an adhesive region where the zooid is attached to its tube. Zooids are freely movable. They leave the tubes and cling to the coenecium surface while feeding.

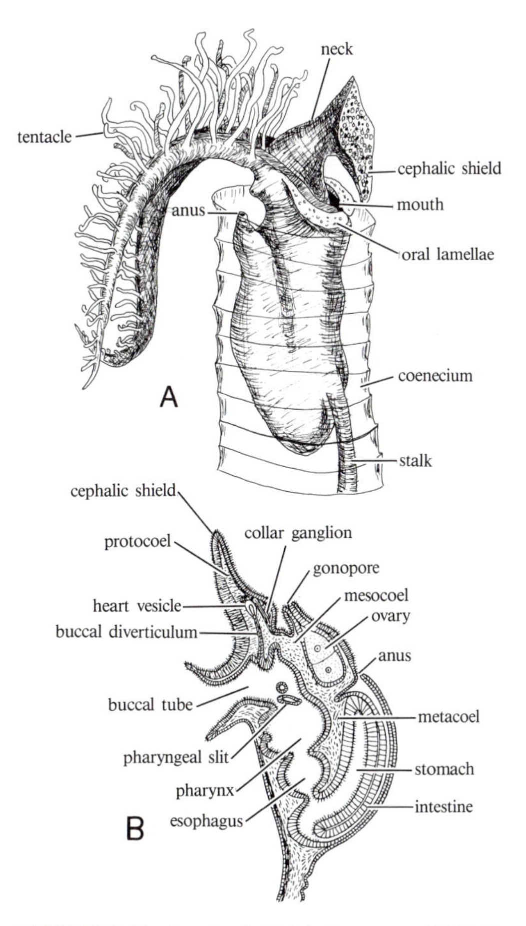

FIGURE 36.6. Pterobranchs. **A.** *Rhabdopleura normani.* **B.** Median sagittal section of *Atubaria.* (**A** after Delage and Herouard; **B** after Dawydoff, 1948.)

Digestive System Pterobranch organ systems closely resemble enteropneust organ systems, and need not be redescribed in detail (Fig. 36.6B). The digestive system is U-shaped instead of straight, as is commonly the case in sessile organisms. It consists of a **buccal tube** in the collar, a **pharynx** sometimes with a pair of **slits** in the anterior part of the trunk, a short **esophagus**, a saccate **stomach**, and an **intestine** that curves dorsally to end in an **anus**.

Circulatory System The main circulatory channels have no cellular walls. A **dorsal sinus** above the esophagus and the pharynx receives blood from the gonadal sinuses and the lacunar plexus of the gut. It runs to a **central sinus** in front of the buccal diverticulum. A contractile **heart vesicle** lies over the central sinus (Fig. 36.6B), but the sinus is also contractile. A large **ventral shield sinus** arises from the posterior tip of the central sinus; it gives off the **peribuccal sinuses** that join to form the main **ventral sinus** at the collar–trunk septum. Folds in the wall of the ventral shield sinus constitute the **glomerulus**.

Nervous System The nervous system of *Cephalodiscus* is wholly epidermal. There is no hollow collar cord as in enteropneusts, but rather there is an indefinite, solid **collar ganglion** that joins middorsal and laterodorsal thickenings of the nerve plexus in the cephalic shield. **Circumenteric connectives** at the collar–trunk boundary pass from the collar ganglion to the **midventral nerve cord**. A short **middorsal nerve trunk** extends from the collar ganglion to the anal region.

Reproductive System The saccate gonads lie in front of the stomach. Short gonoducts open behind the collar–trunk boundary. Sexes are usually separate, but males and females look alike. Hermaphroditic forms are not uncommon. Members of one coenecium may be of the same or a different sex.

BIOLOGY

Rarely encountered, pterobranchs have very little known about them.

Rhabdopleura zooids resemble *Cephalodiscus*, but have no gill slits and only one pair of lophophore arms. The stalk is permanently attached to the stolon, so they are tethered when they emerge from the tube to feed (Stebbing and Dilly, 1972). The dark brown zooids are only about 1 mm long, and have a cephalic shield with a red pigment band but without a marginal notch. In both genera, feeding is achieved by outwardly beating cilia that generate currents to catch particles. Particles that settle on the lophophore surface are quickly swept to the mouth down the gutter of the arm. Particles can move up or down the arm, and apparently unsatisfactory morsels are voided back into the water (Gilmour, 1979). The development of a pharyngeal slit in the cephalodiscans is apparently related to the large volume of water that streams down the numerous arms of the lophophore through the mouth. *Rhabdopleura* laterally deflects excess water by means of the foldlike oral lamella near the mouth.

The stalk epidermis can form buds that receive an extension of the stalk coelom and the dorsoventral mesentery. The cephalic shield of the bud grows large first, and the trunk develops later. The arms arise as buds, the anterior pair first and the rest in serial order. The heart vesicle, different than in enteropneusts, is derived from the protocoel (Fig. 36.6B). The solitary individuals of *Atubaria* apparently cannot reproduce asexually. The members of the other two genera live in tubes embedded in a common case, the coenecium. They reproduce asexually to produce new members of the coenecium, and sexually to form new coenecia.

Little is known of pterobranch embryology. The large, yolky eggs develop indirectly. A tornarialike larva is formed. The primordium of the cephalic shield develops early; its gland cells are thought to secrete the coenecium.

FOSSIL RECORD AND PHYLOGENY

Graptolithina are fossil forms allied with hemichordates. The graptolites had a tube structure virtually identical to that of the pterobranch coenecium, but formed large floating colonies. They are commonly encountered in Paleozoic rocks.

Much confusion exists as to phylogeny within the hemichordates (see, e.g., Dilly, in Barrington and Jefferies, 1975). Some believe that the free-living enteropneusts, with their well-developed organ systems, similarities to primitive chordates, and developmental patterns evocative of echinoderms, are the most primitive hemichordates. Others think that the sessile pterobranchs, with affinities to known fossil forms and similarities to lophophorates are more primitive. The debate is unresolvable until more of the biology of both living groups and the development of pterobranchs becomes known.

Taxonomy

Class Enteropneusta. Solitary, wormlike, with straight gut and many pharyngeal slits; without arms on mesosome. Example: *Saccoglossus* (Fig. 36.2B).

Class Pterobranchia. Aggregative or colonial; living in a zoecium; gut U-shaped, with or without pharyngeal slits; with arms bearing tentacles.

Order Rhabdopleurida. Colonial pterobranchs, each living in tubular zoecium secreted by epidermis and connected to other members by living stolon; with two tentacular arms, and without gill slits. Example: *Rhabdopleura* (Fig. 36.6A).

Order Cephalodisca. Pterobranchs living in a common zoecial mass, but without living stolon connecting members of aggregation; with one pair of gill slits and four to nine pairs of arms with tentacles. Example: *Cephalodiscus* and *Atubaria* (Fig. 36.5A,B).

Selected Readings

Barrington, E. J. W. 1965. *The Biology of Hemichordata and Protochordata.* Oliver and Boyd, Edinburgh.

Barrington, E. J. W., and R. P. S. Jefferies (eds.). 1975. *Protochordates.* Academic Press, London.

Burdon-Jones, C. 1952. Development and biology of the larva of *Saccoglossus horsti. Phil. Trans. Roy. Soc. Lond.* (B)236:553–590.

Dawydoff, C. 1948. Embranchment des Stomocordés. In *Traite de Zoologie,* tome XI (P.-P. Grassé, ed.), pp. 367–494. Masson Cie, Paris.

Gilmour, T. H. J. 1979. Feeding in pterobranch hemichordates and the evolution of gill slits. *Canadian J. Zool.* 57:1136–1142.

Gilmour, T. H. J. 1982. Feeding in tornaria larvae and the development of gill slits in enteropneust hemichordates. *Canadian J. Zool.* 60:3010–3020.

Hadfield, M. G. 1975. Hemichordata. In *Reproduction of Marine Invertebrates,* vol. 2 (A. C. Giese and J. S. Pearse, eds.), pp. 185–240. Academic Press, New York.

Stebbing, A. R. D., and P. N. Dilly. 1972. Some observations on living *Rhabdopleura compacta. J. Mar. Biol. Assc. U.K.* 52:443–448.

Strathman, R. R., and D. Bonor. 1976. Ciliary feeding of tornaria larvae of *Ptychodera flava. Mar. Biol.* 34:317–324.

37

Chordates

Most chordates are vertebrates, and we are not concerned with them in this book; however, two subphyla, Cephalochordata and Urochordata, include invertebrate chordates. Amphioxus, the standard cephalochordate, is so universally studied in vertebrate anatomy courses that it is omitted here, but urochordates are often overlooked, important as they are with over 1200 species.

Three major types of urochordates are recognized and treated separately here: Larvacea or appendicularians, Ascidiacea or ascidians, and Thaliacea or salps.

Definition

Chordates have a notochord present at some stage of their embryonic development, and this structure may persist into the adult. The nervous system has a dorsal, tubular nerve cord, formed by the invagination of a strip of neural ectoderm on the dorsal midline. Pharyngeal slits are present at some time during development, and pharyngeal arches persist in the adults. There is a ventral, pulsating heart. The postanal tail carries extensions of nerve cord, notochord, and body wall musculature.

Larvacea

Larvacea are relatively common plankton organisms known from polar to tropical seas. They are most common at depths of 100 m or less, but have been collected at depths of 3000 m. They are probably neotenic larvae, but may reveal something of the nature of ancestral stocks from which urochordates stemmed.

They are tiny, transparent animals, often scarcely visible unless the gut, gonads, or other parts are pigmented; bright red and violet shades predominate. Their symmetrical, U-shaped body consists of a dilated **trunk** and an elongated **tail**. The latter is attached to the ventral trunk surface that is twisted 90 degrees, effectively bringing the dorsal surface to the left side (Fig. 37.1A). The tail contains a central **notochord**, a **dorsal nerve cord** with swellings for each muscle segment, and two **lateral muscle bands** that tend to be broken into incipient metameres. A large ganglion marks the nerve cord near the base of the tail.

The mouth opens into a dilated **pharynx** with a short, tubular, ventral **endostyle**, containing longitudinal rows of gland cells (Fig. 37.1A). A single pair of simple, ciliated **pharyngeal slits** open into a **branchial sac** or **atrium** that has short, funnel-shaped canals that open ventrally. The rest of the alimentary tract consists of a short **esophagus**, a saccate **stomach**, and an **intestine** that curves anteriorly to a typically midventral **anus**. A large ganglion gives rise to a solid nerve cord that extends into the tail. A richly cilated, tubular process that corresponds to the dorsal tubercle of ascidians is attached to the floor of the cerebral ganglion and connects to the pharynx. A **statocyst** is found on the left side of the brain. Except in *Kowalevskia*, a **heart** lies in front of and below the stomach. There are no blood vessels; blood circulates in a **hemocoel**. Larvaceans are usually protandrous hermaphrodites. The paired or unpaired gonads lie in the posterior of the trunk.

The most remarkable appendicularian feature is the **house**, a transparent structure secreted very rapidly by epithelial gland cells at the body surface (Fennaux, 1985). Houses vary widely in form and extent. In some cases they are little more than a transitory gelatinous bubble expanded from the mouth region for temporary use; in others they are complex. The house of *Oikopleura* is far larger than the animal that makes it, permitting free movement within (Fig. 37.1B). Movements of the tail drive water through the house. Water enters at an incur-

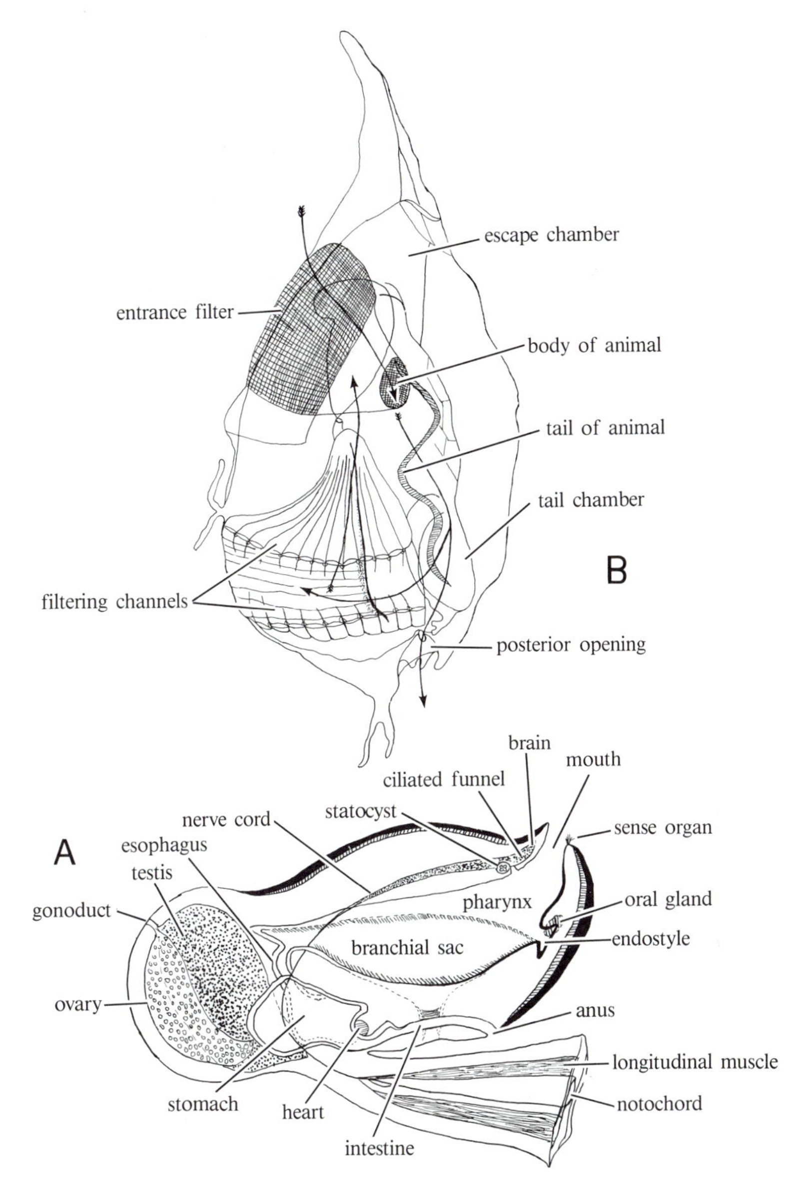

FIGURE 37.1. Larvacean structure. **A**. Section of *Oikopleura*, showing only part of tail. **B**. *Oikopleura albicans*, larvacean, in its house. Animal much smaller than its house and drives water through house by lashing movements of its tail. (**A** after Herdmann; **B** after Lohmann.)

rent pore, guarded by a fibrous mesh that filters out large particles, and circulates through passageways that act as a second filter system. The particles that gather on the inner filters are eaten. An excurrent pore is normally closed by a hinged door that opens suddenly as pressure builds up in the house; the sudden jet of water propels the house and animal through the water. An added feature is the escape hatch or emergency door, through which the animal leaves when the filters are clogged. New houses are secreted frequently, every few hours in some species.

Ascidiacea

Ascidians occur in oceans everywhere, from pole to pole and from shallow to deep water (Monniot and Monniot, 1975). Some are solitary, and others form colonies of considerable size. Some individual ascidians are very small, while others are about 30 cm long. Colonial species usually have relatively small zooids.

Ascidian structure is best studied in solitary species. A soft or tough tunic clothes the body and is attached at a permanent base located at the posterior part of the left side of the body. At the opposite end, an oral siphon marks the anterior tip of the body, and a cloacal siphon indicates the dorsal margin (Fig. 37.2A).

The **tunic** is an epidermal secretion, but amoeboid cells from the mesenchyme invade it and differentiate into pigment cells, stellate cells, and vacuolated cells, and sometimes secrete calcareous spicules that stiffen the tunic. The tunic matrix is composed

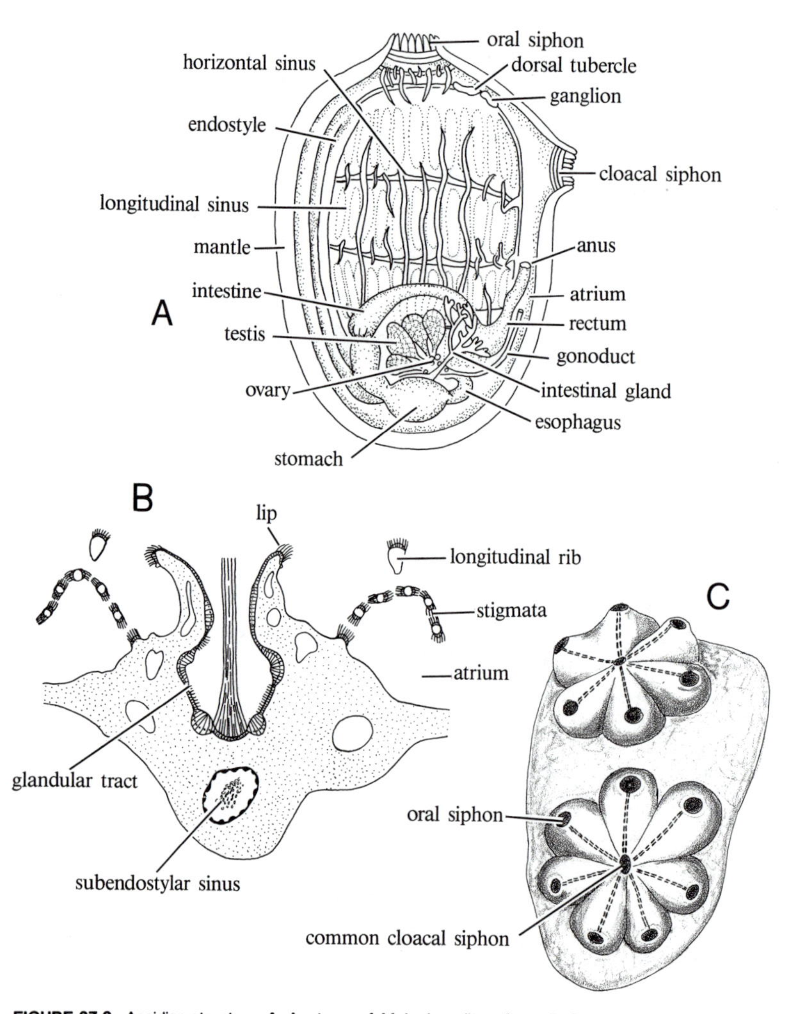

FIGURE 37.2. Ascidian structure. **A**. Anatomy of *Molgula,* solitary form. **B**. Schematic section through endostyle. **C**. *Botryllus violaceus,* colonial form with central common cloacal siphon. (**A** after Van Name, from Kleinholtz, in Brown; **B** after Herdman; **C** after Milne Edwards.)

largely of a cellulose known as tunicin, found in tunicates and some hemichordates (Welsch, 1984). The tunic is often colored, taking its colors from pigment cells and blood corpuscles. Blood sinuses enter the tunic and ramify through it, ending near the surface in terminal knobs. They are abundant enough to aid in respiration, and in some cases pulsate and aid in blood flow.

The mantle or body wall is beneath the tunic. An epidermis covers a thick dermis that contains homogeneous and fibrous connective tissues. Smooth muscles run in all directions; their contraction somewhat modifies body shape. The **siphons** (Fig. 37.2C) have strong sphincter muscles and are lined with tunic and epidermis turned inward at the siphon margins. Tunic stimulation evokes responses, indicating the presence of a body wall nerve plexus. Stimuli spread in all directions and can spread around cuts in the body wall. In addition, the body wall nerve plexus is independent, as destruction of the central nervous system causes relaxation of the body wall muscles but does not destroy their normal reflexes.

Digestive System Water enters the **oral siphon** and leaves through the **cloacal siphon**, having passed through the **pharynx**, the **pharyngeal slits**, and the **atrial cavity** (Fig. 37.2A). The epidermal lining of the oral siphon stops at the whorl of oral tentacles. Neurosensory cells in the oral tentacles evoke reflex contractions of body wall muscles, clearing the siphon of particles or irritating substances by means of forcefully ejecting a stream of water. The pharynx is the body's largest organ. It is lined by tissues derived from larval endoderm, but the lining may arise from tissues derived from other germ layers during the course of asexual budding. The pharynx begins just within the oral tentacles.

The pharynx wall is supported by a system of bars, through which the branchial lacunae of the circulatory system pass (Fig. 37.2A). The bars are arranged in patterns characteristic of species. Internal, longitudinal bars support a delicate grid of transverse and longitudinal bars. The primary gill slits become subdivided, often very intricately, to form many **stigmata**. Stigmata are often so modified that they are lacy or spiral. A deep midventral groove, the **endostyle** (Fig. 37.2B), runs the length of the pharynx, ending in a posterior **caecum**. The pharynx wall is joined at the mantle below the endostyle; everywhere else it is surrounded by the atrium. Four rows of mucous glands lie in the endostyle, as in the amphioxus. The thyroid primordium of the agnath ammocoetes larvae also has four rows of gland cells, indicating that the endostyle is the homologue of the vertebrae thyroid. A median row of long flagella and laterally placed tracts of cilia push mucus from the endostyle up onto the pharyngeal basket.

The pharynx opens into an **esophagus**, usually a short, descending arm of the U-shaped digestive tract (Fig. 37.2A). A dilated **stomach** lies at the turn of the U. The tubular **intestine** ascends to open into the atrium at the **anus**. Circular muscles form a sphincter at both ends of the stomach, but the gut wall is not highly muscular, and food conduction depends on ciliary currents. The stomach wall contains gland cells, mucous cells, and absorptive cells. Digestive glands arise from the posterior end of the stomach.

Excretory, Respiratory, and Circulatory Systems No special excretory organs exist. Respiration is achieved largely by using the water currents passing through the pharyngeal slits.

The tunicate circulatory system is almost wholly open. Blood flows through spaces generally lacking an endothelial lining. The **heart** is unusual from morphological and physiological viewpoints, and the blood has some unique properties.

A **pericardium** and a **pericardial sac** are formed from endodermal tissue during development. The pericardial cavity may be a vestige of the coelom and lies either at the base of the digestive loop or at the base of the postabdomen in species with a long postabdominal region. Its wall is folded to make a more or less U-shaped tube, the heart. The heart has no wall comparable to the heart wall of other chordates, and also lacks a lining.

Three lacunar plexi connect the three principal blood sinuses. The subendostylar sinus arises from the ventral end of the heart and is connected to the ventral side of the lacunar plexus of the pharyngeal basket. The ventral abdominal sinus is connected to the dorsal end of the heart. It opens into an extensive lacunar plexus over the gut and the viscera and issues two branches to the tunic. The median dorsal sinus is connected with the lacunar plexus of the viscera and runs over the dorsal midline of the pharynx, where it is attached to the pharyngeal lacunar plexus. It also receives branches from the lacunar plexus of the tunic.

The **epicardium** is one tube or a pair of tubes arising from the base of the pharynx. When a pair of tubes are present, they usually unite. The epicardium extends along one side of the digestive tract and, in tunicates with a long postabdomen, reaches the base and extends into the stolon. Its function is problematic. It contributes tissue to buds formed in this region, introducing an endodermal derivative into the bud, but many tunicates bud successfully without any endodermal tissue in the bud primordia.

Tunicates apparently never have a separate coelom, but are so evidently related to coelomates that the coelom must have been lost secondarily. Because the epicardium arises from the digestive tube, it may be a vestige of a coelom. This view is somewhat supported by epicardial development in *Ciona.* The two epicardial sacs remain separate, one on each side of the digestive tube. They are greatly enlarged, surrounding the internal organs in the same manner as the coelomic pouches of other animals.

Nervous System A large, usually somewhat elongated ganglion lies between the two siphons (Fig. 37.2B). Nerves arise from each end, passing to the siphons, the viscera, and the gills, where they innervate the languets, dorsal lamina, and dorsal tubercle. The **cerebral ganglion** seems to be of little importance in controlling the body activities. Its removal destroys the crossed siphon reflexes, but other activities are unchanged.

The **subneural gland** is a small mass of gland cells below the cerebral ganglion. A narrow duct connects it with a ciliated funnel located in the dorsal tubercle, a prominence in the dorsal pharyngeal wall. Dorsal tubercles vary greatly in form and are useful in taxonomy (Fig. 37.2A). The subneural gland has been thought to be an excretory organ, a mucous gland, and a lymph gland, but no satisfactory demonstration of a function has been achieved. Extracts of the subneural gland cause effects similar to those caused by pituitary hormones when tested on other animals. This is interesting, as it lies at the same place as the hypophysis of vertebrates. The vertebrate pituitary arises from a double primordium, a depression on the brain floor and a pocket extending upward from the stomodeum. The subneural gland and the dorsal tubercle also arise partly from the pharynx and partly from the ganglion, supporting the idea that they are a pituitary homologue.

Reproductive System Nearly all ascidians are hermaphroditic. Most have a single **ovotestis** in the center of the digestive loop. Each part is served by a **gonoduct**. The parallel gonoducts, sometimes a double tube, run to the atrial wall and open into the cloaca. The lumen of the saccate **ovary** is continuous with the gonoduct. The **testis** is usually composed of many tubular branches. Sperm mature in the tubule wall and enter the tubule lumen, which leads to the gonoduct. *Molgula* and some other ascidians have paired ovaries and testes. The second ovary and testis lie against the right body wall, and have separate gonoducts to the cloaca. Some species are protandrous or protogynous, thus avoiding self-fertilization, but in other species ova and sperm can mature simultaneously.

BIOLOGY

Locomotion Although the tail is functional in larvae in achieving movement, it is lost in adults as they metamorphose into sessile forms.

Feeding and Digestion The pharynx is a ciliary-mucus, food-gathering device (Barrington, 1965). Cilia at the margins of the stigmata pull water into the pharynx and send it into the atrium. Some tunicates pump several thousand times their body volume of water through the pharynx in a day. Mucus from the endostyle moves up across the pharyngeal bars, and food particles are trapped in it. Strands of mucus are delivered to the dorsal lamina, which bends over to form a tube or groove along which food moves to the esophagus.

Water leaving the pharynx goes into the branchial sac or atrium, which is formed by the union of two ectodermal invaginations from the surface and lined with epidermis throughout. Cords of tissue attach the outer wall of the atrium to the pharynx, preventing excessive expansion. The gonoducts and anus enter at the upper end of the atrium, converting it into a kind of cloaca. Water from the atrium departs through the cloacal siphon.

Although ascidian enzymes have not been studied carefully, proteolytic enzymes have been found. Digestion appears to be wholly extracellular and appears to occur in the stomach. The intestine seems to be the primary absorptive part of the gut, based on histological evidence. A prominent dorsal ridge, the typhosole, increases the absorptive surface. Surface is sometimes further amplified by increase in intestinal length and coiling. Food is stored in the intestinal wall, primarily as glycogen.

Respiration The rich lacunar plexus of the pharyngeal wall suggests that most respiratory exchange occurs there. Some respiratory exchange undoubtedly occurs in the tunic lacunae also. Tunicates have no respiratory pigment, but their inactive life requires little oxygen. Less than ten percent of the available oxygen is absorbed from water passing through the gills. Critical oxygen tensions have not been determined; presumably they pass so much water through the body for feeding that respiration is not a critical physiological problem.

Excretion About 90 percent of the nitrogen excreted is in the form of ammonia. Presumably it leaves at the pharynx, tunic, or any other handy surface. A considerable quantity of uric acid and urates are also produced. These are picked up by amoebocytes and incorporated into intracellular concretions. Cells containing such insoluble wastes settle on the surface of various organs, especially the gut and the gonads. Special centers for storage excretion sometimes occur. They are most often delicate, clear-walled vesicles in the intestinal loop. In the absence of definite information, it seems probable that these wastes are end products of purine metabolism. An erroneous impression of the excretory situation would be obtained if only released compounds were taken into consideration. No doubt the same principle applies in many invertebrates where storage excretion is less obvious.

Circulation The circulatory system is essentially triangular. Blood can (and does) flow either way through the triangle. The heart pulsates, each beat a peristaltic wave beginning at one end and passing to the other. There are no valves; blood is milked along by the advancing wave of constricting muscle. For several minutes, perhaps a hundred beats, contractions begin at the dorsal end, forcing blood into the subendostylar sinus, through the pharynx, into the median dorsal sinus, from this into the tunic and the viscera, to return by way of the dorsal abdominal sinus. The beat gradually slows and stops. Then the heart beats again, but in the opposite direction, forcing blood to flow through the system in the opposite direction. This arrangement is unique. Neither its functional significance nor the factors that cause it are understood. The heartbeat appears to be myogenic, originating in the heart muscle itself, for isolated hearts continue to beat and the removal of the brain has little effect in most of the tunicates tested. A pacemaker of ganglion cells may lie in the heart, but discrepant results have been obtained in histological studies of the heart wall. It has been suggested that back pressure causes reversal of beat, but isolated hearts in which back pressures should be missing continue to reverse. When a heart is forced to beat by electrical stimuli, the threshold of excitation rises with each successive beat, suggesting that the pacemaker may undergo changes, gradually becoming less sensitive. In this case the pacemaker at the opposite end of the heart may gain dominance.

Tunicate blood plasma is isotonic or only slightly hypertonic to seawater, but some inorganic ions strangely are regulated (Goodbody, 1974). Vanadium is concentrated enormously, sometimes as much as half a million times. It is accumulated in the blood plasma of *Cionides* and *Diazonides*, and in blood corpuscles known as vanadacytes in some other tunicates. However, the vanadium is not involved in respiratory transport, as it remains in the reduced state within cells. The pyloric glands, which open into the stomach, are thought to excrete calcium. A full analysis of ionic regulation of tunicates is not available.

The blood corpuscles are numerous and strikingly colorful. They contain red, yellow, green, brown, blue, or white inclusions. Some are amoeboid, and some are known to be phagocytic. They move through tissues and apparently pick up uric acid to initiate storage excretion.

Sense Organs The motile larvae have good sense organs, but adults do not. Neurosensory cells of the invertebrate type occur on the body surface, especially in the siphon regions and in papillae on the inner surface of the branchial apparatus. They are probably important in regulating water flow, and in clearing the siphons or particles by evoking ejection reflexes. A light stimulus inside a siphon causes closing of the other siphon. A stronger stimulus causes the closure of the other siphon and body wall contractions that eject a particle in the offending siphon.

Reproduction Little is known of sexual reproduction. Breeding is seasonal, and even in tropical forms it is possible to detect peak larval settling during the fall that might be due to adult reproductive intensity during certain seasons superimposed on an otherwise constant reproductive pattern (Barrington, 1965).

Solitary ascidians do not reproduce asexually, but colonial forms are active budders. Colonial species also regenerate well, whereas solitary forms do not. Tunicates have developed surprisingly diverse methods of budding.

The tissues present in individual buds differ surprisingly. All buds contain epidermis, but other components are varied. Some contain only epidermal tissue, others include mesenchyme, and still others have one or another kind of endodermal derivative, usually the epicardium. In *Botryllus*, only the atrial epithelium and epidermis are used in bud formation. In *Salpa*, on the other hand, buds contain epidermis, part of the endostyle and the mesenchyme; therefore, all germ layers are represented. *Diazona* buds contain part of the intestine, epicardium, and reserve cells in addition to epidermis. Evidently tissue derivation from germ layers means very little in developing asexual buds, however important it may have been during initial embryonic differentiation.

Budding occurs at stolons, along the epicardium in the postabdominal region of elongated ascidians, or from the atrial wall in flattened colonies without a stolon. Regions of bud formation are first seen as centers of rapid growth, which disrupt the overall organization and continuity of the organism. As the growth rate falls somewhat, these areas become morphogenetic fields that gain integrity and independence, differentiating into new organisms. Rapidly elongating parts, such as a stolon or the epicardium of a young ascidian, often divide repeatedly to form a chain of buds (Fig. 37.3A).

Its development is sometimes delayed until the stolon breaks away from the colony or has grown enough to become distant from the parent (Fig. 37.3B), thus gaining the autonomy needed for differentiation.

Colony formation by asexual reproduction is important in tunicates (Berrill, 1975). The relationship between the zooids of tunicate colony varies in intimacy with the nature of the budding process and the extent to which they share a common tunic (Fig. 37.3C). Discrete zooids are attached by a stolon in the simplest colonies. A common tunic partly unites the base of zooids in some stolonic colonies. The zooids are wholly enclosed in a common tunic in other types of colonies. In this case, zooids tend to be regularly arranged, perhaps as an adaptation to prevent the discharge of excreta from the cloacal siphon of one zooid toward the oral siphon of another. This trend results in colonies with a definite shape, with all oral siphons at one side and all cloacal siphons at the other. It is a short step from such an arrangement to the development of a common cloacal chamber for all of the zooids, discharging water through a common cloacal pore (Fig. 37.4B,C). Another type of common atrial chamber is seen in *Bo-*

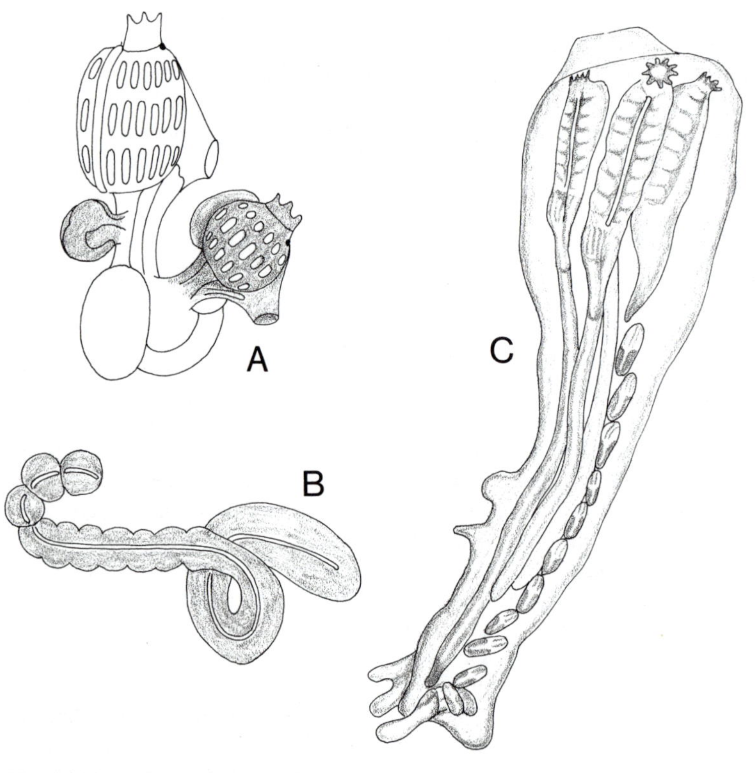

FIGURE 37.3 Budding in ascidians. **A**. Development of thoracic and abdominal buds in Dideminidae that together will make single zooid. **B**. Postabdomen of *Colella* containing epicardial tube, undergoing strobilation. **C**. Strobilation in young colony of *Circinalium*. (**A** after Salfi; **B** after Caullery; **C** after Brien, in Grassé.)

tryllus, where regular systems of small zooids share a common cloacal siphon (Figs. 37.2C and 37.4D). One large colony may have many such systems.

Development Ova differ considerably in size and form among species. As in other groups, brooding species produce large ova with more yolk, undergoing more direct development; and smaller ova show more primitive patterns of development. Cleavage is complete and unequal. The destiny of blastomeres is determined very early, and regions that will participate in formation of some body parts can be distinguished in the ovum! A somewhat flattened coeloblastula invaginates to form a flattened gastrula (Fig. 37.5A). The blastopore shrinks and closes as the embryo elongates. A typical neural furrow forms at the middorsal surface, sinks in as a neural groove, and closes to form a neural tube (Fig. 37.5C). Cords of cells on each side of the archenteron serve as primordia of the mesenchyme. They at no time contain a coelom.

The larva is sometimes called a tadpole because it is shaped like an amphibian tadpole (Fig. 37.5D). The larva is completely enclosed in a tunic and, thus, cannot feed until metamorphosis is completed (Fig. 37.5E,F). The larva is more highly organized than the adult, indicating that adaptations leading to a sessile life have involved the loss of external bilateral symmetry and cephalization, and the reduction of the muscular and nervous systems.

Thaliacea

The free-swimming Thaliacea (Aldredge and Madin, 1982) all have a posterior cloacal opening, resembling some of the colonial ascidians with a common atrial/cloacal chamber. Although organ placement is somewhat modified, the basic anatomy is like that of ascidians.

Pyrosoma, the only genus of the Pyrosomidea, is colonial, forming cylindrical colonies with a common atrial chamber open at one end and closed at the other (Fig. 37.6A,B). The oral siphons are on the

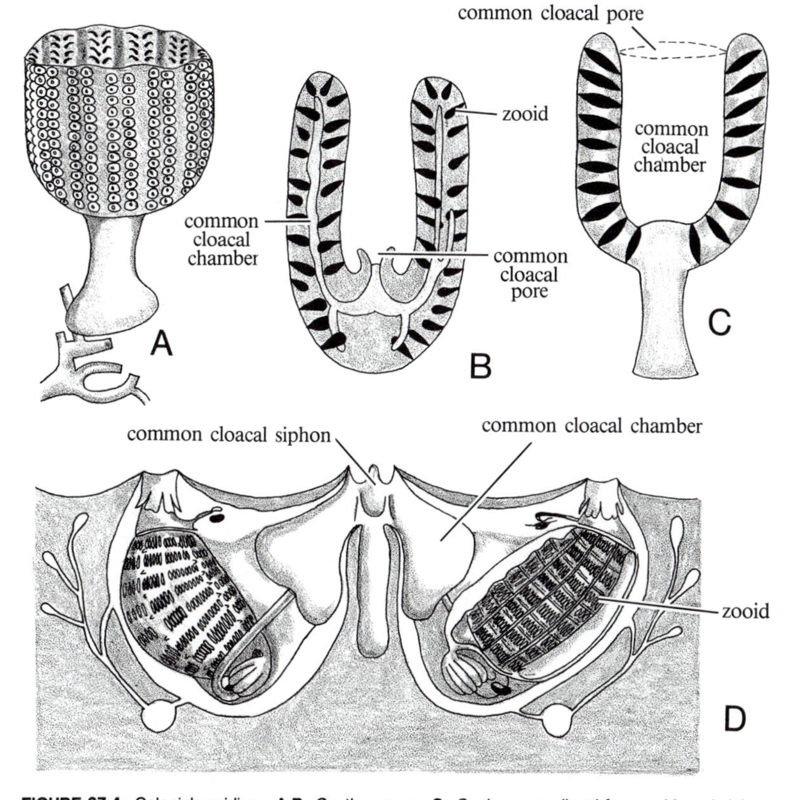

FIGURE 37.4. Colonial ascidians. **A,B.** *Cyathocormus.* **C.** *Coelocormus;* lineal forms with oral siphons on outer surface and cloacal siphons opening into common cloacal chamber. **D.** *Botryllus,* retaining U-shaped body form with common cloacal chamber. (**A,B,C** after Okada in Grassé; **D** after Delage and Hérouard.)

external surface, and the cloacal siphons discharge into the cloacal chamber. The stream of water emerging from the open end of the common cloaca propels the colony slowly through the water. *Pyrosoma* are most common in warm seas and may become very large; some are said to reach 4 m in length. They are brilliantly luminescent. Mosely remarks, "I wrote my name with my finger on the surface of a giant *Pyrosoma* as it lay on the deck in a tub at night, and my name came out in a few seconds in letters of fire." The stimulus spreads; even a flash of light directed on the surface may make a whole colony glow. Rounded gland cells at the end of the pharynx are the luminescent organs. *Pyrosoma* have a pigment cup with retinal cells, evidently a visual organ, embedded in the brain.

Doliolida are solitary, cask-shaped animals with eight or nine muscle bands around the body (Fig. 37.6C). Contraction of the circular muscles forces water out of the posterior cloacal opening, propelling the animal.

Salpidea are similar to doliolids: transparent animals, but without complete circular muscle bands and with two large, open gill slits (Fig. 37.6D,E). Not

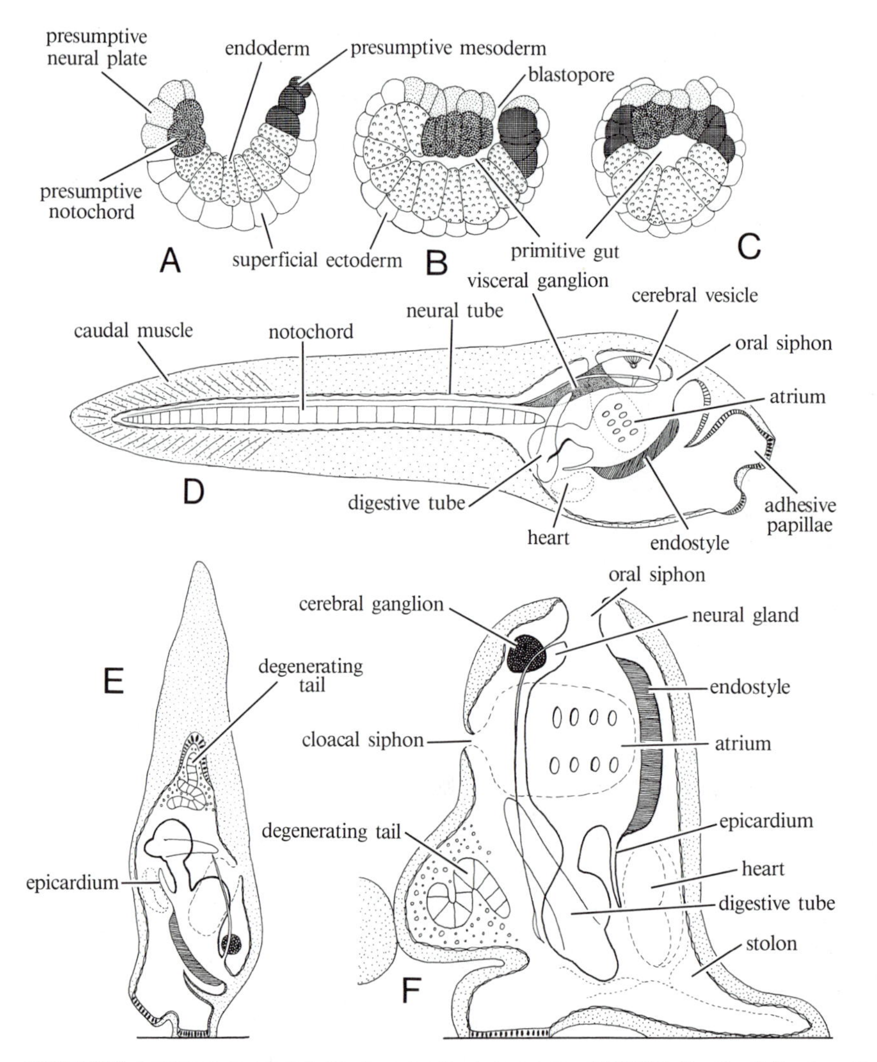

FIGURE 37.5. Ascidian development. **A.** *Styela;* gastrulation by invagination. **B,C,D,E,F.** *Clavelina.* **B.** Very large blastopore closes, shown in sagittal section. **C.** Neural plate cells on right, accompanied by superficial ectoderm, grow over blastopore as neural groove closes, thus forming neurenteric canal. **D.** Fully developed larva with well-developed tail and resembles tadpole. **E.** Larva attaches. **F.** Metamorphosis into typical adult ascidian. (**A** after Conklin; **B,C** after Van Benedin and Julin; **D,E,F** after Seeliger, in Grassé.)

many species of salps are known, but they are widespread, occurring abundantly in cold seas and deep water.

Remarkable life cycles are characteristic of Thaliacea. Sexual reproduction gives rise to a zooid that reproduces asexually (Fig. 37.6E). Asexual reproduction eventually leads to the appearance of sexual individuals (Fig. 37.6D). Sexual and asexual animals are very similar in organization, and the salp life cycle clearly involves alternating generations.

Pyrosoma produces a large, yolky egg that undergoes meroblastic cleavage in the parent atrium. No tailed larva are formed, and the embryo develops into an individual that never completes its development. It forms a stolon that produces four buds. As these appear, the embryonic colony leaves the par-

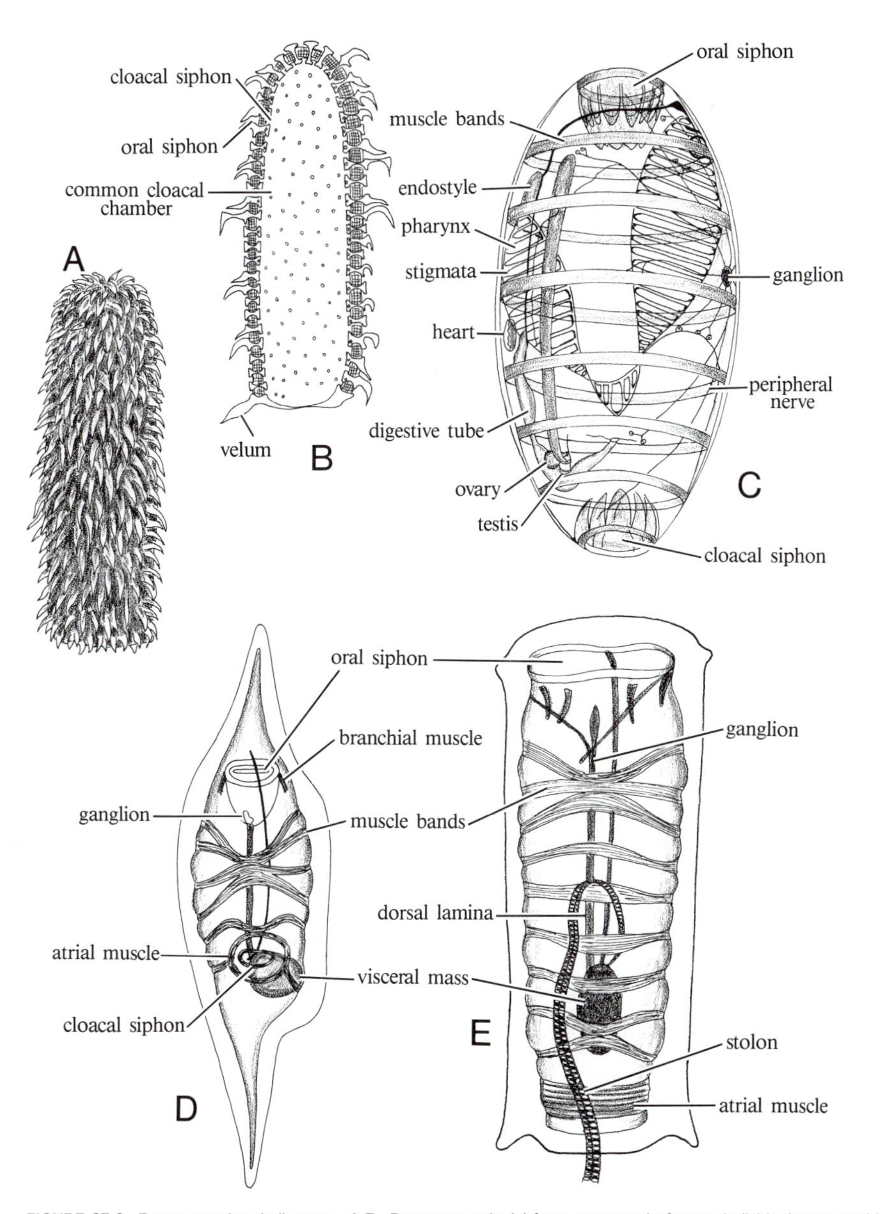

FIGURE 37.6. Representative thaliaceans. **A,B.** *Pyrosoma,* colonial form composed of many individuals arranged in form of open cylinder, propelled by water passing into central cavity from many members of colony, and emerging through opening at one end of colony. **C.** *Doliolum.* **D,E.** *Salpa,* transparent pelagic thaliacean that may occur as single individuals **D** or in chains **E**, depending on stage of life cycle. (**A** after Milne Edwards; **B,D,E** after Herdmann; **C** after Grobben.)

ent. The stolon is a nurse zooid; it digests the yolk and develops a heart and blood vessels that distribute food to the growing buds. The atrium of the stolon becomes very large, eventually becoming the primordium of the common atrium of the colony. When the original buds mature, they in turn bud off new zooids, and the colony continues to grow.

The solitary *Salpa* (Fig. 37.6E) arises from a zygote. When mature, a stolon forms at the end of the endostyle. The stolon constricts to form a series of buds attached to it. These develop while still attached, forming chains of animals with the eldest at the tip (Fig. 37.7B). The chain trails behind the parent salp, and clusters of young buds break off from time to time. Buds are substantially like solitary forms internally, but are rounded and have tubular papillae that attach them together (Fig. 37.6D). They may separate early or remain attached until mature. They eventually produce ova and sperm. The zygote develops in the parent atrium, receiving nourishment through a placental arrangement. When mature, it separates from the parent and eventually produces a stolon of new blastozooids.

Of all Thaliacea, *Doliolum* has the most complex life cycle. Zygotes develop in the sea, becoming typical tailed larvae. The tails are lost as they metamorphose into oözooids. Mature oözooids have a complex posterior stolon and a dorsal process. Probuds form on the stolon and migrate to the dorsal process, where they attach in three rows (Fig. 37.7A). The parent organism becomes modified to support the young, when internal organs are largely lost, but the muscles remain strong and the parent becomes a motile raft on which the young develop. Probuds do not develop directly into zooids. The two lateral rows of probuds develop into nutritive zooids (gastrozooids) and remain permanently attached to the oözooid. Some of the buds in the middle row develop into *Doliolum*-like phorozooids that detach and swim away. While attached, however, other buds from the middle row (gonozooids) migrate to the stalk of the phorozooid and are carried away with it. These buds become sexually mature gonozooids, assume a *Doliolum* shape, and detach from the phorozooid. The controls responsible for modifying bud development in this manner remain unknown.

Taxonomy

Subphylum Urochordata. Chordate characteristics developed only in larvae; usually metamorphosing into highly modified, sessile or pelagic adults without coelom or traces of metamerism; notochord in tail region only; body wall forming secreted case, usually rich in cellulose known as tunicin.

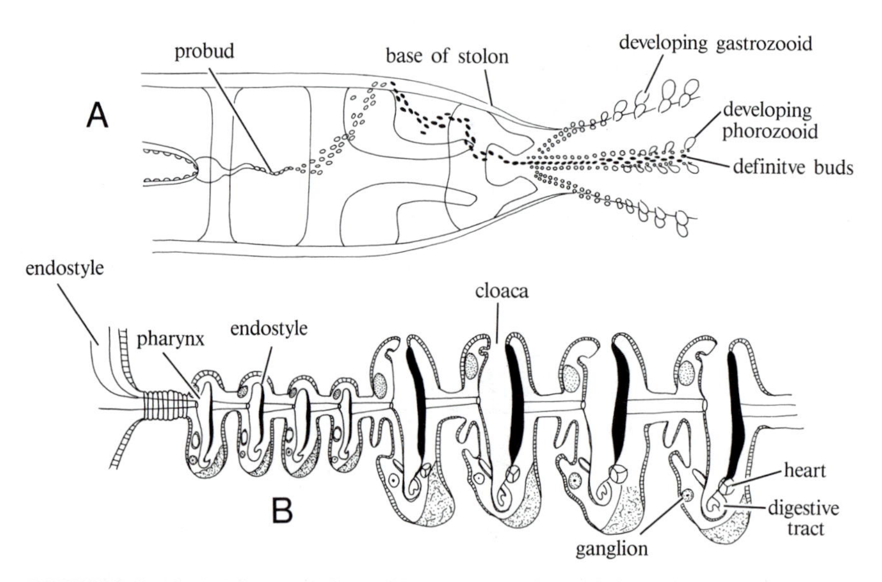

FIGURE 37.7. Budding in thaliaceans. **A.** Stolon of *Doliolum* showing formation of probuds and their migration to stolon where they form three bands of buds, two lateral bands becoming nutritive gastrozooids and center band becoming nurse phorozooids, and definitive buds that will continue their development on stolon of phorozooids. **B.** Scheme of stolon of *Salpa* showing formation of three generations of individuals. Diminutive buds formed at base of stolon develop; at end of stolon they are fully formed, and may break free to take up independent existence or remain in small groups, attached by original stolon. (**A** after Barrois, in Grassé; **B** after Brooks.)

Class Larvacea (= Appendicularia). Larval organization as adults; pelagic, secreting temporary house through which water circulates to capture food (Fig. 37.1).

Order Copelata. Only order.

Class Ascidiacea. Sessile, solitary, or colonial with heavy tunic; adults without tail and with degenerate nervous and muscle systems; pharynx with several gill slits usually subdivided to form pharyngeal basket; atrial cavity receives water from pharyngeal slits, and discharges it through dorsal atrial pore.

Order Apleusobranchia. Gonads nestle in loop of gut (enterogonous); branchial sac simple (Fig. 37.3A).

Order Phlebobranchia. Enterogonous; branchial sac with internal longitudinal bars (vessels) on wall (Fig. 37.2A).

Order Stolidobranchia. Gonads in body wall (pleurogonous); branchial wall folded, with internal longitudinal bars (Figs. 37.2C and 37.4D).

Class Thaliacea. Pelagic, with transparent tunic, and atrial cavity discharging water through posterior aperture.

Order Pyrosomida. Colonies cylindrical (Fig. 37.6A,B).

Order Doliolida. Alteration of generations, sexual gonozooid, asexual oözooids (Figs. 37.6C and 37.7A).

Order Salpida. Solitary asexual oözooids, colonial sexual gonozooids (Figs. 37.6D,E and 37.7B).

Subphylum Cephalochordata. Free living; without well-developed head or brain; notochord extending through entire length of body; segmental musculature.

Order Amphioxi. Single order.

Subphylum Chordata. Free living; well-developed brain and head; with endoskeleton of cartilage and/or bone; pharyngeal apparatus elaborate, often specialized for structural and functional needs other than slit filter feeding.

Selected Readings

Alldrege, A. L., and L. P. Madin. 1982. Pelagic tunicates: Unique herbivores in the marine plankton. *Biosci.* 32:655–663.

Barrington, E. J. W. 1965. *The Biology of Hemichordata and Protochordata.* Oliver and Boyd, Edinburgh.

Barrington, E. J. W., and R. P. S. Jefferies. 1975. *Protochordates.* Academic Press, London.

Berrill, N. J. 1975. Chordata: Tunicata. In *Reproduction of Marine Invertebrates,* vol. 2 (A. C. Giese and J. S. Pearse, eds.), pp. 241–282. Academic Press, New York.

Fennaux, R. 1985. Rhythm of secretion of oikopleurid houses. *Bull. Mar. Sci.* 37:498–503.

Godeaux, J. E. A. 1974. Introduction to the morphology, phylogenesis, and systematics of lower Deuterostomia. In *Chemical Zoology,* vol. 8 (M. Florkin and B. T. Scheer, eds.), pp. 3–60. Academic Press, New York.

Goodbody, I. 1974. The physiology of ascidians. *Adv. Mar. Biol.* 12:1–149.

Monniot, C., and F. Monniot. 1975. Abyssal tunicates: An ecological paradox. *Am. Inst. Oceanogr.* 51:99–129.

Welsch, U. 1984. Urochordata. In *Biology of the Integument.* 1. *Invertebrates* (J. Bereiter-Hahn et al., eds.), pp. 800–816. Springer Verlag, Berlin.

38

Phylogeny

Historic Background

No treatment of the invertebrates would be complete without some attempt at an overview of their evolution. No subject, however, has ever been marked by so much speculation as has that of invertebrate phylogeny. Hardly any workers agree. All of these ideas make for interesting reading. These widely disparate concepts are frequently juxtaposed against each other without comment in lecture presentations and reading assignments, and as a result students all too often become confused. The plethora of rival interpretations of anatomical facts and the confusing array of names applied to all manner of hypothetical ancestors, or paper animals, is intimidating.

Space limitations preclude recounting all these divergent ideas in detail here. The student can consult earlier editions of this book if truly interested. Most of these concepts have their origins in the nineteenth century, and often came from the German school. These scientists were influenced by the eighteenth- and early nineteenth-century German philosophers who dealt with a kind of neoplatonic *Naturphilosophie.* This philosophic school was concerned with identifying archetypes, or hypothetical forms, that were supposed to embody the essence of what it was to be a particular animal kind. These archetypes eventually came to be viewed as actual ancestors. One of the most famous of these was HAM, the hypothetical ancestral mollusc, which so constrained thought that it actually served to suppress consideration of important fossil forms in developing a phylogeny of Mollusca (see Chapter 21).

One of the most productive figures in this regard was Ernst Haeckel. He originated (Haeckel, 1874) several ideas, such as the blastea and the gastrea theories, that have come to dominate phylogenetic speculations about invertebrates. The former is the idea that all metazoans evolved from hollow, ball-like colonies of flagellates. This concept in turn generated two divergent factions within it. One, the planula theory (Metschnikoff, 1886), developed higher invertebrates from an epibenthic, flagellated planula larvalike organism that filled the hollow ball with cells delaminated from the surface layer. This planula organism was then the archetype for the development of cnidarians. An alternative approach (e.g., Hadzi, 1963) assumed that a ciliate stock gave rise to a syncytial, acoeloid stem. A consequence of this school's ideas views cnidarians as derived from platyhelminths.

An alternative to the planula theory is the gastrea theory (favored by Haeckel himself) that saw a spherical blastea give rise to a cuplike, two-layered gastrea. The conceptualization of this animal, which looked much like a cnidarian, took its inspiration from the common gastrula stage shared among higher metazoans. This theory, in turn, gave rise to several contentious subtheories among which some workers speculated that the partitions of the gastrocoel seen in anthozoans prefigured the formation of segmented coelomic septa (Jägersten, 1955; Marcus, 1958; Remane, 1963; Sharov, 1966). In regard to this latter view, all sorts of contending additional ideas were developed that focused on how the coelom might have formed (see Chapter 20). As a result, all too many family trees of the animal kingdom have been produced since the days of Haeckel; and the well-known synthesis of Hyman (Fig. 38.1) is but one example of what these speculations look like.

All these theories, regardless of their details, have certain things in common. First, they attempt to deal with all phyla, but focus on only a few characters in their attempt to arrange the phyla. Second, they often start with *a priori* assumptions about hypothetical ancestors and what genealogical arrangements should be, and then seek facts to back up their position. Third, none of these theories really attempt to make testable statements about animal relationships, but rather prefer to focus on producing definitive stories about animal history.

We can now perceive the error these early natu-

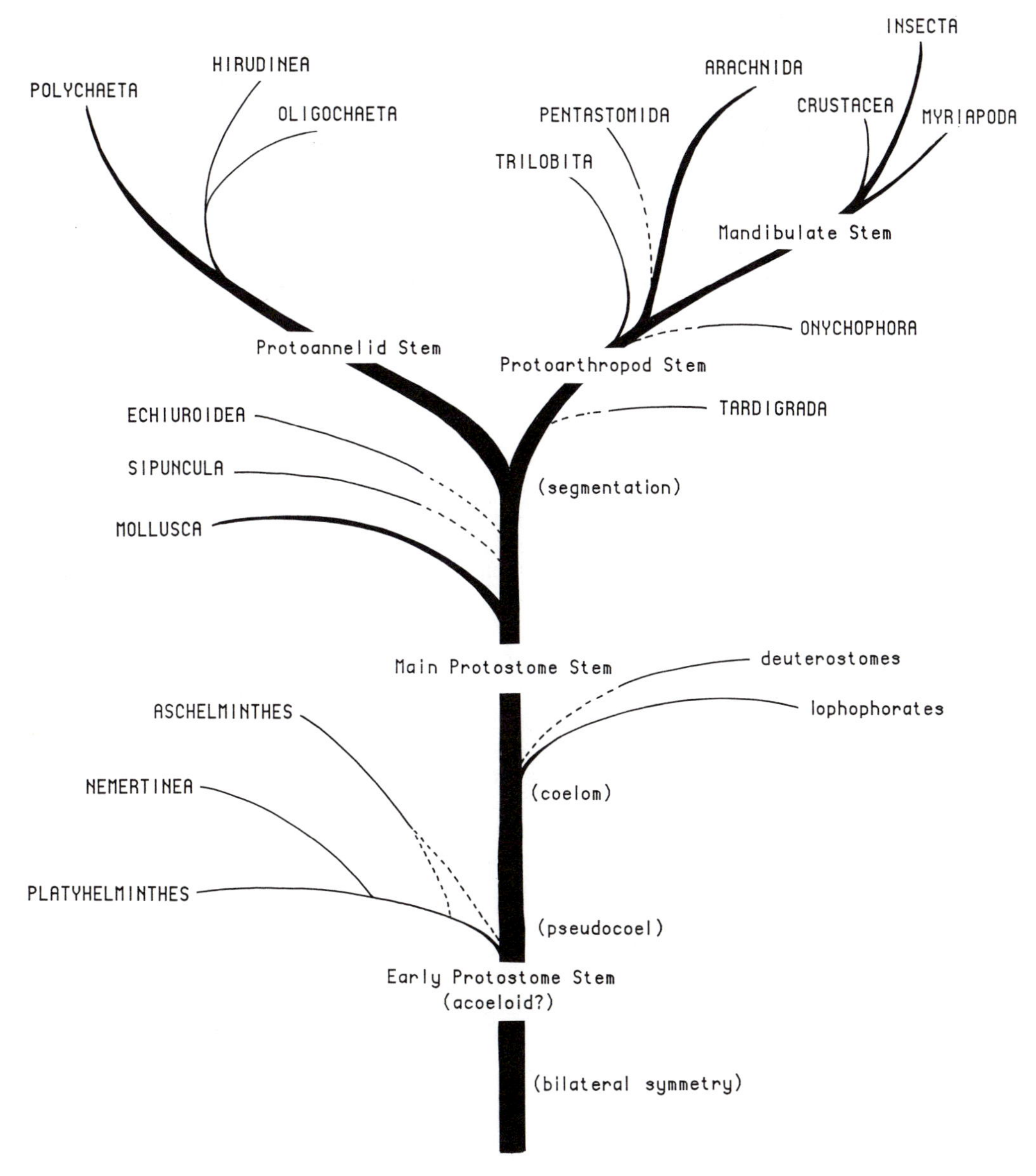

FIGURE 38.1. Phylogenetic tree of animal kingdom based on the writings of Libbie Hyman.

ralists fell into. Many authorities have pointed out that three very distinct phases occur in the course of any phylogenetic analysis. (1) Produce a cladogram. This entails the examination of as many characters as possible, the polarizing of these features by one or more recognized methods of analysis (see Chapter 2), the assembly of a character matrix, and the analysis of this matrix in a manner designed to produce the shortest tree with the least amount of convergence or character reversal and the most amount of congruence of characters in its branches. (2) Produce a phylogram, or evolutionary tree. Some authorities would argue that there is little or no difference between a cladogram and an evolutionary tree, and others would say that there is no need to use anything else but a cladogram. A cladogram, strictly speaking, is only a graphic arrangement of taxa that displays their genealogical relationship. An evolutionary tree is an attempt to qualify the information in the cladogram and interpret its meaning. For example, cladograms can, although not always, be unrooted, that is, have no implicit statements about the direction of evolution within the group in question. A phylogenetic tree is always rooted, and implies a definite sequence of events. (3) Develop a scenario, that is, produce a "story," or evolutionary narrative, about the history of the group based on the information contained in the tree.

Each of these steps entails increasing levels of subjectivity. The first step, production of the matrix and the cladogram, is the most objective in that its assumptions about the direction of character changes and its method of analysis are subject to testing. As we have seen in Chapter 2, however, this method only minimizes subjectivity and does not totally eliminate it. The second step, production of an evolutionary tree, introduces another level of subjectivity into the analysis in that it makes assumptions about the evolutionary events. Those who prefer to use the cladogram directly try to reduce that subjective element. The third step, the outline of a scenario, is the most subjective of the three, and in fact some authorities would not even allow it to pose as science as it lacks clear falsifiable hypotheses inherent in such narratives. Any number of scenarios, or stories, could be developed from the consideration of a single evolutionary tree.

The point to be made here is that the classic, early, traditional speculations on animal evolution start with the scenario, the most subjective element in the process, and then try to create an evolutionary tree and data set to support and match that scenario. In other words, the development of many theories about animal relationships has more often than not been just the opposite from what we would expect of strict scientific methodology. Science, to reiterate lessons from Chapters 1 and 2, is a process of thought in which patterns in nature are recognized, testable hypotheses are produced to explain these patterns, and that we are then free to accept, reject, or modify only in light of new or additional factual information.

Contemporary Views

Recent approaches to invertebrate phylogeny seem to separate into two schools. One group contains those who are marked by a kind of intellectual despair. These workers focus on the differences that separate animal groups and, in so doing, in effect imply that little or nothing can be said about their relationships. The extreme of this school is exemplified by Nursall (1962), who surveys the animal kingdom and concludes that no tree of relationships can be drawn, that all phyla arose independently from various protists, that in some cases the divisions between them extend back to the level of the monerans. A scheme of relationships in this context would resemble a sort of phylogenetic grass rather than a tree.

This school is adamantly opposed to the production of hypothetical archetypes. An example of this is Anderson (1983), who feels that the consideration of archetypes all too often ignores the need to provide functional intermediates between phyla. Although Anderson's views do not go to the extent of Nursall's, in that he does group phyla that he feels are related; he does not hesitate, based mostly on developmental differences, to suggest that metazoans may be polyphyletic. He feels that sponges, cnidarians, and ctenophores may have been independently derived from chaonoflagellates and/or zooflagellates; that the phyla that exhibit true or some modified form of spiral cleavage may have come from ciliates; and that deuterostomes and other minor groups may have had yet other independent ancestries from within protists.

Inglis (1985) does not agree with Anderson's line of reasoning, but nevertheless does come to similar conclusions. He advances a concept he terms evolutionary waves, that is, groups of phyla that arose with multiple, parallel evolutions of major technical improvements. He recognizes five such events: sponges and placozoans, radiates, deuterostomes, protostomes, and what he calls paracoelomates (i.e., mesozoans, pseudocoelomates, and acoelomates).

Investigators, such as Nursall, Anderson, and Inglis, often leave us frustrated in our attempts to understand invertebrate genealogy. As evolutionary biologists, we instinctively believe that one ought to be able to propose hypotheses about invertebrate interrelationships based on shared similarities.

The opposing school to the above approach is dedicated to the old idea of producing evolutionary trees and scenarios in which the node points are various kinds of hypothetical ancestors. Many of these theories are proposed in book-length or near book-length treatments (e.g., Remane, 1967; Siewing, 1969; Jägersten, 1972). Often these workers call attention to the importance of having fully functional intermediates between phyla (e.g., Salvini-Plawen, 1978). Nevertheless, proponents of this school get so carried away with the logic of their assumptions that what is produced is often intimidating to the student in the complexity of their serial, hypothetical ancestors (e.g., Salvini-Plawen, 1980), and are too often riddled with strange terminology (for example, why call Ctenophora "Collaria" as in Salvini-Plawen, 1978). The work of this school, however, is not devoid of useful observations. Many interesting conclusions deserve consideration, for example, equating the coelom of protostomes with the metacoel of the lophophorates (Salvini-Plawen, 1982), recognition of mesoderm in ctenophores (Siewing, 1977), or advancing the concept of archimeric coeloms (Siewing, 1980). Sometimes, though, conclusions based on hypothetical archetypes are indeed strange and quickly brought into question, for example, Siewing's contention that the basic spiralian was a polychaete (Ivanova-Kazas, 1981).

One of the most stimulating papers of this genre is that of Nielsen (1985), although he takes an essentially historical, scenario approach to animal evolution and proposes what is called the trochaea theory. This idea was one of the first effective attempts to integrate larval and adult features into a coherent system. Although several archetypes are proposed in the trochaea scheme, a large number of characters are utilized in the analysis, and an effort is made to define derived character states. The result is something very near to a cladogram, and indeed has been a major inspiration for the analysis in this textbook.

Current Analysis

In light of the various pitfalls outlined above, students may think that analysis of invertebrate phylogeny is an exercise in futility, with little hope of ever reaching a consensus. Yet, if the study of invertebrate phyla and the classification of organisms are indeed sciences, and Evolutionary Theory is the grand paradigm that holds all of biology together, then there is no reason why the analysis of animal interrelationships also cannot be done in a more scientific manner. The analysis here is an effort to move toward that end.

The matrix of characters used (Table 38.1) is one largely based on gross anatomical features derived from the literature, that is, characters that traditionally have been used in discussions of animal relationships. Much new and interesting information is coming to light as a result of studies that employ scanning and transmission electron microscopy. In a few instances, some of these features were utilized in early phases of this analysis (e.g., monociliate versus multiciliate epidermal cells), but these characters were found to be highly convergent, or homoplastic, in different phyla. In addition, it is often not possible to use such characters consistently across all phyla, because not all phyla have been equally examined as yet for all of these features.

Finally, no biochemical data were utilized. Such studies use comparisons of cytochrome c, hemoglobins, ribosomal RNA, and DNA hybrids (e.g., Ghiselin, 1988). Some interesting results have been obtained (e.g., Bergström, 1986). Unfortunately, this data cannot be applied in the kind of analysis discussed here because currently too few groups have been studied with these methods. In addition, the information obtained to date is often contradictory (for example, compare the results of Bergström, 1986, to those of Field et al., 1988). One of the problems with the biochemical analyses is that they are often essentially phenetic, based on total numerical similarity of molecules. Until such time as homology of amino acid or base pair sequences is analyzed, the results of biochemical phylogenies will not be comparable to those obtained by cladistic methods.

The characters that are used in the matrix of Table 38.1 are outlined in Table 38.2.

The data were analyzed using several cladistic computer programs. Both PAUP and PHYSYS, two of the more important programs used today, were employed. Little difference was found between PAUP and parts of the PHYSYS package. Within PHYSYS, two options were used, WAGNER.S and WISS. In comparing results, it was found that PAUP and the PHYSYS WAGNER.S option gave comparable results. The analysis of WAGNER.S is primarily used here. WAGNER.S uses global branch-swapping to select the most parsimonious tree. Reversal of character states is allowed. WISS, on the other hand, produces a tree by not allowing charac-

Table 38.1. Data matrix of characters used in analyses to produce cladograms in Figures 38.2 and 38.3

Taxa	Characters (5 10 15 20 25 30 35 40 45 50 55 60 65 70 75)
Ancestor	00000000000000000000000000000000000000000000000000000000000000000000000000000
Mesozoa	10111000000000000000000000000000000000000000000000000000000000000000000000000
Placozoa	10900000000000000000090000000000000000000000000000000000000000000000000000000
Porifera	11000000110000000000000000000000000000000000000000000000000000000000000000000
Cnidaria	11001111000001100000000000000000000000000000000000000000000000000000000000000
Ctenophora	11000111001111001000000000000000000000000000000000000000000000000000000000000
Gnathostomulida	11010011001910000190010000000010000000000000000000000000000000000000000000000
Platyhelminthes	11010011001110000011000000000010000000000000000000000000000000000000000000000
Gastrotricha	11010011001110010001010100010011001000000000000000000000000000000000000000000
Rotifera	11010111001110010001110101110011110001000000000000000000000000000000000000000
Acanthocephala	10010111001110010001110101010011110110000000000000000000000000000000000000000
Loricifera	11010111001110010909910910090099000000111000000000000000000000000000000000000
Kinorhyncha	11010111001110010901110910010011000000100000000000000000000000000000000000000
Priapulida	11010111001110010001110010010011000000111000000000000000000000000000000000000
Nematomorpha	11010111001110010001111110090000000000000000000000000000000000000000000000000
Nematoda	11010111001110010001111110011100000000000000000000000000000000000000000000000
Chaetognatha	11010111001010010001111001100000000000000000000000000000000000000000000000000
Mollusca	11010111001110010110000000000000000000000110000010001000000001010100000000000
Nemertinea	11010111001110010110000000000010010000000110000000000000000000001110000000000
Sipuncula	11010111001110010110000000000000000010000101000010001000000010000000000000000
Echiura	11010111001110010110000000000000010000000100000010001000000010000000110000000
Annelida	11010111001110010110000000000000000000000110000010001000000001100000000000000
Pogonophora	10010111001110000190000000000010000000000110000010000000000001100010000000000
Pentastomida	11010111001110010190000000000000000000000110000000000000000001100000000000000
Tardigrada	11010111001110010190000000010100000000000110000000010000000001100100001000000
Onychophora	11010111001110010190000000000100000000000110000000000000000001100101001000101
Uniramia	11010111001110010190000000000100000000000110000000000000000001100100001100111
Cheliceriformes	11010111001110010190000000000100000000000110000000000000000001100100009019110
Crustacea	11010111001110010100000000000100000000000110000000000000000001100100000111000
Phoronida	11010111001110010000000000000000000000000101111100101000000000000000000000000
Ectoprocta	11010111001110010000000000000000000000000101111001900000000000000000000000000
Entoprocta	11010111001110010110100000000010000000000101100101000000000000000000000000000
Brachiopoda	11010111001010010000000000000000000000000101111001110000000000000000000000000
Echinodermata	11000111001011010000000000000000000000001000001100110100001100000000000000000
Enteropneusta	11010111001010010000001000000000010000001000001100110111110000000000000000000
Pterobranchia	11010111001010010000000000000000000000000101119100190111000000000000000000000
Urochordata	11011111001010010000000000000000000000000001000010000010110010000000000000000
Cephalochordata	11010111001010010000000000000000000000000110000010001010110010000000000000000

Characters defined in Table 38.2.

Table 38.2. Characters used in the analysis of invertebrate phyla

Primitive state	Derived state
1. No collagen, no acetycholine transmission, microgametes otherwise, reproductive cells (if separate) on surface/exterior	1. Collagen, acetylcholine cholinesterase system, sperm flagellate or ciliate with condensed chromatin and mitochondria, reproductive cells or tissues internal
2. No gut	2. Digestion by means of special cells or body organs
3. General free-swimming larva	3. Nematogen/infusorigen larva
4. No symmetry	4. Bilateral symmetry
5. No alteration of generations	5. Metagenesis of some type
6. No special muscle cells	6. Well-developed muscle cells
7. No basal membranes, nerve cells, or gap junctions	7. Epithelia with basal membranes, nerve cells with synapses, gap junctions between cells
8. No endoderm	8. Endoderm in embryo
9. No special water channels or choanocytes	9. Pores and canals for water circulation facilitated by choanocytes
10. Embryonic cells/layers do not invert	10. Embryonic inversion
11. No mesoderm	11. Mesoderm in embryo
12. Cleavage indeterminate and radial	12. Cleavage determinate
13. Muscles ectodermal in origin	13. Subepidermal muscles
14. No symmetry	14. Radial or biradial symmetry
15. No nematocysts	15. Nematocysts
16. No anus	16. Anus
17. No colloblasts or ctenes	17. Colloblasts and ctenes
18. Cleavage radial	18. Spiral quartet cleavage
19. Ectomesoderm	19. 4d mesoderm
20. Nervous system as poorly polarized nerve net	20. Orthogonal nervous system—anterior nerve ring and several longitudinal cords
21. No body cavity	21. Pseudocoel
22. General free-swimming larva	22. No primary larva
23. Longitudinal and circular body wall muscle	23. Longitudinal body wall muscles only
24. Cleavage radial	24. Aberrant spiral cleavage (monets and duets)
25. Buccal chamber not eversible	25. Buccal introvert
26. Cuticle, if present, solid	26. Cuticularized epidermis with tubules
27. No retrocerebral organs	27. Retrocerebral organs
28. No eutely	28. Eutely
29. No renettes, amphids, or phasmids	29. Renette cells, amphids, and phasmids
30. No molting	30. Ecdysis with ecdysone
31. No special excretory organ	31. Protonephridia
32. Flame cells (ciliate)	32. Flame bulbs (flagellate)
33. No lemnisci	33. Lemnisci
34. No proboscis	34. Proboscis (evertable organ without mouth)
35. Free cilla	35. Cilia ensheathed with cuticle
36. No uterus bell	36. Uterus bell
37. Epidermis solid	37. Epidermal lacunae
38. Cuticle uniform	38. Cuticle with several unit membranes

Table 38.2. Characters used in the analysis of invertebrate phyla (*continued*)

Primitive state	Derived state
39. No scallids	39. Scallids on introvert
40. No lorica	40. Lorica at some stage
41. No caudal appendages	41. Caudal appendages at some stage
42. No body cavity	42. Coelom (metacoel)
43. Mesoderm not segmented	43. Segmented or serial structures derived from mesoderm
44. Gut straight	44. Gut coiled or looped, anus anterior
45. No lophophore	45. Lophophore
46. Downstream particle capture	46. Upstream particle capture in adults
47. Downstream particle capture	47. Upstream particle capture in larvae
48. Larval apical organ does not degenerate	48. Larval apical organ degenerates
49. General free-swimming larva	49. Trochophore or trochophorelike larva
50. No special photoreceptor	50. Photoreceptor cell with tuft of tightly packed cilia
51. Coelom undeveloped	51. Archimeric coelom
52. Coelom, if present, as a schizocoel	52. Enterocoel
53. No special excretory organ	53. Metanephridia
54. General free-swimming larva	54. Tornaria/bipinnaria larval type
55. No pharyngeal slits	55. Pharyngeal slits
56. No buccal diverticulum	56. Buccal diverticulum
57. No dorsal nerve cord	57. Dorsal nerve cord from tubelike infolding of ectoderm
58. Pharyngeal slits simple	58. Pharyngeal slits divided
59. No calcareous endoskeleton	59. Calcareous endoskeleton
60. No water–vascular system	60. Water–vascular system
61. No notochord	61. Notochord
62. Brain not derived from any part of larval apical organ	62. Brain in part derived from larval apical organ, main nerve cord ventral
63. No special cellular derivative for coelom	63. Teloblasts give rise to mesoderm and coelom
64. Coelom large and pervasive	64. Coelom restricted to around circulatory system
65. No rhyncocoel	65. Rhyncocoel
66. No prototrochal lobes	66. Larva with prototroch developed as ciliated lobes (pilum or velum)
67. No hemocoel	67. Hemocoel (unlined mesodermal cavity)
68. Body grows uniformly	68. Apical growth from larva
69. Circular and longitudinal body wall muscle	69. Oblique body wall muscle layer
70. No anal vesicles	70. Anal vesicles
71. No appendages	71. Lobopods or uniramous limbs
72. No antennae	72. Deutocerebral antennae
73. No special excretory organs	73. Segmental, secretory, excretory glands
74. No appendages	74. Biramous limbs
75. No tracheae	75. Tendency to develop tracheae
76. No special excretory organ	76. Tendency to develop Malpighian tubules
77. No special jaws	77. Whole limb jaw

Expression of these characters for each taxon is provided in Table 38.1 and used to generate cladograms in Figures 38.2 and 38.3.

ter reversals, with the result that convergences and parallelisms are common.

Cladogram of Invertebrates

The computer analysis of these characters produced the WAGNER.S cladogram seen in Figure 38.2. Some surprises are noted with respect to where particular phyla ordinarily occur in traditional scenarios of invertebrate relationships.

1. There is but a single stem to this cladogram. The ideas that mesozoans and poriferans may have arisen independently from protists is not supported by this data matrix.

2. The resolution of the relationships of the placozoans and the mesozoans *sensu stricto* (rhombozoans and orthonectids) to each other and the main stem must await more information on structure and modes of reproduction and development, especially for placozoans.

3. Ctenophores and cnidarians are not sister groups within some larger group often referred to as radiates. Rather, the ctenophores are a sister group to all the more derived invertebrates. The recognition of a mesodermal layer and determinate cleavage, along with the beginnings of a subepidermal musculature in ctenophores allies them with advanced invertebrates. The development of radial and biradial symmetry in the Ctenophora is merely a convergence to that seen in the Cnidaria as well as Echinodermata.

4. A surprise is seen in the relationship of the traditional acoelomate, pseudocoelomate, and eucoelomate clades to each other. Traditional views of animal history hold that acoelomates are a grade of organization ancestral to all others. This cladogram (Fig. 38.2), however, reveals that after the essential bilateral body plan is established, with bilateral symmetry and the development of an anus, two main clades can be discerned. One line has a highly organized mode of development, with spiral quartet cleavage and 4d mesoderm as a foundation. This includes the acoelomates and the eucoelomates. The other line, with an orthogonal nervous system, a pseudocoel, and the lack of primary larvae, is composed of the pseudocoelomate phyla. These latter phyla exhibit a variety of divergent monet and duet types of spiral cleavage, as well as the primitive radial form.

5. In the sister lineage to the pseudocoelomates, the acoelomates (platyhelminths and gnathostomulids) can be interpreted as a side branch that reverts to mainly ciliary locomotion, aborts the development of an anus for the gut, and uses protonephridia for excretion and osmoregulation. Interestingly, of the characters used in this analysis, none is unique to Platyhelminthes, and it might be argued that the Gnathostomulida are merely specialized flatworms who have lost a larval stage in their adaptation to the interstitial habitats.

6. Coelomates form a distinct clade with two branches. One of these branches contains many groups that used to be included within the deuterostomes. This is a lineage that is based on the development of a lophophore and looped gut with anterior anus. Many of these groups have sessile or inbenthic life-styles. A transition series is seen in this clade whereby the lophophorate characters are lost and a mobile life-style is regained. Undoubtedly, paedomorphosis played a role in this process, that is, larval and juvenile characters being retained or used in the development of a mobile adult or reproductive stage in the life cycle. Note that the data matrix (Table 38.1) used in this analysis indicates that the lophophorates and the phylum Hemichordata (pterobranchs and enteropneusts) are better viewed as paraphyletic groups rather than distinct monophyletic clades in their own right.

7. The sister group to the deuterostome/lophophorates consists of phyla that are marked by a tendency to develop segmental or serial structures in the tissues derived from mesoderm. This is a feature that became increasingly important as the evolution of this clade occurred. This clade includes the nemertines, the trochophorate phyla, and the annelid/arthropods sensu lato that includes the highly derived pogonophorans.

8. The most highly serial, or segmented, of all groups are the arthropodlike forms. The scheme shown here could imply that either the phylum Arthropoda should be redefined to include groups that are traditionally excluded, such as Tardigrada and Onychophora, or that the traditional phylum Arthropoda is polyphyletic, as Manton (1977) and Anderson (1973) argued.

Students should remember that the above-postulated relationships are dependent on the particular data matrix used here. New knowledge on the structure and development of all the phyla will produce factual information that will alter the matrix and thus this scheme. In addition, different computer programs can and do provide differing views of animal history. An object lesson in this regard can be seen in Figure 38.3. The same data used to generate Figure 38.2 were analyzed using the WISS option within PHYSYS. WISS essentially functions on Dollo's Law, that evolution does not retrace its prog-

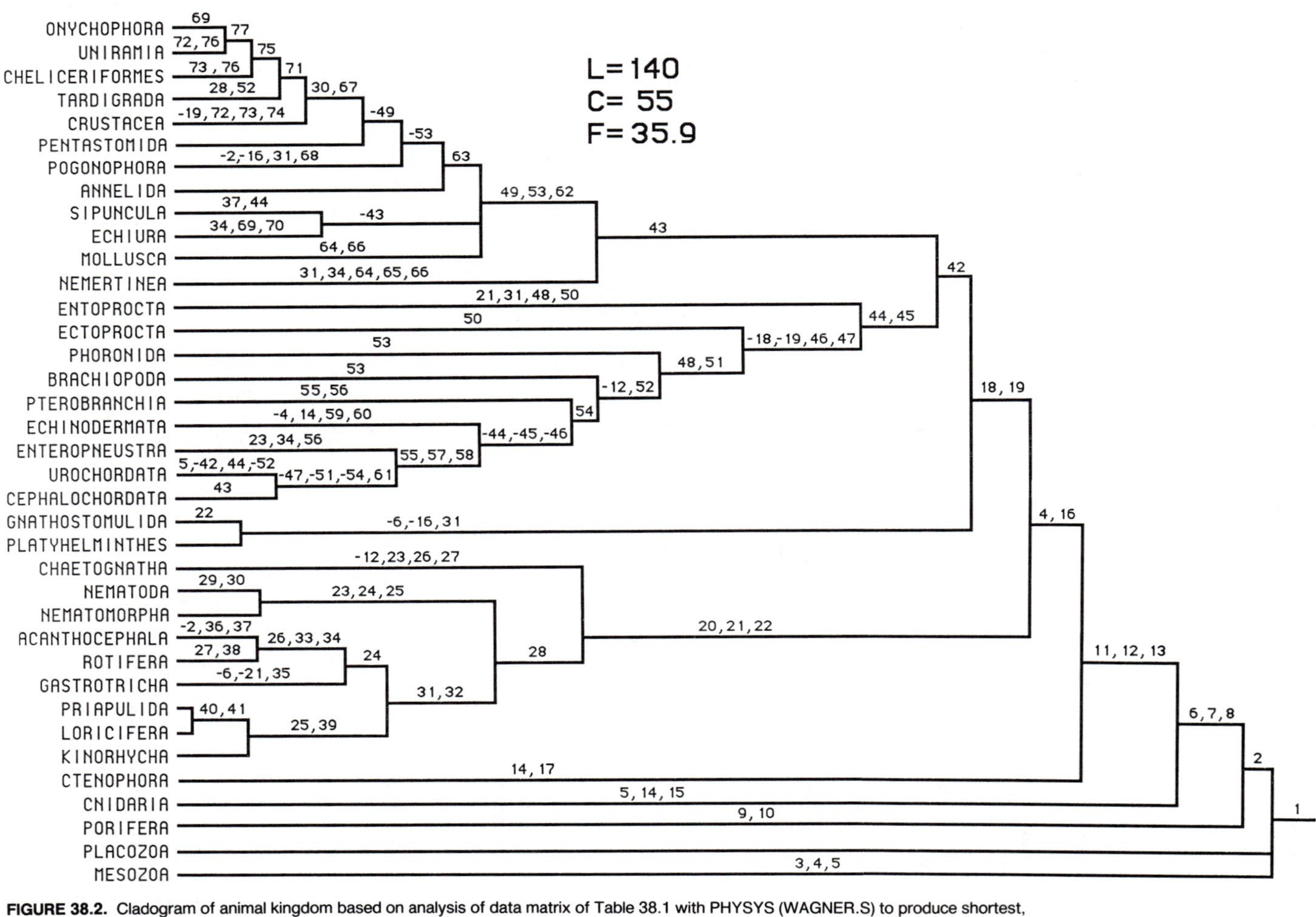

FIGURE 38.2. Cladogram of animal kingdom based on analysis of data matrix of Table 38.1 with PHYSYS (WAGNER.S) to produce shortest, most parsimonious tree. Length, C-index, and F-ratio provided. Characters are those listed in Table 38.2.

ress; WISS does not allow character reversals. WISS does not produce the most parsimonious tree; that is to say, a WISS tree is not the shortest possible (length = 158 versus length = 140), and because it does not allow reversals, it contains a great deal of convergent character expression (F-ratio = 52.2 versus F-ratio = 35.9).

Nevertheless, some interesting points of difference emerge from comparing the WISS tree with that of the WAGNER.S cladogram. Although nearly related phyla in Figure 38.2 reappear in Figure 38.3, the relationships of these groups to each other is unusual. The pseudocoelomate and deuterostome clades are not too different from the PHYSYS WAGNER.S; although in the latter, the Hemichordata emerge as a distinct taxon, the chordates are a sister group to all the rest of the deuterostomes, and these are a sister group to ctenophores and cnidarians. The acoelomates are placed very high in the WISS tree, and the trochophore clade has a variant arrangement of the taxa over that seen for WAGNER.S. It is interesting, however, that the major differences in the placement of taxa is in shifting the position of the major branches rather than a major rearrangement within the branch clades or a reshuffling of the whole tree.

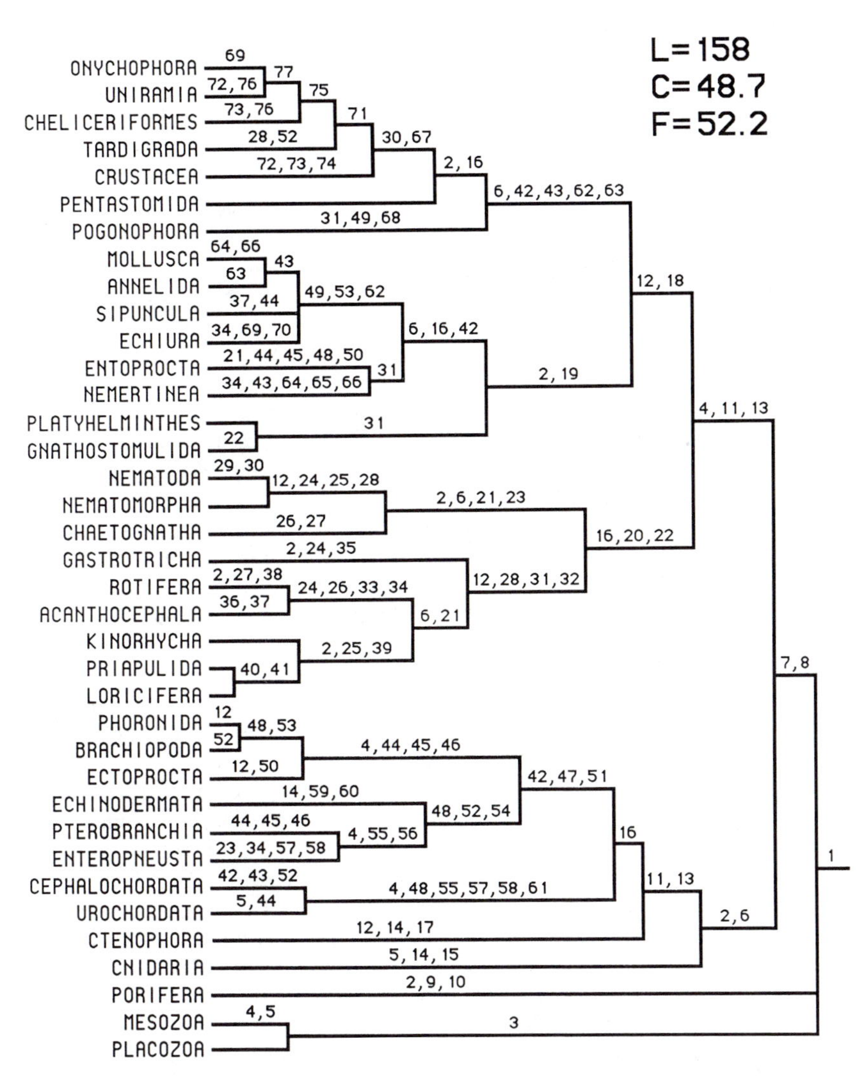

FIGURE 38.3 Cladogram of animal kingdom based on analysis of data matrix of Table 38.1 with PHYSYS (WISS) to produce a tree without any character reversals. Length, C-index, and F-ratio provided. Characters are those listed in Table 38.2

Scenario of Invertebrate Evolution

This cladistic analysis is at least a first approximation toward an understanding of invertebrate relationships, a hypothesis, if you will, that can and should be tested with further observations and data. Accepting this argument for now, we can return to a consideration of the phases of phylogenetic analysis mentioned and develop a plaudible scenario of events that may have occurred in the course of invertebrate evolution based on the cladogram of Figure 38.2. The following might be one possible story.

The early stages in the evolution of the animal kingdom centered on increasing complexity of structure and biochemistry related to facilitating the interaction of cells. This sequence of events began at a protistan grade, wherein individual cells, even those within colonies, were independent entities. A first step in the evolution of metazoans involved the development of an initial loose organization and partial specialization of cell types (at least as far as reproduction is concerned), such as that seen among the mesozoans and placozoans. Next came a stage whereby specific body functions were restricted to specific cells. Sponges reflect this grade, although they lack true tissues as poriferan cells often are capable of performing various functions at different times. The next step in metazoan evolution involved advanced cell specialization to produce true tissues. Cnidarians and ctenophores mark different phases in this stage of evolution, and highlight a sequence of events that involved not only development of tissues (i.e., specialization of cells) but also development of embryonic germ layers (i.e., specialization of a process that could produce tissues).

Animal evolution up to that point was focused, if you will, on internal factors, that is, facilitation of cell interactions, specialization of cell function, and perfection of a process for differentiation of distinct cell and tissue types. An intermediate plateau of animal evolution was achieved whereby moderately complex body forms had evolved that could handle a diversity of relatively sophisticated functions related to survival and reproduction. Animal history then underwent a shift in focus away from internal factors related to basic body organization, toward evolving diverse body plans that could more effectively interact with the external environment. The appearance of bilateral symmetry marks this transition.

Two different clades then evolved, each reflecting different solutions to the problems engendered by this shift. One clade is represented by the pseudocoelomates. This group is often characterized by strange monet and duet patterns of early cell cleavage, and probably, as a result, lack viable primary larval phases in their life cycles. Pseudocoelomates function with nervous systems (termed orthogonal) that are essentially modified from a radial form. Several nerve cords typically occur, none of them really predominating over any of the others, and all of them loosely controlled by a poorly developed nerve ring or brain. The fact that processes of embryonic differentiation seem poorly developed among pseudocoelomates may be responsible for the structural peculiarities we see among them. Their body plans have many parallels to those seen among the more advanced coelomates. There are specialized muscle layers for locomotion, but these are often incomplete (such as lack of circular layers in the nematode/nematomorph/chaetognath line). There is a pseudocoel cavity, but there is a lack of the elaborate specialization of muscles that coelomates have, both in the body wall and around the gut tissues. Some groups, namely priapulids and chaetognaths, even developed linings to the pseudocoel, although these appear to be rather peculiar and rather distinct from the linings of true coeloms. Pseudocoelomates evolved some very special and peculiar sense organs to deal with specific sensory needs, such as amphids, phasmids, retrocerebral organs, caudal appendages (?), and sensory spines, but they lack any effective degree of cephalization.

The other clade, including acoelomates and coelomates, differed from the pseudocoelomate line in having a highly regulated and determinate mode of development, as evidenced by spiral cleavage and 4d mesoderm. These distinctive features of ontogeny were often lost or greatly modified in the course of the evolution of this clade. Nevertheless, the hallmark of this lineage as a whole, undoubtedly related to the specialized mode of development they passed through, is the highly complex tissue types that this mode of development produces. Phyla in this clade are typically characterized by possession of distinct larval stages.

At least three clades resulted from specialization of cleavage pattern and specification of mesoderm formation. One group, the acoelomates, never developed particularly complex body plans, but did perfect increasingly complex life cycles that focused on meiofaunal or parasitic life-styles. Perhaps this emphasis was related to a lack of body cavity.

The other two lineages were based on a coelomic body plan. One of these, the deuterostome/lophophorate line, focused initially on the evolution of a primarily sessile life-style. Many aspects of the original spiralian specialized cleavage and mesoderm formation were lost. These losses perhaps reflected an underlying process operating in the genome of these animals, as the more derived phyla of this clade exhibit paedomorphic processes in the course of their evolution. That is, either juvenile features were held into the adult phase, or gonad function was shifted into a juvenile phase to achieve major morphological changes in body plan. Sustained ontogenetic paedomorphosis, however, brought a return to a mobile life-style among the advanced deuterostome phyla of this lineage.

The evolution of the rest of the coelomate phyla primarily focused on mobile life-styles. In addition, these annulated phyla seem to have possessed a genome that was organized in a way that predisposed these groups to differentiate at least serial, if not truly segmental, body structures. This capacity, when linked in some of the phyla with paedomorphic processes in ontogeny, resulted in an arthropod clade noted for great diversity in body forms as well as an overwhelming number of species.

Thus, we have a story for animal evolution. Whether or not this scenario is a true and accurate description of what really happened in the course of animal history is irrelevant. What is important is that we see that animal relationships are subject to scientific analysis, and that stories based on such analysis need not be pure fantasy, as Libbie Hyman suggested (Hyman, 1959, p. 754).

Selected Readings

Anderson, D. T. 1973. *Embryology and Phylogeny on Annelids and Arthropods.* Pergamon Press, Oxford.

Anderson, D. T. 1983. Origins and relationships among the animal phyla. *Proc. Linn. Soc. N.S.W.* 106:151–166.

Bergström, J. 1986. Metazoan evolution—A new mode. *Zool. Scripta* 15:189–200.

Field, K. G., G. J. Olsen, D. L. Lane, S. J. Giovannoni, M. T. Ghiselin, E. C. Raff, N. R. Pace, and R. A. Raff. 1988. Molecular phylogeny of the animal kingdom. *Science* 239:748–753.

Ghiselin, M. T. 1988. The origin of molluscs in the light of molecular evidence. *Oxford Surv. Evol. Biol.* 5:66–95.

Hadzi, J. 1963. *The Evolution of Metazoa.* Macmillan, New York.

Haeckel, E. 1874. Die Gastria Theorie, die phylogenetsche Classification des Thierreiches und die Homologie der Keimblatter. *Jena Z. Naturwiss.* 8:1–55.

Hyman, L. H. 1959. *The Invertebrates*, vol. 5. McGraw-Hill, New York.

Inglis, W. G. 1985. Evolutionary waves: Patterns in the origins of animal phyla. *Aust. J. Zool.* 33:153–178.

Ivanova-Kazas, O. M. 1981. Filogeneticheskoye znachyeniye sliralnogo drolyeniya. *Biologiya Morya* 1951(5):3–14.

Jägersten, G. 1955. On the early phylogeny of the Metazoa. The bilaterogastrea theory. *Zool. Bidr. Uppsula* 30:321–354.

Jägersten, G. 1972. *Evolution of the Metazoan Life Cycle.* Academic Press, London.

Marcus, E. 1958. On the evolution of the animal phyla. *Quart. Rev. Biol.* 33:24–58.

Metschnikoff, E. 1886. *Embryologische Studien an Medusen.* Holder, Vienna.

Manton, S. M. 1977. *The Arthropoda.* Clarendon Press, Oxford.

Nielsen, C. 1985. Animal phylogeny in light of the trochaea theory. *Biol. J. Linn. Soc. Lond.* 25:243–299.

Nursall, J. R. 1962. On the origin of the major groups of animals. *Evol.* 16:118–123.

Remane, A. 1963. The evolution of the Metazoa from colonial flagellates vs. plasmodial ciliates. In *The Lower Metazoa* (E. C. Dougherty, ed.), pp. 23–32. University of California, Berkeley.

Remane, A. 1967. Die Geschichte der Tiere. In *Die Evolution der Organismen*, vol. 1 (G. Heberer, ed.), pp. 589–677. Fischer, Stuttgart.

Salvini-Plawen, L. 1978. On the origin and evolution of the lower Metazoa. *Z. zool. Syst. Evolut.-forsch.* 16:40–88.

Salvini-Plawen, L. 1980. Was ist eine Trochophora? Eine Analyse der Larventypen mariner Protostomier. *Zool. Jahrb. Abt. Anat.* 103:389–423.

Salvini-Plawen, L. 1982. A paedomorphic origin of the oligomerous animals? *Zool. Scripta.* 11:77–81.

Sharov, A. G. 1966. *Basic Arthropodan Stock.* Pergamon, New York.

Siewing, R. 1977. Mesoderm in ctenophores. *Z. zool. Syst. Evolut.-forsch.* 15:1–8.

Siewing, R. 1980. Das Archicoelomatenkonzept. *Zool. Jahrb. Abt. Anat.* 103:439–482.

Siewing, R. 1969. *Lehrbuch der Vergleichenden Entwicklungsgeschichte der Tiere.* Paul Parey, Hamburg.

Taxonomic Index

Subject Index

Page numbers in **boldface** type indicate first use and/or definition of term.